SPIDER WEBS

Spider Webs

BEHAVIOR, FUNCTION, AND EVOLUTION

WILLIAM EBERHARD

THE UNIVERSITY OF CHICAGO PRESS

CHICAGO AND LONDON

The University of Chicago Press, Chicago 60637
The University of Chicago Press, Ltd., London

For more information, contact the University of Chicago Press, 1427 E. 60th St., Chicago, IL 60637.

Published 2020

Printed in the United States of America

31 30 29 28 27 26 25 24 23 22   2 3 4 5 6

ISBN-13: 978-0-226-53460-2 (cloth)
ISBN-13: 978-0-226-53474-9 (e-book)
DOI: https://doi.org/10.7208/chicago/9780226534749.001.0001

Additional parts of this large project are archived online at press .uchicago.edu/sites/eberhard/; these portions are designated in the text with a capital O followed by the book chapter to which they refer (e.g., section O6.1; table O3.3; fig. O9.1).

Library of Congress Cataloging-in-Publication Data
Names: Eberhard, William G., author.
Title: Spider webs : behavior, function, and evolution / William Eberhard.
Description: Chicago : University of Chicago Press, 2020. | Includes bibliographical references and index.
Identifiers: LCCN 2019050065 | ISBN 9780226534602 (cloth) | ISBN 9780226534749 (ebook)
Subjects: LCSH: Spider webs. | Spiders—Behavior.
Classification: LCC QL458.4 .E24 2020 | DDC 595.4/4156—dc23
LC record available at https://lccn.loc.gov/2019050065

♾ This paper meets the requirements of ANSI/NISO Z39.48-1992 (Permanence of Paper).

Frontispiece: The recently discovered "pseudo-orb" of the New Guinean psechrid *Fecenia ochracea* is a spectacular example of a major theme of this book, evolutionary convergence. Although the spider is not closely related to the "true" orb weavers (Orbiculariae) its web shares with orbs a planar, aerial array of non-sticky radial lines that converge on a hub, frame lines that support the radii, and a uniformly spaced array of more or less circular sticky lines laid on the radii. The recent nature of this discovery, and the mystery of the ribbon-like nature of the sticky lines in the inner portion of the web (such band-like cribellum lines have never been described in any other species) emphasize a second general theme: our knowledge of spider webs (especially non-orb webs) is still profoundly incomplete (from Agnarsson et al. 2012; courtesy Ingi Agnarsson)

To the memory of three great

naturalists of web-building spiders:

REV. H. C. MCCOOK,

MAJ. R. W. G. HINGSTON,

and E. NIELSEN

A NOISELESS PATIENT SPIDER

A noiseless, patient spider,

I mark'd, where, on a little promontory, it stood, isolated;

Mark'd how, to explore the vacant, vast surrounding,

It lauch'd forth filament, filament, filament, out of itself;

Ever unreeling them—ever tirelessly speeding them.

And you, O my Soul, where you stand,

Surrounded, surrounded, in measureless oceans of space,

Ceaselessly musing, venturing, throwing,—seeking spheres, to connect them;

Till the bridge you will need, be form'd—till the ductile anchor hold;

Till the gossamer thread you fling, catch somewhere, O my Soul.

— Walt Whitman

CONTENTS

CHAPTER 5. THE BUILDING BEHAVIOR OF NON-ORB WEAVERS

CHAPTER 6. THE BUILDING BEHAVIOR OF ORB-WEAVERS

CHAPTER 8. WEB ECOLOGY AND WEBSITE SELECTION

CHAPTER 10. ONTOGENY, MODULARITY, AND THE EVOLUTION OF WEB BUILDING

1

INTRODUCTION

1.1 INTRODUCTION

This book is about how spider webs are built, and how and why they have evolved their many different forms. It was conceived during a field trip in a spider biology course, when a student asked me why I had not written a book on spider webs. She noted that such a book would have been useful for the course but did not exist. She was right on both counts, and also that I was well-placed to produce it. I am an evolutionary biologist particularly interested in behavior. I have had the great fortune of living most of my working life in the tropics, and have spent just over 50 years watching a diverse array of spiders and thinking about their webs. Writing this book represents a chance both to contribute general summaries of data and ideas that are currently lacking, and to organize and develop new ideas that grew out of producing these summaries.

Inevitably, the book's coverage is idiosyncratic. I propose more new ideas, give more new data, and discuss more extensively those topics related to behavior and evolution; some discussions in other areas, like the biochemistry and mechanics of silk, are more standard and less original. I focus on behavior and evolution partly because spider webs have the potential to make especially important contributions to resolving central questions in behavior and evolution, just as they did in the past in understanding the characteristic and limits of innate behavior (Fabre 1912, Hingston 1929). An orb web can be easily photographed with perfect precision, and it constitutes an accurate record of a series of behavioral decisions made by a free-ranging animal under natural conditions. And because a web-building spider's sensory world is so centered on silk lines, the orb is also a precise record of some of the most important stimuli that the spider used to guide those decisions; in addition, these stimuli can be experimentally altered by making simple modifications of webs. These advantages offer unusual chances to study some basic questions in animal behavior with especially fine detail and precision.

I hope that this book may help future observers to see how web-building spiders can be exploited to investigate exciting new questions in animal behavior. It is a fragment of a larger project that originally included several additional topics: the role of behavioral canalization and "errors" in the evolution of behavior; the importance of scaling problems in the absolute and relative brain sizes of tiny animals and the limitations in behavioral capabilities that they may impose; the organization and evolution of independent behavior units in the nervous system; the implications of chemical manipulations performed by parasitoid wasps on their spider hosts and of additional experimental manipulations of webs for how behavioral subunits are organized and controlled in the spider; and the possibility that small invertebrates understand, at least to some limited degree, the physical consequences of their actions and the role that such understanding may play in the evolution of new behavior patterns. Other fragments of this larger project are archived online at the University of Chicago Press at press.uchicago.edu/sites/eberhard/; these portions are designated in the text with a capital O followed by the book chapter to which they refer (e.g., section O6.1; table O3.3; fig. O9.1). My hope is that this book will help to project the study of spider webs back onto the forefront of studies of animal behavior.

1.2 A FOREIGN WORLD: LIFE TIED TO SILK LINES

Spiders are ecologically extremely important: for instance, a recent estimate suggests that they capture 400–800 million metric tons of prey each year, or about 1% of global terrestrial net primary production (Nyffeler and Birkhofer 2017). Web spiders live in a world that is ruled by their silk lines. This world is so different from ours, with respect both to their sensory impressions and to the basic facts of what is and what is not feasible for them to do, and they have such exquisite adaptations to this world, that it is useful to briefly describe these differences here, anticipating more detailed descriptions in later chapters (references documenting these statements will be given later). My aim is to help give the reader the kinds of intuition that are needed to understand their behavior.

Orb weavers are functionally blind with respect to their web lines. Their eyes are neither designed nor positioned correctly to resolve fine objects like their lines, and in any case many spiders build their webs at night. Watching an orb weaver waving her legs as she moves is thus like watching a blind man use his cane, with two differences: the spider is usually limited to moving along only the lines she has laid (and thus has to use her "canes" to search for each new foothold); and the lines are in three rather than two dimensions. Slow motion recordings reveal a further level of elegance. Spiders economize on searching as they walk through a web; one leg often carefully passes the line that it has grasped to the next leg back on the same side before moving on to search for the next line. Each more posterior leg needs to make only a quick, small searching movement in the vicinity of the point that is being grasped by the leg just ahead of it.

This dominance of the sense of touch gives observations of orb weaver behavior a dimension that is largely missing in many other animals, because the spider's intentions can be intuited from the movements of her legs. One can deduce a blind man's intentions by watching the movements of his cane; when he taps toward the left just after he hears a sudden growling sound from that direction, he is probably wondering about the origin of that sound. And it is also clear what he has found out; if his cane has not yet touched the dog that growled, he probably does not yet know its precise location. The searching movements that a spider's legs make to find lines, especially at moments during orb web construction, when one can predict the next operations that she will perform, make it possible to confidently deduce both her intentions and the information that she has perceived. For instance, when the orb weaver *Leucauge mariana* is moving back toward the hub after attaching the first radial line to a primary frame while she is building a secondary frame (when she is at approximately the site of the head of the inward arrow from point p in fig. 6.5j), she consistently begins to tap laterally with her anterior legs on the side toward the adjacent radius (o in fig. 6.5j) that she is about to use as an exit to reach the second primary frame. The movements and their timing leave no doubt about what she is attempting to do: she is searching for the adjacent radius. In some cases it is possible to deduce, even in a finished web, what the spider may or may not have sensed at particular moments during the web's construction by taking into account leg lengths and the distances between lines. This access to both the intentions and the information that an animal has (or lacks) offers special advantages in studies of animal behavior.

An orb weaver is also largely chained to its own lines.

In one sense, the spider lives a life similar to that of a blind person who is constrained to move only along preexisting train tracks. And just like a blind man walks around and taps objects in a room to sense its size and the presence or absence of furniture in it, the spider searching for a site in which to build a web must explore it physically, rather than surveying it visually. Much of this exploration must occur by following lines that were already laid previously; the spider cannot simply set out in any arbitrary direction. While spiders sometimes lay new lines by walking along a branch or a leaf and attaching the dragline line, at other times the only way for the spider to lay a new line from one distant point to another is by floating a line on a breeze and waiting until it snags somewhere. The spider has no control over the strength or direction of the wind, nor how or where the line snags. If the line fails to snag firmly on a useful support object, the spider's only recourse is discard it and try again. To evaluate a possible website (is it too large? too small? does it have a projecting branch that would interfere with a web?), the spider probably remembers the distances and directions she has moved while exploring it (sections 6.3.2, 7.8).

And then there is the web itself. Orbs are famously strong, and able to resist both the general loads imposed by wind and the local stresses from high-energy prey impacts. But they are also foreign to some human intuitions. The lines are extremely light, and because air is relatively viscous at the scale of a small animal like a spider, transverse vibrations that would dominate a web at a human scale are strongly damped in spider webs. A more appropriate image is an orb of flexible lines strung under water. Not unexpectedly, spiders rely mainly on longitudinal rather than transverse vibrations of their lines to sense prey and other spiders. And while an orb is strong, it is also delicate in unexpected ways. A few drops of rain, a fog, or even dew can be enough to ruin it: the sticky spiral lines adhere to each other, the radii, and the frame lines, and the sticky material on the lines of most species washes off in water. Wind can cause individual sticky lines to flutter and become damaged when they swing into contact with each other. Some patterns in sticky spiral spacing in orbs may represent adaptations to reduce this type of damage (section 3.2.11).

In sum, web-building spiders live a life that is limited by their own senses, and tightly bound to the silken lines in their webs. To understand these spiders, it is necessary to understand their webs, and vice versa.

Fig. 1.1. Three outstanding observers of spider webs and building behavior from the late nineteenth and early twentieth century (left to right): Rev. Henry C. McCook, Major R. W. G. Hingston, and Emil Nielsen (the last two courtesy of, respectively, Royal Geographic Society, and Nicolas Scharff).

1.3 A BRIEF HISTORY OF SPIDER WEB STUDIES

Spider webs are, of course, common and easily observed. The history of scientific inquiry regarding webs and spider building behavior begins, as with that of many other fields in biology, with descriptions by field naturalists. Important early works, now difficult to obtain and which I have not read, included Quatremére-Disjonval 1792, 1795, Reimarus and Reimarus 1798, Blackwall 1835, Dahl 1885, and Wagner 1894. The American clergyman H. C. McCook (Fig. 1.1) (in addition to providing distinguished service in the American Civil War and writing popular children's books on insects, religion, and leaf cutter ants) described many details of natural history and behavior in unprecedented detail (McCook 1889). The nineteenth-century giant of spider taxonomy, Eugene Simon, also contributed scattered but precise observations of webs (Simon 1892–1903). Also important, with acute observations and the first collections of good photographs of webs, were the books of Emerton (1902), Comstock (originally published in 1912, revised in 1967), and Fabre (1912).

Somewhat later, perhaps the two best naturalist observers of spider orb webs and of building behavior were the Irish medical doctor and explorer Maj. R. W. G. Hingston (Hingston 1920, 1929, 1932) and the Danish school teacher Emil Nielsen (Nielsen 1932) (Fig. 1.1). Hingston parlayed military assignments in India, Pakistan, and

Iraq, and subsequent expeditions to British Guyana and other sites, into opportunities to observe tropical spiders (and other animals) and describe their behavior. His careful observations of the details of web construction, his incisive simple experiments, and his thoughtful analyses are particularly striking in light of his energetic physical undertakings, which included participation in an expedition to climb Mount Everest and scaling tropical trees to collect specimens on the trunks and in the canopy. The other giant contrasted in many ways. Emil Nielsen was a Danish secondary school teacher who did not travel to exotic places to study spiders, but instead made thorough, careful comparative studies of web-building spiders in Denmark. He published an especially thorough book-length study in Danish and, mercifully for English-speakers, combined it with a condensed companion volume in English (Nielsen 1932). Nielsen's work on the parasitoid wasps that induce changes in the web-building behavior of their host spiders was also exceptionally thorough and insightful. Both Hingston and Fabre discussed general questions regarding "instinctual" behavior and the limits of the mental and behavioral capabilities of invertebrates, comparing them with other animals and humans.

Other especially important general studies in the field of webs and their diversity were those by Herman Wiehle (1927, 1928, 1929, 1931) in northern Europe and the Mediterranean, Hans Peters (1931, 1936, 1937, 1939, 1953, 1954, 1955) in Germany, the Mediterranean, and tropical America, and B. J. and M. Marples in Europe and Oceania (Marples and Marples 1937; Marples 1955). Additional laboratory studies of orb web construction behavior during this and the following generation mostly centered on two European araneids, *Araneus diadematus* and *Zygiella x-notata* (e.g., Peters 1931, 1936, 1937, 1939; Tilquin 1942; König 1951; Mayer 1952; Jacobi-Kleemann 1953; Witt 1963; Le Guelte 1966, 1968; Witt et al. 1968), and the American uloborid *Uloborus diversus* (Eberhard 1971a, 1972a). Studies of the webs of isolated species continued, and began to include more tropical species (e.g., Robinson and Robinson 1971, 1972, 1973a, 1975, 1976a).

Up to about 1960, one might say that the history of spider web studies was pretty typical of many topics in zoology. Kaston (1964) presented a pre-cladistic hypothesis concerning spider web evolution, based almost entirely on observations of North temperate zone species. But long after the time had seemingly come for a detailed general, in-depth summary of what was known about webs and construction behavior, it was still missing. Theodore Savory produced a short general popular book (Savory 1952) with an admirable tendency to emphasize general questions, but with many undocumented and sometimes imprecise statements. Peter Witt, Charles Reed, and David Peakall wrote a more scholarly book (Witt et al. 1968), but it was limited almost entirely to laboratory observations of two species, *Araneus diadematus* and *Zygiella x-notata*. An important summary published 24 years later continued this unfortunate tradition, specifically omitting data on any species other than *A. diadematus* (Vollrath 1992). Ernst Kullmann published large numbers of beautiful photographs of the webs of many heretofore unstudied species (Kullmann 1971/1972, 1975; Kullmann and Stern 1981) but did not produce a general overview. An edited book (Shear 1986) provided detailed and thorough summaries of some aspects of the webs and building behavior in particular groups (e.g., Coyle 1986; Lubin 1986) and included a brief summary (Shear 1986), but there was no synthetic, in-depth overview. A more recent general, synthetic book (Craig 2003) emphasized chemical and physical properties of silk, rather than behavior, but was unfortunately largely confined to summarizing the author's own studies (Vollrath 2003; Higgins 2004).

A few large, general papers and chapters giving overviews of webs and building behavior have appeared recently. Some have emphasized their use as taxonomic characters (Griswold et al. 2005; Kuntner et al. 2008a) and have tended to ignore function. Others give more general reviews (Viera et al. 2007; Blackledge et al. 2009c, 2011; Harmer et al. 2011; Herberstein and Tso 2011) but have been necessarily incomplete, especially with respect to behavior and to web designs other than orbs. A thorough, book-length review of spider webs, and of their functions and construction behavior, has yet to be written. This book attempts to fill that niche.

The lack of a general summary on spider webs is having unfortunate consequences. Several aspects of the current academic environment favor neglect of older work: the relatively greater ease of obtaining electronic copies of more recent as opposed to old papers; the widespread emphasis in hiring and promotion decisions on quantity rather than quality of publications; and trends that favor claims to originality over thorough historical research and integration of new data with prior knowledge. Some recent papers on spider web biology have carried the modern trend of neglecting previous work to new heights (or depths), either lacking relevant citations of earlier literature, or simply citing one or two recent papers or a review to "cover" all other previous work. Here are four examples. The clear leader in the field of orb web construction biology in the 1950s and 1960s was Peter Witt; yet three large recent reviews cited only 3, 1, and 0 of his publications. Only one of these three reviews even cited the two old but revolutionary papers, Riechert and Cady (1983) and Wise and Barata (1983), that up-ended most previous attempts to relate web designs to inter-specific competition and the kinds of prey that they capture. In a squabble over whether one group of spiders (theridiids) are capable of the highly coordinated movements that are used in the specialized "cut and reel" behavior (Fig. 6.3) (Benjamin and Zschokke 2002, 2003, 2004; Eberhard et al. 2008a), none of the authors cited the clear, complete older descriptions of this behavior in theridiids (Marples 1955; Bradoo 1972). And none of several recent major reviews of the vexed problem of stabilimentum function even cited the 1932 book of Hingston, in which he devoted 60 full pages to stabilimenta in tropical species.

I believe there is a real danger that much earlier knowledge is being lost. A comprehensive summary may be more important in spider webs than in some other fields. I have thus attempted to minimize citing only recent summaries (which, of course, are often less than complete, and sometimes include imprecise or even erroneous interpretations of previous work). I have made a special effort to be more complete with non-orb webs, because they have been less-studied, and show some previously unrecognized patterns (e.g., Figs. 1.2, 1.3).

Nevertheless, I am certain that I have not covered the sprawling published literature completely. In the best of all possible worlds, with the best of all possible minds and memories, and fluency in other languages (especially German and Japanese), I would have liked to have rendered this service. But I have no illusions of having made a truly thorough review (in any case, it would make very dull reading). I find that even when I read papers a second or third time, I routinely find additional important points that I missed previously; I have had quite different interests and ideas in my head at different stages of this project. My review of recent literature tailed off in late 2016. The MS was delivered to the press in August 2017, and after that I mostly only completed "in press" citations as they were published. I apologize in advance to all the authors whose work I cited only partially or missed completely, and warn the reader of this limitation.

I have also added bits of unpublished information that I have accumulated during nearly 50 years of observing spiders and their webs (cited as "WE"). One result of writing this book was to make me aware of many lacunae in our current knowledge, and the book has launched me on several projects along the way that were designed to fill some of these holes. Several species that are abundant where I live, including the filistatid *Kukulcania hibernalis*, the araneid *Micrathena duodecimspinosa*, the tetragnathid *Leucauge mariana*, and the uloborid *Zosis geniculata* are thus over-represented in the text. I have also pointed out many remaining gaps in our understanding, and compiled outstanding questions and promising research topics, in the online Table O10.1 in the hope that this will be useful to future researchers.

1.4 EMPHASIS ON BEHAVIOR

The emphases here on behavior and behavioral evolution come naturally. Orb web construction behavior has long been considered an iconic example of stereotyped innate behavior, and details of orb construction behavior provide especially valuable traits for higher-level taxonomy (Eberhard 1982; Kuntner et al. 2008a). In addition, orbs offer unusually favorable conditions for studying behavior at especially fine levels of detail and precision and testing general ideas (Ades 1986). This is because an orb web is an exquisitely precise, detailed record of many aspects of the spider's construction behavior (Fig. 1.4). Both the animal's behavior and some of the crucial stimuli that were used to guide that behavior are "frozen" in the pattern of lines in the web (Zschokke and Vollrath 1995a). Behavioral decisions are easily recorded in photographs, because orbs are generally two-dimensional. In addition, the spider is guided largely by the positions of other web lines as she builds, especially during temporary spiral and sticky spiral con-

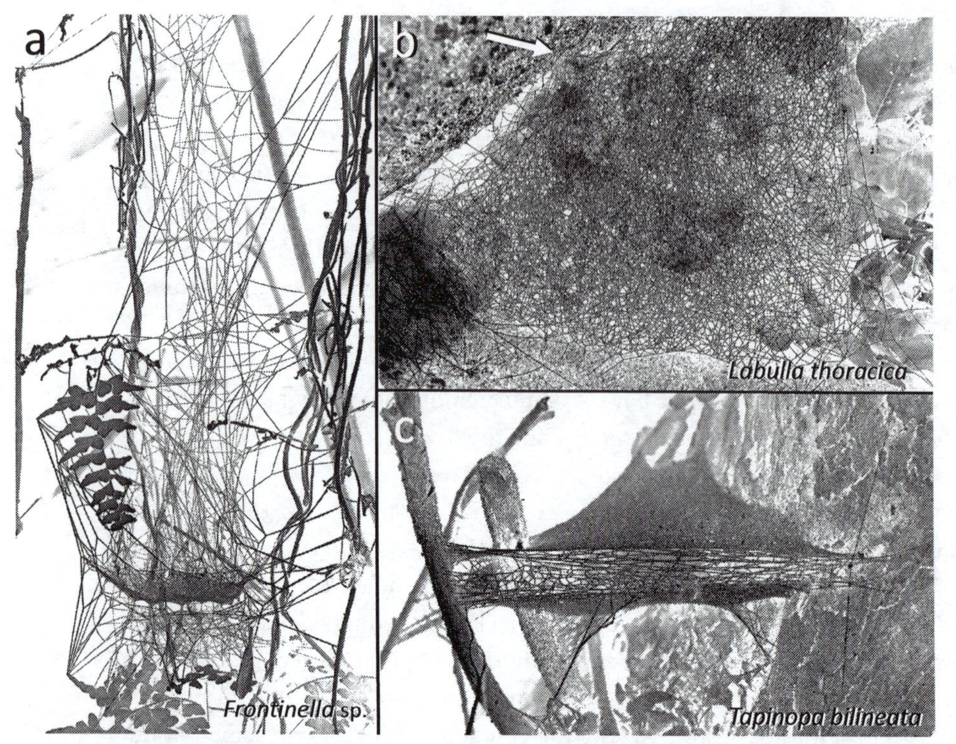

Fig. 1.2. Beautiful complexity and diversity is not limited to orbs, as illustrated by the aerial sheet webs of the linyphiids *Frontinella* sp. (*a*), *Labulla thoracica* (*b*), and *Tapinopa bilineata* (*c*). The *F.* sp. web had an unusually tall knock-down tangle above the dense horizontal sheet where the spider rested, a nearly clear space just below where a few lines pulled the sheet sharply downward into a cup, and a sparse tangle with a hint of a planar structure below this which may have served to protect the spider from predators. The "naked" horizontal sheet of *L. thoracica* (*b*) lacked tangles; it had a retreat at one edge (arrow) and a more open mesh near the edges. The *T. bilineata* sheet (*c*) also lacked tangles, but seen from the side it proved to be a double sheet loosely joined at the edge, forming a cavity that enclosed the spider (photographs by Gustavo Hormiga).

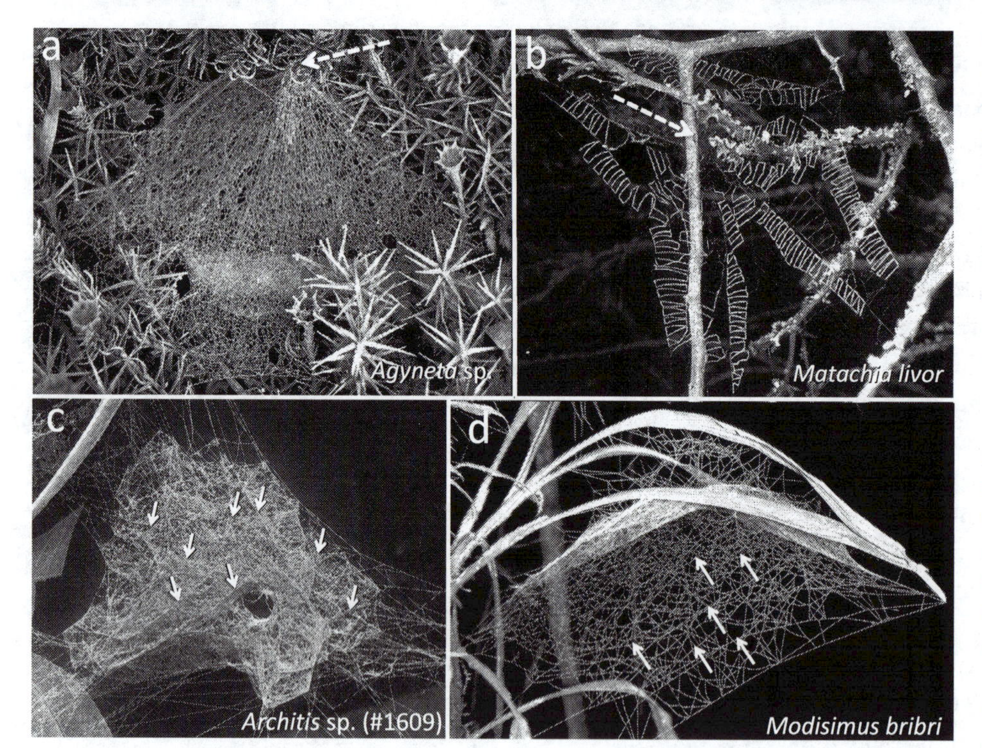

Fig. 1.3. This sampler illustrates some kinds of organization that are as yet only poorly documented in non-orb webs of mature females, including lines radiating from the spider's retreat (dashed arrow) in a linyphiid sheet web (*Agyneta* sp.) (*a*), zig-zag sticky lines centered on a retreat (dashed arrow) in a desid web (*Matachia livor*) (*b*), and skeletons of long, more or less straight lines (arrows) that support the sheet webs and extended beyond some edges in the pisaurid *Architis* sp. (whose web is unusual in having two, approximately vertical sheets, connected by a short tunnel) (*c*), and the pholcid *Modisimus bribri* (*d*) (see also fig. 5.21). The zig-zag sticky lines (strong white lines) in the web of *M. livor* imply two patterns seen in the construction behavior of many non-orb spiders. The sticky lines must have been built after the non-sticky lines (or at least those between which the zig-zags are built). And construction of the sticky lines, must have occurred in discrete bouts in different sectors of the web. Web construction behavior has never been observed directly in this species (or in any other desid), so these inferences remain unconfirmed (*a* photograph by Gustavo Hormiga, *b* courtesy of Brent Opell).

struction, so many of the stimuli that guide her behavior can be deduced from web photos (chapter 7). Other possible stimuli from the web that would be more difficult to quantify from photographs, such as tensions and patterns of vibrations, appear not to provide important stimuli, at least during the latter stages of web construction (section 7.3.2.6). Instead, orb construction is probably largely guided by aspects of an orb such as distances and angles between lines and that can be measured precisely in photos. The precision with which both potential stimuli from an orb web and the spider's behavioral responses can be measured is something that behaviorists studying most other animals can only dream about. And these fine analyses can be conducted on behavior that was performed in the natural context of building its web, rather than some highly constrained or artificial situation.

Orb weavers also offer additional practical advantages for a behaviorist. Because of the limitations of the spider's sensory world and the great importance of stimuli from the lines in the web in guiding her building behavior, simple experimental manipulations of these lines can reveal how they guide the spider. With only moderate care, it is possible to observe a spider closely without alarming it. Spiders tend to build their webs at highly predictable sites and times, and construction behavior often involves many highly repetitive behavior patterns, allowing an observer in the field to observe fine behavioral details over and over, and eventually understand them. Repetition also offers unusual opportunities to evaluate the frequency with which spiders make errors and examine the conditions which promote errors (section 9.7), difficult topics that are presently uncommon in studies of animal behavior but which may be important in understanding how behavior evolves (Eberhard 1990a, 2000a). While these advantages are especially clear in orb webs, non-orbs also offer scope (which has hardly been exploited) for deductions about behavior (Fig. 1.4), including how behavior is organized and the effects that this organization has on behavioral evolution and the mental processes involved in compensating for previous errors (Figs. 5.3, 5.4). Variability is the rule in orbs, even when sequential webs are built by the same individual at exactly the same site (e.g., Fig. 7.41).

In sum, I believe that orb web construction behavior can be in the vanguard as studies of animal behavior move past the current, largely typological stage (focused on "the" behavior of a species), and on to population thinking in which variation and its consequences are described and analyzed (Mayr 1982). For instance, orb webs permit one to employ huge sample sizes to study the sizes and frequencies of true behavioral errors (section 9.7) and possible lapses of attention (Eberhard and Hesselberg 2012), as well as variation in adaptive flexibility in behavior (which in most species has also been treated typologically, such as "the" behavior that is performed under condition *x*, and "the" behavior under condition *y*). Another consequence of the frozen behavior is to facilitate study of the multiple influences on the same decision. For instance, ten or more different variables influence each decision that an orb weaver makes when she attaches a loop of sticky spiral to a radius (section 7.20.1). While the behavioral record offered by an orb is incomplete (for instance, more than one type of leg movement can produce a particular web design) (Vollrath 1992), combining observations of webs with observations of the building behavior itself can nevertheless provide a very unusual opportunity to study behavior in great detail. One objective of this book is to provide a solid base on which a new generation of behavioral studies can be built.

Such "mechanistic" details, which used to be standard fare in animal behavior studies, are currently out of style; many recent studies of behavior emphasize instead ultimate causes—the effects of natural selection on behavior in terms of reproductive payoffs. Sooner or later, however, behaviorists will have to return to the study of behavioral mechanisms in order to understand how selection actually brings about its effects on behavior; in the end, the mechanisms are the traits on which selection acts. The field of morphological evolution has already entered this sort of transition, with a rebirth of interest in the previously neglected effects of developmental processes (Nijhout 1991; Raff 1996; West-Eberhard 2003).

Orb web spiders have often been thought of as small automatons that are tightly programmed to build the same species-specific webs day after day, and that are completely unconscious of their own work (Fabre 1912; Hingston 1920, 1929). It is true that spiders need no previ-

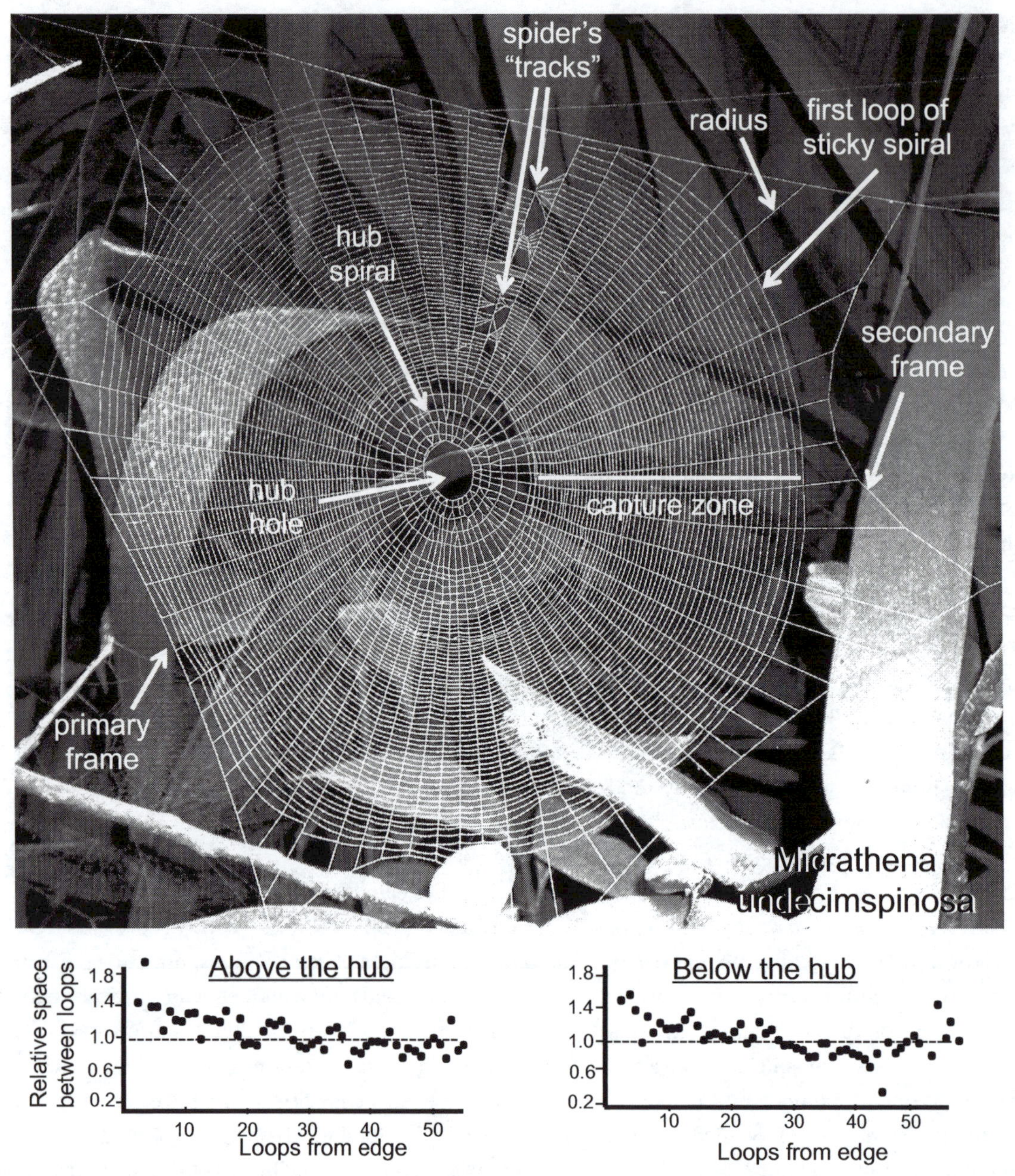

Fig. 1.4. This orb of a mature female *Micrathena duodecimspinosa* illustrates both the beautiful geometric regularity for which orb webs are famous, and subtle patterns of *irregularity*. The graph below shows that although the spaces between the loops of sticky spiral are highly regular, those nearer the outer edges are considerably larger than those nearer the hub (except for the innermost few loops below the hub). This pattern, common in the orbs of several families (Eberhard 2014; see fig. 3.6), leads to the proposal that an orb should be viewed not as a single unit prey trap, but rather as a combination of traps with different prey-catching properties in different portions of the orb (section 4.3). To give an idea of scale, this web of an approximately 1 cm long spider, contains the following: 18.7 m of sticky spiral; 9.6 m of radii (including the set of radii that she broke and reeled up, then later reingested when she removed the center of the hub); 0.4 m of hub loops; 1.6 m of frame lines; and 1.5 m of anchor lines. It has 3840 attachments (319 of non-sticky lines to each other, 3521 of sticky to non-sticky lines). Not bad for about 40 min of work.

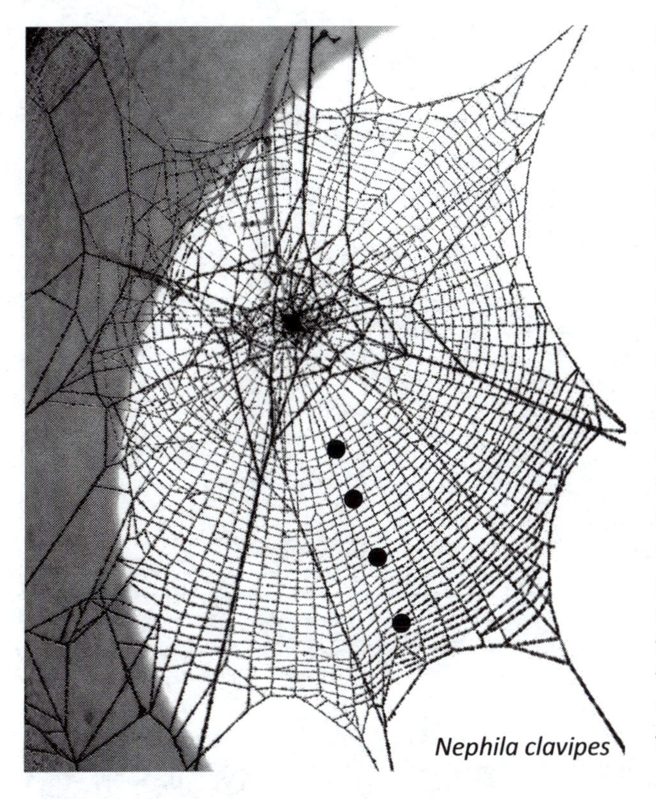

Fig. 1.5. This was the first orb web built by a spiderling of the nephilid *Nephila clavipes* after emerging from the egg sac; it illustrates an additional pair of themes of this book—the apparently minor role of learning in web construction behavior, and the blind spot of adultophilic arachnologists. The black dots indicate intact loops of the non-sticky temporary spiral, an unusual trait that is present in this first orb of the spider's life just as in the orbs of adults (see fig. 4.4). The small tangle in the foreground (which obscures the spider resting at the hub) is another unusual trait also found in adult webs. The web's highly regular design and its similarity to the orbs of mature females show that these spiders need little or no learning to build an orb. The basic similarity between the orb designs of young spiderlings and adults, typical of most orb weavers, is also striking in view of the differences in their sizes: the spiderling weighed on the order of 1 mg, while adult females weighed on the order of 1000 mg. They must confront very different ecological problems such as distributions of prey sizes and abundances, possible websites, predators, and parasites. Arachnologists studying web designs have traditionally analyzed the costs and benefits of orb designs for the adult, and neglected the behavior and ecology of younger stages. This "adultophilic" bias needs to be overcome, because selection must act on web designs throughout a spider's life (see section 3.2.14).

ous learning to produce an orb, and indeed, spiderlings that have just emerged from the egg sac can spin complete webs on their first try (Fig. 1.5); and there is little indication that their behavior improves with practice (section 7.19). As described long ago by Hingston (1920), spiders seem unable to make even what would seem to be simple adjustments to some experimental alterations to their webs: "Introduce difficulties in its circuit . . . , build up obstructions to impede the blind routine, and the spider can do nothing to overcome them; it can only struggle in its course. It can appreciate none of these difficulties; it can understand none of these obstacles; all it can do is but circle on" (p. 134).

This simple automaton view has gradually eroded, however (see summaries in Herberstein and Heiling 1999; Herberstein and Tso 2011; Hesselberg 2015), and one of the themes in this book is that it must now be discarded. Spiders clearly show substantial, multidimensional flexibility in their web construction behavior (see, for example, Figs. 1.6, 6.2, 7.36). It is possible that these adjustments may themselves result from "prewired" alternative behavior patterns, implying that the appropriate change in views is from a simple to a complex automaton. It is also possible, however, that spiders have some sort of primitive understanding of the mechanical consequences of their building behavior (Eberhard 2019b).

The original impression of simplicity and invariability that was associated with the early oversimplified descriptions of orb construction behavior has been followed by a gradual accumulation of exceptions to the behavioral "rules," and an appreciation of previously unnoticed or unreported subtlety and variations. This process is incomplete, and our present understanding is undoubtedly still overly typological. In addition, there is obviously much still to be learned, even about well-studied species (Fig. 1.7). The need to deal with such variation has increased the difficulty of writing this book. I have struggled to avoid two extremes: overly simplistic descriptions of behavior and webs that are clear and easy to understand but seriously underestimate variation; and overly complex descriptions that are more nearly correct but are too convoluted for even the most motivated reader to follow.

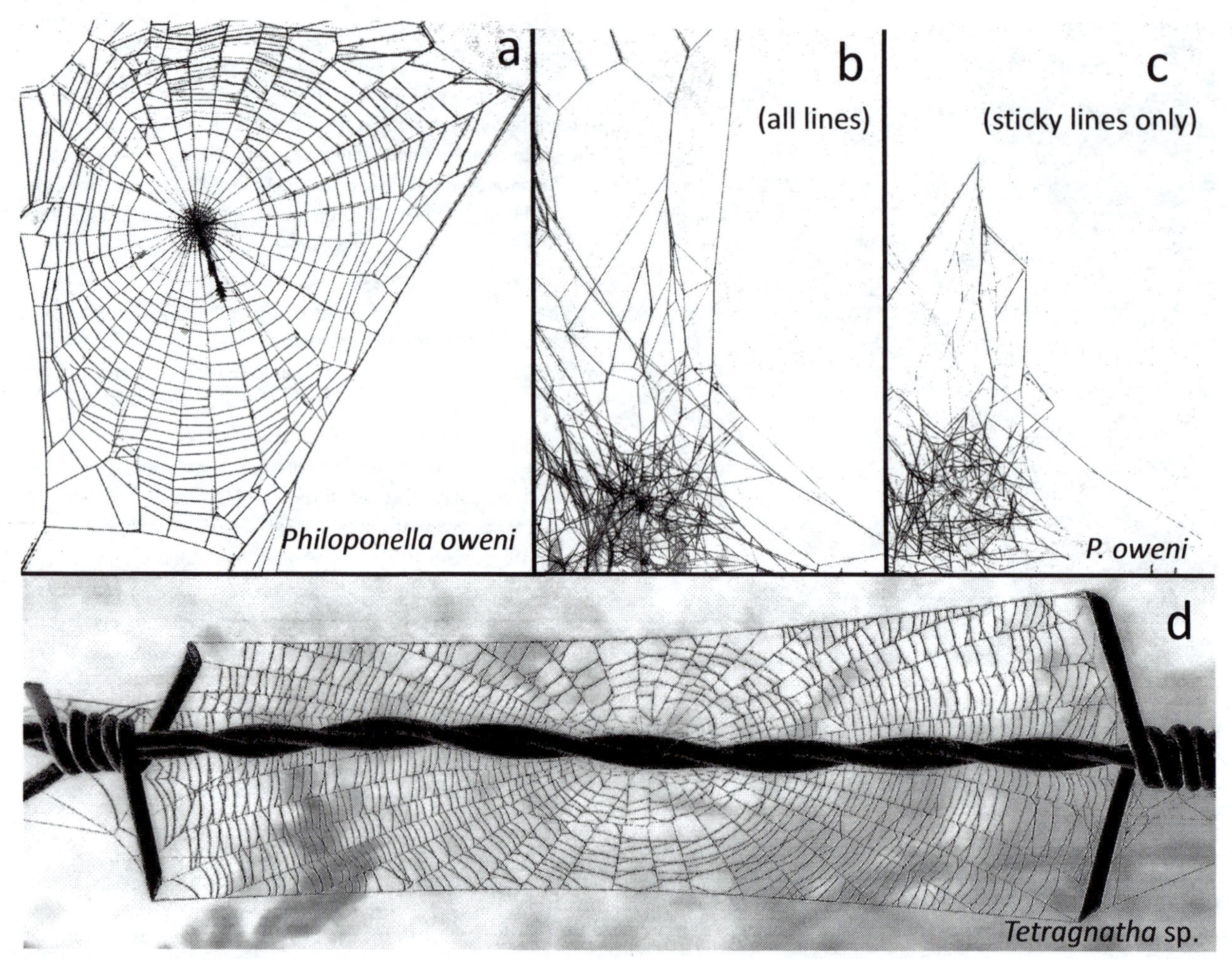

Fig. 1.6. These webs illustrate the behavioral flexibility of orb weavers, a major theme of this book. The two webs built in captivity by *Philoponella oweni* (UL) (*a-c*) illustrate "alternative phenotypes": *a* is a complete horizontal orb with no tangle; *b* and *c* show a tangle web seen from above. In *b* all lines in the web are visible, while in *c* only the sticky lines are visible (the web was coated with white powder and photographed in *b*, and then jarred repeatedly to remove the powder from all non-sticky lines and photographed again in *c*). Other webs of this species included both an orb and a tangle; some individual spiders built all three web types. A second type of flexibility (*d*), in which spiders adapt their web designs to the environments in which they find themselves, is illustrated by the horizontal web of an immature *Tetragnatha* sp. on a strand of barbed wire (the elongate spider was hidden under the wire at the hub).

1.5 THE SCOPE OF THIS BOOK AND TACTICS IN PRESENTATION

A great deal is known about spider webs, and in order to avoid an incomprehensible sprawl of endless lists, I have often emphasized only more general trends and ideas, attempting to avoid giving so much detail that the reader loses view of the most interesting aspects of the forest for the trees. Much detailed data is relegated to tables and online appendices. So much is now known about spider webs that a truly thorough summary of current knowledge would be crushingly dull. I have tried to make the text accessible to a general reader by focusing on particular illustrative examples rather than making exhaustive discussions, and by filling in some details with extensive figure captions. In many places I have avoided giving long strings of citations to document a particular statement, citing instead another section of the book where previous work on the topic is discussed more thoroughly.

I have attempted to evaluate previous studies critically, because there is both imprecision and controversy in some published accounts. Presenting convincing and fair discussions of conflicting ideas sometimes posed

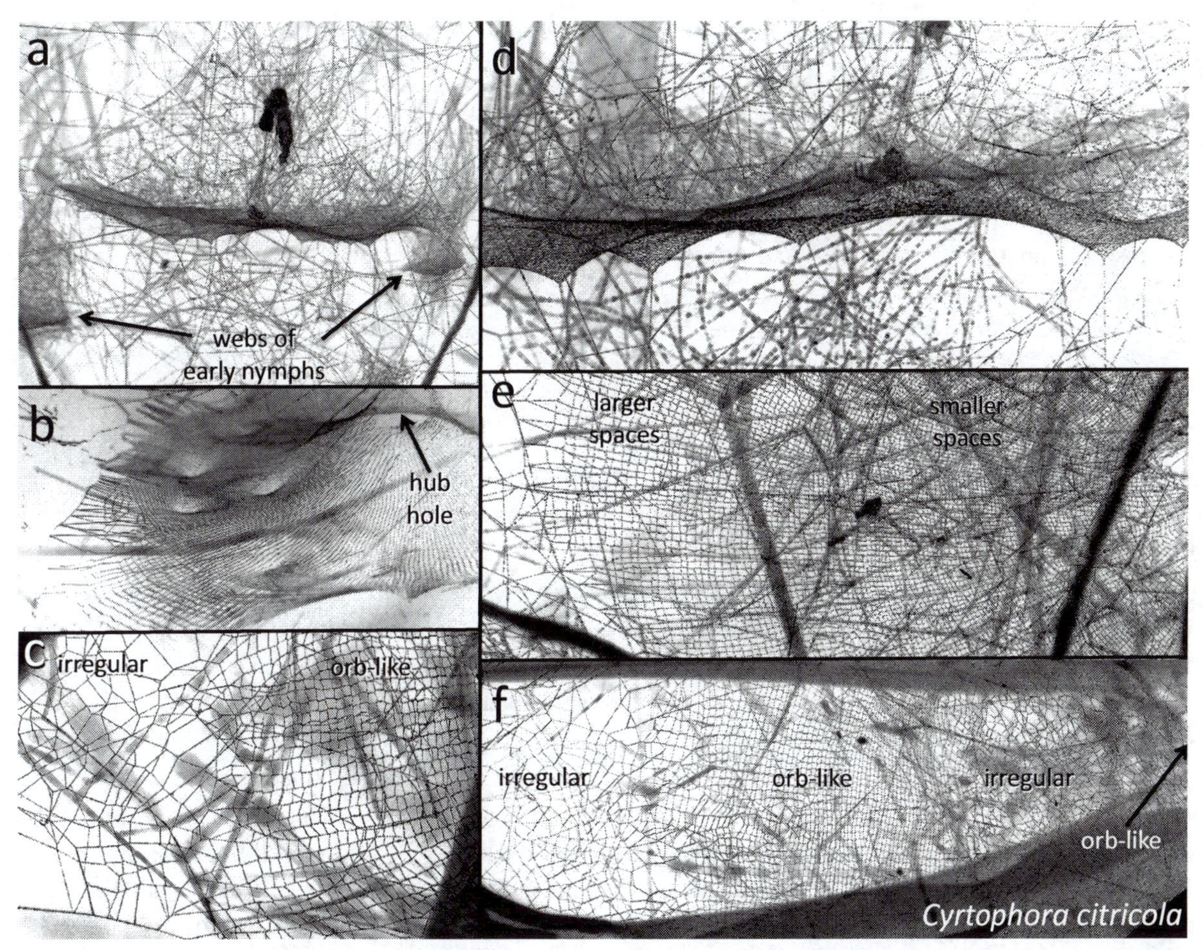

Fig. 1.7. A serious historical problem in spider webs studies has been the tendency to use typological descriptions of web designs that overlooked details and variations. This kind of problem is illustrated in these images of webs of *Cyrtophora citricola* (AR), whose general design was described nearly 100 years ago (Wiehle 1928) and has subsequently been studied extensively (Blanke 1972, 1975; Kullmann 1972; Lubin 1973, 1974, 1980; Sabath et al. 1974; Berry 1987; Peters 1993a; Rao and Poyyamoli 2001). Nevertheless, several previously under-appreciated details reveal further sophistication of these webs as traps. Early descriptions mentioned a dense tangle of lines above and below a horizontal, planar sheet (see *a*, a largely finished, 4-day old web), and a clear organization of the lines in the sheet, with frequently bifurcating radial lines that converged on a central hub that were connected by an uniformly spaced, tight non-sticky spiral. The tangles above and below the sheet were said be connected to its outer edges, and also, in the case of the upper tangle, to the hub. But closer examination adds several details. An image of this same web after only a single night of building (*b*; in this image, in contrast with the others, the web was not dusted), shows that the tangle was built by stages. The spider first built only a sparse tangle above and below the radially organized sheet. The sheet was not planar, but instead had deep, widely spaced downward-directed dimples where a few widely spaced, taut lines connected it to the tangle below; it also had a large hole at the hub, through which the spider climbed to attack prey that were trapped in the upper tangle. The "dimple" lines undoubtedly increased the tension on the sheet itself, but nonetheless left it sufficiently lax for the spider to fold it slightly with her legs, partially enclosing prey as she attempted to grasp and bite them from below the sheet. The spider added to the tangles on subsequent nights, especially above the sheet where many additional, relatively lax lines attached to the top of the sheet scarcely pulled it upward (*d*). The dense array of looser lines just above the sheet may help retain prey on the sheet's upper surface. The sheet itself also showed additional patterns. The circular lines were not evenly spaced as in previous descriptions, but those farther from the hub were farther apart (*e*). In addition, the spider extended portions of the sheet with irregularly-oriented lines attached to its outer edge (*c*, *f*). Rarely, in webs built at sites where there were only elongate open spaces (*f*) this irregular sheet was extended still farther, first by an additional "orb-like" portion with regularly arranged, more or less parallel "peripheral radial" lines and tightly spaced "peripheral spirals," and then by an additional extension with irregular lines (the central, radially arranged sheet and its hub are out of sight on the right). These "peripheral radii" did not converge in a hub-like patterns. The reader should be forewarned that many or probably most web descriptions in this book likely err in being excessively typological. The webs in *a* and *d–f* were of adult females, those in *b* and *c* were of intermediate-sized spiderlings; the arrows in *a* indicate webs of second instar spiderlings built in the lower tangle of an adult female's web.

a challenge, because the data and the arguments were complex and tedious. Nevertheless I wanted to avoid two tactics that are unfortunately common in recent publications: simply accepting all published claims at face value; or dismissing (or even ignoring) some views without giving any justifications. These traits can make for smooth prose, but are unfair to both readers and previous authors. Advances in science often depend on disproving hypotheses, so the reasons for discarding previous ideas are important (Popper 1970). And some imprecise ideas, if not specifically challenged, acquire lives of their own when they are repeated uncritically by subsequent authors. In a few, particularly complex topics I give only simplified discussions in the text, and detailed critical discussions in appendices. Even this tactic has its limits (there is simply not enough room, for instance, for a thorough discussion of every single function that imaginative arachnologists have proposed for the function of silk stabilimenta) (see section 3.3.4).

Another common practice that I have tried to reduce is talking about "the" spider or about "the" web of a species. I usually give the genus and species name of each species that is mentioned. My attempts to be more specific are motivated by discoveries of how much variation exists both within and between species. To help protect readers from drowning in a sea of unfamiliar names, I include the family affiliation of each species mentioned, and use the following abbreviations for the families that are most commonly mentioned (the numbers in parentheses give the approximate numbers of described species and genera [Platnick 2014], and thus indicate some aspects of their biological importance): AN = Anapidae (154; 38); AR = Araneidae (3,045; 169); DEI = Deinopidae (60; 2); LIN = Linyphiidae (4,490; 591); MYS = Mysmenidae (131; 23); NE = Nephilidae (61; 5); SYM = Symphytognathidae (69; 7); TET = Tetragnathidae (967; 47); THD = Theridiidae (2,420; 121); TSM = Theridiosomatidae (106; 18); and UL = Uloboridae (271; 18) (Platnick 2014). I have virtually ignored, however, the large family Linyphiidae, because an extensive review of their webs will be published elsewhere (G. Hormiga and W. Eberhard in prep.). It has not been possible during the protracted process of the production of the book to keep up with the very latest (not necessarily stable) taxonomic opinions; for instance, Nephilidae has been relegated to a subfamily of Tetragnathidae or Araneidae in some recent studies.

When the option was available, I have chosen web photographs of species not illustrated in previous publications, in order to increase taxonomic coverage. I often find that the patterns of web lines are easier to perceive in negative images, so I have used many negatives of web photos.

1.6 EVOLUTIONARY HISTORY AND PHYLOGENY

The probable evolutionary history and the phylogenetic relations between the major groups of spiders form an important general context for much of this book. Given the current uncertainties regarding phylogenetic relations (below), perhaps the most important data concern the evolutionary age of silk use in spiders. Fossils indicate that webs are quite ancient (Fig. 1.8) (Vollrath and Selden 2007). Judging by fossil spinnerets, silk production was already well-developed in the early Devonian (approximately 390–400 million years ago [Mya]). Webs themselves seldom leave fossils, but the taxonomic affinities of fossil groups and the behavior of their modern relatives indicate that prey capture webs (which probably extended the spider's sense of touch near the mouth of the burrow) were present on the order of 300–310 Mya (in Mesothelae). Funnel weavers (hexathelid mygalomorphs) were present about 240 Mya, and possible araneomorphs about 230 Mya (by this time, flying insects are abundant in the fossil record). Sticky (cribellum) lines cannot be confidently attributed to spiders before about 165 Mya, at which time orb-weavers (deinopoids and *Nephila*) were already present (Selden et al. 2011, 2013); the earliest known araneoids date from a little later, about 140 Mya (Fig. 1.8). Dates determined from molecular data are in general accord but somewhat older (Dimitrov et el. 2016). To put these dates into perspective, orb webs had apparently already evolved before or at the same time when birds originated as a branch of dinosaurs (about 150 Mya), the first of the major groups of modern mammals were first appearing (about 150 Mya), and before flowering plants first appeared (about 132 Mya) (Zimmer and Emlen 2013). The ancient nature of webs, and of orb webs in particular, is

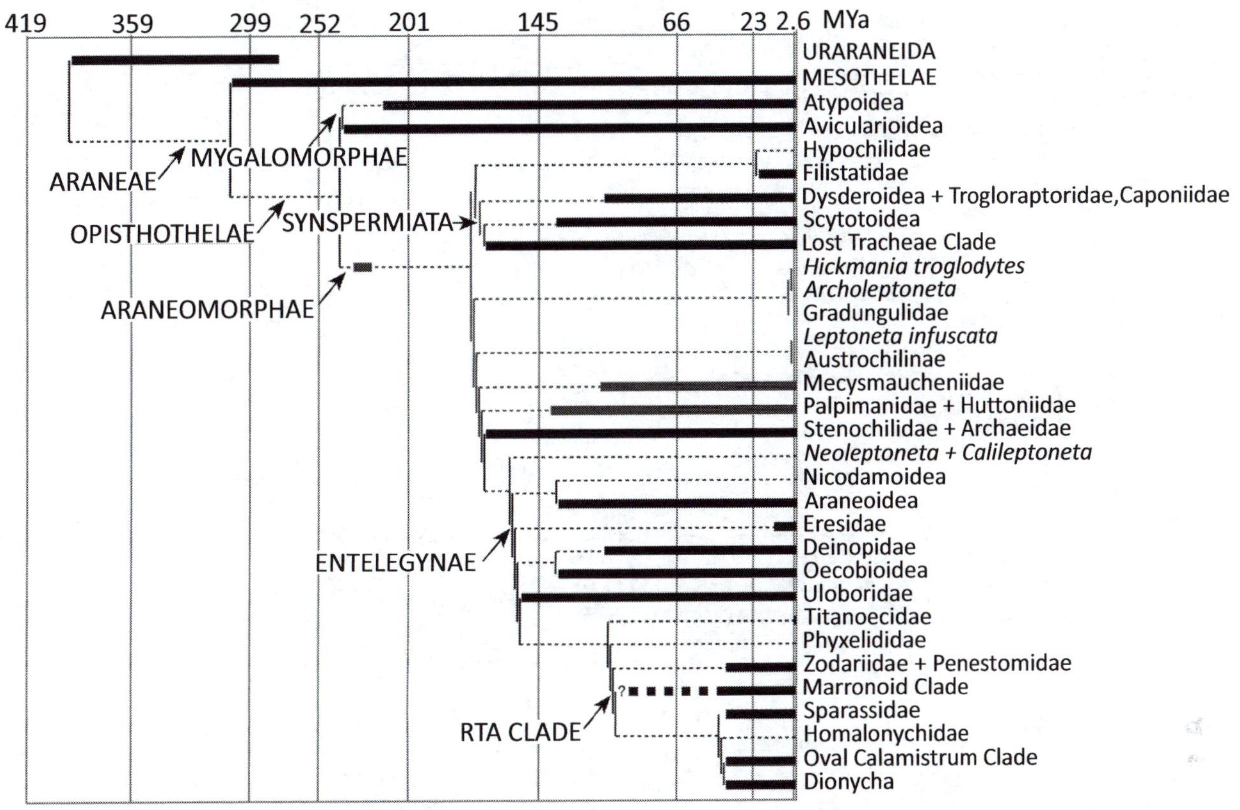

Fig. 1.8. This summary of fossil evidence, placed on a recent phylogeny, emphasizes the ancient nature of spider webs and orb webs in particular ("Orbiculariae") (thick lines give time ranges based on described fossil specimens; dotted lines give estimated time ranges based on ages of fossils of sister groups). Although some details of the phylogeny (and estimated time ranges) may change, several conclusions can be drawn. The Opisthothelae are, as expected, much more ancient than the araneomorphs. Webs with sticky (cribellum) lines arose early (about 230 Mya). Uloborids arose relatively early within araneomorphs, and araneoids were present soon after, thus supporting the ancient nature of orb webs. The RTA clade, in which orb webs were secondarily discarded long ago, may not have become diverse and abundant until much later (no fossils are known for the first half of its proposed existence). Several currently speciose and abundant RTA families, such as salticids, thomisids, and clubionids, are absent from cretaceous amber, where they would presumably have been trapped if they had been common (modified after Vollrath and Selden 2007, courtesy of Paul Selden).

crucial for understanding modern patterns of diversity and evolution (chapters 9, 10).

Unfortunately, although there are many extensive studies of higher-level spider phylogeny, which have used morphological, molecular, and to a lesser extent behavioral traits, many relationships of modern spiders are currently unresolved (Fig. 1.9). "Most deep clades in spider phyogenies are disputed, mainly by molecular results. Not only are new molecular studies incongruent with much 'traditional' knowledge but they are often incongruent with one another" (Agnarsson et al. 2013; p. 83). To date most molecular studies have relied on a small sample of 11 genes (for which primers were easily available). Hopefully, shifts to "next generation" techniques that permit the analysis of many more genes (e.g., Garrison et al. 2016, Fig. 1.10) will improve the quality of the molecular data. Figs. 1.8–1.15 illustrate several recent trees; probably all will change in the future.

In the face of this uncertainty, I will use some traditional groupings for descriptive purposes, without implying strict phylogenetic relationships (unless specifically stated). "Araneoidea" (a well-supported grouping in recent analyses) will refer to the orb-weaving groups Araneidae, Tetragnathidae, Nephilidae, and Symphytognathoidea, as well as to the related but probably secondarily non-orb weaving families Nesticidae, Synotaxidae, Linyphiidae, Pimoidae, and also what at the moment appears to be the sister group for all of these, the "tangle

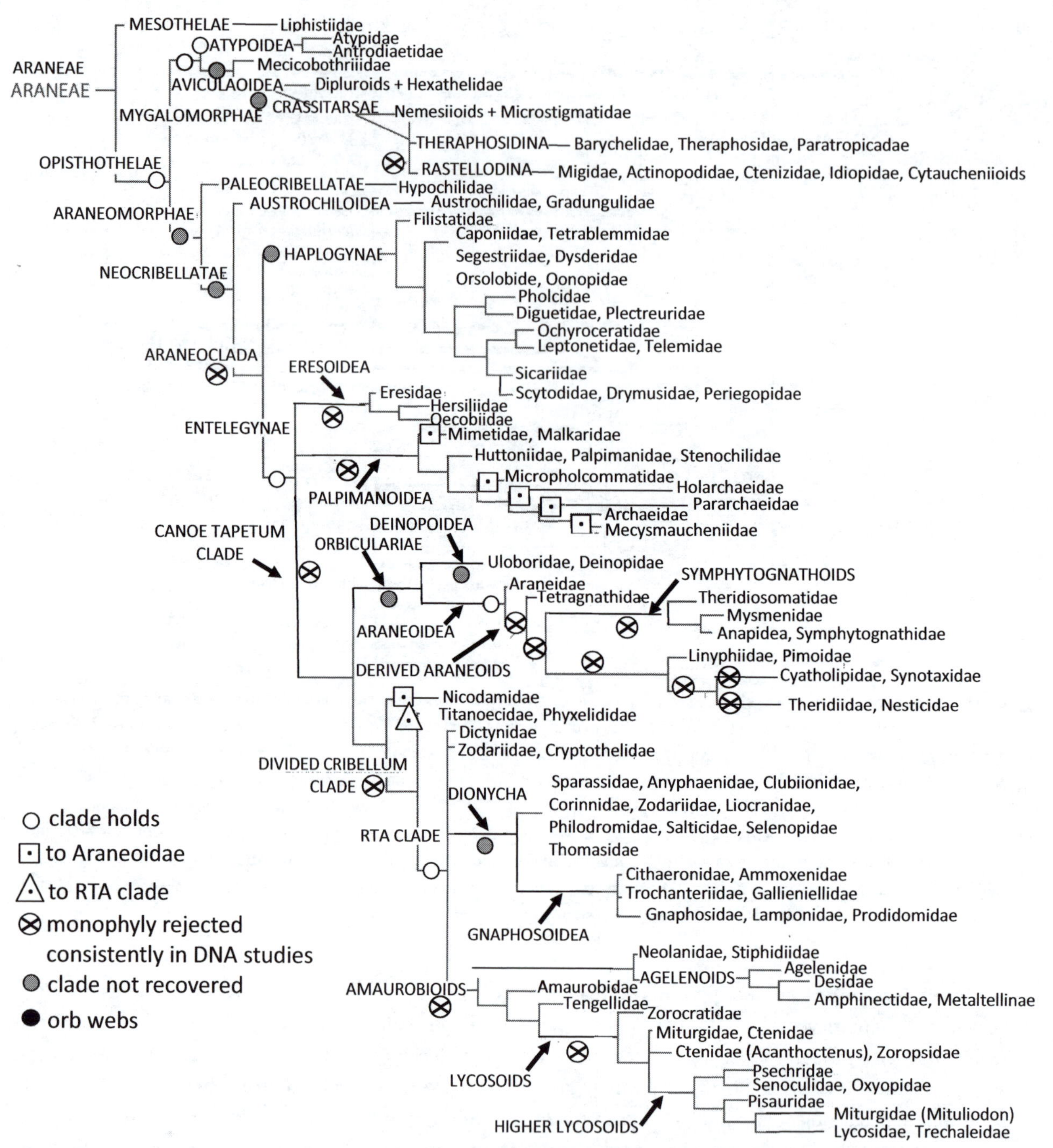

Fig. 1.9. Evolutionary perspectives are important in many portions of this book. Unfortunately, many aspects of the higher-level phylogeny of spiders are as yet uncertain, largely due to conflict between morphological and molecular data (section 1.6). This uncertainty is illustrated with this morphology-based tree published fifteen years ago (Coddington 2005) on which the conclusions from recent molecular analyses have been superimposed. Basal clades that are presently controversial due to the lack of molecular support are indicated by symbols with "x" at branching points. Few of the formerly recognized groups have strong support (white circles). Many groupings in this tree and the others in figs. 1.10–1.15 are only best guesses and are likely to change in the future (after Agnarsson et al. 2013).

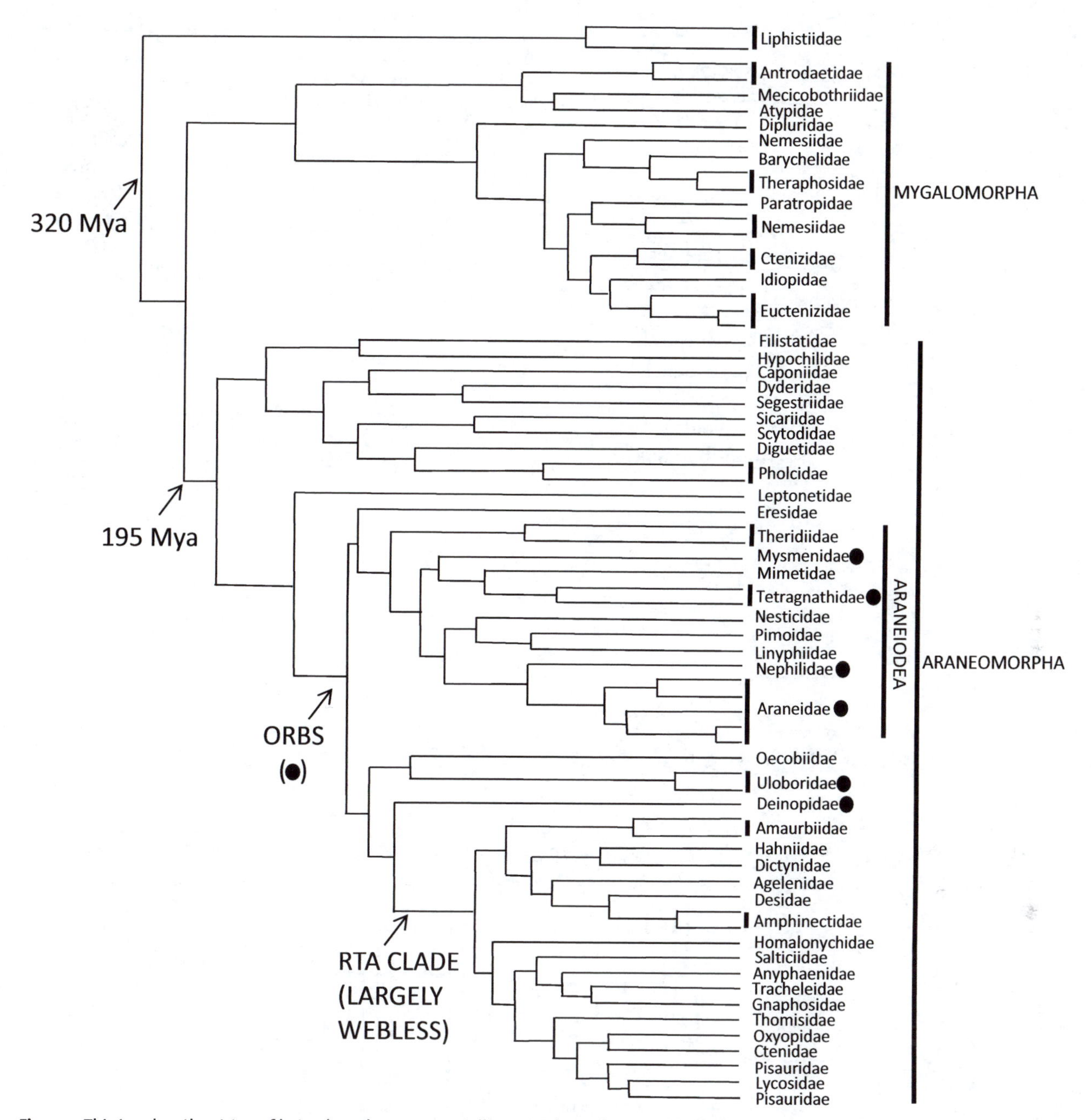

Fig. 1.10. This tree has the virtue of being based on an especially recent, large data set; black dots indicate orb webs or modified orbs (after Garrison et al. 2016).

web" family Theridiidae. "Orbicularia" will refer to the araneoids plus the cribellate orb weaving families Deinopidae and Uloboridae (this grouping appears to be paraphyletic); Deinopoidea will refer to Uloboridae and Deinopidae (as well as the fossil families Mongolarachnidae and Juraraneidae, which may or may not be closely related—Selden et al. 2015); it may also be paraphyletic; the well-supported "RTA clade" (a large set of families that lack orbs and that generally do not build webs at all) is the sister group of Deinopoidea (Dimitrov et al. 2011; Bond et al. 2014; Fernández et al. 2014). I will also employ the term "non-orbicularian" or "non-orb weaver" to designate species whose webs are not orbs; this grouping is surely paraphyletic.

Several additional, less controversial groupings are important to understanding web evolution. Mygalo-

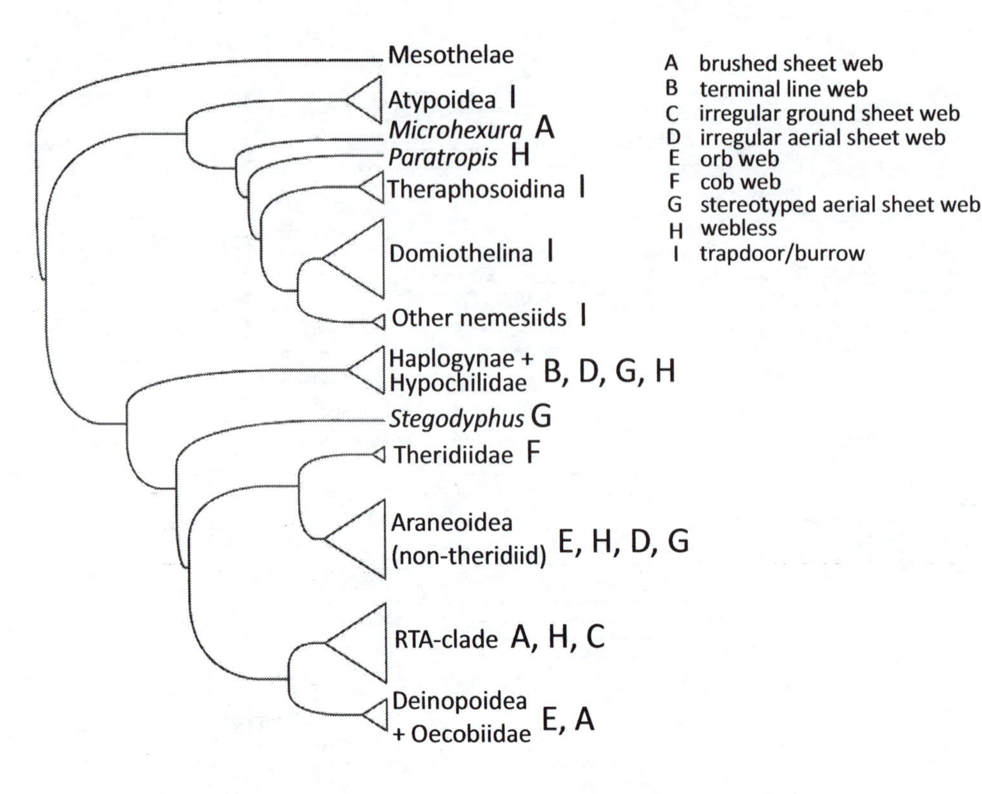

Fig. 1.11. This tentative phylogeny of orb-weaving families and their close relatives in Araneoidea is intended to help the reader place the genus names that are mentioned frequently in the text (after Garrison et al. 2016) (loosely based on Griswold et al. 1998). These characterizations of web types are seriously imprecise (see sections 9.4, 10.5.2).

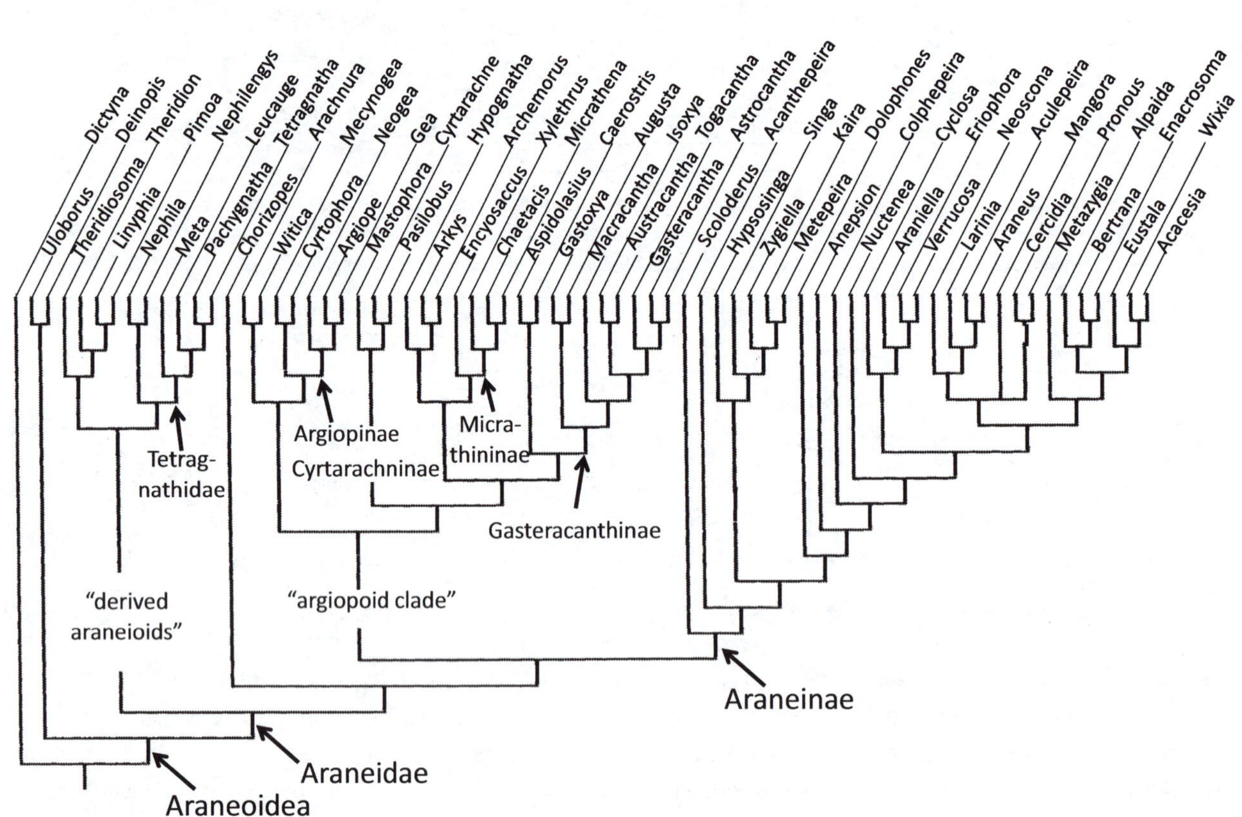

Fig. 1.12. This morphology-based phylogenetic tree has the advantage of including most of the orb-weaving araneoid genera mentioned in the text (modified from Scharff and Coddington 1997), but future data will probably suggest modifications. Probably the lineage of *Zygiella* and closely related genera originated more basally, following the splits that led to Tetragnathidae and Nephilidae; the webless genus *Arkys* may lie outside Araneidae (Gregorič et al. 2015).

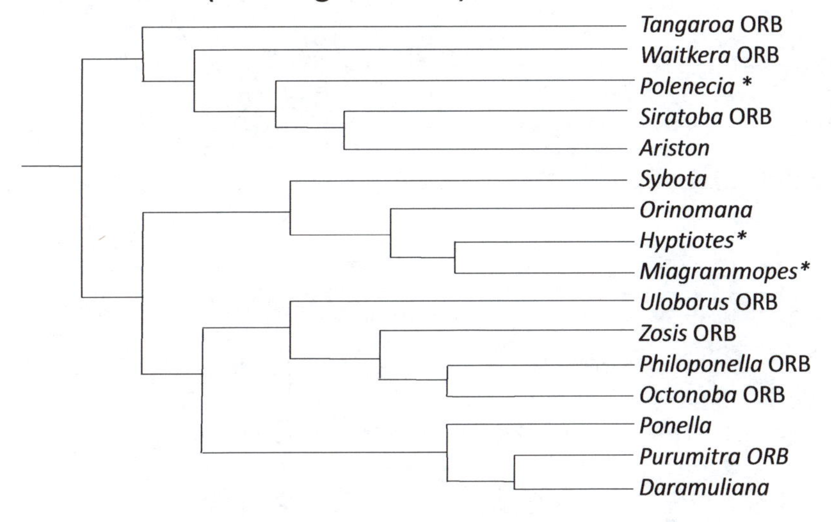

Fig. 1.13. A morphology-based phylogenetic tree for the genera of the orb-weaving family Uloboridae. The genera known to build more or less typical orbs are indicated by "ORB"; genera with substantially reduced webs are indicated with "*" (tree after Coddington 1990, and Opell 2002, with web information on *Purumitra* from F. Soley, pers. comm.).

morpha are the tarantulas and their allies, while Araneomorpha are the "true" spiders (all spiders other than Mygalomorpha and Mesothelae). At the family level, the orb weaver families Deinopidae, Uloboridae, and all of the araneoid families are "probably by and large monophyletic," though the relations between them are as yet uncertain (Agnarsson et al. 2013). Some non-orb weaver families, such as Filistatidae and Pholcidae, are also apparently monophyletic, but others, including Agelenidae, Amaurobiidae, Hahniidae, and Dictynidae, are probably polyphyletic (Griswold et al. 2005; Agnarsson et al. 2013).

1.7 TERMINOLOGY AND OTHER PROCEDURAL MATTERS

Throughout the book, the word "web" with any modifier refers to a prey capture device. I will add a modifier to indicate those silk structures that have other functions, such as egg sac webs, resting webs, silken retreats, silk tangles made by newly emerged spiderlings near egg sacs, sheets made by adult females to protect themselves and their offspring (Sedey and Jakob 1998). These distinctions are admittedly not always clean. For instance, prey that brush against silk retreat lines were sometimes attacked by salticids (Nelson and Jackson 2011; Jackson and Macnab 1989a) and sometimes became entangled there (Hallas and Jackson 1986). Egg sacs sometimes had sticky silk laid on their outer walls, as in the uloborid genera *Uloborus*, *Miagrammopes*, *Philoponella*, and *Zosis* (Lubin 1986; WE), the dictynid *Mallos hesperius* (WE), and the theridiid *Latrodectus geometricus* (Triana et al. 2012). These egg sacs were all at or near the spiders' resting sites, so it is possible that protection for the eggs may have sometimes incidentally served to capture prey. In atypid spiders, the lining of the retreat is also the prey capture web (Coyle 1986); in one atypid, *Sphodros rufipes*, the earth that was dug from the bottom of the tunnel and applied to the wall above ground as camouflage was first broken into small pieces, possibly to facilitate prey capture when the spider strikes through the wall with her fangs (WE). Individuals of the social eresid spider *Stegodyphus dumicola* and the solitary *S. lineatus* responded to the first signs of invasion of their web by ants by laying sticky silk near the point where the ants were about to invade (Henschel 1998; Leborgne et al. 2011). Using silk for other functions is biologically important, and deserves its own discussion. Interesting biological phenomena concerning the silk are associated with egg sacs: when a mature female of the theridiid *Theridion evexum* is attacked in her retreat by small ants, she moves her egg sac outside the retreat and builds a tangle of sticky lines around it (Barrantes and Weng 2007b); a clubionid (in an undescribed genus) makes a delicate, transparent veil over its egg mass that has 13 evenly spaced large holes around the edges (Murphy and Murphy 2000); and females in four different araneid genera that are not obviously closely related, *Pozonia*, *Micrathena*, *Acacesia*, and *Verrucosa*, fold a dead leaf around the egg sac in a tight package and then insert it into the

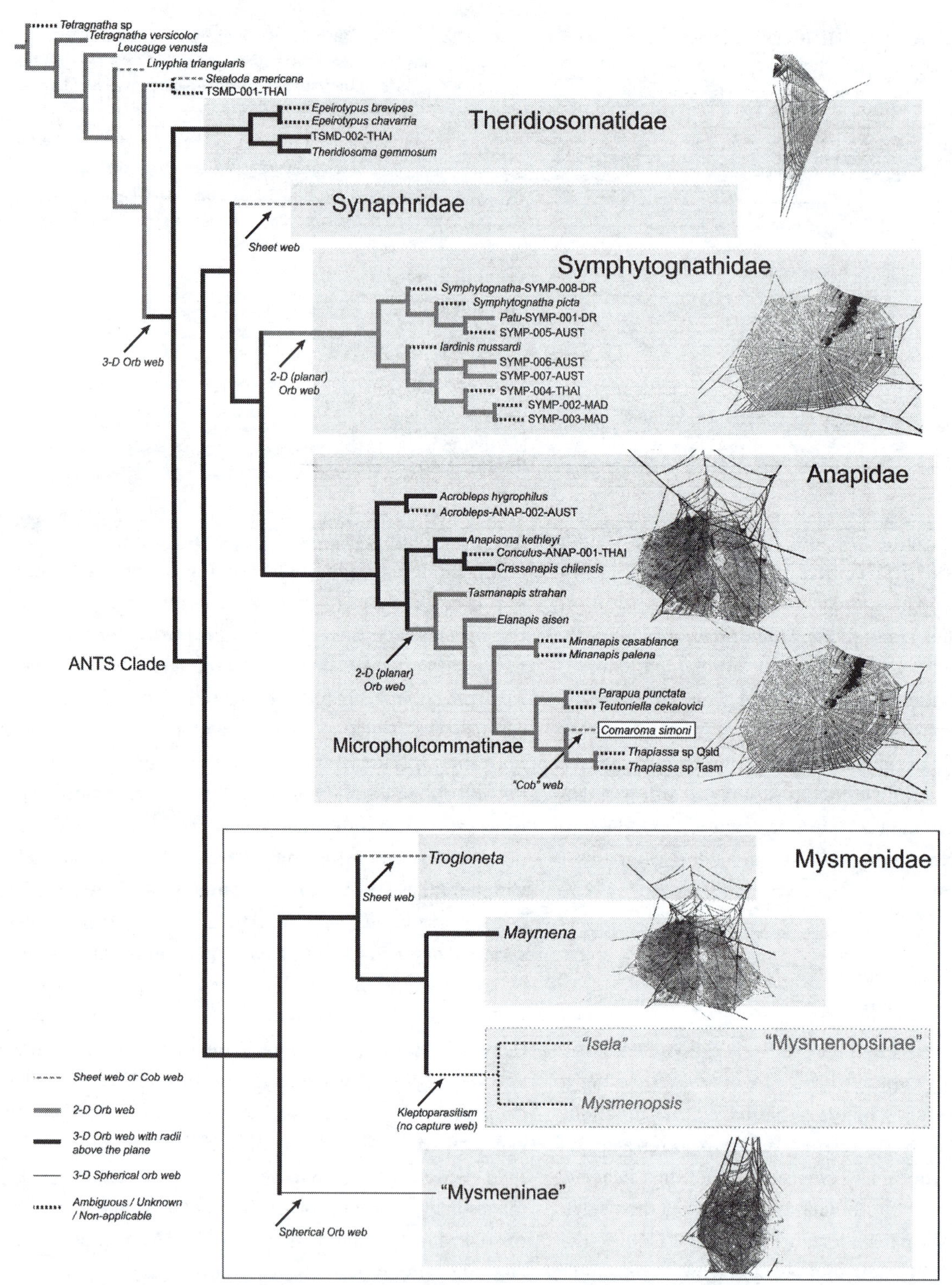

Fig. 1.14. This simplified cladogram shows the complex pattern of web evolution in the four families of "Symphytognathoidea." Planar orbs have arisen twice from three-dimensional orbs in these tiny spiders. "Sheet" webs (see text) have arisen twice from three-dimensional orbs in the species in this cladogram, and two additional times in others (in an undescribed anapid from Madagascar, and in the symphytognatid genus *Anapistula*) (from Lopardo et al. 2011). The monophyly of symphytognathoids is uncertain (Agnarsson et al. 2013).

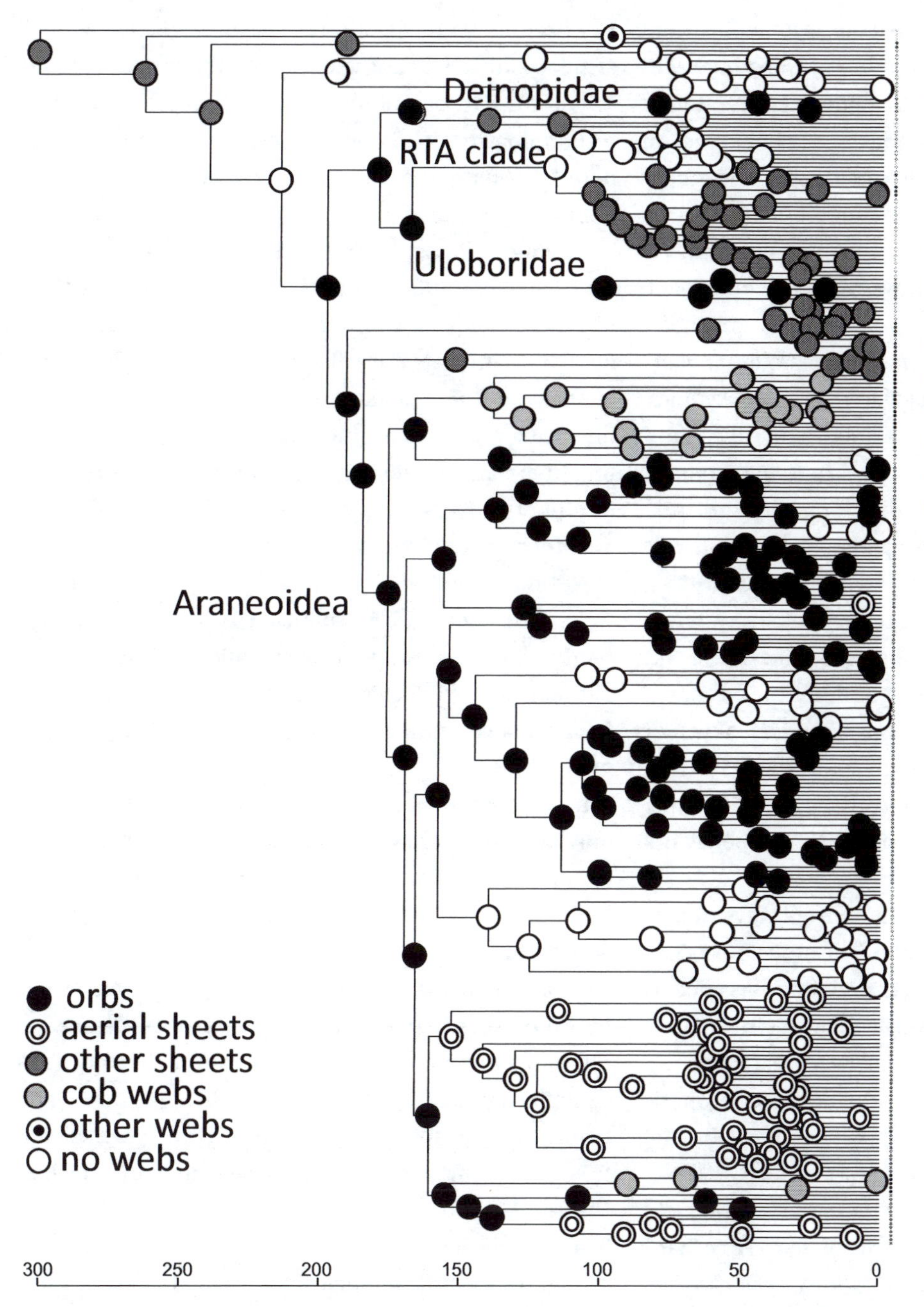

Fig. 1.15. The visual impression in this simplified analysis of the evolutionary relationships of web designs, with only a few clades labeled, shows two typical patterns: web designs are clustered in particular clades; and they have frequently converged in different lineages. The ancient nature of orb webs is also clear. Several of the web classifications are imprecise, however (see sections 9.4 and 10.5.2); the symbols at junctions indicate the most probable state of each ancestor; the scale at the bottom is in Mya (after Dimitrov et al. 2016).

leaf litter (Moya et al. 2010; Eberhard 2015; G. Barrantes pers. comm.). The "silk kisses" of the male pholcid *Modisimus culicinus* are perhaps the most esoteric use of silk: "During the last 30–50 s of each copulation the male bent his opisthosoma [abdomen] ventrally and repeatedly tapped the female's spinnerets with his; a thick, apparently viscid thread became visible between male and female spinnerets" (Huber 1996, p. 239).

In order to clarify behavioral descriptions by avoiding the use of the less specific pronoun "it," I engage in the convenient illusion (as in Spanish, French, German, and others) that all spiders are females. In any case, the overwhelming majority of published behavioral observations have been of mature females. The legs of a spider that is moving around a central point are distinguished as "outside" legs (thus "leg oI" is the first leg that is on the side farther from the hub) or "inside" legs ("leg iI" is nearer the hub). A movement of the spider toward the hub of an orb will be described as "inward," and movements away from the hub as "outward." A radius along which the spider moves outward will be termed an "exit" radius. The word "segment" refers to that portion of a line between two successive attachments of that line (e.g., a segment of sticky spiral); the "current segment" of a line refers to the line being laid between the spider's spinnerets and the line's most recent point of attachment. "Sticky" lines refer to lines that can adhere strongly to smooth surfaces (not to be confused with becoming snagged or entangled on surface irregularities); "non-sticky" lines are those that do not adhere strongly unless they are entangled. Arachnologists often use "viscid" to refer to sticky lines that are coated with droplets of liquid adhesive (thus distinguishing them from "dry sticky" lines that contain instead cribellum silk) (Foelix 2011). Some even distinguish "viscid" from "sticky" (Benjamin et al. 2002). I will adhere (*sic!*) to the simple sticky vs. non-sticky distinction, and add descriptors (e.g., cribellate sticky lines) to distinguish different mechanisms of adhesion. I will often use the past rather than the present tense in referring to behavioral observations ("spiders attached sticky spirals to radii" rather than "spiders attach sticky spirals to radii"), in order to avoid giving the potentially misleading typological impression that these behaviors always occur.

Some terms have been used in unusual ways in previous publications on spider webs. One of the most confusing is that the orbs of a species are often classified as being either "horizontal" or "vertical," when in fact only a small fraction of orb webs are truly vertical (90° slant), and perhaps an even smaller fraction are precisely horizontal (0° slant); to give an idea of this imprecision, the webs of *Leucauge* (TET) are typically described as "horizontal," but the slants of 66 *L. mariana* webs in the field ranged from 8 to 83°, with a mean of 39.9° (Eberhard 1987b; see also Gregorič et al. 2013 on *L. venusta*). As has been customary, I use "horizontal" to refer to webs whose slants seem to be more nearly horizontal than vertical (usually 30° or less with horizontal). I will break with custom in recognizing two types of "vertical" orbs: those whose web planes are very close to 90° with horizontal (85–90°) ("truly vertical"), and orbs whose planes are seldom if ever precisely 90°, but are typically more than about 70° with horizontal ("vertical" or "approximately vertical") (see section 3.3.3.1.1 for reasons why this distinction may be important, and appendix O9.3 for a more detailed discussion).

I use the word "fiber" to designate a single cylinder of silk produced by a single spigot. In many cases a spider produces a pair of fibers, one from each of the corresponding spigots on her bilaterally symmetrical spinnerets; I will use the word "line" to refer to such pairs of fibers (or presumed pairs; the number of fibers was not determined in most descriptions). I avoid one other possible label for multi-fiber lines, "thread," and use "cable" to refer to objects composed of relatively high numbers of fibers. I also use "line" to specify combinations of sticky and non-sticky silk, such as the "sticky spiral lines" of uloborid which include a pair of non-sticky base line fibers and a dense mat of hundreds of fine cribellum fibers. My usage is somewhat arbitrary, and there are other reasonable alternatives (Zschokke 1999).

Several other "arachnological" terms for the forms of webs have been used in confusing ways, and may have hindered understanding web evolution (section 10.5.2). The names for spider web forms often represent attempts to apply abstract human categories, and some webs do not fit these categories; inappropriate use of such names can sometimes inadvertently suggest homologies that may not exist. I do not use "sheet web" in its most encompassing meaning, to refer to any planar array of silk (as, for instance, in Griswold et al. 2005; Blackledge et al. 2009c; Harmer et al. 2011) because I believe that imprecise use of this term has impeded recognition of important diversity (section 9.4). Instead I use "sheet" to refer only to a more or less planar, approximately horizontal array, and whose lines are sufficiently dense and planar that the spider can walk on top of it (see section 10.5.2 for distinctions regarding other categories such as "brushed," "substrate," "aerial," and "stereotyped" used in some recent studies).

I follow the arachnological tradition of not specifying the sex or age of a spider when referring to an adult female. This unfortunate "default" custom of emphasizing the webs of only adult females and neglecting those of males and younger individuals is deeply ingrained (section 3.2.14). Some species for which species and genus identifications were not available are indicated by the numbers of specimens that are deposited in the Museum of Comparative Zoology at Harvard University. These and a second set of specimens that I have placed in the Museo de Zoología of the Escuela de Biología of the Universidad de Costa Rica (including *Allocyclosa bifurca*, *Araneus ex-*

pletus, Argiope trifasciata, Cyclosa jose, C. monteverde, C. turbinata, Diplothyron simplicatus, Dolichognatha sp., *Emblyna* sp., *Epeirotypus brevipes, Kukulcania hibernalis, Leucauge argyra, L. mariana, L. venusta, Mallos hesperius, Mangora* sp., *Metazygia wittfeldae, Metepeira* sp., *Micrathena duodecimspinosa, M. molesta, Modisimus bribri, Neoantisea riparia, Neriene coosa, Nephila clavipes, Philoponella vicina, Physocyclus globosus, Tengella radiata, Uloborus diversus, U. eberhardi, U. glomosus, U. trilineatus, Wagneriana tauricornus, Witica crassicauda*, and *Zosis geniculata*) are vouchers for figures and for the previously unpublished observations in this book that are cited as "WE."

I camouflage some uncertainty regarding the araneid whose behavior Hingston (1920) observed in India by referring to it as *Neoscona nautica* (on the recommendation of H. W. Levi). Hingston referred to it as *Araneus nauticus*, but did not deposit voucher specimens; he observed this species and *Tetragnatha* sp. at the same site, and apparently sometimes lumped them together in his descriptions.

There are several synonyms in English for the terms that I use (Fig. 1.4): *anchor line* = "guy" or "mooring" line; *upper primary frame line* = "bridge frame" or "foundation line"; *radius* = "spoke" or "ray"; *temporary spiral* = "provisional," "auxiliary," "non-sticky," "structural," or "scaffolding" spiral; *sticky spiral* = "viscid," "permanent," "catching," or "capture" spiral; and *hub spiral* = "inner" or "strengthening" spiral (Jackson 1973, Le Guelte 1966, and Zschokke 1999 review terms in English, German, and French). For terms that have been used inconsistently in the literature, I use the following definitions (see also Fig. 1.4). *Primary* frame lines are those frame lines (generally longer and built earlier in construction) that end on (or are continuous with) anchor lines. *Secondary* (= "auxiliary, or "inner") frame lines are those that end on other frame lines. *Primary* radii include those radii that are built during the process of primary frame construction, plus any other radii that were already present prior to initiation of primary frame construction (e.g., radius *y* in Fig. 6.5). *Secondary* radii are all the other radii that originate at the hub and that are built after the primary radii but before the temporary spiral; they include both radii built during secondary frame construction, and others (e.g., Fig. 6.5e). *Tertiary* (or "split" or "subsidiary") radii are those built during temporary spiral construction (Fig. O6.8). Sometimes two secondary radii were fastened together by the temporary spiral near the hub (where radii were especially close to each other), giving the false impression of a tertiary radius (Fig. 6.10). The radial dragline that was laid on the trip away from the hub during radius construction and then immediately broken and replaced on the way back to the hub is a "provisional" radius (see Zschokke 1999 for other terminology). My use of one term rather than another is somewhat arbitrary, and is based largely on descriptive clarity; I have not attempted to trace which terms were used first.

Arachnologists have usually characterized the times at which the webs of any given species are built and operated with the simple terms "nocturnal" and "diurnal," but in fact the topic is substantially more complex. For instance, some nocturnal species build just at dusk, others later at night; some webs that are built at night are removed before dawn, while others are left in place the following day. In addition, there is often substantial intra-specific variation in the timing of web construction. I discuss these complications later (section 8.10, Fig. 8.13, table 8.2), and will generally use the standard oversimplified diurnal/nocturnal classification.

I use "spacing" to refer to the distances between loops of sticky spiral in orbs, and avoid the more common term "mesh." Strictly speaking, the "mesh" of a woven structure refers to both the warp and the woof, not to just one of the two (as, indeed, "mesh" was used by Witt 1971). Substituting the terms "mesh-width" and "mesh-length" becomes confusing (at least for me) when labeling the spaces between lines below and to the sides of the hub. In addition, an orb web is emphatically *not* appropriately characterized as having a single mesh size; the radii are much closer together near the hub than near the outer edge, and these differences have important functional implications (section 4.3).

Arachnologists have generally been lax in attributing particular types of silk lines to particular glands, especially in descriptions of behavior (the careful studies of Hans Peters 1984, 1990, 1993a, 1995b are exceptions). There are several possible but labor-intensive ways to demonstrate the glands that produce different lines. One is to flash freeze the spider and check which spigots were producing lines (Peters 1984), and then trace connections between the spigots and the glands of that species. Others are to check the chemical properties of gland contents and lines (e.g., Townley and Tillinghast 2013), or to demonstrate that particular glands are empty or depleted immediately following production of particular types of lines (Witt et al. 1968). These techniques have been employed in only a few species. And even these techniques may be inadequate: some glands have multiple, possibly different lobes (Řezáč et al. in press; P. Michalik pers. comm.); some produce multiple products (Bittencourt et al. 2010; I. Agnarsson pers. comm.); and some spigots (e.g., of the piriform glands) produce both solid lines and liquid simultaneously (Eberhard 2010a). In addition, because nearly all work on silk has been on orbicularian spiders, glands and spigots of other groups that are thought to be homologous are only assumed to produce lines with similar properties, without experimental tests (section 2.2). Hesselberg and Vollrath (2012) discuss, for example, the difficulties of distinguishing lines from major and minor ampullate glands in *Nephila edulis* webs. For lack of better data, I follow the tradition of assuming that the loosely established associations between glands and lines are correct; the reader should keep these uncertainties in mind.

Different authors have used different names for the early developmental stages of spiders outside the egg; I will follow Foelix (2011): spiders emerge from the egg within the sac as "first instar" spiderlings, then molt once and leave the sac as "second instar" spiderlings (the stage in which most build their first prey capture webs).

1.8 ACKNOWLEDGMENTS

I have benefited immensely from working at the Smithsonian Tropical Research Institute, whose consistent, effective, and generous support has freed me from the frenetic chasing after grants that is an unfortunate aspect of many academic jobs in biology. STRI has asked me to contribute knowledge, and has judged me on the basis of what I have produced, rather than the amount of money I have raised; this is a luxury that many modern biologists cannot afford, and one for which I am truly grateful. I also benefited greatly from working at the Universidad del Valle in Cali, Colombia, for 10 years, and then for >35 years at the Universidad de Costa Rica. Both provided freedom, stimulating colleagues, pleasant working conditions, and inspiring students. I am especially grateful to my colleagues Gilbert Barrantes and Daniel Briceño at UCR for their friendship, collaboration, and sustained, enthusiastic intellectual stimulation.

I had the good fortune to be raised academically in a museum, and I have always felt closer in my spider interests to taxonomist colleagues than to ecologists or even behaviorists. I owe special thanks to the late Herbert Levi for his unstinting identifications of long series of spiders, and his patience in not seeing quick published payoffs for all his work. Others who kindly provided identifications of specimens are Ingi Agnarsson, Brent Opell, Norman Platnick, and Darrell Ubick.

I have also benefited greatly from and been inspired by the high standards set for careful, meticulous, and extensive integration of massive amounts of information by spider systematists like Ingi Agnarsson, Jonathan Coddington, Charles Griswold, Gustavo Hormiga, Bernhard Huber, Matjaž Kuntner, Herbert Levi, Brent Opell, Norman Platnick, and Martin Ramírez. They have been my closest colleagues in arachnology.

More specific thanks go to Ingi Agnarsson, Gilbert Barrantes, Todd Blackledge, Rainer Foelix, Gustavo Hormiga, Sean Kelly, Matjaž Kuntner, Yael Lubin, Yuri Messas, Peter Michalik, Tadashi Miyashita, Jairo Moya, John Murphy, Martin Ramírez, Nicolaj Scharff, Diego Solano-Brenes, I-Min Tso, and Samuel Zschokke for permission to use figures. Paul Selden kindly updated a previously published figure for this book. I benefited from discussions, references, and other additional information provided by Anita Aisenberg, Fernando Costa, Jessica Eberhard, Rainer Foelix, Kyle Harms, Hilton Japyassú, Stano Korenko, Peter Michalik, Brent Opell, Adelberto Santos, Fernando Soley, Rick Vetter, Eric Warrant, and Mary Jane West-Eberhard. Gilbert Barrantes, Fernando Costa, Bernhard Huber, Stano Korenko, Matjaž Kuntner, Peter Michalik, Brent Opell, Martin Ramírez, Diego Solano-Brenes, and Samuel Zschokke allowed me to use other unpublished information. I am proud and fortunate to be part of the often magnificent tradition of sharing that generally characterizes naturalists truly interested in their organisms.

I thank Kenji Nishida and T. Inoue for translations from Japanese, and Angel Aguirre for digging up even the most obscure references. Special thanks to Ruth

Madrigal, who provided much invaluable, enthusiastic assistance in obtaining, compiling, and proofing references, to two anonymous reviewers for perceptive comments on an early draft, the World Spider Catalogue (and its originator, Norman Platnick) for making accurate use of spider names possible even for behaviorists, and to James Toftness for taking on the nightmare process of obtaining permissions to use previously published figures. I especially grateful to Ingi Agnarsson, Mariella Herberstein, Matjaž Kuntner, and especially Thomas Hesselberg, Brent Opell, and Samuel Zschokke for kindly taking the time to make helpful comments and corrections on substantial portions of the MS.

Finally, I am grateful for the support, calm advice, and tolerance that my wife, Mary Jane, has shown over and over in the face of the monomaniacal dedication I have given to this project. I knew when I started that I was getting into a marathon, and that it would impose impediments to living a normal life; but I did not realize it would take nearly so long and be so difficult.

2

THE "HARDWARE" OF WEB-BUILDING SPIDERS

MORPHOLOGY, SILK, AND BEHAVIOR

But how much more wonderful does it all seem when we picture the web as a potential fabric, first woven into an inimitable harmony to lure to death thousands of living creatures, then, tattered and torn in the tragedy, to be again received into the maw of its voracious host, to be repurified in the strange economy of the spider's structure, to emerge again from the spinning-wheel in fine transparent filaments, to be woven again into the same lovely texture, and to repeat day after day the same eternal drama. —Hingston 1920, p. 121

2.1 INTRODUCTION

Web spiders live in a sensory world that is alien to us in many ways. The tactile sense is dominant, and vision and hearing play only minor roles. A web spider's physical world consists largely of her own lines, and her major tasks involve producing, finding, and holding lines. A spider must generally find the lines using her sense of touch. The different types of basic abilities important to a spider in producing a functional web can be divided into two general classes: the properties of the silk lines themselves; and the physical and behavioral equipment that spiders use to build and manage webs of silk lines.

2.2 SILK GLANDS AND SILK

2.2.1 ORIGINS

I will begin by describing spider silk in general, and then turn briefly to silk from different types of glands. This is an active, rapidly moving area of research, and much of what I say will be based on the recent, extensive summaries and reviews of Craig (2003), Harmer et al. (2011), Vollrath (2010), Blackledge et al. (2011), Townley and Tillinghast (2013), Blackledge (2012) and Mortimer and Vollrath (2015). The standard view has been that spiders utilize a relatively consistent array of glands and silks (based on gland morphology and the staining properties of the gland products) to produce a much wider array of web designs (especially in Araneoidea) (Peters and Kovoor 1991). Studies of the silk itself, while available for only a small fraction of species and web designs (e.g., Blackledge et al. 2011, 2012a,b; Mortimer and Vollrath 2015), have already revealed however a "dizzying complexity of different types of silk" (Blackledge et al. 2011). Undoubtedly the descriptions here are overly simplistic.

Silks can be loosely defined as natural proteins in semi-crystalline molecular structures that are extruded outside an organism (Craig 1997); they are defined by their processing from a gel-like dope to a dry fiber (Mortimer and Vollrath 2015). In contrast to other biopolymer proteins such as collagen and keratin, silk lines are not grown inside or as part of organisms, and are used exclusively outside the body. Silk has also evolved independently in more than 20 other animal groups, including other arachnids (pseudoscorpions, mites) as well as in numerous insects. Silk has different proper-

Table 2.1. Types of silk glands, the spinnerets where their lines emerge, and the functions of their products. Spinnerets are designated as follows: AMS = anterior median; ALS = anterior lateral; PMS = posterior median; PLS = posterior lateral). Most data are from Blackledge et al. 2011 and also Peters and Kovoor 1991, Wolff et al. 2014, and Řezáč et al. 2017; those on loxoscelids are from Knight and Vollrath 2002.

Gland	*Spinnerets where spigots are located*	*Functions of the lines*	*Numbers of spigots Many (M) or few (F)*
Major ampullate	ALS	Dragline, supporting lines in webs	F
Minor ampullate	PMS	Bridge and balloon lines, temp. spiral in orbs, sometimes added to draglines[1]	F
Aciniform	PMS, PLS	Wrap prey, stabilimentum, sheet webs, distal tip of bridge line	M
Flagelliform	PLS	Baseline for araneoid sticky lines	F (1 pair)
Pseudo-flagelliform	PLS	Baseline support mat of cribellum fibrils	F (1 pair)
Cribellum	AMS (fused in a plate)	Sticky prey capture lines (both fine fibrils and nodules)	M
Aggregate	PLS	Liquid adhesive, salts on flagelliform lines (adhere to prey, maintain aqueous medium for flagelliform line)	F
Cylindrical	PMS, PLS	Egg sac	M
Piriform	PMS, PLS	Attach dragline to substrate or other lines (both fine fibrils and liquid produced)	M
Epiandrous[2]	Ventral surface of abdomen	Support sperm droplet (males only)	M
Venom (Scytodidae)	Chelicerae	Glue and line that are shot onto prey to wrap it and fasten it to the substrate	F
? (Loxoscelidae)[3]	ALS	Sticky thin bands for prey capture	F

[1] Of nine radii from five orbs of three mature female *Araneus diadematus* (AR), seven radii included both major and minor ampullate lines, but two had only major ampullate lines (Peters and Kovoor 1991).

[2] Epiandrous spigots are lacking in mature males of many spiders, especially those of very small species such as erigonine linyphiids and theridiids (Marples 1967; Agnarsson 2004; Miller 2007). The functional significance of this pattern is unknown.

[3] On the basis of the size and the shape of the gland and the location of its spigot, the gland is said to be homologous with the major ampullate gland, though the proportions of the gland and the duct are markedly different (Knight and Vollrath 2002).

ties in different groups (Sutherland et al. 2010), but in perhaps no other group is it used in such diverse ways as in spiders.

Silk production originated at least 390 million years ago in spiders (Shear et al. 1989), and is a defining character of the entire order Araneae. Some modern species produce up to eight different types of silk (table 2.1). The silk glands are located in the spider's abdomen, and ducts connect them to the abdominal appendages near the rear end of the abdomen called spinnerets. Small silk glands lie within the spinnerets themselves, while larger silk glands are located more anteriorly. Ancestral spiders had eight spinnerets (anterior median and anterior lateral; posterior median and posterior lateral) that corresponded to two pairs of abdominal appendages (Vollrath and Selden 2007). In some spiders the anterior median pair of spinnerets has fused to form the plate-like cribellum, while in many other groups this pair of spinnerets has been lost. A few haplogyne spiders have lost all but one pair of spinnerets.

Each spinneret has a number of hollow, approximately cylindrical spigots that are open at their tips (fig. 2.1, 2.7). The duct of each silk gland runs through the spigot and opens at its tip (fig. 2.1c,d). Exquisitely preserved fossil fragments of spinnerets have been found that bear recognizable spigots, showing that silk production arose in the early Devonian (Shear et al. 1989); silk was thus present soon after spiders moved from the sea onto land (or perhaps even before) (Vollrath and Selden 2007). Current speculation (see chapter 10) supposes that silk arose from nitrogenous waste products, and that its first function was either to cover the eggs (a

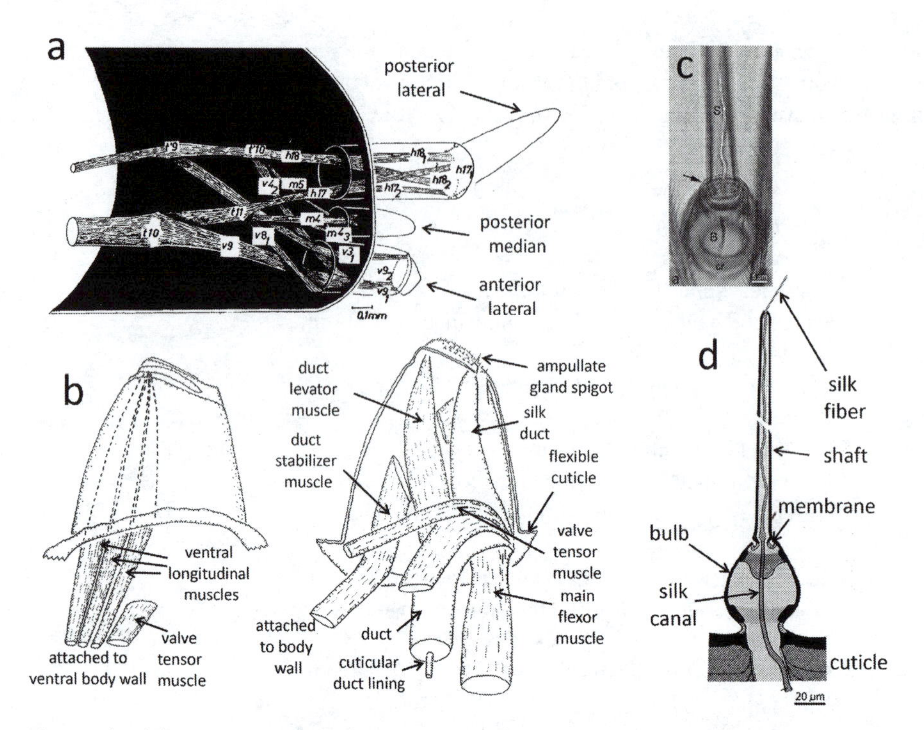

Fig. 2.1. The elaborate arrays of muscles associated with the long spinnerets of the sheet-building agelenid *Eratigena atrica* C. L. Koch (*a*) and also with the relatively short anterior lateral spinnerets of the orb-weaving *Araneus diadematus* (AR) (*b*) reveal that these spinnerets are capable of complex movements. The thin cuticular silk canal is visible through the wall of the hollow spigot of the mygalomorph *Brachypelma albiceps* (Theraphosidae) in *c* and *d*. The silk canal begins as a funnel near the exit of the silk gland, and runs all the way to the spigot tip. It is sometimes filled with silk, and sometimes empty. There may be a valve at the junction between the bulb and the shaft, where there is also a "joint membrane" that probably allows a certain amount of additional movement of the shaft with respect to the bulb. Silk gland ducts in other species have similar silk canals (R. Foelix pers. comm.) (*a* from Peters 1967, *b* from Wilson 1962a, *c* and *d* from Foelix et al. 2013b).

function of silk in nearly all extant spiders), or to trap a plastron or air bubble under water that the spider could use for respiration (Vollrath and Selden 2007). Silk was probably also used to line the tunnels or retreats of early terrestrial spiders (Coyle 1986). Thus, although nearly all spiders are predators, silk lines probably only secondarily came to be used in webs to capture prey.

Although prey capture webs have come and gone repeatedly in the evolution of spiders (chapter 9), silk itself never seems to have been completely abandoned. The silk glands constitute up to approximately 5% of the weight of the orb-weaving araneid *Araneus diadematus* (Peakall 1969 in Prestwich 1977). The molecular genetics of silk production and the physical properties of different silk lines are tightly linked to the foraging ecology of spiders, and understanding their evolution illuminates how molecular and organismal processes interact during evolution (Craig 1992, 2003; Blackledge et al. 2011, 2012; Mortimer and Vollrath 2015).

2.2.2 MECHANICAL PROPERTIES AND HOW THEY ARE DETERMINED

Spider silk is famous for its outstanding strength. The lines from different glands range from rubber-like non-adhesive lines (flagelliform glands) to gluey liquids (aggregate glands), from extremely fine and highly adhesive fibrils (cribellum glands) to perhaps the most tenacious natural fibers known, which combine high extensibility and resistance to breaking (ampullate glands). Dragline silk's toughness (= work/volume; a measure of the energy a material can absorb before rupturing) is about five times that of Kevlar®, and its high strength/weight is about 5 times that of steel; these qualities plus its immunological compatibility with living tissue, and its production under environmentally benign conditions, make it interesting for a variety of practical uses (Agnarsson et al. 2009a). The toughest biomaterial yet known is the major ampullate gland silk of the araneid *Caerostris darwini* (Agnarsson et al. 2010).

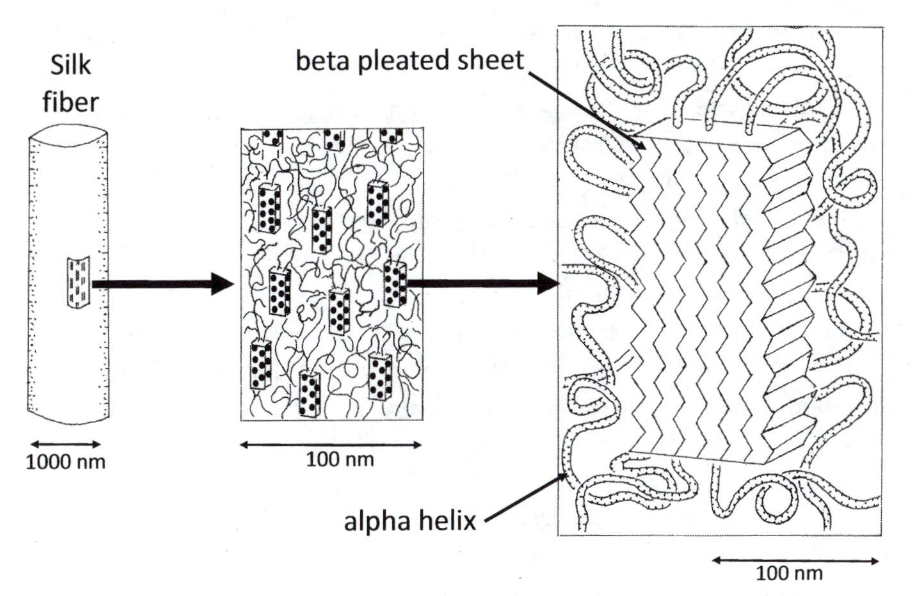

Fig. 2.2. Views at increasing magnification of the alignment of silk molecules in β sheets (conferring stiffness) and loose α helices (conferring elasticity) (from Foelix 2011, after Vollrath 1992).

These properties have inspired repeated attempts to take practical advantage of spider silk—from early French attempts in Madagascar to culture *Nephila* spiders like silkworms (Savory 1952), to recent insertion of silk genes into other organisms (Allmeling et al. 2013) in order to produce silk in large quantities. There are hopes that the products of the next generation of silk studies will be useful in military, industrial, and medical applications (Blackledge et al. 2011; Allmeling et al. 2013). Possible uses include cartilage, bone, and nerve regeneration, as well as more stressful uses such as bullet-proof clothing. Synthetic polymer threads with soft and hard molecular segments that were inspired by spider silk, can have "memories" of shapes that can be recalled on demand, and promote self-healing (Li and Shojaei 2012). The variation among different types of spider silk can also help orient searches for bioinspired fibers by demonstrating which properties are feasible (Blackledge and Hayashi 2006a; Elices et al. 2009).

Nearly all known spider silk proteins are coded by a single family of genes, the spidroins, which have repeatedly duplicated and then diverged over evolutionary time. Only about 5% of silk-gland specific transcripts encode fibroins, however, and some of the numerous other proteins which as yet have no annotations are likely to include novel silk-associated proteins (Chaw et al. 2015). Most spidroins are quite large, 200–350 kDa molecules (Blackledge 2013) that consist mostly of a large central region of repetitive modules ranging from about 50 to 200 amino acids in length. This central region is flanked by an N-terminus and a C-terminus. The N- and C-termini are very similar in the spidroins of different types of silk, and thus provide useful indications of deep evolutionary history (Garb et al. 2010; Dimitrov et al. 2011). In contrast, the modules in the central region, which strongly influence the physical properties of the silk, are very similar within a given molecule but sharply different between different silks; in some cases they are so different that they are difficult to homologize (Gatesy et al. 2001). The modules in the central region are typically dominated by amino acids with short residuals (especially alanine and glycine), and these motifs are thought to become highly organized and form secondary structures such as β-sheets and β-spirals in the silk (fig. 2.2). The crystalline stacking of these long repeats, especially the β-sheets, is thought to be important in explaining how relatively weak hydrogen bonds can make silk so strong. In "amorphous" regions, in contrast, spidroins connect the crystal-forming domains; the hydrogen bonding and entanglement in these regions provide strength and rigidity, but they are easily disrupted as the silk is stretched, making silk both strong and stretchy. Most silks are composed of one or two types of spidroin. The major ampullate silk of orb weavers contains both major ampullate spidroin 1 (MaSp1) and major ampullate spidroin 2 (MaSp2), and its material properties vary with the proportions of these two spidroins. Much of the inter-specific variation among Orbiculariae in major ampullate silk properties, such as extensibility and energy damping, correlates with differences among species in the proline composition of the silk (Blackledge et al. 2012a). A complete sequence for one aciniform silk gene (AcSp1) showed less divergence in three species of *Argiope* (AR) in the highly repeated subunits of the central region than in the terminal regions; purifying selection was deduced to be stronger in the N-terminal than the C-terminal region (Chaw et al. 2014).

Spidroin paralogs evidently originated prior to the divergence of mygalomorphs and araneomorphs (figs. 1.9, 1.10); this divergence was followed by a much greater expansion of the spidroin family in the araneomorphs, which paralleled the divergence in their silk glands (Garb et al. 2007). The total number of spidroins is currently unknown. Previous counts gave a total of 19 for all spiders; but the especially sensitive technique of constructing a relatively complete, annotated genome for a single species (*Nephila clavipes*) (NE) recently gave a total of at least 28 and perhaps up to 35 spidroins (Babb et al. 2017). There was also impressive diversity in the basic construction blocks of silks, including hundreds of unique repeated coding motifs and unique cassettes in *N. clavipes*. Current knowledge of silk diversity may thus represent only the tip of an iceberg.

The processes that occur in the duct from the gland to the spinneret act in combination with the chemical nature of a gland's products to determine the physical properties of the silk line; these can vary quite substantially even in a single gland type such as the major ampullate in a given individual. Silk is converted from a highly concentrated aqueous liquid or gel-like dope inside the gland to a semi-crystalline form as it emerges from the spinneret by a combination of water resorption, secretion of K+ and H+ ions, absorption of Na+ ions, a drop in pH, and shear flow forces as it is moves down the duct. The physical force of the pull causes the silk molecules to align themselves with each other, resulting in the formation of hydrogen bonds that transform the modules in the central portions of the molecules from liquid to semi-crystals (fig. 2.2); the physical properties of the resulting silk line are affected by details of this alignment. Even gross morphological differences, such as the long, winding ducts of major and minor ampullate glands (lacking in all other glands), probably affect the mechanical properties of lines. The tensile strength of a line is also positively correlated with the rate at which it is drawn (Denny 1976).

Some spiders appear able to control, perhaps by means of valves in the major ampullate ducts (Wilson 1962), the shear force applied to the silk line as it is drawn down. This force can strongly affect the degree of alignment of the silk molecules, and thus how stiff and extensible the lines are. This factor can have substantial effects, and variation in some mechanical properties reaches almost 50% even within an individual orb weaving spider (Hesselberg and Vollrath 2012). A variety of spiders, including even the relatively basal araneomorph *Hypochilus gertschi* (Hypochilidae), do not hold their draglines with any legs while they descend at the end of the line, but can nevertheless descend at different velocities and can even stop without grasping the dragline (Eberhard 1986a). This suggests that their silk lines are already at least partially crystallized while still inside the spider's body, and that friction of the silk with the walls of the duct itself is used to "brake" production of silk lines (also section 2.3.5). It is possible that this mechanism differs in other species, and that the fluid is converted into semi-crystalline material only at the mouth of the spigot itself, as in the sicariid *Loxosceles laeta* (Knight and Vollrath 2002).

There were also surprising differences in the physical properties of the gumfoot lines of the black widow *Latrodectus hesperus* and the scaffolding web lines to which gumfoot lines were attached (Argintean et al. 2006), as well as the major ampullate gland silk lines that spiders produced as they fell (Lawrence et al. 2004). Another theridiid, *Parasteatoda tepidariorum*, made thicker lines after encountering larger prey (Boutry and Blackledge 2008). Direct behavioral observations suggested that the gumfoot lines and "scaffolding" lines were continuous with each other (Eberhard et al. 2008b) even though the scaffolding line differed in being more resiliant when under only a slight strain (0–3%) (this difference improved the line's ability to function as a spring to lift prey up off of the substrate) (Argintean et al. 2006).

The mechanical properties of major ampullate silk lines have also been shown to be determined by both the relative amounts of MaSp1 and MaSp2, and how these proteins are assembled (Sponner et al. 2005 in Liao et al. 2009). Spiders themselves may be able to control assembly adaptively. Female *Cyclosa mulmeinensis* (AR) produced stronger lines without altering their protein composition when they were subjected to wind (Liao et al. 2009). It seems to be generally true that silk lines pulled forcefully from restrained spiders differ in their properties from those in the same spider's webs (Work 1977, 1978). One possibility (among others) is that spiders altered the physical properties of their silk by using spidroin-specific antibodies to alter MaSp1 and MaSp2 interactions that affect crystalline formation (Liao et al. 2009).

A second factor affecting the orientation of the amorphous regions of spidroins is humidity. When water infiltrates a silk line after it has been drawn, it disrupts hydrogen bonding, allowing the spidroins to move irreversibly to more disordered states ("supercontraction") (Work 1977). The degree of supercontraction in major ampullate silk varies in different species (Boutry and Blackledge 2010, 2013). It can produce shrinkage of an unrestrained line by up to 50% in high humidity, and can generate substantial forces in web lines. It also softens silk, making it more compliant. The functional implications of the increased tensions and softening of the silk due to supercontraction in natural webs in the field are uncertain. Proposed consequences include the following: reduced sagging when a web is weighted with rain, mist, or dew (fig. 8.10) (Work 1977; Agnarsson et al. 2009b); an improved ability to absorb the kinetic energy of the prey, observed in prey impact with intact orbs of *Nephila clavipes* (NE) and *Argiope trifasciata* (AR), due to the softening effect (Boutry and Blackledge 2013); and facilitation of the alignment of molecules along the fiber axis during the spinning of silk from liquid dope (in this case the water responsiveness of dry threads might be a "non-adaptive by-product" for webs) (Guinea et al. 2005). A distinct process of shortening, "cyclic contraction," also occurs in the major ampullate gland silk of *Nephila clavipes* in response to decreases in humidity (Agnarsson et al. 2009b). Again, the consequences for natural webs are uncertain.

A silk line is a "viscoelastic" polymer, meaning that its material properties change as it is stretched (fig. 2.3). A line thus manifests a suite of physical properties, depending on how far it has been stretched. As a line begins to be extended, the changes that occur are generally reversible; later, just before failure, the line suffers permanent plastic deformation. Many silks show high

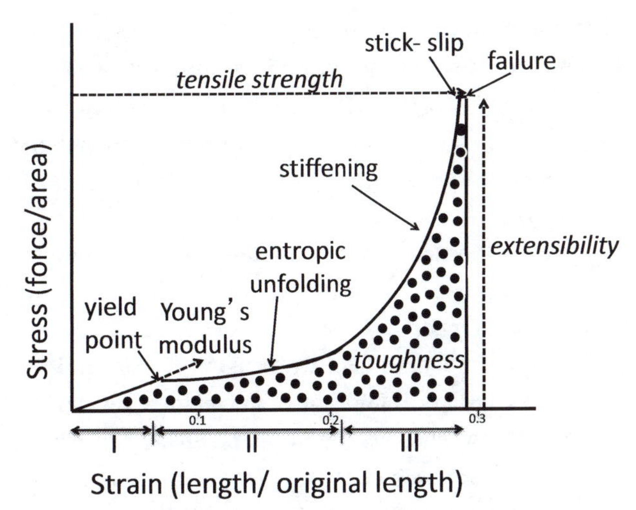

Fig. 2.3. The non-linear stress–strain curve from the major ampullate silk of the araneid *Argiope keyserlingi* illustrates several key details related to the biomechanical properties of spider silks. Stress is calculated as force divided by the cross-sectional area of the fiber ("engineering stress"); true stress is calculated by multiplying engineering stress by L/Lo (length of stretched fiber/original length). This approximates the instantaneous cross-sectional area of the fiber, an important variable for elastic materials. Strain measures the change in the length of a fiber relative to its original length ("engineering strain"). It is usually converted, as in the graph, to "true strain" by taking the natural log of L/Lo. Tensile strength is the stress at the breaking point of a material under uniaxial loading. Extensibility describes the stretchiness of a fiber, for example the percentage that a fiber's length is increased when compared with its original length at the moment it breaks. Stiffness is a measure of the ability of a fiber to resist deformation; it is defined by Young's modulus, calculated from the slope of the initial elastic region of the stress–strain curve. Yield is the point where a fiber transitions from elastic (and reversible) deformation to plastic deformation. Toughness (the energy required to break a thread) is calculated as the area under the stress–strain curve. Hysteresis is the proportion of energy lost as heat during a loading–unloading cycle (the energy required to stretch a silk thread is greater than that required to return it to its natural state). At the molecular level (see fig. 2.2), region I corresponds to a stiff initial response with homogeneous stretching, region II to an entropic unfolding of the alpha-folding of semi-amorphous protein domains, and region III to a stiffening as molecules align and the load is transferred to the beta-sheet crystals; a stick-slip deformation of the beta-sheet crystals occurs at failure (after Harmer et al. 2011 and Cranford et al. 2012).

"hysteresis" during the reversible stage, a property that is functionally important in prey capture. Hysteresis is a measure of how much of the kinetic energy that produced the extension is converted to heat, rather than being stored internally (as in a spring). Major ampullate silk converts about 60% of its loading energy to heat. Thus when the load is removed, the line's return to its original length is slow rather than immediate. In prey capture, high hysteresis means that prey that makes a high-energy impact with the web (higher velocity and/or larger mass) will not simply be bounced forcefully backward, as if the web were a trampoline. Instead much of its impact energy is converted to heat, and the rebound is slower and weaker.

Spiders appeared to modify both the composition of the line itself ("material properties") and the structural properties (e.g., the diameters or numbers) of the lines laid when faced with different problems (e.g., Blackledge and Zevenbergen 2007; Tso et al. 2007; Ortlepp and Gosline 2008). Mechanical properties were also altered (for the worse) when silk was forcibly extracted from *Nephila* spiders instead of being collected from lines built by the spider (Madsen and Vollrath 2000). The magnitudes of the changes produced by even a single individual spider are sometimes quite substantial (Herberstein and Tso 2011) (see section 3.2.9).

As mentioned above, all types of silk (excepting only the independently evolved "silk" that is produced in the venom glands in the spitting spiders in the family Scytodidae) share the property that the gland contents cannot be ejected forcefully by the spider, but must instead be pulled from the spider's body. Thus most spiders cannot match the cartoon character Spider-Man's ability to "shoot" lines. This detail probably had important consequences in the evolution of spinnerets and their spigots, and in building behavior (sections 2.3.2, 7.1.1). Another detail with important consequences for web evolution is that silk lines are apparently coated with a sheath that covers the internal nanobrils and elongate cavities within a core. The sheath may be related to the suggestion from scanning electron microscope (SEM) images that even fibers of non-adhesive silk are apparently slightly "wet" when they first emerge from the spigot. Even non-adhesive fibers from ampullate glands appear to adhere to each other (fig. 2.4a); and the aciniform lines of myglomorphs and araneomorphs often adhered at least weakly to objects that they contacted soon after they were produced (e.g., Barrantes and Eberhard 2007).

The following introduction to silk glands and their products is taken largely from the more detailed and authoritative accounts of Blackledge et al. (2011) and Blackledge (2012, 2013). As they emphasize, and is discussed in section 3.2.9, there is much more variation than is indicated here. The diversity of silk types has only begun to be explored, especially in non-orb weavers, and promises to be exciting (Blackledge et al. 2011; Blackledge 2013). Even in some mygalomorphs that build only simple silk structures, spigots and glands differ morphologically (e.g., Glatz 1972 on the trap door spider *Nemesia caementaria*). I will only briefly discuss epiandrous and tubuliform (= cylindrical) silk glands here, because they are involved in making tiny webs to support sperm and egg sacs, rather than in prey capture. I will also omit the controversial possibility of tarsal silk for locomotion (see Gorb et al. 2001; Pérez-Miles et al. 2009; Pérez-Miles and Ortíz-Villatoro 2012; and Foelix et al. 2013a).

2.2.3 MAJOR AMPULLATE GLANDS

The best-studied glands by far in terms of the chemical composition and physical properties of their products are the large, major ampullate glands, which produce the draglines that most spiders leave behind as they walk (much of the previous section applies to major ampullate silk). The major ampullate gland of *Nephila* (NE) consists of a sac or ampulla with an apical "tail" that produces some part of the silk feedstock and a funnel that leads to the duct. The long duct is folded back on itself in an elongated "S" that progressively narrows to form a hyperbolic die (a nozzle profile capable of generating constant extensional strain rates) (Mortimer and Vollrath 2015). The flow of droplet inclusions within the silk feedstock confirms that elongational flow at a slow shear rate assists in defining axial molecular orientations before the final draw-down. The combination of wall shear and elongational flow, together with subtle changes in the pH, induce beta-sheet transitions in the proteins as they travel through the duct (Vollrath and Knight 2001). The forms and branching patterns of major ampullate glands vary substantially in different taxonomic groups (P. Michalik pers. comm.). Small lobes develop in juveniles just prior to molts, and produce the draglines when molting occurs (Townley et al. 1993).

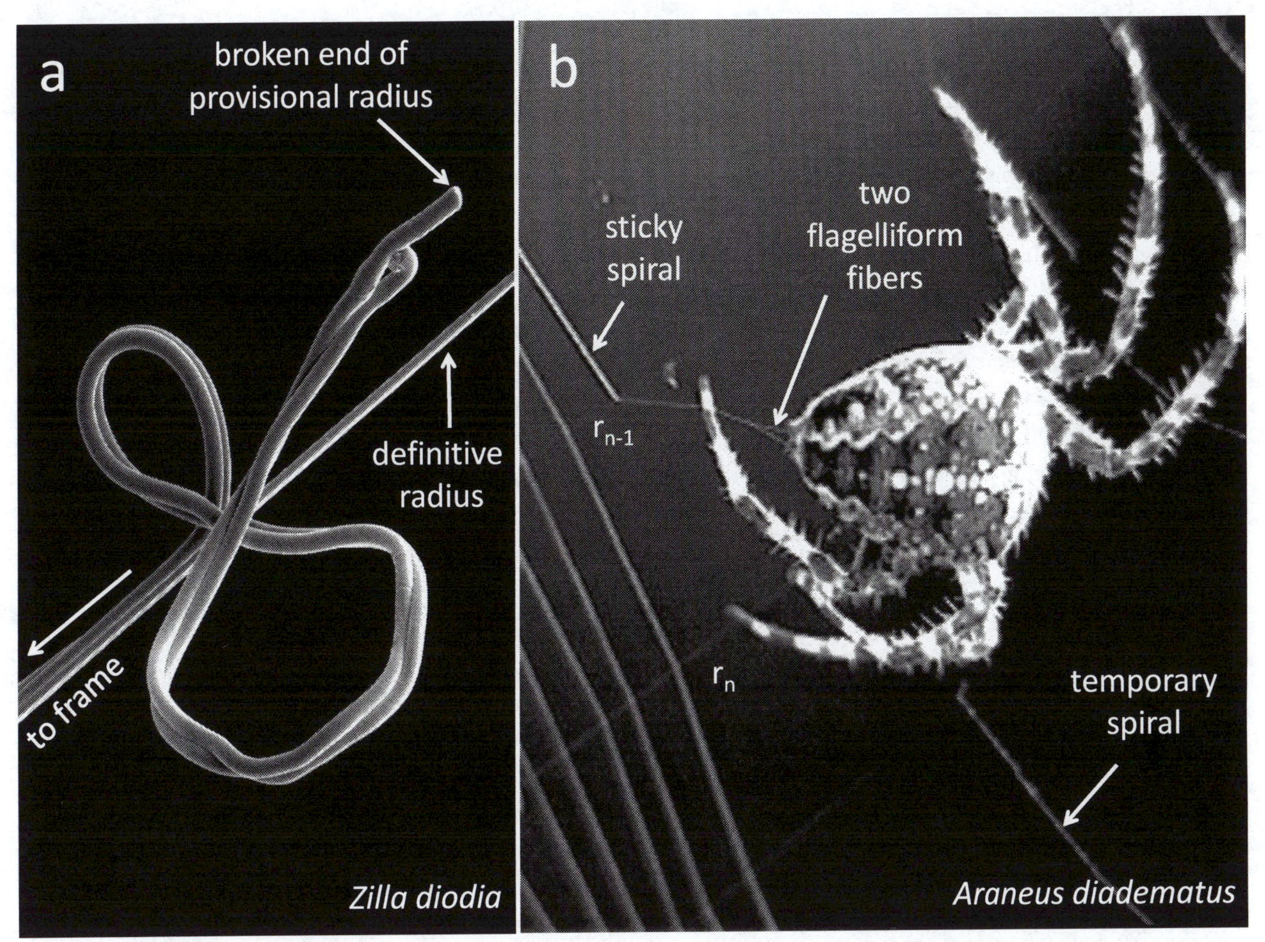

Fig. 2.4. Details of webs and behavior can illuminate silk production. During radius construction, *Zilla diodia* (AR) broke the provisional radius (*a*) and replaced it with a new dragline as she moved to the hub. As shown in the figure, the spider left the two ampullate fibers of the broken provisional radius loose. They adhered tightly to each other and to the fibers of the definitive radius, and thus were apparently sticky when first produced (from Zschokke 2000; courtesy of Samuel Zschokke). A detail in the image of *Araneus diadematus* (AR) engaged in sticky spiral construction from a video recording (*b*) suggests a possible but heretofore unappreciated role for spinneret movements during orb construction. The two flagelliform fibers of the sticky spiral baseline were separated when they emerged from the spider's posterior lateral spinnerets, and merged farther from her body, forming a single line. There was no separation between these fibers, however, at points where the sticky line was attached to a radius (fig. 2.10e). This suggests that the posterior lateral spinnerets probably moved medially, bringing the flagelliform fibers together at the moment the attachment was made (photo courtesy of Samuel Zschokke).

The major ampullate lines do the heavy physical work in the webs of the araneomorphs (figs. 1.9, 1.10), forming the scaffolding on which other lines are placed and which absorbs the major part of the stress exerted by both the spider and her prey (I will follow general usage in attributing draglines to the major ampullate glands, even though more than one pair of fibers often occur in a single dragline; see, for instance, Hingston 1922a on the nephilid *Nephila pilipes*). Ampullate gland silk is strong and tough even in basal lineages, but it is especially strong and stiff in Orbicularia (Swanson et al. 2006) (it is especially extensible in *Deinopis*, however); this difference correlates with the presence of the derived spidroin MaSp2, which may form "nanosprings" that increase mobility within the amorphous region of the silk and provide increased toughness. The ratio between MaSp1 and MaSp2 varies among taxa, and possibly even within individual spiders (Blackledge et al. 2011) (for the surprising range of variations in this and other types of silk, see section 3.2.9). Early major ampullate silk is thought to have originated about 375 Mya, its tensile strength improving rapidly with increased repetitiveness and certain amino acid motifs, maximizing in basal entelegyne spiders about 230 Mya (Blackledge et al. 2012b). Toughness improved subsequently in orb weavers, coupled with the origin of MaSp2. This outline is overly simplistic, as the major ampullate silk of even the single araneid genus *Argiope* showed a wide range of stress-strain relationships (Blackledge et al. 2012b).

It might have been expected that the diameters of major ampullate lines were fixed by the diameters of their spigots, but this is not the case. A clear demonstration comes from the diameters of presumed major and minor ampullate fibers in the tangle above the sheet in the web of *Cyrtophora citricola* in the two webs of two adult females. Of 52 fibers measured under the SEM, 8% were 1 µm in diameter, 25% were 2 µm, 10% were 3 µm, 6% were 4 µm, 15% were 5 µm, 8% were 6 µm, and 29% were 7 µm (measurements are rounded to the nearest µm) (Peters 1993a). A mature female *Nephila* in the lab caused the two dragline-fibers to adhere to each other by medial movements of the anterior lateral spinnerets after they had been experimentally separated while dragline was being pulled from the spider (S. Zschokke pers. comm.). Conversely, flagelliform fibers produced by *Araneus diadematus* (AR) were separated during sticky spiral construction (fig. 2.4b). To my knowledge, the possibility that spiders use spinneret movements to alter the properties of lines has never been checked.

The numbers and the forms of ampullate glands vary among species; some are double, and others have branches of different forms and sizes (P. Michalik pers. comm.). The agelenid *Agelena labyrinthica*, for instance, has eight similar branches, some with sub-branches of their own, all emptying into a common ampulla, and an additional, smaller ampulla with only one branch; and there is substantial morphological variation within the family Agelenidae (Řezáč et al. 2017). The oecobiid *Uroctea* has "numerous" glands that, despite having different sizes, apparently produce the same product (judging by their staining properties) (Kovoor 1987). The major ampullate glands of the araneids *Gasteracantha*, *Cyclosa*, and *Cyrtophora* possess up to at least three histologically distinct secretory regions (Kovoor 1987). A further mystery is the functional significance of the reduction in the number of glands during development in Araneidae. Spiderlings had two pairs of both major and minor ampullate glands on each side, and one of each pair and its corresponding spigot subsequently degenerated (Peters 1993a).

Many spider groups that do not build webs use major ampullate glands to produce draglines, which are said to function as safety lines that allow the spider to climb back up if she falls from an elevated perch. While I have indeed seen wandering spiders such as salticids, sparassids, and clubionids do just this, I am less than completely convinced of the importance of this function. How often is it important for a wandering ground spider like a lycosid that falls or jumps from an elevated perch to be able return to that same perch?

2.2.4 MINOR AMPULLATE GLANDS

A spider's minor ampullate glands often resemble its major ampullate glands in morphology, but are smaller. The physical properties of their silk lines are similar to those of major ampullate gland lines, though they are somewhat more extensible, weaker, less tough, and thinner (Blackledge et al. 2011). Minor ampullate lines are included in the sticky cribellum lines, but also play additional, diverse roles in ecribellate spiders (table 2.1; Řezáč et al. 2017). Gland morphology varies with respect to numbers of branches and sub-branches (Řezáč et al. 2017). The functional significance of differences in minor ampullate morphology is not known, although in amaurobioid families a bifurcation of the minor ampullate glands was associated with producing sticky cribellum silk (Řezáč et al. 2017). Another mystery is the fact that *Araneus cavaticus* (AR) was able to digest (and thus recycle) her major ampullate but not her minor ampullate lines (Townley and Tillinghast 1988).

When the minor ampullate gland silk lines of *Linyphia triangularis* (LIN) emerge they may have an outer adhesive covering that then hardens so that the lines serve to bind other lines together, replacing to some extent piriform gland attachment discs (Peters and Kovoor 1991). The minor ampullate glands of *L. triangularis* produce at least two different products (both of which stain positively for possibly sticky polysaccharides and acid mucosubstances) (Peters and Kovoor 1991). As with major ampullate glands, there is substantial taxonomic and ontogenetic variation in the forms and branching patterns of minor ampullate glands (Townley et al. 1993; Řezáč et al. 2017; P. Michalik pers. comm.).

2.2.5 ACINIFORM GLANDS

Fossils of spigots suggest that aciniform glands (or something similar to them) were the first to evolve (Shear et al.

1989; Vollrath and Selden 2007; Blackledge et al. 2009c). Aciniform lines are relatively thin, and each aciniform gland is small. Typically there are many small aciniform glands and spigots on the posterior median and posterior lateral spinnerets, and multiple aciniform lines are produced simultaneously in broad swaths. In species with long posterior lateral spinnerets, the aciniform spigots are arranged more or less linearly along the spinneret, and the spinneret makes sweeping and tapping movements while the swath is being produced (Rojas 2011; Eberhard and Hazzi 2013). Some species with shorter posterior lateral spinnerets spread them wide while building, and produce swaths of aciniform lines. Spiders typically use aciniform lines to wrap prey, to form the tips of bridge lines floated on the wind (from spigots on the median posterior spinnerets—Peters 1990; Wolff et al. 2014), to produce silk linings for their burrows, and to make dense sheet webs for prey capture (table 2.1).

The amino acid sequences of aciniform silk are remarkably uniform among the repetitive modules, and they have reduced crystallinity and lower alignment; the lines are thus more extensible than major ampullate lines. A single fibroin gene, aciniform spidroin 1 (AcSp1), was expressed in the aciniform silk gland of *Argiope trifasciata* (AR). This protein was composed of highly homogenized repeats that were 200 amino acids long and that had low representation of the crystalline regions of ampullate silk; they did not display substantial sequence similarity with other spidroins (Hayashi et al. 2004). Recent study of the sequence of AcSp1 in a black widow suggested a long history of strong purifying selection on its amino acid sequence, as well as selection on silent sites and intragenic recombination (Ayoub et al. 2013).

Aciniform glands are sometimes said to be the only type of silk gland in many mygalomorphs (fig. 1.9), but the glands and spigots are much less studied in this group, and homologies are not yet certain. Mygalomorphs are thought to use aciniform silk to line their burrows (Coyle 1986), to construct doors to the burrows (Coyle 1986), to wrap prey (Barrantes and Eberhard 2007), and in one family (Dipluridae) to make prey capture webs with silk tunnels and sheets (Coyle 1986; Eberhard and Hazzi 2013). Phylogenetic analyses of spidroin paralogs indicate that a gene duplication event occurred concomitantly with specialization of the aciniform and tubuliform glands (which produce egg sac silk) (Ayoub et al. 2013). Judging by modern mygalomorphs, the earliest fossil species probably produced aciniform silk in multi-strand swaths; there is a more or less linear array of 19–20 similar spigots along the long axis of a fragment of spinneret of the Middle Devonian fossil *Attercopus fimbriunguis* (Shear et al. 1989; Selden et al. 1991).

SEM images indicate that aciniform fibers are slightly wet and adhesive when they first emerge (Kullmann 1975 on ctenizid and diplurid mygalomorphs; Weng et al. 2006 on an uloborid and an araneid; WE on the araneid *A. argentata*). For instance, the theraphosid *Psalmopoeus reduncus* initiated wrapping silk production by attaching aciniform lines to the substrate by wiping her posterior lateral spinnerets back and forth across the surface (fig. 2.5) (Barrantes and Eberhard 2007). The diplurid *Linothele macrothelifera* also wiped her posterior lateral spinnerets across the substrate to attach lines there (Eberhard and Hazzi 2013). The lycosid *Aglaoctenus castaneus* appeared to attach aciniform lines to each other or to her dragline by clapping her spinnerets together (Eberhard and Hazzi 2017). The aciniform lines of the oecobiid *Oecobius annulipes*, in contrast, appear not to be wet or adhesive when they emerge (Glatz 1967), because the lines were held apart by the long anal setae as they emerged from the spigots during prey wrapping, yet there was no obvious sign (e.g., cleaning movements) that these setae become coated with liquid.

2.2.6 FLAGELLIFORM GLANDS

Flagelliform glands are known only in araneoid spiders. They produce the axial lines to which adhesive glue from the aggregate glands is added (below). Each flagelliform spigot is part of a "triplet" on the posterior lateral spinneret that is formed with a pair of adjacent aggregate gland spigots that pour glue onto the flagelliform gland line as it emerges. The flagelliform silk in orb webs lacks β-sheets; its rubber-like, highly extensible properties are associated with its relatively disorganized molecular structure (Blackledge et al. 2011). The dominant repeats probably form repeated β-spirals, producing a spring-like helix that may be the basis for the high elasticity of this silk (Hayashi and Lewis 1998). Both introns and exons of flagelliform genes have undergone concerted evolution, and the intron sequences are more homogenized intra-specifically than are the exons. This unusual pattern could be explained by extreme mutation and

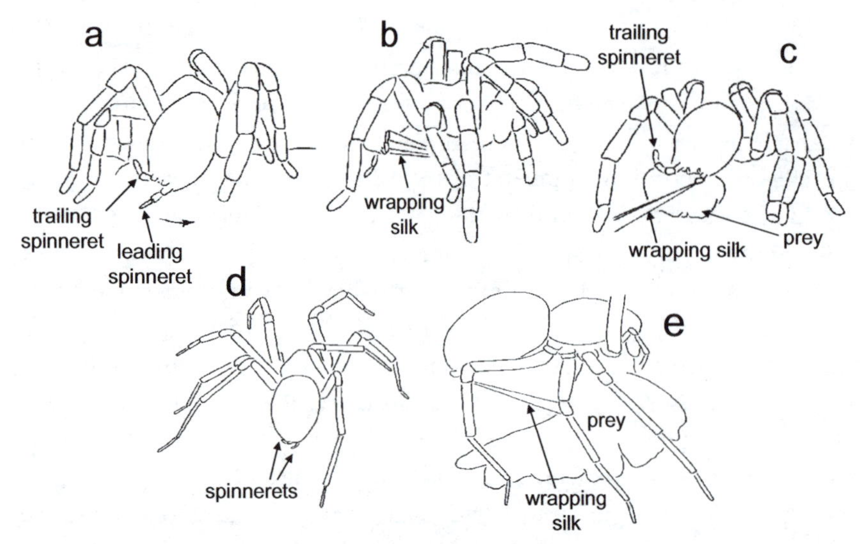

Fig. 2.5. Prey wrapping behavior, shown here in the mygalomorph theraphosid *Psalmopoeus reduncus* (*a–c*) and the zoropsid *Tengella radiata* (*d*, *e*), may have been an important evolutionary source of behavior patterns that enabled spiders to manage swaths of lines during the construction of prey capture webs. Attachments to the substrate by mygalomorphs (which lack piriform silk) were made during both wrapping and web construction by swinging the abdomen laterally, sweeping the leading posterior lateral spinneret and the multiple fine lines it was producing across the substrate (curved arrow in *a*). Typically, both in mygalomorphs (*a–c*) and many araneomorphs (*d*) the trailing spinneret was raised, keeping its swath of fine lines free from the substrate (only a few of the lines in the swath are drawn). By attaching the lines from different spinnerets at different points on the prey and the substrate, the lines in the swaths were spread apart, covering more area on the prey. This same method of spreading swaths of lines was used by the agelenid *Melpomene* sp. and the lycosid *Aglaoctenus castaneus* in building sheet webs (Eberhard and Hazzi 2017) (from Barrantes and Eberhard 2007).

recombination pressures on internally repetitive exons (Hayashi and Lewis 2000, 2001).

Flagelliform lines typically extend several times their resting length before breaking, and the flagelliform silk in orbs is about 1000 times less stiff than major ampullate silk. Pulling on flagelliform silk fibers of the araneid *Araneus* sp. revealed rupture peaks due to "sacrificial bonds" that are characteristic of other self-healing biomaterials (Becker et al. 2003). Flagelliform silk is approximately 10 times more extensible than major ampullate gland silk, and was also much more extensible than the pseudoflagelliform baselines of cribellate sticky silk (fig. 2.6). Extensibility varied by a factor of three among different orb weaver species (Sensenig et al. 2010a). This stretchiness is increased by contact with water, which is normally provided by the liquid in the coating of glue on the line (Sahni et al. 2014), and the effect may be adjusted in environments with different humidities (Opell et al. 2013, 2015; Amarpuri et al. 2015; section 2.2.8.2). This wetting effect also occurs in the major ampullate silk in the gumfoot lines of the theridiid *Latrodectus hesperus* (Blackledge et al. 2005b,c).

The extreme extensibility of flagelliform gland silk has been said to result in part from the sticky spiral baseline being rolled up inside each droplet of sticky silk, and being pulled out when the sticky line was stressed (Vollrath and Edmonds 1989). This "windlass" effect is apparently of little importance in nature, however. The sticky spiral baseline curled on itself when the tension on segments of sticky spiral lines of *Nephila edulis* was reduced by bringing the ends of the line closer to each other (Elettro et al. 2016), but it only seldom curled appreciably in the unmodified webs of one nephilid and six araneid species (Blackledge et al. 2005c; see also Schneider 1995).

2.2.7 PSEUDOFLAGELLIFORM GLANDS

These relatively unstudied glands produce the axial fibers on which sticky cribellum silk is laid (Joel et al. 2015). They are thought to be homologous to flagelliform glands because their elongate spigots occupy similar positions, on the posterior lateral spinnerets. Pseudoflagelliform fibers are stiffer and stronger than flagelliform fibers, but not nearly as extensible (Opell and Bond 2001; Blackledge and Hayashi 2006b). Their lack of extensibility is compensated to some extent because, after the pseudoflagelliform baselines rupture, the highly

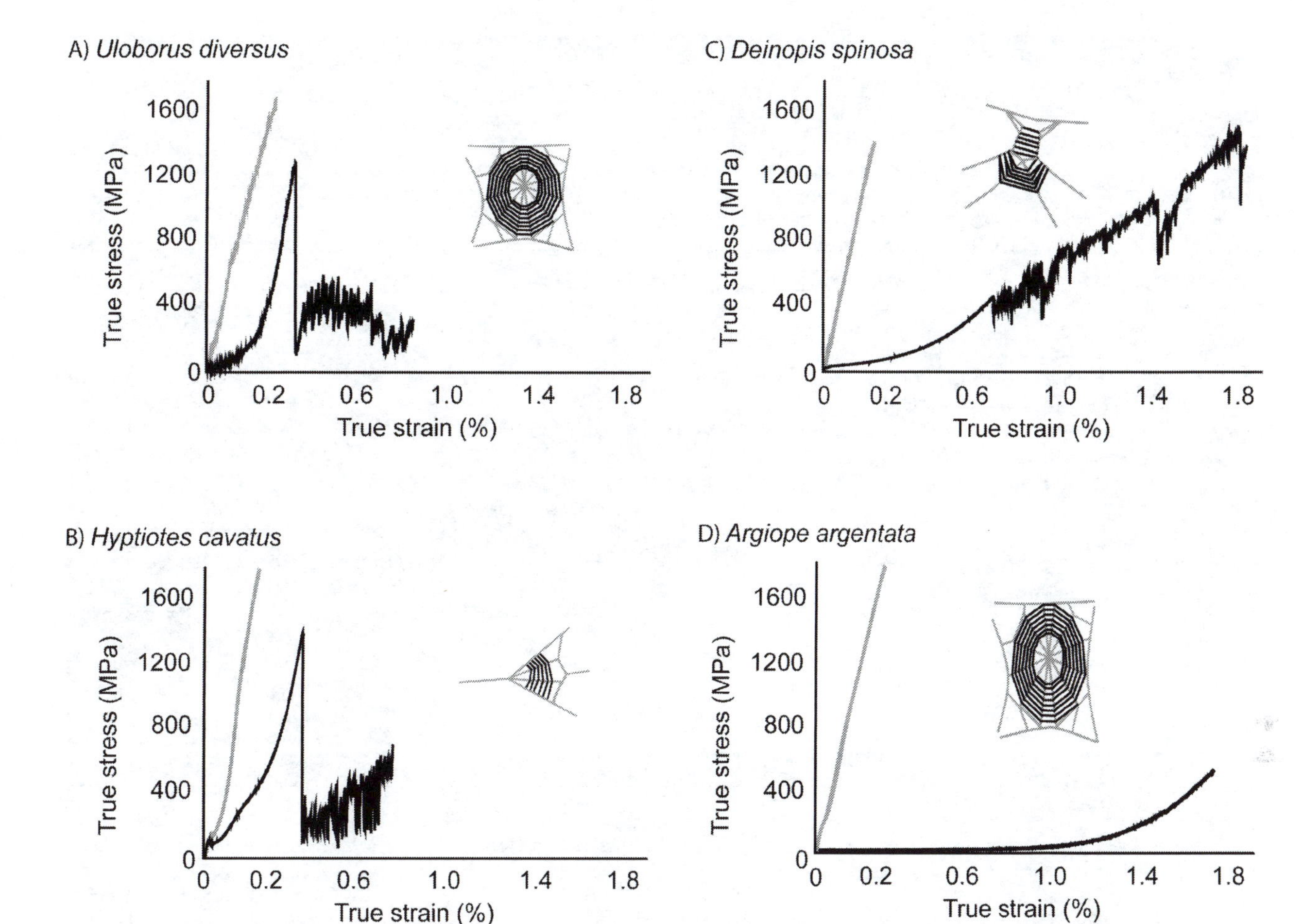

Fig. 2.6. Comparisons showing the strikingly different biomechanical performance of sticky spiral lines (in black) compared with that of the major ampullate gland dragline silk (in gray) of three cribellates (*a–c*) and an ecribellate araneid (*d*) (in the thumbnail sketches of their prey capture webs the sticky lines are black). Major ampullate silk performed similarly in all four taxa, but the sticky spiral of the ecribellate (*d*) differed from that of the cribellates, where most of the stress was generated in the pair of axial fibers; the cribellate axial fibers failed at high peak stresses and moderate strains, and the sticky spiral line then continued to absorb force through the extension and failure of the hundreds of surrounding cribellum silk fibrils until final failure (see fig. 10.18) (the cross-sectional area of these fibrils was not measured, so stress values after failure of the axial fibers can be interpreted only as relative indications of qualitative changes in the forces generated). In the ecribellate, the two axial fibers did not break until the final failure (after Blackledge and Hayashi 2006b).

folded cribellum fibrils unfold (next section). The evolutionary transition from pseudoflagelliform to flagelliform glands in the araneoid lineage is not well understood (Blackledge et al. 2011; Opell 2013; Sahni et al. 2014; section 10.8.5).

2.2.8 STICKY SILK

2.2.8.1 Cribellum glands

The most ancient of the four different types of gland that produce sticky silk are the cribellum glands, which originated in the Triassic (250–200 Mya). They lie just below a ventral abdominal plate (or pair of plates), called the cribellum, which is the vestige of the anterior median spinnerets. The best summary on cribellum silk is that of Opell (2013). There are still major mysteries associated with cribellum silk, including its molecular structure and the question of how cribellate spiders avoid adhering to it (section 2.6).

The cribellum is covered with large numbers of small spigots; there are up to >10,000 in the desid *Badumna insignis*, and nearly 5000 even in the small orb weaver *Uloborus glomosus* (UL) (Opell 1999a, 2002). The spigots are approximately cylindrical, but in most species each spigot has several bulges of unknown functional significance along its length (fig. 2.7). Each spigot produces a single very fine fibril. Fibrils are 25–35 nm in diameter

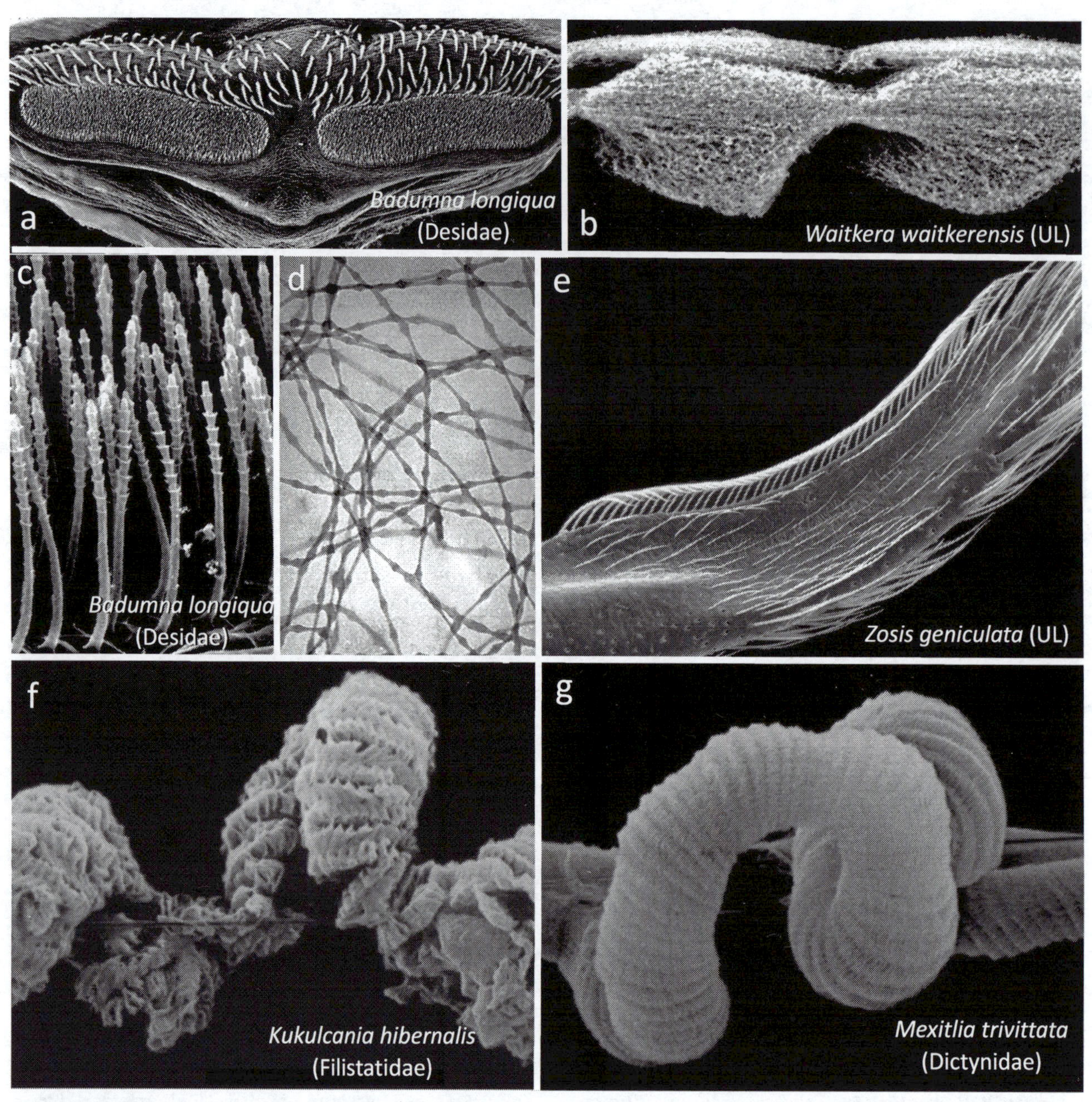

Fig. 2.7. The plate-like cribellum (*a*) has hundreds of tiny spigots (*c*) from which the calamistrum, a comb on tibia IV (*e*), pulls a swath of tiny fibrils that are molded into complex mats (*b*, *f*, *g*) that are composed of hundreds of tiny adhesive fibrils (*d*). The calamistrum brushes posteriorly across the cribellum rapidly, over and over; at the end of each rearward stroke, the cribellum fibrils somehow free themselves from the calamistrum, and the leg moves forward to make the next combing movement. It is unclear how the calamistrum avoids adhering to the treacherously adhesive cribellum silk. The forms of the mats of cribellum fibrils differed substantially among different species (*b*, *f*, *g*); neither the means by which the different-shaped mats were produced, nor the functional significance of the differences (if any) are known (*a*,*c* from Opell 1999a, *b*, *f*, *g* from Opell 2002, *d*, *e* courtesy of Brent Opell).

in most species, but range down to 10 nm. In most cribellate species each fibril has small intermittent swellings that are thought to be hydrophylic. The stickiness of cribellum silk is strongly correlated with the number of cribellum spigots in both orb weaving and non-orb weaving cribellates (Opell 1999a). In orb weavers (but not non-orb weavers) the number of cribellum spigots was positively correlated across species with the spider's weight (Opell 1999a). Spigots with shafts that resemble those of cribellum spigots sometimes occur on the posterior median and posterior lateral spinnerets, producing "paracribellate" fibers; these spigots vary in form, numbers, and locations in different taxa (Griswold et al. 2005).

A special comb consisting of one or two rows of stiff bristles on the spider's hind tibia, the calamistrum, repeatedly brushes rapidly across the cribellum during cribellum fibril production. All species with a cribellum also possess a calamistrum. Recent observations of the basal progradungulid *Progradungula otwayensis* suggest that the cribellum may have evolved before the calamistrum. This species combs out some cribellum silk with the calamistrum, but also produces other cribellate lines without combing; combed cribellum silk was piled more loosely, and was more adhesive (Michalik et al. 2019). A detailed study in the uloborid *U. plumipes* (Joel et al. 2015) supported a model of cribellum silk production that involved a complex, highly coordinated set of leg and spinneret movements. Each rapid rearward brushing movement (about 10/s) combined with a small anterior movement of the cribellum to push and fold the mat of fibrils into an expanded section (a "puff") (fig. 2.7), with a greater density of fibrils where the calamistrum had pushed against the mat. The shapes of puffs vary substantially in different species (Griswold et al. 2005). Individuals of *U. plumipes* in which the clamistra were removed continued to produce mats of cribellum fibrils, but the mats were not formed into puffs. By folding the fibrils into puffs, the spider lengthens the sticky lines and enables them to extend more when stressed by a prey, but the functional significance of different shapes of puffs is unknown.

The spinnerets executed complex, coordinated movements during cribellum silk production. The posterior median and the posterior lateral spinnerets opened laterally and closed and the lateral spinnerets extended posteriorly during each combing movement of the calamistrum. The posterior median spinnerets opened just after the calamistrusm's brushing movement, and paracribellum lines from the long spigots on the posterior median spinnerets became interspersed with the cribellar fibrils; then, when the spinnerets closed medially, the paracribellum lines (and thus, indirectly, the cribellum fibrils) were attached to the pseudoflagelliform axial lines that were emerging from the posterior lateral spinnerets (Joel et al. 2015).

In most species, the axial lines were combined with additional, highly folded "reserve warp" (probably paracribellar) lines that varied both within and between groups in number and shape, from straight to slightly coiled to highly coiled (Eberhard and Pereira 1993; Griswold et al. 2005) (fig. 2.7); in filistatids, for instance, there were large helical fibers, reserve warp fibers, and crinkled axial fibers (Opell 2013). The axial fibers probably serve to sustain the mats; reserve warp fibers may link cribellum fibrils to the axial fibers as in *U. plumipes*, or reinforce the mat after the axial fibers have broken and its cribellum fibrils are being unfolded and extended by a prey.

The unfolding and extension of cribellum fibrils undoubtedly aid in stopping and retaining prey (Opell and Bond 2000; section 3.3). In *Deinopis spinosa* (DEI) up to 90% of the total work performed by capture lines as they were extended could be attributed to these fibrils, rather than to the pseudoflagelliform axial fibers (Blackledge and Hayashi 2006b). The cribellum fibrils were not elastic, and did not return to their initial folded state after being extended (Opell and Bond 2000). Nevertheless, the sticky spiral lines of *Hyptiotes paradoxus* (UL) have been reported to be highly elastic (Marples and Marples 1937) (in contrast to those of other uloborids such as *Uloborus*, *Zosis*, and *Philoponella*) (WE); the sticky lines in a *Deinopis* sp. web also returned to their original lengths after being extended when the spider was induced to briefly expand her web by a nearby noise (WE).

The adhesiveness of cribellum fibrils is thought to result from the combination of three different mechanisms: mechanical interlocking or snagging on irregularities in the surface (e.g., the setae on an insect); van der Waals forces, which are generated when molecules of the fibrils interact directly with molecules on the surface to which they adhere; and hygroscopic or capillary forces, which are generated when water molecules cohere to both a fibril and to the surface. In addition, the epicuticular waxes on the surfaces of insects infiltrate the mat of cribellum fibrils of *U. plumipes*, probably as a result of capillary forces, and increase the strength of adhesion by a factor of eight over wax-free surfaces (Bott et al. 2017). The relative importance of different types of adhesion probably varies. Snagging may be important with irregular surfaces such as setose insect cuticle, but is not necessary for adhesion: cribellum silk adhered tenaciously even to smooth surfaces such as glass. All cribellum fibers but those of hypochilioids and filistatids have small, regularly spaced nodules, and at typical environmental humidities these apparently contribute substantially to the adhesion of fibers via capillary forces. The total adhesiveness of a line of *Hyptiotes cavatus* (UL) (with noded fibers) was twice as great at 46% relative humidity than at 1–2 %. Hydroscopic forces basically replaced van der Waals forces as the most important for adhesion at high ambient humidities (Hawthorn and Opell 2003). Although cribellum lines appear to constitute more effective traps in more humid environments, many cribellate species live in quite dry habitats.

2.2.8.2 Aggregate glands

The second type of sticky silk, the aggregate gland silk of araneoids, is the best studied. Most data on aggregate gland silk are from orb weavers, although the glands also occur in other araneoid families (see reviews of Sahni et al. 2011, 2014; Townley and Tillinghast 2013). There are generally two pairs of these ancient glands (which arose at least 165 Mya), though some have been lost in scattered taxa (Townley and Tillinghast 2013). They produce a complex, watery cocktail of low-molecular-weight, hydrophilic molecules that constitute the sticky droplets that occur like beads on the sticky spiral lines in orb webs, and on major ampullate lines in gumfoot webs

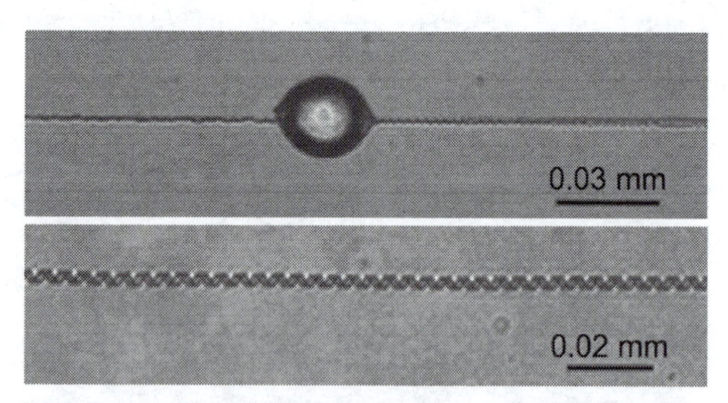

Fig. 2.8. The tightly helical form of these lines and the very sparse droplets of glue in the tropical synotaxid, *Synotaxus* sp., typify the unexplored diversity of properties of silk lines and their functions in different taxonomic groups. The mechanical properties of the silk vary even among closely related species (e.g., Sensenig et al. 2010a), so the previous custom of assuming that lines from a given type of gland have standard properties will have to be abandoned. This will make analyses of the functional significance of different web forms much more challenging (from Barrantes and Triana 2009).

and others (e.g., figs. 2.8, 10.13). These globules constitute the majority of the weight of an orb. Even after the water is removed by evaporation, the water-soluble fraction (largely aggregate gland products) constitutes about one- to two-thirds of an orb's weight (Townley and Tillinghast 2013).

The morphological complexity of the glands, including the densely packed, radially arranged nodules on the ducts (Townley and Tillinghast 2013), suggest a complex secretion process that may include active transport of water and ions (especially phosphate) and possibly organic compounds into the duct lumen (Townley and Tillinghast 2013). Some aggregate gland glycoproteins currently cannot be homologized with other silk proteins.

The crescent-shaped tips of the two aggregate spigots adjacent to the tip of the flagelliform spigot apply the morphologically complex aggregate gland products onto the flagelliform baseline. The viscous liquid glue begins as a cylinder, but as it absorbs atmospheric moisture it almost immediately rounds up spontaneously into a series of morphologically complex ellipsoid droplets, with a thin covering remaining on the baseline between droplets (Townley and Tillinghast 2013). Each droplet contains a core of fibrous glycoproteins (an estimated 5–10% of its volume) surrounded by an aqueous coating (Vollrath and Tillinghast 1991), and a thin surface coat (Peters 1995b). It is the inner core or nodule of glycoproteins that constitutes the principal adhesive (Vollrath and Tillinghast 1991; Townley and Tillinghast 2013). There are also granules of glycoprotein within the core that probably act to anchor the more diffuse region of the droplet to the axial line (Opell and Hendricks 2010), although droplets can slide along the axial lines if sufficient force is applied (B. Opell pers. comm.; WE). There are also hygroscopic salts within the glycoprotein (Sahni et al. 2014). Probably a dynamic equilibrium between the hygroscopic liquid in the glue and the surface coat with the water in the air is dominant in droplet formation (Edmonds and Vollrath 1992; Peters 1995b). Droplet size generally correlates positively with spider size (Craig and Freeman 1991).

The components of the aqueous cocktail surrounding the core (about 80% water) include low-molecular-weight hydrophylic compounds such as amino acids, five different compounds related or identical to known or suspected neurotransmitters (!), and salts (Tillinghast and Townley 1987; Blackledge et al. 2011). These low-molecular-mass (LMM) compounds occur in very different relative amounts in different congeneric species, and even among webs of a single species. Some variation may be due to compounds that the spider is unable to synthesize itself and acquires from gut bacteria and from reingesting previous webs (Townley and Tillinghast 2013). An additional component, potassium dihydrogen phosphate, may prevent bacterial degradation (Schildknecht et al. 1972), though the short length of time most orbs are exposed before being reingested (often 12 hours or less) may not be long enough for this antibiotic function to be important.

The rubbery nature of the baseline and the adhesiveness of the glue are reduced if they are dried out, and one potentially key property of aggregate gland material is the ability to absorb and retain water from the air. The hydrophilic salts and the γ-amino butyramide can increase the adhesive properties of the glue (the volume, contact area, and extensibility of the glycoprotein core) as well as the extensibility and elasticity of the baselines. Droplet volumes increase with greater relative humidity of the air (Sahni et al. 2014; Opell et al. 2015), while viscosity varies with humidity over five orders of magnitude. Adhesion was maximized at very different humidities for five diverse species. Different mixes of proteins and hygroscopic organic salts "tuned" the glue of each species so that it was stickiest in humidity ranges that were typical of their habitats (Amarpuri et al. 2015; see also Opell et al. 2013, 2015). In *Araneus marmoreus* (AR) these differences resulted in biologically significant variation in retention time of houseflies (Opell et al. in prep.). There are probably tradeoffs in the effects of water in increasing the extensibility of the droplets, which allows the formation of "suspension bridges" that recruit adhesion along the thread's length, and decreasing its viscosity, which affects the adhesion of a single droplet. Sharp differences between the water-soluble components of the two *Argiope* species (nine ninhydrin-reactive components were present in the silk of *A. trifasciata*, but only one in that of *A. aurantia*) (Anderson and Tillinghast 1980) hint at further hidden diversity.

A second result of aggregate gland material acquiring water from the air may be maintenance of the spider herself: the water acquired by the spider when she reingests an orb may constitute about 10% of her daily water intake (Edmonds and Vollrath 1992).

A droplet itself can stretch after contacting an object; this is important in providing a "suspension bridge effect" that aids in overcoming the problem in most sticky materials that adhesion during pull off is generated primarily only at the edges of a surface (Opell and Hendricks 2010). As a sticky spiral line begins to peel off an object, the droplets near the edge of the object stretch and remain in contact with its surface, thus contributing adhesive force while more interior droplets become recruited to resist further pull off (Opell and Hendricks 2010).

It has been hypothesized that components of the aggregate gland glue also function as a poison to reduce prey resistence. Such a poison would have to act quite rapidly, however, and in very small amounts, and there are other possible functions for these compounds (Anderson and Tillinghast 1980). Another, perhaps more likely, additional function is for amphiphilic lipids to increase the wettability of insect epicuticles, increasing the glue's contact area with the prey (Townley and Tillinghast 2013).

The watery nature of the droplets of glue might suggest that they are not particularly costly in terms of materials, compared with other types of silk. The aggregate glands are large, however (Townley and Tillinghast 2013), and there are several suggestions that their products are a limiting resource. The amounts of the hygroscopic compound choline in the sticky lines of the araneids *Argiope trifasciata*, *A. aurantia*, and *Araneus cavaticus* decreased when spiders were starved (Townley et al. 2006), and other low-molecular-mass compounds that are costly for the spider to synthesize were also reduced. Experimental manipulation of the amounts of sticky spiral silk available in the spider's glands when orb construction began had strong effects on several aspects of orb size and design (Eberhard 1988a; section 7.3.3.2).

Substantial variations in function may be waiting to be discovered. The surface reflectance (brightness) of sticky lines (and thus the visibility of the orb) correlated positively with droplet size, but the correlation was closer in nocturnal than in diurnal orbs and varied within some genera and also between genera (Craig and Freeman 1991). The adhesion forces of glue on sticky spiral lines differed in different species of *Tetragnatha* endemic to the Hawaiian islands (Alicea-Serrano et al. in prep.).

Differences in aggregate gland silk may be especially strong in non-orb weavers (see Sahni et al. 2011 for contrasts between the araneid *Larinioides cornutus* and the theridiid *Latrodectus hesperus*). The low-molecular-mass compounds in *Latrodectus* spp. (THD) differed strongly from those in araneoids (in some samples their most abundant component was completely unknown in araneids), and there are also differences among different theridiid genera (Kovoor 1977a,b; Kovoor and Lopez 1983). Sahni et al. (2011) found that with respect to stretching, gumfoot glue behaved like a viscoelastic liquid and was not sensitive to humidity differences, while the sticky spiral glue of *Larinioides cornutus* (AR) behaved like a viscoelastic solid and experienced reduced elasticity at higher humidity and peak adhesion at intermediate humidities; they proposed that the differences may be due to different tackifiers.

The more posterior of the two aggregate glands in theridiids apparently produces the glue used in wrapping attacks on prey and has short ducts that lack the elaborate system of nodules typical of araneids and present in the anterior gland, which produces glue for the web (Townley and Tillinghast 2013). The ducts of the aggregate glands of linyphiids, which are apparently used only for web lines, lacked nodules entirely (Townley and Tillinghast 2013).

The glue on gumfoot lines in *Latrodectus hesperus* lacked the internal nodules that are largely responsible for adhesion in araneids. The entire mass of glue at the tip of a gumfoot line may often collapse onto the prey when the break-away attachment to the substrate fails, so design differences are not unexpected. Other species of *Latrodectus* also place glue on other, not easily broken lines in other portions of the web (Eberhard et al. 2008b), but the internal structure of these other droplets is unknown.

Another mystery concerns the tiny, barely perceptible, droplets in the webs of the theridiids *Theridion hispidum* and *T.* nr. *melanostictum* and *Anelosimus* spp. (Eberhard et al. 2008b), the pholcids *Modisimus guatuso* (Briceño 1985) and *Belisana* spp. (Deeleman-Reinhold 1986; Huber 2005) (fig. 9.19), and, the synotaxids *Synotaxus* spp. (fig. 2.8, Eberhard et al. 2008b). These droplets are so small and sparse that they are barely adhesive, and seem likely to be ineffective against all but the smallest, weakest prey. Even tinier droplets that are only visible with magnification occur on tangle lines above the sheet in the webs of species of *Anelosimus* (Eberhard et al. 2008b) and *Nihonhimea tesselata* (THD) (Barrantes and Weng 2006b), and on tangle as well as the sheet lines in some linyphiid webs (Nielsen 1932; Peters and Kovoor 1991; Benjamin et al. 2002). Benjamin et al. (2002) speculated that the droplets serve to fasten the lines of the web to each other, as the webs have few piriform attachment discs (Peters and Kovoor 1991); but the droplets are not restricted to junctions and there is as yet no direct evidence of such adhesion (see, however, section 9.4 on linyphiid sheets). In the synotaxids and *Belisana* pholcids, the droplets are far from other lines (figs. 2.8, 9.19, Eberhard 1977a) and surely do not function in this way.

2.2.8.3 Venom glands that produce contractile sticky "webs" in Scytodidae

A third type of sticky material (its classification as "silk" is preliminary) is part of a spectacular predatory innovation in the small "spitting spider" family Scytotidae. One portion of the venom glands associated with the chelicerae is modified to produce both elastic lines and sticky glue that are shot from the spider's tiny chelicerae onto the prey (Foelix 2011; Stratton and Suter 2009). Raising her basal cheliceral segments and oscillating her fangs rapidly (at about 278–1700 Hz!), the spider *Scytodes thoracica* "spits" at very high velocity (up to 29 m/s!) an orderly, web-like array of zig-zag gluey lines onto the prey and the surrounding substrate (Stratton and Suter 2009). The line is elastic, and is launched in an elongated state, and then contracts during the 0.2 s after it strikes; this contraction probably serves to "wrap" the prey and fasten it to the substrate (Stratton and Suter 2009). Following a successful spit, the spider approaches cautiously, delivers a delicate bite to any small protruding extremity (her small cheliceral size, probably necessary to permit rapid spitting movements, makes her able to bite only small prey structures), and steps back until the prey becomes immobile.

Little is known regarding the composition of these contractile "webs." They are probably not toxic for prey

(Clements and Li 2005), though some prior observers suggested that they contained venom (Nentwig 1985a). Given that the spiders possess genes that code for the silk proteins in their abdominal silk glands and spinnerets, and that there are ultrastructural similarities between the venom gland cells and cells in the abdominal silk glands (Kovoor and Zylberberg 1972), it seems likely that the lines shot from the chelicerae are silk (Suter and Stratton 2009); the glue has not been characterized, however. No sticky material is included in the sparse, non-adhesive tangle webs that some *Scytodes* species build with silk produced by their spinnerets (Nentwig 1985a; Gilbert and Rayor 1985; Eberhard 1973a, 1986a).

2.2.8.4 Ampullate glands in *Loxosceles*

Still another type of sticky line is known only from a small group of poisonous, "brown recluse" spiders in the genus *Loxosceles*, and has scarcely been studied. The highly adhesive sticky silk is not associated with any liquid; it consists of wide, very thin ribbons about 2–4 μm wide that have been estimated in different species to be only 12–40 nm thick (Lehmensick and Kullmann 1956; Coddington et al. 2002; Knight and Vollrath 2002). The ribbons are laid loosely on the substrate near the spider's sheltered retreat under a rock or a trunk (Vetter 2015). The mechanism responsible for their adhesiveness is unknown; speculations include van der Waals and electrostatic forces (Knight and Vollrath 2002) (fig. 2.9). Clumps of loose adhesive silk of *L. arizonica* moved toward objects like fingers, forceps, and a glass slide that were brought near, so electrostatic interactions may be involved (Coddington et al. 2002). The ribbons are unusual in that the liquid gland product is apparently converted to a semi-cristalline state abruptly, at the distal tip of the spigot; the valve that usually occurs within the duct of the ampullate gland of araneids is lacking (Knight and Vollrath 2002). Unique, highly modified toothed setae near the spigots may be involved in silk production (fig. 2.9). The sticky silk of *Loxosceles laeta* is not known to be associated with baselines corresponding to the flagelliform or pseudoflagelliform lines of the other groups (Lehmensick and Kullmann 1957; Knight and Vollrath 2002; Margalhaes in press). The homology with the ampullate gland silk of other spiders is uncertain.

2.2.9 PIRIFORM GLANDS

Both ampullate and piriform silk appeared abruptly with the araneomorph spiders, and are present throughout the modern species in this group (Griswold et al. 2005). The evolutionary sequence may have been complex, however, as some piriform protein products probably had a separate evolutionary origin (below). The association between ampullate and piriform lines makes sense functionally, as fastening a few thicker lines to each other with a mass of fine lines and glue seems more effective than the more ancient technique using swaths of thin aciniform lines using only their weak stickiness when first produced. Piriform silk may have originally evolved to attach lines to the substrate (e.g., in silk linings of tunnels), and only later to attach single lines to each other. Attachments to the substrate are less demanding in terms of precise positioning of the piriform lines. Many wandering spiders, like salticids, repeatedly attach their trail lines to the substrate, and also make attachment discs that attach lines to the substrate but not to other single lines (Li et al. 2002).

The piriform glands probably had especially important consequences for the evolution of webs. Their products fasten silk lines firmly to each other (fig. 2.10, 2.11) and to the substrate (fig. 2.10b,c,e), and also (in uloborids) help wrapping lines from the aciniform glands adhere to prey (Peters 1982). Before these glands evolved, spiders could probably only fasten their lines more weakly to other objects and lines, dabbing or wiping their spinnerets on the substrate to which they wished to attach (section 5.3.8, box 5.2). Adhesion probably resulted from the silk lines being slightly wet when they first emerged (fig. 2.11), or to occasionally snagging lines of silk on the substrate. As with many types of silk glands, piriform glands appeared abruptly with the origin of araneomorph spiders (Blackledge et al. 2009c). The evolution of a combination of strong ampullate lines and piriform attachments that fasten them firmly opened up a new world of possible web forms for spiders (section 10.4).

There are at least two piriform gland products: fibers; and an amorphous, hydrocarbon-rich cement-like substance that includes aligned nanofibrils, lipid enclosures, and a dense, isotropic boundary layer (Kovoor and Zylberberg 1980; Peters 1992; Grawe et al. 2014; Wolff et al. 2015) (see fig. 2.10). The gland products are complex chemically, and the fibers may include at least two

Fig. 2.9. These greatly broadened comb-like setae (some are labeled "c") constitute a morphological reminder of the many remaining mysteries that surround variations in silk constitution and production. Setae of this sort are known only from the posterior lateral spinnerets of the haplogyne fiddle back spider, *Loxosceles laeta* (Sicariidae); presumably they function in conjunction with the very rapid spinneret movements that occur during sticky silk production (Margalhaes in press). Possible functions that have been proposed include such diverse uses as pressing the thin ribbons of silk that are produced by the anterior lateral spinnerets, directing this silk onto the web, pulling it from the anterior lateral spinnerets, or charging it electrostatically as it emerges (scale bar is 20 μm) (from Knight and Vollrath 2002).

unique proteins (Blackledge 2013). Some proteins may be exceptional in not belonging to the spidroin gene family (Hu et al. 2007). The piriform silk in sticky spiral-radius attachments in orb webs of the araneids *Araneus diadematus* and *Argiope aurantia* differed from silk in radius-frame attachments with respect to its affinity for histological stains, and birefringences (Tillinghast and Townley 1987; Townley and Tillinghast 2013), as well as the mechanical properties of the "pulley-like" attachments (below). There may be a great deal of as yet unexplored diversity in piriform gland products.

The adhesion produced by piriform attachment discs, especially those cementing lines to each other, is often strong: when orb webs break under the stress of prey impact, failure is usually due to rupture of the lines themselves rather the junctions between them (section 3.3.2.9). Even attachments to smooth substrates such as Teflon and smooth, waxy leaves are often strong enough to hold the spider's weight (Grawe et al. 2014). The most common type of failure of attachments to a leaf or Teflon film was for the entire disc to peel off. In contrast, when the surface was a glass microscope slide, peeling off never occurred, and the most common type of failure was breakage of the lines that joined the dragline to the area in contact with the leaf (Grawe et al. 2014). The attachments of sticky spiral lines to radii in the orbs of

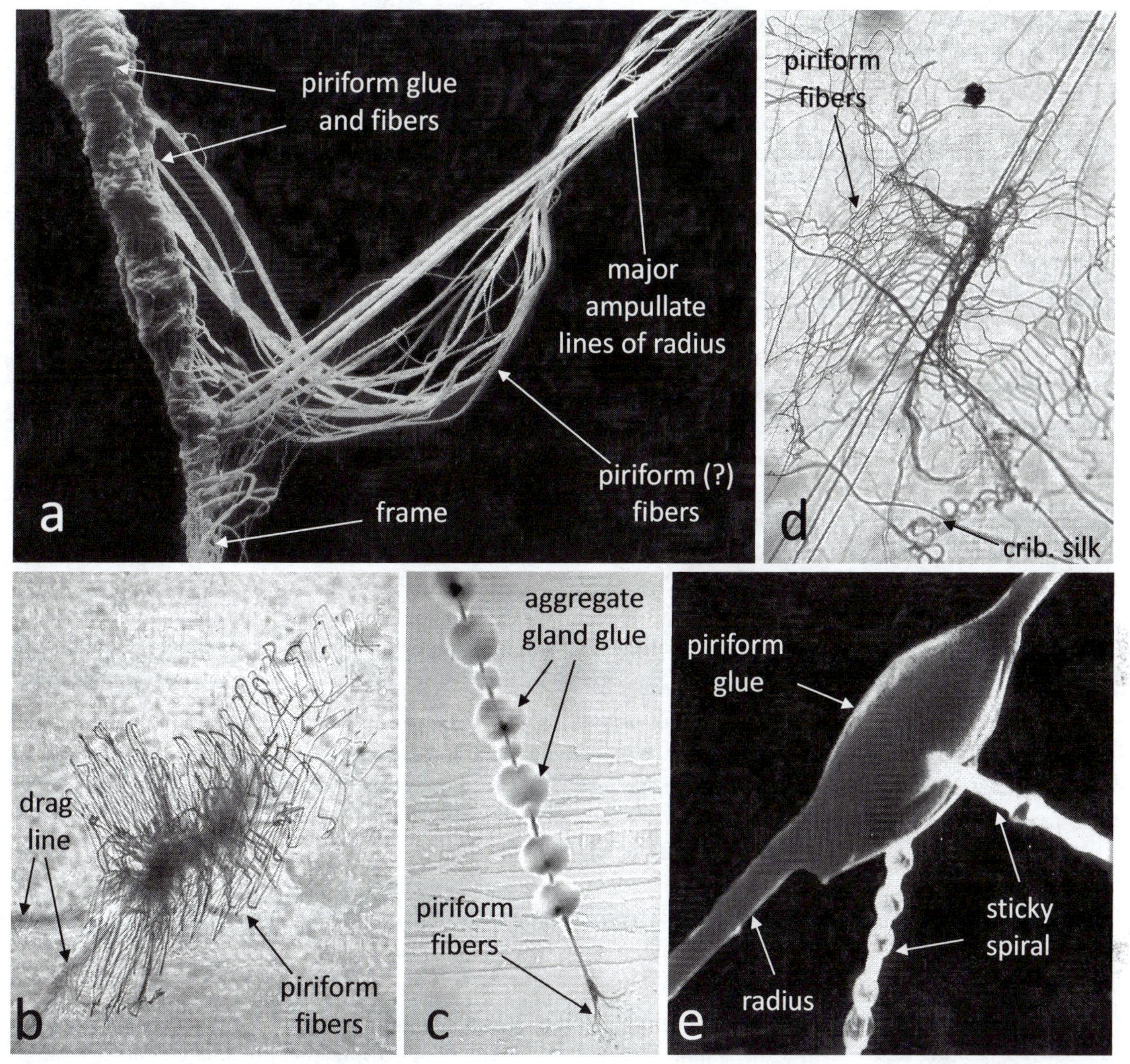

Fig. 2.10. A crucial development in the evolution of spider webs was the piriform disc. Composed of both fine fibrils and liquid glue, piriform discs fasten silk lines to the substrate (*b*, *c*) and to other silk lines (*a*, *d*, *e*). As shown in this sampler, the discs vary in their morphology. *a*) This SEM image shows part of a typical attachment between two non-sticky lines, a radius to a frame line of *Cyrtophora citricola* (AR). Attachments of this sort are typically stronger than either of the lines that they join. *b*) This typical piriform disc attaches a non-sticky dragline of *Latrodectus geometricus* (THD) to a glass slide; the zig-zag pattern of fine lines increases the area of contact with the substrate and thus probably also the strength of the adhesion. *c*) This specialized, reduced piriform disc was built by the same instar of *L. geometricus* as *b* to attach a gumfoot line to a slide. It is designed to be easily broken by tugs by prey that adhere to the glue droplets near the tip of the gumfoot line; when the attachment disc breaks, the tension on the line lifts the prey away from the substrate, where it dangles helplessly. *d*) This piriform disc attaches a line with cribellum silk to the sheet of the zoropsid *Tengella radiata*. Some of the piriform lines are wrapped around strong, straight lines as in *a*, but many are splayed apart on the very fine lines of the sheet (which are not visible at this magnification). *e*) This SEM image shows a coat of glue covering the piriform attachment of the sticky spiral to a radius in the orb of *Nephila madagascarensis* (NE). This special "pulley" attachment breaks partially when force is applied to the sticky spiral line, allowing the sticky spiral baseline to slide past the radius, but maintaining its adhesion to the radius (section 3.3.3.1.6, fig. 3.33) (*a* and *e* from Kullmann 1975, *d* from Eberhard and Hazzi 2017, *b* and *c* courtesy of Gilbert Barrantes).

some araneids and nephilids are unusual in allowing the sticky spiral lines to slide through the attachments ("pulley attachments"), increasing the web's ability to stop and retain prey (section 3.3.3.1.6).

Piriform glue is also quite unusual in how rapidly it hardens (S. Zschokke pers. comm.; Wolff et al. 2015). Most man-made adhesives take seconds or minutes to harden, but piriform attachments made during web construction resist tension almost immediately after they are made; Samuel Zschokke (per. comm.) measured hardening times of around 0.2 s in *Araneus diadematus* (AR). The piriform attachments made during the construction of orbs and other webs when the spider "cinches up" a frame line (sections 6.3.2, 6.3.3, figs. 6.4g, 6.5c) or tightens a line by adding attachments to the substrate or other web lines (fig. 2.10) involve particularly substantial tensions that are applied almost instantaneously.

2.2.9.1 Spinneret morphology

The piriform spigots are on the anterior lateral spinnerets, with their tips often in approximately the same plane on the spinneret's flattened medial-distal surface (e.g., fig. 2.11, Griswold et al. 2005). This flat surface probably permits multiple spigots to touch a flat substrate simultaneously when the spider spreads her anterior lateral spinnerets. In some species, such as *Nephila pilipes* (= *maculata*) (NE) (Hingston 1922a) and *Micrathena duodecimspinosa* (AR) (WE) the spider may facilitate contact of these spigots with the substrate by rocking her abdomen from side to side while she makes an attachment. The positions of the piriform spigots on the anterior spinnerets are appropriate for fastening the ampullate gland lines, which are also produced by the anterior lateral spinnerets, to the substrate or other lines; but they are not appropriate for fastening the aciniform lines, which emerge farther posteriorly, to other objects or lines.

Diverse morphologies, positions, and numbers of piriform spigots are illustrated in the taxonomic literature (e.g., Griswold et al. 2005). The functional significance of the differences is largely unknown, though there are exceptions. The more than two-fold difference in the numbers of piriform spigots in *Poltys illepidus* and *P. laciniosus* (AR) may be associated with the need for a stronger dragline attachment in the heavier *P. illepidus*, which seemed more disposed to jump and dangle on its dragline when disturbed (Smith 2006). Gnaphosid spiders that did not build webs attacked prey by running past and around them, producing a band of sticky silk that immobilized the prey's legs (Bristowe 1958), and they used their enlarged anterior lateral spinnerets with derived piriform spigot morphology to produce the band (Wolff et al. 2017b).

On the other hand, the significance of the loss of piriform spigots in the diguetid *Diguetia* sp. (Platnick et al. 1991), a genus in which spiders make aerial webs with many firm attachments of lines to the substrate and to each other, is a mystery. Web design probably correlates with the relative sizes of piriform glands themselves in some groups. For instance, a typical linyphiid web may have more attachment discs, especially in the tangle above the sheet, than does an orb; correlated with this, a mature female *Linyphia triangularis* has numerous, fairly large piriform glands (Peters and Kovoor 1991). One exceptional species is the sand dune eresid *Seothyra henscheli* (box 9.3). The piriform spigots bristled in different directions at the tip of the anterior lateral spinneret like the spikes of a mace (fig. 9.31d). The spider forms mats of grains by stabbing her anterior lateral spinnerets repeatedly into the sand; the lines from the bristling piriform spigots bind the grains together into a mat to which the spider then attaches her web lines (Peters 1992; fig. 9.30).

2.2.9.2 Morphology of attachment discs

Piriform attachments ("attachment discs") are complex morphologically (fig. 2.10, 2.11) (Schütt 1996; Wolff et al. 2017a). Attachment discs fastening the dragline to a planar substrate had piriform fibers enclosing the dragline that also spread approximately radially on the substrate (fig. 2.10b). This architecture results in a large contact area generated at low material cost, and probably distributes forces over a high number of branches, lowering the stress generated in each single structure. In addition, propagating cracks may be arrested because there are fibers embedded in the cement-like substance. In addition, peeling angles are kept low, making the attachment resist higher forces before detaching (Grawe et al. 2014; Wolff et al. 2015). In *Nephila plumipes* the insertion point of the dragline was shifted toward the center of the attachment disc, forming a flexible tree root-like network of branching fibers around the loading point (Wolff and Herberstein 2017).

Discs attaching draglines to the substrate were morphologically diverse (fig. 2.10, Schütt 1996). One specialized type is theridiid gumfoot line attachments to the substrate, which break easily, allowing the line to lift the prey up, out of contact with the substrate (e.g., Bristowe 1958; see chapter 9). These discs had fewer fine lines contacting the substrate than did other substrate attachments of non-sticky lines of the same spiders (Eberhard et al. 2008b; Sahni et al. 2012, fig. 2.10a,b). The attachments of lines to the surface of tropical streams and lakes by the theridiosomatid *Wendilgarda* spp. had a similar sparse appearance (Coddington and Valerio 1980; Eberhard 2000a, 2001a). In contrast, the substrate attachments of strong radial lines by the filistatid *Kukulcania hibernalis* were compound, including both broad swaths of many fine (presumably aciniform) lines that were applied with rapid waggling movements of the PL spinnerets, and also had smaller masses of possible piriform material (WE) (fig. 10.11). The attachment discs of the pholcid *Pholcus phalangioides* were extreme in being puddles of a liquid silk that lacked fibers (Schütt 1996); the liquid was probably produced by a single pair of very wide spigots that may be derived from piriform spigots (Platnick et al. 1991). The egg sacs of *Zygiella x-notata* (AR) had two types of substrate attachments—typical piriform discs, and more liquid "glue droplets" (Gheysens et al. 2005).

The morphology of piriform discs that attached one line to another involved, in contrast, many fine lines that were wrapped around the two lines being joined, as well as glue (fig. 2.10a,e, 2.11c,d; see also Kavanagh and Tillinghast 1979; Grawe et al. 2014). In *Cyrtophora citricola* (AR) the liquid itself had at least two components, one electron dense and the other electron lucent (Peters 1993a). Details of how these different components emerge from the spigots are not known. Images of at least some piriform lines of the eresid *Seothyra henscheli* suggested they were "wet" when they were produced (Peters 1992). Attachments between lines can differ within a single web (fig. 2.10c–f). In *Cyrtophora citricola* (AR), the attachments of the non-sticky spiral to the radii in the sheet (fig. 1.7) were substantially shorter than the attachments between other non-sticky lines in the upper tangle (Peters 1993a). In the sheet web of the lycosid *Aglaoctenus castaneus* some piriform attachments were much wider and longer than others (fig. 2.10e,f), and appeared to attach the dragline simultaneously to multiple lines in the sheet (Eberhard and Hazzi 2017). The attachments of sticky spiral lines to radii in *Nephila madagascarensis* (NE) (Kullmann 1975) and in several other species of araneids and tetragnathids (WE) differed from the attachments between non-sticky lines in having a relatively greater mass of "glue" and a smaller quantity of fibers (Kullmann 1975, fig. 2.10c). The lines in the dense sheets of *Linyphia triangularis* and *Microlinyphia pusilla* (LIN) were joined by a variety of mechanisms (Benjamin et al. 2002). In sum, the morphology of attachment structures produced by piriform glands is diverse.

2.2.9.3 Different attachment disc morphologies result from spinneret behavior and morphology

Differences in the forms of attachment discs that were made to smooth surfaces by the nephilid orb-weaver *Nephila plumipes* and the webless, trunk-living sparassid hunting spider *Isopeda villosa* (which was 2–3 times heavier) were associated with differences in both spinneret behavior and morphology (Wolff et al. 2017a). The mechanical properties of the dragline attachments of *I. villosa* were strongly "directional"; their highest pull-off resistance was when they were loaded parallel to the substrate and in the direction along which the spider had moved (that is, the direction in which loading often occurred in nature). The substrate attachments of *N. plumipes* were more robust in different loading directions (again, as occurs when webs are anchored to substrates in nature); they resisted pull-off at 90° about twice as well as those of *I. villosa*, despite being produced more rapidly. A larger field of piriform spigots in *N. pilipes*, and rubbing movements of the spinnerets applying piriform material to the dragline, were correlated with these differences (Wolff et al. 2017a).

Peters (1993a) thought that the spinnerets of *C. citricola* "glide along the threads to be connected" (p. 159) as the spider moves. The parallel orientation (on the right in fig. 2.10a) of the piriform fibers and new line that was being laid as the spider moved in the tail end of an attachment, as well as video recordings of the slender posterior section of the piriform field in *Nephila plumipes* (Wolff et al. 2017a), were in accord with this idea. The jumble of piriform lines in the central area of the attachment disc, in contrast, came from the broad anterior section of the piriform field (Wolff et al. 2017a). In general, understanding the spatial relations between the morphology and behavior of the spinnerets, spigots, and lines of different species is complicated by the fact that the area around the spinnerets was inflated in many published photographs (for better visibility); they may thus be in unnatural positions. Probably many details remain to be understood of the processes by which piriform lines emerge from their spigots and are applied to the ampullate or sticky spiral lines that are produced at the same time as well as to other lines already in place (e.g., fig. 2.10a).

The multiple orientations of piriform fibers may be produced by rubbing movements of the spinnerets against each other, such as the brief, rapid anterior-posterior brushing movements by the anterior lateral spinnerets of *Nephila clavipes* (NE) that were observed each time the sticky spiral was attached to a radius (appendix O6.3.6.3). Moving the anterior lateral spinneret anteriorly could bring the new major ampullate gland line into contact with piriform lines that had just been attached to the old line. In a species like *N. clavipes*, the spider's abdomen is more or less parallel to the radius throughout both temporary and sticky spiral construction (fig. O6.4), so new spiral lines will be more nearly perpendicular than parallel to the long axis of the spider's abdomen as they are being attached to a radius; this could bring the major area of piriform spigots on the anterior lateral spinneret closer to the spiral lines.

2.2.9.4 The "piriform queen"—*Cyrtophora citricola*

The piriform attachments of *Cyrtophora citricola* were not morphologically unusual (fig. 2.11), but the speed and agility with which they were produced during web construction was extraordinary: about 4 attachments were made each second, for periods lasting up to many minutes while the non-sticky spiral was being constructed (fig. 1.7) (Kullmann 1958; WE). This speed is spectacular, because each attachment involved complex behavior: finding the next radius; positioning it precisely between the anterior lateral spinnerets and at a precise distance from the previous attachment to this radius; initiating piriform silk production; and then cleanly terminating piriform production. The precisely coordinated processes of initiation and termination probably lasted on the order of only hundredths of a second. The piriform spigots of this species were also unusual (Peters 1993a). There were two distinct sizes, and the smaller spigots were particularly numerous (fig. 2.11a) compared with those of other "typical" araneoid orb weavers such as *Araneus diadematus* and *Argiope bruennichi*. There was also an unusually large U-shaped bald patch on the side of the spinneret distal to the major ampullate spigot; the outer margin of this patch was lined with especially small piriform spigots. The functional significance of these traits is unclear (Peters 1993a).

2.2.10 EPIANDROUS GLANDS

The small, somewhat mysterious epiandrous (or epigastric) glands that occur only on the ventral surface of the abdomen of mature males constitute one final set of silk glands. They are superficially similar to piriform glands (Peters and Kovoor 1991), but they empty externally through a few fine spigots located just anterior to the external male gonopore, far anterior to the spinnerets. Epiandrous glands seem to be present throughout the order (Foelix 2011), though they are absent in scattered species (Agnarsson 2004). The glands and their spigots are so inconspicuous that their absence in other groups could easily go unnoticed.

The mystery of the epiandrous glands is not the function of their silk, but why it is needed. They produce a small mat of fine lines in the "sperm web," onto which the male deposits a droplet of the semen directly from his gonopore; he then takes up the semen into his pedipalps. But why is this tiny mat necessary? For instance, the male of the theraphosid *Grammostola monticola* first makes a larger, dense sheet of fine "aciniform" lines, then adds a small, seemingly superfluous mat of epiandrous silk onto which he deposits a droplet of semen (Costa and Pérez-Miles 2002). The filistatid *Kukulcania hibernalis* also adds epiandrous gland lines to an already dense sheet of lines (Barrantes and Ramírez 2013).

Could it be that the epiandrous silk has some special property that somehow facilitates coherence of the droplet or its uptake by the palps? Perhaps, for instance, it is especially hydrophobic (I. Agnarsson pers. comm.). The fact that the epiandrous glands of *Linyphia triangularis* produce two types of secretion with different staining affinities (Peters and Kovoor 1991) hints at the possible importance of special chemical properties.

Alternatively, it might be that the epiandrous mat is an ancient fail-safe device to assure sperm are not lost. The spider's orientation just prior to sperm deposition may be only approximate, and the male cannot afford to lose track of the site where he will deposit his small but crucial droplet. To my knowledge, sperm deposition always follows immediately after deposition of epiandrous lines; the male moves his abdomen slightly anteriorly and deposits the droplet.

2.2.11 OTHER PRODUCTS ASSOCIATED WITH SILK

There are other chemical products that are associated with silk lines but whose provenance remains to be clarified. One common type are the sex pheromones produced by mature females, which may be lipids or lipid-soluble substances (Gaskett 2007). Another recently discovered and presumably rare product, is an ant repellent pyrrolidine alkaloid that may function to keep ants from invading the webs on the thick non-sticky lines of larger juveniles and adults of the giant nephilid *Trichonephila antipodiana* (Zhang et al. 2011). The possibility of similar effects has very seldom been checked, and there are many species of very small ants which could possibly use thinner lines (ants do occur on some densely meshed sheet webs with accumulated detritus—WE); ants are important predators of *Stegodyphus* spiders (Henschel 1998; Wright et al. 2016). A third recently discovered property is the antibiotic effect of silk from both the retreat and the prey capture web of the eresid *Stegodyphus dumicola* (Keiser et al. 2015) and the agelenid *Tegeneria domestica* (Wright and Goodacre 2012). The functional significance (if any) of this effect for these spiders has not been determined.

2.2.12 CONTROL OF RATES OF SILK SECRETION IN GLANDS

Spiders make substantial adjustments in the rate of silk synthesis. Emptying the accumulated contents of the lumen of the major ampullate gland of *Araneus* (AR) stimulated transport of silk already present in secretory vesicles in the cytoplasm of cells lining the gland's lumen (partially refilling the gland in 30 min), and stimulated synthesis of new silk (Witt et al. 1968; Moon and Tillinghast 2004). Vescicle numbers and lumen contents were replenished in several hours: discharge into the lumen was still heavy at 4 hr, and was nearly complete at 8 hr. Both cholinergic stimulation of the glands (as in stimulation from nerves) and depletion via mechanical pulling induced synthesis (Peakall 1964, 1965). The

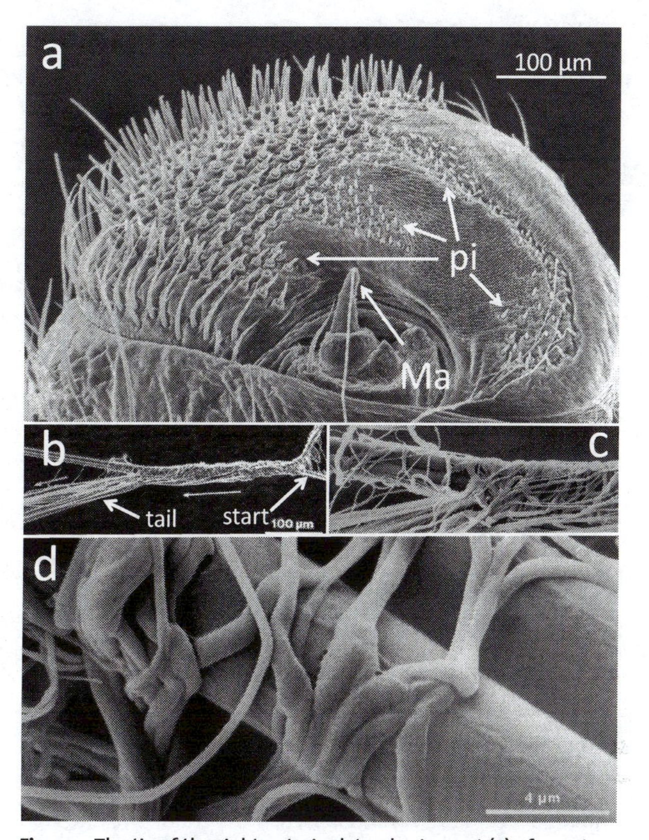

Fig. 2.11. The tip of the right anterior lateral spinneret (*a*) of a mature female of the "piriform queen" of spiders, *Cyrtophora citricola* (AR) is largely covered with small piriform spigots (all of the many spigots are piriform except the single large major ampullate spigot [Ma], which has a line emerging from it). This species makes extraordinary numbers of piriform attachments (>22,000/web) with extreme precision and speed (>4/s) while building its sheet (fig. 1.7, appendix O6.3.5.1) (these spigots took only approximately 0.03 s to produce each attachment). The surface that bears the piriform spigots is approximately planar (as in many other species), probably to facilitate simultaneous contact of many spigots with a surface to which an attachment is being made. There were two size classes of piriform spigots in this species (Peters 1993a), and lines with various diameters were included in each attachment (*b–d*). The smaller spigots are particularly numerous in this species compared with other "typical" araneoid orb weavers such as *Araneus diadematus* and *Argiope bruennichi*. There is also an unusually large U-shaped bald patch near the major ampullate spigot; the outer margin of this patch is lined with small piriform spigots. Judging by the piriform lines in *b*, initiation of an attachment (right side in *b*) was more precise than termination (the left side of the attachment in *b* has a short tail of apparently wasted piriform lines). At greater magnification (*d*), the piriform lines seem to have been covered by a thin layer of liquid. The relationship (if any) between the especially numerous small piriform spigots (which were not specialized for producing glue rather than lines, because some had fibers emerging from them), the large bald patch, and the extraordinarily precise and rapid initiation and termination of piriform silk deposition is not clear (photos and data are from Peters 1993a).

overall rate of production was positively correlated with demand for silk (e.g., web production) (Witt et al. 1968; Reed et al. 1970). As would be expected for an animal that depends on silk to obtain food, the glands persisted in synthesizing silk even in the face of sustained starvation (Witt et al. 1968; Reed et al. 1970). Starved *Araneus diadematus* lost up to 50% of their body weight and still continued to produce silk (Witt 1963).

The ampullate glands are not normally emptied completely when an orb is built, because spiders could be induced to add extra radii if newly built radii were destroyed during radius construction. *Neoscona nautica* (AR), *Araneus diadematus* (AR), and *Leucauge mariana* (TET) added up to at least a complete extra set of radii before the spider abandoned radius construction and began to build the temporary spiral (Hingston 1920; König 1951; Reed 1969; Eberhard 1988c). Thus an orb weaver normally has a small reserve of ampullate gland silk when she finishes her orb. Maintaining a reserve of dragline silk makes adaptive sense, because the spider needs to be able to continue to produce draglines as she moves around on her finished orb (to attack prey, to flee from danger, etc.).

The data on filling and emptying refer only to non-sticky dragline silk. Almost nothing is known regarding the time course and control of filling and emptying the aggregate and flagelliform glands that produce the sticky spiral glue and baseline, other than that the aggregate gland is relatively full at the beginning of orb construction and empty at the end (Witt et al. 1968). This is an important gap, because in at least some species orb designs are more directly affected by the spider's reserves of sticky spiral silk (section 7.3.2.2).

2.2.13 FORMING BRIDGE LINES

Bridge lines are floated on the breeze to establish connections to distant objects. Bridge line construction occurs during exploration and the early stages of web construction, and also when web spiders move long distances. The strength with which the distal portion of a bridge line adheres to objects in the spider's environment varies, and some fail; thus the bridge lines produced by *Micrathena duodecimspinosa* (AR) that contacted large leaves in nature often failed to adhere strongly enough to support the weight of the spider when she walked toward the leaf (WE). The frequency with which bridge lines of *Larinioides sclopetarius* in captivity adhered strongly enough to support the spider's weight varied with the presumed amount of surface area contacted; 17 of 28 contacts with a sheet of cardboard sustained the spider's weight when the sheet was parallel with the air current, but only 3 of 38 when it was perpendicular (Wolff et al. 2014). Adhesion involved both mechanical tangling of the fibers with surface irregularities in the substrate, and adhesive properties of the fibers themselves, and the surface properties of the substrate itself affected adhesion: successful bridges were established to cardboard and glass, but not to aluminum foil, or plexiglass (Wolff et al. 2014).

The bridge lines of subadult female *L. sclopetarius* (AR) were composed of multiple fine aciniform fibers (50–80 nm in diameter) from spigots on the median posterior spinnerets, combined with a pair of minor ampullate fibers (Wolff et al. 2014). The minor ampullate fibers were present in greater numbers in the distal portion of the line (at least 15 are visible in fig. 9.6 of Wolff et al. 2014). Some of the aciniform fibers splayed apart in the wind, in some cases forming loops. They thus presumably increased both the chances of contacting an object, and increased the surface area of the line that contacted the object and its irregularities.

2.3 SPINNERETS AS HIGH-PRECISION INSTRUMENTS

The spinnerets of most spiders are stubby, relatively simple cylindrical structures, and have not inspired studies of their behavioral finesse. Nevertheless, they are probably capable of subtle movements, as indicated by their extensive musculature (fig. 2.1). Even the web-less ctenid *Cupiennius salei* has special mecanoreceptors that probably provide information on the forces pulling on the dragline and its orientation in space (Gorb and Barth 1996). Additional, more "internal" behavior that facilitates the very precise timing in both initiating and terminating lines also occurs in some groups. For instance, the effective and efficient use of piriform attachment discs requires that the piriform lines and glue be initiated and terminated very abruptly and precisely. Similarly, effective use of aciniform lines to wrap prey during attacks requires rapid and simultaneous initiation of many lines, and also spreading these lines apart in a swath to increase contact with the prey. Because spiders cannot "shoot" their lines, initiation of lines from particular glands presumably involves exuding liquid silk at the tips of their spigots, touching these tips against particular substrates (other lines or objects), and then pulling away (see below). Because the spigots themselves are apparently not mobile, these processes call for finely controlled spinneret mobility, and strategic placements of the spigots on the surfaces of the spinnerets.

2.3.1 ANCESTRAL MORPHOLOGY AND BEHAVIOR

The ancestral morphology and behavior of the spinnerets are relatively clear in some respects. As noted above, the Devonian fossil *Attercopus fimbriunguis* apparently had at least some relatively long spinnerets (Vollrath and Selden 2007). Present-day mygalomorphs serve to suggest several details. Each spigot on the spinnerets of a theraphosid consisted of a hollow proximal bulb and a distal shaft from which a silk line was emitted (fig. 2.1), and there were probably spigots on several segments rather than only on the terminal segment, as in araneomorphs (Foelix et al. 2013b). Modern spigots are modified setae or at least derived from similar precursors, but they lack the multiple innervations that are typical of setae, so ancient spigots probably did not serve as tactile organs. The long ancestral spinnerets were probably also quite mobile, and while attaching lines to the substrate the spider probably made crouching movements with her body, dabbing movements with her abdomen, and waving or tapping movements with her posterior lateral spinnerets, as occur, for instance, during web construction by the diplurid *Linothele macrothelifera* (Eberhard and Hazzi 2013), tube construction by the atypid *Sphodros rufipes* (WE), and prey wrapping by the theraphosid *Psalmopoeus reduncus* (Barrantes and Eberhard 2007). Dabbing and side-to-side movements of the abdomen to produce and attach draglines are presumably ancient, and occur in various families, including Salticidae, Anyphaenidae, and Pisauridae, and Sparassidae (Li et al. 2002; Eberhard 2007a; Wolff et al. 2017a).

These probable ancestral movements are conserved in egg sac construction. The abdomen swings and bobs while a swath of lines is produced without any legs being involved, even in groups such as araneoids and uloborids in which the spiders grasp lines and bring them to the spinnerets with their legs in other contexts. Typically the spider dabbed her abdomen repeatedly against the egg sac, touching her spinnerets to the sac with each dip, then pulling out additional silk as she pulled her abdomen away. Dabbing behavior occurred in *Miagrammopes* (UL) (Akerman 1932), *Hyptiotes paradoxus* (UL) (Marples and Marples 1937), *Zosis geniculata* and *Uloborus* sp. (UL) (WE), and the araneids *Cyrtophora citricola* (Kullmann 1958), *Mastophora dizzydeani* (Eberhard 1980a), *Pozonia nigroventris*, *Micrathena* sp. (Moya et al. 2010), and *Argiope argentata* (WE), the austrochilid *Hickmania troglodytes* (Doran et al. 2001), and the diguetid *Diguetia albolineata* (WE) (in the diguetid, dabbing movements of the abdomen also occurred during construction of the long tubular retreat). In *A. argentata* (AR), *Metazygia wittfeldae* (AR), and *Latrodectus geometricus* (THD) the spider used her legs IV to pull out silk lines and push them onto the sac (WE, G. Barrantes pers. comm.), but did not grasp the egg sac lines to which she was attaching these new lines. The newly emerged lines were at least briefly sticky, and adhered to the substrate and to other lines without any piriform attachment discs.

2.3.2 STRATEGIC PLACEMENTS OF SPIGOTS ON THE SPINNERETS OF ARANEOMORPHS

Spinneret morphology provides many useful higher-level taxonomic characters in spiders, and the distributions and forms of the spigots of different silk glands are known for many species (summaries in Coddington 1987; Platnick et al. 1991; Agnarsson 2004; Griswold et al. 2005). Surprisingly, however, there has been little discussion of the possible functional significance of the locations of spigots on spinnerets, despite the fact that the sites of different types of spigots relative to each other are likely to influence the abilities of spiders to initiate and coordinate the production of different types of lines.

2.3.2.1 General considerations

Apparently the only general discussion of the possible functional significance of spigot placement is relatively recent (Eberhard 2010a). The basic argument is that the anterior-posterior order of the spinnerets on the spider's body and of the sites of the spigots on the spinnerets is likely to have important consequences for how the spider will be able to initiate and attach different types of lines to each other and to the substrate. The logic is based on the fact that silk lines can only be produced by pulling them from the spider's body (the spider cannot "shoot" them) (Witt et al. 1968; Foelix 2011). Thus new lines must be initiated by pressing the tips of spigots against some other object and then pulling them away. Two different basic mechanisms have been proposed for line initiation. The best known, "direct contact initiation," occurs when the spider presses her spinnerets against an object (e.g., a branch, or a silk line) and then pulls away (e.g., Kullmann 1975). If the spinnerets are held in appropriate positions, and if the spigots are at appropriate sites on the spinnerets to bring their tips into contact with an object or another silk line, then liquid silk at the tips of the spigots will adhere to the object and the spider can initiate a new line when she pulls her spinnerets (or the silk line) away. Similarly, lines could be initiated by direct contact between spigots on different spinnerets, or of spigots with lines being drawn from other spigots (see below).

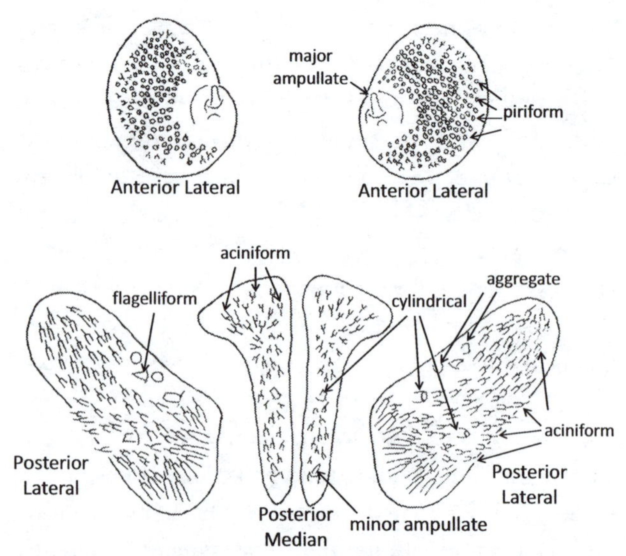

Fig. 2.12. The spigots of different silk glands are located at strategic sites in this generalized diagram of the three pairs of spinnerets of an araneoid spider; posterior, or "downstream," is down (a few of the multiple piriform and aciniform spigots are indicated by arrows). The spigots for the aciniform fibers, whose initiation often depends on their making contact with other lines, are downstream from other spigots such as those of the piriform and major ampullate glands whose fibers are often initiated without contact with other lines (Eberhard 2010a) (after Coddington 1989).

One example of this "dragline initiation" technique occurs when *Nephila clavipes* and *Argiope argentata* wrap their prey (Eberhard 1987c). When a large female *N. clavipes* begins to wrap a prey, the lines in the swath of wrapping lines from the aciniform spigots begin where they are attached to her dragline. Initiation of these wrapping lines presumably occurred when the spider applied the wet tips of the acinoform spigots to the dragline she was producing as she moved toward the prey. Dragline initiation is also presumably used in many species (Eberhard 1987c) to produce airborne lines; the glandular origins of the airborne lines are generally not known for certain; Hingston (1922a) believed they emerged from the posterior lateral spinnerets in *Nephila pilipes*.

Dragline initiation is feasible only if the tips of the spigots of the lines to be initiated can touch the dragline after it has emerged from the major ampullate spigots on the anterior lateral spinnerets. The possibility of such contact will depend on the movements and positions of the spinnerets, and the sites of the spigots on the spinnerets. So long as the spider moves forward rather than backward while spinning draglines (which is generally the case, except during sticky spiral construction in araneoids; section 2.3.2), the general rule would be that spigots that are located farther posterior ("downstream") on the spider's body (fig. 2.12) can initiate lines by touching their tips to lines that are produced by more anterior ("upstream") spigots, but not vice versa (except in unusual cases in which the spinnerets themselves are long and are flexed forward).

A second possible mechanism for initiating lines, "clapping initiation," involves touching spinnerets together and then spreading them apart (Blackwall cited in McCook 1889; Nielsen 1932; Eberhard 1987c). For example, the spider could clap or rub a pair of spinnerets together and then spread them laterally. The spider could then elongate the short lines between the spinnerets by pulling out additional silk with her legs or by using friction with the air. There are reports that spinnerets are spread widely during the initiation of some airborne lines (Blackwall in McCook 1889; Eberhard 1987c), but I know of no confirmed demonstration of clapping initiation. The giant nephilid *Nephila clavipes* rubbed her spinnerets briefly together when she made each attachment of the sticky spiral, presumably initiating piriform lines (WE). Hesselberg and Vollrath 2012 speculated that

Table 2.2. Morphological details of spinnerets that may have possible functional consequences (Eberhard 2010a) (see the text for more complete explanations): the sites of spigots of different glands on different spinnerets in different groups of spiders (data for araneoids from Coddington 1987, for araneomorphs from Griswold et al. 2005) (ALS = anterior lateral spinneret; PLS = posterior lateral spinneret; "—" = gland not present).

Detail	*Araneoids*	*Other araneomorphs*	*Possible consequences*
Aciniform spigots downstream of ampullate spigots	Yes	Yes	Initiate aciniform lines on the dragline
Piriform spigots limited to the tip of the ALS	Yes	Yes	Spinneret movement determines site of attachment disc
Tip of ALS where piriform spigots occur is flat or nearly so	Yes	Yes	Many spigots contact surface
Piriform spigots upstream of flagelliform and aggregate spigots	Yes	—	Lateral orientation of sticky line with respect to abdomen needed to make attachments to radii
Flagelliform spigots downstream in theridiids with respect to ampullate spigots	Yes	—	Initiate sticky wrapping silk on dragline (if this occurs—see text)
Aciniform downstream from ampullate or flagelliform	Yes	—	Initiate aciniform lines on sticky silk lines during wrapping in theridiids (Barrantes and Eberhard 2007)
Spigots for cribellum silk are upstream of other lines	—	Yes	Fibrils are initiated with legs IV rather than on other lines, so do not need to be downstream of others
Aciniform in pholcids downstream from sticky (possibly piriform?) spigots	—	Yes	Initiate "dry" wrapping on sticky wrapping lines
Cylindrical gland spigots downstream of piriform spigots	Yes	Yes	Attachments nevertheless possible (though possibly not as strong) because cylindrical lines are adhesive so few piriform discs needed

the rubbing movements of the spinnerets that they observed during temporary spiral construction by *N. edulis* served to twist lines around each other, but this does not seem topologically possible; the rubbing motions more likely also functioned to initiate and apply piriform silk lines.

A second type of deduction derived from "upstream-downstream" locations concerns how lines are fastened to other lines or to the substrate. Generally speaking, the spigots that produce the fastening silk need to be downstream to or at most level with the spigots that produce the lines that are being fastened.

These types of argument show that the locations of the spigots of several glands in spiders can be explained in terms of their functions (Eberhard 2010a) (table 2.2). For instance, the location of the piriform spigots on the flattened, distal-medial surface on the anterior lateral spinnerets is appropriate for producing attachment discs to fasten the spider's dragline both to more or less planar objects and to other silk lines.

2.3.2.2 Special cases involving web designs

Perhaps the most striking example of a functional relation between spinneret morphology and web design comes from the ochyroceratid *Ochyrocera cachote*, which builds small domed aerial sheet webs with sectors in which large numbers of lines are nearly perfectly parallel (fig. 2.13; Hormiga et al. 2007). Each posterior lateral spinneret of *O. cachote* has an unusual row of tightly spaced aciniform spigots (similar rows of spigots also occur in *O.* sp. and Telemidae and Leptonetidae [Platnick et al. 1991]; *Leptoneta nippora* builds a thin, more or less horizontal sheet web [Shinkai and Takano 1984], but the pattern of lines has not been observed). The similarity between the number of spigots and the numbers of parallel lines in the swaths in *O. cachote*

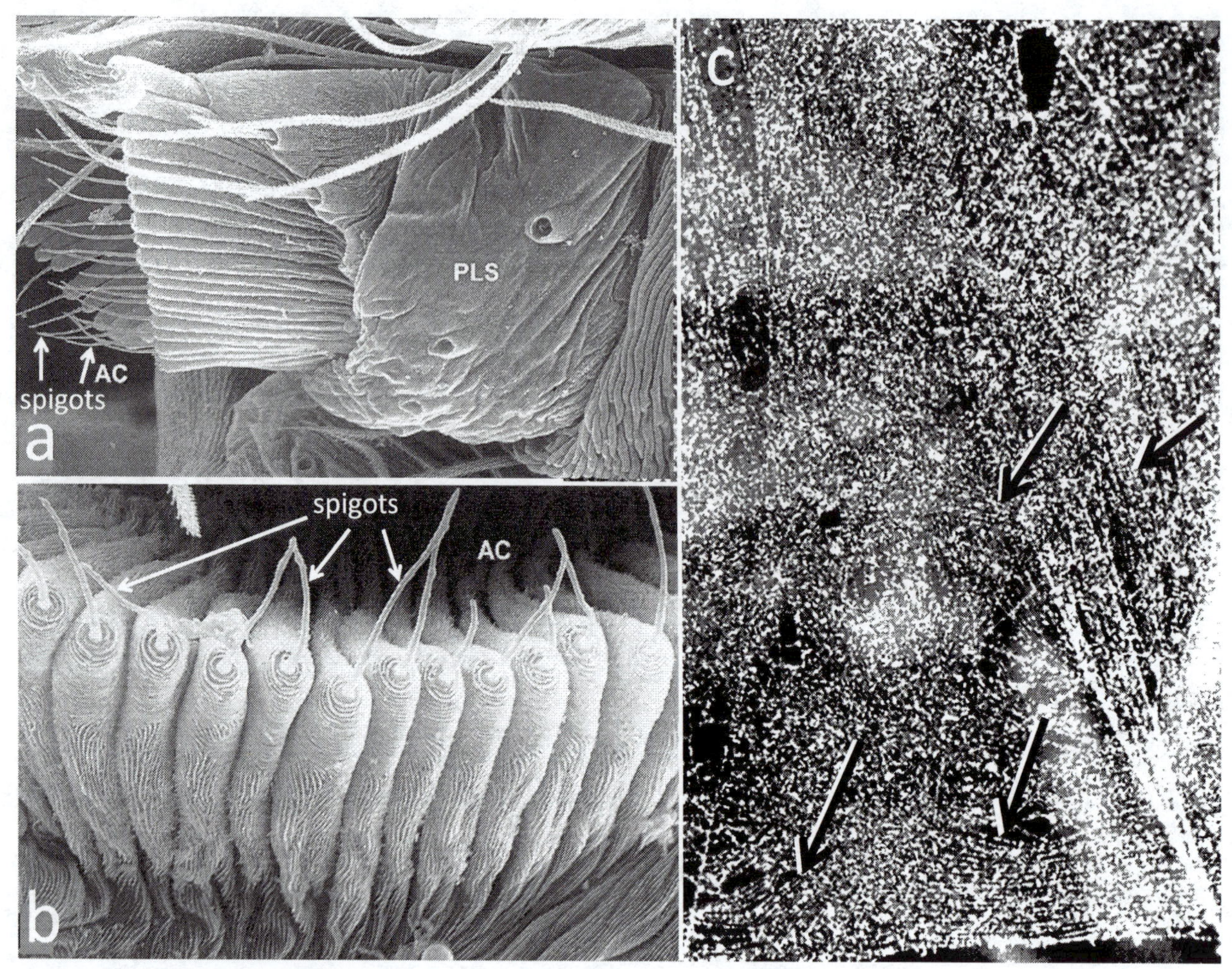

Fig. 2.13. The unusual morphology of the posterior lateral spinnerets of the ochyroceratid *Ochyrocera cachote* (lateral view in *a*, postero-lateral view in *b*), each of which had a tightly packed row of small aciniform spigots mounted on an accordion-like folded membrane, is probably linked to the unusual arrays of nearly parallel lines in the sheets of its web (arrows in *c*). Web construction has never been observed directly, but the correspondence between the number of parallel lines in a swath and the approximate number of these spigots suggests that each swath of parallel lines is produced with a single pass of the spider's abdomen. Recent observations of an unidentified Colombian ochyroceratid showed that the distances between parallel lines in swaths were substantially greater than the distances between spigots (M. Ramírez pers. comm.); perhaps the highly folded tissue associated with the spigots (*b*) is inflated, spreading them apart as the spider is producing the lines (from Hormiga et al. 2007).

(about 20), and the highly parallel orientations of these lines, leave little doubt that each swath was produced during a single pass of spider's spinnerets (fig. 2.13), though it appears that the lines in webs are farther apart than are the spigots on the spinnerets (M. Ramírez pers. comm.). It may be that the highly folded membranous areas where the spigots occur are inflated and cause the spigots to fan apart during web construction; such spreading occurred in wrapping attacks on prey in the piriform spigots on the anterior lateral spinnerets of several gnaphosid species attacking prey by wrapping (Lissner et al. 2016; Wolff et al. 2017b), and in the aciniform spigots on the posterior lateral spinnerets of *Oecobius annulipes* (fig. 2.14).

A second clearly functional morphological detail involves the position of the tips of the two aggregate gland spigots immediately adjacent to the tip of the flagelliform gland spigot in orb-weaving araneoids; the aggregate spigots pour their liquid contents directly onto the sticky spiral flagelliform line as it emerges (Coddington 1989). The less consistent positions of this triad of spigots in "derived orbweavers" such as theridiids (i.e., aranoids

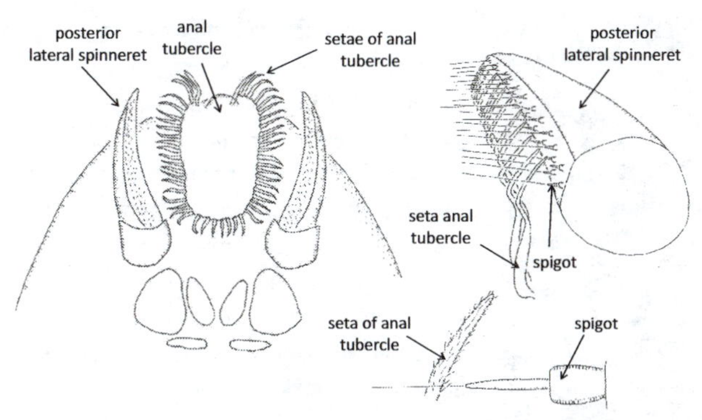

Fig. 2.14. The oecobiid *Oecobius annulipes* uses specialized morphology and behavior to spread aciniform fibers into a wide swath of wrapping silk. Each elongate posterior lateral spinnet has a row of spigots; the long setae on the nearby anal tubercle hold the fibers apart while the spider runs rapidly over and around the prey (after Glatz 1967).

that no longer build orbs) (I. Agnarsson pers. comm.), suggest that perhaps this association is less important (perhaps sticky aggregate silk is applied to ampullate rather than to flagelliform lines?).

One other unstudied consequence of possible general importance occurs in the funnel webs of agelenids and lycosids, in which swaths of lines from the posterior spinnerets (especially the posterior lateral spinnerets) form a dense sheet (Rojas 2011; González et al. 2015; Eberhard and Hazzi 2017). These lines have been said to be "brushed" onto other lines in the sheet without piriform attachments (Blackledge et al. 2009c). This lack of strong piriform attachments is seemingly an inevitable "upstream-downstream" consequence of the location of the piriform spigots on the anterior spinnerets. Similar "brushing" attachments occur in the diplurid *Linothele macrothelifera* (Eberhard and Hazzi 2013). In fact, however, the situation is more complex (at least in the sheet web lycosid *Aglaoctenus castaneus*), because the abdomen is inclined laterally when attachments are made and the lines from the posterior spinnerets on the lower side are apparently sometimes attached with what appear (under the microscope) to be piriform attachments (Eberhard and Hazzi 2017).

2.3.2.3 Additional complications

The upstream-downstream logic just applied to questions regarding attachments to other lines (Eberhard 2010a) depends on whether the dragline or the sticky spiral line that is being attached runs more or less parallel to the spider's own longitudinal axis. There are two clear exceptions. One is when the sticky spiral is being attached to a radius during araneoid sticky spiral construction (see, for different families, figs. O6.2–O6.4). The longitudinal axis of the spider's abdomen is more or less parallel to the radius, and nearly perpendicular to the sticky line. The sticky spiral baseline from the flagelliform gland spigot on the posterior median spinneret passes close enough to the anterior lateral spinneret that the piriform spigots contact the baseline when they make their quick anterior-posterior rubbing movements at the moment of attachment (see description above of *Nephila clavipes*).

A similar exception occurs during hub spiral and temporary spiral construction when the long axis of the abdomen is more or less aligned with the radius but is more nearly perpendicular to the spiral line. This could bring the minor ampullate line emerging from the median posterior spinneret of *Cyrtophora citricola* (AR) (Peters 1993a) close to the field of piriform spigots on the anterior lateral spinneret.

The sensory structures at the cephalothorax-abdomen articulation constitute a further possibly important morphological trait that gives the spider information on the position of her abdomen. As noted by Agnarsson et al. (2008), aerial web spiders probably flex their abdomens at this articulation more, and more precisely, than other spiders. Lifestyle correlations with specialized setae at this articulation appear to be weak, however (Agnarsson et al. 2008).

2.3.3 PHYLOGENETIC INERTIA?

An alternative, "historical constraint" hypothesis for spigot placements needs to be considered. Perhaps the spigots originated by chance at their present locations, and then remained there because no alternative forms arose. The major ampullate spigots apparently evolved only once (Griswold et al. 2005), and the only small variations in major ampullate spigot placement across spiders indicates that their placements in different species represent (at least in some senses) only a single evolutionary event. This skepticism regarding functional interpretations does not explain why, by chance, the present favorable locations originated; clearly the spig-

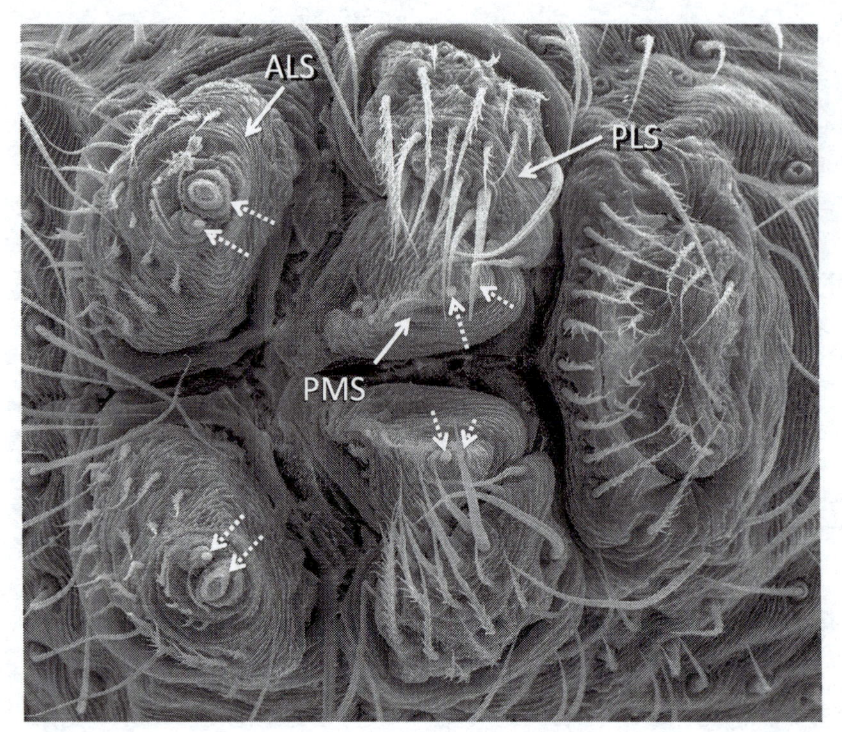

Fig. 2.15. The number of spigots on the spinnerets of the pholcid *Leptopholcus guineensis* has been reduced to only 8 (dotted arrows) (ALS = anterior lateral spinneret; PMS = posterior median spinneret; PLS = posterior lateral spinneret). Reductions in spigot numbers have evolved repeatedly in this family; their selective significance is unknown (from Huber and Kwapong 2013, courtesy of Bernhard Huber).

ots are not randomly scattered over the spinnerets. Nor can it explain why their locations did not subsequently change more substantially. Clearly, the constraint imposed by phylogeny is not severe, as the numbers and locations on a given spinneret do vary to some extent.

In one group, the Pholcidae, there has been some evolutionary flexibility in spigot numbers (though not their locations). The numbers of spigots have been reduced repeatedly from 18–20 to only 8, as in *Leptopholcus guineensis* (e.g., fig. 2.15). The same 12 spigots on the anterior lateral spinnerets have always been lost. The reason for this consistency is not clear; the webs of species with fewer spigots looked much the same, at least superficially (B. Huber pers. comm.).

Further tests of the historical constraint versus the functional design hypothesis could be performed with additional studies of spinneret behavior. For instance, there should be a correlation between how the anterior lateral spinnerets are moved when attachments are made and the site and slant of the planar field on this spinneret where the piriform spigots are located. There are also some spigot placements in cribellate spiders that are not yet intelligible (Eberhard 2010a), also offering scope for future research.

2.3.4 BEHAVIOR OF THE SPINNERETS

Recent studies have begun to describe heretofore neglected spinneret behavior. I will describe a few preliminary glimpses. High-speed video recordings of the giant females of *Nephila plumipes* (NE) making attachment discs to attach the dragline to a glass slide revealed a complex two-step process (Wolff and Herberstein 2017). First the distal tips of the anterior lateral spinnerets (where the piriform spigots are located) were rubbed repeatedly against each other, rolling the dragline over the spigot fields and wrapping it in piriform silk. The second step (often immediately following the first, but sometimes after a minute or more), began when these spinnerets were spread, pulling piriform strands from the spigots, and pressing them to the glass to form the attachment disc. The spinnerets swept simultaneously both anteriorly and posteriorly (in opposite directions); each movement produced an approximately 3 mm long track. In between sweeps, the spinnerets rubbed along each other in a tilted position, presumably strengthening the bridge between the dragline and the disc. The spider made several sets of sweeps while moving her abdomen slightly anteriorly, and a few more while moving posteriorly, producing an approximately circular disc. Tracing the lengths of the paths of the spinnerets and counting the piriform spigots, Wolff and Herberstein (2017) calculated that the spider laid a total of 3.5–8 m of nanofibers in each attachment disc (!).

Video recordings of sticky spiral construction by female *N. clavipes* (WE) gave a more fragmentary glimpse of spinneret behavior during orb construction. At the moment when the sticky spiral line was attached to a radius, the spider's abdomen was aligned very precisely, so that the radius was in the cleft between the right and left spinnerets and its long axis was approximately parallel to the radius (last drawing in fig. O6.4). The tips of the two anterior lateral spinnerets rubbed against each other briefly about twice with a brief anterior-posterior motion that lasted about 0.07 s. During this brief period, the piriform spigots initiated and then terminated production of the lines and glue that fastened the sticky spiral to the radius. Similar rubbing movements with the anterior lateral spinnerets occurred in one bout of prey wrapping, just as wrapping began; later, during wrapping, the anterior and posterior lateral spinnerets clapped together.

Video recordings of construction behavior by the funnel web wolf spider *Aglaoctenus castaneus* (Eberhard and Hazzi 2017) revealed a previously unappreciated variety of spinneret movements (visible in this species because of its large body and spinnerets and its unhurried movements). The posterior lateral spinnerets were frequently spread wide laterally while the spider walked, with each emitting a swath of fine lines. Sometimes the anterior lateral spinnerets clapped or rubbed against each other before making an attachment; the posterior lateral spinnerets also sometimes clapped together before or after attachments. One posterior lateral spinneret was consistently lifted while the other was extended posteriorly when the spider attached lines to other silk lines (fig. 5.17). Spiders also produced long aerial bridge lines that were carried on the weak wind while the spider walked along under another line early in construction, but the spinneret movements involved in initiating these lines (claps?) were not observed. Similar asymmetric raising of only one posterior lateral spinneret while making an attachment also occurred during sheet construction by the diplurid *Linothele macrothelifera* (Eberhard and Hazzi 2013), and was common during prey wrapping in various mygalomorph and araneomorph families (fig. 2.5, Barrantes and Eberhard 2007; Hazzi 2014). Clearly there is a great deal yet to be learned about spinneret behavior in sheet building spiders.

Small rhythmic medial movements of the posterior lateral spinnerets during sticky spiral construction probably had a different function in the uloborids *Uloborus walckenaerius* (Peters 1984) and *Zosis geniculata* (WE). The spinnerets moved medially following each brushing movement of leg IV that pulled fibers of cribellum silk from the cribellum; the movements may function in molding the mat of cribellum fibers around the sticky spiral baseline (fig. 2.7, Peters 1984).

One recent discovery, with no known parallel in other groups, is that the elongate anterior lateral spinnerets of the sicariid *Loxosceles laeta* made rhythmical and very rapid movements (8.5–13/s) while spinning the very thin, ribbon-like, and highly adhesive major ampullate lines (section 2.2.8.4). Highly modified toothed setae (fig. 2.9) on the posterior sides of the posterior lateral spinnerets may hold the ribbon lines while the anterior spinnerets are moving forward, pulling lines from the major ampullate spigots, and then the posterior median spinnerets (which are also moved rapidly after each anterior lateral movement) may free the lines from the setae for the next pull (Margalhaes in press).

These accounts omit at least one further behavior that occurs within the spider's abdomen—the opening and closing of a valve in the duct of the major ampullate gland (Wilson 1962) (I am not aware of valves having been found in the ducts of any other araneomorph silk glands, or in any mygalomorph ducts). Closing this major ampullate gland valve (or perhaps some other structure in the gland duct?) increases the force necessary to draw silk (Work 1978). In a female *A. diadematus* (AR) a pull of only about 5 mg was usually needed to pull dragline silk from her body; but the spider could hang motionless at the end of her dragline even when her legs were not touching it, thus supporting her own weight (about 60 mg) (Work 1978). The force of this "holding brake" is sometimes even substantially greater (Work 1978). Work (1978) hypothesized that braking resulted when the spinnerets were closed medially on each other, and other (piriform) spigots and perhaps setae on the anterior lateral spinnerets produced friction with the line, much like the friction produced by the interdigitating "tension gates" used in textile machines that produce artificial lines, but he did not specify specific structures. Another indication of an internal brake is that even when the strain energy capacity of the dragline is insufficient to absorb the potential energy lost by a spider as she falls, failure of the dragline is avoided by dissipating energy by friction in the duct as dragline silk is drawn (Brandwood 1985). In some species the spider also manipulates the dragline with leg IV as she falls, allowing the line to slip through her tarsal claw (Eberhard 1986a).

2.3.5 HOW ARE LINES TERMINATED?

Very little is known concerning the mechanisms that spiders use to terminate lines, and most concerns only major ampullate lines. When a dragline is being forcibly pulled from an immobilized spider, the anterior lateral spinnerets sometimes move medially from a more spread position, and break the line by rubbing quickly forward and rearward against each other (Work 1978). The tension gate mechanism proposed by Work (1978) could theoretically be responsible. Similar morphological arrangements that would allow termination by rubbing against other homologous spigots do not occur, however, for piriform or aciniform spigots.

Breakage apparently occurs in some other contexts when there is a lack of liquid silk in the duct to the spigot, or the duct itself closes valve-like; the line breaks there while it is being pulled (Work 1978). Presumably subsequent initiation of another line occurs when liquid silk is pushed out the duct (past the broken end?) to the tip of the spigot.

Several types of silk are mostly produced during long uninterrupted periods, and rapid, precise termination may not be especially important, but piriform lines are exceptions. Piriform lines terminate and then rapidly reinitiate production hundreds or thousands of times during the construction of a single orb, with truly amazing, split-second coordination. In building the orb illustrated in fig. 1.4, for example, the female *Micrathena duodecimspinosa* made 832 attachments of her hub and temporary spiral to radii in the space of only 428 s (a new attachment every 0.52 s) (WE). For each attachment, the radius was contacted by the spinnerets for only about 0.07–0.09 s, so the piriform silk must have been initiated and then terminated in even less time. In addition, the sticky spiral was attached to the radii on the order of 3000 additional times, and each of these attachments also lasted only approximately 0.07 s. Some other species have even more spectacular numbers: *Cyrtophora citricola* (AR) makes about four piriform attachments/s, and a total of >22,000 in a single web (Kullmann 1958; Peters 1993a; WE).

How terminations and initiations are accomplished so rapidly and precisely in the short ducts and spigots of the numerous small piriform glands is unclear. Microscopic examination of the piriform silk at major and minor ampullate line attachments in *C. citricola* (AR) webs (fig. 2.11) indicated that initiation was typically more abrupt than termination (Kullmann 1975), as termination involved the production of a short "trail" of fine lines as the spider moved away from the attachment (fig. 2.10a; Peters 1993a). These trails of lines (which also occurred in radius-frame attachments in other araneids—WE) appeared to have less glue on them (fig. 2.10a), and appeared to represent wasted silk. It is possible that selection for ease of termination may also affect spigot placement. For instance, when female *Nephila clavipes* finished wrapping a prey, they often seemed to "scrub" the spinnerets together and to scrape their surfaces with tarsus IV (WE). If such scrubbing is needed to terminate wrapping lines, positioning the spigots so they can be scrubbed could facilitate termination. The apparent need for this cleaning behavior in *N. clavipes* makes the precise, rapid terminations of piriform silk during orb construction seem even more extraordinary.

2.4 LEG MORPHOLOGY AND BEHAVIOR: GRASPING LINES PRECISELY AND SECURELY

2.4.1 GRASPING LINES IN A WEB; TARSAL MORPHOLOGY AND LEG MOVEMENTS

As long as spiders built webs very near to the substrate (e.g., the linings of burrows; section 10.3.1), walking on the web probably did not require radical new abilities. For instance, the tarsi of the filistatid *Kukulcania hibernalis* consistently walked on the substrate rather than on the lines of the web itself, at least in the less dense parts of the web (e.g., fig. 5.10) (WE). When spiders began constructing aerial lines above the substrate, however, new morphological structures and behavioral abilities became advantageous. Walking along an aerial line poses several problems: the spider needs to find lines to grasp; she needs to grasp them securely; she needs to release them quickly and cleanly; and she needs to hang under the line rather than walking on top of it.

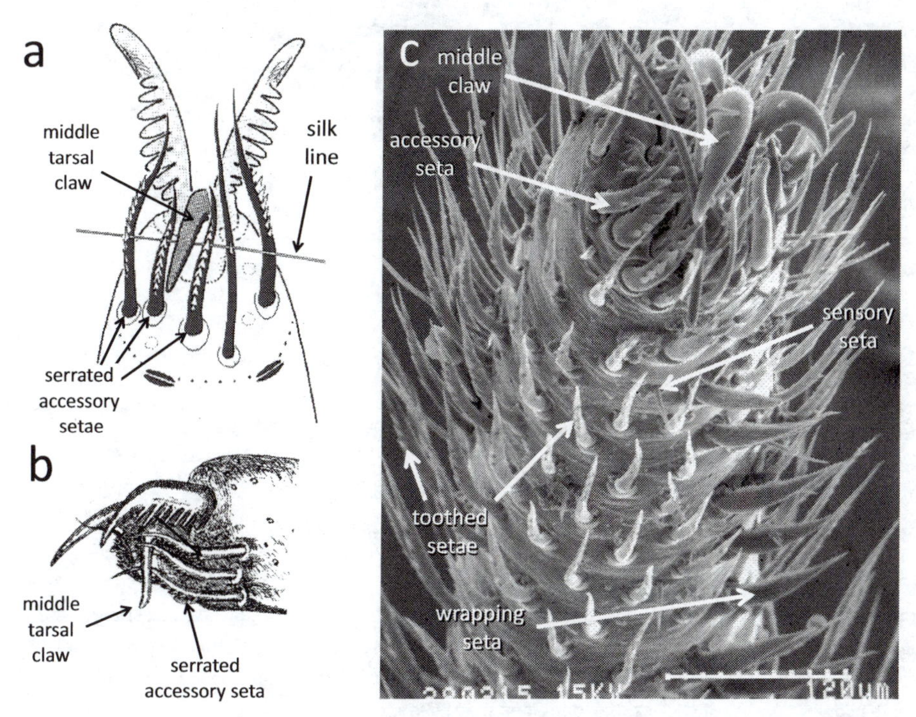

Fig. 2.16. Drawings of *Araneus diadematus* (AR) (*a* and *b*) illustrate how the middle claw and the serrated accessory setae just ventral to it represent crucial evolutionary developments in managing silk lines. By flexing the claw ventrally (*a*), the spider seizes a silk line tightly by trapping it against the teeth on the serrated setae. The tarsus I of the araneid *Araneus diadematus*, seen in anterior view (*b*), illustrates another subtle but important detail: the accessory setae and the median claw are located asymmetrically, on the antero-lateral sides of tarsi I and II and on the postero-lateral sides of tarsi III and IV. The anteriorly directed median claw and the accessory setae on the anterior side of the tarsi I and II facilitate grasping new lines when legs I and II swing medially ahead of the spider. A SEM image (*c*) of the left tarsus IV of a mature female *Gasteracantha cancriformis* (AR) illustrates a veritable tool kit: the asymmetrically placed serrate accessory setae; the median claw (directed to the same retrolateral side where the accessary setae are located); a row of stout setae that are presumably used to wrap prey; presumed chemosensory setae; and a forest of toothed setae that aid in minimizing the spider's adhesion to her own sticky silk (section 2.6) (*a* from Foelix 2011, *b* from Nielsen 1932).

Modifications of the tarsal claws and the associated serrate accessory setae at the tip of the tarsus have long been known to enable spiders with three tarsal claws to firmly grasp and then release single silk lines. The middle claw moves as a unit with the lateral claws (to which it is fused) when trapping the line against the teeth on one (or perhaps more) of the stiff but nevertheless somewhat flexible serrate setae nearby (fig. 2.16) (Nielsen 1932; Wilson 1962; Foelix 1970, 2011). The three claws are supplied with only two tendons, one that can produce a dorsal movement of the unit and the other that can produce a ventral movement; there are no tendons that could produce lateral movements (Labarque et al. 2017). When the claws are flexed ventrally, the middle claw bends the line sharply against the serrated accessory setae and achieves a firm grip; when the middle claw is lifted slightly, the grip is loosened (Wilson 1962) (fig. 2.16a); and when it is lifted completely, the grip on the line is released. At the moment of release, the tension on the line itself as well as the straightening movements of the setae may propel the line away from the claw (Nielsen 1932).

A functional association between the middle claw and the serrate accessory setae is supported by their similar asymmetrical positions on the tarsus: on legs I and II the middle claw is directed prolaterally, and most of the setae are on the prolateral side of the tarsus; on legs III and IV, the middle claw is directed retrolaterally, and most of the setae are on the retrolateral side of the tarsus (fig. 2.16a,b, Nielsen 1932). These asymmetrical positions of the median claw and the serrated accessory setae are correlated with the positions in which the spider often holds her legs (Nielsen 1932); legs I and II are often directed anteriorly and legs III and IV are directed posteriorly. The asymmetric positions of the median

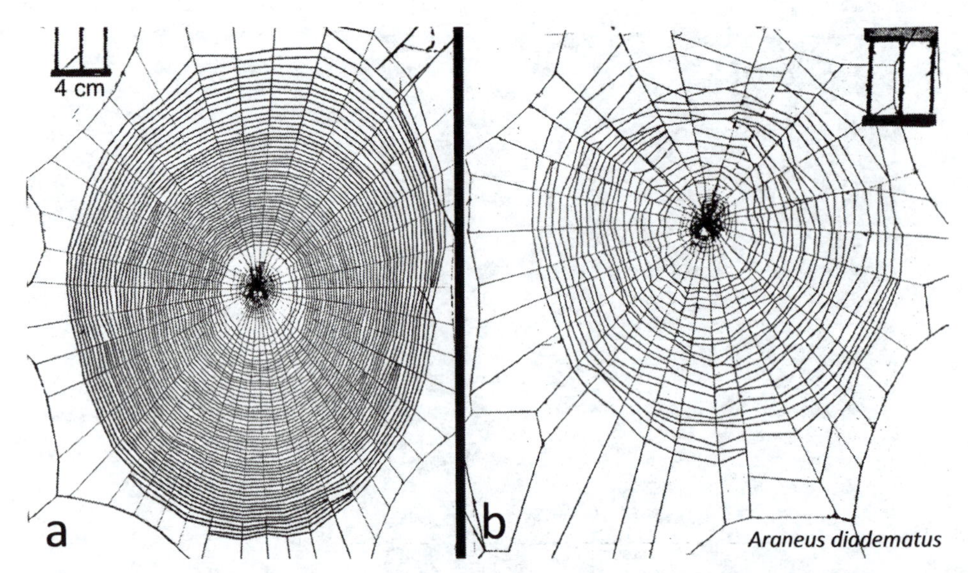

Fig. 2.17. Webs that were built by *Araneus diadematus* before (*a*) and after (*b*) the serrated accessory tarsal setae were removed suggest that reducing the spider's ability to grasp lines firmly affects orb construction. Surprisingly, the effects seem to be concentrated on sticky spiral construction, rather than on earlier stages. The regularity of the hub spiral spacing was hardly affected, even though precise grasping of lines and highly coordinated leg movements also occur during hub construction (section 6.3.3.4, appendix O6.3.4). The lower numbers of radii and hub loops in *b* are associated with a lower number of sticky spiral loops, and may have resulted not from problems grasping lines but rather from an earlier decision during construction to lay a smaller length of sticky line (see sections 7.3.3, 7.3.4). Further experiments will be needed to document differences quantitatively (from Foelix 1970).

claws enable the spider to snag lines by executing medially directed searching leg movements (section 2.4.2). Experimental evidence of a grasping function for the serrate accessory setae was obtained by removing them from all of the legs of *Araneus diadematus*; the spider experienced difficulties in both web construction and in climbing vertical lines (fig. 2.17) (Foelix 1970). Comparative morphological data from other species also support this interpretation. The amaurobiid *Amaurobius ferox*, which lacks the setal teeth, slipped while attempting to climb vertical lines (Nielsen 1932). In the triangle web uloborid, *Hyptiotes paradoxus*, which holds a line to her web under substantial tension with her legs I for long periods while she waits for prey, both the middle claws and their serrated accessory setae are especially stout and have numerous teeth (Nielsen 1932). The most dramatic modification in *H. paradoxus* is on legs III, where there are three serrated accessory setae on each side of the median claw (Nielsen 1932), suggesting an important role in silk manipulation. These legs do not hold the tight line, but instead hold the mass of lax line that the spider releases the moment a prey strikes her web, after which these legs grasp the line (it is not clear whether having extra serrated accessory setae could aid in managing this silk). The serrated setae were stronger and more prominent in the juveniles of *Pachygnatha* spp. that build orbs than in conspecific adults that do not build prey capture webs (Martin 1978). The agelenid *Eratigena atrica* and the amaurobiid *Amaurobius ferox* lack serrated accessory setae, but have tufts of smooth setae associated with both the medial claw and the two lateral claws that may be bent when the tarsus contacts silk lines, and that may thus aid in pushing the line clear of the claws when the spider is releasing it (Nielsen 1932). The eresid *Eresus niger* has a pair of smooth accessory setae associated with the median claw that are presumably bent when it grasps a line.

Even spiders lacking these refined grasping structures are able to walk in webs. Small spiderlings in the mygalomorph families Ctenizidae and Atypidae walked under aerial lines to elevated dispersal sites several m from their mother's burrows (Eberhard 2006; Cox 2015), even though they are normally substrate-bound, burrow-dwellers and lack middle claws (Main 1976; Coyle 1986). Video recordings of other mygalomorph spiderlings, including the atypids *Atypus affinis* (unpublished video by Janni Louise Cox) and *Sphodros rufipes* (WE) and a ctenizid (perhaps *Ummidia* sp.), showed that spiderlings

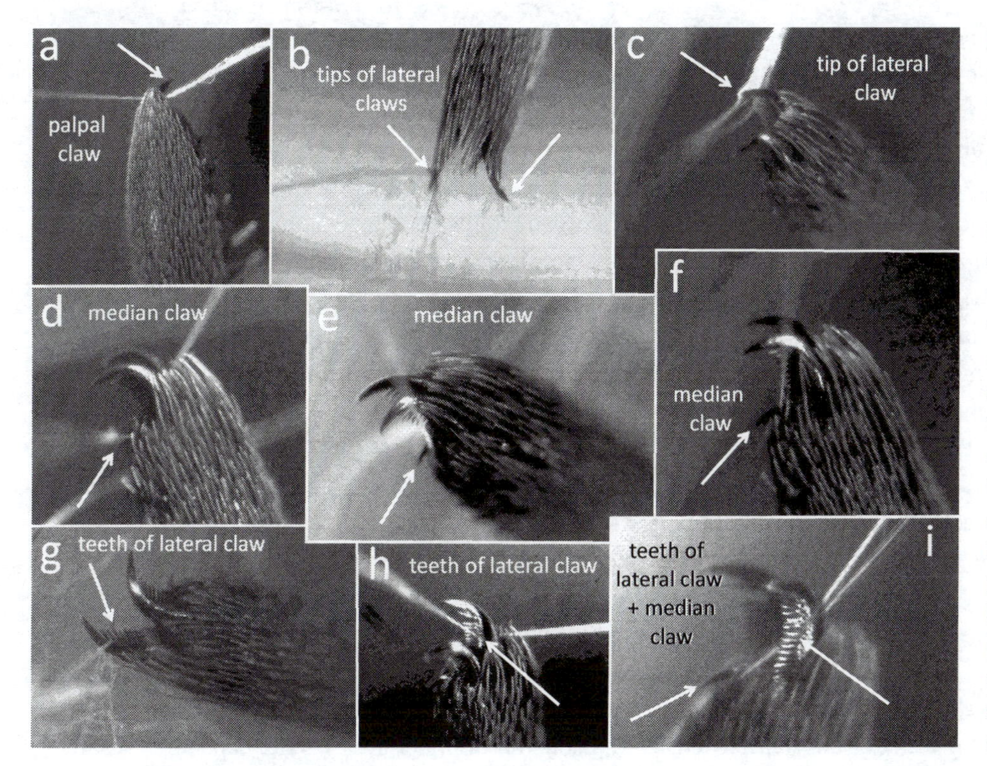

Fig. 2.18. This array of images illustrates the lack of consistency with which a mature female of the filistatid *Kukulcania hibernalis* grasped lines with different portions of her tarsal claws while the spider rested in her retreat (each arrow indicates the portion or portions of the claw that exerted force on the line) (in *b* the claws have snagged some but not all of the fibers in a multistrand cable of lines near an attachment to the substrate). The variety in the orientations of the lines, ranging from more or less perpendicular (*a, c, d, f*) to more nearly parallel (*i*) to the long axis of the tarsus, and in the portions of the claw that snagged the lines, contrasts with the more nearly perpendicular orientations and the contacts with the median claw that were typical of orb weavers (figs. 2.16, 2.20).

gave no indication of having trouble grasping or releasing lines, though their progress was slow and jerky (WE). Many of the lines that they walked along were made up of multiple fine strands. Other groups such as jumping spiders, whose tarsi are also designed for walking on the substrate and not on silk lines (Hill 2006), can also walk in spider webs and even hunt their owners there (Jackson 1990a,b; Li et al. 1997).

Preliminary observations of a mature female of the basal haplogyne filistatid, *K. hibernalis*, which lacks serrate accessory setae, revealed that the details of how a mature female in her retreat grasped lines with her tarsal claws were variable and inconsistent (fig. 2.18) (Eberhard 2017). This spider was only resting quietly in a sparse retreat, and was not exerting special force on any lines; it is possible that spiders show more consistently strong, less variable grips on lines in other contexts. Of course, walking slowly along a line like the *Ummidia* spiderlings, or hanging in a sheltered retreat, is not as difficult as running rapidly across a web to attack a prey.

There are many additional details and variations in this basic morphological scheme whose functional significance is unclear. The position of the middle tarsal claw is also strongly asymmetrical in the basal cribellate non-orb weaving web spider *Thaida*, as well as the non-web builder *Senoculus* (here the middle claws on all four legs are directed prolaterally). In contrast, the middle claws are not oriented asymetrically in the orb weaver *Uloborus*, or in the dictynid *Dictyna*. The possession of a third claw is not always associated with webs, as third claws occur in some non-web spiders such as ground living lycosids (Nielsen 1932; Ramírez 2014). Here they presumably function such as spurs that brace or hook against the substrate as the spider moves.

Serrate accessory setae are also an apparently old trait that is widely distributed in araneomorphs. They occur in spiders both with and without webs (for example *Megadictyna*), from austrochilids to orbicularia (Griswold et al. 2005). There is no clear difference in the serrate accessory setae of austrochilids compared with those of other araneomorphs. Perhaps the most puzzling combination of traits occurs in Psechridae, which have 3 claws, but claw tufts rather than accessory setae (Griswold et al. 2005), and walk under lines in their webs (Bristowe 1930; Eberhard 1987a); perhaps here the median claws serve to allow spiders to climb vertical lines. There are also further subtleties in the median claws of other web builders (concave surfaces on the retrolateral sides of the teeth of the median claw (M. J. Ramírez pers. comm.) whose functional significance is not clear.

Finally, many spiders sometimes allow a line to slide smoothly under the middle claw. Presumably they accomplish this by raising the tarsal claw slightly so that the line remains snagged under the tarsal claw and on the serrated accessory setae, but is not bent sharply. In an orbweaver like *Nephila clavipes* (NE), both tarsi III and IV slid along lines during radius construction (WE, sections O6.3.2.2, O6.3.3.2). I do not know the taxonomic distribution of this ability, and suspect that it could harbor interesting taxonomic information (section 5.3.3). It appears that sliding one tarsus IV along the dragline as it is produced is widespread among orbicularians (WE), while some mygalomorphs and other basal groups such as filistatids (Eberhard 1986a; WE) and the web-building wolf spider *Aglaoctenus castaneus* (Eberhard and Hazzi 2017) apparently never slide the dragline through any claws (proving the absence of a capability is difficult, however).

The ability to clamp lines firmly with leg IV probably enables the spider to lay lines under substantial tension. When the orb weaver *Micrathena duodecimspinosa* attached anchor lines under substantial tension to the substrate, she held the new line she was producing with one leg IV while she pulled herself forward on the substrate with her more anterior legs. This movement on the substrate was accompanied by increased tension on the dragline, as indicated by the angles formed with the web lines to which it was attached and, in some cases, by the bending or twisting of a flexible substrate such as a leaf. The spider's ability to clamp her dragline firmly with leg IV allowed her to raise the tension on the line substantially above the tension that was needed to pull silk from her spinnerets.

One final puzzle regarding tarsal claw morphology, recognized long ago by Hingston (1920), is that sheet weavers are able to run rapidly across sheets that consist of dense arrays of fine lines laid at widely varying angles without snagging their strongly curved claws (e.g., Rojas 2011; Eberhard and Hazzi 2013, 2017). Lifting the tarsal claws as the spider begins to execute a stepping motion is probably a very ancient trait, as it occurs in a variety of groups with and without webs, including liphistiids, and occurs across the entire diversity of foot configurations in spiders (Labarque et al. 2017). Perhaps the tarsal claws of funnel web spiders remain flexed dorsally as they move on their webs? Strangely, Hingston (1920) found that the lycosid *Hippasa olivacea* lost her ability to avoid being snagged when her pedipalps were removed.

An additional derived feature of the tarsi of web builders is subdivision of the podotarsite that confers additional mobility and mechanical flexibility (Labarque et al. 2017). Similar subdivisions also evolved independently in other webless lineages. Both may have been derived from weak areas of the cuticle associated with the sensory organ (the foot slit sensilla) (Labarque et al. 2017).

2.4.2 COMPLEMENTARY SEARCHING AND GRASPING BEHAVIOR

2.4.2.1 The blind man's cane and the art of following

An appropriate image for a spider finding her way around her web is that of a blind man using his cane to explore his surroundings (Vollrath 1992; sections 6.6.1, 7.1.1). A web spider seldom if ever sees her lines during web construction: her eyes are not capable of resolving such fine objects (Foelix 2011); they are directed dorsally while the lines are usually ventral to her body while she builds; and most spiders build in darkness. The spider relies instead on tapping movements of her legs (usually legs I and II) to find new lines. Use of tactile cues was confirmed by the spider's extensive use of apparently exploratory tapping behavior with the legs and small imprecisions in how lines were contacted (sections 2.4.2.2, 6.3.3, appendix O6.2.1). A major, but heretofore generally unremarked, trend is for the spider's more posterior legs to follow "in the footsteps" of her more anterior legs. An anterior leg like leg I explores an area by waving, contacts a line and grasps it; then the next leg on the same side (e.g., ipsilateral leg II) reaches forward directly and seizes this line with little or no exploratory waving near the point held by leg I. Leg I then releases its hold on the line, and searches farther ahead for another line to grasp. Often a line is then passed to more posterior legs in the same way (III follows II; IV follows III). Occasionally a leg I on one side follows the leg I on the other (Briceño and Eberhard 2011 on *Leucauge mariana* [TET]).

Following behavior occurred in many contexts, including exploration and construction of the radii, the hub, the temporary spiral, and the sticky spiral construction (section 2.4.2, table O2.3). The functional significance of following behavior seems clear: it economizes on time and on the leg movements needed while moving on the web. By reducing the need for her more posterior legs to wave and explore like a blind man's cane, the spider can move more quickly, and expend less energy.

Following behavior was first described in sticky spiral construction behavior by *Nephila pilipes* (= *maculata*) (NE) by Hingston (1922b), then later in hub construction behavior in *Uloborus diversus* (UL) (Eberhard 1972a), and sticky spiral construction in at least eight araneid and three tetragnathid genera (Eberhard 1981a), and recently at several other stages and in other genera (Eberhard 2017). I anticipate that further studies will show that the tendency for more posterior legs to follow more anterior legs is taxonomically widespread beyond the orbicularians (section 5.3.7), and represents an ancient adaptation that has been used in many different groups to accomplish many different tasks in web construction. Following behavior appeared not to occur in the *Ummidia* mygalomorph spiderlings mentioned above when they walked under lines (WE), but legs IV of the diplurid *Linothele macrothelifera* tended to follow legs III while the spider walked on the substrate or the sheet of silk already present during sheet construction (Eberhard and Hazzi 2013). Following is thus probably an ancient trait, and was probably already present when early spiders evolved to walk under aerial lines. Similar following behavior has evolved convergently in the emesine bug *Stenolemus giraffa*, whose posterior legs follow the anterior legs, solving the same problem of finding lines to grasp when the bug stalks its spider prey in its web (F. Soley pers. comm.).

One further detail is that following behavior is sometimes facultatively omitted. Sometimes a spider alters her usual "explore and then follow" behavior when she finds herself in an area where the lines are so dense that her leg is likely encounter a line nearby wherever she places it. The large psechrid *Psechrus* sp. used her leg III to tap to find the next line in her sheet to which she was going to attach her sticky line, rather than following her more anterior legs, which held lines far from the attachment site and were moved less frequently (Eberhard 1987a). Some spiders may facultatively abandon following behavior when the lines in the web are dense. For instance, *Leucauge mariana* used following behavior in most parts of her web, where lines were sparse (e.g., Eberhard 1987d hub construction), but when she was turning rapidly at the hub of her orb during prey attack behavior, she seized new lines more or less directly without following other legs (Briceño and Eberhard 2011). An interesting implication of these behavioral adjustments is that the spider seems to "know" the likely density of lines in her immediate vicinity.

2.4.2.2 Asymmetric searching movements that match asymmetric tarsal morphology

Video recordings make it possible to add a behavioral dimension to the old morphological story just described. There are behavioral asymmetries that fit beautifully with the tarsal asymmetries and that facilitate grasping and holding individual silk lines (table 2.3). The small searching movements made by following legs to find and grasp the line held by a leading leg had consistent orientations: legs I and II moved prolaterally; and legs III and IV moved retrolaterally (table 2.3). These movements, which were necessary because the following leg never grasped the line at exactly the same site as the anterior leg, thus brought the asymmetrically placed middle claws and their associated serrate accessory setae into contact with the lines (Eberhard 2017). They may also occur in non-orb weavers, as leg I of *Chrosiothes* sp. nr. *portalensis* (THD) also searched in a prolateral direction during sheet construction (WE). It is not known whether asymmetry in searching movements occurs in other spider taxa, or whether (as predicted by the interpretation here) behavioral asymmetries in the searching behavior of legs are reduced or lacking in those groups in which tarsal tips are morphologically symmetrical (Ramírez 2014).

An exception seen in these same video recordings lent further support to the association between asymmetric behavior and asymmetric morphology. During sticky spiral construction, the spider used one leg to tap to encounter lines, but not to grasp them ("inner loop localization behavior") (figs. O6.2, O6.3, section 6.3.5, appendix O6.2.5). Tapping movements in this context tended to be directed dorso-ventrally rather than laterally in the araneids *Araneus expletus*, *Cyclosa monteverde*, *Gasteracantha cancriformis*, and the uloborid *Zosis geniculata* (leg oI) (WE). In the nephilid *Nephila clavipes* the inner loop localization movements were exploratory

Table 2.3. A tentative list of possible uniformities in two details of leg movements performed during orb construction: following; and prolateral vs. retrolateral short-distance searching movements. The species are those whose construction behavior I happen to have filmed (not all stages are represented for all species). All recordings were made in the field except those of *Cyclosa monteverde*, *Nephila clavipes*, and *Zosis geniculata*. In most species only a single individual was filmed, but in all cases the activity was repeated many times. Sticky spiral construction was characterized in the outer rather than the inner half of the web, and in vertical webs usually included behavior above as well as below the hub. One leg was characterized as "following" another if it consistently moved to and grasped a line near to the point where that same line was being held by the other, leading leg, and in which the leading leg then quickly released its hold on the line (typically moving forward to find and grasp another line). "Short-distance searching" movements were the generally small amplitude movements executed by the tip of the leg approximately 0.1 s before it grasped a line; they were especially clear when one leg was following another (probably because the approximate location of the line was already known by the spider). They differed from the repetitive, larger-amplitude "long distance searching" leg movements made when a spider explored an empty space by waving or tapping with her legs (see text). Legs are indicated by "o" and "i" to indicate their positions during construction: "outer" legs were those directed away from the hub while the spider was spiraling around the web building hub, temporary, and sticky spiral lines; "inner" legs were directed toward the hub (e.g., fig. O6.1). Many of the behavior patterns (both following and short-distance searching) were not absolutely constant, and the characterizations represent the most common types of movements rather than exhaustive lists of all movements. Inconsistent behavior and behavior of species that were too small or moved too rapidly for me to interpret the leg movements are indicated with "—". In sum, this table does not provide final characterizations of all leg movements, but rather illustrates two apparently general trends in the more consistent and easily observed types of leg movements: legs often followed the immediately anterior ipsilateral leg; and short distance searching movements by legs I and II tended to be prolateral in direction, while those by legs III and IV tended to be oriented retrolaterally.

Behavioral operation and spider species	*Legs that followed other legs (leading leg–following leg)*	*Legs NOT follow any other leg*	*Direction of short-distance searching: Prolateral*	*Retrolateral*
Secondary radius construction				
Leucauge mariana TET	I-II; II-I; I-I (contralat.)[1]	I; II[1]	I; II	—
Micrathena duodecimspinosa AR	I-I[2]	—	—	—
Zosis geniculata UL	oI-iI[2]	—	—	—
Hub construction				
L. mariana TET	oI-oII; oII-oIII; oIII-oIV	oI; iIII[3]	—	oIII[4]; oIV
M. duodecimspinosa AR	oI-oII; oII-oIII; oIII-oIV	oI; iIII[3]	oI; oII; iI; iII	oIII; oIV
Z. geniculata UL	oII-oIV; oIV-oIII[5]	oIII; oII	—	—
Temporary spiral Construction				
L. mariana TET	oII-oI; oIII-oII; oIV-oII[6]	—	—	—
Cyrtophora citricola AR[7]	oI-oII; oII-oIII	—	—	—
Nephila clavipes NE	oII-oIII; oIII-oIV; oI-oII[8]	oI; oII; iI; iII; iIII	oI; oII; iI; iII	oIII; oIV
Z. geniculata UL	oI-oII; oII-oIII[9]; oIV-oII/oIII	iII, iIII[10]	oI?; oII?	oIV
Sticky spiral construction				
L. mariana TET	oI-oII; oII-oIII; oIII-oIV?[6]	—	—	oIV (?)
M. duodecimspinosa AR	oII-oIII; oIII-oIV	oII	oI; oII	oIII?; oIV[11]
Gasteracantha cancriformis AR	oI-oII[12]; oII-oIII[13]; oII/oIII-oIV[6]	oII[11]	oI[14]; oII[15]	oIII[14]; oIV
Araneus expletus AR	oII-oIII; oIII-oIV; iI-iII; oI-iI	oII; iI	oI (?); oII (?); iI; iII	oIII; oIV
Cyclosa monteverde AR	oII-oIII; oIII-oIV	iII; iIII	oII	oIII; oIV; iIII; iIV
N. clavipes NE	oIII-oIV[11]	oIII; oII; oI	—	oIII; oIV
Z. geniculata UL	oI-oII[16]; oII-oIII[13]; oIII-oII[13]; oI-iI[17]; iI-oI[17]; iIII-oIII[18]	—	oI; oII[15]	oIII[15]

Table 2.3. Continued

[1] All three following sequences were common; some other times these legs did not follow each other.
[2] As grasped successive possible exit radii.
[3] Did not move at all during construction of the first loop.
[4] The movements were very rapid, so this characterization is tentative.
[5] oIV grasped r_n first, then oIII grasped it nearby. But oIV did not then immediately release its grip and move on; instead both legs held the radius as the sticky line was attached between them.
[6] Leg oII often left the radius one or two frames of the video recording before oIV arrived; during this time oIII (which had followed oII) remained holding the same radius. The site grasped by oIV was closer to that grasped by oII than to that grasped by oIII. It is thus not entirely clear whether it should be said that oIV followed either oII or oIII.
[7] Construction of non-sticky spiral that may be homologous with hub spiral in dense horizontal sheet.
[8] Behavior occurred while turning from temporary spiral construction to lay a tertiary radius.
[9] Both oII and oIII were on r_n.
[10] If following occurred, it was at best inconsistent.
[11] The tarsus often appeared to slide (probably making contact on its retrolateral side) along r_n before gripping it.
[12] oII followed oI to first seize r_n, but later did not follow oI during inner loop localization behavior when it occasionally grasped r_n briefly during its tapping movements to locate the inner loop of sticky spiral.
[13] Hand-over-hand movements reeling in or walking out r_n, with each leg grasping the radius outside the other; often only 1–2 steps.
[14] When reached to grasp r_{n+1}, but not while it tapped to locate the inner loop.
[15] Especially clear as oII and oIII pulled in or walked out r_n hand-over-hand.
[16] Except on the first sticky spiral on r_n, when oI held the temporary spiral and oII grasped r_n.
[17] Infrequent.
[18] Only during the first step of leg oIII following an attachment.

extensions with leg oIV, and were oriented to produce dorsal rather than lateral contact (Eberhard 1982; WE). These legs generally touched the inner loop only briefly (usually in only a single frame in the 30 fps video recordings), and did not seize it.

One further possible dimension is as yet unexplored: the dorso-ventral movements of the tarsal claws themselves (section 2.4.3). Perhaps ancestral spiders lowered the claws when the tarsus was about to contact the substrate, and raised the claws when lifting the leg from the substrate. Such movements would resemble those needed when walking under a line, and may have preceded them evolutionarily. Some other tasks while walking on the substrate may also resemble those while walking under a line, such as coordinating leg movements so that several legs are always in contact with the substrate at any given moment and are thus able to support the spider's weight (e.g., Foelix 2011).

In sum, the tarsi of web-building spiders tell a beautiful story of functional morphology. The forms of the median claws and the serrated accessory setae, their orientations and positions on the tarsi, the movements of the spider's legs, and the results of experimental removal of the setae all argue strongly for their importance in grasping and clamping lines. I have to admit, however, that I was left thinking one sunny morning that there may still be much to be discovered about how spiders grasp and walk on lines after watching a young sparassid, which has only two claws and a dense claw tuft rather than serrated accessory setae, literally sprint up its dragline to hide under the leaf from which I had dislodged it (WE).

2.4.2.3 An additional detail: rotating legs to grasp lines

One further aspect of grasping lines has only recently been studied. While the median claw can clamp a line against the serrated accessory setae when the line runs perpendicular to the direction in which the claw is flexed ventrally (fig. 2.16a), this is not possible when the median tarsal claw grasps a line that is more or less parallel to the leg's longitudinal axis. This orientation is very common in natural contexts, however, as illustrated by fig. 3.28 of legs I and II of a *Leucauge mariana* (TET) and fig. 2.19 of legs I and II of a *Metabus ocellatus* (TET) at the hubs of their webs. Nielsen (1932) noted this problem in *Hyptiotes paradoxus*, in which legs I and II normally hold the line to the orb that is oriented with the spider's longitudinal axis while the spider waits for prey; he believed that the spiders solved this problem with the stronger prolateral displacement of the median claws and the serrated accessory setae on legs I and II of this species, and by bending the line at the point where the claw grasped

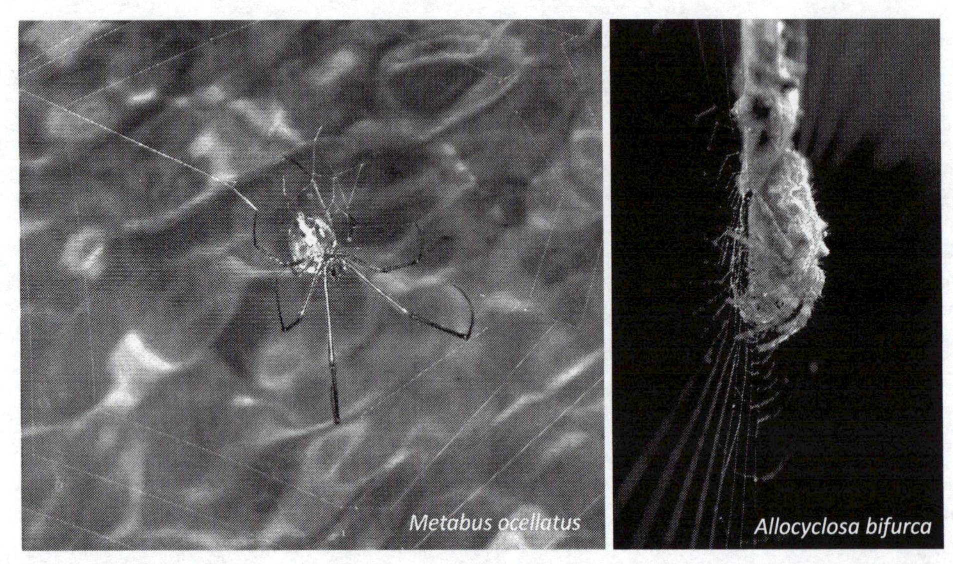

Fig. 2.19. Two species illustrate contrasting extremes in hub designs. The "minimalist" open hub of the horizontal orb of *Metabus ocellatus* (TET) just above the surface of a tropical stream (*a*) is just large enough to be grasped by the spider's legs III and IV (one leg III and one leg IV hold the outermost hub loop). They grip the hub directly above the spider's heavy abdomen, while her long legs II and I hold radii in the free zone (they occasionally even extended into the capture zone). The anterior legs probably support little of the spider's weight, and provide little motive force when the spider turns at the hub to attack prey. The hub of *Allocyclosa bifurca* (AR) (*b*) represents the opposite extreme. The numerous, tightly spaced loops extend far beyond all of the spider's legs, even when she abandons her daytime cryptic crouch and spreads her legs at night (see also fig. 3.27; the object just above the spider is part of an egg sac—see section 3.3.4.1).

it (fig. 2.21). Such bending was not seen, however, in several species in three families of orb weavers that were checked (Eberhard 2017). Instead, the spiders appear to have solved this problem in two different ways. Orb weavers and their allies used a previously unrecognized movement, rotating the distal portion of the leg on its longitudinal axis, thus orienting the middle claw so that it was more or less perpendicular to the line (figs. 2.20, 2.21). The precise site (or sites) where rotation occurs remain to be determined, though the limited dorso-ventral mobility of the claws themselves (see Ramírez 2014) indicates that the it occurred at more basal articulations. A second, clumsier solution, is illustrated in fig. 5.17b of the lycosid *Aglaoctenus castaneus*: the spider's legs on one side were aligned approximately perpendicular to the line, and she used only these legs to walk along it, letting the others dangle in air below. Similar behavior also occurred in spiderlings in the mygalomorph families Atypidae, including *Atypus affinis* (unpublished video recording of Jenni Louise Cox) and *Sphodros rufipes* (WE), and Ctenizidae (possibly *Ummidia*) (WE). Further observations will be needed to determine the taxonomic extent of these types of behavior.

This implies that an orb weaver moving across her web probably constantly adjusts the rotation of each of her legs to align its middle claw perpendicular to the lines that it grasps. It may be that the information acquired during following behavior (preceding section) includes the orientation as well as the location of the line that is being held by the leading leg. This could allow the spider to adjust the degree of rotation of her following leg before it contacts the line, making it easier to grasp the line when the leg encounters it. When *Z. geniculata* moved one leg II moved to grasp a line that was already being held by the other leg II, the moving leg II was already rotated so that its median claw was approximately perpendicular to the line (WE).

It is not known how a spider determines how much she needs to rotate her leg in order to seize a newly encountered line (one not already being held by another leg). Perhaps the direction in which the setae at other sites on her tarsus are deflected by the line just before she grasps it provide this sort of information. But this is only speculation; high-speed recordings of the process of grasping a new line, and lines already held by other legs, could test these ideas. It is theoretically possible for

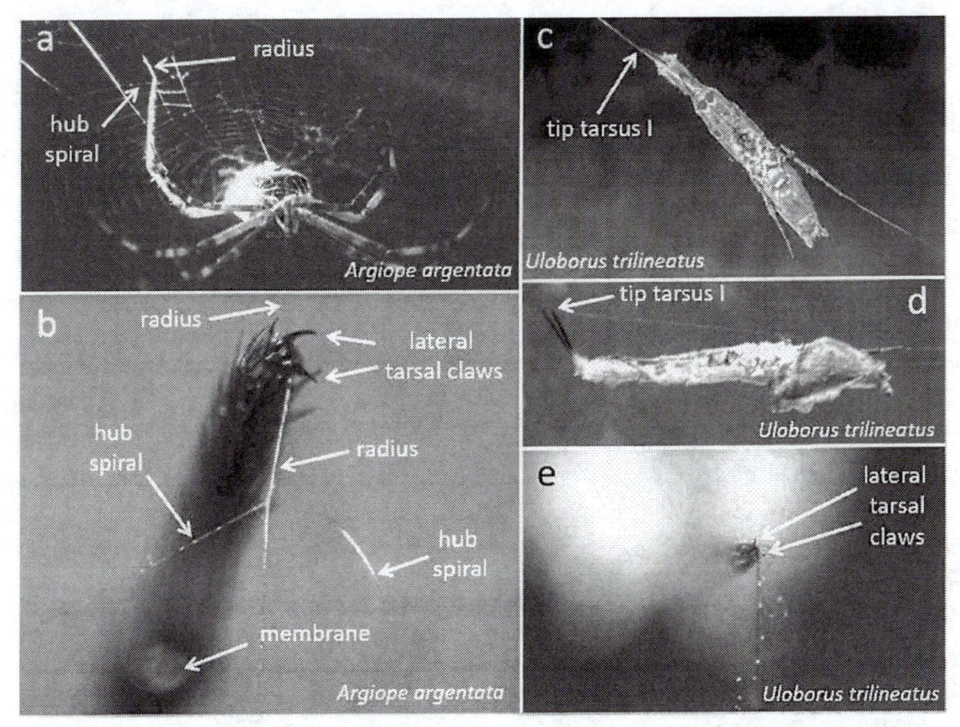

Fig. 2.20. A subtle (and previously unnoticed) behavioral detail allowed this mature female *Argiope argentata* (AR) (*a*, *b*) to grasp a radius with one leg IV at the hub of her orb even though the line was nearly parallel to this leg's longitudinal axis. By rotating a distal portion of her leg on its longitudinal axis (compare the orientations of her lateral tarsal claws with those of the tibia-metatarsus intersegmental membrane in *b*) she positioned the median claw (not visible) approximately perpendicular to the line, allowing it to clamp the line firmly against her serrated accessory tarsal setae (see fig. 2.16). A mature female *Uloborus trilineatus* UL (*c–e*) used similar behavior to grasp a frame line that was nearly parallel to the spider's long axis with both of her legs I, as demonstrated by the orientations of her femur, patella, and tibia (*c*, *d*) compared with that of her tarsus (*e*) (after Eberhard 2017).

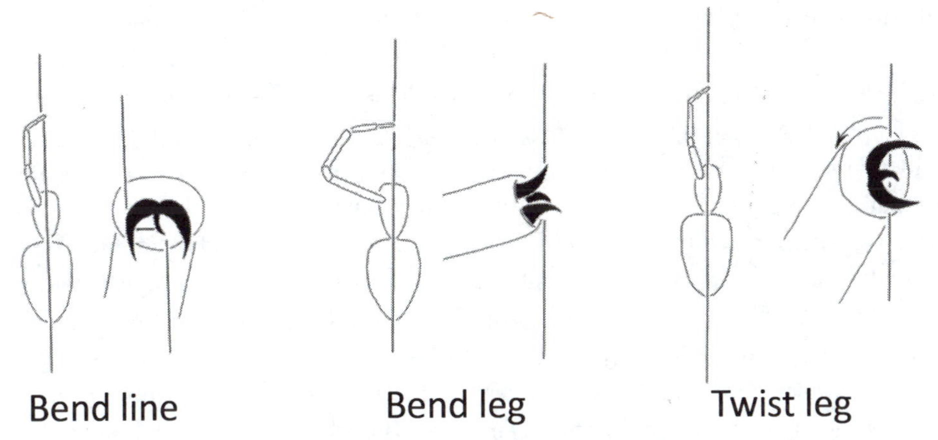

Fig. 2.21. Three theoretically possible ways in which a spider could orient her leg to use her middle tarsal claw to grasp a line that was more or less parallel to the long axis of her leg: *a*) bend the line so that it is perpendicular to the long axis of her leg at the point where the middle claw grasps it (proposed by Nielsen 1932); *b*) bend her leg so that the long axis of her tarsus is perpendicular to the line; and *c*) twist the leg on its long axis so that the middle claw is perpendicular to the line. Orb-weaving spiders apparently use technique *c*.

a spider to sense the orientation of a line that she has grasped, via either the rotation of her legs or the stresses exerted on her leg by the line.

2.4.3 GRASPING A LINE PRIOR TO ATTACHING THE DRAGLINE

Additional motor abilities are needed to attach lines to each other. The spider must grasp lines and hold them in precise alignment against her spinnerets in order to attach lines to them at precisely determined points, and then to release these lines. Mygalomorphs have never, to my knowledge, been observed to use any of their legs in making attachments. The few detailed studies of their spinning behavior suggest that these spiders apply lines to the substrate and other lines only imprecisely, by lowering the abdomen and tapping or sweeping their spinnerets on the substrate without contacting the lines with their legs (e.g., Barrantes and Eberhard 2007; Eberhard and Hazzi 2013; in prep.). It is still possible, however, that more elaborate mygalomorph attachment behavior may yet be discovered in the tangles of strong lines high above the sheets of some diplurids (e.g., Paz 1988).

Grasping lines before attaching to them may have evolved before aerial webs, judging by how the substrate web filistatid *Kukulcania hibernalis* consistently used one leg III to grasp the lines to which she attached cribellum silk (Eberhard 1987a). The legs used and the sites that they grasped varied in different groups: one leg III anterior to the spinnerets in *K. hibernalis*, the pholcid *Physocyclus globosus* (WE), and (probably) the lycosid *Aglaoctenus castaneus* (Eberhard and Hazzi 2017); one leg IV (or sometimes none) behind the spinnerets in the scytodid *Scytodes* sp. (Gilbert and Rayor 1985); and one leg III in front and the ipsilateral IV behind in orb weavers and perhaps all orbicularians (Eberhard 1982; WE).

The precision needed to make such attachments has seldom been appreciated. In typical orbicularians, for instance, the line is grasped with ipsilateral III and IV as just described, and these legs then bring it to the spinnerets, aligning it very precisely in the tiny cleft between her two anterior lateral spinnerets. Only in this position can it be contacted on either side by her piriform spigots (fig. 2.12, section 2.3.2). The spider achieves this precision despite the fact that her abdomen has only limited mobility, few tactile sensory structures, and the spinnerets themselves are short and ill-equipped to sense other lines.

Some spiders achieve the necessary precision in other ways, as illustrated in the building behavior of the pisaurid *Thaumasia* sp. This spider made relatively precise attachments between lines (fig. 5.11) despite never using her legs to explore or hold them (her tarsi consistently rested on the surface of the leaf just below, rather than on the web) (Eberhard 2007a). Perhaps the relatively small distances between the lines, the radial design of the web, or the spider's frequent returns to the center of the web during construction allowed her to make attachments without seizing any lines with her legs.

2.5 CUTTING LINES AND RECYCLING SILK

2.5.1 CUTTING LINES

Cutting lines that are already in place is an important aspect of the construction of an orb. Provisional lines that were laid during preliminary exploration are often shifted or removed, allowing much more consistency in the design of the final product. Exploration usually involves placement of lines that are not organized around the future hub of the web (e.g., fig. 6.1) and are later removed. Cutting lines is also an integral part of the "cut and reel" behavior (sections 5.3.6, 6.3.3) used to adjust tensions on lines, to shift points of attachment (section 6.3.2), and to remove the temporary spiral. The mechanisms used to cut lines have not received much attention. Many authors (myself included) have limited themselves to simply noting that spiders cut lines by bringing them to the vicinity of their mouths. But many spiders' mouthparts are not properly designed to cut lines mechanically: the chelicerae cannot act effectively either as a scissors, or as a knife working against a chopping block (below). The mechanism used to cut lines remains somewhat mysterious; I suspect that it often involves special digestive enzymes (Tillinghast and Kavanagh 1977; Tillinghast and Townley 1987).

The ability to digest dragline silk is probably a derived trait in spiders, because special enzymes in addition to those for digesting food are required; the molecular structure of dragline silk makes it largely immune to the action of many general proteases, which attack alpha helices rather than pleated sheets (Tillinghast and Kavanagh 1977). An enzyme that can digest silk molecules occurs in the digestive fluid of the araneid *Argiope aurantia* (Tillinghast and Kavanagh 1977). The fact that fluffy, cotton-like masses of loose silk at the center of the hub (see section 6.3.6) and entire collapsed webs (section 6.3.9) are often ingested by several araneids in a matter of seconds (or less) (Peakall 1971; Carico 1986; WE), with no sign of masticating movements of the mouthparts (WE) is in accord with rapid extra-oral chemical breakdown of the silk. The fine filtration of food that occurs in the spider's pharynx argues against the possibility that the silk lines are ingested intact and degraded deeper in the spider's digestive tract.

Perhaps the context in which the ability to break lines originated was in freeing prey that had been immobilized from the web in order to carry them to a safer site to feed on them with a minimum of damage to the web (other early uses may have included cutting debris or unwanted prey free). The taxonomic distribution of the ability to cut immobilized prey free is not known. Some spiders such as the theraphosid *Psalmopoeus reduncus* and the diplurid *Linothele macrothelifera* ripped prey free of entangling lines by brute force (Barrantes and Eberhard 2007; Eberhard and Hazzi 2013), and the filistatid *Kukulcania hibernalis* also pulled subdued prey through her web by force as she backed toward her retreat (WE). In a variety of other haplogyne families, including Hypochilidae (Shear 1969), Diguetidae (Eberhard 1967; Bentzien 1973), Scytodidae (Gilbert and Rayor 1985), and Pholcidae (Eberhard and Briceño 1983; Barrantes and Eberhard 2007; I. Escalante in prep.), spiders cut prey free with little or no tugging. The diguetid *Diguetia canities* also sometimes broke lines forcefully while freeing prey (and also during web construction): holding the line near her mouth with a leg II or III, the spider pumped her body up and down several times; the moment the line broke, her body sprang back (Eberhard 1967; WE). Cutting lines is a standard part of attack behavior in orbicularians (e.g., Robinson and Mirick 1971; Robinson and Olazarri 1971; Robinson et al. 1971; Barrantes and Eberhard 2007), as well as in non-orb web-building entelegynes such as psechrids (Robinson and Lubin 1979b).

A second context in which cutting could have arisen is sperm web construction. The clumsy way in which mature theraphosid males used their mouthparts and/or oral secretions to slowly produce a notch in the edge of the sheet of the sperm web by "masticating" it (the process lasted up to several min) (Costa and Pérez-Miles 2002) contrasts sharply, however, with the nearly instantaneous and precise breaking of lines by orb weavers like *Nephila* and *Argiope* spp. during attacks on prey (e.g., Robinson and Mirick 1971; Robinson and Olazarri 1971; Robinson et al. 1971) and during orb construction (chapter 6).

Closeup video recordings of the large, slow moving orbicularian *Nephila clavipes* (NE) showed that cutting did not involve tugging or any other obvious exertion of force; the line was brought to the mouth region by the tarsus of a leg or palp, and often though apparently not always grasped and pulled briefly closer to the mouth with a cheliceral fang; it then "fell apart" without visible indications of strong mechanical force having been applied (WE). Morphological details of the chelicerae of *N. clavipes* are also not appropriate for cutting lines mechanically (WE). The basal segment has no sharp, flat edge that could mesh like a scissors with the fang; its scattered, conical teeth are not appropriately designed for a scissor-like function. An alternative, chopping block hypothesis was also not supported. The internal surface of the fang is not sharp, but is rather slightly rounded and has no sharp edge or ridges. The floor of the fang furrow, against which the fang closes, is flat; it also lacks any structure that would be appropriate for mechanically cutting lines.

Behavioral details also argued against the chopping block hypothesis, because lines making a wide variety of angles with the chelicerae were all cut immediately; the spider made no attempts to align them in any particular way with the chelicerae beforehand. The chelicerae also did not move in ways, such as "sawing" or "snipping" at the line, that might be expected to accompany mechanical cutting. Rather the line seemed to simply fall apart when it reached the spider's mouth area; sometimes the ends fell apart with a gentle "snap," with no perceptible previous tightening (WE).

I hypothesize that selection to reduce damage inflicted on the web by prey removal, possibly combined with selection to economize on the effort needed, probably favored the ability to cut lines chemically. A further, probably originally incidental advantage was the ability to recycle their webs, reducing the costs of moving to new websites (sections 2.5.2, 8.2).

2.5.2 RECYCLING SILK

Orb weavers routinely gather the previous web together and ingest it before they build a new one. Labeled material from a radioactively labeled web began to appear in the ampullate glands of *A. diadematus* as soon as 30 min following ingestion (Peakall 1971; Witt et al. 1968). A large fraction (about 80–90%) of the ingested labeled material ended up in the ampullate glands as opposed to the rest of the spider's body (Peakall 1971). This study suggested an extremely high efficiency of recovery of ingested silk protein in new silk (about 80–95%), but later studies (using another species of *Araneus*, and with some methodological differences) gave recovery rates of around 30%, which is more typical for proteins ingested by other animals (Townley and Tillinghast 1988). In any case, recycling of web protein surely occurs.

Blackledge et al. (2011) stated that web recycling in orb weavers occurs primarily in araneoids that produce viscid silk, rather than cribellate orb weavers or more derived orbicularians that have lost orbs. If this is true (they did not cite supporting data), it could suggest that the primary target of recovery is the viscid glue droplets (Blackledge et al. 2011), perhaps the essential amino acids that constitute 39% of the proteins in aggregate gland secretions (Townley and Tillinghast 2013). Some other otherwise-puzzling observations suggest that this may be an oversimplification. The linyphiid sheet weaver *Diplothyron* sp. produces abundant lines with sticky droplets, but abandons rather than removing old webs (WE). Perhaps it is recovery of particular components of aggregate gland secretions, not the recovery of the protein per se, that has most selective importance in silk recycling. Uloborids, which do not add accessory gland products to their orbs, may be less likely to reingest their old webs. For instance, *Zosis geniculata* (UL) discarded large masses of loose silk when replacing and repairing major sectors of their orbs (I have seen up to 33 substantial masses of fluff accumulated under a single orb), and also discarded egg sacs from which spiderlings had emerged (WE), even though they reingested silk while removing the small mass of loose dragline silk that accumulated from cut and reel behavior during exploration and early radius construction during proto-hub construction (appendix O6.3.2.3), as also occurred in *Lubinella morobensis* (UL) (Lubin 1986). Other species of deinopids and uloborids are known to cut lines easily during web construction, but nevertheless to sometimes discard balls of loose silk from previous webs or abandoned webs without reingesting them (Akerman 1926; Marples and Marples 1937; Eberhard 1972a; Nentwig and Spiegel 1986). Another uloborid, *Hyptiotes paradoxus*, however, consistently reingested silk (Opell 1982). Some tetragnathids and nephilids also sometimes abandoned webs without reingesting them (Buskirk 1975; WE). In sum, the ability to sever and digest lines enzymatically is widespread in araneoids (section 2.5; Carico 1986; Higgins 1987, 1992b), but the present incomplete knowledge of taxonomic distribution of web recycling in other groups does not follow clear patterns.

Another puzzling observation is that, at least in *Nephila pilipes* (= *maculata*), ingestion of the old web sometimes took many hours. Hingston (1922c) saw that an individual that rolled up her web (after he had broken all the anchors except one at the top) fed for approximately 12 hours (from 10 AM to "late at night"). Because of the spider's large size, he was able to observe "the yellow ball" of silk on her mouthparts. Why did this spider take so long to ingest silk, while recordings of *Araneus diadematus* (AR) (Peakall 1971) and *Micrathena duodecimspinosa* (AR) (WE) showed that ingestion of an old web or a mass of fluff took on the order of one second?

One previously unremarked problem in reingesting webs is that the presence of water droplets on lines from rain, fog, or dew could dilute and probably reduces the effectiveness of her proteolytic enzymes (section 2.5). Thus, an araneoid orb weaver may be obliged to discard a web that has accumulated drops of water rather than recycle it (fig. 8.10). Perhaps breaking the lines in an old, water-laden web but not cutting them entirely free (as in fig. 8.10) allows spiders to later rescue at least part of a web by waiting until it dries off, but this is only speculation.

The protein in silk lines may sometimes be discarded because it is not valuable nutritionally to the spider. After all, silk may have evolved from the nitrogenous waste products of these predators, whose food is rich in protein (section 2.2.1). Glatz (1967) used this reasoning to explain the frequent abandonment of apparently intact and functional webs by the oecobiid *Oecobius annulipes* only two to four days after building them; the pholcid *Physocyclus globosus* also frequently abandoned intact webs (Eberhard 1992a). Nevertheless, at least for some spiders in some situations, ingesting silk is apparently a significant source of nutrition. Many orb weavers routinely ingest their previous webs (Shelly 1984; Carico 1986), and some spiders, including the theridiid *Argyrodes lanyuensis* (Tso and Severinghaus 1998) and mature males of the tetragnathid *Leucauge mariana* (WE), steal and ingest the silk from the webs of other spiders.

One ecological consequence of the ability to digest silk, whether or not the major payoff is from the silk lines themselves or from aggregate gland products, is that it probably enables web builders to search for prey more efficiently. Most orb weavers, for instance, can ingest the old web before abandoning a site which has proven to be unsatisfactory, thus reducing the cost of moving away. Section 8.6 discusses possible ecological consequences of recycling silk.

2.6 HOW SPIDERS AVOID ADHERING TO THEIR OWN WEBS: A MYSTERY PARTLY SOLVED

Finally, an important ability that has received puzzlingly little attention is how spiders that make sticky webs avoid becoming stuck to their own webs. Perhaps the most obvious problem is that when a spider attacks a prey, she needs to move rapidly across sticky portions of her web without becoming entangled; she could also benefit from minimizing the damage she produces in her web as she moves. One possible solution was first outlined by the pioneering naturalists Fabre (1912) and Hingston (1920). Working with adult females of *Argiope bruennichi* (AR) and *Nephila pilipes* (NE), they found that when the tips of the spider's legs were washed in a non-polar organic solvent, the spider apparently adhered more tenaciously than usual when she subsequently contacted sticky spiral lines. Hingston (1920) also found that *N. pilipes* also appeared to "recoat" her legs by moistening her tarsi in her mouth, and rubbing them on other parts of her body such as her legs, her abdomen on and near her spinnerets, and the sides of her abdomen, which Hingston thought were especially likely to come into contact with sticky lines. In contrast, she neglected other large expanses of her body such as the ventral surfaces of her abdomen and cephalothorax that do not contact lines. Hingston gave anecdotal evidence that these areas adhered more to sticky lines when he experimentally brought lines into contact with them. Both Fabre and Hingston concluded that orb weavers probably use some substance, perhaps some sort of oil, to avoid adhering to their sticky spiral lines.

Most subsequent discussions simply repeated these accounts, though Vollrath and Tillinghast (1991) argued that the anti-adhesive story is a myth, and that orb weavers do not need a defense against adhesion because they tend to avoid walking on the sticky lines in their webs (as also noted by Fabre and Hingston), moving instead along the non-sticky radial lines. The preference for walking on non-sticky radial lines could also, however, be due to selection to avoid disturbing the uniform spacing between sticky lines by not walking on them (fig. 4.19), or to the fact that radii are much less extensible than the sticky spiral lines, and thus offer more rigid support for the spider's weight (section 2.2).

More importantly, the avoidance hypothesis fails to take into account that at other moments during sticky spiral construction araneoid orb weavers do not avoid contact with sticky lines, but rather repeatedly touch and pull forcefully on them. Just before attaching the sticky spiral to each radius, the spider pulls out additional sticky line from her spinnerets with her legs IV (section 6.3.5; Eberhard 1982; Briceño and Eberhard 2012). In *Nephila clavipes* (NE) and *Gasteracantha cancriformis* (AR), for instance, the spider pulls on the sticky spiral line approximately 1000 to 1500 times with each hind leg during the construction of a single orb (Briceño and Eberhard 2012).The pulls lengthen the sticky line, reducing the tension on it and probably improving its ability to stop and retain prey. Pulls of this sort apparently occur in all species of araneoid orb weavers (Eberhard 1982). Spiders also repeatedly contact sticky lines during inner loop localization behavior just before each attachment to a radius (section 6.3.5). Still further contact often occurs during prey attack in *Araneus diadematus* (AR) (Peters 1933a) and *Nephila clavipes* (WE) when the spider pushes against the orb with the distal portions of her legs I and II to position her body while wrapping the prey at the attack site. In short, orb weavers repeatedly contact sticky lines; in no contexts, however, does the spider

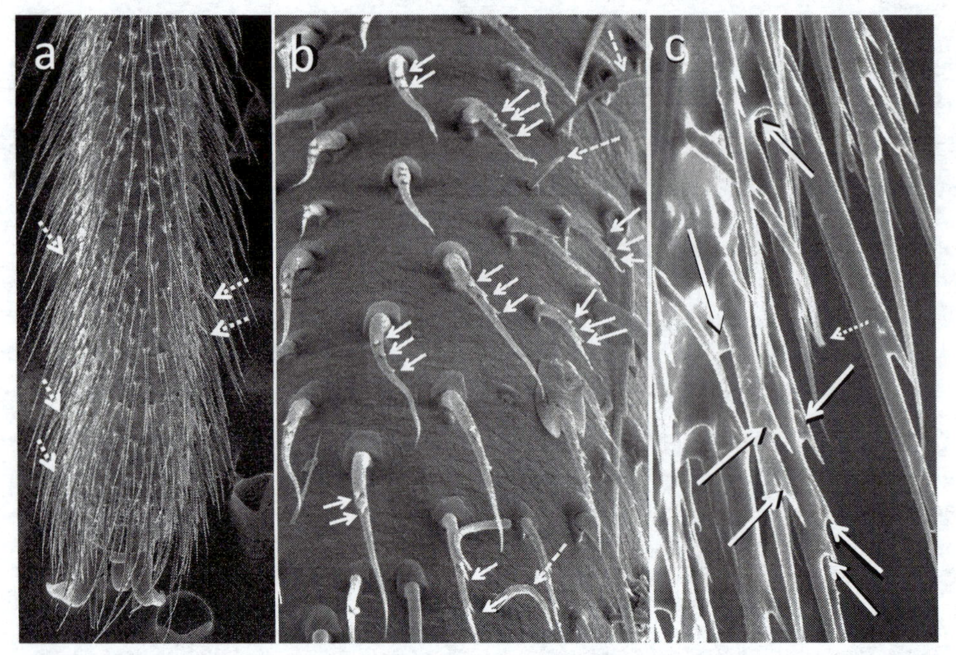

Fig. 2.22. The tarsal setae of a mature female *N. clavipes* (NE) illustrate a morphological mechanism used by some spiders to reduce their adhesion to their own webs. In dorsal view (*a*), the tarsus is densely clothed with setae, a few of which were blunt-tipped and oriented more nearly perpendicularly to the surface of the tarsus, and which probably function as chemo-sensors (dotted arrows). At greater magnification (*b*), small barbs (solid arrows) are visible near the bases of the other, more common setae, always on the surface of the seta farthest from the surface of the tarsus (most setae in *b* are projecting approximately toward the viewer). When the tarsus was pressed experimentally against a sticky spiral line, mimicking a common behavior during sticky spiral construction, the line slid along individual setae until it was arrested by the barbs near their bases; this minimized the area of contact by preventing extensive contact with the tarsal surface below. In *c*, a tarsus IV of a mature female *N. clavipes* that was pressed 400 times against a sticky spiral line illustrates how liquid (solid arrows; presumably glue from the orb) accumulated on the setae. Dotted arrow indicates a blunt-tipped seta. Additional defenses of the spider against adhesion to her own web included details of how she moved her legs, and an oily, anti-adhesion coating.

adhere more than very weakly to the sticky lines (Briceño and Eberhard 2012).

Two recent studies gave further support to the Fabre and Hingston anti-adhesive idea with *Nephila clavipes* (NE) and *Gasteracantha cancriformis* (AR) (Briceño and Eberhard 2012), and with the araneids *Araneus diadematus* and *Larinioides sclopetarius* (Kropf et al. 2012). Washing isolated legs with either hexane or carbon disulfide increased the force of adhesion by sticky spiral lines in standardized tests (washing with water gave inconsistent results). Two additional anti-adhesion traits that were revealed included cleverly oriented leg movements, and barbed, drip-tip setae. Spiders pulled sticky lines from their spinnerets with thrusting movements of leg IV that brought the exterior surfaces of the setae on the tarsus into contact with sticky lines. The direction of the leg's thrust caused the droplets of adhesive on the sticky line to slide basally on these setae until the baseline snagged on other setae or on the basal branches or "barbs" on the exterior surfaces of these setae (fig. 2.22); snagging on the barbs prevented the sticky droplets from reaching the surface of the tarsus where they would have made more extensive contact and thus adhered much more strongly. When the leg then pulled away from the sticky line, it moved more or less parallel to the long axis of its tarsus (Briceño and Eberhard 2012), thus causing the sticky droplets on the line to retrace their paths, sliding distally along the setae (fig. 2.22); at the moment the leg finally pulled the setae free, the area of contact between each sticky droplet and a seta had become reduced to just the tip of the seta, minimizing the force of adhesion. Because the setae are scattered along the length of the tarsus rather than being arranged in rows (Briceño and Eberhard 2012), different droplets pulled free from different setae one after the other rather than simultaneously, reducing still further the strength of adhesion.

The possibility that the anti-adhesion traits documented in these studies occur in spiders in general is raised by a recent observation of the salticid *Thyene imperialis* walking on sticky lines of *Cyclosa deserticola* (AR) "without any problems" (Jäger 2012).

The tests of adhesion to isolated legs also revealed an additional, previously unappreciated problem that orb weavers confront. When an isolated hexane-washed leg was touched over and over to a sticky line, the force with which it adhered gradually increased. The likely reason for this increase was that sometimes when the leg pulled away, a small amount of adhesive substance remained on the leg; this glue gradually built up on the leg. When washed legs that had touched a sticky line 400 times were checked under the SEM, their setae had substantial accumulations of viscous material (fig. 2.22c). This number of contacts is biologically realistic (or even an underestimate) for the construction of a single orb. For instance, the *Micrathena undecimspinosa* web in fig. 1.4 had over 3500 attachments (57 loops of sticky spiral and 63 radii), and the spider pulled sticky line with her IV one or more times prior to each attachment to a radius (section 6.3.5). Orbs like those of *Acacesia* and *Deliochus* (fig. 3.5) and *Pozonia* (fig. 3.35) have even higher numbers of sticky spiral attachments. The anti-adhesive coating (washed off in these experiments) presumably helps protect against such accumulations of glue.

Several mysteries still remain. The largest puzzle is that cribellate spiders also avoid adhering to the treacherously adhesive cribellum silk in their webs, but none of the defensive mechanisms just described will work in their webs. In the first place, the webs of many cribellates, such as those of filistatids, *Stegodyphus*, and dictynoids (see chapter 9; Griswold et al. 2005), have dense, disorganized arrays of cribellum silk that often extend into and around the spider's own retreat, so neither the "tip toe" tactic nor the carefully oriented withdrawal movements of the legs of orb weavers that were just described is a feasible defense. The non-orb weavers routinely contact sticky lines with many parts of their bodies as they move in their webs, but show no sign of adhering. Secondly, an oily substance is not likely to provide a defense against the hygroscopic and van der Waals forces responsible for the stickiness of cribellum silk (see Opell 2013; section 2.6.). If anything, oil might be expected to increase adhesion: the cribellum silk of the filistatid *Kukulcania hibernalis* adhered tenaciously to the tarsi of *Nephila clavipes* (D. Briceño pers. comm.). A possible clue comes from the unusually "hairy" setae on the tarsi (fig. 2.23) of mature male *Kukulcania hibernalis* (Filistatidae), which may have an especially demanding challenge to avoid adhesion, because they grasp and tug strongly on sticky lines in female webs as part of courtship. Just how cribellate spiders avoid adhering to their own sticky silk is not known.

An even greater mystery occurs in all cribellate spiders—the lack of adhesion when the calamistrum pulls sticky lines from the cribellum. The comb-like row of thick setae on leg IV (the calamistrum) snags and pulls swaths of lines from the cribellum during huge numbers of rapidly repeated brushing movements across the cribellum (fig. 2.7, section 3.3.3.3.2.1.2); for instance, a mature female *Zosis geniculata* [UL] brushed posteriorly over her calamistrum on the order of 69,000 times in the course of building a single orb (WE). With each posterior brushing movement an additional length of cribellum silk is snagged by the calamistrum, pulled from the cribellum spigots, and then somehow released. Following each brushing movement, the leg reverses direction moves anteriorly to prepare for the next brushing movement. At this critical moment, the bristles of the calamistrum must somehow free themselves from the highly sticky lines that they have just pulled from the cribellum. The only force that would seem likely to pull the cribellum fibers from the calamistrum bristles is their air resistence.

How do the calamistrum setae free themselves immediately and smoothly from the cribellum lines? I have not found any study addressing this question. A calamistrum is apparently present in all spider species with a cribellum, and the combing movements are very ancient (above; Griswold et al. 2005). Perhaps the complex microsculpture of tiny ridges and teeth on the calamistrum setae of some groups (Ramírez 2014), which do not seem to have any known function, is involved. But the calamistrum setae of some other groups, like *Uloborus* spp., are relatively smooth (R. Foelix pers. comm.). Brent Opell has suggested (pers. comm.) that perhaps the fine cribellum lines are not sticky the instant that they are pulled from the cribellum, and only become adhesive a little later, perhaps after the tiny nodules on the fibrils are formed (Opell 2013).

There are also some puzzling reports of *lack* of ad-

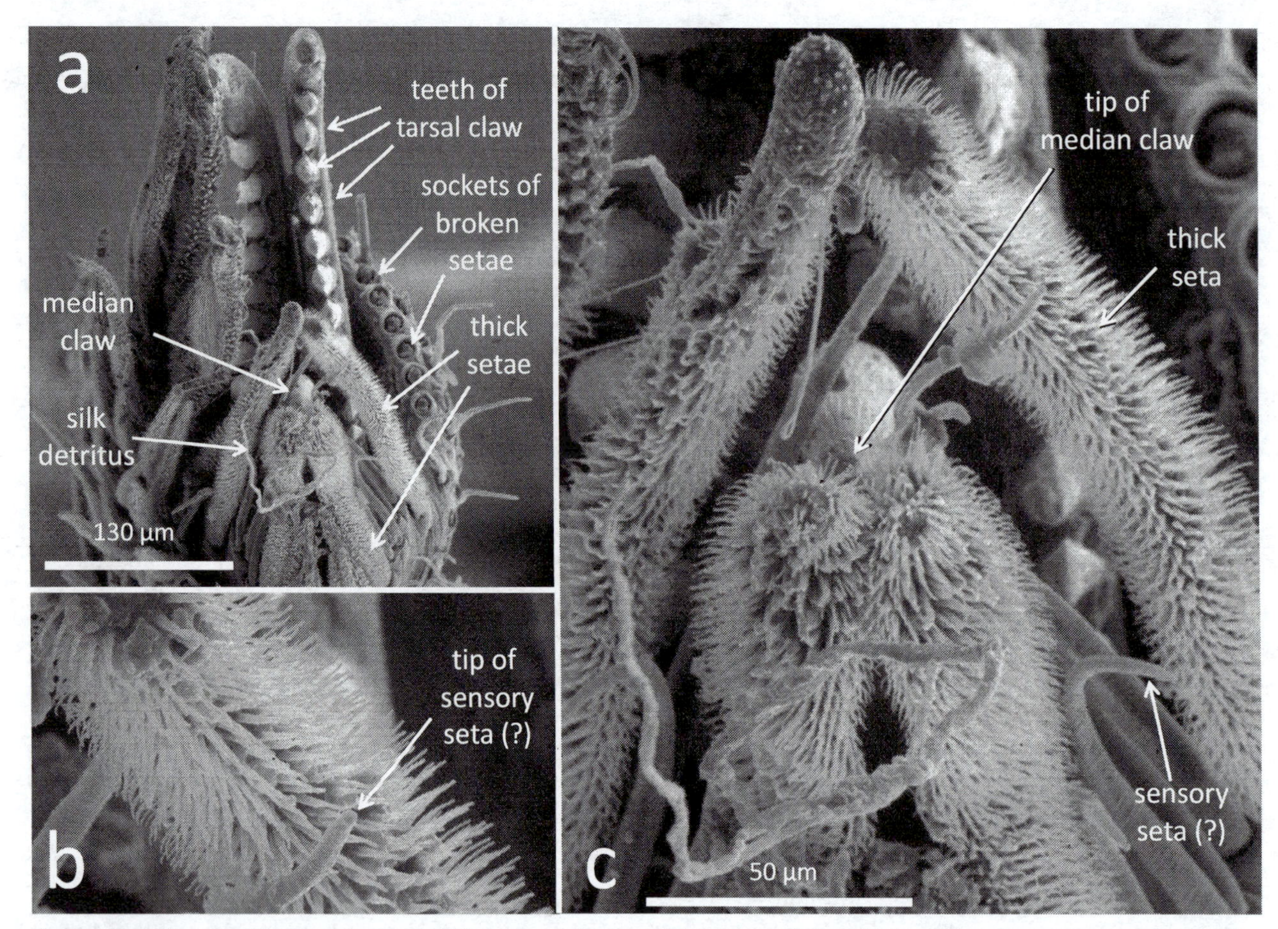

Fig. 2.23. Ventral views of the tip of the tarsus I of a mature male *Kukulcania hibernalis* (Filistatidae) at different magnifications highlight a small cluster of thick setae resembling hairy caterpillars that may reduce the adhesion of the male to the female's highly adhesive cribellum silk (most of the other setae on the right side of photos *a* and *c* were removed). The tips of the male's tarsi made strong contact with the adhesive cribellum lines on the webs of females during courtship, when he jerked and pulled on her web, but they nevertheless adhered weakly if at all. Female tarsi have more reduced versions of these setae (G. Barrantes pers. comm.). This strange male setal morphology supports the idea that processes on setae are somehow important in avoiding adhesion to cribellum lines, and argues against the importance of mechanical entanglement in the adhesiveness of cribellum lines (photos courtesy of Gilbert Barrantes).

hesion by cribellum silk. The eresid *Seothyra henscheli* shook sand grains off the cribellate lines in their webs after wind blew sand onto them (Henschel and Lubin 1992)! Pentatomid bugs and earwigs living in the abandoned web of the eresid *Adonea* moved in the web without adhering to it (Nentwig 1983c), and the salticid *Portia fimbriata* moved through the webs of the social cribellate eresid *Stegodyphus* sp. without perceptibly adhering to them (Jackson 1986). The scytodid *Scytodes globosus* also moved freely in the web of its prey, the amphinectid *Metaltella simoni*, without any signs of adhesion to the abundant cribellum lines there (Escalante et al. 2015; I. Escalante pers. comm.).

The highly adhesive web of *Deinopis longipes* was thrust rapidly toward the flat substrate by the spider's legs I and II, where it captured walking prey but never adhered to the substrate (Robinson and Robinson 1971); the authors say that "the net is held in such a way that the tarsi may touch the substrate in advance of the net," but the published photos do not show the tarsi projecting beyond the lines; indeed the morphology of the spiders, with the tarsal claws that presumably grasp the lines located at the tips of their legs, would seem to preclude this possibility.

Similar mystery surrounds the long, thin ribbons of major ampullate silk that sicariid spiders *Loxosceles* spp. lay in apparently disorganized clumps around their retreats (Lehmensick and Kullmann 1956; Vetter 2015). The disorganized arrangement of the sticky silk precludes defending against adhesion by using careful orientation of the spider's legs or body. The adhesion of this silk may come from electrostatic properties (Coddington et al. 2002). If so, this poses the question of whether these spiders use electrostatic charges as defenses.

2.7 CENTRAL NERVOUS SYSTEM BASIS FOR WEB CONSTRUCTION

Some volumetric measurements suggested that the "central body," an arc of specialized neuropil in the supraesophageal ganglion, is of particular importance in the nervous control of web-building behavior. But comparisons among four species with different lifestyles (only one of which was an orb weaver) showed little difference in the relative fraction of the volume of this ganglion that was occupied by the central body (approximately 3.1 to 5.1 %). Much remains to be learned about which portions of the central nervous system are involved in which aspects of web construction.

2.8. SUMMARY

Silk production is a defining feature of spiders, and this chapter describes the multiple morphological, secretory, and behavioral traits that permit spiders to produce silk and build their webs. Spider silks are proteins that were probably originally derived from nitrogenous waste products. Nearly all have evolved from a single ancestral "spidroin" gene that subsequently duplicated and diverged repeatedly, flowering into an assortment of different types of silk products that are synthesized in at least ten different types of glands. These silks have a variety of properties that adapt them to perform many different functions. Silk is used in forming the linings for burrows, protective enclosures for eggs, protective walls of retreats, safety lines, attachments of lines to the substrate and to other lines, platforms onto which males deposit droplets of semen, thin lines that float free in gentle breezes to establish bridges to distant objects (and also lift spiders to fly), sticky lines that adhere tenaciously to prey, and strong lines to resist stress from the impacts of prey that hit the web and that prevent them from pulling away.

The strongest spider silks are justly famous for their extraordinary combination of extensibility and toughness. Their potential for medical, military, and industrial applications has stimulated much current research on their molecular and mechanical properties. The viscoelastic nature of dragline silk allows an orb web to absorb sudden stresses from prey without breaking; the lines cede with the impact and do not immediately spring back, thus absorbing momentum and avoiding a trampoline effect of launching the prey in the opposite direction. The conversion of silk from an aqueous liquid (in the lumen of the gland) occurs at the tips of the spigots or within the gland ducts, and happens only when the line is pulled. This dependence on pulling has had important consequences for the morphology of the spinnerets and their spigots, and for web-building behavior in general; the spider cannot shoot lines to distant objects, and thus must depend on her own movements, movements of her spinnerets, or air movements to initiate and extend bridge lines.

Recent work has revealed a previously unsuspected diversity in the chemical and mechanical properties of lines in different species, even in the lines from a given gland in an individual spider. Spiders apparently facultatively manipulate the abilities of lines to resist stress and their stickiness (e.g., Gosline et al. 1999; Madsen et al. 1999; Vollrath et al. 2001). A new generation of studies is now beginning to document the extent of this inter- and intra-specific diversity and its relations to other factors such as web designs, prey types, humidity, and environmental stresses (see for example Blackledge et al. 2011; Sensenig et al. 2011; Blackledge 2013; Opell 2013; Boutry and Blackledge 2013; Amarpuri et al. 2015; Opell et al. 2015). It is already clear that uptake of water from the atmosphere (promoted by inclusion of proteins and organic salts in glue droplets) is important in maintaining the stickiness and high extensibility of sticky spiral lines. It appears that the hygrosopic properties of different species' sticky lines are "tuned" to the humidity of their natural habitats to produce maximum adhesion. Silk from different species differs in the degree to which it "supercontracts," and high humidities that cause nonsticky lines to supercontract can alter an orb's ability to stop prey. The degree of supercontraction is correlated with the proline composition of the silk molecules, and the silks of different species contain different proportions of proline (Liu et al. 2008). Study of evolutionary adjustments of silk properties to different environmental conditions is only in its infancy, and exciting developments seem sure to emerge in the future.

Another newly appreciated complication is the structural complexity of the small droplets of adhesive on the sticky spiral lines of araneoid orb weavers and the functional significance of the different components. It is also likely that some types of silk lines (e.g., aciniform, major ampullate, and cylindriform gland lines) are at least somewhat adhesive when first laid, but later become non-adhesive.

Morphological structures that allow a spider to control how and where her silk lines are deposited also affect web construction. These abilities were probably especially important as spiders evolved from solving the mechanical problems associated with webs by "brute force" (for instance, by simply adding more and more lines) to using more precise positioning of fewer, stronger lines to capture prey (see also section 10.3). One key set of structures is the spinnerets. Early in evolution they were long and highly mobile, but they are much shorter in most modern groups. The complex musculature of spinnerets suggests that even short spinnerets are capable of various types of movements, and limited direct observations of their behavior have confirmed that they perform rapid, precise movements that are of crucial importance in processes such as initiating and terminating lines rapidly, and fastening lines solidly to each other and to the substrate. Spinneret movements may be more elaborate and important during web construction than is currently appreciated. The positions of the spigots on the spinnerets probably represent adaptations for initiating lines and fastening them to each other and to the substrate.

Structures on the tarsi and subtle details of leg movements enable spiders to grasp, manipulate, and release silk lines precisely. Early web-building spiders probably never grasped individual lines with their legs, and relied instead on movements of the entire abdomen and their long, mobile spinnerets to place lines. Evolutionary innovations that were important in building prey capture webs include the following: acquisition of a middle tarsal claw and strong accessory tarsal setae that have notches enabling the spider to seize lines firmly when the middle claw flexes ventrally; following behavior by more posterior legs that economizes on searching behavior, by utilizing information on the locations of lines from more anterior legs; asymmetric lateral searching behavior by the legs (which, in conjunction with the asymmetric placement of the middle claw and its accessory setae, facilitates grasping newly encountered lines); rotating the leg on its longitudinal axis so that the middle tarsal claw is perpendicular to the line that is being grasped; grasping a line with tarsi III and/or IVand holding it precisely aligned against the spinnerets to facilitate making attachments by using piriform silk (facilitated in turn by the flat field of spigots on the medial side of the tip of the anterior lateral spinneret); and the combination of branched setae, careful tarsal movements, and a protective chemical coating that allows some spiders to reduce adhesion to their own sticky lines. The mechanisms used by cribellate and *Loxosceles* spiders to avoid adhering to their own sticky silk, however, are still unknown.

Finally, the ability to cut lines is also important in web construction. Cutting is probably accomplished enzymatically rather than mechanically, using special enzymes (typical proteolytic digestive enzymes are not effective with dragline silk). Cutting makes it possible for the spider to make fine adjustments in the tensions on lines and their attachments to each other, and to eliminate the preliminary lines that were laid during the exploratory phase of web construction (these preliminary lines, in turn, are necessary because the spider is unable to shoot new lines, and must instead produce them by pulling).

3

FUNCTIONS OF ORB WEB DESIGNS

The difference between utility and utility plus beauty is the difference between telephone wires and the spider's web.
—Edwin Way Teale, 1953, Circle of the Seasons: The Journal of a Naturalist's Year

3.1 INTRODUCTION

The function of a spider's web is simple: increase the spider's ability to capture prey. But judging the functionality of a given web design is complicated. A web can contribute to prey capture in several ways, and a design feature that increases the performance of one sub-function can decrease that of another. For instance, an orb like that in fig. 1.4 will intercept more prey if it covers more area; but it needs physical strength to absorb the kinetic energy of fast-flying prey like a bee. Given a limited supply of silk, increasing the stopping ability by making thicker lines or spacing them closer together will increase the web's stopping abilities, but will reduce the area it covers and thus the number of prey that it intercepts. Once a prey is intercepted and stopped, retaining it long enough for the spider to reach it requires still other design properties, which trade off in complex ways against the first two functions (table 3.1). A further complication is that not all web functions are of equal importance.

This chapter will focus on the design implications of the functions of orbs rather than non-orbs, for the simple reason that there are very few experimental studies of how other types of webs function; undoubtedly many of the same functions performed by orbs must also be performed by other web types, but their relative importances and trade-offs are probably quite different.

3.2 CORRECTING COMMON MISCONCEPTIONS ABOUT ORB WEBS

To begin, it will be helpful to discard some common misconceptions about how orb webs function. I will start with relatively simple problems and progress to more difficult topics. The discussions here are incomplete; logical and empirical reasons for some omissions are given in appendices O3.1 and O3.2.

3.2.1 ORBS ARE NEITHER SIEVES NOR SOUND DETECTORS

The analogy between an orb and a sieve or fishing net is common, but inappropriate (Sensenig et al. 2012). Firstly, the individual lines of nets and sieves are, in contrast with the lines in orbs, essentially unbreakable for the objects they are designed to retain. Secondly, some lines in orb webs are sticky. An orb can capture a small, weak prey that contacts only a single sticky line, while the net or sieve analogy leads to the mistaken idea that a prey can only be captured if its diameter is greater than the space between lines (e.g., Witt et al. 1968; Denny 1976; Riechert and Łuczak 1982; Watanabe 2000b; Aoyanagi and Okumura 2010; Cranford et al. 2012). In addition, different lines in an orb have different mechanical properties. A radius can absorb about ten times more kinetic energy than can a sticky spiral line in the same web (Denny 1976; Eberhard 1986a). Sticky spiral lines, in contrast, are more important in determining the web's ability to adhere to the prey and thus retain it (see section 3.3.3). The only possible exception to the non-sieve function that I know is the untested hypothesis (Blackledge et al. 2011; J. Coddington pers. comm.) that some small, very tightly meshed symphytognathoid orbs are designed to sieve pollen or fungal spores from the air. Several types of data, including the sites where webs are built, the general lack of abundance of pollen and fungal spores in the air, and the adhesion of these objects to single sticky lines (rather than by sieving) argue against the importance of sieving (see section 3.2.14).

A second old idea is that orbs might detect or convey acoustic signals from prey flying in the neighborhood. This hypothesis was seemingly put to rest by the combination of theoretical calculations of transmission properties and resonant frequencies (Frohlich and Buskirk 1982), and direct measurements that showed virtually no web movements in sound fields <60 db SPL (Finck et al. 1975). Prior to the prey making contact with her web, it is imaginable that the spider might sense the prey's presence via sound-induced vibrations of the radii near the prey and respond by jerking that area of the web before the prey contacts the web, making it harder for the prey to avoid the web. But I know of no reports of such behavior, and high-speed video recordings showed that even the very rapid spider *Leucauge mariana* (TET) turned and jerked only after contact had occurred (Briceño and Eberhard 2011). Once contact is made, direct movements of web lines generated by movements of the prey will induce much higher amplitude movements of the lines than any sound that the prey makes.

Interest in the sonic properties of webs was recently revived by Mortimer et al. (2014), who proposed that silk fibers are "tuned" to a resonant frequency that the spider uses, through "plucking" behavior, to locate both prey and structural damage. But this study concentrated on measuring responses in single lines rather than in intact webs, and thus did not take into account the important effects of damping and dissipation of energy by other lines connected to radii (Landolfa and Barth 1996). In addition, it misinterpreted a crucial behavior pattern. When a prey strikes the web, an orb weaving spider does not pluck lines like a guitar player plucking a string. Video recordings of *L. mariana* (TET) and *Micrathena duodecimspinosa* (AR) (which confirmed direct observations of these and other species), showed that instead the spider jerked lines sharply. The tarsi remained in contact with the lines (Briceño and Eberhard 2011; WE), thus damping vibrations in the line.

Sensitivity to airborne sounds generally seems to result from direct stimulation of sense organs such as trichobothria on the spider's body (Barth 2002), and to be used by orb weavers to avoid predators, rather than to capture prey (Frings and Frings 1966). In several derived groups, including deinopids, theridiosomatids, and bolas spiders, however, spiders apparently use airborne vibrations to trigger attacks on prey (chapter 9).

3.2.2 ORB WEBS ARE NOT THE PINACLE OF WEB EVOLUTION

The sophisticated and subtle morphological and behavioral capabilities that spiders employ to build orbs have tempted arachnologists to conclude that orb webs are somehow the most highly evolved pinnacle of spider web evolution. It has also long been clear, however, that this idea is untenable: some lineages in orb-weaving families have evolved reduced webs or have discarded webs entirely (fig. 3.1). Several other more recently documented cases of this sort include the theridiosomatid genus *Wendilgarda* (fig. 10.29; Coddington and Valerio 1980; Coddington 1986a), and the araneids *Poecilopachys australasia* (Clyne 1973), *Pasilobus* sp. (Robinson and Robinson 1975), *Paraplectana tsushimensis* (Chigira 1978) (fig. 9.23), *Ocrepeira ectypa* (fig. 9.2), and *Kaira alba* (Stowe 1985, 1986).

Recent fossil data have pushed the origin of orbs back in time (Vollrath and Selden 2007), and phylogenetic studies have substantially increased the numbers of descendant groups that appear to have lost orb webs (Blackledge et al. 2009c; Bond et al. 2014; Fernández et al. 2014; Garrison et al. 2016). These include not only

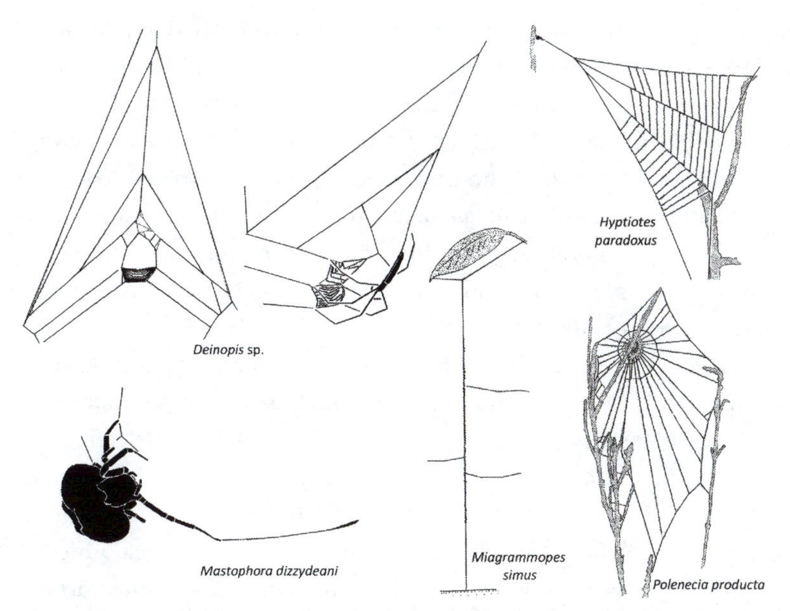

Fig. 3.1. These venerable cases of species descended from orb-weaving ancestors in which orb webs have been lost and replaced by non-orb web designs were all discovered more than 80 years ago. They indicate that the orb is not an "evolutionary pinnacle"; the excuse for the mistaken pinnacle idea ran out long ago. The new contribution from recent phylogeny studies (e.g., Bond et al. 2014; Fernández et al. 2014; Garrison et al. 2016), is not that orb webs have repeatedly been lost; rather it is that a greater number of non-orb and webless modern groups may be descended from orb-weaving ancestors than was formerly believed. The drawings, in which sticky lines are indicated by thicker wavy lines, are based on the following: *Deinopis* (Akerman 1923 and Coddington and Sobrevila 1987); *Miagrammopes* (Lubin et al. 1978); *Polenecia producta* (Wiehle 1931 and Peters 1995a); *Hyptiotes* (Comstock 1967; Wiehle 1927 and Marples and Marples 1937); and *Mastophora dizzydeani* (Hutchinson 1903; Eberhard 1980a) (in this species the sticky silk forms a mass at the tip of the line; the modified, flattened silk line is tightly folded in the drop of glue, and the sticky mass elongates spring-like when, as in this drawing, the spider swings it at a nearby moth).

orb-less members of orb weaving families (e.g., fig. 3.1), but also a major fraction of all modern species that belong to numerous other families whose modern species lack orbs or even webs altogether. It now seems that orb web construction is an ancient (though still successful) adaptation, and that orbs have repeatedly proven inferior in different lineages, and have been replaced by other prey capture strategies.

Why would such beautiful, efficient, and effective traps have so often been disadvantageous? There are numerous possible explanations. In some groups the reduction was accompanied by a clear "compensatory" adaptation, including the chemical prey attractants of the bolas spiders *Mastophora, Celaenia* (AR) and their allies, and *Kaira*, or the ability to make the web collapse around a prey that strikes it in *Hyptiotes* and *Deinopis*. In some *Tetragnatha* spp. (Crome 1954; Gillespie 1999; Blackledge and Gillespie 2004) and the araneid *Chorizopes* sp. (Eberhard 1983) the spiders use quick reactions or powerful weapons to attack their prey directly. Others, like the symphytognathid *Curimagua bayano* (Vollrath 1978) and the mysmenid *Mysmenopsis tengellacompa* (Eberhard et al. 1993; see also Lopardo et al. 2011) have abandoned orbs to parasitize other web-building spiders.

I speculate that one important problem with orbs is that they are fragile, high cost structures. They are often in tatters only a few hours after they are built (fig. 3.19), and most species build a new orb every day (and some regularly build three or four). While a spider generally reingests her silk when replacing an orb (section 2.5), her recycling efficiency is only about 30% (Townley and Tillinghast 1988), so each renovation represents a substantial loss. In addition, construction behavior often involves considerable exertion. For instance, a mature female *Araneus diadematus* (AR) sometimes walked more than 63 m (!) in the process of exploring a site and then building an orb (Zschokke 1996); on a human scale, building a typical orb would involve running on the order of 8 km in the space of about 30 min (section 7.1.1).

In contrast, many non-orb spiders in the field generally make only small repairs and additions to their webs each night (table 5.1). Probably because of an orb's coherent, unitary design, the common non-orb tactic of adding small increments to the web over many successive nights is not feasible (though a few orb weavers have evolved "repairs" that have this result—section 6.3.8).

Another limitation is that orb webs can capture only airborne prey. At least three groups that abandoned orbs for alternative designs, the theridiosomatid *Wendilgarda* spp. (fig 10.29; Coddington 1986a; Eberhard 1989a, 2001a), the araneid *Ocrepeira ectypa* (fig. 9.2, Stowe 1978, 1986), and the highly successful gumfoot webs of theridiids (figs. 9.7, 9.22, 10.30; Eberhard et al. 2008b), gained access to the rich array of walking or floating prey as a result.

One further possible reason for abandoning orbs has hardly been explored—the development of defenses by insect prey against being captured by orbs (see section 4.11). For instance, *Drosophila* flies perform precise flight maneuvers in the near vicinity of orb webs (Craig 1986, 1988) that enable them to avoid being captured. Whether this kind of behavior is common, and whether it might have made orbs unprofitable in some contexts, remain to be determined.

In sum, there has clearly been "life after the orb" for spiders. Orb web construction is more stereotyped and complex in some respects than the construction behavior used to build many other types of webs; but this did not mean that orb weavers always outperformed other spiders that utilized different predatory tactics.

3.2.3 ORBS HAVE NEVER BEEN DEMONSTRATED TO BE "OPTIMUM" STRUCTURES

Optimality in a structure that performs multiple functions is determined by the trade-offs among the different functions. There is a general consensus (Witt 1965; Chacón and Eberhard 1980; Eberhard 1986a; Craig 1987a, 2003; Bond and Opell 1998; Blackledge et al. 2011) that an orb web must perform several tasks: 1) it must be large enough to extend into the insect's flight path; 2) the insect must fail to sense and avoid it; 3) the biomechanical properties of its lines and the connections between them must be strong enough to absorb the insect's momentum on impact, stopping it without rupturing; 4) it must retain the prey long enough for the spider to arrive to attack; 5) the web lines must have survived previous stresses more or less intact, resisting or adjusting to both localized stresses that resulted from previous impacts of prey and detritus, and to generalized stresses resulting from wind and from movements of the objects to which the web is attached; and 6) the lines must provide a path for both the vibrations that will be transmitted from the prey to the spider and for the spider herself as she moves to attack the prey.

Not only are there many functions and trade-offs between different orb web functions, there are many design features that can affect these tradeoffs (table 3.1 gives seven candidates). There are also "fabricational constraints" imposed by both geometry (Coddington 1986b; Zschokke 2002) and the biomechanical properties of the lines of a given type (which are not constant inter-specifically, intra-specifically, or even in some cases within a single orb; see section 3.2.9). Clearly, there is a veritable minefield of variables that can affect the chances that a web will capture a prey. The selective regimes under which spider web designs have evolved are probably very complex.

This complexity means that actually demonstrating that a given design is optimal for a particular species is very difficult. Nevertheless it is common to see statements claiming optimal designs for orbs, even in recent papers: "Spider webs . . . are characterized by a highly organized geometry that optimizes their function" (Cranford et al. 2012; p. 72); or "Orb web spiders, which rely on their webs being optimized for particular conditions," (Harmer and Herberstein 2009; p. 499); "Thus, a suite of bio-mechanical traits interact with web spinning behaviors in optimizing orb web function" (Blackledge et al. 2012a; p. 4). I believe that claims regarding optimality are not well founded, and that optimality has never been demonstrated empirically in any species. Although I was formerly an enthusiastic participant in speculations about optimal orb designs (Eberhard 1986a), I am not not sure that convincing tests are even feasible. In order not to get bogged down in details but nevertheless discuss studies of this sort fairly, I have relegated most of this critical discussion to appendices O3.1 and O3.2. In short, I believe that different lines in an orb serve so many different functions, and that accomplishing the different functions so often involves multiple trade-offs (table 3.1), that it is extremely difficult to determine exactly

Table 3.1. The basic lesson from this undoubtedly incomplete list of the possible trade-offs between advantages and disadvantages of alternative orb web designs and tensions on web lines is that the consequences of building orb webs with different designs is a complex tangle (see also Miyashita 1997; Blackledge and Eliason 2007). The trade-offs are based on the supposition that a spider has a fixed amount of resources (material and energy) that she can invest; they thus underestimate the array of possibilities if the sizes of silk glands relative to body size change over evolutionary time. The advantages and disadvantages marked with "+" are likely to be accentuated for "high energy" prey (that are larger or move more rapidly), many of which will be of greater nutritional value.

Advantages	*Disadvantages*
Larger sticky spiral spacing (thus greater capture area)	
Increased interception (larger areas)	+ Reduced ability to stop and retain prey
Increased interception (web less easily perceived by prey)	Reduced ability to survive environmental stresses such as wind[1]
Horizontal (rather than vertical)	
Less energy expended in construction	Reduced interception[2]
Reduce loading by wind[3]	+ Reduced retention (prey encounter fewer lines as they struggle and fall free from the impact site)
More rapid attack[4]	
Greater portion of web can be close to some prey-rich horizontal habitat (e.g., surface of water)[5]	
Less distortion of web by spider´s weight while building (more precise positions of lines)	
Increased stopping and stability (more equitable distribution of the spider's weight on radii)	
Thinner lines	
Increased interception (larger area covered) and/or improved retention (+greater density of lines)	+ Reduced ability to stop and retain prey
Increased interception (web less visible for prey)	Reduced ability to survive environmental stresses (e.g., wind)
	Reduced ability to support spider
Greater amount adhesive on sticky lines	
+ Increased retention time	Reduced sticky line length reduces numbers of prey intercepted
Increase prey attraction to "sparkly" droplets[6]	Web easier for prey to perceive
Decrease dehydration (and loss of stickiness) in wind[7]	
Tighter web	
Increased stability (survive in wind because web flaps less)	+ Reduced ability to stop and retain prey[8]
Less distorted by spider's weight during construction (more precise positions of lines)[9]	Reduced interception (?) (less web movement in light wind[9])
Improved transmission of vibrations[10]	
More radii	
+ Increased stopping	Reduced interception (web smaller)
Increased stability (resist environmental stress such as wind)	Reduced interception (web more easily perceived visually by prey[11])
Nearer to approximately vertical substrate	
More difficult for prey to perceive web	Fewer interceptions (probably most prey arrive from one side)
+ Increased stopping because of lower prey velocities	

Table 3.1. Continued

[1] Judging by the facultative adjustments of *Cyclosa mulmeinensis* (AR) to windy conditions (Liao et al. 2009; Wu et al. 2013), wind damage will be reduced by larger spaces between sticky spiral loops, but increased by larger web size and longer total sticky spiral length.
[2] This prediction assumes mostly horizontal flight paths of prey. This trend in flight paths is surprisingly uncertain, as horizontally oriented sticky traps did not consistently capture fewer prey (see section 3.2.4).
[3] Assumes wind direction is predominantly horizontal.
[4] This is expected at least comparing prey impacts to the side of and above the spider in vertical orbs.
[5] Greater prey concentration just above the surface of streams was documented by Buskirk (1975) near stream surfaces in Costa Rica for *Metabus ocellatus*; a similar trend was clear in artificial traps hung just above a stream near Cali, Colombia (el. 1050 m) (WE). Importance of the ability to place horizontal orbs closer to the ground is implied by the vertical distribution of horizontal and vertical webs in the araneid *Paraneus cyrtoscapus* (Edmunds 1978). Contrary advantages of vertical orbs would be placing them closer to more or less planar vertical substrates such as tree trunks.
[6] Only a single species of prey (a stingless bee) was tested for attraction to the sticky spiral "sparkle" (Craig and Freeman 1991).
[7] *Cyclosa mulmeinensis* (AR) facultatively increased droplet size in windy conditions (Wu et al. 2013), but this in turn would deplete the spider of sticky silk more rapidly.
[8] The hypothesis that lower tension increases prey retention is supported by the independently evolved active reduction of web tension by several spiders when prey hit their webs (section 4.4.2). For a discussion of the physics involved, see Craig 2003.
[9] This prediction is based on the hypothesis that irregular movements of an orb web in the wind increase its prey interception (the "encounter model" of Craig 1986 that is discussed in detail in section 3.2.11.3; this hypothesis is likely to apply in some circumstances but not others).
[10] This is surely true in the extreme case: slack lines scarcely transmitted the longitudinal vibrations that were transmitted by lines under tension and that are used by spiders (see Landolfa and Barth 1996).
[11] Craig and Freeman (1991) state that the sticky spiral is largely responsible for determining an orb's visibility; but, at least to the human eye, the glint of some radii that are illuminated at an appropriate angle is also visible.

which design would be optimum under any given set of conditions.

3.2.4 THE TRAJECTORIES, DIAMETERS, AND VELOCITIES OF PREY ARE DIVERSE AND POORLY KNOWN

Many models and tests of how webs function mechanically to capture prey (e.g., Craig et al. 1985; Eberhard 1986a; Craig 1987a; Aoyanagi and Okumura 2010; Sensenig et al. 2010a; Cranford et al. 2012; Zaera et al. 2014) assume that prey strike orbs at 90° to the plane of the orb. The idea that theridiosomatids derive an advantage over planar orbs from pulling their orbs into cones because it makes them more difficult for prey to see because all parts of the orb are not in focus at once (Craig 1986) also assumes perpendicular encounters. But prey probably only occasionally strike an orb while moving perpendicular to its plane. Striking an orb at an angle of less than 90° will tend to reduce the effective spaces between lines, especially between sticky spiral lines. It will also alter the patterns of stress on radial, hub, and frame lines in ways that are not contemplated by some models. Some prey appear to sense orbs at a distance and attempt to evade them (fig. 3.2), reducing their velocities and perhaps changing their angles of impact. The angle of impact with the orb's plane could have important consequences for interception (section 3.3.1), stopping (section 3.3.2.1), and retention (section 3.3.3.1), and the trade-offs in these functions (table 3.1). For instance, an orb could be less likely to intercept prey with flight angles oblique to the web plane, but better able to stop and retain them.

A further complication is that "the" diameter of the prey, and thus the number of lines that it can be expected to contact, will be influenced not only by its general body shape at rest, but also by the positions of its legs, antennae, and wings while it flies, and also its flight path (Craig 2003). An insect flying in a straight line has a smaller effective diameter than one flying erratically in a bobbing path (such as, for example, that of many crane flies and mecopterans—Craig 2003; Blackledge and Zevenbergen 2006). Direct observations of the slow, hovering flight of some familiar insects reveals that some, perhaps many species extend their legs in flight. The legs of the paper wasps *Polistes* spp. and the honey bee *Apis mellifera* dangle ventrally; the legs of *Panorpa* scorpion flies and of several groups of nematoceran flies are spread, with legs I projecting anteriorly and legs III laterally and posteriorly (Blackledge and Zevenbergen 2006; WE). High-speed photography (see for instance, andresmorya.photoshelter.com/) reveals similar spread and extended positions in orthopterans, flies, and lacewings. The antennae of many of these insects are long and are extended during flight. In addition, some prey

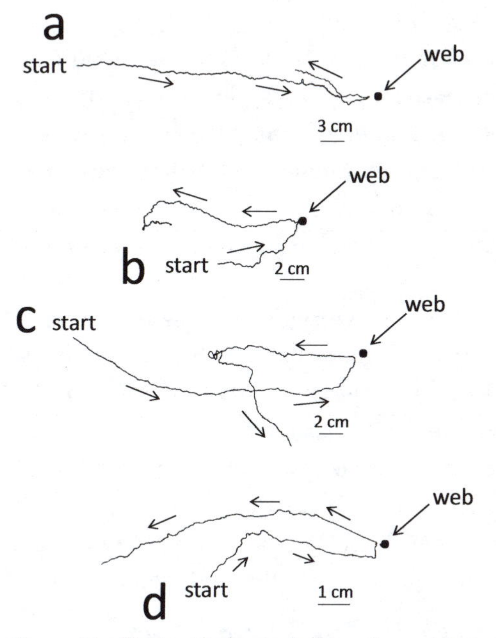

Fig. 3.2. The flight paths (arrows) of four *Drosophila melanogaster* flies as they approached a web line of *Theridiosoma gibbosa* (TSM) (dots) show how three of the four turned away sharply just before contacting the line; the fourth (*b*) touched the web but did not adhere strongly, and then also moved rapidly away (after Craig 1986). These paths have potentially important implications for understanding how orb webs function. At least for prey such as these flies that have slow, hovering flight patterns, the assumption in some attempts to model orb web function that prey have high speed, linear trajectories (see fig. 3.23) may be misleading. And, perhaps more seriously, attempts to quantify those prey that are truly available to spiders in nature by employing artificial traps (e.g., adhesive sheets, sampling insects in the air with suction devices) are imprecise (sections 3.2.5.4, 8.9.2, appendix O3.2.3).

may alter their flight direction just as they reach an orb, altering the areas of their bodies that contact the web. The positions of a flying insect's legs and antennae could have large effects on whether it was stopped by an orb and how long it was retained. In general terms, it appears that it is easier for an insect to free itself of sticky lines that adhere to its legs than to its body, and to distal rather than proximal portions of its legs (Suter 1978; Blackledge and Zevenbergen 2006).

The important point is that insects often extend their appendages during flight, and the details probably vary widely among species. Combining this variation with the effects of different velocities and angles of impact, ignorance regarding differences in the magnitudes of the effects on different species (especially in view of the possibility that only a small fraction of prey are crucial to determining the spider's reproductive success—Venner and Casas 2005; section 4.2.2), and our current ignorance of the shapes and the slopes of curves describing the effects of these variables on spider nutrition, makes even moderately precise modeling of the effects of different orb designs on prey capture extremely difficult (sections 3.2.3, 7.15). Parenthetically, understanding species that build orbs against tree trunks and across single curled leaves (figs. 4.3, 3.34d,e, 9.17a) may be simpler, because impacts should tend to be slower and closer to 90°.

3.2.5 MOST DIFFERENCES IN ORB DESIGNS ARE PROBABLY NOT SPECIALIZATIONS FOR PARTICULAR PREY

A glance at the figures in this book will show that spider orb webs differ in many design aspects, including the numbers and patterns of the spacing of sticky spiral loops and radii, the relative sizes of hubs, deviations from symmetry, and the numbers and relative lengths of frame and anchor lines. Different orb web designs are often said to represent specializations to capture different prey; statements to this effect (among many others) have been made by Łuczak (1970), Uetz et al. (1978), Chacón and Eberhard (1980), Olive (1980), Eberhard (1986), Stowe (1986), Craig (1987a), Uetz and Hartsock (1987), Opell (1997), Blackledge and Zevenbergen 2006, Gregorič et al. (2010), and Blackledge et al. (2003, 2011). Prey specialization offers an appealing explanation for the diversity of different orb designs, especially if designs have evolved as adaptations to avoid competition for prey with closely related species (section 3.2.6). I will argue here, however, that current data are not sufficient to support typical prey specialization arguments.

The word "specialized" can be interpreted for an orb in different ways. At one, inclusive extreme, the term implies only that the orb's design imposes a non-random bias in the subsample of all of the potential prey present in a given habitat that a web captures. That webs have such general biases is undeniable: aerial webs tend to capture flying or jumping prey, while substrate or gumfoot webs tend to capture walking prey; neither type ever captures eagles, cladocerans, or amoebae. At the opposite, "strict" extreme, specialization implies much stronger biases. Closely related spider species could, due to relatively small design differences in their orbs, capture different arrays of prey in the same environment.

There is obviously a continuum of intermediate interpretations of the word "specialized," and interpretations undoubtedly varied among the authors I just cited. I suspect, however, that many of them used the word in a relatively strict rather than relaxed sense (I know that I did). I will discuss this more strict sense here because of its important implications regarding the evolutionary diversification of web designs (see appendix O3.2 for further discussion).

3.2.5.1 Long lists of prey captured seem to argue against strong specialization

One serious problem for the prey specialization hypothesis is that the lists of prey that are captured by orb weavers typically include many different types of insects (e.g., table 3.2; Robinson and Robinson 1970b, 1973a; Nentwig 1985b; Higgins and Buskirk 1992; Blackledge and Wenzel 1999; Venner and Casas 2005; Calixto and Levi 2006). Aerial non-orb webs such as linyphiid sheet webs also captured long lists of diverse prey species (Turnbull 1960; Herberstein 1997a). It is of course true that occasional strays will show up if the sample is large enough (e.g., walking prey like worker ants or mites are occasionally captured in aerial webs), so simple lists of prey species can be misleading. Nevertheless, from the perspective of a spider, the longer the list of prey, the smaller the selective importance of capturing any given species of prey. Although prey lists often have several weaknesses (below), the typically large variety of prey observed (table 3.2) speaks against prey specialization.

One comparative study (Craig 1987a) found patterns in prey capture that matched differences in orb designs in five species in a tropical forest understory; the two species with denser, more radius-rich orbs captured prey that were estimated (indirectly) to be heavier and to have higher impact energies. This association with differences in orb design per se, however, is uncertain because the same two species with "high-energy" webs are larger than the others, and thus produce thicker lines (Craig 1987a). The spiders' biases with respect to higher-energy prey may thus have been due to their own sizes (or to differences in subhabitats where their webs were built—next section).

3.2.5.2 Strong habitat effects

More tellingly for the prey specialization hypothesis, the habitat of a web rather than its design seems to have special importance in determining which prey it captures. Thus, spiders with similar web designs but in different habitats captured widely different arrays of prey (e.g., the summary in Eberhard 1990a on ten studies of *Argiope* spp.). And species with quite different web designs (orb webs, tangles with gumfoot lines, funnel webs, and lampshade webs) that were built in the same habitat captured surprisingly similar arrays of prey (fig. 3.3, Riechert and Cady 1983; see also Wise and Barata 1983). Both Riechert and Łuczak (1982) and Wise and Barata (1983) concluded, from their own data and reviews of other studies, that differences between microhabitats influence prey captures more than differences in web structure.

These conclusions must be treated cautiously. Some differences did exist between web types (fig. 3.3); prey identities were poorly resolved (generally only to order); and prey numbers were not adjusted for the undoubtedly different nutritional values of prey with different weights (e.g., Venner and Casas 2005; Blackledge 2011) and taxa (Toft 2013) (see next section). But it is also true that the relative uniformity occurred even though there were probably differences in details of the subhabitats where the different web types were built. In sum, the lesson is not that there were no differences between web types, but that the differences were less than might have been expected. The relative uniformity in prey captured by these very different web types reduces the expectation that minor changes in orb design, such as the number of sticky spiral loops or the number of radii, will have substantial impacts on which prey are captured.

The study of Yamanoi and Miyashita (2005) on two sympatric araneids, *Neoscona punctigera* and *N. mellotteei*, which build unremarkable orbs with only moderate structural differences in the same general habitat, is especially appropriate for the question of prey specialization. They found a trend for the two species to capture different prey (in this case, the fraction of weevils, elaterid, and carabid beetles as opposed to other insects). I believe, however, that even their data are open to alternative interpretations. The two spider species were said to inhabit the same microhabitat, but this similarity was measured only in terms of percent plant cover and

Table 3.2. This small, incomplete sample of studies of the prey captured by species that build non-orb and orb webs illustrates the wide diversity of prey types that are typically captured by any given type of web (see Blackledge 2011 for many other studies of orbs).

Taxa	*Web type*	*Range of prey*	*Most common prey*	*Reference*
Hypochilidae				
Hypochilus thorelli	Substrate with lampshade sheet	7 insect orders, spiders, myriapods, opiliones	Spiders	Riechert and Cady 1983
Filistatidae				
Pritha nana	Substrate	13 orders	Nematocera	Nentwig 1982b
Theridiidae				
Theridion evexum	Aerial tangle	7 insect orders, centipedes	(not specified)	Barrantes and Weng 2007b
Parasteatoda tepidariorum	Gumfoot	7 insect orders, spiders, myriapods, opiliones[1]	Coleopters, spiders[1]	Riechert and Cady 1983
Achaearanea globispira	Gumfoot	Mostly ants	96% ants	Henschel and Jocqué 1994
Agelenidae				
Coelotes montanus	Funnel	6 insect orders, spiders, myriapods, opiliones1	Lepidoptera, Diptera[1]	Riechert and Cady 1983
Pisauridae				
Hygropoda dolomedes	Sheet	9 insect orders, spiders	Diptera	Cerveira and Jackson 2002
Dendrolycosa sp.	Sheet	9 insect orders, spiders	Diptera	Cerveira and Jackson 2002
Araneidae				
Araneus cavaticus	Orb	6 insect orders[1]	Diptera[1]	Riechert and Cady 1983
Argiope argentata	Orb	8 insect orders	Orthoptera	Robinson and Robinson 1970b
Micrathena gracilis	Orb	4 insect orders	Diptera	Uetz and Hartsock 1987
Nephilidae				
Nephila pilipes	Orb	8 insect orders	Coleoptera, Lepidoptera	Robinson and Robinson 1973a
N. edulis	Orb	3 insect orders	90% leafhoppers[2]	Austin and Anderson 1978
Tetragnathidae				
Leucauge globosa	Orb	(not given)	Cecidomyid flies[3]	Craig 1987b
Theridiosomatidae				
Epilineutes sp.	Orb	(not given)	Cecidomyid flies	Craig 1987b

[1] All prey were captured in the same overhanging rock ledge habitat.

[2] The mix of prey was determined by checking stabilimenta (strings of dead prey hung near the hub). The possibility that this sample was biased by a preference by spiders to put certain types of prey in the stabilimentum was not checked.

[3] Prey importance was determined from supposed relative abundance in the habitat, rather than from data on prey captured, and thus is not compelling.

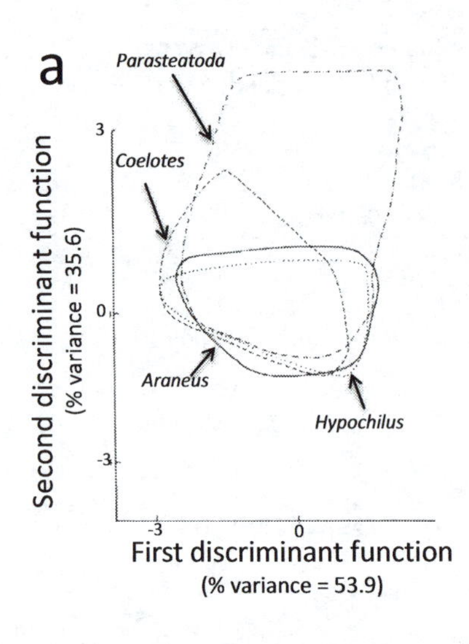

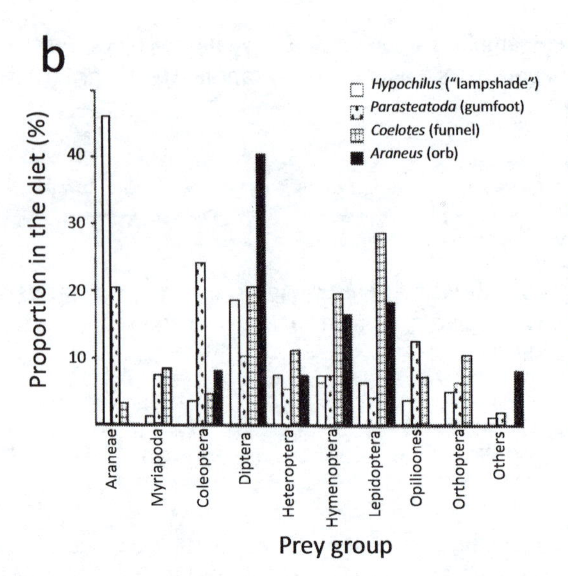

Fig. 3.3. The prey captured in the same habitat (rocky ledges) by four spider species that had very different web designs (an orb in *Araneus*, an approximately cylindrical sticky lampshade attached to the substrate in *Hypochilus*, a dense planar sheet in *Coelotes*, and an aerial tangle with gumfoot lines in *Achaearanea*) were nevertheless surprisingly similar. The overlaps in prey are illustrated in two ways: a principal components analysis (*a*); and the relative proportions of different prey taxa (*b*) (prey were identified only to Order) (from Riechert and Cady 1983).

web height, and several other aspects of their habitats that could have been important may have differed between the two species (e.g., species of plants, living vs. dead plants, the sizes of the open spaces in which webs were built). Their samples were also not large.

This is not to say that important biases in the lists of prey captured by orb weavers do not occur. For instance, both *Metabus ocellatus* (Buskirk 1975) and *Metleucauge* spp. (Yoshida 1989) built their orbs just above the surfaces of streams, and captured mainly midges, mayflies, plecopterans, and other weak-flying insects that are associated with water. The orbs of *M. ocellatus* have relatively widely spaced sticky spiral lines, especially near the outer edge of the web), and only moderate numbers of radii and sticky spiral loops (fig. 7.43). These design details are appropriate for capturing slow, weak-flying, light bodied insects (see sections 3.3.2.1, 3.3.3.1). But it is difficult to disentangle the effects of web designs from those of their websites.

In sum, field data on the prey captured by different orb designs generally include large numbers of prey types, and show strong associations with differences between habitats, arguing against relatively strict prey specialization. Prey specialization resulting from small design details has not not been demonstrated convincingly, despite many confident statements that it occurs. Additional studies might help resolve these questions. For instance, spiders with different web designs that were induced to build webs in moveable supports (e.g., wooden or wire frames) (as was done with portions of webs by Miyashita and Shinkai 1995 and Opell et al. 2006) could be moved experimentally in the field, allowing interchanges of web types between exactly the same sites.

3.2.5.3 Data from prey counts generally have serious flaws

It is not easy to determine the significance of the prey lists that were just cited for the evolution of different web designs, because it is often difficult to translate counts of prey (the most common type of data) to more biologically relevant variables such as nutrition for the spider. One major problem concerns prey weight, which varies enormously and surely has important effects on nutrition (e.g., Tso and Severinghaus 1998; Venner and Casas 2005; Blackledge 2011; sections 3.2.5.3, 4.2.2, appendix O3.2.3). A second major problem is that the nutritional values of prey vary substantially even after prey weight has been taken into account; a spider can even lose rather than gain weight from feeding on some kinds of prey (Toft 2013); spiders can starve in spite of an apparent abundance of prey, just as herbivores can starve in a green world (Toft and Wise 1999a,b; Toft 2013). This consideration may be important, as web spiders in nature rejected some prey they captured (e.g., Turnbull 1960 on *Linyphia triangularis*; Robinson and Robinson 1973a on *Nephila pilipes*). The overall conclusion is that the diversity of the prey species that are most responsible for a spider's growth and reproduction is probably generally less than that suggested by prey lists.

One recent proposal (Sensenig et al. 2011; Blackledge 2011; Evans 2013) that was based on nutritional considerations linked to prey size, is that orb weavers are generally selected to capture the largest prey available because these prey represent disproportionately large nutritional payoffs (Venner and Casas 2005). In effect, the idea is that orbs in general are specialized to capture high-energy prey, and that the capture of many other, smaller prey is incidental and of little selective importance. This "large rare prey" hypothesis may be true in specific cases, but there is evidence against its generality (section 4.2.2; Eberhard 2013a).

One further complication is the possibility that the most important selection on orb designs does not involve the usual, day-to-day captures of prey, but rather the very occasional but nevertheless crucial, life-altering captures of one or a few very large prey; these could determine whether the spider will survive, or will be able to reproduce (e.g., Venner and Casas 2005; Blackledge 2011; Evans 2013). This would imply that nearly all of the many insects that are routinely counted in field studies are, in fact, only incidental by-products of web designs that were favored not by selection to capture these prey, but by selection to capture a few, rare, life-altering prey. The impression of generality derived from the day-to-day counts could thus be a unreliable guide to the selection that has molded these web designs.

This rare, life-altering prey hypothesis for selection on orb designs was shown to be feasible in a modeling study (Evans 2013). Finding that something is feasible, however, is not equivalent to showing that it exists in nature. Several important simplifications and assumptions were made in the model, so it is not clear whether its results provide a reliable guide regarding what happens in nature. Unfortunately, the lower the frequency of the presumed crucial, life-altering events, the more difficult it is to empirically test whether they actually occur. The need for data to be from natural, unaltered habitats in which spiders evolved (rather than anthropogenic habitats) makes testing even more difficult. Empirical proof for the life-altering prey hypothesis is lacking; my own intuitive bias is to be guided by the data showing that orb weavers capture wide varieties of prey, but I have no confident, final solution.

In sum, these problems with prey counts shake my confidence in drawing confident conclusions from the large majority of studies of the prey of web spiders. One general lesson is that the apparent ease with which biologically relevant and precise quantification of spider feeding can be obtained in the field by checking spiders and their webs may be an illusion. The sizes, qualities, and numbers of different prey are important, and these variables require additional data that are difficult to obtain and generally unavailable. In the end, the available data do not allow confidence in either accepting or rejecting general conclusions regarding the consequences of different details of orb designs for prey specialization.

3.2.5.4 Measuring "available" prey is also difficult

Another way to check for prey specialization is to compare the prey captured in webs with representative samples of the prey that are available in the spider's habitat. At first glance this would seem relatively easy to accomplish by simply obtaining a reference sample of the prey that are in the air at a web's site using sticky traps, vacuum samples of the air column, etc. This has given rise to a large literature, but, as discussed in detail in appendix O3.2, "available" is a very slippery concept; it is technically difficult to distinguish prey that are "available" in biologically appropriate sense. I believe all of the many published data of this sort are seriously flawed. The basic problem, discussed in a preliminary way here (see also section 8.9.2 and appendix O3.2.3), is that all known sampling methods take only biased samples of the prey that are present, that both spiders and their orbs are also biased, and these biases are seldom (if ever) the same (Castillo and Eberhard 1983; Eberhard 1990a). Some biases could be overcome if, as occurred in a study of social *Anelosimus* (THD), multiple sampling techniques were used and all the techniques showed the same trends (Guevara and Avilés 2007) (and even here, prey were lumped into such general categories that differences in both numbers and quality could be obscured). I know of no conclusive studies for different orb designs, and see little hope of precise quantification of specialization via comparisons with available prey.

3.2.5.5 Ontogenetic changes in web design (and lack of such changes) can introduce noise

Ontogenetic changes (and lack of changes) present further questions regarding strict prey specialization. The prey that an individual orb-weaver captures during her lifetime undoubtedly change dramatically; but in general, these changes are accompanied by only relatively minor changes in web design. Individual orb weavers generally increase in weight by 1.5–2 orders of magnitude; in a giant species like *Nephila* spp. the change spans >3 orders of magnitude (Babu 1973; Quesada et al. 2011). It is simply physically impossible for a recently emerged spiderling to subdue the same prey that are used by adult conspecifics; and conversely, prey captured by small spiderlings are surely too small to be nutritionally significant for an adult. The sizes of webs and the diameters of their lines increase with the spider's size (Le Guelte 1966; Craig 1987a,b; Blackledge et al. 2009c), but the changes in the designs of orbs often appear to be relatively minor. The orbs of first and second instar *Argiope bruennichi* (AR) averaged 22.0 radii, while those of adult females averaged 29.4 (Tilquin 1942), while *Araneus diadematus* (AR) showed a trend in the opposite direction, from an average 36.6 to 29.4 radii (Witt et al. 1968); but the basic designs in both species changed little. For instance, the orb of a spiderling of *Nephila clavipes* (NE) (fig. 1.5) has many of the same design features of an adult web (fig. 4.4), including a tangle next to the orb, tertiary radii that are held nearly parallel to each other, intact temporary spiral lines, tight and relatively uniform spacing of the sticky spiral, and strong vertical asymmetry (fig. 1.5). The designs of the orbs of early instar *Zygiella x-notata* (Le Guelte 1966), *Allocyclosa bifurca* (AR) (Eberhard 2007b; WE), and *Leucauge mariana* (TET) (Maroto 1981) were also similar in many respects (though not identical) to those of adults.

While further studies of web ontogeny are needed, it is surely true that ontogenetic changes do occur in all orbs, if nothing else in the diameters of the fibers in the lines (Craig 1987a). The physiological demands on spiderlings probably also differ from those on adults, including greater problems with water loss due to their higher surface-to-volume ratios, and a greater need for energy due their disproportionately large nervous systems (Quesada et al. 2011; Eberhard and Wcislo 2011). Ontogenetic changes also occur in silk. In *Neoscona arabesca* (AR), silk investment increased isometrically with body size, but glue production was greater than expected in larger spiders, and the strength and toughness of sticky capture spiral thread increased during ontogeny (Sensenig et al. 2011). Larger spiders spun larger webs with smaller radii; but the increased volume of all silk types and the greater toughness of the capture spiral silk resulted in an isometric scaling of stopping potential. The material properties of dragline silk did not change over ontogeny.

This topic is worthy of mention even in the absence of some crucial data, because some lifetime consistencies in web designs seem to be so clear. At a given site, it is often possible to confidently identify the webs of young spiderlings of many species, using their similar "overall design" to those of adults. Perhaps the major contribution I can make at the moment is to emphasize the neglected topic of the webs of immature spiders (see section 3.2.14).

3.2.5.6 Ecological settings of studies need to be evolutionarily realistic

Looming behind attempts to study the prey specificity of spider webs is the probability that the conditions in most modern-day ecological settings differ from those under which the spiders and their webs evolved. This problem of "ecological realism" also crops up at other places in this book (e.g., section 3.3.4.2.2.2, appendix O3.2.4), so I will examine it in detail here. Understanding the selective advantages that caused a given trait to evolve is a subtle, demanding task, and requires careful thought about sometimes difficult-to-know conditions in the past (Williams 1966; 1998). The organisms that we observe today evolved in the past, under conditions that may have been quite different in some respects but quite similar in others compared with those in the present. It has been common to extrapolate from observations of the selection that occurs in the present, and assume that similar selection occurred in the past. This assumption of uniformity is appropriate and powerful for some traits, but less so for others.

For instance, it is reasonable to presume that the advantage of having a median tarsal claw for gripping silk lines (section 2.4.1) and of having fewer sticky spiral loops to reduce the drag of an orb in windy conditions (Liao et al. 2009) selected for these traits in the past. Both advantages can be observed in present-day conditions, and can confidently be assumed to have occurred in the past. But data from some other types of ecological variables, such as relative numbers of potential prey or potential predators, provide less convincing arguments. Take, for instance, the fact that some orb designs are relatively better than others at capturing prey that fly faster and are larger. These traits involve trade-offs with capturing prey of different sizes (see table 3.1, section 4.2). The overall payoffs from any given design will depend on the mix of numbers and nutritional payoffs from the different types of prey that are present at a given site. Present-day mixes are not necessarily the same as those in the past.

To illustrate with a personal example, there was a particular mix of prey in the large open field of grass and weeds (set among large pastures and sugar cane fields) that was mowed once or twice a year, where my students and I studied a large population of *Metazygia gregalis* (AR) and its prey (Chacón and Eberhard 1980; Castillo and Eberhard 1983). Formerly there was extensive forest at this site, and *M. gregalis*, a species that lives in early second growth habitats, probably only occurred in small, scattered, largely ephemeral patches of early second growth at recent tree falls, landslides, and along river banks. These patches probably resembled the modern field of grass and weeds in some respects, but they undoubtedly differed in others: for instance they surely lacked certain exotic species of grasses, and exotic insects such as honey bees, and they undoubtedly received potential prey that migrated or strayed from other, nearby forest or aquatic habitats not do not presently exist near this field. The fact that *M. gregalis* persisted in this field year after year is evidence that it can survive and reproduce successfully under modern conditions. But its persistence at this site is *not* evidence that it evolved under this particular set of conditions, nor that its orb design represents an adaptation to the modern conditions. Perhaps it survived at this site because of the particular design of its orb; perhaps it survived there in spite of its design.

The limits of reasonable ecological realism are difficult to determine. The large majority of the earth's surface has been modified by humans during the last 10,000 years. Truly "undisturbed" sites are rare or entirely gone in many areas of the world. In addition, the sweeping historical changes that accompanied glacial and inter-glacial periods probably resulted in substantial changes in the species of prey and predators that have co-occurred with a given spider species. Judging by the extensive shuffling of plant species species during the recent transition from glacial to inter-glacial conditions (Colinveaux et al. 2000; Betancourt et al. 1990), changes in community composition were likely substantial.

Another species from the same field, the bolas spider *Mastophora dizzydeani*, illustrates how the problems of ecological realism do not have the same effects on other types of deductions. This spider preyed exclusively on males of two species of noctuid moths. The behavior of the moths suggested that the spiders were luring them chemically. It was reasonable to conclude that these spiders used a pheromone mimic that evolved to trap male moths (Eberhard 1977b; Stowe et al. 1987); it would not be justifiable, however, to conclude that the particular mix of moth species observed in this field was the mix of prey under which the pheromone mimicry evolved.

I emphasize that I am *not* saying that I know what the ecological conditions were for different species in the past; I *am* saying that it cannot be confidently assumed that present-day conditions are the same as those under which the traits of the spiders evolved. This means that hypotheses that depend on certain ecological variables, such as the relative numbers of potential prey insects that have different sizes, nutritional values, and flight speeds, cannot always be tested convincingly by observing modern-day populations. I realize that some ecologically oriented colleagues will see these arguments as nihilistic surrenders to hopeless ignorance. But intellectual rigor is useful; little is gained (and much is lost) by ignoring the limitations of data to support or discard particular conclusions.

3.2.5.7 Flexible construction behavior

The wide range of intra-specific and even intra-individual facultative changes in the details of orb designs that are elicited by variables other than prey (e.g., see Herberstein and Tso 2011 and chapter 7) constitutes still another reason to doubt the prey specialization hypothesis. For instance, the correlation between silk supplies and the substantial intra-individual variations in designs of orbs built at the same site on the same day by *Leucauge mariana* (TET), *Metabus ocellatus* (TET), and *Micrathena sexspinosa* (AR) (figs. 7.18, 7.43; Eberhard 1988a), and by *Zosis geniculata* (UL) in the same portion of the same web from one day to the next (fig. 7.19), suggest that the amount of silk available in the spider's silk glands has major effects on several aspects of orb design. One would have to accept the seemingly unlikely (though not impossible) supposition that when a spider has larger or smaller reserves of silk in her glands, the types of prey that are most advantageous for her to attempt to capture change, and that the altered web designs are appropriate to capture them. Many different factors induce alterations in orb designs (sections 7.3, 8.2.1), making the array of suppositions of this sort that would be required by the prey specificity hypothesis even more unlikely.

3.2.5.8 Possible exceptions: relative prey specialization

There are a few extremely modified orbs that apparently represent specializations for certain types of prey. The most dramatic are the convergent "ladder" webs of *Tylorida* sp. (TET) in New Guinea (Robinson and Robinson 1972), and *Scoloderus* spp. (AR) in the Americas (fig. 3.4) (Eberhard 1975; Stowe 1978; Traw 1995), which are probably specialized to capture moths. The elongate web form, their perfectly vertical orientations, construction late at night, and the prey records from *S. cordatus* (68% of 212 prey were moths; the percentage was apparently lower in more typical orbs of other species in the same habitat, though quantitative data were lacking) (Stowe 1978), make it likely that both represent adaptations to capture moths. Moths are difficult prey for orb weavers, due to their covering of detachable scales that adhere to the orb's sticky spiral lines and allow the moth to fall free (fig. 4.20). Other vertically elongate tree trunk ladder webs, built by the nephilids *Herennia* and *Clitaetra* (Robinson and Lubin 1979a; Kuntner et al. 2008a,b) and the araneids *Cryptaranea atrihastula* (Forster and Forster 1985), *Telaprocera maudae* (Harmer and Herberstein 2009), and *Eustala perfida* and *conformans* (Messas et al. 2014), constitute a further striking convergence in design; these webs represent adaptations to build at similar sites (tree trunks) (Harmer 2009). These sites may result in prey specialization, or other advantages such as camouflaging the web or the spider.

Less dramatic vertical elongations in one species of *Eustala* (fig. 3.10) and in *Spintharidius rhomboidalis* (AR) (Levi 2008) may represent similar moth specializations. Other convergences on orbs that have extremely dense and numerous sticky spiral lines but are less elongate, such as *Acacesia* and *Pozonia* in the New World (figs. 3.5, 3.35; Eberhard 1976a) and *Deliochus* (fig. 3.5) and *Poltys* in Australia (Kuntner et al. 2008a; Smith 2006), may also represent specialization on moths.

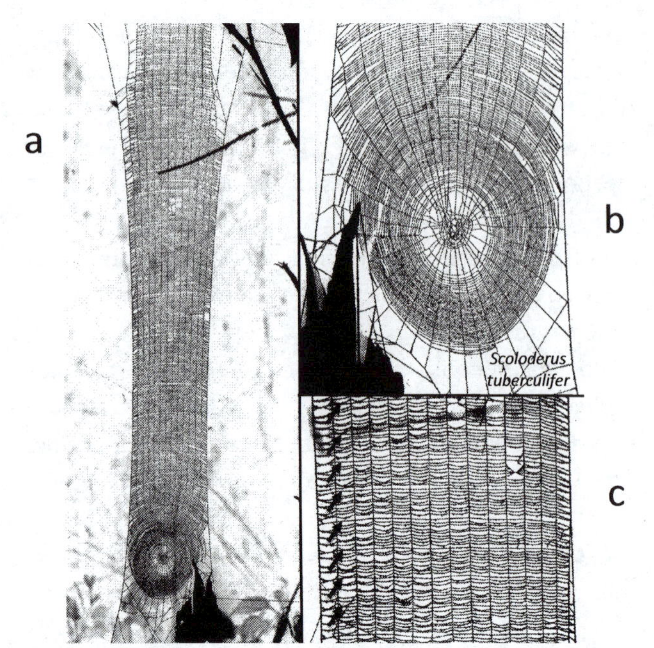

Fig. 3.4. The "true" ladder web of a mature female *Scoloderus tuberculifer* (AR), built late at night and removed at dawn, is probably a specialized trap for moths. These webs were not associated with tree trunks, as are the elongate "trunk" orbs of *Herennia* spp. and *Clitaetra* (NE) (fig. 4.3) and *Telaprocera maudae* (AR). The web's elongate form and perfectly vertical orientation probably increase retention of moths, whose covering of detachable scales on the wings and body makes them difficult prey to retain (the scales adhere to the web but the moth falls free). A moth in a ladder web will shed scales (fig. 4.20), but then fall into additional web below until it is finally stripped of its scales and held fast. This web has several additional design features that are probably also associated with its vertical elongation. Most of the long upper radii are nearly perfectly vertical and parallel to each other for most of their length (*a*, *b*), only curving at their lower ends to converge at the hub (*b*). These radii are held in these usual positions by the temporary spiral, much of which is left intact in the finished web (arrows in *c*). In a spectacular convergence, the orb of the distantly related *Tylorida* sp. is similarly asymmetrical, but has the lower rather than the upper portion elongated (Robinson and Robinson 1972) (from Eberhard 1975).

Four additional extreme specializations are the *Ocrepeira* (= *Wixia*) (AR) "asterisk" webs (fig. 9.2) to capture walking prey (Stowe 1978), the *Wendilgarda* (TSM) webs with sticky lines attached to the surfaces of streams (fig. 10.29; Coddington and Valerio 1980; Eberhard 2001a) that capture prey on or just above the surface of the water (in one species 82 of 85 prey were water striders), the very sparse webs of *Cyrtarachne* (AR) and its close relatives

Fig. 3.5. These beautiful "sticky spiral-rich" nocturnal orbs of mature females of the Australian *Deliochus* sp. (AR) (*a*) and the neotropical *Acacesia hamata* (AR) (*b*) are probably convergently specialized to capture moths. The webs of both species are built in large open spaces, and have relatively few radii compared with the large number of closely spaced sticky spiral loops (for still other probably convergent "moth webs," see figs. 3.4, 3.35) (from Kuntner et al. 2008a and Kuntner unpub., courtesy of Matjaž Kuntner).

that are also probably moth specialists (fig. 9.23, Stowe 1986) (they are made with especially strong and adhesive lines) (Cartan and Miyashita 2000), and the open-meshed design of webs that *Parawixia bistriata* (AR) builds only on days when flying termites are available (Sandoval 1994).

There are also some non-orbs that capture highly specific types of prey. There are several ant specialist groups in Theridiidae that build their gumfoot webs over ant foraging trails (Cushing 2012). In the gumfoot building theridiid *Achaearanea globispira*, for example, 96% of the prey were ants (Henschel and Jocqué 1994), and *Dipoena banksii* may capture nothing except *Pheidole* ants with its single gumfoot line webs that were placed near the entrances of ant nests (Gastreich 1999). Other specialist theridiids are *Argyrodes attenuatus*, which appeared to capture only insects and spiders that used its long, sparse non-sticky lines as resting sites (Eberhard 1979a), *Chrosiothes tonala* that also built only non-sticky lines that it used to explore for termites foraging on the floor of the forest that it had sensed chemically (Eberhard 1991), and *Thymoites melloleitaoni* that has been found only inside termite nests (Bristowe 1939). Two preliminary impressions are that prey specialization in non-orbs seems to be associated with reduction of the complexity of the webs, and with building webs at very specific sites.

In sum, prey specialization does exist; but it appears to be rare, and large confined to highly modified orbs or non-orb webs.

3.2.5.9 A summary regarding prey specialization in orb webs

Summarizing the data from different techniques, the trends that have been identified in studies of typical orb webs tend to argue against rather than in favor of relatively strict prey specializations in orb designs. Although the techniques that have been used in attempts to quantify both prey that are available and those that are captured are often flawed, it is clear that orbs typically capture a wide diversity of prey. In addition, webs with quite different designs that were built in the same habitat captured rather similar arrays of prey. And webs built by different ontogenetic stages of the same species often have quite similar designs but undoubtedly capture very different arrays of prey (though the data here are scarce). A few studies that suggested that particular designs of more or less typical orbs are associated with captures of particular types of prey are weak because the biases in captures could be due either to biases in the prey present at the sites or to differences in the spiders' sizes, rather than to differences in the orbs' designs. In sum, a modified version of the conclusion of Riechert and Łuczak (1982) is most appropriate for orbs: differences in habitats where webs are built, rather than differences in design, are probably responsible for many differences in prey captured.

3.2.6 INTER-SPECIFIC COMPETITION FOR PREY IS PROBABLY NOT COMMON

Prey specialization, if it occurs, could result from competition with other species for prey. It is difficult to demonstrate, however, that differences in orb design are the result of selection due to such competition, as has frequently been supposed (see Wise 1995). While it is routine to find that different species in the same habitat differ with respect to the prey that they capture, there are reasons to be skeptical of the frequent supposition that web differences evolved to avoid competition for food with other species. The most convincing data come from experimental manipulations of the relative population densities in the field by David Wise and his colleagues. Careful field experiments showed that two congeneric araneids, *Argiope trifasciata* and *A. aurantia*, which built webs with similar designs at similar old field sites (and whose small differences in website heights had in fact been attributed earlier to competitive displacement—Enders 1974), did not compete for prey in this habitat (Horton and Wise 1983). Other removal experiments led to similar conclusions. For instance, removal of individual species in the group of four species in different families that captured largely similar arrays of prey in the same subhabitat also failed to result in competitive release (Riechert and Cady 1983). A thorough analysis of published data led Wise (1995) to the conclusion that web-building spiders seldom experience inter-specific competition for prey. Even intra-specific competition for prey may be weak in some species. In at least some habitats, orb weaver populations are apparently limited by predators (Toft and Schoener 1983; Spiller and Schoener 1994) or parasites (Bianchi 1945), not by intra-specific competition for prey (Miyashita 1992) or space (Colebourn 1974).

It should be noted that careful experimental tests of inter-specific competition are available, however, for only a few species, and some evidence from other, "inadvertent experiments" in which invasive species of web builders have displaced related native species appears to be contradictory. The clearest data come from a pair of linyphiids in North America. In plots in a recently invaded area in Maine (USA), the presence of the European and Asian *Linyphia triangularis* was associated with a reduced tendency of the native species *Frontinella communis* to build in the same plot, and also with an increased tendency for *F. communis* to abandon the plot (Bednarski et al. 2009). It also appears that the recent arrival of the invasive, anthropophilic African brown widow *Latrodectus geometricus* has been associated with the near disappearance of the native black widow *L. "mactans"* (species identification is uncertain) from similar habitats in northern Florida, USA (G. B. Edwards pers. comm. using data from both personal collecting and calls from residents regarding spiders in and around their houses). A similar, especially dramatic displacement of *L. "mactans"* by *L. geometricus* was described over 50 years ago in Hawaii (Bianchi 1945).

At least in the case of the linyphiids, however, it appears that the displacement of the native species was due to direct aggressive interactions, not to competition for food (Bednarski et al. 2009). The invasive *L. triangularis* has similar competitive interactions in its native habitats, where it attacks a smaller congener (Toft 1987, 1989). These observations do not rule out the possibility that the selective advantage of the aggressive behavior is to win out in competition for food; but other possibilities, such as excluding other individuals from the web so as to avoid interference in attacks on prey, have not been eliminated. Nothing seems to be known regarding the mechanism of displacement in the *Latrodectus* species, and possible competition for food has not to my knowledge been tested.

A second, less powerful technique to test for inter-specific competition is to look for inverse correlations in the distributions or population densities of different species in the field, and differences in the types of prey captured by different species (e.g., Enders 1974; Blackledge et al. 2003b). Though there are cases of such correlations and differences, reasonable alternative explanations were generally not eliminated, so these studies do not provide convincing demonstrations of competition for prey. *Wendilgarda galapagensis* (TSM), which is endemic to the isolated and depauperate (Hogue and

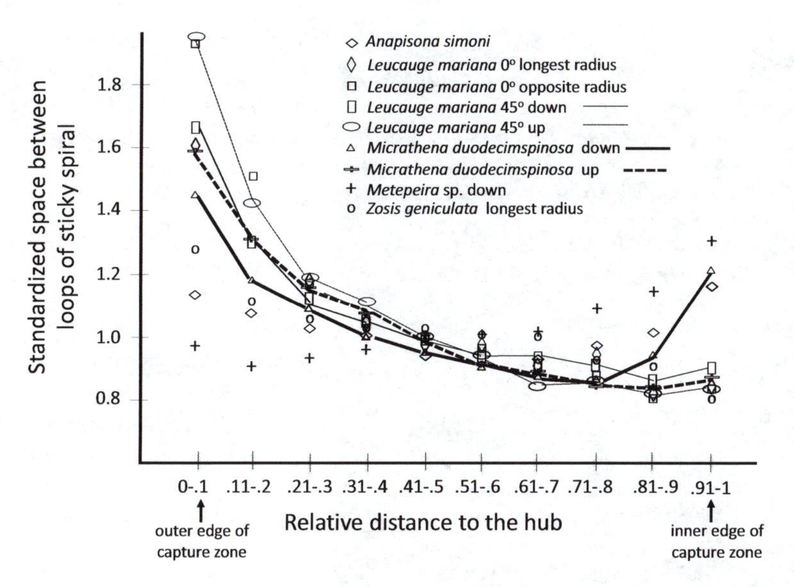

Fig. 3.6. The edge-to-hub patterns of the mean standardized spaces between loops of the sticky spiral (distance/median space) in the orbs of six species of orb weavers in five families illustrate a common pattern: sticky spiral spacing tends to be larger near the outer edge of the web (left side of graph). The exception to this pattern, *Metepeira* sp., is also exceptional (along with *Nephila clavipes*—see fig. 4.4) in capturing prey flying at relatively low velocities; the pattern and the exceptions favor the "prey stopping" explanation (see fig. 3.7 and text).

Miller 1981) oceanic island Isla del Coco, offers especially suggestive evidence favoring the importance of inter-specific competition: in addition to building genus-typical webs attached to the water in small streams (e.g., fig. 10.29), these spiders built quite different, apparently derived webs away from the water, where no other theridiosomatids and indeed few orb weavers were present, and these had quite different designs (fig. 8.1) (Eberhard 1989a). But again alternative explanations were not eliminated.

In sum, careful field experiments with pairs of web-building species revealed that they did not compete for prey: increasing or reducing the numbers of one species had no perceptible effect on the predatory success of the other. It is possible that the few species in these experiments are not representative, but the fact that some species pairs were especially selected as likely candidates for showing inter-specific competition argues otherwise. Given the low densities of many orb weaver populations in nature, the huge numbers of insects in flight, the wide spectrum of species that orb weavers prey on, and the probably largely stochastic encounters of prey with orbs, this lack of inter-specific competitive effects is not really surprising. Far-reaching consequences of the probable lack of importance of inter-specific competition for prey for understanding the evolution of webs are discussed in section 9.9.

3.2.7 STICKY SPIRAL SPACING IS NOT UNIFORM

One of the most striking geometric regularities in an orb web is the uniform spacing between the sticky spiral lines. Many authors refer to "the" sticky spiral spacing (or mesh size—see section 1.7) of a given orb (appendix O3.2.1). Closer examination, however, shows that neither the distances between sticky spiral lines nor those between the radii are uniform in an orb (e.g., figs. 1.4b, 3.6; Nentwig 1983a); more importantly, there are consistent patterns in these differences, and the differences themselves appear to be part of the orb's design (section 4.3).

Radii obviously converge near the hub. The distances between adjacent radii in the typical orbs were on average 4–5 times greater at the outer edge of the capture zone than at its inner edge (Eberhard 2014). In addition, there were also clear, consistent differences in sticky spiral spacing (figs. 1.4, 3.6). In a given orb, the largest spaces between sticky spiral loops was on the order 2–3 times the smallest (figs. 1.4, 3.6) (Peters 1939; Le Guelte 1966; Nentwig 1983a; Heiling and Herberstein 1998; Sensenig et al. 2012).

These differences in sticky spiral spacing are not random, but show a pattern: larger distances generally occur in the outer portion of an orb (fig. 3.6) (Peters 1939; Le Guelte 1966; Nentwig 1983a; Eberhard 2014). This pattern of larger distances near the outer edge appears to be very general, although to my knowledge there never has been a wide-ranging review of within-web variation in the distances between sticky spiral loops in different species. It occurred in published photos of the orbs of 14

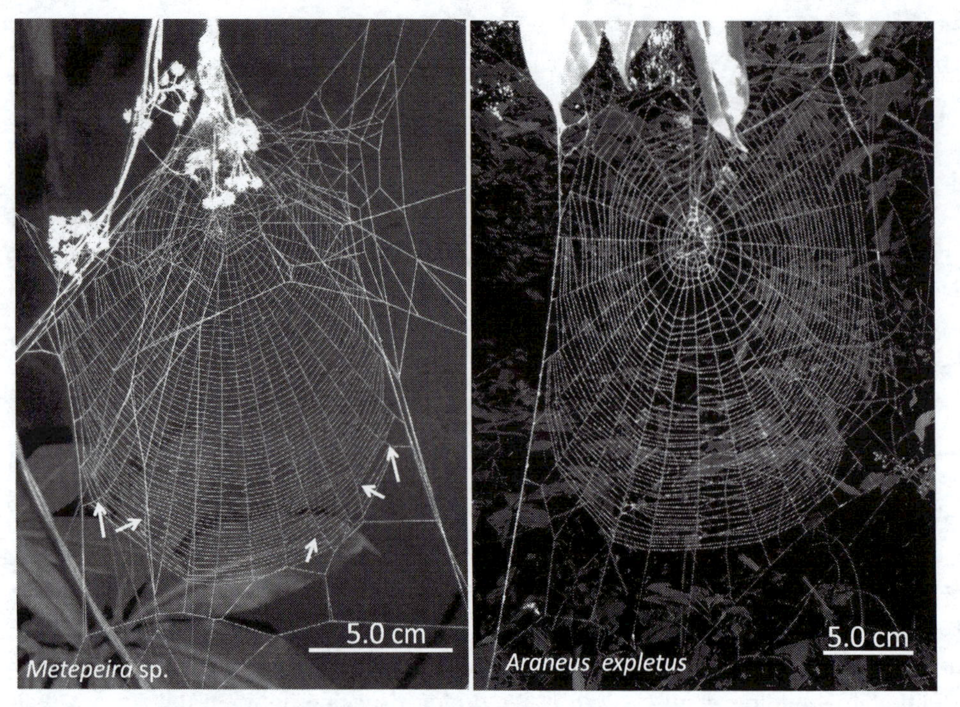

Fig. 3.7. These orbs of mature female *Metepeira* sp. (AR) and *Araneus expletus* (AR) are unusual in having smaller spaces between the sticky spiral loops near the lower edge of the orb (compare with figs. 1.4, 3.8). In both species there is also reason to suspect that prey velocities may be relatively low. This association fits the predictions of the "prey stopping" hypothesis (section 4.3.2). As in many other species, however, the *A. expletus* sticky spiral lines in the upper portion of the orb were consistently farther apart near the edge. The arrows mark the *M.* sp. radii that curved to fill larger inter-radial spaces (from Eberhard 2014).

genera of araneids, 3 tetragnathids, 2 uloborids, a theridiosomatid, and an anapid (Eberhard 2014) (generally there was only one photo/species, however, so further data are needed). There also appear to be inter-specific differences. For instance, the distances between loops below the hub in the web of the araneid *Metepeira* sp. and the nephilid *Nephila clavipes* showed the opposite trend, increasing gradually from the edge to the hub (figs. 3.6, 3.7, 4.4). Other species showed more complex patterns. For instance, in the vertical orbs of the araneids *Zygiella x-notata* (Le Guelte 1966) and *Micrathena duodecimspinosa* (figs. 1.4, 3.6) the spaces between loops below the hub tended to be somewhat larger near the outer edge of the orb, smallest midway toward the hub, and largest very near the hub; above the hub, in contrast, the spaces tended to be greatest near the edge but then decreased monotonically nearer the hub, with little or no increase near the hub.

Another pattern in some species is for the sticky spiral spacing near the lower edge of the capture zone to be especially small, while that near the upper edge of the capture zone is relatively large (see also section 3.3.3.3.1) This pattern occurred in several *Cyclosa* and *Allocyclosa* (AR) species (fig. 3.27) (Zschokke and Nakata 2015; WE), *Metepeira* sp. (AR) (fig. 3.7) (Eberhard 2014), *Epeirotypus brevipes* (TSM) (fig. 3.8), and *Nephila clavipes* (NE) (fig. 4.4) (Eberhard 2014). As suggested by Champion de Crespigny et al. (2001) and Zschokke and Nakata (2015), the tight spacing at the bottom of the orb may be due to selection to prevent prey from escaping by tumbling down the web (see section 5.8).

Still another pattern that was documented in *Uloborus diversus* (UL) (Eberhard 1969) and that seems to occur in other orb weavers was for sticky spiral spacing to be reduced at the point where it doubled back. This reduced the size of the hole in the array of stick lines at the turnback point (fig. 3.9).

The existence of these patterns in the spacing suggests that the differences are not accidents, but part of the orb's design. There are several possible explanations for why these patterns occur, and no single explanation works for all cases (section 4.3.5). For the moment, the point is that sticky spiral spacing is not uniform in orbs, and that the variations in spacing are not simply mistakes.

3.2.8 ORB DESIGNS ARE PROBABLY NOT TAXON-SPECIFIC

3.2.8.1 Species-specificity

A common question asked about orbs is whether their designs are species-specific. Many arachnologists with extensive field experience at particular field sites might

Fig. 3.8. The upper portion of this orb of *Epeirotypus brevipes* (TSM) illustrates the common edge-to-hub pattern, with a gradual decrease in the spaces between sticky spiral lines nearer the hub. The spaces below the hub were much less variable, however. Webs of this species are often built near solid surfaces (this web was in the hollow trunk of a standing tree, with bats perched in the darkness above), so the velocities of prey impacts may often be relatively low. Wind velocities were also very low. On balance the prey-stopping hypothesis offers the best explanation for the edge-to-hub patterns. The two white objects are egg sacs that are hanging on the spring line that connected the hub to the trunk.

be tempted to answer yes; it is often possible to identify most if not all of the orb weavers at a site on the basis of their webs, without seeing the spider (e.g., Shear 1986). Bristowe (1941) noted that "the typical webs of most if not every species [of British araneids and tetragnathids] can be recognized without seeing the weaver herself" (p. 250–251). Similarly, species-specificity may occur in other web designs; for instance, Bristowe (1941) distinguished the funnel webs of four species of *Tegenaria* on the basis of the density and flatness of the sheet, the relative diameters of the tunnels, and silk color.

Perhaps field experience of this sort is responsible for the common published claims that orb webs are species-specific (one chapter of the Witt et al. 1968 book, for example, was titled "Specificity of the web"). But discrimi-

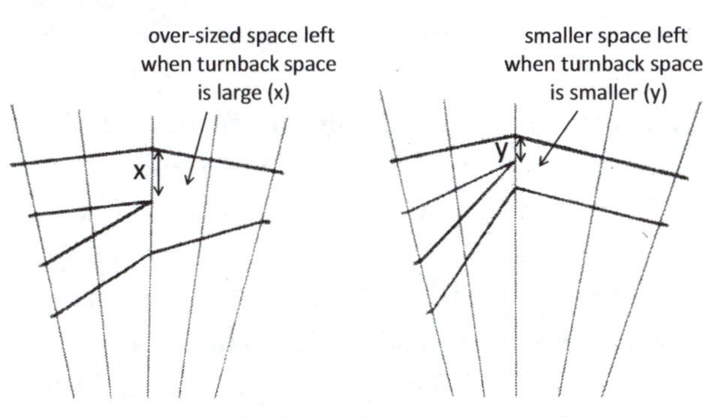

Fig. 3.9. The probable function of the typically reduced spacing at turnback sites (y) in uloborid webs was to reduce the sizes of the open spaces lacking sticky lines (from Eberhard 1969).

nation of local species usually involves comparing only a single species in each genus or, at best, only a few congeneric species. The term "species-specific" means much more; it means that the webs of a species can be distinguished from those of even closely related congeneric species. Are the orb designs of closely related species truly sufficiently different that they do not overlap, despite their usual substantial variation (e.g., fig. 4.12, chapters 7 and 8)? I will argue that the data needed to give a confident general answer are not available, and that the probable answer is "no."

To illustrate with my own personal experience, I can confidently distinguish the orb webs of mature female *Leucauge mariana* from those of the approximately 30 other orb weaving species that are common in the Valle Central of Costa Rica. This, however, is the only common *Leucauge* species, and when I am faced with photos of the webs of other *Leucauge* species from other sites, I often cannot distinguish them from those of *L. mariana*. I experience the same combination of confident local web identifications and inability to make confident distinctions between congeneric species in several other genera with which I have ample experience at this site, such as *Araneus*, *Eriophora*, *Wagneriana*, *Cyclosa*, *Argiope*, *Nephila*, *Uloborus*, and *Philoponella*.

Thus, at least on the basis of the "gestalt" traits of orbs that I have learned to use, the webs of *L. mariana* are not species-specific. The possibility still exists that other more subtle traits will be found to distinguish species in the future, but there are presently few if any data that clearly document species-specificity. This would be a major undertaking. To be convincing, such a study would need to include substantial samples of the webs of each species, because of the considerable adaptive flexibility of orb weavers (Scharf et al. 2011; Herberstein and Tso 2011; chapters 7, 8). It would also have to include similar samples from an ample number of other species in the same genus. Due to the likelihood of design adjustments to environmental conditions (Sensenig et al. 2010b), comparisons would need to standardize the contexts in which orbs were built. There has never to my knowledge been a single study that convincingly tests web species-specificity in any orb-weaver genus. One preliminary attempt using radius number in six species in the large genus *Araneus*, did not give encouraging results. Values for different species varied widely within each species, and overlapped broadly: 25–42 in *A. diadematus*; 15–22 in *Nuctenea umbricata*; 11–26 in *A. quadratus*; 26–66 in *A. diodius*; and 15–22 in *Gibbaranea bituberculata* (Savory 1952) (see also Wiehle 1927, 1928, 1929, and 1931 on *Araneus*, and fig. 3.10 for diversity in the genus *Eustala*). A preliminary test in the *republicana* group of *Philoponella* (UL) also failed to reveal clear species-specific differences (Opell 1979).

Probably the most complete study of possible species-specificity in orb designs came to a similar conclusion. Many species of the genus *Tetragnatha* are endemic to the Hawaiian islands; about 35 species that build orbs are probably all descendants of a single colonizing ancestor (Blackledge et al. 2003b; Blackledge and Gillespie 2004). Other species of this large genus (>100 species) are distributed worldwide, and most make flimsy, more or less horizontal orb webs with open hubs (fig. 3.18c) (see fig. 3.11j for an exception); they also have relatively wide spaces between sticky spiral loops (for an extreme, see fig. 9.10), and often but not always occur near bodies of water (Kaston 1948; Comstock 1967; Dąbrowska-Prot and Łuczak 1968; Shinkai 1979; Shinkai and Takano 1984; Eberhard 1986a; Gillespie and Caraco 1987; WE). The Hawaiian species of *Tetragnatha* are the only native nocturnal orb weavers on the islands, and have undergone a dramatic radiation. The orbs of up to four species were analyzed on each of three islands. On all three islands, the overall web architectures of the different sympatric species differed from one another (using a principal components analysis [PCA] of four web variables); but comparing different islands, there were three pairs of species on different islands that built similar orbs (at least partial overlap in mean ± 95% standard error in PCA analyses) (Blackledge and Gillespie 2004). It is possible that intra-specific variation (and thus overlap between species) was underestimated. The samples of webs were modest (10–28 individuals/species), and the studies were performed in small habitat patches, some only on the order of only 1 ha (Blackledge et al. 2003b). These limitations may be important, given the substantial intra-specific flexibility of orb designs in general (Herberstein and Tso 2011; Scharf et al. 2011; Sensenig et al. 2010b; chapters 7, 8). On the other hand, it is possible that inter-specific overlaps were overestimated, as

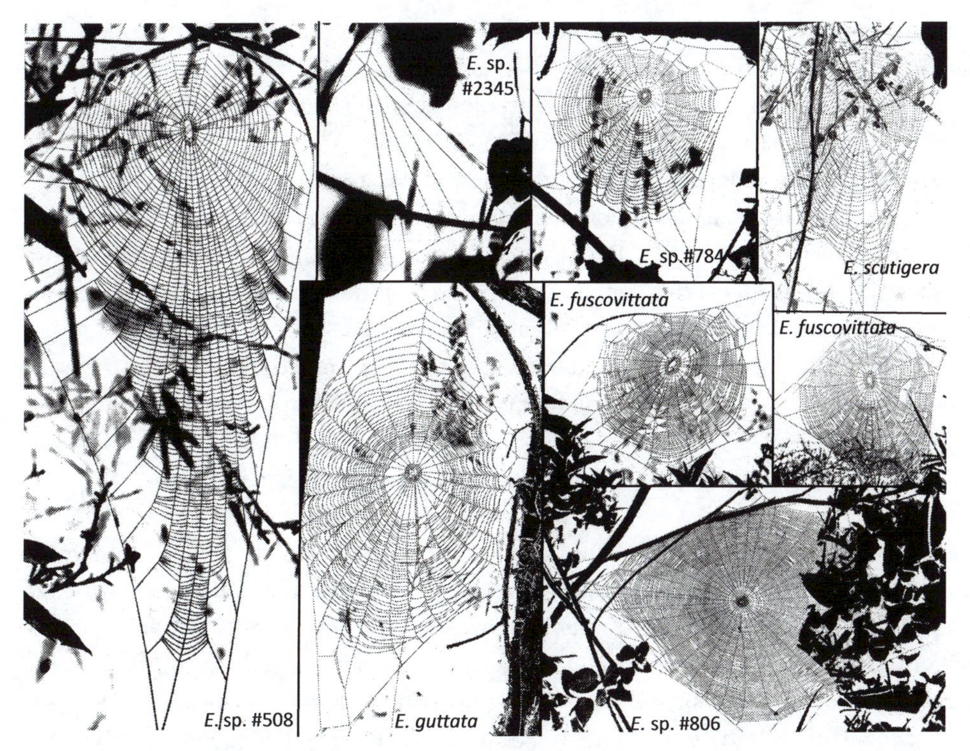

Fig. 3.10. This sampler illustrates the orbs of mature females in the medium large, New World araneid genus *Eustala* (Platnick 2014 listed 85 named species). Although web forms are known for only about 10 species, great intra-generic variation has already been documented. In contrast with others, the web of *E.* sp. #806 was nearly horizontal, and had many radii and many, closely spaced loops of sticky spiral. The great vertical asymmetry of the web *E.* sp. #508 (which was not associated with a trunk) was much greater than that of other species, but it was nevertheless much less than the asymmetry of the trunk webs of *E. conformans* and *E. perfida* (see fig. 4.3c). The webs of *E.* sp. #2345 lacked a sticky spiral, and sticky lines were laid instead on radii and frame lines. The webs of *E. fuscovittata* differed in being built in especially open areas, often with tall grass (two webs are included to give an idea of intra-specific variation). It is not clear whether *Eustala* is atypical in showing such diversity.

many web variables, such as hub and free zone size and shape (see section 3.2.8.3 and figs. 3.10, 3.15) were not included in the analyses.

Two other intra-generic comparisons, between *Zygiella x-notata* and *Z. keyserlingi* (of *Zygiella* sensu strictu) (Gregorič et al. 2010), and between *Leucauge argyra* and *L. mariana* (TET) (Triana-Cambronero et al. 2011; Barrantes et al. 2017), also showed differences in variables such as the number of primary radii, the number of horizontal sticky spiral loops, the size of the orb relative to the spider's size, frequencies and sizes of tangles near the orb, and the rate at which orb area increased with spider size. Only two species were compared, however, in these studies.

In sum, species-specificity in orb design has seldom been studied. The preliminary analyses available argue against species-specificity. It is possible that future studies that include additional web variables may discover as yet unappreciated subtle species-specific differences. There is no reason to expect, in any case, that all genera will show the same pattern. Given the present absence of data documenting specificity, and the extraordinarily large number of variables known to affect orb designs (table 7.3; section 7.21), the best summary regarding species-specificity is that it is yet to be demonstrated. My own intuition agrees with that of Levi (1978; p. 2): "I would expect . . . that in many [genera] the variation of individual webs of a species is too great and the webs of related species [are] too similar to separate."

3.2.8.2 Effects of intra-specific genetic differences

Little is known regarding how much of the substantial intra-specific variation that is commonly observed is due to genetic differences among individuals, and how much is due to differences in the conditions in which the webs were built. While environmental effects can be large (e.g., fig. 7.42), the only systematic study I know of (Rawlings and Witt 1973) concluded that genetic effects are also important. The differences between spiderlings from eight different egg sacs laid by different females were clearest in measurements of aspects of the sizes of their webs, and least evident in measurements of the "regularity" of their designs (e.g., standard deviation of inter-radial angles, standard error of median spacing between sticky spiral loops). More data will be needed to judge whether these are general trends.

3.2.8.3 Genus-specificity?

It is also possible that orb web designs differ at higher taxonomic levels. Craig (2003), for instance, speculated that orb designs are "often specific to a particular spider genus" (p. 65) (but gave no supporting data). The field experience of distinguishing the orbs at given sites (Shear 1986) is relevant here; I *can* distinguish the orbs of *L. mariana* webs from those of the species in the other 15–20 genera that are common in the Valle Central. But a local sample of this sort is again much too limited to justify strong conclusions, and a given site rarely has representatives of sister genera. The data needed to quantify both intra-specific and intra-generic variation are generally lacking (Craig 2003).

Kuntner and colleagues (Gregorič et al. 2010) made a first step in genus-level characterizations by specifying web characters that may be sufficient to diagnose each of the four genera in the family Nephilidae (the webs of some species are not known, however, and the data were not all quantitative). Another recent pioneering quantitative study of Gregorič et al. (2010) also gave preliminary indications of genus-level specificity in the webs of five species in four newly proposed genera that were created by splitting up the old araneid genus *Zygiella* (mostly on the basis of differences in genital morphology) (Wunderlich 2004). They found web differences that characterized each species, and that 7 of the 16 web characters (selected from an unspecified larger number as being useful) showed statistically significant differences among species.

Even though they used an admirably large sample (278 webs in the field from 25 localities), I would argue (as the authors themselves noted) that genus-specificity cannot be tested confidently until the range of intra-generic variation has been determined more completely; this would involve adding further species in each of the three genera that has multiple species. I would also note the prominent variation in the Gregorič et al. data. For instance, the web trait said to distinguish the entire group (an open sector in the orb with a signal line running from the hub to the spider's retreat that is at or beyond the web's edge) was sometimes absent in the webs of all four genera (ranging from 6% to 59% of the webs). As Gregorič et al. noted, it is difficult to deal taxonomically with traits that are inconsistent.

It is useful to note that some of the traits that were found to differ between species are almost certainly influenced by the site at which the web was constructed, rather than by orb construction behavior per se (Gregorič et al. 2010). For instance, *Leviellus stroemi* was the only species in which "non-circulating" sticky spiral lines (an estimate of the relative number of turnbacks in the sticky spiral) were common above as well as below the hub (they occurred in 56.5% of the webs, as opposed to 0–5.5% in the others species) (Gregorič et al. 2010). Turnbacks are used throughout orb weavers to fill in asymmetric spaces with more or less spiral lines, and this *L. stroemi* trait probably resulted from the choice of website and retreat locations. An illustration of how important website shape can be comes from the elongate, ladder web araneid *Telaprocera maudae*, which built a typical, oval web when obliged to build in a non-elongate space rather than on a tree trunk (Harmer and Herberstein 2009). This emphasizes the importance of documenting orbs in nature as well as in captivity (Sensenig et al. 2010b).

The hubs and free zones of orbs, which may be more isolated from habitat-imposed variations, seem to be the most promising aspects of orbs for taxon-specificity. Preliminary surveys show clear variation between different genera in a given family (figs. 3.11–3.15). They were highly consistent in one genus, *Micrathena* (AR) (fig. 3.14), and moderately consistent in *Eustala* (AR) (fig. 3.10); but they differed within others such as *Alpaida* (AR) (fig. 3.15), and *Tetragnatha* (figs. 3.11, 9.10, 10.24, 10.26; see also Blackledge and Gillespie 2004). Careful measurements to quantify intra-specific and intra-generic variation in

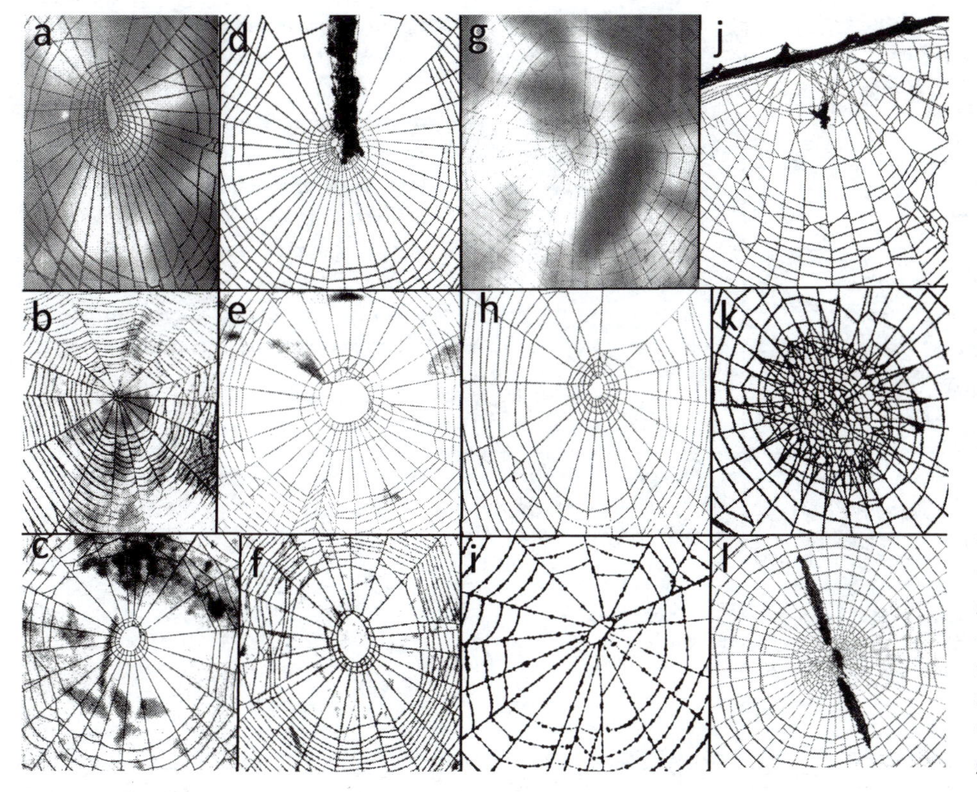

Fig. 3.11. The substantial variation in the designs of the hubs of genera in Tetragnathidae (*a–j*) and Uloboridae (*k-l*) suggest that there may be taxonomic patterns, perhaps at the level of families or genera, but the studies necessary to test this hypothesis have yet to be performed (to different scales; all are field webs except that of *U. glomosus*) (the spider was removed in all but *U. glomosus*; *k* courtesy of Brent Opell). *a–Chrysometa alboguttata*; *b–Cyrtognatha* sp.; *c–Tylorida* sp. #1962; *d–Dolichognatha* sp.; *e–Leucauge* sp. D.; *f–Opas* sp.; *g–Prolochus* (= *Dolichognatha*); *h–Meta* sp. #1232; *i–Pachygnatha* sp. #563; *j–Tetragnatha* sp. #2047; *k–Waitkera waitakerensis*; *l–Uloborus glomosus*.

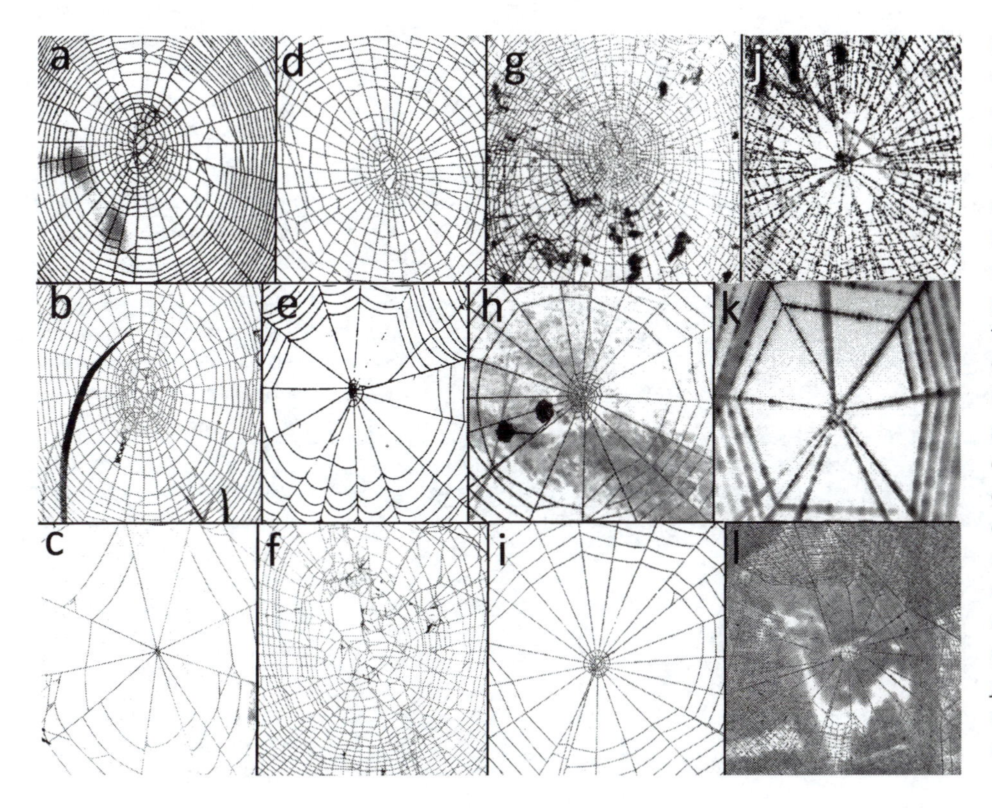

Fig. 3.12. The substantial variation in the designs of the hubs of Araneidae (*a–e*), Nephilidae (*f–g*), Theridiosomatidae (*h–i*), and symphytognatoids (*j–l*) suggest that there may be taxonomic patterns, perhaps at the level of families or genera, but the studies necessary to test this hypothesis have yet to be performed (to different scales). *a–Allowixia* sp.; *b–Argiope argentata*; *c–Cyrtarachne* sp.; *d–Cyclosa* sp.; *e–Hypognatha mozamba*; *f–Nephila clavipes*; *g–Nephilengis malabarensis*; *h–Epeirotypus brevipes*; *i–Naatlo splendida*; *j–Anapogonia* (?) sp.; *k–Anapisona simoni*; *l–Maymena* sp. #2168.

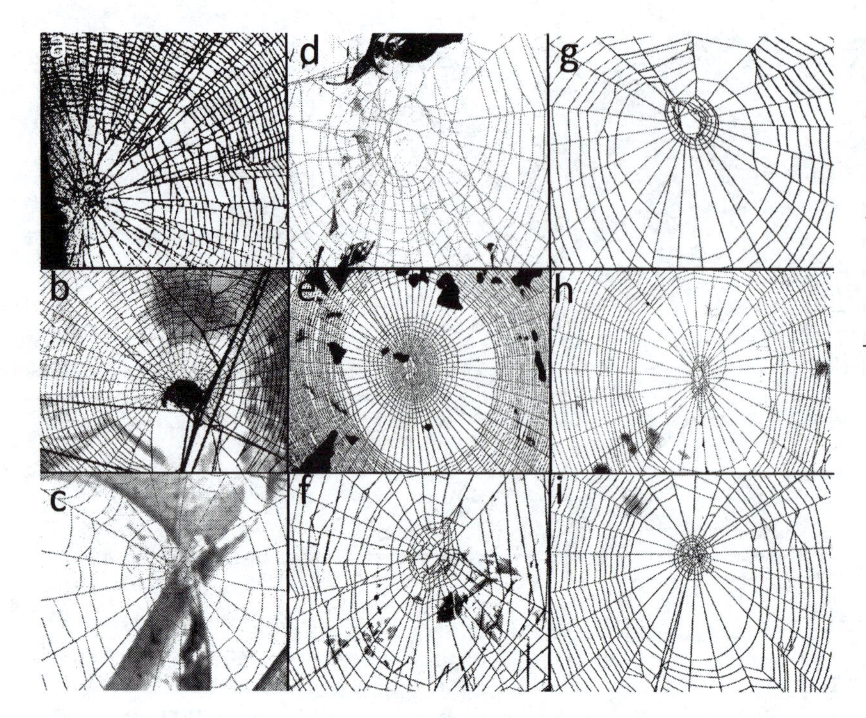

Fig. 3.13. The substantial variation in the designs of the hubs of different genera of Araneidae (*a–i*) suggest that there may be taxonomic patterns, but the studies necessary to test this hypothesis have yet to be performed (to different scales). *a–Enacrosoma anomalum*; *b–Spilasma duodecimguttata*; *c–Bertrana laselva*; *d–Eriophora nephiloides*; *e–Mangora* sp.; *f–Araneus expletus*; *g–Metazygia wittfeldae*; *h–Gasteracantha* sp.; *i–Neoscona* sp. #2002.

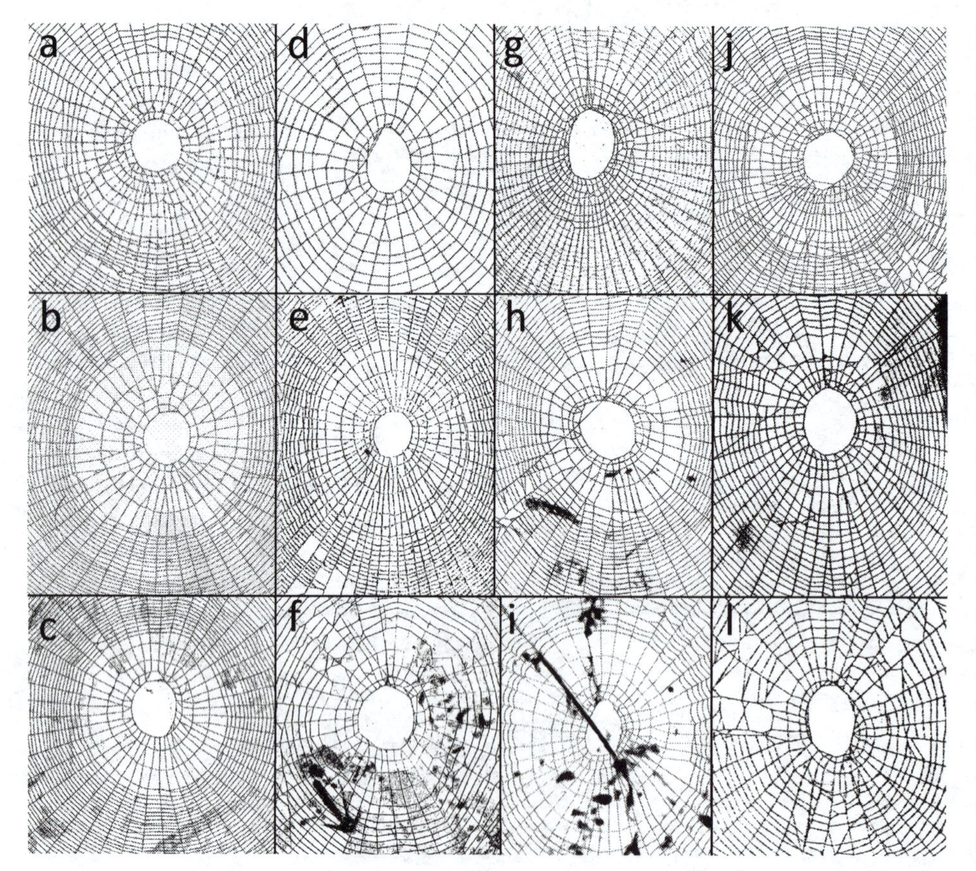

Fig. 3.14. The hub designs are very similar in 11 different species in the araneid genus *Micrathena* (*a–k*) (see, in contrast, the intra-generic variation in *Eustala* in fig. 3.10, *Alpaida* in fig. 3.15). Although quantitative comparisons remain to be made, the hub holes were large, there was little if any free zone between the outer edge of the hub and the innermost loop of sticky spiral, and the hub lines spiraled outward slowly in a relatively circular path after one or a few tight loops at the edge of the hole. Even the hub of a species in the sister genus *Chaetacis* was similar (*l*). In all of these species the spider rested at the hub, spanning the hub hole with her legs and tightening the web (see fig. 3.37) (to different scales). *a–M. molesta*; *b–M. horrida*; *c–M. gracilis* (?); d–*M.* sp. nr. *lucasi*; *e–M. plana*; *f–M. parallela*; *g–M. schreibersi*; *h–M. sexspinosa*; *i–M. bimucronota*; *j–M. duodecimspinosa*; *k–M. guerini*; *l–Chaetacis cornuta*.

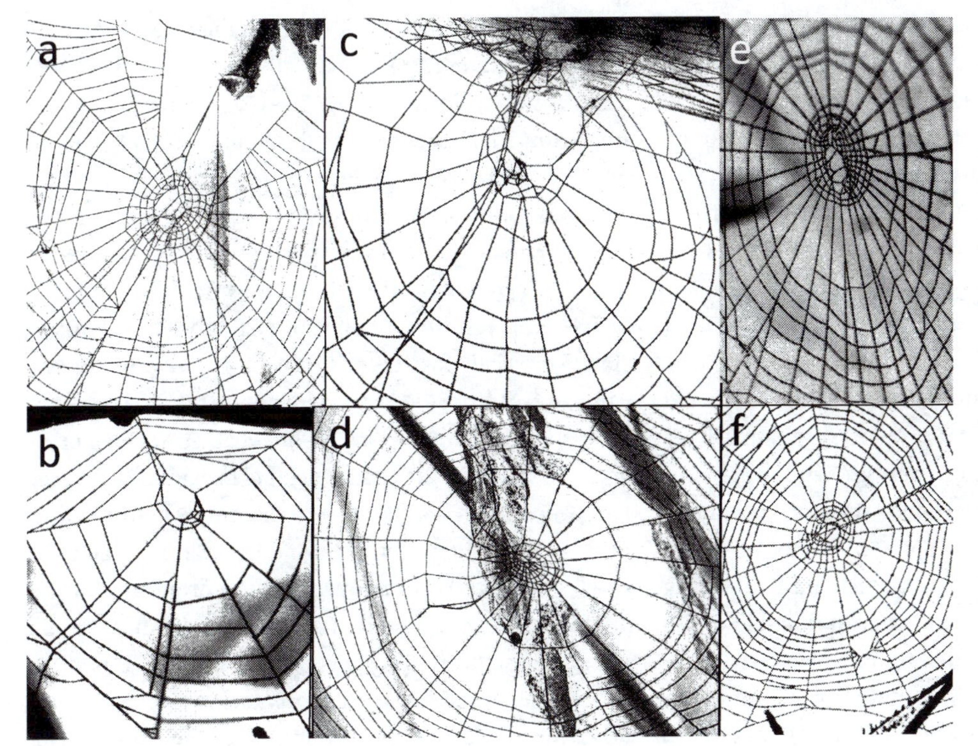

Fig. 3.15. There is substantial intra-generic variation in the hubs and free zones of these six species of the large Neotropical genus *Alpaida* (AR) (with 141 described species—Platnick 2014). Nevertheless, certain traits, such as locating the hub contiguous to or very near an adjoining support, may be typical of at least some species. It is not clear whether this genus is unusual in the degree to which hub and free zone designs vary. *a–A. truncata*; *b–A. anchicaya*; *c–A. acuta*; *d–A. niveosagillata*; *e–A. septemmammata*; *f–A. leucogramma*.

larger samples of species will be needed to test these possibilities. The incomplete nature of current data can only overemphasize the lack of overlap.

Summing up, I estimate from Platnick's The World Spider Catalogue (http://research.amnh.org/iz/spiders/catalog/INTRO1.html) that the webs are entirely unknown, without even a single published photo or description, in about 56% of the approximately 170 genera of Araneidae. Probably the coverage is similar in Tetragnathidae, though it may be better in the smaller families Theridiosomatidae (Coddington 1986a), Uloboridae (fig. 1.13), Nephilidae (Kuntner et al. 2008a), and the symphytognathoids (Lopardo et al. 2011). Web traits are mostly quantitative and vary continuously rather than qualitatively; thus substantial samples of species will be needed to confidently discuss intra-generic variation and whether there is overlap between genera.

3.2.8.4 Differences at higher taxonomic levels and a summary

The possible importance of taxonomic constraints is illustrated by the small, simplified resting webs built by some mature male orb weavers (fig. 9.4). These webs serve not to capture prey, but presumably to protect the males from predation. Not surprisingly, the hubs of these webs resemble the orbs of conspecific females and immatures. The vertical orientation of webs of male *Argiope argentata* and their horizontal orientation in *Mangora* sp. seem unlikely to be related to differences in function, but rather to the orientations of the prey capture webs of conspecifics. Similarly, the lack in araneid species of the effective defense provided by the spring lines in theridiosomatid species is probably due to the lack of spring lines in the prey capture orbs of araneids.

At the taxonomic level of families and subfamilies, I expect that there may be several distinguishing traits. For instance, uloborid orbs have several traits that are apparently widespread in this family and absent or nearly absent in the orbs of araneoids (see figs. 3.31, 7.24, table in Lubin 1986). Nevertheless, the careful quantitative studies with substantial samples of species and genera needed to demonstrate this are lacking.

Summarizing taxon-specificity in orb designs, it appears that species-specificity is rare. Genus-level specificity may be more common, and there are probably also family-level specificities. But we are a long way from documenting these specificities, if they exist.

3.2.9 THE PROPERTIES OF HOMOLOGOUS LINES ARE NOT INVARIABLE

3.2.9.1 Differences between species

Recent findings have confirmed early hints (e.g., Breed et al. 1964) that there is substantial variation in the mechanical and chemical properties of silk from homologous glands in different species (fig. 9.25, sections 3.2.9, 9.2.6). It is now clear that previous studies (e.g., Eberhard 1986a) were incorrect in assuming that the properties documented in early studies of a few species (e.g., Denny 1976) are generally uniform throughout araneoid orb weavers. The mechanical properties of some lines from the same gland types are now known to vary substantially in different species (table 3.3). One particularly striking example is the sticky spiral baselines of *Cyrtarachne bufo* and *nagasakiensis*, which have breaking strengths that are 7–10 times greater than those of the baselines of other araneids that build typical orbs (Cartan and Miyashita 2000). This difference illuminates the design of their webs, in which the sticky lines are so widely spaced that even large prey contact only a single line.

Many differences between taxa are now known in scattered cases in temperate zone species. The greatest difference in the adhesive efficiencies of sticky spiral lines (μN of adhesive force/dry volume) in a sample of five different species was that *Leucauge venusta* (TET) was 18 times greater than *Araneus marmoreus* (AR) (Opell and Schwend 2009). The number of central nodules of glycoprotein in the droplets of glue on sticky spiral lines was double in *Argiope trifasciata* (AR) and single in *Araneus diadematus* (AR) (Vollrath and Tillinghast 1991); Opell (pers. comm.) found that in another population of *A. trifasciata* they were single, while in *Larinioides cornutus* (AR) one web had 2–3 nodules/droplet but others had only 1. The mechanical properties of the sticky silk in the gumfoot lines of the theridiid *Latrodectus hesperus* differed from those of the sticky spiral silk in the orbs of the araneid *Larinioides cornutus* (Sahni et al. 2011). The stress-strain mechanical properties of major ampullate silk differed strikingly even among species in a single genus, *Argiope* (Blackledge et al. 2012b).

Some species differences are correlated with environmental differences. There was a correlation between the chemical properties of the glue and the spider's environment in the webs of *Neoscona crucifera* (AR), which builds in high humidity forest edge habitats at night, and *Argiope aurantia* (AR), which builds diurnal orbs in exposed, low humidity habitats. The glue droplets of *A. aurantia* had hygroscopic components (possibly glycoproteins or other low-molecular-mass compounds) that were not present in *N. crucifera* glue that made them more hygroscopic, and thus promoted sticky line function at lower humidities. At higher humidities (>60%) the *A. aurantia* droplets became saturated and adhesion decreased, possibly because their glycoproteins became over-lubricated, extended too readily, and released their adhesion to a surface (Sahni et al. 2011; Opell et al. 2013). In contrast, the energy needed to extend a *N. crucifera* droplet that had adhered to an object continued to increase at higher relative humidities (Sahni et al. 2011; Opell et al. 2013). Comparisons of the properties of the glue of five diverse araneoid species showed that the mix of organic salts and proteins in the glue was "tuned" to the typical humidities of their habitats to give maximum adhesion under natural conditions (Amarpuri et al. 2015).

Presumably some of these differences in mechanical properties stem from differences in the chemical compositions of silk. The composition of even homologous lines from taxonomically close relatives sometimes differs strikingly. GABamide (4-aminobutyramide) is the predominant low weight organic compound in the sticky spiral glue in *Argiope aurantia* (and in *Araneus* spp. and *Nephila clavipes*), but in *Argiope trifasciata* the principle amine was N-acetylputrescine (N-acetyl-1,4-diaminobutane) (Townley et al. 1991). In three araneid species (*A. trifasciata*, *A. aurantia*, and *Araneus cavaticus*) that varied substantially in four of the principal low weight water-soluble organic molecules in the sticky spiral glue, there were no differences in hygroscopic properties (which are key to maintaining stickiness and extensibility of the base line at different relative humidities because hydration of the droplet maintains both the supercontraction of the axial line and the extensibility of the glycoproteins in the glue droplet—Vollrath and Edmonds 1989; Townley et al. 1991; Opell et al. 2013, 2015). The differences may represent adaptations enabling lines to function at different temperatures.

Correlations with environmental conditions can be complicated because of the large range of temperatures and humidities under which even a single web built at

Table 3.3. The differences in the means of biomechanical properties of the major ampullate dragline silk lines of different species of spiders emphasize that the variability of this silk showed some logical patterns, but also some surprises. The species that do not build webs all have relatively less tough draglines, while the species with the most striking physical demands on its silk, *Caerostris darwini* (whose webs have bridge lines up to >20 m long), has the toughest silk. The next toughest silk, however, comes from *Scytodes* sp., a genus in which prey are restrained by spitting adhesive silk onto them and the spiders have no web or at most a loose tangle near the substrate (Eberhard 1973b, 1986a). The relatively high toughness of *Kukulcania hibernalis*, whose substrate web (fig. 5.10) never has to absorb the impact of prey, is also unexpected. Some species also showed substantial intra-specific variations in mechanical properties (section 3.2.9), and further surprises may be in store (after Agnarsson et al. 2010 with a few corrections of web classifications; references are the following: A = Swanson et al. 2006; B = Swanson et al. 2009; C = Agnarsson et al. 2008; D = Agnarsson et al. 2010).

Taxa	*Stiffness (Gpa)*	*Strength (Mpa)*	*Extensibility (In mm/mm)*	*Toughness (MJ/m³)*	*Web (reference)*
Liphistiidae					
Liphistius murphyorum	5.0	130	0.05	10	Burrow (B)
L. mayalanus	3.5	100	0.1	10	Burrow (B)
Theraphosidae					
Davus fasciatum	1.5	50	0.3	10	Burrow (?) (B)
Grammastola rosea	1.0	20	0.65	15	Burrow (B)
Aphonopelma seemanni	1.5	90	0.35	25	Burrow (B)
Cyriopagopus paganus	3.5	190	0.25	30	Burrow (?) (B)
Pterinochilus murinus	2.5	110	0.3	35	Burrow (?) (B)
Phormictopus cancerides	3.0	110	0.3	45	Burrow (?) (B)
Poecilotheria regalis	4.5	210	0.35	70	Burrow (?) (B)
Dysderidae					
Dysdera crocata	8.0	545	0.177	48	None (A)
Lycosidae					
Schizocosa mccooki	4.6	553	0.242	60	None (A)
Hypochilidae					
Hypochilus pococki	10.9	945	0.17	96	Lampshade (A)
Agelenidae					
Agelenopsis aperta	12.1	958	0.183	101	Sheet (A)
Oxyopidae					
Peucetia viridans	10.1	1089	0.178	108	None[1]
P. viridans	10.9	1064	0.181	111	None[1]
Nephilidae					
Nephila clavipes	13.8	1215	0.172	111	Orb (A)
Plectreuridae					
Plectreurys tristis	16.1	829	0.241	112	Tangle[2]
Amphinectidae					
Metaltella simoni	8.6	765	0.281	114	Tangle[3]
Pholcidae					
Holocnemus pluchei	14.3	1244	0.153	115	Sheet+tangle (A,WE)

Table 3.3. Continued

Taxa	*Stiffness (Gpa)*	*Strength (Mpa)*	*Extensibility (ln mm/mm)*	*Toughness (MJ/m³)*	*Web (reference)*
Salticidae					
Phidippus ardens	14.2	975	0.189	116	None (A)
Deinopidae					
Deinopis spinosa	10.4	1345	0.191[c]	124	Modified orb[4]
D. spinosa	13.5	1329	0.185[c]	136	Modified orb[4]
Uloboridae					
Uloborus diversus	9.1	1078	0.234	129	Orb (A)
Filistatidae					
Kukulcania hibernalis	22.2	1044	0.222	132	Substrate (A)
Araneidae					
Argiope trifasciata	9.2	1137	0.215	115	Orb (C)
A. argentata	8.2	1463	0.184	116	Orb (A)
A. argentata	5.3	1371	0.214	119	Orb (C)
Metepeira grandiosa	10.6	1049	0.235	121	Orb (A)
Mastophora hutchinsoni	9.4	1137	0.268	140	Bolas (A)
M. hutchinsoni	9.8	1152	0.280	161	Bolas (C)
M. phrynosoma	11.3	698	0.340	156	Bolas (C)
Araneus gemmoides	8.3	1376	0.224	141	Orb (A)
A. gemmoides	8.6	1414	0.237	164	Orb (C)
Gasteracantha cancriformis	8.0	1315	0.301	178	Orb[5] (A)
G. cancriformis	7.3	1199	0.331	193	Orb[5] (C)
Caerostris darwini	11.5	1652	0.52	354	Orb (D)
Tetragnathidae					
Leucauge venusta	10.6	1469	0.233	151	Orb (A)
Theridiidae					
Latrodectus geometricus	10.2	764	0.310	117	Gumfoot (C)
L. hesperus	9.5	959	0.224	132	Gumfoot (C)
L. hesperus	10.2	1441	0.303	181	Gumfoot (A)
Scytodidae					
Scytodes sp.	10.7	1179	0.357	230	Tangle[6] (A)

[1] Immature individuals have webs.

[2] Small tangle webs at sheltered sites such as under rocks (WE).

[3] Web form according to Griswold et al. 2005.

[4] The non-sticky lines in the "hand-held" webs of this species are extremely elastic, so the lines whose elasticity was measured must not have been from webs, but rather from the draglines. The lines in the web may be from other glands opening on other spigots on the anterior spinnerets (see section 10.9.3).

[5] Some webs have very long frame lines, up to 5 m or more.

[6] One tropical social species has more or less extensive tangles, but most *Scytodes* have at most very sparse tangles.

a sunny site will be called upon to function from dawn to mid-day (Townley et al. 1991). Because droplet extensibility affects the line's ability to sum up the adhesive effects of multiple droplets when adhering to prey, daily changes in humidity may alter prey capture abilities of orbs significantly (Opell et al. 2011a). At high humidities, a sticky spiral line of *A. diadematus* can increase its glue volume by >40% by absorbing water from the atmosphere (Edmonds and Vollrath 1992). Higher humidity at the moment at which an *A. diadematus* spider produces a sticky line can also increase the size and the subsequent ability of its sticky droplets to absorb atmospheric water (Edmonds and Vollrath 1992). This raises the heretofore unrecognized possibility of effects on the timing of web construction during the day (table 8.2, section 8.10). The stickiness of the sticky spirals of the araneids *Argiope* spp. also varied seasonally (Opell et al. 2009). Environment-induced changes in stickiness also occur in cribellates. The stickiness of the cribellum silk of the uloborid *Hyptiotes cavatus* doubled when humidity was raised from 1–2% to 45% (Hawthorn and Opell 2003).

The upshot of the variations in silk properties is that the ability of a line to retain prey can be affected by the habitat and the time of day when the prey strikes the web. Possible correlations between different web designs and differences in the mechanical properties of homologous lines merit further exploration. Further integration of silk properties and differences in habitats in discussions of web designs, following the pioneering work of Miyashita (1997), Opell and Schwend (2009), Harmer et al. (2011), Sensenig et al. (2011), Blackledge et al. (2012a), and Amarpuri et al. (2015), is likely to be highly rewarding.

3.2.9.2 Differences within species

Even more surprising is the substantial degree of intraspecific "plasticity" (variation in the material properties of silk obtained from a single spider through different spinning methods) in the properties of lines from the same glands (Boutry et al. 2011). I will list a few scattered examples of this as yet largely unexplored field to give an idea of the surprising magnitudes and kinds of variation (see also section 3.2.9 and Herberstein and Tso 2011).

The vertical major ampullate gumfoot lines of *Parasteatoda* (= *Achaearanea*) *tepidariorum* were more compliant than the horizontal major ampullate lines in the tangle used for support (Boutry and Blackledge 2009). There was also variation in the properties of dragline silk spun by spiders in three spider clades when they were in three different situations (line produced while dropping, while walking, and when forcibly extracted from an immobilized spider) (Boutry et al. 2011). Variation was compared between members of the RTA clade and orbicularians with respect their anatomy, silk biochemistry, and phylogenetic relations. The orbicularians seemed less dependent on external spinning conditions, and probably used a valve in their spinning duct to control friction forces and speed during spinning (Boutry et al. 2011). This plasticity probably resulted more from differences in processing the silk dope in the spinning duct than from differences in silk biochemistry.

Regarding stickiness, in *Cyclosa turbinata* (AR), stickiness varied between individuals on the basis of the amount of silk available (Crews and Opell 2006), while the sticky spiral loops in the outer portions of orbs of *Argiope* spp. and *C. turbinata* were stickier than those near the hub (Blackledge et al. 2011; Crews and Opell 2006; Opell et al. 2009). After being exposed to continuous moderate wind (1.1 m/s) for seven days, *C. mulmeinensis* in still air built sticky spiral lines that had sticky droplets whose diameters were larger (had both greater volume and surface area) (Wu et al. 2013). The stickiness of the line remained unchanged, but its tendency to dehydrate was probably reduced, and the glue had greater amounts of hydrophilic low-molecular-weight compounds (Wu et al. 2013). There may even be individual differences in the amount of edge-to-hub change in diameters of the sticky spiral base line. When three webs were measured for each of eight individuals of *Argiope argentata* (four lines in each web), the mean diameter of the outermost sticky spiral lines averaged 233% of that of the innermost lines in one individual, but only 128% in another (Blackledge et al. 2011). The diameters of the sticky droplets on sticky spiral lines were greater below the hub than above the hub in the same webs of *Araneus diadematus* (AR) (Edmonds and Vollrath 1992). The density of puffs of cribellum silk also varied in the same web of *Uloborus diversus* (UL) (fig. 4 of Langer and Eberhard 1969) and *Thaida peculiaris* (fig. 19F of Griswold et al. 2005), so some cribellates may also modulate the adhesiveness of their lines.The draglines that the araneid *Argiope trifasciata* produced when climbing vertically resisted larger loads than those produced when the spider moved horizontally (Garrido et al. 2002). *Araneus ventricosus* silk had higher tensile properties in webs higher above the ground (Pan et al. 2004). Exposure to constant wind during the preceding seven days induced the araneid *Cyclosa mulmeinensis* to produce dragline silk of greater strength, extensibility, toughness, ultimate tension, and breaking energy (Liao et al. 2009). Even the properties of the line produced during a single descent on a dragline can vary, because the speed at which a dragline was drawn correlated positively with its strength (Ortlepp and Gosline 2004; Pérez-Rigueiro et al. 2005). The diameters of the radii in orbs built in captivity by mature females of the araneid *Larinioides cornutus* averaged only 49.6% of the mean diameter of radii in field webs (Sensenig et al. 2010b). Hesselberg and Vollrath (2012) noted that in *Nephila edulis* (NE) "all of our reported silk mechanical properties showed large variations between different webs and spiders"; strain at breaking, breaking energy, initial Young's modulus, and point of yielding all varied in this species with both rate of silk production and temperature (Vollrath et al. 2001). The diameters of non-sticky lines (presumably ampullate gland silk) also varied substantially (by a factor of >2 within a web and by about 6x between the webs of only 8 different individuals) in *Latrodectus hesperus* (THD) (Blackledge et al. 2005c). There was intra-web variation in the theridiid *Parasteatoda* (=*Achaearanea*) *tepidariorum*; the supporting web lines were thicker, able to bear higher loads before deforming permanently and before breaking, and less elastic than other lines in the same web that bore droplets of sticky material (Boutry and Blackledge 2009).

Feeding history sometimes affects the properties of lines. *Latrodectus hesperus* spiders that enjoyed a higher rate of feeding built webs in which the mean diameter of lines was double that of lines in the webs of poorly fed spiders (Blackledge and Zevenbergen 2007). A spider's weight correlated with a line's mechanical properties. The stiffness (Young's modulus) of draglines produced by heavier *Araneus diadematus* (AR) was lower than that of lighter individuals (Vollrath and Köhler 1996). The diameters of dragline fibers in the orb of *A. diadematus* correlated with the spider's weight (Christiansen et al. 1962; Vollrath and Köhler 1996; Blackledge et al. 2011). Appendix O3.3 discusses the additional possibility (somewhat uncertain in my opinion) that orb weavers alter silk properties adaptively on the basis of the mechanical demands made by recent prey impacts (Tso et al. 2007).

As might be expected from this variation, there are also intra-specific differences in silk chemistry. There was substantial variation in the low-molecular-weight organic compounds associated with sticky spiral glue in a single population of *Argiope aurantia*, and low-molecular-weight components of the sticky spiral lines also varied among different populations of *Nephila clavipes* (NE) (Higgins et al. 2001). The silk in the thicker dragline threads produced by heavier individuals of *A. diadematus* also had altered mechanical properties, with larger diameter lines having lower stiffness (Young's modulus) (Vollrath and Köhler 1996).

And even a single line, once it is produced, can show significant changes in its mechanical properties under different environmental conditions. For instance, the adhesive properties of sticky spiral lines changed when temperature and humidity changed. In *Argiope aurantia*, which lives in open weedy habitats, the energy to extend a droplet on a sticky spiral line (an index of its ability to transfer adhesive force and to dissipate energy) was greatest under afternoon conditions (hot and dry), less under morning conditions (cool and humid), and least under hot and humid afternoon conditions (Stellwagen et al. 2014). Comparing several araneid species, changes induced by exposure to UV radiation reduced the energy to extend a droplet in the sticky spiral lines of a species with a nocturnal orb (*Neoscona crucifera* [AR]) and a diurnal forest species (*Micrathena gracilis* [AR]), but not in species that build diurnal webs in open habitats (*Argiope aurantia* [AR], *Verrucosa arenata* [AR], and *Leucauge venusta* [TET]) (Stellwagon et al. 2015). Environmental factors also induced changes in the mechanical properties of dragline silk that differed between species in apparently adaptive manners. Ecologically realistic intensities and durations of UV irradiation resulted in mechanical strengthening of the dragline silk of three diurnal araneoid orb weavers during the operating lifetimes of their orbs, but in weakening of the dragline silk of two nocturnal species (Osaki and Osaki 2011).

In sum, orbs may be much more subtle traps than might have been supposed. It seems likely that spiders may fine-tune the properties of the materials in their orbs to be maximally effective under the environmental conditions at the time of day when their webs will be up, and perhaps even when particularly important prey are active.

3.2.9.3 Consequences for understanding orb web designs

The variations just described sound a strong cautionary note for attempts to understand the functional significance of different orb designs. Differences in the silk itself will clearly have to be taken into account in future attempts to understand the functions of different orb designs. The most promising step in this direction is the pioneering study of Sensenig et al. (2010a) of 24 species in 19 genera and 5 orbicularian families. The authors presented extensive data on the physical properties of the lines in their orbs corrected for the sizes of the adult spiders, and also attempted to correct for the possible effects of phylogeny. They concluded that evolutionary shifts to larger body sizes have been repeatedly accompanied by improved abilities to stop prey, due to changes in both silk material and web designs. Large spiders produced silk with improved material properties, and also used more silk, to make webs that have superior stopping abilities. For spiders of comparable size, those spinning "higher quality" silk build webs with more sparse arrays of lines. These conclusions all seem intuitively reasonable, but they are difficult to evaluate for several reasons (see appendix O3.3).

The run-on lists in this section are designed to emphasize the possible importance of the hitherto unexpected extent of inter- and intra-specific variations. One logical consequence of all this variation would seem to be that spiders experience both mechanical constraints and trade-offs, with greater costs being associated some particular mechanical or chemical properties. Otherwise one might have expected them to consistently make the strongest and stickiest lines possible, something they are obviously not doing. These costs and their trade-offs remain to be quantified. Just as the study of spider venoms has been revolutionized by the discovery of striking, unexpected diversity (up to a thousand different toxin molecules in a single gland!) (G. Binford pers. comm.), so the chemical diversity of silk molecules promises to revolutionize understanding of the functional significance of differences in the designs of different webs.

3.2.10 CORRELATIONS BETWEEN ORB DESIGN AND DETAILS OF ATTACK BEHAVIOR ARE INCONSISTENT

A spider and her web obviously act as a team in capturing prey (Robinson 1975), and it is reasonable to expect that the overall designs of the spiders' webs and attack behavior and morphology should coevolve. Some spider groups provide beautiful examples of evolutionary adjustments of attack behavior and morphology to webs and prey. Atypids strike from within their tubular web at prey on the outer surface, using especially long, stiletto-like cheliceral fangs (Bradley 2013); tube-dwelling *Segestria* fold the dangerous mandibles and abdomen tip of stinging hymenopteran prey away from themselves by pulling the prey into the tunnel (Bristowe 1958); *Oecobius annulipes* utilized the open surfaces where they built webs to rapidly run over and around dangerous ant prey, spreading a swath of wrapping lines using special setae on the anal tubercle (Glatz 1967). The expectations of web design–attack behavior coevolution are only met partially, however, in orb webs. I will begin with some of the numerous cases that do not seem to fit predictions.

3.2.10.1 Inconsistent relationships

Two obvious variables in attack behavior are the speed with which the spider reaches the prey, and her ability to immobilize larger and more dangerous prey more quickly and securely by wrapping them as opposed to just biting. Both of these variables vary widely among orb weavers (e.g., Robinson and Olazarri 1971; Robinson and Mirick 1971; Robinson 1975; Suter 1978; Heiling 2004; Briceño and Eberhard 2011). Olive (1980), for instance, contrasted the effectiveness of attacks on a variety of prey by the araneids *Araneus trifolium* and *Argiope trifasciata*, which co-occur in different subhabitats in old fields; he found that each species was especially effective in attacking the kinds of prey that were more common in its own subhabitat. *Araneus trifolium*, which has large fangs and relatively short legs, were slower to reach prey; they built open-meshed webs that were higher above the ground, and specialized on innocuous but rapidly escaping prey such as Diptera and Lepidoptera. In contrast, *A. trifasciata* had relatively longer legs, smaller fangs, and more densely meshed orbs built near the ground, and specialized on dangerous, slowly escaping insects such as grasshoppers.

Olive attempted to generalize from these correlations in his tiny sample of species, but the associations that he found between leg length, cheliceral size, and web design are not consistent in other species. For instance, the tetragnathid *Leucauge mariana* has relatively long, thin legs and attacks very rapidly (Briceño and Eberhard 2011), but has robust chelicerae. Using relative numbers of radii and sticky spiral loops as a guide (fig. 4.12), its webs have moderate to open sticky spiral spacing and moderate numbers of radii. *Cyclosa* spp. and *Allocyclosa bifurca* have moderately short legs, moderately small chelicerae, and moderate attack times, but have dense sticky spiral loops and proportionally large numbers of radii (Peters 1954; Suter 1978; WE) (figs. 6.21, 7.41). The short-legged, heavy-bodied *Gasteracantha cancriformis* has moderately-sized chelicerae, and executed slow attacks preceded by repeated web shakes (Eberhard 1989b), but it built webs with only moderate densities of radii and sticky spirals (figs. 3.22, 4.12) (WE).

Zschokke et al. (2006) also checked for correlations between orb properties and attack behavior. They found that the mean retention time for the fruit fly *Anastrepha ludens* in the webs of three orb weavers, *Nephila clavipes* (NE), *Verrucosa arenata* (AR), and *Leucauge venusta* (TET), correlated with the mean time needed by the spider to bite or to begin to wrap it. Again, the sample of spider and prey species was small. Common retention times in the webs of the first two species were also more than an order of magnitude greater than the attack times, and retention times for other prey species were very different, so it is difficult to evaluate the general significance of these patterns.

In general, the biological significance of means of the times to attack prey is uncertain, because attack speed varies widely with different prey (e.g., Robinson and Olazarri 1971; Robinson and Robinson 1973a on *Argiope argentata* and *Nephila pilipes*, Zschokke et al. 2006 on *N. clavipes*, *L. venusta*, *V. arenata*, and *Nihonhimea tesselata*), with sites of the prey in the web, and also with differences in previous feeding success (Herberstein et al. 1998 on *A. keyserlingi*). One simplifying step could be to use the maximum attack speeds, and ignore the long delays sometimes associated with dangerous prey or satiated spiders. Even using maximum speeds, however, it is clear that some predictions do not match observations. Spiders such as *Nephila clavipes* (Landolfa and Barth 1996) and *Mangora* spp. (WE) with sticky spiral-rich webs (figs. 4.4, 4.12) are very quick to arrive at trapped prey. In *N. clavipes* 35% of attacks took <1 s, and the first response at the hub occurred in <40 ms in 58% of the trials (Landolfa and Barth 1996). These very rapid attacks are particularly dramatic in view of the very slow movements of *Nephila* spp. in other contexts. At the other extreme, attacks were consistently very slow in some species with sticky spiral-poor orbs such as the stubby-legged theridiosomatids *Theridiosoma* spp., *Ogulnius* sp., and *Epeirotypus* spp. (McCook 1889; Wiehle 1929; Coddington 1986a; WE).

The araneids *Micrathena gracilis*, *M. duodecimspinosa*, and *G. cancriformis* were also slow (Uetz and Hartsock 1987; WE) (*M. gracilis* apparently never reached prey in less than 3 s), but they had intermediate orb designs (figs. 1.4, 3.22, 4.12). In contrast, the tetragnathid *Leucauge mariana* also built intermediate web designs (figs. 4.12, 7.14d), but executed extraordinarily rapid attacks that often took <0.5 s even when the spider had to turn >90° before running to attack (Briceño and Eberhard 2011). The fact that *Arachnura melanura* (AR) employed a jerky gait during attacks (probably as camouflage from predators—Robinson and Lubin 1979a) indicates that speed is not always of overriding importance.

Another possible attack variable is the behavior used by the spider to restrain the prey: bite it with the chelicerae (common for most prey in tetragnathids, nephilids, and some araneids) (Robinson 1975; Kuntner et al. 2008a; WE); or wrap it in silk (common in uloborids and many araneids) (Robinson 1975; Lubin 1986; WE). But the web designs of species that attack prey by biting range widely, from the dense orbs of *Nephila* spp. (NE) that are rich in both radii and sticky spiral loops (Kuntner et al. 2008a), to the dense sticky spiral but only moderate number of radii in *Dolichognatha* spp. (TET) (figs. 3.43a, 4.12) (Eberhard 1986a), to the flimsy, open-meshed orbs of *Tetragnatha* spp. and *Metabus ocellatus* (TET) (figs. 4.12, 7.43). Likewise, the designs of the orbs of species that attack prey by wrapping them range from webs with dense sticky spirals araneids such as *Acacesia hamata* and *Mangora* spp. (Carico 1986; Eberhard 1986a), to the more or less typical designs of *Araneus diadematus* and *A. expletus* (Witt et al. 1968; figs. 2.17, 3.7), to the sticky spiral-poor orbs of *Uloborus* spp. (UL) (fig. 3.31).

A final possibility is that spiders respond more rapidly

to prey in orbs with "higher regularity" (Zschokke and Vollrath 2000). This trend was documented only with two unpublished studies, however; further details will be necessary to evaluate this possibility.

In sum, these failures to find correlations with attack behavior do not falsify the idea that retention times will tend to correlate positively with web design traits such as the relative density and strength of sticky lines. They suggest however that additional factors have also had important roles in the evolution of these designs. Uncertainty regarding cause and effect may also be involved. (Is the web of a spider that attacks slowly expected to be especially well designed regarding relative radius numbers and sticky spiral spacing for retaining prey? Or to other factors, such as thicker or more adhesive lines, or to weak, easy to retain prey that enable spiders to invest less in rapid attacks?) Perhaps with larger samples and corrections for other factors, predicted trends with attack behavior will be discovered. One possibility, a correlation between attack speeds in different portions of an orb and the degree of vertical asymmetry in orbs, is discussed in section 3.8. This topic is clearly complex and poorly understood.

3.2.10.2 More likely correlations

One correlation is very likely to be consistent, even though only few published data are available at the moment: spiders with relatively stubby legs will tend to attack prey more slowly. Relatively long legs, in contrast, allow high running speeds, as in *Nephila clavipes* and *Leucauge mariana* (though they do not guarantee rapid attacks, as illustrated by the slow cautious attacks of the long-legged theridiid *Theridion evexum*—Barrantes and Weng 2007b). The median times to reach prey in the webs of three orb weavers with long, medium, and stubby legs were, respectively 1 s for *L. mariana* (TET), 4 s for *Metazygia yobena* (AR), and 8 s for *Gasteracantha cancriformis* (AR) (Eberhard 1989b). I know of no general study documenting attack time in stubby-legged species, but in my own incidental observations, the relatively short-legged araneids *G. cancriformis*, *Micrathena duodecimspinosa*, *Hypognatha* sp., theridosomatids in general, the uloborid *Hyptiotes*, and the theridiids *Phoroncidia* spp. all approached prey slowly. Some, like *M. duodecimspinosa*, routinely began attacks by turning at the hub and strongly jerking the sector of web containing the prey, rather than running toward it; perhaps the jerks increased the entanglement of the prey. It might be interesting to trace the evolution of stubby legs (presumably they are derived in some of these lineages) and to check for concomitant changes in web design that might make jerks more effective in entangling prey. In general, web traits favoring retention must be especially important in these groups.

While short legs impede rapid attacks, they facilitate holding the web tense while waiting for prey to arrive (the legs of the theridiosomatid genera *Epeirotypus* and *Theridiosoma* and the uloborid *Hyptiotes cavatus*, which tense their webs for hours at a time, are especially short and thick—Shinkai and Shinkai 1985; Coddington 1986a; Opell and Konur 1992). Tensing the web this way is undoubtedly costly in terms of energy and materials. In a sample of four species, the first anterior leg depressor muscles had denser arrays of mitochondria and tracheae in those species with more active web-monitoring behavior (Opell and Konur 1992). Increased strength to hold the web may also be favored in *Hyptiotes*. Perhaps reeling in each radius during sticky spiral construction by *G. cancriformis*, instead of walking outward along the radius (fig. 3.25; WE), also favors strong legs.

Another seemingly consistent association related to attack behavior concerns orb symmetry. In those species in which the spider rests off the orb in a retreat, the hub is consistently displaced toward the side nearest the retreat (fig. 3.16). An arbitrary sample includes *Zygiella x-notata* (AR) (Le Guelte 1966 and Endo 1988), *Araneus legonensis* (Edmunds and Edmunds 1986), *Larinioides sclopetarius* (Heiling 2004), *Metazygia wittfeldae* (WE), *Metepeira* spp. (AR) (WE), *Araneus expletus* (AR) (WE), *Dolichognatha pentagona* (TET) (WE), and *Chrysometa* spp. (TET) (WE). Peters (1937), Le Guelte (1966), and Heiling (2004) found that when they experimentally moved the position of the retreat relative to that of the orb by rotating the frame in which the spider was housed, *Z. x-notata* and *L. sclopetarius* responded by building orbs in which the hub was displaced toward the new site of the retreat. These asymmetries shorten the spider's path in attacking prey, because spiders attacked prey by first going from the retreat to the hub and then to the prey. In *L. sclopetarius*, the median attack time for a spider resting at the hub was 2 s, three times faster than than when the spider was in a retreat at the edge of the web (Heiling 2004).

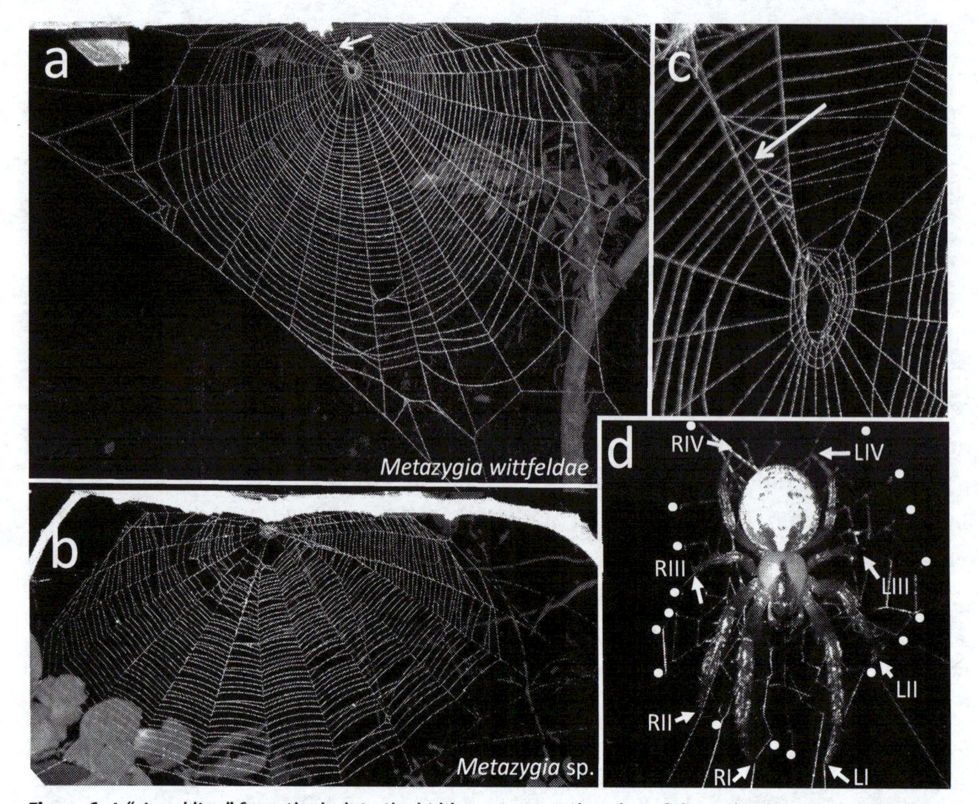

Fig. 3.16. A "signal line" from the hub to the hidden retreat at the edge of the web where the spider rests during the day has been used successfully as a taxonomic character in *Zygiella* spp. (AR) (Gregorič et al. 2010), but the night-time orbs of mature female *Metazygia* (AR) illustrate the difficulty of distinguishing signal lines in other groups. The web of *M. wittfeldae* (*a*, *c*) has a relatively small hub (*a* is a view perpendicular to the web plane; *c* is from slightly from the side, with the focus on the signal line from the hub to the spider's retreat that is slightly out of the plane of the orb, just beyond the upper edge of the web (thick arrow). Some sticky spiral lines were attached to the signal line as if to a radius, where it was closest to the plane of the orb (*c*). In another species of this genus (*M.* sp.) signal lines were often lacking when hubs were immediately adjacent to the spider's retreat on a twig (*b*) (similar to the twig webs of *Tetragnatha* [TET] and *Uloborus* [UL] in figs. 10.24, 10.26). In other genera such as *Zygiella* and *Aculepeira* (AR) (fig. 3.24), spiders build in the early morning and frequently rest in retreats during the day, and the sticky spiral lines are not attached to the signal line, probably improving its ability to transmit prey vibrations to the spider. The legs RI and LI of a mature female *M. wittfeldae* (*d*) also illustrate unusual positions, grasping lines at or just beyond the edge of the hub in the free zone. In contrast to many other araneids (e.g., figs. 2.19b, 3.35), legs RII, RIII, and LII grasped the outer hub loop, and the other legs grasped lines inside the hub (the arrows mark the positions of tarsi, the dots mark the outer edge of the hub). This spider readily fled to her retreat when disturbed.

Given the short retention times for the majority of prey stopped by orbs in the field (section 3.3.3.1.1), lowering the time needed to reach the hub could have a substantial impact on prey capture rates. A second likely correlation with these asymmetries is a reduction in the attenuation of vibratory signals that are transmitted along radii from the prey to the hub and then to the retreat; for instance, some failures of *L. sclopetarius* spiders in retreats to respond to prey may have been due to failure to sense the its presence in the web (Heiling 2004).

Finally, there was a trend in the attack behavior of species with highly reduced webs (eight different groups in table 3.4) to manipulate web tensions when they attack prey. The spider did not run rapidly to the prey, but collapsed the web onto or around it before she began to bite or wrap it.

In sum, a spider's ability to capture prey is undeniably influenced by an intimate interaction between the design of her orb and her attack behavior, which is in turn influenced by her morphology. While attack behavior surely needs to be added to the biomechanical properties of silk lines and the phylogenetic relationships of different groups in comparative studies such as that of Sensenig et al. (2011) to understand orb designs, many details of how to do this are not clear. Some details of orb designs correlate with aspects of attack behavior such as

Table 3.4. Derived attack behavior that has apparently coevolved with derived web designs.

Spider	*Web*	*Attack behavior*	*References*
Gradungulidae			
Progradungula carraiensis	Small patch of sticky lines held by legs II and III (fig. 10.9)	Scoop prey upward into web with anterior legs	Gray 1983
Theridiosomatidae			
Wendilgarda sp.	Sparse vertical lines with sticky tips attached below to a water surface (fig. 10.29)	"Snow plow" attacks—gather together and move a cable of several sticky lines, sliding their tips on the water	Coddington and Valerio 1980
Theridiosoma spp, *Epeirotypus* spp., *Naatlo splendida*	Ray web with spring line out of web plane that spider reels in, pulling orb into cone	Release reeled up spring line, causing orb to spring backward when sense vibrations of nearby prey	McCook 1889, Comstock 1967, Shinkai and Shinkai 1985, Coddington 1986a, WE
Araneidae			
Scoloderus tuberculifer	Ladder web that is elongate above the hub (fig. 3.4)	Run past prey to get above it, then cut radii there, causing the prey to sag into contact with additional sticky lines below[1]	Eberhard 1975
Ocrepeira (= *Wixia*) *ectypa*	"asterisk" web of a few, short non-sticky lines that radiate from a central point spanning a fork of a branch	Run to substrate and wrap prey that brushes against a line by rapidly circling it and the branch over and over	Stowe 1978
Mastophora, *Ordgarius*, *Dichrostichus*, *Cladomelea*, *Agatostichus*	One or a few large sticky droplets, each hanging free at the end of a single line	Swing droplet toward nearby insect when sense vibrations	Stowe 1986, Yeargan 1994
Uloboridae			
Hyptiotes spp. and *Miagrammopes* spp.	Spring line from apex of sector of orb or from end of one or a few sticky lines	Release and reel up spring line repeatedly with legs IV, causing the web to collapse repeatedly around the prey	Akerman 1932, Marples and Marples 1937, Comstock 1967, Lubin et al. 1978
Miagrammopes sp.	Single vertical sticky line or lines	Pack up long sticky line into sticky mass as descend and then apply it to the prey	Lubin et al. 1978
Deinopidae			
Deinopis spp., *Menneus* spp.	Small rectangular array of parallel sticky lines on a network of highly elastic, non-sticky lines held at the corners by legs I and II	Spread the web with legs I and II and lower it onto prey walking below, or expand it rapidly dorsally to trap prey flyng nearby; also repeatedly expand and partially collapse it after the prey is in contact to further enswathe it[2]	Akerman 1926, Kaestner 1968, Robinson and Robinson 1971

[1] Similar behavior that causes prey to sag into additional sticky lines occurred independently in some uloborids with more or less vertical orbs (Lubin 1986), and in the non-orb theridiid *Tidarren sisyphoides* (G. Barrantes pers comm.).

[2] Robinson and Robinson hypothesized that the behavior served in addition to "straighten the net" so that it could be reused.

speed or how prey are immobilized, but in general the expected correlations between orb design and attack speed (or leg length) occur only inconsistently. Attack behavior probably has further effects, but these are yet to be clarified.

3.2.11 ORB MOVEMENTS IN WIND MAY NOT BE GENERALLY SIGNIFICANT IN INTERCEPTING PREY (BUT MAY AFFECT ORB DESIGNS)

3.2.11.1 Web movements and the encounter model

Erratic oscillatory movements in the wind could enhance the probability that a web will contact and capture nearby insects. Even if a potential prey can sense an orb web at a distance and take evasive action (see fig. 3.2, section 3.11), it might still have difficulty avoiding contact. Craig et al. (1985) argued that an orb's ability to capture prey increases as a result of orb movements. Web movements are undoubtedly important in a few groups, including theridiosomatids (e.g., McCook 1889; Wiehle 1929; Coddington 1986a), *Hyptiotes* (UL) (Wiehle 1927; Marples and Marples 1937), and deinopids (Akerman 1926; Robinson and Robinson 1971; Coddington and Sobrevila 1987), in which spiders actively launch their webs rapidly toward nearby insects. The webs of many other species move passively in the wind. For instance, the widely spaced lax sticky lines of *Ogulnius* sp. (TSD) (fig. 10.22c), when observed glinting in patches of sunlight in a tropical forest, swung erratically up to 0.8–1 cm in gentle, nearly imperceptible air movements (WE). Movements of this sort could increase the chances that the lines intercept insects. Craig et al. (1985) extended this idea to typical orbs, whose lines sag little or not at all. They argued that the passive movements when an orb sags and flutters in the wind are important, when combined with small fluctuations in an insect's flight path, in increasing the likelihood that prey will be intercepted (the "encounter model"). Some movements of orbs that they measured were quite large, even in the gentle winds in tropical forest undergrowth; for example, the orbs of the relatively small (2.7 mg) tetragnathid *Leucauge globosa* moved up to 8 cm in winds of 0.05—0.1 m/s.

Some subsequent reviews (e.g., Blackledge et al. 2009c, 2011) ignored the encounter model. I believe that there are serious problems with the data that were used to support the model, but that the basic idea is nevertheless important. The major empirical problem was that some (or perhaps many) of the web displacements that Craig et al. (1985) filmed and measured were artifacts. The webs were coated with white powder to make them visible in order to film their movements, and this powder surely caused the webs to sag (apparently substantially, judging by fig. 4.17 in Craig 2003). Long experience photographing powdered webs in the field leaves no doubt in my mind that greater sagging leads to larger movements in the wind. Nevertheless, further observations (below) showed that web movements were not entirely artifacts: uncoated web lines were also displaced in wind, though their movements were smaller and less consistent than those reported by Craig et al. (1985).

An independent reason to wonder about the encounter model is that many spiders manipulate their orbs in exactly the opposite manner predicted by the model. The spiders tense rather than loosen their webs while they wait at the hub for prey to arrive (table 4.3) (section 3.4.2.3) (see figs. 3.34–3.37); then, at the moment that a prey strikes the web, they release the tension. Tensing the web prior to prey impact would reduce the amplitude of its fluttering movements in the wind at the stage when the encounter model predicts that the spider would profit by increasing them.

In sum, orb webs do at least sometimes move in the wind, but the distances are poorly documented, and web tensing behavior by the spider complicates interpretations; insect flight behavior near webs, especially under windy conditions, is also poorly known. These uncertainties led me to make a few additional, though only roughly quantitative field observations on how orbs move; the topic is complex (see also section 8.4.3.2.12), and what follows is only an exploratory sketch, not a finished study.

3.2.11.2 Different types of orb web movement in the wind

Preliminary observations in second growth in Costa Rica in the late wet season showed that orb movements in the wind occurred at four general levels: the web as a whole moved, including all the frames and anchors, due to movements of the substrate to which the anchors were attached ("substrate movements"); the web sagged and moved as a whole with respect to the supports to which the anchors were attached ("whole orb movement"); the capture area and hub moved with respect

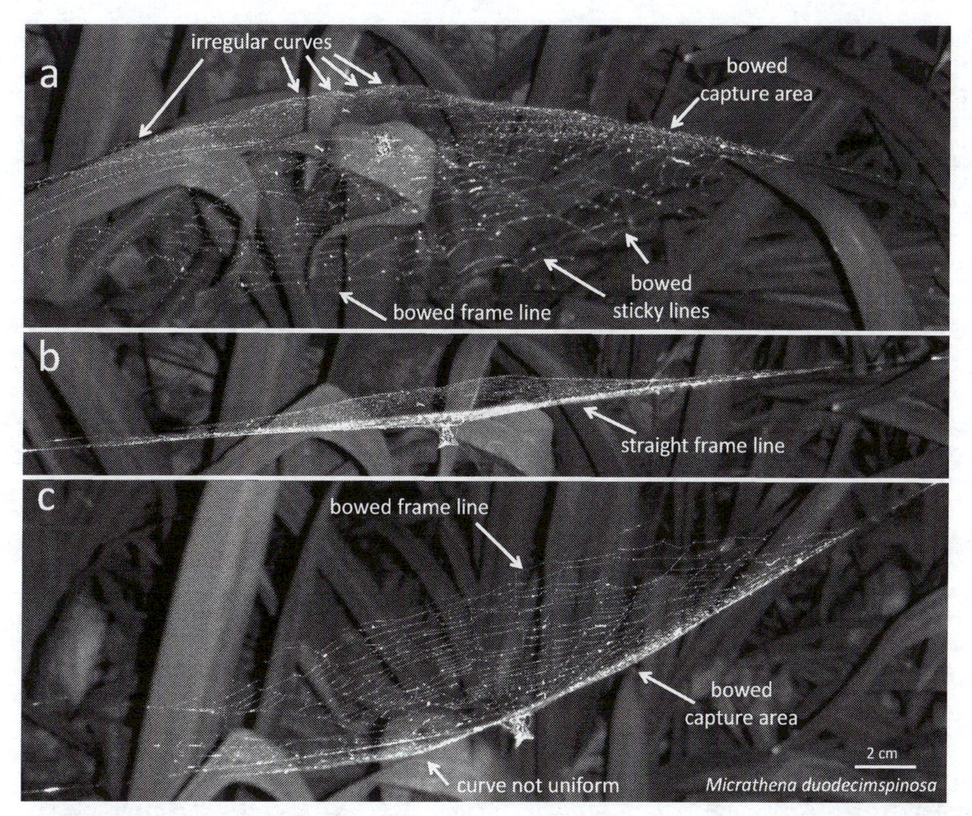

Fig. 3.17. These three photos looking down on the same uncoated, approximately 70° web of a mature female *Micrathena duodecimspinosa* (AR) were taken in the afternoon when the web had accumulated detritus, making it easier to photograph, illustrate different types of web movement (only the upper frame line is clearly visible in *a* and *c*). The moderate, erratic breeze of up to about 1–1.5 m/s first blew in one direction (*a*); then was still (*b*); and then blew briefly in approximately the opposite direction (*c*). The bowing of the frame lines in *a* and *c* resulted from either whole web or substrate movements; it was greater than it appears because the anchor lines extended substantially beyond edges of the photos. The capture zone tended to curve more strongly than did the frame lines, due to "capture zone movements." Places where the curve of the capture zone was not smooth in *a* and *c* were perhaps due to irregularities in air flow(?). Curves in some individual sticky spiral lines that were moving are also evident in *a*.

to the more rigid frame lines, like a sail with respect to a boom ("capture zone movement") (see fig. 3.17); and the individual sticky spiral lines moved with respect to the more rigid radii to which they were attached ("sticky line movements") (fig. 3.18; Triana-Cambronero et al. 2011). I will term, somewhat imprecisely, the first three categories as "entire web" movements. Recognition of these differences helps interpret some of the evidence mentioned above. The measurements of Craig et al. (1985) were of "entire web" movements, combining substrate, whole web, and capture area movements; sticky line movements were ignored. Spiders that tensed radial and hub lines while resting at the hub increased mainly radial tensions, and thus mainly reduced only capture area movements. Tensing behavior contradicted predictions regarding capture area movements, but (as will be seen below) had little relation to other web movements.

Setting aside for the moment sticky line movements, gentle breezes (gusts of only about 1 m/s) displaced uncoated orbs of adult female *Micrathena duodecimspinosa* (AR) up to 1–2 cm. The orbs of this species generally have only three, relatively long anchor lines (typically each was approximately 1–2 times as long as the diameter of the capture zone). In stronger winds with gusts of about 5 m/s, webs moved up to about 5 cm (fig. 3.17; WE). The web of a small, probably second instar spiderling (the first instar outside the egg sac) of this species moved up to 2–3 cm in weaker gusts of up to approximately 0.6 m/s. In addition to capture zone movements, these web movements included substrate and whole web movements: frame and anchor lines were curved in photos (fig. 3.17); and in some cases leaves and other objects to which anchor lines were attached moved. When adult females were induced to walk away from the hub

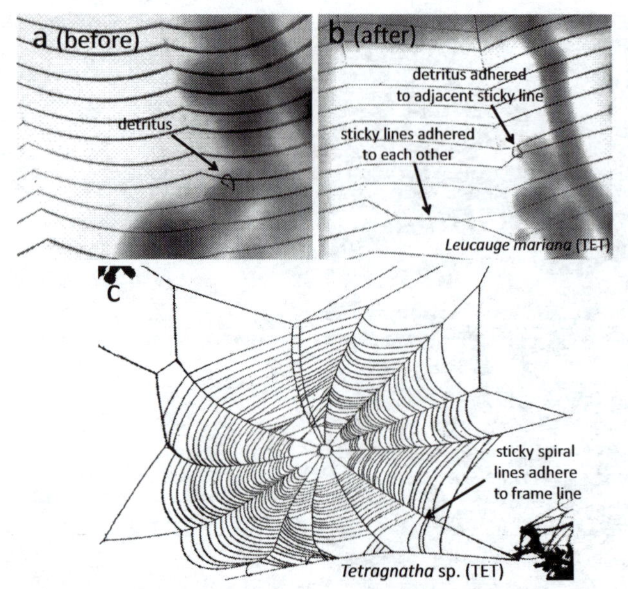

Fig. 3.18. The photos in *a* and *b* illustrate effects of gentle wind on uncoated lines near the edge of a newly constructed orb of a mature female *Leucauge mariana* (TET). The sticky spiral lines were bowed slightly by a weak (estimated <1 kmph) breeze (*a*); damage occurred at two points where sticky lines adhered to each other or to a small piece of detritus when the same lines were blown somewhat more strongly (*b*) (a weaker wind was blowing in a different direction when photo *b* was taken). The photos illustrate several points: the sticky spiral lines bowed much more strongly in the wind than did the radii; wind damage resulted when the fluttering movements of two segments of sticky line brought them into contact in their middle portions (a common pattern in relatively long segments in webs in the field); the irregular nature of the "fluttering" movements of the sticky lines in the wind introduces an element of chance in determining which lines will adhere to each other (note that the longest two segments did not stick to each other); small bits of detritus can increase the chances of adhesion damage, both because the detritus can project toward an adjacent line and because the weight of the detritus will increase the amplitudes of fluttering movements; and weak winds are often irregular in direction (compare *a* and *b*). Still another type of damage occurred in the horizontal orb of *Tetragnatha* sp. (#517) when multiple sticky lines adhered to a frame line (*c*). Wind sometimes produced damage of this sort when the frame line sagged briefly as a result of the supporting structures to which it was attached moving toward each other.

(where they had been tensing their webs), the distances that their webs moved in the wind did not change perceptibly. These wind velocities were 10–100 times higher than those in the encounter model study (Craig et al. 1985); uncoated *M. duodecimspinosa* orbs were relatively immobile at lower wind velocities.

The orbs of other species at the same site moved in different ways under similar wind velocities. The orbs of mature females of *Cyclosa jose* (AR) and *Allocyclosa bifurca* (AR), which had more numerous and relatively shorter anchor lines (much less than the diameter of the capture zone) and that were consistently attached to relatively rigid supports, scarcely moved (though the egg sac + detritus stabilimenta [see figs. 3.41, 7.41] of these species vibrated with low amplitudes). Similarly, the orb of a mature female *Nephila clavipes* (NE), which also had relatively short anchor and frame lines, moved very little in similar breezes. The orbs of mature females of still another species, *Leucauge mariana* (TET), which were slanted 30–60° with horizontal, moved with intermediate amplitudes. In gusts of about 1 m/s, mature females resting at the hubs of orbs moved up and down an estimated 5–10 mm. Some of these webs were attached to leaves that moved, so substrate movement may have been responsible for some of these movements. The webs of *L. mariana* were nearly immobile in low velocity breezes (est. <0.3 m/s).

Sticky spiral movements occurred in lower wind velocities, including even very light, nearly imperceptible wind (<0.3 m/s) (fig. 3.18). The magnitudes of the displacements of the central portions of the longer segments of sticky lines in the webs of *L. mariana* were on the order of the distances between attachments of these loops to the radii in estimated 0.6–1.1 m/s winds. The amplitudes of sticky spiral line movements relative to the spaces between them were smaller in the other species.

In sum, under some conditions in nature, the orbs of some species moved erratically up to a cm or more. Webs also moved at more usual wind conditions (at least at this particular site and season), but the amplitudes of entire orb movements were on the order of only one mm. Sticky spiral movements were greater in *L. mariana* than in the other species, and occurred even at low wind velocities.

3.2.11.3 Orb movements in the wind: are they important?

3.2.11.3.1 PREY CAPTURE

The significance of the different web movements just described for intercepting prey probably depends on several presently little-studied factors, including the sizes, flight paths, and speeds of important prey insects in the field (section 3.11). The focus probably needs to be on possible evasive movements by large prey that may be rendered ineffectual by web movements, rather than on the less nutritionally important small prey that are capable of flying through the openings of the orb (Craig et al. 1985) (section 3.2). The magnitude of entire web movements per se may not be as important as their velocities relative to the velocities of potential prey as they maneuver in the wind to avoid the web. Similarly, it is not the magnitude of web responses to occasional gusts of wind (as in web damage—see next section), but the mean magnitudes of movements of uncoated webs, and not what happens on a single windy day but what occurs on average days. Data are apparently lacking on these topics, so it is uncertain whether entire web movements increase prey capture.

The possible consequences of fluttering sticky line movements for prey capture have never to my knowledge been examined. Sticky spiral movements could play an important role in prey capture in some species but not others: they would seem especially likely to have effects on slowly flying prey, for smaller and more maneuverable prey, and in the outer portions of the orb where the sticky lines are longer. In still air and with well-lit webs, small flies turned away at distances of up to 5 mm from the web, or slowly approached closer, touched the web with only a single extended leg, and then rapidly accelerated away (Craig 1986) (fig. 3.2). Thus the behavior of both insect prey and sticky web lines is such that the encounter model could apply, but further information is needed to judge the possible biological importance of the encounter model for sticky line movements.

Incidentally, sticky line movements of a different sort may sometimes increase the ability of lines to intercept prey. The sticky spiral lines of the araneid *Araneus diadematus* were deflected toward several types of insect prey (and water droplets) that had been experimentally charged with static electricity (Ortega-Jiménez and Dudley 2013), thus increasing the web's interception and adhesive properties. At least some insects become charged (up to several hundred volts) when they fly, due to friction with air molecules (Colin et al. 1991; Vaknin et al. 2000), so this effect may be biologically significant. Again, this effect is most likely to be important in longer segments of sticky lines, and ones that are under lower tensions

In sum, the possible importance of the encounter model for prey capture, both for entire web movements, and for sticky line movements, is still uncertain; but it cannot be ruled out, and it could be important for a variety of orb traits.

3.2.11.3.2 WEB DAMAGE

Orbs gradually accumulate several types of damage in the field (fig. 3.19). Substrate movements, even though brief, can cause multiple adhesions in large sectors of an orb, and thus alter the geometrically regular patterns of the original orb (fig. 3.18c). Damage can result when the attachment point of an anchor line moves inward toward the rest of the orb, and reduces the tensions on many lines (Triana-Cambronero et al. 2011), or when sticky lines swing toward other lines and adhere to them when they flutter in the wind. The apparent preferences that some species show for more rigid attachment sites (section 8.4.3.2.2) are probably adaptations to reduce substrate movement damage. The tangle built above the horizontal orbs of juvenile *Leucauge argyra* (TET) that tenses the orb with lines connected to the hub (fig. 10.3) (Triana-Cambronero et al. 2011) could have a similar function.

A second type of damage can occur during whole orb and capture area movements,when some lines move relative to others, and sticky lines contact other lines such as frames (figs. 3.18*c*, 10.4). Still another type of damage can result from sticky line movements when individual lines flutter in the wind. Fluttering sticky lines can contact other web lines and adhere to them (e.g., the "central adhesion damage" when sticky spiral lines make contact in figs. 3.18, 3.19) and even when tensions on other web lines are unaltered.

The probabilities of these different types of contact depend on whether different lines in the orb move independently of each other, and on the amplitudes of such independent movements (which will in turn depend on the tensions on the lines and on their lengths). The danger of adhesions between web lines due to wind could thus affect selection acting on the distance between sticky spiral and frame lines, sticky spiral tensions, the lengths of sticky spiral segments (the distances between radii), and the spaces between adjacent loops. The general topic of damage from adhesions has received very little empirical attention (see, however, the "sticky spiral entanglement" hypothesis in section 4.3.2.4), and I can give only a preliminary sketch of possibly important points.

Fig. 3.19. It is useful to look at orbs from different angles. In contrast with nearly all of the other photos of freshly built and undamaged webs in this book (and in most of the published literature), this uncoated web of a mature female *Micrathena duodecimspinosa* was photographed late in the day (5:30 PM), after it had been up for about 11 hrs on a moderately windy day (with occasional gusts of up to perhaps 3 m/s). The web (seen from above) had numerous holes and contained many small bits of detritus. At least some of the holes were made by the spider herself as she removed larger bits of detritus that had blown into her web (she removed another piece soon after the photo was taken). Several types of damage can be seen: *a*) "central adhesions" between adjacent segments of sticky spiral (see also fig. 3.18); *b*) adhesions of sticky spiral lines to radii; *c*) breakage of sticky spiral lines; *d*) breakage of radii; *e*) lateral displacements of radii to contact other radii; and *f*) loading of the web by detritus. How well a web stands up under such a beating will obviously affect its ability to trap prey, but the patterns of web damage in nature are virtually unstudied, and even very basic questions remain to be answered. How common are the different types of damage, and what causes them? Are the different types of damage randomly distributed in different parts of orbs in the field? Do they vary with the slant of the web? with the rigidity of the supports to which the anchors are attached?

The sticky lines are slack in some species that build orbs or derivatives of orbs, and form more or less shallow catenaries. These lines would be especially susceptible to damage from sticky line movements. Catenaries occured, for example, in the finished orb webs of *Anapisona simoni* (AN) (figs. 7.9, 9.4), *Conculus lyugadinus* (Shinkai and Shinkai 1988), and *Metabus ocellatus* (TET) (fig. 7.43) (WE). Slack sticky lines also occur in some reduced webs derived from orb, including those of *Pasilobus* sp. (Robinson and Robinson 1975), *Poecilopachys australasia* (Clyne 1973), *Ogulnius* (TSM) (fig. 10.22), and *Mysmena* (MYS) (fig. 10.21) (the webs of *Ogulnius* and *Mysmena* were coated with powder for photography, but uncoated sticky lines also sagged in both species [WE]). Only the outermost sticky spiral loop of an uncoated *Epeirotypus brevipes* (TMS) orb sagged perceptibly in still air, but other loops also sagged in very weak, scarcely perceptible air movements (WE). On the other hand, published photographs of uncoated orbs show no perceptible catenaries in several other species, including *Nephila edulis* (NE) (Hesselberg and Vollrath 2012), *Araneus diadematus* (AR) (except under the influence of methamphetamine) (Hesselberg and Vollrath 2004), *Eustala illicita* (AR) (Hesselberg and Triana 2010), and *Cyclosa caroli* (AR) (Hesselberg 2013); the segments of the sticky spiral formed straight lines between attachments to radii. Catenaries were also lacking under windless conditions in even the longest sticky spiral segments of adult females of the Costa Rican species *Leucauge mariana* (TET), *Allocyclosa bifurca* (AR), *Cyclosa jose* (AR), *Gasteracantha cancriformis* (AR), *Argiope argentata* (AR), *Micrathena duodecimspinosa* (AR), and

Nephila clavipes (NE) (WE). Thus, despite the fact that low tensions on sticky lines would seem likely to be advantageous in both stopping and retaining prey (they would increase the abilities of the lines to absorb stress without breaking) (table 3.1), distinctly slack sticky lines are probably the exception rather than the rule among orb weavers, perhaps to reduce damage from sticky line movements in the wind.

Nevertheless, tensions on the sticky lines were low enough in all of the Costa Rican species that even straight sticky lines sagged and moved in light breezes on the order of 0.6 m/s or less (fig. 3.18a). Similar bowing in a light wind occurred even in short segments of the sticky spiral in the webs of *Cyclosa turbinata* and *L. venusta* (WE). Qualitative evaluations indicated that in comparable winds, the sticky lines moved more (relative to the distances between them) in the webs of *L. mariana* and *G. cancriformis*, less in those of *M. duodecimspinosa*, *A. argentata*, and *A. bifurca* (AR), and least in those of *N. clavipes*.

The sticky spiral segments in an orb tended to all sag in the same direction and in synchrony at low wind velocities (fig. 3.18), and did not contact other lines; but at higher velocities (perhaps accompanied by increased turbulence, which was not measured), adjacent sticky lines often fluttered independently. When a segment of sticky line was long enough, and the bowing was substantial enough, the fluttering movements eventually brought adjacent lines into contact with each other, resulting in adhesion to an adjacent frame or radius ("adhesion damage") (Fig 3.18b). Larger numbers of sticky lines suffered adhesion damage at higher wind velocities. These adhesions opened up holes in the array of sticky lines, reducing the geometric regularity that is probably a key trait of orbs favored by natural selection (section 3.3.3).

Adhesion damage from wind has several important traits. Because a single touch is sufficient to cause a sticky line to adhere to another, it is not the mean wind velocity, but rather sudden changes in direction and the maximum velocities in irregular gusts (the "threads of stronger wind among lesser air movements"—Geiger 1965; p. 120) that determine the amount of damage done. The importance of extremes for damage contrasts with the effects of web movements on prey interception, where mean velocity is probably more important. A second difference is that winds that are more nearly parallel to the plane of the orb may be more likely to produce adhesion damage (though experimental data are lacking). Assuming that wind direction tends to be approximately horizontal, this would mean that more nearly horizontal orbs may be more susceptible to adhesion damage. Again, this contrasts with damage from movements of the entire web or most of it caused by drag in the wind, which seem likely to be greatest when the wind direction is perpendicular rather than parallel to the orb plane and to be greater in more nearly vertical orbs (Craig et al. 1985). One further complication is that a small piece of detritus that adheres to a single line can increase the likelihood of contact with adjacent lines in the wind (fig 3.18b), but probably has little effect on prey capture.

Although there are no quantitative studies of variation in these sorts of wind damage, preliminary observations suggest that even moderate winds can produce adhesion damage under natural conditions (*a* in fig. 3.19). About 29% of 191 *U. diversus* webs spun on a given morning had at least an estimated 10% of the surface missing by late in the afternoon (Eberhard 1971a; wind damage was not distinguished, however, from other types).

Many factors seem likely to affect the frequency with which adhesion damage will occur, including those in the following, probably incomplete list: tension on the sticky lines; the lengths of the segments of sticky lines (which of course vary widely in a given orb—see section 4.3); the distances between sticky lines; the angles of the wind with the plane of the orb; turbulence in air flow patterns (greater turbulence might increase the likelihood of sticky line contact); small scale irregularities in air velocities (which are common in nature) (Geiger 1965); the sizes of the glue droplets on the lines (Wu et al. 2013) (larger droplets tend to produce greater friction with the air, and thus greater bowing of lines) (Zaera et al. 2014); the extensibility of the lines themselves (greater extension under loading from the wind can result in reduction of the distances between lines and in additional friction); the time of day (for example, wind speeds near the ground tend to increase during the morning, and tend to be lower at night) (Geiger 1965); height over the ground (in the first several meters above the ground, wind speeds tend to increase as an exponential function of the height above the ground) (Geiger 1965); the presence of small pieces of detritus adhering to individual sticky lines (fig. 3.18); and the rigidity of the supports to which the web is attached (greater flexibility and displacement of the supports will increase the chances of damage). At a larger scale, different environments also surely differ with respect to wind velocities (see Liao et al. 2009 on *Cyclosa mulmeinensis* and *ginnaga* in seashore vs. forest habitats, Tew and Hesselberg 2017).

One further aspect of damage is the web maintenance behavior of spiders, including repair of damaged orbs to apparently avoid further damage. One aspect of web maintenance is removing fallen detritus such as leaves, flower petals, etc. from the web (Robinson and Robinson 1973a; Pasquet et al. 2007). Detritus removal entails the metabolic cost of the activity, the danger of breaking crypsis or emerging from a sheltered retreat, and the damage to the sticky spiral (the "tracks"—fig. 4.19) made by the spider as she moves across her orb. Removal of detritus is apparently important, as it often occurred within a few seconds after an object hit the web in *Z. x-notata* (AR) (Pasquet et al. 2007) and *Leucauge mariana* (TET) (Briceño and Eberhard 2011). The likely benefits include increased physical stability for the web, especially from stess as the detritus swings under windy conditions (Langer 1969), and possible defense against some predators (Pasquet et al. 2007 on the generalist vespid wasp predator, *Vespula germanica*). Detritus is probably more often a problem in more nearly horizontal webs, but I know of no comparative empirical measurements.

A preliminary list of possible examples of different aspects of orb design that affect an orb's susceptibility to wind damage and are modified by spiders in response to wind includes the following. Captive *Cyclosa mulmeinensis* (AR) spiders that had experienced moderate, persistent wind (about 1.14 m/s) built orbs that were smaller and had fewer but stronger major ampullate lines; their drag in 1.14 m/s wind was reduced (Liao et al. 2009). The orbs of similarly wind-exposed spiders also had smaller capture areas, and shorter and more widely spaced sticky spiral lines with larger glue droplets (Wu et al. 2013). Wind during construction also induced *Araneus diadematus* (AR) to build smaller and rounder webs, with fewer, more widely spaced sticky spiral loops (Hieber 1984; Vollrath et al. 1997) (wind speeds in the two studies were 1.2 m/s and 0.5 m/s, respectively). Wind induced no changes in web design, however, in *A. gemmoides*, other than a tendency to orient their approximately vertical orbs more nearly parallel to the wind (which could also reduce damage) (Hieber 1984).

Field comparisons between *Uloborus diversus* (UL) spiders of different sizes showed that smaller individuals (with weaker webs) built more nearly horizontal webs, that they were closer to the ground, and that similarly-sized spiders in more windy habitats built smaller and more nearly horizontal webs (Eberhard 1971a). Field data in another species, however, showed few correlations with wind. When several aspects of orb designs (including web slant, orientation, area, height above the ground, the vertical asymmetry of the capture zone, and the vertical diameters of the capture zone, the hub, and the free zone) were compared between a windy and a less windy site (mean wind velocities were 1.30 ± 0.13 and 0.03 ± 0.13 m/s) in large samples of *Metellina mengei* webs (N = 430, 279), there were no differences in mean values or in variances, other than that webs were built higher above the ground in the sheltered habitat (Tew and Hesselberg 2017).

There is other, less direct evidence that also suggests that wind can be important: the common limitation of the webs of some symphytognathoids, which have strongly sagging sticky lines, to microhabitats in the litter layer of forests (WE); the choice of relatively windless nights by *Pasilobus* sp. (Robinson and Robinson 1975) and *Poecilopachys australasia* (Clyne 1973) to build their webs (which also have long, strongly sagging sticky lines) (Robinson and Robinson 1975). Avoidance of wind damage to sticky lines may also help explain the general trend in orb web designs for the spaces between adjacent loops of sticky spiral to be greater near the outer edge of the orb and thus avoid adhesion damage even though they are longer (the "sticky line entanglement" hypothesis in section 4.3.2.4).

In sum, the effects of web movements in the wind on prey capture and web damage are poorly documented. Selection to reduce wind damage undoubtedly varies greatly in intensity, and could affect numerous web traits. The absolute values of variables such as wind speeds, the distances that lines are displaced, the spaces between lines, and the heights above the ground likely affect the relative importance of these different factors for different species. The current lack of quantitative studies correlating these factors and wind damage makes it impossible to judge their relative importance in nature, and of wind damage in general; but wind damage is probably often substantial (fig. 3.19). More precise measurements under more controlled conditions are needed, as are further field surveys including measurements of the frequencies of different types of damage in the field.

3.2.12 PREY ARE NOT DEFENSELESS: PROTECTION FROM THE SPIDER'S OWN PREY

I include the topic of the danger to the spider from her own prey (e.g., fig. 3.20) not because it has been well-studied, but because it seems to have hardly been discussed at all. There are nevertheless abundant data showing that danger from prey affects spider attack behavior, and it is reasonable to ask whether this danger has also influenced their web designs.

The starting point is that it is very common for spiders to refrain from attacking some prey, or to make less than maximally rapid attacks after prey have been stopped and are entangled in the web. The frequencies of non-attacks are surprisingly high in a wide variety of different species. In riparian populations of the funnel web agelenid *Agelenopsis aperta*, 41% of the prey that struck webs were not attacked (Riechert and Łuczak 1982). In the aerial sheet web linyphiid *Linyphia triangularis* in the undergrowth in a mixed deciduous wood in England, the corresponding figure was 17.7% of 581 prey that entered webs (Turnbull 1960). In three species inhabiting standstone rock faces, the funnel web agelenid *Coelotes montanus*, the gumfoot tangle web theridiid *Parasteatoda* (= *Achaearanea*) *tepidariorum*, and the orb-weaving araneid *Araneus cavaticus*, >50% of the individual prey of a given taxonomic group that entered the webs were not attacked in 9 of 20 comparisons that were made for the different taxonomic groups of prey that were important in the spider's diet (Riechert and Łuczak 1982). Of 25 large acridid grasshoppers (with large mandibles and spiny posterior legs) that were placed in the webs of *Nephila pilipes* (= *maculata*) (NE), nearly half (12) escaped during the spider's slow, cautious attacks (Robinson and Robinson 1973a) (in some cases the spider's first response was to flee). Bristowe (1941) noted that caution in attacks was especially common in tetragnathids and linyphiids. Given that large prey are rare and thought to be especially important in determining a spider's survival and fitness (Venner and Casas 2005; Blackledge 2011; Evans 2013), these losses of large prey are especially dramatic.

These dangers are not just hypothetical. The most commonly reported observation in this respect is the frequent reluctance of web spiders to attack large, dangerous prey (fig. 3.20*b*). In *N. pilipes* smaller, less well-armed prey were generally attacked more quickly (Robinson

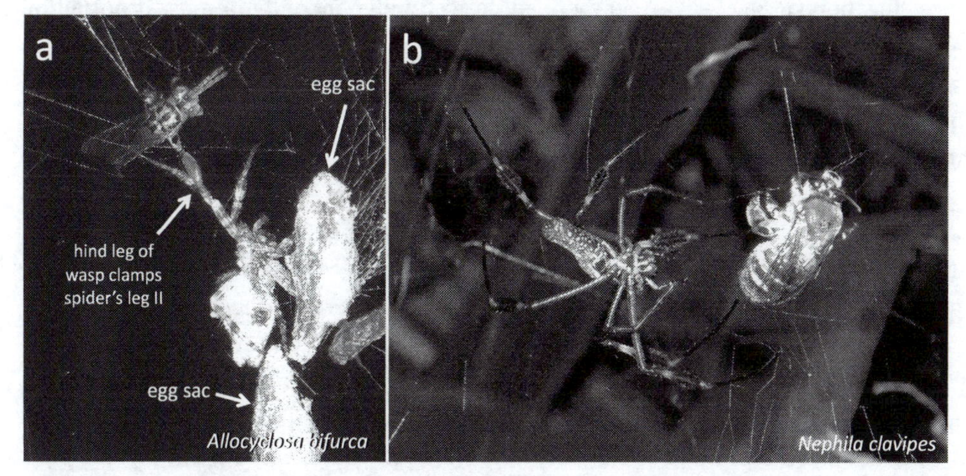

Fig. 3.20. Prey in webs can be dangerous. The predicament of a mature female *Allocyclosa bifurca* (AR), whose leg II was trapped in the powerful grip of the hind leg of a chalcidid wasp (probably *Conura*) (*a*), illustrates the dangers that an orb weaver must confront when she attacks a prey. The largely blind spider must make intimate contact with potentially dangerous insects that are in unpredictable orientations in order to bite and wrap them. The general importance of dangers from prey is suggested by the frequent hesitancy and apparent caution with which orb weavers attack some insects that is illustrated in *b*. This penultimate female *Nephila clavipes* (NE) approached and repeatedly tapped a large solitary bee that had been stopped and retained by her web; but after >8 min of tapping, she returned to the hub without attacking it.

and Robinson 1973a). In general, attacks by orb weavers on more dangerous (larger or more heavily armed) prey tended to include slower, more hesitant approaches, more frequent pauses during the approach with apparent attempts to gather further information (e.g., jerk the web), raising the anterior legs away from the prey when biting it, biting and then backing away from the prey, and initiating wrapping from a greater distance (e.g., Robinson and Olazarri 1971; Robinson and Robinson 1973a; Lubin 1980). *Argiope* spp. (AR), which are at less risk because they wrap prey from a distance rather than biting them, attacked large prey more rapidly than did *Nephila* spp., which attacks prey by biting (Robinson et al. 1969; Robinson and Mirick 1971; Robinson and Robinson 1973a; Robinson 1975).

There are, in addition, scattered observations of direct damage from prey, such as a vespid wasp that stung an attacking *Nephila edulis* (NE) (Austin and Anderson 1978) (see also fig. 3.20*a*). In some cases additional details confirmed that danger from the prey was responsible for the ineffectiveness of some attacks. In *A. aperta*, attack frequencies were higher in a more poorly fed desert population that received only about one-third as many prey, while spiders in a better-fed riparian population were less likely to attack aposematic prey (Riechert and Łuczak 1982). In *L. triangularis*, failures to attack were more common for prey species with which the spider had no previous experience (Turnbull 1960).

If you put yourself in the spider's place, it is easy to understand why orb weavers attack some prey cautiously. For an effectively blind animal, a web would be a particularly frightening place to have to attack a large, powerful prey, even if the spider can distinguish the prey's size through the amplitude of the web vibrations produced on impact (Suter 1978 on *Cyclosa turbinata*). The positions of a prey's weapons, such as the sting of a wasp, the powerful mandibles of a katydid, or the spined hind legs of a grasshopper, are unpredictable when it is struggling in a web. This uncertainty contrasts, for instance, with the fact that when a spider attacks prey walking on a substrate, the prey's dorsal (often less defended) side has a predictable position with respect to the substrate. In sum, the evolution of aerial webs may have been accompanied by selection on spiders to defend themselves more effectively from their prey.

Greater vertical asymmetry, with more of the capture surface below than above the hub, could reduce danger to the spider. This would allow the spider to attack prey from above, reducing the chances that the prey might fall toward the spider as it struggled during an attack (section 4.8). Another aspect of orb design that could reduce the danger for the spider would be for the orb plane to be slanting rather than vertical.

Because spiders are at greater risk of being preyed on themselves while they are attacking prey (Robinson 1975), the reductions in the duration of a spider's attacks on large, dangerous prey that result from attacking prey by wrapping as opposed to biting (Robinson 1975) could potentiate selection favoring orb designs that are better able to stop and retain high-energy prey (Blackledge 2011). Perhaps the reduced danger from prey in slanting and horizontal orbs has also facilitated the evolution of more rapid, less cautious attacks in some groups (table 3.1). These are only speculations, however.

One non-orb design that could reduce the danger to spiders from their prey is a tightly meshed horizontal aerial sheet, with the spider moving on the undersurface while the prey is on the upper surface. This has clearly been a successful web design, both in terms of numbers of species and numbers of independent evolutionary origins (table 9.1). There are probably thousands of species with this basic type of web, and it has apparently been derived independently in most if not all of the following families: Austrochilidae, Araneidae, Theridiidae (a least three times), Linyphiidae and Pimoidae, Synotaxidae, Pholcidae, Ochyroceratidae, Scytodidae, Plectreuridae, Diguetidae, Psechridae, and Oxyopidae. Close observation of spiders capturing prey that fell onto the finely meshed sheet of *Cyrtophora citricola* (AR) (fig. 1.7) showed that this barrier was sometimes almost too effective, because it apparently impeded the spider's attempts to grasp and hold prey from below: some prey escaped repeatedly and the spider had to chase it across the sheet. In some cases she folded the sheet around the prey with her legs during an attack (WE). The barrier formed by the sheet probably explains the greater prevalence of the more dangerous biting as opposed to wrapping attacks in *Cyrtophora* (Lubin 1973; Robinson 1975): it protected the spider from the prey; but it impeded wrapping silk from reaching the prey.

Recognition of potential dangers from prey and the

fact that spiders fail to attack some larger prey raises the possibility that there may be an upper limit on the selection favoring the ability of an orb to stop especially large prey (the "rare large prey" hypothesis—section 4.2.2). Having a web that is able to stop prey that are larger than the size that the spider is willing to attack effectively could represent a waste of resources.

3.2.13 DESIGN DETAILS ARE LIKELY TO BE SELECTIVELY IMPORTANT

Previously (Eberhard 1990a) I argued that the largely stochastic nature of prey capture would make natural selection unable to discriminate among different design details of webs. Many small details of how a prey strikes an orb (for example, whether it strikes the orb near its outer edge, perpendicular to the orb, above or below the hub, with a leg or a wing, etc.) will affect whether the prey will be stopped and retained long enough for the spider to capture it. Because many of these details will have little or no relation to the design details of a particular orb, I thought that this variation would be so great that minor details of an orb's design could not be expected to have significant effects on the spider's feeding success. Selection on design details might thus be so weak that it could not be expected to account for the evolution of detailed differences in orb designs (Eberhard 1990a). Craig (1987b) also doubted the importance of selection on the details of the designs of some orbs, though for other reasons (see section 10.9.4).

I believe that this argument fails to take into account one further point. While measurements of retention times suggest a huge role for stochasticity, even for the same prey that contact the same web (section 3.3.3.1.7), the effects of a given detail in the spider's building behavior on prey capture are not limited to the impacts with a given web. Instead they are summed over the entire set of webs that the spider builds in which this detail is expressed (in most cases, over all the webs built during the spider's lifetime). Furthermore, many aspects of the building behavior of an orb are repeated many times during the construction of a single web (chapter 6). For instance, a bias in the distance at which one sticky spiral loop is laid with respect to the previous loop can affect hundreds of attachment points in a single orb. These types of summation will increase the probability that natural selection will be able to distinguish between even small behavioral differences. The finding that spiders do indeed make fine adjustments in orb designs (e.g., sections 4.3, 4.8) support this conclusion.

3.2.14 ADULTOPHILIA: A SERIOUS ARACHNOLOGICAL PROBLEM

I have relegated critical discussions of several common methodological problems in studies of the functional significance and evolution of different orb designs to appendix O3.2. One problem, however, is so general and potentially important that I include it here, despite the paucity of relevant data. Arachnologists have usually emphasized adults rather than immatures in studies of behavior and ecology (for recent exceptions, see Gregorič et al. 2013; Barrantes et al. 2017). There are several explanations for this bias (below), but it nevertheless results in some basic problems. Just as in other fields such as community ecology (e.g., Green et al. 2014), careful study of juvenile stages may be crucial to understanding adults.

A detailed review of ontogenetic changes in silk composition and geometric designs of the orbs of *Neoscona arabesca* (AR) (Sensenig et al. 2011) showed how complex the mix of change and lack of change during ontogeny can be. Silk investment increased isometrically with body size, but glue production was greater in larger spiders; larger spiders spun larger webs with smaller radii, but the increased volumes of all silk types and the greater toughness of the capture spiral silk resulted in an isometric scaling of stopping potential. The strength and toughness of sticky capture spiral thread increased with diameter and hence also during ontogeny, but dragline thread material properties did not change. The relative importance of sticky spiral lines in dissipating a prey's kinetic energy to stop it was greater in smaller orbs of *Verrucosa arenata, Neoscona domiciliorum*, and *Larinioides cornutus* (AR), with aerodynamic damping being more important (Sensenig et al. 2012); the sticky spiral lines of smaller spiders were also stiffer, increasing their contribution to dissipating prey energy at lower extensions (Sensenig et al. 2010a).

Although both orb and non-orb web designs tend to be similar in juveniles and adults of a given species, some species show consistent ontogenetic changes. For instance, smaller spiderlings of both *Leucauge argyra* and *L. mariana* (TET) built orbs with relatively greater areas and higher densities of sticky lines, and more extensive nearby tangles; assuming that diameters of lines changed in proportion to size, smaller individuals apparently made greater web investments, with designs that increased their abilities to capture relatively large prey (Triana-Cambronero et al. 2011; Barrantes et al. 2017) (see also Higgins 1992b, 1995 on *Nephila clavipes* NE—section 10.2). The ecological variables that change as an individual grows include distributions of prey sizes, identities, and velocities, susceptibilities to environmental perturbations such as wind and dessication, predators, and expenditures on basal metabolism. For instance, a prey that is too small to be of nutritional importance for an adult could be an important resource for intermediate-sized juveniles, and be too large to be utilized by very small individuals. Perhaps not surprisingly, even closely related species can differ in their web design adjustments for size. For example, the number of sticky lines increased with capture area in *L. mariana*, but decreased in *L. argyra*, and the density of sticky lines decreased faster with web area in *L. argyra* (Barrantes et al. 2017). To the extent that web designs remain the same during ontogeny, there may often be a complex interplay between facultative adjustments to deal with the problems of juveniles and those of adults; obviously, a spider will only reach the adult stage if the webs that she builds earlier in her life enable her to survive and grow.

In sum, the ecological problems of juveniles are not necessarily just scaled down versions of those of adults. Some juvenile-adult differences are predictable. I will list a few of the likely problems for smaller individuals, just to illustrate their variety. Web damage due to the impacts of insects that have too much kinetic energy to be stopped seems likely to be greater for smaller spiders, because of their thinner lines and thus weaker webs (e.g., Craig 1987a) (unless small spiders locate their webs in different microhabitats where larger prey are less common). The quantities of falling debris are likely to be similar for larger and smaller individuals, but they likely produce greater web damage in smaller individuals. The relative viscosity of the air will be greater, increasing the relative importance of the role of sticky lines in absorbing impacts (Lin et al. 1995; section 3.3.2.1), reducing problems from trampoline-like rebounds, and increasing likely damage from given wind velocities. Size/volume relationships indicate that the loss of water from sticky spiral glue is likely to be greater, perhaps necessitating different frequencies of web renewal, greater concentrations or mixes of hygrophyllic compounds in the glue to maintain adhesiveness (Amarpuri et al. 2015), or greater relative investments in glue. Smaller spiders themselves will be more susceptible to desiccation, imposing different habitat limits. Juvenile spiders probably have proportionally larger energy demands than conspecific adults, because in spiders (Quesada et al. 2011) (as in many other animals—Eberhard and Wcislo 2011), there is a negative relation between body size and the relative size of the energy-expensive central nervous system.

Another likely difference is that smaller insects tend to be more abundant than larger insects (e.g., Schoener 1980), so prey that are large enough to be nutritionally important probably tend to be more abundant for smaller spiders (see section 6.2.4). Differences of this sort have been documented empirically for some spiders. For instance, the size class that constituted 68% of the captures (by number, not mass) by young *Linyphia triangularis* (Linyphiidae) in May formed only 8% of the captures of adult females in October (Turnbull 1960). In *L. triangularis* less than 10% of the total number of species that were preyed on during the entire year were captured during any given week (Turnbull 1960). Seasonality in prey also had large effects. More than 90% of their prey insects were adults, and adults of >80% of the species were captured only during periods of less than four weeks. A further difference is that the juveniles of some species live in different subhabitats from adults. For instance, adult female *Nephila pilipes* (= *maculata*) (NE) built their orbs above the herb layer, where prey velocities may be high, while juveniles more often built within the herb layer (Robinson and Robinson 1973a), where velocities are almost certainly lower.

A concrete illustration of possible size-related consequences for juvenile web designs comes from pollen ingested when the sticky spiral was reingested (section 3.3.3.3.5). Pollen had a stronger effect on the survival of second instar spiderlings of *Araneus diadematus* (AR) than on third instar spiderlings (at least at certain sites and times of the year in which wind-blown pollen was abundant) (Smith and Mommsen 1984). Pollen is easy to stop and retain, even on single lines in weak webs, and the electrical charges that it accumulates in the air may increase its adhesion to silk (Ortega-Jiménez and Dudley 2013). Pollen capture could tend to favor longer, thinner sticky spiral lines with smaller glue droplets, and fewer radii in smaller spiders. Even a short burst of pollen production in some seasonal environments might be enough to get spiderlings past a particularly difficult life stage (Smith and Mommsen 1984).

One seldom-considered consequence of these juvenile-adult differences is that a given web trait could be highly advantageous for juveniles, but just a sub-optimal "hangover" for adults. Given the frequent general similarity between the web designs of juveniles and adults (e.g. fig. 1.5), it seems likely that the web design at one stage of a spider's life is influenced to some extent by the web designs at other stages. In sum, a given trait may occur in the webs built at one stage of the spider's life not because it is selectively advantageous at that stage, but because it was favored by the conditions at some other stage. The relative importance of the factors associated with different spider sizes are not known, and I cannot offer any quick fixes. My objective here is to point out that the problems are complex and do not have easily intuited solutions, and that future studies of both intra- and inter-specific correlations with body size hold great promise for improved understanding of web designs.

Arachnologists have several good excuses for our bias against studying juveniles. Smaller animals are generally more difficult to observe than larger ones, and consistently using mature animals allows one to eliminate some kinds of variation. Sometimes an emphasis on adults is a practical necessity in the field because juveniles (especially very young juveniles) are difficult to identify to species. A focus on adults can facilitate comparisons with previous studies of adults. But none of these excuses is a justification for continuing to ignore juveniles.

3.3 HOW ORBS FUNCTION

Having cleared away some common misconceptions, we can now turn to how orb webs function. There is general accord that many spiders in nature are hungry—that they could grow more and faster and reproduce more if they could capture more prey (e.g., Anderson 1970; Spiller 1992; Wise 1993). In addition, both lab and field studies consistently show that a substantial fraction of the prey that strike an orb web escape (Nentwig 1982a); means for escape frequencies ranged from 10–82%, with a median of 50% in 12 species in 3 families in table 3.5. The two major causes of prey escape are the inability of the web to absorb the prey's momentum, or failure to retain it long enough for the spider to reach it once it has been stopped (e.g., Uetz and Hartsock 1987) (fig. 3.21). For larger (and thus on average nutritionally more important) prey, escape rates can be quite high (e.g., 75% for prey >9 mm long in the araneid *Micrathena gracilis*). Escapes can be biologically important, because prey impacts are not always abundant. For example, in the space of three nights 26 (35.1%) of 74 individuals of the tropical araneid *Eriophora edax* failed to capture a prey, and 41 (55.4%) captured only one (Ceballos et al. 2005).

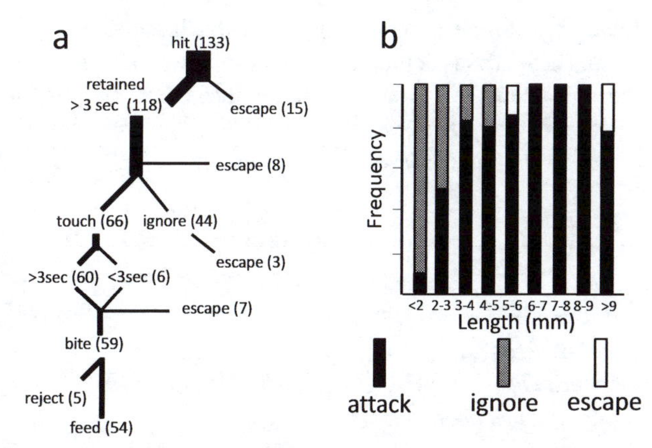

Fig. 3.21. The histories of 133 prey that struck the orbs during 77 web hours of patient observation of *Micrathena gracilis* (AR) in the field, and that were retained for at least 3 s are portrayed in *a*; the frequencies of different outcomes for these prey are shown in *b* according to the sizes of the prey. Once a prey of other than tiny size had been retained for 3 s, the likelihood was high that the spider would reach it and bite it. These data emphasize the possible importance of a variable ignored in many analyses of prey captures—the selectivity of the spider in ignoring or rejecting some prey. These observations do not include prey that broke through the web or that were not retained for at least 3 s, as occurred for most of the larger prey (75% of prey >9 mm escaped). "Ignore" = the spider did not respond, or turned to face the prey and jerked her web, but failed to contact it; "attack" = the spider contacted the prey (usually with her legs I or II); "reject" = the spider contacted the prey and then actively discarded it from her web (after Uetz and Hartsock 1987).

These observations imply that there is natural selection on webs to improve prey capture, and that this in-

Table 3.5. A sample of the frequencies with which prey escaped after encountering webs before the spider was able to attack them. The methods of introducing prey into the web were the following: "manual" = prey grasped in forceps and placed in the web; "flown" or "jumped" = prey induced to fly or jump toward the web (e.g., released from a vial held near the web with a light on other side of web to attract them);"blown" = prey blown gently from aspirator into web; "natural" = prey flew into the web spontaneously with no inducement.

Taxon	*Frequency of escape*	*Method introduction*	*Prey*	*Reference*
Tetragnathidae				
Metabus ocellatus	53%	Natural	Various	Buskirk 1975
Metleucauge (3 spp.)	20–50%	Manual	Various	Yoshida 1989
Tetragnatha elongata	40%	Natural	Various	Gillespie and Caraco 1987
T. praedonia	61%	Manual	Various	Yoshida 1987
Tetragnatha elongata	40%	Natural	Various	Gillespie and Caraco 1987
T. praedonia	61%	Manual	Various	Yoshida 1987
Metleucauge (3 spp.)	20–50%	Manual	Various	Yoshida 1989
Araneidae				
Araneus trifolium	58%	Flown or jumped	2 types	Olive 1980
Argiope trifasciata	63%	Flown or jumped	2 types	Olive 1980
Argiope aurantia	18%[1]	Flown or jumped	4 spp.	Blackledge and Zevenbergen 2006
Cyrtophora moluccensis	58–82%	Flown	Blowflies	Lubin 1973
Metazygia sp.	18–35%[2]	Flown	Sepsid flies[2]	Eberhard 1989b, WE
Metepeira sp.	33%	Natural	Various	Uetz 1986
Micrathena gracilis	17%	Natural	Various	Uetz and Hartsock 1987
Nephilidae				
Nephila pilipes (=*maculata*)	10%	Flown	Blowflies	Lubin 1973

[1] Percent that escaped in less than 1 s from control webs that lacked spiders; presumably this underestimates the rate of escapes from spiders.
[2] 18 and 35% of prey that were retained for at least an instant in the vertical webs of adult and penultimate instar spiderling took <5 s to escape; the median time taken to reach prey that flew naturally into webs was 5.5 s. These percentages surely underestimate escape rates: the flies were relatively small (on the order of 2–4 mg) compared with the spiders (adult females were estimated to weigh on the order of 20–30 mg); and those flies that brushed against the web without being detained at least momentarily were not counted.

cludes selection to perform other functions in addition to simply intercepting prey. These functions are described in the following sections. Because a trait that is likely to improve one web function will often reduce the web's ability to perform others (table 3.1), perhaps the most important general conclusion is that different orb designs probably often have different trade-offs between costs and benefits. These trade-offs make it challenging to understand the selective forces responsible for the evolution of different orb designs (see appendix O3.2).

3.3.1 INTERCEPTING PREY

To capture a prey, the first task for an orb is to intercept it with sticky silk in the "capture zone" (fig. 1.4). Interception is affected by several different orb web design properties (e.g., table 3.1). Covering a greater total area with sticky lines ("prey capture area") is the first, most obvious way to intercept more prey. Given that a spider has only a finite amount of sticky silk available, she can increase the prey capture area by increasing the spaces between sticky spiral lines, or by making the lines thinner or less sticky. Increasing the prey capture area by increasing the spaces between lines will increase interceptions until the distance between sticky lines is equal to the diameter of the largest prey; from this point onward, interception will remain constant as web size increases (Chacón and Eberhard 1980; Lubin and Dorugl 1982). An orb designed especially for interception of prey that are easy to stop and retain would have very widely-spaced sticky lines: possible examples include *Alpaida tuonabo* (AR) (Eberhard 1986a), *Metabus ocellatus* and *Tetragnatha* spp. (TET) (figs. 7.43, 9.10; Shinkai 1988b), and *Ogulnius* spp. (TSM) (fig. 10.22; Coddington 1986a). Extreme cases are *Ulesanis* spp. (THD) (Marples 1955), *Phoroncidia* spp. (THD) (Eberhard 1981b; Shinkai 1988a; WE), *Miagrammopes* spp. (UL) (Akerman 1932; Lubin et al. 1978), and some *Wendilgarda* sp. (TSM) (Eberhard 2001a), whose webs are reduced to a single sticky line (fig. 8.1).

A second important trait affecting interception concerns the geometric regularity of the pattern of lines. If too many lines are concentrated in a small sector, coverage of other sectors will be reduced, and these lines will be "wasted" for intercepting prey. Vollrath (1992) illustrated this idea by projecting photos of different web designs onto an array of random points (a map of stars in the sky). He found a surprisingly strong effect: 87% of the stars rested on a web line within the capture zone in a normal orb web design, but only 61% (nearly a third less) when the web had the same total length of lines but the lines were arranged "irregularly."

A second obvious factor contributing to interception is an orb's durability in the face of environmental stresses such as wind. Greater durability increases the likelihood that the web lines will still be in place when a prey arrives. Mechanical resistance to general stresses like wind will depend on several orb design traits, including the diameters, lengths, and chemical compositions of lines, and tensions (section 3.2.11.3.2). When wind direction changes erratically, lower tensions and longer lines are disadvantageous because lines will snap tight suddenly each time the web flaps, and longer, looser lines billow and snap more (Langer 1969). An additional variable likely to be important is the degree to which lines are reinforced.

The likelihood of interception will also be affected by where the orb is built. Some aspects of an orb's design will affect where it can be located. For instance, longer anchor lines will increase the distance of the orb from surrounding objects (fig. 3.22), altering its chances of capturing insects that are flying in different areas and perhaps also at different velocities. In addition, if certain flight directions of the prey are more common than others, orienting the plane of an orb so that it is more nearly perpendicular to prey flight paths will increase the number of prey intercepted. One likely non-random aspect of insect flight paths is that they are often, though apparently not always, more nearly horizontal than vertical (Chacón and Eberhard 1980; section 3.2.4). Other factors can also influence interception. Opell et al. (2006), for instance, mentioned web visibility (section 3.3.3.2), attractiveness to insects, spider visibility and attractiveness, and the presence of stabilimenta (see section 3.3.4.2).

The most important point is that prey interception is a complex function, and that it can be affected by various orb traits. No single design will be best for intercepting all prey under all circumstances. Possible patterns in the interception payoffs of different designs and their costs are described in section 4.2.1, and limitations of these analyses are discussed in appendix O3.2.

3.3.2 FUNCTIONS FOR NON-STICKY LINES (RADII, HUB, AND FRAME LINES)

3.3.2.1 Stop prey

To stop a prey that it has intercepted, an orb must absorb the prey's kinetic energy (its momentum) without breaking. The momentum (E) of an insect with mass m that is travelling at velocity v is $E = mv^2/2$. This equation emphasizes the importance for web designs of the prey's velocity, which is raised to the second power (see section 4.3.3.1). Two important properties of silk, the toughness (= the energy of mechanical deformation per unit volume prior to rupture) of dragline silk, and the adhesiveness of glue droplets of araneoids, both have especially high values when strains are applied rapidly (Blackledge et al. 2011), thus making them especially good at absorbing the rapid stresses associated with prey impact.

In stopping a prey of mass m, a web must apply a force equal to ma, where a is the rate of deceleration. The major ampullate lines in the radii, frame lines, and anchor lines of an araneoid are on the order of 1000 times stiffer than the sticky spiral lines, and they are generally thought to be of special importance in stopping prey (Denny 1976; Craig et al. 1985; Sensenig et al. 2010a, 2012; Blackledge et al. 2011) (contra Qin et al. 2015). As expected, orbs were

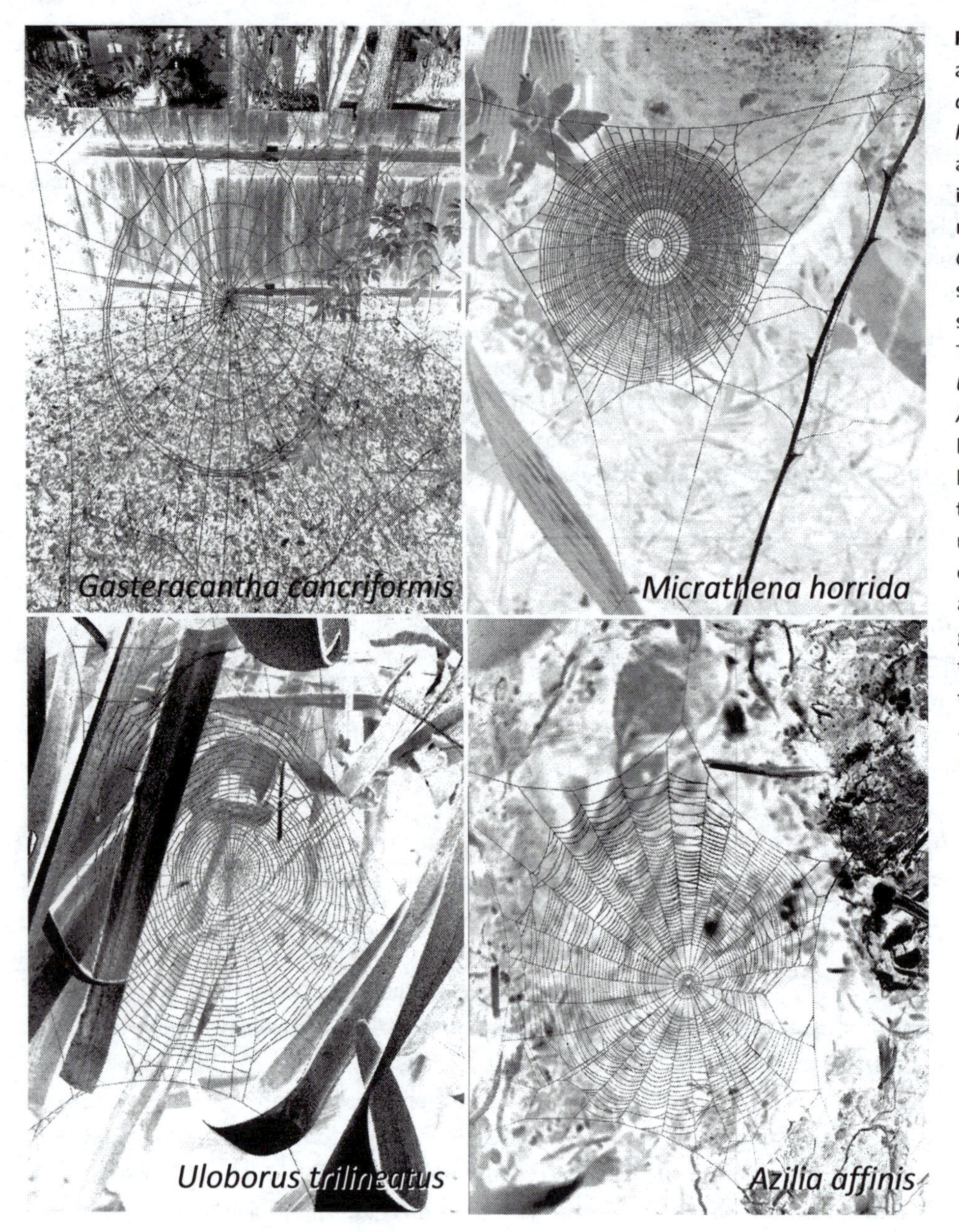

Fig. 3.22. The orbs of two araneid species, *Gasteracantha cancriformis* and *Micrathena horrida*, have relatively long anchor lines and presumably intercept prey flying relatively rapidly across open spaces (the *G. cancriformis* was interrupted soon after she had begun sticky spiral construction). They contrast with the orbs of *Uloborus trilineatus* (UL) and *Azilia* sp. (AR), both of which have short anchors and are built in confined spaces, where they presumably intercept more slowly flying (and at least occasionally falling) prey. Long-anchor webs may tend to show greater symmetry (a result of the spider not having to adjust the web's form to the available attachment sites?), but this has never been quantified.

able to absorb more energy when a 300 mg simulated prey struck more radii (Sensenig et al. 2012). The low initial resilience of radial lines ensures that they dissipate energy (Denny 1976). The shorter the distance in which deceleration occurs (the shorter the line or the greater its original tension), the greater the force that must be exerted to stop the prey (and thus the greater the likelihood of rupture). An idea of how the diameters of lines and the connections between them in an orb influence stopping was obtained from the simple experiment of dropping small weights from equal heights onto single horizontal radial and sticky spiral lines in intact orbs of *Leucauge mariana* (TET) and *Allocyclosa bifurca* (AR): the maximum weight that a single radius could stop was on the order of 10x that which could be stopped by a single sticky spiral line in the same orb (Eberhard 1986a). Although impact stresses were distributed in variable amounts and patterns to other web lines (Sensenig et al. 2012), the lines that failed when an object broke through the web were almost always only those in the immediate vicinity of the impact.

Much of a prey's momentum is converted to heat

(>50%—Blackledge et al. 2011; Sensenig et al. 2012) when the molecules in the silk lines are deformed. These deformations of the dragline silk in radii mean that a radius does not recover its initial length immediately after being extended, but only gradually returns to its original length after absorbing an impact. This visco-elastic property prevents the web from having a trampoline-like effect on the prey; the lines sag briefly, rather than rebounding immediately and launching the prey back in the direction from which it came (Denny 1976). Friction with the air may also reduce the rebound of an orb (Zaera et al. 2014), further reducing the trampoline effect. The elastic modulus combined with the extensibility of ampullate silk of the araneid *Araneus sericatus* (1.25) was close to the optimum (1.40) for resisting a force applied perpendicular to a line (Denny 1976).

The biomechanical properties of the orb webs of uloborids differ from those of araneoids principally in that the sticky spiral axial lines are much stiffer, and may thus play a greater role in stopping prey (Opell and Bond 2001). The sticky spiral lines also lack the sliding connections with the radii that occur in the webs of some araneoids (Eberhard 1976b; section 3.3.3.1.6), reducing their ability to cede without breaking under sudden loads.

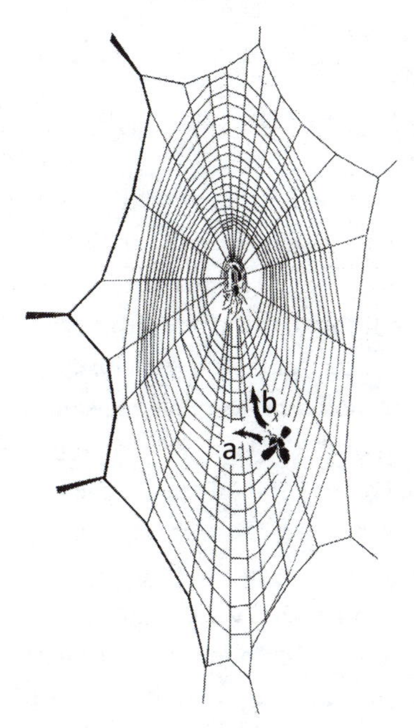

Fig. 3.23. The drawing illustrates how an insect flying toward an orb web at an acute angle with the web plane (*b*) has a greater likelihood of encountering multiple lines (and thus being stopped) than if it strikes the orb while moving perpendicular to its plane (*a*). Although glancing blows are probably more common than perpendicular impacts in nature, most studies of prey interactions with orbs have focused on perpendicular impacts.

At the scale of silk lines, with their small weights and diameters (a few micrometers or less—Blackledge et al. 2005a; Sensenig et al. 2010a, 2012), air is relatively viscous. Much as tiny insects are best imagined as "swimming" rather than flying, spider silk lines may be better imagined in human terms as thin ropes immersed in water. One possible consequence revealed by two modeling studies is that the friction of both radii and sticky spiral lines with the air may play an important role in stopping prey (Lin et al. 1995; Zaera et al. 2014). Air friction would be especially great with the abundant sticky spiral lines. The relative importance of web friction with the air in stopping prey has been challenged, however, by other simulations and by calculations based on high-speed video recordings of objects colliding with real orbs (Sensenig et al. 2010a, 2012). Under the most generous estimates of aereodynamic drag in these studies, air friction was responsible for less than about 10% of the total work performed in stopping the prey. The sticky spiral lines were so easily extensible that they dissipated little of the impact energy. In sum, both factors contribute to stopping prey, but, at least for medium to large spiders (Blackledge 2013), stopping appears to be mostly due to energy absorption by the radii (up to 98% of the work performed) rather than to energy absorption by the sticky spiral lines or by air friction (Blackledge et al. 2011; Sensenig et al. 2012). Aerodynamic drag probably plays a larger role in stopping prey in the orbs of smaller species and younger individuals (Sensenig et al. 2012).

The angle that the prey's path makes with the plane of the orb represents a further complication. In general, a path that is less nearly perpendicular to the orb's plane will bring the prey into contact with more web lines, both sticky and non-sticky. This effect will be especially pronounced for radii (and thus for stopping) when the prey's path is more nearly perpendicular to the radii in its immediate vicinity (e.g., path *a* rather than *b* in fig. 3.23). One aspect of orb design that the spider can easily alter, and which may often change the impact angles of prey, is the web's slant with respect to gravity (see section 4.9.4).

3.3.2.2 Transmit vibration cues for arousal and orientation

3.3.2.2.1 LONGITUDINAL VIBRATIONS AND THEIR AMPLITUDES

Using silk lines to expand the area within which the spider can sense prey vibrations is undoubtedly ancient; it has arisen (probably independently) in such diverse spider families as liphistiids (fig. 10.5; Bristowe 1930), segestriids (Griswold et al. 2005), hersiliids (Williams 1928; fig. 9.1), oecobiids (Kullmann and Stern 1981; Solano-Brenes et al. 2018; fig. 10.5) and thomisids (Jackson et al. 1995). A flexible line can vibrate both perpendicular to the line (transverse vibrations) and along the axis of the line (longitudinal vibrations). The lines are so fine that friction with the air is expected to rapidly damp transverse vibrations, and empirical measurements of vibrations and the responses of spiders to different types of vibration in several araneoid species have confirmed that longitudinal vibrations are probably most important for spiders; they attenuated less and gave more directional information (Liesenfeld 1956; Suter 1978; Masters and Markl 1981; Klärner and Barth 1982; Landolfa and Barth 1996; Mortimer et al. 2014). Longitudinal vibrations are transmitted more effectively by lines under greater tension and are less extensible. Even the ratio of MaSp1 to MaSp2 in the major ampullate silk can influence vibration transmission (Blackledge 2013). In araneoid orbs the radii are both under greater tensions and made of a less extensible type of silk than are the sticky spiral lines, and radii are thought to be much more important in vibration transmission (Liesenfeld 1956; Suter 1978; Landolfa and Barth 1996) (fig. 3.24).

Web-building spiders depend on vibration signals to accomplish at least four functions in prey capture: alert the spider that a prey has arrived in the web; orient the spider toward the prey; provide information about the prey's size and perhaps its identity; and convey vibrations from the spider (jerks on the web) to the prey that elicit further informative vibrations from the prey and that may further entangle it in the web. Web traits that could affect these different types of vibration transmission overlap widely, but in some cases they can be at least partially separated.

The amplitudes of vibrations received by spiders appear to be especially important for the alerting function. Many species rest off of the web or at other hidden locations. Special lines often connect such off-of-the-web resting sites directly to the hub, and these signal lines often run through open sectors or out of the plane of the orb (fig. 3.24, table 9.1), thus reducing the damping effects of air friction with sticky lines. The spider usually holds the signal line with one tarsus while wait-

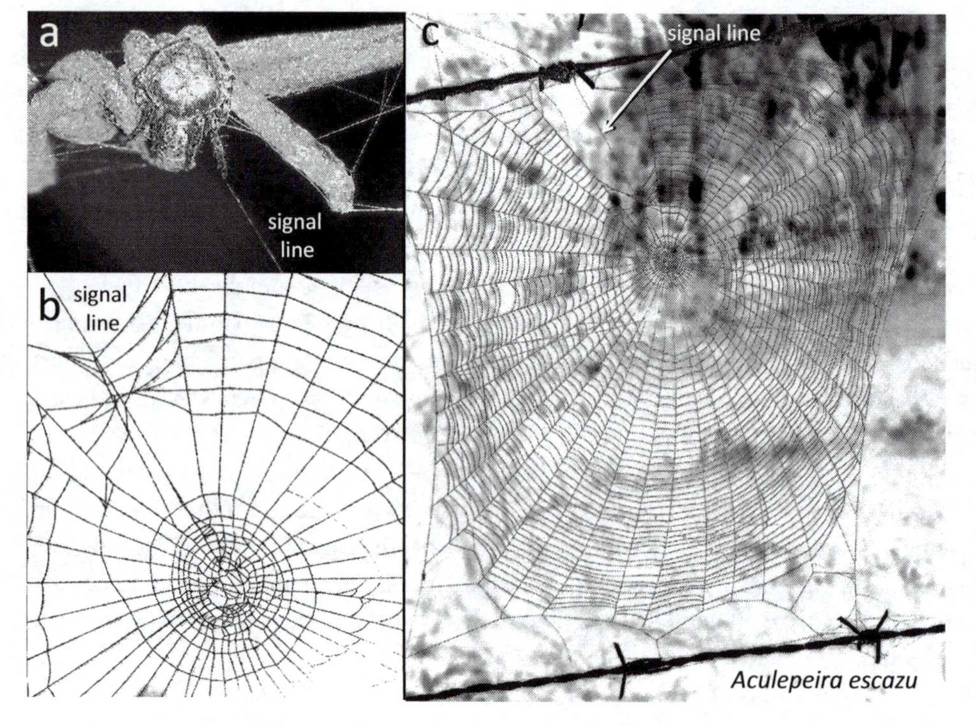

Fig. 3.24. *Aculepeira escazu* (AR) and her web illustrate tradeoffs between the increased security from predation that results from the reduced visual conspicuousness of the spider when she rests beyond the edge of the orb (*a*), and reductions in the web's ability to capture prey (open sector around the signal line in *c*) and in the speed with which she can attack. Her tarsi hold the signal line to the hub (*b*, *c*), which transmits vibrations of lower amplitude to her than she would have received at the hub; she must also traverse the signal line before reaching the hub and then must orient toward prey.

ing in the retreat to receive these vibrations (fig. 3.24). These anti-attenuation devices are important because the amplitude of vibrations is probably crucial in these species. Longitudinal vibrations transmitted through the hub from a radius that is more or less perpendicular to the signal line will produce much smaller longitudinal displacements in the signal line than will longitudinal vibrations in a radius which is more nearly parallel to the signal line. This emphasizes the importance of web design traits that reduce attenuation at other stages of transmission. In *Z. x-notata* the smallest displacements of the tarsus that elicited attacks were 10^{-2} to 10^{-3} mm (Liesenfeld 1956). The frequent association of open sectors that lack sticky lines with these signal lines could improve transmission (fig. 3.24; Landolfa and Barth 1996); they may also allow spiders to execute more rapid attacks by providing a clear pathway to the hub, but I know of no data on this point.

Experimental tests also indicated that vibration amplitudes are important for the alerting function. *Araneus pinguis*, which monitors its orb from a retreat via a signal line, was more likely to attack intermediate-sized prey when they were closer to the hub (and thus presumably produced stronger vibrations on the signal line) (Endo 1988). Araneids in the genus *Cyclosa* responded more rapidly to prey when vibration amplitudes were greater (Suter 1978), and when radii had been experimentally tensed (thus improving vibration transmission) (Nakata 2009). Mature female *L. mariana* (TET) in the field responded more rapidly during the day to small 1.2 mg weights dropped onto radii than weights dropped onto single sticky spiral lines ($chi^2 = 9.3$, $p = 0.0092$, $N = 51, 48$) (WE). Spiders also preferred to hold tense rather than artificially slackened radial lines while resting at the hub (Briceño and Eberhard 2011; WE), suggesting (less directly) the importance of improved transmission. At the level of nerve physiology, the rapidity of the nervous response in *Araneus diadematus* legs was greater when vibration amplitude was higher (Finck 1972).

There may be a previously un-remarked inverse relation in orb designs between signal lines and hub holes that is also related to attenuation of vibrations. Combining a wide hub hole with a signal line to a retreat off the orb may be less functional, because the hole would reduce transmission of vibrations from many radii to the signal line. Among the species listed in table 9.1 that build signal lines to retreats off the web (categories #6 and 7 of orbs: 11 genera with open sectors, and 5 lacking them), the hub hole is either absent (as in *Lubinella*), or is immediately filled in nearly all. The only exceptions are *Chrysometa* (Eberhard 1986a), *Dolichognatha pentagona* (Eberhard 1986a), both of which have small hub holes, and perhaps "*Meta*" (Hingston 1932) (genus identification is uncertain: Hingston's stylized figure is very different from the small, sparse webs of *Meta* spp. illustrated by Wiehle 1927 and Shinkai and Takano 1984). Filling in the hole in the center of the hub almost certainly improves vibration transmission to a spider in a retreat at the end of a signal line. To the best of my knowledge, signal lines are lacking genera with relatively large open hub holes such as *Pronous* (WE), *Micrathena* (fig. 1.4), *Tetragnatha* (Blackledge and Gillespie 2004), and *Metabus* (figs. 2.19, 7.43) (not counting the trail line left when the spider flees from the hub when disturbed). Wiehle (1927) noted that three species of *Meta* have open hubs, but that in *M. merianae* the hub hole is relatively small, and that this species also differed in having more pronounced beginnings of a signal line; there are similar lines with small hub holes in the araneids *Metazygia wittfeldae* (fig. 3.16), *Meta* sp. (fig. 3.11h), and *Eustala illicita* (T. Hesselberg pers. comm.). This explanation is only speculation at the moment, and the numbers of evolutionary derivations are unknown.

Several design properties of orbs, such as greater numbers of radii and lower numbers of sticky spiral lines, could contribute toward increasing the amplitudes of vibration at the hub. Greater tensions on radial, signal, and hub lines would also improve transmission. Nevertheless, the very common behavior of removing the center of the hub and thus reducing the tensions on some radial and hub lines after the orb is otherwise complete (section 6.3.6) argues that the disadvantage of reduced transmission effectiveness may generally be over-balanced by the advantage that a slacker web has in stopping high-energy prey (table 3.1).

3.3.2.2.2 PRECISION OF ORIENTATION AND THE IMPORTANCE (?) OF RADIAL ORGANIZATION

The vibrations of radii received by the spider at the hub can inform her of a prey's location in her orb. One design property, the number of radii, will clearly correlate positively with the amount of information available to guide orientation. Some orb weavers turn at the hub extremely accurately in response to the stimuli from prey impacts. The mean errors in orienting toward prey were only 7.0 ± 8.2° in *Nephila clavipes* (NE) and 3.6 ± 7.7° in *Zygiella x-notata* (AR) (Landolfa and Barth 1996). The especially rapid turns of *Leucauge mariana* (TET) (in 0.1–0.2 s) were also accurate (43 [89.6%] of 48 responses to impacts behind the spider on webs in the field were toward the correct radius) (Briceño and Eberhard 2011). The cues used to execute these orientations are presumably differences in the amplitudes of the vibrations received by different legs. Another possible orientation cue, differences in the times at which vibrations arrive at different legs, is unlikely. The high velocities of propagation of web vibrations would require that the spider's temporal resolution be very high (Landolfa and Barth 1996)—substantially higher than the maximum resolution of 2–4 ms observed in the non-web ctenid *Cupiennius salei* (Hergenröder and Barth 1983; Barth 1993).

The importance of radial orientations of lines for facilitating orientation to prey is uncertain. Many spiders with dense sheet webs that have little if any radial organization nevertheless run quickly and accurately to prey on the sheet (Baltzer 1930; Holzapfel 1934; Görner 1988; Barrantes and Weng 2006a; Barrantes and Eberhard 2007; Eberhard and Hazzi 2017). Even mygalomorphs, such as *Antrodiaetus unicolor* with their legs spread apart gripping inside the soft collar of detritus and silk at the tunnel entrance, or *Ummidia* spp. lurking under a trapdoor that is barely open (about 1 mm), strike accurately at passing prey (Buchli 1969; Coyle 1986). Another detail that suggests that it is relatively easy for spiders to orient precisely to prey is that wind had no effect on the time needed by *Araneus diadematus* (AR) to orient toward prey that struck their webs (Turner et al. 2011), even though it presumably generated vibrations that obscured prey vibrations. Both Szlep (1965) and Lamoral (1968) noted some orientation mistakes, however, by *Latrodectus* spp. (THD), and the linyphiid *Neriene coosa* sometimes had to correct its path one or more times while running toward a *Drosophila* fly that had landed 5–8 body lengths away on its sheet (WE). Szlep (1965) suggested that *L. pallidus*, whose "sheet" has a stronger radial organization, made fewer errors than *L. revivensis*, but provided no quantitative evidence. Perhaps the more open meshes of the *Latrodectus* sheets make precise orientation more difficult.

3.3.2.2.3 TYPES OF PREY VIBRATION

The vibrations from a prey typically begin with the sudden, high amplitude jerk of its impact, followed by the vibrations caused by its struggles to free itself and, in some cases, by resonant swinging movements whose frequency and amplitude are related to the prey's weight. Struggling movements such as wing buzzing and rubbing and kicking with the legs vary among prey species (Liesenfeld 1956; Suter 1978). Suter found that the vibrations produced by five species of prey of *Cyclosa turbinata* (AR) (two flies, a bee, a leafhopper, and a moth) also varied considerably within each species, and overlapped broadly between species; he concluded that prey discrimination via vibrations by this spider was unlikely. Responses by *Z. x-notata* to web vibrations did not depend on the frequencies of the vibrations (Liesenfeld 1956). There is, however, other behavioral evidence that some orb weavers can discriminate the size and to some extent the identity of a prey on the basis of the vibrations that they receive from a distance (Peters 1931; Robinson 1969b; Robinson et al. 1969; Robinson and Mirick 1971; Robinson and Robinson 1976a; Suter 1978). It is unlikely that different web designs are "tuned" to transmit the vibrations of certain preferred prey more effectively (Suter 1978; Landolfa and Barth 1996).

Some prey waited motionless in the web for up to several minutes before beginning to struggle, thus separating the jerk of impact from the other vibrations; this separation reduced the chances of a quick attack by some spiders. In the araneid *Z. x-notata*, which rests in a retreat that is connected to the hub by a signal line, standardized web vibrations were more likely to elicit attacks when they began suddenly rather than gradually (Liesenfeld 1956). In contrast, one moth, whose deciduous scales enabled it to escape rapidly, always struggled immediately after impact (Suter 1978). Some combinations, such as wing buzz plus body swing, elicited more attacks (Liesenfeld 1956; see also Blackledge and Zevenbergen 2006). Different stages of spider attack behavior, such as turn at the hub vs. run to the prey, may differ in their responsiveness to different types of prey vibration (section 3.3.2.6). The possibility that prey responses to being tangled in spider webs have evolved to reduce the effectiveness of spider attack behavior merits further study (section 4.11).

In sum, transmission with minimum attenuation of longitudinal vibrations along the radii from the prey to the spider is an important function of an orb. Several aspects of orb designs, including radius tension, numbers of other lines attached to the radius, special signal lines to retreats, and gaps between lines such as the hub hole can all affect attenuation. Vibrations resulting from impact with the web differ from those produced by prey struggles, and different combinations had different effects in eliciting spider attacks. Prey struggling behavior merits further study.

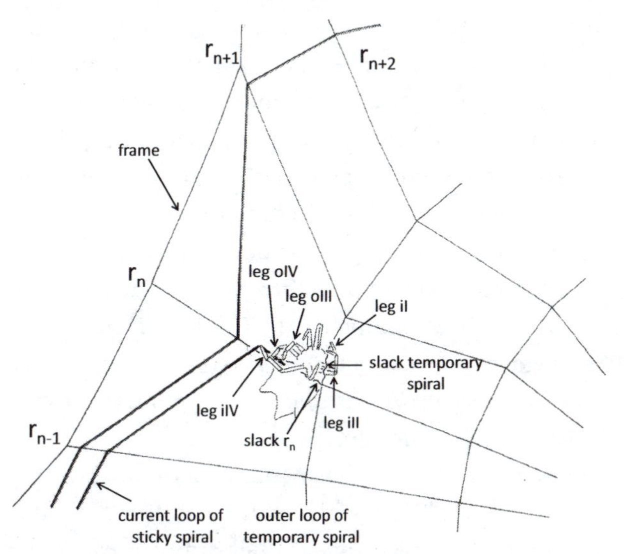

Fig. 3.25. Both the weight of this short-legged, heavy-bodied adult female *Gasteracantha cancriformis* and the strong pulls she made on lines with her legs altered the tensions and positions of lines in her near vicinity when she was just about to attach the sticky spiral (heavier line) to radius r_n (the web was slanted approximately 60° with horizontal in this schematic drawing from a video recording). The spider caused portions of both r_n and the temporary spiral to go slack as she held the outer loop of temporary spiral with her legs iI and iII after having reeled in radius r_n (note the inward displacement of the previous sticky spiral loop on r_n). She grasped r_n with legs oIII and oIV just before making the attachment, and the radius was tight between them.

3.3.2.3 Support the spider and facilitate her movements

Providing physical support for the spider is important in several contexts. In the first place, sustaining the weight of the spider at the hub applies stress to lines above the hub, making them less able to absorb prey impacts; but it lowers the tensions on radii below the hub, increasing their ability to stop prey. This effect of vertical asymmetry, perhaps combined with more rapid attacks in the lower sector because the spider is aided rather than hindered by gravity as she moves toward prey, may explain the general trend in orb webs for lower sectors of the web to be larger (section 4.8). Small distances between lines at the hub facilitate rapid leg movements as the spider turns at the hub to attack prey (section 3.3.2.6). On the other hand, the general trend for there to be fewer and shorter radii above than below the hub (section 4.8, Peters 1936, 1937; Reed 1969) argues that supporting the spider's weight at the hub has only minor influence on orb design.

The radii facilitate access to different parts of the capture zone during attacks on prey. The non-sticky radii are relatively rigid compared with the sticky lines, and probably much easier for the spider to walk along; many authors have noted that orb weavers consistently walk along radii rather than along sticky lines (e.g., Vollrath and Tillinghast 1991). By adding more radii to an orb, a spider will increase her ability to arrive rapidly within attack distance.

Supporting the spider's weight during construction is also important. The spider needs to avoid distorting the orb so much that it becomes difficult for her to make precise measurements of angles and distances and thus attach lines at appropriate sites (fig. 3.25). Distortion is likely to be more important in vertical than in horizontal orbs (because the spider's weight causes her to sag into contact with other web lines) and in webs or portions of webs with more sparse non-sticky lines. Distortion reduction may be an important function of the temporary spiral (section 3.3.2.8).

3.3.2.4 Primary frame lines: adapt to variable spaces, increase extensibility, and avoid resonant vibrations (?)

3.3.2.4.1 THEORETICAL EXPECTATIONS OF BENEFITS AND COSTS

An orb weaver is faced with several options with respect to frame lines. One extreme is no primary frames at all, as in *Tetragnatha lauta* (TET) (Shinkai 1988b) and *T.* sp. (fig. 9.10). Frame lines were nearly absent in the webs of

Meta menardi (TET) in caves (mean 1.1 ± 0.6 per web), with the sparse set of radii (17.0 ± 4.3) nearly all attached directly to the substrate (Yoshida and Shinkai 1993). A second design is low numbers of primary frame lines each supporting many radii or secondary frames. This "long frames-few anchors" design occurs, for example, in the araneids *M. duodecimspinosa* (which usually has the geometric minimum of three primary frames) (fig. 1.4), *Aspidolasius branicki* (AR) (which lacks secondary frames) (Calixto and Levi 2006), *Hypognatha* sp. (Eberhard 1986a), *Gasteracantha cancriformis* (Eberhard 1986a), and *Acacesia* spp. (fig. 3.5); all of these webs are typically built in large open spaces (I will distinguish "primary frame" from "anchor" lines in discussing design properties, even though they are physically part of the same line). A third extreme option is high numbers of primary frame lines, each supporting only one or a few radii. Examples of this "short frames-many anchors" option occur in several families, including *Uloborus* sp. #241 (UL) (Eberhard 1986a), nephilids (NE) (fig. 4.4; Witt et al. 1968; Kuntner et al. 2008a,b), *Anapisona simoni* (AN) (fig. 7.9; Eberhard 2007b), and *Glenognatha* sp. (TET) (Eberhard 1986a). These webs are typically built in more constrained spaces, though I know of no systematic measures to confirm this. The extreme asterisk webs of *Ocrepeira* (= *Wixia*) *ectypa* (AR) (fig. 9.2) had a small frame line for each radius (in this case the frame lines may have the additional function of intercepting prey walking on the substrate) (Stowe 1978).

There are several possible reasons for this variation. A spider can gain at least four advantages from having frame lines to which multiple radii are attached. She can increase silk economy, because many radii extend only to a frame line, rather than all the way to the substrate. She can make radius construction quicker, because she does not need to search along the substrate for an appropriate object to which she can attach each radius (surely a difficult task in many natural situations). It is also possible to build a more standardized design, even in sites that have irregularly positioned (or missing) supports, because frame lines provide attachment sites. This standardization may also reduce the impact of idiosyncratic features of websites on the cues used to direct construction behavior (e.g., Witt 1965 and references). Finally, she increases the ability of her web to absorb the impacts of prey without breaking, because frame and anchor lines effectively increase the length of the set of lines that absorb the impact, and thus increase their overall extensibility.

One more difficult-to-evaluate factor is the strength of the web's mechanical coupling to supporting structures. In general, additional anchors should make the web as a whole less likely to be ripped free due to general stress such as wind. But they will make individual portions of the web less extensible, and thus less able to absorb local stress of prey impact. My impression from web damage in the field and watching orbs in the field in the wind is that resisting local stress is more important but this is not certain. The advantages of mechanical coupling are related to mechanical isolation of the web from possible movements of supporting structures (e.g., leaves or blades of grass that move in the wind); isolation of this sort is likely to be more important where the supports are less rigid or there is more wind (section 3.2.11.2). Lack of data on these variables makes it difficult to discuss these important topics further, other than to note that some species seem to prefer more rigid supports (section 8.4.3.2.2).

A final possible advantage of attaching radii to frame lines rather than directly to the substrate is that attachments to frames reduce possible resonant vibrations of lines in the web that might complicate detection and localization of prey using web vibrations (Landolfa and Barth 1996). The amount of resonance depends on the impedance mismatch between a line and its end supports: the greater the mismatch, the stronger the expected resonant vibrations. The ends of the radii attached directly to a rigid substrate would have high mismatches. In contrast, radii in an orb that are attached to frame lines are generally not rigidly fixed in space; they will thus have relatively low impedance mismatches, and reduced resonant vibrations (Landolfa and Barth 1996). One prediction of this idea, that attack behavior will be less precise or rapid in species or portions of the orb where radii are commonly attached directly to the substrate, has not to my knowledge been tested.

Balanced against the advantages of primary frames is a clear "space utilization" disadvantage: frame and anchor lines reduce the fraction of the available space in which the web is built that can be covered with sticky lines. For spiders building in large open spaces, this is probably not a problem (fig. 1.4), but it is probably im-

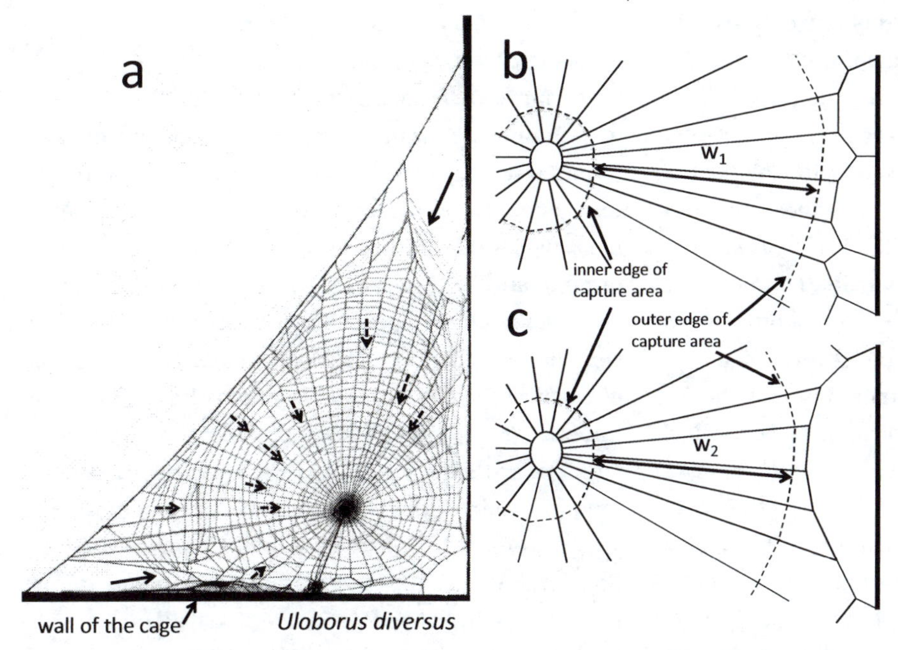

Fig. 3.26. (*a*) This web of *Uloborus diversus* (UL) in the corner of a rectangular cage illustrates the correlation between the relative distance of the hub from the nearest supports and the lengths of frame lines and the numbers of anchors. On the two sides where the hub is nearest the walls, the frames are short and there are many anchor lines; on the other two sides the frames are long and there are few anchors. The drawings in *b* and *c* illustrate a possible selective advantage of the "short frames–many anchors" design when space is limited. When there are numerous, short frames (*b*) the width of the capture zone is greater than when there is a single, long frame (*c*) ($w_1 > w_2$). In effect some of the space where sticky silk could be deployed is "wasted" when frame lines are longer (*c*). A second "space-saving" technique used by *U. diversus* is laying sticky lines beyond the frame lines (solid arrows in *a*) (intact segments of temporary spiral are indicated by dashed arrows).

portant for species that build in small spaces (see next section, figs. 3.26, 10.25).

3.3.2.4.2 TESTS OF PREDICTIONS

The sizes of these advantages and disadvantages undoubtedly vary according to physical characteristics of the environment in which the web is built, such as the sizes of available spaces, the rigidity of supports, and the relative importance of stopping high-energy prey; some observed variations match some expectations. Differences between the frame line designs of different species seem to be associated to some extent with the relative sizes of the spaces used as websites. Thus, for instance, some species of *Cyclosa* (AR) tend to have relatively large numbers of anchors and build in relatively small spaces (fig. 3.34, Peters 1955). Similarly, in 15 webs of mature females of the closely related *Allocyclosa bifurca*, which also often builds in somewhat restricted spaces such as between leaves of bromeliads (Eberhard 2003), the mean number of anchors was 8.3 ± 0.9 (WE). In contrast, the webs of *Micrathena duodecimspinosa* (AR), which were always built in relatively large open spaces (the mean distance between the points where adjacent anchors were attached to the substrate was 81 ± 31 cm), averaged less than half as many anchor lines (3.6 ± 0.8) (WE). The *M. duodecimspinosa* webs, which spanned larger spaces and may intercept prey moving faster, and which might thus seem to need more rather than less mooring, had close to the geometric minimum of three anchors.

In accord with the space utilization hypothesis, strong reductions in the number of radii/frame line were elicited experimentally by confining *Leucauge argyra* (TET) and *Zosis geniculata* (UL) (fig. 7.42) in small spaces (Barrantes and Eberhard 2012; Eberhard and Barrantes 2015). A possibly similar pattern occurred when *Eustala illicita* (AR) was obliged to build in relatively small spaces (Hesselberg 2013; T. Hesselberg pers. comm.).

A second intra-specific test, comparing different sectors of the same orbs, also seems to support the space limitation hypothesis. Many orbs in which the hub is located asymmetrically have a mix of two designs, with shorter frame lines and more anchor lines on the side where the hub is closer to supporting objects (usually above the hub in vertical orbs—see section 4.8). Examples include *Araneus diadematus* (AR) (Witt et al. 1968), *Zygiella x-notata* (AR) (Witt et al. 1968), the horizontal orbs of *Dolichognatha* spp. (TET) (fig. 3.43a, Eberhard 1986a), the reduced web araneids *Pasilobus* and *Paraplectana* (fig. 9.23), and the horizontal orbs of *Uloborus diversus* (UL) built across the corners of rectangular frames (with many short frame lines on the two sides nearest the walls and a single long frame to which most of the radii on the other two sides are attached) (fig. 3.26).

This predicted trend seems clear, though I know of no systematic comparisons. One possible weakness of the space conservation explanation of anchor lines is that some orbs have shorter frame lines on one side even though space per se is not obviously limiting. Another, non-exclusive alternative explanation is that there are more prey closer to the substrate, or that the spider's greater speed of attacks on prey near the hub make it advantageous to place more sticky lines there.

Another apparently consistent (but not systematically documented) pattern is for radii attached directly to the substrate to be relatively short, as in *Meta menardi* (TET) (Yoshida and Shinkai 1993), *Cyrtognatha* sp. #537 (TET) (Eberhard 1986a), *Leucauge argyra* (TET) (Barrantes and Eberhard 2012), *Alpaida tuonabo* (AR) (Eberhard 1986a), *Anapisona simoni* (AN) (Eberhard 2007b; WE), and *Ogulnius* spp. (TSM) (Coddington 1986a; WE) (fig. 10.22). At least some of these species, such as *M. menardi* build in deep sheltered sites, and *A. simoni* occurs in the interstices of the leaf litter and tends to build in small, constrained spaces. These comparisons are again only impressions, and remain to be quantified more precisely.

The unfortunate custom of arachnologists has been to publish web photos in which most of the anchor lines (and even sometimes parts of the frames) are trimmed off (to illustrate other details). This tradition (continued in many photos in this book) makes it difficult to gather data on anchor lines from previous studies. There has also been little published in the way of analyses of anchor and frame designs. Improved, more quantitative tests are needed.

3.3.2.5 Secondary and tertiary frame lines: increase extensibility

Secondary and tertiary frame lines are built "inside" primary frame lines, usually across the corner between two primary frames (fig. 1.4). A survey of published web photos (generally one/species) revealed secondary frames in 17 species in 4 families (Soler and Zaera 2016). I know of no quantitative survey study, but the frequencies of secondary frames apparently vary substantially. They were ubiquitous in the orbs of *M. duodecimspinosa*; all of >100 webs had at least one secondary frame (most had three secondary frames and three primary frames, as in fig. 1.4). Secondary frame lines were also relatively common (4.1/web) in the lower (larger) halves of *Metepeira* sp. (AR) orbs, which had a mean of 5.5 primary frames (fig. 3.7). At the opposite extreme, secondary frames were absent in all *Aspidolasius branicki* (AR) (71 webs observed) (Calixto and Levi 2006) and *Caerostris darwini* (AR) (Gregorič et al. 2011); they were rare in *Epeirotypus brevipes* (TSM) (0.67/web in 30 webs), *Uloborus diversus* (UL) (average of 1.3/web in 21 webs), and *Zosis geniculata* (UL) (0.09/web in 23 webs) (the mean numbers of primary frames in the first two species were 5.7 and 11.1/web) (WE). Strikingly, species of both extremes, *M. duodecimspinosa*, *A. branicki*, and *C. darwini*, all built webs with the same outline—usually three frames anchored by three anchor lines (Calixto and Levi 2006; Gregorič et al. 2011; WE; fig. 1.4).

The functions of secondary and tertiary frame lines are less obvious than those of primary frames, and have seldom been discussed. Gregorič et al. (2011) thought they served to lower the tensions on radii, but gave no supporting data. In the only detailed treatments that I know of, Zschokke (2000b, 2002) and Soler and Zaera (2016) emphasized the increase that these lines would have on the orb's extensibility ("flexibility" was used in the sense of extensibility—S. Zschokke pers. comm.). By adding a secondary frame line, the spider increases the effective length of line between the rigid supports and any impact site in the prey capture zone; a longer line will, all other things being equal, be more extensible and thus more able to withstand the stress of prey impact. Some radial forces will be absorbed by lines running in other directions. If, for instance, radius Y in Fig 3.27a pulls inward toward the hub when a prey strikes it, much of this force will be absorbed by changes in the lengths of the two segments of the secondary frame, and between the secondary and primary frames. In effect, these changes lengthen radius Y. Simulations of prey impacts using finite element models showed that inclusion of secondary frame lines increased an orb's ability to absorb prey impacts without breaking because they produced a more even distribution of stiffness among radii that could eventually lead to failure to stop the prey (Soler and Zaera 2016). Greater extensibility would also increase the orb's ability to adjust to the generalized stresses imposed by wind or movements of the web's supports. The lack of secondary frames noted in the previous paragraph in orbs that had only three frame lines and three long anchor lines was related to the fact that webs with this design are likely to be relatively extensible.

A second, non-exclusive possible function for secondary frame lines is to economize on dragline silk (e.g., fig. 3.27a). Measurements of secondary frame lines and the radius segments that they eliminated (dashed lines in fig. 3.27a) showed (assuming that radius and secondary frame silk lines have a similar composition) that when more radii ended on a given secondary frame in *M. duodecimspinosa* (AR), savings of this sort from secondary frames were greater (fig. 3.27b). The araneids *Allocyclosa bifurca* and *Metepeira* sp. did not show, however, consistent savings from secondary frames, no matter how many radii were attached to them (fig. 3.27b).

A third possibly beneficial consequence of a secondary frame line is that it distributes the tensions exerted by its radii to two primary frame lines rather than only one. In terms of fig. 3.27a, if radius X had been attached directly to the primary frame line 1, its tension would have to have tightened the primary frame 1 to the right of point *m*; because radius X is attached to the secondary frame, its tension is partially taken up by primary frame 2 (the effect of radius X on the anchor remains the same). The sharper the angle between a radius and the frame on which it ends, the more unequal the distribution of tensions on the frame line on either side of the radius attachment; the tendency to place secondary frames where radii would have made more acute angles with the frame (e.g., figs. 3.27, 6.10) makes sense in reducing the chances of overloading portions of the frame line. Under this tension distribution hypothesis, the unusually strong silk of *C. darwini* might explain their lack of secondary frame lines. Of course, these different functions of secondary frame lines are not exclusive, and a spider could reap more than a single type of benefit from a given secondary frame line.

Secondary frame lines were generally displaced more by the radii that were attached to them than were the primary frames in the same webs of *Micrathena duodecimspinosa* (AR) (e.g., fig. 1.4), *Leucauge mariana* (TET) (e.g., fig. 7.30), and *Allocyclosa bifurca* (AR) (e.g., figs. 3.27, 7.41) (WE); this indicates that secondary frame lines were under lower tensions than primary frame lines. The lower secondary frame tensions are in accord with the increased extensibility function mentioned by Zschokke (2002). Lower tensions on secondary frame lines also make sense with respect to the expected advantage of uniformity of stresses in web lines (section 4.4), because secondary frame lines probably tend to be thinner: behavior during secondary frame construction does not involve laying any more lines than during radius construction, while the primary frame lines of at least some species are reinforced with additional lines (section 6.3.3; Eberhard 1972a; Denny 1976).

A possible disadvantage of secondary frame lines is that they reduce the area available for the prey capture zone (see the "space utilization" hypothesis in the previous section).

3.3.2.5.1 TESTS OF HYPOTHESES

One test of the extensibility hypothesis would be to compare the designs of related species in which expected prey velocities differ. One small, preliminary comparison of this sort supports the prediction of fewer secondary frames when prey velocities are lower. *Allocyclosa bifurca* (AR) builds orbs in nature at moderately exposed sites, such as between the large leaves of bromeliad and agavaceous plants. A member of the sister genus *Cyclosa*, *C. jose*, builds its orbs just above the upper surfaces of large, more or less flat leaves, where prey probably tend to be moving slower (section 4.3.3.1). As predicted, the frequency of secondary frames in *C. jose* was lower (2.6/web in 22 webs) than that in *A. bifurca* (5.1/web in 30 webs). Much more extensive comparisons will be needed to test the prey velocity idea.

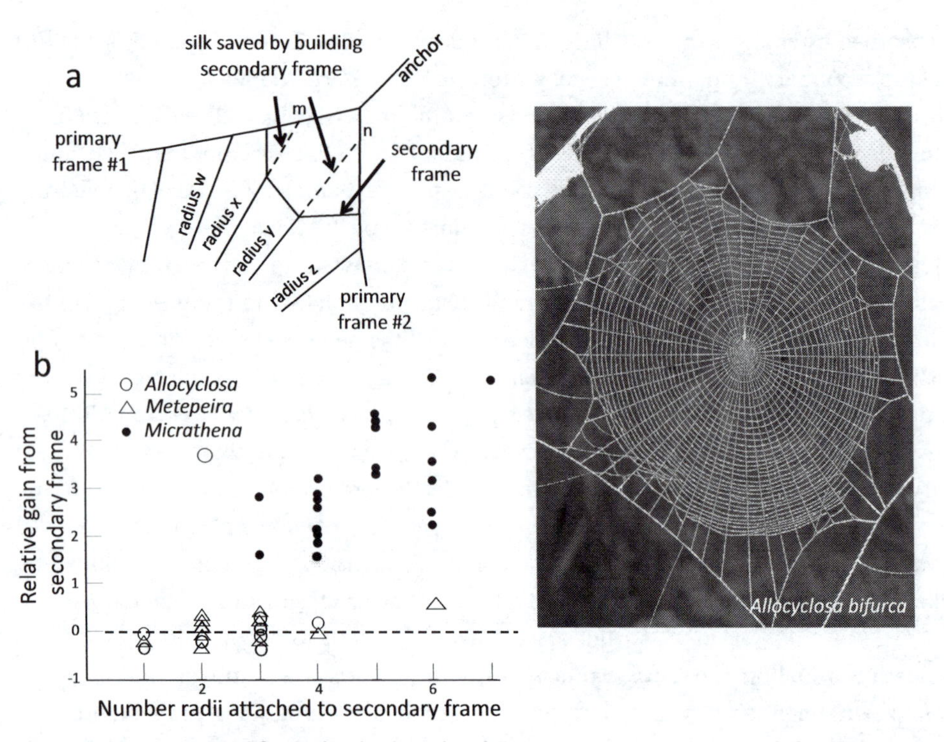

Fig. 3.27. Measurements of finished webs show that different species may experience different amounts of benefit in silk economy from secondary frame lines. The dotted lines in the drawing *a* show that by building a secondary frame line, a spider can reduce the lengths of the radii. The relative investments in radii and secondary frame lines in *b* were quantified by comparing the additional lengths of radii that would have been built if no secondary frame line had been built with the lengths of the secondary frame lines in photos of the webs of three araneid species: *Allocyclosa bifurca* (photo in *c*); *Metepeira* sp. (fig. 3.7a); and *Micrathena duodecimspinosa* (fig. 1.4) (the calculations assumed that radii and secondary frame lines were made of the same silk, which seems reasonable because during construction they seemed to be continuations of the same line—section 6.3.3). The combined relative gain for each secondary frame was quantified on the y axis as follows: $[x-y]/z$, where x = total lengths of radii not laid (dotted lines in *a*), y = length of secondary frame, and z = lengths of the radii that were attached to the secondary frame line). Secondary frame lines did not always save silk (some points were below the dotted line in *b*), though there was a tendency for secondary frame lines to which more radii were attached to save more. As discussed in the text, secondary frame lines also alter the distribution of tension on frame lines; secondary frame lines distribute radial tensions (including tension from prey impact) to larger numbers of frame lines. For instance, part of the tension on radius X in *a* that would have been exerted on the part of the primary frame #1 to the right of point m is transferred to primary frame #2 via the secondary frame. Redistribution of this sort may reduce the chances of over-stressing segments of the frame lines, especially in places where radii make acute angles with the frames. The fact that some spiders built secondary frame lines in places where radii would not have made particularly acute angles with the frame (at 2:00 and 10:30 in *c*) argues against the general importance of tension redistribution and, by elimination, in favor of the importance of increased overall web extensibility (and thus probable increases in web stability when it is subjected to a general stress).

A second test would be to look for exceptional groups that do not build secondary frames. The giant bark spider *Caerostris darwini*, which builds especially large orbs (using unusually strong dragline silk) over bodies of water such as rivers, constitutes such an exception (Gregorič et al. 2011). The extremely long frame lines in these webs (up to 25 m long) mean that the web as a whole is unusually extensible; perhaps, in agreement with the extensibility hypothesis, this makes the additional increases in extensibility that would result from secondary frame lines unnecessary.

Secondary frames are rare in the orbs of some other groups, such as the uloborids *Uloborus diversus* and *Zosis geniculata* (figs. 3.26, 7.24, 6.11). This omission is in accord with the space utilization hypothesis, at least in the better understood *U. diversus*, because orbs in nature are built at sites such as among the debris of pack rat nests and in dense vegetation where open spaces are limited (Eberhard 1973a). Both species are capable of building secondary frames, and the construction behavior per se is very similar to that of primary frames, so secondary frame omission is facultative. A second feature of *U. diversus* webs that also indicates that space is limited is that sticky lines are often laid beyond the frame lines (arrows in fig. 3.26), taking advantage of spaces beyond the frame lines. Restricted spaces are also probably associated with reduced environmental demands on webs, because they are likely to be more sheltered from wind, and because prey velocities may be lower. The balance between costs and benefits of secondary frames may thus be further tilted against their benefits, due to the lower benefits from mechanical stability at such websites.

In sum, the little-studied topic of the costs and benefits of secondary and tertiary frame lines merits further study. There are several possible benefits, and they are likely to have different relative weights in different species. Secondary frames may be omitted by some species that build in confined spaces because the cost of reduced prey capture area may outweigh their mechanical and silk economy advantages.

3.3.2.6 The hub: mechanical stabilizer, information center, and launching platform

3.3.2.6.1 ATTACK BEHAVIOR AND HUB DESIGNS

The hub is one of the most distinctive portions of an orb. Preliminary, largely qualitative analyses have suggested that hub designs often vary relatively little intraspecifically (Japyassú and Ades 1998; Barrantes and Eberhard 2012; Hesselberg 2013) but differ substantially among taxonomic groups (e.g., figs. 3.11—3.15, chapter 9) (Ades 1986; Craig 2003). The reasons for these patterns (if they exist) (samples are small and quantitative measurements lacking) are as yet unclear.

The hub has been thought to have at least four different functions (Witt 1965): to improve the web's physical stability by connecting the radii; to adjust the tensions on the radii (other things being equal, more hub loops will increase radial tensions, while removal of the central portion of the hub will reduce these tensions and thus improve the orb's stopping and retention properties); to provide physical support for the spider and sites for her tarsi to grasp; and to transmit prey vibrations from the radii to the spider's tarsi or to a signal line to the spider's retreat off the web (greater geometrical regularity of the hub may aid the spider in localizing sources of vibrations from prey) (Liesenfeld 1956; Witt 1965; Suter 1978; Landolfa and Barth 1996). Hub-like structures have arisen convergently in other lineages, such as *Latrodectus* spp. (THD) (fig. 10.30) (Szlep 1965; Eberhard et al. 2008a), *Theridion evexum* (THD) (Barrantes and Weng 2007b), *Parasteatoda* (fig. 10.31), and the pisaurid *Thaumasia* sp. (figs. 5.11; Eberhard 2007a), testifying to the importance of these functions. Presumably the hub designs of orbs differ because of differences in the relative importance of the various functions; but little is presently understood regarding these functions.

With respect to the vibration transmission function, it is significant that even with incomplete information, *Leucauge mariana* (TET) was able to orient accurately toward prey in her web. In experiments in which 3 radii had been broken near the hub in the 180° sector of the web opposite to the direction the spider was facing and a *Drosophila* flew into one of these three radii in the capture zone, the spider nevertheless turned to face the correct sector of the web 20 of 20 times, with her two legs I grasping the radii on either side of the broken radius on which the prey rested (Briceño and Eberhard 2011; WE). Turning behavior in this species is probably "ballistic," in the sense that no further orientation cues are gathered once the turn begins (fig. 3.28b). The information available to guide the turn can thus be deduced from the positions of the spider's legs as she rested at the hub. Assuming the mean angles between III and IV legs reported by Briceño and Eberhard (2011) and that there were 30 radii/web (Eberhard 1988a), there were about 4.5 radii in each of the three sectors between adjacent legs III and IV. The spiders oriented correctly despite the likely alterations of the vibratory signals received by their legs III and IV due to the broken radii. The implication is that hub lines do not need to be especially orderly to provide sufficient information for precise orientation by the spider.

3.3.2.6.2 LINES TO PULL, PUSH AGAINST, AND GRASP WHILE TURNING

High-speed video recordings of *L. mariana* (TET) (Briceño and Eberhard 2011) suggested the possible consequences of several details of hubs for the speed of attacks on prey. Attacks by this species are very quick indeed. When a prey struck her orb 90–180° behind her, the spider routinely turned to face the prey in only about 0.15 s (fig. 3.28), and sometimes arrived at the prey when it was 4–5 body lengths from the hub in as little as 0.21 s after it struck her web (Briceño and Eberhard 2011). And the spiders' turns were precise: the spider turned to hold

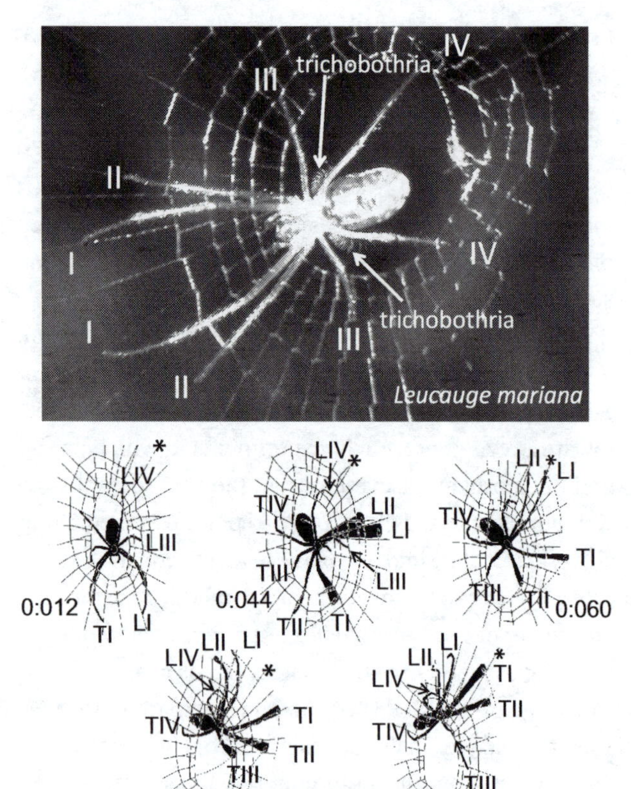

Fig. 3.28. The photo shows a *Leucauge mariana* (TET) resting at the hub of her horizontal orb seen from above; the spider had just turned 90° to face a prey that had struck her web. Legs I and II held radii in the free zone, while legs III and IV held hub lines (the arrows indicate rows of trichobothria on femur IV that may sense approaching predators). The lightning-quick leg movements as a spider turned in response to the impact of a prey are shown in drawings taken from high speed video recordings; the radius nearest the fly is marked with "*"; movements of legs from previous positions are indicated with dotted lines; time is in seconds. During the first part of the turn, legs III and IV maintained their grips, and probably pushed against hub lines while the other legs and body turned laterally; legs I and II on her leading side (LI, LII) swung laterally and rearward (0.044). These movements of legs LI and LII were probably "ballistic," in the sense that the spider received no further information from the web (and certainly no information via these legs) while they were swinging rearward. On arrival in the sector with the prey (just before 0.060), each leg made a quick, small medially directed searching movement to bring each tarsus into contact with a radius (typically only one or two small movements were needed). Soon after legs LI and LII began to move, trailing leg I (TI) swung laterally (0.044), and grasped a radius near the prey radius (0.080). TIII took a large, quick step across the hole in the center of the hub (dotted line in 0.070) to grasp a region of dense lines, while the spider's body axis was turning rapidly. Leg LIII and both legs IV maintained their original holds during this turn, and LIV became sharply twisted (0:080) before finally releasing its grip. Note that all of these literally split-second, highly coordinated movements were done "blind"; the spider could not see any of the hub lines (all of her eyes are on the opposite side of her body). Leg LIII is absent in some drawings because its position was not clear in the recordings, possibly because it was directly above the spider's body (the intersections between hub spirals and radii are imprecise in the drawings, and the sharp zig-zags visible in the photo are omitted) (after Briceño and Eberhard 2011).

the radius closest to the prey with one leg I in 43 (89.6%) of 48 responses that occurred <1 s after a prey-mimicking object struck her orb (WE).

A moment's reflection on the difficulties facing a spider when she turns illuminates both the problems that such a rapid attack entails, and some possible functions of the hub. The positions of the spider's legs relative to her probable center of gravity (probably in the anterior portion of her abdomen) indicate that most of her weight is supported by her legs III and IV (fig. 3.28) in her horizontal web (in a vertical orb the support would be more concentrated on legs IV). In order to make a large turn, the spider pushes and pulls lines with her posterior legs (III and IV) to turn her body. Greater resistance by the hub lines held by these legs will enable her to turn more rapidly. As a general rule, more connections to nearby lines will make these lines better able to resist the pushes and pulls and allow the spider to turn more quickly. While not all orb weavers are as quick as *L. mariana*, speed is probably often important, as testified by the high percentages of prey escape after being intercepted by orbs in the field before the spider arrives to attack them (section 3.3.3.1.1).

The idea that legs III and IV provide the motive force in turns was supported by the temporal coordination of leg movements in *L. mariana*: legs I and II were lifted first from the web, and they and the spider's body were apparently propelled as the spider turned by pushes and pulls on the web exerted by legs III and IV (fig. 3.28). Also as expected, legs III and IV grasped shorter, more highly inter-connected hub lines before and during the turn, while legs I and II grasped single radii in the free zone both before and after the turn.

In addition to aiding in quick propulsion as the spider turns, the hub must facilitate finding the new lines the spider will grasp when she turns. She is effectively blind with respect to where the lines are in her hub, and in other contexts spiders find new lines to grasp by tapping and waving with their legs, like a blind man with his cane (e.g., sections 2.4.2.1, 7.1.1). But exploring this way takes time. By having a dense network of hub lines in just those areas where the spider needs to grasp, the searching time is reduced (fig. 3.28). This factor will favor both a denser hub and one that is large relative to the spider's size.

The hubs of numerous orb weavers seem appropriate for these functions. The diameter of the hub of many species is large enough that even the spider's longest legs (I and II) do not extend beyond its margin, and the hub is also relatively dense so that none of her legs needs to search far to find a line to grasp as she turns (e.g., figs. 2.19b, 3.16d, 3.35, 3.43, 3.45). Providing footholds of this sort may be the function of filling in the hole at the center of the hub immediately after removing it, as occurs in many araneids (fig. 4.17). And the removal of the accumulation of lax lines, which might entangle the spider's tarsi as she turns at the hub, might be the function of removing the center in the first place.

There are, however, many other hub designs less easily explained by these functions (figs. 3.11–3.15). Many tetragnathids have open holes at the center, and in some araneids, such as *Micrathena* spp. (figs. 1.4, 3.36, 3.37), the hole at the center is especially large and the spider holds only its inner edges while waiting for prey. In some tetragnathids and araneids the diameter of the hub is relatively small compared with the lengths of the spider's legs I and II, and these legs typically grasp radii that are beyond the edge of the hub while legs III and IV grasp the hub (figs. 2.19, 3.28). Some other species, such as *Metazygia wittfeldae* (AR) are intermediate; the hub is more densely meshed but small enough that the spider's legs I grasp its edge or on the radii just beyond the edge (fig. 3.16d).

Why do so many hubs have designs that would seem to make turning slower? At present there are only hints. The hub of *L. mariana* (fig. 3.28) shows that a relatively sparse design is enough for quick turns if the movements of legs III and IV are rapid and precise (Briceño and Eberhard 2011). Perhaps the differences in the designs of the hubs of other species are associated with different leg movements and agility as they attack prey. Perhaps grasping radii rather than hub lines with legs I and II somehow facilitates other aspects of the attack beside the turn (sensing prey vibrations? choosing which radius to use to run to the prey?). Perhaps the designs of the hubs are related to the fact that some spiders, including the araneids *Allocyclosa bifurca* and *M. duodecimspinosa* and the uloborids *Uloborus diversus* and *Zosis geniculata*, rely more on the web to hold the prey, and only arrive many seconds after impact, and usually only after having turned rapidly toward the prey and jerked the web repeatedly (section 3.2.10). This behavior makes the prey swing in the orb, and may serve to cause it to contact further sticky lines (Robinson and Olazarri 1971; Suter 1978). But these are only speculations, and the uniformity of hub designs in the genus *Micrathena* (fig. 3.14) despite the differences in spider positions at the hub (fig. 3.37) suggests that the explanations are complex rather than simple; further data (especially high-speed movies) are needed.

The intra- and inter-specific correlations between the number of loops of hub spiral and the number of radii in different species (fig. 3.29) present a possibly related puzzle. A larger number of loops could augment any of the four possible functions proposed above, so the (unanswered) question posed by the correlation is why any particular hub spiral function would be less important in webs with fewer radii.

3.3.2.6.3 TENSING (AND RELAXING) FUNCTIONS OF THE HUB

In many species, the hub spiral is attached twice to each radius it crosses. Double attachments have the effect of shortening or cinching up the radius, increasing its tension (section 4.4, fig. 4.8a): the sharper the radius-hub spiral angle, the greater the tension on the hub spiral compared with that on the radius (fig. 4.9). Preliminary analyses reveal several patterns. It seems to be common for the zig-zags produced by the hub spiral to be larger in the inner portions of the hub, and to be reduced or even disappear in the outer portions of the hub (figs. 3.30, 4.9). In species that remove the center of the hub at the end of orb construction such as *Azilia affinis* (TET) (fig. 4.16), this gradient may result from the more central segments of hub spiral being less extensible due to their shorter lengths, and thus producing larger zig-zags in the radii when radial tensions decrease when the center of the hub is removed. In other species in which

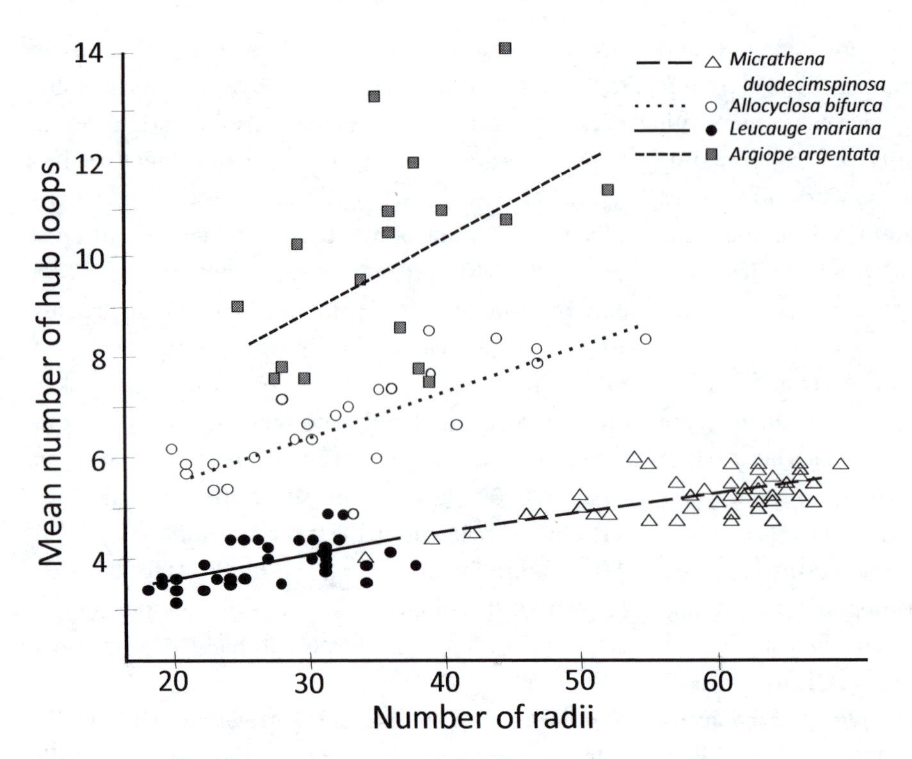

Fig. 3.29. There was a significant relation between the mean number of hub loops (above, below, and to the sides of the center of the web) and the number of radii in orbs of *Micrathena duodecimspinosa* (AR) ($p < 0.0001$), *Leucauge mariana* (TET) ($p = 0.0004$), *Allocyclosa bifurca* (AR) ($p = 0.0001$), and *Argiope argentata* ($p = 0.047$) (loops were not counted above). The relations differed, however, between species.

the center of the hub is not removed, as in *Philoponella vicina* (UL) (fig. 4.9), the gradient suggests (if tensions on hub spiral lines are approximately constant) that there is a gradual cinching up of the radial tensions early in hub construction, and then later hub loops do not increase them further. In *Leucauge mariana* (TET) hub spirals increased the tensions on radii during orb construction but not afterwards. The radial tensions in this species are cinched up somewhat by double hub spiral attachments (fig. 3.30), and were then sharply reduced when the spider removed the center of the hub following sticky spiral construction. The final tensions on the radii were reduced so sharply that the zig-zag deflections of the radii in the finished web were approximately 90° (fig. 3.30b). The likely function of this ephemeral cinching up of radial tensions is to support the spider during temporary spiral and sticky spiral construction with a minimum of web distortion. Reducing web distortions might increase the spider's precision in making attachments at sites and under tensions that more precisely approximate the sites and tensions that they will have when the orb is finished (I know of no data on this point, however).

In some groups the hub center is removed, but the spider then immediately fills in the open space that she has just created with additional lines (section 6.3.6). Hubs that are left open were formerly taken to be a trait that distinguished Tetragnathidae from other orb weavers (Kaston 1964), but they also occur in many araneids (table 9.1). Some araneids utilize the hub hole to manipulate web tensions immediately after prey impact, probably to ensnare it. The spider tightens the orb by grasping and pulling inward on the inner edges of the hub hole (especially with her legs III) while waiting for prey to arrive (fig. 3.37), then releases this tension and jerks the web when a prey strikes the web. The sudden sag in the web may increase its ability to absorb the prey's momentum; the subsequent jerks may also cause the prey to swing into contact with additional lines and increase retention times. Again, however, these ideas are speculative and need further testing.

3.3.2.7 Functions of the tertiary radii

Some radii originate not at the hub, but part way out another radius. As far as I know, all such "tertiary" radii (Le Guelte 1966) are always built during temporary spiral construction. Although I know of no systematic survey, it appears that most tertiary radii are located in the larger sector of asymmetric orbs, and tend to fill spaces between radii that are unusually far apart (e.g., figs. 3.7, 3.31, 4.4, O6.8) (Hingston 1922a; Le Guelte 1966; Kuntner 2007; Kuntner et al. 2008a).

The frequencies of tertiary radii vary substantially in

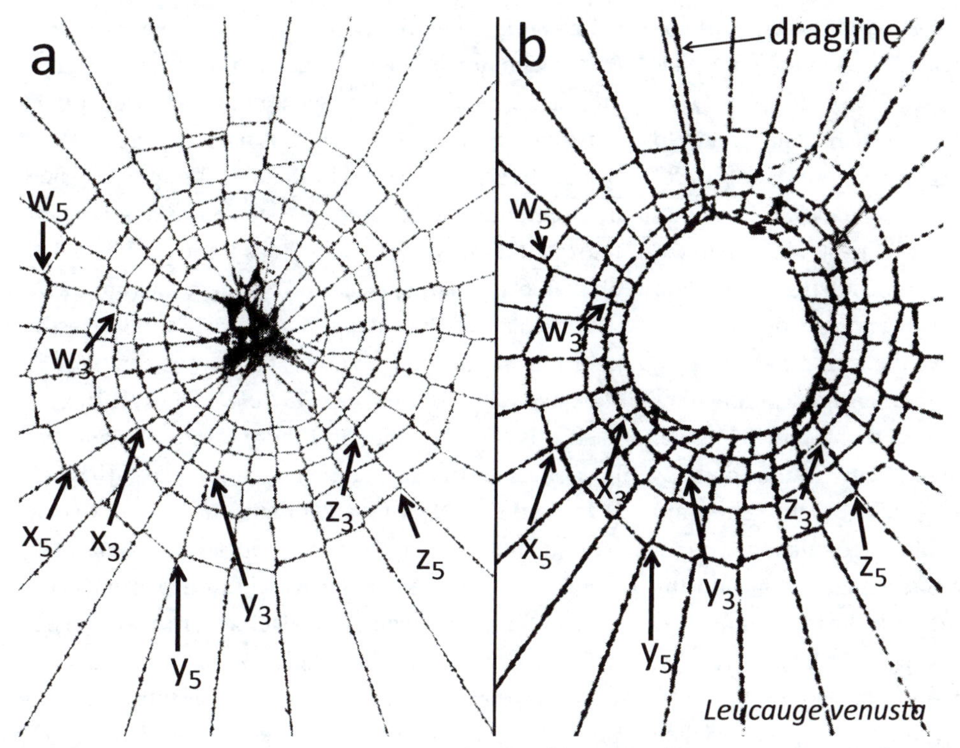

Fig. 3.30. The changes in the zig-zag patterns of radii before (*a*) and after (*b*) *Leucauge venusta* (TET) removed the center of the hub revealed differences in radial tensions (the hub was powdered for the first picture when the sticky spiral was partially complete; after the powder was then knocked off, the spider finished the web and removed the center of the hub, and the hub was then powdered again for the second photo). The increased angles of deflection of the radii at some points (w_3, w_5, x_3, y_5, z_3, z_5) after the hub center was removed reflected decreases in the relative tensions on radii compared with the hub spiral. Removal of the center of the hub (along with the upper portion of the innermost loop of hub spiral) caused the hub to expand, mostly upward and downward. A second pattern was that both before and after the center of the hub was removed, the outermost loop of hub spiral tended to deflect radii less strongly than the previous loops (as mentioned in the text, radii were probably under lower tensions nearer the center of the hub). Similar patterns of increased deflections of radii closer to the center of the hub are apparent in the hubs of *Azilia affinis* (AR) (fig. 4.16) and *Philoponella vicina* (UL) (fig. 4.9).

different groups that make more or less typical orbs. In a sample of species in which I have access to multiple web photos, tertiary radii appear to be entirely absent in the orbs of several species, including *Epeirotypus brevipes* (TSM) (N = 30 webs), *Argiope argentata* (N = 10 webs), *Allocyclosa bifurca* (AR) (N = 23 webs), *Zosis geniculata* (UL) (N = 23 webs), and *Micrathena duodecimspinosa* (AR) (N >100 webs) (direct observations of construction behavior revealed that the occasional apparent "splits" of radii in this species were actually pairs of radii that were bound together by hub attachments). On the other hand there was an average of 1.6 tertiary radii/web (all in the lower half) of 27 webs of *Metepeira* sp. (AR), and they also occurred (0.2/web) in orbs of *Uloborus diversus* (UL) (N = 21 webs). My qualitative impression is that tertiary radii are associated with orbs that have especially large numbers of radii such as *Nephila* spp. (see section 4.3.3.2) and *Cyrtophora* (Kullmann 1964, 1975). More extensive data will be needed to make further comparisons.

One likely function of filling with tertiary radii rather than additional secondary radii is material and energetic economy: the spider fills in open spaces between radii in the outer portion of the orb without "wasting" radial lines near the hub where the density of radii is already high (Hingston 1922a; Eberhard 1972a), and the subsequent tasks of attaching temporary and sticky spiral lines to radii are behaviorally easier (Hingston 1922a), and the spider also expends less piriform silk.

One type of evidence supporting the filling-in hypothesis is that tertiary radii tend to be built in sectors of the web where radii are especially widely separated.

In 70 of 83 cases in 43 webs of six adult female *Uloborus diversus*, the tertiary radius originated in the outer half of the orb; and 45 (64%) of these tertiary radii were adjacent to anchor lines, sectors in which radii tended to be farther apart (Eberhard 1972a). The tertiary radii of *Nephila* spp. and *Nephilengis* (= *Nephilengys*) *cruentata* were bent sharply away from the closest radius in a way that could serve this "filling-in" function (fig. 4.4, section 4.3.3.2). The webs of *Cyrtophora* and its close relatives *Mecynogea* and *Manogea* have the greatest abundance of tertiary radii; the sticky spiral is absent, and the temporary spiral is very tightly spaced and remains intact in the finished web. The large majority of radii (perhaps around 90%) at the outer edge of the web are tertiary radii (Kullmann 1975). Their filling in function is especially important in these webs, which depend on the densely meshed sheet of non-sticky lines to retain prey long enough for the spider to attack (fig. 1.7).

A second probable function of tertiary radii in orbs with large numbers of radii (there were up to more than 100 radii at the periphery of a *N. pilipes* web) is to reduce the number of radii at the hub. This would facilitate hub construction, and simplify behavioral decisions such as where to lay new radii during web construction (Hingston 1922a) and which radii to use in launching attacks on prey.

Hingston (1920) also proposed that tertiary radii may serve to correct mistakes made by the spider when the she failed to sense and fill in holes between radii during secondary radius construction. This "mistake" hypothesis seems unlikely, at least in *U. diversus* (UL), because the spider often laid secondary radii in sectors smaller than those in which she later built tertiary radii (Eberhard 1972a); but it cannot be confidently discarded.

3.3.2.8 Functions of the temporary spiral

3.3.2.8.1 PATTERNS IN TEMPORARY SPIRAL SPACING

The temporary spiral (also known as the auxiliary spiral) has been largely neglected in studies of the designs and functions of orbs (Gotts and Vollrath 1991). This non-sticky spiral is usually ephemeral, as the name implies: it is built just before sticky spiral construction, but is then gradually removed, segment by segment, a few minutes later during sticky spiral construction (section 6.3.4). There are thus few published photos of temporary spirals. It is generally thought to be composed of lines from the minor ampullate glands, rather than the major ampullate glands that are used for other non-sticky lines in the orb, though the evidence is only scattered (Peters 1993a; Hesselberg and Vollrath 2012). The pair of fibers in a temporary spiral line had smaller diameters than those of the major ampullate lines (indicated by spigot diameters in many species and by direct measurements of lines in a few) (Blackledge and Hayashi 2006a; Hesselberg and Vollrath 2012) (see also Peters 1993a on the non-sticky spiral of *Cyrtophora citricola*) (fig. 1.7). The silk itself was slightly stiffer and more extensible, and had lower tensile strength than major ampullate silk (Blackledge and Hayashi 2006a). Perhaps the smaller diameters and inferior mechanical properties of temporary spiral lines are associated with selection favoring economy in lines that need not resist high stresses. The distinction between the hub and the temporary spiral in webs in which it is not removed, such as those of *Cyrtophora* and its allies and *Nephila* spp., is not always clear (Hingston 1922b).

The fraction of species whose temporary spiral spacing has been documented is small (except among the nephilids, which leave the temporary spiral intact in the finished web). It appears, however, that there are patterns in the spaces between temporary spiral loops within a given web, and that these patterns differ in different species. For instance, the temporary spiral lines in *Micrathena duodecimspinosa* (AR) orbs were farther apart in the upper than in the lower portions of the orb, and in portions farther from the hub (figs. 6.10, 7.34). In *L. mariana* (TET) the spaces between temporary spiral loops differed in being larger rather than smaller in the lower sector of webs built at 45°, but also gradually increased farther from the hub (Eberhard 1987d); this hub-to-edge pattern also occurred in *L. mariana* webs built at 0°. Still another adjustment by *M. duodecimspinosa* was that distances between loops were smaller on shorter radii (fig. 6.10; WE). The figures in Gotts and Vollrath (1991) suggested that *Araneus diadematus* (AR) resembled *M. duodecimspinosa* and *L. mariana* in having large spaces between temporary spiral loops farther from the hub (fig. 7.34; Vollrath 1985; Gotts and Vollrath 1991), but there was no clear pattern above vs. below the hub. The significance of this variety of spacing patterns is unclear (see below), but it demonstrates that temporary spirals do not follow a clearly logarithmic pattern (Vollrath and Mohren 1985; Hesselberg and Vollrath 2012) (for an extreme example of the lack of a logarithmic increase in spacing, see fig. 3.42a of *Enacrosoma anomalum* [AR]).

3.3.2.8.2 PROBABLE FUNCTIONS OF THE TEMPORARY SPIRAL (HAND RAIL AND OTHERS)

Despite their ephemeral nature, temporary spiral lines have numerous possible functions. It has long been clear that the temporary spiral provides bridges that spiders use to move from one radius to the next while building the sticky spiral (e.g., McCook 1889; Fabre 1912; Hingston 1920). The spider economizes on the distances traveled, and also avoids pulling out more sticky spiral line than the distances between adjacent radii; she thus avoids laying lax sticky spiral lines. Judging by the paths of most orb weavers (e.g., figs. 6.1, 6.25, Zschokke and Vollrath 1995a), the bridge function is more important in some species than others, and is most important in the outer portion of an orb (when nearer the hub, the spider often simply reached from one radius to the next and moved directly to the next attachment site). In the inner portion of the orb of at least one species, *Uloborus conus* (UL), the temporary spiral does not have a bridge function, because no sticky lines are built there (Lubin et al. 1982).

A second possible function is that the temporary spiral holds the radii in place with respect to adjacent radii during sticky spiral construction: the temporary spiral reduces the amount that the radii sag under the spider's weight, and thus keeps the distances between radii nearer to those that they will have in the finished orb (Hingston 1920). The possible importance of the anti-sagging function is confirmed by details of sticky spiral construction behavior of certain short-legged, heavy species that reel in radii rather than walking along them during sticky spiral construction (fig. 6.24); it is probably less generally important for lighter and longer-legged species, and of limited or no importance in horizontal webs. The anti-sagging effect also occurs during attacks on prey in the finished webs of *Nephila clavipes* (NE), where the temporary spiral is left intact; the relatively small "tracks" that this species leaves when walking across the orb (section 4.9.4; WE) are probably due to the intact temporary spiral.

The bridge and anti-sag hypotheses for temporary spiral function are supported in several species in which the temporary spiral has been reduced or lost, and in which either the bridge or the anti-sag function is not needed. In some bridges are not needed because the spider is huge with respect to the radii and the spaces between them as in *Deinopis* sp. (Coddington 1986c), or because she does not lay sticky lines between radii, at least in the outer portion of the web as in *P. producta* (UL) (Peters 1995a), and *Eustala* sp. (AR) (Eberhard 1985). In still others, including theridiostomatids (Eberhard 1982; Coddington 1986a) and cyrtarachnines (AR) (Stowe 1986), the anti-sag function is not needed because the sticky lines are lax in finished orbs. The elimination of the temporary spiral in the symphytognathoids (which was presumably present in a reduced form in their common ancestor, as in theridiosomatids) may have had a different function, to allow spiders to attach sticky lines that were continuous with the sticky spiral to supporting lines above the horizontal plane of the orb (Eberhard 2011).

There are several additional possible functions. A hitherto unremarked but important consequence of the temporary spiral is that it holds the radii in place, often standardizing the spaces between radii by bending them into oversized spaces and thus filling in spaces through which prey might otherwise pass without encountering a radius (fig. 3.31); it also reduces the lengths of segments of sticky line and thus the dangers that they will stick together due to wind movements (section 3.2.11) (e.g., fig. 3.31; Eberhard 2013a; see also figs. 4.4 of *Nephila* and 3.4 of *Scoloderus tuberculifer*). In other groups, in which the temporary spiral is removed during sticky spiral construction, the radii have similar curved forms (though in somewhat smoother curves) that were originally produced by the temporary spiral. There are species, however, in which bending the radii toward larger inter-radial spaces appears not to occur (e.g., *M. duodecimspinosa*; fig. 6.10).

A further function, seen in the scattered set of species in which the temporary spiral is left intact in the orb (including the nephilids *Nephila* and *Nephilengys*, and the araneids *Scoloderus* spp. and non-sticky sheets of *Cyrtophora* and its relatives), is that temporary spiral lines stabilize the web mechanically by distributing local stresses over more lines (Landolfa and Barth 1996). They

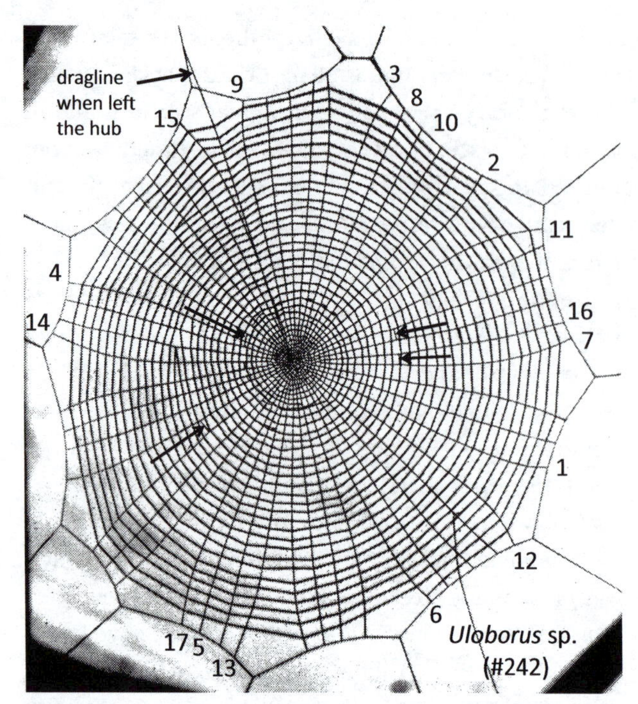

Fig. 3.31. In the petal-like pattern of the radii in this horizontal orb of *Uloborus* sp. (#243), radii tended to bend toward especially large inter-radial spaces in the "corners" between adjacent frame lines. The order in which the last 17 radii were built was deduced from the pattern of lines in the hub (see fig. 6.11) (the radii are numbered in the order that they were constructed, so the last radius built is #17). The earliest radius attached to a frame (probably built as part of frame construction) tended to be attached to the central part of the frame. For example, the first radius to the frame at the upper right was the one between radius 2 and radius 10; for the frame at the lower left, the first radius was between radius 5 and radius 13. This pattern contrasts with that in *M. duodecimspinosa* webs, where the first radius attached to a frame was usually near one end of the frame (fig. 6.10). In most cases, the last radius to a frame in this *U.* sp. web was at the edge of a large inter-radial space; this pattern also occurred in the *U. diversus* web (fig. 6.11). Several other typical uloborid traits are visible in this web: scattered segments of intact temporary spiral lines (some are marked with arrows); a hub in which the center was intact; the hub spiral that was attached twice to each radius it crossed, often pulling the radius out of a straight line; and segments of sticky spiral near the hub that were not attached to every radius they crossed.

also make the web less flexible, and thus reduce the risk that adjacent closely spaced sticky lines will contact each other or radial lines, especially under windy conditions (Hesselberg and Vollrath 2012; see also Eberhard 2014). On the downside, they probably also reduce vibration transmission along radii (Landolfa and Barth 1996).

This stabilizing effect also occurs during sticky spiral construction in some heavy-bodied and short-legged araneids such as *Gasteracantha cancriformis* (figs. 3.25, O6.1) (Eberhard 1982; Briceño and Eberhard 2012; WE), in which the temporary spiral is subsequently removed. The relatively short legs of this species accentuate the distortion caused by the spider's substantial weight during sticky spiral construction because they concentrate her weight in a small area of the web, causing the lines there to sag substantially. Sagging could be especially important when the spider is attempting to locate the inner loop of the sticky spiral and attach the next loop at a precise distance from it. The spider's solution was to refrain from walking outward along the radius and causing it to sag downward. Instead, she stopped at the junction with the outer loop of temporary spiral (which stabilized the radius by connecting it to adjacent radii), and reeled in the radius beyond this point "hand over hand" with her legs oI–III. When her tapping leg oI touched the inner loop, she made the attachment, still without releasing the temporary spiral line (fig. 3.25). In this way the spider could lay sticky lines in a portion of the web where her own weight would otherwise have produced severe distortions.

Still another probable function for the temporary spiral, that it guides the spider like a "hand rail" while she lays further loops of temporary spiral, has not been discussed previously. This hypothesis is brought into focus by realizing that neither the bridge nor the anti-sagging function is likely to be important in the inner portion of an orb. Why does the spider not just omit those inner loops? After finishing the hub, she could walk out along a radius to where bridges begin to be useful, and start to build the temporary spiral there; this would save the behavioral and material costs of the inner loops. The temporary spiral construction behavior of present-day spiders suggests an answer: that each previous loop of temporary spiral is used as a guide for building the next loop (section 7.4.1).

The final result of this "hand rail" function is the form of the outermost loop of temporary spiral, and this loop is then also employed as a hand rail that guides the spider as she lays the first loop of sticky spiral (Zschokke 1993). Both correlations within webs and responses to experimental modifications show that the placement of the first loop of sticky spiral is strongly influenced by the site of the outermost loop of temporary spiral. This effect occurs in uloborids, araneids, and tetragnathids (section 7.3.6). This effect may have important consequences for prey capture, because it determines the size and shape of the prey capture area.

One more probable function of intact temporary spiral lines, in a few groups, such *Nephila* and *Cyrtophora*, is to increase the tensions on radii. The temporary spiral is laid under substantial tension in these species (Hesselberg and Vollrath 2012); and because it is attached twice to each radius that it crosses, it pulls the radii into zig-zag patterns (figs. 1.7, 4.4, 7.31b). In effect each radius is shortened, thus raising its tension. Hesselberg and Vollrath (2012) estimated that the stress on *N. edulis* radii was increased from 50 MPa to 145–193 MPa by a single temporary spiral loop. A further pattern was that the length of the zig-zag (h/w in fig. 4.9d) was relatively greater in attachments farther from the hub in *N. edulis* webs, increasing the amount that they pulled the radius out of line (the "zig-zag index" of Hesselberg and Vollrath 2012: $[\alpha_1 + \alpha_2]/180$ in fig. 4.9d). Because temporary spiral loops farther from the hub were laid after the tensions on radii had already been raised by earlier loops, Hesselberg and Vollrath (2012) estimated that the tensions on temporary spiral lines increased by a factor of about 3 moving from the hub to the edge of *N. edulis* webs. The longer lengths of the zig-zags will also result in an accentuated tightening effect on the radii farther from the hub. An alternate hypothesis for the double attachments of the temporary spiral to radii is to avoid the "slippage . . . seen in other types of junctions" (Hesselberg and Vollrath 2012). This function is unlikely, however, because the only junctions that have been reported to slip this way are of of radii with sticky spiral lines, not radii with temporary spiral lines (Eberhard 1976b). The functional significance of greater temporary spiral and radial tensions farther from the hub (Hesselberg and Vollrath estimated that they differed by a factor of about 1.5–2x in *N. edulis* webs) is not clear.

3.3.2.8.3 PATTERNS IN TEMPORARY SPIRAL SPACING IN ORBS AND THEIR POSSIBLE SIGNIFICANCE

Because of the hand rail effect on the outermost loop of sticky spiral, patterns in the spacing of loops of temporary spiral can affect the size and shape of the orb's capture zone. This realization suggests the hypothesis that the adjustments of temporary spiral spacing function to adjust the placement of the outermost loop of the temporary spiral. For instance, the larger spaces above the hub than below it in *M. duodecimspinosa* lead to a capture zone with an oval outline that is somewhat elongated above the hub.

One way to test this "outer loop form" hypothesis would be to alter the shape of the orb by enclosing spiders in containers with different shapes (e.g., Ades 1986, 1991; da Cunha-Nogueira and Ades 2012; Ades et al. 1993; Hesselberg 2013), and check whether the pattern of temporary spiral spacing changes accordingly. In webs built in a short but wide space, the temporary spiral spacing on approximately vertical radii should decrease in comparison with that on approximately horizontal radii; the changes should be inverted in a tall and narrow space. I know of no detailed study of this possibility. Hesselberg (2013) found no significant evidence of the predicted changes when he performed these two experiments with *Eustala illicita* (AR), but, he checked only the space between the first and second loops of temporary spiral. It is possible that the "outer loop form" hypothesis could help explain differences in temporary spiral spacing with respect to gravity and radius length, but there are several remaining mysteries. Why, for instance, does the temporary spiral spacing increase gradually farther from the hub in both horizontal and vertical orbs?

In summary, the temporary spiral probably performs a surprisingly large number of functions: it is used as a bridge between radii during sticky spiral construction, to avoid making overly long segments of sticky spiral and to save time and effort in moving from one radius to the next; it reduces the sagging of radii during sticky spiral construction, thus keeping the sticky spiral lines from adhering to each other and to the radii; it bends radii to fill oversized holes in the web and thus improves its stopping ability; and it determines the form of the capture zone, because its outer loop serves as a "handrail," guiding the spider as she lays both successive loops of temporary spiral and also as she lays the first loop of sticky spiral. This handrail function can explain some of the patterns in temporary spiral spacing. In some species in which the temporary spiral is left intact in the finished orb, it tightens the tensions on radii via double, zig-zag attachments; in one such species the temporary spiral tensions were gradually higher in the outer portions of the orb. The temporary spiral also stabilizes these orbs mechanically. The bridge, anti-sagging, and handrail functions are supported by experiments and comparative evidence. Further mysteries remain, and our present ignorance of the temporary spiral designs in most species suggests that further surprises may be in store.

3.3.2.9 The other side of the coin: how best to fail

Most discussions of the functional significance of orb designs have focused, quite reasonably, on their effects on capturing prey. But a web in nature also has to minimize the damage that it suffers when it is broken by overpowering forces of two general types: the localized stress near the point of impact ("local loading") from an object whose momentum is too great for the lines to absorb (e.g., falling leaves, large, rapid insects); and general stresses spread over a larger fraction of the orb ("global loading") such as wind. Global loading stresses may be especially severe when a web snaps back and forth due to changes of wind direction that produce transient sharp changes in stress (Langer 1969). Questions of how webs fail mechanically have only recently been discussed in detail (Ko and Jovicic 2004; Alam and Jenkins 2005; Aoyanagi and Okumura 2010; Cranford et al. 2012; Zaera et al. 2014). The finite element method of analysis of structural mechanics, which models a structural system using a set of connected finite elements with known properties that are connected at nodes, has been used to calculate displacements and changes in variables like tensions, stresses, and overall transverse web stiffness that result when particular elements are broken. One basic approach, asking how a web's design affects the damage to other web elements from large local forces that rupture lines, has yielded some surprises.

Cranford et al. (2012) found, using both simulations and experimental local loading of *Araneus diadematus* orbs, that a mechanical property of ampullate silk, the lack of linearity in the responses to stress (fig. 2.3), is crucial in limiting the extent of deformation and damage from local loading. The silk's relative rigidity under low stress up to the yield point (regime I in fig. 2.3), and its subsequent softening (regime II) that becomes more pronounced at higher stresses (regime III) aided in resisting initial loading, but then resulted in ceding before other web lines were damaged. When either radial or sticky spiral lines were over-loaded and failed to simulate overpowering local loading, the only lines that failed were those that were contacted directly (similar, exclusively local damage also occurred in simple empirical tests of *Allocyclosa bifurca* [AR] and *Leucauge mariana* [TET] orbs—Eberhard 1986a). In essence, the damage resulting from overpowering local loading was minimized.

Zaera et al. (2014) noted that in cases in which the web is stressed but does not fail, it is important to distinguish elastic strain (from which the web will recover) from inelastic strain due to higher stresses that do not break the lines but permanently degrade the web's ability to resist future loads (inelastic strain). While inelastic strain may be important in wind loading, it is probably not important for prey impact loads, because the spider immediately cuts the lines in the near vicinity as she removes the prey from the orb. In essence, locally damaged lines that are still intact after the impact of a prey that is stopped are likely to be removed when the prey is cut free.

Additional simulations showed that damage was less extensive when the radial lines had elastic properties similar to those of spider ampullate silk. Under strong loading (immediately prior to failure), most of the radii other than the radius that was directly stressed exhibited deformation states corresponding to regime I in the stress-strain graph (fig. 2.3); the semi-amorphous regions of ampullate silk molecules permitted reversible entropic unfolding due to their alpha helices (fig. 2.2). In contrast, in simulations using radii that failed at the same stress values but that had either linear elastic properties or elastic-perfectly plastic properties (similar to the properties of egg sac silk from the cylindrical glands), more extensive areas of the web were damaged. The reason for this difference was that the radius that was contacted suddenly softened at the yield point; its initial modulus was reduced by about 80% (regime II), ensuring that only the directly loaded radius ("the sacrificial radius") eventually failed. The effects that the stress-strain curves of the material in the lines had on localizing damage showed a trade-off, however, with the prey-stopping ability of the orb. When the lines were elastic-perfectly plastic, the maximum strength of the web increased by 34%.

Tests with global loading of the web to simulate a

wind perpendicular to the orb plane showed, in contrast, that there was little difference between radii that had different stress-strain relations for winds up to about 10 m/s (in regime I, where ampullate silk is relatively stiff). Eventual damage under stronger winds (in terms of % of lines broken) was up to six times greater when lines were elastic-perfectly plastic rather than ampullate lines. The general implication is that the relative costs and benefits of having radii with different elastic properties could differ, depending on the frequencies of different magnitudes and types of stress. There is one extraordinary account of how an accidental sharp knock against the branches supporting the orb web of *Poltys* sp. nr. *laciniosus* (AR) resulted in the whole web "disintegrating as if exploding, leaving nothing but a few frame lines" (Smith 2006; p. 278) (I have never witnessed anything remotely similar to this).

A separate modeling study of sticky spiral lines showed that their much lower spring constant than that of radii (the ratio is about 1:10), along with their geometric placement, resulted in stresses being dispersed in the orb when spiral lines were broken, instead of being concentrated at certain points as is common in other elastic systems (Aoyanagi and Okumura 2010). Greater dispersal translates into an improved ability to survive without breaking.

These data relating material properties of silk to web designs do not permit one to distinguish evolutionary cause and effect. Do orbs have the designs that we see today because ampullate silk has the non-linear properties shown in fig. 2.3? Or did the non-linear responses evolve to improve the mechanical properties of orbs? Comparative data (e.g., the stress-strain properties of ampullate silk in taxa that are not derived from orb weavers, or that have secondarily lost orbs) could help distinguish between these hypotheses. The general topic of damage limitation offers many additional tantalizing questions (table O10.1).

A final aspect related to how webs fail is the relationship between the strength of lines versus that of the attachments between them. I know of no careful study of this point, but it is clear from incidental observations that radii almost always break under stress before their attachments to frame or hub lines (section 3.3.2.9). It might thus seem that the spiders "over-invest" in attachment materials (theoretically, the most efficient use of resources would be a web in which a line and its attachments broke simultaneously). But breaking a line may do less damage to a web with respect to future captures than breaking attachments would when a high-energy object crashes through it.

3.3.3 FUNCTIONS FOR STICKY LINES

3.3.3.1 Retain prey

3.3.3.1.1 SELECTION FAVORS LONGER RETENTION TIMES

Once an orb has stopped a prey, it must retain the prey long enough for several events to occur: the spider must sense its presence; locate where it is in the web; evaluate whether or not to attack it (prey that are too large or dangerous are generally not attacked); run to the impact site; and forcefully immobilize it by biting or wrapping (Robinson 1975). It has long been known that many prey are restrained only briefly by orbs after being stopped, and escape before the spider arrives (e.g., Lubin 1973; Nentwig 1982a; Uetz and Hartsock 1987; Miyashita and Shinkai 1995; Blackledge and Zevenbergen 2006; Zschokke et al. 2006). For example, various species of prey remained in the orbs of *Araneus sericatus* (= *Epeira sclopetaria*) for an average of only 5 s, shorter than typical spider response times (Barrows 1915). In *Micrathena gracilis*, 75% of the prey that were >9 mm long escaped in <3 s, more quickly than the spider was able to attack (Uetz and Hartsock 1987). Even moderately small sepsid flies (an estimated 10–25% the weight of the spider) often escaped quickly after being stopped in sectors of vertical webs of mature female *Metazygia yobena* from which the spider had been removed (Eberhard 1989b): from 12 to 20% escaped in <5 s, approximately the time needed by this species to reach flying prey (mean of 5.5 ± 4.0 s, median = 4 s) when various types of prey were induced to make presumably "normal" impacts with webs of this nocturnal species by attracting them with a light held near the web.

These arguments, plus the likelihood that the high degree of uniformity in sticky spiral geometry and spacing in orb webs is the result of selection, make it logical to conclude that there is probably strong selection favoring longer retention times in orb webs. The variations in retention times are typically very large, however, making the retention time of a given orb design difficult to characterize (section 3.3.3.1.7, appendix O3.2). The combination of great variation and the likely selection on retention is illustrated by the following examples. Greater numbers of lines adhering to prey tended, as expected, to produce longer retention times with several prey types in the webs of *Argiope aurantia* (AR) (Blackledge and Zevenbergen 2006). In apparent contradiction, the variation in retention times when sepsid flies were induced to fly into the orbs of *Leucauge mariana* (TET) was so great, however, that there was no significant effect on the mean retention time comparing webs in which the uniformity of sticky spiral spacing had been substantially altered by previous impacts and struggles of prey (WE). Nevertheless, an additional behavioral detail in these spiders indicates that selection on sticky spiral spacing is indeed strong (and complex—see section 4.3, Blackledge and Zevenbergen 2006). Orb-weavers often leave "tracks" of displaced sticky spiral lines when they walk across their orbs (see figs. 1.4, 4.19). Even though *L. mariana* builds more or less horizontal orbs and produces less extensive tracks than others such as *Allocyclosa bifurca* (WE), the spiders often returned to the hub following an excursion into the capture zone by using a behavior whose only likely function was to reduce tracks. The spider attached her dragline at the hub before moving out; when she returned, she released her hold on the web and fell, swinging free under the web and then climbing quickly back up her dragline to the hub. By executing such a "pendulum return," she avoided leaving tracks (presumably at the cost of the increase in energy needed to climb up to the hub). Similar pendulum returns also occurred in species with moderately slanting webs, such as *Nephila clavipes* and *N. pilipes* (NE) (Robinson and Robinson 1973a; WE), and the araneids *Araneus diadematus* (Peters 1933a), *Argiope argentata* (Robinson and Olazarri 1971), *Micrathena duodecimspinosa* (WE), and occasionally *Allocyclosa bifurca* (WE).

In sum, there appears to be strong selection on sticky spiral spacing, despite high variation in retention times. As mentioned above, selection probably acts on this (and other aspects of web construction) not only with respect to the payoffs from a particular prey and a particular web, but on the sum of all of the payoffs from all of the webs that are built by that individual during her lifetime (section 3.2.13).

3.3.3.1.2 MEANS BY WHICH PREY ARE RETAINED: ADHESION, EXTENSION, AND RESISTANCE TO BREAKING

In a relatively sparse web like an orb, the ability to retain a prey probably depends mostly or exclusively on the tenacity with which its lines adhere to the prey; physical tangling is probably of little importance (though I know of no data on this score). The length of time a prey will be retained before it falls or struggles free will thus depend on the spaces between sticky lines, their effective lengths, their extensibility, their strength, and their adhesiveness.

The adhesiveness of aggregate gland glue varies substantially across spider sizes, and was generally greater for larger spiders (Opell and Hendricks 2009; Blackledge et al. 2011). There is presumably a upper limit to the advantage of increasing the stickiness of a line: it would be disadvantageous to "overbuild" by making a line's adhesiveness stronger than the line itself. Measurements of strength and stickiness (using fine sand paper surfaces) showed that 17 species of araneids, tetragnathids, and nephilids consistently avoided this type of design error: the force of adhesion was substantially less than the force needed to break the spiral line (the range was from 3% to 33%, with a median of 15%) (fig. 3.32) (Agnarsson and Blackledge 2009). The authors estimated that the forces of adhesion to prey cuticle are likely to be greater, bringing the adhesive force closer to but still below the force needed to break the line. Thus when a prey contacts a line at a single site, the prey will be likely to break free of the adhesion before the line itself breaks. One possible advantage of this "safety factor" is that by detaching from the prey instead of breaking, the segment of sticky spiral could remain intact and available to snag the prey as it continues to struggle, as well as to retain other prey (Agnarsson and Blackledge 2009).

There was a tight correlation in inter-specific comparisons between the stickiness of the glue and the strength of the sticky spiral baseline ($R2 = 0.91$), and also a significant correlation between stickiness and the extensibility of the line ($R2 = 0.49$) (Agnarsson and Blackledge 2009) (neither correlation was significantly affected when the phylogenetic relations between the species were taken into account with an independent contrasts analysis). Agnarsson and Blackledge concluded that the biomechanics of sticky spiral baseline silk may have affected

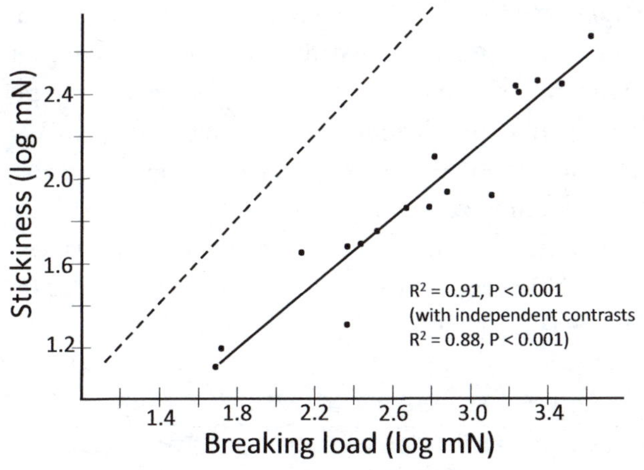

Fig. 3.32. The positive relationship of the stickiness of a sticky spiral line and its own breaking strength illustrate an elegant adjustment between the stickiness and strength of web lines. If the adhesive on a line were stronger than the breaking strength of the line itself, adhesive material would be wasted; but the adhesive strengths and breaking loads of sticky spiral lines in 17 species of araneoids in three families showed that spiders do not make this mistake. Stronger lines tended to be more strongly adhesive, but their adhesive forces did not exceed their own breaking loads (dotted line). Species are *Araneus trifolium*, *A. marmoreus*, *Argiope aurantia*, *A. trifasciata*, *Cyclosa conica*, *Eustala* sp., *Gasteracantha cancriformis*, *Larinioides cornutus*, *Leucauge venusta*, *Mangora gibberosa*, *Metepeira labyrinthea*, *Micrathena gracilis*, *Neoscona arabesca*, *N. crucifera*, *Nephila clavipes*, *Tetragnatha* sp., and *Verrucosa arenata* (from Agnarsson and Blackledge 2009).

the evolution of the stickiness of sticky spiral lines, and that glue properties may also have affected selection on baseline mechanics, a conclusion supported by Opell et al. (2011c). The tensile strength of the glue itself also exceeds its adhesion (Opell et al. 2011b). These linkages between the properties of the glue and the baseline may have had important effects on the evolution of orb web designs.

Another physical property that probably increases prey retention times in araneoids is the high extensibility of the sticky spiral axial fibers (in *Araneus sericatus* they can extend to approximately 300% of their resting length before breaking, as opposed to about 125% for dragline silk) (Denny 1976; Craig 2003; Blackledge et al. 2011). When an insect struggling in a web pushes against a sticky spiral line, it will meet only weak resistance as the line extends. The insect will thus find little support against which it can push or pull, making it harder to scrape or pull off the lines that are adhering to its body (Denny 1976). The ability of a partially extended, highly extensible line to contract and then extend again may also have another positive effect on prey retention. Distortion of the orb in the immediate vicinity of a struggling insect will probably momentarily relax tensions on sticky lines, especially in more or less vertical webs where the weight of the insect will cause the web to sag. The great extensibility of sticky lines would make them able to survive this momentary shortening without sagging into contact with other lines; they would thus still be available to retain the struggling prey.

Sticky spiral lines are often laid in a partially extended state in an orb (Opell and Bond 2000). Most species actively reduce the tension on sticky lines by pulling out additional silk with their legs IV during sticky spiral construction (section 6.3.5), but they nevertheless generally attach these lines under tension (section 3.2.11), in a partially extended state. For example, Denny (1976) found that each segment of sticky spiral was stretched to about 110% of its resting length in the web of *A. sericatus* (AR). There are a few exceptional groups, such as some theridiosomatids and anapids, in which the resting length of a sticky line is often greater than the distance between radii, and the sticky lines hang in catenaries (section 3.2.11, figs. 9.4f, 10.25, Shinkai and Shinkai 1985). Any increase in extensibility is presumably favorable with respect to the ability of a sticky line to absorb the impact of a prey (at least if "pulley conversions" are ignored—see section 3.3.3.1.6). The situation may be more complex for retention (the principal function of sticky spiral lines), because sticky lines will be both tensed and relaxed by the prey's own weight and its struggling movements. A further complicating factor is that the extensibility of a sticky line is also positively related to its length. The longer sticky spiral lines nearer the edge of the orb (section 4.3) are more extensible and likely to be better at retaining prey. The stiffer sticky spiral baselines of uloborids would provide less advantage in this respect (Opell and Bond 2001), though the sticky spiral lines of *Hyptiotes paradoxus* (UL) are highly elastic (Marples and Marples 1937). I know of no quantitative comparisons of the effects of stiffness on the retention of struggling prey.

3.3.3.1.3 MEANS OF ESCAPE: PREY BEHAVIOR

There are surprisingly few empirical data available on details of how prey escape after being stopped by an orb (see section 4.11), and I can present little other than general ideas. The two forces that pull a prey free from an orb are gravity and (less often) the flying force generated by the prey itself (Suter 1978; Blackledge and Zevenbergen 2006) (section 3.3.3.1.5).

Additional forces from the prey's own pushing, kicking, scraping, etc. movements can break lines or their adhesions to its body and free it sufficiently that it is pulled free by gravity or its own flying power. It seems likely that details of prey struggling behavior vary according to the sites on its body contacted by web lines ((Blackledge and Zevenbergen 2006; see Suter 1978 on the order of freeing different appendages, section 4.11.2), and perhaps also its tendency to "play dead" in the web (Suter 1978); but these topics are nearly completely unexplored.

The postures of insects as they jump or fly prior to contacting a web probably also have important effects on retention. For instance, some insects fly with their legs dangling, while others fold them tightly against their bodies (Dalton 1975). Some postures may represent adaptations to defend against being trapped in spider webs. When the long, extended legs of scorpion flies contacted orb webs with only with their distal tips, the insects were often able to maintain both wings free and beating, enabling them to pull free more rapidly than other insects; if, however, their wings contacted the web, they were retained for long times (Blackledge and Zevenbergen 2006) (section 4.11). Drosophilids (Craig et al. 1985), and also mosquitoes that fly with their long thin legs extended (WE), turned away sharply when a leg touched a web line (fig. 3.2). Recognition that the sites where lines contact the prey probably influence retention times introduces caution in interpreting the results of the common technique of measuring retention times by forcefully launching prey or dropping them into webs using forceps (e.g., Nentwig 1982a), rather than allowing them to fly or jump into the web with their normal flight postures (e.g., Blackledge and Zevenbergen 2006).

3.3.3.1.4 SPACES BETWEEN STICKY LINES

One orb web trait that undoubtedly influences retention times is the distance between adjacent loops of sticky spiral. Tighter spacing will presumably tend to result in the prey contacting more lines and being retained for longer. The only direct test of this supposition involved experimentally increasing the distances between sticky spiral loops by breaking some lines in the webs of *Argiope aurantia* (AR); there was a general reduction in retention times as expected, though the effects varied for different insect species (increased spacing lowered retention times for deer flies and large grasshoppers, but not for scorpionflies or small grasshoppers) (Blackledge and Zevenbergen 2006). This experiment may have underestimated the effects of larger spaces between lines, however, because glue from the broken lines remained on the radii.

3.3.3.1.5 AN ORB'S SLANT

An orb's inclination with respect to gravity undoubtedly affects retention times. In an orb close to vertical, gravity tends to pull the prey downward so that it "tumbles" into contact with other web lines directly below it; in an inclined orb, in contrast, gravity tends to pull the prey free from the web. In experiments comparing retention times of sepsid flies in portions of webs of mature female *Metazygia yobena* (AR) that were held in vertical and horizontal positions (most orbs of this species in nature were close to vertical), approximately twice as many prey escaped from horizontal webs in less than the mean time needed by these spiders to reach trapped prey (5.5 s) (Eberhard 1989b). One untested prediction is that the adhesive properties of inclined orbs will tend to be especially tenacious (e.g., denser sticky lines, stickier or more abundant glue). Increases in retention times due to "tumbling" in vertical or nearly vertical orbs are well documented (Eberhard 1989b; Blackledge and Zevenbergen 2006; Zschokke et al. 2006; Nakata and Zschokke 2010; Zschokke and Nakata 2010, 2015). Tumbling may favor vertical orbs that are narrower and more elongate vertically, and may also explain the apparently common pattern of greater edge-to-hub decreases in sticky spiral spacing above the hub than below it (Zschokke and Nakata 2010, 2015; Nakata and Zschokke 2010; section 4.3).

The effects that a small deviation from perfectly vertical (say a 80° or 70° web versus a 90° web) have on the likelihood that the prey will tumble have never, to my knowledge, been carefully studied. At least some tumbling occurred at slants as low as 70° in webs of mature female and penultimate female *Nephila clavipes* (Zschokke et al. 2006; WE). Four of the six possibly independently derived groups with probable moth-specialist webs—the ladder webs of *Tylorida* sp. (Robinson and

Robinson 1972), *Scoloderus tuberculifer* and *cordatus* (fig. 3.4) (Eberhard 1975; Stowe 1978, 1986), *Pozonia nigroventris* (fig. 3.35), and *Deliochus* sp. (fig. 3.5) (Kuntner et al. 2008a), build webs with orientations close to 90°. Moths are tumbling specialists, leaving behind a trail of their detachable scales that adhere to sticky lines as they fall (fig. 4.20; section 4.11.2), so these orbs may be anti-tumbling specializations. This leaves unexplained, however, why the slants of the orbs of two other apparent moth specialists, *Poltys* spp. (Smith 2006) and *Acacesia hamata* (Eberhard 1976a), were not especially close to 90° (table O3.1).

3.3.3.1.6 "PULLEY" ATTACHMENTS OF THE STICKY SPIRAL TO THE RADII

The strength of the attachments of sticky spiral lines to radii is another material property that affects prey retention. These connections usually ceded partially when sticky spiral lines of araneids and nephilids were stressed moderately (Eberhard 1976b): the sticky line remained attached to the radius but slid through the connection, thus converting the sticky spiral–radius junction into a sliding connection (a "pulley"). This allowed sticky spiral line from adjacent sections of the orb to be added to the segment of sticky line that was being stressed, lengthening the stressed segment of sticky line and thus increasing its ability to resist breaking (fig. 3.33). Sliding connections (which were apparently rediscovered by Craig [1987a], who used the term "fiber slip"), may depend on a special piriform gland product found in these junctions (Kavanagh and Tillinghast 1979). Pulley connections were rare or absent in the webs of all four genera of tetragnathids and in one uloborid that were tested (Eberhard 1976b), so there may be taxonomic differences.

It is not known how the components of the attachment discs themselves (which are applied when the radius makes an unusual "sideways" angle with respect to the spinnerets—section 2.3.2.3) lead to slipping. Occasional slips also resulted when a radius near a radius-frame attachment was pulled parallel to the direction of the frame (6 of 22 attempts with two webs of two adult *Allocyclosa bifurca* females); the radius-frame attachment slid along the frame line without the radius breaking free (WE). Stresses of this sort are probably very rare in nature, however (radii in intact orbs pull more nearly perpendicularly to frame lines), and the pulley conversions in this context may not have any biological significance. They do, however, illustrate that radial and frame lines are not twisted or knotted around each other at attachment points. Experimental pulls on radii that were more or less perpendicular to the plane of the orb (and thus exerted stress on the radius-frame attachment that was more or less perpendicular to the frame line) usually

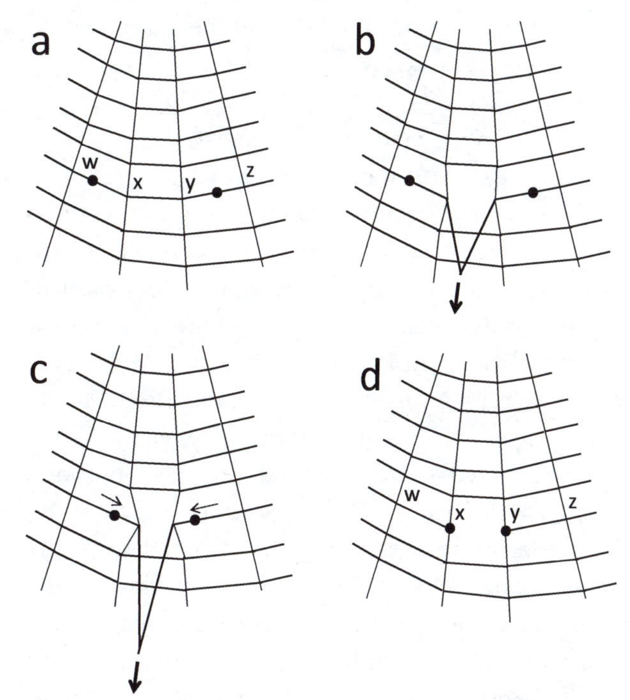

Fig. 3.33. The "sliding connections" of sticky spiral lines to radii illustrated diagramatically here make the orbs of araneid and nephilid spiders even more treacherous traps for prey. When marks (black dots) were placed on two segments of a loop of sticky spiral (*a*), and the segment between them was then pulled away from the plane of the orb (arrow in *b*), the stress on the connections with the radii on either side (*x*, *y*) gradually increased (note radius displacements in *b*). Before the stress became strong enough to break the spiral or to pull it away from the radius, however, the attachments to the radii failed "partially"; the marked portions of the line moved toward the radii (dotted arrows in *c*). Although the sticky line remained firmly attached to both radii, the axial line of the sticky spiral slid through the attachments at *x* and *y*. In effect, these attachments were converted into pulleys, lengthening the portion of the spiral line that was being pulled and thus increasing its ability to absorb further force without breaking. Eventually, as the pull continued, the sticky line broke (*d*). Sometimes it did not break until further nearby attachments (*w*, *z*) had also been converted into pulleys. Sliding connections of this sort are absent, as far as is known, in the orbs of tetragnathids, theridiosomatids, the symphytognathoid families, and uloborids (from Eberhard 1976b).

caused the radius itself rather than the attachment to break (18 of 22 cases) (WE).

3.3.3.1.7 VARIATIONS IN RETENTION TIMES AND THEIR CONSEQUENCES

Studying the effects of web traits on retention times is complicated by the fact that retention times in orb webs typically show huge variation, even for a particular species of prey and spider (e.g., Robinson and Robinson 1973a; Eberhard 1989b; Miyashita and Shinkai 1995; Blackledge and Zevenbergen 2006; Zschokke et al. 2006). For instance, the ranges of retention times for moderate-sized wasps (20–25 mm) that were dropped onto webs of *Nephila clavata* and *Argiope bruennichi* encompassed factors of 75x and 15x respectively, despite very small sample sizes (N = 10, 13) (Miyashita and Shinkai 1995). The same conclusion is suggested by the fact that substantial changes in the physical arrangement and spaces between sticky lines in sectors of orbs of *Leucauge mariana* (TET) and *Argiope aurantia* (AR) had no perceptible or only erratic effects on the retention times (Eberhard 1989b; Blackledge and Zevenbergen 2006).

Many factors are probably responsible for producing this huge variation. The number of sticky lines adhering to the prey presumably varied with the site in the orb where the prey hit and the angle of the prey's path with the plane of the orb, and also according to which sites on the prey's body were contacted. Even apparently minor details could be important. A line contacting an area of the prey whose surface has longer setae or is more oily will presumably result in weaker adhesion (Nentwig 1982a; Opell and Schwend 2007). Adhesion to the prey's wings could be especially important by impeding their use to pull the insect free (Blackledge and Zevenbergen 2006). The strength of adhesion to the cuticle of different prey species varies widely (e.g., Eisner et al. 1964; Opell and Schwend 2007). Variation in the developmental stage of the spider can affect the strengths and stickiness of her lines (section 3.2.14). For example, the webs of penultimate spiderlings of the araneid *Metazygia* sp. lost sepsid flies within 5 s about twice as often as did those of mature females (Eberhard 1989b). Even the humidity and the season of the year can have effects. The webs of the araneid *Argiope trifasciata* were less sticky late in the fall, possibly due to lack of silk reserves or to changes in investment preferences (Opell et al. 2009). The stickiness can even vary within a single orb; the outer loops of *Argiope aurantia, A. trifasciata,* and *Cyclosa turbinata* were stickier than the inner loops (Opell et al. 2009; Crews and Opell 2006). Although some of these effects are only weakly documented, the general message is clear: many different factors must affect retention times.

3.3.3.1.8 SUMMARY

Many different factors can influence retention times, and they can have potentially important effects on prey capture. As yet it has not been possible to demonstrate how evolutionary adjustments have occurred in particular cases, or to explain the sticky spiral spacing in any particular species in terms of measured retention times. Studies of retention constitute an important frontier for future studies.

3.3.3.2 Reduce the web's visibility

3.3.3.2.1 SOME INSECTS CAN SEE ORB WEBS

Documenting that a fly can see and avoid orbs can be as simple as watching empidid flies flying back and forth alongside, around, and through the orb webs of *Tetragnatha* sp. and *Metabus ocellatus* over a stream, repeatedly landing on small insects that are trapped in the webs but never becoming entangled in or apparently even brushing against the webs (Nentwig 1982a; WE). Other insects also maneuver easily in flight near orbs. Several species of stenogastrine wasps regularly pluck small prey from orb webs by hovering close to the web (Turillazzi 2013), and there are many other incidental observations of insects avoiding orb webs in nature (Lubin 1973; Chacón and Eberhard 1980; Rypstra 1982, 1985; Craig 1986, 1994a,b; Gregorič et al. 2011). As might be expected, some insects also apparently use vision to avoid capture by other types of webs such as the aerial sheets of the linyphiid *Linyphia triangularis* (Turnbull 1960).

When insect flight paths were filmed in well-lit glass wind tunnels that contained an orb web with one of two different designs (dense orbs of the araneid *Mangora pia*, or sparse orbs of the smaller theridiosomatid *Theridiosoma globosum*) (Craig 1986), *Drosophila melanogaster* and the culicid mosquito *C. quinquefasciatus* performed five different types of defensive responses that reduced their chances of being captured when they came near an orb: turning away from the web at a distance; approaching and then hovering between 5 and 0.5 cm from the web's surface before flying away; approaching the web, touching its surface, and rapidly accelerating away (fig. 3.2); rapidly turning and flying away from near the web but without touching its surface; and steering to and flying through holes in the web caused by previous web damage. Sometimes these flies extended their front legs as they approached an orb to touch it (Craig 1986), probably thus confirming the web's location and also minimizing the area of contact with it. The fruitflies (but not *C. quinquefasciatus*) began to avoid the dense *M. pia* webs at a greater mean distance (7 cm) than the webs of *T. globosum* (2 cm). The flies' abilities to avoid webs were so good that Craig concluded that "it seems surprising that orb-webs capture any prey at all" (p. 54).

Of course, not all insects have fine control over their flight trajectories or encounter orbs in well-lit flight tunnels. The rapid "turn-away" responses of these relatively slow-flying flies (the mean air speed of *D. melanogaster* in still air was about 11.6 cm/s—Craig 1986) are probably difficult or impossible for many other more clumsy and rapid fliers such as grasshoppers, beetles, neuropterans, leaf hoppers, flying termites, etc. In addition, not all web lines are easily seen. Different types of silk differ in their visibility. Much of the light intercepted by non-sticky orb web threads is not reflected, but the glue droplets on the sticky spiral lines reflect, disperse, and scatter light; the sticky spiral is probably largely responsible for an orb's visibility to prey (Craig and Freeman 1991). Araneoid ampullate silk also reflects less UV, perhaps making it less visible to insect prey (Craig 1994a,b).

3.3.3.2.2 DOES VISIBILITY AFFECT PREY CAPTURE IN THE FIELD?

Have orb weavers evolved web designs whose function is to reduce the web's visibility to prey? Answering this question is technically difficult. The visibility of a web in the field is undoubtedly affected by complex interactions among the spider's size, the sizes of the sticky droplets, the reflectance properties of silk lines, web architecture, ambient light, visual backgrounds, the angle of approach of the prey to the web, and insect visual capabilities (Craig 1990; Craig and Freeman 1991; Craig and Lehrer 2003a,b). Measuring the relative impact of the visibility of a orb design on prey capture under ecologically realistic conditions is not simple. In addition, a well-lit, windless lab setting with a simple visual background is appropriate for testing the *capabilities* of an insect, but not for determining whether these capabilities are sufficient and are used to avoid orbs under natural conditions. A second technical difficulty concerns the ecological realism of the insect's own behavior. It is not certain, for example, that a fly struggling to escape from a container, or one that was just blown from a tube toward a web, uses the same cues or flies in the same ways that it would while foraging in nature.

Several studies have concluded, on the basis of evidence of varying quality, that an orb web's visibility has important effects on its trapping abilities. Perhaps the extreme is the claim that even in orbs that are only up at night there is a negative correlation between web visibility and prey capture; "web visibility is an important factor affecting the foraging performance of *both* nocturnal and diurnal spiders" (Craig and Freeman 1991; p. 251) (emphasis in the original). My own assessment is that it is too early to draw confident conclusions regarding the effects of visibility on the evolution of different orb web designs. Because several problems involve technical questions related to the techniques that were used, I have relegated this critical discussion to appendix O3.1, and will only summarize here by saying that problems with the data are sufficiently numerous and serious to cast serious doubt on the conclusions.

Inter-specific comparisons offer a second type of data. When the sizes of the droplets of glue on the sticky spirals and the silk brightness of species were compared in 23 species in 16 genera of araneids and tetragnathids, the traits that increased visibility were relatively greater in species with nocturnal rather than diurnal orbs. In addition, the nocturnal orbs appeared to tend to have relatively greater numbers of sticky spiral loops (Craig and Freeman 1991). These differences, however, have a second (non-exclusive) explanation: greater adhesion is advantageous at night because moths are abundant at night and are especially difficult to retain because of their deciduous scales (Eisner et al. 1964).

A futher possible indication of the importance of web visibility is the relatively sticky spiral-richness of trunk orbs. Visibility is probably a minor problem for these webs, because (at least for humans) they are very difficult to see against the nearby trunk. Other differences in the prey likely to be trapped by trunk webs, such as their expected lower velocities, complicate these comparisons, however.

In sum, there is clear evidence that some insects, under certain conditions, can see orb webs, and use this information to avoid them. Because of limitations in the current data (appendix O3.1), however, the conclusion that visibility is "a key factor affecting spider foraging success" (Craig 1988, p. 279) is probably too ambitious. The additional possibilities that prey see non-orb webs at night and are attracted to them (Lai et al. 2017), is also discussed in appendix O3.1.

3.3.3.2.3 YELLOW SILK

One way to alter a web's visibility is to alter the color of its lines. Strikingly, yellow silk has evolved independently at least six times in orb weavers (table 9.1) (and has apparently never evolved in any other type of web). The yellow color, which in *Nephila clavipes* orbs results from the presence of four different pigments (Holl and Henze 1988), was stronger in webs built at sites with greater ambient light intensity, and when the illumination was richer in wavelengths between 500 to 650 nm (yellow and orange) (Craig et al. 1996). Spiders like *N. clavipes* may have evolved to produce pigments only in open sites where the background makes the yellow color less visible (Craig and Lehrer 2003a). Similarly, the yellow color of dragline silk is stronger in *N. clavata* later in the season when more leaves are yellow or brown (October as opposed to July and August in Japan) (Osaki 1989). But yellow webs in the tropical areneids *Cyclosa nodosa* and *Araneus expletus* are built during the wet season, when nearly all the nearby leaves were green (WE), and those of *Micrathena schreibersi* occurred low in the understory of older forests, where most of the background was brown or green (WE).

Another possibility is that yellow lines are attractive to insects (Craig et al. 1996). In *N. clavipes*, *M. schreibersi*, and *A. expletus* the color is also more intense on the sticky lines, which at least for humans, are the most visible (Craig and Freeman 1991). The possibility that groups other than *Nephila* also modulate the yellow colors of their lines has not been checked. In sum, the functional significance of the repeated convergences on yellow lines in orbweavers is not yet clear.

3.3.3.3 Other functions

3.3.3.3.1 SURVIVE ENVIRONMENTAL INSULTS

As illustrated by the tattered web in fig. 3.19 that was photographed near the end of a moderately windy day, orbs can take a beating during their short working lives. Most of the photos in this book are of freshly built orbs (see fig. 1.4 of this same species), but environmental damage to orbs is undoubtedly common in nature, so these photos are not entirely representative of the structures with which spiders capture most of their prey in nature. Assuming that the design features of undamaged webs are those which natural selection has favored, it seems reasonable to assume that web traits that allow an orb to minimize the typical damages resulting from exposure to their environments have been favored. In general, sticky lines that are longer, under less tension, or closer together are all probably more susceptible to being damaged by adhering to other lines. Sticky lines are probably also more easily and more often broken than non-sticky lines. But in the absence of data from nature, these are only guesses.

It is often difficult to distinguish the damage due to prey capture, prey escape, detritus, rain, wind, and the tracks made by the spider herself. In the only systematic attempt that I know of, Craig (1989) concluded that among five tropical forest species, the webs of three larger species (5.3–147 mg) suffered less overall damage, and had proportionally more of this damage due to prey, than the webs of two smaller species (0.8–2.7 mg). This study had several weaknesses, however. Most importantly, it lacked clear descriptions of criteria to distinguish some of these types of damage (see appendix O3.1). In addition, wind damage was apparently assayed on the basis of observations of experimental damage to orbs that had been coated with powder (e.g., fig. 2 in Craig 1989) rather than unloaded webs; the sample was small; and many other factors in addition to spider size varied in these taxa. In addition, some damage resulted from lines being combined rather than broken, making their effects difficult to anticipate. For example, the common central adhesion pattern in adjacent loops of sticky line (fig. 3.18b, section 3.2.11) produces a stronger, compound double line that has larger amounts of glue on it, but it opens a space with no sticky lines in the web. The increase in thread diameter could have two simultaneous effects, increasing the web's drag while also increasing its ability to resist drag. The magnitudes of the effects that these different types of damage have on the web's ability to capture prey are unknown; different types of damage likely have different effects on different web functions such as intercepting, stopping, and retaining prey.

Most of an orb's air friction is probably due to its sticky lines, both because the total length of sticky line in an orb is substantially greater than that of the non-sticky lines (in a typical orb of *M. duodecimspinosa*, for example, the relation between their lengths is around 2:1), and because the droplets of sticky material result in greater friction with the air (Zaera et al. 2014). The consequences of friction for orbs may be mixed, however. In general, shorter, fewer, and thinner lines have less drag. Under generalized loading by wind, greater drag is probably disadvantageous because it increases the danger of lines breaking (Langer 1969; Zaera et al. 2014). Under local loading by prey impact, however, greater friction with the air may help dissipate the kinetic energy of the prey and thus improve the orb's stopping ability (Lin et al. 1995; Zaera et al. 2014) (see, however, Sensenig et al. 2012 and section 3.3.2.1). Friction may also reduce the rebound of the orb when stressed by an impact and thus improve its retention ability. Traits that could conceivably affect the likelihood of damage include spaces between sticky spiral loops, the spaces between radii (and thus the lengths of sticky spiral segments), tensions on lines, and the degree of rigidity of objects to which the orb is attached. In short, the factors affecting frequencies of different types of web damage and their consequences for prey capture are likely to be complex.

One pioneering study documented facultative adjustments to different wind velocities by *Cyclosa mulmeinensis* (AR) (Liao et al. 2009). Some spiders lived near the seashore, where winds were relatively stronger than farther inland. When spiders were experimentally subjected to windy conditions in captivity that were similar to those in the seashore habitat, they made orbs that were smaller and had fewer, more widely spaced sticky spiral loops than when they were kept in windless conditions (Liao et al. 2009). The sticky spiral loops in wind webs were also relatively low in number and more widely spaced in inter-specific comparisons with other *Cyclosa* species (fig. 3.34), and direct measurements showed that the drag of these webs was reduced. In addition to these changes in orb geometry, there were changes in the silk lines: the major ampullate gland silk of *C. mulmeinensis* was stronger in windy conditions (Liao et al. 2009). The mechanism by which this improvement in strength was achieved was not clear. Neither the protein composition nor the diameters of the lines changed. In sum, these *Cyclosa* spiders apparently sacrificed some trapping abilities of their webs to improve their webs' abilities to survive under windy conditions.

3.3.3.3.2 REDUCE CONSTRUCTION COSTS AND PHYSICAL CONSTRAINTS

There are two types of energetic cost of a web: energy expended as building behavior; and energy expended in the silk itself. The total energetic costs of a web (Lubin 1986) will be $C = [M + (S - Sr)]\ p + [M + (S' - S'r)]\ q$, where M = the metabolic cost of construction behavior; S and Sr are the energetic content of the total silk and of the portion of the silk that is recycled when a web is replaced; S′ and S′r are the energetic content of the total silk and the portion of the silk that is replaced when the web is repaired; and p and q are the frequencies with which webs are replaced and repaired. Constructing an orb involves a period of sustained physical exertion that lasts on the order of half an hour or more and that can have substantial metabolic costs (e.g., Peakall and Witt 1976; Sherman 1994; Opell 1998). The additional maintenance costs of the silk glands are probably comparatively small (Prestwich 1977).

3.3.3.3.2.1 Behavioral costs?

Nakata and Mori (2016) argued that more complex behavior has greater costs in orb weavers. They found that *Cyclosa argenteoalba* and *Eriophora sagana* (AR) took more time to build more asymmetrical webs, and argued that this was linked to the greater behavioral complexity needed to build the sticky spiral in such webs. This particular measure of the difficult concept of "complexity" is open to discussion, and there is also no certainty that greater complexity is necessarily linked to greater metabolic costs (for instance, greater web symmetry might require more extensive or thorough exploration behavior, and be more costly). Nevertheless,

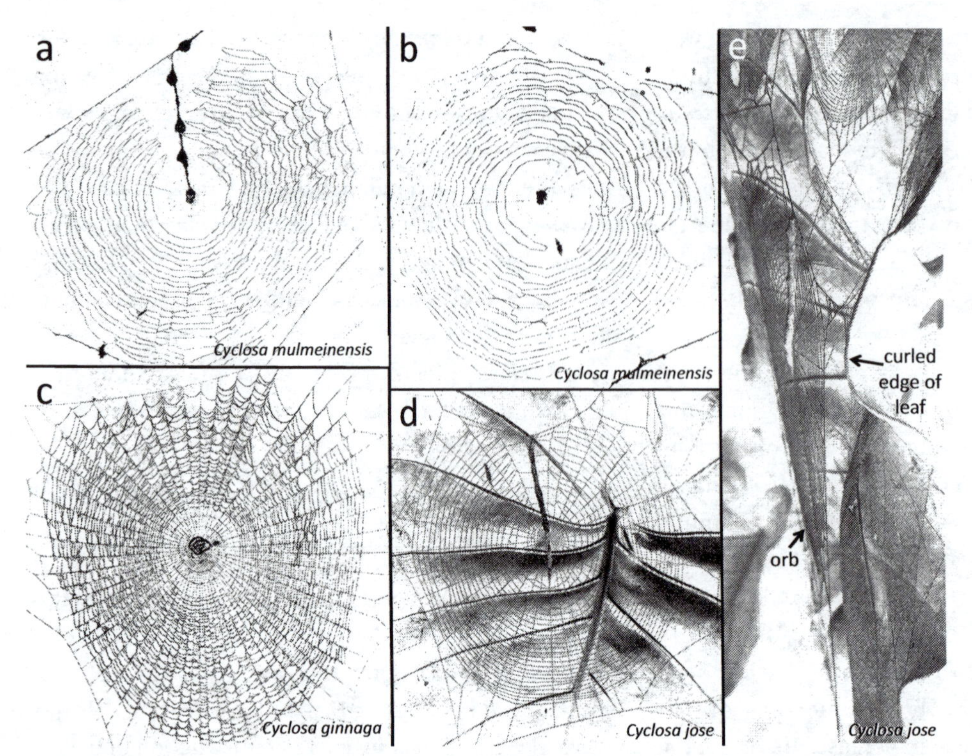

Fig. 3.34. These images illustrate the probable effects of the wind on orb design in the araneid genus *Cyclosa*. Webs of *C. mulmeinensis*, which builds at windy seashore sites, have sharply lower numbers of sticky spiral lines and the sticky spiral lines are relatively widely spaced in webs both with (*a*) and without (*b*) stabilimenta; the maximum numbers of loops were only 23 and 20 in these orbs. In contrast, the orbs of several other species of *Cyclosa*, which are often built in forest habitats, had relatively large numbers of radii, and closely spaced sticky spiral lines, as illustrated in the web of *C. ginnaga* with 47 loops (*c*) (see also figs. 3.27 of a related species). The relative exposure to the wind of the orbs of *C. jose* (*d*, *e*), which were also somewhat sparse, was uncertain; they occurred at open sites, but often spanned large leaves that were curled at the edges (arrow in the lateral view in *e*), that may have sheltered them somewhat from the wind. When *C. mulmeinensis* spiders were experimentally exposed to wind, they made smaller orbs with lower numbers of sticky spiral loops (Liao et al. 2009). Presumably the lower numbers and wider spacing of the sticky spiral loops served to reduce damage from the wind (the "sticky spiral entanglement" hypothesis in section 3.2.11.3.2) (*a* and *b* courtesy of I-Min Tso, *c* courtesy of Daiquin Li).

there may be a link between costs and complexity in the sense that increased analytic abilities may be needed to perform more difficult behavioral tasks. Building and maintaining the neural equipment with which to perform behavioral tasks undoubtedly has costs. Possibly, for instance, the portion of the nervous system of an orb weaver that is responsible for web-building behavior is relatively larger overall and more costly than that of a similarly sized mygalomorph. The only data I know of on this score, however, do not support this predicted cost of web complexity. The relative sizes of the "brains" (the supra- and sub-esophageal ganglia) did not differ in the orb weavers *Mysmena* sp. (MYS) and the related anapid *Anapisona simoni* (AN) (correcting for body size) from those of the kleptoparasitic mysmenid (*Mysmenopsis tengellacompa*) that has secondarily lost webs completely or another kleptoparasitic theridiid *Faiditus elevatus* (Quesada et al. 2011). These considerations do not mean, however, that the time needed to build a particular web trait (whether or not the behavior is more or less "complex") does not entail a cost, both in terms of energy spent and risk of predation (previous section).

3.3.3.3.2.1.1 Orb weavers

Orb weaver metabolism during web construction has, to my knowledge, never been measured directly. Indirect data on nephilids in the genus *Nephila* and the sheet-orb building araneid *Cyrtophora moluccensis* suggested that behavioral costs are a much lower fraction of the total cost than are the material costs (2.6% and 4.2% respectively in the two species) (Lubin 1973). These estimates were made on the basis of the time spent building and measurements of metabolism while spiders rested and moved about (to an undetermined extent) in respirometry flasks, rather than on measurements made during web construction, and may be underestimates (the metabolic rates of two non-orb weavers increased by about 50% and 100% while building—see section 3.3.3.3.2.1). One conservative, speculative estimate that attempted to assign weights to several factors placed the behavioral costs for building a typical orb at 20% of the material costs (Eberhard 1986a).

An empirical study estimated the metabolic costs (but not the material costs) of building an orb indirectly in the moderately large (115 mg) araneid *Araneus diadematus.* Counts of the steps taken by certain legs from movies were combined with literature accounts of the metabolic costs of a "step" in other animals (Gold 1973). Unfortunately interpretation of these data is difficult: movements of the spider's legs vary widely (chapter 6); there was no description of how the complex paths of spiders during sticky spiral construction (e.g., McCook 1889; fig. 6.4) were translated into "steps"; some assumptions regarding the paths that spiders follow during construction of frame lines were incorrect (see sections 6.3.2, 6.3.3); and the uncertain supposition was made that the spiders' movements corresponded to "running" rather than "walking" in other animal groups.

The combined behavioral and material costs of orbs were estimated in another study by determining weight losses in the araneid *Zygiella x-notata* (Venner et al. 2003). During the experiment some spiders were kept for 10 days in glass tubes, where they could not build orbs but presumably moved around to some (undetermined) extent; others were placed in frames where they built one or more orbs. All individuals were apparently kept without food or water during the 10 days prior to the experiment as well as during the 10 day experimental period. Possible technical problems again make it difficult to evaluate the results: the inherent imprecision of estimating metabolism from fresh body weight (which is influenced, for example, by the degree of dehydration, the time since the most recent defecation, etc.); uncertainties associated with assuming that silk lines are uniform (see section 3.2.9); and uncertainties regarding the cost of non-sticky silk material and building behavior (above). In general, determining the metabolic costs of moving is difficult in spiders, and requires subtle analyses (e.g., Moya-Laraño et al. 2007, 2008, 2009).

In sum, the behavioral costs of orbs probably tend to be lower than the material costs, but the fraction of total costs may vary substantially. The preliminary nature of the available data precludes making precise evaluations of the effects of many types of alternative geometrical designs using equal amounts of silk. All other things being equal, the behavioral costs of building vertical orbs should be greater than those of similar horizontal orbs.

3.3.3.3.2.1.2 The special case of uloborids

The cribellate family Uloboridae has special behavioral costs (Lubin 1986). Uloborids pull sticky silk from their cribellum using high numbers of rapid combing movements with their hind legs. For instance, a videotaped mature female *Zosis geniculata* at about 20°C in the middle portion of her web made a mean of 5.82 combing movements/s, and produced one segment of sticky spiral every 20.2 ± 5.7 s (WE). The length of the sticky spiral line in one normal-sized orb was 1281 cm in direct measurements from a photograph; there were 585 segments of sticky spiral line, and the mean length of a segment was 2.19 cm. This gives an intimidating estimate of about 69,000 combing movements/orb.

Approximate calculations for *Uloborus conus* gave a similar result. A spider produced about 7.2 cm of sticky spiral/min (about 8 s/cm) during the 210 min required to build its orb (Lubin 1986). Assuming that about 180 min of this was spent in sticky spiral construction, and that the combing rate in *U. conus* is similar to that in *Z. geniculata* (this is conservative, because *U. conus* is smaller), this gives about 54,000 combing movements/web of *U. conus*. In sum, the energy expended in uloborids in sticky spiral production may constitute a much larger fraction of the total energetic cost of building an orb than in araneoids. I know of no direct measurements of the metabolic cost of combing behavior, but the rapidity of the movements and their huge numbers indicate that it is substantial.

Several additional, hitherto unremarked factors may be associated with reducing the costs of sticky spiral lines in uloborids. They are unusual among orb weavers in skipping radii (not attaching the sticky spiral to each radius crossed) (figs. 3.26, 3.31, 7.24). Skipping reduces the uniformity of the spaces between loops of sticky spiral, but increases the speed of sticky spiral construction, and reduces the energy needed; the advantages may have been enough to compensate for the costs of reduced uniformity of spacing. Other energy savings may come from approximately horizontal webs, and a relatively low ratio of the numbers of sticky spiral loops to the numbers of radii (fig. 4.12, section 4.5).

3.3.3.3.2.2 Energetic constraints?

Some patterns in orb webs may be due to variables that are energetic constraints on the spider, rather than design features that are favored by natural selection to increase prey capture. For instance, Herberstein and Heiling (1999) argued that the asymmetrical catching areas (smaller above than below the hub), and the wider spaces between sticky spiral lines above the hub in the araneids *Argiope keyserlingi* and *Larinioides sclopetarius*, are consequences of the greater effort needed by the spider to raise her abdomen to attach sticky spiral lines above as opposed to below the hub. Younger spiders of both species made webs that had a greater up-down symmetry, and, as predicted, the tendency for wider spacing above the hub was accentuated when weights were added to the abdomens of *L. sclopetarius* spiderlings (Herberstein and Heiling 1999).

I am not certain, however, of the importance of these constraint arguments. More energy will surely be expended to climb upward and to raise the abdomen to make each attachment, but upward and downward movements occur both above and below the hub, and it is not obvious that the up-down distances moved by the abdomen are greater above than below the hub (figs. 6.1, 6.23). Another reason to doubt the constraint argument is that experimental addition of weights to *Araneus diadematus* (AR) gave different results: the vertical asymmetry of the orb did not change, and there was no correlation between body weight and vertical asymmetry in control spiders (Coslovsky and Zschokke 2009). Two different estimates of the costs of adding sticky spiral lines above and below the hub that were derived from videotapes of building behavior in *A. diadematus* showed different patterns. The "specific-energy-cost" (estimated on the basis of the up-down distances moved divided by the length of the spiral line laid) was greater in the upper portion of the web, while the "time cost" (length of time needed to build a given length of sticky line) was greater in the lower portion (Coslovsky and Zschokke 2009). The wider spacing above the hub that resulted from experimentally increasing the spider's weight could also be explained if heavier spiders are relatively slower in attacking prey above rather than below the hub than are lighter spiders. Reduced investment in the sector that is likely to have lower likely payoffs could have evolved in the context of the weight gains resulting from large meals or the accumulation of eggs prior to oviposition. The energetic constraint idea suggests that the wider spacing is, in contrast, an inadvertent consequence of the spider's lack of flexibility in the amount of effort she spends to make an attachment.

One further behavioral cost constraint that has been proposed is that the wider spacing between sticky spiral loop near the edges of orbs of some species (see section 4.3.2) is disadvantageous for prey capture, but is nevertheless favored because it approximates more closely a logarithmic spiral and that this spiral conserves energy because it represents the shortest distance between radii (Burgess and Witt 1976). This idea can be more confidently discarded, because in fact spiders do not move directly from one attachment to the next (e.g., figs. 6.1, 6.23, Zschokke and Vollrath 1995a), so a logarithmic spiral for the line does not imply the shortest distance moved by the spider. The non-logarithmic patterns in the spacing of both temporary and sticky spirals also argue against this constraint model (section 3.3.2.8).

3.3.3.3.2.3 Material costs

The costs of the different types of silk in a web presumably depend on their lengths, their diameters, and their chemical compositions. In araneoid orbs, a substantial fraction (up to just over half in *Nephila clavipes* [NE]) of the dry weight of a web is in the water-soluble fraction (presumably mostly the glue on the sticky spirals) (Fischer and Brander 1960 in Tillinghast and Christenson 1984; Tillinghast and Christenson 1984). Measured by wet rather than dry weight, the aqueous glue undoubtedly constitutes an even larger fraction of the web's weight. Venner et al. (2001) argued that sticky silk is "a limiting factor" in web construction, because of the combined costs of production and of building; this belief was supported by the major changes in web design that were induced by experimentally altering the amount of sticky silk available to the spiders *Leucauge mariana* (TET) and *Micrathena sexspinosa* (AR) at the beginning of construction (Eberhard 1988a). I know of no studies of the fibrous, non-soluble portion of an orb in which the respective amounts and costs of all of the different types of line (draglines, attachment disks, sticky spiral baselines) were distinguished. The trypsin-digestible portion of these fibers in a *N. clavipes* web (the sticky spiral baseline) constituted 25% of the total weight of the fibrous material (Tillinghast and Christenson 1984). The cost of the synthesis of this flagelliform baseline material, compared with that of other lines, is apparently not known. The 25% figure may be an underestimate, because small cages in which the webs in this study were built (Tillinghast and Christenson 1984) may have resulted in atypically small orbs, which tend to have a higher ratio of the length of non-sticky to sticky lines (Eberhard 1986a).

Designs of non-sticky webs that are more durable should, again other things being equal, be cheaper over the long term (Lubin 1973; Ford 1977). This expectation was met in a comparison of *Cyrtophora moluccensis* (AR) with *Nephila clavipes* (NE) webs (Lubin 1973). A mature female *C. moluccensis* may build only a single complete web once in her adult lifetime of about 3.5 months; typically she needs to renew only approximately 10% of it every 18 days (Lubin 1973). In contrast, material in the orb portion of a *N. clavipes* web lasted only 2–3 days (Lubin 1973; Higgins 1987), and the webs of many other orb weavers were even more ephemeral (see section 8.7).

Many species, at least among araneoids, recycle their webs (section 2.5). Recycling recovers about 30% of the material costs (Townley and Tillinghast 1988), and must increase the importance of behavioral in relation to material costs. In general, selection that favors energetic efficiency in construction behavior seems likely to be stronger in orb weavers than in non-orb weavers (Burgess and Witt 1976). The costs of the silk in a web have generally been quantified in terms of calories/mg (e.g., Lubin 1973; Tanaka 1989). This technique, based on bomb calorimetry, may overestimate costs, as the spider presumably synthesizes silk from energy-rich precursor molecules. But I know of no empirical data on this point.

3.3.3.3.2.4 The (unknown) costs of vigilance

The preceding discussion has followed the published literature in concentrating on the metabolic costs of building behavior per se. This is surely only a part of the metabolic cost of using a web to capture prey. Spiders often respond rapidly to avoid losing prey, and it seems likely (though I know of no direct measurements) that remaining "alert" and ready to respond quickly to prey entails metabolic costs above the resting metabolic rate. A metabolic cost is obvious in some species in which the spider actively tightens lines in her web by flexing her legs; examples include the uloborids *Hyptiotes* and *Miagrammopes* (fig. 3.1); the theridiosomatids *Theridiosoma*, *Epeirotypus*, and others (table 3.4); the theridiid *Phoroncidia* (table 9.1); *Pozonia nigroventris* (AR) (fig. 3.35); and araneids in the genera *Mangora* and *Micrathena* (figs. 3.36, 3.24). In *Hyptiotes cavatus* the spider's thick legs, their especially rich tracheal supply (Opell 1987a,b), and their especially dense populations of mitochondria (Opell and Konur 1992) confirm that holding the web tight for hour after hour is metabolically costly. The relatively thicker anterior legs of the theridiosomatids *Theridiosoma* and *Epeirotypus*, which also tense their webs, compared with the thinner anterior legs of other theridiosomatids that do not tense their webs (Coddington 1986a), also suggest increased metabolic costs of tensing the web. The likely payoff in these species is from increased stopping and retention of prey when the spider sags her web immediately following prey impact.

At least some spiders that build typical orbs also appear to maintain high levels of vigilance while waiting at the hub. Consider, for instance, the reaction times of a human athlete and the tetragnathid *Leucauge mariana*. In the 0.1–0.2 s that it takes an Olympic human sprinter, hyper-alert and tensed in the starting blocks for a 100 m dash, to begin to move perceptibly after the starting gun has fired, the spider (with no prior warning) has already located the source of vibrations, turned, and run several body lengths to reach a prey that struck her web (fig. 3.28) (Briceño and Eberhard 2011). Even the large, otherwise slow *Nephila clavipes* can reach a prey more than a body length away in a few tenths of a second after impact

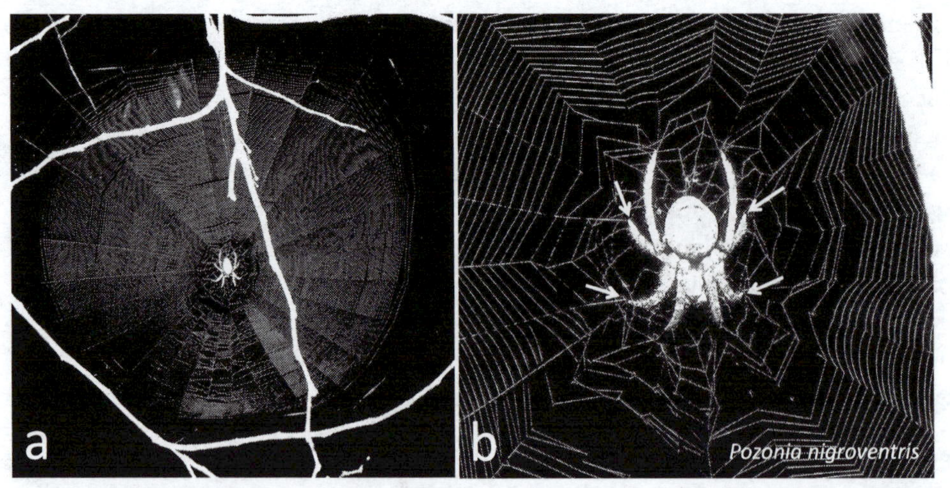

Fig. 3.35. This mature female *Pozonia nigroventris* (AR) tightened the lines at the hub as she waited at the hub (the arrows in *b* indicate the lateral displacements of lines held by legs II and III) of her previously unknown, sticky spiral-rich nocturnal web (*a*). Tensing a planar web while waiting for prey and then relaxing it the moment that a prey strikes has evolved convergently in several araneids (figs. 3.5, 3.36) and in uloborids (fig. 10.23), and may aid in prey retention. Tensing the web probably reduces its billowing movements in response to air movements, arguing against the idea that these erratic movements are important in increasing prey interception (section 3.2.11) (photos courtesy of Gilbert Barrantes).

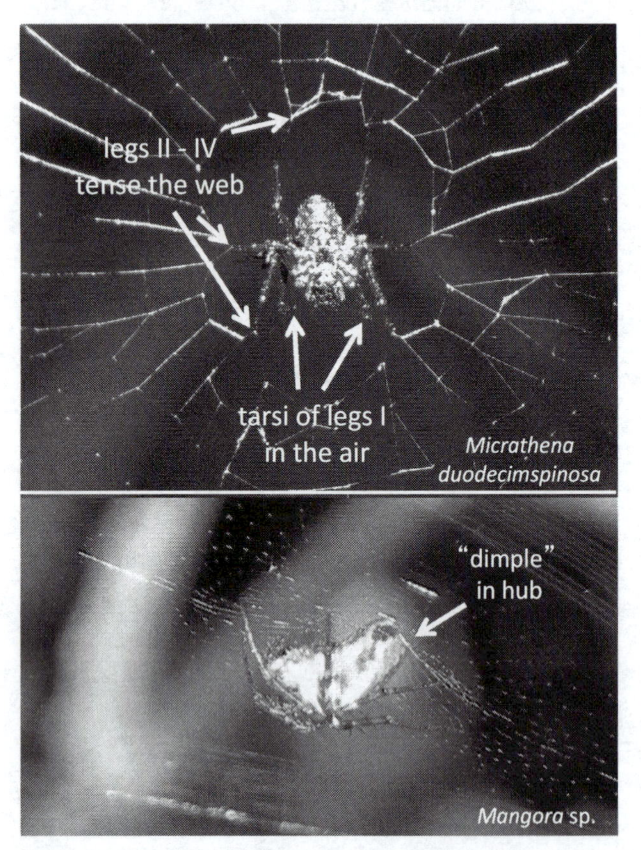

Fig. 3.36. Mature females of the araneids *Micrathena duodecimspinosa* (AR) (*a*) and *Mangora* sp. (AR) (*b*) tensed their orbs while they waited for prey at the hub, a behavior that must be energetically costly. The stubby legs II-IV of *M. duodecimspinosa* tensed lines at the edges of the open hub, while her legs I were held out of contact the web. The spider's response to the impact of a prey was to release this tension abruptly, turn rapidly to face the prey, and jerk the web. *Mangora* sp. probably produced the conical dimple in her approximately horizontal orb just above the tip of her abdomen (arrow in *b*) (also seen in *M.* sp. #1472) by either contracting her legs or pushing upward with her abdomen (or both). The dimple disappeared the moment a prey struck the orb during the rapid attacks on prey. The many zig-zag attachments of hub loops to radii of *Mangora* hubs reduce the tension on lines in the innermost portion of the hub (section 4.4.2). By actively holding the web tense increasing web tension until prey impact, these spiders decrease some types of web movement in the air (section 3.2.11), arguing against the importance of these web movements in promoting prey impact (Craig's "encounter model"). A sudden sag when the prey strikes the web probably improves the web's abilities to retain it, and possibly (if the spider's response is quick enough) it may also help stop the prey. Lines under higher tension also transmit vibrations more effectively, and some spiders also respond more rapidly to prey vibrations on tense radii (Nakata 2010a), so there are other possible payoffs for tensing orbs (section 4.4). The linyphiids *Diplothyron* sp. and *Florinda coccinea* also tense their flat sheet webs by flexing their legs (WE). The functional explanation for why *M. duodecimspinosa* held their legs I out of contact with the web is not known; legs I of three other species of *Micrathena*, *M. bimucronota*, *M. horrida*, and *schreibersi*, gripped hub lines (fig. 3.37).

(WE). Nakata (2009) argued, that when a *Cyclosa octotuberculata* tensed the radii above and below the hub with her legs as she rested at the hub, this was equivalent to her focusing her attention on these sectors; she increased the transmission of vibrations in these sectors, and thus her speed of response to prey that strike the web. In another species with a modified orb, the bolas spider *Mastophora dizzydeani*, the arrival of a moth repeatedly veering rapidly through the night air toward and away from a spider that has cocked her leg and is ready to sling her ball can be tense for both the observer and for the sometimes visibly quivering spider.

Rapid responses to prey are not limited to orb weavers. *Kukulcania hibernalis* sometimes reaches a prey more than a body length away in her tangled web (fig. 5.10) in as little as 0.5 s (WE). And the webless ctenid *Cupiennius salei* can turn and leap into the air to snag an insect flying nearby (Barth 2002).

One minor puzzle may be illuminated by the realization that vigilance at the hub may be metabolically costly. *Allocyclosa bifurca* (AR) often removed portions of its orb early in the evening, long before it replaced them just before dawn the following day (fig. 6.21). *Argiope aurantia* were said (no details were given) to capture "virtually no prey" during the night (Enders 1975). Why did these spiders not wait until just before building the next orb, and thereby take advantage of prey intercepted at night? Even more extreme cases are *Micrathena duodecimspinosa* (AR) and *Wagneriana tauricornis* (AR), which routinely removed the entire orb (early in the evening in the first species, late in the morning in the second), and did not replace it until much later (dawn the following morning in the first, the early evening in the second) (WE). Early removals might be part of "rest" strategies designed to reduce the costs of sustained vigilance (or could promote recycling of web material). Perhaps there are diurnal variations in the degree of alertness and its attendant metabolic costs, or even variations that are related to how long since a given web was built, but these are only speculations.

There are preliminary indications of phylogenetic patterns in metabolic expenditures (which in spiders in general are low, compared with those of other poikilotherms of similar sizes) (Anderson 1970). Comparisons of the basal metabolic rates of representative species from different families showed that the filistatid *Kukulcania hibernalis* and the mygalomorph *Ummidia* sp. had especially low rates compared with other poikilotherms of the same weights (30% and 19% respectively), while *Parasteatoda* (= *Achaearanea*) *tepidariorum* (THD) had higher rates (117%). The resting metabolic rates of wandering species showed diurnal cycles, but those of four web-building species (three theridiids and the filistatid) did not show diurnal cycles when they were on their webs. Kawamoto et al. 2011 found that during daylight hours the metabolic rate of the nocturnal orb weaving araneid *Metazygia rogenhoferi* was about double that of the more diurnally active uloborid *Zosis geniculata* while the spiders rested off their webs. They speculated that this difference was associated with the greater frequency of web repair and movement to new websites by *M. rogenhoferi* that is permitted by their cheaper construction behavior due to the loss of the cribellum (section 10.5.1). One further hint of a taxonomic pattern in differences in metabolism was the much higher rates of accumulation of biomass in the field by several orbicularian species compared with the sheet-weaving zoropsid *Tengella radiata* and the non-web lycosid *Tasmanicosa godeffroyi* (fig 8.4; Santana et al. 1990). The rates of food ingestion and metabolic expenditures were not measured separately, but the data have the advantage of being based on normal behavior and expenditures in the field (as opposed to sitting or struggling in respirometry flasks). No conclusions can be drawn regarding associations with web types; but it is clear that the relative rates of prey capture varied substantially in different species, even when corrected for body size. Differences in metabolic rates may need to be taken into account in understanding web designs.

3.3.3.3.3 NON-ORB WEAVERS

The most complete and direct energy analyses I know of concern a small linyphiid, *Lepthyphantes zimmermanni*, which builds a horizontal sheet on the upper surface of the leaf litter (Ford 1977), and two funnel web builders, the large lycosid *Sosippus janus* (Prestwich 1977) and the agelenid *Agelena limbata* (Tanaka 1989). The energy content of the silk in the web of a 4 mg mature female *L. zimmermanni* was 1.16 J, while the energy expended in building it was 0.72 J, or 38% of the total cost of 1.88 J. The corresponding behavioral cost for *S. janus*, which was nearly 100 times heavier than the linyphiid, was only 18% of the total web cost during the first night.

The total cost of building *L. zimmermanni* webs (energy in the silk plus increase in metabolic rate) was nearly constant at different temperatures, and represented, respectively, 8.2, 5.2, and 4.0 times the total daily maintenance cost for the spider at 4.8, 10, and 15°C (Ford 1977). The combined material and behavioral cost of building a web of this species is thus huge (up to more than a week of body maintenance costs). In the funnel web spiders, the total costs were even higher (23.6 and 9–19 times the daily cost of basal metabolism) (Prestwich 1977; Tanaka 1989). Because *S. janus* adds progressively to her web over many days, this comparison may be overly conservative (the frequencies of repairs and replacements of *L. zimmermanni* and *A. limbata* webs in the field are apparently not known).

An indirect measure of the cost of web construction was obtained in the lycosid funnel web builder *Aglaoctenus lagotis* by estimating the reduction in the spider's immune response immediately after web building. The immune response, estimated by measuring amount of melanic encapsulation of a sterile nylon filament inserted into the spider's body, was stronger in the control immature and mature male individuals that had been held in small 1.5 × 5 cm tubes for 48 hr preceding the test than in others that had built a web during the previous 48 hr (González et al. 2015). The results were inconsistent, however, as there was no significant difference between mature females that had and had not built webs recently.

3.3.3.3.4 DEFENSE AGAINST PREDATORS

A spider resting at the hub of her orb during the day is at risk of being attacked by many different types of predators, nearly all of which orient visually. Birds, lizards, frogs, monkeys, damselflies, sphecid wasps, pompilid wasps, ichneumonid wasps, preying mantids, emesine bugs, asiliid flies, and salticid and oxyopid spiders have all been observed preying on orb weavers (summaries in Bristowe 1941; Higgins 1992b; Eberhard 1973a, 1990a). Defense against predators is one function for several traits of orbs, including tangles built near the orb ("barrier webs") (figs. 1.5, 1.6, 3.7, O9.1, table 3.6), components of the silk itself that repel ants (Zhang et al. 2011), retreats that are built at protected sites beyond the web's edge, and "decorations" of detritus or silk on the web (section 3.3.4.2). Choosing websites farther from objects from which emesine bugs can strike at the spider resting at her hub can also provide defense (Soley et al. 2011). Sticky lines in the sparse webs associated with the egg sacs of the uloborids *Miagrammopes* sp. 1 nr. *unipus* (Lubin et al. 1978), *Uloborus diversus* (fig. 10.16), *U.* sp. nr. *penicillatus* (WE), and *Zosis geniculata* (fig. 10.35) may defend against enemies of eggs, as well as capturing prey. In *U.* sp. nr. *penicillatus* egg sac webs were present only at night and removed in the day (WE), suggesting protection against some undetermined nocturnal enemy.

The hub of an orb probably also sometimes provides a mechanical defense for the spider. Several species of *Argiope* (Hingston 1927, 1932; Frings and Frings 1966; Robinson and Robinson 1970a) and *Clitaetra* (NE) (Kuntner 2006) responded to threatening stimuli by rapidly shuttling to the opposite side of the hub. The defense provided by the hub may also be visual when the hub has a stabilimentum. This function is controversial for some stabilimentum designs (fig. 3.37, section 3.3.4.2), but seems relatively obvious for dense "disc" stabilimentum that obscure the spider's outline (fig. 3.38). Ichneumonid wasps parasitic on *Allocyclosa bifurca*, which has a dense hub, shuttled to the spider's side of the hub when they attacked spiders, suggesting a defensive function for dense hubs (Eberhard 2018). In contrast, the small, open-meshed hubs of the tetragnathids *Leucauge argyra* and *L. mariana* did not provide effective defense against attacks by the ichneumonid wasps *Hymenoepimecis argyraphaga* and *H. tedfordi*, which grabbed the spiders through the web as they rested at the hub (Eberhard 2000b, 2018, WE).

One defense that has evolved repeatedly is to build a tangle of lines adjacent to the orb; it can serve both as a mechanical barrier to some predators (Fincke 1992 on damselflies), and as a preliminary warning system (WE on the ichneumonid wasp *Polysphincta gutfreundi*). Several types of evidence support this defensive function hypothesis, including direct observations of predators avoiding tangles, consistent association of the side of the orb where the spider rested with a greater development of the tangles, and spiders that moved into tangles

Table 3.6. A summary of studies and the types of data used to determine the probable functions of tangles near orbs ("protect" = protect spider from potential predators or parasites; "stabilize" = increase web's ability to resist wind stress; "prey" = attract prey or knock them onto the approximately horizontal orb). The general association with orbs operated during the day (when many predators such as birds, lizards, damselflies, and wasps are active), and the tendency to be more elaborate on the side of the orb where the spider rested, generally support the protective function hypothesis, but there are exceptions.

Species	*Orb diurn. or noct.?*[1]	*Site spider rests*	*Slant of orb*	*Num. sides orb with tangle*[2]	*More tangle on side of spider?*[3]	*Prob. funct.*	*References*
Uloboridae							
Conifaber spp.	D	Hub	Horiz	1 (B)	Yes	Protect[4]	Lubin et al. 1982, Grismado 2008
Uloborus conus, bispiralis, trilineatus, kerevatensis (= albolineatus), quadrituberculatus	D	Hub	Horiz	1 (B)	Yes	Protect[4]	Lubin et al. 1982
U. barbipes	?	?	Horiz	1 (A)	No	Prey?	Lubin 1986
Philoponella variabilis	?	Hub	Horiz	2[5] (A,B)	?	Prey[6], protect?	Lubin 1986
P. oweni	D	?	Horiz[7]	2[5] (A,B)	?	Prey[6]	WE
P. spp.	D	Hub	Horiz[8]	2 (A,B)	?	Prey? protect	Peters 1953, Lubin 1986, Opell 1979, Lahmann and Eberhard 1979, WE
Araneidae							
Arachnura spp.	D	Hub	Vert-horiz	1	Yes	?	McKeown 1952, Robinson and Lubin 1979a, Lubin 1986, WE
Cyrtophora spp.[9,10]	D,N[11]	Hub[12]	Horiz	2 (A,B)	No[13]	Prey[14], protect[15]	Lubin 1973, Blamires et al. 2013
Witica sp.	D	Hub	horiz	1 (B)	Yes	protect[16]	WE
Metepeira spp.[17]	D	Ret	vert	1(2)	Yes	protect	McCook 1889, Emerton 1902, Comstock 1967, Viera 1986, 2008 WE
Phonognatha spp.[18]	D?[19]	Ret	Vert	?	Yes	Stabilize, protect?	McKeown 1952, Hormiga et al. 1995, Kuntner et al. 2008a
Thelacantha brevispina[20]	D?[21]	Hub	Vert	1	?	Protect? lure prey?[15,22]	Tseng et al. 2011
Allocyclosa bifurca	D	Hub	Vert	1(2)[23]	Yes	Protect (?)[16,24]	Eberhard 2003, WE
Araneus omnicolor	?	Ret[25]	Vert	1	Yes	Protect[19]	Gonzaga and Sobczak 2007
Argiope argentata	D		Vert			Stabilize[26]	Lubin 1975
A. trifasciata	D	Hub	Vert	1(2)[27]	Yes	Protect[16]	Edmunds and Edmunds 1986, WE
Tetragnathidae							
Leucauge mariana[28], *L. venusta*	D[29]	Hub	Horiz	1 (B)	Yes	Protect	Zschokke et al. 2006, WE
Leucauge argyra[28,30]	D[29]	Hub	Horiz	1(2) A(B)	No	Stabilize, prey?	Triana-Cambronero et al. 2011
Nephilidae							
Nephila clavipes	D[19]	Hub	Vert	1(2)[27]	Yes	Protect[15,16]	Higgins 1990, 1992b
Nephila clavata	?	?	Vert	?	?	Protect	Blamires et al. 2010a

Table 3.6. Continued

Species	*Orb diurn. or noct.?* [1]	*Site spider rests*	*Slant of orb*	*Num. sides orb with tangle* [2]	*More tangle on side of spider?* [3]	*Prob. funct.*	*References*
Nephila pilipes (= *maculata*)	D[19]	Hub	Vert	1	Yes	Protect[16]	Hingston 1922c, Robinson and Robinson 1973a, WE
Nephilingis spp.	D[19]	Retreat	Vert	1[31]	—[31]	Protect	Edmunds 1993, Kuntner et al. 2008a, WE

[1] "Diurnal" = built in the early morning before dawn, operated during the day (table 8.2).

[2] "A" = tangle above approximately horizontal orb; "B" = tangle below.

[3] "Yes" = in webs with tangle on only one side of the orb, it was the side on which the spider rested; in species in which there was a tangle on both sides and the web was vertical, the tangle was denser on the side of the orb on which the spider rests.

[4] Deduced from defensive behavior of spider.

[5] There was sticky silk in the tangle above as well as in the orb itself.

[6] Deduced from the presence of sticky lines.

[7] Often dome-shaped, with hub at the top of the dome.

[8] Some were dome-shaped with the hub at the top of the dome; many species in this genus had tangles above and below the orb (Lubin 1986).

[9] Orb was a dense non-sticky horizontal sheet of radii and temporary spiral lines, and lacked sticky lines.

[10] Very similar designs occurred in other species of *Cyrtophora*, and in the closely related genera *Mecynogea*, *Kapogea*, and *Manogea* (WE).

[11] Web built at night, functions both day and night.

[12] Spider was at the hub or under detritus such as fallen leaf near edge of orb, where tangle was also present below orb.

[13] At least in *C. citricola* (Kullmann 1958, WE).

[14] Blowflies induced to fly into the web were sometimes stopped by the tangle above and fell onto the sheet, where the spider attacked them; less commonly, they were retained in the tangle and the spider climbed up to attack them there (Lubin 1973).

[15] Deduced from direct observations of predators and experimental removal of the tangle.

[16] Deduced from the fact that the tangle was on the same side of the orb as the spider and lacked sticky lines; Hingston (1922c) argued (less convincingly for me) that the tangle aided in prey capture in *Nephila*.

[17] Deduced from the position of the spider's retreat. The tangle was more dense or occurred exclusively on the same side of the orb as the retreat, and the retreat was in the midst of the tangle; includes *Metepeira labyrinthea*, *M. spinipes*, and *M.* sp. from Costa Rica (McCook 1889; Emerton 1902, Viera 1986; WE).

[18] Tangle was sparse, and largely (or exclusively?) associated with sustaining the curled leaf near the hub (inside of which the spider rested during the day); perhaps this species is not properly said to have a "tangle."

[19] Orb sometimes built at night (sometimes long before dawn, sometimes near dawn); probably best characterized as diurnal.

[20] Tangle also had stabilimentum tufts.

[21] Deduced from the fact that the observations of prey arrival at webs were made during the day.

[22] The tangles may have protected against birds (though the total number of attacks was only 4 in 1066 hours of recordings). But the data are not conclusive for the combined data for two wasp species, *Eumenes* sp. (Vespidae) and *Batozonellus* sp. (Pompilidae), that hovered near webs. *Eumenes* sp. belongs to a group that preys on lepidopteran larvae, not spiders (e.g. Evans and Eberhard 1971) so they were probably only inspecting the spiders and would not have attacked them even if no tangle were present. When the tangle (with its silk stabilimentum tufts) was present, more prey struck the orb than when the tangle (and the stabilimentum tufts) was removed in both years of the study; the difference was statistically significant, however, in only one of the years.

[23] Usually the tangle was on the same side as the spider, which nearly always rested on the lower rather than the upper side of the slanting orb; when the tangle was on both sides of the orb, it was more extensive on the spider's side (WE).

[24] Only a minority of orbs had tangles.

[25] The tangle of lines surrounded and sustained the curled leaf within which the spider rested during the day.

[26] Deduced from the greater frequency of tangles in windy as opposed to sheltered sites. The possibility that spiders at sheltered sites were less well fed (and thus perhaps less likely to build tangles—see Baba and Miyashita 2006 on another species) was not explored.

[27] Only a small fraction had tangles on both sides. Direct observations of the foraging behavior of predatory pseudostigmatid damselflies confirmed that the tangles of *N. clavipes* deterred attacks (Fincke 1992).

[28] Tangles were below the orb and had no sticky lines.

[29] Also built orbs during the day and in the early evening (made several webs/day).

[30] Tangles were much more frequent in orbs of immature than mature spiders.

[31] Tangle was around the retreat, not the orb; it was more extensive in older spiders, which had more elaborate retreats (Edmunds 1993).

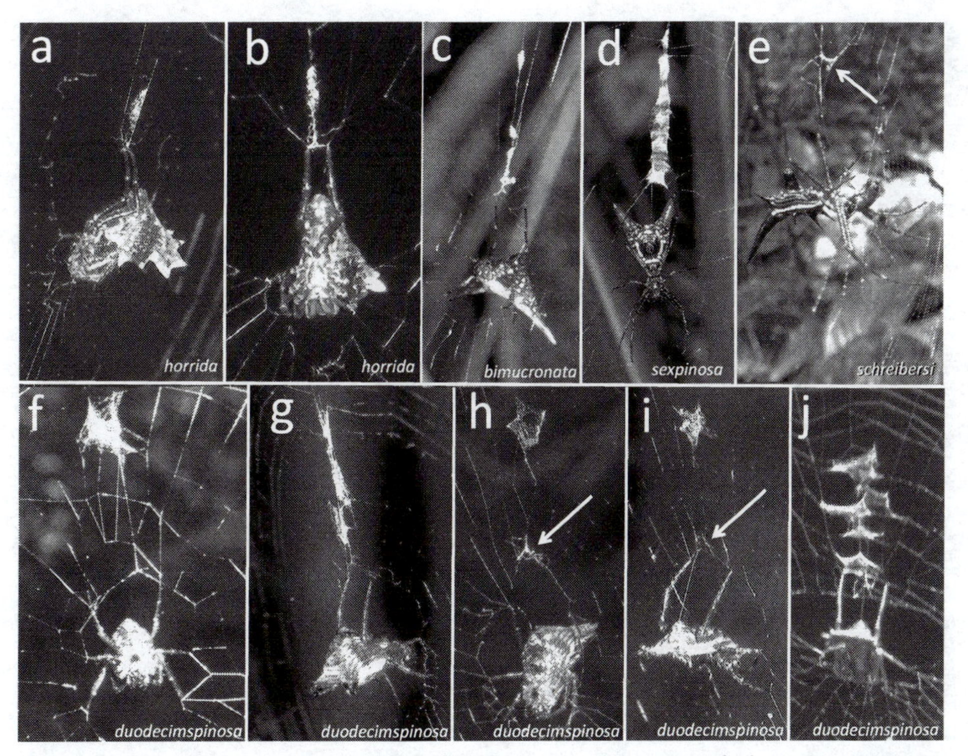

Fig. 3.37. This gallery of views of the adult females of five species of *Micrathena* (AR) seen from different angles at the hubs of webs that had stabilimenta illustrates two points. First, as is typical in other groups, the stabilimenta of *Micrathena* vary substantially both within and between species: some of the intra-specific variation in *M. duodecimspinosa* is illustrated in *f–j*; in addition, 28% of 47 webs of this species lacked stabilimenta entirely (WE). In other species, inter-specific variation ranged from small, barely noticeable accumulations of white fluff in *M. schreibersi* (*e*) to clear, jagged swaths of white silk in *M. sexspinosa* (*d*), and *M. duodecimspinosa* (*j*); Hingston (1932) (on "*Acrosoma*") illustrated still other designs of *Micrathena* stabilimenta. The intra-specific inconsistency suggests that the advantages and disadvantages of stabilimenta are nearly equal (Edmunds 1986). A delicate balance of this sort could make it difficult to obtain clear results from experimental manipulations, perhaps explaining some of the contradictory findings in tables O3.3–O3.5. Secondly, the genus *Micrathena* provides a taxonomic microcosm with evidence both for and against against the visual defense hypothesis. The stabilimentum of the *M. horrida* spiderling (*a*, *b*) blended with the constrained position of its legs pressed against its body, making its visual outline even less spider-like and thus supporting the visual defense hypothesis. In *M. bimucronata* (*c*), one could make at least a strained argument that the white stabilimentum was a visual continuation of the light color patterns of the spider and provides visual camouflage. But there was no resemblance or visual fit between *M. sexspinosa* (*d*), *M. schreibersi* (*e*), or *M. duodecimspinosa* (*f–j*) and their stabilimenta; nor did the spiders modify the positions of their legs or abdomens to merge their outlines with those of their stabilimenta. All legs except legs I were partly extended; the abdomen was inclined sharply away from the web in *M. bimucronata* and *M. schreibersi*, but much less in *M. sexspinosa*. The position of *M. schreibersi* at the hub (*e*) might be thought to be designed to camouflage her resemblance to a spider, or perhaps to emphasize the defensive properties of her large, stiff abdominal spines (their black color does not fit this hypothesis, however); but this would not explain the stabilimentum's function. Just as the fact that the constrained positions and coloration of many other species that make their visual outlines merge with those of their stabilimenta constitute strong evidence that favors the camouflage hypothesis (figs. 3.41, 3.45. 3.46), the *lack* of any visual fit with the stabilimenta in *M. sexspinosa*, *M. duodecimspinosa*, and *M. schreibersi* constitutes strong evidence against it. Perhaps their stabilimenta serve to warn off large animals and reduce web damage (Jaffé et al. 2006). Evidence from *M. sexspinosa* suggested (though weakly—see table O3.3) that their stabilimenta may attract prey (Gálvez 2011).

when disturbed (summarized in table 3.6). In a few cases tangles seem to have other functions, however (table 3.6), and attacks by one stealthy predator, an emesine bug, are aided rather than impeded by a nearby tangle (Soley et al. 2011). A nearby tangle must often be a mixed blessing, as it would deflect some prey from encountering the orb; presumably this negative effect is more than compensated by its benefits, but I know of no data testing this hypothesis.

Another common defense is to rest away from the hub at a more hidden site that is connected to the hub by a signal line. This tactic has arisen repeatedly (table

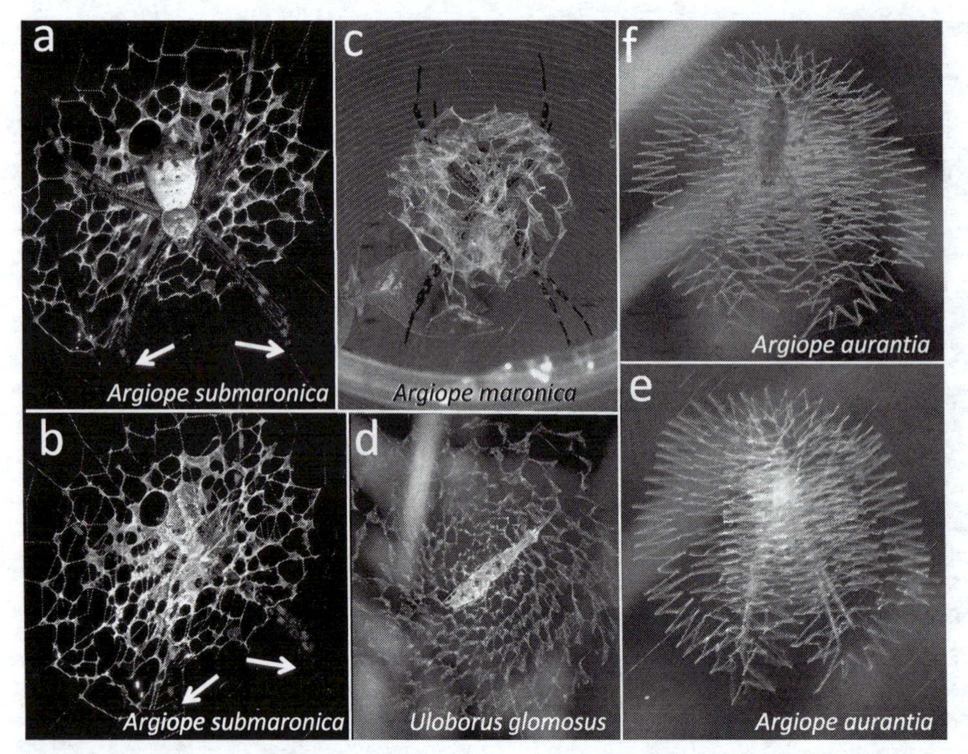

Fig. 3.38. Disc stabilimenta hide spiders visually. Dorsal (*a*) and ventral (*b*) views of a large *Argiope submaronica* (AR) spiderling at the hub of her orb in the field show how a disc stabilimentum can hide the spider visually with a dark background. From the ventral side (*b*), the most likely angle of view for large predators such as birds, given the spider's position on the underside of a slanting orb built near the ground, the spider was essentially invisible to the human eye; the disruptive coloration of her legs made them especially difficult to distinguish (arrows in *b* mark legs LI and LII). Another individual of this species built a small disc stabilimentum on a small tangle of short, non-sticky lines in a small plastic cup (8.8 cm dia. at the mouth) the day after she molted; she was also largely hidden, even against an intermediate-tone background (*c*) (incidentally, building a stabilimentum on a tiny network of non-sticky lines is also evidence against the prey attraction hypothesis for stabilimenum function). The disc stabilimentum (*d*) on the horizontal orb of *Uloborus glomosus* (UL) seen from above (also near the ground) gave less convincing though not entirely neglible visual shelter to the spider. Dorsal (*e*) and ventral (*f*) photos of an immature *Argiope* sp. (probably *aurantia*) in the field also suggested visual camouflage, but with a different contribution from body coloration: the spider's shiny white color with small blotches made it difficult to distinguish her body from the zig-zag disc stabilimentum, even when seen from the dorsal side (photos *a–d* were taken with natural lighting, *e* and *f* with a flash).

9.1). This behavior apparently usually functions as a defense against diurnal enemies, as these spiders typically move to the hub as soon as night falls. There are, however, groups with nocturnal orbs, such as *Wixia abdominalis* (AR), that utilize signal lines (Xavier et al. 2017), and some individuals of *Araneus expletus* (AR) and *Metazygia wittfeldae* (AR) remained in their retreats during at least part of the night (WE) (fig. 3.16).

Finally, the possibility of chemical defenses by web-weaving spiders against predators is little-explored. Bristowe (1941) noted that many linyphiids were rejected as prey by various other spiders, sometimes after only being touched and others after being bitten; he did not give species names, however, or elaborate further. Both *Mastophora dizzydeani* (AR) and *Leucauge mariana* (TET) have disagreeable odors when they are held close to the nose (WE) (the smell from *L. mariana* was weak, but many *Leucauge* species have apparently aposematic coloration—Shinkai and Takano 1984; Murphy and Murphy 2000; Bradley 2013). Further work on chemical defenses might be rewarding.

3.3.3.3.5 OTHER POSSIBLE VARIABLES AND FUNCTIONS

Attack behavior may affect selection on some web traits. Many orb weavers shake or jerk the web one or more times after turning at the hub to face toward a prey before reaching it. The function of web shaking may be to entangle the prey, and could favor any orb trait (such as lower tensions on radial and sticky lines) that would promote the prey swinging into contact with additional lines. In *Hyptiotes* spp., for instance, spiders lower web tensions abruptly by releasing line when a prey strikes the web (fig. 10.23) (Marples and Marples 1937; Opell et al. 1990); attacks sometimes included jerks on a radius that further entangled the prey (Opell et al. 1990). In some species, such as *Micrathena duodecimspinosa* (AR) and *Allocyclosa bifurca* (AR), shaking occurred in all or nearly all attacks (WE); in others such as *Leucauge mariana* (TET) and *Nephila clavipes* (NE) shaking was omitted in many rapid attacks.

The entangling function of shaking has, however, never to my knowledge been documented carefully. This is an important gap, because it may be incorrect; in several species, including *Uloborus diversus* (UL), *A. bifurca* (AR), and *L. mariana* (TET), web jerking sometimes shakes a prey free, rather than further ensnaring it (WE). In fact, shaking by spiders in several families, including Diguetidae (Eberhard 1967), Linyphiidae (WE) and Theridiidae (Eberhard 1972b) and the araneid *Cyrtophora* (Lubin 1973; WE), caused prey to fall free from tangles of lines and onto the sheet below.

Web shaking in orb weavers may have another, sensory function. It causes the prey to swing in the web and generate vibrations that the spider may use to judge its weight (heavier prey would generate larger amplitude vibrations). In some species shaking was more frequent when the prey was larger (and thus potentially more dangerous) (section 3.2.12, Robinson and Olazarri 1971; Robinson 1975), favoring this sensory function hypothesis.

Sticky lines trap airborne "plankton" such as pollen; in some sites such as near a maize field, or in forests of temperate-zone wind-pollinated trees, appreciable amounts of pollen can accumulate on sticky lines during brief periods during the year (Smith and Mommsen 1984; Carrel et al. 2000; Ludy and Lang 2006; Eggs and Sanders 2013). Some spiders benefited from nutrients in the pollen from maize fields when they ingested old webs and replaced them (Ludy and Lang 2006; Eggs and Sanders 2013). The relative nutritional importance of pollen on webs in nature is still uncertain, however. Some types of pollen do not provide nutritional benefits (Carrel et al. 2000), and monocultures of maize, where some studies were performed, are obviously unnatural. Stable isotope analyses with Bayesian mixing models of *Araneus diadematus* spiders collected in May in the vicinity of pine (*Pinus sylvestris*) and spruce (*Picea albies*) trees in Switzerland suggested that a large portion (about 25%) of the diet of immatures was pollen rather than insects (Eggs and Sanders 2013). The likely imprecision of the models used in the analysis, however, which were derived from supposing that the insect prey was mainly small dipterans and hymenopterans (sampled with sweep netting and suction sampling), leaves this estimate in doubt (see sections 3.2.5.4). I have observed a *Leucauge venusta* (TET) behave as if the pollen on her web lines in a temperate forest when oak trees (*Quercus* spp.) were shedding pollen in large amounts was not nutritionally useful. A spider that was removing her web, which carried a substantial load of pollen that made the lines much more visible, prior to replacing it in the early afternoon discarded three of five sectors that she removed with quick flicks of her legs I or II immediately after cutting them free, and packed the other two sectors into small masses, held them at her mouth for less than a minute, and then discarded them also; they floated away rather than falling, suggesting that she had not wet them with digestive enzymes.

There is one report of apparent chemical attraction of the males of one species of diurnal moths by the webs of the araneids *Araneus trifolium*, *Argiope trifasciata*, and *A. aurantia* (Horton 1979). I know of no further attempts to confirm or extend these observations. Bolas spiders and *Kaira* do attract male moths chemically using female pheromone mimics (Stowe 1985; Yeargan 1994), and the high frequency of moth captures by species in several genera (*Cyrtarachne*, *Pasilobus*, *Poecilopachys*) related to bolas spiders hints at possible chemical attraction. Nevertheless, the diversity of moth species that three species of *Cyrtarachne*, *C. inaequalis*, *C. bufo*, and *C. nagasakiensis*, captured in the field does not fit easily with this idea (Miyashita et al. 2001). For *C. inaequalis*, from which most prey were collected, there were 19 moth species in at least 17 genera and 4 families. The other two species captured moths from seven additional subfamilies not included in the *C. inaequalis* list. In addition, the spiders captured moths of both sexes (the ratio of male:female was 0.8:1). Further research is needed to clear up this mystery.

Fig. 3.39. The non-planar orb of a mature female *Mangora* sp. (AR) built in thick grass (above) emphasizes a construction problem that orb weavers usually solve, but which has escaped detailed discussion: choosing attachment sites that are all in the same plane. Lack of planarity (as in this web) is probably disadvantageous, because, as illustrated in the drawing below, the trajectories of some prey (numbered arrows) will be intercepted at two different sites on the web (prey #1). This represents, in effect, a waste of web area. A more planar orb could have intercepted some prey that were missed (e.g., prey #4).

3.3.3.4 Planar and non-planar orbs

Although a few orb-weavers build orbs that are normally pulled into cones (figs. 7.9, 10.3, 10.25; see table 9.1), the large majority build perfectly or nearly perfectly planar orbs. Nothing is known regarding the cues that are employed to achieve planarity, and the causes of deviations from strict planarity have never to my knowledge been investigated systematically. In *Araneus diadematus* (AR) the first web that a spider built at a site (in captivity) tended to be less perfectly planar than later webs (Zschokke and Vollrath 2000). *Herennia* trunk webs tend to have a small curvature (produced by their pseudoradii) that keeps them close to the trunk on which they are built (Robinson and Lubin 1979a; Kuntner 2005), while in *Anapisona simoni* (AN) the degree of deviation from a plane (the sharpness of the angle of the cone) was greater when the spider had less space in which to build her approximately horizontal orb (Eberhard 2011). In a few species, such as *Uloborus diversus* (UL) (Eberhard 1971a) and *Mangora melanocephala* (AR) (fig. 3.39), it appears that substantial deviations from planarity are common (WE). At the opposite extreme, as in *Micrathena duodecimspinosa* (AR) and *Gasteracantha cancriformis*, achieving a perfect plane is often automatic because the orb has only three attachment sites (e.g., fig. 1.4) (three points determine a plane).

In most species, however, there are additional anchors and planarity is not automatic. The advantage of planarity may be related to improving prey interception. If all prey were moving in a given direction, understanding the advantage of planarity would be simple: placing a planar web perpendicular to their paths would maximize prey interception, and non-planar orbs would cover less area and intercept fewer prey. Most orbs in nature, however, probably intercept prey moving in a large variety of directions (section 3.2.4). The consequences of this variation for planar vs. non-planar orbs are not immediately obvious. Bending a web away from a planar configuration could cause it to fail to intercept some prey that it would have formerly intercepted, but would enable it to intercept others that were flying more nearly parallel to it.

Further understanding comes from considering an extreme case (fig. 3.39). Suppose that the distribution of directions in which prey move is perfectly random, that the density of prey is uniform in the space where the orb is built, and that the orb is strongly non-planar (say it is curved into a dome shape as in fig. 3.39). This non-planar web will "waste" some of its silk by being in position to intercept some prey twice (prey #1 in fig. 3.39), rather than once, and thus lowering its interception efficiency (number of prey intercepted/cm of silk line). The more sharply bent the web, the lower its interception efficiency. In other words, a planar orb has the largest projection area/web area (Zschokke and Vollrath 2000). This implies that selection favors more nearly planar orbs, and that webs like that in fig. 3.39 represent "mistakes" (perhaps resulting from a lack of suitable attachment sites or open space in the dense grass and weeds where the web was built? or economies in exploration prior to web construction?).

These ideas run counter to the fact that some orb weavers, such as *Spilasma artifer* (AR) (Eberhard 1986a), *Anapisona simoni* (AN) (fig. 7.9 (Eberhard 2007b), and spiderlings of *Leucauge argyra* (TET) (fig. 10.3) (Triana-Cambronero et al. 2011) regularly build non-planar, cone-shaped orbs (in all cases the central axes of the cones were vertical). The contradiction would disappear if the flight paths of prey were non-random, and tended not to be horizontal, but this is only speculation, as no data are available.

3.3.4 THE FUNCTION(S) OF STABILIMENTA

Finally, we come to the stabilimentum, the orb web structure whose function is easily the most controversial of all, and whose mention can raise the blood pressure of some otherwise calm arachnologists. The term "stabilimentum" is usually employed to designate an object or an easily visible accumulation of silk or other material that is added to an orb web, usually but not always near the hub (fig. 3.40). The name comes from the early hypothesis (now largely discredited—table O3.4) that they serve to stabilize the web; they are sometimes instead given the more neutral name "decorations." The objects in stabilimenta vary, and include silk (apparently from the aciniform glands), egg sacs, plant detritus, and prey remains. Many different functions have been proposed (see Herberstein et al. 2000b, table 3.7). The most extensive reviews to date are those of Herberstein et al. (2000b) and Walter and Elgar (2012). I will begin the present, more complete and critical discussion with the simpler case of egg sac and detritus stabilimenta, and then discuss the more controversial silk stabilimenta.

3.3.4.1 Egg sac and detritus stabilimenta

Egg sac and detritus stabilimenta have evolved independently multiple times (figs. 3.41, 3.42, table 9.1). They have long been thought to function as camouflage for the spider as it rests nearby (e.g., Hingston 1927, 1932; Marson 1947a,b; Edmunds 1986; Chou et al. 2005; Gonzaga and Vasconcellos-Neto 2005; Zschokke and Bolzern 2007; Blackledge et al. 2011; Herberstein and Tso 2011). The only competing hypotheses that I am aware of are that the detritus stabilimenta of *Nephila edulis, N. plumipes,* and *N. clavipes*, which include corpses of prey, function either as caches of prey (Champion de Crespigny et al. 2001; Griffiths et al. 2003), or as prey attractants for carrion flies (Bjorkman-Chiswell et al. 2004; Hénaut et al. 2010). Neither of these alternatives is convincing. The supposed advantage of food caches was that they give "a relatively constant prey ingestion rate" (Champion de Crespigny et al. 2001; p. 43). It is not clear, however, what advantage this might give over the usual spider feeding tactic of ingesting all that it can, and storing nutrients in its own body (where they are not subject to degradation and loss due to decay, to the leaching effects of rain, or to kleptoparasites such as those frequently associated with the webs of these and other *Nephila* species [Vollrath 1979; Agnarsson 2003a; Griffiths et al. 2003]). In addition, the conditions of the experimental treatment in the Champion de Crespigny study (10 days with no food) were probably ecologically unrealistic and extreme, and may have provoked unnatural responses from the spiders; furthermore, the spiders' responses were inconsistent in different sized individuals.

The fly attraction hypothesis also has serious problems. Some prey of *N. clavipes* were mistakenly identified as carrion feeders (elaterid beetles, sciarid flies). The detritus stabilimenta of *N. edulis, N. clavipes,* and *N. plumipes* often included vegetable debris, which sometimes substituted entirely for prey carcasses (Robinson and Robinson 1973a; Griffiths et al. 2003; Bjorkman-Chiswell et al. 2004; WE). Experimental removal of carcasses from the webs of *N. clavata* increased rather than decreased prey interception rate (Blamires et al. 2010a). The prey carcasses in the webs of *N. clavipes* in Costa Rica were

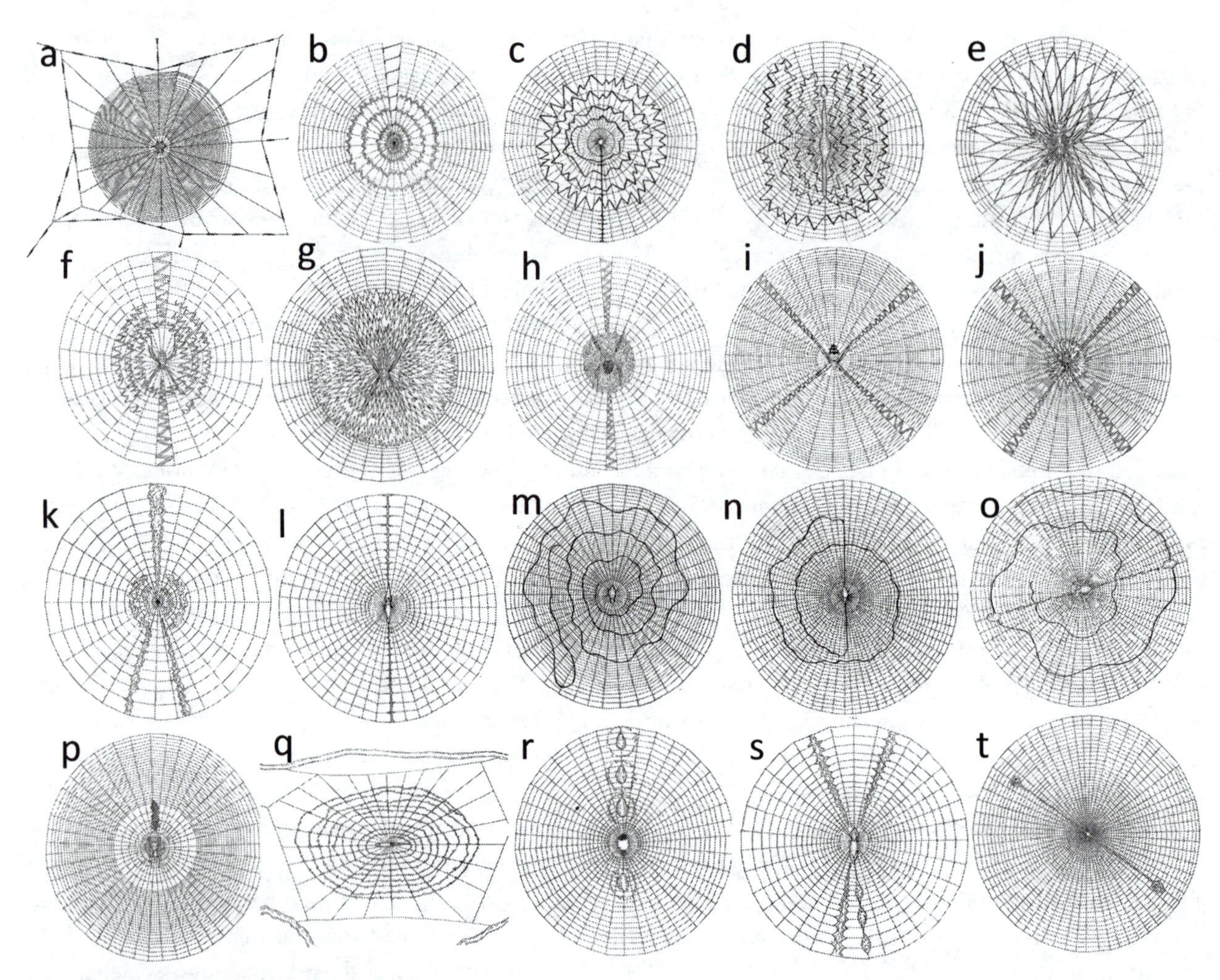

Fig. 3.40. The diversity of patterns in silk stabilimenta has long intrigued arachnologists. This sampler of stylized drawings illustrates some of the 34 different types that were recognized by Hingston more than 80 years ago. The categories include stabilimenta on the frame lines (*a*), various patterns of circular or zig-zag lines on or in the vicinity of the hub (*b-e*) (including a vertical line in *c* and detritus pellets in *d*), circular lines at the hub combined with lines along radii (*f*), a thick mat at the hub (*g*), a mat combined with vertical lines (*h*), an "X" of slanting lines (*i*), a mat combined with an "X" (*j*), slanting lines combined with zig-zag circular lines (*k*), a vertical line with which the spider aligns her own linear body (*l*), wandering thin circular lines around the hub (*m*), a vertical line combined with wandering circular lines (*n*), a slanting line with expansions at either end combined with circular lines (*o*), a short, linear mat (*p*), a long spiral on a web entirely lacking sticky lines (*q*), circular lines around pellets on radii rather than at the hub (*r*), slanting lines combined with a string of egg sacs (*s*), and a slanting line combined with pellets of detritus at its ends (*t*). The species are: *a–Gasteracantha cancriformis* (= *cancer*); *b–Uloborus glomosus* (= *mammeatus*); *c*–probably *Argiope* sp.; *d–A. argentata*; *e–A. argentata* (= *cuyunii*); *f–A. trifasciata* (= *argyraspis*); *g–A. argentata* (= *filiargentata*); *h–A. aurantia* (= *cophinaria*); *i–A. argentata*; *j–A. argentata*; *k–U. glomosus*; *l–U. filidentatus*; *m–Cyclosa fililineata*; *n–C. fililineata*; *o–C. tapetifasciens* (= *filioblicua*); *p–Micrathena militaris* (= *Acrosoma armatum*); *q–U. filinodatus*; *r–C. tremula* (nominum dubidum); *s–U. plumipes*; *t*–near *C. tapetifasciens* (= *filioblicua*) (all are araneids except *Uloborus*; note that some designs varied intraspecifically) (after Hingston 1932).

first sucked dry and fragmented (Robinson and Robinson termed them "prey debris") (fig. 3.41d), and they were sometimes left in place for days and weeks (WE). They are thus seemingly inappropriate for attracting organisms like carrion flies, which are drawn only to carrion in the early stages of decomposition (Gennard 2012). The original observations also have drawbacks. The possible importance in the diet of *N. edulis* of the possibly non-native flies (*Lucilia cuprina*) that were attracted was not quantified in the Bjorkman-Chiswell et al. 2004 study of *N. edulis* study; this and other species of *Nephila* are known to be extreme prey generalists (Austin and

Table 3.7. Descriptions of the four most widely cited hypotheses for the function of silk stabilimenta on orb webs, with a list of other less popular hypotheses, explanations for why some versions of more popular hypotheses are not discussed further, and representative citations for all. Data supporting and contradicting the four most popular hypotheses are discussed in the text and tables O3.3–O3.5 (see also Herberstein et al. 2000b and Blackledge et al. 2011 for strong reasons to doubt the general importance of the less popular hypotheses). Most of the functions are non-exclusive, and more than one function could occur even in a single species (Herberstein et al. 2000b).

A. Major hypotheses

1. **"Stabilize"** (= "mechanical" hypothesis of Herberstein et al. 2000b). The stabilimentum adjusts an orb's mechanical properties in an advantageous way. As noted by Robinson and Robinson (1970a), the exact mechanical function that is presumed to provide greater "stability," and just which portion of the web is stabilized, have never been specified. The original and most frequently cited version (and the one I examine in detail) is that the stabilimentum somehow provides, by unspecified means, mechanical support that makes the web better able to resist damage from prey or other mechanical stress (Robinson and Robinson 1970a; Neet 1990). One variation on this hypothesis is that tensions exerted by the hub spiral (but not linear) stabilimenta of *Octonoba sybotides* webs "tuned" the web, raising tensions on some radii and thus improving the spider's ability to sense smaller prey, and that "spiral stabilimenta at the hub may support the distortion of the radial threads by the hub threads" (Watanabe 2000b)1. Because it is generally true that in uloborids as well as in other groups the stabilimentum silk lines are connected to webs in ways that preclude such tension changes or mechanical support (Eberhard 1973a; Herberstein et al. 2000b), I do not discuss this variant hypothesis in the text or other tables. Another variation based on the mechanical consequence of two particular types of stabilimenta, "disc" (e.g., Hingston 1932; Herberstein et al. 2000b) and "circles or spiral at the hub" (e.g., Hingston 1932; Herberstein et al. 2000b; fig. 3.40) (*Argiope versicolor* and *C. ginnaga*), is that the stabilimentum provides a physical shield against predators: it increases the difficulty of breaking through the hub to gain access to the spider when she is on the opposite side, because there are additional lines that connect the hub lines to each other (Tolbert 1975). This physical shield stabilization effect is not easily separable from the visual shield effect of the same stabilimenta (see #4 below), and is contradicted by the many other designs in which the stabilimentum has an interruption rather than a shield exactly where the spider rests. The possible visual function of disc and circle stabilimenta will thus be emphasized, though the possibility that sometimes disc and circle stabilimenta shield the spider physically cannot be ruled out.
2. **Prey attraction** (= "foraging" hypothesis of Herberstein et al. 2000b). The stabilimentum attracts prey to the web, increasing the spider's foraging success (Ewer 1972; Craig and Bernard 1990; Craig 1991, Watanabe 1999a,b, Watanabe 2000a). An additional, related possibility (which appears not to have been mentioned previously) is that the stabilimentum, by incidentally visually hiding the spider, also increases the likelihood that prey will approach near enough to become entangled in the web (*prey attraction + camouflage* hypothesis). This modified hypothesis could explain, in the context of prey attraction, the widespread and otherwise unexplained trend for spiders to rest in constrained positions that merge their outlines with those of the stabilimenta during the day but not at night. Both the simple and the compound versions are considered in Tables O3.3–O3.5.[2]
3. **Warning** (= "web protection" hypothesis of Herberstein et al. 2000b). The stabilimentum helps reduce mechanical damage to the web by warning large animals away from the web (Horton 1980; Eisner and Nowicki 1983). The most likely intended targets of the warnings were said to be birds, because orbs with stabilimenta survived intact in the early morning when birds are especially active (Eisner and Nowicki 1983); it is possible that other large visually orienting animals could also repond. Learning rather than innate avoidance seems likely (Robinson and Robinson 1970a), though either could be involved.
4. **Visual Defense** (= "anti-predator" hypothesis of Herberstein et al. 2000b). The stabilimentum defends the spider against visually orienting predators by one of at least four different means that involve the predator's vision. The most complete presentations of this hypothesis are those of Hingston (1927, 1932), who discussed 34 different types of stabilimentum patterns and four means by which they could provide visual defense (for a similar, though slightly different classification of visual defenses, see Robinson and Robinson 1970a, 1973b): form a visual barrier ("visual shield" hypothesis) behind which the spider can hide from view (e.g., figs. 3.38 and 3.49 of the araneids *Argiope submaronica* and *A. aurantia*; blend with the coloring and/or the outline of the spider to form a compound object that does not resemble a spider or any other edible object ("blend" hypothesis) (e.g., fig. 3.49 of the araneid *Argiope submaronica*, and fig. 3.46 of the uloborid *Zosis geniculata)* (in *Cyclosa ginnaga* the compound object may resemble a bird dropping—fig. 3.41, Tan et al. 2010); form a decoy object that resembles the spider and distracts a predator's attention from the nearby spider ("decoy" hypothesis) (fig. 3.41 illustrates this effect with the egg sacs of *Cyclosa monteverdi* and *Allocyclosa bifurca*; similar

Table 3.7. Continued

A. Major hypotheses (continued)

patterns occur in the silk stabilimenta of other *Cyclosa* species—see fig. 3.40, Hingston 1932; Marson 1947b); and distract or "disperse" the attention of the predator from the spider, delaying its attack enough to give the spider a better chance of escaping by other means such as jumping off her web (as occurred in *Cyclosa* spp. with detritus stabilimenta—Gonzaga and Vasconcellos-Neto 2005; Chou et al. 2005) ("distraction" hypothesis). One variant of the blending hypothesis is that by merging with the stabilimentum, the spider becomes part of an apparent object that is too large for a gape-limited predator such as a lizard to attack successfully (Schoener and Spiller 1992). More than a single visual defense effect could employed at a time; for instance, Hingston (1932) argued that the stabilimenta of *Cyclosa tapetifaciens* (= *filioblicua*) and *Argiope argentata* simultaneously have blending, decoy, and dispersion functions. The distinctions between different stabilimentum designs can also become blurred. For instance, the term "disc" stabilimentum has been used to designate several designs, including solid mats of white silk (e.g., fig. 3.46) that could function as visual shields, zig-zag lines that were clumped at the hub (figs. 3.38, 3.40) and seemed more likely to blend with the spider to form a compound visual object resembling an object such as a bird dropping (Tan et al. 2010), and intermediates (fig. 3.40). The likely selective advantage of visual defenses is to reduce the danger to the spider of resting at the hub during the day, allowing her to remain at the hub and capture more prey by responding more rapidly when they strike her web.

B. Minor hypotheses

5. **Love path.** Males use stabilimenta as paths to locate conspecific females (Wiehle 1927).
6. **Shade** (= "thermoregulation" hypothesis of Herberstein et al. 2000b). The stabilimentum casts a shadow on the spider as she rests on her web, reducing her temperature and thus her metabolic costs (Humpreys 1992).
7. **Silk reserve.** Silk is deposited in the stabilimentum to be ingested later and used for other tasks (Walter et al. 2008).
8. **Non-functional results of stress.** Stabilimenta are incidental byproducts of stress, and are not selectively advantageous (Nentwig and Heimer 1987).
9. **Regulate silk production** (= "silk regulation" hypothesis of Herberstein et al. 2000b). Deposition of stabilimentum silk stimulates production of wrapping silk in the aciniform and piriform glands (Peters 1993b).
10. **Aposematic signal.** The white tufts on *Gasteracantha cancriformis* orbs are part of the aposematic signals used by these brightly colored, heavily spined spiders to deter predators.
11. **Drinking site.** The spider drinks water from hub portions of the stabilimentum (in the araneid *Argiope bruennichi*) (Walter et al. 2009).

---

[1] Watanabe (2000, p. 568) also cited the stabilization hypothesis to explain the linear stabilimenta of *O. sybotides*, supposing that they "may strengthen the short radial threads."

[2] A further variation on the idea of prey attraction is that the coloration of the spider herself as she rests at the hub serves to attract prey (e.g., Craig and Ebert 1994; Hauber 2002; Tso et al. 2002; Bush et al. 2008; Tso 2013). I will not discuss this possibility in detail. Some weaknesses in its support: estimates of the numbers of prey captured were not convincing because of flawed techniques (only prey left in the web and not attacked by the spider were counted, or counts of holes in the web (which may be subjective—footnote in table O3.4, Blackledge et al. 2011) were used to estimate numbers of prey captured; there was no control for possible differences in prey abundances at websites when webs at different sites were compared; differences in insect size (and thus probable nutritional value) were not taken into account when comparing the numbers of individuals captured; and some sample sizes were small.

Anderson 1978; Robinson and Robinson 1975; Higgins 1987). Nor were there controls for possible differences in the numbers of prey present at the two sites in this study (at one site stabilimenta were removed and fewer prey were subsequently captured).

Data from several other groups with detritus stabilimenta also argue against the carrion attractant hypothesis. Stabilimenta did not attract prey in field or lab experiments with *Cyclosa* spp. (Gonzaga and Vasconcellos-Neto 2005; Chou et al. 2005; Blamires et al. 2010a). Many groups include vegetable rather than animal detritus in detritus stabilimenta, including the araneids *Cyclosa* spp. (Hingston 1932; Tan et al. 2010), *Allocyclosa bifurca* (Eberhard 2003), *Azilia vachoni* (Sewlal 2016), *A.* sp. (Hingston 1932), and *Enacrosoma* spp. (G. Hormiga unpub.; WE) (fig. 3.42), as well as the tetragnathids *Dolichognatha pentagona* (fig. 3.43) and *Landana* sp. (WE). Similar inclusions of plant debris in non-orb webs such as those of theridiids, which generally rest out of sight under the debris (summary in Eber-

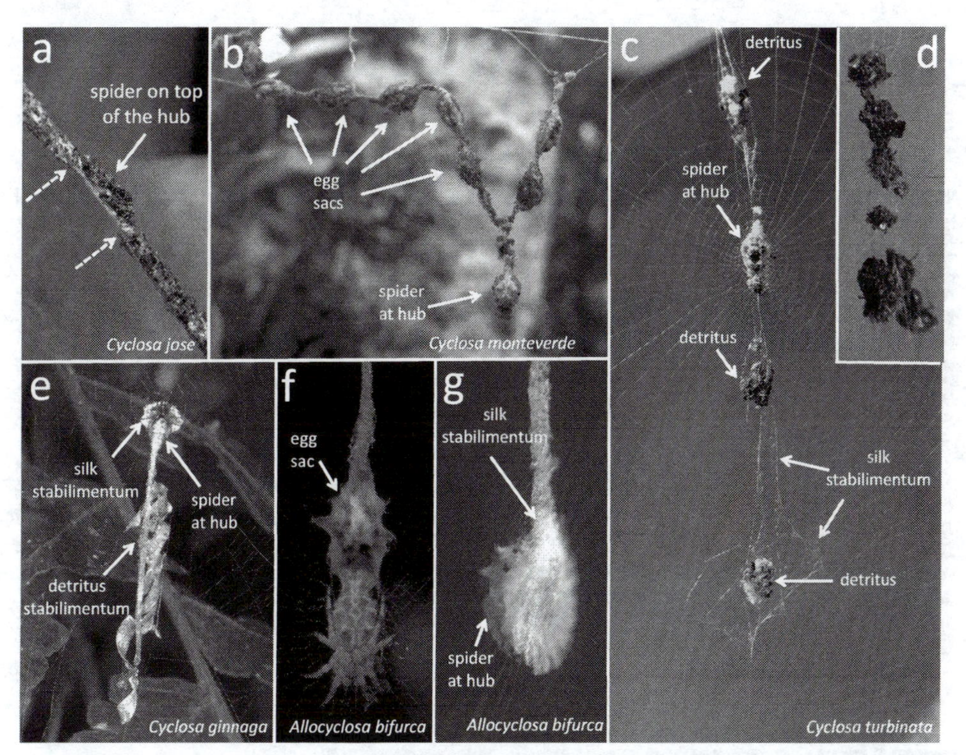

Fig. 3.41. Egg sac and detritus stabilimenta (which are generally thought to function to camouflage the spider at the hub) and silk stabilimenta (of less certain function) are sometimes linked in the araneid genus *Cyclosa* and its allies. In some species, such as *C. jose* (*a*) and *C. monteverdi* (*b*), the visual resemblance between the color and form of the spider and her detritus (*a*) or egg sac (*b*) stabilimentum is so strong that there is little doubt that they function as camouflage. The spider just fills the space between the two linear cylinders of detritus in *C. jose* (dotted arrows in *a*) (this species is unusual in resting on the upper rather than the lower surface of the hub). The webs of *C. turbinata* (*c*) also included pellets of detritus that were close to the size of the spider, but they were sometimes combined with white stabilimentum silk that connected and encircled them (but not the spider). The pellets of prey remains that constituted the stabilimentum of *Nephila clavipes* (NE) were only dry, crumbled masses (*d*), inappropriate for attracting carrion flies as supposed by the prey attraction hypothesis. The web of *C. ginnaga* (*e*) presented a combination of silk and detritus: a long dead leaf extended from a silk disc at the center of the hub where the spider rested; the silvery color of the spider's abdomen blended with the disc. Still another type of visual defense occurred in *Allocyclosa bifurca* (AR). The female's color changed gradually from brownish to gray-green when she reached maturity, and this color change matched a change in stabilimenta; the stabilimenta of immatures were made of brownish detritus, while those of mature females were gray-green egg sacs (*f*). The likely camouflage function of these egg sac stabilimenta was also linked to silk stabilimenta, because when the egg sacs were removed, spiders built silk stabilimenta (*g*); when the egg sacs were returned to the web, the spiders ceased building silk stabilimenta. It thus seems likely that the silk stabilimenta had the same camouflage function of the sacs and debris (*b* courtesy of Adair McNear; *d* courtesy of Daiquin Li).

hard et al. 2008b), are also thought to function to hide the spider. The inclusion of detritus stabilimenta in the resting webs of *Cyclosa oculata* (AR), which lacked sticky spirals and were thus not designed to capture prey, also contradicted the fly attraction hypothesis (Zschokke and Bolzern 2007).

Support for the alternative, visual defense hypothesis for egg sac and detritus stabilimenta comes from several types of data. Spiders have a variety of other morphological and behavioral traits that appear to function as defenses against visual predators. These include special, constrained postures that hide their legs, colors that blend with the background or disrupt the spider's visual outline, and retreats in detritus that is in the web or in which the spider shelters at or beyond the edge of the web (Hingston 1927, 1932; Ewer 1972; Edmunds and Edmunds 1986). Visual predators thus appear to exercise strong selection on spiders resting on their webs. Some spiders have such elaborate, beautiful morphological modifications to produce visual camouflage (e.g., Akerman 1932 on the uloborid *Miagrammopes*; Smith 2006 and Xavier et al. 2017 on the araneids *Poltys* and

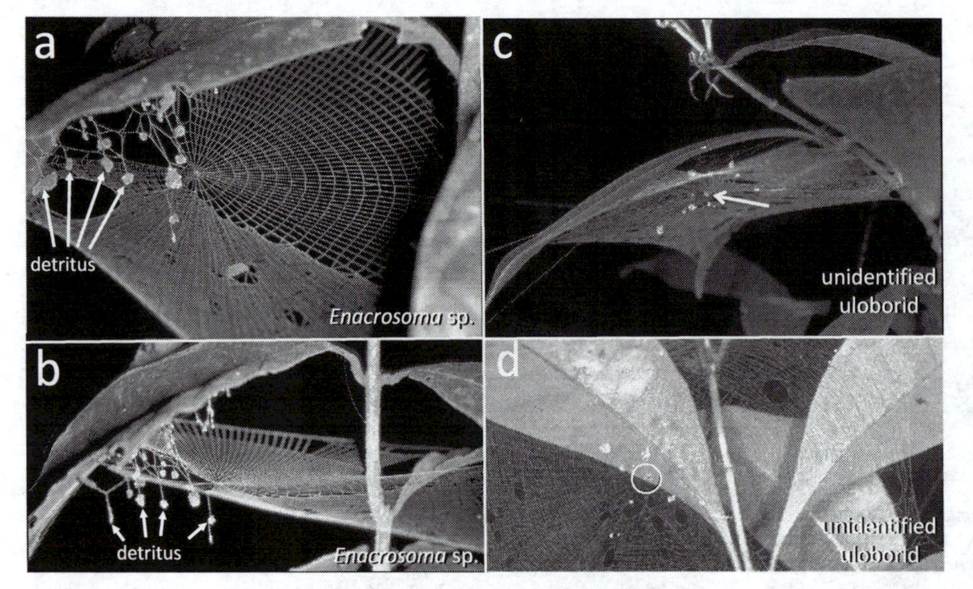

Fig. 3.42. The orbs of *Enacrosoma* sp. (AR) (*a*, *b*) from Ecuador and an unidentified uloborid from Australia (*c*, *d*) have converged on several unusual traits (a large portion of the *E.* sp. orb was being repaired, and the spider had only built the temporary spiral). Both species built especially dense horizontal orbs near the underside of a large leaf. Both species hung multiple masses of detritus from their webs, making it difficult to distinguish the crouching spider (arrow in *c*, circle in *d*) (observations of other, finished webs of *E. anomalum* showed that the spider hung from the orb on a line, but was connected to the hub by a second line—WE). Both species built numerous, closely spaced loops of sticky spiral (photographs by Gustavo Hormiga).

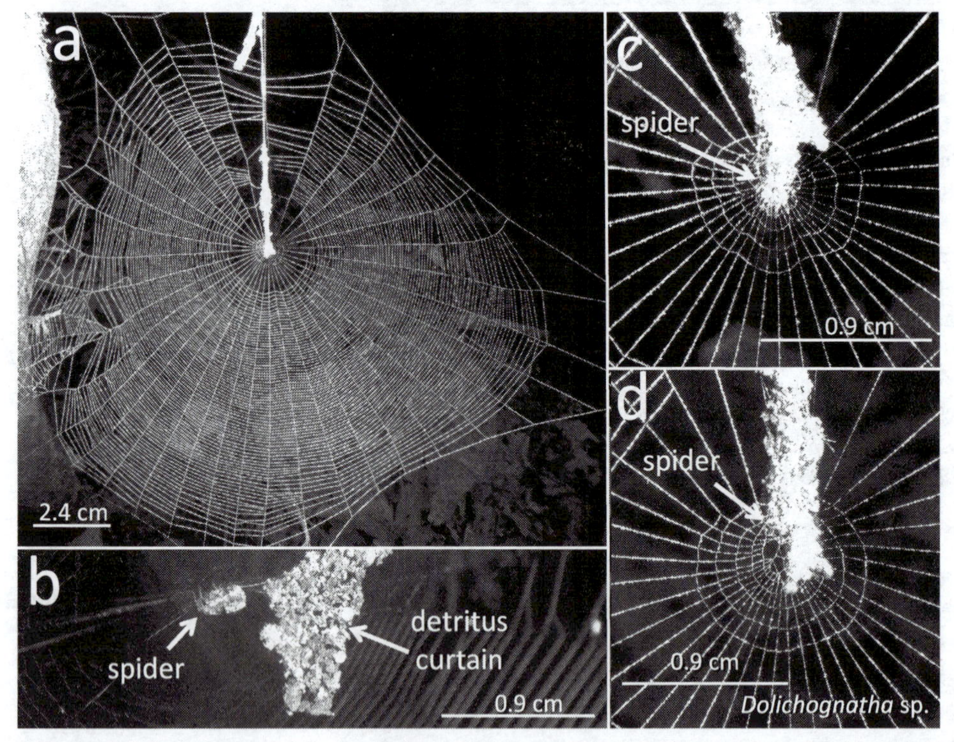

Fig. 3.43. Both the behavior and the morphology of this mature female *Dolichognatha* sp. (TET) from Costa Rica support the hypothesis that her detritus stabilimentum functions to provide visual camouflage. The spider rests with her legs tightly folded by day (*a*, *c*) and with them more extended at night (*b*) at the hub of her more or less horizontal orb. Here the profile of her lumpy abdomen merges with the edge of the similarly colored stabilimentum, a free-hanging curtain of irregular lumps of plant debris fastened together with silk. When disturbed (e.g., when a support for her web was jarred), the spider turned 180° from her usual position (*c*), and grasped the curtain and pulled it partially around her (*d*), thus obscuring her outline even more. Similar use of curtains of detritus near the hub occurred in *Dolichognatha* (= *Epeira*) *lodiculafaciens* (Hingston 1932) and *D.* sp. from New Guinea (Levi 1981); *Cyclosa conica* and *C. turbinata* (AR) also turned to rest on detritus stabilimenta when disturbed (WE).

Wixia; Getty and Coyle 1996 on the deinopid *Deinopis*) that there can be little doubt that there has been intense selection in nature favoring visual crypsis in web-building spiders.

The frequently dramatic correlations between the form of the spider's abdomen and the form of the egg sac or detritus stabilimenta that it builds also favor the visual camouflage hypothesis. Species with long cylindrical abdomens build long, thin egg sac and detritus stabilimenta, those with bumpy abdomens build bumpy egg sacs, and those with globular abdomens build globular egg sacs of a similar size (fig. 3.41) (Hingston 1932). Several species of *Cyclosa* leave a space that is just the size of the spider's body in a string of detritus objects, and that is exactly at the center of the hub where the spider sits; the spider appears to be just one more mass in the string (fig. 3.41). The color of the spider also often resembles that of the detritus stabilimentum, at least to the human eye (fig. 3.41, Hingston 1932; Marson 1947a,b). In some species the timing of color changes in the spider's abdomen matches the timing of similar color changes in the egg sac and detritus stabilimenta that she builds (Edmunds 1986; Eberhard 2003). Behavioral details tell the same story. When disturbed at the hub, *Dolichognatha* sp. (TET) turned and pulled some of the loose-hanging mat of detritus, which matched her own general coloration, around herself (fig. 3.43). *Cyclosa turbinata* also turned onto its detritus stabilimentum when disturbed (WE).

There are even direct observations in *Cyclosa* of predator attacks on parts of detritus stabilimenta away from the spider resting at the hub (Gonzaga and Vasconcellos-Neto 2005; Chou et al. 2005; Tseng and Tso 2009). When I have asked inexperienced students to pick out the *Cyclosa* spider in webs in the field with a detritus stabilimenta, they frequently make mistakes, again supporting the camouflage hypothesis. Pseudostigmatid damselflies (specialist predators of spiders in their webs) may also experience difficulty perceiving *Cyclosa* sp. spiders with detritus stabilimenta, as they did not attack them in the field on their webs unless the spider moved (Fincke 1992).

The camouflage hypothesis is actually more complex than it first appears, however, because it depends on understanding the sensory worlds of a variety of potential predators and their abilities to learn (or fail to learn) under natural conditions. Unfortunately, the behavior of potential predators when they encounter detritus stabilimenta in the field is poorly studied. Two important questions are whether predators in nature learn to associate stabilimenta with food, and whether they distinguish the spiders from their stabilimenta. Two studies of the deflections of attacks by potential predators of *Cyclosa* species (a vespid wasp and an unknown insect, presumably a wasp) (Gonzaga and Vasconcellos-Neto 2005; Chou et al. 2005) documented that, in fact, predators can and do attack detritus stabilimenta in the field. The attacks of *Vespa affinis* tended to be misdirected toward the stabilimentum rather than the spider itself (Chou et al. 2005). The presumably negative experience that a potential *Cyclosa* predator has as a result of such misdirected attack not only saves the life of the spider on that particular web, but may also result in the predator learning not to make future attacks on such objects. Such learning could occur if the first strike was somehow disagreeable for the predator or elicited an effective defensive behavior by the spider such as jumping from the web (Hingston 1927, 1932; Edmunds and Edmunds 1986; Cushing and Opell 1990a) (at least one predatory wasp, however, has the ability to follow spiders to the ground below when they jump defensively—Eberhard 1970a; and parasitic wasp sometimes wait for extended periods for spiders to return to their webs (see review in Eberhard 2018c). The fact that the predators in the *Cyclosa* studies of Gonzaga and Vasconcellos-Neto (2005), Chou et al. (2005), and Fan et al. 2009 did not persistently hover near the web and repeatedly attack different portions of the stabilimenta until hitting the spider suggests that they had failed to learn how to deal with these prey, supporting the visual defense hypothesis.

In sum, several types of evidence demonstrate that detritus and egg sac stabilimenta function to camouflage spiders against visually orienting predators. Evidence for alternative functions is weak.

3.3.4.2 Silk stabilimenta

3.3.4.2.1 THE HYPOTHESES

In contrast to the tidy camouflage story for detritus and egg sac stabilimenta, the function of silk stabilimenta ("stabilimenta" in what follows) has a long history of controversy that shows no sign of a pending resolution. This controversy has furnished drama in otherwise rather staid meetings of arachnologists, where a speaker is sometimes hardly able to finish a talk before being attacked by a supporter of a competing hypothesis. The question continues to be actively discussed, and the literature is sprawling. The objective of this overview is not to argue that there is one single function of silk stabilimenta (I believe, in fact, that there is probably no single answer); rather I aim to present critical evaluations of the available evidence (see especially tables O3.3–O3.5), much of which I think is flawed. I have emphasized the probable importance of some variables that have often been neglected in studies of stabilimentum function, in the hope of aiding future work. While the discussions are tangled, one general lesson that emerges, echoing the findings in previous reviews (Hingston 1927, 1932; Herberstein et al. 2000b), is that the evolution of silk stabilimenta presents both rampant diversity, and widespread convergences.

Tables O3.3–O3.5 constitute the most extensive compilation of evidence and the most thorough examination of arguments regarding stabilimentum function ever assembled (for previous reviews, see Herberstein et al. 2000b; Walter and Elgar 2012) (I will not deal here with the separate and likewise controversial question of the functions of the sometimes bright colors of spiders that build stabilimenta—see Craig and Ebert 1994, Théry and Casas 2009, and section 3.3.4). In an attempt to avoid swamping the reader with more information about stabilimenta than he or she ever wanted to know, I have placed these tables and my evaluations of their implications in the online appendices. The following discussion is a general overview of these tables.

Stabilimenta have evolved many times. Scharff and Coddington (1997) listed 22 genera in which they occur, and Herberstein et al. (2000b) proposed 9 probable independent origins. Recent discoveries of stabilimenta in the araneids *Caerostris darwini* (Gregorič et al. 2011), *Molinaranea* (Levi 2001), *Neogea nocticolor* (Murphy and Murphy 2000), *Plebs eburnus* (Bruce et al. 2004), and *Metepeira gressa* (WE) may represent additional convergences. The multiple evolutionary origins of silk stabilimenta were said by Herberstein et al. (2000b) to imply that they have multiple functions, but this argument is not logical (Eberhard 2003): convergent evolution implies different historical origins, but not necessarily different selection pressures. For instance, the many convergences on the flattened appendages that animals use to swim do not mean that flippers, fins, and webbed feet do not all function to increase mobility in the water.

The limits of what constitutes a silk stabilimentum are not entirely clear, as in the dense array of white tufts of silk associated with the resting place against the underside of a leaf of the pholcid *Cantikus halabala* species group (Huber et al. 2016), or the white silk added to the exposed walls of the silk retreats of *Araneus expletus* (AR) (fig. 3.49c,d) (Eberhard 2008). Stabilimentum forms vary widely, and there have been many convergences on particular designs, such as zig-zag or saw-tooth lines on radii near the hub, circles at the hub, and solid discs at the hub (fig. 3.40, Hingston 1932; Herberstein et al. 2000b).

At least 14 different hypotheses have been proposed for silk stabilimentum function, and they have also been said to have no function at all (Nentwig and Heimer 1987; Nentwig and Rogg 1988; Peters 1993b), or to represent the currently functionless vestiges of former adaptations (Robinson and Robinson 1973b). Careful discussion of all the hypotheses is not feasible here—it would require an entire chapter or perhaps an entire book. I will give an overview of the four most commonly cited (and I believe the most likely) hypotheses (defined in table 3.7): prey attraction; web stability; web advertisement; and visual defense. Reasons for rejecting many of the other, less commonly cited hypotheses are discussed by Herberstein et al. (2000b). Most of these four hypotheses have multiple variants (table 3.7), which increases the difficulty of testing them. The different possible functions are not exclusive, and a given stabilimentum could incidentally confer more than a single benefit, especially in different environments (see below, and for example, the discussion of the stabilizing effect of disc stabilimenta in table 3.7).

During the past 20 years, one major breakthrough was the discovery of a serious flaw in one type of evidence that appeared to strongly favor the prey attraction hypothesis. The webs of *Argiope* (AR) with stabilimenta in the field had been found to capture more prey, but it had not been appreciated that some websites in the field have consistently more prey than others, and that the better-fed spiders at such websites were better able to afford the cost of adding a stabilimentum to their webs (Blackledge and Wenzel 1999, 2001a). It appears that the amount of aciniform silk available may influence stabilimentum size, because selectively depleting these glands in *A. aetheroides* reduced stabilimentum size (Tso 2004). When the amounts of prey captured at different websites were taken into account, the stabilimenta in *A. aurantia* actually *reduced* prey capture (Blackledge and Wenzel 1999, 2001a; Blackledge 1998a,b). These *A. aurantia* data were given little importance, however, in some subsequent reviews (e.g., Herberstein and Tso 2011), perhaps because the trends they showed did not occur in another species, *A. keyserlingi*, in which the same individuals at given websites captured more prey on days in which they built more elaborate stabilimenta (Herberstein 2000).

An important distinction that has sometimes been lost in the dust and smoke of the stabilimentum wars is that the possible reasons why stabilimenta are maintained in current populations (their current costs and benefits) are not necessarily the same as the reasons why they originally evolved. This is especially important when, as is usually the case, the environments where they have been studied have been substantially altered by human activities, and thus have densities and diversities of both prey and predators that may differ from those of the environments in which the spiders evolved. Most experimental studies have failed to take this need for "ecological realism" into account, and have emphasized only present-day costs and benefits (section 3.3.4.2.2.2 below, also discussions in table O3.4).

3.3.4.2.2 PROBLEMS INTERPRETING THE DATA

3.3.4.2.2.1 Inconsistent support and behavior

Perhaps the most important source of heat in the stabilimentum wars is that all hypotheses are strongly contradicted by at least some data. No hypothesis escapes unscathed (Tables O3.3–O3.5). A second cause for conflict is that data from different studies, even of closely related species, not infrequently contradict each other. Table O3.3 lists twelve different topics on which different studies have produced contradictory results. For example, the stabilimenta on *Argiope* webs have sometimes increased prey capture (Craig 1991; Craig et al. 2001; Herberstein 2000; Tso 1996), sometimes decreased it (Blackledge and Wenzel 1999), and sometimes had no effect (Bruce et al. 2001).

Further problems result from 1) published arguments that are internally contradictory (e.g., evidence of types #12, 17, and 21 in table O3.3); 2) defenses of hypotheses that are confusingly convoluted (see #8, 26, and 54 in table O3.3); 3) failures to attempt to distinguish incidental uses made of stabilimenta (e.g., sources of shade, rain water) from the evolved functions for which they are appropriately designed (see Williams 1966 and Alcock 2013 for discussions of this difference); and 4) incomplete arguments that give considerations of the implications of some but not other trends for the hypotheses (for example, Herberstein et al. 2000b and Starks 2002 do not consider the implications of the strict limitation of stabilimenta to diurnal orbs for any hypothesis other than prey attraction). A few publications simply ignored the controversy, discussing almost exclusively work supporting one hypothesis (e.g., Craig 2003); others miscited observations from the published literature in ways that favored a particular hypothesis (see Eberhard 2003 regarding data on *Gasteracantha* in Herberstein et al. 2000b).

In many species stabilimenta are highly variable intra-specifically in both presence and form (e.g., fig. 3.40f,g), even in the successive webs of a single individual (Hingston 1932; Marples 1935, 1962, 1969; Marson 1947a,b; Robinson and Robinson 1970a, 1978; Ewer 1972; Lubin 1986; Nentwig and Rogg 1988; Peters 1993b; Watanabe 1999a; Herberstein 2000; Herberstein et al. 2000b; Seah and Li 2002; Uhl 2008). For instance, better fed individuals in two *Argiope* species built larger stabilimenta, and in one species they also built them more frequently (Blackledge 1998a,b). This variability suggests that the costs and benefits of having stabilimenta, and of having stabilimenta of different types, may often be nearly equal (Edmunds 1986), and may thus be difficult to measure precisely. To cite another complicating environmental variable (among several) that has never to my knowledge been taken into account in a field study, the density of the population of a stabilimentum-building species may affect the balance between costs and payoffs that individual spiders can expect to derive from building a stabilimentum. This is because predators would be more likely to learn to associate stabilimenta with food if the spiders are more abundant (Edmunds 1986; Starks 2002). Learning by predators may play an important role in determining the advantages of stabilimenta (Craig 1994a,b), but has been little studied in this context.

3.3.4.2.2.2 Difficulties with direct measurements I: ecological realism

One logical tactic in attempts to test the hypotheses for stabilimentum function, such as prey attraction and predator defense, which depend on the spider's interaction with other organisms in its habitat, is to measure these interactions empirically. But, despite the substantial number of studies of this type, this tactic has several problems. In the first place, stabilimenta did not evolve under the altered ecological conditions that currently occur at most sites in the world today. Ecological realism (direct measurements of possible functions such as prey attraction or predator defense under conditions similar to those under which stabilimenta evolved) is crucial, but very difficult to attain. In fact, I believe that complete, convincing weightings of this sort of ecological importance for different prey and predator types have never been achieved in any study (tables O3.4, O3.5).

This problem is especially serious for the hypotheses that include interactions with large numbers of other species. If, for instance, some prey are attracted to stabilimenta, some are not (e.g., Tso 1998a,b), and others are repelled (Blackledge and Wenzel 1999), then the relevant overall payoff will depend on the relative numbers (and sizes) of these different types of prey in the habitats in which the species evolved. Similarly, convincing direct tests of a defensive function for stabilimenta would require data on the effects of stabilimenta with each type of potential predator; and then the different effects of stabilimenta on potential predators would have to be weighted appropriately, on the basis of the importance of the different species of predators in killing spiders (not on their abundances!) at different sites and seasons, and in natural (not anthropogenic!) environments.

Nature can be complex. For instance, direct observations of the attack behavior of the wasp *Polysphincta gutfreundi* that parasitizes *Allocyclosa bifurca* (AR) showed that these highly specialized wasps were not distracted by the stabilimenta of these spiders (Eberhard 2018c, WE). Nevertheless, coloration changes from brown to gray-green in *Allocyclosa bifurca* that occur at the same time in the life cycle when the spider changes from making brown detritus stabilimenta to gray-green egg sac stabilimenta, and the contrast between spider's daytime crouching posture and its spread-legged posture at night argue strongly that these stabilimenta function to camouflage the spider, presumably against other, more generalized predators (Eberhard 2003). The same stabilimentum could even attract some predators but defend against others (Li 2005), and these costs and benefits must be measured with enough quantitative precision to permit determination of the balance between them (Bristowe 1941). These are difficult tasks indeed!

Ecological realism problems also affect interpretations of laboratory studies. For example, experimental demonstrations that a stabilimentum does or does not defend against a given predator in captivity are often mute regarding the possibility that the attractiveness of spiders as prey in simplified captive conditions will differ from that in the much more complex environments in the field (Craig 1988). In addition, they often beg the question of the relative importance (if any) of that particular predator among all the other predators in the natural habitat or habitats in which the spider species evolved. Simple co-occurrence in the same modern habitat (e.g., Blamires et al. 2007b) does not imply that the predator actually preys on the spider; nor does it guarantee that the two species co-occurred in previous habitats where the spider evolved. The effects of stabilimenta on different predators are very likely to vary greatly, given the wide variety of likely predators on web spiders that, as noted above, include monkeys, birds, lizards, frogs, sphecid wasps, pompilid wasps, asilid flies, preying mantids, emesine bugs, damselflies, and salticid and oxyopid spiders. The current population densities of some predators at many study sites undoubtedly differ from those at the sites and times when the spiders evolved stabilimenta (for example, many primates are reduced or absent in many modern tropical study sites). The contexts in which some experimental data were obtained are so unnatural (e.g., #22 in table O3.3) that their relevance to selection in nature is highly questionable.

In sum, several problems related to ecological realism make it very difficult to use some types of direct modern measurements to evaluate the possible evolutionary significance of stabilimentum functions that involve interactions with diverse arrays of other organisms. Lack of ecological realism reduces the usefulness of many published studies for evaluating the hypotheses of stabilimentum function. Other, less direct kinds of tests of the hypotheses related to the behavior of the spiders (below) may provide more reliable evidence.

3.3.4.2.2.3 Direct measurements II: behavioral contexts, defensive behaviors, species differences

The visual cues that potential prey find attractive or that they avoid when flying in the vicinity of orb webs during their normal, undisturbed activities in the field (as opposed to during tests in artificial settings in captivity) are poorly known. Even under relatively standardized conditions in a greenhouse, Rao et al. (2008) showed that the likelihood that the stingless bee *Trigona carbonaria* would be captured in the web of an *Argiope keyserlingi* (AR) was lower when it encountered a web during its foraging activities than when it was returning to its nest. This illustrates the importance of the behavioral contexts, and brings to the fore very difficult, unexplored questions regarding the relative frequencies with which webs are encountered in different behavioral contexts by prey and predators in nature.

A further complication concerning behavioral contexts is related to the possibility that a spider's defensive behavior functions in combination with the stabilimentum. Juvenile *A. versicolor* (AR) tended to shuttle to the opposite side of their disc stabilimenta when disturbed (Li et al. 2003), in accord with the visual defense and physical shield hypotheses for the stabilimentum (but not with the prey attraction hypothesis). Adults that built cross stabilimenta (similar to fig. 3.40i) tended, in contrast, to swing or pump the web rapidly back and forth when similarly disturbed. Swinging behavior could make their bodies more difficult to distinguish from the stabilimentum (in accord with the visual defense hypothesis, but not explained by the other three leading hypotheses) (Li et al. 2003). Antipredator responses in *A. keyserlingi* (dropping, pumping the web, or moving to the edge) also varied in different settings (lab vs. field) as well with time of year, but were not correlated with investment in cross stabilimenta (Blamires et al. 2007a). The defensive behavior of *Uloborus glomosus* (UL) also varied with the type of stabilimenum on the web (in accord with the visual defense hypothesis); it also varied, however, with the age of the spider, and even the time of day for less obvious reasons (Cushing and Opell 1990b). Presumably the variations were due to differences in this species' susceptibility to attacks from different predators, but no data are available.

Other aspects of the behavior of potential predators on orb weavers that might be influenced by stabilimenta are also very poorly studied. It is not clear, for example, which stabilimentum traits might confuse visually orienting prey or predators during their foraging activities. In the absence of such data, the visual distraction variant of the visual defense hypothesis (table 3.7) can be stretched to accommodate almost any pattern of white silk by claiming that it "disperses" the predator's attention (Hingston 1932). Another crucial question with no answers is how much learning occurs in different groups of predators in nature, and which aspects of a stabilimentum might facilitate or impede learning?

There are similar difficulties in documenting the degree of attractiveness of stabilimenta for different possible prey. It appears that different taxa of insects respond (and fail to respond) to stabilimenta in different degrees (Tso 1998a,b; Blamires et al. 2008), so lab trials of only a single species of prey are not very convincing. Ideally, the importance of the degree of attraction for each prey taxon should be weighted with respect to its importance in the diets of spiders in nature (including both numbers of individuals and their sizes), but such weighting is seldom even mentioned, much less attempted. The technique of Tso and his colleagues of using video cameras in the field to count interactions with prey and predators seems particularly promising in this respect, though sample sizes for different species of prey and for predation events are small despite heroic sample sizes (e.g., >1000 hr taped by Tseng et al. 2011) (and there were remaining uncertainties regarding ecological realism).

The relative importance of different factors probably varies over time as well as space (see section 3.2.14). As a spider grows during ontogeny, the payoff for a given stabilimentum design is likely to vary substantially (e.g., Robinson and Robinson 1973b). Both the prey and the predators most important for small spiderlings are undoubtedly different from those that are most important for adults (section 3.2.14). Even for a single predator like a lizard that could prey on multiple sizes of spiders, stabilimentum functions might differ: the visual extension of an individual *Argiope argentata*'s legs provided by its stabilimentum may inhibit attacks by gape-limited lizards when the spider is moderately large, but not when she is smaller (Schoener and Spiller 1992). The common finding that stabilimentum patterns change ontogenetically in form and frequency (e.g., Marples 1962; Ewer

1972; Lubin 1974, 1986; Robinson and Robinson 1978; Nentwig and Rogg 1988; Seah and Li 2002; Eberhard 2003; Rao et al. 2007; Uhl 2008) suggests that these balances may differ for individuals of the same species at different times of the year. Furthermore, there was individual variation in the ontogenetic timing of changes in stabilimenta in *Argiope flavipalpis* (Ewer 1972), suggesting intra-specific variation on the balances between costs and benefits for different individuals.

3.3.4.2.2.4 *Direct measurements III: inappropriate measurements*

Another problem concerns the precision or appropriateness of measurement themselves. Take, for instance, the attempts to quantify prey impacts in nature by counting the holes and places where sticky lines adhere to each other in orbs that were built several hours previously (e.g., Craig and Freeman 1991). These data ignored other possible causes of web disturbances, such as detritus, the struggles of prey, the movements of the spider in her web, and movements of the web and its supports in the wind (fig. 4.19, section 3.2). It is not obvious to me, for example, how many prey struck the orbs in figs. 1.6d, 3.5b, 3.7b, 3.16b, or 3.43, even though they are relatively fresh. Counting problems become much more difficult as the web ages—see for example figs. 3.19 and 3.34c (the reader is invited to try counting) (for other reservations regarding this technique, see also Blackledge et al. 2011, appendix O3.1, and footnotes in tables O3.4 and O3.5). The data using this technique were not taken blind, so inadvertent, unconscious biases in the difficult process of quantifying prey impacts could have occurred.

Finally, as noted by Robinson and Robinson (1973b), simple lab studies that have sometimes been used to demonstrate that the individuals of species are physiologically capable of perceiving a stimulus are not sufficient to show that the animals do indeed attend to this stimulus under natural conditions. Thus, for instance, comparing stabilimentum reflectances with the visible spectrum of colors or the color contrasts for the eye on an insect or a bird (e.g., Craig and Bernard 1990; Blackledge and Wenzel 2000; Tan et al. 2010) leaves the question of whether the animals actually use this information in nature unanswered. Ethological studies have repeatedly shown that many animals attend only to relatively simple stimuli ("sign stimuli") even when they are capable of discriminating more complex ones (Alcock 2013).

There are still further complications that are likely to make direct measurements of the costs and benefits of stabilimenta difficult. It is possible that some insects' responses to the sight of a stabilimentum vary depending on their distance from the web. For instance, Rao (2010) found a lack of attraction to the stabilimenta of *Argiope keyserlingi* when potential prey (bees) were 1 m from the web, and possible avoidance of stabilimenta at especially small distances, but mentioned the possibility that nevertheless the stabilimenta might be attractive at longer ranges. Only when data summing these different effects are weighted appropriately for their frequency of occurrence in nature would it be possible to give biologically relevant indications of a stabilimentum's effects on prey or predators.

3.3.4.2.2.5 *The importance of UV reflectance*

The discovery that the aciniform silk in stabilimenta reflects light strongly in the UV (Craig and Bernard 1990), in combination with the fact that insects perceive and respond to UV wavelengths, has provoked extensive research on the reflectance of different types of silk. Determining the implications of these studies for the stabilimentum function hypotheses is not easy. To avoid getting bogged down in details, and because several of the weaknesses mentioned in previous sections also apply here, I have relegated most of the discussion to appendix O3.1, and will give only a summary of the data that lead to the conclusion that UV reflectance does not give convincing support to the prey attraction hypothesis.

The combination of the high sensitivity of insect eyes to UV, the high UV reflectance of stabimentum silk, and the generally low UV reflectance of foliage (Craig 2003) probably makes stabilimenta stand out clearly from their visual backgrounds in the eyes of many insects. But the physiological capabilities of the sense organs of an animal do not necessarily translate directly to the animal's behavior. For instance, a test for the use of the stabilimenta of *Argiope aurantia* (AR) for orientation by the bee *A. mellifera* by Blackledge and Wenzel (2000) found a surprising result. Although physiological studies have shown that bee eyes clearly respond to wavelengths reflected by this species' stabilimenta, the bees turned out to be less rather than more able to learn when a food source was marked with stabilimentum silk. In simple terms, high visibility is not equal to high attractiveness.

It is also well-established that some insects use "nectar guides" on flowers involving UV reflectance to guide them when they are near a flower (Wehner 1981). Again, however, applying these findings to the possible function of UV-reflective stabilimenta (Craig and Bernard 1990; Craig 2003) is not simple. Nectar guides include both UV-bright and UV-dark patterns (e.g., Barth 1985), so UV reflectance per se is not necessarily attractive. Secondly, several studies of the prey of orb weavers that build stabilimenta (see Robinson and Robinson 1970a on *Argiope argentata*, Blackledge and Wenzel 1999 on *A. aurantia*, Cheng and Tso 2007 on *A. aemula*, Gregorič et al. 2011 on *Caerostris darwini*) have not reported a bias toward flower-visiting insects (though it is difficult to determine which prey in a habitat are "available"—appendix O3.2.3). Some insect groups that seldom visit flowers, such as odonates and grasshoppers, constituted major portions of the diet of some *Argiope* and *Caerostris* (AR) species, and many non-flower visiting insects were captured by the stabilimentum building *Octonoba sybotides* (UL) (Watanabe 1999b).

Most stabilimenta are placed on or near the hubs of orbs (figs. 3.40, 3.41), which are inappropriate sites to trap insects that are using them as nectar guides. For instance, the classic "X" stabilimenta of several *Argiope* species is not centered on sticky lines, where it could trap prey that were guided to the intersection of the arms, but rather on the non-sticky hub lines (where, of course, the spider rests) (figs. 3.45–3.47). In sum, there are several reasons to question the simple generalization that the high UV reflectance of stabilimentum silk functions to make stabilimenta attractive to prey in the field.

3.3.4.2.2.6 *The hypotheses are not mutually exclusive*

The four major hypotheses are not mutually exclusive. In fact it is possible to construct a plausible "hybrid" hypothesis that combines two of the most popular hypotheses: the adaptiveness of a stabilimentum in attracting prey could be enhanced if it were to "incidentally" afford visual defense to the spider (Tan and Li 2009) (or, vice versa, a visual defense structure could incidentally attract prey). A combination of this sort is not unreasonable, as some insects are known to perceive spiders resting on flowers and avoid them (Bristowe 1958; Heiling and Herberstein 2004; see also Morse 2007), so camouflaging the spider might increase prey capture (though prey capture in webs of *Argiope aurantia* was reduced by inclusion of a stabilimentum) (Blackledge and Wenzel 1999).

Some comparative data that are unexplained by the simpler version of the prey attraction hypothesis could be explained by this hybrid. For instance, the constrained daytime orientations and positions of the spider and her legs that make her outline merge with that of the stabilimentum, and the almost unvarying association between the site where the spider rests during the day and the site of the stabilimentum (#1, 3, and 4 in table O3.3) are not easily explained by the prey attraction hypothesis, but are compatible with this hybrid (though this reasoning also has its limits—figs. 3.37, 3.50).

Or consider the possible learning experiences of potential predators of *Zosis geniculata* (UL). At least to a human eye, it was very easy to see the stabilimentum on each web under certain kinds of illumination (fig. 3.46), but the stabilimenta sometimes made it hard to distinguish occupied from unoccupied hubs (when the spider has dropped, run to the edge of the web, or abandoned the web) (WE). If a bird or lizard strikes at an unoccupied hub, it may learn a negative lesson: a stabilimentum is not associated with food, but instead with an unpleasant episode of having its head entangled in web threads. Some species of spiders with stabilimenta leave their webs more readily than others, and some predators are probably more likely to frighten the spider and thus give it a chance to escape before striking, so the importance of such negative lessons could differ among predators as well as among spider species.

3.3.4.2.2.7 *Some crucial behavioral phenomena are poorly understood*

The importance of learning per se probably also varies between different predator species, different seasons, and different habitats (Robinson 1969a). Showing that a predator is *capable* of learning (e.g., Robinson and Robinson 1970a) is not sufficient to answer the crucial questions of whether learning does occur in nature, and how frequently it occurs. As emphasized by Craig (1994a,b), learning by predatory insects may play an important role in determining the advantages of stabilimenta.

Even within a given group of potential predators, there is probably important diversity in responses to stabilimenta. Take birds, for example. Some hummingbirds seek out webs and steal silk threads to make their nests; they also prey on small orb weavers that they snatch from their webs, in some cases by poking their beaks through the hub (Robinson and Robinson 1970a). On the other hand, some other potential bird predators such as butcher birds (Robinson and Robinson 1976b) and blue jays (Horton 1980) appeared to be very uncomfortable after blundering into the webs of larger species such as *Argiope* and *Nephila*, and spent up to half an hour cleaning themselves afterward. If bird species such as these learn to use stabilimenta as visual cues, they may use them to learn to *avoid* spider webs.

The general hunting tactics sometimes vary even among different species of a given taxon (Blackledge and Pickett 2000 and Blackledge and Wenzel 2001a on mud-daubing sphecid wasps). In general there is a dearth of direct observations of predator foraging behavior under natural or semi-natural conditions, and most data regarding the effects of stabilimenta on potential predators are seriously incomplete in this respect. Indeed, the use of visual cues by predators is complex, and some important aspects are difficult to study in the field (for an early, extensive summary, see Robinson 1969a).

3.3.4.2.2.8 Comparing many apples with many oranges

The data on stabilimentum functions involve so many, diverse types of data that it is difficult to keep them all in mind, much less balance them appropriately against each other (tables O3.3and O3.4 include >60 types of data, so >240 different evaluations must be made with respect to the four major hypotheses). Some implications are much more important for some hypotheses than others; and some involve much more extensive data and clearer patterns than others. Weighing these differences objectively is difficult.

Stabilimenta may also have multiple, simultaneous but conflicting effects on spider fitness. In the araneid *Argiope aurantia*, for example, stabilimenta seem to both reduce prey capture and to favor the survival of the spider and/or her web (Blackledge and Wenzel 1999). Confusingly, in *A. argentata* these same effects may be reversed: stabilimenta were said to increase rather than decrease prey capture, and to decrease rather than increase spider survival (Craig et al. 2001) (the measures of survival were inconclusive, however—see table O3.4). Is the effect on prey capture in either species large enough to override its effects on survival (whatever they may be)? To answer such a question, one would need to have accurate, quantitative measurements in appropriate habitats (section 3.3.4.2.2.2) of the relative *sizes* of the different effects (not simply the usual "effect" vs. "lack of effect" discrimination that is typical in behavioral ecology studies).

3.3.4.2.2.9 Summary of weaknesses of direct measurements

The previous sections seem like a litany of frustration. For numerous reasons, the hypotheses for stabilimentum function are very difficult to test directly. The implications of many (most) studies that have attempted to measure the effects of stabilimenta are seriously limited by these practical problems. There are also large gaps in critical areas such as the visual stimuli used by prey and predators during their normal, undisturbed movements through natural environments. There is not a single case in the entire literature on stabilimenta with convincing, ecologically realistic direct measurements of all or even most of their costs and benefits. Perhaps it is not surprising that there are so many cases in which studies of the same or closely related species give conflicting data and conclusions (#36–45 in table O3.3).

One possible solution to these problems is to use comparative studies (e.g., table O3.3). In this case, past selection on different species is assumed to have performed the otherwise very difficult tasks of having summed up, averaged, and weighted their costs and benefits appropriately. But these comparisons have their own problems. In the first place, they document only correlations, with no guarantees regarding cause-and-effect relations. In addition, comparative studies generally assume that similar functions occur in different taxonomic groups. Because there are data contradicting all of the hypotheses, this supposition is not convincing.

3.3.4.2.3 FURTHER COMPLICATIONS: ANGLES OF VIEW, ILLUMINATION, AND BACKGROUND

Visual stimuli from stabilimenta are crucial to three of the four hypotheses. The angle from which a spider and

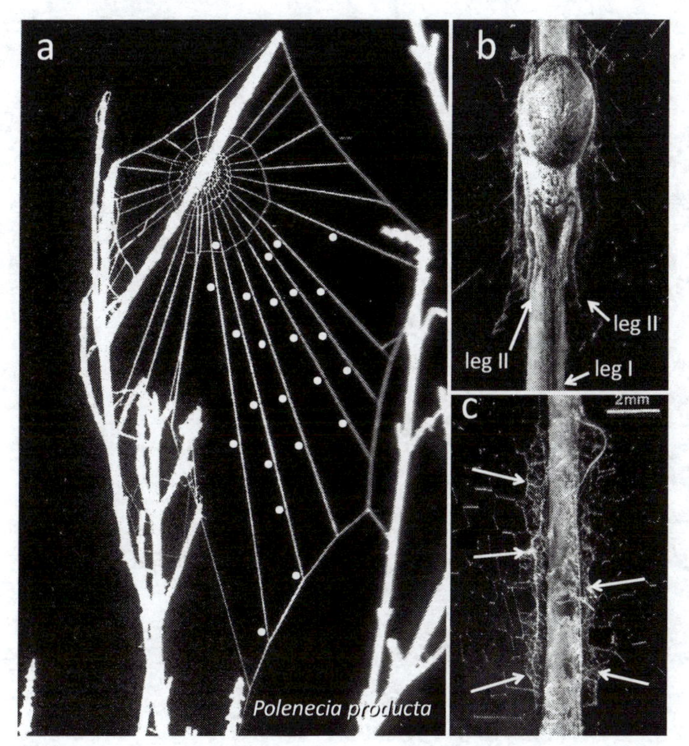

Fig. 3.44. The web of *Polenecia producta* (UL) (*a*) is thought to be derived from an orb. The hub, radii, frames, and temporary spiral were all similar to those of uloborid orbs, but the sticky cribellum silk (thicker lines in *a*) was laid along the radial and frame lines, rather than in a spiral. The white dots indicate the points where the mat of cribellum silk was interrupted briefly where it was presumably attached to the radii (similar interruptions occurred at attachments of the sticky spiral in orbs of *Zosis geniculata*—see figs. 7.1, 7.19, 7.45). The spider rested at the hub, on the undersurface of a twig (*b*) (photo taken from below); her legs I were hidden against the twig, while the distal portions of legs II were slightly spread and grasped hub lines and probably monitored the web for prey vibrations. There was a small "fringe" stabilimentum of white silk (arrows in *c*) in the area where tarsi II grasped hub lines. The placement of this tiny stabilimentum exactly in the area where the spider's legs grasped the web is in accord with the visual defense hypothesis (from Peters 1995a).

her stabilimentum are viewed, the visual background, and the angle of illumination, all of which surely vary substantially in the field, can all have major effects on the stabilimentum's perceived form and its ability to obscure the spider's outline. To take an extreme case: the disc stabilimenta of *Argiope submaronica* (fig. 3.38) and *Allocyclosa bifurca* (Eberhard 2003) almost completely obscure the view of the spider when viewed from the opposite side of the web; she is much more visually obvious, in contrast, in a "same side" view. Supposing that prey or predators approach from both sides, this would give the stabilimentum a chance of hiding the spider in some cases (as in, for instance, the possible shielding effect of the small sheet web of the salticid *Spartaeus spinimanus*—Jackson and Pollard 1990). Another extreme case favoring the visual camouflage hypothesis would be *Polenecia producta* (UL), in which all of the spider but the tips of her legs II is hidden by a twig, and a small mass of stabilimentum silk is placed just where these tips are exposed when they grasp hub lines (fig. 3.44, Peters 1995a).

I will not try to guess at the likely angles of view that prey and predators have as they approach webs with stabilimenta. But for a subset of predators, especially larger animals such as monkeys, birds, damselflies, and perhaps lizards, some viewing angles are more likely than others. This is because most orbs are inclined rather than perfectly vertical, and spiders always rest on the lower sides of inclined webs. For instance, the orbs of the uloborids *Uloborus diversus* and *Philoponella herediae*, which are built deep amongst tangled debris of pack rat nests (Eberhard 1973a) and in low undergrowth and near the leaf litter (Opell 1987c), are more likely to be seen from above (and thus in "opposite side" views) by larger predators. The same may be true for birds approaching slanting orbs of *Argiope aurantia* (AR) built near the ground, in which the spider consistently rests on the lower side of the web (fig. 3.45). In contrast, the orbs of *Zosis geniculata*, which are often built in sheltered sites, for instance near the ceiling of a building, seem likely to be both illuminated and seen more often from below, in "same side" views by birds, wasps, and damselflies (fig. 3.46). Barrier tangles built on one side of an orb (e.g., in the silk stabilimentum-building nephilid *Nephila clavipes* and the araneids *Argiope* spp., *Thelacantha brevispina*, *Metepeira gressa*, and *Allocyclosa bifurca*) (table 3.6), are also likely to cause predators to have "opposite side" views, because these tangles beside orbs are consistently placed on the same side as the spider and must cause the predator to reach the spider from the other

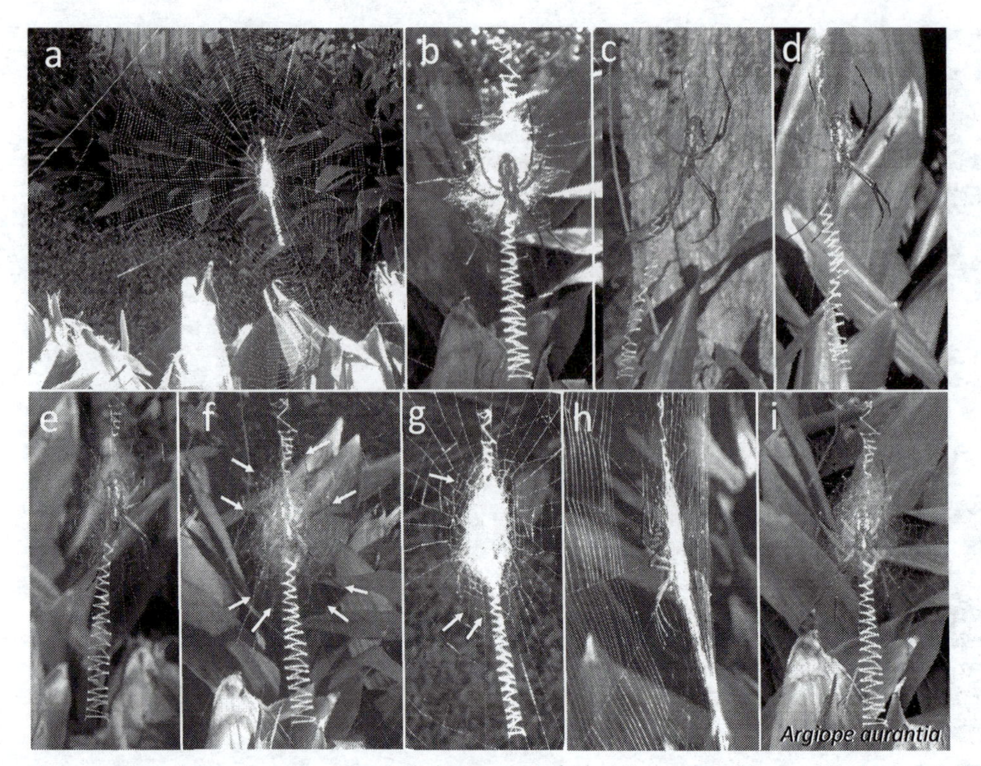

Fig. 3.45. These photos of a mature female *Argiope aurantia* (AR) resting under a "disc + vertical arms" stabilimentum on an unpowdered web in the field (*a*) show that the same stabililmentum can have very different effects on the spider's visibility, depending on illumination and visual background. Orb weavers nearly always rest on the undersides of their slanting orbs, and *A. aurantia* webs are often built in low vegetation (Enders 1974; Horton and Wise 1983), so the ventral and ventral-lateral views of the spider depicted here are probably typical of those that large flying predators like birds or wasps would usually have of spiders at the hubs of their orbs. When the stabilimentum was in direct sunlight at 9 AM in the morning, the spider's legs were nearly invisible, but their shadows were visible on the bright white disc (*b*). At 4:30 in the afternoon in direct sunlight, her visibility varied dramatically when viewed from different angles with different backgrounds. Seen from the web's right side against a brightly lit tree trunk (*c*) or against brightly lit (*d*) or darker leaves (*e*), the stabilimentum (especially the disc) had little or no effect on the spider's visibility. Viewed perpendicular to the orb (*f*) the disc had a clear blurring effect on the spider's body and the dark distal portions of her legs were inconspicuous where they projected beyond the edge of the disc (arrows) (contrast this with the disruptively banded legs of *A. submaronica* and *A. argentata* that, as shown in figs. 3.38 and 3.47, are held in constrained positions next to white bands of stabilimentum silk; this combination of differences in leg positions and coloration supports the visual defense hypothesis for stabilimenta). Viewed at an angle of about 45° to the left (*g*) the spider was completely obscured, with only the distal portions of her legs visible (arrows); the obscuring effect persisted with more acute angles until her body became visible below the stabilimentum when seen nearly directly from the left (*h*). A little later, when the spider and the stabilimentum were in the shade and were viewed perpendicular to the hub (*i*), the blurring effect of the disc was weaker (compare with *f*). Clearly, at least the disc portion of the stabilimentum of this species obscures the visual outline of the spider under certain circumstances; just as clearly, in other circumstances it does not, at least for a human. Summing up the possible camouflage effects of stabilimenta in the field in a quantitative manner, taking into account illumination throughout the day and different angles of viewing, would be a daunting task even for a single "predator" species like humans (much less for wasps or birds in general).

side. The denser barrier webs near the lower side of the orb probably force "opposite side" approaches by potential predators (see Higgins 1992b on damselfly predators), so spiders would be able to hide behind discs. The coordination between ontogenetic changes in barrier webs and stabilimentum design in *Argiope argentata* is interesting in this context. Younger spiders differed in being more likely to build both disc stabilimenta and barrier webs (Uhl 2008).

Webs built against tree trunks offer somewhat similar conditions that also fit with the visual defense hypothesis. *Clitaetra* spp. (NE) presents a combination of a somewhat less dense disc stabilimentum (Kuntner et al. 2008a,b) and the spider shuttling to the side of the hub near the trunk of the tree when she is disturbed (Kuntner 2006); the spider's body is difficult to see in this position, at least for a human (M. Kuntner pers. comm.). The linear stabilimenta that are common in other *Argiope* species are rare in *A. ocyaloides*, which builds exclusively against tree trunks, and the spider's cryptic coloration makes it blend very effectively against the trunk, at least for a human (Robinson and Lubin 1979a).

Illumination and background are further complex variables, and it seems appropriate to end this litany of complications regarding stabilimentum function by introducing one further set of equivocal data. Craig (2003) argued convincingly that both the illumination and the visual background of an orb can have major effects on its visibility for approaching prey; the same factors must also affect the visibility of a spider and her stabilimentum. Photographs of *Zosis geniculata* (UL) and *Argiope aurantia* (AR) with their stabilimentua under different conditions revealed a surprising variety of images (at least in the visible spectrum) (figs. 3.45, 3.46). Under some combinations of lighting and background, the stabilimentum obscured the spider's outline, while under others it had little or no camouflaging effect. Concealment was particularly strong when a well-lit spider

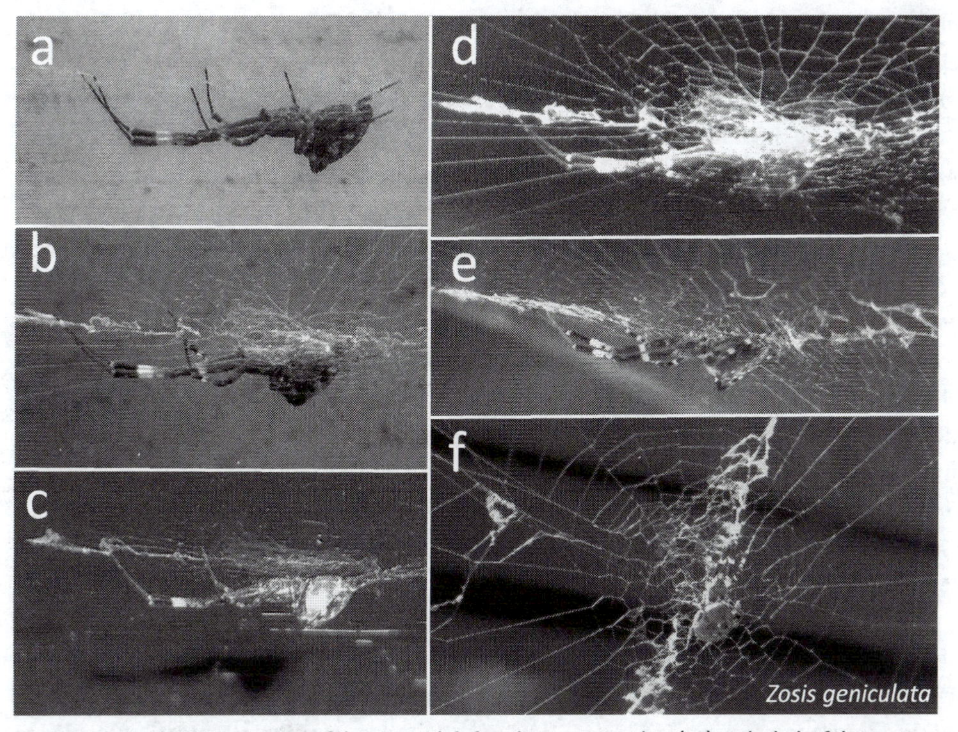

Fig. 3.46. These photos in captivity of the same adult female *Zosis geniculata* (UL) at the hub of the same unpowdered orb with a silk stabilimentum illustrate the importance of illumination, background, and the angle of view in determining a stabilimentum's effect on the spider's visibility (*a–c* and *e* are lateral views; *d* is a lateral view from slightly above the spider's approximately horizontal web). In *a*, with a light background and indirect illumination, the stabilimentum was nearly invisible and the spider's body stood out clearly. In *b*, with the same background but a brighter, more direct illumination from above and to the side, the stabilimentum silk was more visible and the disc portion of the stabilimentum partially obscured the spider's abdomen and cephalothorax. In *c*, with a black background and a strong, direct illumination from the direction of the viewer (the camera's flash), the spider's light colored abdomen and cephalothorax and the stabilimentum (especially the linear arm) stood out clearly. In *d*, with bright lighting from above (no flash) and a dark background, the white disc obscured the spider's abdomen and cephalothorax nearly completely. In *e*, a lateral view looking toward a window, both the spider and her stabilimentum were easily visible, though the coloration of her legs disrupted their outlines (see also *b*). In *f*, also taken near the window with natural illumination but from below, the spider's body and legs were very difficult to distinguish. The prevailing direction of illumination in *c* and *d* (from the viewer toward the spider), and the dark backgrounds in *d* and *f*, could be common conditions for predators and prey peering in from a more open space toward the typically sheltered sites where *Zosis* built their webs in nature (WE). The bright white of the stabilimentum silk in *d* and *f* was due to its being over-exposed;the camera based its exposure on the entire field, most of which was dark, as might happen with the eye of a predator peering into a sheltered site.

and stabilimentum were in front of a dark background (a similar "overexposed" camouflaging effect occurred in the photograph of *Argiope flavipalpis* (AR) on her stabilimentum in Ewer 1972). Do the UV reflecting properties of stabilimentum silk (Craig and Bernard 1990) and the high UV sensitivity of the eyes of some predators and prey result in such "overexposed" images? In some photographs of *Argiope argentata* that used filters to exclude all but UV light, the stabilimentum was bright and the spider virtually disappeared (e.g., fig. 2F of Craig and Bernard 1990) (nevertheless the body of *A. versicolor* reflects UV—Li and Lim 2005). If some predators form such "overexposed" images, then the visual defense hypothesis could accommodate some evidence that otherwise seems contradictory. For instance, the vertical stabilimentum lines in the webs of several *Argiope* species (figs. 3.40, 3.45; Cheng et al. 2010) might obscure the outline of the spider even when the spider failed to align her legs with vertical stabilimentum lines but instead held them in the form of an "X" (#36 in table O3.3). The black distal portions of the legs of *A. aurantia* fit this idea, nearly disappearing in bright sunlight with a dark background (fig. 3.45). Similar situations, in which human vision provides a misleading indication of the biological significance of colors and forms, may explain some apparent "imperfections" in mimicry in other animals (Kikuchi and Pfennig 2013).

3.3.4.2.4 CONCLUSIONS

Having hacked our way through the dense thickets of hypotheses and spiny data here and in tables O3.3–3.5, what general conclusions can we draw? To start with, I believe that it is wise to avoid the temptation to quantify the data in table O3.4 by summing up the numbers of types of evidence for and against the different hypotheses. Any sense of security that might come from avoiding the opprobium of some colleagues for making "merely verbal" rather than numerical analyses would be an illusion. The data truly represent apples and oranges: they vary widely in the generality of their implications for the hypotheses, in the quality of the data and the techniques used, and in their sample sizes. Some findings have much more powerful implications than others.

What are the most reasonable conclusions? First, I agree with several previous reviewers (Robinson and Robinson 1970a; Edmunds 1986; Herberstein et al. 2000b; Walter and Elgar 2012) that probably no single function holds for all stabilimenta. The hypothesis that seems to be least strongly contradicted in table O3.4 is visual defense; but it is unable to explain the numerous tufts of stabilimentum silk on the frames and anchor lines in the orbs of *Gasteracantha* spp., *Thelacantha brevispina*, and their close relatives, and in *Cyclosa argenteoalba*, or the small patches of silk near the hub that, at least in some populations, have no apparent visual relation with the spider in *Witica crassicauda* (AR) (fig. 3.50) and several species of *Micrathena* (fig. 3.37) (see 44b in table O3.3). It is worth noting that most of these (*Gasteracantha* and relatives, *Micrathena*, *Witica*) build webs in large open spaces, where the web advertisement hypothesis has a better chance of being important (see Jaffé et al. 2006).

There are, nevertheless, some truly general patterns that have strong implications. For instance, the unanimous association of stabilimenta with diurnal rather than nocturnal orbs (item 2 in table O3.4) is a strong reason to abandon the web stability hypothesis as a general explanation. After all, nocturnal orbs could also benefit from stabilization, yet not a single one has a stabilimentum. This conclusion is especially strong in light of the large number of evolutionary derivations of stabilimenta. There is also a uniform (though less completely documented) trend for the constrained daytime leg postures that cause the spider to blend more with the stabilimentum to be abandoned at night (fig. 3.47) that is not predicted by the web stability hypothesis. Of course the elimination of this hypothesis (or any other) as a general explanation does not logically imply that there will never be a particular case in which it applies. I would also, however, echo the caution voiced by Blackledge et al. (2011) against facile appeals to multiplicity of functions and avoidance of generalizations.

Because of some strong contradictory evidence, I am not as convinced as some colleagues that the commonly mentioned prey attraction hypothesis is the most strongly supported. Most telling for me is the fact that stabilimenta occur, in a variety of species (in at least 18 species in eight genera and three families), on webs that are built in contexts in which it is clear that the web is neither designed nor used to capture prey (figs. 3.47–3.49) (#8 in table O3.3) (see also Walter and Elgar 2012). In *Nephila clavipes* and *Argiope argentata* the contradiction is even stronger, because stabilimenta were *more*

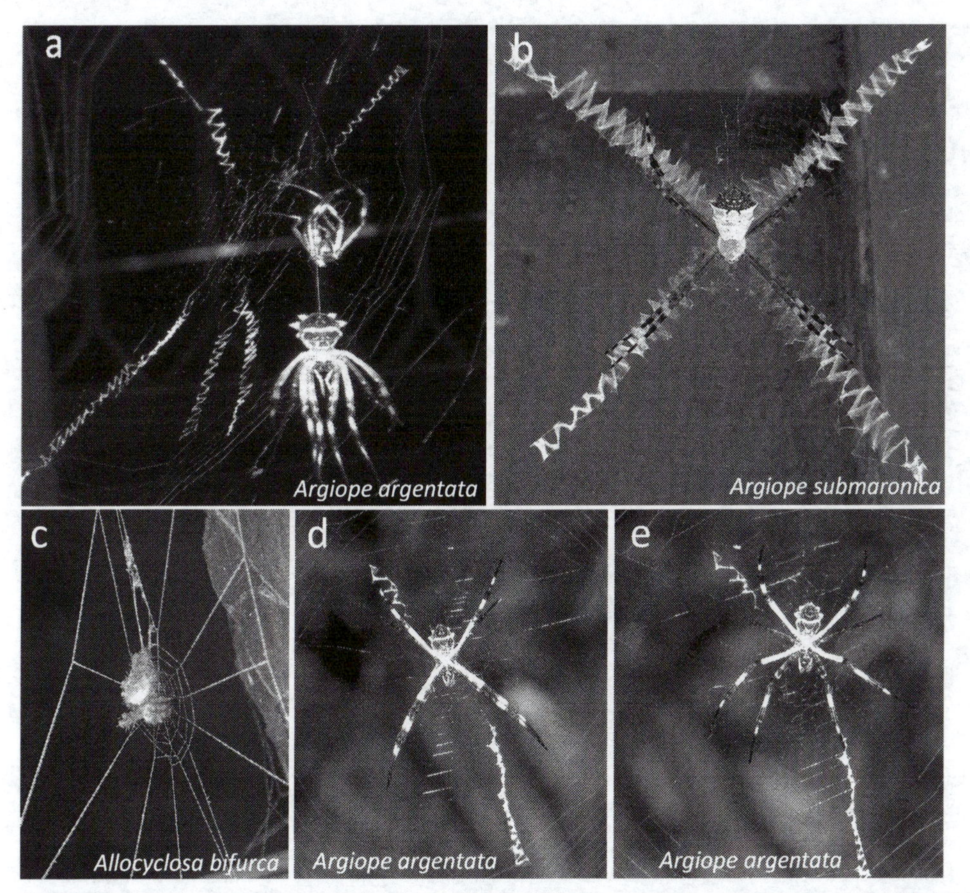

Fig. 3.47. These webs summarize three types of evidence against the prey attraction hypothesis for silk stabilimentum function. The molting web of *Argiope argentata* (AR) (*a*), where the penultimate instar spiderling had rested for three days and nights prior to molting, almost completely lacked a sticky spiral, but it nevertheless had an unusually elaborate stabilimentum including five different lines (the silk in these lines is partially collapsed due to rain); unusually elaborate stabilimenta were especially common on molting webs in this species (Nentwig and Rogg 1988, Robinson and Robinson 1973b); attracting insects during and immediately following the process of molting, when the spider is soft and defenseless, would be dangerous for the spider. Similarly, the inclusion of a small stabilimentum on the resting web of a mature female *Allocyclosa bifurca* (AR) (*c*) that entirely lacked sticky silk also argues against a prey attraction function. A second type of evidence is the disruptive black and white coloration and the positions of the legs of a mature female *Argiope submaronica* (AR) (*b*) that made them merge visually with the zig-zag white bands in the X-shaped silk stabilimentum on her orb; disruptive coloration is not predicted by the prey attraction hypothesis, but supports the visual camouflage hypothesis, though it less convincing when *Argiope* do not hold their legs aligned with their silk stabilimenta (*d*). Still another type of data favoring the camouflage hypothesis was the change by a mature female *A. argentata* (AR) from the usual constrained daytime position at the hub (which reduced her resemblance to a typical 8-legged spider outline) (*d*) to a spread-leg position at night (*e*) when there was no danger from visual predators, presumably improving her ability to sense prey in her web.

common on resting webs than they were on prey capture webs; in *A. argentata* those on resting webs (fig. 3.47) were also larger (Robinson and Robinson 1970a, 1973b, 1978; Nentwig and Rogg 1988). Stabilimenta were so strongly associated with molting webs in *Argiope bruennichi* that Tilquin (1942) called the stabilimentum of this species the molting platform ("le lieu de la mue"). These stabilimenta are associated with moments in the spider's life when she is weak, fragile, and relatively defenseless, and when the presence of insects nearby is likely to be *damaging*. This contradiction of the prey attraction hypothesis is especially striking because it includes five species of *Argiope*, the genus in which the majority of the experimental data supporting the prey attraction hypothesis have been compiled (Walter and Elgar 2012; tables O3.3, O3.5).

Setting apart *Gasteracantha* and its allies, *Micrathena* spp. and *Witica crassicauda*, the hypothesis that seems

Fig. 3.48. These silk stabilimenta offer evidence against the prey attraction hypothesis for silk stabilimentum function, because the webs lacked functional sticky silk. An intermediate-sized spiderling of *Gasteracantha cancriformis* (AR) built a resting web (*a*) after its orb was destroyed following an early morning rain; the web had stabilimentum tufts on several radii, but lacked sticky lines. The resting web that a mature *Zosis geniculata* male (UL) built in captivity (*b*) had a "disc and line" silk stabilimentum, again despite the absence of sticky silk. The web on which a *Nephila clavipes* (NE) spiderling built a stabilimentum (*c*) was an old prey capture orb that was not renewed following a rain and that had a sector missing. The following night the spider added more silk to the stabilimenum, but did not repair the orb. She stayed on this web for two more nights before finally molting, and then built a complete orb that lacked a stabilimentum. Silk stabilimenta also occurred on *N. clavipes* resting webs that completely lacked sticky lines (Robinson and Robinson 1973b). The *Z. geniculata* web was dusted with white powder and photographed with a flash; the others were unmodified and photographed with natural illumination.

the least strongly contradicted as a general explanation in table O3.4 is visual defense. Several lines of evidence favoring visual defense seem particularly compelling: evidence from detritus stabilimenta that visual defense is very important for spiders resting on orb webs; the constrained body positions that cause their outlines to merge or become lost in that of the stabilimentum; the (less well-documented) relaxation of these postures at night (fig. 3.47, #1 in table O3.4); some patterns of coloration that blend with stabilimenta (figs. 3.38, 3.46, 3.47); the visual barrier that is constituted by some stabilimenta (figs. 3.38, 3.49); the correlation between certain behaviors by the spider (shuttling behind the hub, web pumping or vibration) that are almost certainly defensive in function and the presence of stabilimenta (#13 in tables O3.3, O3.4) that provide visual barriers (or, in some, a mechanical defense); and the apparent equivalence suggested by the behavior of several species linking stabilimenta and other structures that likely provide visual defense (egg sac and detritus stabilimenta), as evidenced by the induction of stabilimentum construction by removal of egg sacs and detritus, and its suppression in their presence (#19 in table O3.3) (see also fig. 3.49b). The size and site of the vestigial stabilimentum of *Polenecia producta* (UL) also argue strongly for a visual camouflage function: the spider is completely hidden under a twig, with only the tips of her legs II that hold hub lines in view; and a small mass of stabilimentum silk is placed just where the tips hold the web lines (fig. 3.44, Peters 1995a).

On the other hand, in *Gasteracantha*, *Micrathena* (figs. 3.37, 3.40, 3.48b), and some *Witica crassicauda* (fig. 3.50) the visual defense function seems to be precluded, and web advertisement seems likely. In addi-

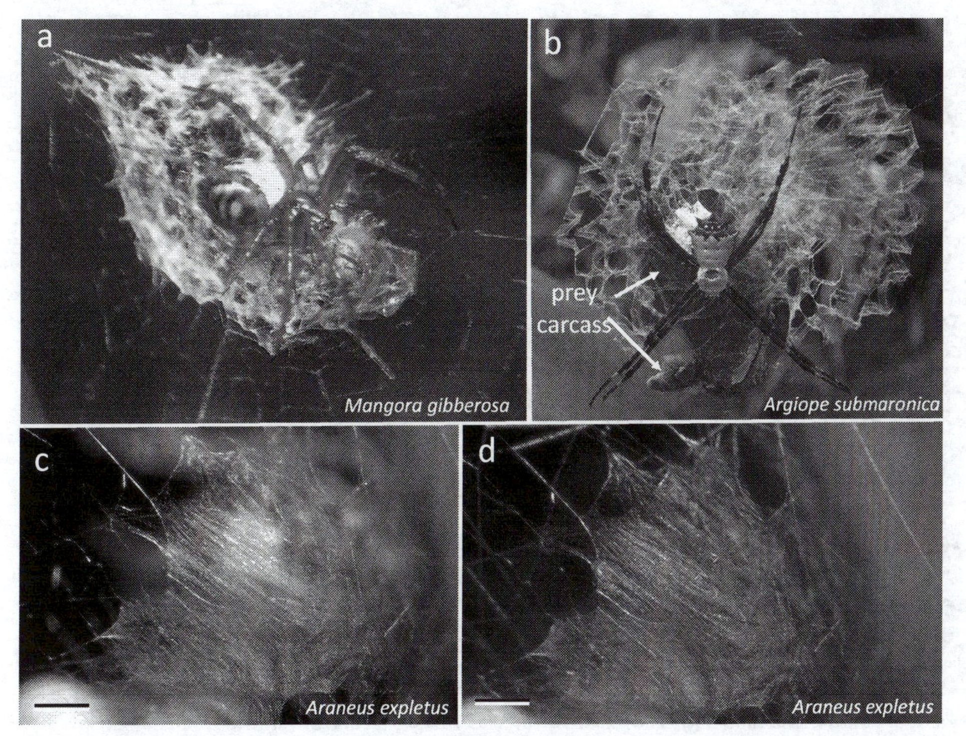

Fig. 3.49. Some webs provide particularly strong evidence regarding particular hypotheses for stabilimentum function. The disc stabilimentum of a mature female *Mangora gibberosa* (AR) (*a*) argues strongly against the idea that stabilimenta function to shade the spider. When direct, early afternoon sunlight fell on the hub, the spider rested above rather than below the disc stabilimentum at the hub of her nearly horizontal orb, where her abdominal white coloration blended visually (photo with natural lighting). When I shaded the web, she quickly moved to rest on the underside of the hub. The stabilimentum of *Argiope submaronica* (AR) (*b*) offers strong evidence against the prey attraction hypothesis, because it was built on a resting web that lacked any sticky line and thus could not capture prey. The spider incorporated the body of a butterfly captured the previous day in the stabilimentum, providing herself with additional visual camouflage; she was virtually invisible when viewed from the other side (see fig. 3.38b). *Araneus expletus* (AR) rested in a silk-walled retreat at the web's edge, rather than at the hub (see fig. 3.7b); in accord with the visual defense hypothesis (but not the others), the spider added a white "stabilimentum" to the outer wall of the retreat rather than to the hub, obscuring the spider while she rested in the retreat (*c*); for comparison, the retreat is shown empty in *d* (*c* and *d* from Eberhard 2008).

tion, multiple studies, utilizing multiple techniques with numerous species (mostly *Argiope*) in which visual defense as well as other functions are feasible, have found that some insects are attracted to stabilimenta (#37, 39, 42, 43 in table O3.3). I believe that viewed as a group, these studies make the strongest case for the prey attraction hypothesis. I have thus made a closer analysis of the clearest studies (table O3.5). The result, unfortunately, was not conclusive: the studies favored the prey attraction hypothesis more often than not, but were inconsistent. For what it is worth, the numerical totals were that 17 studies in table O3.5 favored prey attraction, and 6 contradicted it; and none of them escaped completely from possible criticisms (though the criticisms were more serious for some than others) (the exercise of searching for weaknesses in published studies of this sort is a little like shooting fish in a barrel; I suspect it is usually possible to find imperfections in studies of this sort).

For me, the studies of *Plebs eburnus* and *A. keyserlingi* (Herberstein 2000; Bruce et al. 2001, 2004) are among the most convincing of these studies; but they also gave inconsistent results regarding prey attraction (see footnotes g and q in table O3.5). Another strong study was that of Cheng and Tso (2007), whose innovative use of video recordings to document prey interception and predation on spiders resolved some previous methodological problems that plagued many earlier studies. In the end, these inconsistencies emphasize the prudence of maintaining a certain degree of skepticism regarding these types of study.

In sum, I do not think that it is possible to derive a

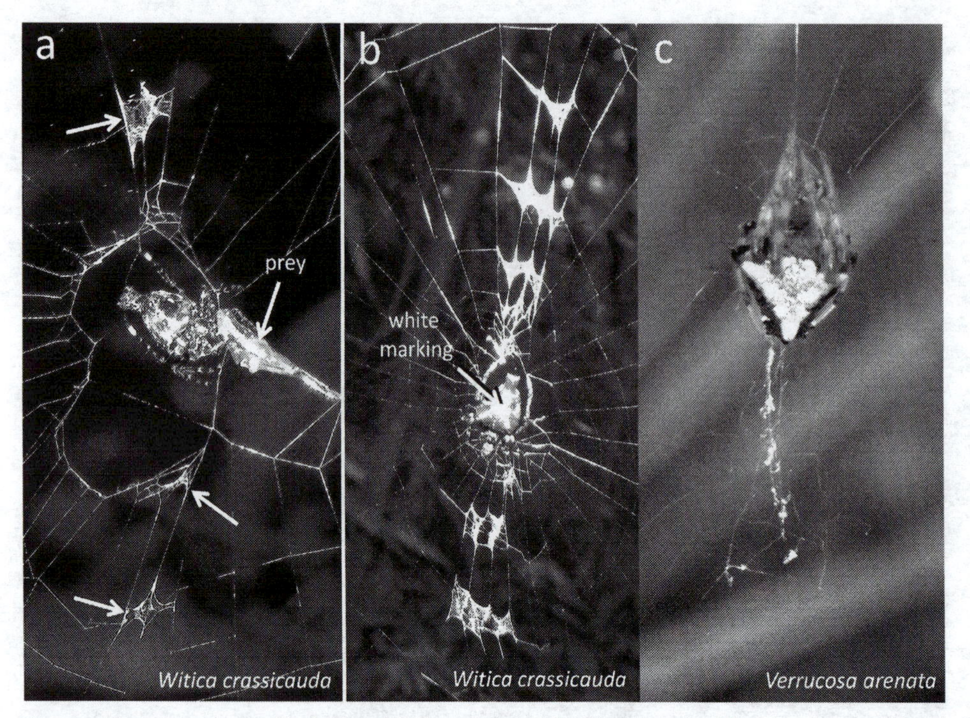

Fig. 3.50. These uncoated vertical webs of *Witica crassicauda* (AR) in Costa Rica (*a*) and Puerto Rico (*b*) and of *Verrucosa arenata* (AR) in Louisiana (*c*) have both negative and positive implications regarding the visual defense hypothesis for stabilimentum function (see table O3.3). In a Costa Rican population of *W. crassicauda* tiny patches of white stabilimentum silk were common near the hub (arrows in *a*) (the spider is seen in ventral view, feeding on a prey; usually spiders crouched facing downward at the hub). Small patches of stabilimentum silk also occurred on lateral frame lines, far from the spider (not shown). These stabilimenta had little or no obvious visual connection with the spider. In contrast, more extensive stabilimenta were common near the hubs of the same species in Puerto Rico (*b*) (S. Kelly pers. comm.), and the bright white area on the dorsum of the spider's black abdomen blended visually with the masses of white silk. *Verrucosa arenata* crouched facing upward at the hub during the day on orbs sited in large open spaces (where the web advertisement function might be important). Some webs had small masses of stabilimentum silk, and the spiders' abdomens also had similar light patches. But it requires an uncomfortable stretch of the imagination to say that these stabilimenta obscured the spider's outline by blending visually with these markings (*a* and probably *b* were illuminated with a flash, *c* was in natural light; to different scales) (*b* courtesy of Sean P. Kelly).

firm conclusion regarding prey attraction in the field; the studies with the most direct evidence tend to support it, but not always or conclusively. If these findings that prey are attracted to stabilimenta are to be believed, I do not see how to reconcile them with the fact that molting spiders in several of the same species nevertheless tend to build as many or even more and larger stabilimenta on their resting webs.

3.4 SUMMARY

Prey capture webs perform many different functions, and several general aspects of webs are designed to accomplish different functions. The focus in this chapter was by necessity on orb webs, due to the current dearth of data and analyses regarding other designs; but non-orbs must accomplish many of the same functions, so in some sense they were covered also. To understand how an orb web functions, it is necessary to leave behind a number of common misconceptions. In the first part of the chapter I reviewed the evidence against several of these. In the second part I discussed how the different functions performed by orbs—especially intercepting prey, stopping prey, retaining prey, and facilitating attack behavior—are likely to affect selection on different design aspects of orbs. I paid special attention to how difficult it is to gather really convincing data on many of these points. To some extent this negative emphasis was a reaction to previous models (my own included—Eberhard 1986a) that attempted to explain orb designs in relatively simplistic terms. Perhaps the most important points in the chapter were the following. Orb webs do not function like sieves or fishing nets, because their lines can be broken by the prey that they are designed to capture. For small prey, contact with only a single sticky line in an orb is sufficient, but for larger (and generally more nutritionally profitable) prey it is generally necessary to contact a prey with several lines. There is as yet little evidence that orb designs tend to be species-specific, and the kinds of data and analyses needed to test this possibility are not yet available. Some basic web traits such as the spaces between sticky lines, the numbers of radii, and the area covered by the web are unlikely candidates, as they can vary strongly from one web to the next built by the same spider. Web specificity at higher taxonomic levels such as the genus is more likely than is species-specificity, but data needed to test this idea are still very scarce.

It does not appear that differences in aspects of orb designs such as different spaces between lines, hub sizes, radius numbers, and frame and anchor numbers and lengths represent specializations to capture different types of prey. The prey-specialization hypothesis fails on several counts. Most importantly, the orbs of a given species typically capture a wide variety of prey in the field, even at a given site and season. Differences between prey of different species often seem best explained by differences in the subhabitats where webs are built. Furthermore, many aspects of orb design are typically relatively constant throughout a spider's lifetime, despite the fact that her size often changes a hundredfold or more between when she emerges from the egg sac and when she reaches maturity, and her nutrition thus depends on completely non-overlapping arrays of prey during her lifetime. This ontogenetic consideration has not been fully appreciated in some discussions, perhaps due to the unfortunate tendency of arachnologists to focus exclusively on adults and ignore earlier developmental stages.

Possibly related to the lack of prey specificity is the likelihood, revealed by careful field experiments and analyses of data (Wise 1985), that orb weaver species generally do not compete for prey. Thus many differences in orb designs probably did not evolve to avoid competition with other species for prey. In fact, as will be discussed in the next chapter (section 4.3), a given orb is probably not appropriately seen as a trap with a single design, but rather as a combination of multiple designs that have quite different prey capture properties. The relative lengths of an orb weaver's legs provide a useful estimator of her attack speed (species with stubby legs do not run fast), and preliminary analyses have revealed a lack of clear correlations of orb designs with attack speeds, or with different attack tactics such as wrapping vs. biting prey.

Another factor that is only beginning to be studied concerns the differences between some species in the mechanical properties of silk lines from the same gland type. The mechanical properties of dragline silks clearly vary, even though only a small number of related species have been studied. Surprisingly, the properties of the draglines and sticky spiral lines produced by the same individual spider, even in the same web, sometimes also vary substantially. The correlations between these variations, the geometric designs of webs, and environmental variables such as relative humidity suggest the exciting possibility that spiders adaptively manipulate the properties of the lines while they are spinning their webs; these possibilities are only beginning to be studied.

New data also suggest the possible importance of wind, both with respect to the previously neglected idea that movements by typical orb webs in air currents increase prey interception rates and also regarding certain types of damage to orbs in the field. Different aspects of orb design (e.g., the lengths of frame lines, the lengths and spacing of sticky spiral lines, the rigidity of the supports used) can affect different aspects of how webs move in the wind. Some designs may make it more difficult for prey near the web to avoid contacting it. Web designs can also affect the frequency and types of web damage.

Orbs in the field gradually become tattered from damage inflicted by prey impacts, detritus, wind, and by movements of the spider herself on her web. During most of its short lifetime in the field an orb is at least partially tattered, rather than in the virgin state that is typically illustrated in scientific publications. Further study of which kinds of damage are the most common in nature, of the effects that these types of damage have on different web functions, and of the ways that different web designs can affect the likelihood that damage will occur, may yield further insights into the functions of different orb web designs.

Minor details of orb designs may affect their effectiveness as traps. An incomplete list includes the following.

The widespread trend for frame lines to be reduced (and in some cases omitted) on the side of the orb where the hub is closest to the frames or substrate increases the web's capture area (area covered by sticky lines) when the website is near the substrate. The sliding or "pulley" connections of the sticky spiral to the radii allow elongation of sticky lines under stress, making it more difficult for prey to break free. The close spacing of lines in the hub give the spider resistance against which to push and pull as she turns rapidly to begin an attack on a prey, and also increases the speed with which she can find and grasp new lines as she turns and then runs toward the prey. The modulation of the distances between loops of during temporary spiral construction probably serves to adjust the outline of the outer loop of sticky spiral and thus the size and shape of the capture area. Deviations of orbs from strict planarity probably represent "errors" (presumably in the choice of a website or of anchor points) that reduce the web's effectiveness in intercepting prey.

The topic of insect defenses against spider webs has scarcely been broached, but already there various possible anti-web traits known. These include slow, tentative flight, long thin legs or antennae that are extended in flight and can act as long-range sensors with small contact areas, "playing dead" immediately after impact, careful timing of wing buzzing to reduce attraction of the spider, special positions of the wings to reduce the area of contact with the orb during struggling behavior, and reorientation of the body to slip through holes between lines. Further studies will undoubtedly uncover further tricks by insects, and perhaps also responses in spider web designs and attack behavior to counter these ploys.

Another field ripe for further advances is that of metabolic costs, including both that of the material in a web and the building behavior per se, the spider's vigilance once the web is built, and also other efforts, such as tensing the web.

Undoubtedly the most thoroughly studied aspect of orb web function concerns the silk ornaments (stabilimenta) that are added to some orbs after construction is finished. After presenting the most thorough and critical review of the evidence yet made, I concluded that the original hypothesis of stabilizing the web mechanically should be rejected as a general explanation, because of the total lack of stabilimenta on orbs built and operated at night (when stabilization would be also be useful). There are data from at least some species that strongly contradict every one of the several other hypotheses that have been proposed for stabilimentum function, so stabilimenta almost certainly perform multiple functions. Experimental tests of the hypothesis that stabilimenta attract prey have given mixed results, though they favor it more often than not. Many of these studies have methodological weaknesses, however, and they frequently lack ecological realism, making interpretation of the data uncertain. The strongest trend contradicting the currently popular prey attraction hypothesis is the repeated finding that stabilimenta are built more frequently or are more elaborate on the reduced webs (which often completely lack sticky lines) that spiders build on which to molt. In these cases it would be advantageous to *not* attract prey. The hypothesis that stabilimenta provide visual defense for the spider received most support overall, but it was also strongly contradicted in several species.

I will postpone summarizing several other topics discussed here until the end of the next chapter, where I will attempt to integrate the different possible orb web functions more completely.

4

PUTTING PIECES TOGETHER

TRADE-OFFS AND REMAINING PUZZLES

4.1 INTRODUCTION

This chapter continues the focus from the previous chapter on the functions of different parts of an orb, and attempts to further integrate understanding of different aspects of web design. The largest challenges are to combine the multiple effects that a given design trait can have on the different functions of an orb. While many problems remain unsolved, some progress has been made.

4.2 "OPTIMAL" ORB DESIGNS: TRADE-OFFS BETWEEN FUNCTIONS ARE DIFFICULT TO MEASURE

Claims that orb web designs are "optimal" are common (section 3.2.3) (e.g., Witt 1965; Craig et al. 1985; Eberhard 1986a; Blackledge et al. 2009c; Sensenig et al. 2011). But none of these claims are documented quantitively, and to actually demonstrate optimality quantitatively is very difficult in a complex structure like an orb web. An orb has multiple design traits, each of which has trade-offs between costs and benefits whose magnitudes will vary under different environmental circumstances (table 3.1). Some traits of an orb improve some functions while simultaneously damaging others, so it is not possible to maximize all functions with any particular design (table 3.1) (e.g., Eberhard 1986a, 2013a; Blackledge and Eliason 2007; Blackledge et al. 2011). In general terms, trade-offs are inevitable, because a spider has only a finite amount of material and energy to invest in her web, and different designs are associated with biologically significant differences in the payoffs from these investments. Any particular orb web trait is likely to result in both advantages and disadvantages.

I believe that disentangling these complex interactions to determine an optimum is not presently possible, for lack of sufficiently precise quantitative knowledge. I have relegated a detailed discussion of this complicated topic to appendix O3.2, and will only summarize a few points here. Briefly, I will discuss several types of empirical problems: the large number of possible web functions and the frequently conflicting effects that any given variable has on different functions; the large variations in the effects of web design on several key variables (and thus the practical difficult of quantifying their effects precisely); difficulties in determining the magnitudes of the effects under evolutionarily realistic conditions (e.g., habitats similar to those where the spiders evolved, rather than highly altered, anthropogenic conditions); and the lack of a common, readily quantifiable currency for measuring all of the costs and benefits.

4.2.1 TRADE-OFFS BETWEEN FUNCTIONS

Perhaps the most serious problem in understanding orb web designs is posed by conflicting effects (table 3.1). It is not sufficient to simply determine the direction ("polarity") of the effect on a given variable (e.g., increase vs. decrease prey retention times), as is often the case in other studies in behavioral ecology. Instead it is necessary to quantify multiple effects with enough precision to accurately determine the balances between costs and benefits: to decide which is larger, and by how much. This requires accurate answers to difficult-to-solve quantitative questions. *How much* does reducing the spaces between sticky spiral loops by 0.2 mm increase retention? *How much* does an increase in retention time increase the probability of capture? *How much* does this trait simultaneously reduce prey interception? *How much* more material does it cost? The precision needed to answer these types of questions, especially under evolutionarily realistic conditions, is much greater than that which we behavioral ecologists are accustomed to providing, and it is generally lacking in current studies.

One tactic for understanding complex situations in which multiple potentially interacting variables affect an outcome (such as nutrition) is to construct models in which the multiple variables are all included, and determine how their different effects balance against each other quantitatively under different conditions. Claims of optimality have often been associated with discussions and models that attempt to determine the advantages of different orb web designs. Although I formerly participated enthusiastically in modeling efforts, I have since become more skeptical, and have arrived at an uneasy balance. I am still convinced that it is possible to determine the polarities (costs vs. benefits) of particular consequences of particular traits on particular functions (e.g., table 3.1); but I believe that the models that have attempted to determine the quantitative balances between different effects are of only limited usefulness in understanding why different orb designs evolved (appendix O3.2). The basic problem is that models are based on certain assumptions regarding empirical relationships, and they can be evaluated only by comparing their predictions with empirical observations. It seems to me that in constructing biologically relevant models, "the devil is in the details," and that the quality and quantity of the data currently available are not up to these challenges.

Fig. 4.1. Capturing a rare large prey can be life-altering experience. This single large meal will go a long way toward providing this slim *Nephila clavipes* (NE) spiderling with enough reserves to molt to the next instar.

4.2.2 THE "RARE LARGE PREY HYPOTHESIS": A DOMINANT ROLE FOR THE STOPPING FUNCTION?

One recently popular way to simplify the tangle of different functions has been to argue that two web functions, intercepting prey and stopping them, are so crucial that they overshadow all of the others (Blackledge 2011). The argument is as follows. A spider's basic payoff from an orb is nutritional, and large organisms are nutritionally more valuable than small prey organisms. So web traits that promote the capture of large prey (especially the ability to absorb the momentum of high-energy prey) will prevail. Under certain relative frequencies of large and small prey, the benefits from occasional very large prey items will be enough to outweigh those from the more frequent captures of smaller prey (fig. 4.1). In the extreme case, small prey will be nutritionally irrelevant to the reproductive success of the spider.

Although small prey are generally more abundant in nature, the "rare large prey hypothesis" proposes that their abundance in nature is not enough to compensate for their lower nutritional values. One recent study of the araneid *Zygiella x-notata* in which the nutritional value (mass) of prey was estimated carefully on the basis of each prey's length (Venner and Casas 2005) supported this hypothesis. A subsequent meta-analysis of data on the prey of 31 species in 18 genera led to the same conclusion (Blackledge 2011). Blackledge (2011) concluded that "the 'rare, large prey hypothesis' can apparently be generalized across orb weavers" (p. 205).

This conclusion could have great potential importance for understanding the evolution of spider webs, because it can cut through the complex tangle of interacting factors (table 3.1). It implies that selection will consistently favor those design properties of orbs and the physical properties of their lines (table 3.1) that promote capture of especially large as opposed to smaller prey (see Sensenig et al. 2010a; Blackledge 2011; Blackledge et al. 2011). This in turn implies that one can focus on only two functions—the abilities of the orb to intercept prey and to absorb high kinetic energy impacts without breaking—and can safely ignore most of the complex array of other possible orb functions. A simulation study of foraging, growth, and survival that incorporated assumptions regarding the biomechanics of prey capture, the frequencies of prey, the relation between prey length and weight, and silk thread and orb design properties (Evans 2013) showed that it is indeed possible that selection favoring the ability to stop high-energy prey could be very strong. The relative simplicity that results from accepting the rare large prey hypothesis is appealing, and it formed the basis for some recent analyses of orb web evolution (Opell and Bond 2001; Sensenig et al. 2010a; Evans 2013). In a study of 23 species in four different families, in which extensive comparisons of differences in the biomechanical properties of the lines in the webs were included in analyses of orb designs, an orb web's "performance" was used to mean its ability to stop high-energy prey (Sensenig et al. 2010a).

There are several reasons, however, to doubt that the support for this hypothesis is convincing (Eberhard 2013a). In the Venner and Casas study, the weights of especially long prey were probably strongly overestimated, and the assumption made that long prey consistently produce high-energy impacts with orbs is incorrect (many of the especially long prey in this study [crane flies] often fly quite slowly and tentatively). It is very likely that the longer prey that were captured by orb weavers in this and other studies represent a strongly biased subsample of species of this size, favoring those with relatively low weights and low flight speeds when compared with other prey species of the same length (most of the heavier, faster prey of these lengths will have escaped) (Eberhard 2013a). The data that were used (prey captured) are not sufficient to test ideas regarding natural selection on orb designs in the absence of additional data on prey escapes. And the habitats in which many of the studies were conducted are highly altered (e.g., cotton fields, or the windows of a building in the midst of sports fields); the mixes of large and small prey at these sites may not have been similar to those in the natural habitats where the spiders evolved. In sum, several theoretical reasons suggest that the data are not sufficient to support Blackledge's confident conclusion that the rare, large prey hypothesis is generalizeable across orb spiders.

A second reason to doubt the generality of the rare large prey hypothesis comes from the multiple lineages in which designs have evolved to be *less* capable of stopping prey (their orbs have much fewer radii and much more widely spaced sticky spiral loops than did those of their ancestors). A list (limited largely by current uncertainties regarding the phylogeny of the details of orb designs in most lineages) includes the following: *Alpaida tuonabo* (AR) (Eberhard 1986a); the facultative daytime orb web design with few radii and more widely spaced sticky spiral loops of *Parawixia bistriata* (Sandoval 1994); an undetermined species of *Eustala* (fig. 7.25, Eberhard 1985); some species in the tetragnathid genus *Tetragnatha* (fig. 9.10, Shinkai 1988b); the group of related araneids that includes *Poecilopachys australasia* (Clyne 1973), *Paraplectana tsushimensis* (Chigira 1978 in Stowe 1986), *Pasilobus* sp. (Robinson and Robinson 1975), and *Cyrtarachne* spp. (Stowe 1986); and theridiosomatid genera such as *Ogulnius* (fig. 10.22c), *Wendilgarda* (Eberhard 1986a; Coddington 1986a), *Epeirotypus* (fig. 10.22a), and also *Theridiosoma gemmosum* (fig. 59 of Coddington

1986a). The facultative reduction in dragline strength by the araneid *Cyclosa mulmeinensis* in response to lack of wind stress (Liao et al. 2009) is also not easily reconciled with the idea that rare large prey are overwhelmingly important in webs built under non-windy conditions. Because it is conceivable that additional data on line diameters and physical properties in the webs of these species will fit the rare large prey hypothesis, it necessary to leave these cases still open; but present indications do not clearly favor it.

Further reason to doubt the overwhelming importance of the stopping function in orb web evolution comes from within-orb differences in sticky spiral spacing in different araneoid species (section 4.3.3). There are various within-orb patterns, including larger spaces below rather than above the hub in some species, and clear within-orb patterns in species that build horizontal orbs, which argue that these patterns are probably the result of selection regarding the retention (rather than interception and stopping) abilities that are determined by spacing between sticky loops.

In sum, the biologically important variable for an orb-weaver is not the number of prey she captures, but the total amount of resources that she receives from these prey; a larger prey usually provides more nutrition than a smaller one. This reemphasis on prey size is important, and renders previous work of mine (e.g., Castillo and Eberhard 1983; Eberhard 1986a) and others that emphasized prey numbers of dubious value for understanding orb evolution. But extrapolating from this insight to deduce that selection on orb web designs to stop high-energy prey has dominated over other functions such as retention and damage reduction is not obviously correct. It thus seems wise to keep in mind the many different functions of orbs, and avoid emphasizing only the interception and stopping functions. Probably it will eventually turn out that rare large prey are of overwhelming importance in some cases, but that they are not in a nontrivial number of other cases. In short, it is prudent to take into account the multiple trade-offs between different orb functions (table 3.1).

4.2.3 INVESTMENTS IN FORAGING

Quantifying the foraging efforts made in the field by a flycatcher, a damselfly, or a house fly can be technically daunting. Quantifying the investment in foraging by an orb weaver seems like child's play in comparison. Measures of energy in the silk of the webs, the energetic costs of building, and recycling are all accessible. In a recent example, Venner et al. (2006) took advantage of these advantages to analyze how the araneid *Zygiella x-notata* modulates web size during its adult lifetime, using the technique of dynamic optimization over an infinite-time horizon. They asked how adult females balance the costs and benefits of different web designs in determining the designs of successive webs in order to gain energy, grow, and lay their first clutch as fast as possible, while at the same time limiting the risk of death from starvation or predation (death by starvation was assumed to occur at 25 mg, oviposition at 80 mg; approximate values for both weights had been documented in previous laboratory analyses).

The model took into account previous studies of the metabolic costs of building behavior and quantitative estimates of several other aspects of foraging by this species in the field and the lab (Venner et al. 2003; Venner and Casas 2005). The basal predation risk was determined in a population on and around the windows of a building in an outdoor sports complex, and it was assumed that the larger the web the longer it would take to build and thus the greater would be the predation risk (the spider spends the rest of the day in a more sheltered retreat). The two most fundamental results were that 1) spiders reduce their web size as they gain weight due to body mass-dependent costs of web construction behavior; and 2) under greater predation risks they begin this reduction in web size at lower weights.

It is not clear how much trust to place in these conclusions, however, because the model had several weaknesses. Rates of both prey capture and predation were measured in a population living under conditions that are surely not representative of those under which the species evolved (Eberhard 2013a). In determining predation rates, it was assumed that when a spider disappeared from a site in the field where she had built webs previously that she had been preyed upon, thus ignoring the possibility that she had migrated to another site. It was apparently supposed that predation risk varies directly with the duration of web construction behavior, despite that fact that both pompilid and sphecid wasps have been observed hunting other orb-weaving spiders in their retreats during the day (e.g., Eberhard 1970a), and that this species usually builds its orb in the dark, pre-dawn hours (Venner et al. 2000; table 8.2); it is uncertain whether any of its predators would be drawn by the spider's building behavior (some visual predators like insectivorous birds, wasps, flies, odonates, etc. would probably not). Again, the devil is in the details.

4.2.4 OVERVIEW

Summarizing, the problems just reviewed (see also appendix O3.2) imply that the results of theoretical analyses of the functional consequences of different design traits are seriously limited (table 4.1 gives a pessimistic summary). Synthetic analyses will apparently have to take into account many variables, including phylogenetic effects, the strength, toughness, stickiness, and extensibility of lines, and prey abundance and energy, in order to sort out the likely costs and payoffs from different adjustments in or combinations of these variables. Several variables almost certainly interact with others, further complicating analyses. Perhaps the best general analysis is that of Sensenig et al. (2011), which I believe points the way forward but which nevertheless has several serious methodological problems (appendix O3.2).

Models can be useful in exploring the multiplicity of different effects of particular design features, and in demonstrating that a given feature may improve a web's ability to perform some functions while simultaneously reducing its ability to perform others. These models can be seen as starting points, and can reveal important weaknesses in available data. These can then guide future empirical research. For example, the weaknesses of the strategic investment models of the effects of predation risk on foraging decisions of *Z. x-notata* call attention to the potential usefulness of further studies of predation on these spiders. But, in my opinion, the currently available models are affected by too many variables that are too poorly known in terms of empirical measurements for them to be very useful in explaining why some orb designs rather than others are built by spiders (table 4.1).

4.3 "MULTIPLE TRAP" DESIGN: A NEW WAY TO VIEW ORB WEBS

This section extends the pessimistic theme of overwhelming complexity from the previous section, but takes us around a corner of sorts to open up a new vista on orb webs. Instead of attempting to typify "the" type of prey that a given orb is designed to capture, or "the" design properties of an orb, I will argue that any given orb includes a variety of designs that are appropriate for capturing a variety of prey.

This change in perspective grows out of recognition that prey specialization is weak (section 3.2.5), that the distances both between radii and between sticky spiral lines vary substantially within an orb, that there are clear patterns in both types of within-web variation, and that the stresses from stopping and retaining prey are largely local phenomena in an orb and cause only nearby lines to break (section 3.3.2.9). Instead of seeing a given orb as a unitary trap for particular types of prey (e.g., Eberhard 1986a; Vollrath 1992; Craig 2003; Blackledge 2011), the systematic variations in spacing between radii and sticky lines suggest viewing an orb as a suite of different traps. The differences in spacing in different areas of an orb have profound consequences for the payoffs from an orb, because both stopping and retaining prey depend mostly on the lines in the immediate vicinity of where the prey strikes, not on the overall properties of the orb. I will argue that some of the differences in the designs of different parts of an orb are probably adaptations that exploit these different properties (Eberhard 2014; Zschokke and Nakata 2015).

4.3.1 UNEQUAL SPACING OF RADII IS UBIQUITOUS

The spaces between the radii in an orb are anything *but* uniform: they are closer together near the hub than near the edge. The average distances between radii at the outer edge of the capture zone ranged from 4.4 to 5.4 times greater than those at its inner edge in the same web in five species in four families that build typical orbs (Eberhard 2014). Radial patterns of lines are undoubtedly very ancient in spiders; they occur even in the webs of *Liphistius*, the sister group to all other spiders, as well as in other distantly related groups like desids (*Matachia*) (fig. 1.3b) (Griswold et al. 2005), segestriids (Griswold et al. 2005), titanoecids (Griswold et al. 2005), hersiliids (fig. 9.1, Williams 1928), pisaurids (fig. 9.1, Eberhard 2007a), oecobiids (figs. 9.6, 10.5, Crome 1957; Glatz 1967; Solano-Brenes et al. 2018), the psechrid *Fecenia* (Frontispiece, fig. 9.18), and filistatids (fig. 5.10) (Comstock 1967; Griswold et al. 2005). On this basis, it seems likely that

Table 4.1. A pessimistic (and undoubtedly incomplete) summary of factors that can complicate attempts to make convincing determinations of optimum web designs. The word "must" is used here to distinguish a priori suppositions for which I know of no direct data from empirical findings. These factors probably reduce the strength of correlations between web geometry and function. This list could serve as a "to do" list to make future models more convincing.

Web traits

1. The angle with which a prey strikes an orb, which presumably varies greatly in nature (the distributions of impact angles in nature are unknown), must have a large effect on whether a prey will be stopped and retained.
2. The direction from which a prey approaches a web must have large effects on the visual background, and thus on the web's visibility; there is little knowledge of backgrounds experienced by prey in the vicinity of diurnal orbs in nature; presumably they vary greatly (in contrast, visual background probably has no importance in nocturnal webs, offering opportunities for comparisons).
3. The mechanical properties and diameters of lines from a given gland vary within a single web, within a species, and between species (section 3.2.9).
4. The mechanical properties of lines can change dramatically with weather conditions such as humidity (section 3.2.9.1) and temperature (Denny 1976).
5. Distances between lines (especially radii) vary dramatically within a web; supposing that an orb has "a" mesh size is an illusion (section 4.3, appendix O3.2.1).
6. Many web traits must have multiple functions, and many modifications of a trait must result in both costs and benefits (table 3.1).
7. The "currencies" of the costs and benefits regarding different functions are different (survivorship, metabolic costs, material costs, web survivorship, prey capture); this impedes making precise calculations of the balances between costs and benefits.
8. The correlation between web design and prey capture may be quite weak; some comparative studies found a lack of significant differences between even quite different web designs (e.g., Riechert and Cady 1983; Wise and Barata 1983).
9. The correlation between web design and retention time may be quite weak; some substantial alterations in sticky spiral geometry had no significant effect on retention times (Eberhard 1989b).
10. There is substantial variation in spider attack behavior due to, for instance, learning (e.g., Turnbull 1960) and satiation, and this must alter payoffs from different prey.
11. Variation in feeding behavior (due to, for instance, hunger, digestive enzymes) alters payoffs from given prey (Collatz 1987).
12. The physical properties of viscid glue droplets vary within and between webs (e.g., Benjamin et al. 2002; Herberstein and Tso 2011; Amarpuri et al. 2015)
13. Both the design of the orb (Harmer and Herberstein 2009) and the mechanical properties of homologous lines (Christiansen et al. 1962) vary with the spider's weight.
14. General web designs are relatively constant throughout a spider's ontogeny, but the ecological problems that a spider and her web face as she grows from spiderling to adult (often the weight changes by a factor of 100 or even more) must be quite different (e.g., relative abundances of prey of different sizes relative to that of the spider, types and abundances of different potential predators, relative metabolic costs, and danger of desiccation). The costs and benefits of web traits need to be determined throughout a spider's lifetime, not (as is currently done) just for adult females.

Prey traits

15. The diversity of prey captured varies within and between geographic sites both within and between species.
16. The relative payoffs from prey with different characteristics (e.g., different sizes, with and without defensive chemicals) are difficult to determine under natural conditions.
17. For some prey species, context-dependent traits (e.g., flight speed) must strongly influence the likelihood of their being captured.
18. The importance of retention varies widely, depending both on the rapidity with which spiders arrive at impact sites (from a few tenths of a second to many seconds [section 3.2.10.1]), and differences in the rapidity of responses to different prey species (e.g., Suter 1978).

the radial pattern in orbs was inherited from an ancestor that spun radially organized non-orb webs. A radial pattern in which non-sticky supporting lines converge on the spider's resting place has a number of advantages in vibration transmission, rapid access to the web for the spider, and (perhaps) mechanical stability (Witt 1965).

4.3.2 EDGE-TO-HUB PATTERNS IN STICKY SPIRAL SPACING ARE ALSO COMMON—WHY?

Although the sticky spiral lines in orbs are obviously arranged in highly regular arrays, and despite the occasional direct statements and simplifying assumptions in models that the spaces between loops are constant (e.g., Hingston 1920; Vollrath and Mohren 1985; Eberhard 1986a; ap Rhisiart and Vollrath 1994), the sticky spiral spacing typically varies substantially within a given orb. In five species in four families with typical orb designs, the mean spacing in the tenth of the capture zone in which the sticky lines were farthest apart was between 1.2 and 2.4 times greater than the spacing in the tenth in which the sticky lines were closest together (fig. 3.6). In a large sample of *Zygiella x-notata* (AR) webs, the mean space near the edge in the upper part of the web was on the order of twice that near the hub, but was nearly the same (or slightly smaller) than spaces very near the hub in the lower portion of the web (data in graphs of Le Guelte 1966) A preliminary survey of published photos of the webs of species in 19 genera in four families (larger samples are needed, as many species were represented by only a single web) showed that this same pattern was common in both vertical and horizontal orbs: sticky lines tended to be spaced farther apart near the outer edge of the catching zone than near its inner edge (Eberhard 2014). Similar patterns have been noted in *Araneus diadematus* (AR) (Peters 1936, 1937; ap Rhisiart and Vollrath 1994; Zschokke 2002), *Cyclosa* spp. (Zschokke and Nakata 2015), and *Wixia abdominalis* (Xavier et al. 2017).

Four hypotheses that have been proposed might explain this common "larger-near-the-edge" pattern in sticky spiral spacing.

4.3.2.1 Constraint by cues

Peters (1939) noted a correlation between the distances between adjacent radii and the spaces between adjacent loops of sticky spiral attached to these radii in *Araneus diadematus* (AR) orbs, and proposed that the cue used by the spider to determine sticky spiral spacing is the inter-radius distance. Sticky spiral spacing, however, often shows complex, non-linear changes with the distance from the hub in this and other species (figs. 1.4, 2.17a, Eberhard 2014). In addition, the spacing sometimes changes abruptly in the same area of an orb (e.g., fig. 7.19), arguing against consistent use of inter-radial distances as a cue. The constraint by cue argument is thus not convincing as a general rule.

4.3.2.2 Speed of attacks

Spiders are generally able to attack prey nearer the hub more rapidly (Masters and Moffat 1983), so an investment in sticky silk to retain prey that strike the orb and are retained nearer rather than farther from the hub is likely to yield a larger payoff to the spider (Heiling and Herberstein 1998).

4.3.2.3 Stopping prey

Different areas of the orb have different abilities to stop prey, because whether or not a prey will be stopped by an orb is determined by the lines that the prey encounters at or near its point of impact, not by the orb as a whole. The tensions on radii near the hub are lower (section 4.4.2), increasing their ability to withstand the stress of an impact. In addition, the smaller distances between radii mean that the prey is more likely to encounter multiple radii. These spacing effects will be accentuated when prey strike the web at acute rather than perpendicular angles (figs. 3.2, 3.23) (Eberhard 2014), as is likely to often be the case (section 3.2.4). The overall result is that a high-energy prey is more likely to be stopped if it strikes the web nearer the hub. If one assumes, as seems likely, that larger prey are more difficult to retain after they have been stopped, and that contact with a larger number of sticky lines increases the retention time, then an orb weaver could gain greater payoffs by improving subsequent retention of the prey by investing sticky silk in the inner portion of the capture zone, where the ability to stop prey is higher, than investing it far from the hub. This is the "radius density" hypothesis of Zschokke (2002) and Zschokke and Nakata (2015); I will refer to it as the "prey stopping" hypothesis, to focus on the function.

4.3.2.4 Sticky spiral entanglement

Longer segments of sticky spiral swing in greater arcs, so they are more likely to become entangled with or adhere to other web lines (section 3.2.11.3). The segments are longer near the edge of the orb because the radii are farther apart, so they need to be spaced farther apart to reduce the danger of entanglement.

4.3.3 ILLUMINATION FROM EXCEPTIONS

Evaluating these hypotheses is complicated. They are not mutually exclusive, and a given design could result in multiple benefits in the same web. Comparative data indicate that each of the hypotheses can be ruled out in particular cases (table 4.2; Eberhard 2014), but ruling out a given hypothesis for one species does not mean that it cannot apply in others. Analyzing the effects of the different factors is thus likely to be difficult. I will, however, describe a few tests of the prey stopping hypothesis, because it seems likely, on the basis of several types of comparative data, to be of especially general importance (see Eberhard 2014 and Zschokke and Nakata 2015). As is often true in biology, exceptions to patterns can be used to test general hypotheses; I will discuss two types of exceptions whose consequences for the hypotheses are summarized in table 4.2.

4.3.3.1 Low prey velocities

Because the kinetic energy of a moving object is proportional to the square of its velocity (kinetic energy = [1/2] mv^2 where m is the mass and v is the velocity), a prey's velocity is especially important in determining whether an orb can stop it. Larger insects will tend to fly faster; Denny (1976) calculated that the kinetic energies of six insect species whose weights varied by a factor of about 500 had typical kinetic energies in flight that varied by a factor of about 700 at presumed maximum flight velocities. There are several araneid species in which it is likely that the flight velocities of their prey tend to be unusually low. As expected under the prey stopping hypotheses, these species also build orbs that have atypical sticky spiral spacing patterns: spacing between loops is smaller rather than larger near the orb's edge. For example, the approximately vertical orbs of *Gea heptagon* (AR) were built close to the ground in open grassy areas; the hubs were mostly only 15–20 cm above the ground. The lower edge of the orb was deep in the grass, only a few cm above the ground, while the upper edge was near the upper tips of the grass blades (fig. 4.2, Sabath 1969; WE). The speeds of insects that strike the lower portions of these orbs, where the sticky spiral lines are closer together (fig. 4.2), are thus probably especially low. The tighter spacing at the bottom of the orb might be explained by the fact that prey in a vertical orb often "tumble" or fall into lines directly below as they struggle (fig. 4.20). Tighter spacing of the sticky spiral at the bottom edge of the prey capture zone could reduce escapes via tumbling (Eberhard 2014; Zschokke and Nakata 2015) (the "prey tumbling" hypothesis). The sticky spiral lines in the upper portions of the same orbs, where prey are likely to be moving more rapidly, showed the opposite (typical) larger-near-the-edge pattern of spacing.

In a second exceptional species, *Metepeira* sp. (AR), orbs are built high above the ground (WE), but their prey are also likely to be moving relatively slowly, because the spider builds a more or less extensive, strong tangle immediately adjacent to the orb. Both those prey that arrive at the orb from the side with the tangle, and those that pass through the orb but then bounce back from the tangle in ricochet fashion (Uetz 1989), are likely to be moving relatively slowly. The spaces between sticky spiral loops were smaller near the outer edge of the web in the lower portion of the orb (which was by far the largest) (fig. 3.7) (this "smaller-near-the-edge" pattern also occurred in a photo of a *M. labyrinthea* web) (Emerton 1902; Comstock 1967).

Finally, there is the probable effect of nearby vegetation or other objects on reducing prey velocities in webs built high above the ground. The orbs of *A. expletus* are often inserted among leaves and branches in dense vegetation, sometimes so close to nearby leaves that the leaves touch and adhere to the web when they move in the wind; I have never seen this kind of problem in open space orbs like those, for instance, of the araneids *Cyclosa caroli*, *Micrathena* spp., or *Gasteracantha cancriformis* (WE; see also Craig 1988 on *M. schreibersi*). Similarly, the orbs of *Cyclosa jose* and, at least sometimes, those of *C. divisa* were built across a single, large, slightly folded leaf, and those of *Epeirotypus brevipes* (TSM) were near large objects such as tree trunks. The flight velocities of the prey of *A. expletus*, *C. jose*, and

Table 4.2. A preliminary summary of the support (Y) or lack of support (N) that different patterns in the data provide regarding the predictions of different hypotheses to explain within-orb patterns in the spaces between loops of sticky spiral. Webs were classified as "horizontal" or "vertical" only approximately. Although the implications of some types of sticky spiral patterns that have been observed are less than certain, no single hypothesis (at least in the simple versions described in the text) is favored by all types of evidence.

	Patterns in the data							
Hypotheses	*Spaces were larger near the outer edge in vertical webs*[1]	*Spaces were larger near the outer edge in horizontal webs*[2]	*Diffs. in spacing were smaller when radii were denser*[3]	*Diffs. in spacing were smaller when radii were more nearly parallel*[4]	*Diffs. in spacing were smaller below the hub*[5]	*Spaces were larger near the hub in vertical webs*[1]	*Spaces were larger near the hub in horizontal webs*[2]	*Exceptions that occurred when prey were likely slower*[5]
Radius density	Y	Y	Y	Y	N[6]	N	N	Y
Sticky spiral entanglement	Y	Y	Y	Y	N	N	N	N?[7]
Prey tumbling	Y	N	N	N	Y	?[8]	N	Y
Attack speed	Y	N	N	N	N	N	N	N
Danger from prey	N[9]	N[9]	N	N	N	Y	Y	N[10]

[1] Many species show this pattern; the number of derivations is uncertain.

[2] There are probably at least four different derivations of this patten in approximately horizontal orbs: *Mangora* spp.; *Azilia affinis*; tetragnathids such as *Leucauge* spp., *Tetragnatha* spp., *Dolichognatha* sp.; and several uloborid genera.

[3] In *Cyclosa* spp. (AR), *Enacrosoma anomalum* (AR), *Patu* (MYS), uloborid (UL) (fig. 4.5).

[4] In *Nephila clavipes*.

[5] Probably independently derived cases of this pattern include the araneid species *Gea heptagon*, *Metepeira* sp., and *Araneus expletus* (and perhaps several trunk web groups; the data on their web designs are of poor quality).

[6] This is N if prey velocities below the hub similar, but it is Y if prey velocities below the hub are lower.

[7] Some of these sites (low in thick grass, against tree trunks) probably have lower wind velocities, making closely spaced sticky spiral lines less likely to adhere to each other, in accord with the entanglement hypothesis; some species in this category, however, such as *Metepeira* sp., built orbs in relatively exposed sites (WE).

[8] This pattern could favor the hypothesis when it occurs below the hub, but perhaps not when it occurs above the hub (if radii also help retain tumbling prey however, then the greater density of radii near the hub might make wider spaces between sticky lines advantageous for reasons of economy of material).

[9] Prey impacts in the outer portions of the orb would pose no threat to the spider.

[10] There seems no obvious reason to associate lower impact velocities with changes in the danger from the prey.

E. brevipes were thus likely to be low; again, as predicted, the spacing between sticky spiral loops was smaller near the outer edge of the orb below the hub (figs. 3.7, 3.8).

Another group of orb weavers whose prey likely have low velocities are the trunk orb species (fig. 4.3), whose prey are presumably usually either slowing down to land, or just taking off. The stopping abilities of these orbs may thus also tend to be less severely tested. In addition, the radii in the trunk orbs of the nephilids *Clitaetra* and *Herennia* resemble those in *N. clavipes* webs in being more nearly parallel and less spread apart at the outer edge of the orb than those in typical orbs, due to extensive splitting and to the addition of parallel vertical lines (figs. 4.3, 9.17). Once again, as predicted by the prey stopping hypothesis, the sticky spiral spacing appears not to follow the usual larger-near-the-edge pattern, at least in the larger, lower portion of these orbs (fig. 4.3a). This confirmation of the prey stopping hypothesis is tentative, however, as it is based only on single photos of these species' webs. And, just as clearly (and just as tentatively, because only a single photo is available), a trunk web of *Eustala conformans* appears not to conform to the pre-

Fig. 4.2. This exception may prove a rule (but which one?). The lower portions of the vertical orbs of *Gea heptagon* (AR) are unusual in having an inverted edge-to-hub pattern of sticky spiral spacing: the spaces between sticky spiral loops at the bottom edge are smaller rather than larger than those near the hub. This pattern may be explained by the sites where this species builds its webs near the ground in grassy areas (many blades of grass were cut away to take this picture in a pasture). Insects flying through the dense "forest" of grass blades near the ground probably move slowly, making them unusually easy to stop. Similarly, wind velocities near the ground are likely to be especially low; and struggling prey working their way down through the lower portion of the orb could be retained by the dense array of sticky lines at the bottom. In sum, the prey stopping, the web damage, and the tumbling hypotheses could all account for this exception.

dictions, as it has larger sticky spiral spacing near the edges of both the upper and lower portions of the orb (fig. 4.3c).

It should be noted that, except for *Metepeira*, wind velocities may be lower at the protected sites (e.g. near the ground in *Gea*, the tree trunks in *Herennia* and *Epeirotypus brevipes*) so the differences in sticky spiral spacing might also be predicted by the sticky spiral entanglement hypothesis.

4.3.3.2 Tightly and widely spaced radii

Other exceptional orb designs in which radii are especially close to each other provide additional tests of the prey stopping hypothesis. The orbs in the genus *Nephila* are especially interesting, and relatively well-known (fig. 4.4; Wiehle 1931; Witt et al. 1968; Shinkai 1985; Kuntner et al. 2008a, 2010b; Hesselberg and Vollrath 2012). The radii are much more nearly parallel to each other than in typical orbs, because the spider adds many extra, tertiary radii while building the temporary spiral (fig. 4.4). In addition, the temporary spiral is left intact in the finished orb, and it bends each tertiary radius sharply away from the radius on which (or near which) it originated and holds it nearly parallel to its neighbors for the rest of its length (fig. 4.4). The distances between adjacent radii in *Nephila* webs are thus much more nearly uniform throughout most of the capture zone than in more conventional orbs (fig. 4.4c). *Nephila* webs are also commonly accompanied by a tangle of strong lines on at least one side of the orb (Hingston 1922c; Robinson and Robinson 1973a; Higgins 1993; Kuntner et al. 2008a). This tangle probably protects the spider from predators (Higgins 1993), but it undoubtedly also has the additional effect of reducing the velocities of at least some of the prey that strike the orb (Hingston 1922c). The prey stopping, tumbling prey, and sticky spiral entanglement hypotheses (but not the attack speed hypothesis) predict that the larger-near-the-edge pattern should be reduced or absent in *N. clavipes* webs. The prediction is confirmed: there is little edge-to-hub difference, and what differences do occur are reversed (sticky lines farther from the hub are slightly closer together) (figs. 3.6, 4.4; Eberhard 2014).

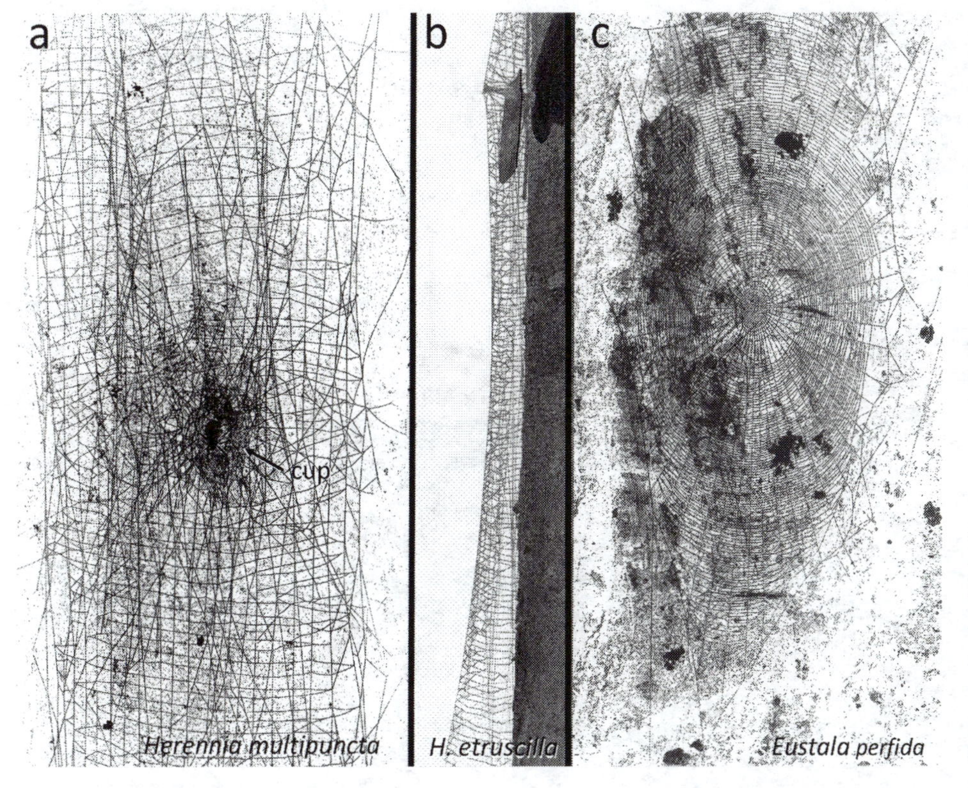

Fig. 4.3. The nephilids *Herennia multipuncta* (*a*) and *H. etruscilla* (*b*) and the araneid *Eustala perfida* (*c*) convergently evolved vertically elongate trunk orbs built near the surfaces of standing tree trunks. The *Herennia* webs are more strongly modified in two respects. The hub was altered to form a cup that was attached directly to the surface of the trunk where the spider rested (*a*); in contrast, the hub of *E. perfida* was aerial. In addition, the *Herennia* webs were curved (*b*) rather than planar; the sticky spiral lines were attached to "pseudo-radii" that ran directly from the upper to lower edges of the web and did not run through the hub. They allowed the orb to fit the curved surface of the trunk. This curve was accentuated in *b*, because the web was attached to the head of a protruding nail and was thus unusually distant from the trunk (*a* and *b* from Kuntner et al. 2010a, courtesy of Matjaž Kuntner; *c* from Messas et al. 2014, courtesy of Yuri Messas).

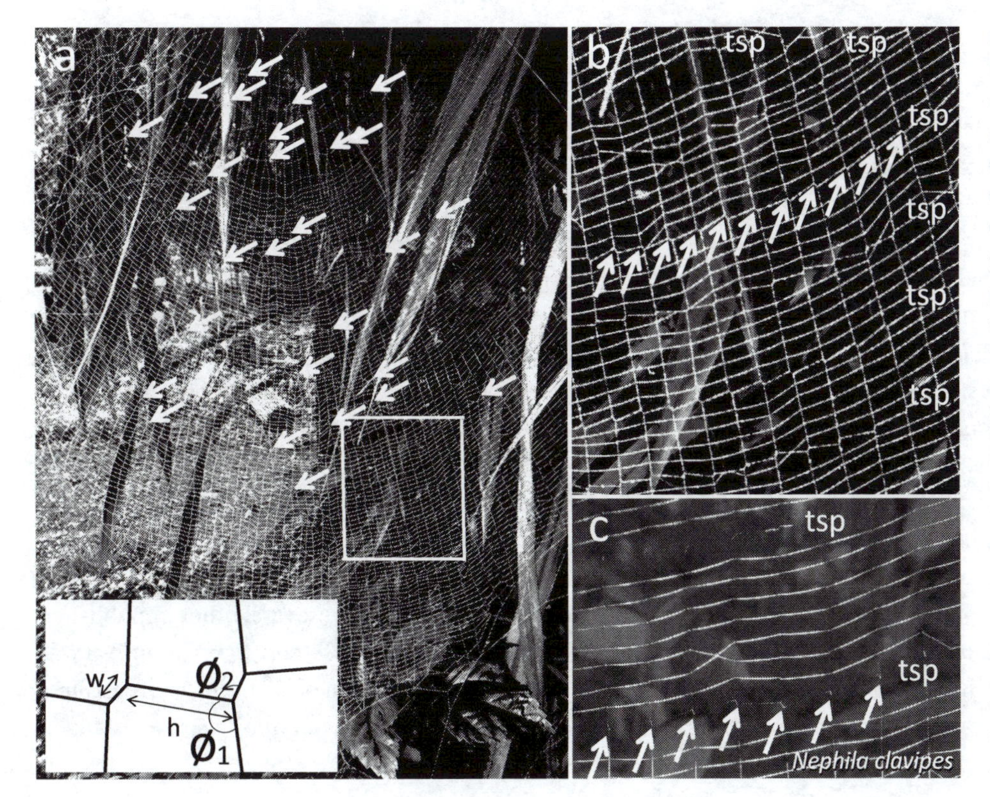

Fig. 4.4. The "rectangular" mesh in this web of *Nephila clavipes* (NE) has secondary and tertiary radial lines that are nearly parallel to each other in large areas of the web. In the overall view (*a*) the arrows mark points where tertiary radii originated between preexisting secondary radii during temporary spiral construction. The closeup view of the portioned indicated by the rectangle shows the nearly uniform spaces between radii (*b*). In a closeup of a small section of a different, unpowdered orb (*c*), the sticky spiral lines (more clearly visible) are easily distinguished from the non-sticky radial and temporary spiral lines. The arrows in *b* and *c* mark double attachments of the temporary spiral to the radii, which tighten the radii by pulling them out of line (inset in *a*).

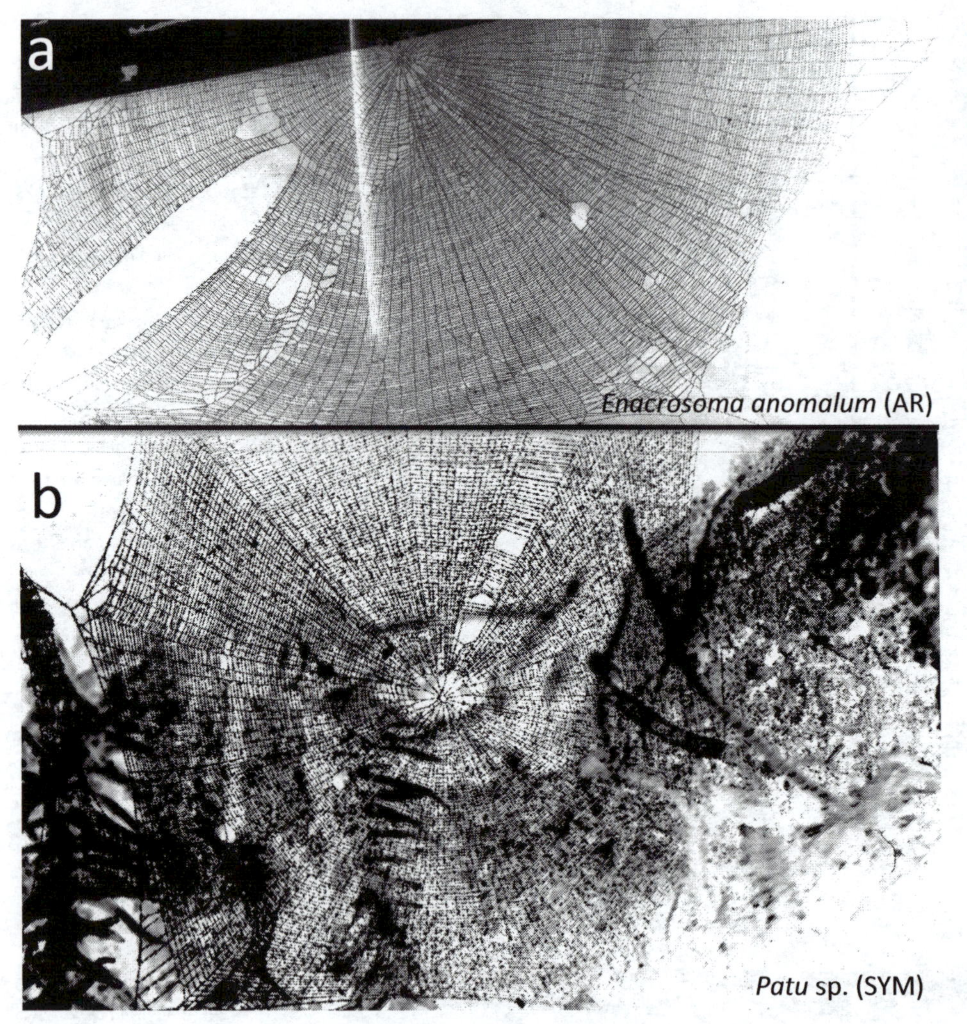

Fig. 4.5. The dense, horizontal webs of mature females of *Enacrosoma anomalum* (AR) (*a*) and *Patu* sp. (SYM) (*b*) support the hypothesis that edge-to-hub decreases in sticky spiral spacing are due to the wider spaces between radii near the edges of more typical orb webs. Both orbs are exceptional in having hardly any edge-to-hub change in the sticky spiral spacing, and in also having relatively small spaces between the radii near the edge. There are 75 radii at the edge in the half of the *E. anomalum* web that is visible, and a total of about 175 in the *Patu* web.

Several other apparent exceptions in which radii are very closely spaced and the sticky spiral spacing is also not perceptibly greater near the edge of horizontal orbs (not predicted by the tumbling hypothesis) occur in the symphytognathid genus *Patu* sp. (fig. 4.5b; Griswold and Yan 2003), the anapid *Elanapis aisen* (Ramírez et al. 2004), an unidentified Australian uloborid (fig. 3.42), and in *Enacrosoma anomalum* (AR) (fig. 3.42). In a small sample of *E. anomalum* webs in Costa Rica, webs were built just under the surface of a large leaf, and may thus also capture prey with lower velocities; one web of the congener *E. quizarra* had considerably fewer radii, and differed in showing a clear pattern of larger sticky spiral spacing farther from the hub (WE).

One further group in which the usual edge-to-hub sticky spiral spacing pattern appears to be inverted in the lower portion of the orb is the araneid genus *Cyclosa* and its monotypic sister genus *Allocyclosa* (Levi 1999). Although the data are incomplete, there is published documentation of smaller spacing near the outer edge of the lower sector of the orb in four species, *C. argenteoalba*, *C. confusa*, *C. ginnaga*, and *C. octotuberculata* (Zschokke and Nakata 2015), and additional unpublished observations (all with samples that included ≥10 webs) showed the same pattern in *C. caroli* (Hesselberg 2013; T. Hesselberg pers. comm.; WE) and *C.* nr. *atrata*, *C. insulana*, *C. jose*, and *Allocyclosa bifurca* (WE). The pattern was absent in one published web of *C. conica* (Bradley 2013), but present in several other unpublished photos of this species (T. Hesselberg pers. comm.). It was also lacking in a single photo of *C. mulmeinensis* from a windy site (Liao et al. 2009), and may also be lacking in

C. fililineata (Viera et al. 2007). The orbs of at least some of these same *Cyclosa* species were especially radius-rich (fig. 4.12) (one certain exception is the *C. mulmeinensis* web from a windy site).

Low prey velocities are unlikely to explain the patterns in *Cyclosa*, because the websites of most species of *Cyclosa* are not exceptional with respect to expected prey velocities (though no measurements are available). Some *Cyclosa* species, such as *C. caroli* (Hesselberg 2013, pers. comm.; WE) and *C. insulana* (WE) build at relatively exposed sites where prey velocities do not seem likely to be especially slow. Two possible exceptions are *C. jose* and *A. bifurca*, where the velocities of the prey of may be relatively slow because the webs tend to be built next to the surfaces of large, slightly curled leaves (*C. jose*) or against walls, windows, or tree trunks, and in the spaces between the large, stiff leaves of bromeliads and *Agave* plants where prey velocities may be low, and where they sometimes also have a small tangle adjacent to the orb (WE). In sum, the designs of *Cyclosa* webs often do not fit expectations from the prey stopping hypothesis.

A complementary prediction of the prey stopping and sticky spiral entanglement hypotheses is that the larger-near-the-edge pattern will be accentuated in orbs that have especially low numbers of radii. Some tetragnathid spiders have especially few radii; and, as predicted by the prey stopping and sticky spiral entanglement hypotheses, it appears that they have especially pronounced edge-to-hub differences in their more or less horizontal orbs (fig. 4.6). Further data are needed to test this possibility.

As noted in table 4.2, the sample sizes documenting some of these patterns are small, and the conclusions derived from them are only suggestive. Another possible (non-exclusive) explanation that was omitted in this analysis is phylogenetic inertia (Harvey et al. 1991). This possible constraint seems unlikely to be important, however, because the degree of variation seems easily large enough for natural selection to work on.

4.3.3.3 Other patterns, other explanations

4.3.3.3.1 ABOVE VS. BELOW THE HUB

Although data and analyses are only preliminary, there are probably at least two additional common patterns in the variation in sticky spiral spacing. The first pattern is that the larger-near-the-edge pattern seems to be more consistent in the upper than the lower portions of vertical orbs (Eberhard 2014). The prey tumbling hypothesis mentioned above could explain this pattern. The prey tumbling hypothesis fails to explain, however the larger-near-the-edge pattern both above and below the hub in the approximately vertical orbs of *Micrathena duodecimspinosa* (WE), or the weaker larger-near-the-edge pattern in the spacing below the hub in *Wixia abdominalis* (AR) when the orb had a free sector above the hub than when a free sector was lacking (Xavier et al. 2017). More importantly, it does not explain the clear larger-near-the-edge pattern seen in many horizontal orbs.

There are additional reasons to expect that the radii below the hub may be better than those above it at absorbing the impacts of high-energy prey, thus linking tighter spacing below the hub to the prey stopping hypothesis. In the first place, there are typically more radii below the hub, so they are closer together, and the prey is more likely to strike multiple lines. Secondly, the radii below the hub are usually longer; other things being equal, a longer line can absorb more momentum than a shorter one. Finally, the radii are under less tension, which also increases their ability to absorb more momentum. The lower tensions result from the spider's own weight at the hub, which will relax the radii below the hub, and tighten those above it, and from the larger numbers of radii below it (the force/radius will be lower where there are more radii) (fig. 4.7, Denny 1976; Wirth and Barth 1992).

4.3.3.3.2 INNER EDGE OF THE CAPTURE ZONE, "FLIMSY" ORBS, TURNBACKS

A second probable pattern is that in several species the larger-near-the-edge pattern disappears in the inner-most loops, which are relatively widely spaced (fig. 3.6, Eberhard 2014). One possible, though not especially convincing, explanation for this pattern is that by reducing retention of large and potentially dangerous prey that strike the web very near the hub, spiders reduce the dangers from their struggles (section 3.2.12, Eberhard 2014) (the "danger from prey" hypothesis). While this hypothesis could explain the "free zone" with no sticky lines near the hub (section 4.9.3), it seems less convinc-

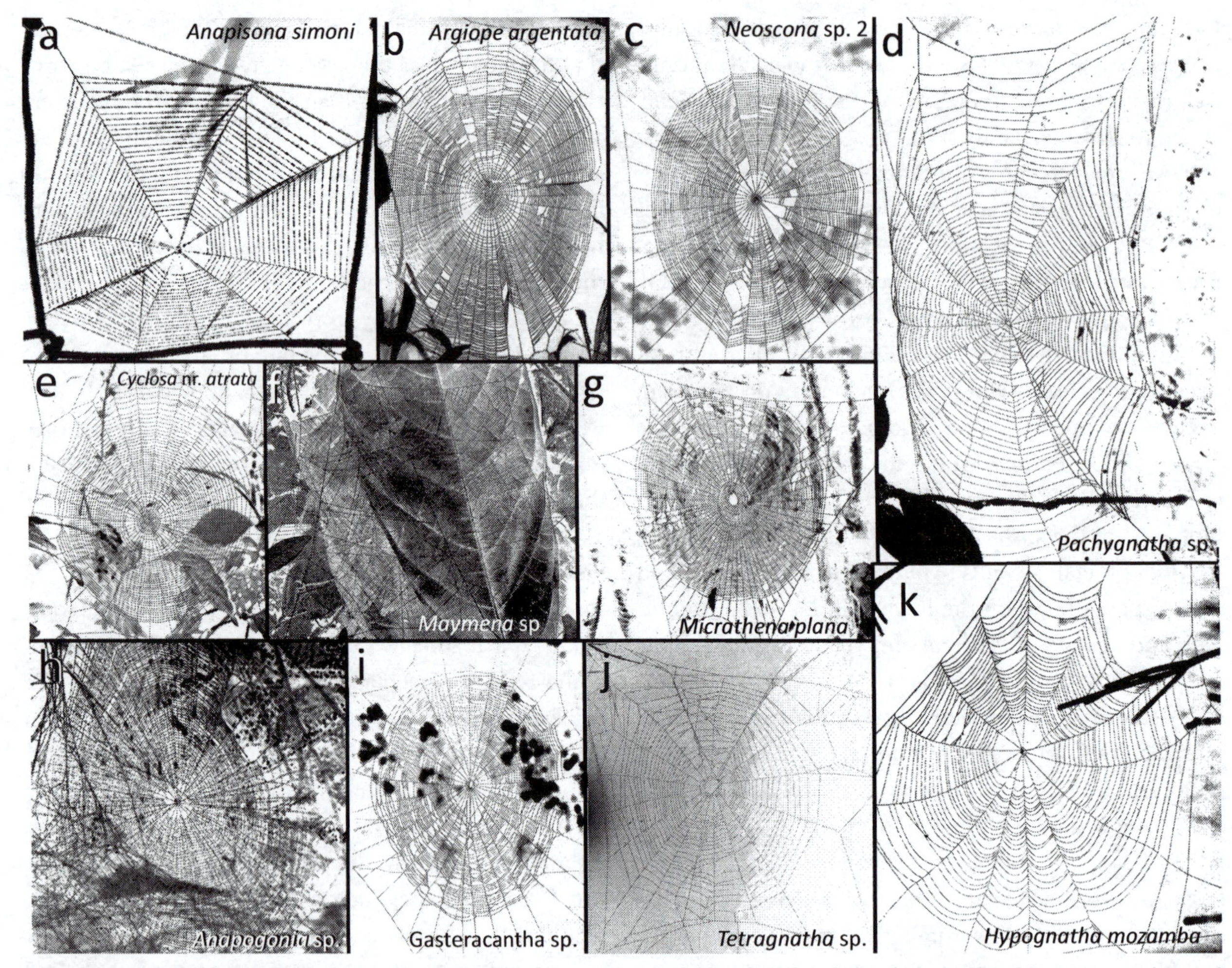

Fig. 4.6. The webs of this sampler of 11 species in 3 families intimate that the strength of the edge-to-hub pattern of decreasing sticky spiral spacing may be more dramatic in flimsy webs with low densities of radii (*a*, *j*, *k*), least dramatic in sturdy webs with numerous radii (*f*, *g*, *h*), and intermediate in others (*b*, *c*, *d*, *e*, *i*). In several intermediates (especially *b* and *e*), the edge-to-hub pattern is clear in the upper but not the lower portion of the orb. The slants of the webs were as follows: *a* about 0° (a shallow cone); *b* about 90°; *c* 77°; *d* 77°; *e* 89°; *f* 59°; *g* 76°; *h* about 0°; *i* 83°; *j* 26°; and *k* 82°. The outer spiral loops on the left side of the *Pachygnatha* sp. (TET) web adhered to each other and to the frame (*d*), illustrating a typical pattern of damage that occurs when the supports for anchor lines move (fig. 3.18c). Further work will be needed to quantify sticky spiral patterns and the densities of radii in different species, and to test the significance of possible associations.

ing for sticky spiral spacing; in vertical orbs prey above rather than below the hub would seem most potentially dangerous as they could tumble downward onto the spider, but the increase in spacing was larger just below the hub (fig. 3.6).

Fig. 4.6 intimates that a further, as yet unquantified dimension—the flimsiness or robustness of the web—may be important. Orbs that seemed more flimsy and tended to become damaged more quickly in the field had more pronounced edge-to-hub patterns of sticky spiral spacing. This pattern (if it exists) might be associated with the sticky spiral entanglement hypothesis, but further work will be needed to quantify these subjective impressions and test for possible associations.

An additional fine adjustment in sticky spiral spacing emphasizes that careful examination of patterns of sticky spiral spacing is biologically relevant. In the uloborids *Uloborus diversus* (Eberhard 1969) and *Zosis geniculata* (WE) the distance of the attachment from the inner loop was reduced at points where the spider turned back, and this adjustment in spacing at a turnback point reduced the open space not covered with sticky lines (fig. 3.9). Such fine adjustments testify to the apparent selective importance of even small details of sticky spiral spacing.

4.3.4 A LIMITATION OF CURRENT DATA: HETEROGENEITY OF LINES

There may be substantial inter- and intra-specific variation in the properties of sticky lines (section 3.2.9) as well as their spacing. For instance, Sensenig et al. (2010b) found, when they compared 13 webs of the araneid *Larinioides cornutus* built in captivity with 11 built in the field, that the field webs had major ampullate silk lines that were 50% thicker (the field webs were also less symmetrical, had 14% more radii, and had sticky spirals that were 58% longer). An individual of *Cyclosa turbinata* (AR) tended to reduce the stickiness of sticky spiral lines when it had fed less (Crews and Opell 2006). Even within a given orb of *Argiope argentata* (AR), the diameters of the sticky spiral baselines were 130% larger near the outer edge of the orb than near the hub (Blackledge et al. 2011). In the araneids *Cyclosa turbinata, Argiope aurantia*, and *A. trifasciata* the outermost loops of sticky spiral were thicker or stickier than the two or three innermost loops (Blackledge et al. 2011; Crews and Opell 2006; Opell et al. 2009). The difference in stickiness was attributed to the spider running out of sticky material (especially phosphorylated glycoproteins) in the later stages of sticky spiral construction (Opell et al. 2009). Alternatively, this flexibility may be adaptive. Because the sticky spiral loops are farther apart near the edge of the orb, perhaps they need greater amounts of sticky material to be able to retain prey effectively than do loops very near the hub where there are many other lines nearby. The possibility that there are differences in the stickiness of sticky spiral lines in different parts of the orb needs to be explored in the other species.

In *A. trifasciata* the within-web differences in glue droplets were even larger. The volumes of the individual droplets on the outermost loops below the hubs of field webs were about twice the volumes of those on the outermost loops above the hub when sectors of orbs were placed in a uniform environment (Stellwagen et al. in prep.). The compositions of the droplets also varied. Those near the lower edge below the hub had a lower ratio of aqueous to glycoprotein volume (2:1) than those above (3:1). Droplets from the middle regions of the top and bottom sectors were larger than those from the top, but maintained the same ratio of aqueous to glycoprotein volume as the top droplets. Droplets from the innermost spirals tended to be similar in size to those at the top, but had four times more aqueous than glycoprotein volume, the highest ratio of droplets from any region of the web (Stellwagen et al. in prep.).

These differences in size and material properties of the droplets were correlated with differences in trapping properties (Stellwagen et al. in prep.). Larger droplets and those with a higher ratio of aqueous to glycoprotein volume had improved glycoprotein function. Bottom droplets, which had the lowest aqueous to glycoprotein ratio, also absorbed the least amount of energy per glycoprotein volume. Their possibly lower viscosity, suggested by a lower force/glycoprotein volume, permitted them to extend further but greatly reduced the work required to extend them. Inner spiral droplets, which had the greatest proportion of aqueous material, were the most viscous, and exhibited the greatest relative toughness.

These differences between different parts of the web were probably not simply due to gradual depletion of the spider's glands, but more likely to the spider having some control over the glue deposited on the axial threads. Their functions are not entirely clear. The toughness per glycoprotein volume was high in top droplets and low in bottom droplets, but the larger total glycoprotein volumes of bottom droplets inverted this difference: overall, the droplets in the bottom spirals were tougher. However, the greatest relative toughness of all occurred in the center of the top and bottom web regions. Thus the spiders apparently allocated resources in a way that optimized the retention abilities of the central web region, which is nearer the spider's resting position at the web's hub, where she can presumably attack more rapidly and which is also more likely to be able to stop larger prey. The lower region of the web, with sticky lines of lower quality, may serve as a "last ditch" area for securing insects that would otherwise tumble down the web and escape (Stellwagen et al. in prep.). A further complication is the probably somewhat greater humidity near the bottom of *A. trifasciata* webs, which are often built in low vegetation near the ground; this could enhance the performance of the lower portion in the field (Amarpuri et al. 2015).

4.3.5 CONCLUSIONS REGARDING WITHIN-WEB PATTERNS OF STICKY SPIRAL SPACING

The question of why there are clear patterns in the variation in sticky spiral spacing in orb webs is not completely resolved. The evidence for the different hypotheses is mixed (table 4.2), though in general it favors the prey stopping and the sticky spiral hypotheses over the others. The hypotheses are not mutually exclusive, and have different degrees of importance in different species. Heterogeneity in different parts of the sticky spiral, which seems to be common, may also offer further explanations. The growing realization of the magnitude and subtle nature of the variations in the physical properties of sticky lines in a given orb, and in the orbs of different species (see also section 3.2.9) makes me fear that much of the discussion in this section (4.3) may be premature, and that it must be treated with extreme caution until further data on these properties are available. There is much yet to be learned about the patterns in sticky spiral spacing in orb webs.

4.3.6 CONSEQUENCES FOR UNDERSTANDING HOW ORBS FUNCTION

The clear patterns of variation in both radius and sticky spiral spacing signal the need for substantial changes in common interpretations of how orbs function. Although it may be reasonable to consider an orb as a unitary device for some functions, such as supporting the spider and transmitting vibrations (Japyassú and Ades 1998), it is best seen as a combination of designs that have different prey capture properties in different portions of the web, rather than as a single unit trap. Because orb weavers have conserved the ancient pattern of lines radiating from a central point where the spider waits, different portions of the same orb have very different densities of strong, non-sticky support lines. And, perhaps due in large part to these differences, orb weavers usually adjust the densities of sticky lines substantially in different parts of their orbs. While analyses of orbs as unitary traps have improved knowledge of the functional significance of different orb web designs, more complete understanding will require analyses that integrate the stopping and retention properties of different portions of the same orb. In one sense, this abandonment of a unitary vision of an orb represents one more skirmish in the long battle of biologists to resist making overly simplified, typological analyses of natural phenomena (Mayr 1982).

4.4 TENSIONS AND STRESSES

4.4.1 THEORETICAL EXPECTATIONS OF UNIFORMITY IN HOMOLOGOUS LINES ARE NOT CONFIRMED

Table 3.1 outlines the expected consequences of differences in tensions for six different web functions; increased tensions likely improve some functions while interfering with others. Changes in tensions on non-sticky supporting lines in orb webs are likely to have both advantages and disadvantages. The greater the tension on a line in a web, the less its ability to resist mechanical stresses imposed by external agents such as prey, wind, etc. The crucial variable is the stress on a line, which is directly proportional to the line's tension, and inversely proportional to its cross-sectional area (stress = force/area). Greater stress on a line in the web at rest will make it less able to absorb additional, externally applied stress before it breaks. Theoretically, in a minimum volume network of uniform lines for resisting non-localized loads such as wind forces, all lines should be of similar materials, equal diameters, and under equal stress ("Maxwell's lemma") (Denny 1976; Craig 2003).

These expectations are not met in orb webs. With respect to tensions, it is well established that not all homologous lines in a any given orb are equally tense (fig. 4.7). In the first place, the radii in most orbs are not equally distributed in all directions around the hub. Usually there are more radii on the side that is larger (usually the portion below the hub in vertical orbs). Since the sum of the forces exerted by the radii on one side of the hub is precisely equal to the sum of the forces exerted by the radii on the opposite side of the hub when the hub is at rest (otherwise the hub would be moving), the tensions on radii on the more radius-poor side must be greater (the total force is distributed among fewer radii) (fig. 4.8c). Tensions on the radii in smaller portions of an orb are thus generally greater (Eberhard 1981a). Calculations based on measurement of the tension in one anchor line and on the angles between this and all the other lines in the orb of a mature female *Larinioides sclopetaria* (= *Araneus sericatus*) confirmed that radius tensions varied by a factor of about 8 (from 0.42 to 3.38 N ×

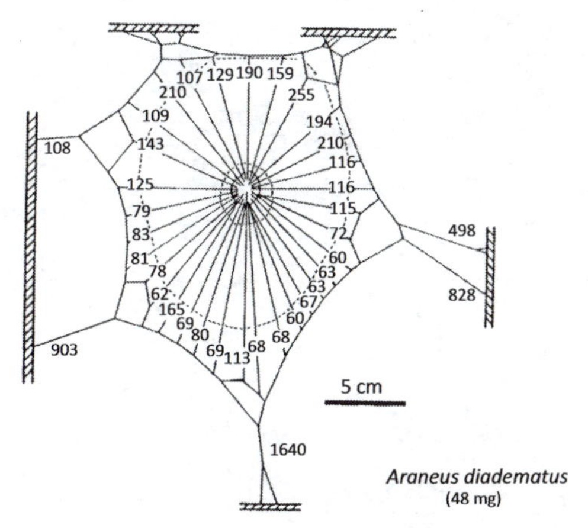

Fig. 4.7. The pretensile forces in μN (μNewtons) on radii and anchor lines in the orb of a mature female *Araneus diadematus* (AR) illustrate several general patterns: anchor lines tended to be under greater tensions than radii; radii above the hub tended to be under greater tensions; and tensions on radii varied widely (from 60 to 255 μN in this web). To put these forces in perspective, this spider's mass (48 mg) would exert 471 μN at sea level, increasing the forces on radii above the hub and lowering them on radii below the hub. The tensions on radii were measured where they were highest, near their intersections with frame lines (the dotted line marks the outline of sticky lines; the forces exerted by the sticky spiral would reduce tensions on radii nearer the hub) (see fig. 4.8b) (from Wirth and Barth 1992).

10^{-4}) (Denny 1976). Radii built earlier and in the upper portion of the orb of *Zygiella x-notata* (AR) were also deduced, on indirect grounds, to be tighter than others (Le Guelte 1966).

Direct, non-destructive in situ measurements by Wirth and Barth (1992) confirmed and extended these patterns in the araneid *Araneus diadematus*. The radii in the upper half of the orb had pretensile forces that were two to three times those of the lower half. In addition, primary radii built during the early stages had greater forces than secondary radii built after the major frame lines had been built. Heavier spiders had greater radial pretensions, and the ratio of tension:mass was approximately constant. Because a longer radius can better absorb the increased stress associated with the impact of a prey than a shorter radius under the same tension, the effect of the trend toward higher tensions on shorter radii on reducing their ability to absorb prey impacts is expected to be even greater than the differences in tension would have suggested.

In addition, the diameters of the radial lines also varied substantially, even in a particular orb: in four species of araneids, the coefficient of variation in fiber diameter in a single orb varied from 15% in *Eriophora fuliginea* to 34% in *Argiope aurantia* (Work 1977); in one orb of *L. sclopetaria* the largest diameter of a radial fiber was 2.9 times of that of the smallest (Denny 1976). Denny (1976) calculated that the stresses on radii in an orb of *L. sclopetaria* ranged by a factor of more than 6, from 8 to 54 MN/m^2 (comparable ranges for frame lines and anchors were 9 to 125, and 15 to 105 MN/m^2).

These data may overestimate the differences, at least in some species of araneids. The outer portion of a radius is under more tension, due to the inward forces exerted by the loops of sticky spiral (fig. 4.8b; Eberhard 1972a; Zschokke 2000). In addition, the radii of some species have twice as many fibers near the edge of the orb, due to the spider breaking and replacing only the inner portion of each new provisional radius during radius construction (section 6.3.3, Eberhard 1982, 2019a; Zschokke 2000). Partial doubling of radii would reduce the stress on the doubled portion of the radius, which included up to >50% of its length in *Zilla diodia*, and also occurred to a lesser extent in *Neoscona nautica* (AR), *Tetragnatha* sp. (TET), and *Leucauge* sp. (TET) (WE). Doubling was absent in other species, however, including *Araneus diadematus* (AR) (Zschokke 2000), *Witica crassicauda* (AR), and *Micrathena gracilis* (AR) (WE). This heterogeneity in radius diameter is another deviation from expectations under Maxwell's lemma.

Tensions and stresses also varied between the different types of non-sticky lines. Anchors were under greater tensions than frames, and frames were under greater tensions than radii. In an orb of *L. sclopetaria*: the ratios for tensions on the three types of line were approximately 10:7:1 (Denny 1976). Similar ratios occurred in direct measurements in the orbs of the araneids *A. diadematus*, *Z. x-notata*, and *Nuctenea umbratica*, while the anchor lines of the nephilid *Nephila clavipes* were under proportionally even greater tensions (Wirth and Barth 1992). Differences in stresses on these different types of line were less dramatic in the *L. sclopetaria* orb because lines under greater tensions tended to have greater cross-sectional areas, but they were still appreciable: the stresses on radii averaged about half of those on frame and anchor lines (Denny 1976). The temporary spiral of *A. diadematus* was also under relatively high tension compared with that on the radii (Wirth and

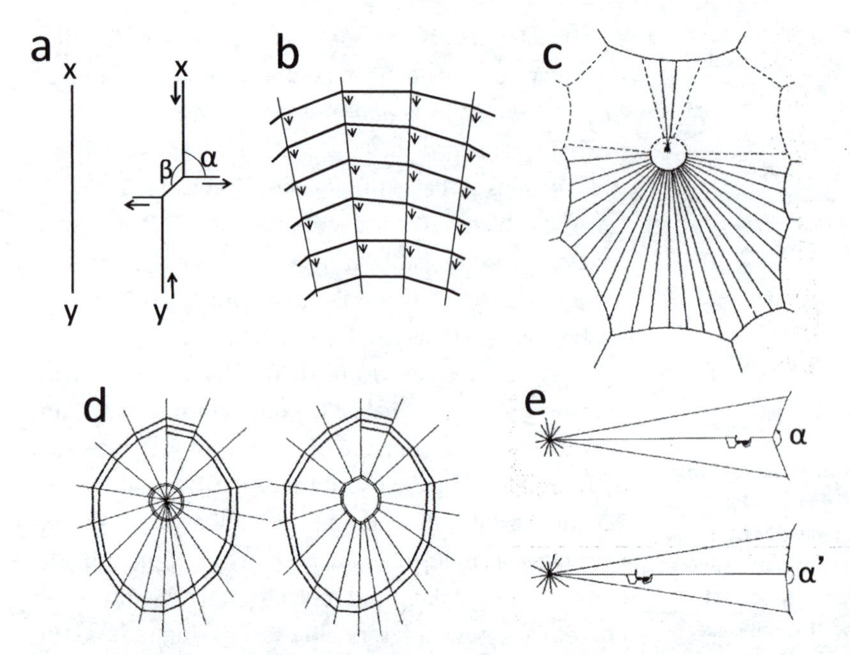

Fig. 4.8. These schematic representations illustrate several effects on tensions on lines in orb webs that are mentioned in this chapter. *a*) When a straight line between two points (x and y) is distorted (right) into a zig-zag conformation by connections with another line (e.g., hub or temporary spiral loops attached to a radius), the tension on the x-y line will be increased (vertical arrows), due to tension on the other line (horizontal arrows). The size of angle β indicates the relative increase in tension on the radius. One extreme would be a large increase on a radius that was under a very low relative tension (β = 90°); at the other would be a radius that was under a very high relative tension and whose tension was not increased (β = 180°) (see also fig. 4.9). *b*) At each point where a sticky spiral line is attached to a radius, the tension on the spiral generates a force on the radius directed toward the hub (arrows). These forces are summed along each radius, so greater relaxation of radial tensions occurs nearer the hub. *c*) When a spider resting at the hub of her orb pulls or jerks on a pair of radii above the hub with her front legs, the increased tension on these radii (the two solid lines on one side of the hub) will be balanced by increased tensions on all the radii in the other half of the orb (solid lines below the hub). During a jerking movement, the tips of her front tarsi will move posteriorly while her body moves a shorter distance anteriorly. If the radii are under similar tensions in the portion of the orb in front of her, the spider could sense their lengths on the basis of their extensibility (how strongly she pulled and how far her tarsi moved posteriorly with respect to her body). *d*) When a spider removes lines from the center of a hub, the diameter of the hub expands, and the tensions on the radii tend to be reduced. *e*) When a spider breaks and reels up a new radial line while she returns to the hub, she usually releases more new dragline than she reels up, and thus lowers the tension on the new radius. This tension reduction results in a smaller deflection in the frame line by the radius (α < α') (modified from Eberhard 1972a, 1973a, 1987d).

Barth 1992), as is often true in *Nephila* (NE) (Hesselberg and Vollrath 2012).

Another factor altering the distribution of tensions on radii in an orb is the weight of the spider herself. When she rests at the hub of a more or less vertical orb, her weight will increase the tensions on the radii, frame, and anchor lines above the hub, and lower the tensions on the lines below the hub. These effects will tend to accentuate the differences due to having fewer radii above than below the hub. In more nearly horizontal orbs, her weight will increase tensions on most of the non-sticky lines in the orb (except the innermost hub lines).

In sum, the failure of orb webs to fit the expectations of Maxwell's lemma, despite the fact that spiders have ample opportunities to adjust tensions during construction (next section) suggests that selection may have acted more intensely on properties that promote resisting the localized loads that are produced by prey impacts than on those that resist general wind loading. When strong, localized loads are applied to an orb, the radial lines that break are always those very near the loading site; other radial lines, frame, anchor, or hub lines are seldom affected; in addition, radii typically break near the impact site, and not near their outer ends (Cranford et al. 2012; WE). Thus, the stresses on the lines that are in the immediate vicinity of the impact appear to be more important than the balance of stresses throughout the web in determining whether a prey will be stopped. In gen-

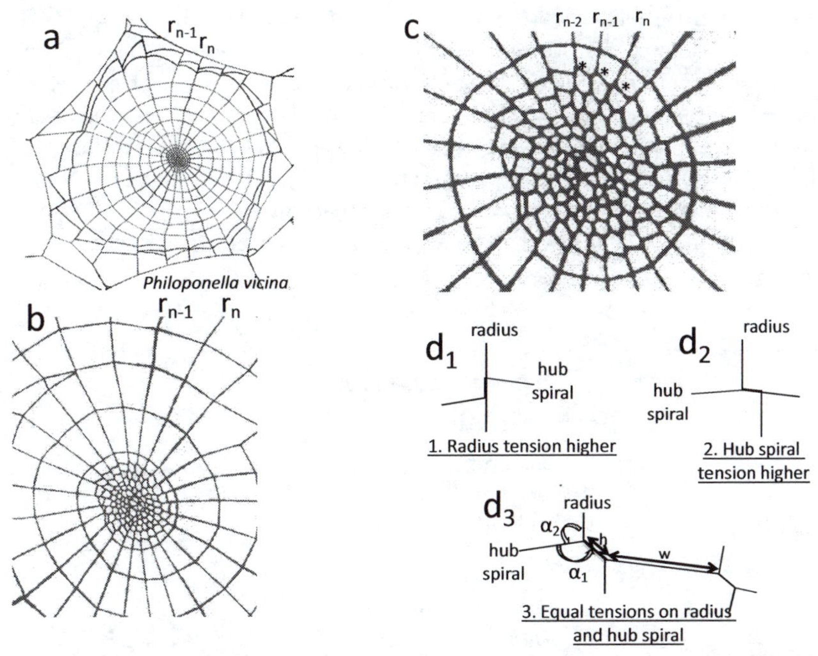

Fig. 4.9. Increasingly magnified views (*a–c*) of lines in the hub of an incomplete orb of a mature female *Philoponella vicina* (UL) in which 5 segments of the second loop of temporary spiral were broken at 11:00–2:00 during temporary construction, causing the next loop of temporary spiral to be displaced inward, toward the hub. Schematic representations of double attachments of the hub spiral to radii in *d* show how the relative tensions on radial and hub spiral lines in a *P. vicina* hub can be deduced (see also fig. 4.8). When the tension on the radius was relatively high compared with that on the hub spiral, the radius was not displaced by the double attachment (d_1). When the tension on the loop of hub spiral was relatively high compared with that on the radius, the radius was pulled into a zig-zag configuration (d_2). In some cases the zig-zag angle approached 120° (d_3), which indicates equal tensions on the radius and the hub spiral. The sizes of the displacements of the radii by the hub spiral lines varied in this *P. vicina* hub. For instance, the asterisks mark sites where the next to last loop displaced r_{n-2} less than r_{n-1} or r_n, indicating that r_{n-2} was under greater tension. Tensions on radii and hub spiral lines undoubtedly changed while the hub was being built, because construction of the outer hub loops reduced the tensions on lines in more inner portions of the hub.

eral, when lines are under lower stress in the resting orb (longer and under lower tensions), they will have greater abilities to absorb a prey's momentum and stop it. This is not to say that mechanical stability under wind stress has not been favored (see section 3.2.11.3.2) and that equalizing stresses on lines throughout the orb would tend to increase this stability. Instead it suggests that fine precision in balancing stresses throughout an orb web seems not to have been of overriding importance.

4.4.2 TENSIONS VARY DESPITE ABILITIES TO ADJUST THEM

4.4.2.1 Tensions on non-sticky lines in finished orbs

Figures 4.8 and 4.9 illustrate how web design traits that can alter the tensions on non-sticky lines in an orb, including double attachments of hub spiral to radii, greater numbers of sticky spiral loops (next section), differences in the numbers of radii on different sides of the hub, removal of the center of the hub, and break and reel behavior performed while laying a radius. Tension changes also occur at other moments during construction: lines are both tightened and loosened during exploration, and during primary radius and frame line construction; secondary radii are sometimes added on alternating sides of the web, thus balancing tensions. Some "false starts" during secondary radius construction clearly reduced the tension on a radius (appendix O6.3.2.3). In some symphytognathoids spiders systematically broke and loosened each radius by lengthening it after the web was otherwise complete (Eberhard 1987e). Still other spiders tensed their finished orbs while waiting at the hub (Table 4.3).

In at least one case, tension alterations apparently resulted in greater consistency in tensions. When *Araneus diadematus* (AR) was forced to anchor its webs to thin flexible rods that bent toward the web when web

Table 4.3. An incomplete list of species of spiders that do (A) and do not (B) actively tense their orb webs by reeling up lines or contracting their legs while waiting at the hub for prey to arrive (in some cases deduced from checking the angles between lines and the spider's tarsi in published photos or living spiders). "Tense" is used here not in the sense of leg position (e.g., crouching vs. extended, as in Scharff and Coddington 1997), but to indicate exertion of tension on web lines in addition to that exerted by the spider's own weight. In all species that I have observed tensing their webs (WE), the spider released the tension either at the moment a prey struck the web or, possibly in some cases, in the instant just before impact. These preliminary data suggest that tensing orbs may be confined to certain taxonomic groups. There are also species in other, non-orb families such as the theridiid *Phoroncidia* spp. in which tensing the web while awaiting prey evolved independently (Marples 1955; Eberhard 1981b; Eberhard et al. 2008b).

Spider	*Reference*
A) Tense the orb	
Araneidae	
Cyclosa octotuberculata	Nakata 2009
Cyrtophora citricola[1,2]	Blanke 1972, WE
Gasteracantha cancriformis[3,4]	Peters 1955, WE
Mangora sp (?)[1,2]	WE
Micrathena spp.[3]	Peters 1955, WE
Poltys illepidus	Shinkai and Takano 1984
Plebs (= *Zilla*) *astridae*	Shinkai and Takano 1984
Pozonia nigroventris	G. Barrantes pers. comm., fig. 3.35
Wagneriana spp.[3]	WE
Theridiosomatidae	
most genera[3,5]	Comstock 1967, Coddington 1986a, WE
B) Do not tense hub lines	
Uloboridae[6]	
Uloborus spp.[1,2,3]	WE
Philoponella spp.[1,2,3]	WE
Zosis geniculata[1,2,3]	WE
Araneidae	
Allocyclosa bifurca[1,3]	WE
Araneus ventricosus	Shinkai and Takano 1984
Argiope spp.[1,3]	WE
Cyclosa spp.[1,3]	Peters 1955, WE
Tetragnathidae	
Azilia affinis[7]	WE, fig. 4.16
Leucauge spp.[1,2,3]	Briceño and Eberhard 2011, WE
Metabus ocellatus	WE, fig. 2.19
Tetragnatha spp.[1,2,3]	WE
Nephilidae	
Nephila clavipes[3]	WE

[1] Observations of the distortion (or lack of distortion) of the hub lines that are grasped by the spider's tarsi (WE); Blanke (1972) also showed the web being tensed in "Hitzstellung (heat posture) I."

[2] Web was more or less horizontal.

[3] Direct observations of behavior before prey arrived and the spider's response.

[4] Another possibility might be that spider flexed her legs to camouflage her outline instead of to tense the web; but flexing could have been accomplished without tensing the web.

[5] Spiders spring their webs in response to sudden humming noises, so presumably at least sometimes they spring the web just prior to rather than following prey impact.

[6] Uloborids that have reduced webs, including *Miagrammopes* spp. (e.g., Akerman 1932; Lubin et al. 1978) and *Hyptiotes* spp. (McCook 1889; Marples and Marples 1937; Comstock 1967) but not *Polenecia* (Peters 1995a), tense their webs.

[7] When resting at the hub during the day, the spider's legs adopted constrained positions (probably to camouflage her outline) and legs III were flexed and pulled the distant inner edges of the hub hole closer to her body. At night, however, web lines were not tensed while the spider waited at the hub (fig. 4.16), so flexing the legs III may have functioned to improve camouflage.

lines pulled on them, the tensions on the web lines remained in their normal range in finished webs, indicating that the spiders had actively corrected tensions when the rods were bent by the spider's lines (Wirth and Barth 1992).

Two types of behavior also support the idea that lower radius tensions are adaptive (presumably to promote stopping and retention): the lowering of the tension on each new radius just before it is attached at the hub (section 6.3.3); and the destruction of the center of the hub at the end of sticky spiral construction by most orb weavers (in some groups the entire hub is removed) that reduces and equilibrates tensions among radii (section 6.3.6). On the other hand, the fact that the tensions were lowered at the end of orb construction, and that the temporary spiral is under relatively high tension (Wirth and Barth 1992) support the idea that there is a trade-off (table 3.1) with the advantages of greater tensions during construction, which reduce the degree to which the web is deformed by the spider's weight during construction.

In sum, orb weavers have ample opportunities to adust tensions on lines in their webs. As noted in the previous section, however, both deductions from the angles between lines and direct measurements of the tensions on non-sticky lines and their diameters show that these multiple adjustments do not result in uniform tensions. Thus the differences in tensions noted in the previous section, which showed that the tensions and stresses on homologous lines in real orbs are not even approximately equal, cannot easily be attributed to inability of the spider to adjust them. Perhaps it is sufficient for tensions to be only approximate equal?

4.4.2.2 Different tensions along the length of a single radius

4.4.2.2.1 WEB MODIFICATIONS THAT PRODUCE TENSION CHANGES

The tension along the length of a radius will vary as a result of the forces exerted by the sticky spiral lines that are attached to it (fig. 4.8b). The net force typically applied by each spiral loop is inward (toward the hub), so the overall effect is to lower the tension and the stress on the inner portions of the radius, and to raise the tension and stress near its outer end (Eberhard 1972a; Aoyanagi and Okumura 2010). This difference will increase the ability of a radius to stop high-energy prey in the inner portion compared with the outer portion of the orb (section 4.3).

A similar "looser near the center" pattern results from the double rather than single attachments of hub spiral and temporary spiral lines to radial lines. As with sticky lines, the net effects of these spirals is to pull inward on the radius; but these spiral lines are less extensible than the sticky spiral, and thus probably have stronger effects on radial tensions. In addition, when the attachments are double, the tension on the spiral line often pulls the radius into a zig-zag configuration, effectively shortening it and thus increasing the tension on it (fig. 4.8a). The same segment of hub spiral sometimes pulled adjacent radii out of line to different degrees (see for example, the junctions marked with "*" in fig. 4.9c). This pattern demonstrates that tensions on adjacent radii were not aways equal when the hub line was laid, and also that the zig-zag attachments have the effect of reducing tension differences between radii. Because both hub and temporary spirals are built from the hub outward, they gradually raise the tensions on the outer portion of radii compared with those on the inner portions as the web is being built.

Zig-zag patterns in the hubs of uloborids occur in *Uloborus* spp., *Philoponella* spp., *Octonoba sybotides*, and *Polenecia producta* (e.g., figs. 3.11, 4.9, 6.11) (Wiehle 1931; Eberhard 1972a; Opell 1979; Lubin 1986; Peters 1995a; Watanabe 2000b), *Cyclosa insulana* (AR) (Wiehle 1928), *Mangora* sp. (AR) (fig. 3.13; WE), and *Tetragnatha* sp. #2047 (TET) (fig. 3.11). Zig-zag patterns also occur in the temporary spiral-radius attachments in the webs of several nephilids (Hesselberg and Vollrath 2012) (figs. 4.4, O6.8), and *Cyrtophora* (AR) and allied genera (fig. 1.7; Kullmann 1958, 1975; WE). It is striking that spiders achieved relatively uniform temporary spiral tensions in vertical orbs despite the substantial displacements of web lines during temporary spiral construction that were caused by the spider's own weight.

One other possible pattern, documented in only a preliminary way as yet, is that the frequencies of double hub-radius attachments may tend to be greater in the central portion of the hub and absent or nearly absent at the outer edge of the hub (see figs. 3.11—3.15) (WE). Early hub loops increased the tensions on the radii more than later hub loops did.

4.4.2.2.2 FUNCTIONS OF ALTERED TENSIONS

Changes in tensions are likely to have both positive and negative consequences that trade off against each other (table 3.1). Double attachments of the hub spiral to radii represent costs in terms of extra silk and special behavior; presumably they confer selective advantages. One possibility is that they represent "fail safe" devices to increase the strengths of attachments between lines, but the lack of damage from prey impacts or wind stress to hub attachments (section 3.3.2.9) argues against this hypothesis. In addition, zig-zag double attachments have the inevitable effect of tightening lines, which I will explore here. Adjusting line tensions using zig-zag attachments has the virtue that the magnitude of the adjustments can be flexible. For instance, if the spider maintains a constant tension on different segments of the temporary spiral, she could cinch up slack radii more than adjacent, tighter radii (fig. 4.9c).

It may be that increased tensions result in quicker attack times, at least for small prey. Female *O. sybotides* on orbs that had been artificially tightened (by expanding the frames in which they were built by 1.5–2.0 mm) "responded" more quickly to small prey (0.7–1.2 mg *Drosophila*) (time between spider turning at the hub and beginning to move away from the hub toward the prey; this was only part of the time needed to attack) than did spiders on control webs (Watanabe 2000b) (the prey's vibrations were not typical of captures in nature, however [section 3.3.2.2.3], because the flies were chilled beforehand, and only gradually began to struggle after being placed in the web). In contrast, response times for larger prey (2.1–2.9 mg *Drosophila*) did not change when the web was tightened.

The sizes of the zig-zag deformations of radii by hub spiral lines in *O. sybotides* (UL) hubs were greater in webs to which the spiders subsequently added spiral stabilimenta (the zig-zag deviations reduced by 2.9% of the straight line distance from the edge of the hub to its center) as compared with webs to which they added linear stabilimenta (where the corresponding value was 1.5%) (Watanabe 2000b). Assuming that tensions on hub line were equal (I know of no data), and that the orbs were of equal sizes (orbs with spiral stabilimenta were larger), this suggested that radii in orbs with spiral stabilimenta were under more tension. More poorly fed spiders tended to build spiral stabilimenta and better-fed individuals built linear stabilimenta (Watanabe 1999a,b), and the spacing between loops of sticky spiral was smaller in orbs with spiral stabilimenta (Watanabe 1999a). Watanabe argued that these differences suggested that more poorly fed spiders (for which capturing smaller prey might be more crucial) built more tense orbs that were better designed to capture smaller prey (Watanabe 2000b). Due to the unnatural and incomplete measurements of response times, and untested assumptions regarding hub spiral tensions, I cannot evaluate these arguments.

To date, only species that are known to make zig-zag temporary spiral attachments to radii also leave the temporary spiral intact in the finished web. This suggests that the zig-zag attachments do not function during sticky spiral construction, but rather after the web is finished. The function of these zig-zag attachments could be that by progressively raising the tension on the radius with the temporary spiral attachments, the spider avoids having the inner portions of the radius eventually go slack as the result of the many attachments of temporary spiral lines to it. Additionally, in none of the species in which zig-zag hub or temporary spiral attachments are especially common (*Nephila* spp., uloborids, and *Cyrtophora* and its allies, a few other taxonomically scattered groups) does the spider remove the center of the hub. This association was especially dramatic in *Tetragnatha* sp. #2047 (in which a twig runs through the hub—fig. 3.11j); the unusually strong zig-zag patterns in the hub were combined with a lack of hub removal (which is otherwise the rule in this genus). The functional significance of this association is unclear.

Because of the large numbers of double attachments of temporary spiral loops in the webs of *Cyrtophora* spp. (AR) (fig. 1.7), their radii must have especially high tensions at their outer ends (there were approximately 100 temporary spiral loops in one *C. citricola* web—appendix O6.3.5). Similarly, tension increases must occur, to a lesser degree, in *Nephila* and *Nephilengis* (NE) due to their intact temporary spiral lines (Hesselberg and Vollrath 2012); there can be up to approximately 20 temporary spiral loops attached to each radius in the lower portion of the web (fig. 4.4, Hesselberg and Vollrath 2012). The substantial changes in radial tensions in *Cyrtophora* probably do not have consequences for stopping prey, because the orb is horizontal, and has a tangle above that presumably absorbs most of the momentum of prey.

The larger amplitude zig-zags in the outer compared with the inner temporary spiral lines in *N. edulis* indicated that the tensions on the outer loops of temporary spiral were higher, by a factor of about 3, than on the inner temporary spiral loops (Hesselberg and Vollrath 2012). Again, the function of this difference is not clear.

Craig (1987b) found that higher tensions were associated with orbs designed to stop high-energy prey. This counter-intuitive conclusion (greater radius tension implies greater stress, and less ability to absorb the prey's momentum) was based, however, on a sample of only five species. Adult body size, which is also positively related to web tension (Wirth and Barth 1992), was greater in the species with greater tensions, so it is possible that the higher tensions were simply incidental consequences of the larger bodies of some species.

In sum, it is clear that the tensions on different types of non-sticky lines in an orb, and even those on different radii in the same web, vary substantially. Uncertainties regarding possible differences in diameters (Work 1977), the numbers of fibers in each line, and the physical properties of their silk (section 3.2.9) make conclusions regarding stresses less certain; but the general impression is they also vary substantially. The consequences of the inequalities in tensions and stresses on radii, both comparing different radii and along the length of a given radius, are apparently not large enough to be of general functional significance in stopping high-energy prey. When strong, localized loads were applied to orbs by simulated prey impacts, the damage was nearly always only to radial lines very near the loading site; the radii did not break far from the impact site, nor did hubs, frames, or anchor lines (Cranford et al. 2012; WE, section 3.3.2.9).

4.4.2.3 Some spiders manipulate tensions in finished orbs

Many spiders tense their finished orbs by flexing their legs while they wait at the hub. Judging by photos in which the spider's tarsi visibly pull on and displace radii or hub lines (figs. 3.35, 3.37), it appears that tensing lines at the hub is taxonomically widespread, occurring in the araneid genera *Araneus* (Nielsen 1932 vol. 2), *Cyclosa* (Nakata 2009), *Eriophora* (Robinson et al. 1971), *Micrathena, Pozonia*, and *Wagneriana* (WE), the family Theridiosomatidae (Coddington 1986a), and in genera with derived, reduced webs such as *Hyptiotes* and *Miagrammopes* (Lubin 1986) (but not in the orb weavers *Waitkera, Octonoba, Philoponella, Tangaroa, Uloborus*, and *Zosis*—Opell and Eberhard 1984; Opell 1992; WE) (see an exception below). In additional cases, the spider tensed the tightly meshed hub in a horizontal orb so that its central part (which may be under especially low tension, due to the large numbers of hub or temporary spiral loops—see preceding section) was pushed into a small cone around the tip of the spider's abdomen in *U.* sp. (#2072) (UL), and the araneids *Mangora simplicatus* (fig. 3.36) and *Cyclosa turbinata* (WE). I know of no cases, however, of similar active web tensing by tetragnathids or nephilids. The linyphiids *Diplothyron simplicatus, Neriene coosa*, and *Frontina coccinea* also routinely tensed their sheets by flexing their legs while waiting on their nearly horizontal sheets (WE).

The contexts in which spiders actively manipulate tensions on their webs offer evidence that supports the importance of some of the trade-offs listed in table 3.1. Nakata (2009) argued that when *Cyclosa octotuberculata* (AR) tensed radii at the hub by flexing her legs, she was able to respond more rapidly to prey that contact the tightened radii. In tensing some radii, she must have loosened others, and it is not clear why, before a prey had struck the web, a spider would be interested in paying more attention to some sectors of her orb than others. If higher radius tension improves the spider's response time, why does she not simply build all radii with higher tensions? Tightening radii while they were being built would seem much more efficient energetically compared with flexing her legs for long periods while waiting for prey.

A more likely function for tensing a web built under lower tensions would be to obtain the double benefit of increased mechanical stability (and perhaps vibration transmission) at higher tension while waiting for prey, and an increased ability to stop and retain prey by releasing the tension at the moment of prey impact (Eberhard 1981a). Web tensing followed by a quick sag at the moment of impact occurs in several genera (*Cyclosa, Micrathena, Gasteracantha, Wagneriana*) (Peters 1955; Nakata 2009; WE). The sag may also cause the web to move with respect to the prey and thus increase the number of sticky lines that contact the prey, as observed directly in the multiple quick sags by *Hyptiotes paradoxus* (UL) (Marples and Marples 1937). None of these explanations is satisfactory, however, for the sudden sagging behavior by some theridiosomatids such as *Ogulnius* sp., which combined long-term tensing with quick sags of a non-sticky line that had little or no connection with the rest of the web (WE).

A final source of tension changes is the phenomenon of "supercontraction," the reponse of dragline silk to humidity changes. When a dragline of *Nephila clavipes* was subjected to a rapid increase from about 10% to 80% relative humidity, a rapid increase in stress was triggered at about 70% (dashed lines in fig. 4.10a,b) due to supercontraction. Unrestrained draglines supercontracted up to 50% of their original length, and restrained lines (as

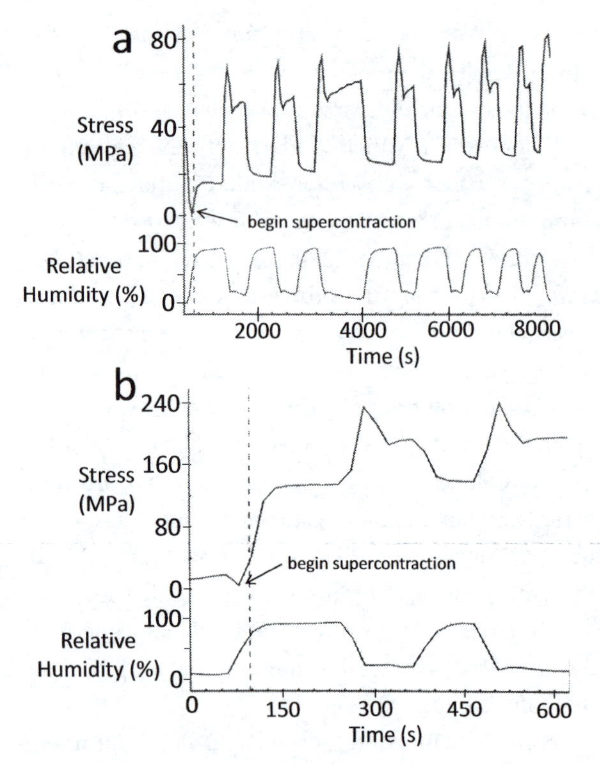

Fig. 4.10. The effects of humidity on the mechanical properties of the draglines of the nephilid *Nephila clavipes* (NE) complicate understanding of the tension relations in their webs. When a loaded line was subjected to slow (*a*) and rapid (*b*) oscillations of relative humidity between high (80%) and low (20%) values, the stress increased under high humidity and decreased under low humidity. During the first increase from low to higher humidity (dotted lines in both graphs), there was also a sudden increase of stress due to "supercontraction" when the humidity reached about 70% (dotted lines in *a* and *b*). Supercontraction was stronger when the change in humidity was more rapid (*b*) (after Blackledge et al. 2009b). The significance of these changes for how webs function in nature (where at least slow changes in relative humidity surely occur) is not clear.

in a web) generated substantial stresses during supercontraction; these responses occurred at levels of humidity that commonly occur in nature (Work 1981; Work and Morosoff 1982). Supercontraction results when water infiltrates the silk, disrupting the hydrogen bonding within the amorphous region of the proteins, increasing molecular mobility so that entropy can drive the proteins into less organized configurations. Supercontraction was greater when the increase in humidity was more rapid (over a period of a few seconds; compare fig. 4.10a with b) (Agnarsson et al. 2009b). Such rapid changes are probably rare in nature. Repeated oscillations of humidity (again, an unnatural condition) resulted in gradual increases of supercontraction at higher humidities, and decreases at lower humidities (fig. 4.10) (Agnarsson et al. 2009b).

The consequences of these effects of humidity on the functional properties of orb webs in nature are not clear. The different parts of a web must experience similar humidities, so the tensions in web as a whole would be expected to increase or decrease simultaneously in its different component lines as humidity changes. The draglines of *N. clavipes* in nature (e.g., in lowland tropical forests) must often experience >70% humidity the moment that they emerge from the spider's body in the early morning (i.e., they may not start from low humidities, as in the experiments in fig. 4.10), and then gradually lower humidities later in the morning, except when it rains; the consequences (if any) of such a history for supercontraction are not clear. The draglines of other species undoubtedly experience different humidity regimes; for instance, webs built in the leaf litter of the same forests (for instance, many anapids, some linyphiids and pholcids) probably never experience anything other than very high humidities. Perhaps supercontraction changes are only incidental consequences of the molecular structure of silk lines rather than adaptations; the possibility of inter-specific and intra-specific variations in responses that confer adaptive advantages remains to be investigated.

4.4.2.4 Tensions on sticky lines

The highly extensible sticky spiral lines of araneoids are generally thought to function in retaining prey, and to play little role in transmitting vibrations or in absorbing the momentum of prey (section 3.3.2.1). All other things being equal, lower tension on a sticky line would seem to improve retention of a struggling prey; it would be less likely to break when the prey pushed and kicked.

Araneoid spiders building sticky lines are faced, however, with the problem that a certain amount of tension is apparently needed to pull the sticky silk from the spider's spinnerets; the sticky spiral line is under appreciable tension when it emerges from the spinnerets (fig. 2.4b). Araneoids commonly lowered the tension by pulling out extra sticky line by pulling on it repeatedly with one or both legs IV before attaching it to a radius (figs. O6.2–O6.4) (Eberhard 1982); pulling behavior has been recorded in 21 genera in four families (Peters 1954;

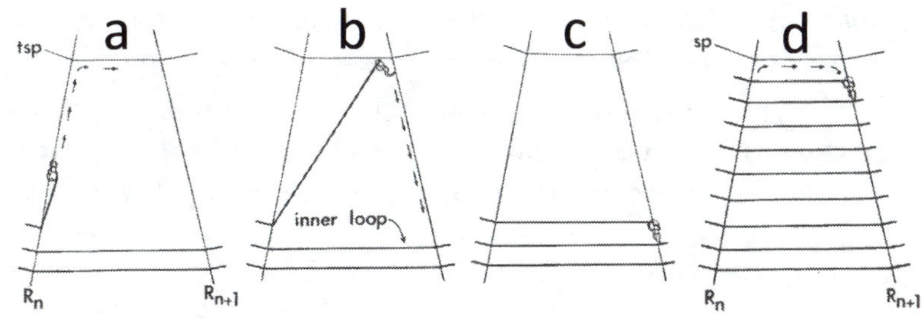

Fig. 4.11. This diagrammatic representation of sticky spiral construction behavior by *Naatlo splendida* (TSM) illustrates an exception that supports the tension adjustment hypothesis to explain sticky spiral pulling behavior with legs IV. It also testifies to natural selection acting on even small details of orb construction. In contrast to most araneoid orb weavers, *N. splendida* makes more pulling movements on the emerging segment of sticky line when she is laying sticky spiral lines near the hub (*d*) than when she is far from the hub (*a–c*). These spiders also differ from most other araneoids in that the temporary spiral is very distant from the outer loops of sticky spiral; during construction of each segment of the sticky spiral far from the hub, the spider thus produces a line that is substantially longer than the distance between the points where it is attached to radii (R_n and R_{n+1}) while she is moving along the temporary spiral (*b*). Thus extra pulls with legs IV just before attaching are less necessary to lengthen the line and lower its tension when she is near the outer edge of her web (for other apparently adaptive adjustments in behavioral details, see fig. 6.26) (after Eberhard 1981a).

Eberhard 1981a). In *Gasteracantha theisi* (AR) the spider pulled up to 16 times before each attachment (Robinson and Robinson 1975). Because the axial line is so elastic, it generally does not become slack even after these extra pulls.

The hypothesis that the pulls with legs IV function to reduce the tensions on sticky spiral lines in the web is supported by the tendency seen in seven different araneid species to pull more times on the new sticky spiral line when the spider is ascending than when she is descending during sticky spiral construction in a more or less vertical orb (Eberhard 1981a). The sag that the spider's own weight produces in the more horizontal radii pulls her body away from the previous attachment of the sticky spiral when she is descending (this pulls additional sticky line from her spinnerets and makes fewer additional pulls necessary); when she is ascending, the same sag keeps her body closer to the latest attachment, making more pulls necessary.

A second adjustment that fits the tension adjustment hypothesis is the reduction in the number of pulls/attachment when the spider is nearer the hub; this difference has been observed in 12 araneid and tetragnathid species (Robinson and Robinson 1975; Eberhard 1981a). Sticky lines nearer the hub are shorter, and need less elongation to achieve a given tension. Changes in the pulling behavior at the moment of attachment (Eberhard 1981a) also support the tension adjustment hypothesis.

A few exceptions to these patterns occur, but they also fit the tension adjustment hypothesis because they involve species (the theridiosomatid *Naatlo splendida* and the anapid *Anapis* sp.) in which the outer loops of sticky spiral are very far from the bridge that is used by the spider to go from one radius to the next (the temporary spiral in theridiosomatids, the hub in anapids) (fig. 4.11). These spiders pulled more rather than less with their legs IV when laying sticky lines near the hub. Near the outer edge of the orb the length of the sticky line drawn as the spider walked to the bridge was substantially longer than the final distance between the attachments to radii (fig. 4.11), so less extra line was needed to achieve a reduced tension (Eberhard 1981a).

Four genera of uloborids, whose sticky spiral silk is less extensible (Opell and Bond 2000), also lowered the tension just before attaching sticky lines (Eberhard 1981a). After arriving at an attachment site the spider continued to comb out further cribellum silk for one or more seconds before she attached it to the radius in *Uloborus diversus* and *Zosis geniculata*; the elasticity of the line took up the slack (Eberhard 1972a; WE). The non-orb cribellate eresid *Stegodyphus* sp. and the filistatid *Kukulcania hibernalis* also reduced tensions on sticky lines by continuing to comb out cribellum silk at the attachment site before attaching it (Eberhard 1987a; WE)

The deinopid *Deinopis* sp. is extreme in that both the sticky and the non-sticky lines in its small web are unusually elastic (Coddington and Sobrevila 1987). This allows the spider to hold the web in a collapsed state without the lines becoming entangled with each other, and then to greatly expand her web very rapidly to capture prey.

During radius construction this spider performs the otherwise unique behavior of pulling multiple times on her non-sticky dragline before attaching it (Coddington 1986a); these pulls probably have the same function as pulling on sticky silk in araneoid orb weavers.

Turning this discussion upside down, one could ask why it is, if lower tensions on sticky lines make them better able to retain prey, that orb weavers generally stop pulling additional sticky line before it becomes slack (and why do a few species, in contrast, cause it to go slack)? One likely factor is the advantage of keeping the sticky spiral under some tension so that adjacent sticky lines do not swing into contact with each other in the wind (some exceptions with lax sticky lines are built in the leaf litter, where wind velocities are low) (fig. 7.9) (section 3.2.11.3). A second contributing factor involves the costs of materials. Presumably natural selection favors a balance that produces a maximum payoff from prey retention corrected for the costs of the material invested. This cost-benefit trade-off is likely to vary with different prey and in different environments, but to the best of my knowledge, this topic has never been explored quantitatively.

4.4.2.5 Experimental manipulation of tensions

When tensions on the orbs of *Octonoba sybotides* (UL) were experimentally increased, spiders responded more quickly to small flies, but with equal delays to large flies (Watanabe 2000b). This result seems to fit the expectation of improved vibration transmission by tighter line, as smaller prey would produce lower amplitude vibrations. But the results were more intriguing. The response time that was measured was the delay (on the order of 2–6 s) between the moment the spider turned to face the prey, and the moment when she began to move toward it. The spider's turn demonstrated that she had already detected the vibrations produced by the prey, so the response time that was measured seems likely to be related to some sort of subsequent decision, rather than vibration transmission per se. The significance of the delay before attacking is suggested by the fact uloborids often jerk and bounce the web after they turn at the hub, prior to launching an attack (Lubin 1986; WE). During this time the spider may be attempting to judge the prey's size and/or identity (perhaps indicated by the amplitude or the frequency of its resonant vibrations in the web—see Landolfa and Barth 1996, section 3.3.2.2), and/or to further ensnare it (Robinson and Olazarri 1971; Suter 1978; Lubin 1986). Explaining why the delay was shorter for smaller prey when the web was tightened will require further work.

4.5 RELATIVE NUMBERS OF RADII AND STICKY SPIRAL LOOPS

There is a strikingly widespread trend throughout orb-weaving spiders, apparently first noted by the taxonomist M. Emerit (1968) in *Gasteracantha versicolor* (AR): orbs that have larger numbers of radii tend to also have larger numbers of loops of sticky spiral. This relation occurred in intra-individual, intra-specific, inter-generic, and inter-familial comparisons (fig. 4.12, Eberhard 1972a; Craig 1987b; Opell 1997). The relation also occurred in intra-specific ontogenetic changes in *Leucauge argyra*, where both the number of radii and the number of sticky spiral loops decreased with spider size (in *L. mariana* neither decreased) (Barrantes et al. 2017). The correlation is not geometrically necessary (see, for example, fig. 3.5). One possible explanation stems from the fact that the number of radii also consistently correlates positively intra-specifically with the area of the web (e.g., Peters 1936, 1937; Eberhard 1972a; Craig 1987b; Barrantes and Eberhard 2012; Eberhard and Barrantes 2015). Greater numbers of radii could function to avoid sticky spiral entanglement, holding the longer segments of sticky lines apart in the outer portions of larger orbs.

Whatever the explanation may be for the correlation, a graph of the numbers of radii vs. sticky spiral loops (fig. 4.12) invites comparisons of the functional properties of different orb designs. Radius-rich designs (e.g., below the dotted line in fig. 4.12) are likely to be relatively good at stopping high-energy prey, while radius-poor designs are likely to be relatively good at retaining prey (Craig 1987b; Eberhard 1986a) (Craig used equal numbers sticky spiral loops and radii as a reference criterion). For instance, the uloborids *Uloborus diversus* and *Zosis geniculata*, and *Cyclosa* spp. (AR), build radius-rich webs compared with other orb weavers as a whole (fig. 4.12); the araneids *Acacesia hamata*, *Pozonia nigroventris*, *Poltys* spp., and *Deliochus* sp. build sticky spiral-rich designs.

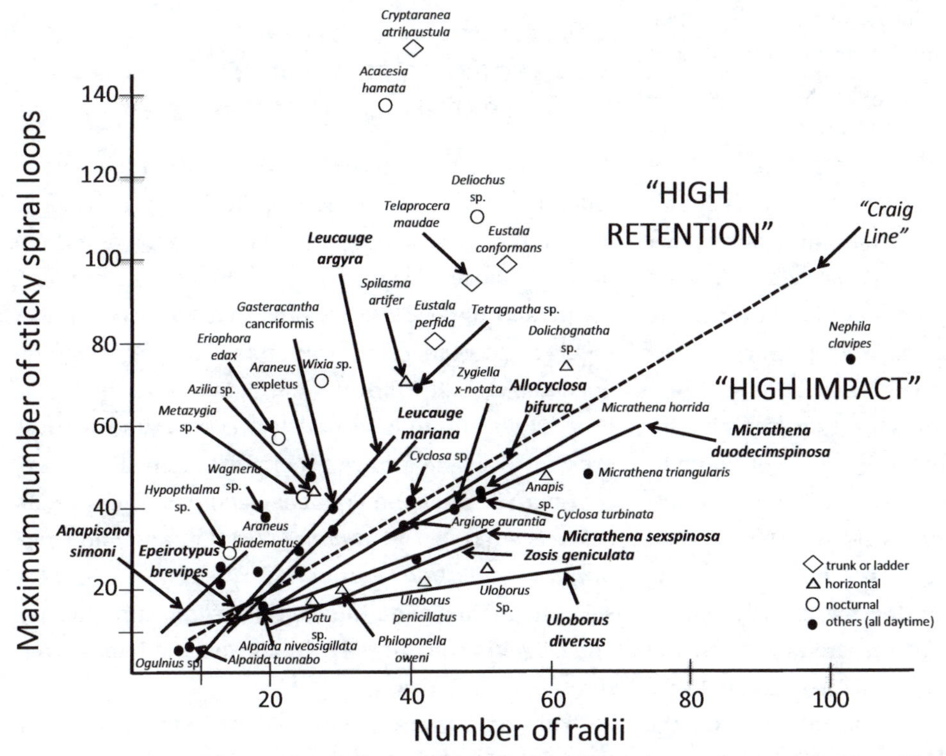

Fig. 4.12. This sampler of the relationships between the number of radii (counted at the outer edge of the orb) and the maximum number of sticky spiral loops in the orbs of mature females of species in five different families illustrates a strong, positive correlation throughout orb weavers, although the slopes and intercepts differ substantially between groups. Some taxa build relatively radius-rich "high impact" orbs, while others build sticky spiral-rich "high retention" orbs. The species were chosen to represent a variety of orb orientations, designs, spider sizes, websites, and taxonomic groups, so the wide dispersal of the points is not representative of orb weavers in general (most would fall in the central area). Individual webs are indicated by points (those with unusual orientations and websites, and those at the edges of the distribution are labeled with species names). Intra-specific correlations are indicated by solid lines and names in bold type; the lengths of these lines were determined by the maximum and minimum values observed (linearity was not tested statistically). The sticky spiral-richness of highly asymmetrical (trunk and ladder) webs is overestimated by the use of the maximum number of sticky loops, because other portions of the webs had relatively fewer loops. The dashed line marks equal numbers of radii and loops, the criterion that Craig (1987a) used to distinguish between orbs that she termed "high energy" and "low energy" ("high impact" and "high retention" here). There is a possible trend for nocturnal orbs to be relatively sticky spiral-rich; in addition, some taxonomic groups, such as uloborids, *Micrathena* spp., and *Leucauge* spp. appear to cluster on the graph. Further analyses are needed, however. Individual points are from photos in Witt et al. (1968), Coddington (1986a,b), Eberhard (1976a, 1986a), Kuntner et al. 2008a, and WE (for *Azilia* sp.). The sources of data and sample sizes for the lines are the following: *U. diversus* (UL) (N = 36) (Eberhard 1972a); *A. bifurca* (AR) (N = 27) (Eberhard 2007b, 2011); *L. argyra* (TET) (N = 38) (Barrantes and Eberhard 2012); *A. simoni* (AN) (N = 115) (Eberhard 2011); *E. brevipes* (TSM) (N = 28) (an undescribed congeneric species had a very similar line—Eberhard 1986b); *L. mariana* (TET) (N = 139) (WE); *M. duodecimspinosa* (AR) (N = 55) (WE); *M. sexspinosa* (AR) (N = 229) (at least 2 webs/spider) (WE); and *Z. geniculata* (UL) (N = 39) (WE).

Correlations of other variables with the radius-rich vs. radius-poor contrast are not all as might be expected, however. Craig (1987b) thought that larger adult body size correlated with building radius-rich orbs, but she based this conclusion on a sample of only five species, and there are many species that do not fit this idea. For instance, the orbs of the small araneids *Cyclosa* spp. and its relative *Allocyclosa bifurca* are neither radius-rich nor sticky spiral-rich; the large araneid *Eriophora edax* has a radius-poor orb; and the medium-sized araneids *Acacesia hamata* (which is only slightly smaller than the largest species in Craig's study) and the relatively large *Pozonia* and *Poltys* all have extremely sticky spiral-rich orbs (figs. 3.5, 3.35, 4.12). Another variable that is also apparently not correlated with radius- or sticky spiral-rich designs is the web's slant, despite the expectation that prey generally fly more rapidly horizontally than vertically (Opell 1997; Bond and Opell 1998) and that vertical orbs should thus be more radius-rich to stop prey (I know of no data on horizontal vs. vertical speeds of prey, however). In contrast, at least in the sample of species in fig. 4.12, it may be that nocturnal orbs tend to be more sticky spiral-rich. I have not attempted to perform an analysis that corrects for phylogenetic inertia in these data, because the substantial intraspecific variation implies that the designs are not strongly constrained by phylogeny. Further analyses will be needed to test this.

Another possibility is that orbs that span larger spaces are designed to stop prey that are flying more rapidly. But simple inspection again shows that there are multiple exceptions to the predictions: *Acacesia hamata* and *Deliochus* sp. orbs are more or less vertical and have long anchors (e.g., fig. 3.5), but both are extremely radius-poor. The orbs of *Uloborus diversus* are horizontal and have short anchors (they are built in relatively sheltered sites, such as among the debris of packrat nests or in the bases of thickly branched, bushy plants), but they are clearly radius-rich. The especially long-anchored orbs of the araneids *Gasteracantha cancriformis* (Peters 1955; WE) and *Hypognatha* sp. (WE) are relatively ordinary, and slightly radius-poor rather than being radius-rich. The araneid *Caerostris darwini* builds its nearly vertical orbs in especially open sites, with upper frame lines that sometimes span small rivers (Gregorič et al. 2011), but its orb is unremarkable with respect to the relative numbers of radii and sticky spiral loops. At the opposite extreme, the orbs of the tiny symphytognathid *Patu* sp. are horizontal and are built very close to the ground in small confined spaces in the leaf litter where the velocities of flying prey might be expected to be low, but nevertheless have radius-rich designs (figs. 4.5, 4.12).

In sum, several simple predictions based on the relative numbers of radii and sticky spiral loops do not fit currently available data well. Perhaps differences in the effects of phylogeny (which I have ignored in this discussion), in prey (moths, for instance, which are especially difficult to restrain), or in the biomechanical properties of the silk of different species, such as those documented by Sensenig et al. (2011) and Amarpuri et al. (2015), are at least partially responsible. Further analyses of deviations from Craig's 1:1 reference ratio could be revealing.

4.6 TESTING VISIBILITY AND STOPPING FUNCTIONS: THE EXTREME CASE OF TRUNK ORBS

The tangle of hypotheses concerning the functions of different orb web designs is simplified in species whose natural histories render particular functions of little or no importance. For example, as discussed in section 3.3.3.2 and appendix O3.1, orbs that are used only at night can serve to test the importance of selection to reduce web visibility. The orb webs built against the trunks of trees constitute a second test case. Data on trunk orbs are available for five genera: the araneids *Telaprocera* spp. (Harmer and Framenau 2008), *Eustala perfida* (Messas 2014), *E. conformans* (WE) (fig. 4.3), and *Cryptaranea atrihastula* (Forster and Forster 1985); and the sister nephilid genera (fig. 4.13) *Herennia* spp. (fig. 4.3) and *Clitaetra* spp. (fig. 9.17) (Robinson and Lubin 1979a; Kuntner et al. 2008a,b).

Trunk orbs have apparently evolved independently at least four times. In the absence of a phylogeny for the large New World genus *Eustala*, I will assume conservatively that *E. perfida* and *E. conformans* result from a single derivation of trunk webs. *Eustala* is apparently not closely related to *Araneus* (fig. 1.12), which is evidently close to *Cryptaranea* (this spider was called *Araneus atrihastula* in the original web description). The taxonomic affinities of *Telaprocera* in Araneidae are not known (Harmer and Framenau 2008), but it likely represents another independent derivation of trunk orbs, because numerous species in *Eustala* (fig. 3.10) and *Araneus* build typical, non-trunk orbs. Nephilid phylogeny, in contrast, has been studied intensively (Kuntner et al. 2008a), and the sister genera *Herennia* and *Clitaetra* represent a single derivation of trunk orbs.

The webs of all these groups are only a cm or so from the trunk of a tree, and are (at least for humans) very difficult to see against the irregular background of the bark (Robinson and Lubin 1979a; Harmer 2009; Messas et al. 2014; WE). In addition the webs of *E.* spp., *T.* spp., and *C. atrihastula* are built at night; *T.* spp. did not even respond to prey entangled in their webs during the day (*Herennia* webs, judging by their relative lack of damage in the morning, are probably built at night or in the early morning—WE); *C. atrihastula* orbs remained in

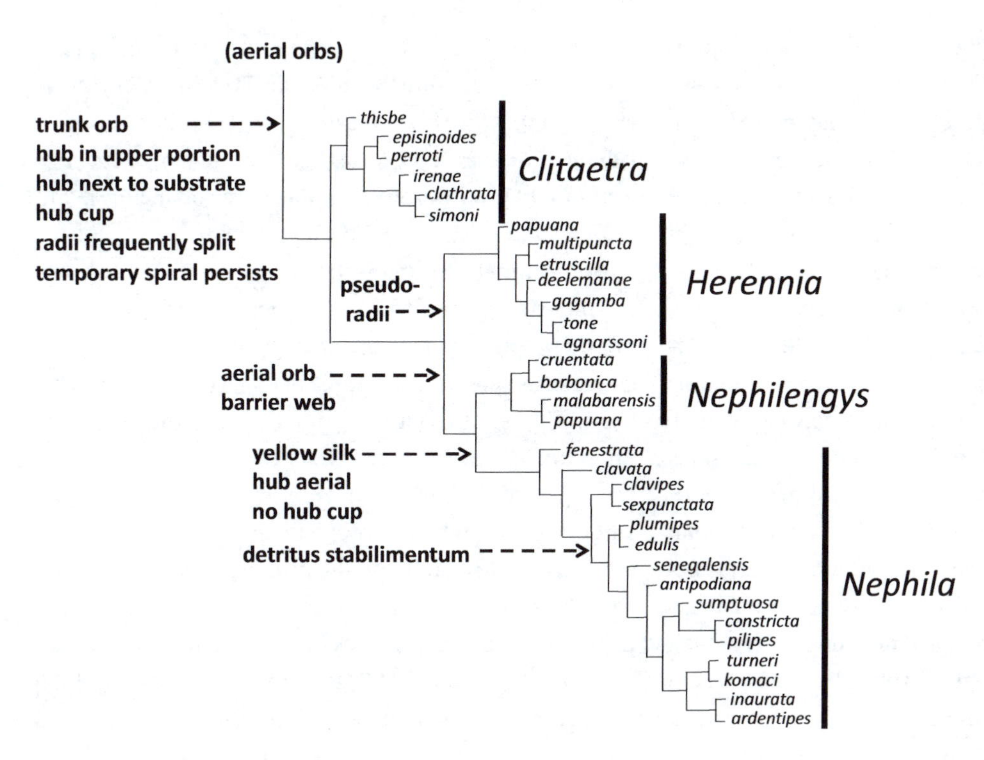

Fig. 4.13. The distribution of web-related traits on the preferred phylogeny of Nephilidae of Kuntner et al. (2008a) illustrates the inconsistent nature of web evolution, and how apparently highly derived traits may turn out to be ancestral in some groups. The two more basal genera, *Clitaetra* and *Herennia*, build tree trunk orbs; the derived pseudo-radii of *Herennia* aid in curving the orb to keep it close to the cylindrical surface of the trunk. The hubs of *Nephilingis* orbs are near a large object such as a trunk, but the rest of the web is in open space rather than against trunks. The orbs of the other more derived genus, *Nephila*, are built in open spaces, a reversion to the designs in many other non-nephilid groups.

place during the day, with the spider resting nearby, connected to the hub by a signal line (Forster and Forster 1985). In sum, web visibility probably has little effect on prey interception in the trunk webs.

The stopping function is also probably of relatively little importance in the trunk orbs. Prey are likely to be either moving slowly in preparation to landing on the trunk, or just taking off from the trunk, without having had time to acquire high flight velocities. Hopping insects jumping away from the trunk might be an exception but, at least in *Telaprocera*, such prey were rarely captured: the most common taxa in a sample of 169 prey were Diptera, Hymenoptera, and Coleoptera, without a single member of a typically hopping group like Orthoptera or Homoptera (Harmer and Herberstein 2010) (Coleoptera and Hymenoptera probably contributed the greatest biomass). In sum, trunk webs are probably less often called upon to absorb high-energy prey impacts.

These special traits of trunk orbs lead to the prediction that they will be sticky spiral-rich. If selection is less intense to reduce visibility, then the numbers of sticky spiral loops could be larger and the spaces between them smaller. Relatively large numbers of sticky-spiral loops compared with numbers of radii are also predicted by the reduced selection to absorb prey momentum. The designs of trunk orbs differ in the predicted ways from those of the orb webs in general: the webs of all four trunk orb groups are distinctly sticky spiral-rich compared with orb weavers in general (fig. 4.12) (the data for *Telaprocera* are from a drawing rather than a photo, and thus may be less precise). Comparisons with close relatives in the two *Eustala* spp. (with other *Eustala* species), and in *C. atrihastula* (with *Araneus* spp.), also show greater sticky spiral richness in the trunk web species (fig. 4.12). Comparisons in the nephilids, which must be made with more distant relatives because all species of *Herennia* and *Clitaetra* whose webs are known build trunk orbs (fig. 4.13), suggest similar confirmations. The webs of *Nephila clavipes* are radius-rich, rather than sticky spiral-rich (fig. 4.12).

These confirmations, though tentative owing to small sample sizes, are encouraging. There are, however, "constructional constraints" that may have also been important in the evolution of trunk orbs. Several additional traits may also be unusual in trunk orbs: their supports are rigid and immobile (the anchor lines are usually attached to the trunk itself); the supports are close to the orb (allowing multiple anchors and relatively short frame lines); and the wind speeds that they experience are probably often reduced compared with those in nearby open areas (Geiger 1965). Thus the danger of sticky spiral entanglement damage due to substrate movements and fluttering sticky lines (section 3.2.11.2) may be reduced. This could allow sticky lines to be more closely spaced, and make sticky spiral-rich designs more favorable.

In addition, as orbs evolved to be narrower to fit on a trunk, they would of necessity have gradually come to have more and more loops below and above the hub, and fewer and fewer loops to the sides of the hub (assuming that the total length of the sticky line remains constant and a lower limit on the spaces between loops). In effect, the sticky spiral lines that had been at the sides of the hub will be displaced to be located above or below it. If, in addition, it was advantageous to limit the total distance from the hub to the web's outer edges (for example, to allow prompt attacks on prey, or because appropriate cavitites in trunks have only limited lengths), then the spaces between sticky spiral lines above and below the hub will become smaller.

In sum, the sticky spiral-rich designs of trunk webs offer confirmations of the importance of both orb visibility and stopping abilities. This support is tentative, however, because of small sample sizes and of possible complications from constructional constraints.

4.7 CORRELATIONS BETWEEN SPIDER SIZE AND ORB DESIGN?

Several studies have tested the effects of spider size by comparing the designs of the adult webs of different species (e.g., Peters 1936, 1937; Witt et al. 1968; Craig 1997; Sensenig et al. 2011). But other species differences may also be involved (websites, web slants, attack behavior, phylogenetic inertia, etc.), so the functional significance of these inter-specific differences (some of which are quite substantial) is not easy to judge. A more powerful way to isolate the effects of size is to compare the orb designs of individuals of a given species as they grow from spiderlings to adults. Many such differences have been documented in ontogenetic studies. Some differences are intuitively "obvious" and not surprising. For instance, larger individuals tended to have webs with larger areas, larger spaces between loops of sticky spiral, and (in at least some cases) larger diameters of the lines in the following: *Argiope bruennichi* (AR) (Tilquin 1942), *Zygiella x-notata* (AR) (Le Guelte 1966), *Araneus diadematus* (AR) and *Neoscona vertebrata* (AR) (König 1951; Witt et al. 1968), three species of *Cyclosa* (AR) (Miyashita 1997), *Allocyclosa bifurca* (AR) (Eberhard 2007b), *Nephila clavipes* (NE) (Higgins 1990, 1992b), *Leucauge mariana* (TET) (Maroto 1981; Barrantes et al. 2017), and *Anapisona simoni* (AN) (Eberhard 2007b).

Some other variables that are less trivially linked to spider size, such as the numbers of radii, numbers of sticky spiral loops, eccentricity, and web slant, also changed as spiders grew. Some of these relations showed sharp differences, but their functional significance was often unclear. For instance, to cite species with particularly clear differences, the number of radii increased smoothly throughout ontogeny in *L. mariana* (TET) (Maroto 1981); in contrast, the number of radii increased sharply in the webs of *A. bruennichi* (AR) at about the third molt, but was nearly constant thereafter (Tilquin 1942); in *N. vertebrata* (AR) the number of radii was high in early instars, and then decreased (Witt et al. 1968); and in *Z. x-notata* (AR) the number of radii was lower in early instars, increased in fourth and fifth instar spiderlings, and then decreased again in the adults (Le Guelte 1966). The functional significance of these different patterns of change in radius number is not known, but they argue against consistent effects of spider size on these design traits.

Miyashita (1997) took ontogenetic analyses to the next level, comparing the patterns of change in three closely related species in the large araneid genus *Cyclosa*. The ontogenetic changes were similar in all three, showing positive slopes in graphs of body size against area, web weight, bridge size (an indicator of the size of the space in which the web was built), and sticky spiral spacing (fig. 4.14); none of these changes is particularly surprising. There were also inter-specific differences which were more interesting. The species with the largest adults, *C. octotuberculata*, differed from the other two (*C. argenteoalba* and *C. sedeculata*) in having the spaces between sticky spiral loops increase more rapidly as the spider's body size increased (fig. 4.14b). Web area, in contrast, increased in a similar manner in this species, but was always proportionally smaller at any given body size (fig. 4.14a). For a given body size, *C. argenteoalba* consistently chose larger spaces in which to build (fig. 4.14c). Miyashita found that silk properties varied with body size and also between species. For instance, for a stan-

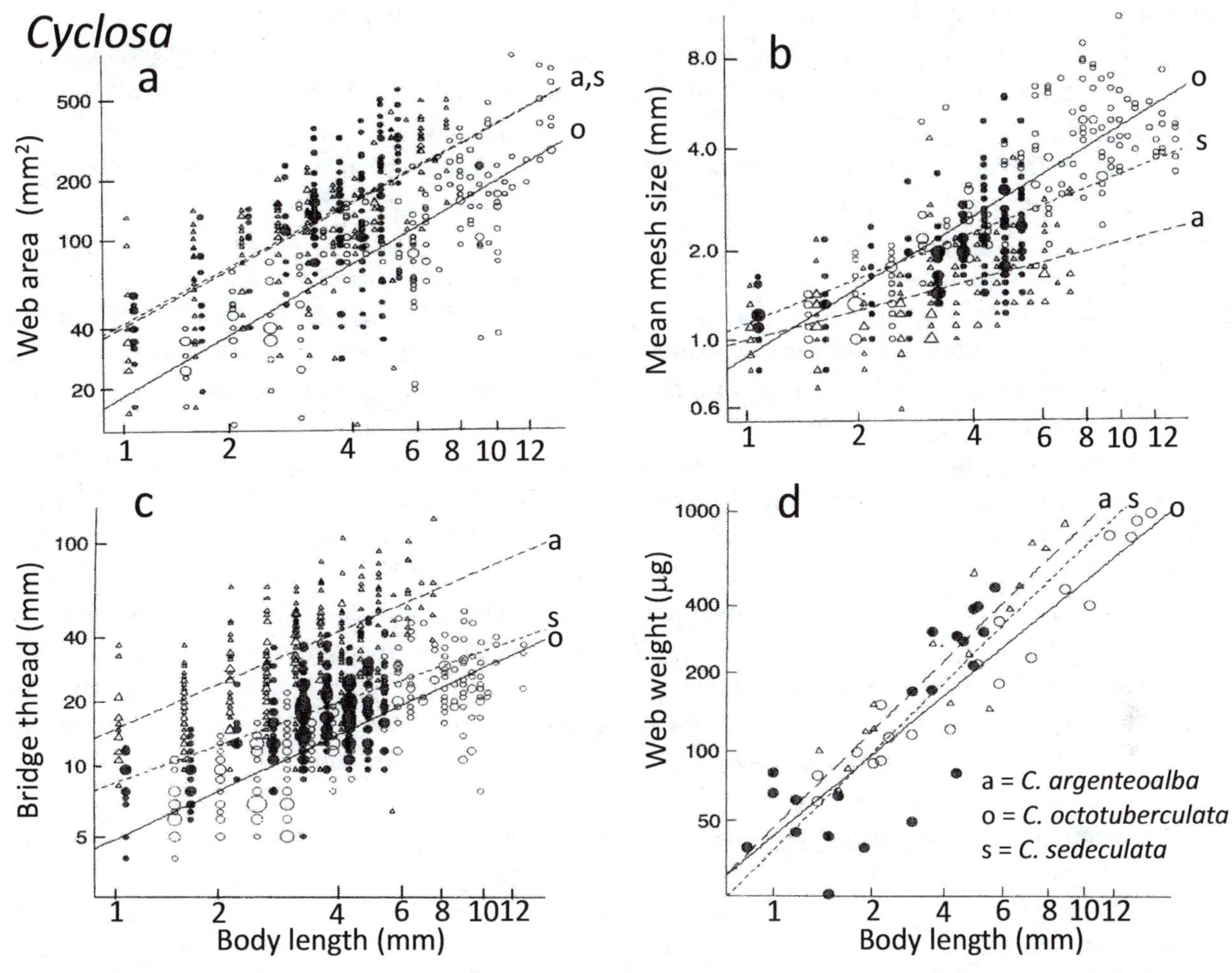

Fig. 4.14. The lines in log-log plots of several properties of orb webs in the field against body length in three species of *Cyclosa* (AR) document species-specific differences: *C. octotuberculata* webs had relatively smaller areas (*a*); *C. argenteoalba* webs had relatively longer bridge lines (*c*). Whether there are phylogenetic patterns in such correlations, with correlations being more similar among more closely related *Cyclosa* species, remains to be determined (after Miyashita 1997).

dard body size (5 mm long), the same length of sticky line had an estimated 15 times greater volume of adhesive substance in *C. octotuberculata* than in *C. sedeculata* (Miyashita 1997). Further work will be needed to determine whether these trends occur in other groups, and to formulate and test adaptive explanations.

In sum, correlations between spider size and orb web design abound in both inter- and intra-specific comparisons, but explanations of their adaptive significances are often only tentative. It is clear that the relations between body size and some aspects of web design can differ, even among closely related species, and that, as recognized previously (Witt et al. 1968), there is no simple relation between body size and orb design. The possibility that the designs of the webs of younger spiders may tend to be more ancestral (less derived) than those of older spiders (section 10.2, table 10.1), and that the adult spider's weight as well as her length affect her web's design (e.g., Witt et al. 1968; Eberhard 1988a; Herberstein and Heiling 1999) need to be included in future analyses.

4.8 SPIDER POSITIONS, ATTACK BEHAVIOR, AND UP-DOWN ASYMMETRIES IN ORBS

4.8.1 ORB DESIGN AND ATTACK SPEED

In species that built non-horizontal orbs, the radii below the hub generally tended to be longer than those above the hub, while those to the sides of the hub tended to be intermediate (e.g., Witt et al. 1968; Herberstein and

Heiling 1999; Nakata and Zschokke 2010; Gregorič et al. 2013; Zschokke and Nakata 2015). The classic explanation for this web size asymmetry is that spiders can run more rapidly downward than upward to attack prey (Masters and Moffat 1983 on *Larinioides sclopetaria*, ap Rhisiart and Vollrath 1994 on *Araneus diadematus*). In *L. sclopetaria* the spider took longer to reach a source of vibrations above the hub than when it was below the hub, even though the first turning response at the hub occurred equally rapidly. This difference in attack speed correlated with the design of *L. sclopetaria* orbs: the distances from the hub to the top and the bottom of the orb were proportional to the difference in the time taken to reach the sources of vibration in the two directions (Masters and Moffat 1983). Further data have confirmed and extended this attack speed hypothesis to explain vertical asymmetries in orbs, though there were some complications.

A pair of "upside down" exceptions to the usual vertical web size asymmetry also support the attack speed hypothesis. Two species of *Cyclosa*, *C. ginnaga* and *C. argenteoalba* (AR), built nearly perfectly vertical orbs that were unusual in that the capture zone was larger above than below the hub; the spiders also faced upward while resting at the hub (Nakata and Zschokke 2010; Zschokke and Nakata 2015). These species (for reasons that are not clear) ran equally fast upward and downward in their webs (time required to move from the hub to the attack site). In a larger "control" species, *C. octotuberculata*, spiders built orbs that were larger below the hub, faced downward, and ran more rapidly downward than upward. A fourth, intermediate species, *C. confusa*, sometimes (when the spider was smaller) built orbs that were larger above than below the hub and faced upward; other, on average larger individuals that built orbs that were larger below the hub, faced downward and ran downward more rapidly (Nakata and Zschokke 2010).

These interpretations of *Cyclosa* orb designs made the seemingly reasonable assumption that the densities of potential prey are similar in different portions of the orb, but this needs to be confirmed; for instance, orthopteran interception rates were consistently higher in the lower half of the webs of the araneid *Argiope trifasciata* (Tso 1996). Asymmetries of this sort seem more likely when (in contrast with *Cyclosa* spp.) the orb is near the ground or the top of the herbaceous layer just above the ground, as, for instance, in *Argiope* spp., *Gea heptagon*, and *Pronous* spp. (fig. 4.2, Enders 1974; Sabath 1969; WE).

The attack speed hypothesis also made other, more risky assumptions.

The attack speed hypothesis depends on the assumption that spiders run to attack prey. Although this is often true for some groups such as *L. sclopetaria*, *Argiope* spp., *Leucauge mariana* (Briceño and Eberhard 2011), and *Nephila clavipes* (WE), it was clearly not true for others such as *Cyclosa* spp. (Nakata and Zschokke 2010; WE), *Gasteracantha cancriformis* (Muma 1971, WE), *Verrucosa arenata*, *Micrathena* spp. (AR) (WE), *Theridiosoma* and *Naatlo splendida* (TSD) (Coddington 1986a; WE), and many uloborids (Lubin 1986; WE). These species often turned rapidly at the hub and jerked the web when a prey struck the web, but only then moved slowly on their relatively stubby legs to attack the prey. The orbs of some of these species are strongly asymmetrical (e.g., fig. 4.15). The prediction that stubby-legged species, which do not attack prey rapidly, have reduced vertical asymmetry in their orbs has not been checked.

Another context in which spiders did not reach prey rapidly occurred in some species that initiated attacks on a large or dangerous prey by apparently attempting to increase the retention time, cutting the radii just above prey and causing it to fall into contact with additional lines below. This behavior has been seen in *Scoloderus tuberculifer* (AR) (Eberhard 1975), uloborids in the genera *Uloborus*, *Philoponella*, and *Zosis* (Lubin 1986), and an *Araneus expletus* (AR) spiderling attacking a bee approximately its own size (WE). This type of attack provides a second explanation for vertically elongate orbs.

A second supposition of the attack speed hypothesis is that orb spiders are equally responsive to prey in different sectors of their orbs. This supposition was supported in *Argiope keyserlingi* (AR), in which prey above or below the hub were equally likely to be attacked (Champion de Crespigny et al. 2001). In mature female *L. mariana*, however, tests during the day in the field by dropping small 0.7 mg lengths of fine wire in the outer quarter of more or less horizontal webs (mean 35 ± 11°, N = 70) while the spider rested at the hub revealed that prey in front of the spider elicited more responses than those behind her (WE). She was more likely to run "immediately" (in <0.5 s) toward the "prey" when it landed in front rather than behind her ($chi^2 = 5.9$, $p = 0.015$, N =

39, 50); and she was also more likely to reach the prey "immediately" when it was in front of her ($chi^2 = 10.3$, $p = 0.002$, N = 29, 45). Spiders were also more likely to respond very rapidly at night than in the day ($chi^2 = 11.0$, $p = 0.012$, N = 182, 71).

The reduced responses of *L. mariana* when the prey was behind the spider were apparently not due to problems of detection, because increasing the weight of the wire to 2.5 mg had no effect on the speed of responses to wires dropped behind her ($p = 0.36$, N = 38, 36). Instead, attention or responsiveness may have been involved: spiders that were feeding were more likely to take >3 s to arrive at the prey ($chi^2 = 11.6$, $p < 0.001$ in 117 cases); and when they attacked, they were much less likely to have turned quickly to face the prey in <1 s ($chi^2 = 69$, $p < 0.0001$, N = 23, 91). It is a puzzle why spiders were less disposed to attack prey in some portions of their orbs than others (if they were less willing to launch attacks in those sectors, why build them?).

Elongation of the web could help solve the problem presented by prey in nearly vertical orbs that "tumble" downward after being stopped (section 4.3.3.1, fig. 4.20). A prey that strikes the orb above the hub and then tumbles may require a somewhat shorter excursion by the spider than a prey that hits the web at the same distance below the hub. Tumbling itself is a complex phenomenon. It is likely less common in insects with long legs and wings (e.g., Blackledge and Zevenbergen 2006), and more common in those that are covered with deciduous scales or hairs (Eisner et al. 1964) (fig. 4.20). Its importance probably varies with the abundances and sizes of different types of prey.

The asymmetry of *Leucauge venusta* (TET) orbs was also not in accord with the attack speed hypothesis, in that webs with relatively steep slants (61–90°) were not more asymmetrical than those with lower slants (0–30°, 31–60°) (Gregorič et al. 2013). Webs of adult female *L. mariana* in the field also failed to show a correlation between slant and vertical asymmetry (the mean top/bottom asymmetry was 0.85 ± 0.15, with a range of 0.59 to 1.41 in 74 webs) (WE) (in neither species, however, was the speed of attack shown to be greater when the web was more steeply slanted). Perhaps the extremely rapid attacks of *Leucauge* even when the prey strikes the web above (behind) the spider (Briceño and Eberhard 2011) make such a correlation less likely.

A further prediction (untested as far as I know) concerns species with vertical webs in which the spider rests in a retreat off the web rather than at the hub. Their orbs should be less vertically asymmetrical, because the weight of the spider would not enhance the abilities of radii below the hub to stop prey by reducing their tensions.

Still another orb trait that may be linked to vertical asymmetries is the typical "larger-near-the-edge" pattern often stronger above the hub than below it (section 4.3.2, figs. 1.4, 3.6) (careful quantitative studies with reasonable sample sizes are lacking, however). Data from four species of *Cyclosa* (AR) suggested this same pattern, though analyzed in a different manner (Zschokke and Nakata 2015). This asymmetry was clear in the only moderately slanted orbs of *Leucauge mariana* (TET) (WE) (the mean slant of 66 webs in the field was 39.9°), so it is probably not an adaptation to tumbling. In this species the lower portion of the orb tended to be larger both in captivity (Eberhard 1987b) and in the field (the lower 45° sector of the orb was larger than the opposite 45° sector in 27 of 35 webs with a slant of >30°) (WE).

This tangle of relations makes separating cause-effect relations from incidental correlations difficult. In *Cyclosa* there were several additional web traits that also show general up-down asymmetries: below the hub, the number of sticky spiral loops tends to be greater, the spaces between them smaller, and the angles between radii smaller (Zschokke and Nakata (2015). Furthermore, the radii below the hub are probably better able to absorb the impacts of high-energy prey because of their greater lengths and lower tensions (section 4.4). Using the known overall vertical asymmetry in orb deisgns, and intra- and inter-specific variation in up-down asymmetries in size ("web-extent asymmetry") in four species of *Cyclosa* and other orb weavers, Zschokke and Nakata (2015) concluded that up-down asymmetries in the suite of other web traits were linked both to gravity per se, and also to web-extent asymmetry. Understanding the vertical asymmetries in *L. mariana* orbs is more difficult.

Perhaps a combination of the advantage of smaller inter-radial angles in larger (lower) portions of the orb (in order to maintain minimum distances between radii in the outer portions of the orb), along with the "prey stopping" and the "sticky spiral entanglement" hypotheses, offers the best explanation for the tighter sticky spiral spacing where radii are closer together (section 4.3). The tendency to build longer radii in the lower sectors may be explained by downward attacks being more rapid in the direction in which the spider consistently faces while resting at the hub (next section). The reason for the unequal responsiveness of this species to prey in different sectors of the orb remains to be explained.

4.8.2 SPIDER ORIENTATION AT THE HUB

A further asymmetry that is apparently associated with gravity's effects on attack speed and vertical asymmetries in orbs is the nearly universal tendency for orb weavers to face downward while waiting at the hub for prey. In two species of *Cyclosa* (*C. ginnaga* and *C. argenteoalba*) that built nearly vertical orbs that were "upside down," with the portion above the hub larger than that below it, the spiders were also unusual in facing upward at the hub (Nakata and Zschokke 2010; Zschokke and Nakata 2015). A "control" species of *Cyclosa, C. octotuberculata,* had both the more typical vertical web-extent asymmetry (larger below than above the hub) and spider position at the hub (facing downward) (Nakata and Zschokke 2010; Zschokke and Nakata 2015) (both traits also occur in the closely related *Allocyclosa bifurca* [WE]). Another species, *Cyclosa confusa,* (true to its name) sometimes faced downward and sometimes upward. Heavier individuals were more likely to face downward, presumably because they moved upward more slowly due to their greater weight (Nakata and Zschokke 2010).

Spiders also tended to face downward in some species that build less nearly vertical orbs. Of 82 adult female *Leucauge mariana* on orbs with slants of at least 20° with horizontal, 81 (98.8%) faced toward the lower half of the orb, and 65 (79.9%) faced within 20° of the most nearly downward directed radius (corresponding figures for 17 *L. venusta* were 17 [100%] and 13 [76.5%]) (WE).

Body weight, in turn, was correlated with web design in some species. An intra-specific correlation between body weight and the degree of vertical asymmetry in the orb also occurred in *Nephila pilipes* (NE), *Argiope keyserlingi* (AR), and *Larinioides sclopetarius* (AR), which all faced downward (Robinson and Robinson 1973a; Herberstein and Heiling 1999; Kuntner et al. 2010a; M. E. Herberstein pers. comm.). The amount of vertical web-extent asymmetry increased when weights were added to *L. sclopetarius* females (Herberstein and Heiling 1999), and was correlated more strongly with body weight than with cephalothorax or leg length in *N. pilipes* (Kuntner et al. 2010a).

4.8.3 FURTHER FACTORS INFLUENCING SPIDER POSITIONS AT THE HUB

Several species in the araneid genus *Verrucosa* posed further questions regarding orientation at the hub. They built apparently typical daytime, vertical orbs that were distinctly larger below the hub (fig. 4.15), but rested at the hub facing upward (Levi 1976; Bradley 2013; Dinish

Fig. 4.15. The orb of *Verrucosa arenata* (AR) presents an unsolved mystery. As is typical in many orb weavers, it is vertically asymmetrical, with a substantially larger capture zone below than above the hub; nevertheless, the spider, in contrast to most other orb weavers, faces upward rather than downward while resting at the hub (fig. 3.50c) (Levi 1976; Rao et al. 2011). The orbs of other species in which the spider faces upward, *Cyclosa ginnaga* and *C. argenteoalba* (AR), have the opposite asymmetry, and are larger above than below the hub (Nakata and Zschokke 2010). The orbs of *V. arenata* tended to be close to perfectly vertical (the mean angle with horizontal was 88 ± 2° in five webs) (WE).

Rao unpub.; WE). *Verrucosa* orbs were typically built in large open spaces, and tended to be nearly perfectly vertical (the mean angle with horizontal in five webs of *V. undecimvariolata* was 88 ± 2°) (webs of *V. arenata*, which were also built in the early morning, were also nearly perfectly vertical—B. Opell pers. comm.). Spiders crouched at the hub with their legs pressed against the body. Perhaps their unusual orientation was related to the colorful markings on the dorsal surface of the shiny, triangular abdomen (but not on the ventral surface). One *V.* sp. performed an unusual defensive behavior when disturbed at the hub: she repeatedly flexed her abdomen dorsally to a horizontal position and then slapped it briskly ventrally, vibrating the web; after two or three slaps she rested with her legs I and II tilted into a crab-like position (WE).

Defense against predators is an additional factor that appears to affect the orientations of some orb weavers. Free-ranging adult female *Zosis geniculata* (UL) on orbs built indoors most often faced nearly directly upward (0 ± 20°) (57% of 126 observations of 10–15 unmarked individuals that were checked intermittently on their moderately slanting orbs [20–40° with horizontal] over a period of 34 days); nevertheless, their webs were generally larger below than above the hub. One defensive behavior of *Z. geniculata* when disturbed was to run to the edge of the web (WE) (as also occurred in the uloborids *Uloborus glomosus* and *U. diversus* [Cushing 1989; Eberhard 1973a]), so facing upward may improve the speed with which they exit their webs.

A second link with defensive behavior comes from *Azilia affinis* (AR), which builds highly asymmetrical orbs at protected sites such as at the base of a large tree trunk or between buttress roots. The orbs were approximately horizontal ($x = 34.5 \pm 12.6°$, range 9–57°) in 11 adult female *A.* sp. (probably *A. affinis*) (WE), and they consistently slanted downward away from the sheltering trunk or buttress root, with the hub nearer the upper, more sheltered side (G. Barrantes pers. comm.; WE). During the day *A. affinis* spiders usually faced upward, toward the small side of the orb, with their legs pressed tightly together (fig. 4.16a); they responded to even minor disturbances by running quickly to the upper, trunk side, where the spider's dark brown color made her difficult to see (G. Barrantes et al. in prep.). At night, however, spiders rested at the hub with their legs spread, facing outward and downward toward the larger side of the orb (fig. 4.16b), presumably facilitating rapid attacks on prey (G. Barrantes et al. in prep.). The association between the spider's preferred escape direction and the direction in which she faced at the hub during the day was not consistent, however, in other species. In *L. mariana* (TET), spiders consistently tended to face downward on their orbs (above), but in a sample of 48, most fled upward when they were disturbed during the day by a puff of air from above: 38 (79%) turned and fled toward the upper half of the web, and 10 (21%) fled toward the lower half; the mean angle of the direction she fled with directly upward was 48 ± 39° (in 18 other cases the spider dropped from the web on her dragline to the vegetation below) (WE). Perhaps these lightning fast spiders respond so rapidly to predator attacks that they can afford to have slightly less than maximum escape speeds (several times I have observed a spider to fail to flee, however, from ichneumonid parasitoid wasps [*Hymenoepimecis tedfordi*], that attacked from above the hub before the spider could respond).

The direction of the sun's rays may also occasionally affect the spider's orientation at the hub. On warm days, mature females of *Z. geniculata* (UL) in full sunlight assumed atypical positions, facing away from the sun and away from the nearest edge of the orb (WE). This aligned the spider's long axis with the sun's rays so only the posterior surface of her abdomen was directly illuminated. This same orientation with the tip of the abdomen toward the sun was assumed by *Nephila pilipes* (Robinson and Robinson 1973a) and *N. clavipes* (Robinson and Robinson 1978) (NE), and it reduced heating under direct illumination (Robinson and Robinson 1978).

Still another complication is presented by the apparently unrelated araneid *Acacesia hamata* and perhaps also *Deliochus* (only a single spider and its web have been described), that built nocturnal sticky spiral-rich orbs in which the usual vertical web-size asymmetry was inverted (fig. 3.5); the spiders, nevertheless, faced downward at the hub. These orbs are thought to be designed to capture moths (e.g., Bradley 2013), and the upper, more sparse extension may function to cause the moth to tumble into denser parts of the orb. The orbs of *A. hamata*, however, were not well-oriented to promote tumbling because they were not consistently close to vertical: the mean angle with horizontal in the field was

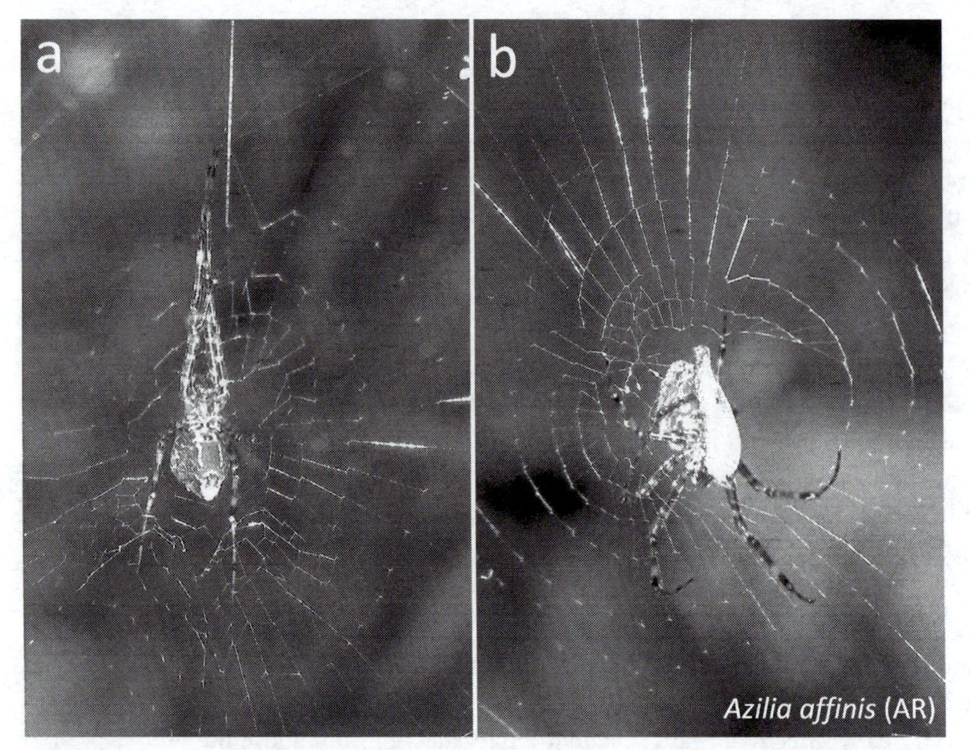

Fig. 4.16. The day-time (*a*) and night-time (*b*) positions of this mature female *Azilia affinis* (AR) emphasize the probable importance of defense against diurnal predators rather than attack speed in the orientations of some orb weavers at the hub. The spider faced upward during the day, toward the smaller, more protected side of her highly asymmetric, gently slanting orb, which was built near the base of a large tree. She responded to even minor disturbances by running rapidly to that protected edge, where she was difficult to see. Her legs I and II had disruptive coloration near their tips, extended anteriorly, and were pressed tightly together (*a*), additional defensive traits that probably camouflaged her outline against visually orienting predators. At night (*b*) (holding a recently captured prey), her position was more appropriate for capturing prey, facing toward the larger portion of her orb (downward and away from the trunk) and with her legs spread.

75 ± 12° (WE; table O3.1), so their inverted vertical asymmetries are thus left unexplained.

4.8.4 SUMMARY

In sum, two vertical asymmetries are widespread in orb webs: the lower portions of vertical orbs tend to be larger than the upper portions; and spiders tend to face downward at the hub. Both asymmetries are probably often associated with greater attack speeds downward than upward, and exceptions to these trends fit predictions of this attack speed hypothesis in species in the genus *Cyclosa*. But many species with typical vertical web asymmetries perform an alternative attack behavior that appears to reduce or eliminate the importance of running speed in capturing prey. Defensive behavior and temperature control also appear to affect the spider's orientation at the hub in some species. It appears that defensive behavior and alternative attack behavior tactics, as well as attack speed, prey tumbling, and the "prey stopping" hypothesis for sticky spiral spacing, will all probably need to be taken into account to understand spider orientations at the hub in different species, and their possible links to vertical asymmetries in orb web designs.

4.9 REMAINING PUZZLES

4.9.1 THE PUZZLE OF TEMPORARY SPIRAL REMOVAL

Most orb weavers carefully break all segments of the temporary spiral while they build the sticky spiral (section 4.3.5.1). At first glance, temporary spiral removal seems paradoxical: by breaking these strong, ampullate gland lines, the spider probably weakens the ability of her orb to absorb the momentum of the large, high-energy prey that have been thought to be especially important (Robinson and Robinson 1973a; Venner and Casas 2005; Blackledge et al. 2009c; Sensenig et al. 2010a; Evans 2013). There are several possible explanations, though none is especially appealing.

One possible function is to make the tensions along each radius more nearly equal. The tensions on the temporary spiral lines appear to be relatively high (section 4.4.2.2.2), so each junction of the temporary spiral with a radius will exert an appreciable net inward force on the radius; this will increase the tension on the portion of the radius beyond the junction, and decrease the tension on its inner side. By breaking the temporary spiral lines, the spider reduces the stresses on radii in the outer part of the orb, and thus increases their abilities to absorb the impacts of high-energy prey. Breaking the temporary spiral will increase the radial tensions in the inner portion of the orb, increasing their abilities to transmit vibrations. Their abilities to stop high-energy prey will be reduced, but the distances between radii are smaller nearer the hub, making it more likely that high-energy prey will be stopped by striking several radii rather than just one.

A second possible explanation takes these considerations of tension into account, but includes reductions in the overall stability of a web under stress resulting from intact temporary spiral lines. This hypothesis was based on models of orbs that take into account the spring constants of spiral lines ("k") and those of radii ("K") (typical values of K:k are 10:1 for sticky spiral, and 1:1 for temporary spiral) (Aoyanagi and Okumura 2010). The maximum force in any line in the orb was reduced when K was substantially greater than k, as compared with when K = k. The lines with the greatest stress were the peripheral portions of radii. This is important because of the "weakest link" argument: an orb's strength can be defined by the maximum stress that any single line is called upon to resist (Aoyanagi and Okumura 2010). In addition, when K/k = 10 rather than 1, the change in the distribution of forces in the web that resulted when a spiral line was broken was reduced. Most importantly, stresses resulting from breaking spiral lines were not concentrated at particular sites. Dispersing rather than concentrating stresses enhances an orb's ability to absorb the impact of high-energy prey (it also potentially increases the extent of damage that is inflicted by an unstoppable object such as a falling leaf, a large insect—see section 3.3.2.9). Whatever advantage may accrue from an intact temporary spiral via better vibration transmission by more tense inner portions of radii must be weighed against the increased stopping abilities of the less tense inner portion of the orb, and the concomitant decrease in the stopping abilities of the more tense outer portions of the same radii.

A simple experiment called into question, however, the importance of the K/k ratio for overall web stability. In webs of *N. clavipes*, in which temporary spiral lines are left intact (fig. 4.4), the only lines that were broken by experimental impacts of high energy objects (pebbles tossed into the webs at intermediate distances from the hub) were lines (both sticky and non-sticky) near impact sites; lines near the outer ends of the radii (and the other parts of the orb) remained intact (WE).

Still another explanation for temporary spiral removal is that temporary spiral lines reduce the ability of radial lines to transmit vibrations because they divert vibrations from the radii. Vibrations transmitted by radii were increased by 2.2–5.8 dB by cutting the temporary spiral lines in *N. clavipes* webs (Landolfa and Barth 1996). There are reasons, however, to suspect that vibration transmission is not a crucial trait in orbs (section 3.3.2.2).

An alternative focus is to ask the opposite question: why are temporary spiral lines left intact in a few orbs? In both nephilid (fig. 4.4) and *Scoloderus tuberculifer* (AR) webs (fig. 3.4), temporary spiral lines bend the radii and hold them apart, and bend them into larger interradial spaces. In effect they spread the especially closely spaced radii in these webs more uniformly (Eberhard 2014), probably making these orbs more effective in stopping and retaining prey. In *Nephila* the temporary spiral lines also "cinch up" the tensions on radii by means of double attachments (fig. 4.4), suggesting the additional function of adjusting tensions on the radii (section 4.4). In some species such as the uloborids *U. diversus* and *Zosis geniculata*, a few segments of temporary spiral are often left intact in finished webs (figs. 3.26, 7.24), but I see no pattern in the sites where they are intact, and cannot discard the hypothesis that they represent mistakes during sticky spiral construction (Eberhard 1990c).

In sum, the puzzle of the selective advantage temporary spiral removal does not stem from a lack of hypotheses, but from their general weakness, and a lack of good reasons to choose among them.

4.9.2 THE PUZZLE OF HUB REMOVAL AND OPEN HUBS

Hub modification varies more than many other aspects of orb construction, and the differences among different taxa cloud understanding of its functional significance. Many orb weavers immediately break and remove the lines in the central portion of the hub as soon as they have finished the sticky spiral. Many of these species immediately lay new lines across the hole they have created. Generally it appears that the lines that fill in the hole are under lower tensions (fig. 4.17, appendix O6.3.7, Eberhard 1987d). Other species leave the hub open. The theridiosomatids and the related symphytognathoid families remove not only the center but the entire hub; they either relax each radius (by lengthening it), and then reattach it individually to another radius, or in some cases they build an entirely new hub. All of these types of hub modification lower the tensions on radii (appendix O6.3.7).

One hypothesis to explain hub modifications is that they serve to lower radial tensions that, earlier in orb construction, it was advantageous to have higher. Lower tensions will improve the web's ability to absorb the impacts of high-energy prey, but they might be disadvantageous earlier, during construction, because they would increase the local distortions produced by the spider's own weight. The result of the hub center being removed and then replaced in *Argiope argentata* (AR) was that tensions on radii decreased, more so on vertical than horizontal radii (fig. 4.17). In species such as *Micrathena* spp. (fig. 3.37), the spider immediately grasped the inner edges of the hole she had just created and pulled them together as she rested at the hub, increasing the tension again. Possibly this increases the mechanical ability of the orb to resist wind stress (it will sag and flap less—section 3.2.11), and then increased its ability to stop prey by releasing the tension at the moment the prey struck the web.

Another, non-exclusive alternative possibility is that removal of the numerous small tufts of loose silk that accumulate as a result of the break and reel behavior during radius construction (section 6.3.3) reduces the possibility that the spider's tarsi might become entangled in this fluff. This "clean up" hypothesis is in accord with the fact that both of the major groups in which hub removal does not occur (uloborids and nephilids) also lack tufts of loose silk in the hub because they not break and reel lines during radius construction (appendix O6.3.2).

These proposed advantages and disadvantages could be tested experimentally, but to date none of them have been studied, so these explanations remain speculative.

4.9.3 THE PUZZLE OF THE FREE ZONE

Another, largely unremarked puzzle in orb web designs is why sticky spiral lines are usually omitted close to the hub, where the radii are closest together and under lower tensions due to the inward forces exerted on the radii by the sticky spiral lines. This is the portion of the orb with the greatest ability to stop high-energy prey, but, paradoxically,it is apparently incapable of trapping them. There are several possible explanations.

One possible function for the free zone is that it frees the spider's legs from adhering to sticky lines as she grasps lines as she turns at the hub in response to prey impact (Briceño and Eberhard 2011). This hypothesis seems unlikely, however, because many groups have a hub that is so large that the spider always grasps lines within the hub or at its edge but there is nevertheless a very clear free zone—e.g., *Zosis geniculata* (UL) (fig. 3.46), *Dolichognatha* sp. (TET) (fig. 3.43) and *Azilia affinis* (TET) (fig. 4.16), and the araneids *Metazygia wittfeldae* (fig. 3.16), *Pozonia nigroventris* (fig. 3.35), *Deliochus* sp. (fig. 3.5), and *Argiope* spp. (figs. 3.45, 3.47b,d,e).

A second possible function of the free zone is to allow the spider to shuttle from one side of the orb to the other and defend herself against predators, as occurs, for example, in *Argiope* spp. (Hingston 1932; Robinson and Robinson 1970a) and *Cyclosa jose* (WE); *Araneus diadematus* (AR) also sometimes changed sides when repairing web damage (Tew et al. 2015). This hypothesis, however, cannot explain the existence of large free zones in many other groups like the tetragnathids *Leucauge* spp., *Dolichognatha* spp., *Tetragnatha* spp., and many uloborids that build horizontal orbs and thus never change sides (e.g., figs. 3.11, 3.28, 8.10).

Two additional speculations, which are as yet untested, are the following. Perhaps large prey are physically dangerous for the spider as they struggle in the web near the hub, and the free zone serves to reduce the dangers they present. Or perhaps the removal of prey tangled in the web very near the hub would involve cut-

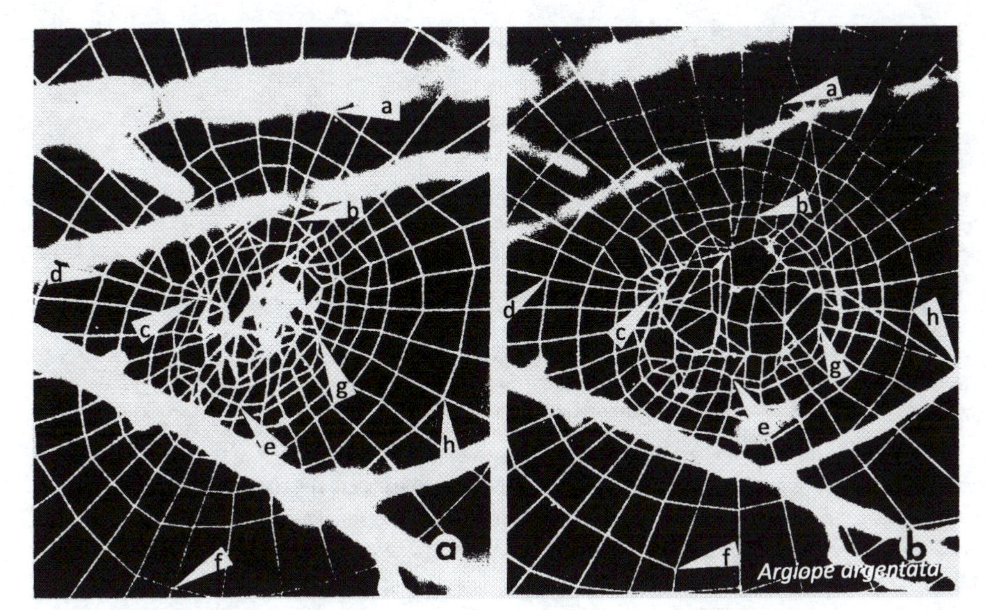

Fig. 4.17. Views of the hub of a mature female *Argiope argentata* (AR) before (*a*) and after (*b*) the spider removed and replaced the central portion of the hub suggest substantial changes in tensions on some lines (photos are at the same magnification; the arrows indicate some corresponding junctions). The horizontal distance across the inner edge of the hub increased by 30%, and across the outer edge by 2%. The corresponding increases in the inner and outer vertical distances were 16% and 5%, so vertical radii were relaxed more than horizontal radii. Other hub replacements in this species showed different patterns (from Eberhard 1981a).

ting so many lines that the damage to the web would be greater than the benefit from the prey. Neither of these hypotheses seems very attractive.

There is a great deal of variation in the relative size of the free zone compared with the hub (figs. 3.11–3.13), and in a few groups the free zone is entirely lacking, including *Micrathena* spp. (AR) (figs. 1.4, 3.14), and several nephilids such as *Nephila clavipes* (fig. 4.4), and the trunk orbs of *Herennia multipuncta* (fig. 4.3) and *Clitaetra episinoides* (fig. 9.17) (but not the trunk orbs of the araneid *Eustala perfida*—fig. 4.3). The pseudo-orb of *Fecenia ochracea* (see the frontispiece) also lacked a free zone. Perhaps these groups hold the key, but at present I see no obvious associations between these differences and any hypothesis to explain free zones.

In summary, the free zone puzzle lacks plausible general explanations.

4.9.4 THE PUZZLE OF NON-VERTICAL ORBS

The angle at which a prey moves with respect to the plane of an orb influences the web's abilities to intercept, stop, and retain prey. Aligning an orb so that more prey impacts are more nearly perpendicular to the orb plane could improve the interception function. If, as studies of web designs have often assumed (e.g., Eberhard 1986a), the paths of flying insects tend to be more nearly horizontal than vertical, then a more nearly vertical orb would intercept more prey. In addition, the retention function of a more nearly vertical orb is greater, because as prey gradually work themselves free they tend to fall into additional web lines (section 3.3.3.1.5). If vertical orbs have these advantages, then why are the orbs of so many species consistently non-vertical, often more nearly horizontal than vertical (table O3.1)? And why are the slants of some species so variable intra-specifically?

As illustrated in table O3.1, there is much inter- and in some cases intra-specific variation in web slants. I distinguished four somewhat arbitrary categories there: spiders that consistently built nearly perfectly vertical webs (mean > 80°, "strictly vertical") (these were all araneids); spiders that built inclined orbs with less steep slopes (about 60–80°) and moderate to high standard deviations ("approximately vertical") (these were mostly araneids, all nephilids, and a few tetragnathids and uloborids); spiders that built somewhat horizontal orbs (20–60°) with moderate to high standard deviations ("approximately horizontal") (these were tetragnathids and uloborids, and a few araneids); and spiders that built close to perfectly horizontal orbs (0–20°) with moderate to small standard deviations ("strictly horizontal") (these were tetragnathids, uloborids, and an anapid). Arachnologists typically label both the strictly vertical and approximately vertical categories as "vertical", and both the approximately and strictly horizontal categories as "horizontal." (Two alternative classifications are 0–45° and 46–90° of Kuntner et al. 2008a, and 0–30°, 31–60°, and 61–90° of Gregorič et al. 2013). The amount of variation in web slants (measured with either standard de-

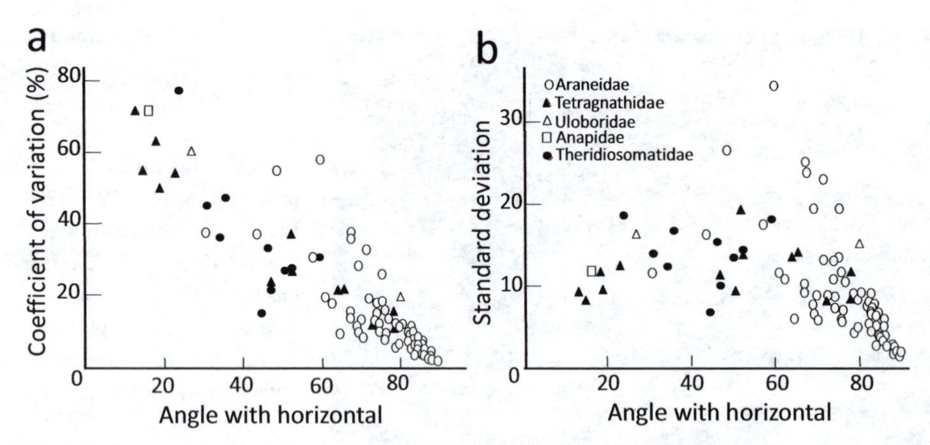

Fig. 4.18. The amount of variation in the slants of orbs (their angles with horizontal) may provide indirect measures of the selectivity of spiders when they choose websites, because lower variation could imply greater selectivity with respect to the slant. Relative variation, indicated by the coefficient of variation for each species (*a*), was similar for different slants; when measured by the absolute value of the standard deviation of the mean, however, it was especially low in webs with higher slants (*b*) (data from table O3.1). The substantial inter-specific variation in the species of the tetragnathid genus *Leucauge* (the tetragnathid points in the graphs) suggests (though does not prove) that phylogenetic constraints play only a minor role (most *Leucauge* were not identified to species, but most of the unidentified species were distinguished from other sympatric species during extended stays at two distant sites [near Puerto Lopez, Meta, Colombia, and Ayanar Falls, Tamil Nadu, India] that presumably do not share *Leucauge* species).

viations or coefficients of variation) was greatest in the approximately vertical and approximately horizontal categories (fig. 4.18). To some extent this is simply a mathematical artifact (an extreme mean nearly equal to 90° is not possible when there is substantial variation). But the spiders that built more nearly vertical orbs were more consistent than those that built nearly horizontal orbs.

One hypothesis explaining the widespread existence of non-vertical orbs is that horizontal orbs are ancestral and there have been subsequent phylogenetic limitations (Bond and Opell 1998). The orbs of tetragnathids and uloborids are generally more nearly horizontal than those of araneids, and the evolution of vertical orbs in araneids was taken to be a key innovation responsible for the evolutionary success of araneids. But there are multiple exceptions to these trends in all three families (table O3.1), and additional evidence suggests that the role of phylogeny in constraining the evolution of web slants is limited. For instance, the webs of the eleven species of *Leucauge* (TET) in fig. 4.18 spanned a wide range of slants ranging from approximately horizontal to approximately vertical, and there was also substantial intra-specific variation in the uloborids *Philoponella* sp. nr. *fasciata* and *Zosis* sp. (Diniz et al. 2017) (see also section 10.9.1). In addition (and more importantly), there is substantial intra-specific variation in the web slants of many species (table O3.1). Orb web slants thus tend not to be strictly limited by the orientations inherited from ancestors: there is usually ample variation in slants on which natural selection could act. This suggests that associations between slants and phylogeny may be indirect, rather than cause-effect.

Another possible explanation for non-vertical orbs is that the assumption that prey consistently move more horizontally than vertically is incorrect. Surely many insects must have vertical as well as horizontal components to their trajectories (compare, for instance, hopping vs. flying insects, or insects like ants falling from trees), but I know of no direct studies of the flight paths of different groups in the field where orbs are present. An indirect indication of prey paths was available, however, for orb webs built in the large tangles of lines above the funnel webs of two *Aglaoctenus* species; orbs at the edges of the tangle would be more likely to receive impacts of prey traveling more nearly horizontally than those in the central portions of the tangle, where prey would be more like to be falling vertically. As predicted by the prey-trajectory hypothesis, the orbs of both *P.* sp. nr. *fasciata* and *Z.* sp. were more nearly horizontal in the central area of *A.* spp. tangles than those of conspecifics near the edges (Diniz et al. 2017).

An additional, unexplored possibility is that prey do tend to fly horizontally, but that the increase in the stopping function of an orb outweighs the decrease in its interception function. When a prey's angle of impact is less nearly perpendicular with the orb's plane, the numbers of lines that the prey contacts are likely to increase.

One test of ideas regarding insect flight is to place webs or web-like traps in horizontal and vertical orientations in the field. Such experiments have produced, however, a confusing mix of results. In a large, uniform open weedy field of grass at night, standardized planar sticky traps placed at the same heights above the ground, at the same angle with the wind, and at the same time of night gave data favoring the assumption that prey tend to fly horizontally. Horizontal traps captured only 32% the number of insects that were captured by vertical traps, and 45° traps captured intermediate numbers (Chacón and Eberhard 1980) (the traps' abundant adhesive and strong lines probably retained all but perhaps the largest insects that contacted them).

In sharp contrast, however, horizontal traps of similar though not identical design that were set for 72 consecutive hours in three tropical forest habitats captured approximately double the number of prey when they were horizontal than when they were vertical, and at two of the three sites traps at intermediate angles (56° and 64°) captured even more (Bishop and Connolly 1992). The logical possibility that some prey fell from trees overhead seems to be ruled out by the fact that the prey in these traps were "overwhelmingly small diptera." A third experimental study (Opell et al. 2006) at temperate sites (an open field and a forest) that used sectors of unoccupied webs of three species orbs mounted on artificial supports produced still different results. They were checked every three hours during the day, and the prey that they intercepted were evaluated by counting the small insects retained in the webs as well as the damaged areas in the webs (see section 3.2.5 for drawbacks to these techniques). Both the webs of *Uloborus glomosus* (UL) and *Leucauge venusta* (TET) (species that build orbs that are more nearly horizontal than vertical) intercepted more prey when oriented vertically than when oriented horizontally (about 220% and 136% respectively) (data read from graphs); but there was no significant difference between the vertical and horizontal webs of *Micrathena gracilis* (AR) (the orientations of webs built by *M. gracilis* varied from nearly horizontal to nearly vertical in nature—B. Opell pers. comm.). The only lesson I can see from these contradictory results is that there may be substantial complexity in the factors that affect interception success of webs with different slants.

Another possible explanation for horizontal orientations is that horizontal orbs permit spiders to place their orbs especially close to horizontal surfaces where prey are more common. The tetragnathids *Metabus ocellatus* and *Glenognatha heleois* place their orbs very close to the surfaces of streams and marshy ground, where prey are thought to be more abundant (fig. 2.19) (Buskirk 1975; Dimitrov and Hormiga 2010). Nevertheless, the orbs of *M. ocellatus* are less consistently near the water surface than would be expected (Buskirk 1975; WE), and some webs over streams had slants of up to 60° or more (WE; fig. 7.43). In addition, the approximately horizontal orbs of many other groups of orb weavers (table O3.1) are not located next to horizontal surfaces where prey are obviously abundant (e.g., fig. 3.39). In sum, this hypothesis is not likely to be generally applicable.

On the other side of the coin, slanted orbs have possible advantages. Building a regularly spaced sticky spiral in a strictly vertical orb may require special adjustments, because the spider's own weight causes her to sag toward previously built lines as she works. One individual of *Allocyclosa bifurca* (AR), building a close to perfectly vertical orb, held her body out of the plane of the orb, rather than hanging below the web lines, while she built sticky spiral lines (WE). A spider at the hub of a slanting orb may also be more protected from the attacks of some enemies, such as salticid spiders (e.g., Tolbert 1975), large damselflies (Young 1980; Fincke 1992; Esquivel 2006; WE), and parasitoid wasps that have limited maneuverability, and must attack from above rather than below when the web is slanted (Eberhard 2000b).

Another advantage of more nearly horizontal webs may stem from reduced damage by the spider to her own web. In nearly vertical orbs, the spider disrupts the sticky spiral spacing every time she moves across her web (she leaves "tracks" on her orb). When webs of *A. bifurca* built in captivity were experimentally oriented horizontally and a prey was added, the tracks made by the spider moving to the prey and returning to the hub were less substantial than when the orb was vertical (fig. 4.19) (WE). The fact that several orb weavers utilize a special

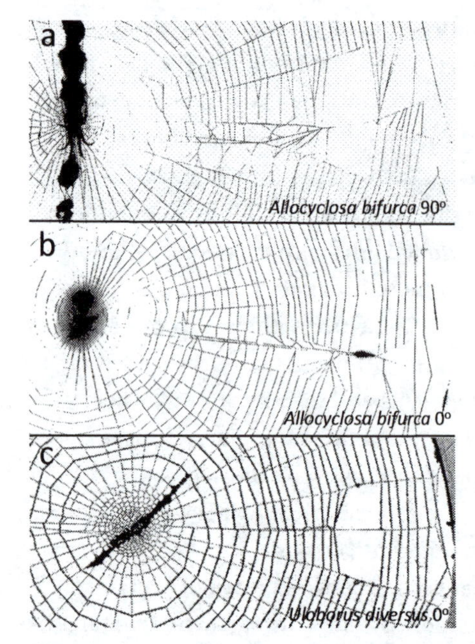

Fig. 4.19. Three typical patterns of disturbances ("tracks") produced as spiders ran out from the hub to attack a *Drosophila* prey in the outer third of the orb, and then returned to the hub (webs are unpowdered, allowing discrimination of the sticky (brighter) lines from non-sticky lines). The tracks in *a* and *b* were made by *Allocyclosa bifurca* (AR) when the web was nearly vertical (90°) (as is typical) (*a*), and when it was horizontal (0°) (*b*) (in *b* the spider left the fly wrapped at the attack site). Disturbance of the sticky spirals was usually greater when the web was 90° (WE). It also tended to be greater in the area below the radius along which the spider had moved than above it in 90° webs (presumably because the spider's body was below rather than above the radius). These attacks included "rotisserie wrapping" behavior, in which the spider rotated the prey on the axis of the radius, often pulling at least one other radius into contact with the exit radius and causing further disturbance to both radius and sticky spiral spacing. Attacks by *Uloborus diversus* (UL) on 0° webs (this species typically built approximately horizontal orbs) produced, in contrast, only very small tracks or none at all, except for the hole left where the prey was cut free (*c*).

behavior to return to the hub that avoids leaving tracks, the "pendulum" returns of *Araneus diadematus* (AR) (Peters 1931), *Argiope argentata* (AR) (Robinson and Olazarri 1971), *Nephila pilipes* (NE) (Robinson and Robinson 1973a), and *Leucauge mariana* (TET) (WE), confirms that such tracks are selectively disadvantageous.

The large intra-specific variation in slants in the field (table O3.1) suggest that attachment sites may sometimes be limited, and that the differences in payoffs from orbs with different slants are small enough that it is advantageous for the spider to adjust her choice of a website to local conditions. A narrow tolerance could impose the cost of searching more extensively to find an acceptable site. In general, larger standard deviations and coefficients of variation were associated with slants less than about 70° (fig. 4.18). This may be associated with wider tolerances in website choice (section 8.4.3.2), though it is also possible that appropriate attachment sites for orbs with more nearly vertical slants are easier to find because of gravity.

4.9.5 THE PUZZLE OF PROVISIONAL RADII

In the large majority of orb weavers (all except nephilids, uloborids, and a few araneids associated with *Cyrtophora*), the spider breaks each new radial line (the "provisional radius" of Zschokke 2000) immediately after she attaches it to the frame; instead of doubling it as she returns to the hub, she reels it up and replaces it with the "definitive" radius (section 6.3.3). Why would the spider deliberately weaken by half the very lines that are most responsible for her orb's ability to stop prey? I know of two possible explanations, but am convinced by neither.

Breaking and then reeling up the provisional radius is accompanied by substantial tension changes on the definitive radius; typically the tension decreases soon after the provisional radius is broken, and then again while the spider is nearing the hub, just before she attaches the new radius there. Perhaps breaking the provisional radius functions to allow the spider to adjust tensions (Eberhard 1981a). But even if this is correct, the details of provisional radius replacement remain paradoxical. By breaking the provisional radius near the frame instead of waiting until she is near to the hub, the spider loses the benefits of doubling the number of fibers in most of the new radius. One araneid, *Zilla diodia*, was exceptional in doubling at least the outer portion of each radius (which are probably under more tension—section 4.4); as expected, she doubled a larger proportion of the radii above the hub, which are under greater tensions (Zschokke 2000). Why is the behavior of *Z. diodia* the exception rather than the rule; and why does even this species typically double less than half of each radius (Zschokke 2000)?

A second possible explanation, proposed by Zschokke (2000), is that the stiffness of the radius would be increased where it was doubled, and that this increase would reduce its ability to absorb the kinetic energy of prey impacts. But the stiffness or Young's modulus is a material property of the silk itself; it is not known to vary in this way. It is the stress on a line (tension/area of its cross-section) (Denny 1976) that will largely determine whether the line will break when subjected to the impact of a prey. A doubling of the total cross-sectional area of the fibers in a radius should halve the stress that is produced on that radius by a prey's momentum, greatly increasing the ability of the orb to stop the prey without breaking.

In sum, only a minimum replacement of the innermost end of the provisional radius might be the most advantageous behavior, yet not a single orb weaver is known to do this, even though a few trend weakly in this direction. The puzzle remains.

4.9.6 THE PUZZLE OF OPEN SECTORS FOR DETRITUS STABILIMENTA

Several groups of orb weavers place linear strings of detritus and/or egg sacs in their orbs. In vertical webs these strings are usually on a radial line above and/or below the hub (see section 3.3.4.1, fig. 3.41). In at least some of these species there is an open space that lacks sticky spiral loops on either side of this string; these include *Arachnura melanura* (AR) (Robinson and Lubin 1979a), *A. logio* (Shinkai 1989), *Cyclosa* spp. (Hingston 1932; Shinkai 1989; Gonzaga and Vasconcellos-Neto 2005), *Acusilas coccineus* (AR) (Shinkai 1989), and *Allocyclosa bifurca* (AR) (Eberhard 2003). I know of no previous attempts to explain this pattern, which presumably requires special behavior on the part of the spider (turn back during sticky spiral construction just before encountering the string). Perhaps prey that are intercepted by this area of an orb tend to escape more easily, pulling themselves free by climbing onto the string, so sticky silk invested in this area may be less likely to pay dividends; but I know of no data to test this hypothesis.

4.9.7 EXCEPTIONS TO TRENDS

Many readers will have been impressed, as I have been, by the frequency in the discussions above with which expectations based on clear, apparently reasonable hypotheses are confirmed in some groups, but are nevertheless contradicted in others. The function of silk stabilimenta is a particularly well-documented illustration, but there are many others (see table 9.3). I believe that often these exceptions do not disprove the hypotheses, but rather result from the effects of additional variables that also have important effects.

This "contrariness" of spiders is, I suspect, related to topics that are discussed in chapters 9 and 10: web-building spiders' great success in terms of large population numbers, large numbers of taxa, wide diversity, and their very long persistence through evolutionary time. Silk webs away from burrows, for example, arose at least 245 million years ago, and orbs at least 165 Mya (Vollrath and Selden 2007); for comparison, the entire group of primates arose only about 65 Mya. This long evolutionary history and great evolutionary success has led to spiders making a relatively thorough evolutionary exploration of the niche space available to predators using silk traps to capture prey.

4.10 NON-ORB WEBS

Non-orb webs need to accomplish the same functions as orbs (intercept, stop and retain prey, etc.; table 3.1), but the effects of their designs on these functions have seldom been studied, and I can give only small snippets of information here.Presumably there are trade-offs between different functions in non-orbs. For instance, a larger, denser tangle above an agelenid's sheet web will probably intercept and knock more prey down onto the sheet, but will also cost more to build, be less mechanically stable where there is wind, and be more visible and more often avoided by prey in some environments. I know of no data to quantify such trade-offs in non-orbs. The one thing that seems certain, judging by the diversity of web designs in non-orb builders such as theridiids (Eberhard et al. 2008b) and linyphiids (Hormiga and Eberhard in prep.), is that the balances between different factors must vary quite dramatically.

Some aerial non-orbs have combinations of functions not known in orbs. For instance, the long legs of the pholcids *Pholcus phalangioides* (Maughan 1978; Nentwig 1983b; Kirchner 1986; Jackson and Brassington 1987; Lopez 1987), and *Physocyclus globosus* (Eberhard 1992a) enable them to run very rapidly, and to execute effective wrapping attacks at a distance, facilitating the capture of walking prey that brush against web lines, as well as airborne prey. A few walking insects were also captured by the aerial sheet weaver *Linyphia triangularis* (Turnbull 1960), but these may have fallen into the webs from above, rather than being attacked on the substrate.

Lai et al. (2017) proposed the novel idea that the aerial, relatively sparse sheets and abdominal markings of the psechrid *Psechrus clavis* attract moths visually at night. These conclusions must be considered tentative, as the study had small samples of factors known to influence moth flight (weather, phase of the moon, and websites); in addition, when the spider was present on the web (as in nature), reducing the web's visibility had no significant effects on moth arrivals (they increased slightly). Nevertheless, further research could be interesting.

Retaining prey by entangling them in loose, slack lines is another tactic that may be widespread in non-orb spiders but absent in orb weavers. The slack, "screw threads" in the webs of the pholcid *Pholcus phalangioides* are thought to entangle prey (Kirchner 1986). The diplurid *Linothele macrothelifera*, the lycosid *Aglaoctenus castaneus*, and the agelenid *Melpomene* sp. often laid fine, slack lines in the sheet by keeping one PL spinneret raised while attaching lines during sheet construction (Eberhard and Hazzi 2013; 2017). The theridiid *Nihonhimea tesselata* and the diguetid *Diguetia canities* sometimes broke lines and released them during tangle construction above the horizontal sheet, thereby causing these and other lines to become slack (WE) (see also fig. 1.7). Tangling may be combined with intrinsic stickiness in the piles of loose sticky silk in some loxoscelids and filistatids (Griswold et al. 2005).

One mysterious aspect of some non-orb webs is the inclusion of tiny droplets of adhesive that are at or just below the lower limit that can be resolved with the naked eye. Tiny droplets occur (presumably convergently) in several groups, including pholcids (Briceño 1985; Deeleman-Reinhold 1986; fig. 9.19), ochyroceratids (Hormiga et al. 2007; fig. 2.13), agelenids (R. Foelix pers. comm.), zoropsids (Eberhard and Hazzi 2017), synotaxids (Eberhard et al. 2008b), theridiids (Barrantes and Weng 2006b; Eberhard et al. 2008b; Madrigal-Brenes and Barrantes 2009), and linyphiids (Peters and Kovoor 1991; Benjamin et al. 2002; WE). They were on lines in tangles, in more or less dense sheets, and in sparse planar arrays that also included non-sticky lines (chapter 9, table 9.1). Some droplets were spaced far apart, and casual, non-quantitative observations suggested that they provide only very small adhesive forces. Their function is unclear.

Another special trait is the high extensibility and elasticity of sticky (cribellum) and non-sticky lines in deinopid webs. These spiders rapidly expand and collapse both sticky and non-sticky lines. I speculate that the unusual multiple "ampullate" gland spigots on the AL spinnerets of deinopids (Griswold et al. 2005) are related to the production of both normal and highly elastic ampullate silk, but know of no data to test this idea.

Some non-orb weavers undoubtedly manipulate tensions while they are building their webs. For instance, the synotaxids *Synotaxus* spp. make double attachments that draw lines into zig-zags, and thus increase their tensions (fig. 5.3, section 4.4.2.2.1); the zoropsid *Tengella radiata* builds very slack cribellum lines just above the edges of the horizontal sheet, where they hang as caternaries and presumably function to retain prey that are about to escape (Eberhard et al. 1993). The spring-like contraction of theridiid gumfoot lines lifts prey free when the prey breaks the weak attachment to the substrate (fig. 2.10); the lift apparently comes from the gumfoot line itself and from tensions on the tangle lines to which it is attached (figs. 5.9, 9.14a, Argintean et al. 2006).

The biomechanical properties of the draglines of the theridiid *Parasteatoda tepidariorum* were affected by the prey the spider had eaten (Boutry and Blackledge 2008), so non-orb spiders may facultatively manipulate silk properties, as in orb weavers (section 3.2.9).

I suspect that in many groups that build non-orb webs around sheltering retreats, the most crucial function was not even listed in table 3.1 for orb webs: to protect the spider by helping to form a retreat that is more or less impregnable—not too large or too small, not too hot or too cold, not too wet or too dry. Once a spider has found a satisfactory retreat, she may adapt her web design to its immediate surroundings (e.g., fig. 10.7), and then employ the ancient "hunker down and endure" strategy of spiders, documented for instance in the sand dune eresid *Seothyra henscheli*, which did not abandon webs even in suboptimal sites (Henschel and Lubin 1992, 1997) (section 8.6). Building more aerial webs (or building retreats constructed from leaves or other detritus) probably represented an important evolutionary escape from dependence on such retreats, allowing web designs to become both more effective traps, and to have more consistent forms (section 10.3; Blackledge et al. 2009c). Dense tangles can also defend spiders (Higgins 1992b; Blamires et al. 2013), and Blackledge et al. (2003) argued that the evolutionary replacement of orbs by 3-D tangle webs in araneoid families such as Theridiidae and Linyphiidae was responsible for their subsequent evolutionary success, due to reduced danger from mud-daubing wasps (assuming that these wasps are major predators of web spiders, and that the lack of inclusion of these families in the prey captured by these wasps was due to web defenses).

As with orbs, better understanding of prey behavior may help us understand non-orb designs. For instance, the recent finding that the gumfoot theridiid *Latrodectus geometricus* preferred to attach its gumfoot lines to smooth rather than rough surfaces, and to attach them to the crests rather than to the valleys of small surface irregularities (Gutiérrez-Fonseca and Ortiz-Rivas 2014), suggests that studies of the choices of paths by walking insects (e.g., the routes preferred by ants) would be rewarding.

4.11 EVOLUTIONARY RESPONSES BY INSECTS? A NEGLECTED ASPECT OF PREY CAPTURE

Web-building spiders are widespread, ancient, and abundant predators, and they capture large numbers of prey (up to about 40% of the spider's body weight/day—Eberhard 1979b; Craig 1989). It is thus very likely that their insect prey have evolved defenses against encountering webs, against adhesion and entanglement in webs that they have contacted (fig. 9.27), and against eliciting attack by spiders once they are entangled in a web. The study of anti-web adaptations holds great promise, but has been largely neglected. Improved understanding of insect countermeasures is likely to illuminate understanding of how natural selection acts on web designs, website selection, and construction times (chapter 8).

4.11.1 AVOIDING OR REDUCING CONTACT WITH WEBS

The ability to see orb webs at a distance has been demonstrated in several insect species (Craig 2003; appendix O3.1.1), and spiders may thus be under selection to reduce web visibility in order to increase prey impacts (section 3.3.3.2). Web illumination, the thicknesses of lines and their adhesive coatings, and the visual background all affect the visibility of a web. The generality of these effects remains to be tested for many prey groups and for many web types. Many non-orb webs are more easily seen, at least by humans, than orbs, so visibility to prey may be more important for them. Nocturnal captures may be relatively more important in non-orb species with more visible webs, but I know of no comparative studies.

Many insects hold their legs and antennae in extended, wide-spread positions during flight (section 3.2.4; Dalton 1975), and these positions probably affect the probability of being captured in a spider web due to two counteracting effects. Extending the legs increases the insect's effective diameter. In a species that flies clumsily and spreads its legs in flight, as for instance a mantid, a tettigoniid, or a pentatomid bug, the extended legs and antennae probably increase the danger that the insect will contact web lines (presumably these insects extend these appendages to stabilize themselves as they fly, to avoid collisions, or to sense landing sites). If the insect moves slowly and has enough control of its flight direction, its extended legs and antennae could also serve as early warning devices for webs, allowing it to reduce the area of its body that contacts the web. For instance, the slow flight and extended long legs of *Hylobittacus* sp. often resulted in the insect contacting the web with only the distal portions of its legs; this left the wings free to beat vigorously, and the insect avoided further contact while pulling free by flying (Blackledge and Zevenbergen 2006). This species had the shortest retention times of the four insects tested by Blackledge and Zevenbergen. Mosquitoes and *Drosophila* gave quick and immediate avoidance responses after briefly contacting a line with an extended leg (fig. 3.2, Craig 1986). Another aspect of insect flight that has not been explored with respect to defense against spider webs is the angle of the long axis of the insect's body with horizontal. Craig (1986) noted that at the moment in which a *Drosophila melanogaster* flew through a hole in a vertical orb, the fly oriented its body to nearly perfectly horizontal, thereby apparently reducing the likelihood of contact with the lines of the orb. Simple experimental tests could test some of these hypotheses (table O10.1).

Still another possibly important aspect of insect flight (and jumping) concerns the effects of velocities on the insect's chances of being stopped and then retained in spider webs. A possible strategy for some clumsy, more or less ballistic fliers (such as scarab beetles and jumping grasshoppers) may be to move as fast as possible and burst through webs rather than avoiding them. Higher velocities might also be favorable due to shorter exposure to visually orienting predators, and being a more difficult target to hit.

The insects whose effects on web spiders are easiest to understand may be those like wingless and immature homopterans and orthopterans, collembolans that can only jump and those that fall from elevated perches (e.g., arboreal ants). Jumping and falling, in contrast with flying, must always involve at least some vertical displacement (thus favoring more horizontal web orientation). Fine flight path adjustments are precluded (see, however, Yanoviak et al. 2005), making the spread legs and antennae of some jumpers (Dalton 1975) useless as web sensors. This kind of prey would presumably not impose selection against having a more visible web.

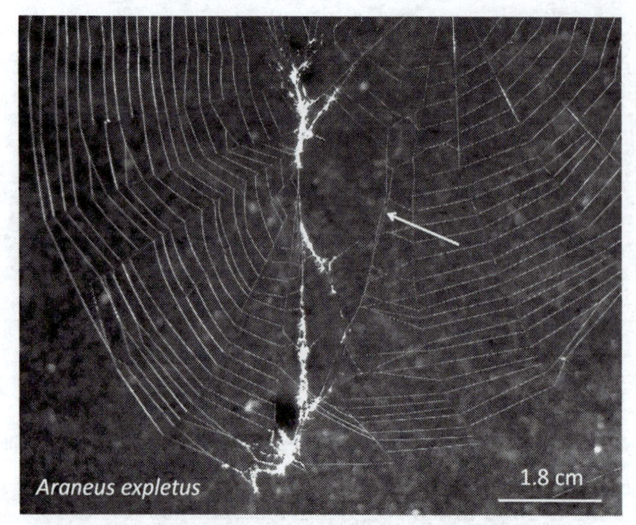

Fig. 4.20. This glistening trail of moth scales in a recently built, uncoated vertical orb of a penultimate instar *Araneus expletus* (AR) spiderling tells a dramatic tale. A large moth was intercepted near the hub, and then fell downward ("tumbled") through the orb, leaving behind a swath of deciduous scales that adhered to the sticky lines. The large holes in the web to the right of the trail were probably made by the spider as she descended, trying to seize the moth (the arrow marks her probable dragline). The moth apparently escaped at the bottom (the web was probably built around 5–6 AM, but the spider did not have a prey package in her retreat when observed at 8 AM).

4.11.2 REDUCING RETENTION TIME AFTER HAVING BEEN STOPPED BY A WEB

Another type of insect countermeasure is to work free of the web before the spider arrives. It often takes a spider several seconds to arrive after a prey has struck the web (section 4.8), so reducing retention time can save a prey's life. One morphological trait that dramatically reduces retention time is a covering of easily detachable setae or scales like those covering many moths (Eisner et al. 1964) as well as some beetles, trichopterans, and flies. The scales adhere to the web, leaving the insect free to pull away from the line (fig. 4.20). This prey defense has probably favored, in turn, especially dense arrays of sticky spiral lines in some spider species (fig. 3.5), lines with special adhesive and mechanical properties (Cartan and Miyashita 2000), and attack behavior that features rapid biting rather than wrapping in many araneids (Robinson 1969b, 1975; Robinson and Olazarri 1971).

Some prey also have behavioral tactics that can reduce retention times, though they are less studied than their probable biological significance would merit. Suter (1978) found that the moth *Sitotroga cerealella* immediately struggled energetically to escape (as would be expected, given its defensive scales), but vibrated the wings in an unusual way. The wings were not extended laterally, so the moth maintained a more compact form with less surface area in contact with the web; presumably this reduced the length of time that the moths were retained.

4.11.3 REDUCING THE PROBABILITY OF BEING ATTACKED IMMEDIATELY

Not all insects behave the same way after they have become trapped in an orb (Suter 1978; Nentwig 1982a; Blackledge and Zevenbergen 2006; Zschokke et al. 2006), and some differences probably represent adaptations to increase the chances that they can work free without eliciting an attack by the spider. In contrast with the scale-covered moth *Sitotroga cerealella*, the bee *Halictus* sp. remained more or less motionless ("played dead") on first impact with the web (Suter 1978). These bees also postponed beating their wings (which produced vibrations that were highly attractive to the spider) until they had freed both of their anterior wings by pushing away lines with their legs, thus giving them a better chance of pulling free from the web by flying. Incidental observations show that mosquitoes almost always became immobile for extended periods (up to many minutes) when introduced into an orb (though often abundant in the field, they make wretched subjects for eliciting prey capture behavior of spiders) (WE). In contrast, some other insects such as *Drosophila melanogaster* flies and *Paratrechina* ants (e.g., I. Escalante in prep.) usually struggle continuously as soon as they are snared. Bad tasting aphids remained motionless in linyphiid webs and were eventually rejected by spiders, while the non-defended isotomid collembolans struggled continually (Alderweireldt 1994). By failing to struggle immediately, an aphid may reduce the level of arousal of the spider when she eventually encounters it, thus lowering the probability of being bitten. Some prey seemed less likely to elicit attacks by the web-building pisaurid *Architis tenuis* when they moved more slowly (Nentwig 1985c).

Attacks by the araneid *Zygiella x-notata* were more likely when vibrations from the prey's impact were followed immediately by additional vibrations from its struggles than when either type of stimulus was presented alone (Liesenfeld 1956). Mature female *Leucauge mariana* in the field attacked simulated prey (short, 2.5 mg strands of fine wire) less rapidly when they were gently introduced into the web rather than being dropped from about 2 cm above it ($chi^2 = 16.7$, $p = 0.00098$, $N = 102, 182$) (WE). Thus when *Halictus* bees separated the two types of vibration by playing dead (Suter 1978), they may have increased their chances of escaping without the spider attacking.

These considerations have practical consequences for studies of web function. Simply measuring retention times (e.g., Eberhard 1986a,b, 1989b) without taking into account playing dead and its effects on spider responses will not always give reliable indications of the probability of capture by the spider. The possible importance of the fine details of ensnarement (which parts of the prey's body touch which web lines) also argues that care must be taken with respect to exactly how prey are brought into contact with webs in experimental studies of prey retention. Simply placing a prey in an orb with a forceps, or dropping or otherwise propelling it into the web (probably the most common techniques in the large literature on attack behavior), may not give biologically realistic data. Even inducing insects to fly into webs under their own power might produce different results if they are frightened and fleeing from a disturbance, rather than foraging undisturbed.

One apparent evolutionary response by web spiders to the defensive tactic of prey flying slowly with their legs extended is to launch the orb web actively toward prey. This occurs in theridiosomatic ray spiders (McCook 1889; Comstock 1967; Coddington 1986a), the deinopids *Deinopis* (Akerman 1926; Robinson and Robinson 1971), and the bolas spiders (Yeargan 1994): all can be induced to lauch their webs by the sound of humming nearby (WE). As far as is known, the primary prey of theridiosomatids are nematocerous flies and other weak-flying insects (Coddington 1986a), whose slow, tentative flight and long, extended legs probably help protect them against other orbs. Increased investment in vigilance and rapidity of response to prey (section 3.3.3.3.2.4) could also counteract some prey defensive tactics.

In sum, few of the many possible adaptations by potential prey to avoid being captured by web-building spiders have been explored; as exemplified by fig. 9.27, there are many mysteries still to be solved. Further studies of the behavior and morphology of insects when they are ensnared by spider webs will almost certainly reveal additional anti-web traits. A better understanding of prey defenses against spider webs promises to illuminate understanding of web designs, silk properties, attack behavior, and website selection behavior.

4.12 SUMMARY (INCLUDING PART OF CHAPTER 3)

The list of functions that webs perform in prey capture is long and varied: intercept prey; absorb their momentum and stop them; adhere to them and retain them until the spider arrives to attack; maintain the web's physical integrity in the face of environmental perturbations such as wind; break, when the web is unable to stop an object, with a minimum of damage to the rest of the web; transmit vibrations from the prey to the spider to alert her to its presence and location; protect the spider from predators and parasitoids; minimize the web's visibility to prey; provide a path by which the spider can reach the prey and that will support her weight and minimize the damage she produces as she moves; minimize the cost (in both material and behavior) of building the web; and (possibly) attract prey.

The most important point is that it is not possible to optimize all these functions for different prey with a single design. A design that is likely to improve one function is usually also likely to impair at least some of the other functions (table 3.1). For instance, larger amounts of glue on sticky spiral lines should improve a web's retention capabilities, but will cost more to produce; they will thus result (for a spider with only finite resources) in shorter sticky lines that cover less area, or that are spaced farther apart. Larger amounts of glue may also make orbs more easily visible during the day and thus reduce the numbers of prey that are intercepted. Although there are a few extreme web designs that probably represent specializations for particular functions (e.g., retaining moths, whose covering of detachable scales makes retention more difficult), untangling the complex set of selective factors acting on the overall design of an orb is generally difficult. An important recent advance is a renewed emphasis on the nutritional payoffs from different prey rather than simply the total numbers of prey captured. Under conditions in which especially large prey are of overriding nutritional importance (that is, when their nutritional payoff is sufficiently great and they are also sufficiently common), traits that are important in capturing especially large prey (in particular the stopping function) will be more important than all the others, and may drive orb evolution. I argue, however, that, contrary to some recent claims, such conditions have not been demonstrated to be generally prevalent in orb weavers, and that the available data thus do not justify an exclusive or overriding emphasis on stopping abilities. I review here (and in appendix O3.2) the reasons why modeling orb web functions is difficult, and sketch some improvements that could help make new models more realistic.

These discussions of possible functions based on the mechanical properties of silk provide a basis with which to reexamine several empirical patterns in the designs of orb webs. For instance, the spaces between loops of sticky spiral and between radii vary substantially in a single orb, and in most species there is a clear edge-to-hub pattern in the spaces between radii and between loops of sticky spiral. Radii are nearly always farther apart at the outer edges of an orb. The most widespread pattern in sticky lines is for them to also be farther apart near the outer edge, but the details of patterns differ in different species. I argue that although there is no single reason for all of the patterns, the differences in the spaces between radii in the orbs of most species probably play an important role, because they result in differences in the prey stopping properties of different portions of an orb, and because they also result in greater needs to hold sticky lines apart from each other in the outer portions of an orb. The orbs of *Nephila* spp. are especially interesting in this respect, because their "radial" lines are bent so that they are largely parallel to each other (rather than divergent near the outer edge); as predicted, the sticky spiral spacing largely lacks the common pattern of larger distances farther from the hub (fig. 4.4).

The spacing between loops of temporary spiral in a given orb varies even more substantially than the sticky spiral spacing, and these differences show both edge-to-hub and above versus below the hub patterns. Because the placement of each temporary spiral loop is guided by the loop immediately preceding it, and because the first loop of sticky spiral (and thus the outline of the capture zone) is guided by the outer loop of the temporary (chapter 7), the most likely explanation for the modulations of temporary spiral spacing is to adjust the overall shape of the capture area of the orb.

A third, very widespread trend in orb webs is the approximately 1:1 correlation between the number of radii and the number of sticky spiral loops (fig. 4.12). This correlation occurs in comparisons both within and between species, and deviations from the 1:1 relationship show taxonomic consistencies. For instance, the orbs of some groups such as *Micrathena sexspinosa* (AR) and *Uloborus* spp. (UL) tend to be relatively radius-rich, while those of others such as the araneids *Acacesia, Deliochus*, and *Pozonia*, and perhaps nocturnal orb weavers in general are sticky spiral-rich.

Because the basic functions of radii and sticky spiral lines differ, it is tempting to consider radius-rich orbs as specialized to stop high-energy prey, and sticky spiral-rich orbs as specialized to retain difficult to hold prey. Inter-specific comparisons, however, gave only mixed support to these characterizations. These included tree trunk orbs, in which the stopping function is probably relatively unimportant, horizontal orbs in which the stopping function is likely to be less important (though empirical data are lacking), and species whose webs are built in large open spaces, in which the stopping function might also be likely to be more important. These contradictions do not justify abandoning ideas regarding the mechanical properties of radius-rich and sticky spiral-rich orbs, but rather suggest that other factors (including, perhaps, inter-specific differences in the mechanical properties of silk lines) are also important. More data are needed to test these ideas.

Another very widespread trend is for non-horizontal orbs to show vertical asymmetries in their designs. Orbs tend to be larger below than above the hub. This top-bottom asymmetry is probably related to two effects of gravity: spiders probably generally run more rapidly downward than upward when they attack prey; and prey struggling to escape from an orb often fall downward gradually ("tumble") through the web. Neither of these factors can explain, however, the common asymmetries in more nearly horizontal orbs. A second widespread up-down symmetry, a tendency for spiders to face downward when resting at the hub of a non-horizontal orb, is correlated with this above-below hub web asymmetry, but a few exceptional species suggest that defense against predators probably also affects some spiders' orientations on their orbs.

The designs of hubs and their functional significance are relatively unstudied. This is unfortunate, given the apparent trend (yet to be carefully documented quantitatively) that hub designs seem to be more consistent intra-specifically, and to be more distinctive in different taxonomic groups such as genera, than are many other orb traits. The likely multiple functions of hub loops include distributing radius tensions and the vibrations from prey and providing footholds for the spider as she turns to face toward prey and launches attacks. The removal of the central portion of the hub following completion of the sticky spiral, which occurs in all orb weavers except uloborids and nephilids, appears to lower tensions on radii and frame lines, and may thus serve to increase the web's ability to absorb the momentum of high-energy prey. It probably also tends to equalize the tensions on radii, which may improve the orb's ability to absorb more generalized stresses such as wind.

There is a general positive correlation, in both intra- and inter-specific comparisons, between the number of hub loops and the number of radii in an orb. Only in the orbs of some theridiosomatids and a few other scattered taxa, whose orbs have very low numbers of radii, are hub loops completely missing. Differences in the spacing of hub loops in different parts of the hubs are obvious in some species, but remain to be investigated systematically. The functional significance is not clear for the relative intra-specific consistency of hub designs, the within-hub differences in the spacing of hub loops, or the differences in hubs between different genera.

There are multiple possible, non-exclusive functions for secondary and tertiary frame lines, including silk economy, equalizing tensions on primary frame lines, and increasing the general extensibility of the orb. The relative importance of these different functions appears to vary in different species. In some, the disadvantage of secondary and tertiary (and even primary) frames in terms of wasted space within the primary frame lines appears to have resulted in their reduction in portions of the orb where the radii are particularly short.

Several new, previously neglected variables promise further insights into the functional significance of different details of orb designs. One concerns expected prey velocities, and the likelihood that they differ according to the distance of the orb from nearby objects. The ability of an orb to stop a prey depends on the prey's momentum, and its momentum is especially sensitive to its velocity (momentum = mass × velocity^2). Thus differences in velocity will have greater effects on the stopping function than comparable differences in prey mass. Objects in the near vicinity of an orb such as tangles of lines, vegetation, tree trunks, etc., probably tend to reduce prey velocities (though direct measurements are lacking). Attention to details of where orbs are built may aid in understanding some details of orb designs.

Another advance comes from close observations of the movements in the wind of individual strands of sticky line, and the realization that wind can damage an orb when fluttering segments of sticky spiral stick to other web lines. Selection to avoid such damage could affect several basic aspects of orb design, including the spacing between adjacent loops of sticky spiral, the distances between adjacent radii (and thus the lengths of segments of sticky spiral), and tensions on sticky lines. One example of the payoff from this insight is the realization that the explosion of diversely arranged and often extremely low tension sticky lines, and their sometimes very close spacing in tiny symphytognathoid spiders was probably made possible by their being built in a nearly windless habitat—the interstices of leaf litter on the floor of humid forests. A second possible application of web damage ideas is in explaining the widespread trend for the loops of sticky spiral to be farther apart near the edge of an orb than in its more interior portions. In general, more data on how and why orbs become damaged in nature could provide useful insights into the functional significance of the details of their designs.

Another factor that may illuminate a suite of orb traits in intra-orb, intra-specific, and inter-specific comparisons concerns the effects on orb designs when there is only limited open space in which to build. Orbs at such sites share several design traits: relatively large numbers of anchor lines; relatively short frame lines; relatively few secondary and tertiary frame lines; relatively small distances between the outer loop of sticky spiral and the frame lines (in some species sticky lines are even laid outside the frame lines); and relatively small spaces between loops of sticky spiral. This combination of traits may constitute a syndrome of space limitation in species that build in constrained spaces such as the leaf litter, and it is also expressed facultatively in some species when they are experimentally confined in small spaces. The same combination is often manifested within the same orb on the side that is closer to the substrate.

Adaptations by insects to avoid being captured in spider webs have hardly been studied, but promise to improve understanding of different web designs. They include avoidance or minimization of contact with the web, adaptations to reduce retention times, and modulation of struggling behavior to reduce the chances that the spider will attack.

In sum, a great deal is now known regarding the functions that orb webs perform, and the significance of differences in the mechanical properties of different silk lines and of differences in orb designs. Perhaps not surprisingly, given the diversity of orb designs and frequent convergences on particular designs (see chapter 9), there does not seem to be any single factor or simple combination of factors that is of overriding importance in understanding orb web evolution. Several factors may contribute to the lack of consistent support for particular hypotheses: there are multiple, more or less interrelated functions, and alternative hypotheses are often not mutually exclusive; there are surprising differences in the mechanical properties of homologous silk lines, and their patterns across different taxa and different orb designs remain to be elucidated.

5

THE BUILDING BEHAVIOR OF NON-ORB WEAVERS

The snow-ball rolling over the carpet of white grows enormous, however scanty each fresh layer be. Even so with truth in observational science: it is built up of trifles patiently gathered together. . . . the collecting of these trifles means that the student of spider industry must not be chary of his time. —*Fabre 1912, pp. 230–231*

5.1 INTRODUCTION

Although a substantial amount of information has now accumulated on the designs of the webs of non-orb weaving spiders (chapter 9), only scattered fragments of their web construction behavior have been observed. Tables 3.A and 3.B are the first summaries ever compiled, and represent reasonably exhaustive summaries of current knowledge of certain details. There is not a single thorough detailed description of the construction behavior from start to finish for such common non-orb designs such as funnel webs, aerial linyphiid sheets, or 3-D tangles. Experimental manipulations that could elucidate the cues that guide their behavior are still completely lacking. The reason for this neglect is probably related to the slow pace of web construction in the many species in which web construction usually lasts several nights (table 5.1), to the readiness of some species to interrupt construction for long periods with even a tiny disturbance (e.g., Tretzel 1961; Szlep 1965; Eberhard 1977a; Lopardo et al. 2004), and, probably most importantly, to the difficulty of understanding and describing behavior that has only a low degree of stereotypy and that often involves movements in three dimensions. It is often very difficult to keep points of reference clearly in mind while the spider is working "aimlessly" (e.g., Benjamin and Zschokke 2002, 2003; Jörger and Eberhard 2006; Eberhard and Hazzi 2017). The contrast with orb web construction is dramatic. Watching a *Nephila clavipes* (NE) interrupt orb construction to add lines to the tangle at the side of her orb and then resume orb construction (Eberhard 1990b) felt like moving from bright sunlight into a black tunnel and then coming out into the light again on the other side.

Because of the scarcity of studies of how non-orb webs are built, this chapter combines descriptions of the construction behavior itself (corresponding to chapter 6 on orbs), what little is known regarding the cues that guide this behavior (corresponding to chapter 7 on orbs), and how some behavior patterns may have changed through evolutionary time (discussed further in chapter 10). I have set apart two "boxes" to serve as illustrations of behavioral descriptions of two species that have been studied in somewhat more detail than usual. I have divided construction behavior into two levels of detail: the order in which lines are added to the web;

Table 5.1. Aspects of higher-level patterns (e.g., the order of placement of lines) in the building behavior of non-orb building spiders deduced from direct observations and patterns of lines in webs. ("n.a." = not applicable because spider does not build this type of line; "?" indicates lack of information, and does not imply "no.")

Taxon	Flexible order of stages	Const. Single/mult	Stages?	St. only after non-st	Skeleton web then fill in: Non-st	Skeleton web then fill in: Sticky	Partial repairs	Interrupt to return to retreat (repeated)	Reg. space betw. st. lines	Sticky from edge move in	Later lines obscure regularity	Pattern in filling in sheet non-sticky lines	Zig-zag sticky lines	Extend radically/ edge	Reference
Non-Orbiularia															
Dipluridae															
Linothele sp.	?	Mult.	?	n.a.	?	n.a.	?[1]	Yes	?	n.a.	?	?	n.a.	?	Paz 1988
L. macrothelifera	?	Mult.	Yes[2]	n.a.	No[2]	n.a.	Yes	Yes	n.a.	n.a.	Yes	No[2]	n.a.	No[2]	Eberhard and Hazzi 2013
Gradungulidae															
Progradungula carraiensis	?	Single	?	Yes[3]	?	Yes[3]	?[4]	?	?	Yes	No?	n.a.	Yes	?[5]	Gray 1983
Austrochilidae															
Thaida peculiaris[6]	?	Mult.	Yes	Yes	Yes[7]	Yes	Yes[8]	No	Yes	Yes	Yes[9]	Often	Yes[9]	Edge	Lopardo et al. 2004
Hypochilidae															
Hypochilus thorelli[4]	?	Mult.	Yes	Yes[10]	Yes[10]	Yes[10]	?	?	?	No	?	n.a.	No	Edge[10]	Comstock 1967, Shear 1969
Filistatidae															
Kukulcania hibernalis	?	Mult.[11]	Yes[11]	Yes	No	?	?	Yes	No[12]/ Yes[5]	Yes[12]	?	n.a.	No	Radial[13]	Comstock 1967, Eberhard 1987b, WE
Pholcidae															
Modisimus guatuso	Yes	Single	Yes[14]	Yes	Yes	Yes	?	No[15]	No[16]	No[16]	Yes	No[17]	No	Edge[18]	Eberhard 1992b
Dictynidae															
Dictyna sp.	?	Mult.	?	Yes	?	?	?	?	Yes[5]	Yes	?	n.a.	Yes	?	Eberhard 1987b
D. sublata[3]	?	Mult.	?	Yes[3]	?	?	Yes	?	?	Yes	?	n.a.	Yes	?	Comstock 1967
D. volucripes	?	Mult.	?	?	?	?	?	?	?	?	?	n.a.	Yes	?	Blackledge and Wenzel 2001b

Table 5.1. Continued

Taxon	*Flexible order of stages*	*Const. Single/mult*	*Stages?*	*St. only after non-st*	*Skeleton web then fill in:* *Non-st*	*Sticky*	*Partial repairs*	*Interrupt to return to retreat (repeated)*	*Reg. space betw. st. lines*	*Sticky from edge move in*	*Later lines obscure regularity*	*Pattern in filling in sheet non-sticky lines*	*Zig-zag sticky lines*	*Extend radically/ edge*	*Reference*
Psechridae															
Psechrus spp.[19]	?	Mult.	Yes[20]	Yes	Yes[7]	Yes	Yes	Yes	Yes[5]	Yes	Yes	No[21]	No[21]	Edge[13]	Robinson and Lubin 1979b, Eberhard 1987a, Zschokke and Vollrath 1995b
P. singaporensis	Yes[22]	?	Yes	Yes	?	Yes	?	Yes[23]	Yes	Yes	No	n.a.	No	No[5,23]	Zschokke and Vollrath 1995b
Fecenia sp. nr. *angustata*	?	Mult.?[24]	?	Yes	?	Yes?[3]	?	?	Yes[25]	Yes	?	n.a.	?	?	Robinson and Lubin 1979b, Agnarsson et al. 2012
Oecobiidae															
Oecobius (?)[26]	?	Single	?[26]	?	No	?	?	Yes	—	—	?	?	?	Edge	Hingston 1925
Zoropsidae															
Tengella radiata adult	?	Mult.[11]	Yes	Yes	?	?	Yes	?	No	No	?	?*	No	Edge[27,28]	Eberhard et al. 1993, Barrantes and Madrigal-Brenes 2008, WE
Agelenidae															
Agelenopsis naevia[4]	?	Mult.	?	n.a.	?	n.a.	Yes	?	n.a.	n.a.	?	?	n.a.	?	Comstock 1967
Melpomene sp.	Yes	Mult.[11]	Yes	n.a.	Yes[7]	n.a.	?	Yes	n.a.	n.a.	Yes	No	n.a.	Edge[29,30]	Rojas 2011
Coelotes terrestris	?	Mult.	Yes	Yes	Yes	Yes	Yes	?	?	No	?	Yes	No	No	Tretzel 1961
Eresidae															
Seothyra henscheli	Yes[31]	Mult.	Yes	Yes	No	?	?	Yes	No?[31]	?	?	?	No	Edge[31]	Henschel and Lubin 1992, 1997, Peters 1992

Table 5.1. Continued

Taxon	Flexible order of stages	Const. Single/mult	Stages?	St. only after non-st	Skeleton web then fill in: Non-st	Sticky	Partial repairs	Interrupt to return to retreat (repeated)	Reg. space betw. st. lines	Sticky from edge move in	Later lines obscure regularity	Pattern in filling in sheet non-sticky lines	Zig-zag sticky lines	Extend radically/ edge	Reference
Stegodyphus nr. *sarasinorum*	?[32]	Mult.	Yes[32]	Yes	?	?	Yes	No[32]	Yes	Yes	Yes	n.a.	Yes	Edge[32]	Eberhard 1987a
Titanoecidae															
Titanoeca albomaculata	?	Mult.	Yes	Yes[33]	?	Yes	?	?	Yes?[34]	Yes	?	Yes?[34]	?	?	Szlep 1966a
Diguetidae															
Diguetia albolineata, D. canities[35]	?	Mult.	Yes[36]	n.a.	Yes[37]	n.a.	?	Yes[38]	n.a.	n.a.	?	No[39]	n.a.	No?	WE
D. imperiosa	?	Mult.	Yes	n.a.	Yes	n.a.	?	?	n.a.	n.a.	?	?	n.a.	?	Bentzien 1973
Ochryoceratidae															
Ochryocera cachote	?	Mult.	?	n.a.?	Yes[4,40]	n.a.?	?	?	n.a.?	Yes[41]	?	?	n.a.?	?	Hormiga et al. 2007
Lycosidae															
Aglaoctenus lagotis[42]	Prob.	Mult.	Yes	n.a.	Yes[43]	n.a.	?	Yes	n.a.	n.a.	?	?	n.a.	?	González et al. 2015
A. castaneus	?	Mult.[44]	Yes[44]	n.a.	Yes	n.a.	?	Yes	n.a.	n.a.	?	No	n.a.	Edge	Eberhard and Hazzi 2017
Hippasa olivacea	?	Mult.	Yes	n.a.	Yes	n.a.	?	?	n.a.	n.a.	?	No	n.a.	?	Hingston 1920
Sosippus sp.	?	Mult.[11]	?	n.a.	?	n.a.	Yes	?	n.a.	n.a.	?	?	n.a.	Edge[30]	WE
Pisauridae															
Archititis tenuis	?	Mult.	Yes	n.a.	?	n.a.	Yes	?	n.a.	n.a.	?	?	n.a.	?	Nentwig 1985c
Thaumasia sp.	n.a.	Single	No	n.a.	No	n.a.	?	Yes	n.a.	n.a.	No	n.a.	n.a.	Edge	Eberhard 2007a
Orbicularia															
Theridiidae															
Latrodectus mactans[45]	?	Single	Yes	Yes	Yes	?	?	?	?	?	Yes	n.a.	No	Yes[45]	Lamoral 1968
L. geometricus[46]	Yes	Mult.	Yes	Yes	Yes	Yes	Yes	Yes[47]	Yes[48]	Yes[48]	Yes	n.a.	No	No	Lamoral 1968, Eberhard et al. 2008b, Barrantes and Eberhard 2010
L. hesperus	?	Mult.	?	Yes	?	?	?	Yes[47]	?	?	?	?	?	?	Salomon 2007
L. variolus	?	Mult.[49]	Yes	?	?	?	?	?	?	?	?	n.a.	No	?	Szlep 1966b
L. tridecimguttatus	Yes	Mult.	Yes[14]	Yes	Yes	?	Yes	?	?	Yes	Yes	No	No	Edge[50]	Szlep 1965
L. pallidus	?	Single	?	?	?	?	?	?	?	?	?	no[51]	No	Radial[52]	Szlep 1965

Table 5.1. Continued

Taxon	Flexible order of stages	Const. Single/mult	Stages?	St. only after non-st	Skeleton web then fill in:		Partial repairs	Interrupt to return to retreat (repeated)	Reg. space betw. st. lines	Sticky from edge move in	Later lines obscure regularity	Pattern in filling in sheet non-sticky lines	Zig-zag sticky lines	Extend radically/ edge	Reference
					Non-st	Sticky									
Steatoda triangulosa	Yes	Mult.	Yes	Yes	Yes	?	?	?	?	?	Yes	n.a.	No	Edge	Benjamin and Zschokke 2002, 2003
S. lepida	?	Single	Yes	Yes	Yes	?	?	?	?	?	Yes	n.a.	No	Radial	Lamoral 1968
S. nr. *hesperus*	?	Mult.	?	?	?	?	?	Yes	No	No	?	?	No	?	Barrantes and Eberhard 2010
Nihonhimea tesselata	Yes	Mult.	Yes[14]	?	Yes[7]	?	Yes	Yes	?	?	Yes[53]	No	?	No[54]	Jörger and Eberhard 2006
Chrosiothes sp. nr. *portalensis*	Yes	Mult.	Yes	n.a.	Yes[55]	n.a.	Yes	Yes	n.a.	n.a.	No	Yes[53]	n.a.	Edge[53]	WE
Tidarren haemorrhoidale	Yes	Mult.	Yes	?	Yes[7]	?	Yes	Yes	No	?	No	No[56]	?	No	Madrigal-Brenes and Barrantes 2009
Theridion evexum	?	Mult.	Yes	?	?	?	Yes?	?	?	?	?	n.a.	No	?	Barrantes and Weng 2007
T. purcelli	?	?	Yes	Yes	Yes	?	?	?	?	?	?	n.a.	?	?	Lamoral 1968
Cryptachaea saxatilis	?	?	Yes	Yes	?	?	?	?	?	?	?	n.a.	n.a.	?	Freisling 1961
Phoroncidia spp.[57]	?	Single	?	No[58]	?	?	Yes	No	?	n.a.	No	n.a.	No	?	Eberhard 1981b, WE
Nesticidae															
Gaucelmus sp.	?	?	?	Yes?[58]	?	?	Yes	?	?	?	?	n.a.	No	?	Barrantes and Eberhard 2007
Linyphiidae															
Linyphia hortensis	Yes[59]	Mult.	Yes	Yes	?	Yes?	?	Yes?[59]	No	No	Yes	Yes[59]	No	No	Benjamin and Zschokke 2004
L. triangularis[60]	Yes	Mult.	Yes	Yes	?	Yes?	?	Yes?	No	No	Yes	Yes	No	No	Benjamin and Zschokke 2004, Benjamin et al. 2002

Table 5.1. Continued

Taxon	Flexible order of stages	Const. Single/mult	Stages?	St. only after non-st	Skeleton web then fill in: Non-st	Skeleton web then fill in: Sticky	Partial repairs	Interrupt to return to retreat (repeated)	Reg. space betw. st. lines	Sticky from edge move in	Later lines obscure regularity	Pattern in filling in sheet non-sticky lines	Zig-zag sticky lines	Extend radically/ edge	Reference
Frontinella pyramitela	?	Single	?	?	?	?	?	?	?	?	?	?	?	?	Comstock 1967
Synotaxidae															
Synotaxus spp.	No[61]	Single	Yes	Yes[62]	Yes[63]	Yes[63]	?	No	No[64]	Yes	No	n.a.	No	Edge[65]	Eberhard 1977a, 1995

[1] Large sheet webs often lasted several months.
[2] The sample was very small, so characterizations are only tentative.
[3] Behavior deduced from web, not observed directly.
[4] Judging by the description of prey capture, the "ladder" portion of the web with sticky silk is probably usually destroyed during many prey captures, and is then built anew; the upper tangle ("retreat network") may be much longer-lived and perhaps added to gradually. Web structure implied that the pattern in sticky silk placement was either working upward from the lower edge, or downward toward the lower edge.
[5] Sticky lines.
[6] Some details also apply to *Austrochilus forsteri*.
[7] Preliminary sheet was later filled in; the tangle above the sheet in *A. tesselata* and T. *sisyphoides* was also filled in later.
[8] Non-sticky lines had "homogenous, uniform, open scaffolding," but building behavior that produced this pattern was not observed. The lines of the skeleton sheet were farther apart near the lateral and distal edges of the web in fig. 1 of Lopardo et al. 2004.
[9] At least early in sheet construction.
[10] Assuming that the "foundation lines" mentioned by Comstock were non-sticky, and that many but not all lines that filled in the lampshade were sticky. Judging by the photos of Comstock, the lower edge of the lampshade was gradually extended.
[11] Web was usually extended during subsequent sessions. At the level of individual movements in *K. hibernalis*, there were two stages (lay non-sticky line as leave retreat, lay non-sticky plus cribellum line as return); but at a higher level no stages were clear.
[12] Non-sticky lines; the radial orientation of the non-sticky lines on which most sticky lines were placed imposed a regularity in spacing between sticky lines.
[13] On different nights.
[14] Stages were not competely discrete, and activities intergraded somewhat, with a somewhat flexible order: extending the skeleton web and filling in the skeleton web; and filling in the sheet.
[15] Returns to the general area of the central dome were frequent, but there was no specific center.
[16] "Wander" across sheet, with no apparent pattern except two details: wandering was more concentrated in central dome area where spider sometimes rested between bouts of activity; and there were abrupt turn backs when the spider reached the edge of the sheet.
[17] Sticky lines being laid, but further confirmation is needed.
[18] Skeleton of sheet was extended laterally, but was subsequent filled in without a clear pattern.
[19] Uncertainties regarding species identification made it unclear whether one or more species were observed in the different studies.
[20] Alternated between laying sticky and non-sticky lines.
[21] Lines alternated in direction at a large scale, but not in sequential attachments.
[22] Used same web in captivity up to several weeks.
[23] Y. D. Lubin pers. comm.
[24] Radii followed by sticky "spiral" lines were all built first on one side of the web, then on the other (Zschokke and Vollrath 1995b).
[25] Radius construction.
[26] The species was identified only as a member of the family Urocteidae (Hingston 1925). The description of the web as a star-shaped tent with a surrounding mat of very fine lines suggests that of *Oecobius*. The association with human buildings and its abundances suggests the widespread anthropophilic *O. annulipes*. Probably all that Hingston witnessed was the construction of the tent portion of the web.
[27] Tangle above orb was nearly entirely missing after only one night (the tiny sample size precludes confident conclusions).
[28] Sheet was sometimes extended at farthest edge from the retreat on subsequent nights.
[29] This pattern was not clear during the first night, but gradual extension was clear over several nights.
[30] Sheet that was added on successive nights in captivity gradually became more elevated away from the substrate and previous sheet lines; this may have been an artifact, however, because sheets in the field had few "layers" of this sort.
[31] Variation in whether sticky silk was added on the first night suggests flexibility. Nevertheless, some sequences of stages seemed to be rigid. Construction of the preliminary mat by walking on the surface occurred at the initiation of the web, and at least the initiation of the mat always preceded deposition of sticky silk; the mat was extended at the edge, though the sticky silk was apparently not extended (details were lacking).
[32] Social species; additions on successive nights were probably not by the same individual, giving clear indication of modular behavior.

Table 5.1. Continued

[33] Behavior was described as being very similar to that of orb weavers, so this easily observed behavioral detail, though not specifically mentioned in the text, was probably similar and was probably used to produce the regularly spaced sticky lines.
[34] Could not be certain from photograph.
[35] No differences were noted, but only two webs were observed for both species.
[36] Very early in construction, the spider ceased returning directly to the mouth of the retreat following each trip away, and began to move and attach lines to other sites in the web. The returns resulted in a poorly defined small platform of lines around the mouth of the retreat.
[37] On the horizontal sheet.
[38] "Rests" of 3–540 s (mean = 100 s, N = 29) facing into the retreat occurred before the spider resumed activity by moving into the mouth, turning 180°, and emerging to continue construction.
[39] In some cases paths resembled (approximately) arcs of circles centered at retreat (and were thus approximately perpendicular to the more radial lines).
[40] Four SEM micrographs of attachments showed some sticky lines attached to some non-sticky lines. There may be many other types of attachment, however.
[41] Pattern of parallel adjacent sticky lines may result from positions of spigots on spinnerets rather than from regularly oriented behavior (see text); but some adjacent swaths also are more or less parallel, so approximately parallel movements of the spider's body may also occur.
[42] Observations were made in relatively small containers in captivity, and spiders may have failed to perform some types of behavior seen in the congeneric *A. castaneus* in the field (Eberhard and Hazzi 2017). The possibility that the spiders made swaths of very fine (difficult to observe) lines instead of (or in addition to) draglines early in construction was not checked.
[43] The movements were the same for building the tube as for building the sheet.
[44] Bridge lines were also floated on the breeze very early during skeleton web construction, so the distinction between stages is not clear. Although the sheet was apparently built as a single unit, and not subsequently extended (at least over the span of three days in the field), the spider apparently added the tangle above it on subsequent days, and may have also added more lines to the sheet itself.
[45] This is *L. indistinctus*, according to Kaston 1970; approximately radial lines were built among very early primary lines in the upper tangle, but at no other stage.
[46] The observations by Lamoral were in South Africa, while the others were in Central America where *L. geometricus* has recently invaded; the unsettled state of *Latrodectus* taxonomy leaves uncertainty as to whether they were the same species.
[47] Return to central disk between bouts of gumfoot line production (especially in young spiderlings).
[48] Pattern for the attachments of several gumfoot lines to each "support" line; this pattern did not occur in successive support lines.
[49] Web is sometimes complete after a month or more!
[50] The edge that is extended is the lower edge of the tangle; the middle and lower layers followed construction of the upper layer.
[51] *L. pallidus* contrasted with *L. tridecimguttatus* in that radial movements that attached the central disc to surrounding lines were common, as were circular movements that may have added lines to the outer edge of the disc.
[52] During construction of central disc and associated gumfoot lines.
[53] In the sheet portion of the web; in *Chrosiothes* sp., expansion of the web at the edges occurred both during a single burst of sheet construction, and from one burst to another or from one day to the next.
[54] In the sheet portion of web; filling in the sheet, however, sometimes began in the center and gradually moved outward, and both tangle and sheet were sometimes extended somewhat on subsequent nights.
[55] Filled in the tangle gradually, but with no apparent pattern; in the sheet, the space between a pair of long, more or less parallel lines was filled in with a highly uniform, more or less hexagonal mesh that was extended crochet-style toward the outer edge.
[56] Drawings of paths suggest that there may be a pattern in which the spider walked more or less straight for 10–30 s and then turned; but the sample is too small to be certain.
[57] An unidentified species in Parque Estadual Intervales, Sao Paulo, Brazil performed identical behavior (WE) to that described for *P. studo* from Colombia.
[58] Sticky line(s) added immediately after prey capture.
[59] The authors describe sticky lines as being laid in "semi-circles," returning to a central area of the sheet; but they also state "there was no clear pattern" (p. 123).
[60] Described as performing similar behavior to that of *L. hortensis*, but precise points of similarity were not specified; I have assumed similar behavior with respect to all of the details in this table.
[61] The order in which unit webs were built was flexible, but not the order of operations during the construction of a single unit; this order was highly stereotyped as a unit, but varied according to whether 1, 2, or 3 sticky lines were produced.
[62] Strict alternation of wet and dry lines that were built in a given sector.
[63] First some non-sticky lines, then alternating sticky and non-sticky lines.
[64] Fill in between parallel lines from top (near the retreat) downward.
[65] Built by adding to lower edge of given rectangular sector; new sectors added on the side of a previous sector.

and the movements of the different legs and spinnerets and how the lines are manipulated while the spider lays lines. These two levels overlap to some extent, but in the better-studied orb weavers these two levels show different overall patterns of evolution (chapter 6) so I discuss them separately here.

It is important to keep in mind the paraphyletic and thus non-evolutionary nature of the category "non-orb" webs (figs. 1.9, 1.10, 1.15). Searches for phylogenetic and functional patterns will have to await additional behavioral observations and better resolution of phylogenies (section 1.6).

5.2 ORDER OF LINES AND OTHER HIGHER-LEVEL PATTERNS

5.2.1 MODULARITY

One theme in this book is that orb web construction is organized in semi-independent combinations of behavioral "modules" (chapter 4), and that this modular organization has had important consequences for their evolution (chapter 10). It is thus of interest to see that several types of data also suggest modular organization in non-orb webs (table 5.1).

Construction by non-orb weavers was often organized into "stages" in which one type of behavior was repeated over and over (table 5.1). A clear example is the pholcid *Modisimus guatuso*, illustrated in figs. 5.1 and 5.2, and described in detail in online box O5.1. Some of this temporal ordering was of physical necessity; for instance, when skeleton web lines supported lines in the sheet, they had to be in place before the sheet filling-in behavior could be performed (table 5.1). But ordering seemed not to be a physical necessity in many other cases (though published descriptions were sometimes unclear on this detail). For the species in table 5.1 that I have observed personally, stages occurred (and often alternated) despite apparent physical independence in the diplurid *Linothele macrothelifera*, the diguetid *Diguetia albolineata*, the psechrid *Psechrus* sp., the lycosid *Aglaoctenus castaneus*, and the theridiids *Nihonhimea tesselata* and *Latrodectus geometricus*. Tretzel (1961) was struck how the entire funnel web of the agelenid *Coelotes terrestris*, which took up to seven days to build, was sketched out at first with a few strong lines.

Perhaps the clearest behavior modules occurred in the synotaxids *Synotaxus turbinatus* and *S. ecuadorensis*. There were several indications (see figs. 5.3, 5.4) that the web of *S. ecuadorensis* was homologous with a single long rectangular "unit" of *S. turbinatus*, and a third species, *S.* sp., also had multiple units like those of *S. turbinatus*) (Eberhard 1995).

Construction behavior also appeared to be modular in other families (table 5.1). Preliminary observations of an undetermined species of *Ochyrocera* (near Yotoco, Valle, Colombia) by Martin Ramírez suggested behavioral modularity: the spider began by making a skeleton web, then filled in several smaller dome-shaped sheets with bands of closely parallel lines, starting from the top of each dome and working downward (M. Ramírez pers. comm.). "Modules within modules" also occurred in the behavior of *Tidarren sisyphoides* (THD): while filling in the dome-shaped sheet, the spider worked in one small section for 1–3 min, then moved to another section (sometimes nearby, sometimes on the opposite side of the web) to fill in some more (Madrigal-Brenes and Barrantes 2009).

The order in which different behavioral modules occurred often appeared to be flexible (Benjamin and Zschokke 2004). For example, *Linyphia hortensis* and *L. triangularis* (LIN) (Benjamin and Zschokke 2004) and *Latrodectus revivensis* (THD) (Szlep 1965) switched back and forth between laying sticky and non-sticky lines; the same may have been true of the psechrid *Psechrus* sp. as she filled in the sheet (Zschokke and Vollrath 1995b) (figs. 5.20, 9.29). Some pseudo-orb webs of the psechrid *Psechrus singaporensis* were built in two blocks of multiple modules—first the radii followed by the sticky spiral on one side of the hub, and then the radii followed by the sticky spiral on the opposite side (fig. 9.29) (Zschokkve and Vollrath 1995b); other sequences have also been observed (W. Piel pers. comm.). The pholcid *M. guatuso* is an apparent exception in that the spider apparently added further lines to the skeleton even after she had begun the "fill in the sheet" stage (Eberhard 1992b). Organizing behavior into modules is probably an ancient trait in spiders; burrow construction was organized in blocks in mygalomorphs (e.g., Coyle 1971, 1981a; Coyle et al. 1992; Coyle and Icenogle 1994).

Modular patterns seemed to be lacking in some fine behavioral details. For instance, no perceptible spatial patterns were noted when spiders filled in sheets in the diplurids *Linothele* sp. (Paz 1988) and *L. megatheloides* (Eberhard and Hazzi 2013), the pholcid *Modisimus guatuso* (Eberhard 1992b), the lycosid *Aglaoctenus castaneus* (Eberhard and Hazzi 2017), the theridiid *Nihonhimea tesselata* (Jörger and Eberhard 2006), or the linyphiids *Linyphia hortensis* and *L. triangularis* (Benjamin and Zschokke 2004). Given our current fragmentary knowledge, however, this conclusion is only tentative.

5.2.2 BEHAVIOR DEDUCED FROM PATTERNS OF LINES

In freshly built sheets of some species, especially those in which lines are relatively widely spaced (figs. 5.2, 5.5),

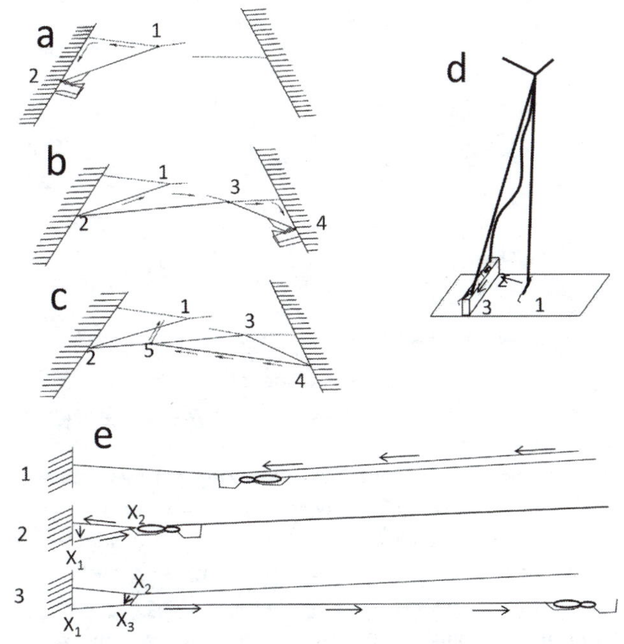

Fig. 5.1. The schematic representations in *a–c* (seen from above) show the order in which a mature female of the pholcid *Modisimus guatuso* attached lines while extending the edge of a sheet that she was building between the buttresses of a large tree (additional lines in the sheet that were closer to the tree trunk are omitted for clarity). After attaching her trail line to a line already present at the web's edge (point 1 in *a*), she walked to the end of this line and then more or less horizontally along the buttress to point 2, where she attached her dragline by bending her abdomen ventrally to touch her spinnerets to the substrate, holding her dragline with one extended leg IV as she did so. The spider then returned along this newly laid line (*b*), attaching her dragline to it or to other web lines (point 3) one or more times as she went. Often, especially in the early stages of sheet construction, the spider immediately went to the other side of the web and extended the edge in a similar manner (point 4). Occasionally the trip to the opposite side was abbreviated, and the spider turned back before reaching the substrate and returned to the first side to extend it further. A series of extensions usually ended (*c*) when the spider attached her trail line part way across the new outer edge of her web (point 5), and walked inward, toward the area nearer the central trunk where the top of the dome of the sheet would be located (see fig 5.2a). The spider seemed to walk more slowly during web extension than during other activities, but in no case did she break and reel up lines (see fig. 6.3). Two web construction behavior patterns in *Diguetia canites*, a member of the little-known family Diguetidae, are illustrated in *d* and *e*. After the spider descended to the floor of the cage early in web construction (1 in *d*); she climbed onto an object there which brought her somewhat closer to the point where she had started (2), thus causing her dragline to go slack. Instead of attaching her dragline immediately, she moved along this object until she was farther from her point of origin and her dragline had become tense again before she attached it (3). A second behavioral detail early in construction (*e*) was unusual in that the spider briefly moved her abdomen rearward (from x_2 to x_3). As she began to return along a new line which she had attached to the substrate (x_1), she attached her dragline to it at x_2 and then reversed direction briefly, moving her abdomen a short distance rearward toward the substrate and attaching her dragline to the line that she had just laid (at x_3). She then returned to the central portion of her web (arrows indicate directions moved by the spider) (pholcid drawings from Eberhard 1992b).

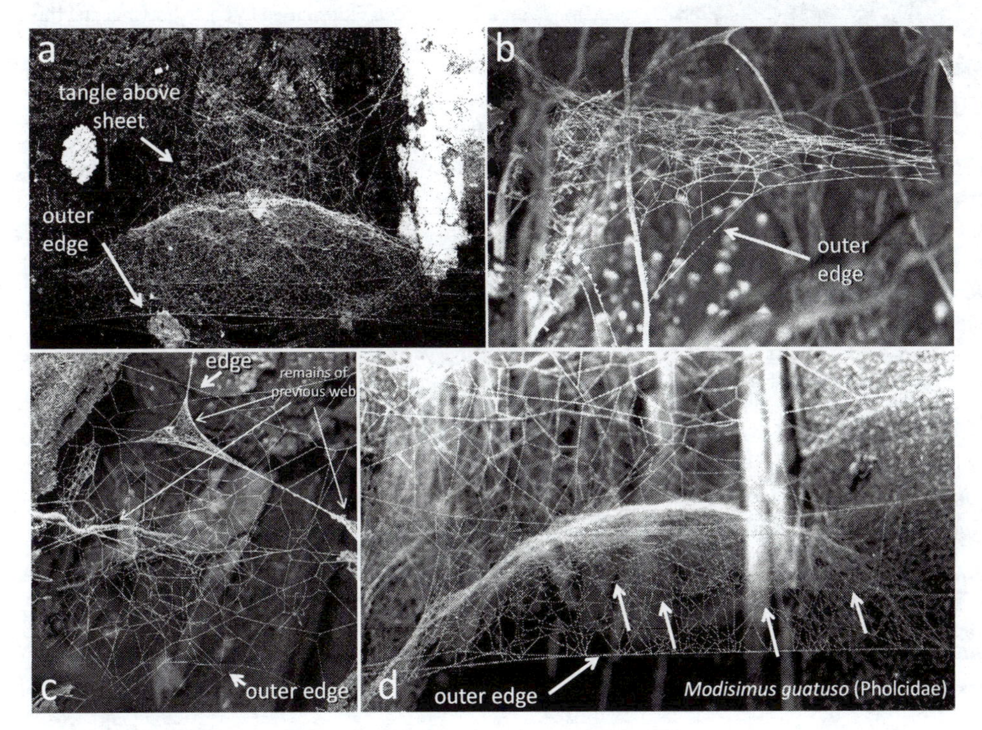

Fig. 5.2. These photos illustrate different stages of construction of the domed sheet webs of the pholcid *Modisimus guatuso*. In a finished web (*a*) the sheet was less dense near the outer edge than near the top of the central dome. Replacement webs still in the skeleton stage an hour or so after the first web was destroyed are seen in lateral (*b*) and dorsal (*c*) views, and a finished replacement web is seen in lateral view in *d*. The replacement sheet was less dense than the original. The long, straight lines in *d* (arrows) were presumably part of the skeleton web (after Eberhard 1992b).

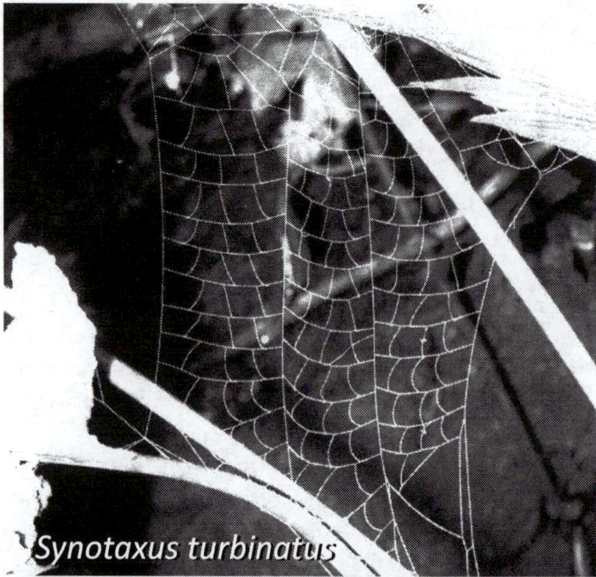

Fig. 5.3. These newly-built webs of two species of the synotaxid genus *Synotaxus* were both vertical, planar, and open-meshed, with small, sparse patches of stickiness on some of their lines; they differed sharply, however, in overall organization. The *S. turbinatus* web consisted of repeated rectangular modules; but no modules were evident in the *S. ecuadorensis* web. Nevertheless, direct observations of construction behavior (fig. 5.4) showed that the differences in design resulted from a relatively minor reorganization of modular construction behavior (after Eberhard 1977a, 1995).

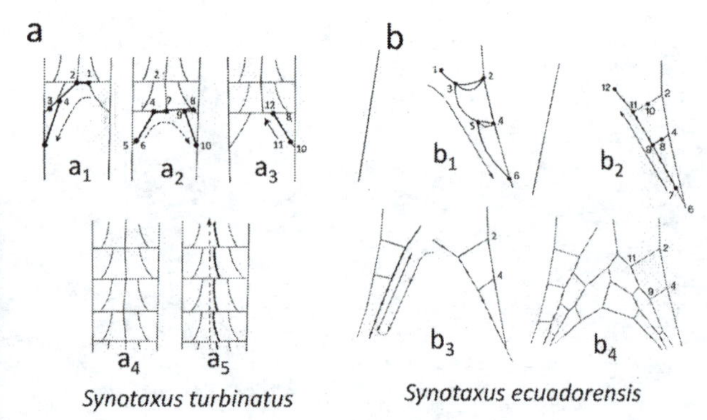

Fig. 5.4. These schematic representations of sticky line construction behavior for a single module of the web of the synotaxid *Synotaxus turbinatus* (*a*) and an entire web of *S. ecuadorensis* (*b*) illustrate that even though their web designs differed (fig. 5.3), their webs were produced using similar sequences. Sticky lines were laid descending along a more or less vertical non-sticky line at one side, then descending along a similar, approximately parallel line at the other side, and finally breaking lines and laying more sticky line while climbing up the middle. In contrast with many spiders, these spiders added lines in a crochet-like fashion, rather than with a radial organization. They also shared the unusual detail (unique among araneoid spiders that have been studied) of reversing direction for a short distance (e.g. attachments #3–4 in a_1, attachments #2–3 in b_1). The implication is that *S. ecuadorensis* web construction is homologous with the construction of a single module by *S. turbinatus*. This, in turn, implies that the spider's nervous system is functionally organized in modules of behavior (after Eberhard 1977a, 1995). Sticky lines are represented with small dots and short thickened segments; numbered dots and thick lines indicate the lines and the attachments in the order in which they were made during the period represented by each drawing; and the spider's path is indicated by arrows.

it is possible to use the web's structure to deduce the spider's path while laying different types of lines. Several of these details can reveal patterns in the spider's movements during web construction. Large numbers of straight lines radiated from one corner of the sheet where the spider rested in several species of the linyphiid genus *Agyneta* (Hormiga and Eberhard in prep.), implying repeated trips to and from this corner. There were "V" shaped patterns at the edges of the sheet that implied that the spider turned back abruptly when she encountered the edge of the sheet during web construction in the pholcid *Modisimus guatuso* (fig. 5.2d, Eberhard 1992b), *Nihonhimea tesselata* (THD) (Jörger and Eberhard 2006), the lycosid *Aglaoctenus castaneus* (Eberhard and Hazzi 2017), and the linyphiids *Walckenaeria* sp. and *Diplothyron* sp. (WE). Direct behavioral observations in the pholcid confirmed that the inferred behavior occurred (Eberhard 1992b). Repair of a hole near the edge of the sheet of the zoropsid *Tengella radiata* (fig. 5.6) suggested that stronger lines were attached to the edges of the hole and then finer lines filled in the space. In contrast, the patterns of lines in some other linyphiid species (*Neomaso patagiatus*, and a mature male of *Po-*

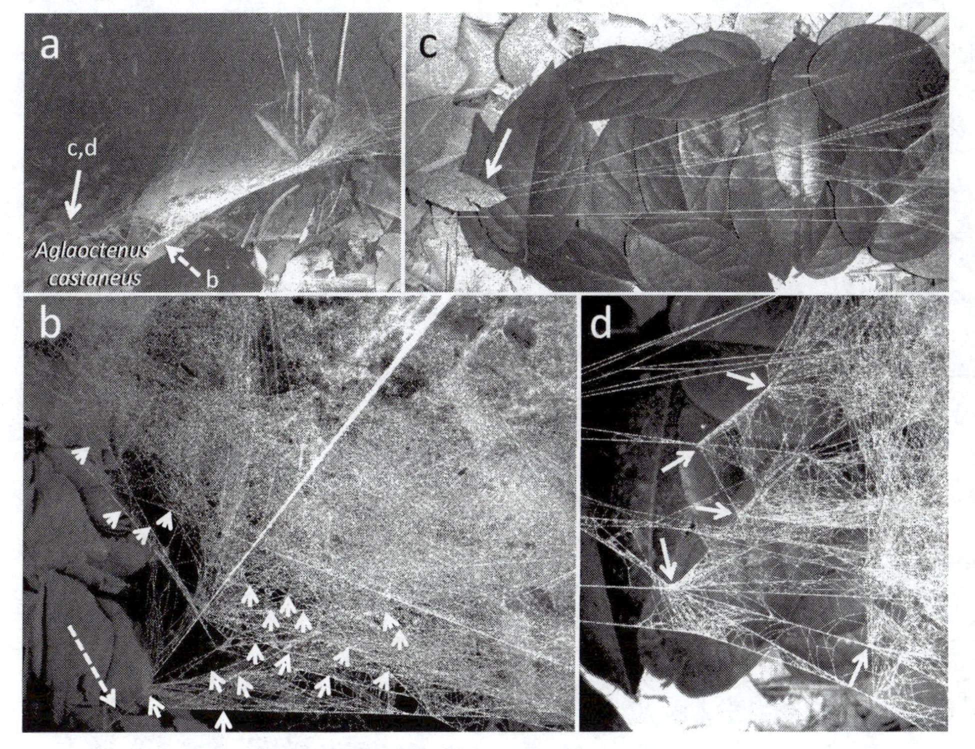

Fig. 5.5. The lines at the edge of this newly built funnel web of a mature female lycosid *Aglaoctenus castaneus* reveal that building behavior was divided into skeleton web and fill-in construction (the photos were taken the morning after the web was built from scratch). Most of the web is shown in *a*, with the areas depicted in closeups indicated by *b–d*. Long straight lines (short arrows in *b*) were laid early in web construction, forming a skeleton web that was anchored to the substrate (dotted arrow in *b*, arrow in *c*) (at least some of these apparent "lines" were actually cables of many fibers). Later the spider connected lines in the skeleton, and filled in with swaths of fine lines to form the sheet. In some places large numbers of fine fill-in lines radiated from attachments to skeleton lines laid earlier (arrows in *d*). In places where the spider did not fill in the skeleton all the way to the edge (*c*, *d*), the distal ends of skeleton lines were easier to distinguish (from Eberhard and Hazzi 2017).

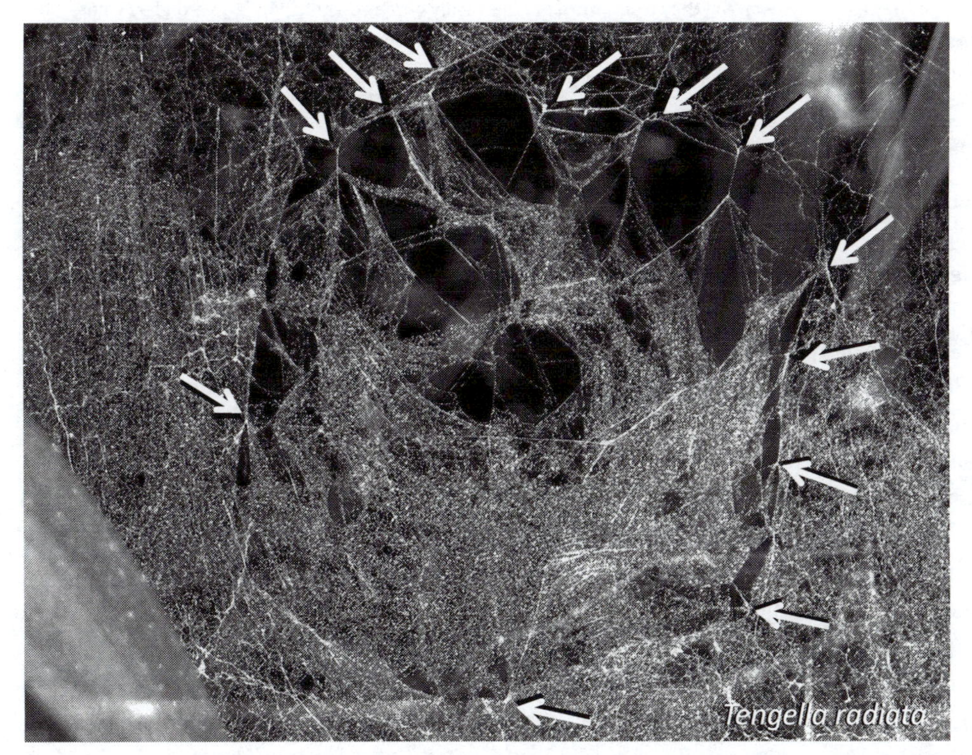

Fig. 5.6. This incomplete repair of a small 2–3 cm diameter hole near the outer edge of the sheet of a mature female of the zoropsid *Tengella radiata* shows two types of line—a skeleton of thicker lines (some of their points of attachment to the edges of the hole are marked by arrows), and swaths of finer lines (the web was damaged the day before, and was lightly dusted with white powder just before being photographed). The upper edge of the hole in the photo was near the edge of the sheet; the greater ease of distinguishing lines in this portion of the repair zone probably resulted from a less complete repair. This non-orb weaver can find holes and adjust her building behavior to the particular characteristics of the damage (see also fig. 7.49).

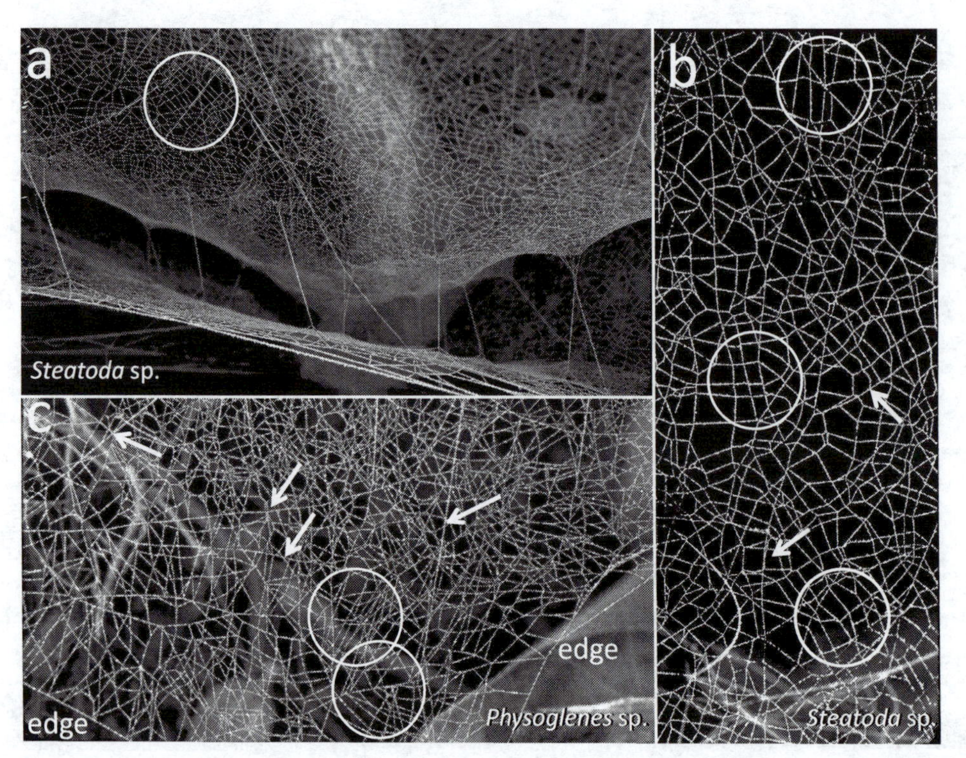

Fig. 5.7. Patterns in the lines of a freshly built web can reveal details of the spider's construction behavior, but care is needed in searching for patterns. The lines in the moderately dense sheets of *Steatoda* sp. (THD) (*a* and *b*) and the physoglenid *Physoglenes* sp. (*c*) offer tantalizing hints. Some relatively straight lines appear to run for long distances (arrows in *b* and *c*). Perhaps these lines were part of a sparse skeleton that was subsequently filled in (as was confirmed for similar patterns of lines in the sheet of the pholcids *Modisimus* spp. and the lycosid *Aglaoctenus castaneus*—see figs. 5.2 and 5.5). In other areas groups of lines seem to be more nearly parallel than would be expected if they were oriented randomly (circles in *a–c*). Did they result from geometrically organized behavior by the spider? Or are they only illusions of regularity among randomly organized lines in the sheet? Observations of web construction behavior are needed (photographs by Gustavo Hormiga).

cobletus sp. 3) suggested that the spider did not tend to turn back sharply from the edge (Hormiga and Eberhard in prep.).

In some sheet web species, such as *Frontinella pyramitela* (LIN) (Suter 1984), *Tidarren sisyphoides* (THD) (fig. 9.5) and the ochyroceratid *Ochyrocera cachote* (fig. 2.13, Hormiga et al. 2007), the densities of lines in sheets were relatively uniform from edge to center. Qualitative inspections indicate that the lines in the sheets of many others, in contrast, were less dense near the edges, including the pholcids *Modisimus* spp. (figs. 5.2d, 5.21, 8.7, Eberhard 1992b), several linyphiids (Dimitrov et al. 2016; Hormiga and Eberhard in prep.), an undescribed cyatholipid from Australia (Dimitrov et al. 2016), the physoglenids *Pahora* and *Runga* sp. (Dimitrov et al. 2016), the lycosid *Aglaoctenus castaneus* (fig. 5.5; Eberhard and Hazzi 2017), and the pisaurid *Architis* sp. (fig. 1.3c). More quantitative analyses confirmed this pattern in the linyphiids *Neriene coosa* and *Diplothyron simplicatus* (WE). Direct behavioral observations also showed that the spiders spent more time filling in central areas in *M. guatuso*.

The sheets of *Diplothyron* sp. (LIN) had sets of curved lines that were nearly parallel to each other and that may have been laid when the spider repeatedly swung her abdomen laterally while walking during construction (Hormiga and Eberhard in prep.). The multiple small subsheets in the webs of *O. cachote* had sections with many parallel lines (fig. 2.13); these were probably laid simultaneously using special arrays of spigots on the posterior lateral spinnerets (Hormiga et al. 2007). The sheets of others such as *Steatoda* (THD) and *Physoglenes* (Synotaxidae) had zones where lines seemed to run perpendicular to each other (fig. 5.7c), implying still further sequences of movement (see Hormiga and Eberhard in prep. for these and additional patterns).

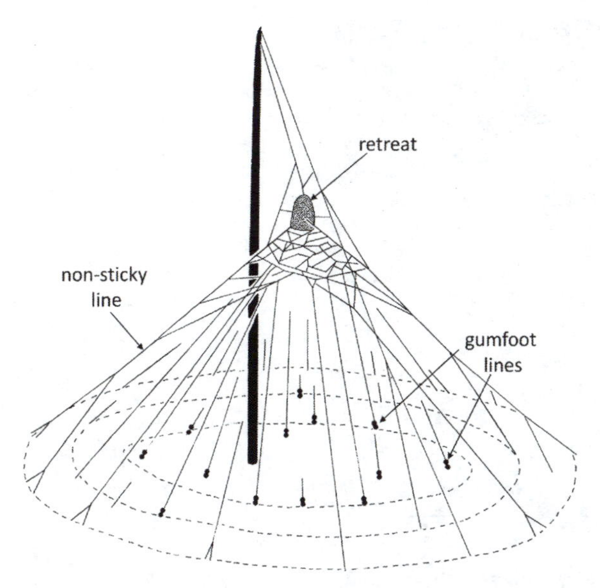

Fig. 5.8. This schematic drawing of a *Cryptachaea saxatilis* (THD) web illustrates a possible general pattern in gumfoot webs: a central location for the spider's retreat, an outer array of non-sticky supporting lines, and an inner array of gumfoot lines more nearly directly below the retreat and more nearly perpendicular to the substrate. The non-sticky lines were apparently under greater tension and were often forked at their ends, possibly to better sustain the tension (redrawn from Freisling 1961).

In the gumfoot+tangle web of *Cryptachaea saxatilis* (THD), some non-sticky lines appeared to be under greater tension (fig. 5.8); they supported the retreat, and enclosed an inner array of gumfoot lines that were laid later (Freisling 1961). Although he did not measure tensions directly, Freisling noted that web lines sometimes deflected plant stems and leaves. He believed that tensions played important roles in guiding construction behavior: that higher tension on a line being laid could trigger construction of a bifurcation; that loosening the tensions in one portion of the web could trigger repairs that raised them again; and that an inability to balance tensions on lines supporting the retreat could trigger abandonment of a site. Tensions on the non-sticky lines that supported gumfoot lines, as well as those on the gumfoot lines themselves, also played a role in lifting captured prey above the substrate (Freisling 1961). Another pattern observed directly in *Parasteatoda tepidariorum* and *Latrodectus geometricus* (Eberhard et al. 2008a), was that series of several gumfoot lines were built while walking along a single more or less horizontal line; the patterns of gumfoot lines in finished webs suggested similar behavior in other species (fig. 5.9).

Fig. 5.9. Several long lines connected to the curled leaf retreat (one is indicated with an arrow) each supported a group of widely-spaced gumfoot lines, suggesting that this Australian *Achaearanea* sp. (THD) builds sequences of consecutive gumfoot lines. Direct behavioral observations of another theridiid, *Latrodectus geometricus*, confirmed that this spider built several gumfoot lines, one after the other along a single approximately horizontal line while she moved toward the central portion of her web (Eberhard et al. 2008a) (photograph by Gustavo Hormiga).

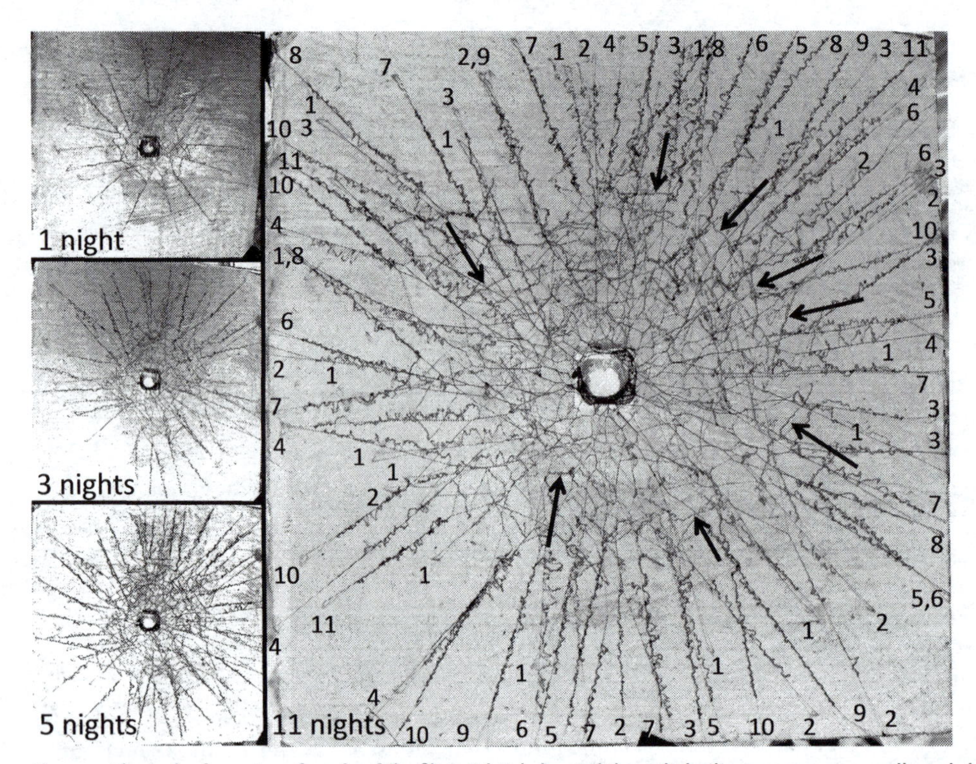

Fig. 5.10. The web of a mature female of the filistatid *Kukulcania hibernalis* built on a 21 x 21 cm cardboard sheet was photographed after 1, 3, 5, and 11 nights of construction (the numbers on the 11 night web indicate the night on which each major radial line was built). Each major radial line consisted of a non-sticky radial line that was laid starting from the central retreat area and attached peripherally to the substrate (thinner, straight lines); on her way back to the retreat, the spider added sticky cribellum silk piled in loose, curled masses (wiggly lines) along the radial line. Near the retreat, the spider occasionally stretched non-radial sticky line (arrows) by extending her legs IV. Several patterns in where radial lines were added are apparent: the first night's web was remarkably symmetrical, and included both long radial lines and connections between them near the retreat; on later nights, the attachments of the radial lines to the substrate tended to be farther from the retreat; lines were added each night to several different sides of the web, rather than being concentrated in a given sector; later radial lines were added in the spaces between earlier radial lines (they thus differed from the radii in orb webs, which were laid at "final" angles that were not subsequently subdivided—see fig. 6.5, section 6.3.3.2); and later radial lines tended to originate on others (including radial lines) that were already in place, rather than at or near the edge of the retreat.

5.2.3 OTHER PATTERNS

One behavioral sequence may be universal. In species in which sticky silk was included in the web, the sticky silk was always laid after other, non-sticky lines had been laid (table 5.1). This pattern was striking for its consistency (it is one of the only behavioral patterns in this entire book with no known exceptions!). Other patterns (e.g., laying sticky lines first and then connecting them and further supporting them with non-sticky lines) are eminently feasible, so it is not a "frabricational constraint" (sensu Coddington 1986b). Perhaps, if sticky silk is relatively expensive compared with non-sticky silk, then it may be advantageous to lay exploratory lines first with non-sticky lines, to sketch out the area that will later be covered with sticky silk.

A second pattern, seen in the sticky lines of a variety of cribellate spiders, including austrochilids (Griswold et el. 2005), phyxelidids (Griswold et al. 2005), dictynids (Emerton 1902; Nielsen 1932; Comstock 1967; Blackledge and Wenzel 2001b; WE), desids (Griswold et al. 2005; Opell 1999), titanoecids (Szlep 1966; Griswold et al. 2005), and eresids (Eberhard 1987), is for a sticky line to zig-zag back and forth between two more or less parallel non-sticky lines, forming a "ladder" (fig. 1.3b).

Another pattern was that non-orb webs were usually built over an extended period of time (often multiple days), rather than in a single burst of activity as occurs in orb weavers; usually the web increased gradually in size during this process (fig. 5.10). Extension over multiple days occurred, for instance, in the genera *Linyphia* (LIN) (Wiehle 1929), *Latrodectus* (THD) (Szlep 1965), *Pholcus* (Pholcidae) (Duncan 1949), *Coelotes* (Ageleni-

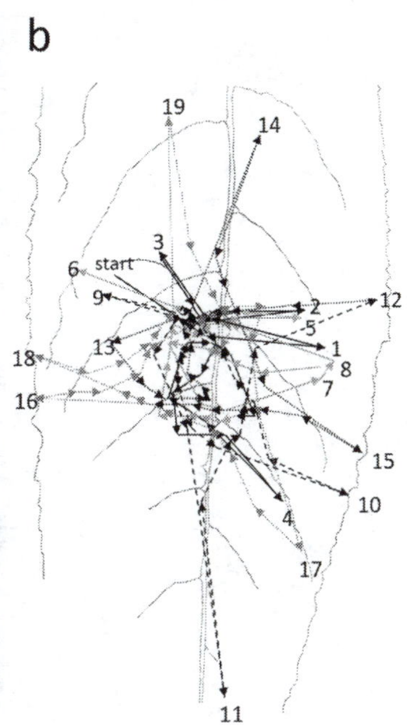

Fig. 5.11. The photo (*a*) shows a finished web of non-sticky lines of the pisaurid *Thaumasia* sp., with the spider resting in a central, cup-like pocket. The web apparently functioned to extend the spider's sense of touch. The diagram (*b*) gives the sequence in which a smaller web was built on a leaf; arrows mark the path of the spider during construction and numbers indicate the order in which radial lines were built. Several patterns were apparent: the spider returned to the central area after building each radial line; consecutive radial lines tended to be laid in opposite sectors of the web; and early radial lines were shorter (from Eberhard 2007a).

dae) (Tretzel 1961), *Diguetia* (Diguetidae) (WE), and the dictynids *Dictyna* and *Mallos* (Blackledge and Wenzel 2001b, WE). Several factors might promote this strategy: gradual accumulation of silk in the silk glands; relatively durable, long-lived webs, in which long-term accumulation of lines is feasible; and web designs in which adding more lines can increase the web's ability to capture prey. Tretzel (1961), noting the less frequent renewal of the thin adhesive layer just above the sheet in captivity than in the field, believed that hunger induced more frequent additions in *Coelotes terrestris* (another possibility would be more frequent destruction in the field). A few taxonomically scattered species are exceptional, building their webs in a single burst of construction behavior: the gradungulid *Progradungula carraiensis* (fig. 10.9) (Gray 1983); species in the synotaxid genus *Synotaxus* (figs. 5.3, 5.4); *Phoroncidia* sp. (THD); the pholcids *Modisimus guatuso* and *M. bribri* (figs. 5.2, 5.21, 8.7); and the pisaurid *Thaumasia* sp. (fig. 5.11).

Some patterns with which web lines accumulate may necessitate changes in ideas about web evolution. As a spider gradually extends the periphery of her web (e.g., figs. 5.10, 5.12a, 5.15), she continues to come into contact with the substrate at the edge. She thus does not gradually become isolated from stimuli from the substrate, as occurs in orb weavers (Witt 1965). Nevertheless, as illustrated by the geometric regularity of the webs of the filistatids *Kukulcania hibernalis* (fig. 5.10) and *Misionella mendensis* (fig. 10.12), the pisaurid *Thaumasia* sp. (fig. 5.11), and the oecobiids *Uroctea durandi* (fig. 10.5) and *Oecobius concinnus* (fig. 9.6), sustained, intimate interactions with the substrate do not preclude highly organized geometrical web designs, as is suggested by some general web classifications (Blackledge et al. 2009c; Fernández et al. 2014; Bond et al. 2014; Garrison et al. 2016).

Another possible general pattern is that many spiders periodically returned briefly to a retreat or a central area and then resumed web construction (table 5.1). Returns of this sort occurred during sheet web construction in the funnel-web agelenid *Melpomene* sp. (Rojas 2011), the dipurid *Linothele macrothelifera* (Eberhard and Hazzi 2013), and the lycosid *Aglaoctenus castaneus* (Eberhard and Hazzi 2017), and during sheet+tangle web construction by the diguetid *Diguetia canities* (WE), and several theridiids (Jörger and Eberhard 2006; Madrigal-Brenes and Barrantes 2009), including *Chrosiothes* sp. (WE). Some spiders paused at the retreat for several minutes before resuming building behavior (the mean duration of 29 pauses in *D. canites* was just under 2 min) (WE).

The returns themselves may function to reset the spi-

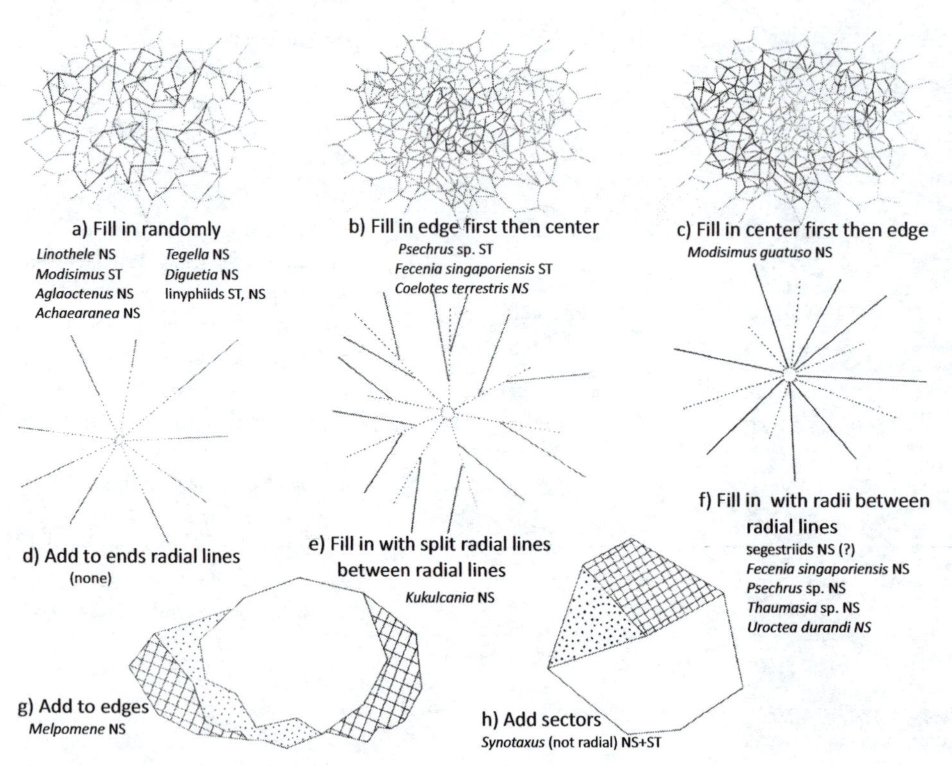

Fig. 5.12. Sketches of different patterns in which new lines (solid) could be added to preexisting lines (dotted) in non-orb webs, with names of species that have been observed building in each pattern (ST = sticky; NS = non-sticky; agelenids were assumed to produce non-sticky silk). Most of these classifications are tentative, because most descriptions were only qualitative and preliminary, and the "random" category is especially uncertain because searches for patterns have been only preliminary at best (cross-hatched and stippled portions in *g* and *h* indicate additions following construction of the original blank portions). Some species used different patterns for adding sticky and non-sticky lines (e.g., the pholcid *Modisimus guatuso*, the psechrid *Psechrus singaporensis*). Some species combined patterns; for instance, the pisaurid *Thaumasia* sp. filled in with radial lines by moving radially away from the central resting area (*f*), but routinely split the radial lines on the way back (*e*). References are as follows: *Achaearanea* (Jörger and Eberhard 2006); *Aglaoctenus* (Eberhard and Hazzi 2017); *Coelotes* (Tretzel 1961); *Fecenia* (Zschokke and Vollrath 1995b); *Kukulcania* (WE); *Linothele* (Eberhard and Hazzi 2013); linyphiids (Benjamin et al. 2002; Benjamin and Zschokke 2004); *Melpomene* (Rojas 2011); *Modisimus* (Eberhard 1992b); *Psechrus* (Eberhard 1987a; Zschokke and Vollrath 1995b); *Synotaxus* (Eberhard 1977a); *Tengella radiata* (WE); *Thaumasia* (Eberhard 2007a); *Uroctea* (Kullmann and Stern 1981).

der's memory of the location of her retreat. Path integration ("dead reckoning"—see section 7.3.2) is probably an ancient trait in arachnids, enabling animals to return to burrows or other retreats after attacking prey in various groups, including ctenids (Seyfarth et al. 1982), mygalomorphs (Coyle 1986), scorpions (Polis 1990), and other invertebrates (Shettleworth 2010). A second possible advantage of repeated returns to the retreat was that they produced radial lines converging on the retreat; these could improve the spider's ability to rapidly locate and then reach prey in her web. In *N. tesselata* (THD) the spider arrived at and left the mouth of the retreat along the same few lines that radiated from the retreat mouth to the nearby sheet (Jörger and Eberhard 2006). These lines may have subsequently increased the spider's ability to locate prey on the sheet while she was still in her retreat (Barrantes and Weng 2006a).

The functions of the pauses at the retreat are less certain. Two speculations are to refill the silk ducts or glands, and to rest and recover from oxygen debts due to activity. Many pauses were short (nearly a third of pauses in *D. canites* lasted less than 20 s) (WE), favoring the rest hypothesis.

Spiders used a variety of patterns to extend non-orb webs (fig. 5.12). Some made radial extensions around a central site, as in the filistatid *Kukulcania hibernalis* (fig. 5.10, Comstock 1967), the psechrid *Psechrus* sp. (Zschokke and Vollrath 1995b), and the pisaurid *Thaumasia* sp. (fig. 5.11) (Eberhard 2007a). In some of these species, radial lines built later branched from those built earlier (figs.

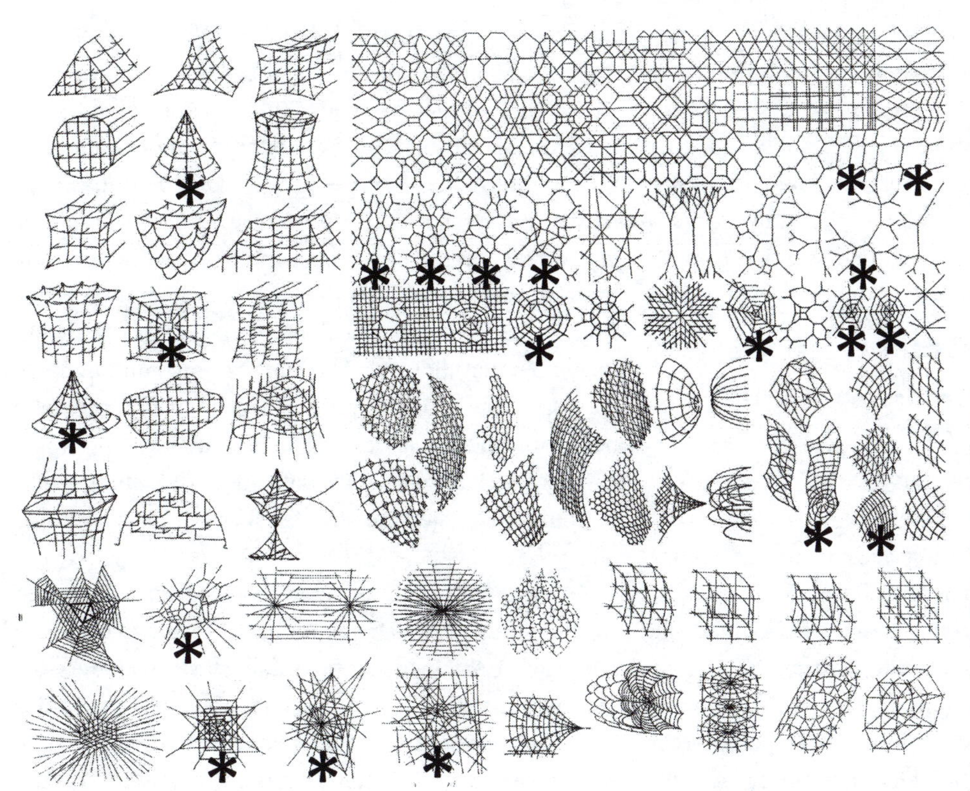

Fig. 5.13. There are many possible ways of organizing lines that are *not* used by spiders. This array of designs, drawn by architects interested in light-weight buildings supported by networks of flexible lines, illustrates designs similar to those of spider webs (marked with "*") represent only a subset of possible arrangements. Notably lacking in spiders are geometrically regular patterns in three dimensions, and patterns that are not centered on a point of convergence (see also fig. 9.28). Patterns that *do* occur frequently in spider webs involve planar, radial arrays; several lineages have evolved to pull a plane into a cone. Perhaps behavioral limitations, such as the difficulty of maintaining precise orientations while moving in three dimensions, have limited web evolution (from Otto et al. 1975).

5.10, 5.11). A mature female of the diguetid *Diguetia canities* began construction of her tangle+sheet web by laying nearly vertical lines extending downward from near the mouth of her tubular retreat. In each case she walked several cm along the substrate before attaching the new line. She tended to use less nearly vertical lines for later descents, and later lines gradually extended the volume encompassed by her growing tangle (WE).

A second pattern was to make repeated additions to the web's edge, causing it to grow outward (fig. 5.12a), as in the linyphiid *Linyphia hortensis* (Benjamin and Zschokke 2004) and the pholcid *M. guatuso* (Eberhard 1992b) (and also in the derived araneid *Cyrtophora citricola*—fig. 1.7). In *Dictyna volucripes* (Dictynidae), relative increases in web size varied according to prey capture rate: all webs of well-fed spiders increased by approximately 50% over a 9-day period; some poorly-fed spiders with small initial webs increased web size by 200–300%, while other large initial webs increased little (Blackledge and Wenzel 2001b). The authors hypothesized that web investment may have been influenced by previous foraging effort and success.

Still another pattern was to make additions throughout an original area that had been blocked out earlier, as in *Steatoda lepida* (fig. 5.19, Lamoral 1968). Filling-in behavior also showed several patterns: gradually filling from a central portion moving outward (fig. 5.12b); gradually filling from the edge moving inward (figs. 5.12c, 5.15) (both patterns also occur in orb construction—see chapter 6); and filling in with no clear pattern (fig. 5.12a). The linyphiids *Linyphia hortensis* and *L. triangularis* were described as moving in semicircles and also as lacking any pattern during sticky line construction (Benjamin and Zschokke 2004). The striking intra- and inter-specific diversity in the finished sheets of webs, including lines and attachments between sticky and non-sticky lines of the linyphiids *Linyphia triangularis* and *Microlinyphia pusilla* (Benjamin et al. 2002), intimates that "filling-in" may be more complex in some groups than direct behavioral observations suggest (see also fig. 5.5 of the lycosid *Aglaoctenus castaneus*).

There are many geometrically regular non-orb patterns that are *not* known to be utilized by spiders (fig. 5.13). In general, those geometrically regular patterns used by spiders have some sort of radial symmetry. A pattern conspicuous by its nearly complete absence was straight lines running from one side of the web to the other (exceptions include webs spanning single leaves built by the dictynid *Mallos hesperius* [WE] and the pisaurid *Hygropoda* sp. #3656 in fig. 5.14). There will likely be additional surprises regarding what spiders are capable of doing, as exemplified by the widely spaced, nearly perfectly parallel lines (much too far apart for the spider to reach from one to the next) used by *Achaearanea* sp. to curl a leaf in which to build its web (fig. 5.16).

5.3 LOWER-LEVEL PATTERNS: LEG MOVEMENTS AND MANIPULATION OF LINES

5.3.1 WALK UPRIGHT ON THE SUBSTRATE OR A DENSE SHEET

Walking on the substrate during web construction is undoubtedly less derived than walking under lines, and Bristowe (1930) argued that the ability to walk under silk lines rather than on top of a solid substrate or a dense sheet was a major evolutionary advance in web evolution. Walking under lines may be associated with other behavioral as well as morphological differences, because leg movements and positions are severely constrained by the positions of the lines themselves (section 2.4).

Unfortunately, the descriptions of how non-orb spiders move during construction or on their finished webs seldom include details of leg movements. There are preliminary indications that behavior may vary. Lycosids in the genus *Aglaoctenus* walked on the substrate while laying early lines and on the upper surface of their dense sheets when attacking prey (González et al. 2015; Eberhard and Hazzi 2017), but walked under individual lines during intermediate stages of web construction (Eberhard and Hazzi 2017). The amphnectid *Metaltella simoni* also walked on top of its sheet when attacking prey, but hung upside down from the ceiling of its more or less tubular silk retreat (I. Escalante pers. comm.). It seems inevitable that diplurids like *Diplura* spp. and *Linothele* spp. that walk on the upper surfaces of their sheet webs but that also built aerial tangles of lines far above their sheet webs (Paz 1988) must walk under these lines while building them; their construction behavior has not been observed.

One particular mystery concerning sheet webs needs further study (table O10.1). As noted long ago by Hingston (1920), the fact that the tarsal claws of spiders are curved ventrally (fig. 2.16) means that when a spider walks on a dense sheet of silk lines, she must run the risk of snagging her tarsal claws on the sheet. To my knowledge, however, such snagging has almost never been observed. The only exception I know is a mystery itself. Hingston (1920) reported that the sheet-weaving lycosid *Hippasa olivacea* normally ran freely with no problems across its dense sheet, but her legs became snagged when her palps were removed (!).

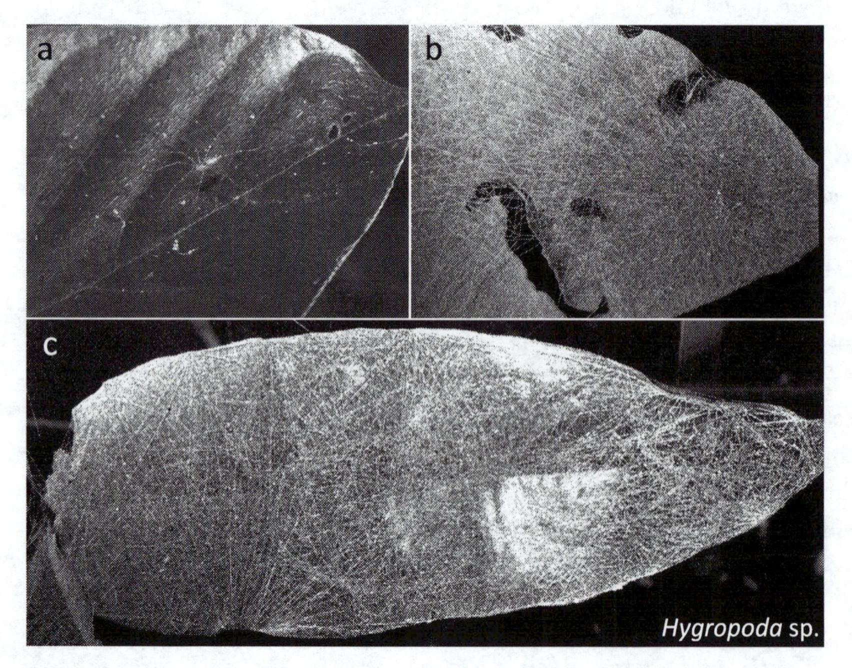

Fig. 5.14. The pisaurid *Hygropoda* (?) sp. (#3656) rested on the upper surface of a small sheet spanning a single slightly curled leaf (*a*). Closeup views of two webs (*b*, *c*) revealed that they were sheets largely made up of numerous long, straight lines (especially evident in the left portions of *b* and *c*), a pattern not seen in the lines in the sheets of other groups (figs. 5.7, 5.21, 8.5, 8.7) (the bright patches in *c* were reflections from the shiny leaf surface below). The webs described for other pisaurids were quite different (figs. 1.3, 5.11).

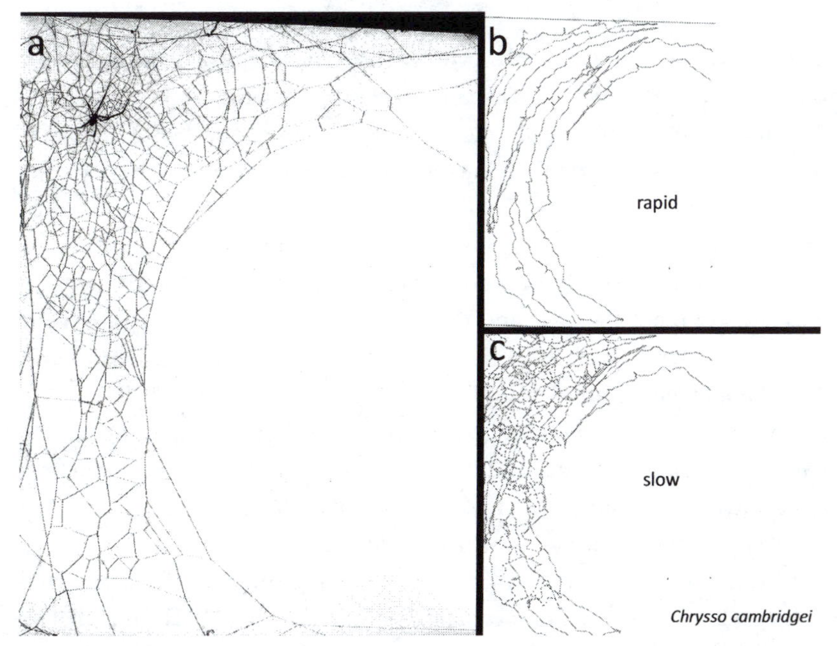

Fig. 5.15. The photo in *a* shows the planar web of a mature female *Chrysso cambridgei* (THD); the drawings *b* and *c* (traced from behavior recordings) show the paths of her body as she built two types of lines in this web. The density of the paths is approximately the same as that of the lines in the web, indicating that the spider laid a single line as she moved (numerous additional paths of the spider walking along the surface of the cage during apparent exploration are omitted). Early during construction (*b*) the spider apparently extended the plane by moving back and forth along the edge, laying new lines as she went, a technique shared with the distantly related pholcid *Modisimus guatuso* (fig. 5.2). She moved relatively rapidly while building these non-sticky lines (mean 3.6 mm/s). Later (*c*), she moved more slowly while moving along the paths indicated by the dotted lines (mean 1.6 mm/s); these were probably sticky lines. Fast and slow movements were intercalated, occurring in alternating bursts. Direct observations are needed to confirm the link between speed and sticky vs. non-sticky lines (from S. Zschokke unpublished data, courtesy of Samuel Zschokke).

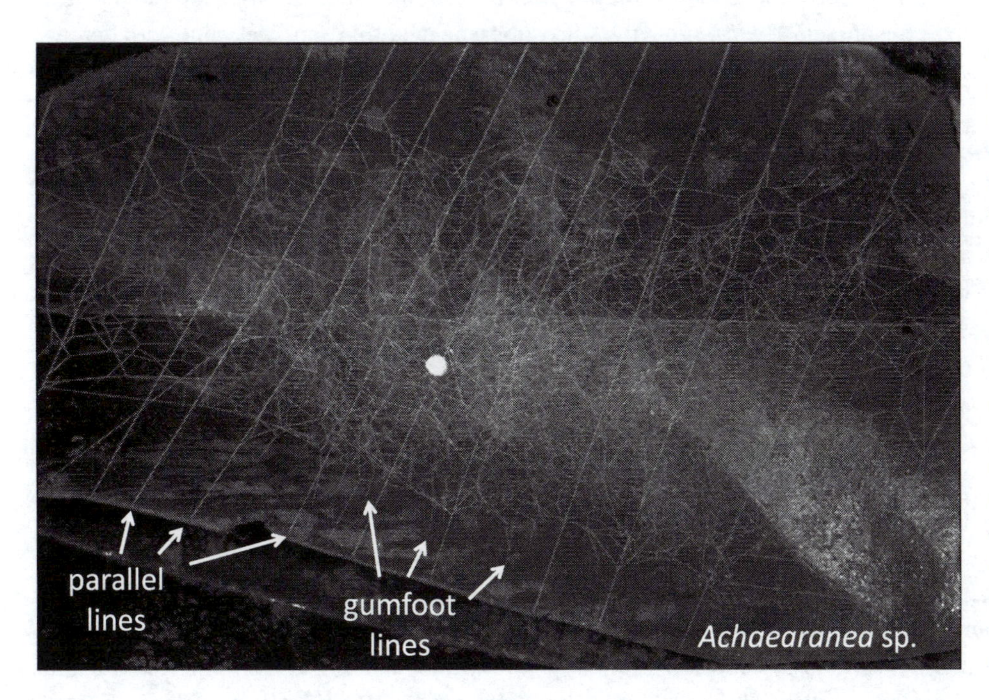

Fig. 5.16. These strikingly parallel, widely spaced lines may have served to curl the leaf in which *Achaearanea* sp. (THD) from Thailand built a tangle web that had short apparent gumfoot lines (a few are labeled; the white sphere is probably an egg sac). This species may thus have the ability (heretofore unsuspected in theridiids) to measure distances that are longer than her own body (photograph by Gustavo Hormiga). The distantly related dictynid *Mallos hesperius* showed a similar ability to build widely spaced parallel lines across curls in leaves (WE).

5.3.2 WALK UNDER SINGLE LINES

Walking under a single line is morphologically and behaviorally challenging, and includes the need to rotate the legs on their long axes (section 2.4; Eberhard 2017). The details of the ways that different taxonomic groups solve these problems may vary (and could provide taxonomically useful characters), but very little is known. One strange behavior was that of the lycosid *Aglaoctenus castaneus*, which built long aerial lines early in the construction of its sheet web. When she returned to her retreat under a long line, she held the line with only the legs on one side of her body while those on the other side hung down in an extended position and were more or less immobile (Eberhard and Hazzi 2017). Apparently similar positions occurred in fragmentary observations of spiderlings of the mygalomorphs *Ummidia* sp. (WE), *Atypus affinis* (Cox 2015; J. Cox pers. comm.), and *Sphodros rufipes* (WE). This asymmetric position contrasts with all descriptions of orbicularians, which used legs on both sides while walking under a single aerial line.

5.3.3 HOLD THE DRAGLINE WHILE MOVING AND WHILE ATTACHING IT

Some groups nearly always hold the non-sticky dragline with one tarsus IV while building; these include orbicularians (both orb weavers and non-orb weavers) as well as some haplogynes such as the pholcid *Modisimus guatuso* and the diguetids *Diguetia canities* and *D. albolineata* (WE). In orbicularians, the dragline appears to slide through the tarsal claw IV as the spider moves forward. But in the haplogyne *D. albolineata*, legs IV usually gripped the line in alternation; these "hand over hand" movements also occurred when the spider descended hanging from her dragline (WE).

The consistency with which legs IV contacted draglines varied in different groups (table 5.2). Some spiders, such as the pisaurid *Thaumasia* sp. (Eberhard 2007a), the filistatid *Kukulcania hibernalis* (WE), the diplurid *Linothele macrothelifera* (Eberhard and Hazzi 2013), and perhaps mygalomorphs in general (e.g., Coyle 1971, 1981a; Coyle et al. 1992; Coyle and Icenogle 1994; F. A. Coyle pers. comm.), may never touch the lines they are producing with legs IV (or any others). Both orb-weaving and non-orb weaving orbicularians spiders, in contrast, seldom moved without holding the dragline with one leg IV. Others, like the lycosid *Aglaoctenus castaneus*, were intermediate, and held it only occasionally. In this species, however, which typically produced swaths of lines rather than single draglines, the contact may have been with the ventral surface of the tarsal segment, rather than the tarsal claws (Eberhard and Hazzi 2017).

Holding the dragline with one leg IV just at the moment it was being attached to another line or to the substrate was common in the pholcid *Modisimus guatuso* (fig. 5.1) and the diguetid *Diguetia canities* (Eberhard 1967; WE), and also occurred in *A. castaneus* (Eberhard and Hazzi 2017). This grip probably allows the spider to raise the tension on the new line. If, instead, her leg did not hold the line, an increase in tension would presumably only be possible if valves in the silk gland ducts grasped the silk (section 2.2).

An instructive case of tension manipulation occurred in mature female *Phoroncidia* sp. (THD) immediately following prey capture on her single horizontal sticky line web. At other times the spider held her non-sticky dragline with one leg IV, but after an attack she added a new segment of sticky line as she returned to her resting site at one end of the line (Eberhard 1981b) without holding this new sticky line with her leg IV (presumably her claw would have scraped off the glue droplets as it slid along the line). Nevertheless, she then raised the tension on her web by adding a short, non-sticky extension to the sticky line: she held this non-sticky line with her tarsus IV, and when she reached her resting site, she turned 180° to face the sticky portion, tightened the line by reeling in non-sticky line with her legs I, and then reeled in additional non-sticky line hand-over-hand with her legs IV (WE). In sum, she laid the sticky line, which she could not grasp with her leg IV, at a low tension, then raised its tension by adding non-sticky line that she grasped and reeled in.

Holding the dragline with one IV at the moment of attachment in pholcids and diguetids may also aid the spider in finding the new line with her legs just after making an attachment, preparatory to returning along it. This localization function was apparently accomplished in an alternative manner by *K. hibernalis* when laying radial lines. The spider turned nearly 180° while attaching to the substrate, so that each attachment formed a loop (fig. 5.10). In some cases no leg held the dragline, while in others one leg III held it, but then released the line soon after the attachment was made. The spider

Table 5.2. Aspects of lower-level patterns described in the text (e.g., positions of legs at the moment of attachment) observed in building behavior of non-orb building spiders. Nearly all of the studies were only preliminary, so all characterizations with "no" must be taken as uncertain.

				Hold dragline with IV while:			Legs that hold line to which attach						
Taxon	*Attach "around corner" to substr.*	*Explor. descents*	*Break lines at mouth*	*Walking*	*Att. to substrate*	*Att. toline*	*Non-st.*	*Sticky*	*Break and reel*	*Not att. to each line crossed*[1]	*Walk under lines*	*Turn back onto n-s line just attached*	*References*
Dipluridae													
Linothele macrothelifera	No	No[1]	No	No	No	No	none	n.a.	No	Yes	No[1]	No	Eberhard and Hazzi 2013
ARANEOMORPHA													
Non-Orbicularia													
Gradungulidae													
Progradungula carraiensis	?	?	?	?	?	?	?	?	?	Yes[2]	Yes[3]	?	Gray 1983
Austrochilidae													
Thaida peculiaris[4]	?	?	Yes	?	?	?	?	III	?	?	Yes	Yes	Lopardo et al. 2004
Filistatidae													
Kukulcania hibernalis	?	?	?	No	?	?	III	III/None	No	No	No	Yes[5]	Eberhard 1987a, WE
Misionella mendensis	?	?	?	?	?	?	?	None[6]	?	?	No[7]	?	Lopardo and Ramírez 2007
Pikelinia tambilloi	?	?	?	?	?	?	?	None	?	?	?	?	Lopardo and Ramírez 2007
Pholcidae													
Modisimus guatuso	?	Yes	No[8]	?	Yes	?	III	None	No	Yes	Yes	Yes[9]	Eberhard 1992b, WE
Digutidae													
Diguetia albolineata	Yes	?	Yes[4]	Yes[10]	Yes	?	IV[11]	n.a.	No	Yes?	Yes	Yes[12]	WE
Eresidae													
Stegodyphus sp.	?	?	?	?	?	?	ipIII&IV[13]	?	?	?	Yes	?	WE
Psechridae													
Psechrus argentata	?	?	No[14]	?	?	?	?	?	?	?	Yes	?	Robinson and Lubin 1979b
Psechrus sp.	?	?	?	?	?	?	III	?	?	?	Yes	?	Eberhard 1987a, WE
Fecenia sp.	?	?	Yes[14]	?	?	?	?	?	?	?	Yes	?	Robinson and Lubin 1979b
Agelenidae													
Melpomene sp.	?	No	No	No	No	No	No[15]	No[15]	No	Yes[15]	No	No	Rojas 2011

Table 5.2. Continued

Taxon	Attach "around corner" to substr.	Explor. descents	Break lines at mouth	Hold dragline with IV while:			Legs that hold line to which attach		Break and reel	Not att. to each line crossed[1]	Walk under lines	Turn back onto n-s line just attached	References
				Walking	Att. to substrate	Att. toline	Non-st.	Sticky					
Coelotes terrestris	?	?	Yes	?	?	?	?	?	?	Yes	?	?	Tretzel 1961
Pisauridae													
Thaumasia sp.	No[7]	No	No	No	No	No	None	n.a.	No	Yes	No	Yes	Eberhard 2007a
Lycosidae													
Aglaoctenus castaneus	?	Yes	No	No	Occas.	No	III[16]	n.a.	No	Yes	Yes/ No	Yes	Eberhard and Hazzi 2017
A. lagotis[17]	?	?	Yes[18]	No	No	No	None	n.a.	?	Yes	No[19]	?	González et al. 2015, M. González pers. comm.
Orbicularia													
Theridiidae													
Latrodectus geometricus[20]	?	?	Yes	Yes/ No[21]	Yes	Yes	ipIII&IV[22]	III&IV[23,24]	Yes[25]	?	Yes	Yes	Lamoral 1968, Eberhard et al. 2008b
Asagena fulva, Steatoda grossa	?	?	?	Yes	?	?	?	?	Yes	?	?	?	Barrantes and Eberhard 2010
S. triangulosa	?	Yes	?	?	?	?	?	?	No[26]	?	?	Yes[27]	Benjamin and Zschokke 2002, 2003
Nihonhimea tesselata	Yes	Yes	Yes	Yes	Yes	Yes[28]	III&IV[29]	n.a.	Yes[30]	Yes	Yes	Yes	Jörger and Eberhard 2006
Chrosiothes sp. nr. *portalensis*	?	?	Yes	Yes/ No	Yes	?/No	ipIII&IV	n.a.	No	Yes/ No[31]	Yes	Yes	WE
Tidarren haemorrhoidale	Yes	Yes	Yes	Yes	Yes[32]	Yes	ipIII&IV	?	Yes	Yes	Yes	Yes	Madrigal-Brenes and Barrantes 2009, G. Barrantes pers. comm.
Phoroncidia sp.	?	?	Yes	Yes/ No[33]	?	?	?	n.a.	Yes	?	Yes	?	WE
Synotaxidae													
Synotaxus spp.	?	?	Yes?	?	?	?	?	?	Yes	No	Yes	Yes	Eberhard 1977a
Linyphiidae													
Linyphia hortensis	?	?	No?[34]	Yes	?	?	?	?	No[34]	?	Yes	Yes/ No[35]	Benjamin and Zschokke 2004
Deinopidae													
Deinopis sp.	?	?	Yes	Yes[36]	Yes[36]	Yes[36]	ipIII, IV	ipIII, IV	Yes	No	Yes	Yes	Coddington 1986c
Orb-weaving orbicularians[37]	Yes	Yes	Yes	Yes	Yes	Yes	ipIII, IV	ipIII, IV	Yes	No	Yes	Yes	Eberhard 1982, WE

Table 5.2. Continued

[1] Web sites were on the surface of leaf litter, and thus were not appropriate for this type of behavior.
[2] Deduced from statement that cribellum lines are "only loosely adherent" to non-sticky support lines. This weak attachment may be adaptive, as it allows the sticky silk to "thoroughly enfold" the struggling prey.
[3] Web is so sparse that it seems impossible that spider walks on top of the lines; spider also hangs under lines while in prey capture position, and after prey capture.
[4] Some details also apply to *Austrochilus forsteri*.
[5] When building long radial non-sticky lines on which sticky silk was then laid.
[6] The leg III that supports the leg IV that is combing out cribellum silk in fig. 7 of Lopardo and Ramírez 2007 is bent under the body exactly as occurs in *K. hibernalis*, suggesting that just previously leg III had held the non-sticky line to which the cribellum silk was attached.
[7] Deduced from photographs of webs or of spiders building webs.
[8] Did not break lines during construction; broke lines at mouth quickly and smoothly when attacking prey.
[9] During construction of skeleton web only; did not occur during filling of skeleton web or while laying sticky lines (in the latter case, the spider turned back repeatedly at the edge of the web, but did not return along the same line that she had been laying).
[10] Leg IV regripped the drag line repeatedly (or alternated with the other leg IV) as the spider moved forward, and also when the spider descended while hanging from the drag line; thus the tarsal claw may not slide along the line as in orbicularians.
[11] Occasionally leg III.
[12] Occasionally failed to hold the line (see text).
[13] Occasionally only III.
[14] During prey removal from the web (construction behavior has not been observed). Cutting occurred, however, during courtship in *F.* sp.
[15] While attaching swaths of fine, probably aciniform lines to form the sheet. Observations of attaching draglines or cables of line to each other are lacking.
[16] Very rarely ipsilateral IV also.
[17] Ipsilateral legs.
[18] With substantial tugging and exertion of physical force on the lines.
[19] Observations were in a captive situation in which it may not have been possible for spiders to build lines far enough above the substrate for this behavior to occur.
[20] The observations by Lamoral were in South Africa, the others in Central America, where *L. geometricus* recently invaded; the unsettled state of *Latrodectus* taxonomy leaves doubt as to whether they were the same species.
[21] The exception is during production of sticky lines, when no leg touches the newly produced line; leg IV also does not hold the line just after the spider begins a descent to lay a new gumfoot line.
[22] With both legs III and only IV when initiating and finishing gumfoot lines, and also in central area.
[23] Attachment of upper, non-sticky end of gumfoot line.
[24] Ventral movement of laterally twisted abdomen may have had an exploratory function; the pedipalps may also help locate lines.
[25] Clear early in construction and when build gumfoot lines; otherwise rare.
[26] Description probably wrong, at least for some lines (see discussion in Eberhard et al. 2008b).
[27] Building gumfoot line.
[28] Not always; apparently, late in construction of the sheet, the spider sometimes grasps the sheet with both IV and does not hold the dragline.
[29] When building sheet, the spider does not hold line to which attachment is being made with any legs.
[30] During early stages ("exploration") only.
[31] While filling in the sheet crochet-style, the spider attached to each line; while adding lines from the sheet to the tangle above, many lines in the sheet were skipped.
[32] But not always.
[33] While producing non-sticky line; no leg held the line while the sticky portion of the line was lengthened immediately after prey capture.
[34] Failure of same authors to see this behavior in theridiids, where it occurs (see Eberhard et al. 2008), makes this characterization uncertain.
[35] Contradictory accounts were given (on p. 123 in the first and third paragraph of Benjamin and Zschokke 2004). Perhaps the "yes" refers to non-sticky lines, and the "no" to the sticky lines.
[36] Pers. comm. J. Coddington.
[37] See chapter 6.

then used her palps to find and follow the line while she returned toward the retreat, combing out cribellum silk.

This line localization function may be important in species in which the spider frequently returns to the retreat during construction (table 5.1), as it would enable the spider to return directly to the site from which she came. The possible importance of this orientation function was illustrated when a mature female *D. canities* occasionally seemed to become lost in the early stages of construction (WE). She occasionally failed to turn back after attaching a line to the side of the cage, and moved onward along and attached her dragline several more times to the side of the cage without holding her dragline. When she eventually appeared to attempt to turn back, she was out of contact with her web (and her retreat), and she wandered for up to >30 min in the cage; she repeatedly retraced routes she had recently followed, and only occasionally made major new explorations. The cage was closed, and she eventually encountered a line that led back to her retreat; but such an incident in nature might result in loss of the retreat. The retreat in this species is surely valuable, as it is a tube made of many small pieces of leaf and stick attached together where the spider hides during the day, and it often contains one or more egg sacs (fig. 9.16f).

5.3.4 HOLD THE LINE TO WHICH THE DRAGLINE IS BEING ATTACHED

A spider can control the location where her dragline will be attached more precisely by holding the line to which she is about to attach (the "attachment line") with her legs. She may also ensure a stronger attachment by gripping the attachment line, because she can position it precisely between her anterior spinnerets and thus bring her piriform spigots to bear on it (section 2.2.9). Orbicularians (both orb weavers and non-orb weavers) usually grasped the attachment line with ipsilateral tarsi III and IV; leg III was just anterior to the attachment point, and leg IV just posterior to it (Eberhard 1982; WE, appendix O6.2.1). Given the highly conserved details concerning which legs are used to manipulate which lines in orbicularians (Eberhard 1982; Kuntner et al. 2008a), details of attachment behavior by non-orbicularians may provide characters for higher-level taxonomy.

Other species used several other techniques to grasp the attachment line. The pholcid *Modisimus guatuso* usually grasped it with only one leg III just anterior to her spinnerets. This grasp was much less stereotyped than in orbicularians, however; sometimes both legs III grasped the same line (both anterior to the spinnerets), and more rarely ipsilateral leg IV also grasped it just posterior to the attachment site (Eberhard 1992b). Grasping the attachment line also varied in the diguetid *Diguetia canities*; most often one leg IV gripped it during skeleton web construction and filling in the sheet (WE). The sheet-weaving lycosid *Aglaoctenus castaneus* (Eberhard and Hazzi 2017) and the filistatid *Kukulcania hibernalis* (WE) also grasped the attachment line with one leg III prior to making an dragline attachment (and, in the latter species, to attach cribellum silk). Video recordings of *K. hibernalis* showed that the claw of leg III snagged the line, and that the pedipalps often contacted this line before leg III; leg III thus appeared to be guided by the palps. Use of leg III in *A. castaneus* was inconsistent, and was omitted when an attachment was made to a sheet of silk rather than to a single line (WE). At the opposite extreme, the diplurid *Linothele macrothelifera* never grasped any lines to which she was attaching lines (Eberhard and Hazzi 2013). Szlep (1966) mentioned that the titanoecid *Titanoeca albomaculata* attached sticky lines only by means of the spinnerets, with no involvement of leg IV. In *A. castaneus* the anterior lateral spinnerets occasionally grasped the attachment line between them and lifted it toward the other spinnerets to make an attachment. Variation in the legs that *Nihonhimea tesselata* (THD) employed to make dragline attachments was instructive in this context. As the sheet of the web became more dense, the spider ceased to carefully grasp each line to which she attached with her ipsilateral legs III and IV (Jörger and Eberhard 2006). Presumably the need to guide the spinnerets to a given attachment line was reduced when many lines were available, and in addition the likelihood that both tarsi III and IV would grasp the same line in the sheet became smaller. A similar pattern occurred in *L. hortensis* (LIN): the spider did not hold each line to which an attachment was being made while constructing its dense sheet (Benjamin and Zschokke 2004).

While constructing the retreat, a spider that was probably *Oecobius* (see footnote in table 5.1) performed an apparently unique behavior; she walked rapidly on the substrate but raised her abdomen dorsally and built lines dorsal to her body (Hingston 1925).

5.3.5 SNUB LINES

"Snubbing" lines to promote stronger attachments to the substrate, originally discovered in orb weavers (section 6.3), also occurs in the aerial sheet weaving haplogyne diguetid *Diguetia canities* (WE). When attaching a line to a small object such as a twig, the spider applied her spinnerets to the far surface of the object to make the attachment, in effect snubbing the line in addition to physically attaching it. One spider made a loop of 360° around a cylindrical support, thus anchoring the line especially securely. I cannot judge whether the lack of mention of snubbing in behavioral descriptions of other non-orb species is due to lack of attention by the observers to this detail, lack of opportunity for the spider (e.g., observations of web construction in smooth-walled cages), or true absence of this ability.

5.3.6 CUTTING AND RECONNECTING LINES

There are scattered reports of spiders breaking lines as part of web construction. Among mygalomorphs, mature male *Grammostola monticola* (Theraphosidae) broke lines with their chelicerae when preparing the edge of the sperm web (Costa and Pérez-Miles 2002) (section 10.4.1), and the diplurid *Linothele macrotheli-fera* broke lines by pushing her entire body through the blind end of a tube to create an opening (Eberhard and Hazzi 2013). The agelenid *Coelotes terrestris* opened holes in tube walls or sheets after first searching with her front legs for a small site with less dense lines; she produced a small hole by seizing lines with her chelicerae, and then inserted her front legs through the hole and slipped through. The process was so smooth that the animal seemed to "walk though a wall" (Tretzel 1961). These agelenids, as well as the hahniid *Hahnia helveola* and a theraphosid, also used another breaking behavior;the spider used a strong oblique blow with a front leg to tear the thin sheet that closed the mouth of a secondary tunnel in order to enter a tight space in which to hide from danger (Tretzel 1961).

The behavior that Coddington (1986c) called "cut and reel" is ubiquitous during at least some stages in the construction of orbs and the non-orbs in groups directly derived from orb-weavers (see chapter 6) (fig. 6.3): the spider attached its trail line to the line under which she was walking, then broke the line and, as she moved forward, reeled up the line in front of her and replaced it with the trail line emerging behind her (for details, see fig. 6.3). This behavior allowed the spider to remove lines, to shift their points of attachment, and to alter their tensions.

The occurrence of cut and reel behavior has been controversial in the non-orb orbicularian family Theridiidae. Benjamin and Zschokke (2002, 2003) stated that cut and reel behavior did not occur in *Steatoda triangulosa* or *Parasteatoda* (= *Achaearanea*) *tepidariorum*, and did not mention it in descriptions of web construction behavior by *Latrodectus geometricus* and two species of *Theridion*. They noted that although that they were uncertain in some phases of web construction, they were confident that it did not occur during gumfoot line construction (Benjamin and Zschokke 2002). Nevertheless, cut and reel behavior was later established as a consistent part of gumfoot line construction by direct observations and also by examination of accumulations of loose silk at the upper ends of gumfoot lines in the webs of *P. tepidariorum*, *Latrodectus geometricus*, *L. mirabilis*, *Steatoda hespera*, and *S. grossa* (Eberhard et al. 2008a; Barrantes and Eberhard 2010; WE). It has also been described in other contexts in other theridiids by Marples (1955) and Bradoo (1972). It seems likely that the occurrence of cut and reel behavior was simply missed by Benjamin and Zschokke. Some other theridiids, such as *N. tesselata* (Jörger and Eberhard 2006) and *Chrosiothes* sp. (WE), seemed to limit cut and reel behavior to very early stages of construction.

Cut and reel behavior may constitute a synapomorphy of the Orbicularia, as it has never been noted to my knowledge in any non-orbicularian. There are preliminary indications that it does not exist in some taxonomic groups (table 5.2), though it is difficult to give definitive proof of a lack of a behavior. One subroutine of cut and reel behavior, the ability to reel up silk, is probably very ancient, however. A wide array of taxa reeled up and packed together the silk line as they re-ascended the draglines on which they had descended (e.g., Eberhard 1986c).

5.3.7 FINDING LINES AND FOLLOWING BEHAVIOR

Many spiders (both with and without webs) wave or tap their anterior legs in front of them like antennae, and these movements almost certainly serve to gather information on the presence of nearby web lines and other objects. The general lack of understanding of building behavior by non-orb weavers often precludes, however, understanding whether specific types of tapping behavior constitute searches for particular lines. In one situation, however, it was possible to deduce when spiders of several groups were probably attempting to finding a line—the moment when the spider had just finished combing out cribellum silk and was about to attach it to another line. The spider tapped repeatedly with one leg III until it encountered a non-sticky line; she seized it with this leg, and attached the cribellum silk to it. Exploratory tapping behavior of this sort was performed by the austrochilid *Thaida peculiaris* (Lopardo et al. 2004) (this species also sometimes used the other leg III), the psechrid *Psechrus* sp. (Eberhard 1987c), and the filistatid

Kukulcania hibernalis (often just before the attachment made as combing began) (WE). Szlep (1961) noted that the tapping with legs I and II by the amaurobiid *Titanoeca albomaculata* while laying cribellate lines in their "pseudo-orbs" resembled inner loop localization behavior of orb weavers during sticky spiral construction (sections 6.3.5, 7.3.2.1), but did not give details.

Orb weavers employ "following" behavior to guide the movements of more posterior legs by following their more anterior legs (Eberhard 2017; fig. 6.12, section 2.4.2.1). There is no survey of following behavior in non-orb weavers (it generally requires slow motion analysis of recordings). Video recordings showed that it occurred in the pholcid *Physocyclus globosus* (G. Barrantes and W. Eberhard unpub.), the lycosid *Aglaoctenus castaneus* when the spider walked along a single line (Eberhard and Hazzi 2017), and the non-orb orbicularian *Chrosiothes* sp. (THD) (WE).

Following behavior when walking on silk lines may have evolved from walking behavior on the substrate. In the slow walking gait of the webless wandering wolf spider *Alopecosa*, posterior legs followed the ipsilateral immediately anterior leg in time and (in the case of legs III and IV) in space (Ehlers 1939 in Foelix 2011). In the funnel web lycosid *A. castaneus*, legs followed each other in time but not in space when the spider walked on the substrate, but shifted to following in both time and space when the spider walked along a single line (Eberhard and Hazzi 2017).

Finding lines that have already been laid (e.g., to attach a new line to them) appears to present special problems in "substrate" webs, whose lines lie on or near a surface around the mouth of a tunnel or retreat; these webs occur in various families (Blackledge et al. 2009c; Garrison et al. 2016). The lines lie on or very close to the substrate, and, as seen especially clearly in the early stages of construction by the agelenid *Melpomene* sp. (Rojas 2011) and *K. hibernalis* (WE), the spider tends to walk with her tarsi on the substrate rather than on her web lines. Video recordings of *K. hibernalis* revealed a previously undescribed behavior that may be of general importance in facilitating this search in substrate webs (e.g., fig. 5.11). The spider began the construction of a new radial line by moving away from the retreat in a radial direction, attaching her dragline (actually a cable of fine lines) several times to the network of lines already in place near the retreat. She walked with her tarsi on the substrate, but paused several times, bent both palps and legs III ventrally to seize the network of non-sticky lines, briefly lifted them briskly several mm, and then released them and continued onward. The lifting movements had the effect of extending the cribellum silk at sites where it connected the network to the substrate below, and it may have thus raised the non-sticky lines slightly from the substrate and made them easier to locate subsequently. Another, non-exclusive possibility is that lifting may inform the spider regarding the presence or numbers of web lines below her.

The palps of another species with a substrate web, the pisaurid *Thaumasia* sp., tapped actively during much of web construction (WE). They may have helped the spiders sense lines already laid (individual lines were not visible in the videos, however, so this could not be confirmed); these spiders did not lift their webs as in *K. hibernalis.*

5.3.8 RUBBING, BRUSHING, LIFTING, AND CLAPPING MOVEMENTS OF THE SPINNERETS

The spinnerets of most spiders (including all orb weavers) are short, and their behavior is both difficult to observe and seemingly unremarkable (or, at least, largely unremarked). More detailed observations of web construction by non-orb species with longer, more easily observed posterior lateral spinnerets, including the agelenid *Melpomene* sp. (Rojas 2011), the diplurids *Linothele* sp. (Paz 1988) and *L. macrothelifera* (Eberhard and Hazzi 2013), and two other moderately large species, the lycosid *Aglaoctenus castaneus* (Eberhard and Hazzi 2017) and the filistatid *Kukulcania hibernalis* (WE), which have shorter spinnerets but move slowly enough to be observed in detail, give a glimpse into a previously unappreciated world of complex, subtle, and sometimes puzzling spinneret behavior.

Both *M.* sp. and *Linothele* spp. attached the swaths of fine lines they produced by lowering and swinging their long posterior lateral spinnerets laterally with a graceful waving motion, brushing or wiping the ventral spinneret surfaces (where the silk spigots are located) against the substrate; the tip of the abdomen itself approached but often did not contact the substrate. The spigots are arranged approximately linearly along the ventral surface of a spinneret (not just at its tip), so the spinneret produced a wide swath of fine lines (presumably aciniform lines in the agelenid). Similar sweeping or brushing movements of the spinnerets occur during prey wrapping in these and other species that have long posterior lateral spinnerets (Barrantes and Eberhard 2007; WE). The likely function of sweeping movements during web construction is to attach the lines from these spinnerets to the substrate; the spider immediately walked away, drawing out new lines that were attached to the substrate where her spinnerets had brushed. By sweeping her spinnerets across the substrate, rather than just dabbing them, the spider presumably increases the area of silk-substrate contact, and thus the strength of adhesion. It is also likely that the long spinnerets serve as sense organs, informing the spider of objects behind her (Eberhard and Hazzi 2013).

The filistatid *K. hibernalis* has short anterior and posterior lateral spinnerets, but the spider increased the area of contact in each attachment to the substrate by moving her spinnerets. She repeatedly spread her posterior lateral spinnerets laterally and then closed them while she swung her abdomen laterally while pressing her spinnerets against the substrate (see fig. 10.11) (WE).

A second, more puzzling spinneret movement was made during web construction by *L. macrothelifera*, *M.* sp., and *A. castaneus* when they attached lines to other silk lines; the spider raised one of her posterior lateral spinnerets and held it more or less dorsally while making the attachment (fig. 5.17; Eberhard and Hazzi 2013, 2017). This movement raised the lines being produced by this spinneret away from the substrate, and thus paradoxically *prevented* their being attached at the same moment that the other lines were being attached. The spinneret that was raised was on the spider's trailing side rather than her leading side when she swung her abdomen laterally to attach (fig. 5.17). Similar raising of the trailing posterior lateral spinneret was performed by a wide variety of mygalomorph and araneomorph spiders when they wrapped prey that they had placed on the substrate (Barrantes and Eberhard 2007; Hazzi 2014; Eberhard and Hazzi 2013, 2017).

Raising a spinneret while attaching lines during web construction (and prey wrapping) probably functions to widen the area covered by the swath of lines (Eberhard and Hazzi 2017). By refraining from attaching some but not all of the lines that she is producing at a given site, and by then later attaching these lines at other sites, the spider causes her swaths of lines to run in many more directions than those in which she herself has moved (fig. 5.17; Eberhard and Hazzi 2013, 2017). Elongate spinnerets and spreading movement of the spinnerets have the same likely function (Barrantes and Eberhard 2007).

The frequent, medially directed clapping movements of the posterior lateral spinnerets of *A. castaneus* when the spider was moving from one attachment to another and just before and just following an attachment may also affect the width of the swath (fig. 5.17). These movements probably caused the aciniform lines to contact and possibly to adhere to each other and possibly also to the ampullate draglines. Clapping could thus convert swaths into cables. Adhesions of this sort might explain the otherwise mysterious arrays of lines that were not directed toward any attachment the spider had made to the substrate that were evident in video recordings of *A. castaneus* (fig. 5.17). Another possible function of clapping, originally proposed in the context of balloon line production (e.g., Blackwall in McCook 1889; Nielsen 1932; Eberhard 1987c), would be to initiate lines. The tips of spigots would contact other objects (spigots on the other spinneret, other lines), and droplets of silk at their tips would be pulled into lines when the spinnerets moved apart. The two functions are not exclusive. Clapping movements were not seen in the diplurid *L. macrothelifera* (Eberhard and Hazzi 2013).

A mature female of the filistatid *Kukulcania hibernalis* moved both her anterior and her posterior lateral spinnerets rapidly and continuously during the entire process of laying a radial, non-sticky supporting line from the retreat to the edge of her web (probably she rubbed her two anterior lateral spinnerets against each other, and the two posterior lateral spinnerets against each other, but no ventral view of the spider was available to confirm this) (WE). The function of these movements is uncertain. Microscopic examination showed that each radial line was composed of a large number of fine lines (fig. 10.11).

Still another context in which the posterior lateral spinnerets moved rhythmically was during cribellum silk production. Peters (1984) observed clapping movements of the posterior lateral spinnerets of *Uloborus walckenaerius* (UL) and concluded that they functioned to fold or press the mat of cribellum fibers around the

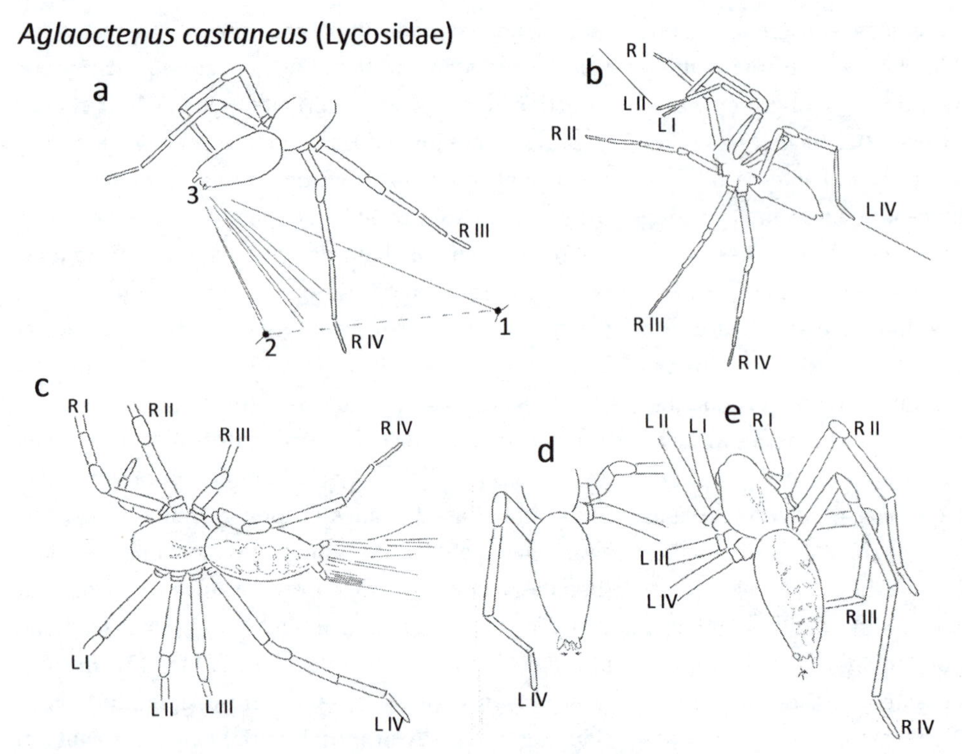

Fig. 5.17. These stylized drawings of the graceful positions of the lycosid *Aglaoctenus castaneus* traced from video recordings of sheet construction illustrate subtle behavior patterns that served to widen the swaths of lines as they were produced, and thus to increase the area that the swaths covered. In *a*, the many fine lines that the spider was laying were visible when light glinted off of them while she made the attachment to a line at point 3 (she held the line against her spinnerets with left leg LIII). Surprisingly, only a subset of these lines had been attached at the previous attachment (point 2). Some others ran directly from the previous attachment (point 1) to point 3, while others had apparently been attached at sites between point 1 and point 2 while the spider was moving between these two sites (represented as a dashed line; it was not visible). In *b* the spider was walking along an approximately horizontal aerial line that she held with only her left legs, allowing her right legs to dangle in the air ahead and below her; this may be an alternative to the behavior of orb weavers and their allies (fig. 6.3) allowing the spider to grasp lines with her asymmetric tarsal claws (section 2.4.2.3). The spinneret movements (*c–e*) that the spider used to selectively attach some lines but not others that they were producing are only partially understood. One way spiders made selective attachments was to raise one posterior lateral spinneret dorsally at the moment when the she was making an attachment (the right spinneret in *a*, the left one in *e*); this ensured that the lines that were being produced by this spinneret were not included in the attachment. The spider also frequently moved her spinnerets while moving from one attachment site to the next. Commonly she spread her posterior lateral spinnerets laterally, widening the swath of fine lines (*c*). In addition, she sometimes clapped the two posterior lateral (and at least sometimes the anterior lateral) spinnerets against each other (*d*), perhaps causing lines to adhere to each other (as may have occurred, for example, between points 1 and 2 in *a*) (from Eberhard and Hazzi 2017).

axial fibers. Similar movements of the same spinnerets also occurred in *Zosis geniculata* (UL) (WE). The filistatid *K. hibernalis* also extended the posterior lateral spinnerets rearward rhythmically while combing out cribellum silk, while the dictynid *Mallos hesperius* spread both anterior and posterior lateral spinnerets and did not close them while combing (WE). There is substantial diversity among cribellates in the forms of the mats of cribellum lines on the baseline: they include "puffs," "coils," and other more complex forms (Gray 1983; Griswold et al. 2005). Perhaps the forms result from movements of the posterior lateral spinnerets during cribellum silk production; their functions are unclear.

The anterior lateral spinnerets of *A. castaneus* also moved during web construction. In occasional lateral views, they were briefly directed ventrally, and appeared to grasp the line between them, and then pull it dorsally, toward the other spinnerets just before the spider made an attachment to another silk line. These spinnerets also made clapping movements just before an attachment was made (to initiate piriform lines?).

These observations of spinneret behavior only scratch the surface, and further work will probably reveal additional types of movements and functions.

5.3.9 DABBING AND SWEEPING WITH THE ENTIRE ABDOMEN

In many spiders, especially those lacking long spinnerets, much or most of the movement when lines are attached is by "painting" them onto the substrate by moving the entire abdomen. One especially dramatic case was the salticid *Pellenes arciger*, which swept her abdomen from side to side while building the approximately vertical sheet of her extraordinary sail-like web (fig. 9.3; Lopez 1986).

5.3.10 OTHER LOWER-LEVEL BEHAVIOR PATTERNS THAT ARE ABSENT IN ORB WEAVERS

It is well known that spiders float lines on air currents, thus establishing bridges to distant objects and balloons for long-distance dispersal (Foelix 2011). Typically, the spider passively allows air currents to pull bridge lines from her spinnerets, either while hanging at the tip of her dragline or while standing at the tip of a projecting object. A mature female of the pholcid *Modisimus* (probably *M. bribri*) produced a bridge line with a different, apparently unique behavior (WE). She dropped from an object in the undergrown of an old secondary forest to dangle on her dragline with a clutch of eggs held in her chelicerae; while holding her dragline with a more anterior leg or legs, she then made rapid prey wrapping movements with her legs IV (e.g., Barrantes and Eberhard 2007), producing a line that gradually extended horizontally downwind in the almost imperceptible breeze. Eventually this bridge line snagged the stalk of a plant 1–2 m away, and spider reeled it in quickly with her anterior legs to tense it, attached her dragline near the mass of accumulated fluff, and walked along the bridge to the stalk. Here she repeated the performance, eventually walking on another bridge line to an object further downwind.

Early in web construction, when establishing the connections to the substrate for her tangle plus horizontal sheet web (fig. 9.16), a diguetid *Diguetia canities* often made short trips "backward." Behavior of this sort, in which the spider makes such an attachment part way back along a line that she has just laid, has never to my knowledge been described in an orb weaver. The only other description of repeated backward attachments that I know occurred in the synotaxids *Synotaxus* spp. building their "rectangular orbs" (fig. 5.4, Eberhard 1977a, 1995). Backward movements in synotaxids were especially complex, because the spider alternated sticky and non-sticky lines.

A further behavior pattern seen in non-orb weavers that has never been described in an orb weaver is breaking lines in the web during construction and simply releasing them and allowing the slack lines to be included in the web. Breaking and releasing occurred during the construction of tangles in *Nihonhimea tesselata* (THD) (Jörger and Eberhard 2006), *Chrosiothes* sp. (THD) (WE), and *Kapogea nympha* (AR) (WE) (see also fig. 1.7). It presumably also occurred in the production of the loose "screw" lines in the phocid *Pholcus phalangioides* (Kirchner 1986). Temporary spiral lines are broken and released during sticky spiral construction, but are so short that they play no role in prey capture. The orb weavers *Leucauge mariana* (TET) and *Micrathena duodecimspinosa* (AR) broke and released lines during orb construction, especially in the late exploration phase; most were attached to the substrate, however, and drifted away and were not incorporated into the final web (WE).

5.4 STEREOTYPED BEHAVIOR IN NON-ORB CONSTRUCTION

Many non-orb weavers show a great deal of unexplained variation as they build their webs. Descriptions of non-orb construction behavior frequently include reports that particular aspects of the spider's behavior showed no patterns (e.g., the "drunken wandering" of *Nihonhimea tesselata* described by Jörger and Eberhard 2006). Lack of patterns in filling in sheets has also been mentioned in the diplurid *Linothele macrothelifera* (Eberhard and Hazzi 2013), the pholcid *Modisimus guatuso* (Eberhard 1992b), the linyphiid *Linyphia hortensis* (fig. 5.18b) (Benjamin and Zschokke 2004), the agelenid *Melpomene* sp. (Rojas 2011), the lycosids *Aglaoctenus lagotis*

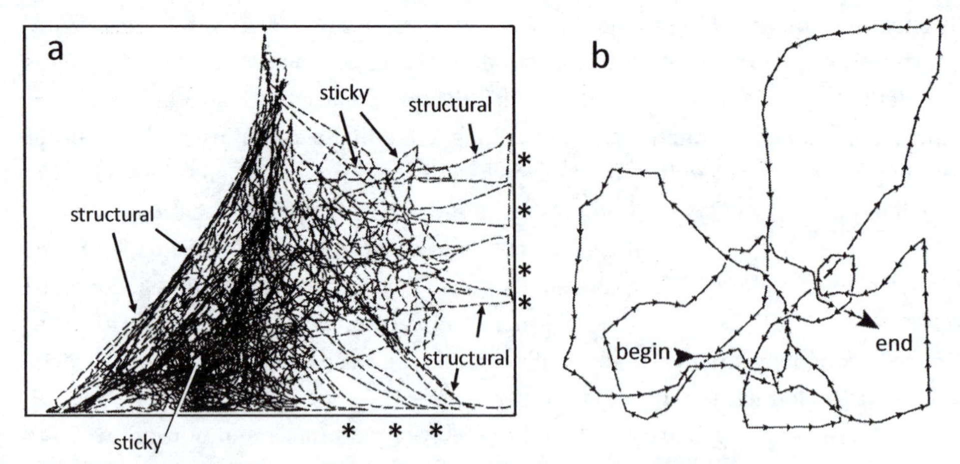

Fig. 5.18. Two patterns are perceptible in the paths traced from video recordings of a *Linyphia hortensis* (LIN) seen from above as she laid "supporting structure lines" and "sticky lines" in a rectangular cage (*a*) (walls of the cage are solid lines), and (in a closer view) as she added sticky lines to the horizontal sheet (*b*). In building supporting structural lines that were attached to the frame (*a*), the spider repeatedly walked to the end of a line, moved laterally along the cage wall ("*"), and then returned to the central area. Judging by the trips to the right and bottom sides of the frame, she attached her dragline after she had walked along the frame and then returned along this newly laid line, rather than returning along the "exit" line that she had produced while moving toward the cage wall (see fig. 10.34 b) (this same pattern is typical in radius construction by orb weavers—section 6.3.3). In laying sticky lines (*b*), she wandered in large looping paths, seldom contacting recent sticky lines. Behavior of this sort could not produce the more or less orderly arrangements of lines seen in the sheets of some other linyphiids such as *Agyneta* spp. (fig. 1.3a), suggesting that there is variation in sticky line construction behavior in this family (after Benjamin and Zschokke 2004).

(González et al. 2015) and *A. castaneus* (Eberhard and Hazzi 2017), and the psechrid *Psechrus* sp. (Zschokke and Vollrath 1995b). Two webs spun by the same individual *L. macrothelifera* at exactly the same site on successive evenings differed sharply, also suggesting a lack of a strict pattern (Eberhard and Hazzi 2013). None of these studies was extensive enough, however, to eliminate the possibility that a pattern was missed due to insufficient data or analysis. Patterns in finished webs can be difficult to perceive when early patterns in construction (e.g., *Psechrus* sp. in fig. 5.20, *Steatoda lepida* in fig. 5.19) are obliterated by later additions.

In any case, some stereotyped patterns of web construction do occur in non-orb webs, as illustrated by the following examples. The pisaurid *Thaumasia* sp. executed a highly repetitive, relatively consistent pattern, in which the spider made many short, radially oriented trips on the upper surface of a single leaf, laying draglines from near the center of her web to the edge and back (fig. 5.11). The synotaxid *Synotaxus turbinatus* made a much more complex, difficult to decipher series of highly stereotyped attachments while building a series of sticky and non-sticky lines in her "rectangular orb" (fig. 5.4). *Chrosiothes* sp. (THD) also built highly orga-

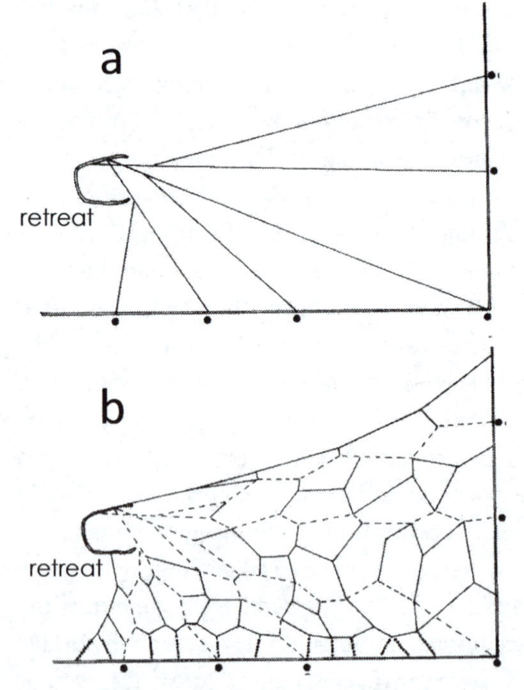

Fig. 5.19. This schematic figure of web construction by *Steatoda lepida* (THD) illustrates how the organization of lines laid early in web construction (*a*, dashed lines in *b*) can later be obscured by additional lines. The early pattern of radial lines converging on the retreat was obscured by later lines (sticky portions of gumfoot lines are indicated by multiple tiny balls) (after Lamoral 1968).

nized webs that included a closely-knit horizontal sheet of non-sticky lines (fig. 9.16); sheet construction behavior was rapid, smooth, highly stereotyped, and very difficult to decipher (WE).

Finally, as expected, the construction of the pseudo-orbs of the psechrid *Psechrus singaporensis* (fig. 9.29) (Zschokke and Vollrath 1995b) and the titanoecid *Titanoeca albomaculata* (Szlep 1966a) showed patterns during construction of sticky and non-sticky lines. The designs of other geometrically regular non-orbs such as that of the gradungulid *Progradungula carraiensis* (Gray 1983) intimate strongly that their construction behavior (which is yet to be described in detail) is also stereotyped.

This combination of stereotyped and non-stereotyped behavior reveals that characterizing the non-orb webs of some major taxonomic branches as "irregular" or as "stereotyped aerial sheets" (Blackledge et al. 2011; Garrison et al. 2016) is a serious oversimplification.

5.4.1 BUILDING TUNNELS

Although tubular retreats are widespread (table 9.1), there are few descriptions of their construction. The most complete description is of the agelenid *Coelotes terrestris* (Tretzel 1961), which built primary tunnels while first sketching out its funnel web. These were stronger-walled and more evenly woven than the secondary tunnels that were added later and apparently received less construction work. At first the interior was polygonal rather than rounded, but subsequent additions smoothed the walls. The spider moved along the tunnel, swiveling her abdomen with a characteristic rotation about its longitudinal axis. The spider also sometimes rotated her entire body and walked in a spiral path along the side or the roof of the tunnel. Even when walking straight, she zigzagged her spinnerets against the tunnel wall by swiveling her abdomen (a movent seen in spiderlings as young as five days old). The tunnel diameter was generally just large enough to admit the spider in a crouched position (much smaller than the puzzlingly large diameter tubes of the lycosid funnel web weaver, *Aglaoctenus castaneus*—Eberhard and Hazzi 2017) (fig. 5.5a), but was expanded in chambers for resting and feeding, and for storing prey remains.

5.5 ADJUSTMENTS TO SUBSTRATE-IMPOSED CONSTRAINTS

Many non-orb webs are physically more tightly associated with the substrate than are orbs (compare, for example, figs. 1.4 and 5.10). Tighter physical linkage might result in substrate differences inducing stronger changes in web design (e.g., Tretzel 1961; Blackledge et al. 2009c) (section 7.20.4). One (as yet untested) prediction would be that non-orbs that are more aerial are less variable. Selection on spiders to adaptively adjust web designs to substrate variations has probably been more intense in spiders with substrate webs (Blackledge et al. 2009c; section 10.3). Although experimental and comparative tests of the effects of particular substrate stimuli are still lacking, apparent substrate-imposed changes in the designs of non-orbs are indeed sometimes quite substantial, as illustrated by the following examples.

The zoropsid *Tengella radiata* builds large, horizontal sheet webs that extend from a tunnel at one edge, with an extensive tangle above (Eberhard et al. 1993). But some individuals in the field omitted both the sheet and the tangle partially or completely; extreme webs had only a small collar with a few sticky lines near the mouth of the tunnel (fig. 10.7). These were built at sites that had strong, impregnable retreats (cracks and holes in large tree trunks) but that lacked attachment sites nearby (WE). The webs of the theridiid *Anelosimus jocundus* were usually built on branches or small plants with multiple forks, and had a dense, durable, concave sheet at the bottom of a tangle of threads (as is common in *Anelosimus*—Eberhard et al. 2008b); but webs built on a single unforked branch lacked sheets entirely (Nentwig and Christenson 1986). The height of the tangle built above the horizontal sheet of *Novafrontina uncata* (LIN) was sometimes up to three times the diameter of the sheet, while in others it was only about two-thirds of the diameter of the sheet; apparently this depended on the availability of attachment sites above the sheet (fig. 7.51). The funnel webs of the lycosids *Aglaoctenus* spp. lacked the large tangle that was built above the sheet in other webs when they were built at sites that lacked suitable attachment sites (González et al. 2015; Eberhard and Hazzi 2017). Webs of the filistatid *Kukulcania hibernalis* built on the vertical walls of a house were elongate vertically when there were vertical shelves in the wall that offered opportunities to elevate lines from the surface of the wall; they were elongate horizontally when there were horizontal shelves (WE). Further examples of substrate effects occur in the agelenid *Coelotes terrestris* (Nielsen 1932; Tretzel 1961), the titanoecid *Titanoeca albomaculata* (Szlep (1966), and the pisaurid *Architis tenuis* (fig. 1.3) (Nentwig 1985c).

I speculate that the cues used by non-orb weavers in general (excluding highly ordered webs like those of *Synotaxus* spp. and *Achaearanea* sp.; figs. 5.3, 9.12) are less tightly linked to the presence and positions of particular lines in the web than are the cues used by orb weavers (section 7.3). In an orb, the imprecise placement of a few segments of sticky line can severely alter the distribution of all subsequent sticky lines (fig. 7.2). Probably similar imprecisions in the filistatid *Kukulcania hibernalis* (fig. 5.10) or the zoropsid *Tengella radiata* would have little effect on the general designs of their webs. A speculative consequence is that non-orb weavers may be under less intense selection favoring sustained attention during web construction (section 7.17.3). Another possible but as yet unexplored consequence is that non-orb weavers show more individual biases in "handedness" (systematic favoring of the animal's right or left side), as seemed to occur in the austrochilid *Thaida peculiaris* (Lopardo et al. 2004).

5.6 MANAGING SWATHS OF FINE LINES

Until recently, there were no detailed descriptions of spiders building webs using swaths of lines, and descriptions of construction referred to "the" dragline and "the" sticky spiral line. It is now clear that some spiders lay wide bands of fine lines while building their webs, and that some behavioral traits probably function to widen these swaths. Widening the swath is probably also the function of elongate posterior lateral spinnerets of some mygalomorphs and other scattered groups such as hersiliids, in which the spinnerets have rows of spigots on their ventral surfaces.

Two types of movements of the posterior lateral spinnerets in the lycosid *Aglaoctenus castaneus*, which has moderately short spinnerets, resulted in wider swaths of aciniform lines during web construction: the spider spread the spinnerets laterally as she moved; and she raised one spinneret away from the substrate while lines from the other posterior lateral spinneret were being attached (fig. 5.17a) (Eberhard and Hazzi 2017). The diplurid *Linothele macrothelifera*, which has long posterior lateral spinnerets, also raised one posterior spinneret while making attachments, and also raised these spinnerets while walking across the web. In addition she swept these spinnerets laterally across the substrate when making an attachment, also resulting in a wide swath of lines being laid (Eberhard and Hazzi 2013).

Two of these movements, raising one spinneret and sweeping long spinnerets across the substrate, also occurred during prey wrapping in a taxonomically wide spectrum of species, where they presumably also serve to widen swaths of lines wrapping prey (Barrantes and Eberhard 2007). Perhaps these two behavior patterns originally evolved as part of attack behavior, and were later incorporated into web construction (section 10.4). Widening the swath of lines is likely to be advantageous both when building a sheet (filling in more area more rapidly), and when wrapping prey (covering more of the prey's surface).

A second consequence of making only intermittent attachments of swaths of lines was to cause them to be lax in the finished web (e.g., Eberhard and Hazzi 2017). The numerous lax lines in the finished sheet of the agelenid *Coelotes terrestris* (Tretzel 1961) probably resulted from this type of behavior. Lax lines, especially numerous fine lines, may facilitate entanglement of prey.

None of these movements has been reported during orbicularian web construction; these spiders all have relatively short spinnerets, and apparently always produce only a few fibers at a time (typically a pair of ampullate fibers in the dragline or a pair of flagelliform fibers in the sticky spiral line) during web construction. The only spreading behavior reported was by the posterior lateral spinnerets during prey wrapping (e.g., Robinson and Olazarri 1971; Robinson 1975).

One additional type of movement, rapid rubbing of the spinnerets against each other, may function to combine multiple fibers into a multi-strand cable. The filistatid *Kukulcania hibernalis* performed sustained, rapid rubbing movements of her posterior lateral spinnerets as the lines emerged from her spinnerets while she built each long radial line (fig. 5.10) (which is a cable made of many fine lines—see fig. 10.11) (WE). Similar intermittent clapping movements by *A. castaneus* may also function to attach fine lines to each other (Eberhard and Hazzi 2017).

5.7 SUMMARY

The most basic message of this chapter is that not enough is known about non-orb weaver construction behavior to make coherent syntheses. The scattered bits and pieces of information on behavior in this taxonomically diverse group are summarized for the first time in tables 5.A and 5.B.

5.7.1 HIGHER LEVELS OF BEHAVIOR

Non-orb webs tended to be built over the space of several sessions (often several days), though with scattered exceptions. Usually webs were extended gradually; the geometric patterns of expansion varied (fig. 5.12). While extending her web, the spider often continued to contact the substrate around the edge of the web, rather than becoming more isolated and behaving in a more stereotyped fashion as occurs in orb construction. Nevertheless, tight physical association with the substrate did not imply, as some have argued, that "substrate-bound" webs were necessarily geometrically less regular (figs. 9.6, 10.12).

The building behavior of non-orb weavers can be organized into blocks of similar behavior, though these modules sometimes intergrade. Another, less well-documented trend, was for the temporal organization of behavioral modules to be flexible, in contrast with the rigid temporal organization of orb web construction. Comparisons of the webs of several species of *Synotaxus* suggests that rearrangements of modules may play an important role in evolutionary changes (Eberhard 1977a, 1995; Agnarsson 2003b).

Making periodic returns to the retreat or resting place during construction was also a very widespread and is probably an ancient pattern. Some returns were only momentary and were immediately followed by further building behavior, suggesting that one likely function of the returns is to refresh the spider's memory regarding the location of the retreat to guide further construction, rather than simply resting from physical exertion. Some returns to the retreat were very direct, implying that the spider had precise information regarding its location, presumably using path integration (memory of distances and directions that she had traveled recently). The most striking probable examples and experimental demonstrations of path integration have been in the context of returning to the retreat after attacking prey (Görner 1988) and searching for females (Henschel 2002), but the same ability is probably also employed during web construction (section 7.20.1). Path integration is probably a very ancient ability in arachnids, as it is known in scorpions, and is probably used by mygalomorph spiders to return to their retreats after attacking prey. This account may be overly simple, as differences in tensions on different lines represent an alternative cue that some spiders may use to find their retreats.

There are tentative indications of several other patterns in some groups of scattered taxonomic affinities: building behavior is nocturnal rather than diurnal; the production of sticky lines is always preceded by non-sticky lines; preliminary, "skeleton" webs often establish the general outlines and supports for the web, and are later filled in with additional lines (fig. 5.21); skeleton webs are consistently made of non-sticky lines, and are later filled in with further similar non-sticky lines, swaths of fine lines, or sticky lines; and filling in with sticky lines is done starting near the edge and working inward (fig. 5.20). Probably the most common characterization of filling-in behavior is that there is no pattern; but most studies have made only superficial attempts to search for patterns.

Many mysteries remain. Nothing is known, for instance, regarding the cues used to achieve the approximately radial symmetry of webs built over the space of several nights (e.g., filistatids) or how web extension is consistently concentrated on particular edges of the web (e.g., agelenids, zoropsids).

5.7.2 LOWER LEVELS OF BEHAVIOR

The ancestral movements used to attach lines involve direct application of silk using ventral and dabbing movements of the abdomen and lateral tapping or sweeping movements of the spinnerets, with no direct intervention of the legs. Dabbing abdominal movements also occur in egg sac construction, even in groups that use more derived attachment behavior during web construction. Using legs to actively manipulate both the dragline as it is produced and the lines to which the dragline will be attached are probably derived traits in spiders.

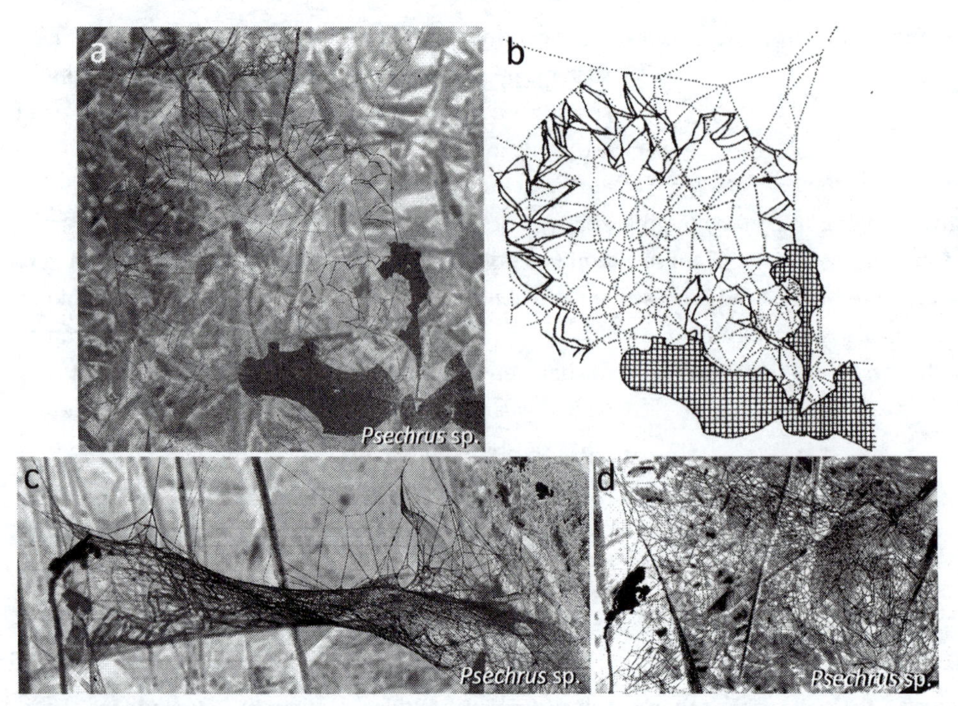

Fig. 5.20. A photo (*a*) and a drawing traced from the photo (*b*) of the horizontal sheet of *Psechrus* sp. (Psechridae) (#2299) show the lines present when the spider was interrupted after having partially replaced a large distal sector of her web that had been experimentally destroyed (the new non-sticky lines in *b* are dotted; the new sticky lines are thicker pairs of solid lines; and the leaves below the web are cross-hatched). The intact portion of the previous web, near the retreat, is at the right. The spider first built the non-sticky lines, then added the sticky lines, beginning at the edge and later working inward. A different, finished web (#1918) is shown from the side with the retreat at the right (*c*), and from above (*d*). The peripheral-first pattern of building is obscured by later lines (*a* and *b* from Eberhard 1987a).

There are three ways in which legs are used in attaching lines. Grasping the line to which an attachment is about to be made and holding it against the spinnerets occurs in even extremely basal branches of labidognaths, but is not known in mygalomorphs (they are, however, poorly studied). The legs used for this task vary among different taxa. Grasping the attachment line probably enables the spider to control the exact point at which she will attach more precisely, and perhaps also to make stronger attachments. This behavior may have arisen in conjunction with the evolution of piriform gland products to fasten lines together.

Grasping the dragline with one leg IV also occurs during construction of non-orbs as well as orbs. It may function to increase the tensions on web lines to values that are higher than the tension needed to pull the line from the spinnerets. Holding the dragline with one leg IV is an integral and necessary part of "cut and reel" behavior (see section 6.3.2.3), which is apparently limited to orbicularian spiders (both orb and non-orb weavers). This powerful technique allows the spider to shift attachment points, and to adjust tensions during web construction. I speculate that holding the dragline with one leg IV first arose in the context of attack behavior, was later used as an orientation device (allowing the spider to return along the line that she had just produced), and later came to be used to control tensions during web construction.

A third type of leg use is wrapping movements to apply sticky silk during non-orb construction. It is known only in two haplogyne families, Pholcidae and (in a modified form) Filistatidae. Outgroup comparisons suggest that this behavior also arose from prey attack behavior.

A newly described suite of behavioral traits is associated with laying a swath or a cable of fine lines (presumably from the acciniform glands) rather than a simple, two-fiber dragline during web construction. Two traits that probably function to increase the width of the swath during sheet construction include lifting one spinneret away from the site where other lines are being attached, and spreading the spinnerets wide while the spider walks across the web. Spreading the swath wider probably functions to distribute the lines more widely,

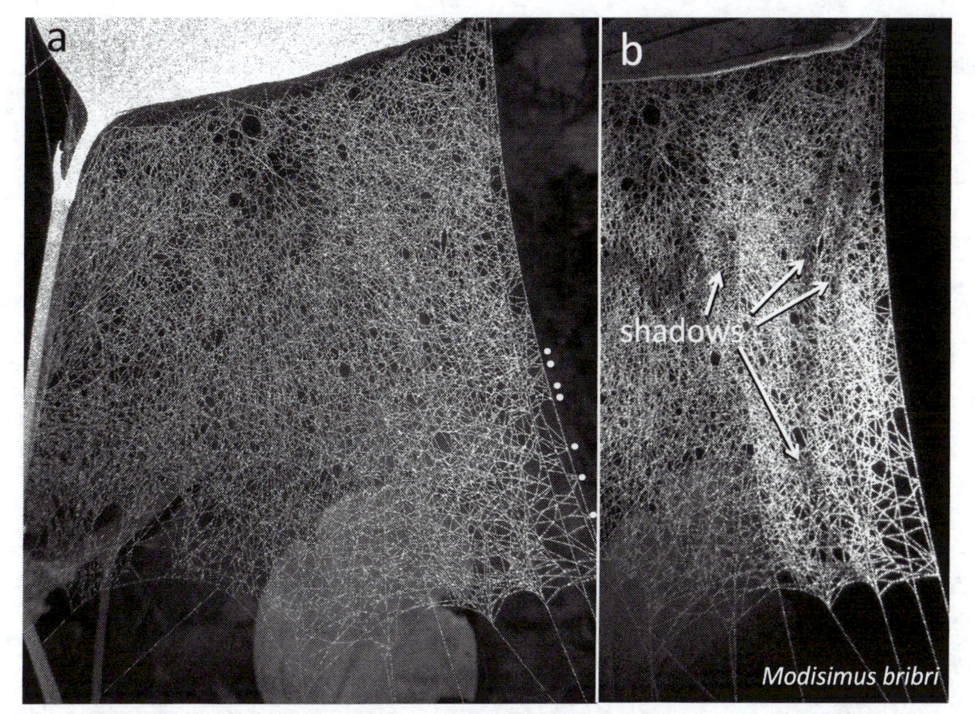

Fig. 5.21. The greater mechanical support provided by the early skeleton lines for the subsequent fill-in lines in the sheet of the pholcid *Modisimus bribri* web was apparent when a web coated with white powder (*a*) was illuminated from the right side (*b*). The shadows directly behind long straight lines in *b* showed that the fill-in lines sagged more when loaded with powder. This mechanical heterogeneity suggests that the early skeleton lines of this species function not only as scaffolding on which the spider lays additional lines, but as mechanical support for these later lines (which bear sticky droplets that probably function in prey capture). A second indication of lower tensions on fill-in lines than on long support lines at the edge of the sheet comes from the lack of deflection of the long lines by multiple attachments of fill-in lines (white dots in *a*).

and thus facilitates filling in the sheet. Both types of movement may have originally arisen as part of prey wrapping behavior, and have later been transferred to web construction. Spiders that were spinning swaths of lines also repeatedly rubbed their posterior lateral spinnerets together, sometimes continually and sometimes intermittently; these poorly understood movements may serve to attach some lines to others.

Still other types of spinneret movements were the pressing or pushing movements of the posterior lateral spinnerets on the cribellum fibrils just after they were combed from cribellum. These appear to mold the mat of cribellum fibrils into different configurations on and around the base line; the functional significance of these configurations is unknown.

Some non-orb web spiders show extreme flexibility in their adjustments to the environment around the web. An extreme illustration is the zoropsid *Tengella radiata*, in which the typical large horizontal sheet and a tangle above were sometimes omitted entirely when webs were built around sturdy retreats that lacked appropriate supports nearby; the webs had only a small tangle with a few sticky lines that surrounded the mouth of the retreat. It is likely that the construction behavior of non-orbs tends to be guided by more general cues, rather than by the presence and orientation of particular lines (as in orb weavers in later stages of construction).

A number of these ideas are only guesses at the moment.

Box 5.1 The funnel web diplurid *Linothele macrothelifera*

The diplurid *Linothele macrothelifera* used different types of spinning movements to build the tunnel and a simple sheet that lay on the upper surface of the leaf litter in a tropical forest (fig. 10.6; Eberhard and Hazzi 2013). When building her tunnel, she generally extended both long posterior lateral spinnerets (PLS) more or less directly rearward, and simultaneously or nearly simultaneously touched or tapped them repeatedly against the substrate (either objects in the litter or a previously laid sheet of silk). The spinnerets contacted the substrate with their basal segments and at least sometimes also their more distal segments; multiple silk lines were laid by each segment. The anterior spinnerets also projected ventrally, and probably contacted the substrate during tunnel construction. The spinnerets were brought into contact with the substrate largely due to anterior-posterior rocking movements of the spider's entire body, combined with minor ventral movements of the abdomen and of the posterior lateral spinnerets themselves. Her short, anterior lateral spinnerets were also lowered when posterior lateral lines were attached, and presumably also deposited lines. Successive taps were closely spaced on the wall of the tunnel. As noted by Tretzel in the agelenid *Coelotes terrestris*, close spacing between tunnel attachments guaranteed that new lines were laid close to the walls of the tunnel rather than across it. The simplicity of this behavior to build what may be the most ancient of spider silk constructs, a silken tunnel, fits with what one would expect for an especially ancient spinning movement.

In contrast, the spider's two long PLS were usually used asymmetrically when building the sheet, which is the more derived portion of her web. Each PLS produced a swath of fine lines, but these swaths were attached at different sites and times, thus increasing the number of directions in which lines were laid in the sheet. When attaching lines (usually near the edge of the sheet), the spider swung her abdomen laterally, and the posterior lateral spinneret on her leading side tapped laterally and swept across the substrate to attach its lines. The trailing spinneret, in contrast, was raised out of contact with the substrate and was often directed more or less dorsally. The lines produced by the raised spinneret were thus held above the sheet, and were not attached until the spider later swung her abdomen toward the other side and lowered this spinneret to sweep across the surface. Each spinneret produced on the order of 20 fine lines (Eberhard and Hazzi 2013) (a similar number of fine lines was produced by what appeared to be a posterior lateral spinneret of another diplurid—Foelix 2011). The spider relied more on the movements of her PLS, and less on the movements of her entire body, to bring her spinnerets into contact with the substrate and attach lines during sheet construction, though lines from the anterior lateral spinnerets (ALS) were also attached in some places. Some details of tunnel and sheet construction movements varied, but the two types were distinct. There was no sign of a preliminary skeleton web such as those made by several other non-orb web weavers such as psechrids, pholcids, agelenids, and theridiids (table 5.1).

In neither type of spinning movement were the legs used to contact, hold, or manipulate lines being constructed. Nor were the legs moved in ways suggesting that they provided sensory information to guide the spider's behavior. Instead, the long spinnerets appeared to be used as sensory organs, waving and tapping in apparent exploration of the areas where attachments were made. This exploratory function and the more elaborate webs of diplurids may explain why the apical segment of the PLS has pseudosegments in this family. Locating sites to which to attach lines would seem not to have been a problem for ancestral spiders building the lining of a tunnel, but longer spinnerets also enable a spider to lay a wider swath of lines with a single movement.

Another possibly primitive trait (as deduced from the likelihood that ancestral webs were built at the mouth of a tunnel) was the tendency of *L. macrothelifera* to make frequent returns to the tunnel between short bouts of sheet construction (table 5.1).

6

THE BUILDING BEHAVIOR OF ORB-WEAVERS

Imagine, it is midnight in the silent woods. Before us is a great vertical wheel with a diameter as tall as a man. It is supplied with a hundred spokes, transparent, slender rays of silk which sparkle where we shine the light. Yet so delicate, they are all firm and strong; yet so numerous, they are all geometrically exact; and they are spread abroad with perfect uniformity over all this immense sheet. Amidst them moves the great architect herself. She looks at first sight an unwieldy creature, groping aimlessly and fruitlessly with her limbs, and scrambling out over the invisible sheet as though she were climbing unsupported in the air. But it is not so. She is a marvelous and consummate architect. She works with a perfect regularity and precision; she weaves with an inimitable skill. Every movement of her widespread limbs has a definite act in view. They are now feeling, now testing, now touching, now measuring . . . —Hingston 1922a, p. 649 on Nephila pilipes

6.1 INTRODUCTION

Building an orb web is truly an extraordinary behavioral feat. Take, for example, the white rat of orb weavers, the north temperate araneid *Araneus diadematus*. Early every morning before sunrise, in the space of about 30 min, the spider makes several thousand attachments (averaging about one every second) as she weaves about 30 m of silk line into a 45 × 45 cm web. On a human scale, this is equivalent to running about 7.5 km to build a web about 70 m in diameter. This web is not built with a rigid, invariable series of behavior patterns. Instead, as will be documented in this and the following chapter, the spider seamlessly adjusts different aspects of the orb's design, such as the numbers of lines and the angles and spaces between these lines, using a probably infinite array of variations in leg and body movements. The behavioral adjustments are elicited by a swarm of factors, including the spider's silk supply, her size and weight, the direction of the force of gravity, and the widely variable positions of support sites. As noted by Cesar Ades (1986), orb construction is simple in terms of repeated behavior patterns, but complex in terms of the adjustments that the spider makes to other variables. For an animal with limited brain power, working in a hurry, in highly variable environments, and in complete darkness, a finished orb web is truly worthy of admiration.

This chapter and its appendices (O6.1, O6.2, O6.3) describe how the spider moves her legs and body to produce an orb. Along with the next chapter on the cues that spiders use to guide orb construction, it constitutes the nuts and bolts of how orbs are built. Because the cues that the spider can sense are largely determined by her own behavior (they are apparently mostly tactile), the descriptions of behavior below are biased toward behavioral details that may help guide construction behavior.

6.2 SIMPLIFICATIONS FOR SMOOTHER READING

6.2.1 SPECIES AND TOPICS

The largest problem in writing this chapter was the opposite of that in the previous chapter: there is a large overabundance of data, due to the extensive published literature describing orb construction, and to additional notes as well as video recordings of several species. To make matters worse, many of these details are crucial, of only parochial significance, and boring. Clarity with respect to certain esoteric details (e.g., does the spider locate the inner loop by tapping laterally with her leg oI, or anteriorly with her leg iI, or posteriorly with her leg oIV?) can be key to answering certain spider-focused questions, such as which cues she might use to guide construction, or which behavioral details can serve as characters for higher-level taxonomy (where they are especially useful—Kuntner et al. 2008a). But a general reader is likely to get lost in a dense forest of details that otherwise lack general biological consequence.

I have dealt with these problems by using simplified accounts of the behavioral data from just two representative araneid species, *Araneus diadematus* and *Micrathena duodecimspinosa*, and only briefly mention other species in the general discussion in this chapter. To compensate, I have added three appendices: O6.1) further discussion emphasizing additional details of the behavior of *A. diadematus* and *M. duodecimspinosa*; O6.2) detailed descriptions of fine-scale movements by *M. duodecimspinosa* (mostly from analyses of video recordings); and O6.3) comparisons with orb construction behavior in different taxonomic groups.

My choice of the "representative" species needs explanation. Choosing only two was in one sense an act of desperation in the face of the profoundly tangled nature of the published descriptions of orb web construction behavior. In the first place, the literature includes observations of >150 species, whose behavior is not uniform (see Eberhard 1982; appendix O6.3). The major descriptive studies of behavior are on the araneid genera *Argiope* (Fabre 1912; Tilquin 1942), *Neoscona* (Hingston 1920), *Araneus* (Peters 1936, 1937; Jacobi-Kleemann 1953; Mayer 1952; Witt et al. 1968; Vollrath 1992; Zschokke and Vollrath 1995a; Zschokke 1996), *Zygiella* (Mayer 1952; Le Guelte 1966), and *Micrathena* (Dugdale 1969; Eberhard 2012; Eberhard and Hesselberg 2012), the tetragnathid *Leucauge* (Eberhard 1987b,d, 1988a,c, 1990b, 2012; Eberhard and Hesselberg 2012), the nephilid *Nephila* (Hingston 1922a,b,c; Eberhard 1990b; Hesselberg and Vollrath 2012), and the uloborids *Philoponella* (Eberhard 1990b) and *Uloborus* (Eberhard 1972a; Lubin 1986). Unfortunately many (probably most) of these descriptions include imprecisions, ranging from clear mistakes in verbal descriptions and diagrams of what occurred, to partial omissions of details stemming from overly simplified descriptions. An attempt to straighten out this tangle of precisions and imprecisions (e.g., which author got which detail wrong for which species, who was the first to get which detail correct, who failed to include which variations and made overly simple claims for which species) would make extremely tedious reading; it simply is not worth the effort (and I would probably commit further sins of commission and omission in discussing these problems). Secondly, orb construction behavior has been strikingly conservative throughout evolution (Eberhard and Barrantes 2015) and is probably monophyletic (section 10.8), so utilizing representative species tells much of the story for other taxonomic groups.

I selected the two species for different reasons. On the one hand, the north temperate *Araneus diadematus* has been the subject of more studies of web construction behavior than any other species, and these have been thoroughly reviewed (Witt et al. 1968; Vollrath 1992); its choice was more or less automatic. The second species, *Micrathena duodecimspinosa*, has several different advantages. Very little has been published on its behavior (all statements below without references are new), but the general layout of their orbs is unusually consistent and geometrically simple. There are nearly always only three anchors, three primary frame lines, and three secondary frame lines, one in each of the corners between the primary frame lines (figs. 1.4, 6.10). The orbs are relatively symmetrical, and both the temporary spiral and the sticky spiral are often laid without a single turnback. In contrast with species like *Leucauge mariana* (TET) and *Nephila clavipes* (NE) (Eberhard 1990b) (and, perhaps to a lesser extent, *A. diadematus*), the order of operations is very consistent; for instance, additional primary frame lines are never laid after the first three frames are in place and radius construction has begun. These simplicities make observations, descriptions, and interpretations of behavioral observations (especially during the earlier stages) clearer because the objectives of the spider's actions are often easier to understand (incidentally, the ability of *M. duodecimspinosa* to adjust to unusual situations, such as the repair of a damaged web [figs. 6.2, 7.36], suggests that the simplicity of *Micrathena* is secondarily derived rather than ancestral). A second virtue of *M. duodecimspinosa* is that it is common where I live and often builds after dawn and is not disturbed by light, so I could make close-up video recordings of all stages of construction. These allowed me to fill in details and give relatively complete and coherent descriptions.

6.2.2 LEVELS OF DETAIL

Describing orb construction is a challenge. A spider has eight legs, and they execute complex, more or less simultaneous movements that have different functions; descriptions can easily become so complex that even the most motivated reader becomes bored and confused. Orb construction behavior is organized hierarchically in nested sets of units or behavior patterns (Vollrath 1992), as in other animals (and in robots, for that matter—Hogan and Sternad 2013); decisions and processes at higher levels are executed via the decisions and actions at lower levels. I have therefore divided construction behavior (somewhat artificially) into two levels: "operations," or the order in which different lines are produced and how they attached to each other; and "detailed movements," the physical movements, especially of the spider's legs, that a spider uses to orient herself, to locate previous lines, and to manipulate the lines in her web while she performs the operations. Only summary descriptions of the movements are given here; more complete descriptions are in appendices O6.1 and O6.2.

The patterns of behavioral repetition sometimes differ sharply at the different levels. For instance, the operations used to build the sticky spiral were highly repetitive: attach the sticky line to a radius, move to the next radius, locate the inner loop of sticky line already laid, attach to this radius, and so on. But the leg and body movements that were used to perform this rigid sequence of operations varied widely, even from one radius to the next (table O6.3, figs. 6.1, 6.23). This variation was itself highly patterned; the detailed movements were adjusted flexibly, according to the particular circumstances, in ways that permitted the higher-level behavior to be rigidly consistent (e.g., section 6.5.1). In effect, there was a roiling sea of variations in the detailed movements underlying the uniformity of several operations in orb construction.

Previous studies have largely ignored the lower, detailed movement level (McCook 1889, Mayer 1952, and Jacobi-Kleemann 1953 are partial exceptions). But attempts to understand the evolution of higher-level operations will eventually have to take into account the underlying movements and the rules determining their adjustments. Just as other evolutionary biologists have recognized that understanding the evolution of morphological phenotypes is illuminated by studies of the developmental processes that produce these morphological traits (West-Eberhard 2003), behaviorists interested in the evolution of high-level behavioral patterns will eventually need to understand the details of the behavioral mechanisms that generate the higher-level patterns (sections 6.5, 10.11).

6.3 BEHAVIOR OF TWO ARANEIDS

6.3.1 HIGHER-LEVEL ORGANIZATION: THE STAGES OF CONSTRUCTION

Orb construction is organized in *Micrathena duodecimspinosa* and *Araneus diadematus* (as in all orb-weaving species and families whose behavior has been observed) into relatively discrete stages that are performed in a consistent, nearly invariant order (fig. 6.1, 6.2): exploration; early radius and frame construction; secondary radius and secondary frame construction; hub construction (sometimes along with a few additional radii); temporary spiral construction (also occasionally with a few additional radii); sticky spiral construction and simultaneous removal of the temporary spiral; and final hub modification (Witt et al. 1968; Eberhard 1982; Vollrath 1992; WE).

Fabre (1912) and Hingston (1920) found that this order was inflexible when orbs were experimentally damaged during construction. For instance, when temporary spiral or radial lines were cut during during sticky spiral construction, *Neoscona nautica* (AR) continued to lay sticky spiral rather than replace the non-sticky lines that had been cut, even though replacement would have been simple, and the absence of these lines caused the spiders substantial difficulties as they moved from radius to radius (Hingston 1920). Subsequent radius elimination experiments with *M. duodecimspinosa* have revealed, however, that sometimes (four of six experiments with as many different individuals) spiders showed flexibility, interrupting sticky spiral construction to add non-sticky lines (fig. 6.2, section 7.10; WE). Related experiments also revealed some (inconsistent) flexibility in the order of operations in *A. diadematus* (Petrusewiczowa 1938;

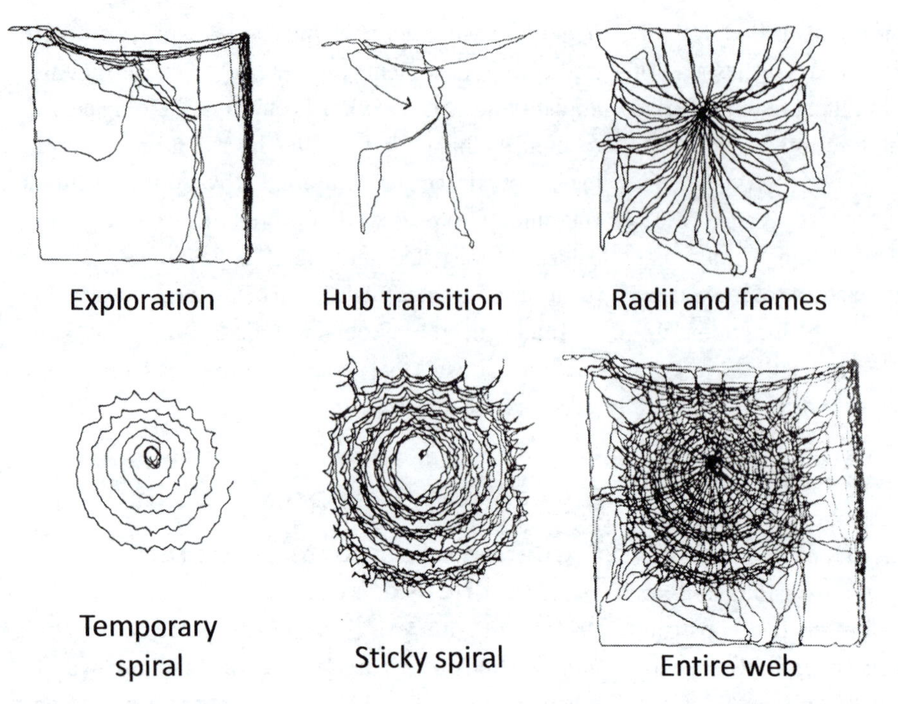

Fig. 6.1. These lines trace the path of an *Araneus diadematus* (AR) during the different stages in the construction of an orb in a rectangular cage. They are probably good representations of the paths of spiders in the field, except for the exploration stage, in which the cage and the lack of air movement probably had several effects: allowing the spider to walk directly along the substrate between attachment points; eliminating her ability to float lines; and reducing the range where she moved (after Vollrath 1992).

König 1951; Reed 1969; Peters 1970). In sum, the construction stages and the order in which they were performed were consistent in both species in normal construction, but showed modest flexibility when webs were modified experimentally.

6.3.2 EXPLORATION AND ESTABLISHING EARLY LINES

6.3.2.1 Use previous lines or start from scratch?

The first steps of orb construction were often determined by previous decisions of the spider regarding which lines of her prior orb to remove. In *M. duodecimspinosa* initiation of an orb typically involved starting nearly from scratch each morning, because the spider removed all or nearly all of the lines in her orb every evening. At night, spiders often rested on only one or two long, isolated lines. Some preliminary exploration may occur at night, because at about 5 AM, most individuals were resting immobile on a few lines, even though they had not yet begun to build (similar field observations of *A. diadematus* are not available). When a *M. duodecimspinosa* lacked such preliminary lines, or was moved away from such lines early in the morning, or when *A. diadematus* was placed in an empty cage, the spider began construction by finding new attachment points. Searching depended on the vagaries of air currents and the distribution of potential attachment sites and connections between them; sometimes *M. duodecimspinosa* failed repeatedly to encounter suitable sites, despite floating multiple aerial bridge lines (appendix O6.2.1.2).

In probably the majority of orb weavers (see Carico 1986), the spider usually replaces the radii and sticky lines of the previous day's orb with a few radii, leaving most of the frame and anchor lines intact, and then reuses these lines in her next orb (of course, in all species individuals also sometimes build orbs from scratch). Reuse of the previous web's lines resulted in a large savings in the distance traveled during exploration in captivity by *A. diadematus*, from a median of about 5.6 m (and a maximum of 63 m!) with no prior lines to about 2 m (to remove the previous web) (Zschokke 1996). I will concentrate here on how orbs are built starting from scratch under field conditions (and thus on *M. duodecimspinosa*) to give a more complete view of the process.

6.3.2.2 Problems in starting from scratch

Choices made at the earliest stages of web construction affect the structures and behavior in later stages and the eventual size and form of the web (Ades 1986; Vollrath 1992). Despite their obvious importance, however, these

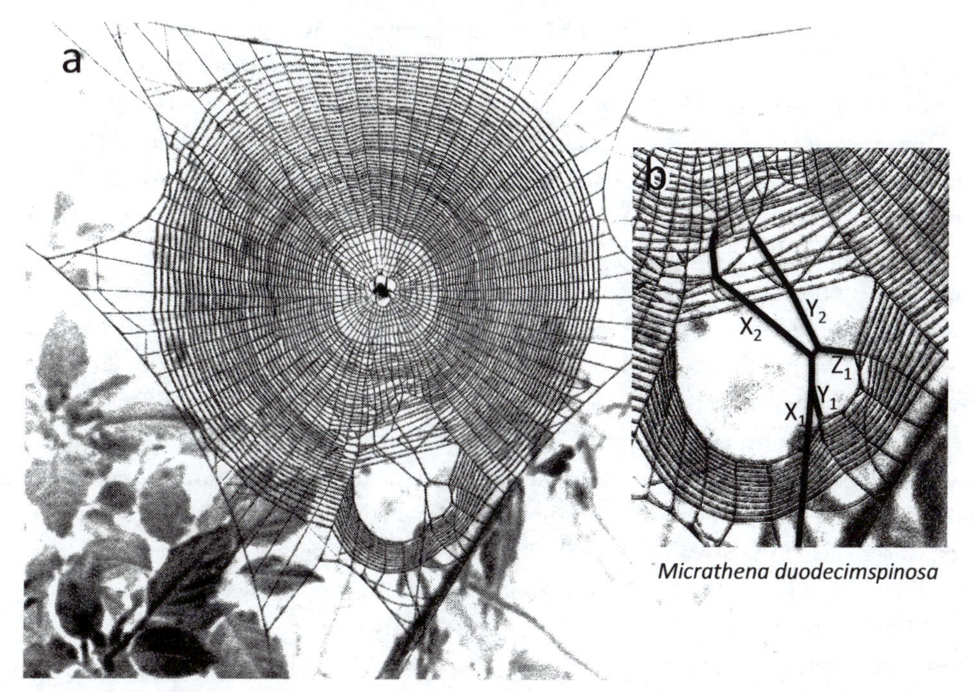

Fig. 6.2. This web of a mature female *Micrathena duodecimspinosa* (AR) (*a*) constitutes an exception to the Fabre-Hingston Rule that orb weavers are inflexible in the order in which they build sticky and non-sticky lines. When ten radii were broken in the lower portion of this web after the spider had laid 13 loops of sticky spiral, the spider paused briefly, then continued sticky spiral construction and laid seven additional loops of sticky line across the large open hole. But then she interrupted sticky spiral construction, and laid repair lines (the heavy lines in *b*), moving downward and then back upward across the broken sector. She then moved to the hub, removed the lines in the center (see fig. 6.16), paused for about 1–2 min, and then resumed sticky spiral construction. When the powder was gently knocked from lines in the repair zone, lines X_1 and X_2 (*b*) were revealed to be neither sticky nor highly elastic (i.e., they were apparently draglines); lines Y_1 and Y_2 were sticky but not highly elastic (perhaps draglines to which sticky lines had adhered); and line Z_1 was sticky and highly elastic (probably a segment of sticky spiral laid before the repair). In sum, the spider interrupted sticky spiral construction to lay non-sticky repair lines, then resumed sticky spiral construction, a flexibility that Fabre and Hingston thought spiders lacked. The repair increased the tensions on the inner portions of the radii to which the repair lines X_2 and Y_2 were attached.

early stages are the least studied and most poorly understood aspects of orb construction. There are several reasons for this relative neglect, not the least of which is the intrinsic difficulty of describing the spider's behavior.

The spider's first major problem is to find supports to which anchor lines can be attached that are neither too close together nor too far apart, and to ascertain whether there is enough open space, free of obstacles, between them in which she can build her orb (to a large extent, these problems did not occur in the cages in which *A. diadematus* behavior has always been observed). The sense of sight in orb weavers in general is poor, many species build in the dark, and experimental blinding in captivity did not impede orb construction (Witt et al. 1968). Thus orb weavers in general can find attachment sites only by encountering them directly. Even in *M. duodecimspinosa*, which often built webs after dawn and thus has visual cues available, spiders sometimes walked close to potential attachment sites without extending any legs toward them, and thus apparently did not use visual information in searching for attachment sites.

The basic technique used by *M. duodecimspinosa* to discover attachment sites in nature was to float lines on the breeze until they snagged on distant objects. Spiders also sometimes walked along possible supports, but usually only for short distances. They also sagged or tensed their draglines and the lines along which they walked, causing them to move through space and contact objects there (see also Zschokke 1996 on *A. diadematus*). Searching for attachment sites by floating lines on the breeze was not always effective, as it depended on the vagaries of air movement; some *M. duodecimspinosa* (and *M. horrida*) in the field spent an hour or more launching lines without finding suitable attachment sites (I could induce them to immediately begin orb construction by snagging a floating line on an appropriate site) (see appendix O6.2.1.2). While a spider was locating possible attachment sites, she needed to also determine whether there was an open planar space between them that satisfied at least three conditions: large enough to harbor an orb with a preferred slant or slants; small enough that the lines supporting her orb would not be overly long; and free of any objects projecting through the plane of the orb or that were too close to it (see Zschokke 1996 for *A. diadematus*). One indication of the importance of checking for possible obstacles is the unusual omission of the exploration stage by *Caerostris darwini* (AR) (Gregorič et al. 2011). These spiders built elevated orbs over rivers, where such obstacles are probably rare.

As suggested in Vollrath's discussion of *A. diadematus* behavior (1992), it seems inevitable (though not yet directly demonstrated) that making these discriminations must involve some sort of kinesthetic memory of the distances and directions the spider moves ("path integration") during the early stages of orb construction; but the stages during exploration when this information is gathered are unknown. The use of path integration to estimate the sizes and shapes of areas traversed has also been implicated in orb weavers at other, later stages of construction (chapter 7; Eberhard 1988c; Eberhard and Hesselberg 2012). In sum, the spider's movements in the area she is exploring probably often have multiple functions: discovering attachment sites; placing and moving lines; and providing information needed to decide whether there is an appropriate sized space for her web (Vollrath 1992).

It is important to note that there are numerous published descriptions of the early stages of orb construction that are partly or completely wrong (McCook 1889; Hingston 1920; Comstock 1967; Savory 1952; Levi and Levi 1968; Dugdale 1969; Forster and Forster 1973; Levi 1978; Foelix 2011). Unfortunately such errors sometimes acquire a life of their own when they are later cited uncritically (see discussions in Coddington 1986b; Eberhard 1990a; Zschokke 1996). A further problem is that even the most precise descriptions generally share the shortcoming of ignoring quite substantial variations (Peters 1933b; König 1951; Mayer 1952; Eberhard 1972a; Coddington 1986b) (for some exceptions, see Tilquin 1942).

6.3.2.3 Basic operations during exploration

Two major operations during exploration in *A. diadematus* and *M. duodecimspinosa* were A) establishing bridging lines between points of attachment; B) fastening these lines together at one or more points. The positions of these lines were often shifted repeatedly by being broken and then reattached to each other at different sites with "break and reel" behavior (fig. 6.3), or (less often, at least in *M. duodecimspinosa*) reattached to the substrate at a different site. These processes included discarding previous lines, replacing lines, shifting points of attachment between lines, and "false starts" (in which apparent trips from the proto-hub toward the periphery were aborted without laying a new radial line), and they varied substantially in their details.

6.3.2.3.1 GATHERING SENSORY INFORMATION AND LAYING THE FIRST LINES

No overall patterns have been discovered in the early stages of exploration of either *M. duodecimspinosa* or *A. diadematus* (fig. 6.1. Vollrath's description of *A. diadematus* in a cage is illustrative: ". . . clambering about crisscrossing the available space. . . . Eventually, sometimes after hours of climbing about, the spider seems to 'settle' for a particular area of the explored spaces" (Vollrath 1992; p. 155) (fig. 6.1). Probably at least some of the variation in exploration is due to differences in the positions of the attachment sites and of possible obstructing objects that the spider discovers. The mean distances that *A. diadematus* traveled while exploring increased dramatically (from 5.6 m to 26.6 m; their respective maxima were 63.2 and 212.5 m!) when three short crossbars were added to each of the two vertical sides of a support (Zschokke 1996). Even the behavior of the same individual at a standardized site sometimes varied substantially; one *A. diadematus* moved 7.3 m while exploring on one day, and 63.2 m on the next). The inclusion of twigs in a cage also caused *Cyrtophora cicatrosa* (AR) to displace its domed sheet web to one side, away from the center of the cage, and to reduce the sheet's size (Rao and Poyyamoli 2001).

Many orbs of *M. duodecimspinosa* were built in open sites, in which there were no nearby objects that could physically interfere with the orb (for a contrast, see fig. 3.7b of *Araneus expletus*). Presumably the consistent choice of open sites was due to information gathered

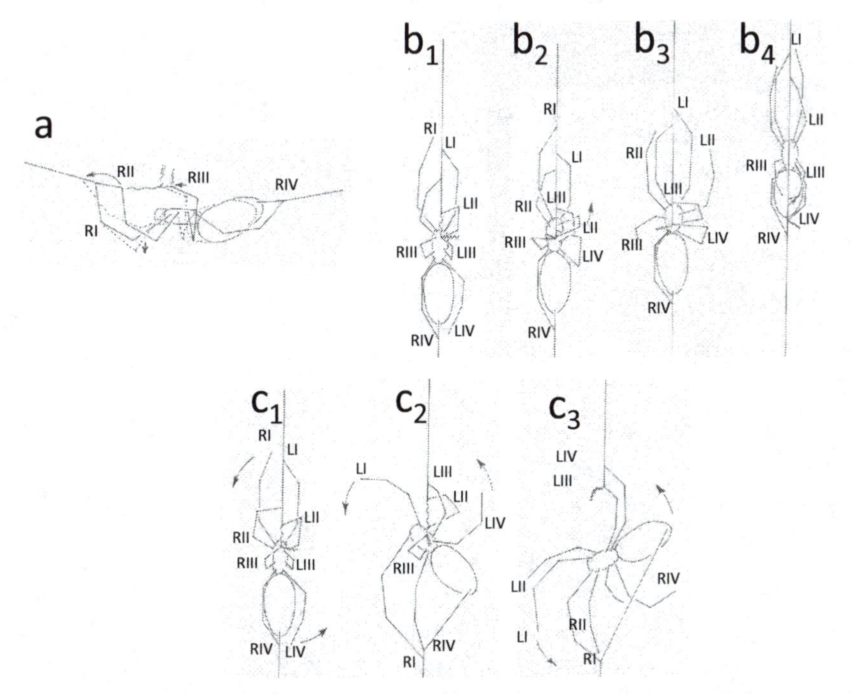

Fig. 6.3. Agility and fine motor control are illustrated in these schematic drawings of the stereotyped "ballet" of leg movements of *Leucauge argyra* (TET) during "break and reel" behavior (*a*), and two methods of reconnecting the broken ends of a line following break and reel behavior (*b, c*). During break and reel (*a*), legs I and II stepped forward with alternate, "hand over hand" movements (seen from the side, with only the legs on the spider's right side are drawn for simplicity) (dotted lines indicate leg and body positions that followed the positions indicated by solid lines). With each step the leg grasped the line just ahead of where the other leg was grasping it, and the spider pulled herself forward by flexing her leg. Thus, leg RII moved forward (dotted lines) to grasp the line in front of leg RI, which had just been flexed to pull the spider forward. Meanwhile, legs III were directed anteriorly and ventrally and, together with the palps, they packed the loose line (wavy line in this and other drawings) that accumulated into a ball of loose fluff (tightly curled lines in this and other drawings) just ventral to her sternum. Periodically, legs III and the palps held the fluff against her mouth, where it disappeared (sometimes it was possible to see that a small wet ball replaced the mass of fluff). Meanwhile one leg IV was held in the air, apparently without any function, while the other (RIV) held the emerging dragline, which slid through its tarsal claws. The dragline was deflected where tarsus IV held it, both while the spider was moving and when she paused, indicating that it helped support the spider's weight. One type of reconnection behavior (b_1–b_4; ventral views of the spider seen from above) occurred without the spider turning around. First she brought one leg II farther posteriorly, so that its tarsus was more nearly ventral to her body than usual (LII in b_1) while its tarsus held the line that had been reeled in as she moved forward. Her ipsilateral III and (a little later) IV then reached forward and grasped this line just anterior to the site being held by leg II (arrow and LIII in b_2), and leg II then stepped forward (LII in b_3 and b_4). These same III and IV legs, which would hold the line (b_3) so that the dragline could be attached to it (in b_4), were on the side opposite the leg IV that held the dragline (RIV). There was a broken stub of loose silk just posterior to where leg IV held the line (tightly curled line in b_3 and b_4). The spider moved forward and dabbed her spinnerets to the line between the tarsi of legs III and IV to attach her dragline (b_4), attaching her dragline to the line ahead. A second type of reconnection (c_1–c_3) involved the spider turning her body 180°. She began by turning her body partially (c_2) and reaching posteriorly with one leg I (RI in c_2) to grasp the dragline beyond the site she held with the ipsilateral tarsus IV. As she turned, the leg III on the opposite side swung anteriorly to grasp the line that was being reeled in just anterior to the site held by her ipsilateral tarsus II (LIII in c_2). As her body turned farther (c_3), the leading IV (LIV in c_3) also swung anteriorly to grasp the reeled line slightly beyond the point being held by leg III (LIII and LIV in c_3); the spider then swung her abdomen laterally and touched her spinnerets to the line between the sites held by legs III and IV, thus attaching her dragline (leg IV sometimes preceded rather than followed leg III). All of these movements were quick and smooth and took only fractions of a second. Careful coordination and precision were crucial, because the spider was simultaneously supporting her own weight with the same legs (from video recordings by Rosannette Quesada).

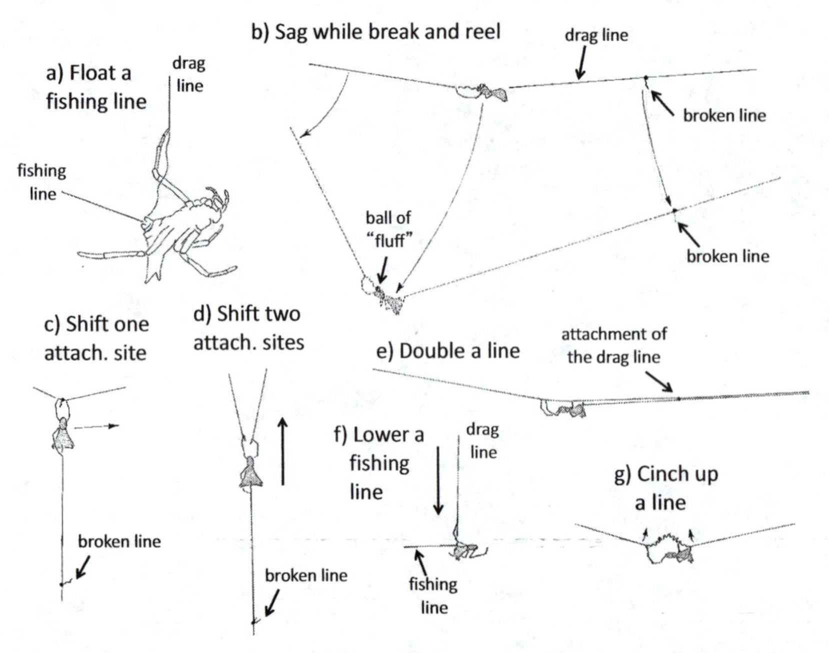

Fig. 6.4. These schematic drawings illustrate different types of exploration behavior by *Micrathena duodecimspinosa* (AR) prior to initiation of the orb proper. *a*) The spider floated a "fishing" line on wind currents by releasing line while hanging on her dragline, searching for possible attachment sites for an anchor line of her new orb. *b*) As the spider moved along a more or less horizontal line, she broke it, and sagged deeply as she released additional drag line, then climbed back upward while releasing little or no drag line. This behavior may function as searching, because any object in the plane that was described by the spider and the line as they sagged downward would be contacted. *c*) The spider shifted the site on a line where a line was attached by first breaking the line and attaching her dragline to its broken end, and then moving along the other line (arrow to the right) and attaching her dragline to it. *d*) The spider shifted two attachment sites by beginning as in *c*, but also breaking and reconnecting the second line. *e*) The spider doubled a line by attaching her dragline to it and laying additional line as she walked along it. *f*) The spider lowered a fishing line that was floating in the breeze (as in *a*) by descending further on her dragline. *g*) The spider cinched up a line (increased its tension) by reeling it in and then attaching her dragline near the points held by her legs; the increase in tension caused the spider to move upward (arrows).

during exploration. The distance of the web above the ground or other substrates may be sensed in *Micrathena duodecimspinosa* (and other species) during occasional long descents from which the spider returned without having made an attachment (Eberhard 1990b; WE). When placed in a previously unoccupied stick maze, *A. diadematus* apparently explored it, then built her web in the only space that was large enough for it. After walking around briefly, she seemed to fix on one particular site, which then formed the focal point of all of her further, sometimes protracted explorations (Chessell unpub. and Zschokke unpub. cited in Vollrath 1992; Zschokke 1996).

One particular type of descent behavior by *A. diadematus* may have a searching function. The spider moved horizontally after attaching her trail line, and then released her hold and fell downward, swinging like a pendulum as she hung from her dragline. By swinging this way, the spider could contact (and thus discover) objects that were not directly below her descent site. Perhaps this behavior represents a search for attachment sites, or for nearby objects that could become entangled in an eventual orb.

Commonly *M. duodecimspinosa* moved or removed and replaced many of the lines that she laid during exploration. These alterations enabled her to produce an orderly array of lines in her orb despite the fact that preliminary, exploratory lines and attachments were often not appropriate for the final orb itself; the spider could evaluate the dimensions of a potential construction site only by actually moving about in it, laying draglines as she went. Shifting and removing lines was generally accomplished by "break and reel" behavior, which involved complex leg and spinneret coordination in breaking, seizing and producing draglines (figs. 6.3, 6.4, appendix O6.2.1.2).

6.3.2.3.2 THE END OF EXPLORATION AND THE "HUB TRANSITION"

The end of exploration occurred abruptly in both *A. diadematus* (Zschokke and Vollrath 1995a) and *M. duodecimspinosa* when the upper frame line and its associated radius were built (fig. 6.5). This marked the end of the period of building tentative lines that would subsequently be removed or shifted; the subsequent new lines were not later removed nor were their attachment sites changed. Similarly, the intermittent, long pauses that characterized exploration ended; for the next 30–45 min, the spider woud pause only once, for 10–20 s just before initiating the sticky spiral. For an observer, this transition marks the abrupt (and welcome) end of behavior that generally seems unintelligible, and the beginning of movements with consistently clear purposes.

After building the upper frame, the spider tightened it (note loose silk at 5 in fig. 6.5d), and then broke and lengthened its radius, thus moving the proto-hub downward to the approximate position that it would occupy in the final web (see fig. 6.5d for *M. duodecimspinosa*, and Zschokke 1996 and Zschokke and Vollrath 1995a for *A. diadematus*). This "hub transition" was immediately followed by the construction of two additional radii and frame lines in *M. duodecimspinosa* (fig. 6.5). At this point, the web typically consisted of a "Y" of three (or occasionally four) radii and their frames (fig. 6.5); *A. diadematus* typically had 4–7 radii (Petrusewiczowa 1938; Mayer 1952; Zschokke and Vollrath 1995a; Zschokke 1996) (not just three, as described by Peters 1937).

In *A. diadematus* the spider sometimes replaced at least the inner portion of some of the other radii immediately after she had lowered the proto-hub, perhaps adjusting their tensions (Zschokke 1996); similar adjustments were not seen in *M. duodecimspinosa*. Comparisons with other araneid species suggest some slight differences with respect to the hub transition process. For instance, the hub's position subsequently moved after the hub transition in 7 of 32 cases in *A. diadematus*, but in 11 of 12 in *Larinioides patagiatus* (AR) and in 4 of 4 in *Cyclosa insulana* (AR) (Zschokke and Vollrath 1995a). The hub transition in uloborids was somewhat different (fig. 6.6, appendix O6.3.2.3).

6.3.3 FRAMES, SECONDARY RADII, AND HUB LOOPS

6.3.3.1 The other "primary" frames

In *M. duodecimspinosa*, the spider almost always built the two other "primary" frame lines immediately following the hub transition, using a sequence similar to that employed in building the first frame (fig. 6.5e-h). In *A. diadematus*, there were more primary frames (up to about seven), and their order of construction was more variable. Some details of primary (and secondary) frame construction are not yet clear (appendix O6.1.1, O6.2.1). The details of many of the different sequences of behavior that are used by other orb weavers to build primary and secondary frame lines varied substantially (Eberhard 1990b; Zschokke 1996; appendix O6.3.3). One consistent pattern was that in all of the large variety of different types of frame line construction behavior, the spider always began by leaving the hub by walking along a radius that was already in place. In *M. duodecimspinosa* there was always a radial line with each frame that was continuous with it, but in *A. diadematus* and others there were exceptions (Zschokke 1996) (appendix O6.1.1, O6.3.1). In *M. duodecimspinosa* and often but not always in *A. diadematus*, the hub was completely surrounded by primary frame lines when primary frame construction ended.

6.3.3.2 Secondary radii

Once the site of the hub had been set and the primary frame lines were in place, the spider continued to add additional, "secondary" radii. Secondary radius construction behavior in *M. duodecimspinosa* and *A. diadematus* (Peters 1936, 1937; Mayer 1952; Vollrath 1992; Zschokke 1996) was similar to that described for many other species. The spider attached her dragline at the hub and grasped two adjacent radii with her legs I (see fig. 6.7 and appendix O6.2.2 for details). She then moved away from the hub along the upper of these radii (the "exit" radius), paying out a dragline. On reaching the frame she moved "laterally" along it a short distance, and attached her dragline to it (fig. 6.8a,b) (sometimes *A. diadematus* moved along the substrate rather than along a frame line). Her dragline formed a radial line to the hub (the "provisional" radius of Zschokke 2000), but she immediately broke this provisional line as she turned and moved back toward the hub; she reeled up the provisional

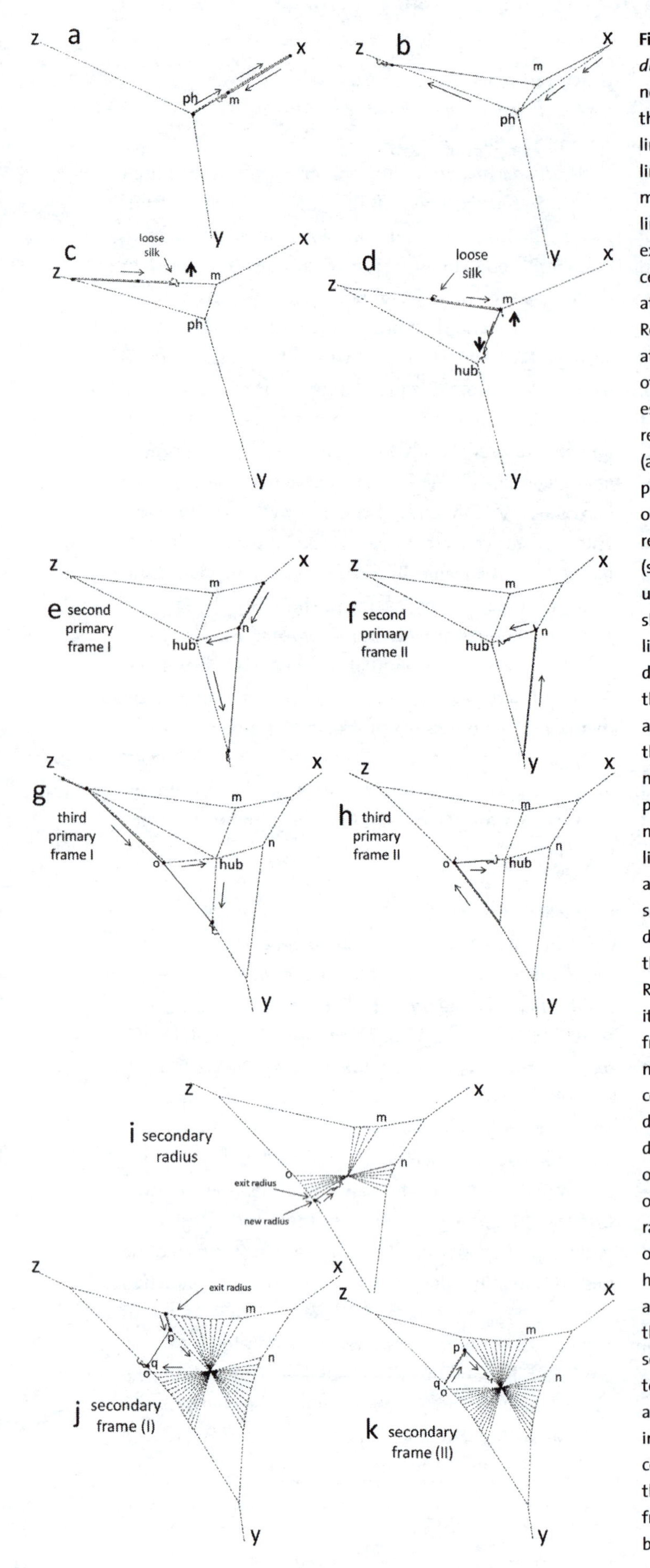

Fig. 6.5. Some basic steps in orb construction by *Micrathena duodecimspinosa* (AR) are depicted in these stylized drawings (they are not to scale and are drawn as if the weight of the spider did not cause the lines to sag; solid arrows depict movements of the spider; solid lines are new lines laid since the previous drawing; dotted lines are lines (single or compound) from before; black circles are attachments made since the preceding drawing; thick arrows are movements of web lines; squiggly lines are accumulations of loose silk). After more or less extensive exploration in a large open space, the spider established a central point, the proto-hub (ph), where three lines (attached to supports at x, y, and z) intersected, and built the first primary frame line (*a–d*). Recognizable orb construction behavior began (*a*) when the spider attached her dragline at the proto-hub, walked out to the end of one of the lines that was above the proto-hub, and made a second, often especially long (up to 12 s) attachment to the substrate (at x). She then returned toward the proto-hub, attaching her dragline at least once (at m) to the line she had laid on the way out. She continued past the proto-hub and out a second radial line (toward z) to attach her dragline one or more times, thus forming the first frame line (*b*). From here she returned along the new frame, pausing to cinch it up along the way (*c*) (see fig. 6.4g); cinching caused the spider and the frame line to move upward a cm or more (heavy arrow). When she reached the point where she had attached radial line to the new frame (m), she broke the radial line and returned to the proto-hub, reeling it up as she went (*d*). As she did so, she relaxed the tension on this radial line considerably, causing the proto-hub to move downward (heavy arrow downward) to the approximate position that it would occupy in the finished web ("hub"); the distal end of this line (m) moved upward (up to a body length or more) (heavy arrow upward). The spider immediately built a second primary frame line (*e–f*). After attaching her dragline at the hub, she moved out the original radial line past its attachment to the first frame line at m to attach farther from the hub (upper dot in *e*) (usually this attachment was to the first frame line, but occasionally it was to the substrate); then she returned along this line to the hub, attaching to her dragline at least once on the way (at n) and then out to near the end of the third radial line where she made another attachment (lower dot in *e*). Returning along this new line, she broke the radial line at n and reeled it up as she returned to the hub (*f*). She next made the third primary frame (*g–h*) with a similar series of movements and attachments, with a new radius attached to the new frame at o. At the end of primary frame construction (*h*) there were four "primary" radii—the three radii built during frame construction and a fourth, more or less vertical line running downward from the hub that constituted the only unmodified remnant of the original "Y" in *a*. The spider immediately began to make a series of secondary radii (*i*). Each new radius was built using a previously built radius as an "exit" to reach the frame. On reaching the frame at the end of the exit radius, the spider moved a short distance laterally, attached her dragline, then returned to the hub along this new radius, breaking and reeling as she went. In nearly all cases the exit radius was above the new radius. The secondary frames were built (*j–k*) after numerous secondary radii had been built, using a sequence of movements similar to those in the construction of the primary frame lines. After attaching a new radial line to the upper of the two primary frames (upper dot in *j*), the spider attached her drag line to the new radial line (p) and continued to the hub without breaking and reeling, and then moved out the adjacent radius and attached the new line to the second primary frame (q). Then she walked along the new secondary frame to p, and broke and reeled up the radius while returning to the hub (*k*).

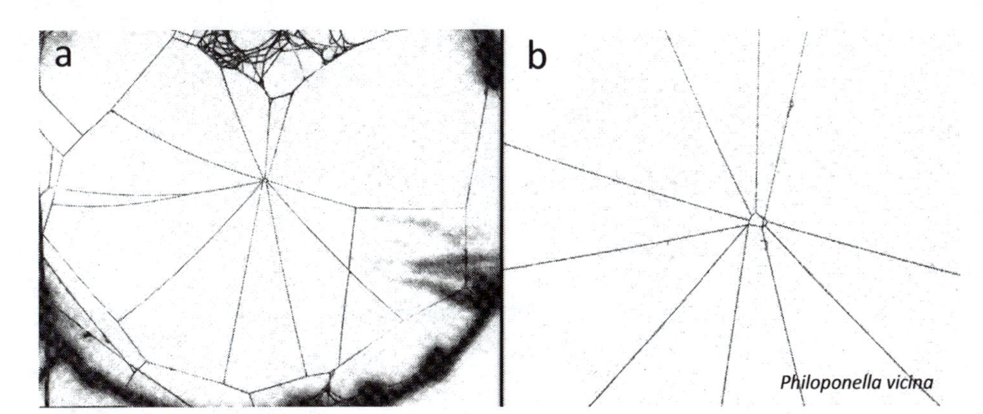

Fig. 6.6. *Philoponella vicina* (UL) built the first hub loop while removing the proto-hub (the central area where the first few radii had been connected) (*a* shows the entire web just after the first hub loop was built, *b* is a closeup of the hub). Similar construction of the first hub loop occurred in *Uloborus diversus* (UL), but no araneoid orb weaver is known to perform this behavior (from Eberhard 1990b).

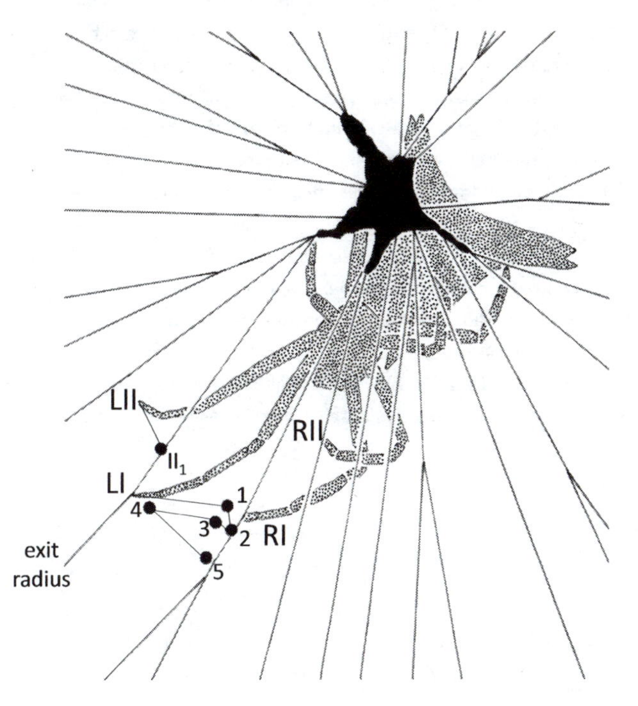

Fig. 6.7. Late in the radius construction this mature female *Micrathena duodecimspinosa* (AR) repeatedly slapped laterally with her legs I in the space in which she was about to build a radius (this was the last radius laid before hub construction began) (the black mass represents loose silk that had accumulated from cut-and-reel behavior during the construction of the previous radii). The lines and dots trace two medial slapping movements with leg LI that occurred immediately after the moment depicted in the drawing (successive dots were 0.033 s apart; the other legs were more or less immobile except that LII moved to grasp a radius 0.033 s later); these movements had been preceded by slaps with both RI and LI, and were followed by three more slaps with RI and then simultaneous slaps with LI and RI before the spider finally began to leave the hub to lay the new radius just below the exit radius; she dabbed her abdomen against the silk mass (presumably to attach her dragline) and moved away from the hub along the exit radius. Extensive explorations preceding radius construction presumably represented searches for previous radii; they only occurred during construction of the last few radii. By grasping adjacent radii with her anterior legs, the spider may obtain information on inter-radial angles.

radius and replaced it with a new dragline (the "definitive" radius). She consistently altered the tension on the definitive radius by releasing more or less dragline as she walked (fig. 6.8a,c) (appendix O6.2.2). Both *M. duodecimspinosa* and *A. diadematus* broke the provisional radius near its attachment to the frame, but *Zilla diodia* (AR) generally did not break the provisional radius until she was about halfway back to the hub (Zschokke 2000). In at least this species, the spider apparently did not attach her dragline to the provisional radius when she broke it (Zschokke 2000); in *M. duodecimspinosa*, however, a white speck was left at this point that, under a compound microscope, proved to be an attachment (and not simply a mass of loose silk). The site where the break occurred in *M. duodecimspinosa* varied, from less than one-tenth to about two-thirds of the distance to the hub. The spider deposited the accumulation of silk from the provisional radius that she reeled up at the hub, where a small white mass of fluffy silk gradually accumulated as radii were built. Zschokke and Vollrath (1995a) reported that both *A. diadematus* and *Uloborus walckenaerius* (UL) tightened the new radial line just before attaching it to the frame, but I have never seen this behavior in *M. duodecimspinosa* (nor in other uloborids) (see, however, fig. 6.9 of *L. argyra*). In some other taxa the provisional radius was not broken (appendix O6.3.2); the details of secondary radius construction constitute useful characters for distinguishing nephilids, uloborids,

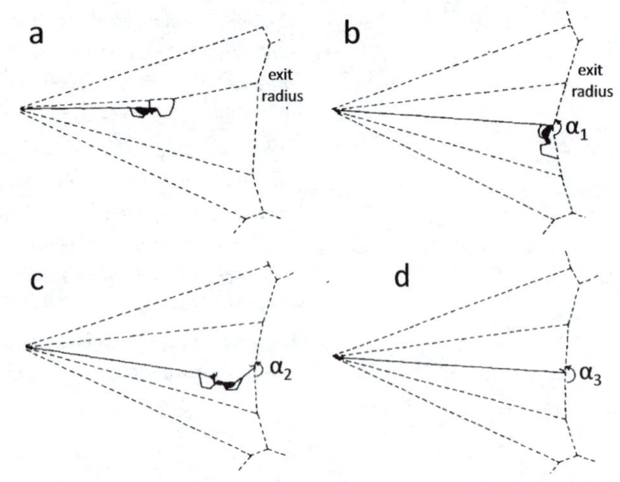

Fig. 6.8. The typical sequence of operations and tension changes during radius construction by araneids and tetragnathids (not to scale; new lines are solid) began with the spider attaching her dragline at the hub and moving toward the edge of the web along one of the radii bounding the open sector (the exit radius), producing a new dragline (*a*). On reaching the frame line she moved laterally a short distance and attached this new dragline to the frame (*b*). She immediately broke this new line (the "provisional radius") (*c*), then moved back to the hub, replacing it with a new dragline that she attached there (*d*). The relative tensions on the new radial line are reflected by the angles α_{1-3} in the drawings. At the moment of attachment to the frame the tension was relatively high (α_1 was smaller); at the moment she broke the provisional radius (*c*) the tension was reduced (α_2 increased), and when she attached the new radius at the hub the tension was even lower (α_3 increased still more) (after Eberhard 1982).

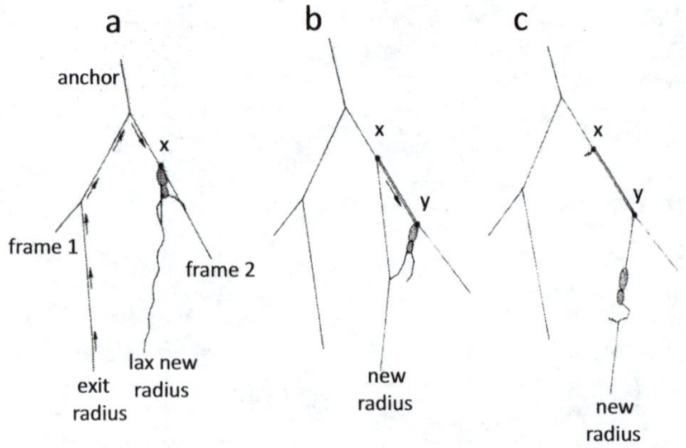

Fig. 6.9. A mature female *Leucauge argyra* (TET) compensated for an especially lax radius by using the subtle variation in radius construction behavior outlined here. When the spider reached the corner between frame 1 and frame 2 and began moving along frame 2 (*a*), her path brought her closer to the hub and thus caused her dragline to go slack (wavy line). She modified typical radius construction behavior to tighten this dragline, attaching her new dragline to the frame at two points (*x* and *y* in *b*) instead of one (as is typical in this and many other species—see fig. 6.5). This extra adjustment involved altered leg movements. This previously undocumented ability to compensate for a lax line, which I missed in a review of radius and frame construction behavior in *L. mariana* (Eberhard 1990b), sounds a note of caution: there may be many additional subtleties of building behavior that are waiting to be discovered, even in relatively well-studied species (drawn from a recording by R. Quesada).

and cyrtophorine araneids (Eberhard 1982; Kuntner et al. 2008a).

One standard pattern in *M. duodecimspinosa* and *A. diadematus* (Zschokke 1996) as well as in all other species in which secondary radius construction has been described in detail (e.g., Eberhard 1982; Zschokke and Vollrath 1995a; appendix O6.3.2), is that nearly all new radii were laid at a "final" angle with respect to the exit radius (e.g., radius 4 in fig. 6.10): only seldom was a subsequent radius laid in the space between a new radius and its exit radius. Analyses of radius connections with hub lines in the hubs of *Uloborus diversus* (UL) (fig. 6.11, Eberhard 1972a) confirmed that this pattern was widely distributed. This pattern gives the spider precise control over most inter-radial angles (appendix O6.1.1.5.1).

Another common pattern in both species was for successive radii to be laid on more or less opposite sides of the web (see radii 2–31 in fig. 6.10). This "alternate sides" pattern occurred in many araneoids, but not uloborids (appendix O6.3.2.3). The pattern was somewhat inconsistent in *A. diadematus* (Witt et al. 1968), however, and also clearly did not occur in one figured web of *Zilla diodia* (AR) in which six successive radii were laid to the uppermost frame (Zschokke 2000). Alternating the sides where new radii were laid probably served to balance the tensions generated by the new radii, so that the hub of the orb did not migrate substantially during orb construction (see fig. 1 of Zschokke 2000) and the tensions on radii did not change substantially as radius construction proceeded.

Often, especially when the next radius was not opposite the previous one, the spider turned slowly at the hub, successively grasping adjacent radii with legs I, as if measuring the angles between them (fig. 6.7, appendices O6.1.1.5, O6.2.2). Finally she ceased turning and laid the next radius between the two radii that she had been

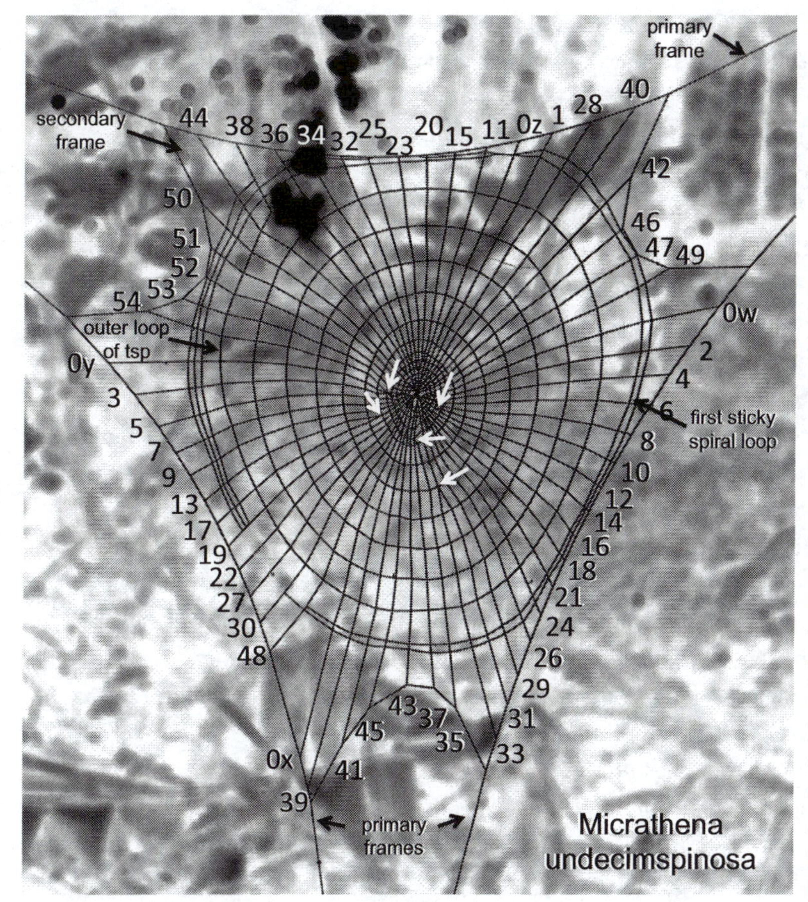

Fig. 6.10. Several typical patterns in radius construction are illustrated in this approximately vertical orb of an adult female *Micrathena duodecimspinosa* (AR) in which radius construction was observed directly, and construction was then interrupted after the spider had completed only 2–3 loops of sticky spiral, before she had broken any of the temporary spiral lines (the sticky lines attached to radii 22, 27, 30, and 48 were accidentally broken). Radii O_w, O_x, O_y, and O_z were the primary radii (see fig. 6.5); the numbers of the other, secondary radii indicate the order in which they were laid. The secondary radii attached to the lateral frames were generally added in strict order from the top moving downward; for example, the successive radii to the left frame (#'s 3, 5, 7, . . .) were built in order moving downward, with each new radius laid just below its exit radius (e.g., the spider used radius 3 as an exit line when laying radius 5, etc.). In contrast, the radii to the upper primary frame were laid on both sides of the primary radius O_z; here again the exit radius was the one immediately preceding it to that frame (i.e., radius 15 was the exit radius for building radius 20). The temporary spiral was a continuation of the hub spiral; the spaces between its loops tended to increase farther from the hub in the first loops and to level off in some parts of the orb, and to be greater above the hub than below it. The white arrows indicate sites at which attachments of the hub or temporary spiral fastened adjacent radii together, giving the false impression that tertiary radii had been added. In webs such as this in which the temporary spiral was intact, it was possible to trace the spider's path and determine whether the distances between different lines that she encountered correlated with her placement of the first and second loops of sticky spiral (see text).

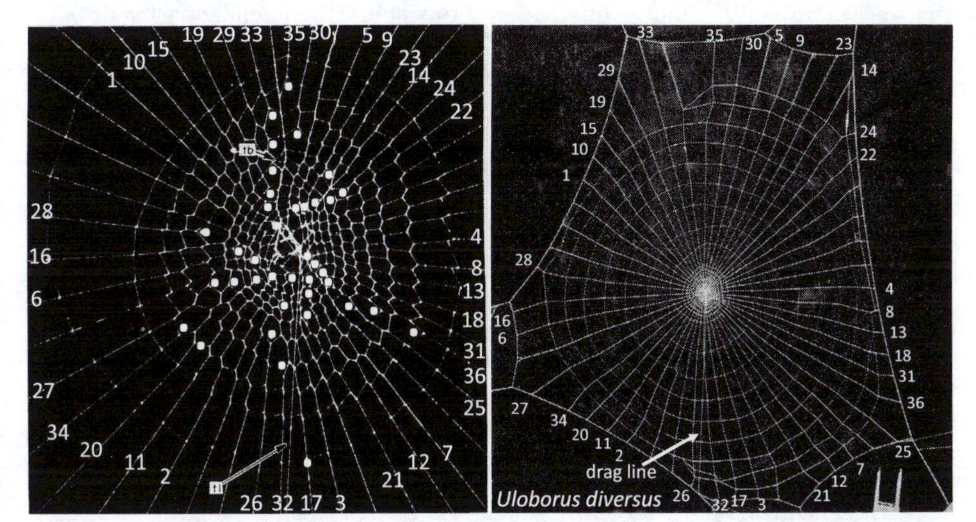

Fig. 6.11. The order in which the radii were built in this horizontal orb of an adult female *Uloborus diversus* (UL) was deduced from the pattern of radius branching in the hub (dots in the closeup of the hub at the left mark the origins of new radii), and radii are numbered in the order in which they were built; the order of the earliest 13 radii, many of which were probably associated with the proto-hub (see fig. 6.6), could not be determined. Nearly all of the radii built following proto-hub destruction were placed at "final angles" (no subsequent radius was laid in the space formed between a new radius and its neighboring exit radius). For instance, the early sector between r_1 and r_{10} was never subdivided by another radius (r_{23} was an exception to this pattern; it was built in the space between r_{14} and its unlabeled exit radius). From r_1 to r_{23} the spider turned in a counter-clockwise direction at the hub; then she turned back (labeled "tb" in the photo on the left), and from then on moved in a clockwise direction. The spider usually moved along the frame in the opposite direction to that she had been moving at the hub. For instance, she moved clockwise along the frame from the exit radius r_{10} to attach r_{15} (from Eberhard 1972a).

holding with her legs I. Before leaving the hub, *M. duodecimspinosa* often tapped repeatedly in the space between these radii with one or both legs I, as if verifying that no radius was already present there (fig. 6.7).

Another radius construction behavior that was very rare in *M. duodecimspinosa* (only one in probably many hundred radii in more than 100 webs) and that has not been reported for *A. diadematus* but that is nevertheless much more frequent in other orb weavers such as *L. mariana* (TET), were "false starts" during radius construction. The spider interrupted hub construction to face outward briefly, or even moved away from the hub as if to lay a radius, but instead turned back, broke the exit radius, and replaced its inner portion as she returned to the hub and resumed hub construction. False starts in *L. mariana* occurred on the order of one every 15 radii during the last two-thirds of web construction (approximately 2 per web) (Eberhard 1990b). Perhaps these brief interruptions represent "mistakes," in which the spider decided to lay a new radius but then changed her mind. Those false starts in which the spider broke and replaced the inner portion of a radius may have functioned to adjust the tension on the exit radius.

6.3.3.3 Secondary frames

Secondary frames (frame lines attached to primary frames—fig. 1.4) were built in corners between the primary frame lines by both *M. duodecimspinosa* and *A. diadematus* during radius construction (figs. 3.27, 6.10). In *M. duodecimspinosa* orbs there were almost always three secondary frames (fig. 1.4); secondary frames were less common in *A. diadematus* (8 of 55 frame lines in 9 orbs—Zschokke 1996). The basic pattern of operations in *M. duodecimspinosa* and *A. diadematus* was similar to primary frame construction (fig. 6.5j,k); a new radius was consistently added as part of the construction of each new secondary frame (fig. 6.5j), or at least before making the first attachment to the frame (point 13 in fig. 6.5). Apparently *M. duodecimspinosa* decided that the new line would be a secondary frame rather than a radius while she was moved away from the hub (or at least prior to making the first attachment to the frame) (point 13 in fig. 6.5), because this attachment point was farther from the exit radius than it would have been if it were a radius (WE, section 7.7.2).

6.3.3.4 Hub loops

During the latter stages of secondary radius construction or just after it ended, the spider began to build the hub, connecting the radii with hub loops as she turned at the hub. Contrary to some descriptions (e.g., Coddington 1986b) hub loops were not added during the early stages of radius construction. The exact moment at which hub construction began varied among species, and, to a lesser extent, between individuals (appendix O6.3.4). In *M. duodecimspinosa* hub construction did not begin until nearly all radii were in place (*L. mariana* [TET] was even more extreme: in only 2 of >50 webs was a single radius added after hub loop construction had begun) (Eberhard 1987d). The hub loop lines in *M. duodecimspinosa* and *A. diadematus* were continuations of the radius lines.The mystery of how the spider achieved the circular shape of the hub and the highly uniform distances between hub loops (e.g., fig. 1.4) even though none of her legs appeared to make contact with the outer loop of hub spiral, may have been solved by video recordings of *M. duodecimspinosa* and the large, slow-moving *Nephila clavipes*. Briefly, these spiders used two techniques (for details, see appendix O6.3.4). The circular outline of the first hub loop was achieved by holding the tip of one extended leg III at a central point in the hub and pivoting around this point. The regular spacing of subsequent hub loops was achieved using a small, quick, almost imperceptible sliding movement of leg oIII along r_n in the instant just before the attachment was made (fig. 6.12). This brought the leg into contact with the outer loop, where it could be used as a point of reference for attaching the next hub loop. Given the substantial taxonomic distance between these two species and the similarity of their behavior, I suspect that similar sliding movements of leg oIII occur in many other species.

6.3.4 TEMPORARY SPIRAL AND TERTIARY RADII

The temporary spiral represented an extension of the hub spiral, in several senses: both were draglines; both were attached to each successive radius; both were laid as the spider spiraled away from the center of the hub (fig. 6.10); and hub spiral construction changed gradually into temporary spiral construction (in contrast with some tetragnathids and theridiosomatids—see figs. 7.7, 7.30, appendix O6.3.4). The temporary spiral differed in

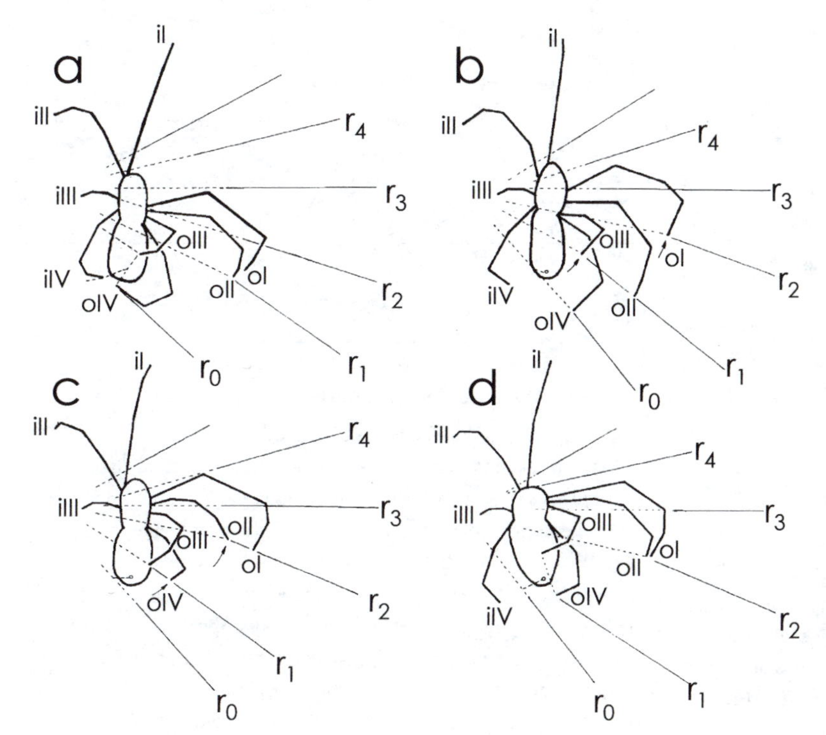

Fig. 6.12. Early hub construction by *Leucauge mariana* (TET) seen from above in a horizontal orb illustrates "following behavior" by legs oII–oIV. After the spider attached the hub line to radius r_0 (*a*), legs oI and oIII each advanced one radius (*b*); oI reached forward to find and grasp r_2, while oIII followed oII to grasp r_1 (arrows). Then (*c*) legs oIV and oII then each advanced one radius each (arrows); oIV followed oIII to grasp r_1, and oII followed oI to grasp r_2. The spider bent her abdomen upward and slightly laterally (*d*) to attach the hub line to r_1 between the sites held by oIII and oIV. Legs on the spider's other side did not move in regular sequences. Leg iIII did not shift its grip at the center of the hub; the other inner legs moved irregularly and only approximately once every 2–4 attachments of the hub loop. Leg iIII gradually extended more and more as the spider spiraled outward, and was extended to close to its full length when hub construction ended (the dashed portions of lines were less clear in the recordings) (after Eberhard 1987d).

that the spaces between successive loops were much larger, and the movements used to build it differed. These movements also varied as the spider moved away from the hub and the distances between radii increased (appendix O6.2.3). The spider nevertheless maintained contact with the outer temporary spiral loop that was already in place at all times, using it as a reference point to determine the attachment site for the next loop (section 7.4), and also (in the outer portion of the orb) as a bridge between radii (Zschokke and Vollrath 1995a; Zschokke 1996). Contrary to some statements, temporary spirals were not logarithmic: the spaces between loops did not increase gradually as the spider moved outward, and the angles with radii were not constant (fig. 6.10).

In *M. duodecimspinosa*, the spider occasionally interrupted temporary spiral construction to build at least one tertiary radius, using movements similar to those used when a new radius was laid during hub construction. Tertiary radii were more common in other species (appendix O6.3.5).

Termination of temporary spiral construction was abrupt. The spider stopped, remained motionless for a brief period (on the order of 30 s in *M. duodecimspinosa*), and then began laying sticky spiral.

6.3.5 STICKY SPIRAL

Sticky spiral construction was the longest phase of orb construction. Even though *M. duodecimspinosa* worked rapidly, sticky spiral construction typically lasted 30 min or more (in contrast, radius and frame construction lasted on the order of 15 min and hub and temporary spiral construction about 4 min). The rate at which *M. duodecimspinosa* made sticky spiral attachments gradually increased, from about 0.7 attachments/s near the outer edge to 1.4 attachments/s in the innermost portion of the orb (fig. 6.13, WE). This was a comparatively brisk rate of attachments. The rate of attachments by *Araneus diadematus* was about 0.3 attachments/s in the outermost part of the web (fig. 6.14). In *A. expletus* it was just under 0.2 attachments/s in a web with about one-third of the sticky spiral complete (WE). A cribellate spider, *Uloborus walckenaerius* (UL), was much slower (.07 to .1 attachments/s) in the outer portion of the orb, because of the time needed to comb out the sticky cribellum silk (fig. 6.14). The velocities with which the spider moved during sticky spiral construction were also slower; adult females of *A. diadematus* reached velocities of up to 6.8 cm/s during radius construction, but averaged only 0.8 cm/s while building the temporary spiral, and

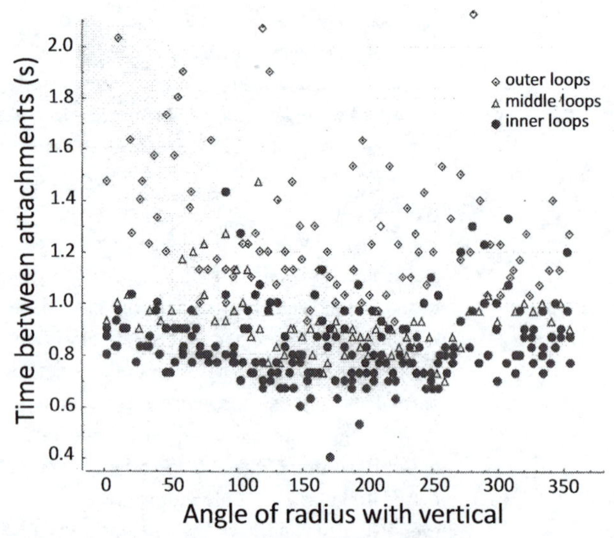

Fig. 6.13. The times between successive attachments of the sticky spiral to radii by a mature female *Micrathena duodecimspinosa* (AR) showed two trends. The times were greater by a factor of about two in the outer portions than the inner portions of the orb, probably largely due to greater distances traveled. In addition, the spider took more time between attachments of the outer loops when descending (angle of radius 0–180°) than when ascending (angle of radius 180–360°). This descending-ascending difference may have resulted from the greater locomotor difficulties when the spider's own weight caused the lines she was holding to sag toward the lines immediately below her, making them more difficult to distinguish from each other. This difference is linked to the variation in behavioral details that were employed to execute the uniform sequence of tasks during sticky spiral construction (see also figs. 6.23, 6.24, 6.25).

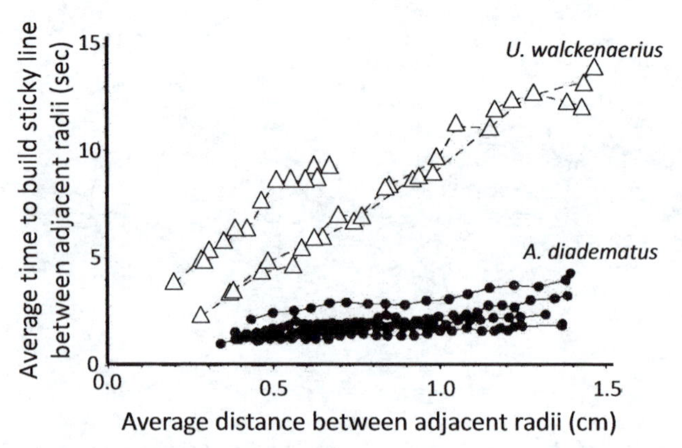

Fig. 6.14. This graph documents a major disadvantage of using cribellum silk in *Uloborus walckenaerius* (UL) rather than viscid silk in *Araneus diadematus* (AR) for the sticky spiral in an orb: cribellum silk takes longer to produce. Each point represents the segment length and the time between attachments for a single segment of sticky spiral between two adjacent radii; each connected series of points represents one complete spiral around the web. It is likely that the metabolic energy expended in combing out cribellum silk is even more important than the time cost: cribellate spiders moved their legs IV rapidly and continuously to comb out cribellum silk, while araneids simply pulled out sticky lines by moving between attachments (from Zschokke and Vollrath 1995b).

0.5 cm/s during sticky spiral construction (Zschokke and Vollrath 1995a).

The operations during sticky spiral construction and the order in which they were performed were relatively rigid (fig. 6.15). The spider began near the outer edge of the orb, and gradually worked her way inward, attaching the sticky line she was producing to each radius that she crossed. The placement of sticky spiral attachments was guided by inner loop localization behavior (section 7.3): the spider searched for and touched the innermost loop already in place briefly with a leg. This was leg oI in *A. diadematus*, as is typical for araneids. Leg oI was extended laterally and tapped (Eberhard 1982) until it touched the inner loop (fig. 6.15); the spider immediately ceased tapping and turned to make the attachment. Inner loop localization by *M. duodecimspinosa* was very unusual among araneids (Eberhard 1982), as the spider

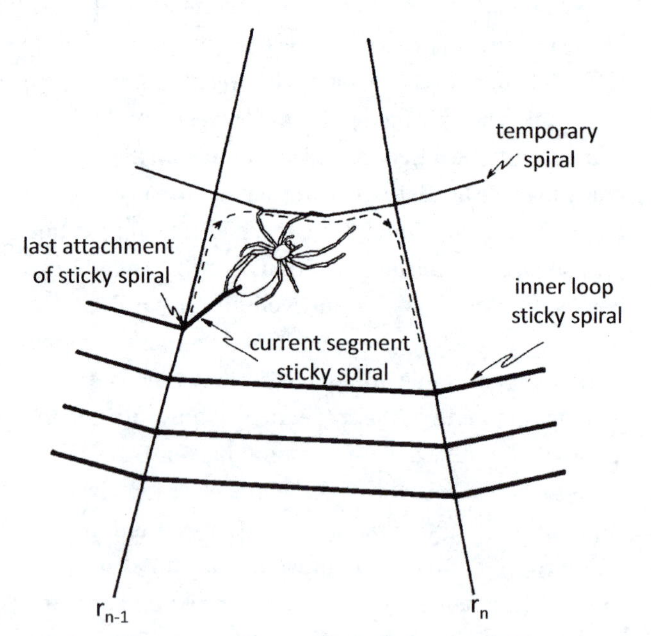

Fig. 6.15. Several terms used in behavioral descriptions are illustrated in this schematic drawing of the path (dotted line) of an araneid spider building the sticky spiral.

mostly used leg oIV to locate the inner loop (appendices O6.2.5, O6.3.6).

The rapid, seamless movements used to build the sticky spiral concealed the fact that each attachment required extraordinarily precise coordination between legs, spinnerets, and silk production from the spigots. The spider accomplished at least seven different operations for every attachment: locate the radius (r_n) to which the sticky spiral line would be attached with her leg oII; locate other non-sticky lines with other legs (especially legs iI, iII, and iIII) to grasp them and thus support her weight; locate the inner loop of sticky spiral already in place; seize the radius with her legs oIII and oIV on either side of the point where the new line would be attached; measure the distance that she moved outward along r_n before contacting the inner loop; align her abdomen precisely with the radius so that the radius was located exactly in the center of the cluster of spinnerets; execute a brief burst of spinneret movements that bound the spiral to the radius (not observed directly in *M. duodecimspinosa* because of its small body size; see Eberhard 2010a, section 2.3, appendix O6.3.6.3 on *Nephila clavipes*); and initiate and then terminate within a few hundredths of a second the production of piriform lines and glue. During this process the spider made decisions regarding the exact site of attachment that were influenced by approximately 10–15 different variables (chapter 7). And these many operations were accomplished very rapidly: as just noted, *M. duodecimspinosa* made more than one attachment/s in some parts of her web, and the spinneret movements lasted on the order of only 0.03 s/attachment (WE).

Sticky spiral construction thus involved dexterity and sustained attention by the spider. Precise leg movements were important, because the exact site of an attachment was physically determined by the sites where legs oIII and oIV grasped the radius (r_n) and the site between these tarsi where the spider touched the radius with her spinnerets (fig. 6.15) (appendix O6.2.5).

6.3.5.1 Break temporary spiral lines

During sticky spiral construction, the spider gradually broke all segments of the temporary spiral, one by one. Temporary spiral cutting is essentially unstudied; not even the leg or palp movements used to bring the lines to the spider's mouth are described for any species. Presumably cutting was triggered when the distance between the inner loop of sticky spiral and the outer loop of temporary spiral (the "TSP-IL distance") fell below some threshold. The threshold in *Neoscona nautica* was the equivalent of about 1–2 sticky spiral spaces according to Hingston (1920), and video recordings of *M. duodecimspinosa* and *A. diadematus* showed that these spiders also usually broke the temporary spiral at similar TSP-IL distances (T. Hesselberg pers. comm.; WE). There are no quantitative data or experimental evidence to test which cues are used. Because the outer loop of temporary spiral is often not parallel with the inner loop of sticky spiral (e.g., fig. 6.10), only a few segments of temporary spiral were broken during the construction of any given loop of sticky spiral, and they were not always adjacent to each other.

6.3.6 MODIFY THE HUB

On finishing the sticky spiral, both species moved to the hub and cut out and ingested the lines in its center, including the mass of fluff that had accumulated there as a result of cut and reel behavior during radius construction (fig. 6.16). In *M. duodecimspinosa* the hub was then left nearly unaltered; at most the spider attached a few broken lines to neighbors while she removed the center. In *A. diadematus*, as is typical in many araneids, the spider immediately "sewed up" the hole that she had created (Witt et al. 1968), laying draglines that connected the lines at or near the inner edges of this hole. In some species spacing was regular, as in *Cyclosa* spp. (fig. 3.12d), and in others such as *Alpaida* spp. it was less regular (fig. 3.15) (see appendix O6.3.7.).

The details of removal and replacement of the center of the hub have never been carefully described, despite the fact that removal is probably very precise to avoid damaging the hub. I give only a summary here for removal by *M. duodecimspinosa* (see fig. 6.16 and appendix O6.2.6 for further details). The spider turned to face downward when she reached the hub, spanning the hub with her legs I and grasping the second or third hub loops on either side. This positioned her mouth slightly above the hub's geometric center. She then began to pull lines to her mouth with her legs II and III and palps, where she broke and ingested them. Removal of lines began at the top of the central area of the hub, and the spider gradually eased downward, packing together and ingest-

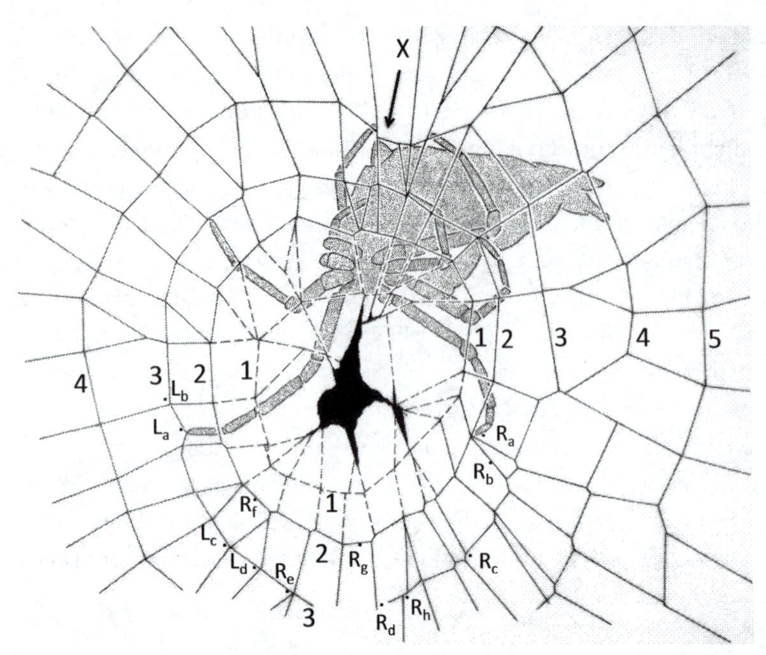

Fig. 6.16. This female *Micrathena duodecimspinosa* (AR) (in ventral view) was in an early stage of the delicate process of removing the center of her orb without damaging the rest of the hub (hub loops are numbered 1 to 5; all dashed lines were gone or at least collapsed against others and no longer visible when removal was finished; the sites grasped by right (R) and left (L) legs I during the process are marked with small dots whose labels indicate (in alphabetical order) the sequence with which lines were grasped). The spider had moved from the inner end of the sticky spiral (not visible) and had more or less finished turning and centering her body under the central portion of the hub at the moment pictured here. She "embraced" the hub with her spread legs I and II, thus accomplishing what was probably the most crucial part of the removal process—positioning her body directly under the central portion of the hub. Legs I and II had repeatedly grasped and regrasped lines, mostly seizing the second or third hub loops or radii that were attached to these lines. In the drawing the spider was just about to initiate her non-sticky dragline by attaching it (at point X) to the line that she held with her left legs III and IV. Subsequently, after repositioning her legs several times and attaching her dragline twice more in the next 7 s, she inclined her abdomen away from the plane of the orb, turned to face more directly downward, and began to break lines at her mouth. She kept these grips with legs I with only minor subsequent movements (from La to Lb and from Ra to Rb) for most of the time as she gradually eased slowly downward, removing lines in the central area of the hub along with the mass of loose silk (irregular black mass). Lines were cut and ingested, beginning at the upper end of the central area and working slowly downward, by pulling them to her mouth with legs II and III and perhaps her palps (their behavior could not be resolved). Legs II and III often grasped and regrasped lines repeatedly in apparent exploratory movements before bringing them to her mouth. Lines often seemed to break just inside (the side nearest the spider) the point where a leg gripped them; the leg then released its hold, and the outer broken end of the line was pulled away by the other lines attached to it, floating free before it curled up on itself or became entangled with other lines. Some lines, in contrast, were apparently held with ipsilateral legs II and III on either side of the mouth and broken between them. One of the first lines broken was an upper segment of hub loop 1, and subsequent breaks removed some other portions of this first loop or caused it to collapse against loop 2; loop 2 thus became the innermost loop in the finished web. Late in the process, when she had reached the lower portion of the central area and was removing the last lines in the lower left hand portion of the hub opening, the spider moved her right leg I repeatedly (first from Rc to Rf, and then back to Rh).

ing the sagging array of lines in the center as she went. It remains to be determined how she chose which lines to grasp with her legs, and which lines to cut.

6.3.7 STABILIMENTUM

Various orb weavers, including *M. duodecimspinosa*, added a silk decoration or "stablimentum" to the orb, usually on or near the hub (section 3.3.4.2) (*A. diadematus* did not make stabilimenta). Stabilimentum silk was laid under little or no tension, using wrapping movements of her legs IV or dabbing movements of her abdomen. The silk formed a white, cottony mass or band of fine lines that often zig-zagged between two radii directly above the hub, or was placed on a single radius (fig. 3.37). The fine lines in stabilimenta are thought to be aciniform gland silk.

6.3.8 ORB WEB REPAIR

Both Fabre (1912) and Hingston (1920) thought that orb weavers were incapable of web repair, and cited this as evidence that spiders do not "understand" the function of their own construction behavior. It is now clear, however, that several species of orb weavers can make quite elaborate repairs to their orbs. Repairs can be conveniently divided into two types: small scale "shoring up" repairs made soon after an orb is damaged (fig. 6.17 of *A. diadematus*); and "replacement" repairs that include sticky lines laid on a substantial fraction of the web's surface (fig. 6.18, 6.20). I will discuss replacement repairs here (see appendix O6.3.8.3 for shoring up repairs). I have not studied replacement repair behavior in *M. duodecimspinosa* other than to determine that it does occur at least occasionally (see fig. 7.36 for a close relative), and I know of no studies of replacement repairs by *A. diadematus*. Because repair behavior offers important opportunities to understand the cues that guide orb construction, however, I will outline replacement repair behavior in orb weavers in general (see also appendix O6.3.8.3).

Replacement repairs typically involved construction

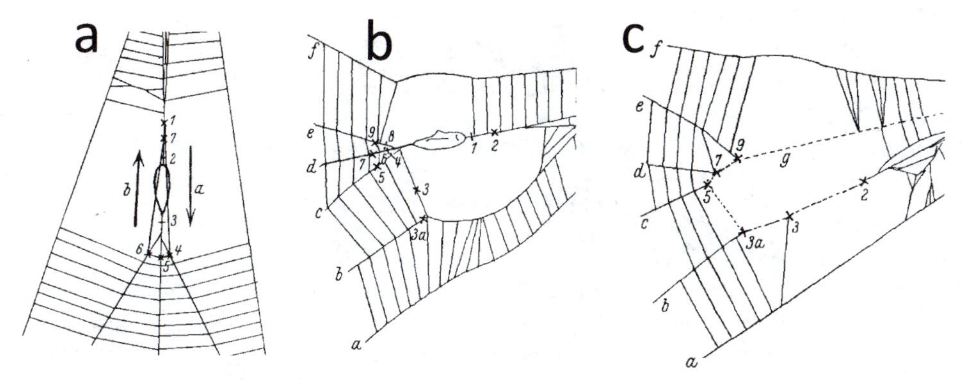

Fig. 6.17. *Araneus diadematus* (AR) repaired their orbs with non-sticky lines after removing captured prey. Each "x" marks a point of attachment of the spider's dragline and the numbers accompanying them indicate the order of the attachments. Some repairs were relatively simple (*a*), while repairs of larger holes tended to be more complex (the prey was still in the web in *b*; the spider had removed it and taken it to the hub in *c*). The spider almost always attached repair lines to non-sticky rather than sticky lines (this was also the case in *Uloborus diversus* [UL]) (WE); other cues used to guide repair behavior have not been determined (from Peters 1931).

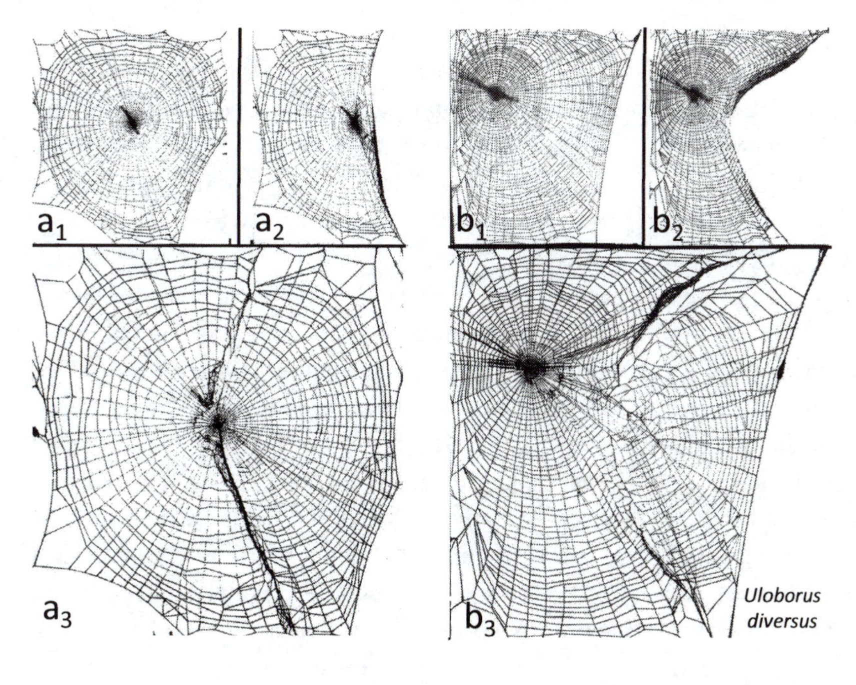

Fig. 6.18. Differences between these two repairs of web damage by *Uloborus diversus* (UL) reveal unexpected flexibility in radius construction behavior. Spiders usually repaired major damage to the orb (a_1–a_3) by building a new, replacement sector that included a new hub. Radius construction in these repairs was very similar to that in normal orb construction (appendix O6.3.2.3). Rarely, however, spiders built a replacement sector without replacing the previous hub, and without building a new one (b_1–b_3). Radius construction behavior in these webs has never been observed directly, but must be quite different. The spider did not build the hub lines that normally form an integral part of radius construction (fig. 6.8), but instead attached new radial lines to different points along the edge of the damaged web (after Eberhard 1972a).

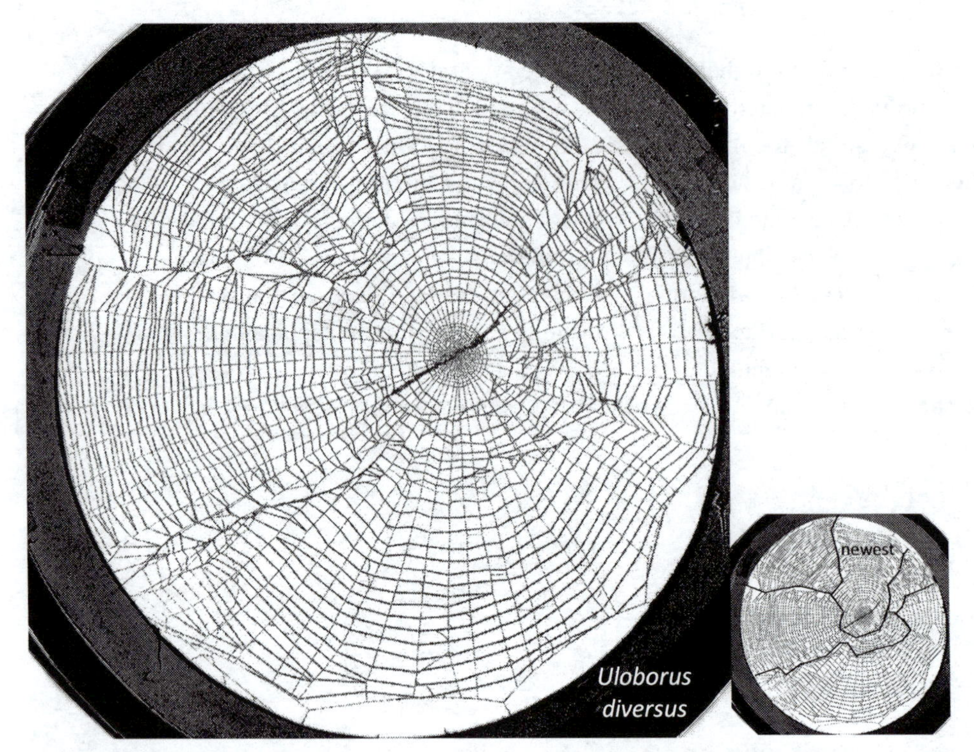

Fig. 6.19. This patchwork of "repairs" allowed a mature female *Uloborus diversus* (UL) to expand her web and make it more dense over the course of >2 weeks, during which she was kept undisturbed in a closed round container (dark lines in the inset at the lower right mark the borders of repair zones). Even though her web suffered no damage from prey or other foreign objects, she replaced different sections in at least four bouts of building, probably each on a different night. The "renewed" portion was sometimes larger than the sector that it replaced, so the overall area of the web gradually increased to fill the available space. In many cases newer sectors also had more tightly spaced sticky spirals (compare the newest, 12:00–2:00 sector with the lower, 2:30–7:00 portion). Web expansion by "repairs" also occurred in other groups that build in sheltered sites, like *Zosis geniculata* (UL) and *Azilia affinis* (AR), and in other species that built especially durable webs, like *Nephila* spp. (NE). Such expansion may enable the spider to have a web that is larger and has more silk than she is able to produce in a single day.

of new radii, temporary spirals, and sticky spirals. They usually occurred long after the damage, typically at the time of day when normal orb construction occurred (in the best studied species, *U. diversus*, early in the morning). Replacement repairs (fig. 6.19) are apparently widespread in the uloborid genera *Uloborus*, *Octonoba*, *Philoponella*, and *Zosis* (Eberhard 1972a; Lubin 1986; Watanabe 1999b; WE), the nephilids *Nephila* spp. (Wiehle 1931; Robinson and Robinson 1973a; Nentwig and Spiegel 1986; Higgins 1987; Kuntner et al. 2008a; WE) and *Herennia papuana* (Robinson and Lubin 1979a), and the araneids *Cyclosa confusa* (Shinkai 1998a), perhaps *Enacrosoma anomalum* (fig. 3.42; WE), and occasionally others such as *Micrathena* sp. nr. *M. lucasi* (WE) (fig. 7.36). Many webs of *O. sybotides*, *Z. geniculata*, *U. diversus*, and *N. clavipes* in the field are patchworks of repairs of previous webs (Wiehle 1931; Eberhard 1971a; Higgins 1987; Higgins and Buskirk 1992; Watanabe 1999b; Eberhard and Barrantes 2015). In *Z. geniculata*, *U. diversus*, and *N. clavipes* in captivity replacement "repairs" were made even though the original orb was not damaged (Eberhard 1971a; WE). They probably serve to extend the area covered by the original orb, or to fill it with more closely spaced lines (see fig. 6.19), allowing the spider to extend her web to cover a greater area than the amount of silk available on a given day would permit. *Nephila clavipes* in the field often repaired about half of the orb each day (Higgins 1987), and individual spiders alternated the sides that were repaired: of 34 repairs by two mature female *N. clavipes*, 30 were on the side opposite the side that the spider had repaired the previous day (and in two of the exceptions, the spider had skipped a day without repairing her orb) (WE). Juveniles were more likely than adult females to replace the entire web rather than only

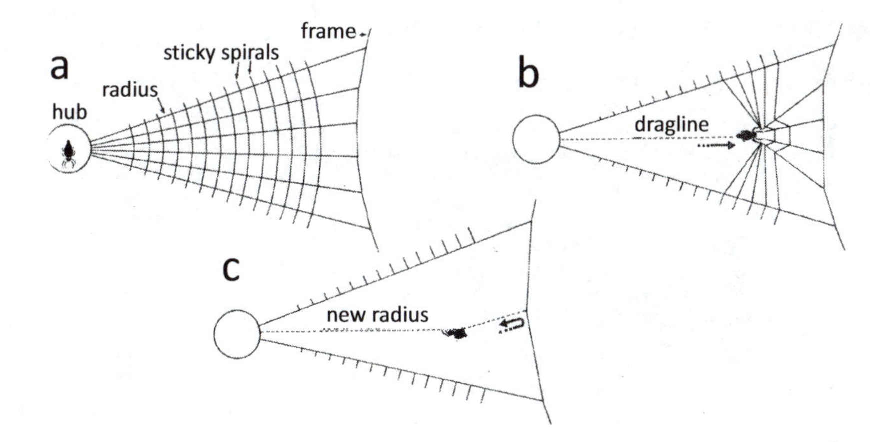

Fig. 6.20. The process of slow, "snow plow" web removal by *Cyclosa turbinata* (AR) began when the spider removed a sector of lines as she was producing a new dragline while moving away from the hub (*b*); she then replaced this dragline with a new radial line on the return trip (*c*) (from Carico 1986).

a part, and the entire web was also replaced more often if it had rained the previous day or night (Higgins and Buskirk 1992). Not enough is known about the relative web investments by juveniles and adults (web size, densities of lines, diameters of lines and glue droplets) to judge whether the less frequent total renewals by adults are associated with their making relatively larger web investments; it is also unclear whether spiders remember recent rains, or can sense rain damage to the web (the non-sticky lines are often intact).

In some respects, repairing a damaged orb may be more difficult than building an orb from scratch. Effective repair requires appropriate adjustments to conditions that vary in different orbs, such as the shape of the damaged sector, the presence or absence of preexisting radii and frame lines, and their positions (Fabre 1912; Hingston 1920). Direct observations of replacement repair behavior in *U. diversus* (UL) (Eberhard 1972a), *Zosis geniculata* (UL) (WE), and *Cyclosa confusa* (AR) (Shinkai 1998a), show that the spider solved these problems by first modifying the damaged sector (see appendix O6.3.8.3), thereby simplifying and standardizing her tasks. Then she used many of the same behavior patterns that she used when building an entire orb (the possibility that uloborids sometimes make "insightful" behavioral adjustments when repairing their orbs is discussed in Eberhard 2019b).

6.3.9 WEB REMOVAL AND RECYCLING

Both *M. duodecimspinosa* and *A. diadematus* re-ingested their webs (Witt et al. 1968; WE), but there are no studies of this behavior in either. Other orb weavers apparently employ two different techniques (described in appendices O6.1.3, O6.3.8): a quick "window shade" removal when speed is urgent (e.g., when it is about to rain); and a more gradual "snow-plow" technique, in which the spider removed sectors in succession (fig. 6.20); in both cases she added radial lines that were used later, in the next web (Nielsen 1932; Carico 1986). Orb removal was often only partial, and some lines were reused in the subsequent orb (figs. 6.20, 6.21, 7.41).

Two cues that may, in combination, elicit orb removal were suggested by the following observation of a mature female *Leucauge venusta* at about 9:30 on a dark morning with impending rain (which began to fall less than an hour later). When I caused a cloud of cornstarch to drift downwind onto the orb in order to photograph it, the spider twice ran to the hub (despite being chased repeatedly to the web's edge) and immediately began "snow plow" removal behavior. No similar behavior was observed in powdering 10–15 other webs of this species and many other webs of *L. mariana* on other, non-rainy days. Perhaps the combination of ambient cues regarding weather (reduced barometric pressure?) plus loading of the web (as would occur with rain, though without the impacts of rain drops) triggered removal (table O10.1).

6.4 SENILITY IN ORB CONSTRUCTION: A NEW FRONTIER?

Behavioral senescence, in which there is a gradual, irreversible decay of behavioral capabilities, is a widespread phenomenon, but both the evolutionary reasons why it occurs and the mechanisms that cause it are only incompletely understood (Monaghan et al. 2008). One re-

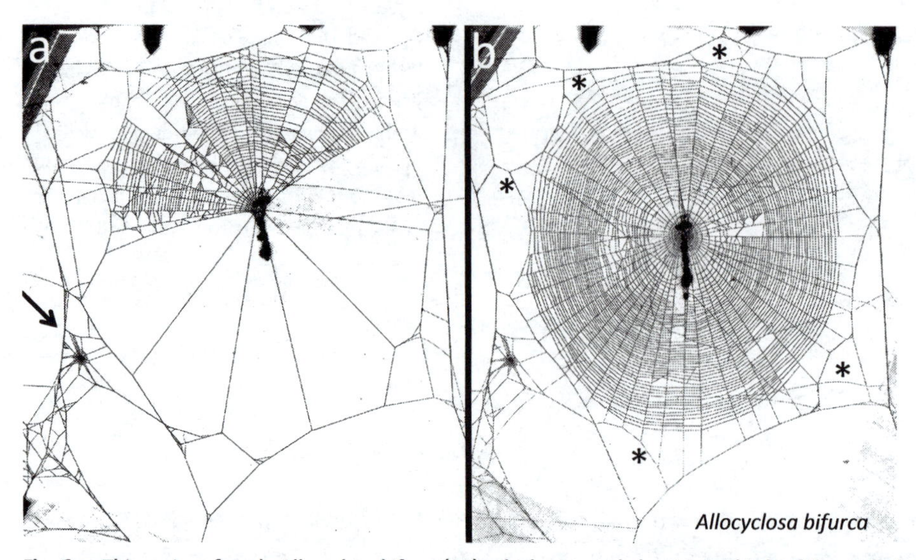

Fig. 6.21. This mature female *Allocyclosa bifurca* (AR), which removed about two-thirds of the sticky lines in the evening between 8 and 9 PM (*a*) and then built a new orb early the next morning (*b*), illustrated both the spider's flexibility and Carico's point (1986) that orb construction is cyclical in many species. Radial lines in *a* that the spider laid as she removed portions of the previous orb (see fig. 6.20) were incorporated in the next orb (*b*); although several frame and anchor lines remained intact (especially the larger frames), some other frame lines changed (sites where frame lines were added, removed, or modified are indicated with * in *b*). This pair of webs also illustrates a puzzle. This species generally removes portions of the orb early in the evening, but does not build a new orb until 3–6 AM; spiders thus often spend much of the night at the hubs of orbs such as that on the left, in which much of the sticky spiral has been removed. Why do they remove sticky lines early in the evening, rather than waiting to remove them until they are about to build the next orb, and thus forego the possibility of capturing additional prey during most of the night? Perhaps orb removal is related to sleep (see text)? An alternative explanation, that they are avoiding the risk of losing silk and glue to rain, is not convincing because this species responded to approaching rain storms, quickly removing sectors before it began to rain. The tiny orb-like web at the bottom left (arrow in *a*) was probably the resting web of a mature male.

warding way to illuminate such widespread phenomena is to compare data from taxonomically distinct groups. Orb-weaving spiders show promise for such comparisons (Anotaux et al. 2012), because they appear to show multiple types of behavioral senescence.

The orb designs of older adult females gradually became inferior (e.g., had reduced geometric regularity) in several araneids (three species of *Araneus*, two of *Zygiella*, and *Meta reticulata*) (Bristowe 1941; Anotaux et al. 2012) and in the uloborids *Uloborus diversus* and *U. glomosus* (fig. 6.22; Eberhard 1971b). The "senile" webs of *U.* spp. resembled those of mature male uloborids (fig. 10.36c) and of spiderlings that had recently emerged from the egg sac (fig. 9.20) in having mats of non-radial lines that were so fine that they were difficult to see, even when well illuminated and viewed against a dark background (appendices O6.3.2.3, O6.3.8.1.1; Szlep 1961; Eberhard 1977d); they lacked (or had fewer) sticky spiral lines. Senile orbs were more likely in virgin compared with non-virgin females of *U. diversus* (Eberhard 1971b). Early spiderlings and mature males are morphologically unable to produce sticky cribellum silk (Wiehle 1927; Szlep 1961), so it may be that the senile orbs of adult females were also associated with problems producing cribellum silk.

Changes in several traits of the senile webs of *Z. x-notata* were different, and were correlated with the time elapsed since the spider's molt to an adult ("age" below): the total length of the sticky spiral decreased; the coefficient of variation in the distances between loops of sticky spiral increased; a measure of the "parallelism" of adjacent sticky lines decreased; and the numbers of anomalies, such as split radii, turnbacks in the sticky spiral, and sticky lines adhering to each other increased (Anotaux et al. 2012). The interpretation of some traits as "anomalies," such as the turnbacks in the sticky spiral, is open to question, as sticky spiral turnbacks are normal features in the orbs of many species (fig. 6.22c; Le Guelte 1966; Eberhard 1969); and the reduced sticky spiral length could be due to a gradual reduction in the sizes

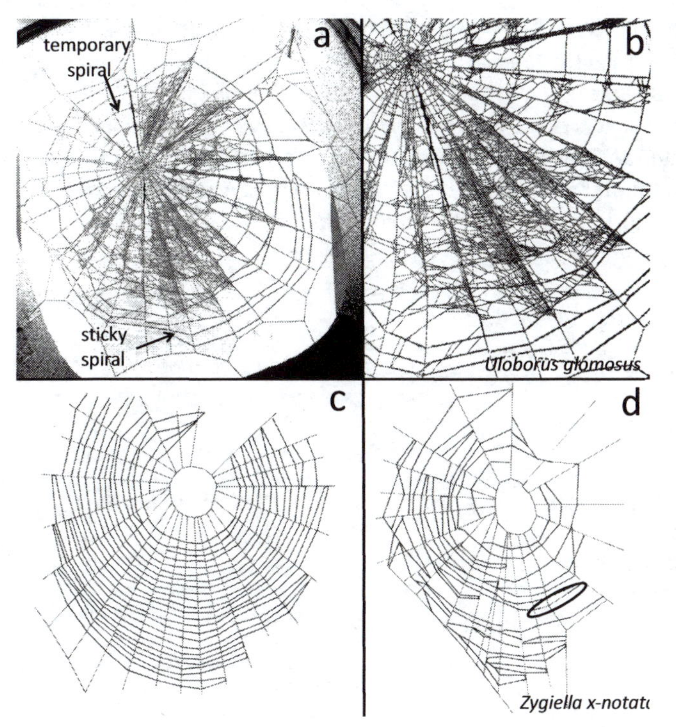

Fig. 6.22. Senility in spiders? The web (*a*, *b*) of a mature female *Uloborus glomosus* (UL) shows several "senile" traits (also seen in the webs of very old *U. diversus* (Eberhard 1971b): the temporary spiral is mostly intact; the sticky spiral is highly reduced; and there are mats of numerous, very thin lines. Similar mats of thin lines are built by uloborid spiderlings in the first instar after they leave the egg sac (fig. 9.20), when the cribellum is not yet functional and the spiderling is thus incapable of producing sticky cribellum silk (Szlep 1961; Eberhard 1977d). This similarity suggests that very old female *Uloborus* spiders have difficulty producing cribellum silk. The drawings of webs of a normal (*c*) and very old female (*d*) of *Araneus diadematus* (AR) (traced from Anotaux et al. 2012; hub lines were omitted because they were not clear in the original photos) provide a second possible case of senility. The regularity of the spaces between sticky spiral lines was reduced in the old female's web, and some segments were apparently only covered with sticky material for part of their length (oval in *d*). "Senility" in both of these distantly related species was manifested in a reduction in sticky lines.

of silk glands due to their lack of use in the small cages in these experiments, rather than aging per se (Witt et al. 1972). Nevertheless, web photos (fig. 6.22d) give eloquent indications that true behavioral differences exist.

Another possible model spider for senescence studies is the linyphiid *Frontinella pyramitela*. Mature males all built normal webs immediately after molting, but routinely lost the ability to construct prey capture webs gradually during the first 1–11 days of their adult lives (Suter et al. 1987). Neither mass at maturity nor adult feeding history were important contributors to the variability in the rate of senescence.

These brief studies open the door on a host of possible questions: Do the changes in senile webs result from sensory, motor, or analytical deficits? Are they associated with physiological changes (e.g., oxidative stress), or genetic changes (e.g., reduced telomere length)? There may be two classes of individuals in *Z. x-notata* that have different patterns of behavioral senescence (some spiders lay their eggs and die late in autumn, while others die in early spring) (Thévenard et al. 2004; Bel-Venner and Venner 2006), making this species especially appealing for further study.

6.5 DETAILED MOVEMENTS

Detailed descriptions of some of the movements of the body and legs of *M. duodecimspinosa* are listed in table 6.1, where possible taxonomic implications are mentioned. More complete discussions are given in appendix O6.2. But these movements show some important general patterns that merit discussion here.

6.5.1 PATTERNS IN VARIATION: HIGH DIVERSITY PRODUCES LOW DIVERSITY

The degree of stereotypy appears to increase progressively in the later stages of orb web construction (Witt 1965; Witt et al. 1968; Eberhard 1972a, 1990a; Vollrath 1992) (precise quantifications of stereotypy are lacking, however). This pattern may result from the spider becoming more and more isolated from the idiosyncratic differences in her surroundings as the orb is built; she is guided more and more by stimuli from the web itself, and her behavior becomes progressively more consistent and predictable (Witt 1965).

This pattern of gradual increases in consistency differs, however, from the pattern at a lower level of analysis: leg and body movements showed apparently endless variation, even in sticky spiral construction (fig. 6.23, ap-

Table 6.1. This undoubtedly incomplete list of the behavioral "bricks" that spiders use to construct orbs and that have been used in phylogenetic analyses of orb weaving spider groups at the familial level and above, and of other traits that show promise of being similarly useful at the genus level and higher demonstrates of the variety of levels of analysis at which behavioral traits can be taxonomically informative. Characters range from the overall order in which operations are performed to the details of which legs grasp which lines when an attachment is made (for a similar array of traits in the family Theridiidae, see Eberhard et al. 2008b). It is possible that other behavioral traits may be useful in grouping species at low levels such as genera (see for example Kuntner et al. 2008a); but in general the extensive sampling within such lower level groups that would be necessary to demonstrate such usefuless has not yet been done (section 3.2.8) (see, however, Gregorič et al. 2010).

Traits	*Comments*	*Representative references on taxonomic distribution*
I. Traits already documented or at least mentioned previously		
A. Patterns in construction behavior		
Order in which construction stages (frames, radii, hub, temporary spiral, sticky spiral) occurred	Several sequences among all orb weavers[1]	Hingston 1929, Eberhard 1982
Inner loop localization behavior during sticky spiral construction	Several types among all orb weavers	Eberhard 1982
Number of attachments of each radius to the frame	2 in Nephilidae, 1 in others	Eberhard 1982
Break and reel radius during construction	Yes/no[1,2]	Eberhard 1982
Legs that were used to hold the radius to which a sticky spiral line was to be attached	oIII and oIV in all orb weavers except uloborids (both IV)[1]	Eberhard 1982
Leg IV pulled out additional sticky line just before sticky spiral is attached to a radius	Only araneoid orb weavers[1]	Eberhard 1982
Maintained contact with temporary spiral while building first loop of sticky spiral	Yes/no[1]	Eberhard 1982
Attached sticky spiral to each radius it crossed	Yes/no[1]	Eberhard 1982
Removed center or hub when sticky spiral ended	Yes/no[1]	Eberhard 1982
Center of hub filled in immediately after being removed	Some araneids[1]	Eberhard 1982
Temporary spiral is 1–2 circles rather than a spiral	Theridiosomatidae only[4]	Eberhard 1982, Coddington 1986a
Omit temporary spiral entirely	Some symphytognathoids only[3,4]	Eberhard 1982, 1987e, Lopardo et al. 2011
Added supplementary radii after sticky spiral ended	Some symphytognathoids only[3]	Eberhard 1987e, Lopardo et al. 2011
Destroyed entire hub, and reconnected radii with new hub after finishing sticky spiral	Symphytognathoids only[3]	Eberhard 1987e, Lopardo et al. 2011
Hub cup against substrate	Some nephilids[1]	Kuntner et al. 2008a
Transition from hub to temporary spiral	Gradual/abrupt[1]	Eberhard 1982
Pseudo-radii	A few nephilids[1]	Kuntner et al. 2008a
Temporary spiral left intact in finished orb	Nephilids, one araneid[1]	Kuntner et al. 2008a
Added tertiary radii during construction of temporary spiral	Facultative in some araneids[1]	Kuntner et al. 2008a
B. Cues used to guide construction behavior		
Location of inner loop used as a reference point during sticky spiral construction	All orb weavers except theridiosomatids and symphytognathoids[3]	Eberhard 1982, Eberhard and Barrantes 2015
Memory of IL-TSP distance used to adjust distance between sticky spiral loop	All orb weavers[3]	Eberhard and Barrantes 2015
Amount of silk in glands used to adjust distance between stick spiral loops	All orb weavers?[3]	Eberhard and Barrantes 2015

Table 6.1. Continued

Traits	*Comments*	*Representative references on taxonomic distribution*
Amount of open space available used to adjust numbers of radii, frames, hub, and sticky spiral spacing	All orb weavers?[3]	Eberhard and Barrantes 2015
II. Less extensively tested possibilities		
Legs used to hold the line to which another line is about to be attached[5]	Possibly also useful for non-orb spiders	Eberhard 1982, WE
3-D tangle "barrier" web beside the orb	Repeated derivations probable[6]	Kuntner et al. 2008a
Held dragline with tarsal claw IV and slid it along the dragline	Only in orbicularians (?) (both with and without orbs)	WE
Built frames in strict order going around the orb	Perhaps exclusive in uloborids (?)	Eberhard 1972a, 1990a
Zig-zag attachments of hub and temporary spiral lines to radii, moving inward along the radius	Almost only in nephilids and uloborids	Lubin 1986, Kuntner et al. 2008a
Replaced proto-hub with tiny circular hub early in radius construction, immediately after frame construction was complete (or nearly complete)	Uloborids only (?)	Eberhard 1990b
Zig-zag attachments of sticky spiral to radii only near outer edge of orb (occasionally also to frame lines)	Uloborids only?	Lubin 1986, WE

[1] Relatively large numbers of species have been studied, so the taxonomic consistency is relatively well established.
[2] Some uloborids cut and reel radii early in construction, but cease to cut and reel as soon as they begin to build the hub spiral.
[3] Only smaller numbers of species have been studied, so taxonomic consistency is less certain.
[4] Probably associated with very precise ability to judge distances traveled.
[5] Non-sticky lines.
[6] See appendix O9.1.

pendix O6.2.5). This variation had its own pattern, however: the details were adjusted according to the local conditions in the web that the spider encountered while she was building (Ades 1986; Gotts and Vollrath 1992). Analyzing sticky spiral construction at a higher level, for example, the spider repeated the same operations over and over: attach the sticky line; move inward and onward to encounter the next radius; move outward, usually exploring until she encounters the inner loop of spiral already laid; turn and grasp the radius with her legs II and IV; and attach the sticky line. Underlying this stereotypy, however, was an extreme diversity of movements of the spider's legs (fig. 6.24, table O6.3) and the path of her body (figs. 6.23, 6.25, Zschokke and Vollrath 1995a). The positions of the spider's legs, their movements, and the functions they performed in sustaining her weight during temporary and sticky spiral construction in a more or less vertical web all differed sharply comparing her behavior near the outer edge versus near the hub, or above the hub, below the hub, or at the side of the hub (fig. 6.26) (appendix O6.2.5). And of course every time the spider reversed direction and a spiral doubled back on itself, all her legs smoothly shifted roles (Hingston 1922b; König 1951).

Thus for the spatial details of leg movements, the order in which they were moved, the forces they exerted, and the coordination between them, the image of an automaton performing the same movements over and over is completely wrong. The relative stereotypy at higher levels in the hierarchy of control was possible because the spider chose appropriately among the different possible movements she could make in different parts of her web, selecting those particular adjustments to the mechanical demands that were imposed by gravity and the locations of lines (Ades 1986). As shown by the orbs built at near-zero gravity by *A. diadematus* (Witt et al. 1977), this adaptive flexibility extends even to evolutionarily novel situations, in which specific hard-wired compensatory abilities are unlikely.

There may be even more flexibility than is currently

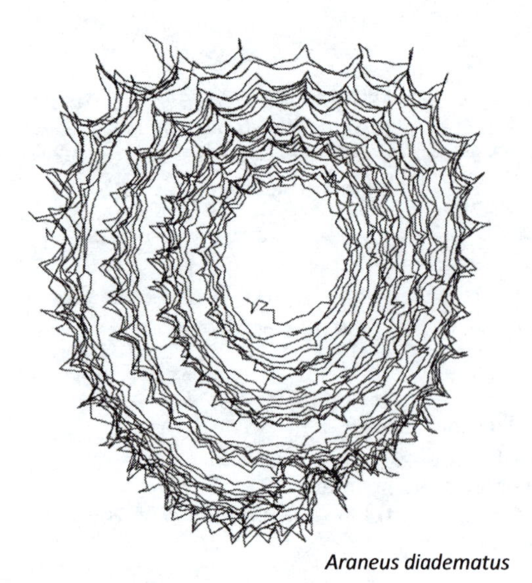

Fig. 6.23. The left-right asymmetry in the paths followed by an *Araneus diadematus* (AR) while she built the sticky spiral in this orb may result from differences in the mechanics of movement when ascending and descending in a orb. The paths (lines) were much more discretely bundled together in the left-hand portion of the web, where the spider was always ascending as she moved in a clockwise direction (except while she was building the outermost bundle). The greater definition of the bundles occurred because the spider moved inward more consistently after each attachment, using the temporary spiral as a bridge to the next radius (the bundles mark the approximate positions of the temporary spiral lines). When descending (on the right side), the spider moved more directly downward from one radius to the next. These differences in bundling (and the intermediate patterns in other parts of the web) illustrate a general pattern in orb construction: there is often a *lack* of uniformity in the lower level details of the behavior patterns with which spiders nevertheless produced regular, highly uniform spaces between radii and sticky spiral lines (see also figs. 6.9, 6.24, 6.27) (courtesy of Samuel Zschokke).

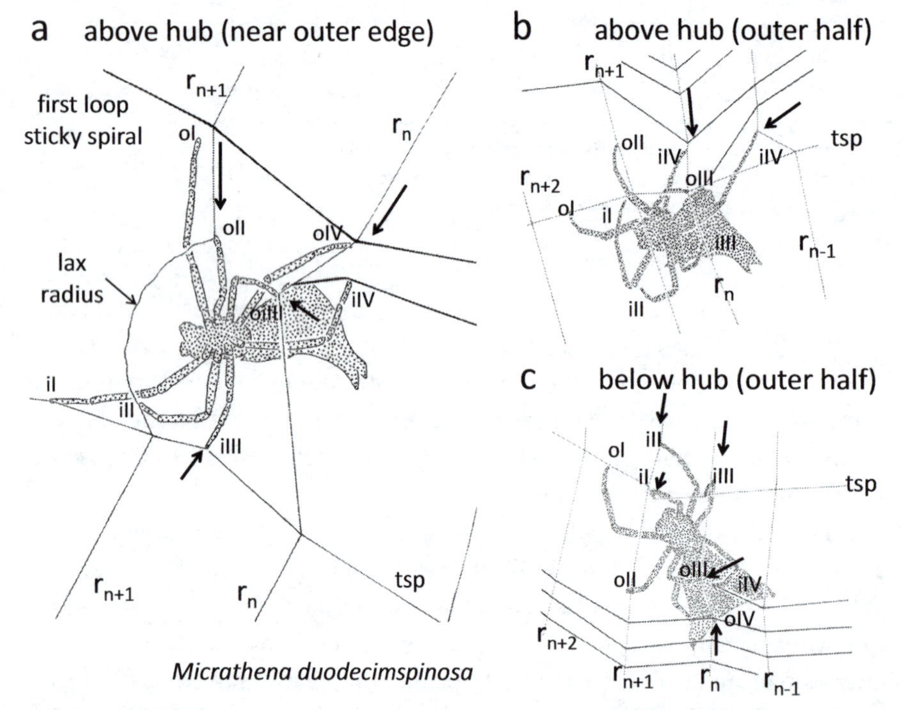

Fig. 6.24. Stylized drawings (traced from video recordings) of how the forces exerted by a mature female *Micrathena duodecimspinosa* (AR) changed when she was in different parts of her web during sticky spiral construction (the directions of the stronger forces on non-sticky lines are indicated by heavy arrows). The movements of legs oIII, oIV, and iIV were relatively uniform in terms of which lines they grasped (section O6.2.5), but the orientations of these legs and the forces they exerted varied widely depending on the spider's site in her web. Distortions of the web caused by the spider's weight, along with variations in her orientation and her position in her web (*a-c*), were associated with changes in the details of leg movements and the forces they exerted. As she made an attachment above the hub (*a*), legs oIV and oII sustained much of her weight; soon after attaching (*b*), leg iIV also exerted substantial force but oII very little. Below the hub just before making an attachment (*c*), in contrast, leg iI sustained much of the spider's weight; legs oIII and oIV pulled on r_n, pressing it against her spinnerets. Some forces were large enough to cause lines to go slack (e.g., leg oII on r_{n+1} in *a*). Not all forces exerted by the legs were downward (note the upward pull by iIII on the temporary spiral in *a*). In sum, the highly stereotyped sequence of "encounter a radius, find the inner loop of sticky spiral, and then attach the new sticky spiral line" involved a highly variable array of forces exerted.

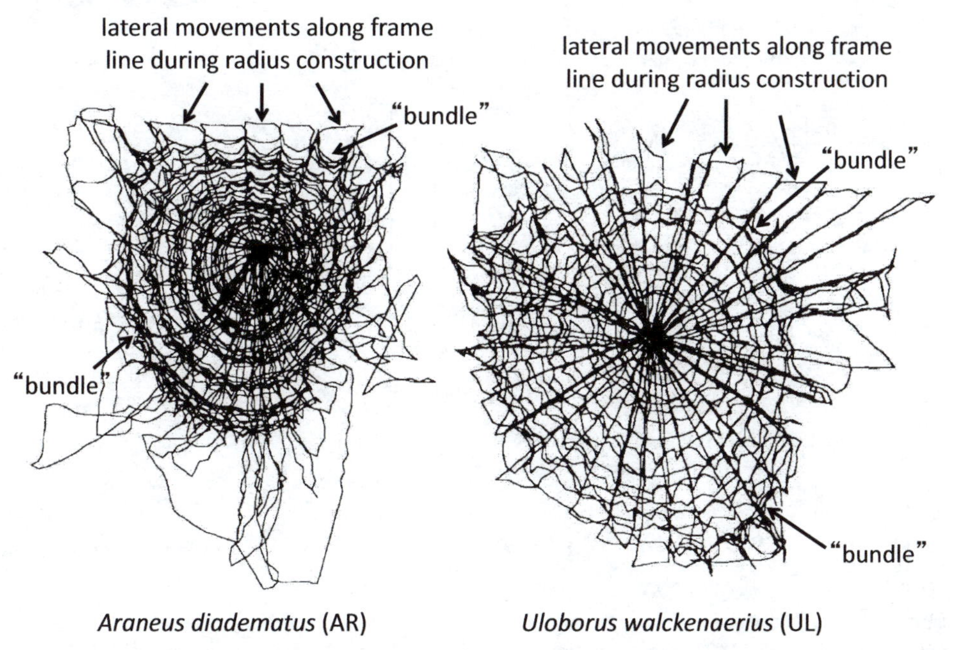

Fig. 6.25. These automated records of the paths of *Araneus diadematus* (AR) and *Uloborus walckenaerius* (UL) as they built radii, and temporary and sticky spiral lines, tell stories of behavioral flexibility. The lateral movements along the frame lines during radius construction were much more distinct for the shorter radii in the upper portion of the orb of *A. diadematus* than in the lower portion; they seemed equally distinct for long and short radii in the *U. walckenaerius* web. The concentrations ("bundles") of paths during sticky spiral construction were especially distinct in the outer portions of both orbs, because the spiders tended to use the temporary spiral as bridges rather than moving more directly from one attachment to the next in areas where the radii were widely separated. In addition, the bundles on the left side of the vertical *A. diadematus* web were more distinct because the spider was circling in a clockwise direction while building the sticky spiral; similar asymmetry in bundling was not apparent in U. *walckenaerius*. There were no differences in sticky spiral spacing corresponding to the different degrees of bundling. These finely graded, detailed behavioral adjustments are "hidden" in the relative uniformity of the spacing of the radii and the sticky spiral in finished orbs (from Zschokke and Vollrath 1995a).

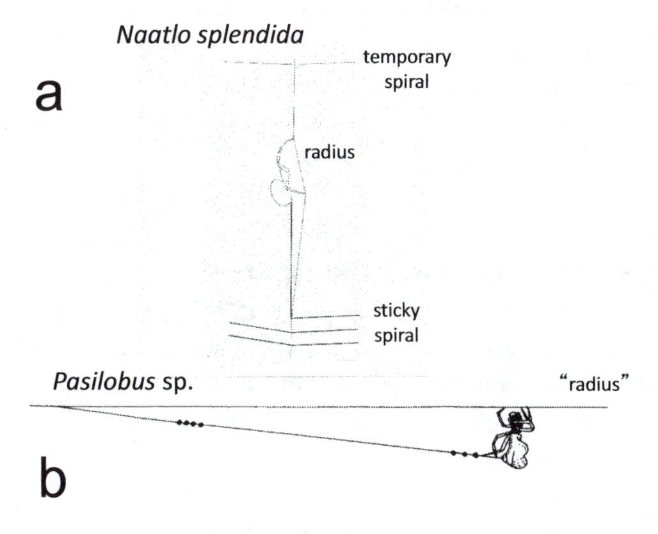

Fig. 6.26. These drawings illustrate convergent fine-tuning of small details of sticky spiral construction behavior. *a*) The theridiosomatid *Naatlo splendida* (#1863) used one leg IV to hold the radius (thin line) away from the new segment of sticky spiral (thick line) that she was producing as she moved inward to reach the temporary spiral. Her relatively small size and the long length of sticky line made it important to hold the two lines apart in order to reduce the danger that her current segment of sticky line might adhere to the radius. *b*) The araneid *Pasilobus* sp. (AR) solved a similar problem of small body size relative to the length of a new sticky line by separating the lines, but in a different way. While walking along a non-sticky line, the spider flexed her abdomen dorsally and held her spinnerets as far as possible from the radius while she pulled out sticky line with her legs IV (drawing from Robinson and Robinson 1975). Still another convergence on holding long sticky lines away from the web while they were being produced occurred in *Hyptiotes paradoxus* (UL) (Marples and Marples 1937).

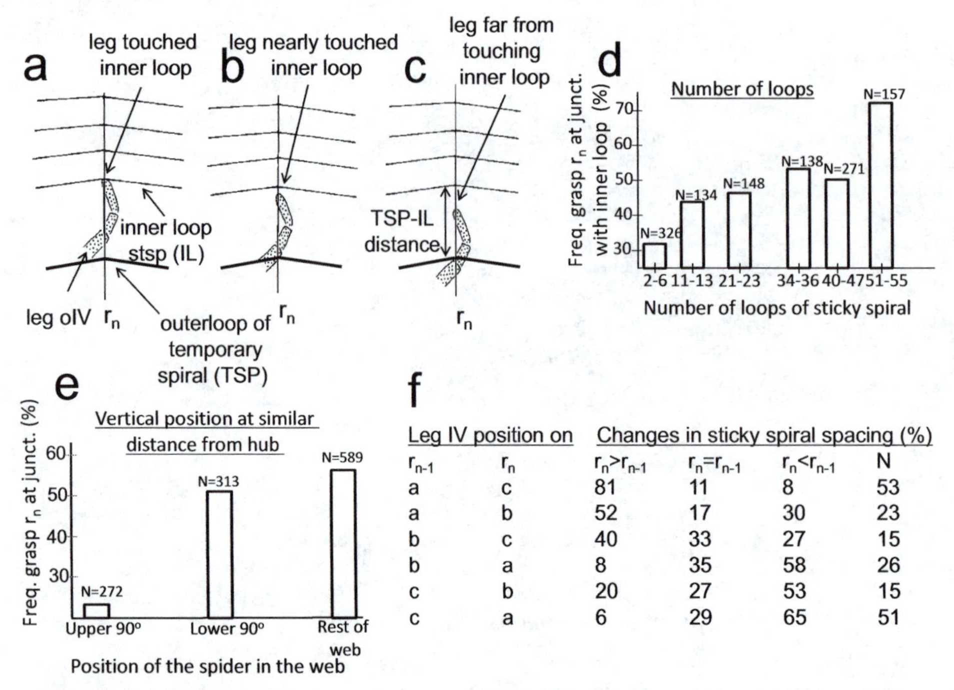

Leg IV position on		Changes in sticky spiral spacing (%)			
r_{n-1}	r_n	$r_n>r_{n-1}$	$r_n=r_{n-1}$	$r_n<r_{n-1}$	N
a	c	81	11	8	53
a	b	52	17	30	23
b	c	40	33	27	15
b	a	8	35	58	26
c	b	20	27	53	15
c	a	6	29	65	51

Fig. 6.27. The stylized drawings (from video recordings) illustrate variations in the positions at which a mature female *Micrathana duodecimspinosa* (AR) grasped r_n with leg oIV just before making an attachment of the sticky spiral (*a–c*). Grasping the radius nearer to the junction was much more common when the spider was nearer to the hub (*d*) and below rather than above the hub (*e*). Changes in relative positions of leg oIV were correlated with decisions regarding where to attach the sticky line to r_n (*f*). When the spider grasped r_n at a site farther from the inner loop than it had on the previous radius (r_{n-1}), the distance between attachment of the sticky spiral from the inner loop on that radius (r_n) also tended to be greater (*f*). For instance, when the leg had grasped the previous radius (r_{n-1}) at position *a*, and then grasped r_n at position *c*, the spacing of the attachment to r_n was greater than that on r_{n-1} in 81% of 53 cases (top line in table *f*). These trends were highly significant; there were many exceptions, however, possibly due to behavioral imprecision (from Eberhard 2012).

recognized. Videotapes of sticky spiral construction behavior showed that the supposedly consistent inner loop localization behavior, which has been used as a taxonomic character for 30 years, is inconsistent: *Leucauge* sp. (TET) occasionally used leg oI instead of leg iI; and *Deliochus* sp. (AR) occasionally used other legs instead of oI (Kuntner et al. 2008a). I anticipate that much additional variation will be discovered using video recordings.

This pattern of greater variation at lower levels of analysis is linked to evolutionary changes in different taxa. To take extreme examples, both *Nephila* (NE) and *Deinopis* (Deinopidae) attach sticky lines to radial lines one after another in strict order as is typical of orb-weavers. But the motor behavior that these spiders used to get from one attachment to the next was highly altered. In both cases the spider is so large with respect to her catching web that her legs I and II move only intermittently or not at all during much of the process (Hingston 1922b; Wiehle 1931; Clyne 1979; Coddington 1986c; WE), in sharp contrast to the probable behavior of their more typically-sized orb-weaving ancestors.

There are exceptions to this pattern. For instance, the tendency of more posterior legs to follow more anterior legs, and for legs I and II to search for lines by moving prolaterally and for legs III and IV to swing retrolaterally to find lines (section 2.4), seems to have been profoundly conservative during evolution, and widely distributed (at least within the orbicularians that have been studied). The movements of legs and spinnerets when one line is attached to another are also apparently relatively invariable (Ades 1986).

6.5.2 VARIATION: A CAUTION AGAINST STEREOTYPY AND TYPOLOGY

The reader should keep in mind that most (if not all) of the descriptions in this chapter and the appendices give overly strong impressions of stereotypy (this is of course always a problem when describing behavior with words). Descriptions of the movements of leg oIV of *M. duodecimspinosa* during sticky spiral construction illustrate this problem. I stated that the tarsus of leg oIV usually contacted the inner loop, but contact sometimes failed to occur; in addition, the exact site where the tarsus grasped r_n was inconsistent (fig. 6.27a). The distance of sticky spiral attachment from the inner loop also varied. It tended to be larger when leg oIV was farther from the inner loop (Eberhard and Hesselberg 2012), but there was also variation here. Sometimes the spider grasped the radius far from the inner loop but did not make a corresponding inward displacement of the attachment site. I expect that similar variations occur in many other details of construction.

Many descriptions of the functions of movements and morphology also suffer from being overly typological. Take, for instance, the way the spider sustains herself in her web, and the problems that her own weight can cause. The forces that the spider's weight exert on her web while she is building often result in substantial distortions of the distances and angles between lines; these are likely to be especially important for spiders with relatively short legs, and for those with more nearly vertical orbs. Take, for instance, a *M. duodecimspinosa* building sticky spiral. Her weight distorts the orb (which is usually 60–80° with horizontal) by displacing downward the radii that she is holding (fig. 6.24). When she is moving downward and more or less level with the hub, her weight strongly reduces the distance between the radius to which she has just attached (r_{n-1}) and the next radius (r_n) while she holds r_{n-1}, but increases as soon as she releases r_{n-1} and holds r_n. In addition, her method of grasping lines (hooking them with her tarsal claws, rather than "standing on them") probably makes it more difficult for her to find and grasp radii with her anterior legs; her legs will tend to approach the lines from above, but they must seize them from below in order to grasp them. The spider's posterior legs will largely sustain her weight, while her anterior legs will push against the orb to keep her body from contacting lines below (counteracting its tendency to hang directly under the line sustaining her). In contrast, just a few seconds later, all these details will change. She will be climbing upward (say moving upward at 9:00); her anterior legs will sustain most of her weight, her posterior legs will be called on to push against the orb, her weight on r_{n-1} will increase rather than decrease its distance from r_n, etc. These distortions make it difficult for the spider to build lines with lengths and spacings appropriate for the final orb, and the functions of her legs in making the adjustments that were necessary to achieve regular spacing changed dramatically.

Abilities to make even greater behavioral modifications were revealed when legs were lost or damaged. For instance, a mature female *Nephila clavipes* that had lost one leg IV compensated during sticky spiral construction by altering the movements of the remaining leg III on that side. When moving across the orb (fig. O6.4) toward the side with the missing leg IV, the behavior of this leg III changed dramatically. It tapped repeatedly much more often before grasping r_n, and sometimes one or more of these taps contacted the inner loop of sticky spiral. Thus this leg III apparently performed inner loop localization behavior that was normally performed only by leg IV (appendix O6.3.6.3). When this same spider moved toward her intact side, her leg III on this side stepped directly from one radius to the next, with little or no exploratory tapping, as in intact spiders (Eberhard 1982; Kuntner et al. 2008a). She also compensated for her missing leg IV by using the leg III on this side to wrap prey, snagging lines near her posterior lateral spinnerets with the tarsus and carrying them to the prey, a movement made only by legs IV in intact spiders (Robinson and Mirick 1971; WE).

Similar substitutions of leg II for a missing leg I occurred in inner loop localization behavior when building the sticky spiral in the araneid *Araneus diadematus* (Jacobi-Kleemann 1953; Reed et al. 1965), and in *Nephila clavipes* (NE) when waiting at the hub (Escalante et al. in prep.) (for a less complete description of the effect of damage to the tips of legs I of a *Neoscona nautica*, see Hingston 1920). These adjustments did not require practice, but began as soon as the leg was lost (possible additional effects of learning in such situations have never to my knowledge been tested, however). Alternative movements and coordinations of movements that compensate for leg loss are pre-programmed in some insects (Gallistel 1980). The important point here is that some of the descriptions here severely underestimate the variations the spider is capable of performing.

Recognition that stereotypy is limited has some important consequences in addition to avoiding underestimates of the spider's behavioral capabilities. The variations in the movements of the spider's legs, the forces they exert, and the distortions that her weight produces in the web vary dramatically during the construction of a single loop of sticky spiral, and make it unreasonable to attempt to simulate web building on the basis of consistent leg positions for each sticky spiral attachment or other simple rules (Eberhard 1969; Krink and Vollrath 1999). At this level of analysis, the behavior of a spider is much more complex than that of such virtual models. The evolution of behavioral traits may also be influenced by variations (Eberhard 1990a,c). Perhaps, for instance, the use of leg oIV for inner loop localization evolved in nephilids and some *Micrathena* species because it was already present in their ancestors as a programmed alternative to compensate for loss or damage to another leg. Or perhaps the adjustment by the spider arose in some cases not as a pre-programmed behavior, but as an "insightful," goal-directed adjustment (Eberhard 2017).

6.6 GENERAL PATTERNS

6.6.1 DEXTERITY, THE BLIND MAN'S CANE, AND FOLLOWING OTHER LEGS

Probably the most revealing image of a spider finding her way around her web is that of a blind man using his cane to explore his surroundings (section 2.4.2.1). Even during the day, a web spider cannot see her web's lines, and relies instead on tapping movements of her legs (mostly legs I and II) to find new lines. Following behavior by more posterior legs reduces the need for them to wave and explore, allowing her move more quickly and efficiently. Following behavior occurred during exploration, and the construction of radii, the hub, the temporary spiral, and the sticky spiral (table 2.3; Eberhard 2017). In some contexts following behavior was so consistent that I could replay a videotape frame by frame, and often predict almost to the frame when each leg would make its next move and where it would go (see appendix O6.2).

The details of following behavior answer an interesting general question that was first raised by Hans Peters (1954) in connection with inner loop localization behavior during sticky spiral construction: does exploratory tapping behavior serve to find the exact point that is to be grasped by a leg; or does it instead provide the spider with general information on sites (and perhaps orientations) of lines that is used in later decisions? The details of following behavior clearly support the second, general orientation alternative: following legs did not grasp the same point that was located (and grasped) by the leading leg, but rather a different point on the same line. Only small, nearly imperceptible searching movements were made by the following leg to find the line, suggesting that a spider senses not only the location in space of the site where her leading leg has grasped a line, but also the line's orientation (perhaps by the angle that her leg was rotated when holding the line?).

Leg movements probably provide both information on lines that the spider can grasp and use to support herself, and also more general information on the presence or absence of lines that guided subsequent construction behavior. These two functions can of course overlap, but in some contexts they did not. For instance, during radius construction, the spider held two adjacent radii that were already present (usually with her legs I), and then waved or slapped laterally with her legs (usually a leg I) in the space between the radii. In *M. duodecimspinosa*, *L. mariana*, and *Z. geniculata* these were not dorso-ventral taps, but instead rapid, large-amplitude side to side slapping movements (WE). Often the spider repeated these movements one or more times, slapping in the hole with one I, and then grabbing a radius with that leg and slapping in the same hole with the other (fig. 6.7, appendix O6.2.2). Presumably she was checking for the possible presence of a radius in the space between the two radii, to avoid laying a new radius where a radius was already present, rather than attempting to walk along a radius.

The dexterity and flexibility displayed by an orb weaver like *Micrathena duodecimspinosa* in modulating the following behavior of her legs was extraordinary. Take, for instance, sticky spiral construction. The spider's two legs I or her iI and iII often moved hand over hand along the temporary spiral; their movements were coordinated in both time and space (see appendix O6.2.2). Meanwhile, legs oII, oIII, and oIV were also following each other in tight coordination, but at a different, more rapid rhythm, as they successively grasped each radius the spider encountered so that the sticky spiral could be attached to it. Each grasping movement of oIV was complex and had a double function after following oIII to the radius: it performed a quick, subtle searching movement to locate the site where the inner loop of sticky spiral crossed the radius; and it then grasped and held the radius precisely against her spinnerets. Leg oIII executed still another highly coordinated movement when the inner loop of sticky spiral was close to the outer loop of temporary spiral, pulling the temporary spiral line to the spider's mouth to be broken as she moved toward the next radius. Meanwhile, another leg, iIV, performed still another task, pulling out additional sticky spiral line to lengthen the new segment just before each attachment. More posteriorly, the spinnerets themselves were probably executing coordinated movements that allowed attachments to be made nearly instantaneously (in *M. duodecimspinosa,* the radius was against the spinnerets on the order of only 0.06 s for each attachment). Even the attachments were complex and precise: the sticky spiral bent sharply, running parallel along the radius (for up to 2–3 mm in some araneids) (Kavanagh and Tillinghast 1979), presumably resulting from fine coordination between leg and spinneret movements. In addition, the spider used different combinations of legs to sustain her weight and adjust for the distortions it produced in her web (above). And all of these detailed movements were repeated at a very high rate (up to <1 attachment/ sec near the hub) and continued for 30 min or more (truly sustained concentration!). One could compare the mechanics of a *M. duodecimspinosa* laying sticky spiral with those of a person playing a long, difficult piece on the piano, with different rhythms and notes for the two hands. Except in the case of the spider, there are eight different legs engaged in different tasks.

6.6.2 PATTERNS OF TENSION CHANGES DURING CONSTRUCTION: A TENDENCY TO RELAX

In a few contexts uloborid and araneoid orb weavers consistently increased the tensions on lines they had already laid during orb construction, but more commonly they reduced the tensions on lines (table 6.2). This bias has apparently not been noticed previously. Perhaps its functional significance is related to the trade-off between the advantages of greater tensions on lines, which can increase the spider's ability to move in her web without producing large local distortions of its form (section 4.4, appendix O6.2), as opposed to the advantages of more lax lines, which can increase the web's ability to stop and retain prey (section 3.3.2.1). The spider may build the first lines in her orb under relatively high tensions to facilitate subsequent movements and construction behavior, and then, when the web is more complete and/or it is less important to be able to move about without distorting it, she can reduce the tensions on lines to make the orb a more effective trap. These ideas (along with the advantages of a more equitable distribution of tensions) seem likely to help explain the tension reductions produced by removal of the center of the web that was followed by tension reductions in radii, and, in many species, the reduction in the tension on each radius during its construction. Reductions in tensions on sticky spiral lines (table 6.2), however, probably had a different explanation—that the tension required to draw sticky lines from the spinnerets was greater than the optimum tension for retaining prey in the finished orb. Possible tests of these ideas are described in table O10.1. Perhaps these considerations will help explain further details that are still not understood (fig. 6.28).

I know of no non-orb building spider that systematically modifies the tensions on lines after they have been laid. Lamoral (1968) thought that *Steatoda lepida* laid early lines in the tangle portion of its web under higher tensions, but made no direct measurements. Differences in tensions in the bowl-shaped sheet of *Frontinella pyramitela* (LIN) were altered by breaking "tensor threads" (Suter 1984), but it was not clear whether these lines were added after the sheet had been built. The occasional breaking of lines, as in the theridiid *Nihonhimea tesselata* (Jörger and Eberhard 2006), obviously lowered some tensions, at least locally.

6.6.3 MISSING DETAILS

One of the strongest impressions I have from writing this chapter is the almost complete lack of attention that has been paid to the "atoms" or building blocks of orb construction behavior. The "roiling sea of variation" in those behavioral units, which I mentioned in the introduction and described superficially in appendix O6.2 and which becomes evident as soon as one studies a slow motion recording of any stage of orb construction, has gone unexplored. The lack of studies at this level in recent meetings of behaviorists and arachnologists and in recent volumes of leading journals in these fields make it clear that this relative ignorance will continue, at least into the near future.

Table 6.2. Changes made by spiders in the tensions on lines during or immediately following construction of orbs or orb-derived webs. The lists of species are undoubtedly incomplete (the groups were chosen to illustrate taxonomic diversity).

Portion of orb or stage in construction	*Behavior*	*Species*	*References*
Increase tension			
Frame lines	Cinch up by going back and forth	Many orb weavers (section 6.3.3)	(section 6.3.3)
Spring line	Reel in with legs I and II after finish orb	*Theridiosoma* and allies	McCook 1889, Comstock 1967, Coddington 1986a
Spring line	Reel in with legs IV	*Hyptiotes* spp., *Miagrammopes* spp.	Marples and Marples 1937, Akerman 1932, Lubin et al. 1978
Bridge line	Zig-zag attachments following prey capture	*Chrosiothes tonala*	Eberhard 1991
Center of hub	Add new lines across hub hole	Many araneoids	Eberhard 1982, 1987d
Decrease tension			
Hub (at start of orb construction	Break and lengthen radius (fig. 6.5d)	*Araneus diadematus*, other araneids	Zschokke and Vollrath 1995a,b
Hub[1]	Early replacement of proto-hub	*Philoponella vicina*, *Uloborus diversus*	Eberhard 1972a, 1990b
Radius construction	Break and lengthen each final radius soon after attach it to the frame	Many araneoids	Eberhard 1981a, Zschokke and Vollrath 1995a,b
Radius construction	Lengthen each radius as arrive at the hub	Many araneoids	Eberhard 1981a, 1982, Zschokke 2000
Center of hub	Remove center of hub after finish sticky spiral	Many araneoids	Hingston 1920, Wiehle 1928, Witt et al. 1968
Sticky spiral	Pull line out of spinnerets "hand-over-hand" with legs IV (or with repeated pulls with single IV)	Many araneoids	Eberhard 1982, section 6.3.5, appendix O6.3.6.1
Sticky spiral	Pull line out with pushing movement of iIV just before attaching to each radius	Many araneoids	Eberhard 1982, WE
Sticky spiral	Continue to comb cribellum silk for sev. seconds just before attaching to radius	Uloborids (including *Zosis geniculata*)	Eberhard 1972a, 1982, WE

[1] Effect on tensions of radii is less certain.

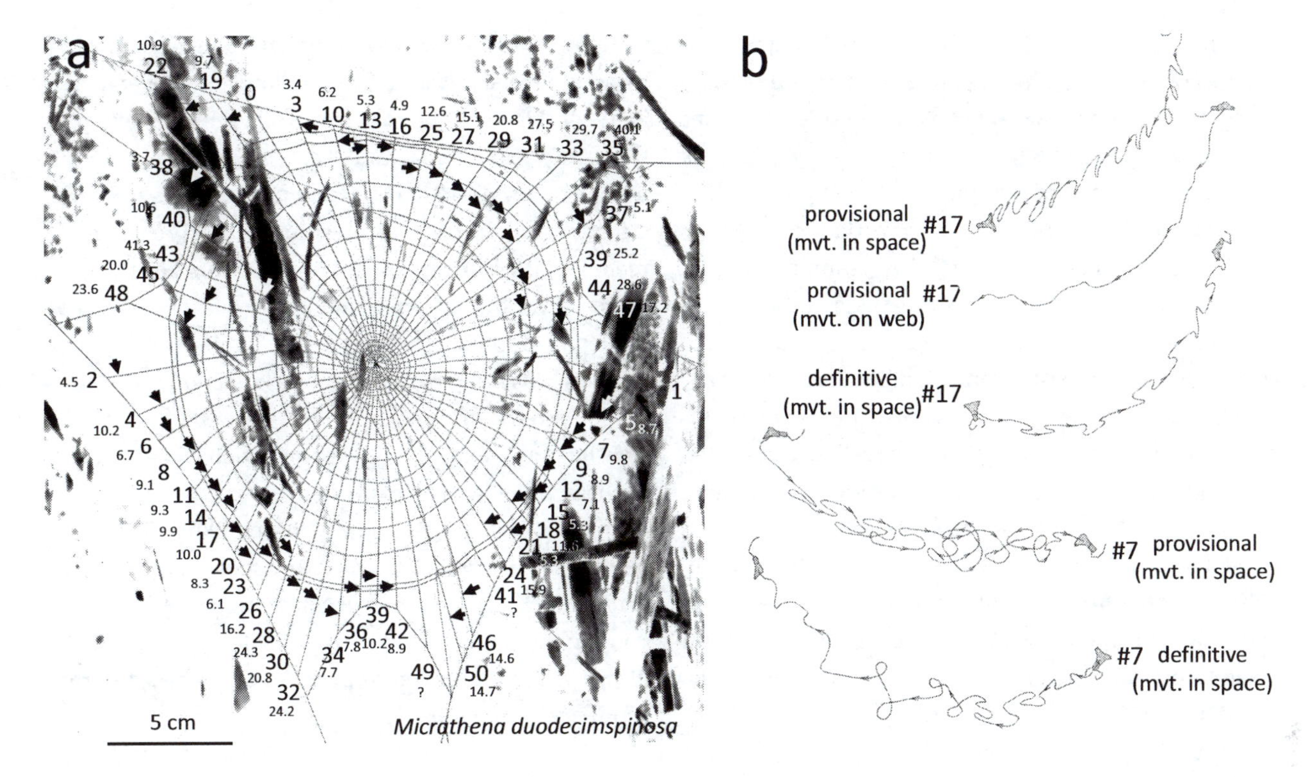

Fig. 6.28. This photo of an orb of *Micrathena duodecimspinosa* (AR) (*a*) illustrates two patterns in the removal of provisional radii laid during radius construction whose functions are unknown. The numbers 3–50 indicate the order in which the radii were added after the "original" radii (0, 1, 2) were built; the smaller number near each radius number indicates the approximate % of the provisional radius that was left intact when the definitive radius was built (measured from video recordings; the points where breaks occurred are indicated with arrows). A larger fraction of the provisional radius was left intact in the last few radii attached to each of the primary and secondary frames; the functional significance of this pattern is unknown. The jerky, bouncing paths that the spider followed while building radii are illustrated for radius #7 (below) and radius #17 (above) in *b*. When the movement of the web was taken into account (middle trace for radius #17), it was clear that much of the spider's movement was due to bouncing of the web itself. Jerky movements of this sort did not occur during exploration prior to radius construction, and ceased after the radii were complete. Perhaps the jerky movements functioned to sense tensions or extensibilities of web lines, but this is only speculation.

Unfortunately, there is reason to suspect that this neglect of behavioral mechanics at lower levels may ultimately limit understanding of the evolution of behavior at higher levels. Just as understanding the ontogenetic building blocks is important in understanding the evolution of morphology, so understanding the evolution of behavioral building blocks is likely to contribute to understanding behavioral evolution at higher levels. For instance, *Leucauge* sp. (TET) usually performed the iI inner loop localization behavior that is typical of other tetragnathids, but occasionally it used the less energetically costly oI behavior typical of araneids when near the hub, where the iI behavior was especially onerous (Kuntner et al. 2008a). Similarly, *Micrathena* sp. (AR) abandoned the typical araneid oI technique for the less costly oIV technique typical of nephilids when near the center of the orb, and *M. duodecimspinosa* used this nephilid technique nearly exclusively. These facultative changes suggest probable pathways in the evolution of inner loop localization behavior.

6.7 SUMMARY

This summary includes information from appendices O6.1–O6.3 as well as this chapter. Orb construction behavior can be analyzed at several levels of organization, and these different levels show different patterns in both intra-individual flexibility and inter-specific diversity. At the very low level of leg and body movements, the behavior of even a single individual is extremely variable, and the old image of an orb weaver as a simple automaton is especially inappropriate. For instance, a spider building the sticky spiral while she is above the hub in a vertical orb makes quite different movements than she does only seconds later when she is below the hub. These lower level variations are highly patterned, however, in ways that enable the behavior at the next level up to be relatively stereotyped: large amounts of variation at a lower level produce smaller amounts of variation at a higher level. Lower level adjustments are made and the individual movements are coordinated smoothly and rapidly to produce the higher-level invariability. For each attachment of the sticky spiral, for instance, the spider *Micrathena duodecimspinosa* (AR) accomplished at least ten decisions and seven different operations/attachment, and often made >1 attachment/s.

Similarly, different levels of analysis of orb construction behavior showed quite different phylogenetic patterns. At the high level of the ordering of different operations (explore, build frames and radii, then hub, temporary spiral and sticky spiral, modify the hub), there has been very little change, and the changes are consistent at high taxonomic levels (families, subfamilies). There are also several patterns at the next lower levels that show great uniformity, both among conspecific individuals and across wide taxonomic ranges. For instance, frame construction was generally associated with simultaneous construction of a radius (usually adding a new radius, sometimes replacing an old one); radii were added at "final" angles, rather than by subdividing the open sectors into equal parts; radius construction (and in particular the decisions to add new radii) was centered at the hub, rather than at the edge (on frame lines); and centripedal activity, working gradually from a central area to the periphery, always occurred in hub and temporary spiral construction, while sticky lines were always laid from the outside working in. One behavioral pattern that is widely used in orb construction is the behavior in which more posterior legs follow more anterior legs on the same side (section 2.4.2.1). Following behavior occurred during radius, hub, temporary spiral, and sticky spiral construction.

I suspect that there has also been similar conservatism at very low levels, such as the "following" movements of legs, the movements used to break, reel, and reattach lines, to float bridge lines, etc. (appendix O6.2), but this is largely speculation due to the dearth of detailed descriptions.

In contrast, it seems that the greatest documented behavioral diversity was at intermediate levels of analysis. Some of this variation showed strong phylogenetic patterns, and some behavioral traits, such as for example inner loop localization behavior during sticky spiral construction and the order of attachments and breaking and reeling of lines during radius construction, have proven to be useful in characterizing and grouping species at higher taxonomic levels such as families and superfamilies. In some cases a relatively large body size compared with the sizes of the spaces between lines in the web has probably selected for new behavior patterns, such as inner loop localization behavior with leg IV, which has arisen convergently in nephilids, deinopids, and some *Micrathena* species. The most variable stage (excluding the poorly known exploration stage) is frame construction; one particularly variable species, *Nephila clavipes*, was estimated to perform up to fifty different sequences (appendix O6.3.3.2).

Descriptions of construction behavior have been plagued by typology. One likely future development will be a greater use of video recordings to study variation in orb construction. Video analyses could also help fill in current gaps in behavioral descriptions between the demonstrations that spiders use certain cues to guide their construction behavior (see chapter 7) and the effects that these cues have on crucial behavioral details. For instance, the location of the inner loop surely influences the choice of the site on the radius for an attachment, but it is not known whether this happens because the sites where legs oIII and/or oIV grasp the radius change, or whether the change occurs in site between these legs where the spinnerets contact the radius. The exact sites that were grasped by leg oIV of *M. duodecimspinosa* correlated with where the sticky spiral was attached, but information on the positions of leg oIII and the spinnerets is lacking. Details of this sort, of course, are what actually determine the traits of finished orbs, and they may have had important effects on the evolution of orb designs. Just as understanding the details of embryological processes helps to understand how and why morphology has evolved, so these kinds of behavioral details are likely to be useful in understanding spider web evolution.

One thing that is already clear, however, is that even in a single individual spider building the sticky spiral in a single orb, there is substantial variation in the movements of legs, the forces that they exert, and the distortions that they produce in the web; it is also clear that the spider must control for these differences. At this level, the virtual models that I and others have made of orb construction are huge oversimplifications.

Some variations in lower-level details of behavior probably represent adaptive solutions to mechanical problems. One such problem is that the spider's own weight sometimes produces substantial local distortions of both the angles and the distances between lines in her web. These distortions, which will be greater for a spider with shorter legs, because her weight will be distributed over a smaller portion of the orb, may make it more difficult for the spider to choose appropriate attachment sites that will give favored spacings and orientations in the finished orb when her weight is no longer distorting it. One behavioral detail that probably represents an adaptation to reduce this distortion problem occurred in the stubby-legged *Gasteracantha cancriformis* (AR). While building the outer loops of sticky spiral, the spider maintained contact with the outer loop of temporary spiral and reeled in the radius to which she was about to attach, instead of walking out toward the frame (see fig. 6.24a, fig. O6.2j). Because she remained near the supporting temporary spiral, the distortion that her weight produced in the radius was reduced. A second possibly adaptive behavioral detail during sticky spiral construction evolved convergently in species in three different groups that built relatively long segments of sticky spiral (for instance, when the temporary spiral was reduced or absent). The spider held the radius away from her body with one leg III, which slid along the radius as she walked inward from the latest sticky spiral attachment (fig. 6.26), thus reducing the possibility that the new sticky line she was producing would accidentally adhere to the radius.

Tensions on lines have been measured in absolute terms in a few species, and have been quantified in relative terms in many more. There is a general trend (with exceptions during the construction of major frame lines) for spiders to lower rather than increase the tension. The reason for this trend is not obvious to me. Perhaps it is as simple as the physically greater ease of making such an adjustment: the spider need only release a little more line while resting immobile, as opposed to backing up or otherwise gathering in line and holding it tight while making a new attachment. There are a few scattered cases of consistent backward movements by spiders: the filistatid *Kukulcania hibernalis* further entangles prey she has seized in her chelicerae by backing toward her retreat; the uloborid *Miagrammopes* tenses the line she rests under; the nephilids *Nephila* spp. back up their draglines while holding prey clear of the web (Robinson and Robinson 1973a). But backing up does not occur during orb construction behavior; interestingly, stereotypic backing up behavior does, however, occur during web construction by two theridiids that build secondarily derived, highly organized webs, *Synotaxus* spp. (Eberhard 1977a, 1995) and *Chrosiothes* sp. (WE).

Behavioral senility may occur in orb weavers, as evidenced by predictable alterations of web designs in old females that appear to be disadvantageous. Some changes may be manifestations of an overall weakening of the spider, but others do not show any obvious correlation of this sort, suggesting that further study of senility in orb weavers may provide useful points of comparison with studies of aging in other organisms.

Other general points made in this chapter include the following. The "vocabulary" of leg and body movements used during orb construction by *M. duodecimspinosa* (AR) was sketched out to illustrate their diversity and provide a basis for future comparative studies (appendix O6.2). The previously overlooked question of how orb weavers are able to produce the highly regular distances between hub spiral loops was answered, noting that different species may use different combinations of mechanisms: *M. duodecimspinosa* (AR) used both leg iIII (early in hub construction) and leg oIII (later); *Leucauge mariana* (TET) used leg iIII; and *Nephila clavipes* (NE) used leg oIII. Previously unappreciated flexibility in the order of sticky and non-sticky line placement in response to web demage, was described. Several hitherto neglected differences in uloborid frame construction and the early stages of radius construction may constitute additional behavioral traits that distinguish this group from araneoids (appendix O6.3): strict sequential ordering of radius construction; strict ordering of frame construction; use of the "leading" radius for an exit during radius and frame construction; proto-hub removal and simultaneous construction of a small circular line joining the earliest radii; shifting the point of attachment of the new radius outward along the frame line just before returning to the hub during frame construction; laying segments of the sticky spiral on radii, creating zigzag patterns in the outer portion of the orb; regularly skipping radii without attaching the sticky spiral to them in the central portion of the orb; perhaps a greater frequency of leaving unbroken segments of the temporary spiral in the finished orb; and (less certain) nearly complete limitation of frame construction to very early in orb construction and a nearly complete lack of secondary frames. The complex, rapid, and precise set of highly coordinated movements of legs, spinnerets, and chelicerae involved in the key "cut and reel" behavior, which allows spiders to change the tensions and locations of lines and are thus critical in exploring websites and in producing orbs, was described in detail (fig. 6.3, appendix O6.2). The potential danger to the orb that could result from mistakes in the widespread behavior of removing lines from the center of the hub at the end of orb construction was recognized, and a preliminary description was given of how *M. duodecimspinosa* (AR) avoids this danger (fig. 6.16). Finally, a rare type of orb repair in *Uloborus diversus* may represent a behavioral missing link with the web construction behavior of ancestral orb weavers (appendix O6.3).

7

CUES DIRECTING WEB CONSTRUCTION BEHAVIOR

A Spider sewed at Night
Without a Light
Upon an Arc of White.
If Ruff it was of Dame
Or Shroud of Gnome
Himself himself inform.
Of Immortality
His Strategy
Was Physiognomy.
—Emily Dickenson

7.1 INTRODUCTION

Yes, but how do they do it? For many people (myself included), the most intriguing questions about spider webs center on how a small, nearly blind animal with limited brainpower is able to build such a precisely organized, geometrically regular, complex web, with not a single lax line in the whole structure. The focus of this chapter is on the internal and external stimuli that guide spiders while they build their webs, and their responses to these stimuli. Aside from their intrinsic interest, these topics also offer a gateway to more general questions concerning the nuts and bolts of animal behavior. I believe that orb webs have the potential to be especially useful in exploring the limits of the behavioral capabilities of small invertebrate animals. Concepts that are otherwise difficult to study, such as "attention," animal "errors," and evolutionary homologies in behavior, are more easily accessible in orb weavers than in most other animals (chapter 10). The summary at the end of this chapter, the first exhaustive review ever made of the cues that guide orb web construction, will reveal some surprising conclusions (section 7.21.1).

7.1.1. BUILDING AN ORB IN HUMAN TERMS

Perhaps it is easiest to appreciate the achievements of an orb weaver like *Micrathena duodecimspinosa* by putting yourself in her place, and imagining what it would be like to build an orb. To start, you must first blindfold yourself, because the spider is surely blind with respect to her own lines and often with respect to potential attachment sites nearby (section 2.4.2.1). The closest human analogy would be that of a blind man who must explore and climb through the tangled branches of bushes and trees on the basis of touch, leaving a silk line behind wherever he goes.

Finding and recognizing an appropriate site in which to build is probably one of the most challenging tasks. The physical supports to which you might attach lines are arranged extremely irregularly in space, and each site is different (the smooth rectangular or circular frames used in most studies of web construction in captivity are NOT fair approximations to natural conditions). And you will not be able to explore any of the spaces by swinging or flapping your lines, because the air is highly viscous at the scale of the spider; an analogous situation on a human scale would be building a web of elastic ropes under water.

You cannot sense the presence or absence of nearby attachment points or obstacles, much less their positions relative to each other, until either you or your lines have contacted them. And even if you could have seen these objects, you are unable to fly to them, to shoot your lines out to reach them, or even, in many cases, to walk to them (imagine trying to lay a line between the widely spaced twigs on different branches of different trees while blindfolded). You must instead depend on the vagaries of the wind; you will have to launch new lines, allowing their tips to float away on irregular air currents, and hope that they will eventually snag on objects that are located appropriately to support your web. As you gradually discover attachment sites in this haphazard way, you will need to recognize whether they border an open site that is sufficiently large to accommodate the orb that you are planning, and whether this space is free of further protruding objects that could interfere with it. You will need to adjust your plan for the area of the open space that your web will occupy in accord with the amount of silk you that have available in your silk glands. The space must have sufficient attachment sites on its perimeter that are not too far apart and that are in the same plane; and this plane must have the particular slant (horizontal, vertical, in between) that your particular species uses for its orbs.

Once you have found a good site, there are many further challenges. You must determine the approximate center where you will place the hub (no peeking from behind the blindfold!). You must clear away all the potentially distracting lines that you produced while you were exploring this area, but must avoid removing the lines to the supports that you will need for the final web. You will have to adjust several of the design features, such as the number of radii and the spaces between the sticky lines, to your silk reserves, and you will have to adjust the overall form of the web to the positions of the attachment sites. In fact, you will have to be quite extraordinarily flexible in your behavior, because many factors are known to influence orb web geometry: climate (especially wind velocity and direction) (e.g., Lin et al. 1995; Vollrath et al. 1997; Liao et al. 2009); previous feeding (Vollrath and Samu 1997; Weissmann and Vollrath 1998); your growth stage (e.g., spiderling vs. adult, proximity of the next molt) (Witt et al. 1968); your reproductive state (e.g., proximity of oviposition) (Sherman 1994); and the presence of conspecifics (Leborgne and Pasquet 1987). Each of the several hundred decisions you will make regarding where to attach the sticky spiral to radii that it crosses will be influenced by ten and perhaps up to sixteen types of stimuli (gravity, your distance from the hub, the distance traveled to reach the current radius, the lengths of your legs, memories of distances traveled on the previous radius and of adjustments to changes in such distances in the recent past, your recent feeding history, whether or not you have built a web at this site previously, etc.). Some decisions will be complex in terms of the variables themselves; for instance, as you will need to have accurate memories of the distances and directions that you have moved recently, and you will also need to sense the distances and angles between your legs when they are holding different lines, and to sense the direction of gravity. You will also need to adjust the tensions on lines as you lay them, and correct for the local distortions in the web that are produced by your own weight, so that lines laid later in the process do not cause earlier lines to become slack or overly tight in the finished orb. Happily, you have eight legs, and can execute several different tasks simultaneously; but you must keep track of and manage all of those legs at the same time!

Most of the thousands of decisions and approximately 3800 attachments that you will make must be performed in a hurry, because the total construction time is only approximately 30–40 min. Some movements must be very precise, for instance bringing lines into the narrow space between your spinnerets (section 2.3). You thus cannot afford either to rest or to lose your concentration at any point along the way; at times you will be making a new attachment of sticky spiral (with its ≥10 decisions) every second or less.

In addition to these mental challenges, there are purely physical demands. Scaling an orb like that of the typical 1.2 cm long adult female *Micrathena duodecimspinosa* (AR) to the size of a human body (about 1.8 m long), you will lay >2.8 km of sticky spiral to cover an area about 45 m in diameter; you will make >2.5 km of supporting lines (radii, frames, anchors) to build a support structure attached to supports that are up to 150 m or more apart. You will have to run >5 km in approximately 30–40 min, and if your orb is vertical, you will have to climb something like 1 km upward, and an approximately equal distance downward.

In sum, building an orb web is not child's play; it involves surprisingly complex mental activities and sustained attention in order to make literally thousands of rapid decisions, and also represents a substantial physical effort.

7.2 CLASSIFYING THE CUES

7.2.1 STIMULI FROM REPEATEDLY SENSED "REFERENCE POINTS" VS. MORE NEARLY CONSTANT "GENERAL SETTINGS"

Two general types of cues have been shown to guide the thousands of decisions made during orb construction regarding where to lay and attach lines. Many variables influence these decisions, and they can be classified with respect to how often the spider perceives the stimuli anew prior to making different decisions. At one extreme, she continually perceives and adjusts to the changes in the positions of the lines in her immediate vicinity as she moves across her web. During sticky spiral constructions for instance, the sites where the innermost loop of sticky spiral and the outermost loop of temporary spiral cross the radius, and the angle of the radius with respect to gravity, are probably sensed anew on each radius that the spider encounters (figs. O6.2–O6.4). I will refer to these repeatedly evaluated cues as "reference point" variables.

At the other extreme are relatively fixed variables that do not change during web construction, such as leg lengths, body weight, previous feeding history, weather conditions, impending oviposition, available space, and sites of prey captures and escapes that occurred in previous webs (summary Zschokke 1996). I will refer to these as "settings" variables, to indicate the possibility that the cues may be sensed only a single time at the beginning of construction, rather than being evaluated anew for each repeated decision (though it is not known how often the spiders actually make these evaluations). Several other possible variables, such as the amount of silk the spider has available in her glands, the spider's relative edge-to-hub position in the orb, and her position relative to the hub (above, below, to the side), are intermediate; they change, but more slowly, and may thus require less frequent re-evaluations.

This classification of cues helps in determining the minimum number of types of stimuli that a spider evaluates from moment to moment, one of the major objectives of this chapter. It is possible (though this is only speculation) that some or all settings stimuli are combined into one overall compound "settings variable" prior to or early in construction, and are not subsequently reevaluated during the process of construction; this would reduce the number of variables that the spider evaluates in making particular decisions (such as where to attach the sticky spiral to a radius), and gives a minimum estimate of the complexity of her decisions (section 7.21).

7.2.2 OTHER INTRODUCTORY NOTES

Many of the discussions in this chapter concern correlations between web traits and other variables: one can explain, for instance, at least some of the variation in sticky spiral spacing by taking into account the spider's positions with respect to the hub and to gravity, her hunger, and her supply of sticky spiral silk. I will leave another type of variable—imprecision in the behavior of the spider or "errors" (Eberhard 1990c)—for discussion elsewhere. I will be taxonomically erratic, and skip from one species to another in order to take advantage of the available behavioral studies. This lack of concern for possible taxonomic differences is justified by the empirical finding that cues and responses to cues seem to be extremely uniform throughout orbicularia, probably as a result of the monophyletic origin of orbs (Eberhard and Barrantes 2015; section 10.8).

Some of the morphological details discussed earlier may affect the stimuli that spiders are able to sense. In the first place, it is obvious that if the spider grasps the same line with two legs, as often occurs when legs follow each other (section 2.4.2.1), the two points of contact will reveal the orientation of the line that they are holding with respect to the spider's own body. Less obviously, a spider may be able to sense the orientation of a line when she grasps it with only a single leg: the amount she has to rotate her leg on its longitudinal axis to bring her tarsal claw more or less perpendicular to the line will depend on the line's orientation (fig. 2.16, section 2.4.2.3). Deflections of the serrate accessory setae or other setae near the distal tip of the tarsus could provide another source of information.

In addition, as an orb weaver grasps the lines in her

web and moves under them, the strengths and directions of the forces exerted on her legs to sustain her weight will correlate with the angles that the lines make with her legs. Spiders' legs carry extremely sensitive slit receptors in a variety of orientations and groupings in the cuticle, and these can sense stresses in the cuticle and thus the positions of her legs and the forces exerted on them; at certain frequencies a spider's metatarsal lyriform organ can respond to displacements as small as about 10Å (Barth 2002). These sensations could potentially give the spider information about the lines themselves. For instance, pulling on a longer line will result in greater extension than pulling with the same force on a shorter line; when the spider walks along a horizontal line, the shorter segment of this line will pull more nearly upward than the longer segment. A line that is more nearly perpendicular to the long axis of a leg will be displaced more when the leg is flexed than a line that is more nearly parallel to it.

In sum, it is quite feasible, in terms of the spider's sensory capabilities, for her to obtain precise information on the positions and orientations of lines that she contacts with only a single leg. In terms of the image of a blind man and his cane, it is as if the cane has delicate sensors that can provide information not only on the presence or absence of an object like a line, but also additional characteristics such as its orientation and length. It is reasonable, however, to wonder whether the distortions caused by the spider's own weight may not make it impossible for her to measure angles in finished webs accurately (S. Zschokke pers. comm.). In some stages of orb construction, such as sticky spiral construction near the outer edge of the web, the spider's weight is applied to only a few lines and causes substantial distortions; these distortions were especially large in a short-legged species (e.g., figs. 6.24, O6.2). Using angles to guide construction behavior seems unlikely in these situations, because the spider would need to make complex, seemingly difficult adjustments to translate the angles she sensed into the angles in the finished, undistorted web.

On the other hand, in a species with very long legs in comparison with the mesh of the web like *N. clavipes*, the stress of her weight is applied over a wider area (fig. O6.4), and the spider's weight would cause much less distortion of the lines where she was building. Similarly, distortions caused by a spider's weight in a horizontal orb (fig. O6.3) would cause more minor alterations of the angles between lines. And, at some stages of construction such as hub construction or removal of portions with many interconnected lines, even a relatively short-legged species would cause smaller distortions (fig. 6.16). In summary, some species in some contexts could obtain useful information on angles between lines, but angles are unlikely cues in others.

Another introductory point concerns what happens after the spider perceives a cue—the often neglected topic of the functional significance of her responses. Many responses provide adaptive flexibility—the spider adjusts details of the orb's design to the particular circumstances in which she finds herself. These circumstances include both her physical and her internal environment. For instance, an immature of the araneid *Argiope bruennichi* used her feeding and molting status to guide her decision whether or not to build a barrier web (Baba and Miyashita 2006). When she had fed poorly, she refrained from building a barrier, thereby sacrificing protection from predators for gaining additional prey (see Blackledge 1998a; Blackledge and Wenzel 1999 for similar arguments regarding stabilimenta). There is no dividing line between cues that are "essential" to building an orb, and those that provide adaptive plasticity. The frequent importance of plasticity is part of the "spiders are not automatons" argument made in other parts of this book (see sections 1.4, 6.5, 6.7, 7.17.2, 7.20).

I will begin by discussing the cues that are used near the end of orb construction, in building the sticky spiral, and gradually work backward in the construction sequence. This is because the later stages are better understood, because they are easier to study because the movements are more repetitive and continue for a longer time, and because the cues are easier to characterize because most of them come from the lines already in place (Witt 1965).

One difficulty in studying orb construction is that the lines are geometrically highly organized with respect to each other, so many different variables are correlated with each other; it can sometimes be difficult to discriminate cause and effect relations from secondary or incidental correlations. I will thus emphasize experimental manipulations, some performed by human observers, and others performed by the spiders themselves (when

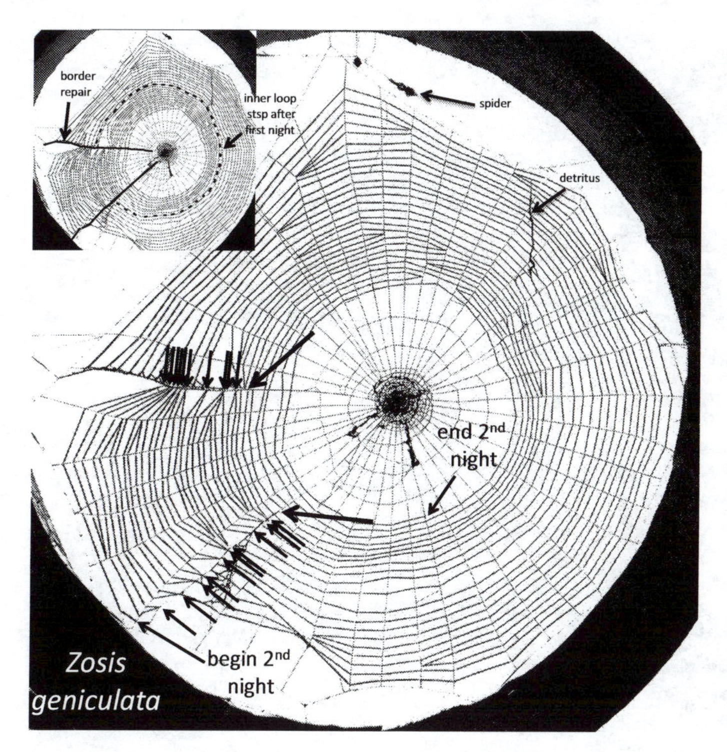

Fig. 7.1. This repaired web of a mature female *Zosis geniculata* (UL) gives an experimental demonstration that the sticky silk in the borders of a repair sector and not the border itself induced the spider to turn back when she was laying sticky lines. The experiment employed a web that the spider had only partially filled with sticky spiral (dotted lines in the inset) on the first night. The next day the 7:30–9:00 sector was removed experimentally (solid lines in the inset); the sticky lines at the edges of the broken sector adhered to each other and to the remaining intact lines along the borders (the innermost extensions of the sticky borders are marked with large arrows in the larger photo). The next night, the spider built new hub lines and radii in the broken sector, and then filled this sector and the central portion of the entire web with sticky silk. While building the sticky spiral, the spider turned back each of the 19 times that she reached a border of the broken sector that included sticky lines (small arrows). In contrast, she did not turn back on any of the 12 occasions in which she reached the more interior portion of the borders which lacked sticky lines.

atypical, altered line placements affected their own subsequent behavior).

Behavioral details were also crucial in deciding which cues spiders used. This problem was illustrated dramatically by a computer simulation study that produced web patterns that were quite similar to the orbs built by spiders, despite the complete omission from the program of the cue that experimental manipulations later showed to be the most likely cue used by the spiders (section 7.15). Even strong correlations between variables in orbs can be misleading. In discussing which cues were used in *Araneus diadematus* (AR) to determine inter-radial angles, Peters (1953, 1954) used the length of the outer segment of sticky spiral. There is a statistical correlation with this variable, but sticky spiral length cannot be a cue determining radius angles, because the sticky spiral is not built until long after the radii are laid.

7.3 CUES FOR STICKY SPIRAL CONSTRUCTION

7.3.1 DISTINGUISHING STICKY FROM NON-STICKY LINES

Spiders distinguished between sticky and non-sticky lines during sticky spiral construction (and other stages of construction). The most dramatic indication of this ability was the abrupt shift from inner loop localization tapping behavior to sticky spiral attachment behavior that immediately followed each contact with a sticky line (section 6.3.5, appendix O6.2.5). Some species also distinguished between sticky and non-sticky lines in other contexts. For instance, uloborids and *Nephila clavipes* (NE) attached new radii to old radii rather than to sticky spiral lines when repairing or replacing sectors of their orbs (fig. 6.17), and also doubled back during sticky spiral construction when they encountered sticky lines in approximately radial orientations, but not when they encountered radii that had similar orientations (fig. 7.1). The effects of deviant sticky lines on the placement of subsequent sticky lines (fig. 7.2, next section) also demonstrated this ability. Perhaps the chemosensory setae on the tarsi (fig. 2.16c, 2.22) were used to perform these discriminations, but I know of no data on this point. The extensibility and elasticity of sticky spiral lines are both greater than those of non-sticky, ampullate gland lines, but details of the spider's behavior during sticky spiral construction (section 6.3.5, appendix O6.2.5) do not suggest that these traits were evaluated. In any case, I distinguish between contact with sticky and contact with non-sticky lines in the following discussions.

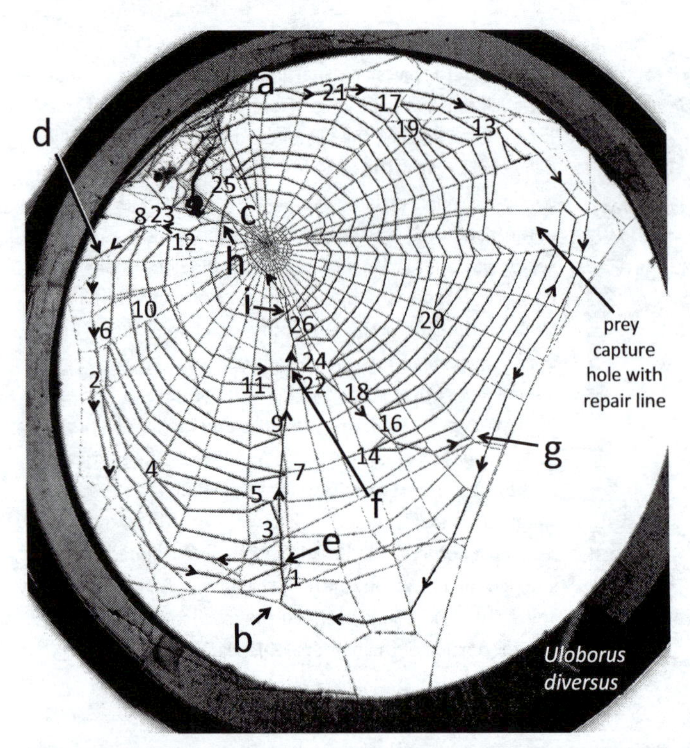

Fig. 7.2. Tracing the unusual path in this web of a mature female *Uloborus diversus* (UL) during sticky spiral construction (indicated by arrowheads on lines and the numbers indicating the order of turnback points) suggests that the inner loop of sticky spiral guided the spider while she was making sticky spiral attachments. The spider began the sticky spiral at *a* and moved clockwise along the edge, but after circling about half of the web she turned at *b* and moved inward toward the hub (perhaps mistaking radii for temporary spiral lines and vice versa?). When she reached the hub she turned outward (*c*) until she encountered the frame (*d*), where she began moving counter-clockwise in the usual circular pattern along the edge. She doubled back (turnback 1) when she encountered the first deviant sticky line (*e*), and continued to zig-zag back and forth between the two deviant sticky lines (turnbacks 1–12). She broke this pattern, crossing the deviant sticky line at *f*, and moved outward and counter-clockwise into the large open sector until she encountered the first loop she had laid at *g*. She then moved in a normal, approximately circular zig-zag pattern, turning back each time she encountered the deviant sticky line between *f* and *g* until she had filled this sector (turnbacks 13–26). The highly abnormal pattern of the sticky spiral apparently resulted from the sticky lines she laid from *b* to *d* while temporarily disoriented. This web also demonstrates that spiders made occasional strategic exceptions to the "do not cross the inner loop" rule (*f*, *h*, and *i*). Exceptions of this sort, which can be induced experimentally, show patterns that are difficult to explain without resorting to attributing higher mental functions to spiders (section 7.3.3.2, Eberhard 2019b).

7.3.2 RAPIDLY CHANGING, REPEATEDLY SENSED REFERENCE POINT CUES

7.3.2.1 Location of the inner loop

The spider began the sticky line near the outer edge of the orb, and spiraled gradually inward toward the hub, attaching her line to each radius that she encountered. Both the spider's own behavior and experimental modifications of webs demonstrated that the location of the inner loop of sticky spiral already in place (the "inner loop" in fig. 7.3) constituted a reference point when she attached the sticky spiral. Inner loop localization behavior (section 6.3.5, appendices O6.1.1.8, O6.2.5, O6.3.6) ceased the moment the leg touched the inner loop (figs. O6.2–O6.4). The results of the classic experiment performed by Hingston (1920) with *Neoscona nautica* (AR) and later repeated with several other araneids and tetragnathids (fig. 7.3a,b; Peters 1939, 1954; Eberhard and Hesselberg 2012) showed that this point of contact served as a reference point for the ensuing sticky spiral attachment. When a segment of the inner loop was removed during sticky spiral construction (in effect moving the inner loop outward in this sector of the orb) the next sticky spiral loop was displaced outward (fig. 7.3b).

A second type of experiment resulted after *Uloborus diversus* (UL) occasionally seemed to temporarily lose track of where she was while laying sticky spiral (for instance, when illuminated by a bright light) and wandered briefly while she produced sticky line. When she

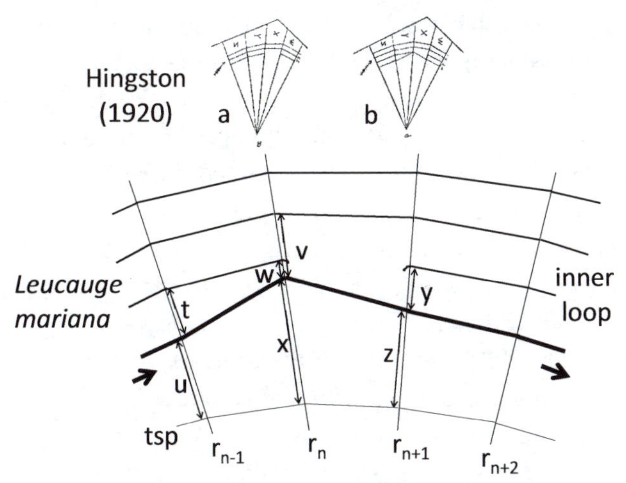

Fig. 7.3. Hingston's drawings (*a*, *b*) illustrate the results of his classic experiment with *Neoscona nautica* (AR). When he broke one segment of the inner loop of sticky spiral while the spider was building the sticky spiral (*a*), the spider placed the next loop of sticky spiral farther outward (*b*) (the arrows indicate the direction in which the spider was circling). This response demonstrated that the site of the inner loop of sticky spiral affected the placement of the next loop of sticky spiral. Analogous experiments with *Leucauge mariana* (TET) demonstrated a similar but less dramatic effect of the site of the inner loop (*c*) (the drawing represents the average relative distances). The spider responded to the experimental damage by attaching farther out on r_n ($w < t$), but the amount of this outward displacement was generally less than that described by Hingston ($w+v$ was greater than rather than equal to t) (*a* and *b* from Hingston 1920; *c* from Eberhard and Hesselberg 2012).

then reoriented and resumed filling in the web with sticky spiral lines in more nearly spiral patterns, her subsequent sticky lines followed the deviant sticky line (fig. 7.2; Eberhard 1972a). Similar patterns of wandering that strongly affected subsequent loops of sticky spiral also occasionally occurred in *Zosis geniculata* (UL) (WE), and *Araneus diadematus* (fig. 1C in Vollrath 1986). This "Joe Morgan" effect (Eberhard 1972a; the first spider seen to manifest this effect was named after a baseball player) showed that even sticky lines with very unusual orientations with respect to the radii were used as guides.

In sum, several independent lines of evidence have demonstrated that the site of the inner loop of sticky spiral is used by orb weavers as a reference point in deciding where to attach the next loop of sticky spiral. Judging by the nearly universal occurrence of inner loop localization behavior in orb weavers (Eberhard 1982; Scharff and Coddington 1997; Kuntner et al. 2008a), this cue is probably nearly universally important. The next sections will show, however, that this is only one among many additional cues that are used to determine the distance from the inner loop at which attachments will be made.

7.3.2.2 Distance from the outer loop of temporary spiral ("TSP distance")

Several types of evidence indicate that spiders used the outer loop of temporary spiral as a second point of reference to guide sticky spiral placement (the "TSP distance") ($v + w + x$ in fig. 7.3, fig. 7.4). This has also been called somewhat inappropriately the temporary spiral to inner loop distance (the "TSP-IL" distance in Eberhard 2012; Eberhard and Hesselberg 2012) (sometimes the spider stopped moving outward before she reached the inner loop, and thus she could not always sense the TSP-IL distance) (I will use "TSP-IL" distance in some places, however, to refer to measurements from web photos).

In webs of *Leucauge mariana* and *Micrathena duodecimspinosa* in which the temporary spiral was still intact and in which it was thus possible to deduce the distances between lines that the spider encountered as she built the first few loop of sticky spiral, there was a trend for sticky spiral spaces to be larger when the TSP-IL distance was larger (fig. 7.5). Experimental manipulation of the TSP distance, by cutting temporary spiral lines during sticky spiral construction by *M. duodecimspinosa* and *L. mariana*, confirmed that the correlation

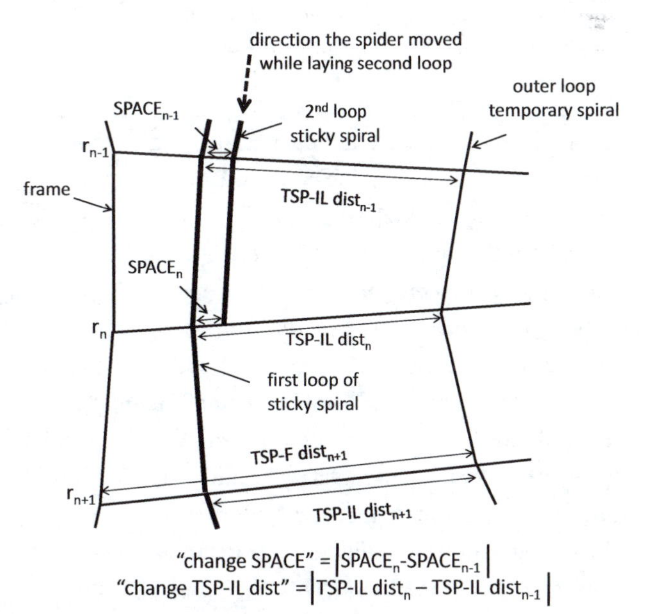

Fig. 7.4. This stylized drawing illustrates variables measured in the orbs of *Micrathena duodecimspinosa* (AR) (see text).

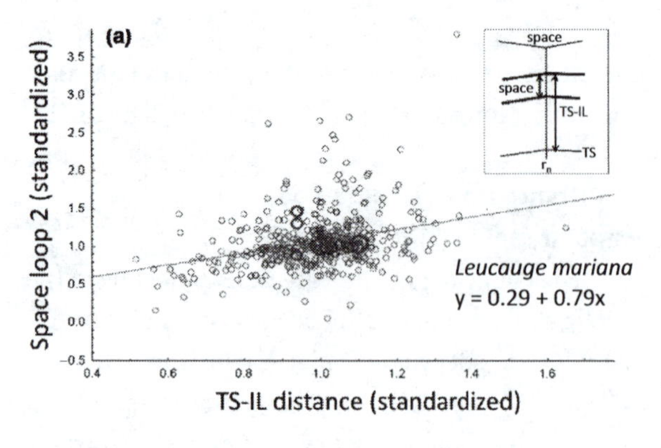

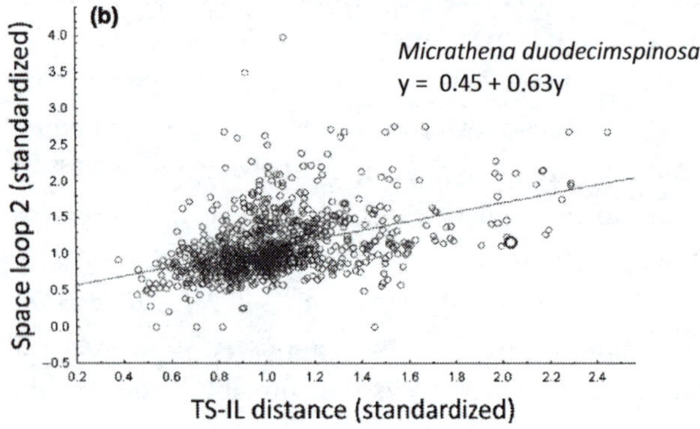

Fig. 7.5. The standardized space between the outermost two loops of sticky spiral (inset above) was greater when the distance between the inner loop of sticky spiral and the outer loop of temporary spiral ("IL-TSP distance") was greater in both *Micrathena duodecimspinosa* (AR) (N = 19 webs) and *Leucauge mariana* (TET) (N = 18 webs) (variables were standardized by dividing each value by the mean for that variable in that web) (from Eberhard and Hesselberg 2012). These correlations imply that the spider used a kinesthetic cue: when she had to move farther, she attached the current sticky loop farther from the inner loop.

resulted from a cause and effect relation: experimental increases in the TSP distance induced greater spaces between sticky spiral loops (fig. 7.6).

Still another indication that the TSP distance affects sticky spiral placement came from species in which sticky spiral spacing was regular but the spider never touched the inner loop of sticky spiral. Although the distances between lines have not been manipulated experimentally in these species, the uniformity of the spaces between sticky lines in intact webs and the lack of other obvious cues indicate that the spider used the distance she traveled along a radius as a cue to guide sticky spiral placement. In the webs of *Hyptiotes* spp. (UL) and *Theridiosoma gemmosum* (TSM), the small size of the spider, the large distance of the outer loop of sticky spiral from both the temporary spiral and the frame (see caption for fig. 7.7), and direct observations of the spiders' behavior (Eberhard 1982) showed that the spider did not contact the inner loop of sticky spiral (Peters 1954; Eberhard 1982). Similar patterns occurred in the webs of other theridiosomatids such as *Epilineutes globosus*, and *Naatlo splendida* (fig. 7.8; Coddington 1986a; WE), and in anapids with radial lines above the orb (fig. 7.9).

One additional, previously unremarked pattern in sticky spiral spacing in some orbs is compatible with the hypothesis that spiders used of the TSP distance cue to guide sticky spiral construction. When a spider made an overly large space between sticky spiral loops on a given radius, the next loop attached to that radius tended to be attached at an unusually small distance from the previous loop; these adjustments were documented in *Leucauge mariana* (TET) and *Micrathena duodecimspinosa* (AR) (Eberhard and Hesselberg 2012). The compensatory adjustments could be due to an effect of the TSP distance: on average, the TSP distance would be smaller the next time around the web following an oversized sticky spiral space, resulting in a smaller spacing of the next loop. Distances from the temporary spiral were not measured in these webs, however (the broken temporary spiral was not visible, because the webs were powdered), so this hypothesis has not been confirmed directly.

The leg positions and movements that spiders presumably used to measure the TSP distance varied in different species. In araneids and nephilids, in which the spider never lost contact with the outer loop of the temporary spiral during sticky spiral construction, the spider probably used her own body and the positions of her legs (figs. 6.24, O6.1). In *Gasteracantha cancriformis* (AR) the spider did not move outward when she was near the edge of the web, but instead reeled in the radius with hand-over-hand movements of legs oI and oII while holding onto the temporary spiral; in this case the amount of reeling could signal the TSP-IL distance. Krink and Vollrath (1999) believed that the TSP-IL distance was constant in *A. diadematus*, but that is surely not the case (Zschokke 1993; Zschokke and Vollrath 1995a; Eberhard and Hesselberg 2012). In other groups, such as tetragnathids, theridiosomatids, and anapids, the spider moved substantially outward beyond the outer loop of temporary spiral

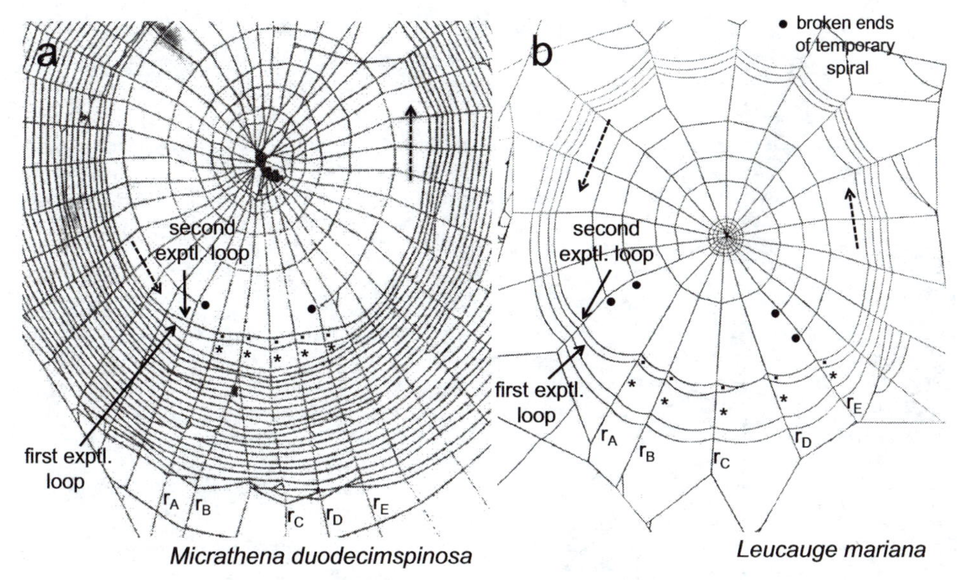

Fig. 7.6. Mature females in two different families, *Micrathena duodecimspinosa* (AR) (*a*) and *Leucauge mariana* (TET) (*b*) gave strikingly similar responses to experimental removal of five segments of the outer loop (*a*) or the outer two loops (*b*) of temporary spiral during sticky spiral construction, thus illustrating the uniformity among orb weavers regarding the cues that spiders use, and their responses to these cues to guide orb construction. The spider sharply increased the sticky spiral spacing ("*" in the photos) when she built the first loop across an experimental sector created by breaking the temporary spiral lines, thus increasing the TSP-IL distance sensed by the spider (dashed arrows indicate the directions in which the spiders were circling their webs). On her next pass across the sector, however, the spaces (marked with "." in the photos) were only slightly larger than normal, despite the still relatively large TSP-IL distances. The memory of the previous adjustments may have had a buffering effect on the spiders' responses when building the second loop.

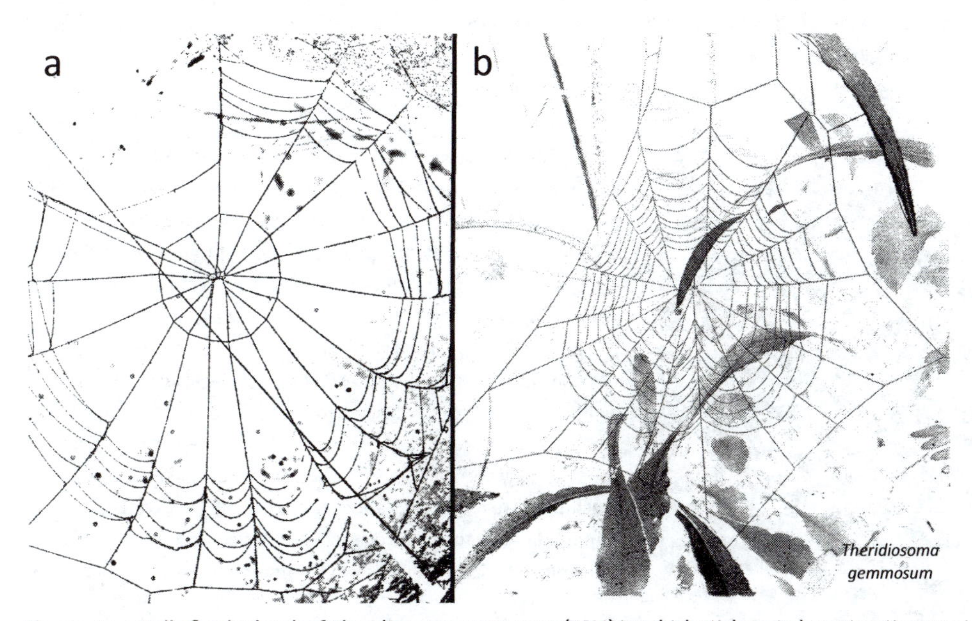

Fig. 7.7. A partially finished web of *Theridiosoma gemmosum* (TSM) in which sticky spiral construction was interrupted (*a*), and a finished web of the same species (*b*), suggest that this spider estimates the distance she has moved along the radius to guide placement of the temporary "spiral" (it is a circle in *T. gemmosum*), and also of the sticky spiral. The spider did not maintain contact, as do many other orb weavers, either with the hub during temporary spiral construction, or with the temporary spiral during sticky spiral construction (sections 3.3.2.8.2, 6.3.4, 6.3.5) (from Coddington 1986a, courtesy of Jonathan Coddington).

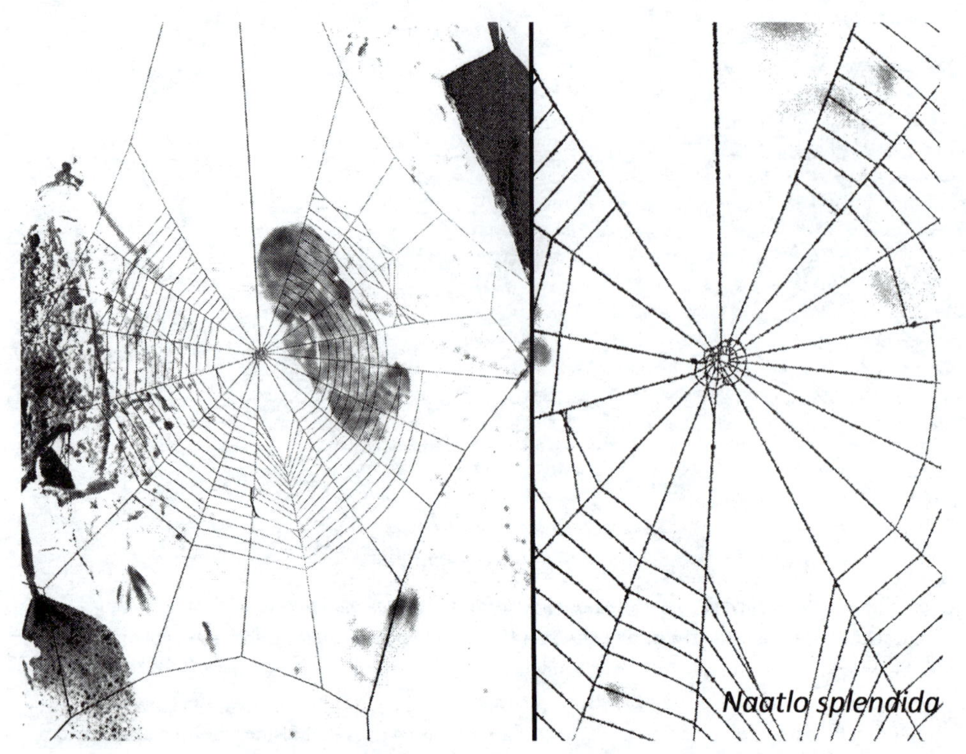

Fig. 7.8. The consistent distances between sticky spiral loops in the web of *Naatlo splendida* (TSM) suggest that the spider probably used the distance she moved from the outer loop of the circular temporary spiral (or the hub) to guide where to attach sticky spiral loops, because direct behavioral observations showed that these spiders never reached the inner loop of sticky spiral during sticky spiral construction. This nearly horizontal web (16° slant) also provides an example of a possible alternative design. In contrast to other webs of this species, it had an open sector, and the spider used the radial line in the open sector as a spring line, reeling it in and holding it tight while waiting for prey (the spring line in other webs of this species was out of the plane of the orb, as in other theridiosomatids—see figs. 3.8, 10.22) (the radius at 5:30 was broken after the web was complete).

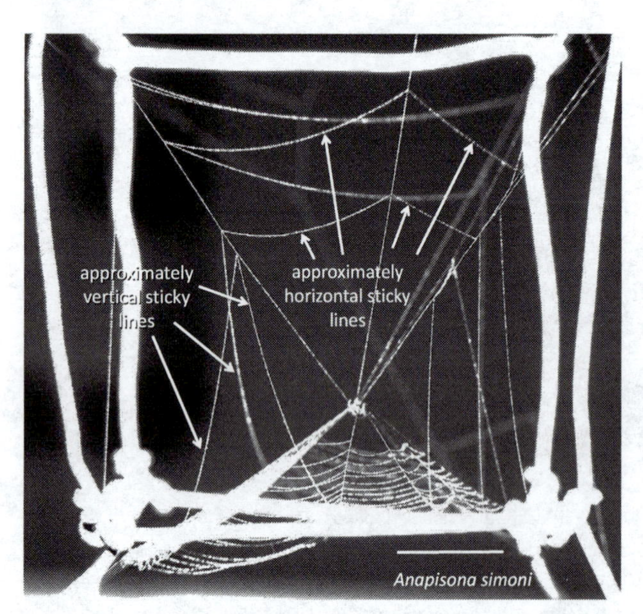

Fig. 7.9. This lateral view of an approximately horizontal conical orb built by *Anapisona simoni* (AN) in captivity provides evidence that spiders used the distance they moved along the radius to guide sticky spiral placement (sticky lines can be distinguished, as they all sag perceptibly in this photo; non-sticky lines are straight). The spider occasionally interrupted sticky spiral construction in the conical portion to attach a sticky spiral line to one of the radial lines above the orb. In these cases the distance of the attachment from the hub was approximately the same as that on the immediately preceding radius, resulting in the sticky line above the cone being approximately vertical. When the spider made two successive attachments to radial lines above the orb plane they were both at approximately the same distance from the hub, resulting in a nearly horizontal sticky line (from Eberhard 2007b).

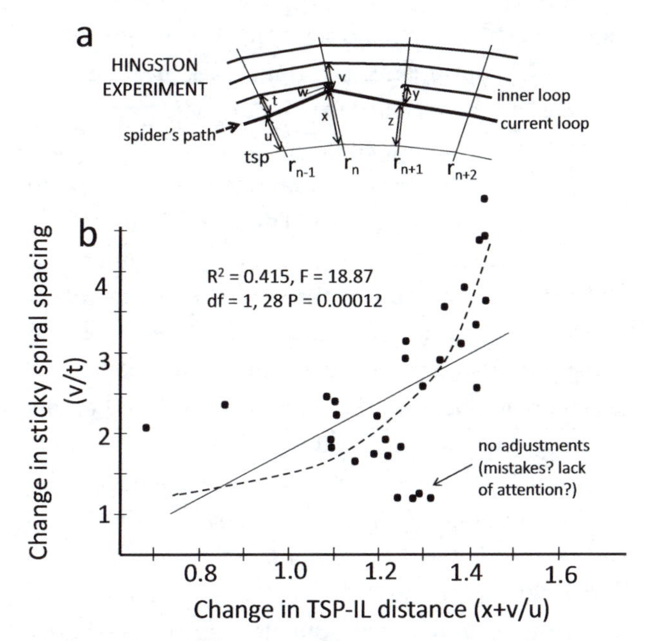

Fig. 7.10. Some possible execution errors in response to Hingston experiments (see fig. 7.3) by *Leucauge mariana* (TET) (*a*) may represent inattention or a lack of response to a cue. The sizes of the adjustments that spiders made in the spacing (comparing *v* with *t*) varied (*b*), ranging from essentially no change to increases of several times the value of *t*. In general, change in sticky spiral spacing was greater when the size of the increase in temporary spiral distance on r_n over that on r_{n-1} was greater. It is not clear whether this effect was linear (indicated by the straight line, and the regression statistics) or curvilinear (indicated by the hand fitted curve). In either case, there was a cluster of points below the line in *b* that did not fit this pattern.

to attach the first sticky spiral loop, and cues presumably came from distances traveled by her entire body.

In sum, several types of evidence indicate that spiders use the TSP distance in deciding where to attach the sticky spiral line, making a larger space when the TSP distance was larger.

7.3.2.3 Memory of the TSP distance along the immediately preceding radius

Another cue that appears to influence sticky spiral spacing is the memory of the TSP distance on the radius preceding the radius to which the sticky spiral is to be attached (e.g., the difference between the distance "t + u" on r_{n-1} and the distance $v + w + x$ on r_n in fig. 7.10). In webs of *M. duodecimspinosa* and *L. mariana* in which all temporary spiral lines were still intact (sticky spiral construction had been interrupted before the spider removed any temporary spiral lines) and TSP-IL distances could thus be measured precisely, a multiple regression of the space between the first and second loops of sticky spiral on the TSP-IL distance for r_n, and on the change in TSP-IL distance (comparing r_n with r_{n-1}), gave significant positive values for both (Eberhard and Hesselberg 2012). The memory effect on the sticky spiral space (fig. 7.10b) was thus independent of the effect of the TSP distance mentioned in the previous section. Gotts and Vollrath (1992) suggested a similar memory effect for *Araneus diadematus* (AR), but did not demonstrate it quantitatively.

Of course, many variables are more or less correlated in the geometrically regular array of lines in an orb, so it is thus possible that the "memory" correlation was not due to a cause-and-effect relationship. Similar memory effects also occurred during temporary spiral construction (section 7.4). They have a smoothing effect on the paths of both sticky spiral and temporary spirals, and tend to reduce or round out irregularities in the path of the sticky spiral (Eberhard and Hesselberg 2012). It is important to note that the spider removed segments of the temporary spiral one by one as she built the sticky spiral and adjacent segments were not always removed on the same trip around the web, so the TSP distance thus sometimes varied abruptly on adjacent radii during normal orb construction.

As predicted by the TSP-IL distance hypothesis, the number of loops removed experimentally also had an effect in *L. mariana*, though it was small. In 75 experiments with mature *L. mariana* females in which the outer two loops of temporary spiral were broken during sticky spiral construction in a sector of three adjacent segments, the mean spaces on the two experimental radii of the next loop in this sector were increased by 100% and 106% over the spaces of the previous loop that had been built just before the experiment began. In contrast, breaking only a single loop of the temporary spiral had a somewhat less dramatic effect. When only one loop was broken in five adjacent segments, the sticky spiral spaces on the experimental radii increased by 98%, 72%, 90%, and 78%.

Another type of evidence that memories of previous TSP-IL distances influence sticky spiral construction has been exploited only qualitatively to date. Hingston observed (1920) that removing one or more loops of temporary spiral altered sticky spiral construction behavior by *Neoscona nautica* (AR), causing the spider to pause when she reached the altered portion of her web ("It shows some hesitation, but does not interrupt its work") (p. 128). Such pauses indicate that the spider sensed that the web was altered, and suggest that she had some sort of expectation regarding the lines that would be present. Given the other evidence implicating recent memories, it seems possible that the spider's expectation was due to memories of recent experiences with other segments of temporary spiral. It is also possible, however, that the pre-programmed instructions for sticky spiral construction include instructions to search in each new sector, and not proceed until she has performed a minimum search for the temporary spiral line, so further observations are needed.

7.3.2.4 Memory of less recent responses to changes in TSP distances

An experiment in which several segments of temporary spiral lines were removed (e.g., fig. 7.6) revealed a further variable that is probably also associated with memory that affected sticky spiral spacing. As just described, the first time a spider building sticky spiral encountered a site where the two outer loops of temporary spiral had been experimentally broken, she increased the space of the sticky spiral loop ("first experimental loop" in fig. 7.6). When she encountered the broken temporary spirals on her next trip around the web, however, her response was different. Even though there was still quite sharp change in TSP-IL distance in the experimental zone (e.g., fig. 7.6b), she nevertheless made much less substantial adjustments, and sometimes (in about 40% of the experiments) failed to make any adjustment at all ("second experimental loop" in fig. 7.6). Both *M. duodecimspinosa* and *L. mariana* showed this same pattern of responses (fig. 7.6). It was as if the spider remembered the site where she had made an adjustment previously, and reduced her response the second time around. In 29 experiments with *L. mariana* in which only one loop of temporary spiral was broken, this memory effect was evident only when the TSP-IL distances were especially large (fig. 7.11) (Eberhard and Hesselberg 2012).

The duration of this memory was much longer than the memory discussed in the previous section (in *M. duodecimspinosa*, it was on the order of 30–60 s as compared with 2–3 s). The function of these longer term adjustments, as with the adjustments involving the shorter term memories, may also be to smooth the outline of the inner loop, and thus promote overall regularity in sticky spiral spacing.

In *L. mariana* it appeared that these longer-term memories of sudden changes in the TSP distance were associated with a memory of where in the web the first encounter with altered TSP distances occurred. In "double" experiments, in which a second set of temporary spiral loops was cut while the spider was laying sticky spiral on the opposite side of the orb where the first set of temporary spiral segments had been cut, the spider usually responded to the second experimental sector with a large change in the sticky spiral place-

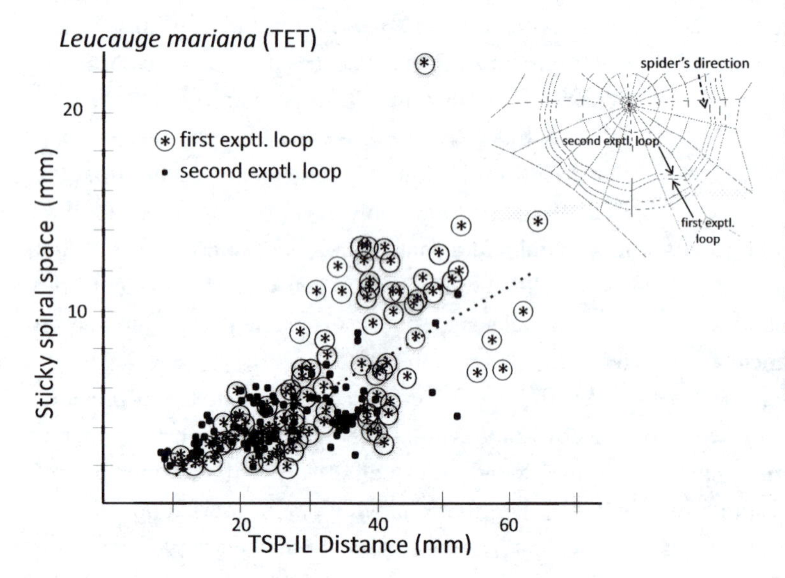

Fig. 7.11. When five segments of the outermost loop of the temporary spiral were broken experimentally in 29 webs of *Leucauge mariana* (TET) (upper right), the sticky spiral spacing of both the first and second loops in the experimental section increased. The increase was proportional to the TSP-IL distance in the first loop. The same relationship (dotted line) obtained in the second loop when this distance was not large (left portion of graph); but at larger TSP-IL distances (right portion of graph), the relationship was steeper in the first loop than in the second (after Eberhard and Hesselberg 2012).

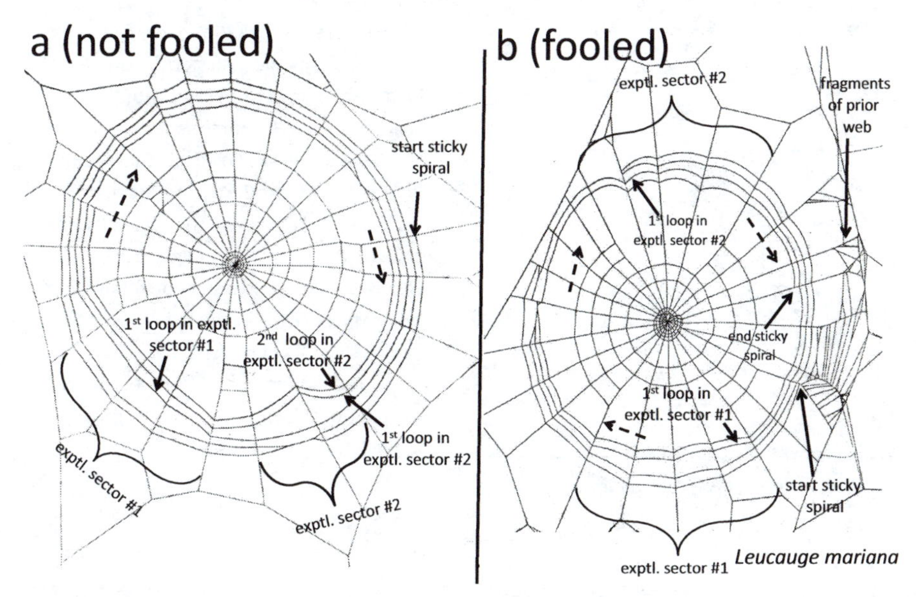

Fig. 7.12. These horizontal orbs of two mature female *Leucauge mariana* (TET) represent mixed success in attempts to "confuse" the spider by creating two experimental gaps in the temporary spiral. The typical "memory effect" (see fig. 7.6) occurred when two loops of temporary spiral were cut in one sector: the spacing of the first sticky spiral loop in the experimental sector was very large (the sticky line veered sharply inward), but the second experimental loop returned to near-normal values. In other words, the effect of the increased TSP-IL distance on sticky spiral placement was suppressed when the spider laid the second loop; the spider apparently "remembered" her first encounter with the experimental sector. I attempted to confuse the spider by removing temporary spiral lines in a second sector of the same web just after she had laid the first loop in the first experimental sector (lines were burned with a brief touch with a hot soldering iron, producing minimal vibrations). If she was fooled, I expected that she would treat her first encounter with the second experimental sector as if it were her second encounter with the first experimental sector, making little or no adjustment in the sticky spiral spacing. Typically, however, she was not fooled (*a*): the first loop in the second experimental sector veered sharply inward (dashed arrows indicate direction spider was circling the web). Occasionally, nevertheless, a spider was "fooled," and the first loop of sticky spiral in the second experimental sector did not veer inward (*b*). Both "fooled" and "not fooled" responses occurred when the experimental sectors were adjacent (as in *a*) and when they were on opposite sides of the web (as in *b*). The reasons for this variation are unknown.

ment (fig. 7.12) (WE). That is, rather than giving only the reduced response that would have been expected if she was remembering having encountered a sudden change in the distance to the temporary spiral and was confusing this second set of experimentally altered distances for the first, she treated this second experimental sector as if it were new.

Occasionally *L. mariana* appeared to be "fooled" in a double experiment, and failed to respond or only barely responded to the increased TSP-IL distance in the second experimental sector (i.e., she treated it as if it were her second arrival at the first experimental sector; fig. 7.12b). As would be expected if this "fooled" interpretation is correct, the frequency of such weak responses was lower when the discrimination of the second experimental sector was presumably easier (when the second sector was about 180° to the first) than when it was more difficult (about 270° to the first) (fig. 7.12): the frequencies of weak or non-existent responses were 2 of 23 for 180°, and 7 of 19 for 270° ($chi^2 = 4.9$, $p = 0.027$) (WE). The low level of significance and the semi-qualitative data make the conclusion that spiders knew the approximate locations of experimental sectors only tentative.

There are no data regarding the stimuli that spiders might have used to establish the locations of experimental sectors. These experiments were performed during the day in an outdoor screened cage, so use of environmental cues such as wind direction and possible visual stimuli was feasible. The spider might also have used path integration (e.g., sensing how far she had gone around the web, using cues from the distances and angles she had traveled). Further experiments under more controlled conditions are needed.

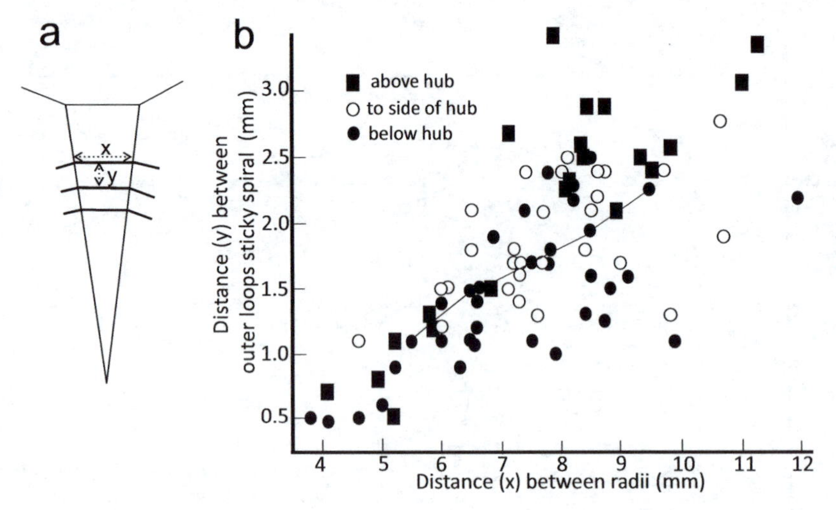

Fig. 7.13. The "sector ratio" pattern of Peters is illustrated by the data from 79 quadrants in 31 webs of 13 *Cyclosa caroli* (AR). The space between the outermost two loops of sticky spiral (*y*) was positively correlated with the length of the outermost segment sticky loop (*x*) (graph *b*). The spider did not perform behavior appropriate to allow her to sense *x* directly (she did not, for instance touch the two intersections of the outer loop with radii while laying the subsequent loop—see sections 6.3.5, 7.3.2, O6.3.6.1), so the relationship of *y* with *x* is presumably indirect (from Peters 1954).

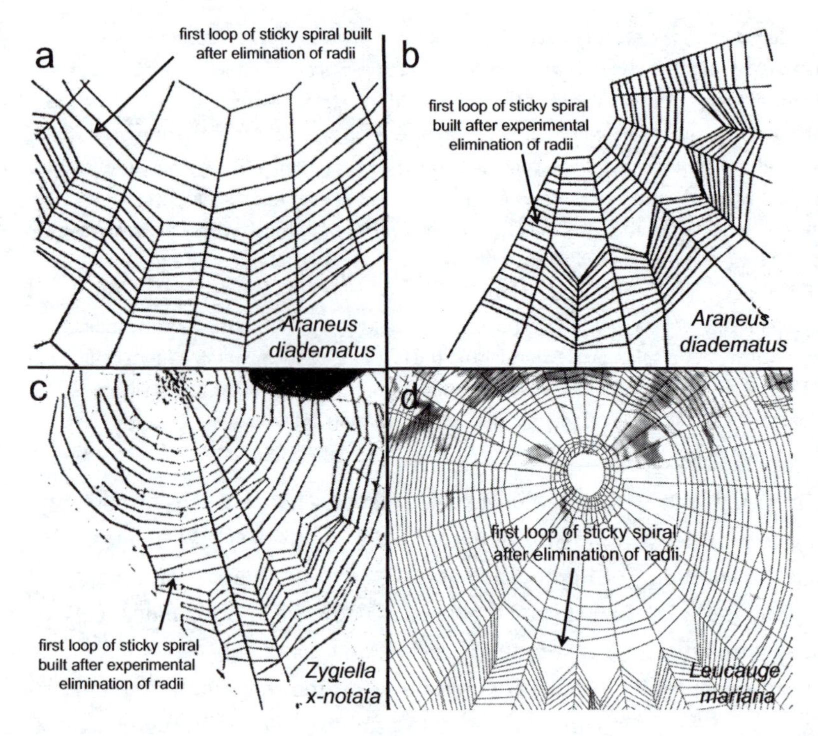

Fig. 7.14. These webs illustrate the inconsistent results of experiments that both favored and contradicted Peters' "sector ratio" hypothesis that sticky spiral spacing increases when adjacent radii are farther apart. When alternate *Araneus diadematus* (AR) radii were broken during sticky spiral construction, creating wider spaces between adjacent radii, the spacing between the next loops of sticky spiral sometimes increased abruptly, as predicted (*a*); but other times it was unchanged (*b*); both *Zygiella x-notata* (AR) (*c*) and *Leucauge mariana* (TET) (*d*) also sometimes responded with increased sticky spiral spacing. The reason for the inconsistency is unknown (*a* and *b* were traced from photos) (*a–c* from Peters 1939).

7.3.2.5 Distance between radii

There was a positive correlation between the space between the outermost two loops of sticky spiral (*x* in fig. 7.13a) and the length of the segment of the outermost loop of sticky spiral (*y* in fig. 7.13a) in the araneids *Araneus diadematus* and *Zygiella x-notata* (Peters 1939), and several other araneids including *Cyclosa caroli* [AR] (fig. 7.13) (Peters 1953, 1954). Peters (1939) posed the question of whether this correlation reflected a cause-effect relationship, or was instead an incidental consequence of other correlations due to the geometric regularity of the orb, by experimentally increasing the distances between radii in webs of *A. diadematus* and *Z. x-notata.* He broke alternate radii both near the inner loop of sticky spiral, and also farther inward toward the hub (he thus also broke the segments of radial lines between the loops of temporary spiral) (see fig. 7.14a–c) while the spider was building the sticky spiral. As predicted, the spaces between the loops of sticky spiral increased abruptly in some webs when the spider encountered the experimental sector where the distances between radii were larger (fig. 7.14). Puzzlingly, however, this response was inconsistent: it occurred clearly in only 2 of 7 experiments with *A. diadematus* (fig. 7.14b,c), and 4 of 5 with *Z. x-notata*. Nevertheless, the clear patterns in some webs, and the fact that they also occurred in similar experiments with some *Leucauge mariana* (TET) (fig. 7.14d), leave little doubt that there was some sort of effect. Further work is needed to explain this variation.

The sensory cues that were responsible for the radius distance effect are not known. The spider was unlikely to be able to sense the distance *y* directly (fig. 7.13) from only touching the inner loop once (section 6.3.5, appendix O6.2.5), but it would have been feasible for her to sense the inter-radial distance in several other ways (Eberhard 1972a).

7.3.2.6 Lack of influence of radius tension

Experiments have consistently indicated that radial tensions do not affect sticky spiral spacing. When one or more adjacent radii were broken while *Uloborus diversus* (UL), *Zosis geniculata* (UL), *Leucauge mariana* (TET), and *Micrathena duodecimspinosa* (AR) were building sticky spiral, the spacing between attachments of the sticky spiral on the broken radii did not differ from that on the adjacent, undamaged radii (fig. 7.15, 7.16) (Eberhard 1972a; WE). Conversely, the dramatic response in the Hingston experiments (fig. 7.3) was produced despite very little change in tensions on lines in the orb (deduced from the essentially unchanged angles between lines).

Behavioral observations of undisturbed spiders also suggest that tensions are not important cues. Sometimes a *M. duodecimspinosa* interrupted sticky spiral construction part way through, went to the hub and removed the center of the hub (and thus reduced the tensions on radial lines throughout the web); then, after a short rest, she resumed sticky spiral construction, with no perceptible before-vs.-after effect on sticky spiral spacing (WE). It is not surprising that radial tensions do not guide sticky spiral spacing because they must be altered substantially by the spider's weight, and the alterations must vary greatly according to the spider's location in her web

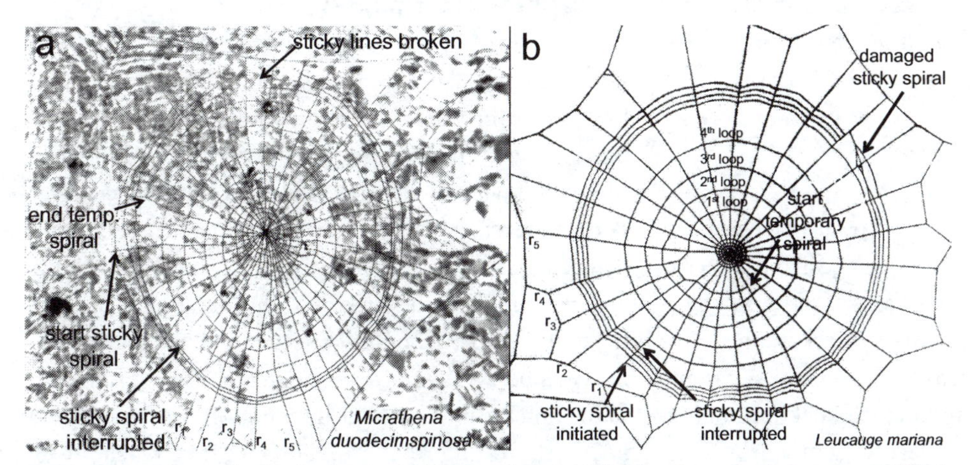

Fig. 7.15. The responses (and lack of responses) in experiments with *Micrathena duodecimspinosa* (AR) (*a*) and *Leucauge mariana* (TET) (*b*) suggest that recent memories of distances traveled during temporary spiral construction but not radial tensions affect the spacing between loops. In both experiments three adjacent radii (r_2, r_3, and r_4) were broken near the hub (just after the first loop of sticky spiral was built in *a*, during construction of the first loop of temporary spiral in *b*); each web was photographed just after the spider had completed 3–4 loops of sticky spiral. The spaces between the two experimental loops of sticky spiral on the broken and thus less tense radii (r_2–r_4) did not differ from those on the nearby intact radii (r_1, r_5). In *b*, the tensions on the r_2–r_4, which analyses of the angles in the web indicated were reduced by about 50%, pulled the first loop of temporary spiral outward. When the spider reached this sector while building the next (second) loop of temporary spiral, she reduced the temporary spiral spacing; the spaces on r_2–r_4 were less than those on the intact radii (r_1, r_5) she encountered immediately before and after. The change in the temporary spiral spacing was gradual: the space on r_3 was less than that on r_2; the space on r_4 was less than that on r_3. In the subsequent (third and fourth) loops of the temporary spiral, however, the spaces on the experimental radii were similar to those on control radii, indicating that the lowered radius tension did not affect temporary spiral spacing (from Eberhard 1988c).

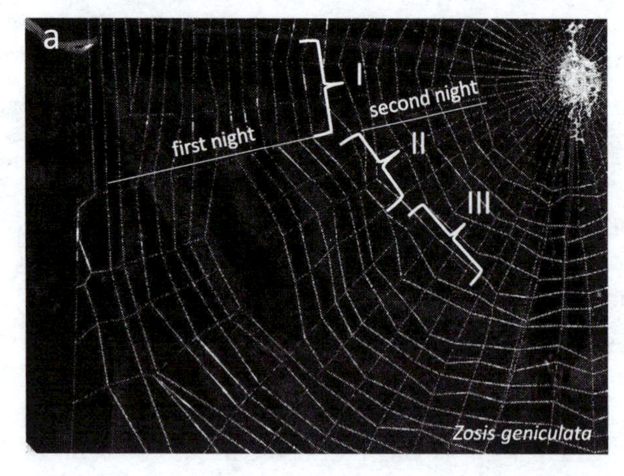

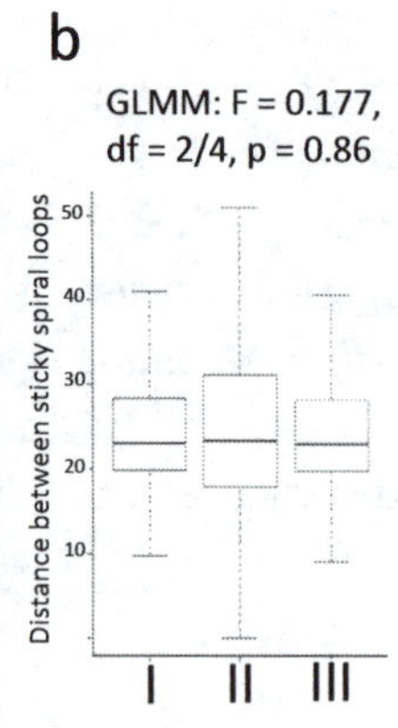

Fig. 7.16. An experiment in a two-night web of a mature female *Zosis geniculata* (UL) (*a*), in which three radii were cut experimentally following the first night and the spider then laid additional loops of sticky spiral on the second night, suggested that radius tension does not affect sticky spiral spacing. Sticky spiral spacing on the second night (*b*) was not different on the less tense, broken radii in zone II than on the adjacent intact radii in zones I and III (from Eberhard and Barrantes 2015).

(see fig. 6.24). For instance, tension on the portion of the radial line where she attaches the sticky spiral will seemingly be substantially increased as a result of her weight when she is above the hub, but will be reduced when she is below the hub.

The lack of effect of radial tensions is one of the justifications (in combination with the inner loop localization behavior of the spider) for the interpretations above that spiders use cues that are based on the physical locations of lines rather than other possible cues such as vibrations, resonant frequencies, etc. These other cues would be strongly affected by changes in radial tensions.

7.3.2.7 Mistakes in discriminating sticky from non-sticky lines?

Some aberrant patterns of sticky spiral placement in *Micrathena duodecimspinosa* (AR) suggested that occasionally the distinction between sticky and non-sticky lines was either ignored or not made successfully. A spider sometimes failed to break one or more segments of temporary spiral while building the sticky spiral (fig. 7.26). In effect, these "errors" left "experimental" non-sticky lines in the web that the spider encountered subsequently when she was building the next loop of sticky spiral. In at least some cases, the spider apparently treated the intact temporary spiral lines as if they were the inner loop of sticky spiral, and used them as reference points in attaching the next sticky spiral loop at a distance similar to the spaces between nearby sticky lines (fig. 7.26).

Perhaps this tendency of *M. duodecimspinosa* to occasionally fail to distinguish between non-sticky and sticky spiral lines resulted from this species using the derived (and perhaps less perfected) inner loop localization technique that employed leg oIV rather than the usual oI, but this is just speculation. Another genus that has evolved to use leg oIV to locate the inner loop instead of oI, *Nephila*, also routinely used both sticky and non-sticky temporary spiral lines to guide sticky spiral construction (appendix O6.3.6.3).

7.3.3 INTERMEDIATE, MORE SLOWLY CHANGING CUES

7.3.3.1 Angle of the radius with gravity

For an animal like a spider that literally spends her life hanging from threads, the direction of gravity seems likely to be an important cue. The ability to distinguish up from down, and to use this as an orientation cue, was demonstrated experimentally in the araneid genera *Araneus*, *Zygiella*, and *Argiope*: when the frame in which the web was built was rotated 180° during attacks on prey (so that the portion of the web that had been above the hub was below the hub), the spider attempted to return to the hub by moving in the direction where the hub had been prior to the rotation (Peters 1937; Le Guelte 1966; Ades 1991). Similar 180° rotation experiments with *Zygiella x-notata* during sticky spiral construction took advantage of the typically larger spaces between loops of sticky spiral above the hub than below it: the sticky spirals that were built below the hub after rotation were more closely spaced than those above it (Le Guelte 1966). Reorienting the orb of *Araneus diadematus* (AR) during sticky spiral construction, so that it was horizontal rather than vertical, strongly reduced the normal difference in spacing above and below the hub (Zschokke 2011), and resulted in slightly decreased regularity in sticky spiral spacing (Vollrath 1986). The up-down sticky spiral asymmetry of a typical vertical orb of *A. diadematus* was lost when, after the radii were finished, the web was laid horizontal (Vollrath 1992). These experiments showed both that gravity was a cue responsible for up-down differences in sticky spiral spacing, and that several other aspects of these orbs (e.g., angles between radii, radius length) that differed between the upper and lower parts of an orb (Le Guelte 1966) were not. These changes in spacing may have been associated with changes in leg positions at the moment the sticky spiral was attached; in *Micrathena duodecimspinosa* (AR) leg oIV tended to grasp the radius farther from the inner loop just before attaching in the upper portion of their more or less vertical webs than in the lower portion (fig. 6.27e); and when this distance was larger, the spaces between sticky spiral loops tended to be larger (fig. 6.27f).

Clever experiments by Vollrath with *A. diadematus* revealed an additional, unexpected complexity: the spider appeared to "expect" that her angle with gravity would change predictably as she circled while building the sticky spiral in certain parts of her orb. Vollrath (1986) rotated the rectangular frame in which the spider was building her orb in a vertical klinostat in total darkness when the spider was midway through sticky spiral construction; he used a set of pulleys that allowed him to produce accurate and vibration-free rotation by hand on an axis perpendicular to the center of the web's hub and moved the frame just enough to hold the spider treadmill-fashion at a certain location in space. When the spider was caused to remain moving horizontally (directly above or directly below the hub), she soon spiraled sharply inward toward the hub after transversing only 180–270° (fig. 7.17). In contrast, when she was held moving vertically (up or down), she did not spiral inward (although the variation in spaces between spiral loops increased, and she often failed to remove the temporary spiral). These treadmill experiments revealed an otherwise cryptic, internal mental process. They suggested that the spider "expected" that when she moved horizontally from radius to radius her orientation with respect to gravity would change; when these changes did not occur, she veered toward the orientation that she was expecting.

In additional experiments with the klinostat, Vollrath rotated the frame with the spider in it at different speeds around an axis that was perpendicular to the plane of the orb while the spider was engaged in producing the sticky spiral. At intermediate rates of rotation (2.3–30 rpm), the spider became disoriented and wandered on the web and left many temporary spiral lines intact (fig. 7.17c–e). In contrast, the spider's behavior was less affected by very slow rotation (0.3–0.7 rpm) (fig. 7.17b) and by very fast rotation (100–150 rpm) (fig. 7.17f) (Vollrath 1988a). When the frame was rotated on an axis parallel to the plane of the orb, there was no effect on sticky spiral construction.

Vollrath (1986, 1988a) proposed several different explanations for the different responses. The lack of an effect on sticky spiral construction at very high rotation speeds could have been due to the spider's perception of the effects of centrifugal force replacing the usual effects of gravity. The ability to use the temporary spiral as a "hand rail" to guide the sticky spiral (see section 3.3.2.8) may have also helped to allow the spiders to correct for moderately high rates of rotation (60 rpm). The lack of an effect at very slow rotation speeds such as 0.3 and 0.7 rpm could have been due to the changes in the direction of apparent gravity being slow enough that the spider could "adjust its internal representation accordingly" (Vollrath 1988a). The lack of effects when webs were rotated on an axis parallel to the plane of the orb (and also when webs built in outer space, free from all of the effects of gravity—Witt et al. 1977) could also have been due to use of the "hand rail" that was provided by the temporary spiral (Vollrath 1986; Zschokke 1993; section 7.3.2.2).

To explain the strong effects of intermediate rates of rotation (and of the treadmill experiment), Vollrath (1988, p. 418) speculated that "the main factor responsible for the disorientation of the capture [sticky] spiral is the animal's confusion about its own position with respect to landmarks of the web (e.g., the hub)."

In the most esoteric experiment of all, in which *A. diadematus* built orbs in the nearly complete absence of gravity in a space capsule, the spider produced apparently normal, uniformly spaced sticky spirals (Witt et al. 1977). These results are not easily compared with the klinostat experiments, in all of which the spider experienced a pull (from gravity or otherwise). There were hints of learning, as the precision of the spider's leg movements improved over time under weightless conditions (Witt et al. 1977).

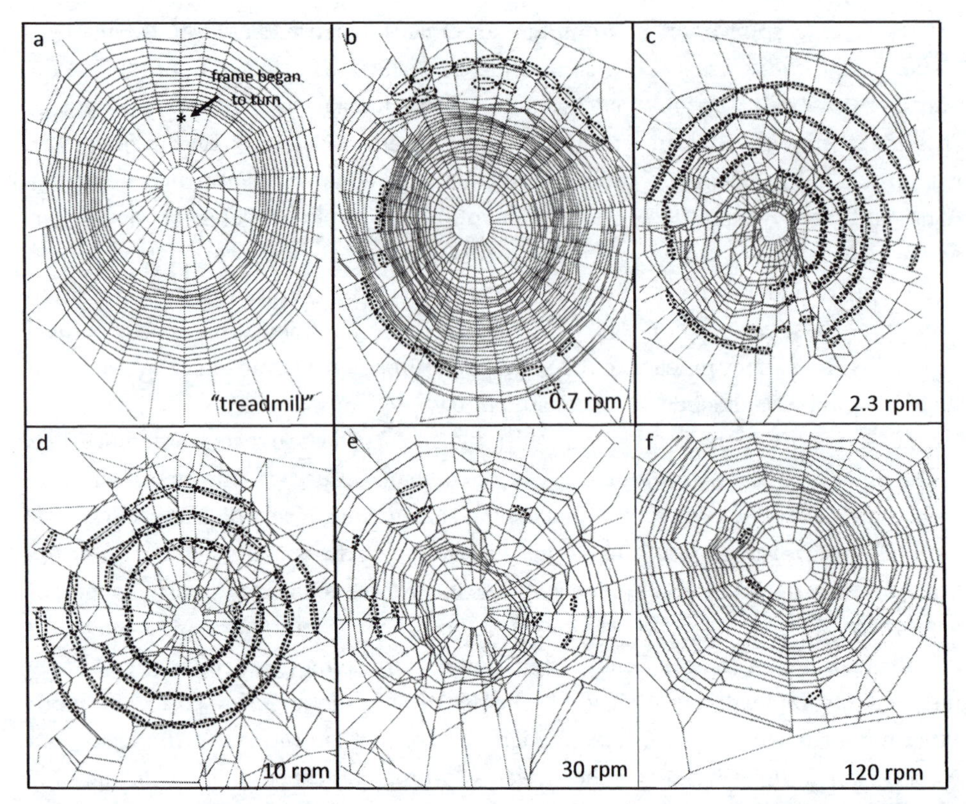

Fig. 7.17. Experimentally mixing gravity and centrifugal forces by rotating the vertical frame in which the spider was building the sticky spiral caused major alterations in both sticky spiral placement and temporary spiral removal in the orbs of *A. diadematus* (AR) (sticky spiral lines are drawn thicker than non-sticky lines; temporary spiral lines that persisted in finished webs are circled with dotted lines). A "treadmill" treatment (*a*), which began while the spider was building the 15th loop in the upper sector and was moving horizontally at about 12:00, produced a severe disorientation. The vertical frame was rotated slowly to the right, so that the spider remained in approximately the same spot as she moved. After the frame had rotated just under 180°, the spider veered inward and back out, and then (about 30° later) she veered sharply inward (starting at about the 5:00 sector in the figure) and continued until she reached the hub and the experiment was concluded. In the webs in *b–f*, the cage rotated continuously at different speeds starting with initiation of the sticky spiral. Sticky spiral construction and temporary spiral removal were relatively unaffected at very slow (*b*) and very fast (*f*) speeds, but were severely disrupted at intermediate velocities (*a* drawn after Vollrath 1986; others drawn after Vollrath 1988a) (the hubs were not drawn because many lines there were unclear).

7.3.3.2 Amount of silk available vs. web area

Several types of evidence indicate that orb weavers modulate sticky spiral spacing (as well as other aspects of their orb designs) on the basis of the relationship between the amount of silk they have available in their silk glands and the area of the web to be covered with sticky silk. Web area may be sensed early, during the exploration stage of construction; the careful analysis of Hesselberg (2015) indicated that orb weavers construct at least rudimentary cognitive maps of where they have walked. It appears that the amount of sticky silk in reserve in their glands had a greater effect than the amount of dry silk. To evaluate the evidence presented below for these claims, keep in mind that both the ampullate and the aggregate glands are relatively full at the beginning of orb construction, and more or less empty at the end (Witt et al. 1968). The contents of at least the ampullate gland begin to be replenished very soon after the web is finished, as the gland cells dump the contents of cytoplasmic vesicles containing silk that had been synthesized previously into the lumen.

Field experiments with *Leucauge mariana* (TET) and *Micrathena sexspinosa* (AR) documented the probable effects of silk supplies on orb design (fig. 7.18, Eberhard 1988a). The silk supplies that spiders had available in different glands when they began an orb were altered experimentally by causing spiders to replace orbs that they had built a few hours earlier the same morning (the

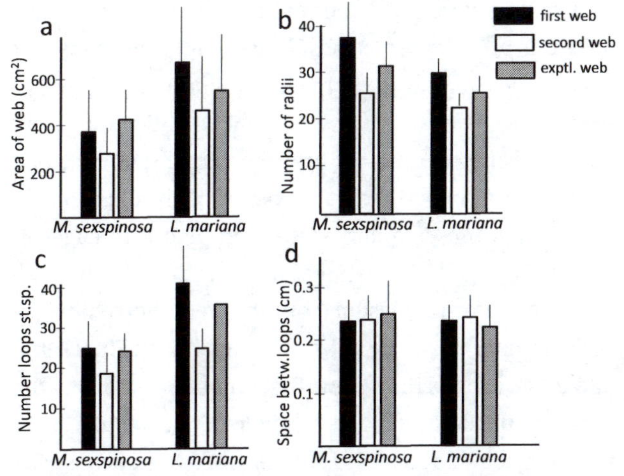

Fig. 7.18. Experimental manipulations of the reserves of sticky silk in the glands of the araneid *Micrathena sexspinosa* (AR) and the tetragnathid *Leucauge mariana* (TET) in the field induced changes in web designs that suggested that these orb-weavers can estimate the amount of sticky spiral silk available in their glands, and that they alter decisions made early in web construction such as the overall web area and the number of radii in the web accordingly. Both species normally built their orbs around dawn (black bars in the graphs). When this web was destroyed about 8–9:00 in the morning, spiders often replaced it 2–3 hrs later (white bars). These control second webs had significantly smaller areas (*a*), lower numbers of radii (*b*), and fewer loops of sticky spiral (*c*) than first webs. The hypothesis that spiders made these changes in response to having lower reserves of silk in their glands was supported by experimental second webs (gray bars) built after the first web was destroyed just when the spider was beginning to build the sticky spiral, presumably leaving her with undepleted sticky spiral silk reserves in her glands (aggregate and flagelliform glands). In both species, the experimental second webs were larger, had more radii, and more loops of sticky spiral than did the control second webs. The mean distance between loops of sticky spiral, in contrast, was unaltered (from Eberhard 1988a).

frame lines and three radii of the first web were left intact). Assuming at least approximately similar rates and details of synthesis to those observed in *A. diadematus*, the reserves of ampullate gland silk that spiders had when they started the second webs were depleted but not exhausted, and they were more severely depleted when the spider was induced to build a third web by destroying the second one. The third webs of *L. mariana* accentuated the differences in second webs in having sharply larger spaces between loops of sticky spiral, as well as fewer radii, fewer sticky spiral loops, and a smaller fraction of the web area that was covered with sticky spiral. These experiments held several factors constant, including spider size, area available for the web, and the effect of web destruction per se.

The relative amounts of sticky as compared with non-sticky silk that the spider had available were also manipulated experimentally. Some first webs were destroyed soon after they were complete ("control" treatment), when all three types of glands were presumed to be relatively empty. Other webs were destroyed just before the sticky spiral was initiated ("experimental" treatment), so that the aggregate and flagelliform glands (which produce sticky lines) were relatively full, but the ampullate glands were depleted. The resulting experimental webs differed from control second webs in not having a reduced area compared with the first web. They had fewer radii and, in *L. mariana*, fewer sticky spiral loops than first webs, but the reduction in both species was significantly less than that in control second webs. The fraction of the web area that was covered with sticky spiral was not reduced, as occurred in the control second webs. Similar results were obtained when *Araneus diadematus* (AR) built four rather than the usual one web in the space of 24 hr: the webs became sequentially smaller and had fewer radii; the sticky spiral was shorter; and the loops were spaced farther apart (Vollrath et al. 1997). In sum, the amounts of silk available in both dragline and sticky silk glands affected web design, and the amount of sticky spiral silk had particularly strong effects.

Lab experiments with the araneid *Araneus diadematus*, in which ingestion of previous webs was either allowed or prevented, gave further evidence that gland contents influenced several design traits, including web size, the number of radii, and the number of sticky spiral loops (Breed et al. 1964; Zschokke 1997). They also revealed additional details. When the spider was prevented from ingesting her web and it was destroyed, the length of time elapsed since the previous web was completed correlated positively with the web area and sticky spiral length (but not spacing) of the second orb up until 20 hr later (perhaps by this time the silk glands were completely filled again?) (Zschokke 1997). In addition, there was a positive correlation between feeding on previous webs and the diameters of threads; these design changes were sequential (the first webs built during the experiment were larger; webs built after longer pauses also had thicker lines).

There was also evidence that *Cyclosa turbinata* (AR)

spiders "assessed their silk resources before they initiated web construction and altered their behavior" to make even more subtle adjustments in silk lines in accord with their silk reserves (Crews and Opell 2006; p. 427). When deprived of food, the diameters of the droplets of glue on their sticky spiral lines decreased gradually in successive webs. The diameters of the droplets did not change from the outer to the inner edge of a given orb, nor did the spaces between loops of sticky spiral.

A remarkably similar association between the contents of entirely different glands and orb design was suggested by observations of unmanipulated webs of *Zosis geniculata* (UL) and *Uloborus diversus* (UL). These data were from spontaneous rather than experimental interruptions of orb construction. Spiders sometimes built an "interrupted web," which had a complete array of non-sticky lines (hub, radii, frames, etc.), but in which only the outer portion was filled in with sticky spiral on the first night; the following night the spider finished the web, filling in the inner portion of the same orb (without adding any further non-sticky lines). The spaces between the first loops of sticky spiral that were laid on the second night, when the contents of the cribellum and pseudo-flagelliform glands that produced the sticky spiral had presumably been replenished, were sharply smaller than those between the last loops laid the previous night (figs. 7.19, 7.20) (Eberhard and Barrantes 2015; WE). A similar but even greater effect occurred when sticky spiral construction by a mature female *Argiope argentata* (AR) was interrupted by a brief rain; the distances between loops were much smaller when the spider resumed construction after the rain ceased (fig. 7.19b).

The hypothesis that *Z. geniculata* spiders can assess the space to be covered and their silk reserves gained further support from the fact that interrupted webs were common when the spiders built in large (50 cm diam) containers (31 of about 90 webs), but never occurred among the several hundred webs that were built by the same individuals in smaller (14.8, 7.8, and 6.5 cm diam) containers (WE; Eberhard and Barrantes 2015). Apparently the spiders always had enough sticky spiral silk to fill the smaller but not the larger webs, and interruptions in larger webs may have been triggered by the lack of sufficient silk. Further work on these surprisingly sophisticated adjustments could be rewarding.

A second type of adjustment in web design was produced as an unexpected result of an attempt to test for the effects of learning on orb construction in *Argiope aurantia* (AR). Reed et al. (1970) produced individuals lacking experience in building orbs by rearing them in small containers in which there was not enough room to spin orbs. When these spiders were then placed in cages in which they could build orbs, they produced orbs with apparently normal designs, but the webs were substantially smaller than the orbs of control individuals that had made orbs as they grew. The ampullate glands of the inexperienced spiders were smaller than the glands of the control individuals. Presumably this was because silk glands (like many other animal tissues) atrophy when they are less active (Witt et al. 1968). These results imply that the spider was able to sense not only the relative degree of filling of her gland (which could, for example, be sensed by stretch receptors in the wall of the gland), but also the relative sizes of these glands themselves.

A final technique to manipulate the relation between the amount of silk in the spider's glands and the space to be filled with sticky spiral was to reduce the available space. Both *Leucauge argyra* (TET) (Barrantes and Eberhard 2012) and *Zosis geniculata* (UL) (Eberhard and Barrantes 2015) gave further evidence of the importance of silk reserves compared with the space to be covered. When spiders were enclosed in tiny spaces as little as about one tenth of the diameter of the smallest spaces they use in nature, the spaces between sticky spiral loops were much smaller than in control orbs in larger containers or in the field.

These data refer to settings-like effects of the gland contents early in orb construction. It remains less clear whether spiders also sensed and responded to the amount of silk in their glands later during the process of orb construction, as the contents of first the ampullate and then the aggregate and flagelliform glands gradually became depleted. Three sets of data give tentative support for this "slowly changing cues" possibility. Reed (1969) attributed the eventual termination of radius construction and the transition to temporary spiral construction when he experimentally removed radii as soon as they were built (fig. 7.37) to the spider sensing that she was running out of ampullate gland silk (Reed 1969); but there are other possible explanations (for instance, memories of previous behavior), so this evidence is not

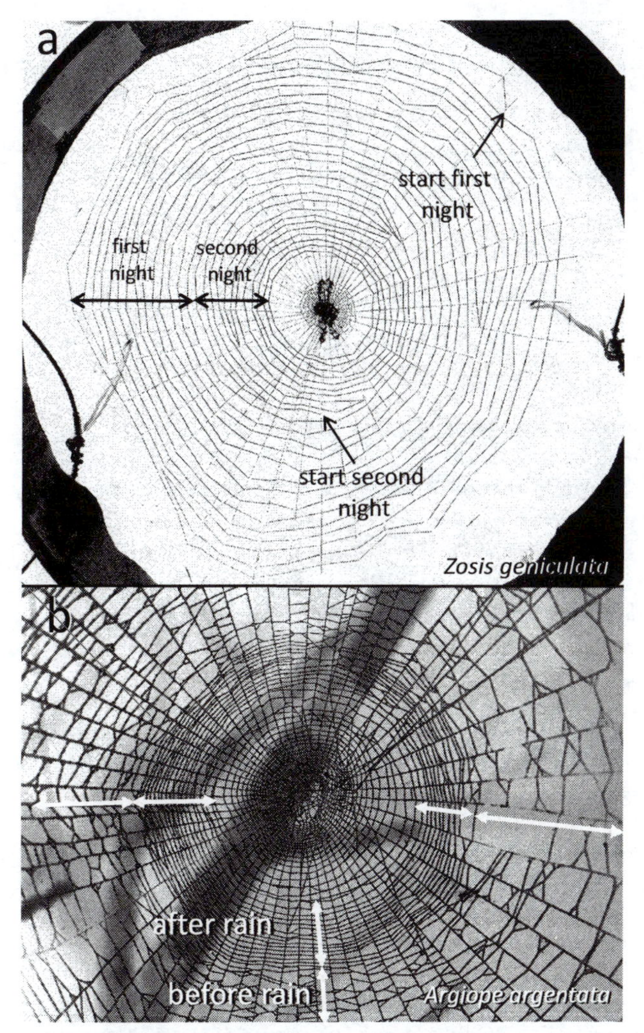

Fig. 7.19. A "two night" web of a mature female *Zosis geniculata* (UL) (*a*) had sharply smaller spaces between the loops of sticky spiral that were built on the second night, supporting the hypothesis that the spiders tended to reduce the spaces between sticky spiral loops when they had more sticky spiral silk in their sticky silk glands (cribellum and pseudoflagelliform glands) compared with the area to be covered with sticky lines. The spider built all radii, hub, frames, and the stabilimentum on the first night, laid the sticky spiral loops labeled "first night," then went to the hub. On the second night, she filled in the rest of the central portion of the web with sticky silk. The spaces between loops of sticky spiral were significantly smaller on the second night (see also fig. 7.20). Interruption of sticky silk production may be triggered when the spider does not have enough sticky silk to cover the area of the web to be filled, because two-night webs of this type were built only in large containers such as the 50 cm diameter container in this photo, and never in smaller 5.8–14.8 cm diameter containers. A similar abrupt reduction in sticky spiral spacing occurred in an unusual field web of *Argiope argentata* (AR) (*b*), in which the outer portion of the sticky spiral was built early in the morning, probably at or prior to dawn, and then damaged by a brief rain. Later in the morning the spider added more, very closely spaced loops of sticky spiral in the small open space between the hub and the inner loop of the previous sticky spiral (double headed arrows span the zones with early and late sticky spiral loops in the photo). The silk gland cells may have began dumping secretion into the gland lumen only minutes after the gland was emptied and commenced further synthesis within 1–2 hours after the web was finished, as occurs in the major ampullates of *Araneus diadematus* (the only glands that have been studied in this respect—Witt et al. 1968), so the abrupt reduction in spacing in this web may have been due to the spider having a large amount silk available to fill the small open space that remained in the center of her orb.

conclusive. In some orbs of the uloborids *Uloborus diversus* and *Zosis geniculata* in which the sticky spiral was interrupted (above), the spaces between the inner loops of sticky spiral increased sharply just before the spider interrupted the sticky spiral, as if she perceived that she was running low on sticky silk and widened them in an attempt to reach the hub (figs. 7.19, 7.21).

A third set of data supporting the "slowly changing cue" from web area hypothesis concerns the way that *Zosis geniculata* and *Uloborus diversus* (UL) changed their use of an otherwise usually dominant cue guiding sticky spiral placement (Eberhard 2019b). When a spider laying sticky spiral found herself "confined" to one part of her orb and excluded from another portion that lacked sticky spiral lines (e.g., after laying the line from *b* to *d* in fig. 7.2, she was excluded from the right half of the orb), she generally altered her behavior: she chose to ignore the usual cue from the inner loop of sticky spiral, thereby gaining access to the unfilled portion (see also fig. 7.45). The hypothesis advanced to explain this change in the cues used to guide behavior (discussed in more detail in Eberhard 2019b) supposes that the spider has a sense of the sizes of portions of her orb that already have sticky lines and those that lack them.

In sum, numerous observations indicate that spiders alter sticky spiral construction, along with several other aspects of web design, in accord with information on their supplies of silk and the area to be covered. The spider's ability to "plan ahead," estimating the demands for different types of silk against her reserves in different glands (both in terms of the absolute size of the gland relative to her body, and the degree to which it is filled), are surprising. These "plans" do not necessarily imply any sort of complex cognitive analysis. They suggest, however, an ability to judge the size of the area

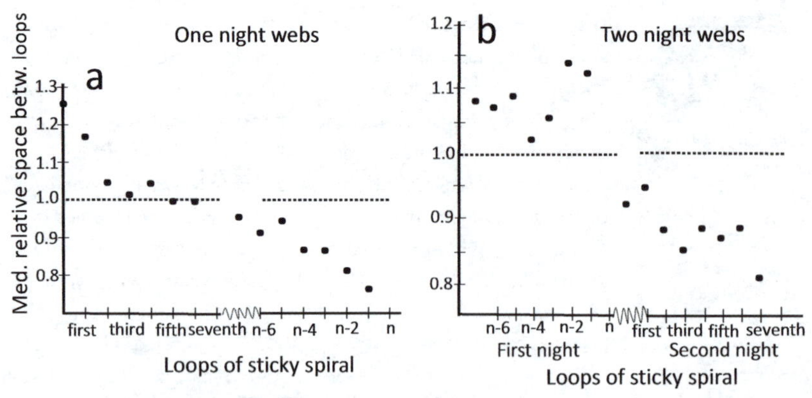

Fig. 7.20. The changes in the spaces between loops of sticky spiral in one- vs. two-night webs of mature female *Zosis geniculata* (UL) (see fig. 7.19) suggest that the amount of silk available in a spider's glands affects her decisions regarding spacing between sticky spiral lines (the spaces were standardized by dividing each measurement by the median space on that radius). In webs built in a single night (the usual case), the spaces between loops decreased gradually as the spider moved from the outer edge toward the hub (*a*). In 31 two-night webs (*b*), in which the hub and all of the radii, frames, and temporary spiral but only part of the sticky spiral were built on the first night, the rest of the sticky spiral was added the second night. In general, the spaces between sticky spiral loops also decreased in these webs from edge to hub; but the last two spaces that were built on the first night, just before the spider suspended sticky spiral construction, increased sharply. When the spider began on the second night (when she had presumably replenished the silk in her glands), the first two loops (built immediately adjacent to these lines, where the spider presumably sensed similar stimuli from the web) were sharply smaller (after Eberhard and Barrantes 2015).

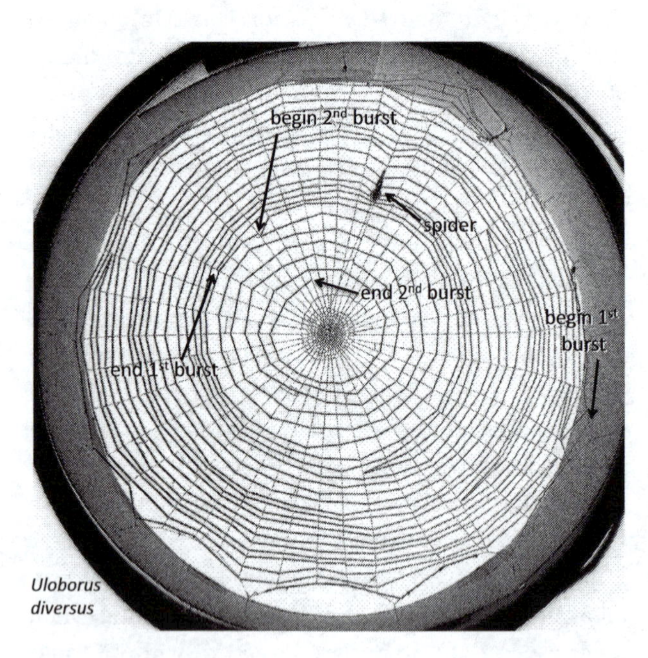

Fig. 7.21. Running low on silk? This mature female *Uloborus diversus* (UL) filled her web with sticky spiral on a single night. She laid most of the sticky silk during a first burst of building, but the spacing became erratic during the last 10 loops of this burst: first the spacing increased substantially for 3–4 loops, and then decreased in 4–5 more tightly spaced loops just before it ended ("end 1st burst"). Later she performed a second burst of construction, laying 5–6 more, widely spaced loops. The wide spaces between these final loops suggest that she may have been running low on silk, and altered sticky spiral spacing in order to be able to cover the remaining space before running out. Whether or not this speculation is correct, webs like this one, with abrupt changes in sticky spiral spacing, demonstrate the effects of other factors in addition to the geometric relations such as the distances between the radii, which change gradually rather than abruptly. Specifically, the abrupt changes argue against Peters' sector ratio hypothesis (fig. 7.14).

to be filled with sticky lines (perhaps involving a cognitive map of the area—see Hesselberg 2015), and several elaborate, finely tuned, and to some extent mutually independent "rules" by which several behavior patterns, including those determining the sites at which sticky spiral lines are attached, are modulated on the basis of information from the silk glands. Presumably these modifications take best advantage of the resources that the spider has available, but the selective advantages of the different adjustments have yet to be explored.

7.3.3.3 Distance from the hub (?)

It has long been known that in *Z. x-notata* the spaces between loops of sticky spiral are different near the edge of the orb as compared with nearer the hub (Peters 1939, 1954; Le Guelte 1966). Different spacings at different distances from the hub also occur in many other species of orb weavers, though the patterns of change vary in different species (section 4.3.2). Probably the most common pattern is for sticky spiral spacing to be larger in the outer portion of the orb and gradually decrease closer to the hub (e.g., figs. 1.4, 3.6). Presumably these changes are induced by stimuli that change gradually as the spider spirals inward from the outer to the inner portion of her orb. Among the cues mentioned above, two possibilities are decreasing distances between radii, and decreasing reserves of sticky silk. Whether one or both, or still another cue is responsible for the edge-to-hub patterns is as yet undetermined.

7.3.4 MORE OR LESS CONSTANT GENERAL SETTINGS

7.3.4.1 Length of the spider's legs

Direct observations of inner loop localization behavior during sticky spiral construction in many species (e.g., Jacobi-Kleemann 1953; Eberhard 1982, section 6.3.5, appendices O6.1.1.8, O6.36), and the disruptive effect of experimental damage to leg I on sticky spiral patterns in finished orbs in species like *Neoscona nautica* (AR) (Hingston 1920) and *Araneus diadematus* (AR) (Reed et al. 1965), leave no doubt that spiders use their legs to guide sticky spiral placemement. It thus seems likely that the sizes of these legs affect sticky spiral placement.

Nevertheless, attempts to understand web designs in terms of the sizes of the spider's own body parts have a checkered history of failure and success. Hingston (1920) speculated early that the spider might use the number of steps from the temporary spiral in the first loop, and later her own legs I as cues to decide where to attach the sticky spiral ("On its body it carries its organs of measurement—the number of its paces, the length of its body, the divergence of its limbs") (p. 106). Subsequent studies were not all in accord with these ideas. Some intra-specific comparisons favored this leg length hypothesis (Witt and Baum 1960; Risch 1977), but in other species such as *L. mariana* the relation was inverse, with larger adult female spiders making smaller spaces (Eberhard 1988a). In *Larinioides* (= *Nuctenea*) *sclopetaria* there was no significant relation between leg length (or any other measure of body size) and sticky spiral spacing in adults, though in juveniles (of unspecified instars) all the measures of body size correlated positively with the spaces between sticky spiral loops (Heiling and Herberstein 1998). Comparative data also showed clearly that a simple version of this hypothesis could not explain inter-specific variations. For instance, there are very large species (e.g., *Nephila*) that space their sticky spirals relatively close together; and there are small species that space them relatively far apart (e.g., *Theridiosoma*) (Peters 1954, 1955).

These comparisons were based on correlations that did not distinguish cause and effect relations from incidental correlations. It was not until the experiments of Vollrath (1987a) that experimental data directly favoring Hingston's leg length hypothesis for intra-specific variation became available. Vollrath took advantage of the fact that when an immature *A. diadematus* loses a leg, this leg is regenerated at the next molt, but the regenerated leg is shorter than the undamaged leg on the other side. He was thus able to create spiders that had a longer leg I on one side than the other. Although he did not observe these spiders building, he found that, as predicted by Hingston's leg length hypothesis, the spaces between loops of sticky spirals in finished webs built by experimental individuals showed a bimodal distribution (fig. 7.22). This experiment has not to my knowledge been repeated with any other species.

7.3.4.2 Previous prey (escaped or captured)

7.3.4.2.1 GENERAL RESPONSES TO PREY

It is well-established that the quantity and the quality of the prey captured in the recent past can affect the frequency of construction, the orb design, and silk composition (e.g., Witt et al. 1968; Sherman 1994; Vollrath and Samu 1997; Tso 1999; Herberstein et al. 2000a; Venner et al. 2000; Watanabe 2001; Tso et al. 2005; Blamires 2010; see the extensive summary of Herberstein and Tso 2011). Thus, for instance, changes in the nutrient content of *D. melanogaster* prey that were reared on different larval media correlated with changes in both sticky spiral spacing and the area covered by the sticky spiral in the araneid *Zygiella x-notata* (Mayntz et al. 2009). These were generalized responses that resulted from the nutrients available from which to synthesize silk, rather than from specific modifications of orb design that were selectively favored because they produced web designs that were especially appropriate for a particular type of

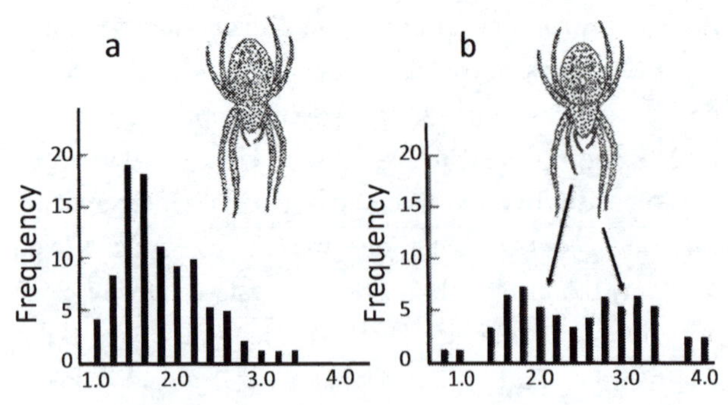

Fig. 7.22. The spaces between sticky spiral loops built by an experimentally altered spider gave an elegant demonstration that the length of the spider's leg I has an effect on the spaces between loops of sticky spiral. *Araneus diadematus* (AR) spiders were created that had a shorter leg I on one side than on the other by experimentally removing this leg in the previous instar; the spider regenerated the leg when she molted, but the replacement leg was shorter (drawing in *b*). The distribution of spaces between sticky spiral loops built by experimental spiders was bimodal, probably reflecting the different lengths of the two legs (it was not possible to deduce for any given sticky spiral loop which set of legs was used to measure, because the spider doubled back repeatedly while building the sticky spiral) (from Vollrath 1987a).

prey. Some of these responses follow the logic of foraging theory (section 8.2).

7.3.4.2.2 PREY-SPECIFIC RESPONSES (?)

A more controversial set of possibilities, for which there are conflicting data, is that the properties of previous prey are used as cues affecting the designs of future webs or the mechanical properties of silk lines—that the spider learns from experience which prey her orb should be designed to capture. Altering her orb design on the basis of the properties of her previous prey might not seem likely to be advantageous, given the likely highly stochastic nature of arrival of different types of prey in orbs in the field and the wide variation of prey captured by a given species (Caraco and Gillespie 1986; see section 3.2.5, appendix O3.2.3). Nevertheless there is evidence (as yet inconclusive) that such alterations may occur.

One suggestion is that *Cyclosa octotuberculata* (AR) adjusted its orb design on the basis of design failures of previous webs (Nakata 2007). In one experiment in captivity, flies were introduced into webs but then removed just as the hungry spider arrived to attack five different times in one day; from the spider's point of view (supposing that she was not simply frightened by this process), the retention function of her web failed repeatedly. The subsequent orbs of spiders that had been "robbed" this way had more closely spaced sticky spiral lines (and thus greater rentention abilities) than did those of spiders that were robbed only once or not at all (Nakata 2007). Both the area covered by the sticky spiral and the total thread length also increased when five prey were robbed.

Several other studies suggested prey-specific responses. Young juveniles of the araneid *Araneus diadematus* built more closely spaced sticky spirals after capturing *Drosophila melanogaster* flies on previous days than after capturing *Aedes* mosquitoes (Schneider and Vollrath 1998). As just mentioned, however, changes in the nutrient content of *D. melanogaster* (reared on different larval media) correlated with both sticky spiral spacing and the capture zone area (Mayntz et al. 2009), so this *Drosophila-Aedes* difference in orb designs could have been due to nutritional differences. The araneid *Argiope keyserlingi* increased the spaces between sticky spiral loops when fed juvenile crickets compared with adult crickets (Blamires 2010); stopping and retaining juvenile crickets would presumably require less densely spaced lines (though this was not demonstrated, and details of the balance of different nutrients were probably not uniform in the two treatments). And the nephilid *Nephila pilipes* built orbs with larger catching areas (covered by sticky spiral) and with more radii when fed with houseflies (*Musca domestica*) than with house crickets (*Achaeta domestica*), a prey that is presumably more difficult to stop and retain (Tso et al. 2005, 2007; Blamires et al. 2011). Again, however, there was no control for possible nutritional differences, and prey stopping and retention were not measured directly. When stimuli from the prey struggling in the web and the type of prey the spider fed on were isolated experimentally (approximately equal wet weights were provided of each prey type) (Blamires et al. 2011), webs built after the combination of fly stimuli and fly food differed from the other three combinations in having a larger catching area and more radii. Changes in silk composition have also been found. Expression of *MaSp1* silk genes in *N. pilipes* increased after feeding on live crickets compared with feeding on live flies (deduced from increases in alanine and glycine in major ampullate silk) (Blamires et al. 2010b). Stimuli from the prey itself seemed not to have an effect, as amino acid compositions of major ampullate silk did not differ when spiders fed on crickets with fly stimuli or flies with cricket stimuli (Blamires et al. 2010b); the authors supposed that these chemical changes resulted in webs whose mechanical properties were more appropriate for retaining crickets.

On the other hand, when the *Argiope keyserlingi* (AR) was exposed experimentally to prey vibrations but prevented from capturing the prey, neither web size nor sticky spiral spacing changed compared with controls (web construction frequency increased, however) (Herberstein et al. 2000c). This species also failed to alter sticky spiral spacing in response to prey nutrient concentrations (comparing cockroaches with crickets) (Blamires 2010). Neither *A. bruennichi* nor *Larinioides cornutus* (AR) altered sticky spiral spacing in response to the presence of different prey types (honey bees vs. damselflies) in the same container, or after having fed on these two prey (Prokop 2006). *Araneus trifolium* (AR) and *Argiope trifasciata* did not alter their web designs when fed equal amounts (dry biomass) of either blowflies or acridid grasshoppers for two weeks (Olive 1982).

These studies of the possible effects of prey on orb design reveal a lack of consistent trends. One problem for the "prey-specific effects" hypothesis is the question of the ecological realism. It does not seem reasonable, for instance, to expect that the Holarctic species *A. diadematus* or the Holarctic-Neotropical *Z. x-notata* (Platnick 2014) should have evolved specific responses to a tramp species from Africa such as *D. melanogaster*. The frequency of drosophilids in the diets of spiderlings of these species under natural conditions is, to my knowledge, unknown. Similarly, *N. pilipes* (= *maculata*) are extreme generalists in the field (Robinson and Robinson 1973a), and because of the positioning of their aerial webs, they are unlikely to capture largely terrestrial crickets like *A. domestica*, so specific, biologically appropriate behavioral responses to these crickets would seem unlikely to have evolved. Judging by habitat preferences, the wingless wood cockroaches used in the *A. keyserlingi* study are also very unlikely prey. In general, the extreme breadth of orb weaver's diets (section 3.2.5) also argues against natural selection favoring such prey-specific responses. A related unresolved problem concerns the number of different categories of prey that a spider can be expected to discriminate and associate with appropriate alterations in web design traits.

Further problems include small sample sizes in some studies (as low as 11), and lack of controls for variables such as spider size, age, and previous feeding history, both in experiments that did and in those that did not find effects. Negative evidence with respect to changes in response to one prey type does not eliminate the possibility of responses to others. In sum the available experiments do not lead to confident general conclusions regarding cues from capturing and consuming particular types of prey.

Another postulated effect regarding previous prey concerned not the prey's identity, but rather the portion of the web where they were captured. The araneids *Larinioides sclopetarius* (juveniles) and *Argiope keyserlingi* (adults) both responded to capturing prey below rather than above the hub by building orbs in which the area below the hub was proportionally larger (Heiling and Herberstein 1999). The araneid *Cyclosa octotuberculata* tensed radii in sectors in which the spider had captured prey on previous days more than radii in other sectors (Nakata 2012). Again the possible lack of ecological realism hangs over these experiments (do consistent biases in the portion of the web where prey are captured occur often enough in nature that natural selection would favor adjustments of this type?).

In summary, the quantity of prey captured in the recent past clearly affects sticky spiral spacing (and other design properties) in orb webs. I believe, however, that it is not clear whether the traits of previous prey that were captured by a spider induce her to adaptively modify the design of her orb to increase captures of these particular types of prey. In the end, my strongest reservation is that natural selection seems unlikely to favor such adjustments in nature. Natural selection on thousands of past generations of the spiders seems likely to have favored the web traits that gave, on average, the best prey capture payoffs. Altering web construction behavior from this "default" program, on the basis of having recently captured, for instance, a few flies, would be advantageous only if natural websites tend to offer sufficiently consistent and different arrays of prey from one day to the next (see table 3.2; also Eberhard 2013a; Blamires et al. 2011). I know of no data that directly address the question of day-to-day, local predictability of particular types of prey in nature; such predictability seems unlikely (appendix O3.2.3; see, however, Sandoval 1994 for an unusual case). The most extensive surveys of orb weaver prey (e.g., Robinson and Robinson 1970b on *Argiope argentata*, 1973a on *N. pilipes*) do not help with this problem, as they grouped captures by different individuals together. It is true, however, that some websites differ, at least in overall quantities of prey (Gillespie and Caraco 1987; Blackledge and Wenzel 1999). And large objects near a given orb could result in flyways that made certain parts of orbs at that site more likely to intercept prey. These are only speculations; we need further data.

7.3.4.3 Presence of predators (?)

Another possibility is that stimuli from potential predators induce spiders to alter the designs of their orbs. Most studies on this topic have concerned the stabilimentum (section 3.3.4.2); the area covered by sticky spiral has also been implicated, though less convincingly. Juveniles of the araneid *Argiope versicolor* reduced the capture area and the length of sticky spiral (but did not alter the sticky spiral spacing) in response to chemical cues from a predator, the salticid *Portia labiata* (controls included chemical cues from another spider not thought to prey on orb weavers) (Li and Lee 2004). It was not clear, however, that reduced web area would constitute an effective defense against a predator such as a salticid that attacks by jumping (a smaller web would seem to be a *poorer* defense). Making a smaller web might save on energy that could be dedicated to performing defensive behaviors such as dropping from the hub, shuttling from one side to the other, or pumping the web (Li and Lee 2004) but the prediction that spiders on smaller orbs perform more defensive behavior has not been tested. The idea that smaller webs make *Argiope* spiders less visible to *Portia* spiders (Li and Lee 2004) is also not convincing.

The araneid *Eriophora sagana* decreased total thread length (though not the capture area or sticky spiral spacing) when exposed to airborne vibrations from a vibrating 440 Hz tuning fork, thought to mimic a predator's flight (Nakata 2008). Both this species and *Cyclosa argenteoalba* (AR) built webs that were more symmetrical vertically when exposed to the vibrations (Nakata and Mori 2016). The hypothesis that the shorter duration of construction of smaller and more symmetrical webs functioned to reduce the danger of predation (Nakata 2008; Nakata and Mori 2016) was somewhat uncertain, however. A second common reaction by *E. sagana* spiders to this stimulus, lifting the anterior legs, indicated that the spiders may have been reacting as if to prey; and it was not clear if either species, which both wait at the hub of the web during the day, builds before or after dawn (for relatives see table 8.2).

In sum, these reservations, which are no less speculative than the original arguments, constitute reason for skepticism but not for rejection. The possibility that spiders vary orb designs on the basis of stimuli from possible predators needs further testing.

7.3.4.4 Wind

Wind can damage orbs in at least two ways. Summarizing the earlier discussion in section 3.2.11, wind can apply a general load throughout an orb as if the web were a sail, stressing the orb's mechanical scaffolding lines (the radii, frames, and anchor lines), and causing them to break. Reducing the density of sticky spiral lines could reduce a web's resistance to the wind, and thus reduce this sort of stress. A second type of damage does not involve breaking lines, but rather their sticking together. If sticky spiral lines swing into contact with each other or with radii, they will adhere and produce holes in the regular spacing of sticky lines. This "sticky spiral entanglement" damage (section 3.2.11.2) can be reduced by spacing sticky lines farther apart, or by shortening the length of each sticky spiral segment by placing radii closer together. Both types of wind damage could favor increased spaces between sticky lines at windier sites.

As predicted, when adult females of *Cyclosa mulmeinensis* (AR), a species that sometimes occurs at windy sites, were experimentally subjected to continuous wind (1.1 m/s) for seven days in captivity, on the eighth day (without wind) they built smaller webs than controls that had significantly larger spaces between the loops of sticky spiral (51% greater than controls), shorter total silk length, and smaller areas (Liao et al. 2009; Wu et al. 2013). The webs of spiders that had experienced wind also had larger droplets on the sticky spirals, which may have helped reduce dehydration in the wind (Wu et al. 2013), and had stiffer radii (Liao et al. 2009). Wind experiments using a more gentle breeze (0.5 m/s) in *Ara-*

neus diadematus (AR), gave similar though less dramatic effects (Vollrath et al. 1997): sticky spiral spacing was unchanged above and below the hub, but was greater to the side of the hub (an 8% increase); web area was also reduced, but only by 13%.

All of these changes could function to reduce wind-induced damage to webs (Wu et al. 2013). Their effects on the orb's ability to stop prey are less certain, though increased radius stiffness could help to absorb prey impact (Wu et al. 2013). Wu et al. (2013) stated that "in strong winds prey fly in multiple directions and at a multitude of speeds," but gave no references to document such differences or their effects on prey stopping and retention. Some prey probably tend to fly against the wind, and if this is common and webs tend to be perpendicular to the wind, prey might be easier rather than harder to stop at windy sites. Additional studies of wind effects on prey behavior and capture could be rewarding.

7.3.4.5 Time of day

In species that build multiple orbs in a single day at the same site (table 8.2), it is possible to compare different webs built by the same individual at different times of the day. In the colonial tetragnathid *Metabus ocellatus*, marked adults and subadults in colonies near tropical streams often built two or three orbs in a single day. Webs built during the last two hours of the day (16:00–17:00), when there was a peak of prey flying just over the surface of the stream (Buskirk 1975), had fewer and shorter radii, and larger spaces between sticky spiral loops (Buskirk 1975, 1986). The designs of late afternoon webs may have been appropriate if the species flying at that time of day were weaker fliers or easier to stop and retain; this has not been checked.

The speed with which spiders of this species moved during sticky spiral construction, an only infrequently measured variable, was greater early in the morning and late in the afternoon than at mid-morning (Buskirk 1986). The higher speeds of movement correlated with prey abundance (there was a second peak of prey abundance around dawn—Buskirk 1975), but the possible functional significance of this correlation is unknown.

7.3.4.6 Season of the year and light rain

One Brazilian species, *Parawixia bistriata*, uses still further cues. In the Brazilian savanna, termites are common, and colonies release swarms of weak-flying winged reproductives from June to September. The flights from different termite colonies were coordinated by local weather and the time of the day: they occurred during the daylight hours on days with light rain. Outside the termite season, the colonial araneid *P. bistriata* built tightly meshed orbs at dusk, just prior to sunset or in the early evening; usually all the spiders in a colony built in the same 60 min period (Fowler and Diehl 1978; Sandoval 1994). About half of the prey (by numbers) were small flies in the phorid genus *Dohrniphora*. But during the termite season, the same individual spiders often built a second web during the day, in addition to the usual evening web. The times of day at which daytime webs were built varied; on each of 13 occasions, however, construction by spiders in a spider colony began within 30 min of when termites in the vicinity of the colony began swarming. On four of these days, construction began before termite flights, so daytime webs were at least sometimes built without the spider having previously interacted with termites that day. Daytime orbs differed sharply in design compared with the evening webs: the web diameter was nearly doubled (182%), the mean space between sticky spiral loops was more than tripled (323%), the number of radii was reduced by a third (66%), and the number of sticky spiral loops was reduced by more than half (42%) (Sandoval 1994). The total estimated lengths of sticky spiral and of radii did not change significantly.

The larger, sparser daytime designs were appropriate to increase the interception function to capture this weak-flying prey (Sandoval 1994). Alternative explanations for the design changes (lack of silk for daytime webs, rain damage) were discussed and discarded (Sandoval 1994).

7.3.4.7 Temperature

Decreasing the temperature from 24°C to 12°C caused *Araneus diadematus* (AR) in captivity to build sticky spirals with fewer loops and wider spaces between them (Vollrath et al. 1997). The overall capture area did not change, while the total length of sticky line decreased. When the temperature was subsequently returned to 24°C, the number of loops returned to control values, but the larger spacing between them persisted. Possible selective advantages of these responses are not clear.

7.3.4.8 Humidity

When the relative humidity was lowered from 70 to 20%, the web and capture area of *A. diadematus* in captivity decreased; the number of loops of sticky spiral decreased, while the spaces between them remained constant (Vollrath et al. 1997). When the humidity was subsequently increased to 70%, web and capture areas returned to control values; the spaces between sticky spiral loops increased, but the number of sticky spiral loops remained reduced. Again, the selective consequences of these changes are uncertain.

7.3.5 ADDITIONAL DECISIONS BY SPIDERS BUILDING STICKY SPIRALS AND CUES TRIGGERING THEM

7.3.5.1 Turn back

Several additional decisions made during sticky spiral construction are also probably influenced by cues from the web. Only one has been studied experimentally, however—whether or not to double back. Many orb weavers adjust their sticky spirals to asymmetrical web forms by doubling back repeatedly in the outer areas of the larger portion of the orb, then spiraling inward once the unfilled space around the hub is more symmetrical (e.g., Hingston 1920; Witt et al. 1968) (figs. 6.22c, 7.1). This implies that the spider decides whether or not she will double back each time she attaches the sticky spiral to a radius. The cues influencing turnback decisions have not been studied extensively, and there are no experimental tests of any particular candidate stimuli. Nevertheless, some patterns in unmanipulated webs give strong indications that one stimulus inducing turnbacks is contact with sticky lines that are oriented more or less radially, rather than in the usual more or less circular orientation.

One type of evidence comes from the occasional brief periods of wandering out of contact with the inner loop of sticky spiral by *Uloborus diversus* and *Zosis geniculata* (UL) that were mentioned above (fig. 7.2). When the spider eventually contacted the "wandering" inner loop of sticky spiral again, she used it to guide subsequent loops of sticky spiral more or less parallel to it. In other cases the wandering line was oriented more or less radially, and encounters with these lines usually resulted in turnbacks (fig. 7.2). Because these turnbacks occurred in a variety of contexts (on long radii, on short radii, near the hub, far from the hub, etc.), and because their locations appeared not to be patterned (they were the result of the spider's temporary disorientation and "wandering"), it is likely that the abnormal location of the sticky line, rather than other stimuli, triggered these turnbacks.

More or less radially oriented sticky lines also appeared to trigger sticky spiral turnbacks when spiders replaced sectors of their orbs. In the early stages of web repair behavior *U. diversus* and *Z. geniculata* (UL) collapsed and condensed sticky and non-sticky lines accumulated at the approximately radial edges of the sector to be repaired (Eberhard 1972a, 2019b). Thus the turnbacks at the edges of repair sectors occurred when the spider encountered sticky silk from the previous web that was approximately radially oriented. Similar repairs with consistent sticky spiral turnbacks at the more or less radially oriented borders of repair zones also occurred in other families, including *Nephila clavipes* (NE) (Kuntner et al. 2008a; WE) and *Enacrosoma anomalum* (AR).

Experiments in both *U. diversus* and *Z. geniculata* demonstrated that it was the sticky silk in these borders that induced the turnbacks, and not the borders themselves. Sectors in "two night" orbs were broken after the first night, when the spider had filled only the outer portion of the orb with sticky spiral. The edges of these sectors thus contained sticky lines in their outer portions, but lacked sticky silk in their inner portions. On the next night, when the spider filled in the repair sector and the rest of the orb with sticky lines, she consistently turned back each time she encountered an edge of the first web that contained sticky lines, but seldom turned back when she encountered edges that lacked sticky silk (fig. 7.1).

A computer simulation of sticky spiral construction (fig. 7.23a,b) gave further, though inconclusive evidence. A pattern of sticky spirals similar to those seen in steep-sided spaces in orbs of this species was produced using a relatively simple set of rules in which more nearly perpendicular angles between the current loop of sticky spiral and the inner loop of sticky spiral already in place triggered turnbacks (simulations in more circular spaces suggested, in contrast, use of the angle between the current loop and the radius; both angles may be used) (Eberhard 1969) (see section 7.15 for limitations in the conclusions that can be drawn from simulations).

Contact with sticky silk that is oriented more or less radially thus seems to trigger turnbacks. How the spider senses the orientation of sticky lines that she encounters

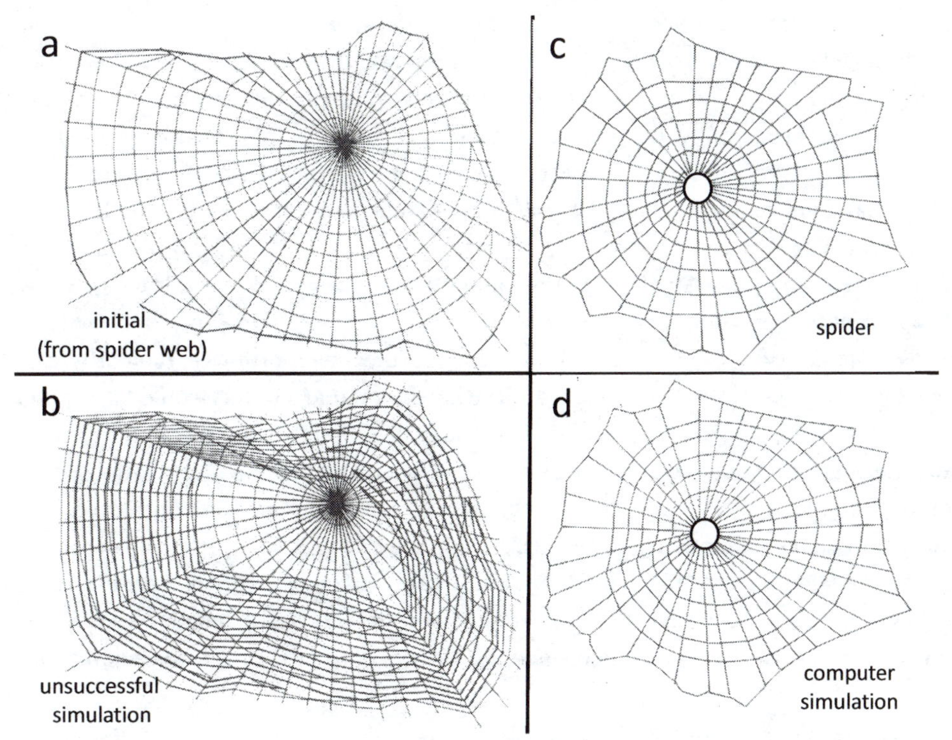

Fig. 7.23. Interpretation of the results of computer simulations of orb construction behavior is not always easy. In *a*, the computer was given a web traced from a photo of a *Uloborus diversus* (UL) web, with radii, frames, and the first loop of sticky spiral in place; the simulation program then generated the sticky spiral (thicker lines) in *b*. This pattern was never seen in spider webs, and the confident conclusion was that this particular program was not used by spiders. In *c* the computer was given the positions of the radial and frame lines traced from a photo of an *Araneus diadematus* web (AR). A program to decide where to attach each loop of temporary spiral generated a pattern (*d*) very similar to the temporary spiral that the spider herself built on the same radii and frames (*c*). This apparent "success" illustrated two important points. There were many small differences between the computer and spider webs (note, for example, the 1:00–2:00 sector). Deviations during orb construction can be cumulative, because early decisions (by both the spider and the computer) influence subsequent behavior. Details of the simulation might differ from the spider's web because of a failure to simulate the spider's program perfectly, or because of imprecisions on the part of the spider herself (Eberhard 1990a). Secondly, contrary to appearances, the program in this particular simulation was probably quite different from that of the spider, because it did not use memory of previous distances moved along radii, a variable that experiments with several species have suggested has a strong effect (section 7.4.1; see figs. 7.30, 7.31, 7.33). Successful mimicry of the spider's behavior does not guarantee that the cues used were the same (*a* and *b* from Eberhard 1969; *c* and *d* from Gotts and Vollrath 1992).

while building the sticky spiral is not certain. There are several possible mechanisms, including multiple contacts with the sticky inner loop (allowing determination of the radius-inner loop angle), and the relative TSP-IL distance (a sharp reduction in this distance from one radius to the next could indicate that the inner loop was directed more radially) (Eberhard 1972a). There are no experiments to discriminate among these possibilities.

Behavioral details also help indicate when the cue is sensed. Direct observations and video recordings of *Araneus expletus* (AR) (WE) and *U. diversus* (Eberhard 1972a) showed that the spider did not move beyond the radius where she made a turnback, suggesting that the turnback cue(s) was probably sensed prior to or at the moment of attachment. This conclusion was in accord with the fact that the spacing of the turnback attachment from the inner loop was consistently smaller than that on other nearby radii in the uloborids *U. diversus* (Eberhard 1969) and *Z. geniculata* (fig. 3.9) (WE).

An alternative was proposed by Hingston (1920), who noted that turnbacks in *Neoscona nautica* (AR) often occurred where the spider was moving from an area with long radii into one with shorter radii; he speculated that the spider may use differences in tensions and vibrations of longer vs. shorter radii as cues. Two types of data argue against this hypothesis: the lack of effects of altering radius tensions on turnback decisions in several species (above); and the induction of turnbacks in *U. diversus* and *Z. geniculata* by both wandering sticky lines and the edges of repair zones, which were not correlated with radius lengths (figs. 7.1, 7.2).

Turnbacks occurred in additional contexts in other species, including on radii near vertical egg sac and detritus stabilimenta in *Allocyclosa bifurca* and *Cyclosa* spp. (figs. 3.34d, 7.41), and on radii adjacent to signal lines in the webs of the araneids *Zygiella x-notata* (Petrusewiczowa 1938; Le Guelte 1966; Peters 1969) and perhaps also *Araneus mitificus*, *Acusilas coccineus*, *Milonia* sp., and *Dolichognatha* sp. (TET) (Gregorič et al. 2015) that have open sectors (in the latter groups these open sectors may have been created by removal of sticky lines rather than turnbacks). There are no data on the stimuli triggering these turnbacks, or whether they differ from those triggering turnbacks in other contexts.

7.3.5.2 Attach to each radius

Another decision made by a few species is whether or not to skip a radius without attaching to it. In general, araneids attach to each radius they encounter. The only exceptions I know of were in *Araneus diadematus* (AR) orbs built under near-weightless conditions in space (Witt et al. 1977), and those built under the influence of d-amphetamine (Jackson 1974; Witt et al. 1968). Some uloborids, however, routinely skip some radii without attaching to them in the inner portions of their orbs (e.g., figs. 7.24). The likelihood that a radius would be skipped was greater nearer the hub in *U. diversus*, and the numbers of radii that were skipped were also larger closer to the hub; the mean number of radii skipped increased from 0.62 ± 0.4 in the first 5 loops to 3.12 ± 1.22 in the last five loops (WE). I know of no experimental studies of the cues affecting these decisions. The length of sticky line laid since the last previous attachment to a radius may

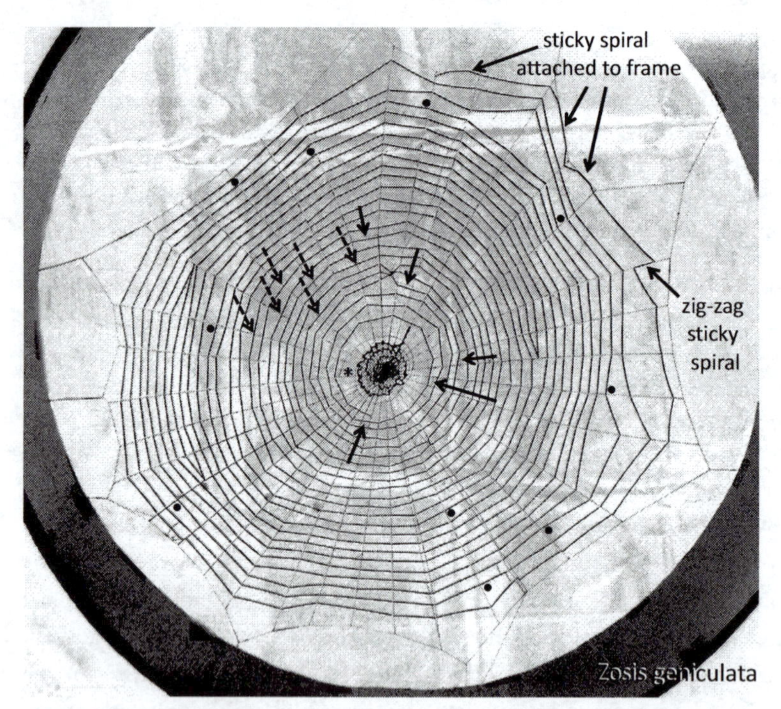

Fig. 7.24. This horizontal orb of a mature female *Zosis geniculata* (UL) shows several details that may distinguish uloborid from araneoid orbs: occasional attachments of the sticky spiral to a frame line; double attachments of the sticky spiral to some radii in the outer portion of the orb that produce zig-zag patterns of sticky silk; sticky spiral lines in the inner portion of the orb that crossed some radii without being attached to them (a few are marked with dashed arrows); and scattered intact segments of temporary spiral (solid arrows). The hub (partially obscured by the stabilimentum), has additional distinctive uloborid traits: an intact central area, numerous double attachments of the hub spiral to radii that produce zig-zag patterns in the radii, and a gradual spiral outward to the temporary spiral ("*") (see also figs. 4.9c, 3.11k,l, 3.31). The taxonomic distributions of these differences, however, have not been tested with wide-ranging, quantitative comparisons in different families. One additional pattern, which uloborids share with araneids (probably as a result of the profound conservatism in the cues used to guide construction behavior—section 10.8.2), is that the spaces between sticky spiral loops tended to be reduced in the next loop following an oversized space (black dots).

be important; the lengths of the segments of sticky spiral in the orbs of *U. diversus* were relatively constant (Eberhard 1972a). There are several possible ways in which the length of the current segment might be sensed, including memory of distance traveled from the previous attachment, the time elapsed since the previous attachment, and the distance the spider swings laterally as she moves (Eberhard 1972a).

7.3.5.3 Number of attachments

Another decision restricted to a small group of orb weavers was whether or not to attach twice to the same radius, and thus produce a zig-zag sticky spiral pattern. Zig-zag patterns were nearly unique to uloborids, and are known in only two distantly related araneids (fig. 7.25); they occurred in the uloborid genera *Uloborus, Philoponella, Conifaber*, and *Hyptiotes*, especially in the outer loops of sticky spiral (figs. 3.22, 7.24, 10.17, 10.23, O9.1) (Lubin 1986). Zig-zag patterns tended to be restricted to only the first loops of sticky spiral. For instance, in the 17 of 36 webs of *Uloborus diversus* in which there was at least one zig-zag pattern, 41 of the 43 zig-zag patterns were in the outermost two loops of sticky spiral, and 29 were in the first loop (WE). A second consistent trend was that zig-zag patterns always occurred on longer radii. In these same 17 webs of *U. diversus*, 33 of the 41 radii on which there was a zig-zag pattern were longer than the median length (WE). There are no experimental tests of stimuli that induce these patterns.

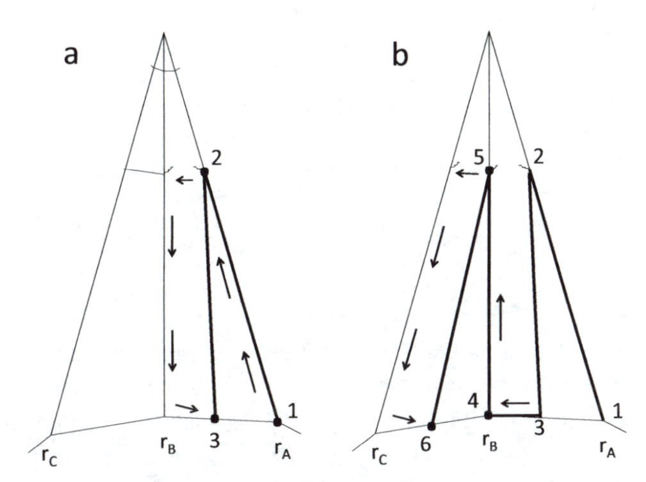

Fig. 7.25. Construction of the sticky "spiral" by *Eustala* sp. (AR) (#2345) (thicker lines in the drawing) was unique in that the spider moved in strictly alternating clockwise and counter-clockwise directions. The spider began laying sticky line (*a*) moving inward (upward in the drawing) along a radius (r_A) until she reached the temporary spiral, where she attached the sticky line (point 2). Then she broke the temporary spiral as she moved clockwise across it, and out the next radius (r_B) to the frame; she then moved counterclockwise along the frame to attach the sticky line at point 3. She reversed her direction again, moving clockwise to r_B and then inward again to repeat the process, attaching at points 4, 5, and 6 (*b*). Other species of *Eustala* built typical orbs (figs. 3.10, 4.3c), so these reversals of direction are relatively recently derived (after Eberhard 1985).

7.3.5.4 Break temporary spiral

A decision that occurred in nearly all orb weavers was whether or not to break the temporary spiral. The segments of the temporary spiral were broken one by one during the process of sticky spiral construction (section 6.3.5.1). Judging by the fact that the temporary spiral was generally broken when the inner loop of sticky spiral had approached it closely (Hingston estimated that it was broken in *Neoscona nautica* when the loop of sticky spiral was about one sticky spiral space from the temporary spiral; in other species it may be closer to twice this distance), the TSP-IL distance may be involved.

Different portions of the temporary spiral were broken at different times, and this probably resulted in another previously unappreciated problem for the spider. There were often isolated segments of temporary spiral in the web (e.g., fig. 7.24), and the spider had to distinguish these while she was building the sticky spiral. In video recordings, *M. duodecimspinosa* appeared to encounter the broken end of the temporary spiral with her legs oI and oII (WE). Apparent mistakes sometimes occurred where short sectors of temporary spiral were left intact (and they apparently induced the spider to use the lingering temporary spiral line instead of the inner loop of the sticky spiral to guide placement of the next loop of sticky spiral, leading to oversized spaces between sticky spiral loops—fig. 7.26). Intact segments of temporary spiral and oversized spaces were particularly common in the innermost portions of *M. duodecimspinosa* webs (fig. 7.26), and in orbs of several uloborids (fig. 7.24).

7.3.5.5 Terminate

Finally, the last decision that the spider makes during sticky spiral construction is to end it. I know of no experimental studies of what stimuli trigger this decision. The criteria for sticky spiral termination decisions may

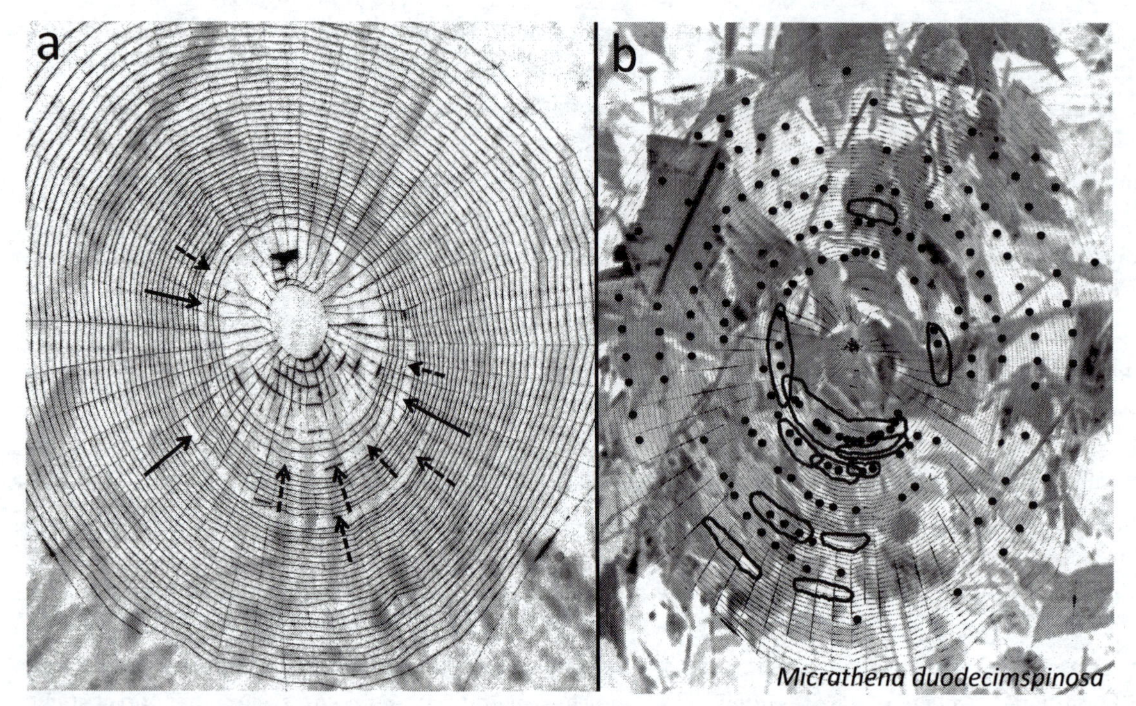

Fig. 7.26. These unpowdered field webs of adult female *Micrathena duodecimspinosa* (AR), in which sticky lines are generally more distinct than the non-sticky lines (except for those that glinted in the camera's flash), demonstrate possible associations between both intact and broken temporary spiral lines and larger spaces between sticky spiral loops. One common pattern (*a*) was that especially large spaces between sticky spiral lines (a few are indicated with solid arrows in *a*) were often associated with "errors" where segments of temporary spiral were left intact (some are indicated by dashed arrows in *a*). It appears that the spider sometimes responded to intact temporary spiral lines as if they were the inner loop of sticky spiral, using them as points of reference to determine attachment sites. The dots in *b* that mark sites on radii in another web where vestiges of the temporary spiral were visible (others could not be distinguished with certainty). It appears that unusually large sticky spiral spacing (outlined with thick lines) tended to occur where the temporary spiral had been present but was broken. Further measurements will be needed to test whether associations between previous temporary spiral loops and larger sticky spiral spacing are consistent.

differ quantitatively in different taxonomic groups. For instance, *M. duodecimspinosa* (AR) and *Nephila clavipes* (NE) continue laying sticky spiral in the area where radii are very close together, and leave little or no open space between the outer loop of the hub and the inner loop of sticky spiral (e.g., fig. 1.4). In many others the sticky spiral ends far from the outer hub loop (e.g., figs. 3.11–3.13).

Two possible stimuli for inducing termination are exhaustion of the supply of sticky silk and the distances between adjacent radii that fall below some minimum. The webs of *L. argyra* spiders confined to very small spaces demonstrated, however, substantial flexibility with respect to both the length of sticky silk spiral and the distance from the hub (fig. 7.42). Witt and Reed (1965) also argued against the idea that running out of silk is the termination cue in *Araneus diadematus* (AR); the central area that was left uncovered with sticky lines (hub plus free zone) was highly correlated with the sticky spiral area ($r = 0.8$ in 103 webs), and they reasoned that if the spider stopped only when she ran out of material, such a relation would not be expected (this argument is not completely convincing, however).

7.3.6 FIRST LOOP OF STICKY SPIRAL: A SPECIAL CASE

Data presented above show that the site of the inner loop of sticky spiral is a crucial reference point that guides sticky spiral placement. This cue is not available, however, when the spider builds the first loop of sticky spiral. Instead, the site of the outer loop of temporary spiral was probably used like a "hand rail" to guide this first loop. In accord with this hypothesis, araneids and nephilids maintained contact with the outer loop of temporary spiral while building the first loop (section 6.3.5, Eber-

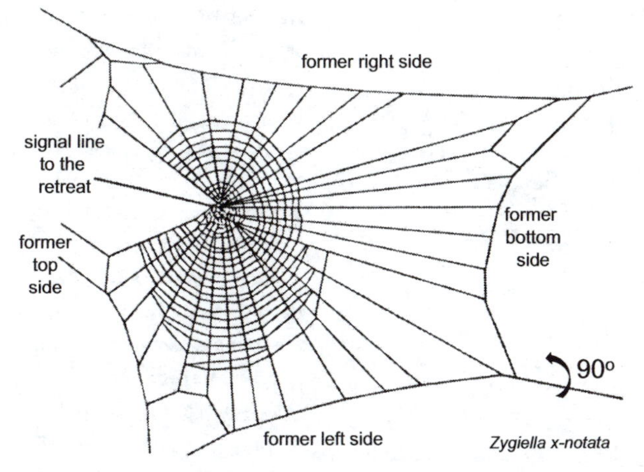

Fig. 7.27. This web of an adult female *Zygiella x-notata* (AR) resulted from Hans Peters' simple, elegant experiment demonstrating that the outlines of both the temporary spiral and the sticky spiral are largely determined by gravity, rather than by the outlines of the frame and radial lines. After the spider had built the frames and radii with a substantially larger area below the hub than above (as usually occurs in this species), Peters rotated the frame in which the web was being built counter-clockwise 90°, just as she began to build the temporary spiral (the former bottom side was now at the right, and the former left side was at the bottom). The outline of the temporary spiral that the spider then built also rotated 90° from normal, with the largest portion toward the former left side (now below the hub), and outline of the sticky spiral had the same 90° change in orientation (after Peters 1937).

hard 1982). In other groups, however, the spider moved beyond the outer loop of the temporary spiral toward the frame before attaching (Eberhard 1982). Hingston (1920) mentioned such behavior in *Neoscona nautica* (AR) and *Tetragnatha* sp. (TET), and thought that the temporary spiral was nevertheless used as a reference point; he said that the spider always moved "three paces" beyond the temporary spiral, but did not specify which leg or define a "pace."

Perhaps the most elegant experimental evidence supporting the hand rail hypothesis comes from the classic study of Peters (1937) with *Zygiella x-notata* (= *litterata*) (AR) that is illustrated in fig. 7.27. By rotating the frame in which the web was being built at the moment when temporary spiral construction began, he showed that cues from gravity prevailed during temporary spiral construction over possible information from the outline formed by the frames and radii; in addition, the outline of the sticky spiral then followed that of the temporary spiral. Additional correlations between temporary spiral and sticky spiral outlines when the temporary spiral was especially small compared with the total web area also occurred in two other species, *Araneus diadematus* (AR) (Peters 1970), and *Uloborus diversus* (UL) (Eberhard 1972a), and when the temporary spiral outline of *A. diadematus* was altered by changing the web plane from vertical to horizontal (Zschokke 1993, 2011) (the two outlines showed a correlation of $r^2 = 0.898$) (Zschokke 1993).

While the outer loop of the temporary spiral thus functions as a hand rail guiding the first loop, an additional cue also influenced attachment sites. When the distance was measured between the first sticky spiral loop and the outer loop of temporary spiral (fig. 7.4) in webs of *M. duodecimspinosa* (AR) and *Leucauge mariana* (TET) in which the temporary spiral lines were still intact (figs. 6.10, 7.12), the distance between the outer loop of temporary spiral and the first loop of sticky spiral was not constant, as expected under the hand rail hypothesis. Instead, there was a positive correlation between the site of attachment of the sticky and temporary spirals and the distance between the temporary spiral and the frame line (TSP-F distance) (fig. 7.28) (Eberhard 2012). This TSP-F distance effect, however, was manifested only when the TSP-F distance was relatively small (fig. 7.28b,e) (perhaps when the spider contacted the frame line). When the frame was more distant (and the spider probably could not reach to touch it), there was no significant TSP-F distance relation (fig. 7.28c,f). Observations of sticky spiral construction behavior by *M. duodecimspinosa* (AR) confirmed that the spider remained in contact with the temporary spiral with her inner legs while she built the first loop of sticky spiral, and that she contacted the frame line only when the frame was not distant from the temporary spiral (r_a, r_b, r_e, and r_f in fig. 7.29).

7.3.7 INTERACTIONS AMONG CUES

Spiders were clearly influenced by multiple stimuli during sticky spiral construction, but little is known about how different cues are combined to influence each attachment decision. Cases in *Micrathena duodecimspinosa* (AR) in which three reference point cues were in conflict with each other provide a glimpse into how cues may be integrated. The resolution of conflicts between the inner loop (IL) site, the TSP-IL distance, and short term memories of this distance apparently varied, de-

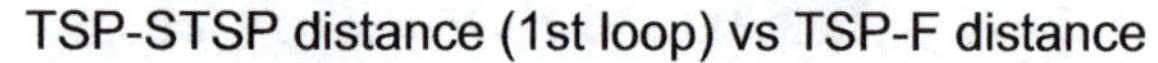

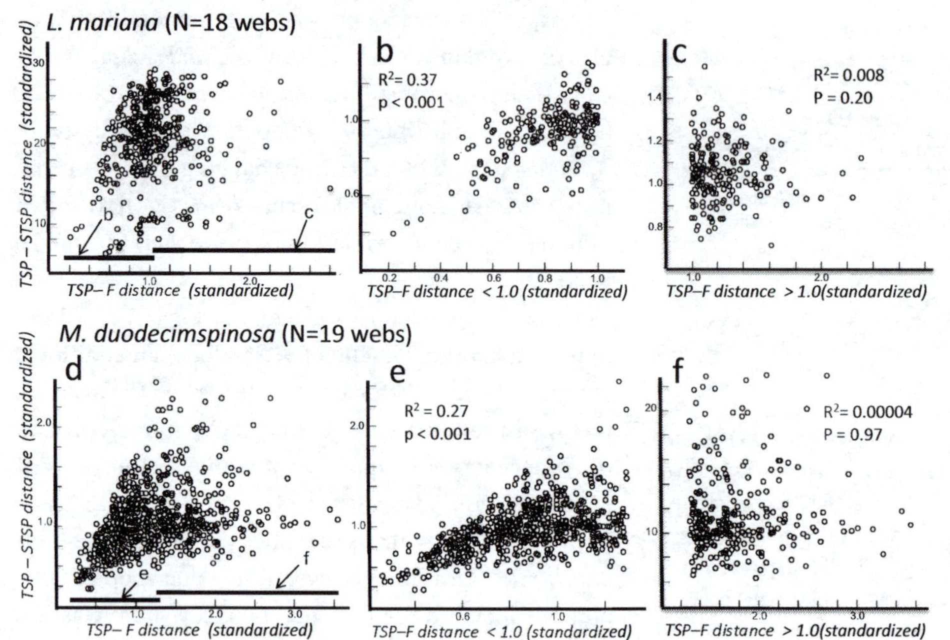

Fig. 7.28. These graphs of standardized distances between the outer loop of temporary spiral and the first loop of sticky spiral from 19 webs of *Micrathena duodecimspinosa* (AR) and 18 of *Leucauge mariana* (TET) show a strong correlation with the distance of the TSP from the frame when the temporary spiral-frame distance was small (<1.0) (*b*, *e*), but a lack of correlation when the TSP-F distance was larger (>1.0) (*c*, *f*). The thick lines in *a* and *d* show the ranges of TSP-F values included in the other two graphs (from Eberhard 2012).

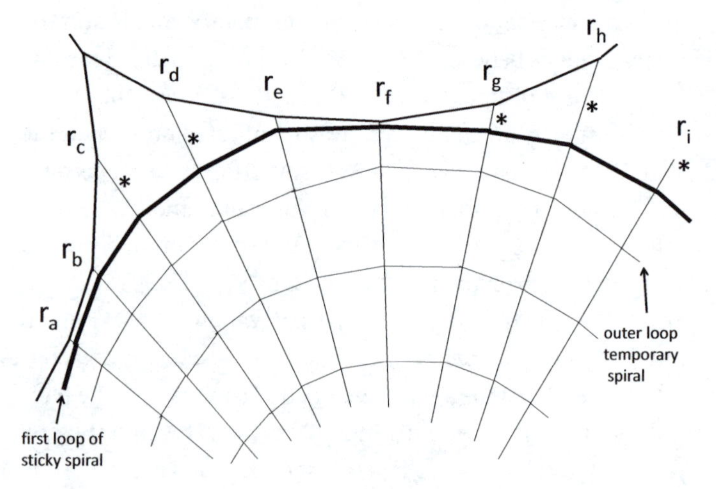

Fig. 7.29. While building the first loop of sticky spiral, *Micrathana duodecimspinosa* (AR) always maintained contact with the outer loop of temporary spiral. This schematic drawing illustrates the places in her orb where the spider could (r_a, r_b, r_e, and r_f) and could not ("*" on r_c, r_d, r_g, r_h, and r_i) contact the frame line while she built the first loop (from Eberhard 2012).

pending on the intensity of the conflict (Eberhard and Hesselberg 2012). At low levels of conflict (the usual case during normal orb construction), the spider contacted the inner loop of sticky spiral using inner loop localization behavior (section 6.3.5), and could thus sense the IL position and the TSP-IL distance, and she used all three cues. When the TSP-IL distance was especially large, however, she sometimes failed to touch the inner loop before she attached the sticky line, thus ignoring both the IL cue and recent memories of the TSP-IL distance (e.g., figs. 7.6, 7.12) (Eberhard and Hesselberg 2012).

The details of this story are currently only speculative. It could be, for instance, that she did not ignore the IL position cue, but used a continuous, non-linear weighting factor that reduced the relative importance of the IL cue when the TSP-IL distance was large. There are other possibilities. What seems clear is that the multiple cues for determining where to attach the sticky spiral may be combined in different ways in different contexts.

7.4 TEMPORARY SPIRAL

Working backward in the sequence of orb construction, the next stage is temporary spiral construction. Begin-

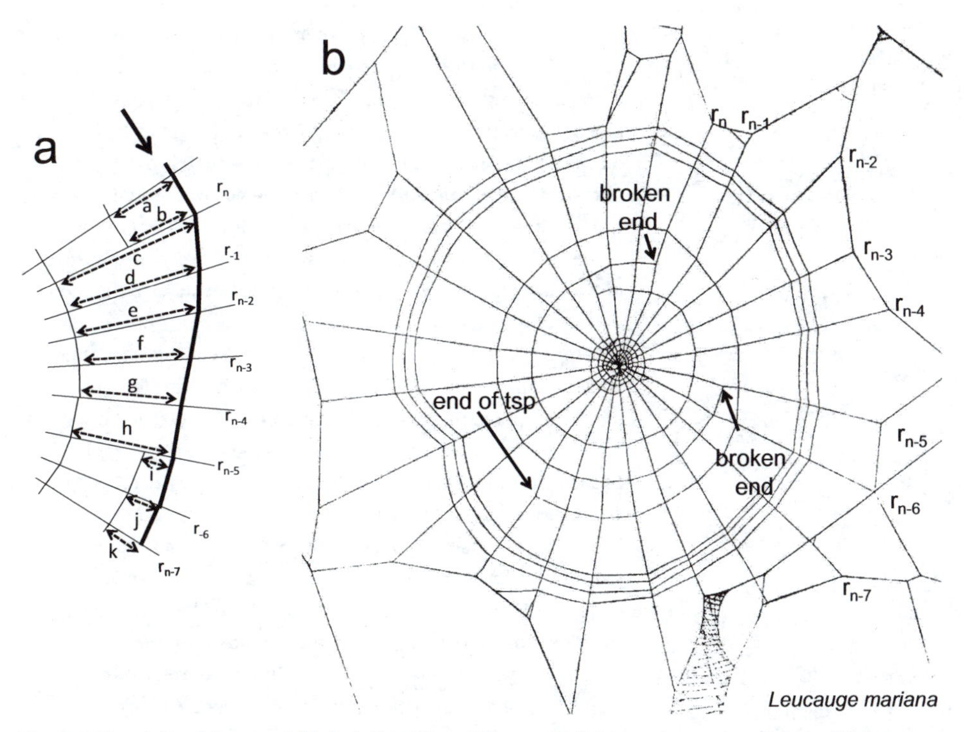

Fig. 7.30. The stylized drawing (*a*) labels the different lines and distances in an experiment with *Leucauge mariana* (TET) (*b*) in which five segments of the outer loop of temporary spiral were broken between radius r_{n-6} and r_n while the spider was laying temporary spiral on the opposite side of the orb. When she arrived at the modified sector (moving clockwise—heavy arrow in *a*), she veered inward as she laid the new temporary spiral line (heavy lines in *a*). The change was gradual (d > e > f > g), rather than abrupt as would be predicted by the simple hypothesis that the new loop was placed a standard distance from the previous outer loop. Recovery of normal spacing outside the experimental zone was also gradual (i < j < k). These gradual rather than abrupt adjustments suggest that the spider remembered the immediately preceding distances to the outer loop, and refrained from changing more than a certain fraction of this distance (section 7.4) (after Eberhard 1988c).

ning at the hub, the spider gradually spiraled outward, maintaining (in most but not all species) contact with the hub and then later with the outer loop of temporary spiral as she attached her dragline to each radius that she crossed. The hub and the outer loop were thus potential points of reference (table 10.3), and the distances from these points seem likely to be measured by the spider's own legs and body (e.g., Hingston 1920). Perhaps the measurements depended more on the body than the legs in *Araneus diadematus* (AR), because the temporary spiral spacing was not affected perceptibly when spiders had longer legs on one side than the other (Vollrath 1987a) (the web measurements reported in this study, however, were not detailed).

There have been few careful descriptions or experiments concerning the stimuli used to guide temporary spiral construction. There is one pair of experimental studies of temporary spiral placement by *Leucauge mariana* (TET) (Eberhard 1987b, 1988c), and one simulation experiment with *Araneus diadematus* (AR) (Gotts and Vollrath 1991, 1992). These studies provide some clues regarding the cues used to guide temporary spiral construction, and a cautionary lesson regarding simulations.

7.4.1 DISTANCES TRAVELED AND PATH INTEGRATION

Experimental cutting of segments of the outer loop of temporary spiral and of radii during temporary spiral construction indicated that the distance from the outer loop affected the placement of the next loop of temporary spiral, as had been supposed long ago (Hingston 1920; König 1951) on the basis of construction behavior. Experimental elimination of a portion of the outer loop resulted in a substantial inward displacement of the next loop in this same area in *Leucauge mariana* (TET) (fig. 7.30, Eberhard 1988c). But the effect was manifested only gradually (figs. 7.30, 7.31), rather than being abrupt

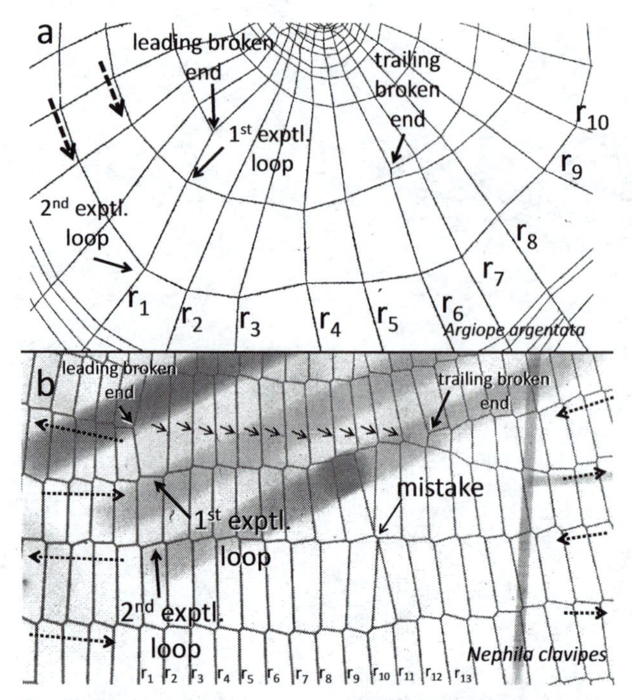

Fig. 7.31. The gradual adjustments in temporary spiral spacing in these incomplete orbs of the distantly related species *Argiope argentata* (AR) (*a*) and *Nephila clavipes* (NE) (*b*) reveal that they used cues similar to those used by *Leucauge mariana* (TET) (fig. 7.30) to guide temporary spiral construction (the dotted or dashed arrows indicate the directions the spiders moved while constructing different loops). When five segments of the first loop of temporary spiral (between radii 1–6) were experimentally cut while a mature female *A. argentata* was building temporary spiral (*a*), the path of the next ("first experimental") loop of temporary spiral veered gradually inward in the experimental zone (r_2 to r_6); by the time she reached r_6 the spider had returned to approximately the original spacing. Then, as she moved farther along where the original loop was still intact, she gradually veered outward (r_7–r_{10}). Another araneid, *Gasteracantha cancriformis* (AR), gave similar responses in a similar experiment. The nephilid *N. clavipes* showed the same pattern of gradual changes when 12 segments of the outer loop of temporary spiral between radii 1 to 13 were cut experimentally while the spider was building temporary spiral to the left of the area in the photo (*b*) (small arrows indicate the stubs of the severed temporary spiral lines). She gradually veered inward toward the hub as she laid the first experimental loop of temporary spiral in this sector; then, when she reestablished contact with the outer loop beyond r_{13}, she gradually veered outward. The second experimental loop did not follow the inward deflection of the first experimental loop. The different spider taxa probably all used the same two cues to guide temporary spiral placement: the site of the outer loop of temporary spiral already in place (which produced the inward displacement of the experimental loops in these experiments); and the memory of immediately preceding attachment sites (which produced the gradual rather than abrupt responses). This similarity is especially striking in *N. clavipes*, which used entirely different legs to contact the outer loop of temporary spiral ("mistake" marks an error in which *N. clavipes* fastened together two radii at the same temporary spiral attachment).

as would have been expected if the spider were simply measuring some standard distance from the outer loop Hingston (1920). Photos of similar experiments in *Argiope argentata* (AR) (fig. 7.31a), *Nephila clavipes* (NE) (fig. 7.31b), and *Philoponella vicina* (UL) suggested similar gradual responses in these species (WE).

A second experiment altered the positions of the temporary spiral in *L. mariana* orbs by breaking three adjacent radii just inside the outer loop of temporary spiral during temporary spiral construction (the outer loop sagged sharply outward on the broken radii, especially the middle radius, as a result) (fig. 7.15). In building the next loop of temporary spiral, the spider also made gradual adjustments: the spacing of the next loop of temporary spiral decreased gradually on the first and second broken radii, then gradually increased on the third broken radius and the first intact radius (figs. 7.15, 7.32, 7.33) (Eberhard 1988c). Similarly, on the third broken radus (r_4 in fig. 7.32), the farther inward the spider went while moving along the temporary spiral to reach r_4 (*y* in fig. 7.32), the greater the increase in the temporary spiral space on r_4. Single webs in which similar experiments were performed in four other species (one uloborid, three araneids) gave qualitatively similar results (fig. 7.33), giving a preliminary indication that the *L. mariana* responses may be typical of orb weavers in general.

Finally, in similar "experiments" that occurred naturally in unmodified webs of *L. mariana*, where the initiation of the temporary spiral occurred abruptly, the spider made similar gradual rather than abrupt changes in the distances between the second and first loops of temporary spiral in the first two or three attachments (fig. 7.12) (Eberhard 1988c).

In sum, these consistent responses to abrupt changes in the position of the outer loop of temporary spiral in different species and in experimental and natural contexts leave little doubt that this line guides the placement of the temporary spiral. In addition, direct observations of the species in these experiments, plus many other araneids, tetragnathids, nephilids, and uloborids showed

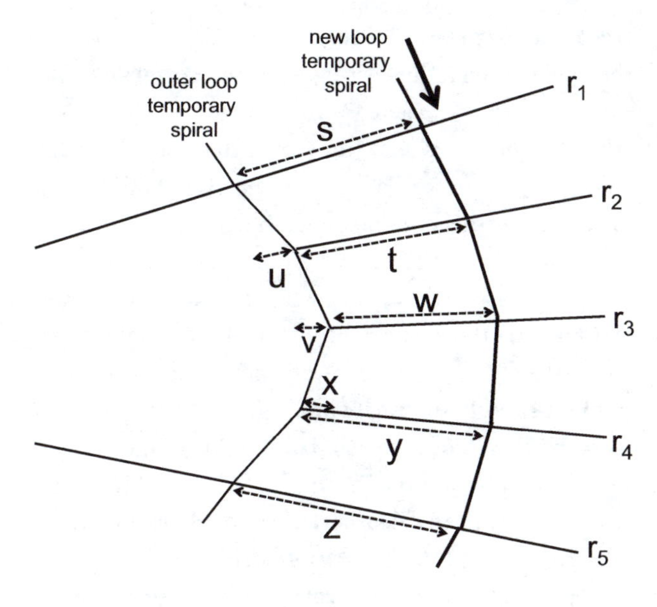

Fig. 7.32. This drawing (based on observations of *Leucauge mariana* [TET]—figs. 7.15, 7.33) illustrates the radial distances that the spider encountered while moving in a clockwise direction (heavy arrow) when she built the temporary spiral in an experimental sector where three adjacent radii (r_2, r_3, r_4) had been broken. The outer loop was deflected outward by the broken radii, so the spider moved outward (*u*) while walking along the outer loop of temporary spiral between r_1 and r_2, and then continued outward (*t*) along the radius r_2 before she attached the new loop. Similarly, in moving onward from this attachment, she moved inward (*t*), then outward along the outer loop (*v*) and farther outward (*w*) along radius r_3. Finally, in going from r_3 to r_4 she moved inward along the outer loop (*x*), and then outward (*y*) along radius r_4. Distance *s* in experimental webs was always less than $t+u$ and correlated strongly with it (r = 0.85); similarly, *t* was nearly always less than $v+w$ and they were strongly correlated (r = 0.84). These correlations support the hypothesis that the spider remembered and tended to standardize the distances that she traveled in radial directions when building the temporary spiral. The distances are described as if the spider walked along the lines; in fact her legs reached across spaces between lines; no part of her body actually moved along the lines (after Eberhard 1988c).

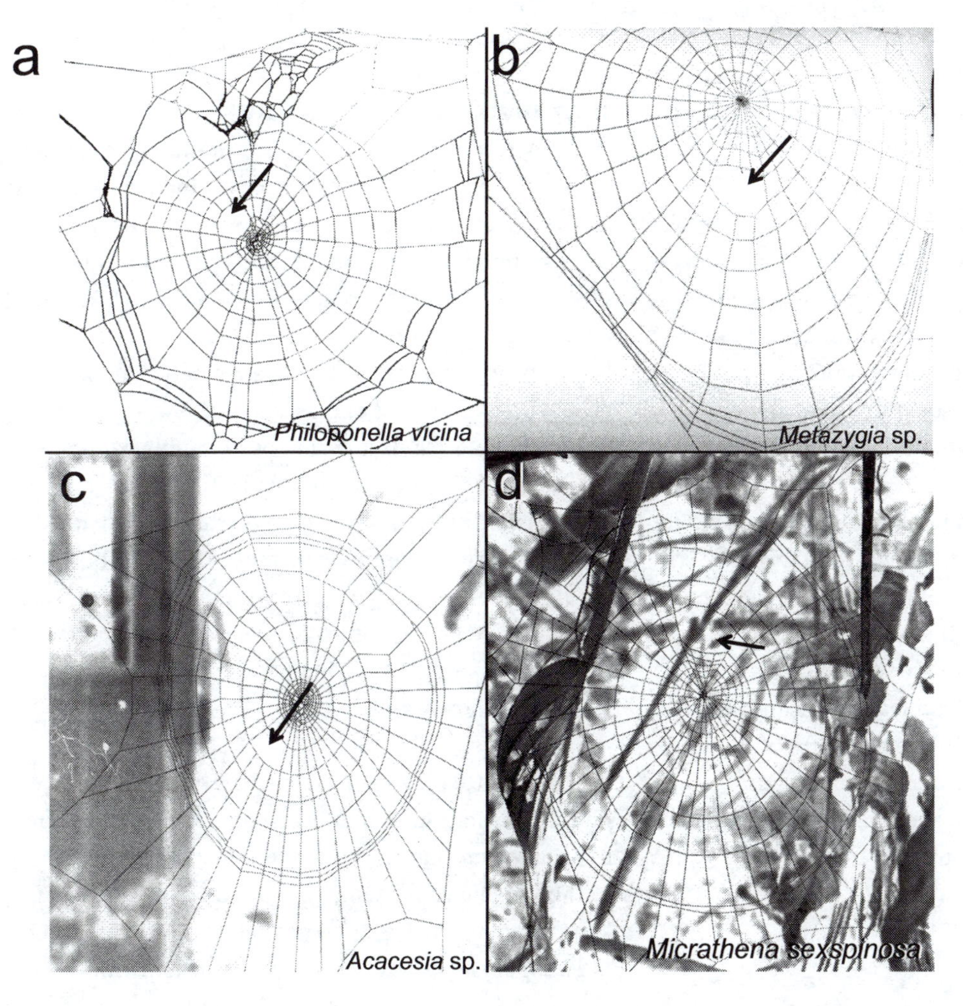

Fig. 7.33. Four other species in Uloboridae (*a*) and Araneidae (*b–d*) showed the same two modifications of temporary spiral spacing in response to cutting adjacent radii during temporary spiral construction as those seen in *L. mariana* (TET) (see fig. 7.30) (arrows mark the spaces where radii were broken). These similar response give further indications of the uniformity of the stimuli that guide orb web construction in different taxa.

that the spider maintained contact with the outer loop during temporary spiral construction (WE). Thus, as Hingston (1920) first proposed, some aspect of the spider's own body size is probably used to measure the distance from the outer loop. But these responses were consistently gradual, rather than abrupt inward or outward displacements of the outer loop (figs. 7.15, 7.31–7.33), implying that some further cue related to the distances traveled toward and away from the hub was involved.

One proposal (for *L. mariana*) was that the spider compared the distance she moved inward to reach the outer loop of temporary spiral following each attachment with some reference distance; if the reference distance was larger, the next space would exceed the inward distance by an amount proportional to the difference; similarly, if the reference distance was smaller, the next space should be proportionally less than the inward distance (Eberhard 1988c). This "sliding reference" hypothesis was supported by the correlations (s with $t + u$, t with $v + w$ in fig. 7.32) mentioned above.

An independent suggestion that spiders used measurements of distances they have traveled along radii to position their temporary spiral loops comes from the extraordinarily precise circular forms of the temporary spirals in some theridiosomatid orbs (fig. 7.7 of *Theridiosoma gemmosum*) (see also *T. epeiroides* in Shinkai and Shinkai 1985, and *Epilineutes globosus* and *Epeirotypus chavarria* in Coddington 1986a). These spiders are much too small to have maintained contact with the hub or temporary spiral lines during temporary spiral construction; probably they instead moved a standard distance outward along the radius to guide placement of the next loop.

Leg length was tentatively ruled out as a "sliding reference" cue in *Araneus diadematus* (AR), because temporary spiral spacing was not perceptibly affected in experimental spiders that had shorter legs on one side (Vollrath 1987a). This lack of an effect, however, did not rule out possible effects of leg length, because the sum of the lengths of the legs on the two sides of the spider as it held the outer loop of temporary spiral remained the same, whether the shorter or the longer legs were directed inward toward the hub.

7.4.2 GRAVITY

The angle of the radius with gravity was correlated with temporary spiral spacing in the approximately vertical orbs of *Zygiella x-notata* (fig. 7.27, Peters 1937) and the moderately slanted orbs of *L. mariana* (Eberhard 1987b). In *L. mariana* the loops below the hub were farther apart than those above the hub when the web slant was 45°, and this asymmetry was lost when the web was built at 0°. Changing the web's slant just before temporary spiral construction began showed that these differences resulted from both gravity and other web traits: when 45° webs were reoriented so that the side that had been below the hub was now above the hub, the spider made larger spaces in the half that was now below the hub (Eberhard 1987b). In *Araneus diadematus* (AR), which normally builds nearly vertical orbs, the vertical "eccentricity" of the temporary spiral (see fig. 7.23) (which was probably due in part to larger spaces between temporary spiral loops below the hub) was reduced when orbs were experimentally laid flat during temporary spiral construction (Zschokke 2011).

7.4.3 DISTANCE FROM THE HUB

In *Araneus diadematus* the spaces between successive loops of the temporary spiral typically increased rapidly as the spider left the hub, became more nearly constant in much of the web, and then decreased slightly near the outer edge (Gotts and Vollrath 1991). Preliminary data from an orb of *Micrathena duodecimspinosa* (AR) also indicated that the distance between loops increased on average as she circled out toward the web's outer edge (fig. 7.34); superimposed on this pattern was the response to gravity mentioned above. There are several possible cues that might guide this change, including kinesthetic memories of distances traveled, but there are no experimental data to test them. There was a significant partial correlation between radius length and temporary spiral (= 0.32) in one web (fig. 7.34) (WE), so radius length may be involved.

7.4.4 LACK OF EFFECT OF RADIUS TENSION

Radius tension did not affect temporary spiral spacing in *L. mariana* (fig. 7.15): there were no further changes in the spacing of subsequent loops attached to the broken radii after the first loop laid (above). Photos of webs of *Philoponella vicina* (UL), *Acacesia* sp. (AR), *Metazygia*

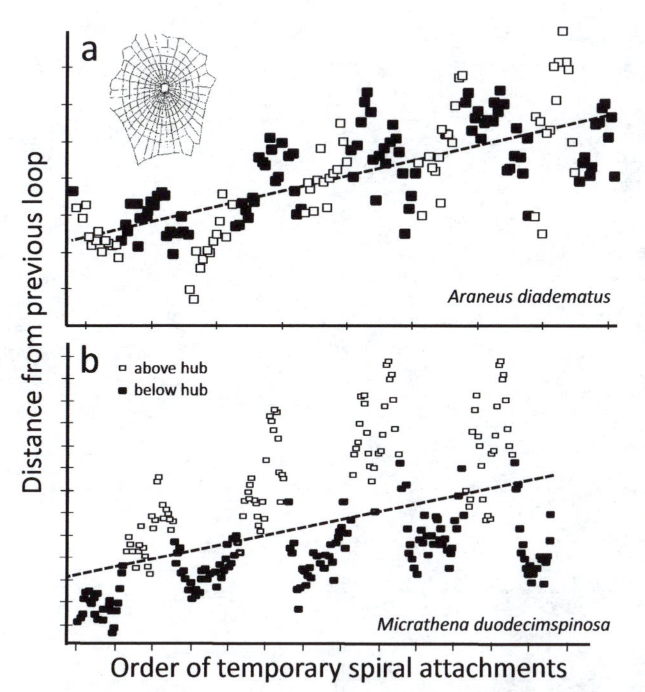

Fig. 7.34. The distances between loops of temporary spirals in webs of mature female *Araneus diadematus* (AR) (inset in *a*) and a mature female *Micrathena duodecimspinosa* (AR) (*b*), graphed in the order in which the spider made successive attachments of the temporary spiral, illustrate two trends. The spacing tended to increase (dotted lines) in both species as the spider spiraled away from the hub into the outer portions of the web, but with substantial scatter. Much of the variation in the spacing in *M. duodecimspinosa* (but not in *Araneus diadematus*) was due to consistently larger spaces above the hub (open rectangles) than below it (solid rectangles).

sp. (AR), and *Micrathena sexspinosa* (AR) suggested a similar lack of effect of radius tension in these species (fig. 7.33). As with sticky spiral construction, the lack of an effect of radius tension implies that radius tensions and vibrations probably do not influence decisions during temporary spiral construction, and that instead cues regarding the positions of lines sensed by direct contact are likely to be important.

7.4.5 ADDITIONAL POSSIBLE CUES

Variant webs that were built by *L. mariana* (TET) and *Micrathena* sp. nr. *lucasi* (AR) in spaces in which the hub was far from its usual central position suggested that spiders may use additional cues that have not yet been documented experimentally. In one of these "inadvertent" experiments the spaces between successive loops in the horizontal replacement web of a mature female *Leucauge mariana* (TET) were substantially reduced in the area where the remains of the sticky spiral of the previous web were especially close to the hub (the lower part of fig. 7.35). A similar pattern of flexibility occurred in a replacement web of *M.* sp. nr. *lucasi* (AR) (fig. 7.36). The normal positions of the body during temporary spiral construction (section 6.3.4) suggest that both of these spiders encountered the remnants of the previous web while they were building portions of the temporary spiral that had reduced spacing; these encounters may thus have been responsible for alterations in the spacing.

Replicated experimental manipulations of the slants of *L. mariana* webs during web construction suggested that stimuli from the web other than gravity (see section 7.4.2 above), also influenced temporary spiral spacing (Eberhard 1987b). When a web that was being built at a 45° slant was reoriented to the opposite 45° slant just after the hub was finished and before the temporary spiral had begun (the portion that had been above the hub was now below it and vice versa), the usual increase in temporary spiral spacing below the hub in 45° webs (control webs) was eliminated. In a different treatment, in which webs being built at 45° were reoriented when hub construction ended so that the slant was 0°, the temporary spacing in the portion that had previously been below the hub was larger, especially near the hub, but the difference was significantly reduced. This implied that both gravity and some other, as yet undetermined cue (or cues) from the web (longer radii below the hub?) were responsible for the usual above-below difference in temporary spiral spacing in *L. mariana* orbs (Eberhard 1987b).

I know of no studies concerning the cues that spiders use to decide to terminate the temporary spiral. In *M. duodecimspinosa*, the spider had usually contacted at least one frame line before she ended the temporary spiral (e.g., upper portion of the web in figs. 6.10, 7.15). When an *Araneus diadematus* (AR) was forced to rebuild the temporary spiral (by breaking all segments of the first temporary spiral she had built), the spider tended to build a smaller temporary spiral the second time (Peters 1970), suggesting that lower reserves of ampullate gland silk may induce termination (the data were from only a single web, however).

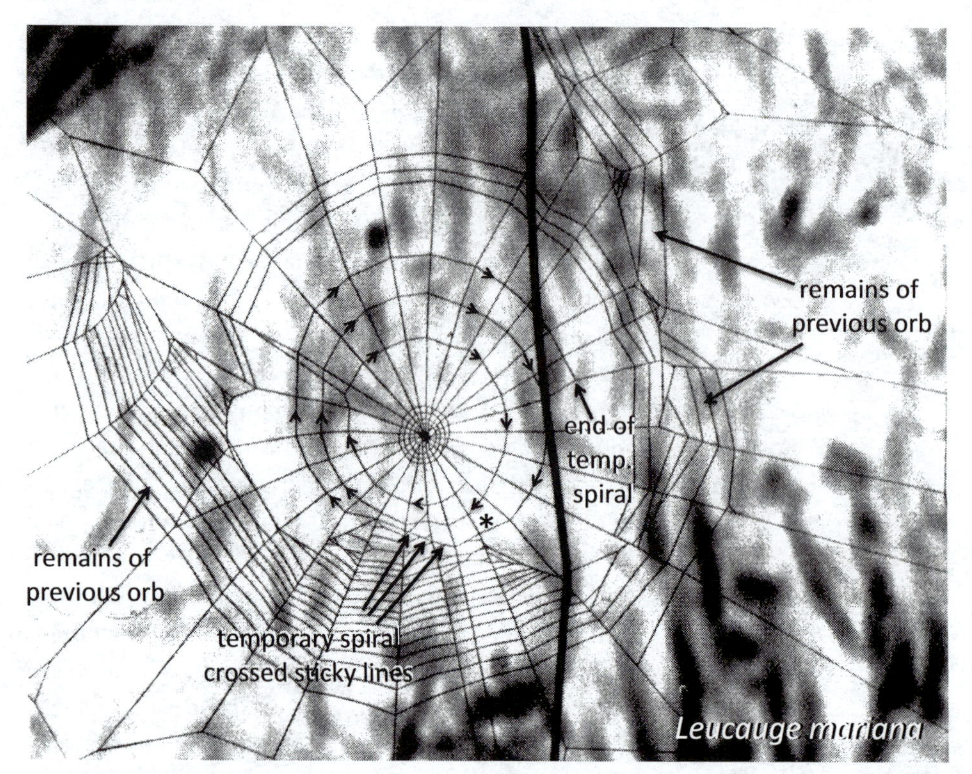

Fig. 7.35. A mature female *Leucauge mariana* (TET) temporarily suspended the usual rules for temporary spiral construction in this partially repaired horizontal orb (construction was interrupted after three loops of the repair sticky spiral were complete). The spider removed only part of her previous web before building the replacement (an unusual behavior in this species). She moved clockwise (small arrows) while building the temporary spiral. When she encountered sticky spiral lines from the previous web ("*"), she abruptly reduced the temporary spiral spacing, and subsequently reduced it even more and in addition repeatedly violated the "do not cross sticky lines" rule (arrows) (see fig. 7.36 for a similar pattern in an araneid, and figs. 7.2 and 7.45 for a similar violations of this rule during sticky spiral construction in two uloborids). These changes in behavior allowed the spider to extend the temporary spiral into a large, open area of her orb. Such flexibility in responding to the stimuli that normally guide temporary spiral construction does not fit previous images of orb weavers as automatons (see also figs. 6.2, 7.41, 7.48).

7.5 HUB

7.5.1 SPACES BETWEEN HUB SPIRAL LOOPS

There are, to my knowledge, no experimental studies of the cues directing hub construction. A few details of hub construction behavior, however, suggest possibilities. Early in hub construction *Leucauge mariana* (TET) held the center of the hub with her tarsus iIII, and she did not release this hold as she slowly turned, laying hub spiral (Eberhard 1987d). Gradually leg iIII was extended farther and farther as the spider spiraled outward; the leg finally released its hold when the hub spiral ended and temporary spiral construction began. In *Micrathena duodecimspinosa* (AR), leg iIII also held a central point in the hub while the first hub loop was built; but the spider then released this hold while she made subsequent loops of hub spiral (section 6.3.3.4). The sustained contact and gradual leg extension suggested that the spider used her leg as a compass to measure the distance from the center of the hub, and to build the nearly round hub loops that gradually spiraled outward.

Another possibility was that the outer rather than the inner leg III is used to guide hub spiral construction after the first loop has been built. During hub construction by *Nephila clavipes* (NE), leg oIII slid inward along the radius until it reached the junction with the outer hub loop, and then immediately grasped the radius and then attached the hub spiral (appendix O6.3.4.2). Once the spider's leg oIII had grasped the radius at or near the junction with the outer loop of hub spiral, the spacing of the next loop would be determined by where she contacted the radius with her spinnerets relative to leg oIII; no data are available to determine whether the spinnerets were positioned consistently with respect to leg

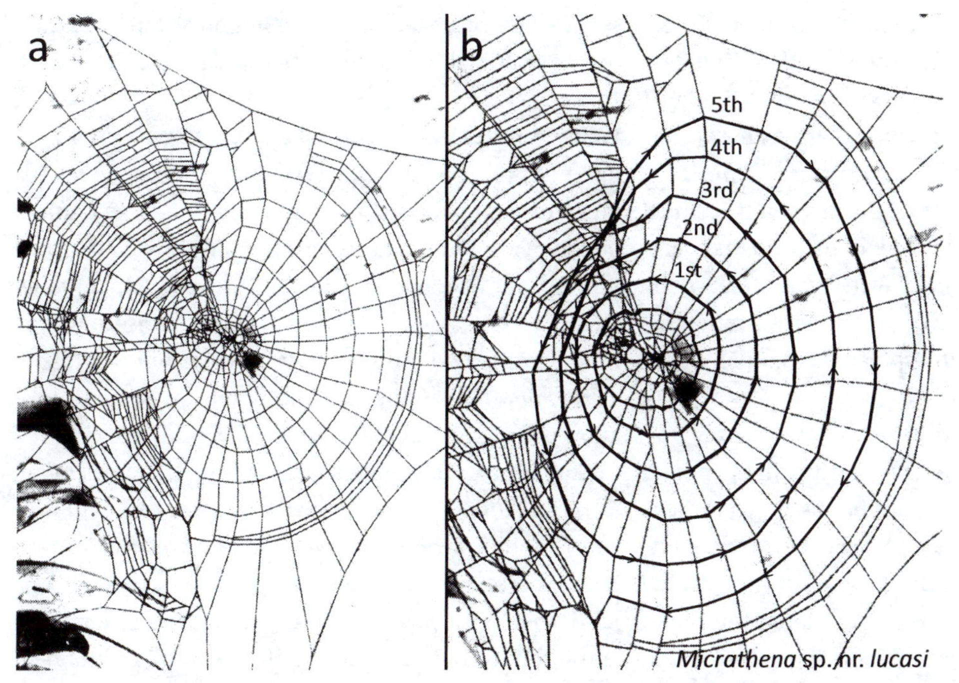

Fig. 7.36. Sometimes an unusual event in which an animal overcomes an impediment is more revealing than is the usual smooth execution of normal behavior. This tangled, partially repaired orb of a female *Micrathena* sp. nr. *lucasi* (AR) (*a*) is an example. The spider's first orb of the morning was damaged by a falling leaf, and later the same morning she replaced the damaged section and the hub (I interrupted her repair after she had finished the temporary spiral and had built three sticky spiral loops). The tracing on the right (*b*) depicts the temporary spiral in bold and the directions that the spider moved with arrowheads; it shows how she crossed the more or less disorganized tangle of sticky and non-sticky lines from the original orb going downward while she built each of the first three loops of the temporary spiral; despite the tangle of lines, she maintained a nearly constant bearing in each loop until she reached the other side of the replacement sector. Finally, she doubled back on her fourth encounter with the original web. The spider's ability to maintain her bearing despite encountering many disorganized lines is not evident in normal orb construction. How was she able to do this? Replacement repairs are rare in the field in *Micrathena* (I have seen only two among several hundred webs), so pre-programmed repair behavior seems unlikely. Alternatively, the web may have resulted from an insightful use of an ability to walk in a circular path (see text).

oIII. Legs oIII of *L. mariana* and *M. duodecimspinosa* were in similar positions and could be performing similar sensory movements during hub construction, but their rapid movements and their positions near the spider's body precluded resolution of this detail in video recordings.

In species in which some hub spiral attachments were double and at least sometimes pulled the radius into a zig-zag shape, spiders must make further decisions to determine whether to make a single or a double attachment to each radius, and how tight to pull the hub spiral line. Given the radius-tightening effects of such attachments, I speculate that cues associated with lower tensions on particular radii elicit double attachments, but no data are available to test this idea.

These simple ideas will probably require further modification to take into account further variations. When adult female *Leucauge argyra* (TET) built orbs in highly constrained spaces, for instance, they reduced the spaces between loops of hub spiral, the area of the hub, and the number of hub loops (fig. 7.42) (Barrantes and Eberhard 2012). Some cue or cues related to the area available for the orb appears to be added to the simple body size variables just discussed, at least under these extreme conditions.

7.5.2 TERMINATION OF THE HUB SPIRAL

In many orb weavers it is difficult to investigate the factors that may induce the spider to terminate the hub construction, because the hub loops do not terminate abruptly, but are instead gradually spaced farther and farther apart as they gradually turn into temporary spirals (e.g., fig. 6.10). In tetragnathids, theridiosomatids, anapids, and symphytognathids, however, there was an abrupt transition from one to the other (figs. 7.6, 7.7). In *L. mariana* gravity probably affects the decision to terminate hub construction, because the hub loop nearly always ended in the upper 45° sector of the orb (webs have not been rotated during or just preceding hub construction, however, so the possibility cannot be eliminated that some other trait of the web that correlates with gravity, rather than gravity itself, affects hub termination) (Eberhard 1987d). In addition, termination almost always occurred as the spider was spiraling upward rather than downward.

The number of hub loops, which is presumably related to termination, correlated positively with the number of radii both intra-specifically and inter-specifically (fig. 3.29, section 3.3.2.6, Savory 1952; Barrantes and Eberhard 2012). It seems unlikely that spiders count either hub loops or radii, so presumably some other, correlated stimulus is responsible for inducing the spider to terminate hub construction.

7.6 STABILIMENTUM CONSTRUCTION

Silk stabilimenta that are placed on or near the hub are generally built after the rest of the orb is complete (Herberstein et al. 2000b). I will discuss three different decisions related to stabilimentum production.

7.6.1 BUILD A STABILIMENTUM OR NOT?

The presence or absence of a stabilimentum is notoriously variable, even in the different webs of the same individual (section 3.3.4.2.2.1). Various cues appeared to influence this decision in different species. A greater amount of silk in reserve may have increased the likelihood of stabilimentum production in *Argiope aurantia* and *A. trifasciata* (AR), as better-fed spiders built more stabilimenta (Blackledge 1998a). Brighter illumination during the night also increased the likelihood that a stabilimentum would be built by *Argiope pulchella* (AR) (Marson 1947a) and *Uloborus diversus* (UL) (Eberhard 1973a) (it had no effect, however, in several other species of *Argiope*) (table O3.3). One study concluded that windy weather may induce stabilimentum construction in *Cyclosa insulana* (Neet 1990), but these data were not convincing (Herberstein et al. 2000b). Experiments in *Allocyclosa bifurca* (AR) showed that the presence of egg sacs or a detritus stabilimentum inhibited the construction of a silk stabilimentum (Eberhard 2003).

7.6.2 WHERE TO PLACE THE STABILIMENTUM?

It seems probable that gravity is used to select the radius on which to build a stabilimentum by some species that build their stabilimenta on vertical radii above and below the hub, as in the araneid genera *Cyclosa* and *Allocyclosa* (Hingston 1927, 1932; Robinson and Robinson 1970a, 1974; Eberhard 2003), and *Argiope* (Hingston 1932), and in *Nephila* (NE) (Robinson and Robinson 1973b), but I know of no experimental test of this hypothesis.

Stabilimentum lines in the more or less horizontal orbs of *U. diversus*, where gravity cues are absent or less consistently available, tended to occur on relatively short radii, and on radii that ended near anchor lines (Eberhard 1973a). One possible cue to achieve this orientation is the reduced extensibility of shorter radii that ended near anchor lines when the spider pulls on them while at the hub, just prior to stabilimentum construction (appendix O6.3.8.2; Eberhard 1973a). Spiders did not leave the hub area during or just prior to stabilimentum construction, so they presumably used only indirect cues to judge the length of a radius and its position in the web. If radius extensibility is used as a cue, this would be the first proof of a tension-related cue being used in orb construction (section 7.11).

The function of placing linear stabilimenta on these radii is probably defensive. During the day the spider generally rests at the hub aligned with a stabilimentum line, with her anterior legs extended anteriorly (as in *Zosis geniculata*—fig. 3.46). Placement of the stabilimentum on a short radius that leads directly off the web would enable the spider to leave the orb and reach the substrate quicky to avoid attacks from predators (Eberhard 1973a).

7.6.3 WHICH STABILIMENTUM DESIGN?

Many species build more than a single stabilimentum design, and in some species the same individual often builds different designs on sequential orbs (Marson 1947a,b; Ewer 1972; Edmunds 1986; Blackledge 1998a; Herberstein et al. 2000b). For example, *Argiope argentata* (AR) built 1, 2, 3, or 4 arms of an "X" of zig-zag lines at the hub (Robinson and Robinson 1970a); the uloborids

Uloborus diversus and *Zosis geniculata* sometimes built linear stabilimenta, sometimes built loops of stabilimentum silk, and sometimes combined the two (Eberhard 1973a, WE).

Several stimuli may affect the choice of stabilimentum design. In *A. aurantia* and *A. trifasciata* (AR) better-fed spiders built more extensive stabilimenta (Blackledge 1998a). In *Z. geniculata* there was a strong correlation between the size of the space in which the web was built and the stabilimentum design: in the large, 50 cm diam. spaces, 26 (58%) of 45 webs had circular loops of stabilimentum silk at the hub, while circular stabilimenta never occurred on any of 71 webs built in small (5.8, 6.5, 7.8, and 14.8 cm diam) containers (WE). The functional significance of this association is unclear (see section 3.3.4.2 for ideas regarding stabilimentum function).

7.7 RADII, FRAMES, AND ANCHOR LINES

Radii are laid in several contexts. Some radial lines are laid, shifted, and removed early during preliminary exploration; only a few of these are incorporated into the final orb (section 6.3.2). After the definitive hub has been established, additional radii are added as part of primary frame construction ("primary" radii), or during the process of secondary or tertiary frame construction ("secondary" radii).

I will begin with secondary radius construction, which is simple and repetitive. The spider turns slowly at the hub until she selects an sector into which she will insert a radius (the "open sector"), attaches her dragline at the hub (or to an intersection between radii), then moves out along one of the two radii bordering the open sector (the "exit radius") laying a new radial line. The angle between the new radius and the exit radius is "final" in the sense that the spider very seldom adds any more radii in the space between the radius and the exit line (section 6.3.3.2, Reed 1969).

The spider makes at least three decisions during each new secondary radius construction: In which sector of the web should she build a new radius? Which of the radii bordering the open sector should be the exit radius? And how far from the exit radius should she attach the new radius to the frame? Additional possible decisions concern whether or not to add a secondary frame as part of radius construction, and under how much tension each radius should be laid. To my knowledge, there are no experimental studies that demonstrate directly which cues are used in these decisions (see, however, Reed 1969). Thus understanding of radius construction is a poor cousin of sticky spiral construction, and strong experimental tests of most hypotheses are lacking. Nevertheless there are several indications based on construction behavior and correlations in finished webs that suggest which cues may be used.

7.7.1 SECONDARY RADII

7.7.1.1 Choosing an "open" sector

Details of the spider's turning behavior at the hub prior to the construction of each radius have suggested three different hypotheses regarding the stimuli used to choose the next sector in which to lay a radius. One possibility, proposed by Reed (1969), is that angles per se are not involved, and that "the sector to be filled by a new thread is determined by the interplay of length, orientation, and mooring of the threads already present." The details of this "interplay" were not specified, however, so it is not feasible to test this hypothesis. A second hypothesis is that the spider sensed the open sector via a break in the rhythm of her turning movements at the hub (Tilquin 1942); when she encountered an oversized space between radii, the rhythm of her turning was broken (presumably because she failed to immediately find a radius to grasp in the oversized space). In *N. clavipes* (NE), whose movements were slower and easier to resolve, the break in the rhythm seemed to occur when the leading measuring leg (oII) moved forward as if to grasp the next radius, and failed to find it because the space was oversized; the spider then waved the leg one or more times in the space before it contacted the next radius (WE). The rhythm hypothesis was eliminated, however, by details in video recordings of *Micrathena duodecimspinosa* (AR) and *Lecauge mariana* (TET) in which successive radii were laid on opposite sides of the web, especially in the earlier half or two-thirds of period of secondary radius construction. The spider hardly paused at the hub between finishing one radius and beginning the next; there was simply no rhythm of turning at the hub to be interrupted (WE).

The third, most popular view supposes that by successively grasping adjacent radii as she turns at the hub (section 6.3.3), the spider measures the inter-radial angles between radii by the separation between her legs (Hingston 1920; König 1951; Savory 1952; Eberhard 1972a). In most araneoids, these "measuring legs" were the legs I; they were oII and oIII in *Nephila* (NE) (appendix O6.3.5.2); and they were oI and oII in uloborids (appendix O6.3.5.3). By comparing this angle with some minimum value, the spider decides whether to add a radius in this sector (whether the sector is "open"). Behavioral details from video recordings of *M. duodecimspinosa* were compatible with this angle measurement hypothesis (fig. 6.7). Spiders always grasped the two radii bounding the sector into which a new radius was about to be laid (the "edge radii"). The legs were moved in a highly stereotyped pattern as the spider turned at the hub, and each successive pair of adjacent radii was almost always grasped at least briefly simultaneously by the two measuring legs. When a spider found an "open" sector in which she would add a new radius, she often grasped the two radii bordering the open sector with her legs I and jerked them gently one or more times. The jerk might constitute a second measurement in addition to the inter-radial angle, but seems more appropriate to measure the extensibility of the two radii, and thus their lengths (longer lines will, other things being equal, extend more under a given force) (see below for the likely importance of radius length).

The spider thus had access to information on inter-radial angles as she turned at the hub. Furthermore, in uloborids and nephilids the spider always turned to face outward away from the hub prior to laying the radius only after grasping the two edge radii with the measuring legs. That is, the spider only made the decision to add a radius after having had the opportunity to sense the angle between the edge radii. Finally, in both *M. duodecimspinosa* and *L. mariana* the spider usually probed several times with legs I and II in the open sector while holding the edge radii with other anterior legs (fig. 6.7). This behavior was appropriate to confirm that there was no radial line already present in the open sector that she had missed. Hingston mentioned apparently similar leg movements by *Neoscona nautica* (AR) when radii were experimentally broken ("search[ed] vainly with its legs for the broken radii. . . . It was most amusing to see the little creature sweeping its leg in the air for the lost radii," Hingston 1920, p. 133–134). Further support for the angle measurement hypothesis comes from the fact that removal of legs I in *Araneus diadematus* reduced the degree of similarity in adjacent inter-radial angles (Reed et al. 1965).

Sensing inter-radial angles may be somewhat more complicated than just described. Measuring inter-radial angles via the angle between the legs that hold them is complicated by the fact that that the separation between radii that is associated with a given inter-radial angle differs according to the spider's distance from the center of the hub. A given inter-radial angle will be associated with a smaller inter-radial distance (and thus with a smaller separation between the spider's legs) when the spider is closer to the geometric center of the hub. This complication could be more serious in some groups like uloborids and nephilids in which the hub is gradually built throughout the period in which secondary radii are added: the spider gradually moves farther from the center of the hub as secondary radius construction proceeds. One untested possibility is that the degree that her leg is rotated to grasp a radial line (section 2.4.2.3) informs the spider regarding the orientation of the radius, and thus permits more precise angle measurements.

Reed (1969) used the powerful technique of experimentally eliminating individual radii repeatedly during radius construction and photographing each replacement radius to demonstrate that in 8 *Araneus diadematus* (AR) webs, one simple version of the angle hypothesis (that the spider fills in the radii by adding a radius to the largest open sector remaining in her orb) was inadequate (Reed 1969). Larger angles (open sectors) were not always filled preferentially. For instance, the 30.8° angle between r_g and r_h in fig. 7.37 was filled on one occasion while the 43.5° angle between r_k and r_l was left unfilled; and the 42.8° angle between r_p and r_a was filled three times during the experiment, but was not subsequently replaced after the experiment ended, while seven other radii were added to smaller open sectors. There was substantial variation in which sector received the next radius (e.g., five different sectors were chosen at different times in the web in fig. 7.37). Similar cases of some smaller rather than larger sectors being filled occurred in other webs (Reed 1969).

The version of the angle measurement hypothesis that Reed tested, however, was based on the unlikely suppo-

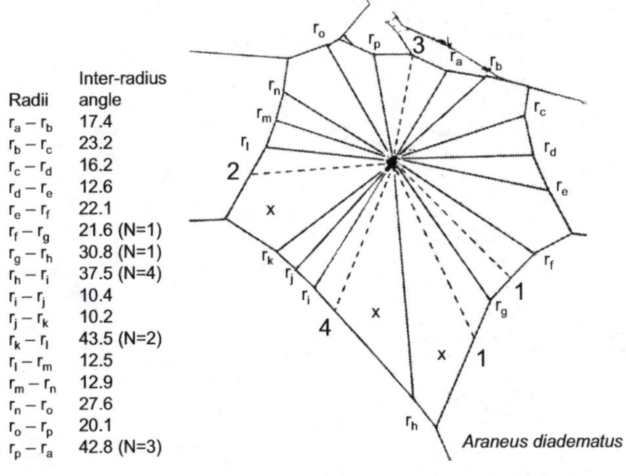

Radii	Inter-radius angle
$r_a - r_b$	17.4
$r_b - r_c$	23.2
$r_c - r_d$	16.2
$r_d - r_e$	12.6
$r_e - r_f$	22.1
$r_f - r_g$	21.6 (N=1)
$r_g - r_h$	30.8 (N=1)
$r_h - r_i$	37.5 (N=4)
$r_i - r_j$	10.4
$r_j - r_k$	10.2
$r_k - r_l$	43.5 (N=2)
$r_l - r_m$	12.5
$r_m - r_n$	12.9
$r_n - r_o$	27.6
$r_o - r_p$	20.1
$r_p - r_a$	42.8 (N=3)

Fig. 7.37. This unfinished web of an *Araneus diadematus* (AR) gives evidence contradicting simplistic versions of the "angle hypothesis" that the spider always chooses the largest open sector for the next radius. The spider had built 16 radii undisturbed (r_a to r_p), but then each additional radius that she constructed was immediately removed experimentally, leaving the web with its original 16 radii. A total of 11 experimental radii were built and then destroyed in five different sectors (dotted lines) (the numbers/sector are shown; they varied from 1 to 4). The spider was then left to complete the orb. She replaced all of the radii that had been destroyed during the experiment except the three in the sector between r_p and r_a, and also added a radius in three other sectors (indicated by "x"). Overall, new radii tended to be placed in the larger open sectors, both in the final web and during the experimental removal phase; but they were not consistently placed in the largest sector available. For instance, sector r_k–r_l was the largest, but only received 2 of the 11 experimental radii; and the second largest sector, r_p–r_a, failed to receive a radius in the finished web, even though it had been filled three different times previously (from Reed 1969).

sition that the spider can sense all radial angles each time she returns to the hub. A more reasonable supposition would be that she senses only the angles between the lines that she grasps with her legs I, and that the appropriate comparisons regarding which sectors are filled with radii includes only those radii that she encountered (grasped with these legs), and then filled or failed to fill. Behavioral details of this sort are available for only one species, *Uloborus diversus* (UL), in which both the order in which radii were constructed and the direction that the spider turned at the hub while she was building them can be deduced from the connections of the hub spiral (fig. 6.11, Eberhard 1972a). An analysis of ten webs demonstrated that it was common for a spider to skip sectors where she would subsequently lay one or more radii; about 30% of the final ten radii laid in an orb, and 60% of the previous ten radii, involved skipping at least one open sector where one or more radii would later be laid; the spider skipped up to six such sectors as she turned at the hub before laying a radius. In sum, a more complex version of the angle hypothesis is needed.

7.7.1.1.1 RADIUS LENGTH

A further variable, radius length, affects decisions where to add radii. The inter-radius angles were larger in smaller webs of given individuals in *Uloborus diversus* (UL) (fig. 7.38a), *Zosis geniculata* (UL), and *Leucauge argyra* (TET) (fig. 10.15d); they were also larger between shorter radii in the same orb in several species (fig. 7.39, König 1951; Mayer 1952; Le Guelte 1966; Eberhard 1972a). Thus the minimum angles that signaled that a sector was "open" and elicited radius construction were probably smaller for longer radii.

The strength of the effect of radius length on radius construction is probably underestimated by the graphs in figs. 7.38a and 7.39. Some inter-radial angles in an orb are not the result of the spider's decision to lay a new radius at such an angle with the adjacent, exit radius, but are "left over" angles between groups of radii that the spider chose not to fill (Reed 1969). For instance, in fig. 6.10, the angle between radius 48 and o_x was "left over," while the angles between 30 and 48, 27 and 30, 22 and 27, etc. were due to active attachment decisions of the spider. If one separates the angles associated with the radii adjacent to anchor lines (triangles in fig. 7.39), the correlation between radius length and angle appears to be tighter. In sum, the length-angle correlation in radii implies that the spider probably sensed both the lengths of adjacent radii (possibly by radius extensibility) and the angles between them (possibly by the angles between her legs as she held them) when she was deciding which open sector of the web should receive the next radius.

Whatever the cues were that spiders used, they sometimes led to quite consistent decisions. Hingston (1920) found that if he cut a newly laid radius just after *Neoscona nautica* laid it (during hub construction, when all or nearly all radii had already been built), the spider replaced it with a new radius in the same sector up to 25 times. Finally the spider simply left the sector with the broken radius open, and proceeded to build the temporary spiral and finish the web. Similar experiments with *Leucauge mariana* gave generally similar results (WE).

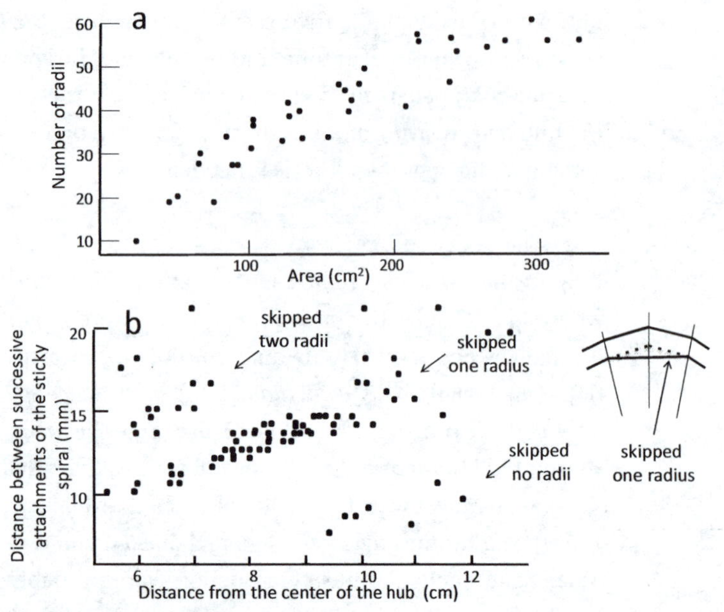

Fig. 7.38. The graph in *a* of data from the webs of a single mature female of *Uloborus diversus* (UL) illustrates two general trends in this species: the number of radii tended to be larger when the estimated area of the web was larger; and there was substantial variation in both variables, even in the webs of a single individual. Graph *b* is from a portion of the orb of a penultimate female *U. diversus* where the angles between adjacent radii were nearly the same, and illustrates another trait that is apparently unique to uloborids—skipping some radii during sticky spiral construction (illustrated in the drawing on the right; see section O6.3.6.4.2). The pattern in the data suggests that the length of the current segment of sticky spiral (or a related variable) affects the spider's decision whether or not to attach her sticky spiral to a given radius. At intermediate distances from the hub (about 7–9 cm), the spider consistently chose to skip alternate radii (as in the drawing), and built segments of sticky spiral that were about 11–13 mm long. At larger and smaller distances from the hub, her choices were less consistent. The consequences of skipping radii included larger spaces between sticky lines and reduced investment in sticky lines (section 7.3.5.2) (after Eberhard 1972a).

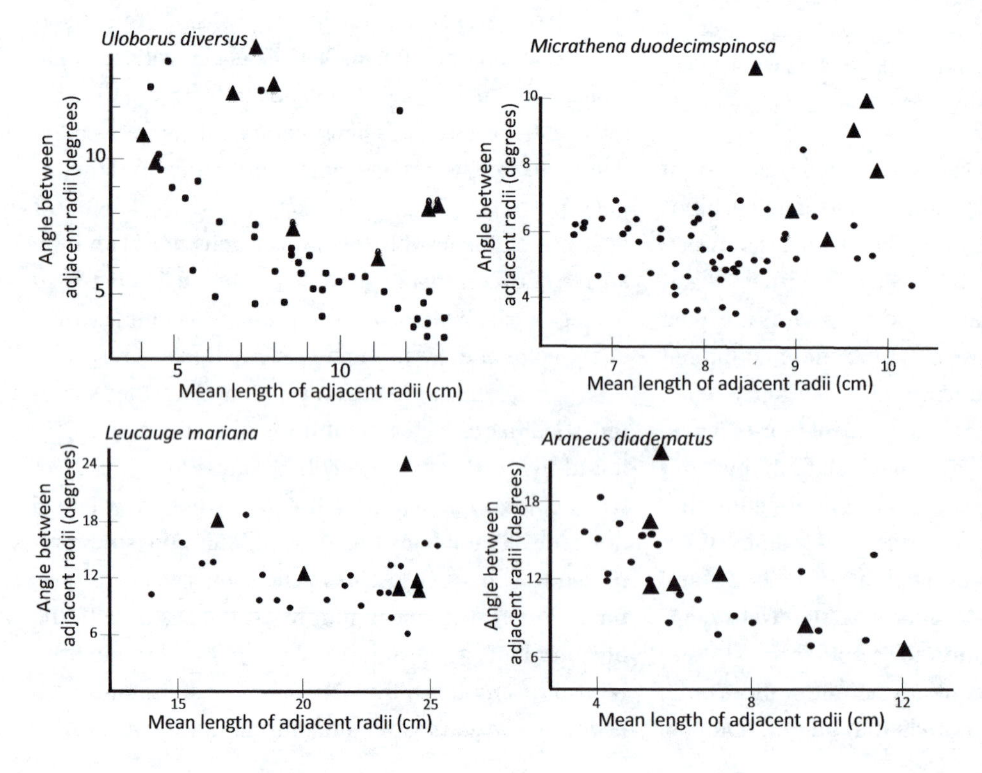

Fig. 7.39. The general negative relationship between radius length and inter-radius angles is illustrated here in typical orbs of *Uloborus diversus* (UL) (from Eberhard 1972a), *Micrathena duodecimspinosa* (AR) (the web in fig. 1.4), *Leucauge mariana* (TET) (the slant was 45°), and *Araneus diadematus* (AR) (from Zschokke 1999). The angles between the radii on either side of an anchor line (indicated with triangles) tended to be larger in all four species (Eberhard 1972a).

In *L. mariana* not all the replacement radii were attached at exactly the same site on the frame, suggesting some level of imprecision (Eberhard 1990c) (see also Reed 1969). It was not possible, however, to cut the radii without leaving a short slack remnant attached to the frame line, so it is possible that stimuli from these loose segments of line were responsible for this variation.

7.7.1.1.2 FALSE STARTS

One further type of behavior that may be related to choosing the sectors in which to add new radii is the occasional "false start" behavior seen in *L. mariana*, *U. diversus*, and *Philoponella vicina* (UL) (Eberhard 1972a, 1990b). False starts may result when the spider originally decided to lay a new radius while at the hub, but then "changed her mind" after starting to move away from the hub on the exit radius. There are other possible explanations, however (e.g., adjusting tensions on radii). The frequency of false starts varied substantially among different species (appendix O6.3.2); their distribution in the web with respect to radius length and inter-radial angles is not known.

Some species showed a tendency to make successive radii on opposite sides of the web (e.g., Savory 1952; Reed 1969; Dugdale 1969; section 6.3.3.2), so another stimulus that probably guided radius construction is a memory of the position of the immediately preceding radius. I know of no studies of this stimulus; the "memory" could be as simple as the position of the spider's body, which faced away from the previous radius when she returned to the hub.

7.7.1.2 Choosing an exit radius

When a spider selected an open sector, she chose one of the two radii bordering this sector to use as an "exit" radius to move to the frame. In vertical orbs such as those of the araneids *M. duodecimspinosa* and *A. diadematus* and many others, the upper of the two was virtually always the exit radius (section 6.3.3). In *U. diversus*, which built nearly horizontal orbs, the spiders showed a weak tendency to use radius length: the shorter of the two radii was the exit in 59% of 133 cases in 10 webs, $p = 0.03$ with chi^2). The occasional gentle jerks just before the spider left the hub might have given the spider information related to radius length (longer radii, if they are under equal tension, would be more extensible). In addition, *U. diversus* showed a stronger tendency ($p < 0.01$) to lay successive radii in an open sector on the same side of the sector (in fig. 6.11, for example, the exit radii for radii 33, 29, 24, 19, and 6 were all on the left side of the sector, while only the exit for radius 12 was on the right side). In another uloborid, *Philoponella vicina*, the radius chosen to be the exit was usually (100 of 127 cases in 31 webs) on the "leading" or far side of the sector that would be spanned (Eberhard 1990b). This bias may explain the consistency in *U. diversus*, because uloborids apparently only seldom turn back while laying hub loops (1–2 times/web in *U. diversus*—Eberhard 1972a); thus the leading edge of a given sector would tend to be the same for much of the construction period. The functional significance of exiting along the leading edge radius is unknown.

7.7.1.3 Choosing a final angle: how far to move along the frame

How does the spider determine where to attach a new radius to the frame, and thus establish the angle that it will make with the exit radius? These decisions affect the distances between adjacent radii and the number of radii in the entire web. These variables, in turn, have important consequences for the ability of the orb to absorb the impacts of high-energy prey (section 3.3.2.1). There were clear patterns in both inter-radial angles and in the total numbers of radii in orbs in several araneids (e.g., Peters 1939, 1954; Mayer 1952) and also *U. diversus* (UL) (Eberhard 1972a). Peters (1939, 1954) discussed one especially clear pattern, the "radius rule" ("Speicheregel" or "Regla de rayos"): the angles between adjacent radii were smaller in webs that had longer radii. Spiders clearly adjusted inter-radial angles in different circumstances.

Nevertheless, a spider is surely *not* able to measure inter-radial angles directly at the moment when she is attaching the new radius to the frame; she is at the edge of her web, while the angle between the new radius and the exit radius is formed at the hub (fig. 7.40a). One variable that would be feasible for the spider to sense and which is tightly linked to this angle is the distance that she moves along the frame before attaching the new radius (*f* in fig. 7.40a). The farther she moves along the frame before attaching, the larger the inter-radial angle will be. Several authors have hypothesized that the spider uses her own body size as a reference (sec-

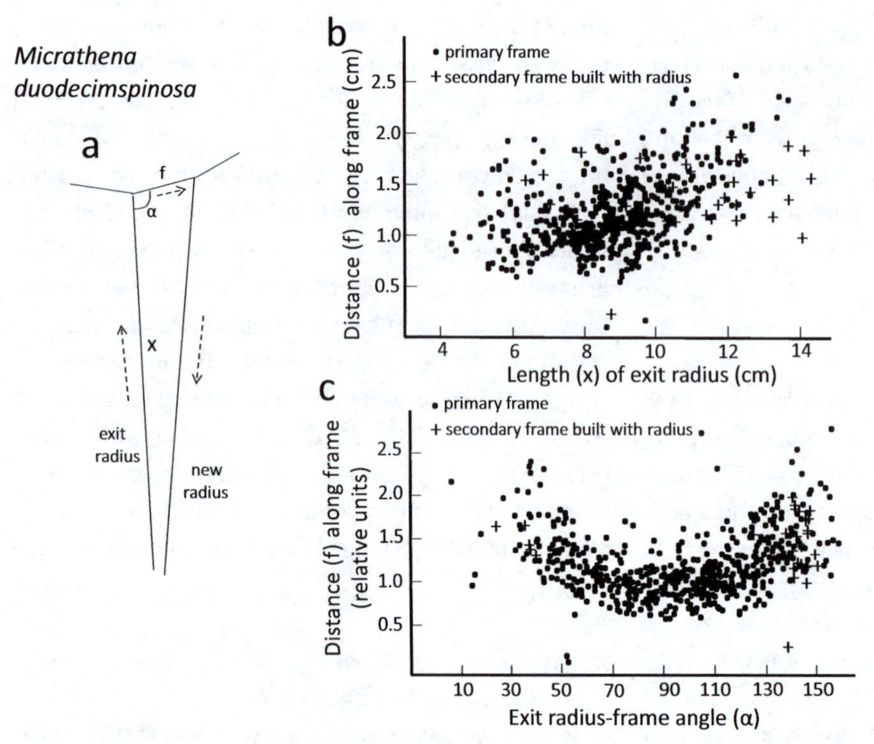

Fig. 7.40. The path of the spider during secondary radius construction is indicated in *a* by dashed arrows: the spider moved away from the hub a distance *x* along the exit radius, laterally along the frame a distance *f* after turning through angle α, and then back to the hub along the new secondary radius. The distances traveled along the frame line in 13 webs built by 13 different adult females of *Micrathena duodecimspinosa* (AR) are illustrated in *b* and *c* (measured from photographs of webs in which the exit radii were known from direct observations of construction behavior). The distance *f* was greater when the length of the exit radius (*x*) was longer (*b*), and when the angle made by the exit radius with the frame line (α) differed more from 90° (*c*). In a multiple regression analysis of *f* on the independent variables *x* and the absolute value of α – 90°, the beta values for both independent variables were significant. Those radii that were built as part of secondary frame line construction (section 6.3.3.3) (crosses) tended to be relatively longer, and to show less correlation with radius length. This suggests that the spider had made the decision to build a secondary frame earlier (on the way out from the hub?), before she attached her dragline to the frame (after Eberhard and Hesselberg 2012).

tion 7.3.4.1), but with less than completely convincing arguments.

Hingston (1920) argued that in *Neoscona nautica* (AR) the "radius rule" (the inverse correlation between radius length and the inter-radial angle) was the result of a very simple mechanism: the spider moved a constant distance ("four paces") laterally along the frame from the exit radius before she attached each new radius. He found that in *Nephila pilipes* (= *maculata*) (NE) a fixed distance was achieved by holding the exit radius with one leg IV while attaching the new radius (Hingston 1922a). Hingston noted that if the distance along the frame is constant, the inverse correlation between inter-radial angles and radius lengths in orbs will result. These ideas are not completely satisfactory, however. It is not clear how he counted "paces" (for instance, which legs were used?). More importantly, the distances traveled along the frame lines are far from constant in many species (Peters 1936, 1937, 1954; Eberhard 1972a; fig. 7.40b,c).

Peters (1937) expressed the radius rule in different terms, noting the positive correlation between the distance of the outermost loop of sticky spiral from the hub and the distance between the two ends of this segment (*x* in fig. 7.13). He believed that this ratio was constant in several araneids, including *Argiope argentata, Araneus diadematus, Zygiella x-notata, Gasteracantha cancriformis*, and probably also *Micrathena gracilis*, and noted that the relation was particularly clear in different webs of the same individual. He also argued that two species of *Cyclosa* are exceptions, in having values of *x* that are

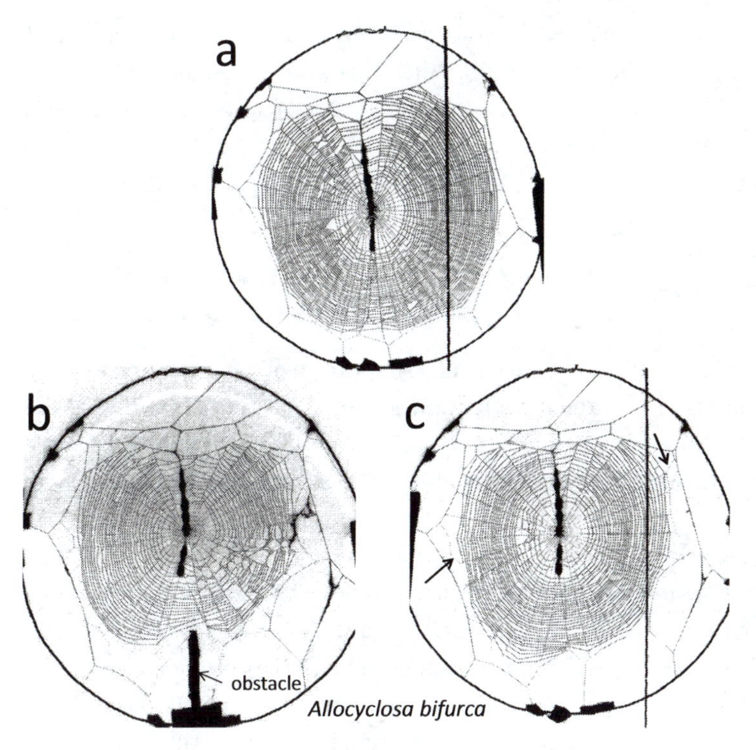

Fig. 7.41. These three successive orbs built by a mature female *Allocyclosa bifurca* (AR) illustrate adaptive behavioral flexibility. The day after she had built an orb in a vertical wire hoop (*a*), an obstacle was introduced at 6:00. On the following night the frames and sticky spiral lines in this sector of the web were altered (*b*); the frame was closer to the hub, there were fewer radii, and the outer loops of sticky spiral turned back repeatedly (the spider apparently left "tracks" by walking subsequently in the 3:00–5:00 sector) (see fig. 4.19). The obstacle was removed the following morning, and the web she built the next night (*c*) was similar to the first web, but still somewhat altered in the area where the obstacle had been, as well as in other details (arrows mark two tertiary frame lines that were added). In sum, the spider altered the second web to adjust to the obstacle; and the frame lines in this second web then apparently influenced the design of the subsequent web. Similar adaptive flexibility occurred in *Argiope argentata* (AR), in which the type of adjustment appeared to depend on the size of the obstacle (Hesselberg 2015) (photos courtesy of Jairo Moya).

more or less constant and not related to the distance from the hub.

Peters (1936, 1937) also attempted to understand radius angles in the orbs of the araneids *Araneus diadematus* and *Zygiella x-notata* in terms of the lengths and positions of the spider's legs while she rested at the hub of a finished web. He found that the ratios of angles between certain legs were similar to the ratios of the angles between the upper six and the lower six radii in the orb. The relationship (if any) between these leg angles and the cues that guide radius construction or the function of the orb in trapping prey is not clear, however, and these ideas have not been followed up subsequently.

A more complete analysis showed that Peters' versions of the radius rule fail to capture the entire picture. In the first place, it is appropriate to include measurements of only variables that the spider might be able to sense. Thus, instead of measuring all the inter-radial angles in a web as Peters did, it is more appropriate to substitute measurements that would be feasible for the spider to sense at least indirectly during construction, such as the angle between each new radius and its exit radius. Similarly, rather than measuring the separation between radii by using the length of the outermost segment of sticky spiral as Peters did, one can measure it as the distance along the frame from the end of the exit radius to the attachment of the new radius to the frame, as experienced by the spider. With these improvements (and correcting for individual differences in spider and web size and in previous feeding by relativizing all distance measurements in terms of the median value for each web) the correlation between exit radius length and radius separation was clear in *Micrathena duodecimspinosa* (AR) (fig. 7.40b). But substantial scatter remained. Some of this scatter was explained by the angle between the exit radius and the frame. The larger this angle, the farther the spider moved along the frame before attaching the new radius (fig. 7.40c).

The strong correlations between inter-radius angles and their lengths suggest that spiders measured radius lengths, but I know of no experimental studies demonstrating how such measurements were made. The spiders's anterior legs may have been used, because their removal affected inter-radial angles (Reed et al. 1965). The spider may have measured radius lengths by pulling on radii while at the hub (longer radii would be more extensible if they are under similar or lower tensions) (Denny 1976). Or she may have sensed the distance that she walked along the exit radius from the hub to the frame, the distance along the frame line, or the angle between the exit radius and the frame line.

The only other quantitative data on the possible cues determining the angles between radii are from *U. diversus* (UL). This species also did not move a standard distance along the frame, and thus seemed to fit the ideas of Peters better than those of Hingston (Eberhard 1972a). Without measuring additional correlations such as those in fig. 7.40, however, it is difficult to confidently distinguish evolved tendencies to make adjustments in the distance from simple imprecision.

In sum, there are several preliminary reasons to suppose that neither the constant distance hypothesis of Hingston (1920) nor the radius rule of Peters (1936, 1937) is generally applicable. Probably spiders sensed the length of the radius and the exit radius-frame angle, but the stimuli used to determine where to attach the radius to the frame are not known.

7.7.2 SECONDARY FRAME CONSTRUCTION

The behavior that determines the angle between a new radius and its exit radius when a spider builds a new secondary frame line (fig. 6.5j,k) is very different. While the angle of other new radii with their exit radii was typically determined by how far the spider moved along the frame before she attached the new radius (fig. 6.5i), the angle of the new radius that is built as part of secondary frame construction is determined by how far the spider walked back toward the hub before she attached her dragline to the radial line that she had laid on her way out (at p in fig. 6.5j), how far this line was then pulled from the exit radius by the new frame line (at q in fig. 6.5j), and (in some cases) by a subsequent shifting of the attachment site when the spider attached the frame to the radius just before returning to the hub (fig. 6.5k). The stimuli affecting these behavioral decisions are not known and, as far as I know, have never been studied.

The only detail that is clear is that in *M. duodecimspinosa* the moment in time when the decision was made to build a secondary frame occurred at or before the moment when the spider made the first attachment to the frame line: the distance of this attachment from the exit radius was consistently less than it would have been if were for a new radius (crosses in fig. 7.40). The inter-radial angles that were produced by *M. duodecimspinosa* as part of secondary frame construction tended to be larger than the other inter-radial angles (fig. 7.39; see also fig. 6.10). The advantage of these differences is not clear.

7.8 EARLY RADII, AND FRAMES AND ANCHOR LINES: DETERMINING WEB SIZE, SHAPE, AND DESIGN

The overall size and shape of an orb are determined early in the construction process; to some degree all of the subsequent stages of construction constitute a simple filling in of the area that was blocked out earlier. There are several indications that orb weavers can sense the size and shape of the area available for a web, and then adjust the designs of their webs accordingly. Very little is known, however, regarding the specific cues, the behavior that spiders use to sense these cues, or their responses to these cues during the early stages. I will speculate more here than in earlier sections, because it seems useful to at least set out some hypotheses that can be tested in the future.

7.8.1 POSITION OF THE HUB

The hub of an orb web is almost never placed at the web's geometric center. In vertical orbs, for instance, the hub is almost always nearer the top than the bottom of the web (e.g., Masters and Moffat 1983; Nakata and Zschokke 2010; section 4.8). The position of the hub is determined at the very initiation of web construction per se (fig. 6.5d, section 6.3.2), and it seems inevitable that gravity is used as a cue, but the manner in which the spider senses its direction is unknown (perhaps by perceiving the direction in which she hangs under lines?). In general, the behavioral mechanisms responsible for hub placement—the cues that the spider perceives and how she responds to them so as to alter the position of the hub—are not known. Nevertheless, a few details can be deduced.

Perhaps the best experiment on how hub location is determined was performed by Tilquin (1942) with *Argiope bruennichi* (AR). When he experimentally varied the rigidity of the two supports of the original bridge line, the spider formed the original "Y" that would be the hub of the orb closer to the more rigid of the two supports.

Determination of the site of the hub also involved other cues. For instance, the hub was consistently displaced toward one edge in the horizontal orbs of some species. *Dolichognatha* spp. (TET) and *Azilia affinis* (TET) built their nearly perfectly horizontal orbs near large objects such as tree trunks (Eberhard 1986a; Álvarez-Padilla and Hormiga 2011), and the hubs of their orbs were usually nearer the edge closest to the large object (fig. 3.43) (Barrantes et al. in prep.; WE). The hubs of *Uloborus diversus* (UL) were also displaced toward larger supports in captivity (Eberhard 1972a; WE). Presumably these asymmetries facilitated rapid defensive runs to the substrate, where the spiders were more difficult to see (e.g., Barrantes et al. in prep.).

Other experiments showed that the position of the hub was altered in the araneids *Zygiella x-notata, Argiope argentata, Eustala illicita,* and *Allocyclosa bifurca* by altering the shape of the available space (Le Guelte 1966; Ades 1986; Hesselberg 2013; fig. 7.41) and, in *Z. x-notata,* the locations of sites of the retreats in which spiders rested during the day (Le Guelte 1966). When *Z. x-notata* spiders were kept in hexagonal frames that could be rotated to change the position of the retreat, the hub tended to be displaced toward the side on which the retreat was located; it was in the upper half of the orb when the retreat was above, and in the lower half of the orb when the retreat was below (Le Guelte 1966). Similarly, it tended to be closer to the side of the cage with the retreat ("horizontal eccentricity") when the retreat was on one side of the cage or the other. These asymmetries did not function to promote escape, because the spider rested in the retreat during the day; instead they probably functioned to enable the spider to reach the hub more quickly when attacking prey. The cues used to produce these biases in the hub location have not been studied.

Another puzzling factor affecting the site of the hub was the spider's previous building experience. When individuals of *Larinioides sclopetarius* (AR) were deprived of prior orb building experience by raising them until maturity in small 3 cm diameter cylindrical cups "filled" with paper strips, they built orbs in which the hubs were nearer to the vertical center of the orb than did spiders from the same egg sacs that had built orbs while being raised in captivity (Heiling and Herberstein 1999). Judging by experiments with other araneids (Reed et al. 1970), the experimental *L. sclopetarius* spiders probably had less silk available in their glands. Indeed, on average the orbs of inexperienced spiders covered only 70% of the area and had a shorter (78%) sticky spiral line (these differences were not significant, but the samples were small). Nevertheless, I see no a priori reason why smaller amounts of silk should correlate with changes in orb symmetry.

7.8.2 SIZE AND SHAPE OF THE SPACE IN WHICH TO BUILD

7.8.2.1 The decisions the spider makes

The size and the outline of the space that will be enclosed by an orb's frame lines are also largely determined early in the construction process. Experiments have demonstrated a great deal of behavioral flexibility (reviewed by Herberstein and Tso 2011; section 8.4.3). Some design adjustments to the website were relatively trivial: for instance, orbs built in larger spaces often had longer anchor lines. Sometimes, however, the adjustments that spiders made were more complex. When the araneids *Argiope argentata* (Ades 1986, 1991), *Eustala illicita* (Hesselberg 2013), and *Allocyclosa bifurca* were placed in cages with different contours, the spiders modified the shapes of their webs to fill the irregularly-shaped spaces (fig. 7.41). Experimentally adding twigs in a cage apparently caused *Cyrtophora cicatrosa* (AR) to build smaller domed sheet webs, and to position them farther from the center of the cage (Rao and Poyyamoli 2001). These adjustments involved changes in the frame lines and the sticky spiral. The cues used by the spiders to distinguish the shapes of the spaces were not determined.

Perhaps the most spectacular flexibility in web design adjustments to the physical space available is the recently discovered ability of *L. argyra* (TET) and *Zosis geniculata* (UL) to build tiny orbs when they were caged in restricted round spaces that were less than a tenth of the diameter of the normal orbs in the field (some cages were as small as about 3 spider lengths) (fig. 7.42) (Barrantes and Eberhard 2012; Eberhard and Barrantes 2015; Quesada 2014). The designs of these tiny orbs differed in multiple ways from those of the "normal" orbs built in the field, and the changes were quite similar in these two distantly related groups. These changes included many aspects, including the anchor lines, frame lines, hub loops, radii, and the sticky spiral (table 7.1, fig. 7.42). Although free-ranging spiders of these species apparently never built such small webs in nature (the evi-

Table 7.1. Summary of the likely independence between the changes in design features that were measured when orbs were built in smaller containers by adult females of *Leucauge argyra* (TET) and *Zosis geniculata* (UL) (see also table 10.3). Criteria to distinguish independence included differences in the timing of decisions during construction, the use of different cues to make decisions, and the physical feasibility of independent variation (see text). Those differences (shorter radii, smaller capture area) that can reasonably be attributed directly to simple physical limitations imposed by a smaller available space, were not included as independent decisions in the analyses in the text, and are marked with "**." Variables that seem likely to be determined by independent decisions are labeled with different letters; the order of the letters reflects the order in which they occur during orb construction ("?" indicates no information available) (from Eberhard and Barrantes 2015).

	Changes of orbs built in smaller spaces			
Web traits	*Zosis geniculata* (UL)	*Leucauge argyra* (TET)	*Probable different independent decisions*	*Same trend in smaller L. argyra webs in the field?*
Relative areas				
Capture area[1]	Smaller	Smaller	**	Yes
Hub area	Smaller	Smaller	**C**	Yes
Free zone area	Smaller	Smaller	**G**	Yes
Overall symmetry	?	Reduced	**D**	No
Radii, frames, anchors				
Number of radii	Smaller	Smaller	**B**	Yes
Length of radii	Shorter	Shorter	**	Yes?
Number of frame lines[1]	Smaller	Smaller	**A**	Yes
Proportion of radii attached directly to substrate	Larger	Larger	**A**	No
Proportion of frame lines with only a single radius attached to them	Larger	Larger	**A**	No
Number of radii/frame	Smaller	Smaller	**A**	Yes
Proportion of radii that end on a "V"	Larger	Larger	**A**	No
Sticky spiral				
Distance from outer loop of sticky spiral to end of radius	Smaller[2]	Smaller[2]	**E**	No
Number of loops of sticky spiral	Smaller	Smaller	**F**	Yes
Space between loops of sticky spiral on longest radius	Smaller	Smaller	**F**	Yes
Distance from outer to inner sticky spiral loop	Smaller	Smaller	**G**	Yes
Consistency of sticky spiral spaces	No change	No change	?	Yes[3]
Hub				
Number of loops of hub spiral	?[4]	No change	?	Yes
Frequency of circular stabilimentum loops	Lower	—[5]	**H**	

[1] In both species, the positive relationship with total web area was significant only in the four smallest containers, and in the largest space the number of frames did not increase.

[2] G. Barrantes and W. Eberhard, unpub.

[3] Same *lack* of a trend.

[4] Inner portions of hub could not be distinguished due to the stabilimentum.

[5] No stabilimentum built in this species.

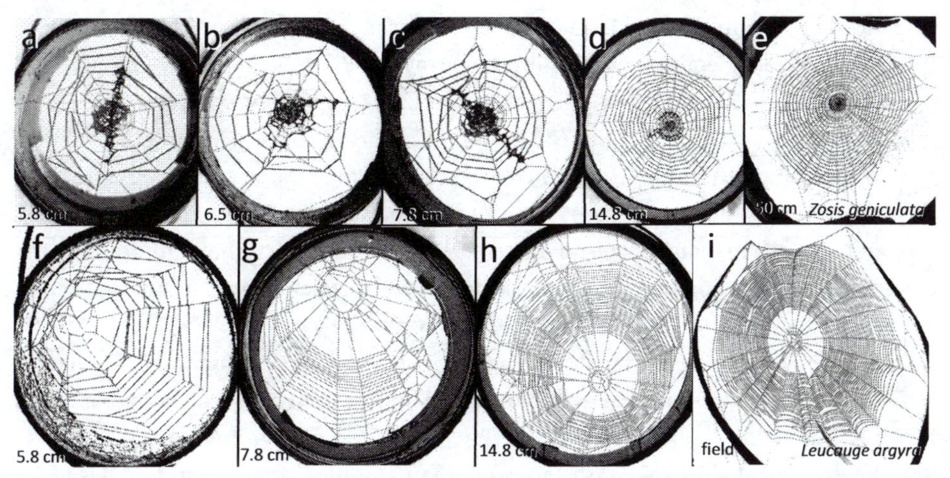

Fig. 7.42. Dramatic changes were induced in the orb designs of mature females of *Zosis geniculata* (UL) (*a–e*) and *Leucauge argyra* (TET) (*f–i*) by caging spiders in spaces of different sizes (cage diameters are given in the photos) (see also fig. 10.15). The spiders altered at least seven different, apparently independent aspects of orb design (table 7.1). The adjustments of *L. argyra* were especially striking, because none of the many hundreds of orbs of mature females seen in the field were nearly as small as the orbs built in 7.8 and 5.8 cm containers; the adjustments made by the spiders to building in these restricted spaces were thus not the direct result of natural selection. The similar reduction and elimination of frame lines by the two species in the especially constrained cages (*a, f*) suggest that the apparently convergent reduction or elimination of frame lines in the normal webs of some other species, such as *Anapisona simoni* (AN) (figs. 4.6a, 9.4f, 10.25a,b), *Pachygnatha* sp. (TET) (fig. 4.6d), *Ogulnius* sp. (TSM) (fig. 10.22c), and *Eustala illicita* (AR) (for radii near the retreat) (Hesselberg and Triana 2010), is probably the result of a single shared, ancestral response to small spaces. The specific cues from the size of the space in which the spiders were caged that they used used to trigger these changes are not known (most of the lines across the central holes in the hubs in *g–i* may have been draglines laid after the web was complete) (*a–e* adapted from Eberhard and Barrantes 2015, *f–i* from Barrantes and Eberhard 2012).

dence on this point is especially certain for *L. argyra*—Barrantes and Eberhard 2012), several of the changes in design were nevertheless appropriate to enable the spider to employ her available silk to build a more effective trap in a restricted space. For instance, the disproportionate reductions in anchor and frame lines avoided their taking up most of the available space, and thus left more space for radii and sticky spiral lines (e.g., fig. 3.26b, Barrantes and Eberhard 2012).

A few especially small orbs of *L. argyra* in the field had minor versions of several of the design changes seen in webs built in constrained captive conditions (Barrantes and Eberhard 2012). For instance, an especially small field web had a few radii attached directly to the substrate (fig. 7.42i). Presumably, such trends in natural contexts evolved under natural selection, and were extended to the extreme contexts in captivity. In accord with this hypothesis, *Eustala illicita* showed a similar tendency for shorter radii to be attached directly to the substrate (Hesselberg 2010; Hesselberg and Triana 2010). Other species also made different design adjustments on the basis of available space that increased the trapping abilities of their webs. The anapid *Anapisona simoni* (AN), when confined in smaller spaces, built more steeply inclined conical webs with a smaller angle at the top (Eberhard 2011); this change allowed the web to have a larger area covered by sticky spiral in the limited space.

These space constraint experiments affected many different web traits, but they may have involved a few "higher order" decisions that subsequently resulted in changes in the stimuli that guide later behavior (e.g., radius length, total area of web, the total area to be covered relative to the amount of sticky silk available). How many types of decisions by the spider were altered independently when she built a web in a restricted space? Table 7.1 gives a conservative estimate of seven, based on physical independence and "common sense" criteria related to the behavioral contexts that would allow decisions to be made (that they occur separately in time, that they are influenced by different cues, and that it was physically feasible for the decisions to be independent of each other) (Eberhard and Barrantes 2015). The independence criterion emphasized the possibility that there can be variation in one decision even after another, previous decision has been made. It thus emphasized the possibility that natural selection can act separately on the two decisions and that they can evolve independently.

Incidentally, it is important to keep in mind that the question of biological independence is different from the question of statistical independence when determining the of numbers of cues and decisions. For instance, the length of the radius of *Leucauge mariana* affects the decision of how far from the end of the radius to place the first loop, and the radius length is tightly correlated with the number of sticky loops ($r = 0.87$, $df = 70$, $p < 0.00001$). But radius length is determined at a much earlier stage of orb construction than is the placement of the first loop, and the two kinds of decision are influenced by different stimuli (sections 7.7, 7.3.6); the attachment site of the first loop is influenced by the site of the outer loop of temporary spiral, a line that is not even present when radii are constructed. Thus these two aspects of design are properly considered to result from separate design decisions, despite statistical correlations (for a similar argument, see Zschokke 2011).

Using these independence criteria conservatively, the experimental confinement of *L. argyra* and *Z. geniculata* spiders to very small containers resulted in the spiders changing several design decisions (table 7.1): (a) whether or not frame lines would be built as part of radius construction and how long the frame lines would be; (b) the angles between adjacent radii during radius construction; (c) the spaces between hub loops (built after radii were finished); (d) the degree of asymmetry in placement of the hub; (e) the distance between attachments of the outer loop of sticky spiral and the outer ends of the radii; (f) the spaces between sticky spiral loops; and (g) the distance from the hub at which sticky spiral construction was terminated. This is a conservative list, because possibly more than one decision was involved in producing the changes in the five different variables that are cataloged as resulting from decision (a).

There is other evidence that these design decisions are independent. Differences in the timing of ontogenetic changes in different web traits suggest the existence of semi-independent modules controlling orb construction in the nephilid *Nephilengys cruentata* (Japyassú and Ades 1998), and there is evidence of similar independence in the ontogenetic changes in other species. Different species of parasitoid ichneumonid wasps elicited quite different arrays of changes in the orbs of their host spiders, in some cases by highly specific stimulation and repression of particular details of orb construction (e.g., elicit one subroutine of frame construction and repress others) (Eberhard 2001b). In addition, the webs of related species of orb weavers show different mixes of similarities and differences (e.g., Coddington 1986b; Eberhard 1986a; Lubin 1986).

Nevertheless, independent subprograms of behavior that are independent may also be linked at higher levels of analysis (Eberhard and Gonzaga 2019). Krink and Vollrath (2000) argued stimuli perceived and analyzed by the spider during preliminary exploration may alter several web construction algorithms. In general, flexibility in both behavior and morphology results from hierarchies of decisions (West-Eberhard 2003).

7.8.2.2 The cues used in decisions

One crucial aspect of a potential website is the presence or absence of objects such as leaves or branches that could touch or interfere with the completed web. Neither the cues used by the spiders to sense the size of the open space nor the behavioral mechanisms that translate these cues into changes in web design have been determined. Sensing the presence of objects in the space cannot be easy, because the spider usually cannot simply survey the scene visually. Some of the dramatic sagging downward, descending and then climbing back up single lines, and swinging movements that occurred during exploration (section 6.3.2, appendices O6.2.1, O6.3.1) may function as searches for potential interfering objects (section 6.3.2.2). It may be more difficult for spiders on horizontal as compared with vertical orbs to explore for nearby objects, because it is simple to explore the vertical plane immediately below a horizontal line by cutting the line and sagging downward. On a horizontal orb, only objects immediately below the web plane but not those above it would be encountered by such maneuvers. Experiments are needed to test this hypothesis for the function of sagging behavior.

I speculate that path integration is probably used to sense the size of the spaces available in which to build webs (fig. 7.43). Path integration, based on the distances and angles that the spider has moved, has been demonstrated experimentally in ctenids that do not build webs

Fig. 7.43. The central portion of this web that has more regularly spaced sticky spiral loops is a replacement web built by a mature female of the colonial tetragnathid *Metabus ocellatus* (TET) that was constructed above a tropical stream at 4:00 in the afternoon, just before the 5:00 peak of insect abundance that is typical of such stream habitats (Buskirk 1975; section 8.10). This replacement web has implications regarding the behavioral mechanism by which spiders may achieve the common correlation between the number of radii and the area covered with sticky spiral lines (the capture zone) (fig. 7.38a) (section 4.5). When this spider removed the central portion of the previous orb and replaced it with a fresh orb, the capture zone of the replacement orb was smaller than that of the older web; as expected on the basis of the radius-capture zone correlation, there were fewer radii in the replacement orb. Each new radius was attached to the broken inner end of an old radius, but three of the 14 old radii (at 1:30, 6:30, and 9:30) were not used (dashed arrows). This reduction in radii occurred despite the fact that the lengths of the radii that the spider contacted while she was at the hub while she built the radii in the replacement web were equally long during both the original and the replacement orb construction. The decrease in the number of radii in the new orb thus appears not to have been made on the basis of radius length (which is usually strongly correlated with capture zone size); instead the spider may have used another indicator of the area of the new capture zonethat she was about to build (from path integration?) (for similar patterns see fig. 7.35 of *Leucauge*, and fig. 6.19 of *Uloborus* in the 8:00–10:00 sector of newest portion of the web).

(Barth 2002), and in sheet web agelenids (Görner 1988). It has also been documented in orb weavers in two other contexts at a smaller scale: determining the spaces between loops in temporary spiral in *Leucauge mariana* (TET) (Eberhard 1988c), and in sticky spiral construction in both *L. mariana* and *Micrathena duodecimspinosa* (AR) (Eberhard and Hesselberg 2012). Memories of distances traveled are also almost certainly used by theridiosomatids and related symphytognathoid families, and also by *Hyptiotes* (UL) in sticky spiral construction; the relatively widely spaced sticky spiral loops (compared to the lengths of the spiders' legs), combined with direct observations in some species, show that the spiders often fail to contact sticky spiral lines already present during sticky spiral construction but are nevertheless able to attach successive loops of sticky spiral at very uniform distances from each other (figs. 7.7, 7.9) (Peters 1954; Eberhard 1982, 2011).

Path integration is generally performed with reference to a point such as the hub or a retreat. I thus hypothesize that during the process of estimating the size and shape of a space while she is exploring it, the spider chooses, and probably periodically resets, a point of reference. The first step might be to choose a provisional reference point (such as the approximate site of the future hub), and then keep track of the distances and angles she moves as she explores in the area around this point (checking for sufficient attachment sites in different directions, and for objects that might protrude through or otherwise interfere with the web). This may be more complicated than it sounds, because intersections between lines move during exploration when lines are added, shifted, or removed, and visual reference points are often not available (section 7.1.1). Periodically the spider may have to choose a new reference point if her explorations reveal that a given site is not appropriate (for instance, that there is a twig protruding through the space). Either she should then extend her explorations into adjacent areas, or should move away and begin again. Reference points might be especially important when building a horizontal orb, because of the lack of useful cues from gravity.

One tantalizing observation regarding possible kinesthetic cues is that removal of the accessory setae from the tarsi of *Araneus diadematus*, which probably increased the difficulty that the spiders experienced in grasping lines, resulted in their building smaller orbs that had fewer radii, and fewer sticky spiral loops that were farther apart from each other (Foelix 1970). If this modification caused spiders to slip as they walked (I am speculating—no direct behavioral observations of construction were reported), the spiders may have in fact walked shorter distances than they "believed" that they had moved during exploration, and may thus have built smaller orbs than they "intended."

The capture areas of *Argiope keyserlingi* webs were greater at higher air temperatures (Blamires et al. 2007b); the relationship, if any, with the spider's sensation of distances traveled is not clear.

7.8.3 SPIDER SIZE AND WEIGHT

Not surprisingly, there is a positive relation between web size and spider size (e.g., Hingston 1920; Witt et al. 1968). More interestingly, there are also positive relationships with the spider's food reserves. Experimental feeding of *Argiope keyserlingi* (AR) and *Larinioides sclopetarius* (AR) resulted in reduced orb size; both treatments resulted in an increase in the vertical asymmetry of the orb (larger below than above the hub) (Herberstein and Heiling 1999). Acute starvation also resulted in decreased web area in (probably) *Araneus diadematus* (Witt 1963). Presumably the spider senses her weight via stresses in proprioreceptors, perhaps the lyriform organs (especially in the femora) that are known to be crucial in obtaining kinesthetic information in the non-web ctenid *Cupiennius salei* (Barth 2002). But body weight per se was ruled out as a cue in the Herberstein and Heiling study, because increases in weight due to the addition of lead weights to the abdomen did not affect web size.

7.8.4 SILK AVAILABLE IN THE GLANDS

Histological studies showed that injection of acetylcholine into the abdomen induced protein synthesis in the ampullate silk gland of *Araneus diadematus* (AR) (Witt et al. 1968). The amounts of silk available at the start of orb construction were altered by administering physostigmine, a cholinergic substance: the amount of silk protein in webs increased by up to 33% in one experiment (Witt 1963). Atropine, an anti-cholinergic substance, slowed ampullate gland synthesis (Witt 1962). Spiders built larger webs after receiving physostigmine, and smaller orbs after receiving atropine (Witt 1962). The relative effects of these drugs on other silk glands was not determined.

Experimental manipulation of the relative silk reserves available in the ampullate, aggregate, and flagelliform glands for replacement webs (Eberhard 1988a) also influenced both web size and design in *Leucauge mariana* (TET) and *Micrathena sexspinosa* (AR). Individuals with depleted reserves built orbs in which the estimated area enclosed by the frame lines was smaller (Eberhard 1988a). Spider weight changes in these experiments were probably very small, so some other sensory cue, possibly from the glands themselves, was probably involved. Still another indication that a larger supply of silk induces a spider to build a larger web is that *A. diadematus* spiders built progressively larger orbs during a period in which they were allowed to reingest their previous webs (Breed et al. 1964). Information on glandular reserves of silk also influenced a building decision, in this case the size of the stabilimentum, in *Argiope aetheroides* (AR) (Tso 2004).

In sum, the overall size of an orb is determined early in construction by the sites of the major frame lines, and is adjusted precisely and harmoniously to the spider's reserves of both sticky and non-sticky silk, to the spider's recent feeding history, and to the size of the available space. Exactly how cues from the glands are translated into the behavior that determines web size is not known, but the translation is apparently relatively precise.

7.9 TO BUILD OR NOT TO BUILD: TRIGGERING ORB CONSTRUCTION AND DESTRUCTION

Finally, in this backward regression we come to the first decisions of all: whether or not to build a new web, and whether or not to move to a new site. Regarding initiation, observations of *Metepeira incrassata* (AR) during a solar eclipse showed that a decrease in illumination elicited decisions to remove the web (when light levels sank) and then build a new one (when light levels increased) (Uetz et al. 1994). Because at least some of the cues from websites that are used in decisions to build are tightly linked to the cues that are used to make choices regarding habitats and websites, I will postpone discussing them until the next chapter on the ecology of websites and website selection.

7.10 CUES THAT TRIGGER TRANSITIONS BETWEEN STAGES OF ORB CONSTRUCTION

Orb construction normally occurs in rigidly ordered stages (section 6.3.1). It is possible that a single cue triggers both termination of one stage and initiation of the next (say, from secondary radius construction to hub and temporary spiral construction), which would make the order of operations invariable. It is also possible that the transition between one stage and the next involves one set of cues that induces the termination of the first stage, and a second set that elicits the initiation of the next. Cases in which spiders show flexibility in the order of stages argue in favor of the second alternative.

Early experiments with *Neoscona nautica* (AR) argued in favor of inflexibility (Hingston 1920). When spiders were returned to earlier stages of construction by breaking some of the lines that they had already laid, they failed to repeat previous stages that they had already finished. When, for instance, the temporary spiral segments were broken after a spider had completed the temporary spiral and was building the sticky spiral, she did not resume temporary spiral construction, despite having to struggle to move from one radius to the next for lack of bridging lines, and produced highly irregular, disorganized arrangements of sticky lines. Similarly, these spiders failed to replace or repair radii that were broken during sticky spiral construction. König (1951) described a similar experiment by Lassen and Toltzin (1940) in which they broke three radii during temporary spiral construction by *A. diadematus*; the spider failed to replace the radii, and instead continued building the temporary spiral.

Another, even simpler experiment of this sort that concerned the end of radius construction and the initiation of hub and temporary spiral construction in several species suggested a degree of flexibility. When radii were repeatedly removed during radius construction as soon as they were built (Hingston 1920; König 1951; Reed 1969), the spider replaced broken radii up to 25 or more times. Eleven tetragnathid and araneid species in six genera gave such responses in König's experiments; the maximum mean number of additional radii was 82 in *Cyclosa conica* (AR). Eventually, however, the spiders always stopped building radii and began hub and temporary spiral construction, leaving the oversized sectors where the broken radii had not been replaced. König (1951) achieved an extreme in spider frustration by leaving only three radii intact during the entire radius construction period, immediately cutting each new radius that the spider added. Eventually, after she had laid a total of two to three times the normal number of radii, the spider began temporary spiral construction and then sticky spiral on the three intact radii (fig. 7.44).

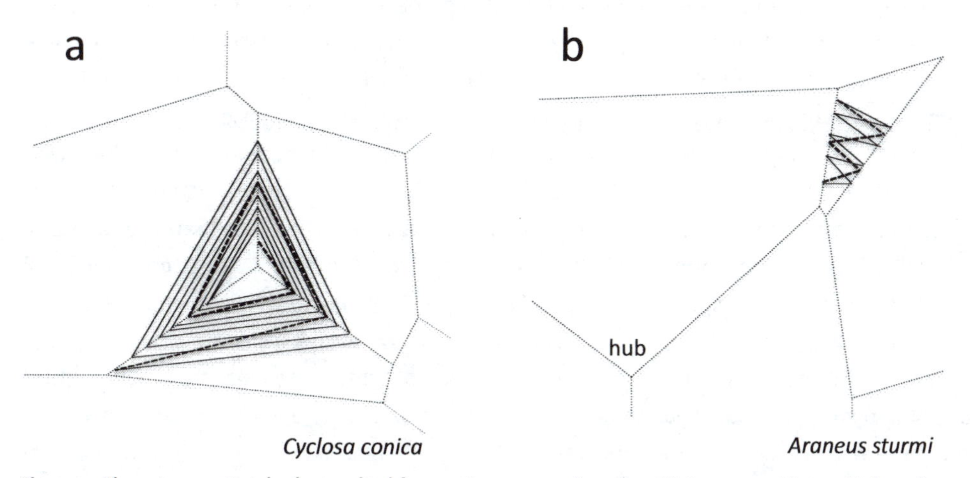

Fig. 7.44. These two vestigial orbs resulted from extreme experiments with two araneids in which each new radius was broken during radius construction as soon as it was laid, leaving the spider with only three intact radii. In each case the spider eventually abandoned radius construction, and built a skimpy temporary spiral and a few loops or zig-zags of sticky silk. The "spirals" in the *A. sturmi* web were built in a corner between two frame lines (the temporary spiral is represented by heavy dashed lines, other non-sticky lines by light dashed lines, and the sticky spiral by solid lines). The fact that temporary and sticky spirals were built in such abnormal circumstances testifies to the strength of the spider's tendency to perform the normal sequence of behavior, even in the absence of the usual stimuli that are associated with the initiation of temporary and sticky spiral construction (after König 1951).

Several conclusions follow from these results: experimental production of holes in the array of radii can prolong the radius construction stage, suggesting that such holes elicit radius construction and inhibit termination of radius construction. The termination of radius construction can be elicited by stimuli other than the lack of holes that have not been filled (section 7.5.2). And normal termination was not due to lack of silk in the ampullate glands. The transition from radius to temporary spiral construction is probably normally influenced by the stimuli that elicit termination of radius construction, but some other stimulus (memory of previous radii already constructed? lack of silk?) also elicts termination of radius construction, at least under some abnormal circumstances.

Additional, more quantitative observations with larger samples have shown that there are further complications. Similar experimental techniques, as well as transfers of individual *Araneus diadematus* (AR) that were interrupted in one stage of construction and transferred to an orb in a different stage, gave more variable results (Petrusewiczowa 1938; König 1951; Reed 1969; Peters 1970). For example, Peters (1970) removed seven spiders from their webs just after they had finished the temporary spiral, broke all of the temporary spiral lines in the web, and then returned the spider to her web: in five of these cases, the spider built a second temporary spiral. This implies that stimuli from a web lacking a temporary spiral were sometimes sufficient to elicit a second round of temporary spiral construction. In the remaining cases, either the memory of having finished the temporary spiral was enough to inhibit temporary spiral construction and elicit instead sticky spiral construction, or the spider failed to perceive the other stimuli. Perhaps differences in the details of these experiments and those of Hingston (who did not remove the spider from the web while he cut the lines), or differences between species were responsible for the differences between these results and those of Hingston. What is clear is that there is more flexibility with respect to how construction stages are ordered than Hingston thought. An incidental observation of *Mysmena* sp. (MYS) (Eberhard 1987e), and experimental damage to *Micrathena duodecimspinosa* (AR) webs (fig. 6.2, section 6.3.1), resulted in the spider interrupting sticky spiral construction to repair damage by adding radii, and then returning to sticky spiral construction; such flexibility may thus be taxonomically widespread.

Peters (1970) also performed web transfer experiments involving pairs of webs that were at similar stages of construction (either radius construction had ended and temporary spiral construction had not begun, or temporary spiral construction had ended but sticky spiral construction had not begun). These experiments did not test whether spiders would return to perform construction stages in the second web that they had already completed in the first web, but rather whether the stage at which a spider was interrupted affected the likelihood that she would remove the second web entirely and replace it with a new web (in some cases re-utilizing some of the frame lines). There was a weak tendency for those that had completed only the radii in the first web to be more likely to remove the second web and start over (4 of 6) than those that had completed radii and the temporary spiral in the first web (0 of 6) ($p = 0.06$ with Fisher Exact Test).

7.11 OTHER STIMULI THAT SPIDERS CAN SENSE BUT THAT ARE NOT (YET) KNOWN TO GUIDE ORB CONSTRUCTION

7.11.1 TENSIONS

Only a few general types of stimuli from the web have been shown to guide orb construction: the direction of gravity with respect to the spider's position in the web; the positions of lines with respect to each other and to the spider's own body; and memories of distances and directions that the spider has moved recently along web lines. Conspicuously absent are cues that are directly related to tensions on lines, such as resonant frequencies, vibrations transmitted from other lines, extensibilities of lines, and tensions. Such tension-related cues have been ruled out for guiding decisions that determine the spacing between loops of temporary spiral and sticky spiral (sections 7.3.2.6, 7.4.4). This evidence does not mean, however, that tension-related cues do not influence orb construction in other contexts.

Several types of relatively direct evidence suggest that uloborids, araneids, and tetragnathids can sense tensions or cues that are related to tensions on radii while they are at the hub, perhaps via stimulation of stress receptors in their leg cuticle (Barth 2002). When alternate radii were cut experimentally in the capture zone of a finished web of *Leucauge mariana* (TET) while the spider was at the edge of the orb, the spider tended to grasp those radii that were intact rather than those that were broken with her legs I and II after she returned to the hub (Briceño and Eberhard 2011). *Araneus diadematus* (AR) responded to more massive web damage, caused by experimentally cutting a major anchor line in the lower half of the web and collapsing between a quarter and a third of the orb, by turning to face the collapsed sector of the web, and then pulled on the lax radii (Tew et al. 2015). When the damaged area was large (on the order of more than an estimated third of the web area) she reeled in the lax radii and fastened them to the hub. When the damage was smaller, she moved across or along one edge of the damaged sector to the frame, and tightened the frame line by adding one or two draglines under tension, and then added a new anchor. The spiders' responses were apparently to tension changes rather than web damage per se, as spiders did not respond when all the sticky spiral lines were experimentally broken in similar-sized sectors but the radii were left intact (and thus remained under similar tensions) (Tew et al. 2015).

Smaller damage to the orbs of *Zosis geniculata* (UL) sometimes elicited repairs, perhaps also triggered by tension differences. When I broke three adjacent radii in the capture zone after the first night in a "two night" orb (fig. 7.16), the next night the spider occasionally laid one or more radial lines across the holes I had made, joining the inner and outer fragments of at least one broken radius on the following night. Presumably the spider sensed the lower tensions on the broken radii, either while resting at the hub or while she laid additional sticky spiral on the second night. In *Uloborus diversus* (UL), spiders tended to place stabilimentum lines on shorter radii that ended nearer to anchor lines (section 7.6.2). The spider's behavior suggested that she may have responded to the low extensibility of the radius, which is affected by the length of the radius and its nearness to an anchor line, as well as the tension itself (Eberhard 1973a). Apparent adjustments of tensions occurred during break and reel behavior during radius construction (fig. 6.3); spiders consistently lowered the tension on the new radius just before reaching the hub (section 6.3.3).

One set of stimuli related to radius tensions are resonant vibrations, which would be of higher frequencies and transmitted with less attenuation when tensions were higher. Are such vibrations used to guide orb construction? I know of no direct tests of this possibility, but can offer one weak, indirect indication that they are not. Resonant web vibrations would be affected by the effective lengths of lines between rigid supports (e.g., Klärner and Barth 1982). One indirect indication that this type of stimulus is not important in guiding orb construction comes from intra-specific variation in two species, *Anapisona simoni* (AN) and *Tetragnatha* sp. (TET), in which spiders built both typical orbs, and alternate, "twig orb" designs (fig. 10.25, 10.26). In the twig orbs the distances between rigid supports, and the locations of the supports themselves, are greatly altered; the hub of the orb is connected directly to a supporting object. In the alternate, "typical" orb designs of these species the hub is free in the air (fig. 10.26). While it is possible that the spiders used quite different stimuli to guide the construction of these two web types, it seems more parsimonious to suppose that they employed relatively similar cues for both.

In sum, spiders apparently sense tension differences on radii in certain contexts. But they did not use tension differences as cues to guide construction behavior, at least in some contexts. It seems very likely that tension-related cues are sometimes used (especially during early stages), but experimental manipulations of tensions have yet to be performed in these stages.

7.11.2 HANDEDNESS?

Lateral behavioral asymmetries (handedness) occur in a wide range of animals, including the scytodid spider *Scytodes globula*, which used its left legs I and II more often when making exploratory taps on its prey (Ades and Ramires 2002). Handedness may be due to the fact that the high complexities of nervous systems make it difficult to produce perfectly symmetrical structures. The orb weaving spider *Argiope argentata* (AR), however, did not show handedness (Ades 1995). Further tests with other species to determine whether part of the evolution of orb webs involved repression of asymmetrical behavioral tendencies would be interesting.

7.12 HINTS OF ABILITIES: FOLLOW CIRCULAR PATHS AND SENSE RADIUS LENGTHS

Other observations of building behavior hint that orb weavers may have an ability to move in a circular path without guidance from particular lines. During orb web repairs by the araneids *Cyclosa confusa* (Shinkai 1998a) and *Micrathena* sp. nr. *lucasi* (fig. 7.36), the spider laid more or less circular temporary spiral (and also sticky spiral in *C. confusa*) lines across a large sector of an orb that already had many sticky spiral lines in place, and that lacked the usual temporary spiral "hand rail" guide (fig. O6.11) (section 3.3.2.8.2). Another possible manifestation of an apparently intrinsic circling ability occurred in *Zosis geniculata* and *Uloborus diversus* (UL) when the spiders crossed intact portions of orb in approximately circular paths when moving from one repair sector to another (figs. 7.2, 7.45; Eberhard 2019b). Still another possible case occurred during the normal construction behavior of the cone below the orb of *U. conus* (UL): "After meticulous, slow sticky spiral construction . . . the spider suddenly began spinning out cribellar silk in a rapid and seemingly reckless fashion" (Lubin et al. 1982; p. 34), spiraling smoothly inward across radii and temporary spiral lines toward the hub (fig. 10.37a,b). In all of these cases the spiders encountered but apparently ignored myriads of potentially confusing lines. The stimuli guiding these circling movements are not known.

A second mysterious behavior concerns a pattern in the temporary spiral spacing in typical orbs of *Micrathena duodecimspinosa*. The spaces between the sticky spiral loops on shorter radii were smaller than those on longer radii (fig. 6.10). These differences did not depend on the spider making contact with the frame lines, because even in the early, inner loops of temporary spiral the spaces were smaller on shorter radii. The implication is that the spider sensed the lengths of the radii,

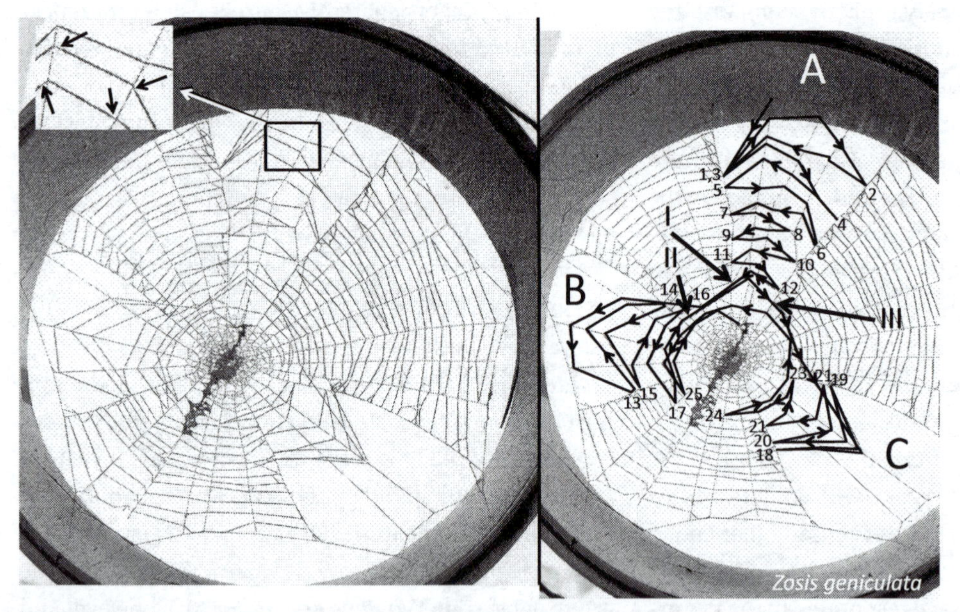

Fig. 7.45. This three-sector repair web of a mature female *Z. geniculata* (UL) constitutes evidence against an automaton interpretation of orb construction (the un-annotated web is at the left; on the right the spider's path while laying the sticky spiral is traced with thicker lines, and the turnbacks are numbered in the order they were built). After a compete orb was built, three sectors (A, B, and C) were experimentally destroyed. The spider laid new radii in all three sectors the next night, and temporary spiral in sector A. She then filled sector A with zig-zag sticky spiral. Following turnback 12, she failed to turn back at the edge of sector A, perhaps breaking the do-not-cross rule (it is not clear whether the lines at point I were sticky or not). She crossed the intact sector between A and B in a more or less circular path until she came to sector B, where she veered sharply outward toward the edge of the web. She filled this sector with zig-zag lines (turnbacks #13–17), and then again broke the rule (at points II and III) as she circled nearly 180° past both a repaired and an intact sector to reach the third empty sector (C). Here she again veered sharply outward, and then laid more zig-zags (turnbacks #18–24) to partially fill it. Finally, she left this sector and made a final loop of nearly 300°, turned back one more time (#25), and ended the sticky spiral. This spider thus broke the "do not cross sticky lines" rule multiple times, giving herself access to areas that were lacking sticky spirals.

and built temporary spiral loops that were farther apart when the radii were longer. The cue used to sense radius length is unknown.

Finally, there is one additional set of observations in which an effect on orb web design is clear but the immediate cue used by the spider is not known. Several different groups of orb weavers have convergently evolved to sometimes build their orbs in groups; in three of these the designs of solitary orbs differ from those colonial orbs (Buskirk 1986). Colonial orbs had smaller diameters in *Cyclosa caroli* (AR), *Tetragnatha elongata* (TET), and *Metabus ocellatus* (TET); in contrast, the diameters were unchanged in colonies of *Leucauge mariana* (TET), *Metazygia wittfeldae* (AR), *Meynogea lemniscata* (AR), or *Nephila clavipes* (NE). Colonial orbs of *T. elongata* were more nearly horizontal, but the slant is not known to be affected in the other species. It is not clear what cues trigger these changes.

The multiple cues used to guide some decisions are presumably integrated by the spider. One type of adjustment that may be especially challenging is the integration of the 10-plus different types of cues that influence the final decision as to where to attach the sticky spiral to the radius, especially when the stimuli conflict with each other. One type of conflict that may be common is between the two reference stimuli, the IL location and the TS-IL distance. When the relative attention paid to these cues was checked, the IL location cues tended to be ignored when their conflict with the TS-IL distance was especially large (Eberhard and Hesselberg 2012). In both the tetragnathid *Leucauge mariana* and the araneid *Micrathena duodecimspinosa* the spider took into account different combinations of stimuli under different circumstances; both the TS-IL distance and the memory of recently encountered TS-IL distances appeared to override the IL location cue when there was strong conflict.

In some other contexts, there may be a hierarchy of attention to different cues. For instance, even if orb weavers have some sort of "circling ability," it was apparently sometimes "over-ruled" by the experimental removal of temporary spiral lines, which induced spiders to follow less-circular paths during both temporary and sticky spiral construction (figs. 7.6, 7.30, sections 7.3.2.2, 7.4.1). Another possibility related to cue switching was illustrated by the question of why *M. duodecimspinosa* failed to respond to experimental changes in radius tension during temporary spiral construction (5.3.3.2), even though the spider uses some aspect of radius length that is likely influenced by radius tension to guide temporary spiral placement. Perhaps spiders switched between sets of cues, normally using the outer loop of temporary spiral to guide temporary spiral placement while building an orb from scratch, but instead depended on her "circling ability" when she encountered an intact sector during orb repair (see Eberhard and Hesselberg 2012 for cue switching in other contexts). A second possibility is that the conclusions drawn from the experiments intended to determine which cues are used may have been inappropriate. For instance, behavioral responses may not be simple, linear functions of tension-related stimuli, as assumed above. During temporary or sticky spiral construction, for instance, encountering a few very lax radii in the midst of others that are under normal tensions may not elicit a response but encountering more gradual changes in radius tension or extensibility may elicit changes. These are only speculations, but they serve to illustrate the limits of present understanding.

7.13 EFFECTS OF PSYCHOTROPIC DRUGS ON ORB CONSTRUCTION

The effects of at least 23 different substances known to affect human behavior, including such powerful drugs as mescaline, LSD, and scopolamine, stimulants such as caffeine and amphetamine, depressants such as chlorpromazine and barbituates, the muscle relaxant diazepam (valium), and others like di-ethyl ether, carbon monoxide, carbon dioxide, and nitrous oxide have been tested on orb web construction behavior, especially in *Zygiella x-notata* (summaries in Witt and Reed 1965, Witt et al. 1968). The variables measured included web size, web shape, frequency of construction, regularity of radial angles (mean of all differences between adjacent angles) and regularity of sticky spiral spacing (mean of all differences between adjacent spaces). All but two substances (imipramine and iproniazid) affected at least one of these variables; all variables were affected by at least some substances but different substances affected different combinations of variables. Larger doses of some substances (e.g., d-amphetamine) produced larger and longer-lasting effects (Witt 1971).

The most general conclusion is that many substances that alter human behavior also alter spider behavior. Evidently the neurophysiology of neural control mechanisms have an overall similarity in the two species. The relative potencies of different substances were similar, though larger dosages (per kg of body weight) were needed in most cases to obtain measurable effects in spiders (Witt 1971). Attempts to go into further detail regarding the mechanisms by which the effects were produced have been hampered by incomplete knowledge of the behavioral mechanisms affecting given web measurements. An exception is the demonstration in *Zygiella x-notata* that the regularity in sticky spiral spacing decreased after d-amphetamine ingestion even though the inner loop localization behavior of leg oI during sticky spiral construction (section 6.3.5) was not affected (Peters et al. 1950). The amphetamine effect apparently occurred during information processing, following information collection and before motor execution (Witt 1971).

Nearly all drug effects on the regularity of orbs were disruptive (Witt and Reed 1965). An exception was the increased regularity in *Z. x-notata* of both inter-radial angles and sticky spiral spacing that was produced by LSD (fig. 7.46, Witt and Reed 1965; Witt 1971). Further experiments with drugged spiders, such as, for instance, experimental removal of segments of sticky and temporary spiral lines or radii during construction (sections 7.3.2.1, 7.4.1, 7.7.1.1), could be rewarding.

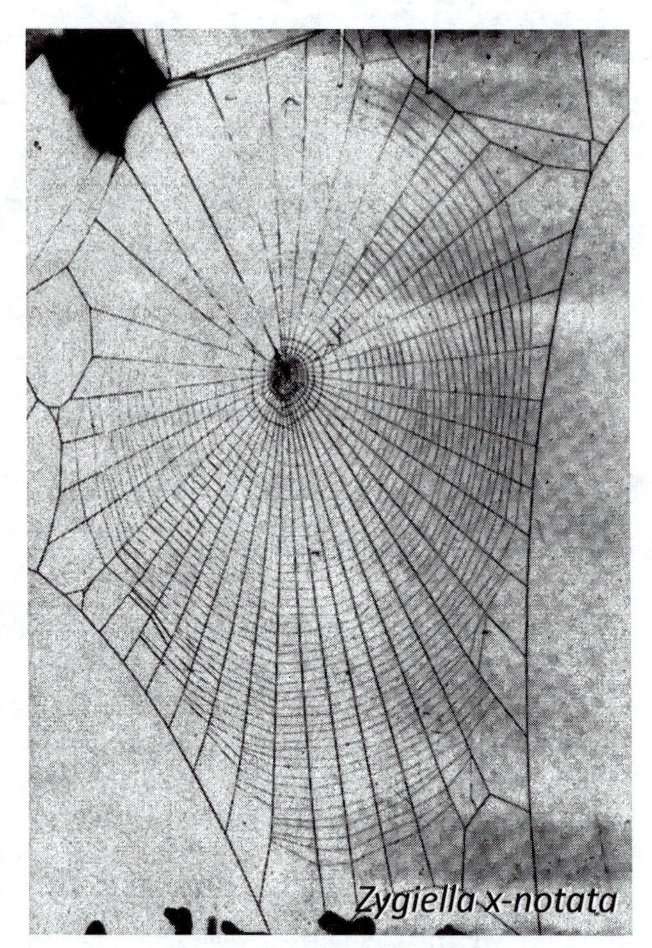

Fig. 7.46. The *Zygiella x-notata* (AR) that built this web had been given a dose of the drug LSD. While many other psychoactive drugs resulted in spiders building less geometrically regular orbs, the effect of LSD was to increase the uniformity of the angles between the radii and the spaces between loops of sticky spiral. The behavioral mechanisms affected by the drug are not known. Perhaps the drug caused the spider to ignore some stimuli (such as radius length, or her position with respect to the hub during radius construction?), or to concentrate more effectively. Using drugged spiders in experiments like those depicted in figs. 7.3, 7.6, 7.12, 7.27, 7.37, and 7.53 might yield useful information about how drugs influence behavior (from Witt 1971).

7.14 COORDINATING DIFFERENT ADJUSTMENTS TO DIFFERENT CUES

One would expect that natural selection would favor responses to different cues that are "harmonious" in the sense of producing adjustments that combine to increase the effectiveness of webs as traps. But it is also possible that such "harmony" is not pre-coded, but instead results from higher-level decisions by the spider. This second possibility can be tested by checking the behavioral adjustments to non-natural situations in which natural selection cannot have favored harmonious coordination of responses.

One such test was to enclose spiders in unnaturally small spaces (Barrantes and Eberhard 2012; Hesselberg 2013; Eberhard and Barrantes 2015). Webs built in tiny spaces had major simultaneous adjustments in the placement of the outer loop of the sticky spiral, the spaces between subsequent loops, the numbers of radii, and the frame and anchor lines to which the radii were attached (and their frequent omission). More minor adjustments also occurred in the size of the hub, the spaces between hub loops, and the area of the free zone. The different adjustments that spiders made to the unnaturally cramped spaces in these experiments were not all "harmonious"; some combinations of adjustments resulted in less than completely functional designs. For instance, the moderate reductions in hub size plus the smaller inter-radial distances at which sticky spiral production was terminated resulted in free zones that were so small that they ceased to function as open spaces where the spider could rapidly locate radii when turning at the hub to attack prey without interference from sticky lines (Briceño and Eberhard 2011). Patterns of this sort suggested that different aspects of orbs can be controlled separately. The tentative conclusion from these limited data is that the hypothesized higher-level analysis did not occur, and that at least some of the different adjustments were made independently.

7.15 THE (LIMITED) ROLE OF SIMULATIONS IN UNDERSTANDING ORB CONSTRUCTION BEHAVIOR

A photograph of an orb web is an exquisitely detailed record from which one can deduce and quantify very precisely major aspects of both the spider's construction behavior and the cues that guided this behavior. It is thus feasible to use web photos to evaluate attempts to simulate orb construction by providing a set of beginning conditions that mimic a real orb at some intermediate stage of construction, and a set of rules for deciding how and where to attach lines. The match between the final form of a simulated construction with the real orb that the spider built can be used as a measure of how well the construction rules in the simulation matched the rules used by the spider.

This tactic was used to study sticky spiral construction in *Uloborus diversus* (UL) (Eberhard 1969), and temporary and sticky spiral construction by *Araneus diadematus* (AR) (Gotts and Vollrath 1991, 1992; Krink and Vollrath 1999). Although the questions asked differed to some extent, and the possible guiding stimuli also differed, the studies all came to the same conclusion: construction of temporary and sticky spirals in an orb required only rather simple behavioral programs.

But these simulations all suffered from two major limitations. It is virtually certain that there is some degree of imprecision in the spider's execution of her own program (as there would be in a human!) (Eberhard 1990c). And these imprecisions will have cumulative effects because previous decisions (e.g., where to attach an early loop of spiral to a radius) affect later decisions. Thus even a simulation that was based on a perfect replica of the spider's own set of instructions cannot be expected to produce a perfect copy of any particular orb built by a spider. The spider herself does not produce identical webs from one day to the next. This leads to the problem of deciding how much a simulated orb can differ from a real web before one can conclude that the program of instructions in the simulation was different from that of the spider. None of the studies answered this question: all used only only the general appearances of the simulated orbs (e.g., fig. 7.23). The proposal to use similarities in "emerging properties" (Krink and Vollrath 1999) suffers from the same limitation.

A second problem, related to the first, was that the geometric regularity of an orb makes it possible to mimic at least approximately a spider's behavioral decisions using more than one particular set of guiding stimuli and decision criteria (Eberhard 1969; Gotts and Vollrath 1992). As Gotts and Vollrath (1991) noted, there is a "vast range of possible models" and the problems of testing which models best mimic the spider "remain unsolved" (p. 510). This problem is illustrated by the fact that the improved virtual model 2 of sticky spiral construction of Krink and Vollrath (1999), which held the distance between the virtual spider's body and the outer loop of temporary spiral constant, improved the fit with certain aspects of finished spider orbs. But behavioral observations show that the body-temporary spiral distance was not constant in *A. diadematus* (Zschokke and Vollrath 1995a; T. Hesselberg pers. comm.), so this model does not help explain the real world of spiders.

Similarly, a simulation study of temporary spiral construction by *Araneus diadematus* (Gotts and Vollrath 1991) did not take into consideration the possibility that spiders use memories of the distances moved inward and outward following previous attachments, variables that had previously been implicated as cues in experiments with *Leucauge mariana* (TET) (Eberhard 1988c). If, as is likely (Eberhard and Barrantes 2015), the cues that are used by tetragnathids and other araneids to guide temporary spiral construction behavior are also used by *A. diadematus*, then the results of a simulation that ignores such memories may lose biological relevance.

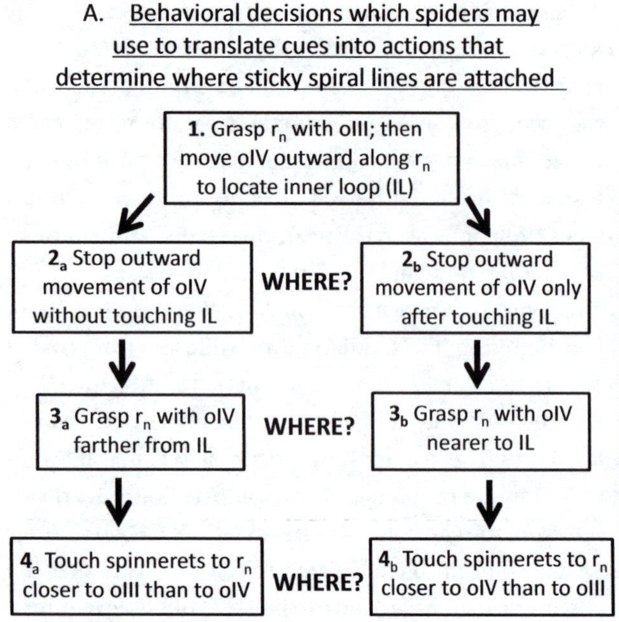

B. Behavioral decisions that are based on different cues

Cues (variables known to influence sticky spiral spacing)	Decsions that they affect
Above/below hub	2,3
Near/far edge	2,3
Weight of spider	?
Spider's hunger	?
Area to cover	?
Supply of sticky silk	?
Length of leg	?
TSP-IL distance	?

Fig. 7.47. This flow chart (left) of the decisions by *Micrathena duodecimspinosa* (AR) that influence her choice of an attachment site on each radius that she encounters while building her sticky spiral includes behavioral mechanisms (left) by which the variables that affect sticky spiral spacing (partial list at right) are translated into the actions that determine the distances between sticky spiral loops. Translations of this sort between variables and physical movements have been little studied (see fig. 6.27), but will probably be important in understanding the evolution of orb web construction behavior.

Models are nevertheless useful in helping to focus attention in future studies on possibly important variables and behavioral details. For instance, one stimulus used to guide the computer simulation of temporary spiral production in fig. 7.23 was the angle of the outer loop of temporary spiral with the previous radius (Gotts and Vollrath 1992). This angle is almost surely not sensed by the spider during temporary spiral construction, because the spider is often out of contact with the previous radius by the time that she reaches the vicinity of the site where the next attachment will be made. The model focuses attention on alternative possibilities. If the spider can remember the orientation of the previous radius from when she held this line, she could combine this with measurement of the angle through which she has moved subsequently and thus sense the angle that was used in the simulation.

7.16 A MISSING LINK: TRANSLATING CUES INTO ATTACHMENT SITES

This chapter has described how various stimuli affect the decisions that determine the positions of lines in orb webs. These data leave a gap with respect to behavioral mechanisms. What were the positions and movements of the spider's legs and body that resulted in the execution of these decisions? For example, a decision to reduce the space between loops of sticky spiral could alter the site where leg oIV gripped the radius, where leg oIII grasped it, or where the spinnerets make the attachment between the two legs (fig. 7.47). Improved understanding of such mechanism questions will be crucial to a more complete understanding of the behavioral process of orb web construction, and of its flexibility and evolution (Peters 1954; Vollrath 1988b). Illumination of behavioral details can clarify the order in which decisions are made and the mechanisms by which particular decisions are executed.

The advent of cheap, portable video recording equipment has made it easier to perform behavioral analyses at the fine scales of time and morphology needed to answer such questions. Although some video studies of arachnid behavior have examined heretofore inaccessible details in spider sexual behavior (Peretti et al. 2006; Aisenberg and Barrantes 2010) and attacks on prey (Barrantes and Weng 2006a), video recordings have rarely been used to study orb construction behavior. Even the most sophisticated of these studies (Zschokke 1993, 1996, 2000; Zschokke and Vollrath 1995a,b) involved recordings made at a distance that documented only large-scale patterns of behavior rather than fine behavioral details.

The only two studies I know of at this finer level, recent analyses utilizing video recordings of the araneids *Micrathena duodecimspinosa* (Eberhard 2012) and *Zygiella x-notata* (Toscani et al. 2012), both concluded that leg positions were important in determining the sites where loops of sticky spiral were attached. The positions at which leg oIV grasped the radius just before the spider attached the sticky line to that radius correlated with variations in the spaces between sticky spiral lines: when leg oIV grasped the radius farther from the inner loop of sticky spiral, the distance of the attachment from this loop was greater (fig. 6.27). Some of the variations in *Z. x-notata* may have occurred at sites where there was a turnback in the previous loop (as in the "spontaneous Hingston experiments" in section 7.3.2.1; Eberhard and Barrantes 2015) (see fig. 2 of Toscani et al. 2012). In *M. duodecimspinosa* two additional variables that were associated with the spider's location in her web (above or below the hub, near the edge or near the hub) correlated with the grasping sites. Thus the tendencies in finished webs of *M. duodecimspinosa* (fig. 1.4) for sticky spiral spacing to be larger near the edge of the web and above rather than below the hub can thus be explained at least in part by where the spider tended to grasp the radius with her tarsus oIV just prior to attaching the sticky line. These correlations were far from perfect, however, suggesting that other behavioral details, such as the site grasped by leg oIII, and/or the positioning of the spinnerets with respect to the sites grasped by legs oIV and oIII, may also be important. Other factors known to influence sticky spiral spacing (fig. 7.47, section 7.3.5) may also correlate with the sites at which legs grasp lines, but no data are available.

Further understanding of behavior at this neglected level of analysis may prove to be useful in understanding the evolution of orb construction. For instance, do all orb weavers use changes in oIV grasping sites as a mechanism to change sticky spiral spacing at different distances to the edge of the web, or do different taxonomic groups use different mechanisms?

7.17 SUMMARIZING THE BEHAVIORAL CHALLENGES MET BY ORB WEAVERS

A summary of the behavioral challenges described in this and the previous chapter that an orb weaving spider meets illustrates the considerable demands on her central nervous system in terms of coordination, agility, decision-making abilities, and sustained attention. Understanding spider behavior at this level may ultimately be important in understanding how and why they have evolved their current forms. The general conclusion of this section is that the complexity of the feats that orb weavers accomplish is greater than has been previously appreciated. This is not to claim that orb weavers are unusual among other spiders or arthropods in their capabilities (I would speculate, conservatively, that they are not; the relative sizes of orb weavers' brains did not differ from those of related non-orb spiders—Quesada et al. 2011). Rather, the special advantages that orb weavers offer for studies of behavior (section 1.4) permit an unusually thorough look into the details of their behavior, and have revealed unexpected complexity.

One initial pattern regarding the cues used to guide orb construction is their great uniformity, both in the cues themselves and in the rules that spiders use for responding to them. This uniformity occurs throughout orb construction, not only during sticky spiral construction, and is even greater than a previous discussion indicated (Eberhard and Barrantes 2015). It is true that the data are far from complete, and two major groups of orb weavers (Nephilidae, and the theridiosomatids and their symphytognathoid allies) are as yet only poorly covered by experimentation. But to date, in each case in which experimental tests of cues and responses have been compared between representatives of different taxonomic groups the cues have proven to be similar. Examples include the following: the dramatic similarity in changes in numerous web traits, including the hub, radii, frame lines, and sticky spiral in *Leucauge argyra* (TET) and *Zosis geniculata* (UL) to building webs in very tightly constrained spaces (figs. 7.42, 10.15); the lack of effect of radius tension on placement of temporary and sticky spiral lines in araneids, tetragnathids, and uloborids (figs. 7.15); the similar effects of the removal of a segment of the inner loop of sticky spiral on the placement of the following loop ("Hingston experiments") in araneids, tetragnathids, nephilids, and uloborids (fig. 7.3); the similar effects of manipulation of relative amounts of sticky silk available to the spider when the orb is begun in tetragnathids, araneids, and uloborids (figs. 7.18, 7.19); and the similar combination of immediate, dramatic responses and then subsequent damping of responses (presumably due to memories of the first response) when the TSP-IL distance was increased by breaking the outer loop of temporary spiral (fig. 7.6) in an araneid and a tetragnathid. The cues and the spiders' responses to them have been very conservative in orb weaver evolution.

In sum, even though these animals have only limited brain power, orb construction entails substantial physical agility and precise, complex coordination.

7.17.1 MECHANICAL AGILITY AND PRECISION

The smooth flow of rapid, precise movements by a spider building her orb gives an impression of simplicity. Sticky spiral construction seems especially uncomplicated: she moves outward, attaches, moves inward and onto the next radius, then outward to repeat this process over and over and over. But the data in the preceding chapter (section 6.5, also appendix O6.2.5) reveal that the mechanical demands with which a spider deals during this process vary widely. I will concentrate first on sticky spiral construction behavior because it is better understood.

Consider the variety of leg and body movements needed to accomplish the multiple, tightly coordinated tasks each time the spider attaches the sticky spiral to a radius; these include coordination between different legs, between leg and spinneret movements, initiation and termination of silk production (e.g., from the piriform glands), and perhaps also modulation of the diameters of lines and the amounts of glue that are added to the lines (Crews and Opell 2006). Each time she makes an attachment, the spider performs at least five basic mechanical tasks: locate radial and sticky spiral lines with her anterior legs; seize the radius with her legs oIII and oIV to make the attachment; measure the TSP distance; align her abdomen with the radius very precisely, so that the radius runs through the center of the cluster of spinnerets at its tip; and execute a brief burst of spinneret movements that attaches the spiral to the radius (in *M. duodecimspinosa* the burst of spinneret movements lasts on the order of only 0.03 s/attachment). Add to this the complication that the movements and forces exerted by the legs in order to accomplish these precise orientations vary substantially in different parts of the web (e.g., above versus below the hub, near vs. far from the hub), so that leg movements and the forces they exert must be continually adjusted to local conditions as the spider spirals around her web. In the course of laying about 60 loops of sticky spiral over a span of only about 20–30 min in a typical orb with about 50 radii, *M. duodecimspinosa* performs about 15,000 tasks, continually adjusting her behavior to different mechanical demands.

7.17.2 ANALYTICAL ABILITIES: MULTIPLE CUES AND DECISIONS

The number of decisions that spiders make while building an orb, and especially while building the sticky spiral, is dramatic. Multiple studies of sticky spiral construction have revealed a surprisingly large number of different cues that affect each sticky spiral attachment decision. I count at least ten and perhaps up to sixteen variables. Two or more "reference point stimuli" are perceived anew at each radius: the site of the inner loop of sticky spiral; the distance between the outer loop of temporary spiral and the inner loop of sticky spiral (the TSP-IL distance); and perhaps the separation between adjacent radii, and the length of the current segment of sticky line are also involved. There are also two memories of recently perceived reference point stimuli: the TSP distance on the immediately preceding radius; and sharp changes in the TSP distance that were encountered 30–60 s previously. The spider probably also remembers the approximate locations in the web of these somewhat less recent encounters. In addition, she uses at least three cues that vary gradually during the process of sticky spiral construction: the supply of silk in her glands in comparison with the area that she needs to cover with sticky lines (she may also sense and respond to her supply of non-sticky silk in her ampullate glands); the direction of gravity versus the direction of the hub (whether she is above, below, or to the side of the hub); and her edge-to-hub position (whether she is in the web's inner versus the outer portion). Finally, she also uses four additional "settings stimuli" that are more or less constant throughout sticky spiral construction: wind velocities; nutritional state; the proximity of oviposition; and (perhaps) memories of particular types of prey or sites in the web where they were captured.

Decisions during sticky spiral construction are also made in addition to those determining the spacing between loops of sticky spiral. The amount of aggregate gland glue applied to the sticky lines may vary on the basis of available silk. The spider also decides while making each attachment whether or not she should turn back. In at least some species this decision is made before the sticky line is attached. Experimental induction of changes in web geometry and changes in the rhythm of her behavior at sites with experimental alterations in the orb indicated that the spider also seemed to have "expectations" in at least two contexts: changes as she moved around the orb in her position with respect to gravity (fig. 7.17) (Vollrath 1986); and the presence of radii and temporary spiral lines while she was spinning sticky spiral (Hingston 1920).

Who would have thought that there is so much going on the brains of these small, "simple" animals as they circle around and around their orbs? I realize that the number of independent cues may be somewhat overestimated in this list. For instance, a spider could sense her edge-to-hub position by measuring the distances between radii; her nutritional status and the proximity of oviposition could both be derived from some basic physiological state. It is also possible that some or all of the four "settings" stimuli are combined into a single variable before sticky spiral construction begins, and that only this single variable is taken into account at each attachment. But most of the cues just mentioned seem likely to be independent.

However it is that the cues are combined, it is clear that these "simple" animals integrate a great variety of information in complex ways. I have not attempted to review the numbers of cues used by other animals to guide particular decisions, but believe that the length of the list for sticky spiral construction behavior is unusual, if not unique, for a single behavioral act, at least in an arthropod. It is possible that other animals also take into account similarly long lists of cues when making decisions, and that the large number of cues documented in orb weavers is simply the result of the ease with which sticky spiral construction can be studied. In any case, the data reaffirm one basic theme of this book, that it is a mistake to conceive of orb weavers as simple automatons that execute the same simple behavior patterns over and over.

Both the cues and the spiders' responses to them are apparently ancient, and they have been very conservative once they arose (Eberhard and Barrantes 2015). The similarity in cues and responses in uloborids and araneoid orb weavers constitutes especially important evidence favoring a single evolutionary origin of orb webs (section 10.8), because the geometric regularity of orbs means that several alternative sets of cues could have provided spiders with sufficient information to guide construction behavior.

7.17.3 SUSTAINED ATTENTION—WHERE ORB WEAVERS TRULY SHINE

Animals receive floods of information from the environment through their sense organs, much more than their brains can process (Dukas 2004). By biasing the input and processing the information appropriately ("paying attention" to a subset of stimuli), an animal can respond more consistently, efficiently, and rapidly to those stimuli that are most relevant to its current behavioral context (Shettleworth 2010). Attention is a well-established phenomenon in vertebrates, and there are indications that other animals such as insects and spiders also present "attention-like" phenomena (Shettleworth 2010; Eberhard and Hesselberg 2012) (there are even studies of "selective attention" by individual neurons; Wiederman and O'Carroll 2013). The study of attention, as well as possible lapses in attention that are associated with behavioral errors, may help understand present-day variations in behavior. Lapses in attention may also influence the directions in which evolution can most readily proceed, because the behavioral variants present in a population will inevitably influence how natural selection can act, and thus future evolution (Eberhard 1990a; West-Eberhard 2003).

As just noted, an orb weaver makes literally thousands of measurements and behavioral decisions while building her sticky spiral. Typically, these occur in the space of the 20–30 min, but in extreme cases the process is much longer: in the giant, slow moving *Nephila pilipes* (= *maculata*) (NE) sticky spiral construction lasted up to 4 hours and 45 min almost without interruption (Hingston 1922b); the smaller, fast-working araneid ladder web spiders *Tylorida* sp. and *Scoloderus tuberculifer* both took approximately 1.5 hours to build the sticky spiral (with multiple short "rest periods" in *T.* sp. (Robinson and Robinson 1972), but none in *S. tuberculifer* (Eberhard 1975) (this species made about two attachments/s). Lapses in attention to the repeated, rapid decisions, and to the rapid and variable but nevertheless very precise adjustments in motor behavior, could result in imprecision in placement of the sticky line. As noted above, *M. duodecimspinosa* made literally thousands of rapid decisions regarding where to attach the sticky line (with

<1 s between attachments in the inner portions of the orb) for a long stretch of time (20–30 minutes), typically with no interruptions and apparently with few "errors" (Eberhard and Hesselberg 2012).

The rapidity, the behavioral precision, and the long, continuous nature of sticky spiral construction together imply sustained and intense attention on the part of the spider. I suspect that such sustained attention in tasks that involve so many rapid decisions and such behavioral precision may be unusual among invertebrates (and perhaps also vertebrates, for that matter). Introspectively, I wonder whether I myself am capable of such a feat of sustained concentration that involves so many decisions made on the basis of so many stimuli and requiring such fine motor control. Even if sticky spiral construction by a spider involves as little attention as does knitting for an expert, knitting a sweater nevertheless requires special kinds of mental abilities (How much yarn do I have left? How tight are the stitches? Have I finished this row? When should I change color?). Seen in this light, orb construction is a surprising product of such tiny brains (section 9.7, Quesada et al. 2011).

7.18 INDEPENDENCE (?) OF THE SPIDER'S RESPONSES

One curse of studying orb webs is that, because of its geometric regularity, there are inevitably strong correlations between many different aspects of an orb's design, only some which are likely to be due to cause-and-effect relationships. For instance, the area of an orb is correlated with the number of radii, the number of radii is correlated with the number of sticky spiral loops, and the number of sticky spiral loops is correlated with the spaces between them. Is the area of the web a cue used by spiders to determine the spaces between loops of sticky spiral, or is the correlation between them only incidental?

Appropriate experimental manipulations and careful statistical analyses (e.g., Zschokke 2011) can help discriminate between incidental correlations and cause-effect relationships. But for many of the variables the necessary experiments and analyses have not been performed. In addition, most studies analyze only a limited subset of the possible variables, and the multivariate statistical techniques that attempt to discriminate between first order and more incidental correlations lose power as the number of variables increases. Probably no study combines all of the possible cues for any given decision.

How many cue-response relationships are there? Or, to rephrase the question, how many types of decisions by an orb weaver are altered independently, and as a result of sensing how many stimuli? I cannot give final numerical answers. I can, however, illustrate one type of analysis that gave a conservative estimate of independent responses in a study in which *Leucauge argyra* (TET) was induced to build orbs in very small spaces (fig. 7.42, table 7.1) (Barrantes and Eberhard 2012).

A total of 12 design variables and 28 relations between these variables were measured and found to have changed using multivariate statistical analyses. But some of the variables and relations between them that changed were incidental products of the physical constraints on the spider in small spaces, so their "responses" cannot be properly considered to be due to choices made by the spiders. For instance, capture area and radius length were necessarily reduced in the small containers. Secondly, it seems highly unlikely that some of the design features in table 7.1 result from independent responses by the spiders. For instance, it is unlikely that the number of loops of sticky spiral in an orb results from any sort of decision by the spider regarding numbers per se; they may be able to count (Rodríguez and Gamboa 2000, Rodriguez et al. 2015), but surely they cannot count that high. More likely, a combination of decisions made during sticky spiral construction affects the number of loops, including how near the end of the radius to attach the outer loop of sticky spiral, how far apart to space the subsequent loops, and (probably to a lesser extent) when to terminate sticky spiral construction.

On the other side of the coin, it is also important to remember that statistical independence is not necessarily equivalent to biological independence (section 7.18). Some aspects that have strong statistical correlations may nevertheless be properly considered to result from separate behavioral decisions by the spider. Two commonsense criteria can help distinguish the degree of biological independence of different web design decisions: that they occur separately in time; and that they are influenced by different cues. Use of only these criteria, however, leaves open the possibility that two decisions that would be considered separate by these criteria might nevertheless be linked because they are both direct, lower-level consequences of a single higher-level decision. For instance, it is not physically possible for the number of radii to be independent of the mean angle between adjacent radii when the spider builds new radii. On the other hand, it is entirely feasible for the spider to build a long radius, and then to either attach the first loop of sticky spiral near its outer end, or to attach it far from its end. This suggests a third commonsense criterion—the physical feasibility of independence in the lower-level decisions. This independence criterion takes into account that there is variation in one decision even after the other, previous decision has been made. It thus emphasizes the possibility that natural selection can act separately on the two decisions, and that they can evolve independently. These behavioral questions are related to discussions of evolutionary "constraints," which are more often focused on morphological evolution (Gould and Lewontin 1979; West-Eberhard 2003; Müller and Wagner 1991).

Using these three criteria conservatively, the 16 different web variables that were affected by experimentally confining spiders in very small containers resulted from at least seven different kinds of design decisions (letters in table 7.1): *a*) whether or not frame lines would be built as part of radius construction and how long the frame lines would be; *b*) angles between adjacent radii during radius construction; *c*) spaces between hub loops (built after radii were finished); *d*) degree of asymmetry in placement of the hub; *e*) distance between attachments of the outer loop of sticky spiral and the outer ends of the radii; *f*) spaces between sticky spiral loops; and *g*) distance from the hub at which sticky spiral construction was terminated. This list is conservative, because possibly more than one decision was involved in producing the changes in the five different variables classified as resulting from decision *a* (table 7.1).

Several other types of data also suggest that different aspects of orb design are to some extent independent of each other. Japyassú and Ades (1998) showed that differences in the timing of ontogenetic changes in different web traits suggest the existence of semi-independent modules controlling orb construction in the nephilid *Nephilengys cruentata*, and reviewed evidence of similar independence in the ontogenetic changes in other species. Different species of parasitoid ichneumonid wasps elicit quite different combinations of changes in the orbs of their host spiders, in some cases by highly specific stimulation and repression of particular details of orb construction (Korenko et al. 2014; Eberhard and Gonzaga 2019). In addition, the normal orbs of related species show different mixes of similarities and differences (e.g., Lubin 1986; Coddington 1986b; Eberhard 1986a). As noted by Japyassú and Ades (1998), uncoupling between such modular behavioral routines enhances the evolutionary plasticity of orb weavers.

It is important to note that sub-programs of behavior that are independent may nevertheless be linked at higher levels of analysis. For instance, as argued by Krink and Vollrath (2000), stimuli perceived and analyzed by the spider during preliminary exploration behavior may alter several different web construction algorithms. In general, phenotypic flexibility in both behavior and morphology results from hierarchies of decisions (West-Eberhard 2003).

7.19 CHANGES IN RESPONSES TO CUES: LEARNING AND MATURATION

This chapter has focused on what stimuli spiders perceive (the "cues") and how they respond to them when building an orb. Some changes in the responses, and in the designs of orb webs, occurred over the lifetime of the spider (table 10.1; see Witt et al. 1968 for an early summary). These changes have two general types of explanation: gradual aging of the spider; and learning. Although there are clear demonstrations that orb web spiders have the ability to learn (Bays 1962; Le Guelte 1967; Ades et al. 1993; Hénaut et al. 2014), ontogeny and changes in the relative development of silk glands (Reed et al. 1970) are thought to be more important in causing many of the changes in orbs than is learning (e.g., Witt et al. 1968; Rawlings and Witt 1973; table 7.2) (see also section 10.2). The review that follows illustrates that there are data, however, from only very few species.

The classic study of Petrusewiczowa (1938) found that learning was not responsible for a typical change in the orbs of *Zygiella x-notata* (AR). Adults and older instar spiderlings tended to leave one sector of the orb open (with no sticky spiral lines in it), with a signal line running through it from the hub to the spider's retreat beyond the edge of the orb. Both open sectors and signal lines were lacking in the orbs of young individuals (Petrusewiczowa 1938; Mayer 1952) (Peters 1967, however, observed that *Z. x-notata* spiderlings built an open sector, even in their first orbs). In earlier webs, the sticky spirals were laid in the open sector and then later removed by the spider when she built the signal line; they were omitted entirely from the sector (by turning back at the edges during sticky spiral construction) later in life. Petrusewiczowa found that when she reared spiderlings to maturity in small containers, and thus prevented them from building orbs, they nevertheless built open sectors with signal lines when they were placed in larger containers. Similar experiments in which lab-raised *Argiope aurantia* were deprived of web-building experience showed that several quantitative changes were also apparently not due to learning, but rather to maturational changes in the central nervous system and to different degrees of development of the spider's silk glands (Reed et al. 1969, 1970).

The results of other studies (summarized in table 7.2) generally indicate that learning does not have a large role in determing orb design. This is not surprising, because there are no obvious payoffs to the spider for learning any of the particular abilities that have been studied. The possible rewards (greater prey capture) for altering orb designs (such as changing the vertical asymmetry) are likely to correlate with orb designs only very loosely, and the rewards will be received by the spider only some time (often hours) after she has built the web (Heiling and Herberstein 1999), making association even more difficult. In addition, immediate behavioral competence, without any learning period, is crucial for a newly emerged spiderling, which depends completely on its web to survive. A young spider probably cannot afford the luxury of only gradually learning how to build a web.

One additional possible reward from learning that a spider might more reasonably be expected to perceive is the possible savings in effort that would result from improved precision and coordination of her construction movements. The rewards could be a reduction in the time or the effort expended in building. This "practice makes perfect" hypothesis has not been supported, however, by the few available data from web measurements. The adjustments of adult *Eustala illicita* (AR) to different-shaped spaces for their webs did not improve with experience (Hesselberg 2015). Early instar spiderlings did not build webs with less consistent spaces between adjacent loops of sticky spiral in *Allocyclosa bifurca* (AR), *Anapisona simoni* (AN) (Eberhard 2007b), or *Leucauge argyra* (TET) (Barrantes and Eberhard 2012; Quesada 2014).

I know of only two cases in which learning seemed to affect web design. One was the gradual improvement in the ability of *Argiope argentata* (AR) to build horizontal orbs when imprisoned in a horizontal space (fig. 7.48, da Cunha-Nogueira and Ades 2012; Ades et al. 1993). This ability is not used in nature, because the orbs of this species are generally more or less vertical (table O3.1); the ability to learn that was demonstrated by this experiment could, however, be used to adjust to other website factors. Similarly, the later webs built by *Araneus diadematus* (AR) when several were built at the same site were more perfectly planar and vertical (Zschokke and Vollrath 2000). Here, however, the possibility of more thorough exploration of the website as opposed to learning was not eliminated. A further possibility (documented under unnatural conditions) was the apparently gradual improvement in locomotory and web-building skills in *A. diadematus* under weightless conditions in space, though the limited sample size precluded confident conclusions (Witt et al. 1977). In none of these cases was it clear what reward the spider might obtain from her improved performance. Claims that spiders used information from the locations in their webs where they had captured prey recently to modify future webs are open to similar doubts (see section 7.3.4.2, table 7.2).

In summary, there are numerous cases in which orb designs change during the lifetime of an individual (table 10.1) that could be due to either ontogenetic changes or learning. Evidence of several sorts indicates that learning has little role in most of these changes, except perhaps in one case, in improving the spider's adjustments to difficult conditions (table 7.2). There are also strong theoretical reasons to expect that learning will seldom be important. The implication is that ontogenetic design changes are often selectively advantageous, though in most cases it is not obvious what these advantages might be. Few species have been studied in sufficient detail, however, to conclusively rule out learning.

Table 7.2. Evidence that learning does not play a major role in producing "normal" adult orb web designs.

1. The basic designs of the first webs built by spiderlings after emerging from the egg sac were similar to those of the orbs of adult females in *Araneus diadematus* (Mayer 1952; Reed et al. 1970), and *Zygiella x-notata* (Petrusewiczowa 1938; Le Guelte 1966; Peters 1969) and *Nephila clavipes* (fig. 1.5).
2. Recently emerged second instar spiderlings of *Uloborus walckenaerius* and *U. plumipes* (UL) that were kept in small vials until they molted to the next (third) instar, without allowing them to build the typical "primary type" sheet-orb webs of second instar spiderlings (fig. 9.20), built normal, adult-like orb webs as is typical of third instar spiderlings (Szlep 1961).
3. Early instars of *Zygiella x-notata* (AR) that had been kept in a constricted space and then released built their first orbs with a free sector, as is typical of more mature spiders (a free sector is sometimes missing in webs of early instars) (Petrusewiczowa 1938; Le Guelte 1966; Peters 1969). The inverse experiment, of removing spiderlings prematurely from their egg sacs, sometimes resulted in very young spiderlings building small three-dimensional tangles of lines that included short sticky lines near silken retreats (Peters 1969). These differences were presumably related to incomplete growth and differentiation of the central nervous system, and/or incomplete maturation of the silk glands (Peters 1969), and cannot be attributed with confidence to learning.
4. When the araneid *Eustala illicita* (Hesselberg 2014) was forced to adjust orb designs to an abnormally confined space, there was no improvement after the first web. Similarly, *Araneus diadematus* did not improve in its ability to overcome disorientation resulting from vertical rotations of the web at 10 rpm during sticky spiral construction (Vollrath 1988a).
5. When spiderlings of *Araneus diadematus* were raised in small tubes from the period preceding construction of the first web for up to 194 days, they built webs whose patterns were unaltered except for being of smaller size (apparently due to poorer supplies of silk because of atrophied silk glands) (Reed et al. 1970).
6. The consistency or precision with which loops of sticky spiral were spaced from each other was just as high in the webs of second instar juveniles as in webs of adult females in both *Allocyclosa bifurca* (AR) and *Anapisona simoni* (AN) (Eberhard 2007b).
7. The adjustments made to building orbs in highly constrained spaces by *Leucauge argyra* (TET) were similar in early instar spiderlings and adult females (Barrantes and Eberhard 2012, Quesada 2014).
8. The conclusions of one study that suggested that web building experience increased the vertical asymmetry of orbs in *Argiope keyserlingi* and *Larinioides sclopetarius* (AR) (Heiling and Herberstein 1999) did not rule out either the likely effects of reduced silk gland development (Reed et al. 1970) in those animals that matured without building webs or the difficulty that inexperienced spiders had in ingesting prey (Heiling and Herberstein 1999) (the study controlled for spider weight, but not relative gland development) (feeding the spider below rather than above the hub showed a marginally significant association with increased vertical asymmetry in both species, however).
9. Learning details of orb construction is theoretically unlikely, because specific reinforcement for "correct" vs. "incorrect" details of construction behavior is unlikely: prey encounters with webs are highly stochastic, and will be only loosely related (if at all) to details of the design of any particular web (section 7.19); and spiders are unlikely to be able to associate particular details of construction behavior (which may have occurred hours earlier) with rewards from prey capture (see discussion in Heiling and Herberstein 1999, Eberhard 2019b).

7.20 CUES GUIDING THE CONSTRUCTION OF NON-ORBS

There are few descriptions of non-orb construction behavior, and an almost total lack of experimental studies of the cues used to guide their construction. Very few studies even mention possible cues, and some studies have concerned variables (e.g., pruned vs. unpruned bushes) that spiders surely did not sense as such, leaving the specific cues still unknown. Our current ignorance surely should not be taken to mean that non-orb weavers do not sense and respond to cues from their webs and their surroundings during web construction. Some behavioral details suggest just the opposite. The diguetid *Diguetia canities*, for example, made probing movements with her anterior legs as she approached her retreat when she was about to take a brief "rest" from web construction, indicating a search for cues (WE). But the experiments that are needed to move beyond descriptions to cause-and-effect conclusions are yet to be performed. Because this is the first review of possible cues ever assembled for non-orb spiders, I will be a little more speculative than for orbs.

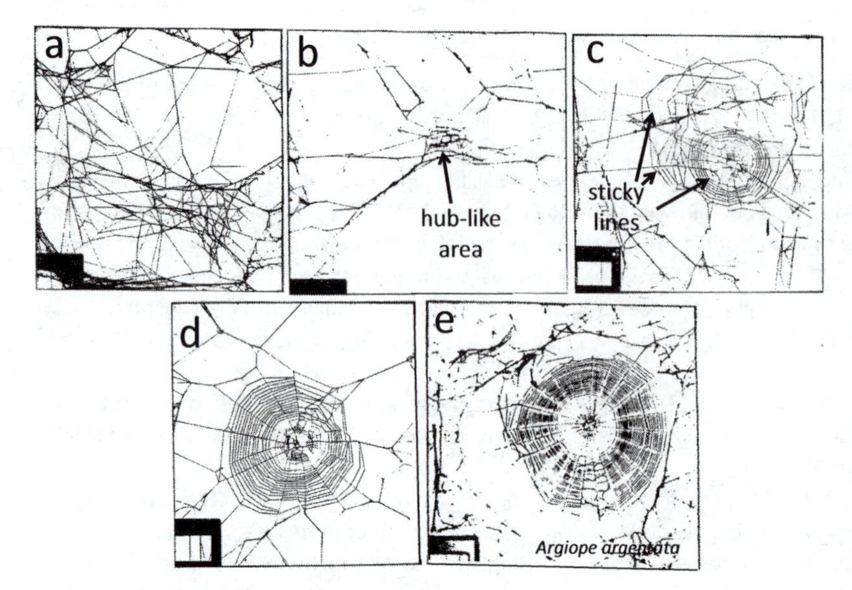

Fig. 7.48. These webs, made by mature female *Argiope argentata* (AR) when they were confined in rectangular frames laid flat, and thus could not build their normal vertically oriented orbs, documented unexpectedly elaborate memories. Spiders built several types of webs in horizontal frames: *a*) a tangle; *b*) a hub in a tangle with no sticky lines; *c*) a rudimentary orb; *d*) a "defective" (small) regular orb; and *e*) a regular orb. Spiders gradually "acclimated" to horizontal frames over the space of several days: they built more readily, and made more nearly normal orbs. Spiders apparently remembered this process of acclimation. When they were moved back to vertical frames for several days, and then re-introduced into flat, horizontal frames, they built more nearly typical orbs sooner (from da Cunha-Nogueira and Ades 2012).

7.20.1 PATH INTEGRATION

Path integration, which involves use of recent memories of the distances and directions the animal has traveled ("ideothetic cues") to remember the location of a reference point such as a burrow or retreat in two or three dimensions (Shettleworth 2010), is probably an ancient ability in spiders. For instance, whip spiders (amblypygids), the possible sister group of spiders, show signs of having a working knowledge of the areas around their retreats, probably formed as the result of previous explorations (Chapin and Hebets 2016). The arboreal tarantula *Avicularia avicularia* takes up predatory positions up to 1 m from the retreat each night, but returns to the retreat during the day (Stradling unpub. in Coyle 1986), and ctenizids in the genus *Nemesia* often leave their burrows in pursuit of prey, and return again without any apparent difficulty (the long legs and small abdomen of *N. dorthesi* [= *ariasi*] are associated with particularly rapid runs of this sort) (Buchli 1969). The prey-sensing extensions of the trap doors of some species are up to 25 cm long in some aganippines (Main 1976). These observations support the likelihood that the common ancestor of spiders could perform path integration, though they do not rule out the possible use of chemical or tactile orientation cues. Experiments have shown that the wandering spiders *Cupiennius salei* (Ctenidae) and *Pardosa amentata* can perform path integration (Seyfarth and Barth 1972 and Görner and Zeppenfeld 1980, respectively). Use of ideothetic cues during web construction in an non-orb weaver has, however, never to my knowledge been tested experimentally. The best experimental information for possible use of path integration comes from studies of how funnel web spiders return to the retreat after attacking a prey, rather than during web construction. In the agelenids *Agelena labyrinthica* and *Agelenopsis aperta* the spider used visual landmark cues, path integration cues, and cues from the web itself to return to its retreat after running onto the sheet to attack prey (Mittelstaedt 1985; Görner 1988).

Presumably path integration is also used during web construction. The agelenid *Melpomene* sp. did indeed repeatedly return more or less directly to her retreat after making short excursions to add silk to the sheet; these occurred even very early in construction, when there was little or no web present that could guide her (Rojas 2011). Fragmentary observations intimated that similar behavior may occur in the diplurid sheet web builder *Linothele macrothelifera* (Eberhard and Hazzi 2013). Path integration is likely to be much more difficult in three-dimensional webs (in contrast to orbs and sheet webs), but may have occurred in the three-dimensional tangles of *Nihonhimea tesellata* (THD) (Barrantes and Weng 2006a).

A further likely possibility is that during website selection non-orb builders assemble information acquired during multiple excursions in different directions from

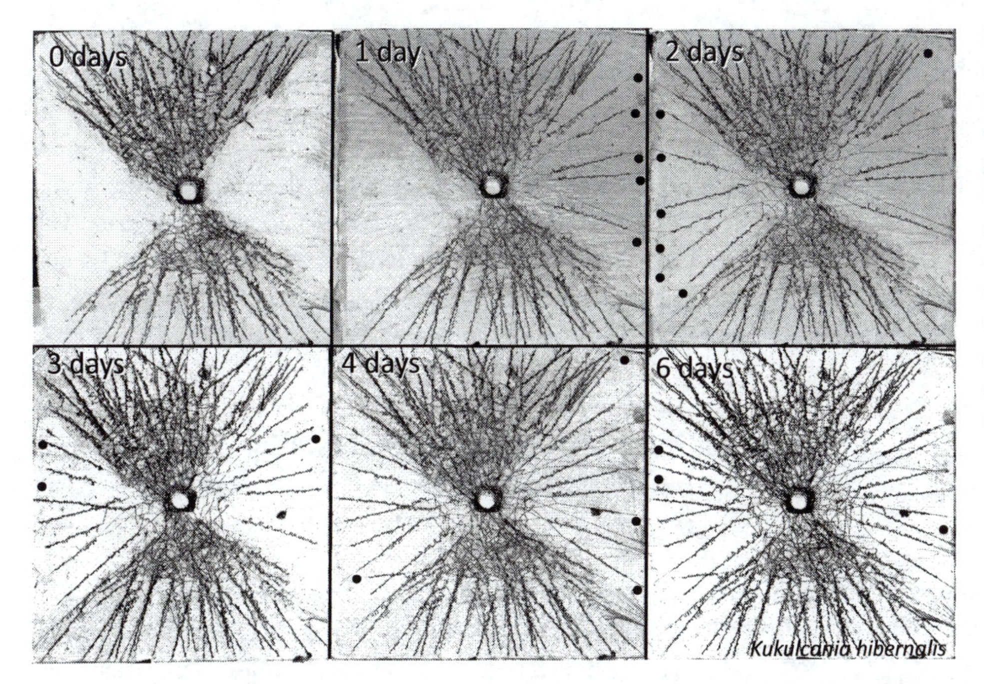

Fig. 7.49. The filistatid *Kukulcania hibernalis* had an impressive ability to sense where damage had occurred, and to concentrate subsequent building behavior in damaged areas. When two sectors of a web were removed (*a*), all of the 22 new radial lines (marked with dots) built during the subsequent six nights were in damaged rather than intact sectors.

some central point to estimate the overall size and/or shape of an open space in which a web could be built. The fact that the webs of different species have minimum and maximum sizes implies that spiders can avoid spaces that are too small and too large. For one example among many, individuals of different sizes of the pholcid *Physocyclus globosus* chose websites of appropriate sizes; when offered similar arrays of sites, mature males and females built webs in larger compartments than did fifth instar spiderlings (I. Escalante in prep.).

Another mystery that may be related to path integration cues is the relatively symmetrical gradual growth of the webs that were built over the course of many nights by the filistatid *Kukulcania hibernalis* (fig. 5.10). When one side of such a web was damaged and the other left intact, additions over the subsequent nights were, in contrast, concentrated on the damaged side (fig. 7.49). Did these spiders inspect their webs by walking around on them each night to check for damage and determine where the next building bouts should occur? Did they distinguish and then remember sites that need additions by using path integration cues? Such questions, which apply even to the rudimentary arrays of signal lines and debris around the mouths of the burrows of some mygalomorphs (fig. 9.26), have yet to be studied.

7.20.2 SMOOTH SUBSTRATES FOR GUMFOOT LINES

Gumfoot lines designed to capture prey walking on the substrate (e.g., ants) occur in theridiids (figs. 9.7, 9.14, 9.22), nesticids, and at least two genera of pholcids (Japyassú and Macagnan 2004; Eberhard et al. 2008b; Escalante and Masís-Calvo 2014). A recent experiment (Gutiérrez-Fonseca and Ortiz-Rivas 2014) showed that the brown widow *Latrodectus geometricus* (THD) preferred to attach gumfoot lines to smoother surfaces. Adult female and second instar spiderlings were kept in cages in which half of the floor area was covered with smooth cardboard strips, and the other half had similar cardboard strips that were corrugated (the positions of smooth and corrugated strips were alternated nightly). The spiders attached more gumfoot lines to the smoother surfaces (fig. 7.50). In addition, the spiderlings tended to attach lines to the peaks rather than the valleys of the corrugated surface. If potential prey of these spiders are more likely to walk along smooth surfaces (where the energy costs of walking are probably lower) (Gutiérrez-Fonseca and Ortiz-Rivas 2014), then the spiders' preference for smoother surfaces may lead to greater rates of prey capture. Similarly, if prey are more likely to walk across the upper portions of a corrugated surface (surely true for relatively large prey), then attaching gumfoot

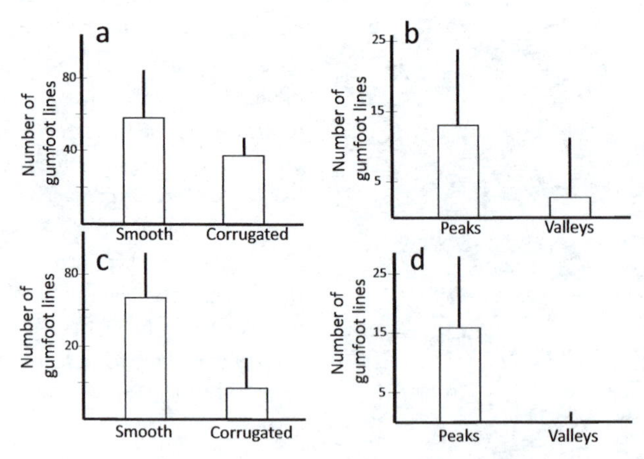

Fig. 7.50. Experiments with captive brown widow (*Latrodectus geometricus*) spiders confirmed the impression derived from many theridiid webs in the field (e.g., fig. 9.7) that spiders prefer to attach their gumfoot lines to smooth surfaces. When alternating, equally wide strips of smooth and corrugated cardboard formed the floor under the spider's retreat, more gumfoot lines were attached to the smooth cardboard than to the corrugated cardboard in four successive webs by adult females (*a*), and most attachments to corrugated were to peaks rather than valleys (*b*) (peak-to-valley depth was 0.45 cm, peak-to-peak distance was 0.75cm). The same trends occurred in webs by second instar spiderlings (the first instar outside egg sac) (*c, d*) with another type of cardboard that had smaller corrugations (depths of 0.15 cm and peak-to-peak distances of 0.25 cm) (after Gutiérrez-Fonseca and Ortiz-Rivas 2014).

lines to the peaks rather than the valleys also makes adaptive sense.

The mechanism used by the spiders to discriminate smooth from rough surfaces was not determined. In some theridiids the smoothness of a substrate may be evaluated early in construction. Tangle lines were built by *Latrodectus* spp. prior to gumfoot lines (Lamoral 1968; Eberhard et al. 2008b). Substrate choice during tangle construction could explain the common field observation (never properly documented, however) that the tangles of theridiid gumfoot webs tend to be located near especially smooth surfaces such as tree trunks or the surfaces of large leaves where their gumfoot lines are attached (fig. 9.7, Nielsen 1932; Eberhard et al. 2008b; WE).

7.20.3 RIGIDITY OF SUPPORTS

Not all potential attachment points offer equally firm and appropriate support for a web. More rigid objects also seemed to be preferred as supports for anchor lines in some orb weavers (Tilquin 1942; section 8.4.3.2.2). *Cyrtophora citricola* (AR), which builds a tangle within which it constructs a horizontal non-adhesive sheet (fig. 1.7), preferred more rigid supports (Madrigal-Brenes 2012). When spiders were offered a series of similar paper strips that were alternately rigid (fastened at both ends) and flexible (fastened only at the upper end) to which they could attach lines to support their tangles, they attached more lines to the rigid strips. Two additional trends were evident: spiders attached more support lines near the corners of the rectangular boxes in which they were housed than near the center of each wall; and they attached more lines to the upper rather than the lower portion of the box (Madrigal-Brenes 2012). The reason for the first of these trends is uncertain, but the second may provide more physical support for the spider and her web. The apparent preferences of non-orb spiders such as *Dictyna* spp. (Dictynidae) for dead twigs (Nielsen 1932, Comstock 1967, Blackledge and Wenzel 2001b) may be linked to their rigidity, but studies testing such preferences have yet to be performed.

7.20.4 LOCATIONS OF SUPPORTING OBJECTS

The stimuli used by spiders to choose attachment points may influence the web's design. For instance, different webs of the linyphiid *Novafrontina uncata* suggested that the form of the tangle above the sheet was adjusted to the distribution of supports (fig. 7.51). A photograph of the web of the amaurobiid *Amaurobius fenestralis* shows the same apparent adjustment, extending along a crack in a wall (Nielsen 1932 vol. 2). The usual approximately circular outline of the horizontal sheet of *Nihonhimea tesselata* (THD) changed to an elongate rectangle when a web was built on a narrow window sill (Jörger and Eberhard 2006). The most extreme case I know of is illustrated in fig. 10.7 in which the zoropsid *Tengella radiata* webs lacked most of the usual structures (sheet, tangle above the sheet) when attachment sites near the retreat were lacking.

The flexibility in the overall form of the substrate web of the filistatid *Kukulcania hibernalis* (fig. 5.10) has a useful implication. The spider attached her lines preferentially to objects that protruded from the surface, so the positions of lines were influenced by the locations of such objects (fig. 7.52). Webs in which the retreat was located in a groove running vertically on the wall of a house tended to be elongated vertically, while webs with the retreat in horizontal grooves in the same wall tended

Fig. 7.51. These two webs (*a*, *b*) of two mature female *Novafrontina uncata* (LIN) illustrate great flexibility in the relative proportions of the tangles above and below sheets that had approximately equal diameters. The differences presumably resulted from fewer available attachment sites for web *b* (photographs courtesy of Gustavo Hormiga).

to be elongated horizontally (fig. 7.52) (WE). Interestingly, the webs of the synanthropic *K. hibernalis* are often thought to be radially symmetrical (e.g., Comstock 1967; Kaston 1965; Griswold et al. 2005; see fig. 5.10), but this symmetry is probably to some extent an artifact of the highly planar surfaces of human structures. In a more natural and irregular environment, the spider would probably take advantage of whatever projecting points were available; as these would tend to be distributed irregularly around the retreat, they would give rise to a less symmetrical web. For many arachnologists like myself, nearly all of our experience with filistatid webs is with those on buildings. The stimuli that guide the spider in adjusting their webs to local substrate contours are presumably perceived as the spider explores near her retreat, but they have never been studied systematically.

7.20.5 RADIAL SYMMETRY

Radial symmetry has evolved convergently many times in non-orb spiders (see, for example, fig. 9.26 for mygalomorphs, also table 9.1), suggesting that the cues needed to produce such symmetry may be readily accessible. In most groups, the spider's burrow or retreat forms a natural point of convergence for radial lines, and the degree of symmetry is determined by the distribution of these lines around the retreat. The symmetry implies that the spider somehow balances the quantity of lines laid on the different sides of her web. Some of these webs are built over a span of many different nights, making the symmetry a more impressive behavioral feat.

For some webs, like that of *Liphistius* (figs. 9.26, 10.5a) for instance, the radiating lines converge on the burrow mouth where the spider rests, and it would be feasible for her to sense the separations between adjacent radial lines that are already in place when she is at the burrow mouth. The decision mechanism for where to lay a new line could be as simple as grasping different pairs of adjacent radial lines with legs I, and sensing the degree of separation between them. I know of no cases, however, in which apparent measuring movements of this sort have been observed. The difficulty of achieving radial symmetry seems much greater in webs such as those of the pisaurid *Thaumasia* sp. (fig. 5.11), the filistatid *Kukulcania hibernalis* (fig. 5.10), the hersiliid *Hersilia* sp. (Wil-

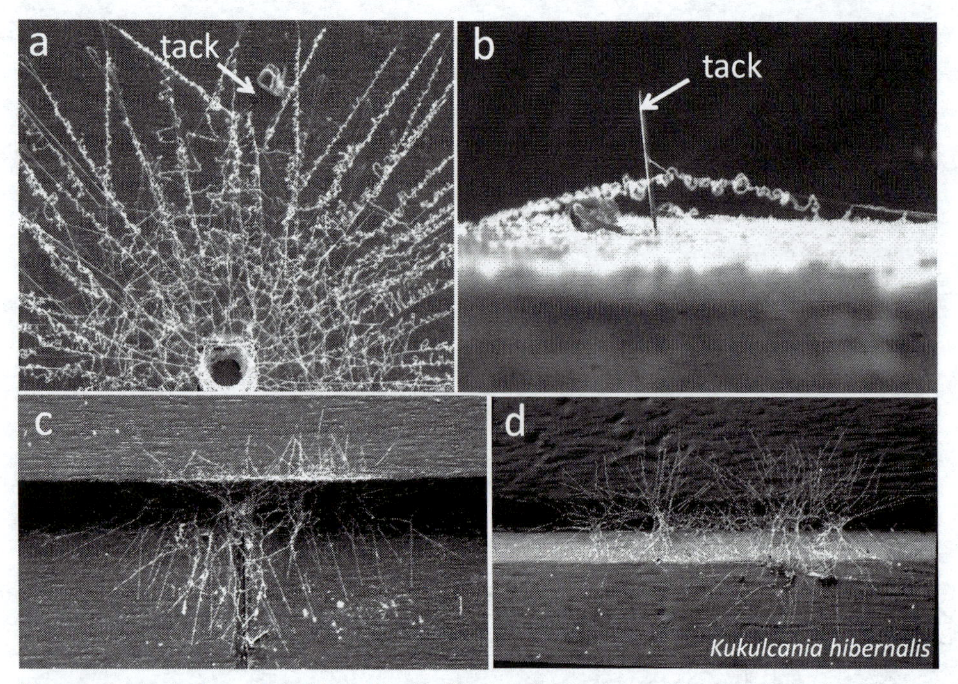

Fig. 7.52. The photos from above (*a*) and the side (*b*) of a radial line in the web of the filistatid *Kukulcania hibernalis* that was attached to a tack, which had been pushed through from the other side of the sheet of cardboard on which the spider's web was built, illustrate a problem with the commonly-used category of "substrate-bound" spider webs (e.g., Blackledge et al. 2009c, Bond et al. 2014, Fernández et al. 2014). Spiders consistently attached their radial lines to objects that protruded above the substrate, so at this fine scale this web might be more appropriately called an "aerial substrate" web. This preference for building elevated lines probably explained variations in the designs of the webs built on a vertical wooden wall of a house (*c*, *d*). In areas where the boards overlapped each other shingle-fashion (*c*), the portion of the web below the retreat (where web lines were elevated) was larger than that above the retreat (where lines generally ran along the surface). The possibility that these up-down asymmetries were due to gravity rather than elevated attachment sites was contradicted by the lack of up-down asymmetry in other webs on the same wall but that were sited in the horizontal grooves between two boards (*d*). By elevating her radial lines from the substrate slightly, the spider presumably increased prey retention by increasing the surface area of the prey that contacted sticky cribellum silk. The overly simplistic "substrate-bound" classification scheme leaves unanswered the question of how far above the substrate a web's lines need to be for the web to be considered an "aerial" rather than a "substrate" web.

liams 1928) (fig. 9.1), the oecobiids *Uroctea durandi* (fig. 10.5) (Kullmann and Stern 1981) and *Oecobius concinnus* (fig. 9.6) (Solano-Brenes et al. 2018), and the hypochilid *Hypochilus gertschi* (Shear 1986) (radial symmetry seems likely independently derived in each of these families). In at least some of these webs the inner ends of the radial lines were not close enough for the spider to be able to sense adjacent radial lines simultaneously while she was at the central point (the hersiliid may be an exception). Thus she cannot use leg positions to judge where new lines need to be added. Some species may use short-term memory, such as a tendency to lay successive radial lines on opposite sides of the central resting place, as observed in *Thaumasia*. Web construction in *T.* sp. was brief (the web in fig. 5.11 took just under 5 min to build), and the spider did not perform any other behavior during web construction that seemed designed to sense the positions of previously laid lines (Eberhard 2007a).

In *K. hibernalis*, the problem of achieving radial symmetry seems greater, because its web was built over the space of many days (fig. 5.10). If spiders used memories of the locations of lines built previously, these memories would need to be stored for long periods of time. Two details of *K. hibernalis* construction behavior (WE) intimate that such long memories may not be necessary. As the spider moved away from the retreat while beginning a radial line, she tapped actively with legs I and II, perhaps sensing lines already in place to enable her to place new lines in spaces between previous radial lines and avoid laying them too close to each other. In addition, the spider did not always return all the way to the retreat between building one radial line and the next.

Sometimes she returned only about half way to the retreat, then moved in a more or less circular path with respect to the retreat on the order of 60–120°, and moved outward to lay another radial line. Circular movements could help spacing successive new lines apart (fig. 5.10). Further observations are needed.

7.20.6 DIFFERENCES IN TENSIONS

As with path integration (section 7.20.1), the best data available regarding the possibility that tension cues might be used during web construction come from behavior that was observed in a different context, attacks on prey. In the funnel web spider *Agelena labyrinthica* the spider's ability to return to the retreat after running onto the sheet was strongly affected by the pattern of tensions on web lines (Baltzer 1930; Holzapfel 1934). Tensions on lines running in different directions in the web were altered by slightly folding the cage from a square to a rhombus, lowering the tensions on lines in one diagonal, increasing the tensions on the opposite diagonal. When the retreat was on the high tension diagonal the spider had no trouble returning; but she was able to return directly in only 3 of 24 trials when the box was folded so that the retreat was on the low tension diagonal (Holzapfel 1934). The spider often walked back and forth along the line of greatest tension in these trials as if searching for her retreat. Holzapfel believed that the denser array of lines in the horizontal sheet that occurred in the vicinity of the retreat were under greater tension when stressed by the spider's weight, but did not measure tensions. A portion of the sheet with more lines would be less extensible under the spider's weight, so differences in extensibility could have been the cue used.

The distantly related bowl and doily linyphiid *Frontinella pyramitela* also built sheet webs with patterned differences in tensions (Suter 1984). The central area (the lowest portion of the "bowl") was under the most tension, and was surrounded by a ring of sheet in which lines were under lower tensions; tensions rose again near the edge of the web (especially near supporting objects). The spider may use the tension differences to find the central area where she rests, as she chose the central, high tension area as a resting site even when the sheet was reoriented 90° so that the sheet was vertical rather than horizontal. The pattern of tensions in the web was affected by 10–15 vertical "tensor" lines that attached the sheet to the tangle below, pulling the sheet into downward-directed "dimples" (figs. 1.7, 9.16) (Suter 1984). Similar dimples associated with vertical lines to the tangle below occur in the webs of several other linyphiids (Hormiga and Eberhard in prep.), and several other distantly related species that build dense horizontal sheets with tangles below, such as the diguetid *Diguetia canities* (Eberhard 1967), the theriids *Nihonhimea tesselata* (Eberhard 1972b) and *Chrosiothes* nr. *porteri* (Eberhard et al. 2008b), and the araneid *Cyrtophora citricola* (Wiehle 1927) (see fig. 1.7).

The fact that the spider senses tension differences in other contexts implies that she could also sense tension differences while she is building her web. Whether or not such cues were used to guide web construction, however, is unknown.

7.20.7 TEMPERATURE

In desert environments a spider resting on her web in the sun can be subject to extreme temperatures (witness the diguetid *Diguetia imperiosa* that in the 47°C summer tempartures in a desert in Mexico emerged from her cylindrical retreat in the tangle [fig. 9.16], and rested in the shade provided by the retreat—Bentzien 1973). In the Negev desert in Israel, the black widow *Latrodectus revivensis* shifted its websiιe preferences in hotter weather to the cooler microclimate provided by the upper branches of taller shrubs (Lubin et al. 1993). Another theridiid, *Parasteatoda* (= *Achaearanea*) *tepidariorum* avoided both high and low temperatures in a gradient from 13 to 27° (Barghusen et al. 1997). When kept for six day periods in chambers with different temperatures this species built larger webs (in terms of total mass) at an intermediate temperature (20°) (Barghusen et al. 1997).

7.20.8 APPARENT SENSORY MOVEMENTS OF LEGS

Given the way that orb weavers use lines that were produced early in construction to guide later construction behavior (sections 7.3–7.8), non-orb weavers probably also use previously-laid lines to guide subsequent construction. I know, however, of no experimental tests that demonstrate such effects directly. Nevertheless, observations of their behavior hint strongly at such cues being used. The colonial dictynid *Mallos gregalis* consistently initiated webs in captivity on the most directly illuminated side of the cage, but if the cage was turned the next day so that another side was illuminated, the spiders continued building on the first side on subsequent nights (Tietjen 1986). Tapping or waving movements of legs, especially legs I and II, are common in some species, and often end at least temporarily when the leg encounters and seizes a line as in the diguetid *Diguetia canities* (WE), the lycosid *Aglaoctenus castaneus* (Eberhard and Hazzi 2017), and the theridiids *Nihonhimea tesselata*, and *Chrysiothes* sp. (WE). One exceptional species that performed few waving and tapping movements during construction was the pisaurid *Thaumasia* sp., which walked on a leaf surface and thus did not need to find lines to support herself. The "pseudo-orb" amaurobiid *Titanoeca albomaculata* tapped laterally with her anterior legs while she was working inward laying sticky lines, apparently in order to locate the inner loop of "sticky spiral" already in place (Szlep 1966a), just as occurs in orb weavers (section 6.3.5). The anteriorly directed tapping movements of the filistatid *K. hibernalis* when beginning a new radial line may also serve to find sticky radial lines already in place (WE). Two other suggestive cases are the apparent use of the edge of the sheet to guide the path of the first cribellum lines in the psechrid *Psechrus* sp. (Eberhard 1987a), and the use of the edge of the skeleton sheet when extending this edge further in both the pholcid *Modisimus guatuso* (fig. 5.2) (Eberhard 1992b) and the theridiid *Chrysso cambridgei* (fig. 5.15) (Zschokke and Vollrath 1995b). I speculate that use of the positions of lines to guide subsequent behavior is common, and that the scarcity of documented examples is due to the scarcity of appropriate studies.

The only experimental manipulation I know of involving a non-orb weaver produced surprising results. When the pedipalps of the sheet-weaving lycosid *Hippasa olivacea* were removed, the spider built "a tangled and shapeless fabric" (Hingston 1920). Even more surprisingly, the movements of the spider's legs seemed to be affected: "It continually catches its feet in the filaments of its web, an act which, in the uninjured spider, never occurs, and I have even seen it tear the snare in the endeavor to free its limbs" (p. 155). These unexpected observations surely merit further study of the role of the palps in web construction in this and other species.

There are also behavioral suggestions that the linyphiid *Frontinella pyramitela* measures tensions on lines. The slow leg flexions performed while returning to the tighter central area of its sheet after being away could sense tensions directly (Suter 1984). Occasionally these spiders also made rapid abdomen flexions in this same context; these could enable her to sense tensions via resonant motions of the web or its transmission properties (Suter 1984).

7.20.9 RESERVES FROM PREVIOUS FEEDING

In some species previous prey captures affect whether or not the spider will alter the design of the next web. The desert-dwelling eresid *Stegodyphus lineatus* makes risk-aversive, conservative decisions. Food-supplemented spiders, both in nature and in captivity, built significantly smaller webs (Pasquet et al. 1999). Web construction in this species is costly; it requires about six hours of work, and results in the loss of 3–7% of the spider's weight.

In contrast, well-fed individuals of two theridiids, the common house spider *Parasteatoda* (= *Achaearanea*) *tepidariorum* and western black widow *Latrodectus hesperus*, added more web during the first 24 hours in a new habitat than did poorly fed individuals (Lien and Fitzgerald 1973; Salomon 2007). The first study did not analyze separately the tangle versus the gumfoot lines that function strictly for prey capture, so the possibility of flexible adjustments in relative investments in the two (changes in web design) was not tested (Salomon 2007). Well-fed *L. hesperus* built lines that were twice as thick as those of poorly-fed individuals, and had fewer lines in the sheets and sticky gumfoot lines but more in the rest of the web ("supporting" lines) (Blackledge and Zevenbergen 2007; Zevenbergen et al. 2008). In sum, the well-fed spiders altered web designs, investing relatively less in structures for prey capture, and more in those for defense against predators. Not surprisingly, the relatively greater investment in sticky gumfoot lines by starved spiders resulted in greater numbers of prey captured when non-flying prey were introduced into the cages; this was true even when well-fed spiders were placed in webs built by starved spiders (Zevenbergen et al. 2008).

The sheet-weavng lycosid *Aglaoctenus lagotis* may make smaller sheets after producing a clutch of eggs than before (González et al. 2015). It was not clear whether the small web size resulted from the spider building a smaller web from the start, or from simply making fewer additions to its edges.

7.20.10 Conspecifics (Gregarious and Social Species) and Possible Constraints Imposed by Orbs on Sociality

The webs of the social dictyid *Mallos gregalis*, which lives in colonies of up to 20,000 individuals, included loosely woven sheets with cribellum silk, "papery" areas, and "spongy" zones that had numerous tunnels and chambers (Tietjen 1986). The optical density of silk deposited in 52 cc Petri dishes by different-sized groups of adult females after five days increased in larger groups, but not proportionately (Tietjen 1986). In arbitrary units of optical density, silk deposition/spider was negatively correlated with group size: for groups of 1, 2, 5, 10, and 20, the densities/spider were 48, 23, 18, 8, and 5 respectively (Tietjen 1986). This implies that in the larger groups some factor associated with other individuals (their presence per se? the lines they laid?) inhibited web production.

The colony members of several social spider species produce unitary communal webs rather than collections of individual webs (for instance, a single compound sheet rather than multiple sheets). Social species with unitary communal webs include the theridiids *Anelosimus* spp. (Agnarsson et al. 2006a, 2006b), *Parasteatoda wau* (Lubin 1986), and the agelenid *Agelena consociata* (Darchen 1965; Riechert et al. 1986). They also differ from other group-living species that build large, more amorphous colony webs that have neither a single unit design nor clear subunits, such as the webs of *Mallos gregalis* (Tietjen 1986) and *Stegodyphus* spp. (Kullmann et al. 1971; Bilde and Lubin 2011).

The cues guiding construction behavior are not known in these species, but it would seem that the stimuli available to guide the construction of a unitary web such as, for example, the single huge sheet in a giant colony of thousands of individuals of *Anelosimus eximius* or of *Achaearanea wau*, must differ from those used in web construction by the close solitary relatives of these species such as *Anelosimus studiosus* or *Nihonhimea tesselata*. Judging by the close taxonomic relations between some social species that build unitary webs and solitary relatives (Agnarsson et al. 2006a on *Anelosimus*), the changes in cues used by social species may not be large.

On the other side of this coin, group living has evolved in orb weavers many different times, but in none of these lineages do spiders build a unitary communal orb web. Bilde and Lubin (2011) list six taxonomically scattered orb-weaver genera, to which two other distant relatives can be added—*Leucauge mariana* (TET) (Valerio and Herrero 1976; WE) and *Cyclosa* (AR) (Wu et al. 2013; WE). In five of these eight groups the subsocial route to more extensive cooperation that was followed in the evolution of increased cooperation in other spider lineages (Bilde and Lubin 2011), would seem to have been open; egg sacs were in or near the colony, and some young spiderlings inhabited the colonies where their mothers were present (Lubin 1980; Uetz 1986; Buskirk 1975; Liao et al. 2009). Nevertheless, contact between adult females and their offspring did not occur in *Parawixia bistriata* (Fowler and Diehl 1978), *Nephila* spp.(Robinson and Robinson 1973a), or *Leucauge* (Valerio and Herrero 1976; Castro 1995).

Perhaps the individualism intrinsic to the design of an orb, with a single hub where signals converge and routes to attack prey originate, but where there is space for only a single spider may have restrained evolution toward more extensive cooperation. This "orb-selfish" argument is supported by the observations of a general lack of prey sharing in several colonial species, including *Philoponella semiplumosa* (Lahmann and Eberhard 1979), *P. republicana* (Lubin 1980), *Metabus ocellatus* (Buskirk 1975), and *Metepeira spinipes* (Uetz 1986). There was also a similar lack of cooperation in the colonial pholcid *Holocnemus pluchei*, which built non-shared individual sheets suspended in a common tangle (Jakob 1991, 2004). Occasional communal feeding did occur, however, in the colonial orb weaver *P. bistriata* (AR), but only on relatively large prey (Fowler and Diehl 1978; Fernández-Campón 2007).

7.20.11 Web Repair

One of the most powerful techniques for documenting the cues that spiders use to guide web construction behavior is to observe how the spider responds to experimental modification of particular lines in her web (see, for example, section 5.3.1). I do not know, however, of a single experimental study of this sort with a non-orb weaving spider. Studies of repair are feasible in many families, and incidental observations (e.g., fig. 7.49) suggest that they could yield much useful information on cues that guide web construction. Given the long lives of many non-orb webs in the field, and thus the likelihood of broken lines (and perhaps fatigue in the lines themselves) (section 8.5.1.4), selection has probably favored the ability to repair damaged portions of webs in many groups. Comparative studies of cues guiding repairs would surely be rewarding.

7.21 SUMMARY

7.21.1 SURPRISING PATTERNS IN ORB CONSTRUCTION, ESPECIALLY STICKY SPIRAL CONSTRUCTION

When a biologist struggles, as I have in this chapter, to make an exhaustive review of particular types of behavior or structures, the result is often disappointing—a mind-numbing list whose only major virtue is its comprehensive coverage. Surely the reader of this chapter can be excused for being tempted to skip, for example, the seemingly endless series of esoteric effects of different variables that guide sticky spiral construction. But the abundant data in this first-ever compilation of more than a century of scientific scrutiny of orb construction, and of sticky spiral construction behavior in particular, reveal some exciting patterns of general significance.

The most general message is one of rampant flexibility: orb construction is not rigid and consistent, but rather changes in diverse ways under the effects of an unexpectedly large number of different cues. For example, up to twenty (!) different variables affect sticky spiral placement (table 7.3) (and this does not take into account additional decisions regarding the thickness of lines or the amount of glue on them, which also clearly vary systematically between webs and even within a single orb) (section 3.2.9.2; Crews and Opell 2006; Sensenig et al. 2010b). While not all of these variables are necessarily independent, I count at least ten that seem very likely to be independent (section 7.18), and the number may well be higher. Orb weaver construction behavior is clearly very flexible. Who would have thought that there is so much going on in the brains of these small, "simple" animals as they plod around and around their orbs building the sticky spiral? I have not attempted to review the numbers of cues used by other animals to guide particular decisions, but believe that the length of this list for orb weavers is unusual, at least for an invertebrate.

Table 7.3. A summary of the surprising number of cues and variables that may affect decisions of where to attach the sticky spiral in an orb (#18 is discussed in the next chapter). While the strength of the evidence varies ("?" indicates less confidence), and a few cues may only be used by a subset of all orb weavers (e.g. #16, #17), the overall message is clear: sticky spiral construction behavior involves surprisingly complex information processing by the spider. At least 4 and probably 5–6 variables are sensed anew for each new attachment, and these are combined with many other "settings" stimuli (probably 5–10 in any given case).

A) Reference point cues that are sensed anew for each attachment

1. Site of inner loop of sticky spiral
2. IL-TSP distance (distance between inner loop of sticky spiral and outer loop of temporary spiral)[1]
3. IL-TSP distance on preceding radius (or radii)[1]
4. Angle of radius with gravity

B) Intermediate cues that vary more slowly and that may not be sensed anew for each attachment

5. Supply of silk in glands
6. Current edge-to-hub distance
7. Current orientation with respect to hub (above, below, to the side)

C) Settings cues that may be sensed only once (before sticky spiral construction begins)

8. Body size
9. Body weight
10. Recent feeding history (total intake)
11. Recent feeding history re. specific prey?
12. Presence of predators?
13. Wind[2]
14. Impending oviposition
15. Sites on web where prey were captured previously
16. Time of day
17. Season of the year
18. Previous orbs at this site ("pilot web" phenomenon; see chapter 8)
19. Temperature
20. Humidity

[1] Sensed via the positions of the legs as they hold lines, plus distances moved.

[2] Affects two decisions: spaces between lines, and sizes or composition of glue droplets from aggregate glands.

A second overall pattern is that there is surprising uniformity both in the cues that orb weavers use to decide where to lay lines and make attachments, and in how they use these cues to guide their behavior. Experimental comparisons of cues and responses to them have shown that different taxonomic groups of orb weavers are very similar. As discussed in chapter 10 (and Eberhard and Barrantes 2015), this uniformity argues strongly in favor of a monophyletic derivation of orb webs. Both the cues and the responses to them are apparently ancient, and have been very conservative among orb weavers. The data give little sign of a gradual evolutionary process of perfecting web construction among orb weavers. In contrast, both cues and responses to cues have frequently been modified in species descended from orb weavers that now build non-orb webs (see section 10.9).

The physical agility and coordination needed to build the sticky spiral also revealed previously unappreciated variation. The uniformity in the verbal description of sticky spiral construction as "walk out one radius and attach, walk out another and attach, . . ." is an illusion; it conceals the quite different leg and body movements and forces that are exerted when orb weavers perform these "repetitive" movements. The intrinsic complexity in coordinating and directing the spider's orb construction behavior brings into focus the additional point that building an orb must require intense and sustained attention to choose appropriately among all the possible leg movement patterns those that will give consistent results at the level of the web designs that are produced.

One possible major evolutionary trend in orientation mechanisms has been to economize on the movements used to obtain cues that guide sticky spiral (as well as temporary spiral) construction behavior. Spiders in the derived theridiosomatoid families have apparently abandoned use of one of the two direct cues, the site where the inner loop is attached to the radius, that is important in other families (Eberhard and Hesselberg 2012). Instead they apparently rely on kinesthetic cues to guide both temporary spiral and sticky spiral construction. A second, similar economy in sensing the location of the inner loop of sticky spiral with small movements of leg oIV instead of larger movements of leg I, has evolved at least six and possibly seven times (see appendices O6.2.5, O6.3.6, section 10.9).

These considerations reaffirm a basic theme of this book, that orb weavers are not simple automatons that execute the same behavior over and over. The pendulum has now swung dramatically away from the conclusions that pioneers like Fabre and Hingston drew regarding the "blind" and "rigid" nature of orb construction behavior. The attribution of rigidity and an inability to adjust to different circumstances has given way, under abundant data of several types, to recognition that orb weavers show striking plasticity (see also Herberstein and Tso 2011). There is still much variation whose adaptive consequences are not understood (e.g., Vollrath et al. 1997), and the explanations for why some responses occur have yet to be worked out. It seems likely that further data will increase rather than decrease the types of adaptive adjustments.

The more subtle question of whether all of this behavioral complexity results from pre-programmed sets of instructions for how and when to gather and then respond to multiple cues, or whether some responses result from a primitive understanding of the tasks being performed, is discussed elsewhere (Eberhard 2019b). It is still possible that an orb weaver is a complex rather than a simple automaton; but it may be that, at least in some respects, she does not completely lack understanding of what she is doing. Further experiments, utilizing newly developed techniques (fig. 7.53), have great promise.

7.21.2 OTHER STAGES OF CONSTRUCTION

The other stages of orb construction have not been studied nearly as extensively with respect to cues and responses, and conclusions cannot yet be drawn regarding numbers of cues and responses. In general it is also too early to describe their evolutionary patterns. Nevertheless, the conservative trend to use the same cues in different groups seems to hold, both in the way that the TSP distance is used to guide temporary spiral construction, and the way inter-radial angles are used in radius construction.

The earliest stages of orb construction are the most mysterious, both because they are the least studied and because they are when the spider probably performs the most demanding analyses. She apparently assesses both the size and shape of the area that her web will occupy (largely determined in nature by the unique array of sites where supports are available and the open spaces between them), and assesses the silk resources that she

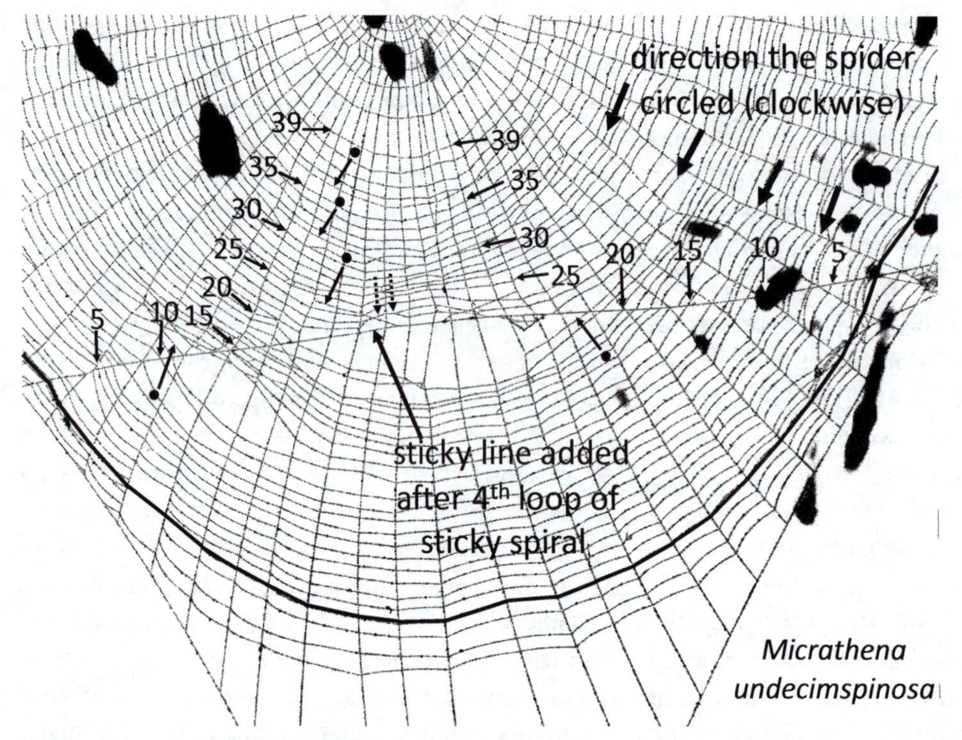

Fig. 7.53. New experimental techniques promise further revelations regarding the cues guiding orb construction. When a long sticky line (the long approximately horizontal line in the figure) was made by collapsing the finished orb of another spider to a single line and was placed on the web of a mature female *Micrathena duodecimspinosa* (AR) early during sticky spiral construction, the spider's subsequent behavior changed in complex ways (the heavy black line traces the innermost sticky loop present at the moment when the experimental sticky line was added; the subsequent loops that the spider added as she moved clockwise are numbered in order; the black dots with arrows indicate sections of temporary spiral that the spider left intact). The first 22 times that the spider encountered the experimental line in the 4:00–5:00 portion of the orb, she responded little or not at all (the spacing of loop 21 was slightly increased). On loop 23, however, she crossed the experimental line but immediately turned and moved along it for about one radius and then used it as a guide, spacing the new loop inward from it. She laid loop 24 approximately parallel to and spaced inward from this loop (the two adhered to each other at the dotted arrows). Subsequent loops (25–40) were normal, though the spaces between loops 25–30 were reduced in the experimental sector. Adding sticky lines to webs during the process of sticky spiral construction makes it possible to control how and where cues are altered, and allows experiments to be replicated, a major improvement over depending on occasional "wandering" by spiders, as in fig. 7.2.

has available. Circumstantial evidence indicates that orb weavers probably use path integration to perceive sizes and shapes of spaces. The spider then translates this information (via unknown processes) into criteria for making various decisions during construction. These include where to build frame lines, how far apart to space radii, where to lay the outer loop of sticky spiral, how far apart to place loops of sticky spiral, and how thick to make the lines and the covering of sticky material on the sticky spiral lines.

7.21.3 NON-ORB WEBS

The cues guiding the construction of non-orbs are generally unstudied. I speculate that a kinesthetic sense ("path integration"), integrating distances and directions that the spider has moved to keep track of the location of her retreat, is probably an ancient arachnid trait. Experiments with artificial nerve networks indicate that a kinesthetic sense is surprisingly simple to produce (see Chittka and Niven 2009). Non-orb weavers are clearly able to adjust the designs of their webs to aspects of the sites at which they build them. In is unclear how much the evolution of orb webs involved the development of new abilities, and how much it relied on using sensory cues and analytic abilities that were already present in their non-orb weaving ancestors. It is also unclear whether the non-orb spiders that have secondarily lost orbs continue to use the cues and analytical abilities that were used by their orb-weaving ancestors.

8

WEB ECOLOGY AND WEBSITE SELECTION

8.1 INTRODUCTION: WHAT IS AND IS NOT INCLUDED

It is clear that web-weaving spiders have substantial ecological impacts. For instance, Robinson and Robinson (1974) estimated that the web spiders in a New Guinean coffee field consumed 160 kg of insects/ha/yr. Taking into account only the individuals of a single species, *Nephila pilipes* (= *maculata*) (for which they had a great deal of direct field data on prey captures), they estimated a yearly catch of 271,755 individual insects/ha each year. Because their searching techniques undoubtedly underestimated webs of spiders near the ground and of smaller web spiders in general, and because smaller spiders tend to capture a larger fraction of their body weight/day (Eberhard 1979b; Santana et al. 1990), these numbers probably represent substantial underestimates (Nyffler and Birkhofer 2017).

This book concentrates on behavior and evolution, not ecology, so I have not tried to review the many studies of the ecology of web-building spiders. I emphasize in this chapter where construction behavior and ecology overlap, because understanding the functional significance of present-day behavioral traits requires understanding their ecological contexts. I will briefly discuss webs and foraging theory, because of the unusual advantages that webs offer in quantifying foraging decisions. I will spend most of the rest of the chapter on the variables that affect a web-building spider's choice of habitat and website, variables that elicit web construction behavior itself. I have omitted the faunistics of web builders (the associations of different species with different general habitat types), and also their community ecology, except for crucial studies relating to intra- and inter-specific competition (see section 3.2.6). I emphasize orb weavers, largely because of the large bias in published studies. I have also omitted other, undoubtedly interesting ecological relations with other species (e.g., the possible effect of prey bouncing off of the horizontal orbs of *Tetragnatha* spp. and falling to the ground where they promote predation by ground-dwelling wolf spiders [Takada et al. 2013]).

A brief glance at some general aspects of spider biology provides a useful introduction (Anderson 1974; Wise 1995). Several morphological and physiological traits adapt spiders especially well to the lifestyle of a predator. Spiders can lower their metabolic rates to withstand long periods without food, and have soft-walled, expandable abdomens that allow them to take up large quantities of food when they do capture a prey (though, strangely, some are more likely to die at molting if they have fed especially abundantly—Higgins and Rankin 2001). The rates at which they capture prey in nature, corrected for the spider's own size, are greater for smaller than larger spiders, and there may also be substantial differences between "fast lane" and "slow lane" taxa (Anderson 1974; fig. 8.4).

Most spiders in nature are hungry, and can benefit from additional food (Riechert and Łuczak 1982; Wise 1995). Prey capture in nature appears not to be generally limited by the presence of other species (sections 3.2.6, 8.8), though the apparent "ecological release" of species such as *Wendilgarda galapagensis* (fig. 8.1) (Eberhard 1989a) and *Tetragnatha* spp. (Blackledge and Gillespie 2004) on extremely depauperate oceanic islands suggests (though does not prove—Wise 1995) that sometimes inter-specific competition may be important. Most (though not all) spiders can move long distances quickly and with little expenditure of energy, either by ballooning, or by producing long, air-borne, super-light lines that float on the breeze to form bridges to distant objects. Because these mechanisms depend largely on the strength and direction of the wind, which are beyond the spider's control, dispersal by spiders to new habitats is probably more fraught with danger than is dispersal by animals capable of self-powered flight. Very long-distance travel is more feasible for spiders, however, than for many self-powered animals.

8.2 WEBS AND ECOLOGICAL FORAGING THEORIES

Spiders are predators, and their webs are physical manifestations of their foraging strategies. Just like other animals, they must invest energy, time, and resources in foraging in order to obtain food. Foraging is risky and also uses up the animal's reserves, so its costs must be balanced against its benefits. An orb-weaver is of particular interest for foraging behavior theory because her foraging investments can be measured readily and precisely in terms of her webs. The investments are to some extent disjunct in time from her decisions regarding her diet breadth (accepting or rejecting prey in the web) (Higgins and Buskirk 1992) and her development (growth per molt, number of instars, adult size) (Higgins 1992a). The choice of a website can represent a compromise among different preferences, including the physical space, supports, microclimate, danger from predators, and available prey (Harwood et al. 2003).

Webs have been analyzed using optimal foraging theory. One general prediction from theory is that both extremely good and extremely poor feeding success are likely to lead to reduced investment in foraging effort, as reflected in the area covered by the web, the amount of silk invested in the web, and the spacing between lines in the web (fig. 8.2). Especially well-fed spiders (e.g., x_5 in fig. 8.2) benefit less from acquiring additional prey, and can dedicate more of their accumulated resources to growth or egg production. Especially poorly fed spiders (x_1 in fig. 8.2) lack the reserves needed to make large investments in foraging. Intermediate levels of feeding are expected to correlate with the highest investments in foraging. In general, spiders seem to conform to these expectations (Higgins 1995; Herberstein and Tso 2011).

Spiders (both orb weavers and non-orb weavers) use cues related to previous foraging success in guiding several decisions regarding foraging effort, such as the area to be covered by the web, the distances between loops of sticky spiral, and whether or not to build at all (Witt et al. 1968; Lien and Fitzgerald 1973; Higgins and Buskirk 1992; Higgins 1995; Pasquet et al. 1999; Venner et al. 2000; Salomon 2007; Blackledge and Zevenbergen 2007; Zevenbergen et al. 2008). Expectations outlined in fig. 8.2 are complex, however, and difficult to test, because they include both positive and negative correlations between feeding success and foraging investment. Thus an increase in foraging success from x_2 to x_3 could lead to an increase in investment, while the same absolute change, but from x_3 to x_4, could lead to a reduction. And some comparisons (e.g., between x_1 and x_5, or x_2 and x_4) may show no correlation at all. Foraging theory thus predicts that the biological contexts in which spiders make some decisions are important. For instance, the decrease in foraging effort observed in well-fed compared with poorly fed individuals that occurred in *Argiope trifasciata* (AR) (Tso 1999), *A. keyserlingi* (Herberstein et al. 2000a), *Octonoba sybotides* (UL) (Watanabe 2001), and the eresid *Stegodyphus lineatus* (Pasquet et al. 1999) suggests that in these cases the poorly fed spiders were

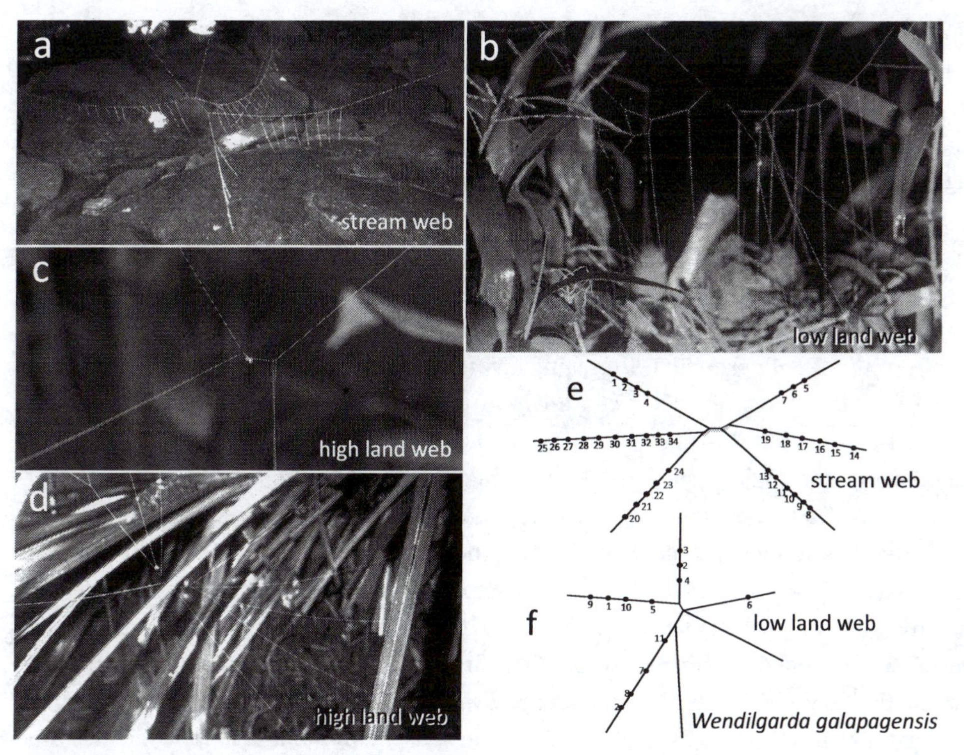

Fig. 8.1. The unusually extensive intra-specific diversity in the webs (*a–d*) built in different habitats by *Wendilgarda galapagensis* (TSM), a species endemic to the isolated oceanic island Isla de Coco where there are few other orb-weaver species, suggests that inter-specific competition among web spiders may be at least occasionally important. Other species in this genus on continental North and South America build relatively uniform "stream" web designs, but *W. galapagensis* varied even in the basic organization of construction behavior: the erratic order of sticky line construction in low land webs (*f*) lacked the edge-to-hub order of adding lines, and one-radius-at-a-time patterns that occurred in stream webs (*e*). Less convincing, but nevertheless possible alternative explanations for the explosion of diversity on this island are lack of predators, and more intense intra-specific competition (from Eberhard 1989a).

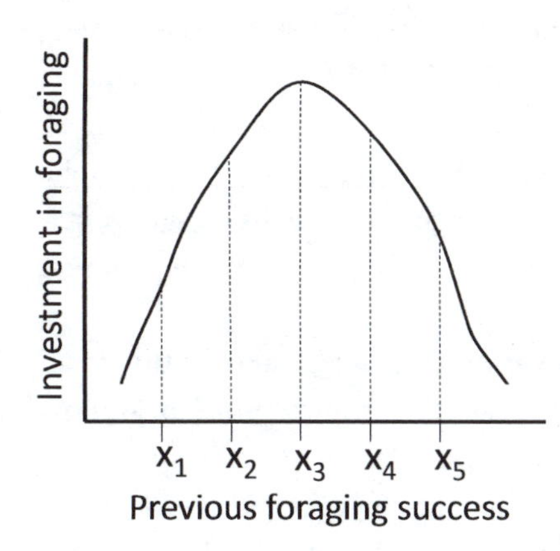

Fig. 8.2. The general prediction from foraging theory is that when an animal's previous foraging success has been extremely poor (x_1) or extremely successful (x_3), its current investment in foraging (such as a spider's web) should be smaller than the investments following moderate success. Following very poor foraging success, the animal lacks sufficient reserves to make large investments in foraging. When foraging success has been especially good, the animal will gain more from utilizing its resources in other ways (e.g., growth, making eggs) than from acquiring further prey. The greatest foraging investments are expected from animals with both adequate reserves and a greater need for food. Details of the shape of the curve are likely to vary in different species in different ecological settings (after Higgins 1995).

nevertheless far from starving to death; both well-fed and poorly-fed spiders may even have been on the right hand side of the curve in fig. 8.2.

Little is known in general regarding the particular cues that are used as indicators of previous foraging success (e.g., body weight, stretching of the abdominal cuticle, fat reserves, memories of prey captures, etc.). Increasing the spider's weight, however, by gluing weights on her body had a negative effect on orb size in the araneids *Araneus diadematus*, *Larinioides sclopetarius*, and *Argiope keyserlingi* (though these results were interpreted as resulting from difficulty in supporting the spider's weight, rather than foraging strategies) (Witt et al. 1968; Heiling and Herberstein 1999). It is generally not clear which decisions during web construction are affected by foraging success. For instance, orb web area might be adjusted by changing the criteria for the distances between attachment sites, or by altering the sites on early lines where subsequent new frame lines are initiated (e.g., fig. 6.5b-g), by terminating the temporary spiral nearer or farther from the frame lines, etc.

One simplifying assumption of the theory sketched in fig. 8.2 is that increased investments in foraging necessarily tend to lead to increased payoffs, but this may not always be true. A larger investment could result, for instance, in a web that is mechanically less stable, that is more visible to prey, or that has some parts too far away from the hub for the spider to reach in time to capture prey. Trade-offs regarding payoffs may explain one common trend in spiders. Spiders tended to reduce their foraging effort or to cease building webs entirely just prior to molting or laying eggs. Pre-molting and pre-oviposition reductions in webs may be due to reduced payoffs or increased costs resulting from physical weakness associated with the degradation of the old cuticle prior to shedding it, or to shifting of resources to produce silk in the cylindrical glands or to the ovaries, thus increasing the physiological costs of investing in a web.

The costs of investing resources in foraging rather than in growth may explain some counter-intuitive observations of responses to different levels of prey capture in juvenile *Nephila clavipes* and *N. pilipes*. When subjected to slightly reduced prey capture rates, they increased their orb sizes (as predicted on the right hand side of the curve in fig. 8.2) (their rates of weight gain decreased), built larger orbs relative to their body sizes, and foraged for longer periods (kept their webs up longer) (Higgins and Buskirk 1992). But when they were subjected to more substantial reductions in prey capture, they decreased orb size as well as the rate of weight gain (as predicted on the left hand side of the curve) (Higgins 1995). At very low food levels (which are typical of some habitats of *N. clavipes*—Higgins and Buskirk 1992; Higgins 1995), webs were small; no spiders lost weight, and a few gained weight. The cost of foraging in *N. clavipes* is exemplified by the negative correlation between the concentration of choline in a spider's webs and that in her cephalothorax (Higgins and Rankin 1999). Choline is an essential B-complex vitamin that is used in cell membranes and neurotransmitters, as well as in silk, and it was limiting when spiders were raised in captivity.

The substantial individual variations among field collected individuals of *N. pilipes* in their responses to food reduction, raise the possibility that differences in previous foraging experience may have induced different individuals to make different short-term foraging decisions. Some data suggest that the interactions between foraging investment, growth, and previous feeding success are more complex than predicted by classic optimal foraging models (Higgins 1995).

Marked individuals of the nocturnal *Larinioides cornutus* (AR) showed somewhat different responses to low food intake in the field. In 27 individuals that captured few prey over 10 consecutive nights, the area of the capture zone and the length of thread were similar when the webs of the first five days were compared with those on the second five days, but the mean spaces between sticky spiral loops decreased; investment in foraging may thus have increased, if thread diameters and amounts of glue were constant (Sherman 1994). Another araneid (*Araneus diadematus*?—three species were listed in the methods section) gave still another response when individuals were starved and prevented from reingesting their webs in the lab (Witt 1963): web area and design remained constant during the first six days, but body weight decreased (thread diameters were said to decrease, though no data were presented). After four more days, web area had decreased but web designs were unchanged. After 20 days, webs were smaller and wider-meshed, and the spider had lost further weight. In all of these studies, spiders sacrificed body weight to continue foraging during the early stages of starvation.

The effects of environmental seasonality on the trade-off between foraging effort and growth are also illustrated in *N. clavipes*. This species proved to be univoltine (producing a single generation/year) in five relatively seasonal sites in Mexico (seasonality was due to dry or cold periods, or both). There was a period early in ontogeny during which web size increased disproportionately with increases in body size: later the relative web sizes became smaller (fig. 8.3 a_1, a_2) (Higgins 2006). In a less seasonal environment in Panama where spiders were present at all times of the year, the relative sizes of webs were constant throughout ontogeny (fig. 8.3b).

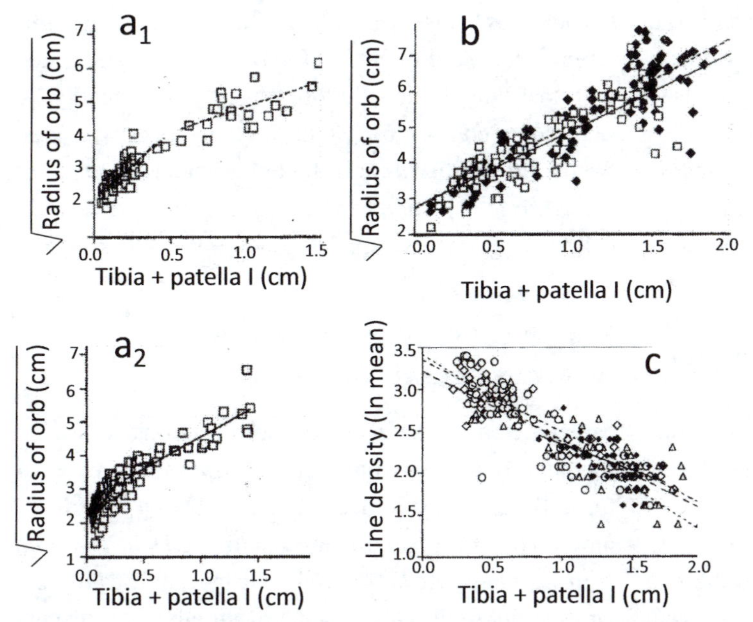

Fig. 8.3. These graphs document variations in how *Nephila clavipes* (NE) adjusted the relative sizes of their orbs (scaled against an indicator of body size) in habitats with different degrees of seasonality. In six highly seasonal habitats in Mexico (two are illustrated in graphs a_1 and a_2), spiders made larger investments in foraging effort (web size) rather than in their bodies early in ontogeny (steep slope early in life); this probably enabled them to reach maturity soon enough to reproduce before dying due to harsh seasonal changes (cold, dry weather). In a less seasonal tropical site in Panama (*b*), where seasonal changes did not cut spiders' lives short, they did not make such disproportionate foraging efforts early in life (black diamonds = early rainy season generation; white squares = early dry season generation). Other aspects of foraging, such as the relative density of lines, did not show similar adjustments in the different Mexican sites (*c*) (different symbols indicate data from different sites) (or in web construction times—data not shown), so the measurements of the relative web size in the other graphs *a* and *b* are probably reasonable indicators of foraging effort (though there are no data on possible changes in fiber diameters and chemical composition of the silk—see section 3.2.9) (after Higgins 2006).

Thus in the more seasonal environments young spiders appeared to engage in the more risky behavior of investing relatively more resources in webs than in their own bodies by making relatively greater foraging efforts, presumably due to a greater need to grow and mature rapidly before the favorable season ended. Two other measures of foraging effort, the relative density of spiral lines in the web, and the rates of orb renewal compared with body size, did not vary with ontogenetic stage in the Mexican populations (fig. 8.3c). The important point is that ecological foraging strategies can vary ontogenetically; typological characterizations of foraging strategies can lead to errors (see also Sensenig et al. 2010b).

The fact that spiders manipulate the diameters and properties of their silk (Higgins et al. 2001; section 3.2.9) challenges the assumption in these studies that silk properties were uniform at different growth stages and in different habitats, so more work is needed. Nevertheless, an important conclusion is that *N. clavipes* may modify their foraging strategies (orb designs) away from short-term optima in order to to achieve reproductive maturity rapidly in strongly seasonal habitats. It was not clear whether these differences between populations were due to genetic differences or to phenotypic flexibility.

Some other aspects of foraging theory appear to be less appropriate for understanding web-building spiders. For instance, a spider at the hub of her web repeatedly leaves and then returns to the hub, just as do many animals that forage away from some central place. Predictions from central place foraging theory failed for *Argiope keyserlingi* (AR). The spider was no less likely to attack prey at the edge of its web than nearer the hub (Champion de Crespigny et al. 2001). The density of prey found trapped in unoccupied orbs of *Araneus pinguis* (AR) increased with the distance from the hub, but this trend did not occur in inhabited webs (Endo 1988), suggesting (contrary to the theory) that spiders attacked prey farther from the hub more readily than prey closer to the hub. These data included prey left in the web by the spider, however, and not those that escaped or that were attacked by the spider, so the results are difficult to interpret. A different modeling study that was based on this theory predicted (not surprisingly) that for orbs built on tree trunks, optimal web form should become more and more elongate as the width of the available space becomes more limited, and the optimal web area decreased with more reduced horizontal space (Harmer et al. 2012). The model had several limitations, however (Harmer et al. 2012), and was not checked quantitatively against real webs.

Still another aspect of foraging concerns decisions whether or not to change sites. Optimal foraging theory requires more precise information regarding website and habitat quality than spiders are likely to be able to gather, due to the high day-to-day variability likely to occur in the mass of prey that is captured (Sherman 1994; Edwards et al. 2009) and their own sensory limitations (section 8.1). For this reason, I believe that the theory is not likely to yield useful detailed predictions, although it helps focus attention on important general topics.

8.3 WHAT IS ENOUGH? "FAST LANE" AND "SLOW LANE" SPIDERS

There are substantial differences in the rates with which different species accumulate prey biomass and grow in the field (fig. 8.4). In the first place, of course, larger species (or larger individuals of the same species) tend to have higher rates of prey capture in terms of absolute biomass (see also Higgins 2006). When the rate is corrected for the spider's body weight, however, two further differences are apparent: first, smaller spiders accumulate prey relatively more rapidly than do larger spiders (note the negative slopes in both intra- and inter-specfic comparisons in fig. 8.4). Craig (1989) found a similar relation in estimated prey capture rates in five other orb weavers. Second, there are also substantial taxonomic differences in the relative rates of feeding. For instance, the "slow lane" *Tengella radiata* accumulated prey much more slowly than the other species in fig. 8.4. Another extreme case was the small linyphiid *Bathyphantes simillimus* that preyed almost exclusively on collembolans; adults in captivity captured only up to four prey items per month (30 prey were provided/week), and were very slow to respond to prey in their webs (mean 220 s) (Rybak 2007). It is not clear why there are taxonomic differences, or whether different web designs are associated with fast lane or slow lane lifestyles. Two species with highly derived web designs, the rectangular orb theridiid *Synotaxus turbinatus* and the bolas spider *Mastophora dizzydeani* (AR) (*i* and *b* in fig. 8.4), captured prey at typical rates for orb-weavers of the same sizes.

8.4 PROCESSES THAT PRODUCE HABITAT BIASES

Anyone who has searched for web-building spiders in the field soon learns that different species tend to occur in particular types of habitats and subhabitats. For example, the orbs of the tetragnathids *Dolichognatha pentagona* and *Azilia affinis* occur in or at the edges of New World tropical forests, only at sites that are adjacent to very large objects (such as the bases of large trees); and only at those sites within this subset of subhabitats, such as the space between two buttresses, where there is an open horizontal space that has potential attachment sites. In addition, *A. affinis* webs are likely to be closer to the ground than those of *D. pentagona*.

The published literature (including many taxonomic studies) is replete with such characterizations of websites, and I will not attempt to make any sort of survey or review. Spiders build webs in a huge variety of ecological settings. Instead I will discuss more general questions concerning how and why spiders decide where to build and where not to build their webs, using only illustrative examples rather than attempting a general review.

Three general types of process can produce the non-random distributions seen in nature. Two involve "pre-building" biases manifested prior to building a web at a given site: philopatry (remaining in the natal habitat); and non-random movements or settlement patterns. The third involves "post-building" biases manifested after a spider has already built a web at a given site, such as expanding the web in one direction rather than another, or leaving vs. staying. The effects of the three processes are combined for any given species, and different patterns can occur at different spatial scales in the same species. Distinguishing among these possible processes on the basis of observations of the final result (a common kind of data in web ecology studies) can be difficult.

Movements between websites probably normally

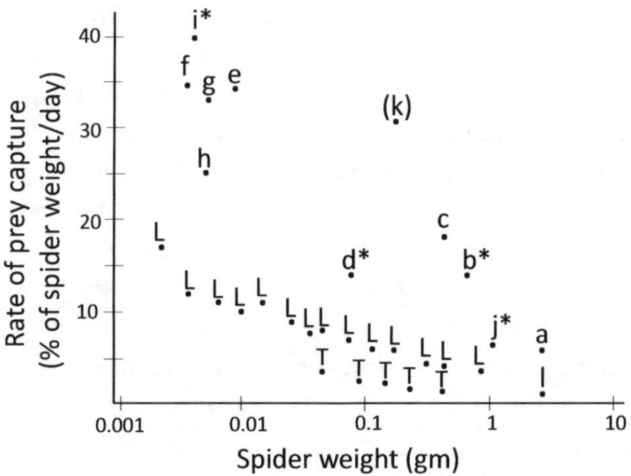

Fig. 8.4. Fast- and slow-lane spiders are indicated by measurements of the mean rates of prey capture per day in the field, expressed in terms of the estimated total wet weight of the prey/day corrected by the spider's own wet weight. In general, rates were negatively correlated with the spider's weight, but they varied widely. Species *a–j* are orbicularians (those marked with "*" build non-orbs); *k* and *T* build sheet webs, while *L* builds no web. All data were from mature females except for species *L* and *T*, in which captures by juveniles of different sizes are also reported. Data from *a–c* were calculated from daily collections of prey remains below the web; others (*d–j*) were from rates of weight gain or egg production in the field, assuming a 33% conversion of prey tissue into spider tissue; *k* may be less precise (as indicated by the parentheses) because it was based on overall acceptance rates for various types of prey (ignoring escape and avoidance behavior of the prey) and placing sticky traps near webs in the field (see Eberhard 1990a and sections 3.2.5.4 and 8.9.2 for doubts regarding the accuracy of such estimates from sticky traps). All of these techniques had the advantage of giving complete, 24 hour summations of prey captured without interventions. Orb weavers and their close kin seemed to capture a larger percentage of their weight than did the others (with the uncertain exception of species *k*), but differences in web designs in this group were not associated with different prey capture rates: the bolas spider (*b*), aerial sheet weavers *d* and *j*, and the sparse, weakly adhesive "rectangular orb" weavers (*i*) all had rates typical of other orbicularians of similar sizes. Species: *a–Nephila maculata* (NE); *b–Mastophora dizzydeani* (AR); *c–Argiope argentata* (AR); *d–Mecynogea lemniscata* (AR); *e–Cyclosa caroli* (AR); *f–Cyclosa* sp. (AR); *g–Dolichognatha* sp. #1 (TET); *h–Dolichognatha* sp. #2 (TET); *i–Synotaxus turbinatus* (THD); *j–Cyrtophora moluccensis* (AR); *k–Agelenopsis aperta*; (Agelenidae); *L–Tasmanicosa godeffroyi* (Lycosidae); and *T–Tengella radiata* (Zoropsidae) (after Santana et al. 1990; the original references are given there; also Higgins 2006 for *a*).

occur at several stages during the life cycle of most web-building spiders. The newly emerged spiderling leaves the egg sac and finds a website of her own. And, as she grows, she probably usually needs to periodically relocate, because websites most appropriate for early instar spiderlings are likely to be inappropriate for later instars or adults. For instance, the young and old spiders of the black widow *Latrodectus revivensis* (THD) occupied the same general habitat (small desert bushes), but larger spiderlings tended to move to larger shrubs, and there were age-dependent preferences for different shrub species (Lubin et al. 1993). Spiders periodically ran the high risks of moving (an estimated 40% mortality vs. <1% mortality when not changing sites) to find a new website. The advantages of website relocation in this species probably stemmed from architectural features that provided support for webs, and protection from thermal extremes and predators (Lubin et al. 1993).

Before discussing the variables that influence website choice, it is useful to review the mechanism by which spiders move between sites, and the kinds of sensory information they are likely to have available while searching for sites.

8.4.1 SEARCHING WITH LINES FLOATED ON THE BREEZE

Web-building spiders use silk lines for long-distance movements, so most movements, except for walking along the substrate, must be downwind. Orientation by sensing chemical cues from an object at a distance (as commonly occurs in insects, for example) would seem to be particularly difficult for them. If the breeze is in a consistent direction, a spider moving downwind will only be able to sense the odor of an object after she has passed it. Theoretically, she might be able to return upwind along the same line if the line passed directly over the object; but if her line passed to the side of the object rather than over it, even this return tactic will be ineffective unless she walks on the substrate. These limitations could be less acute if the wind direction is more variable. Despite these problems, male spiders routinely use pheromones to locate females (Trabalon 2013). I know of no studies of the behavioral details of male searching behavior. Using silk lines for long-distance travel is effective because air movement is essentially continuous in the field (Geiger 1965). The erratic movements of smoke from a cigarette in a "quiet" forest, where one's own senses suggest there is no air movement, testify to this ubiquity. The importance of long-distance chemical cues in website choice remains unclear.

8.4.2 SENSORY BIASES: "SATISFICING" AND SPECIAL PROBLEMS FOR AERIAL WEBS

Selection of a website surely depends on first choosing the general overall habitat, and then distinguishing the sizes and shapes of open spaces and the availability of objects to which the spider can attach her lines (e.g., sections 7.1, 7.4; Szlep 1958; Kullmann 1958). But sensing these aspects of her environment is not easy for a spider, because she probably generally operates without any appropriate visual cues. Orb weavers (and presumably other groups making aerial webs) probably use cues from path integration, but the specific cues used to evaluate possible websites are poorly understood (reviewed by Vollrath 1992; Hesselberg 2015). The spider likely integrates information on both the distances and the directions that she has moved to estimate the size and shape of the area through which she has moved (Vollrath 1992). Hesselberg (2015) mentioned "stigmergy," defined as the expression of complex behavior based on interactions with, modifications of, and feedback from the environment that rely on simple rules. An orb weaver might have simple rules for responses based on information from the silk lines laid during site exploration. Hesselberg contrasted possible predictions of this hypothesis with those of a cognitive map hypothesis, but came to no confident conclusion, as the necessary data are largely lacking.

Even if she uses path integration, a spider has a special problem: the apparent lack of a reference point. Classically, when animals use path integration they sense the angles and distances that they have moved relative to external points of reference such as the entrance to an ant nest or a spider's retreat, or to the position of the sun or some other distant, large object, (Görner 1988; Shettleworth 2010). An orb weaver probably has few or no visual reference points as she explores a new site in the dark; she also lacks the network of radial and spiral lines that she uses as reference points at later stages of construction.

One potential type of reference point would be the particular objects to which she attaches lines, perhaps distinguished by their textures or chemical properties. Direct observations of exploration behavior by the araneid *M. duodecimspinosa* showed, however, that the spider does not behave as if she is using such characteristics as references; typically she contacted an attachment site only once during exploration. On subsequent trips along the line toward this attachment she usually stopped short of the end and turned back; she thus showed no sign of the periodically "refreshing" of her memory (as is typical later in orb and non-orb construction behavior, with repeated visits to the retreat or other central point [section 5.2, figs. 5.1, 5.11]). Only after the decision to build had been taken, and she was in the process of strengthening the eventual anchor lines of her orb and their attachments to the substrate, did *M. duodecimspinosa* return to contact the objects to which her lines were attached (section 6.3).

Another type of potential reference cue during exploration would be the points where different lines intersected. But intersections also seem difficult to use, because the spider often shifted the points of attachment between lines. It is imaginable that the spider also sensed and remembered the distances that points of attachments were shifted, but other unaltered attachments also changed positions substantially when lines were added and removed, so this seems unlikely. Sensing the distances that she had walked (e.g., counting her steps, as proposed by Hingston 1920) was also sometimes only an incomplete measure of vertical distances, when the spider broke the line she was on and released additional line, sagging downward; descents were sometimes 20 cm or more in *Micrathena duodecimspinosa* without taking a single step (section 6.2.3). Again it is possible to imagine a solution (sense the length of dragline released), but the difficulty of the analysis needed by the spider seems substantial.

Even for a human, the process of breaking, shifting, and reattaching lines at different points makes describing (and much less understanding and remembering) exploration behavior very difficult. Seemingly the only constant reference cues available during the exploration of a site would be gravity and (much less reliable) wind direction. Even these cues were not necessary for *Araneus diadematus* (AR), which built complete orbs under conditions of weightlessness in space (Witt et al. 1977) (of course demonstrating that neither gravity nor wind direction is necessary does not mean that they are not used—see Vollrath 1992). One possible solution to this mystery is that spiders use the properties of lines (e.g., their lengths to rigid attachment sites) to identify and remember them, but this is only speculation.

However it may be that spiders sense that the size of a potential website is sufficient and free of nearby objects, the minimum and maximum acceptable sizes undoubtedly vary among species. Previous studies of openness have focused not on website choice per se, but on the consequences of choices for web visibility (e.g., Craig 1988, 2003), presumed exposure to predators (Blamires et al. 2007b), and the likely velocities of prey that strike the orb (section 3.2.11). To my knowledge there are no studies of how spiders judge the openness of a space. The criteria used could conceivably include innate biases, learning, or both.

Another complication is that the preferences of spiders are apparently flexible. If the match between the site and the spider's preferences is especially poor, a free-ranging spider presumably simply moves on to another site. But if the match is close enough, some spiders will adjust the size and design of the webs that they build (see section 7.8.2); particularly extreme adjustments of this type occur in the zoropsid *Tengella radiata* (fig 10.7). In one comparison, *Cyclosa caroli* (AR), whose webs in nature are are suspended by long anchor lines in relatively open spaces, proved to be less flexible in adjusting to spaces with constrained shapes and sizes (spiders more often failed to build an orb) than was *Eustala illicita*, a species whose webs in nature are typically built close to the branches and leaves of *Acacia* trees and are more variable in form (Hesselberg 2013). The fact that individuals of the theridiid *Latrodectus revivensis* that moved farthest from their previous websites tended to achieve greater increases in the quality of their new websites (shrub height) (Lubin et al. 1993) illustrates the possible payoff to individuals that have higher thresholds, and that presumably rejected more of the plants that they encountered along the way.

Spiders' probable sensory limitations suggest an important general conclusion regarding the importance of two different basic searching strategies: locate several possible sites and then choose the best among them; or choose the first site encountered that fulfills certain minimum threshold requirements. Although the comparative tactic is superior because it allows the animal to pick the best among several good options, web-building spiders are probably condemned to use the threshold technique ("satisficing" in Riechert 1985). This is because spiders apparently do not retrace their steps between websites during exploration. I know of no systematic studies on this important point; but I have never seen obvious retracing of paths between sites in extensive observations of uncaged *Leucauge mariana* (TET) and *M. duodecimspinosa* (AR).

One implication of satisficing, combined with the danger inherent in moving between websites (Lubin et al. 1993), is that web-building spiders may often "make do" at suboptimal sites. One evolutionary consequence of such a tactic is that natural selection would act to favor greater physiological tolerances and flexibility in behavior (including web design), reducing the impacts of occupying suboptimal websites (Lubin et al. 1993). These considerations are likely to be especially important for those species in which migration or construction costs are especially high. The limitation imposed by web costs implies that a spider whose web is longer-lived and requires a more substantial original investment (as in many non-orb webs—e.g., fig. 5.10, section 8.6) will be under more intense selection to accurately judge a site, in terms of both prey and physical conditions, before building there (especially if she is unable to reingest her web before abandoning the site). I know of no comparative data to test these predictions.

8.4.3 BIASES IN CHOOSING WEBSITES

8.4.3.1 Philopatry—remain near the natal web

One cue that indicates that a location has provided favorable conditions for survival and reproduction is when it is the website of the spiderling's mother. In addition, at least in the short term, neither competition with the mother for food nor danger from her predators is likely to be important, because neither food nor danger is likely to be the same for young spiderlings. Examples of specific habitat requirements would be soil particle type and drainage for digging burrows in the antrodiaetid mygalomorphs *Antrodiaetus* and *Aliatypus* (Coyle 1986), or hard substrates like rock surfaces or walls where cracks or other cavities occur in which the spider can establish retreats, as in the filistatid *Kukulcania hibernalis* (in this species, however, I have repeatedly observed spiderlings descending on long lines from egg sacs in the ceiling of a room, thus performing at least medium-distance dispersal).

Philopatric biases are widespread in social spiders (e.g., Avilés 1986), but this possibility has not been checked in most other groups. A striking example comes from the African sand dune eresid *Seothyra henscheli* (see box 9.3 for its web) (Henschel and Lubin 1997). In the Nabib Great Sand Sea where these spiders live, the relative physical and biological simplicity of the environment makes it possible to focus on relevant variables. The key traits are exposure to the wind (measured both visually and by sand flow), temperature (which reached up to 65–75°C on the surface!), and food (which was assessed by direct observations of the major prey, two species of ants). Spiders experimentally released in the field readily built burrows and webs nearby soon after release; 67 of 70 stayed for many weeks afterward at the web, which they often built within 0.5 m of the release point on the night of release; their feeding success could thus be followed.

The distribution of control webs was strongly clumped in space, and at some sites clumps lasted for at least seven generations (Henschel and Lubin 1997). Spiderlings generally dispersed only short distances from maternal burrows (they emerged partially grown from the mother's burrow, where they had at first fed on prey provided by the female and later on the female herself; they apparently did not balloon), and tended not to move subsequently. Abandoning one burrow to establish another is especially risky in this species because of predation, the risk of desiccation, and the cost of building a new web. In addition, prey availability may be difficult to predict in both space and time, making payoffs from relocation uncertain. In sum, *S. henscheli* showed an extreme combination of traits, including high risk and cost of relocation, inability to make reliable pre-building distinctions between good and bad sites, and a tolerance of variable conditions, that resulted in highly clumped distributions dictated largely by philopatry, even in a relatively uniform environment.

It is not clear how common this combination of traits is in other spiders. Surely lack of homogeneity in the environment with respect to scarce, special requirements could accentuate the advantage of philopatry. Aggregated distributions like those in *Zygiella x-notata* (AR) (Leborgne and Pasquet 1987) might be thought to be due to philopatry, but probably resulted instead from responses to patchiness of the environment (next section).

8.4.3.2 Disperse and then settle selectively—possible cues

A spider that has moved away from a previous site can use several variables to bias her choice of a website so as to increase her likelihood of surviving and capturing prey. I will discuss these variables one by one. It is important to keep in mind that there are two basic questions: do spiders prefer or avoid potential websites that have particular characteristics ("site bias"); and how do spiders sense these characteristics ("sensory bias")? Obviously site and sensory biases are intimately related (if a spider cannot sense a distinction, she cannot exercise a bias); but they are not the same, and documenting them requires different types of data.

8.4.3.2.1 PROBLEMS QUANTIFYING WEBSITES IN THE FIELD

Some types of webs such as orbs and the aerial sheets of linyphiids and pholcids require open spaces of particular sizes and shapes that are free of projecting objects such as leaves or twigs (WE). For instance, Janetos (1982, 1986) suggested that *Frontinella pyramitela* and *Neriene radiata* (LIN) in the field chose open spaces above forked branches that had overhanging projections to which they could attach the tangle of lines above the sheet (e.g., fig. 7.51). In contrast, other web types, such as funnel webs, often have twigs and other objects projecting through the sheets, as in the lycosid *Aglaoctenus castaneus* (fig. 8.5), the zoropsid *Tengella radiata*, and the agelenids *Melpomene* sp. and *Agelenopsis* sp. (Eberhard and Hazzi 2017). The presence of such obstacles is probably disadvantageous, as they would impede the transmission of vibrations and impede direct access to some prey, reducing the speed of the spider's attacks. It appears that many species choose open spaces of certain minimum sizes, but that some others are less selective regarding the openness of a site; perhaps they are incapable of evaluating this aspect of potential websites, or perhaps the importance of other aspects, such as having a physically secure retreat sometimes overrides the disadvantages of having obstacles in the web (fig. 10.7).

It is difficult, however, to study these choices, partly

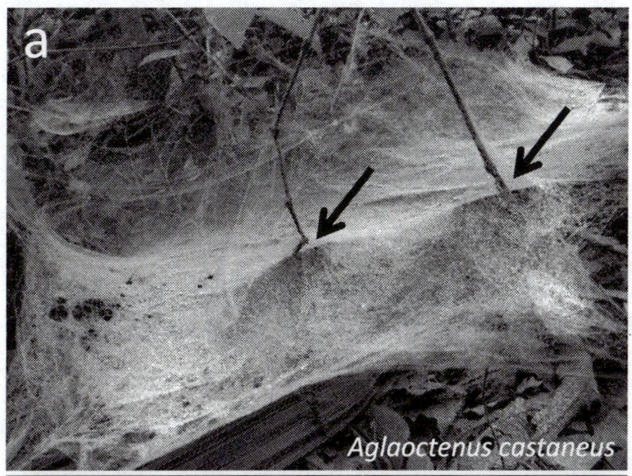

Fig. 8.5. The small plant stems (arrows in *a*) poking through this sheet web of the lycosid *Aglaoctenus castaneus* illustrate a seldom discussed problem for spiders during web construction—avoidance of objects that project into the space where the spider builds her web. In this species, such objects represent obstacles for the spider's rapid runs to attack some prey, as well as potential escape routes for prey (see Eberhard 2018b on a hahniid). It is not known whether reduction or avoidance of such obstacles is a criterion in website choice in this species or other funnel web spiders. The general lack of objects projecting through the orb webs and linyphiid and theridiid sheet webs in the field testifies to the ability of orb weavers and their kin to avoid this problem (from Eberhard and Hazzi 2017).

because it is not easy to quantify the amount of available space. One technique used was to measure the traits of websites in the field. For instance, the mean distance (in the plane of the orb) from the center of the hub to the nearest object was taken as an indication of the "free space" for the orbs of different species: the mean was 1–1.2 m for three species of *Araneus* that built mostly in the upper zone of a habitat rich in alfalfa stalks and umbellifers, while it was 0.5–0.6 m in *Argiope bruennichi*, which built in a lower zone rich in grasses (Pasquet 1984). This measure has the advantage of being simple and easily replicable. But it seems unlikely that a spider could sense this distance during the process of site selection. Another variable that is easily and commonly measured is height above the ground, but height has no necessary relation with the amount of open space (and indeed in the Pasquet study it had no relation to his space measure in three of the four species).

Another possible indicator is the length of the longest frame line (usually the uppermost frame in vertical orbs). This variable can be standardized with respect to the overall size of the web (for instance by dividing by the length of the longest radius), yielding a dimensionless variable, which could perhaps be sensed by the spider. Species like *Gasteracantha cancriformis* (AR) and *Aspidolasius branicki* (Calixto and Levi 2006) presumably have large values, while those like *Gea heptagon* (AR) (fig. 4.2), *Pronous* sp. (AR), and *Anapisona simoni* (AN) (fig. 7.9) small ones. I know of no systematic review of this type of data, however.

These measures can be useful in comparing the websites utilized by individuals of of different species; but they say nothing about possible sites that were *not* chosen, and are thus mute regarding the spider's possible selectivity. Even the comparisons between species just mentioned must assume that the sites that were available to the different species were similar, something that is probably often not true (Enders 1974; Pasquet 1984). What is needed to document preference is a technique for counting unoccupied as well as occupied websites. Measurements of "vegetational complexity," such as counting intersections of plants with randomly placed horizontal and vertical lines (McNett and Rypstra 2000), do not really reflect the combinations of supports and open spaces between them that a spider needs for a web. I have not found any publications with clear answers for these conundrums.

Perhaps the best attempt of this sort was that of Schoener and Toft (1983), who worked on tiny, low-lying island sites with low (0.3–1.0 m) and simple vegetation. They defined a potential site as a location where it as possible to place a web whose design was within the 70% range of heights, spans, depths, and orientations seen in the general population of webs on the island, and that did not overlap with any other similar site. This technique runs the risk of not including other possibly crucial variables (for example, rigidity of possible attachment sites, wind speeds, shade, etc.), and may thus overestimate the numbers of available sites (see Schoener and Toft 1983 for further discussion). Using their criteria, Schoener and Toft (1983) concluded that only about one in ten potential sites was occupied.

This indication that websites were not limiting in nature is especially striking because the population that Schoener and Toft studied was unusually dense (38 large females/54m^2). It fits my strong but admittedly unquantified impression, from observing spiders at the same site for >40 years, that individual websites were occupied in some years but not in others by species like the orb-weavers *Micrathena duodecimspinosa* (AR), *Leucauge mariana* (TET), *Nephila clavipes* (NE), and the sheet-web zoropsid *Tengella radiata*, despite the fact that these species were generally common year after year.

8.4.3.2.1.1 Website choice in simple field situations

One unusually thorough field study took advantage of the sparse populations of plants that constituted the only potential websites for the black widow *Latrodectus revivensis* (THD) in the Negev desert highlands in Israel. Here every individual shrub that could possibly harbor an individual spider could be surveyed, and spiders did not build anywhere except on these shrubs. The webs were more common in taller shrubs, and in *Zygophyllum dumosum* rather than another abundant shrub, *Artemisia herba-alba* (Lubin et al. 1993). When a marked individual changed websites, her new plant tended to be taller than her previous plant. Some of this bias was related to the spider's ontogeny, as movements into larger plants were correlated with increases in body size, a trend also seen just after molts in the orb weaving araneid *Argiope aurantia* (Enders 1975). The fecundity of female *L. revivensis* was also positively correlated with shrub height, independent of the relationships among shrub height, spider size, and survival (Lubin et al. 1993).

Similar kinds of data allowed dissection of the preference of *L. revivensis* for *Z. dumosum* plants over *A. herba-alba* plants. Here the crucial variable may have been the spatial distribution of supports and spaces between them that were available for webs. The *Z. dumosum* plants had more open space than did *A. herba-alba* plants of the same height at the level in the shrub at which the spiders normally built their webs (about two-thirds of the shrub's height). Similarly, larger *Z. dumosum* shrubs had more open space at this height in the shrub than did smaller conspecifics, so moves by spiders to larger *Z. dumosum* plants tended to result in more space in which to build. The exploratory behavior of *L. revivensis* while moving between websites has apparently never been observed directly, so the behavioral mechanism(s) used to choose shrubs is unknown.

Interestingly, food availability did not appear to have any role in the spiders' decisions regarding websites. Neither estimates of prey capture rate (using prey remains below the webs) nor experimental prey supplements for one set of individuals showed any correlation with website tenure in *L. revivensis*. This was despite the fact that, for a given body size (estimated by leg length), the amount of prey consumed had a strong correlation with fecundity. *Argiope aurantia* (AR) and *Nephila clavipes* (NE) showed a similar lack of response in website tenacity in the field to a large range of prey capture rates (Enders 1975; Vollrath 1985).

A likely benefit that may compensate *L. revivensis* for the high risk of predation entailed by changing websites (Lubin et al. 1993) was the possibility of improved thermal conditions and lower predation risk. High daytime temperatures inside the spiders' retreats force them to adopt thermoregulatory positions that exposed them to visually orienting predators such birds and mantids. The generally lower air temperature at greater distances from the soil (and also, possibly greater exposure to cooling breezes in more sparsely branched shrubs) imply lower thermal stresses (Lubin et al. 1993).

8.4.3.2.1.2 Experimental evidence

One experimental tactic is to alter potential websites in the field by trimming away possible supports (Colebourn 1974; Lubin et al. 1993; McNett and Rypstra 2000) and then observe colonization or website tenacity. Some types of physical support were important when individuals of *Argiope trifasciata* (AR) were introduced into plots in old fields from which two-thirds of the dominant plant, the thistle *Cirsium arvense*, had been removed; the numbers that built webs and stayed for the next two weeks were lower than in control plots (McNett and Rypstra 2000).

Over-abundance of supports (e.g., spaces that are too small) can also be important. For instance, Colebourn (1974) concluded that the dense growth form of heather (*Calluna vulgaris*) (that is, the lack of open spaces in the plant) was responsible to some extent for the low numbers of the araneid *Araneus diadematus* that it harbored. He introduced 50 marked adults, 50 marked third instar juveniles, and 50 marked fifth instar individuals into spaces that were 50 cm in diameter and 10 cm deep that he had cut into patches of heather plants; as controls he introduced spiders into unmodified patches of heather where the spiders were naturally uncommon. Significantly more marked adults and instar 5 juveniles had webs the following day in the experimental heather plants into which he had cut spaces than in the control plants. A significant difference persisted four days later.

In other cases adding possible supports resulted in increases in the numbers of spiders occupying a particular site. For instance, adding strings at 25 cm intervals in a 3 m sector of *Inula viscosa* hedge near the edges of ponds resulted in greater numbers of orbs of six araneid and tetragnathid species over control plots that had lower string densities or no string (Ward and Lubin 1992) (the effects on different species were not distinguished). Similar addition of building sites resulted in increases in the numbers of the linyphiid *Floronia bucculenta* (Schaeffer 1978), and the linyphiid *Linyphia triangularis* (Toft 1987). The theridiid *Theridion petraeum* also became more abundant in artificial websites made of 30.5 cm^3 cubes of chicken wire that were placed in an area dominated by sage brush (*Artemia tridentata*) when the cubes were threaded with horizontal bands of jute (a similar increase in density but with vertical bands had no effect); the spiders were also more abundant when the jute bands were more closely spaced (5.1 cm vs. 10.2 cm intervals) (Robinson 1981). Colonization of the artificial sites was relatively slow: after 1 day in the field, only a average of 1.0 spiders of all species was present; after 16 days the average was 5.7. Not all species responded in the same way. When dead stems were added experimentally to grassland sites with limited natural shrubs, the density of some web builders (of small and medium orb webs) but not of others (funnel webs, tangle webs, aerial linyphiid sheets) increased in one study (Gómez et al. 2016). In another study the dictynids *Dictyna* spp. increased but the effects on orb-weavers were mixed: *Aculepeira packardi* (AR) increased but not *Tetragnatha laboriosa* (TET) (Smith et al. 2016).

In sum, many spiders are probably not limited by lack of sites in which to built their webs. Some others occupy experimental websites when they are provided in the field, suggesting that potential attachment sites for their webs had been lacking. It is possible, however, that some or all of the same individual spiders would have eventually found other unoccupied, equally appropriate sites if the artificial sites had not been provided. These studies indicate that the attachment sites that were present previously were less attractive than the artificial substrates that were offered experimentally. In all of these cases, the behavior patterns and the cues used by spiders to make these discriminations remain unknown.

8.4.3.2.2 RIGIDITY, SPACING, AND SURFACE CHARACTERISTICS OF SUPPORTS

There are numerous incidental mentions of orb weavers that seem to prefer to attach their orbs to relatively rigid objects or dead vegetation; examples include McCook 1889 on *Metepeira labyrinthea* (AR) and *Uloborus glomosus* (= *americanus*), Marples and Marples 1937 on *Cyclosa conica* (AR), Tilquin 1942 on *Araneus diadematus* (AR), Opell 1982 on *Hyptiotes cavatus* (UL), Carico 1986 on *Cyclosa turbinata* (AR), Watanabe 2000b on *Octonoba sybotides* (UL), Smith 2006 on *Poltys* spp., Messas et al. 2014 on *Eustala sagana* and *E. taquara* (AR), and Bradley 2013 on *E. anastera*, *Mangora gibberosa*, and *M. maculata*. The araneid *Larinioides* (= *Araneus*) *sclopetarius* even seems to prefer wooden objects over stones (Crome 1956). In contrast, the orbs of some other species such as *Leucauge mariana*, *Neoscona oaxacensis*, and *Eustala fuscovittata* are often attached to flexible supports such as long blades of grass, and *L. mariana* sometimes built orbs while the leaves to which anchor lines were attached were swinging back and forth 1 cm or more in a light wind (WE). Some of these preferences may result from spiders using chemical cues from the substrate, but I know of no experimental studies on this topic.

Preferences for particular substrates also occur in non-orb weavers. For instance, several species of dictynids in the genus *Dictyna* seemed to build most webs at the tips of twigs and on projecting flower heads of plants (Bristowe 1941). Populations of both *D. major* and *D. coloradensis* increased sharply at sites where invasion by the exotic spotted knapweed plant *Centaurea stoebe*, which has prominent, branched flower heads, was simulated experimentally (Smith et al. 2016).

Some orb weaver species seem to prefer the web lines of other, larger species as supports for their webs, as in for example *Philoponella* spp. (UL) on various other species (Bradoo 1989; Fincke 1981; Rypstra and Binford 1995; Eberhard and Hazzi 2017), and *L. venusta*, which preferred to build its ephemeral webs attached to the relatively strong and rigid lines of *Nephila clavipes* (Hénaut and Machkour-M'Rabet 2010). Some species use the lines of conspecifics, and one of the advantages of colonial life in web-builders such as *Metabus ocellatus* (TET), *Cyrtophora citricola* (AR), and *Parawixia bistriata* (AR) is thought to be the opportunity to use conspecifics' lines where supports would otherwise be unavailable (Buskirk 1975; Blanke 1972; Fowler and Diehl 1978). This is additional evidence that appropriate supports at particular sites may sometimes be in short supply.

Rigid supports presumably reduce the rate of web damage (section 3.2.11.2). I do not know, however, of a single conclusive quantitative field test of this prediction. Nor are many of the preferences noted above fully documented. In one more complete study, there were more *L. venusta* webs in an area where *N. clavipes* was present than in one where *N. clavipes* was absent (Hénaut and Machkour-M'Rabet 2010), but this could have been due to other factors (e.g., differences in available prey). Counting the numbers of unoccupied "potential" websites in nature is difficult (see above).

Lab studies have documented preferences for more rigid supports experimentally in *Argiope bruennichi* (Tilquin 1942) and *Cyrtophora citricola* (Madrigal-Brenes 2012). The behavioral mechanism used by *C. citricola* to produce this bias may have been selective cutting of lines that were attached to more flexible supports (R. Madrigal-Brenes pers. comm.), but further observations are needed.

At a finer, within-web scale, adult females of the brown widow *Latrodectus geometricus* favored smoother substrates when building gumfoot lines below an already established tangle web (Gutiérrez-Fonseca and Ortiz-Rivas 2014). *Deinopis longipes* also showed a preference for smooth surfaces building its web, with which the spider snagged insects on the substrate just below (Robinson and Robinson 1971). Smoother substrates may be more often used by walking prey. When the substrate had regular, closely spaced ridges and valleys, *L. geometricus* also attached gumfoot lines to the ridges rather than the valleys, which might also increase the rates of prey encounters by pedestrian prey (Gutiérrez-Fonseca and Ortiz-Rivas 2014).

Many spiders also seem to avoid attaching lines to slick or oily surfaces in captivity, such as sheets of plastic wrapping material, glass, or plexiglass sheets covered with a thin film of oil or vaseline (Hesselberg 2013; Eberhard et al. 2008a). When the spiders *Leucauge mariana*, *L. argyra*, and *Z. geniculata* were observed arriving at a plastic sheet and apparently preparing to attach a dragline, they pawed repeatedly at the surface with their anterior legs, then finally turned away without attaching (R. Quesada unpub.; WE). Apparently they sensed surface characteristics with their anterior tarsi. Presumably this preference functions to guarantee firm adhesion of attachment discs to the substrate (fig. 2.10b-d, section 2.2.9).

8.4.3.2.3 TEMPERATURE

A spider on an aerial web in direct sunlight can be subject to high temperatures, and several species adopt special postures that minimize the area of their bodies receiving direct sunlight on warm days; these include *Frontinella communis* (LIN) (Suter 1981), *Nephila pilipes* (Robinson and Robinson 1973a), *N. clavipes* (NE) (Robinson and Robinson 1974), *Micrathena gracilis* (AR) (Biere and Uetz 1981; Hodge 1987a), and *Cyrtophora citricola* (AR) (Blanke 1972). The location and orientation of the web itself may also be chosen to reduce insolation in warm habitats (Caine and Hieber 1987), or to increase insolation in others (Rao et al. 2007 on *Argiope keyserlingi* [AR]). In the north temperate zone, webs sloping in north-south orientations (giving maximum exposure to the sun) rather than east-west orientations were significantly more common in cool seasons and habitats in *Argiope trifasciata* (AR) (Tolbert 1979), *N. clavipes* (NE) (Krakauer 1972; Carrel 1978), and *M. gracilis* (Biere and Uetz 1981). In the autumn, some cooler sites that had been acceptable for adult female *N. clavipes* earlier in the season became unfavorable, because of temperature dependence in egg development (Rittschof and Ruggles 2010; Rittschof 2012). The webs of the araneid *Mangora gibberosa* in a relatively open habitat (a young pine plantation) in October in northern Florida, USA, showed a tendency to be oriented so that the spider's dorsum was oriented in the direction of the rising sun, perhaps enabling the spiders resting at the hub to warm up more quickly on chilly mornings (Caine and Hieber 1987). In contrast, *Leucauge regnyi* (TET) did not show any trend in the directions in which their orbs were inclined in a lowland tropical forest (250–500 m elevation) (Bishop and Connolly 1992). *Araneus gemmoides* (AR) tended to orient the plane of its orb perpendicular to the direction of illumination (possibly to increase prey interception) but *A. diadematus*, however, showed no response in similar experiments (Hieber 1984). The biased orientations in several of these studies may have resulted from biased abandonment of sites in which the spiders were oriented in different directions, rather than biased building.

Selection of thermally favorable websites may be more critical for *N. clavipes* later in the year, because of greater differences between sunny and shaded sites (Rittschof 2012). The web orientations of the nephilid *Nephila clavipes* in the winter in southeast Texas, USA tended to maximize insolation of the spider at the hub, but their orbs did not show consistent orientations in the summer (Higgins and McGuinness 1991), or at a tropical site (Robinson and Robinson 1974).

It was not clear whether these biases resulted from pre-construction choices by the spiders, or selective abandonment of sites with unfavorable temperature regimes. Selective abandonment seems more likely in view of the cues available to the spiders. Website tenacity in *M. gracilis* was lower at woodland sites where direct sunlight had induced posturing previously (Hodge 1987a), and when spiders were released in a nearby open pine habitat where they normally did not occur (Hodge 1987b).

The funnel-web agelenid *Agelenopsis aperta* chose locations in a thermally stressful desert environment where its body temperature during the day was about 31°C, and avoided websites with direct sunlight (Riechert and Tracy 1975; Riechert and Łuczak 1982). During the hot dry season, spiders tended to inhabit webs whose tubular retreats were oriented in a way that minimized the amount of sunlight entering the tunnel (Riechert and Tracy 1975). The ability to capture prey was more severely limited by the thermal characteristics of sites (which varied by a factor of eight with respect to the duration of the period when prey handling was thermally feasible) than by prey abundance, which varied by only a factor of two (at least according to sticky trap catches; see section 3.2.5.4 for limitations of this technique). It is not clear whether spiders made these discriminations by abandoning unsuitable sites, or biased selection of websites prior to construction.

A further possible effect of temperature was to alter the time of day when the web was built. Biere and Uetz (1981) had the impression that in a cool season, the araneid *M. gracilis* built its orbs later in the day. Diurnal cycles in both temperature and light provided cues in *Araneus diadematus* that elicited web construction at certain times of the day. Experimental coordination of diurnal light and temperature cycles, so that dawn occurred near the daily temperature minimum, elicited the greatest frequency of web construction by this species (near dawn) (Witt et al. 1968).

In sum, temperature has multiple effects on website choice.

8.4.3.2.4 EGG SACS AND RETREATS

Having a detritus stabilimentum or egg sacs appeared to reduce the chances that a spider would move away from a site, perhaps because of the combination of the difficulty of moving the stabilimentum and the protection from predators that it afforded (Gonzaga and Vasconcellos-Neto 2005), and the protection that the spider may be able to afford its eggs. In some species, such as the araneids *Cyclosa turbinata* (Rovner 1976), *Allocyclosa bifurca* and *Metepeira* sp.—WE, and the theridiids *Parasteatoda wau* (Lubin 1986), *Nihonhimea tesselata* (Barrantes and Weng 2006a), *Tidarren sisyphoides* (Madrigal-Brenes and Barrantes 2009), *Theridion evexum* (Barrantes and Weng 2007a), and *Chrosiothes* sp. (WE), it was frequently possible to induce a spider to build at a particular site in captivity by simply hanging the stabilimentum, the retreat, or egg sacs along with the spider at the site. In *C. turbinata* the spider sometimes carried her detritus stabilimentum with her when she moved to a new website (Rovner 1976).

The presence or absence of a suitable cavity for a retreat also affected website selection in some species. An extreme case is illustrated in fig. 10.7 of *Tengella radiata*, in which especially attractive retreats apparently induced spider to inhabit sites that lacked the minimum supports needed to support their usual large sheets.

Simple tests in cardboard containers showed that the brown widow *Latrodectus geometricus* (THD) preferred retreats in corners with smaller angles between the walls, in deeper cavities, in cavities with rougher surfaces, and in refugia that contained silk from a previous inhabitant (Vetter et al. 2016). The brown recluse sicariids, *Loxosceles reclusa* and *L. laeta*, which differ from *L. geometricus* in building their webs immediately adjacent to their refugia, did not prefer to build in tighter cavities in the lab when the walls were separated by 9, 15, 18, and 21 mm, but did prefer rectangular cavities with an open side facing downward over the same cavities oriented so the open side faced laterally (Vetter and Rust 2008). They also preferred previously inhabited retreats that had silk from a previous owner; washing retreats with various solvents did not reduce this preference, so physical rather than chemical cues may be involved (Vetter and Rust 2010). Surprisingly, when offered four similar retreats in captivity and checked daily, individuals of both species switched retreats frequently over a period of 30 days (the maximum was 18 switches in a sample of 20 *L. laeta*); the degree of hunger had no effect in either species (Vetter and Rust 2010). In the field, *Loxosceles gaucho* preferred artificial cavities (produced in wooden blocks) that had more acute angles: triangular cavities were preferred over pentagonal or cylindrical cavities of the same depth (Stropa 2010).

Choosing favorable refuge sites is probably an ancient trait, even in wandering spiders without webs. In general, the cues used to make the discriminations just described and the behavior used to sense these cues are not known; it seems likely that some sensory and analytic capacities used to evaluate retreats at potential websites originally evolved in non-web spiders (and vice versa in non-web spiders that are descended from web-building ancestors).

8.4.3.2.5 LIGHT—ARTIFICIAL AND OTHERWISE

The webs of some spiders consistently occur in dark, sheltered sites, while those of others are built more in the open. As noted by Craig (1986), less well-illuminated websites may be preferred by species with diurnal webs because they are more difficult for prey to see and avoid. Direct illumination of web lines by the sun makes them especially visible to human eyes in forest and second growth habitats; the temporary interruption of hunting behavior by the damselflies *Megaloprepus* and *Mecistogaster* in tropical forest sites each time the sun was covered for a short period by clouds suggested they also use visual cues from the webs of their prey (WE). In a dense population of the pholcid *Modisimus* sp. in a secondary forest where there were flecks of sunlight, only the few individuals in direct sunlight oscillated slowly with their long legs on their small sheet webs (e.g., fig. 5.2) (WE), presumably to make it more difficult for a predator to strike them. A spider could not judge the amount of time her web would be directly exposed to the sun before building a web; but she might be able to judge the overall illumination of a site beforehand, using the brightness of the nighttime sky (at least if she climbs onto an object so that her eyes are directed upward, and it is not cloudy). Greater overall nighttime darkness in captivity increased the likelihood that *Uloborus diversus* (UL) would build an orb and the webs were larger (Eberhard 1971a) (experimental light levels were <l and 4.5 L; moonlit nights can be as bright as 25 L). This species' preference for sheltered sites may thus result from avoiding building webs at sites that are especially exposed to the night sky. Another possible nighttime indicator of sheltered sites could be especially low velocities and frequencies of air movement, but I know of no data to test this hypothesis.

The slanting orbs of *Argiope keyserlingi* (AR) built in large leafy *Lomandra* sp. bushes tended to be placed so that the web's plane and its lower surface were oriented toward light and vegetation gaps, and in the lab the lower surface also tended to be oriented toward light (Herberstein and Heiling 2001). If insect prey move into the vegetation from gaps, this orientation of webs might increase interception rates. Data on flight directions at the *A. keyserlingi* study did not show this trend (Herberstein and Heiling 2001), but there were many complications in sampling prey appropriately (see appendix O3.2.3). Webs of *A. keyserlingi* near *Lomandra* sp. bushes were most commonly built between bushes rather than on or within a single bush, and they also tended to be positioned so that the lower side of the web plane faced southeast (thus orienting the spider's dorsum toward the early morning sun) (Rao et al. 2007).

The diurnal light-dark cycle is apparently used by spiders to trigger web construction behavior. Experimental manipulations of the light-dark cycle so that "dawn" occurred later in the day caused delays in web construction times of several normally dawn-building species of araneids, tetragnathids, nephilids, and uloborids (Witt et al. 1968; Eberhard 1972a, 1990b). Web-building spiders have internal clocks, just as in other organisms, which allow them to anticipate the timing of periodic light cues.

Some night-building orb weavers have been said to be attracted to build their webs near artificial lights, presumably to capture the higher densities of insects that were attracted by the light (Heiling 1999; Adams 2000). Use of this cue seems unlikely, however, because spiders are unlikely to have evolved a response to such an unnatural situation as a light bulb turned on at night in an otherwise dark area. The greater spider densities near light might be due not to attraction, but to a greater tendency to remain at such sites due to higher feeding rates. Heiling (1999) performed a lab test of this idea with *Larinioides sclopetarius* (AR), and concluded that there was a preference for a better-lit chamber, but it was not clear whether this resulted from attempts by the spider to escape (spider preferences were apparently judged by "the presence of silk threads" [p. 45], rather than by orb webs, so interpretation of the data is difficult). Even if the silk threads were all in orbs, an alternative hypothesis ("attempt to escape toward the better-illuminated side, then build an orb wherever you find yourself") cannot be ruled out. More ecologically realistic experiments will be needed to determine the functional significance of possible responses to nighttime illumination.

8.4.3.2.6 THE PRESENCE OF PREY

As noted by Riechert and Łuczak (1982) and Vollrath (1985), it would be advantageous for web spiders to evaluate the prey richness at a website before building a web there, but this is usually difficult because prey are temporally and spatially heterogeneous. Evaluation of hypotheses that spiders sense the prey richness of a website is complicated by the inadequate nature of many of the techniques that have been used to quantify the available prey (section 3.2.5.4, appendix O3.2.3).

Nevertheless, cues from prey sometimes appear to elicit web construction. *Zygiella x-notata* (AR) was more likely to build in the presence of flies in captivity, even when spiders had not captured them (fig. 8.6) (Leborgne and Pasquet 1987; Pasquet et al. 1994; Thévenard et al. 2004). There was a weak trend of the same sort ($p = 0.04$) in a similar experiment with *Argiope keyserlingi* (AR) using two types of fly (Herberstein et al. 2000c), and a trend to build earlier may occur in the tetragnathid

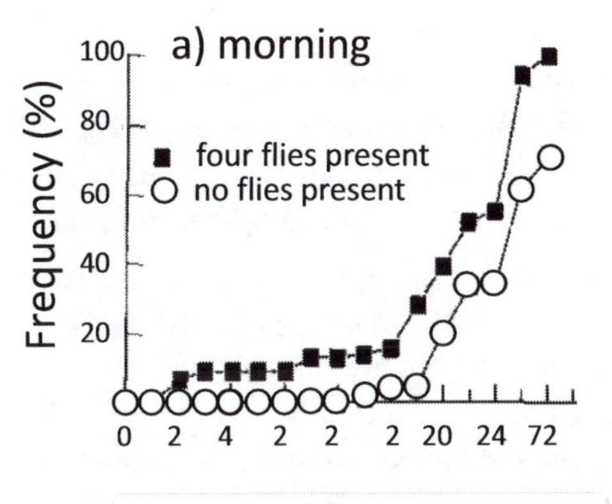

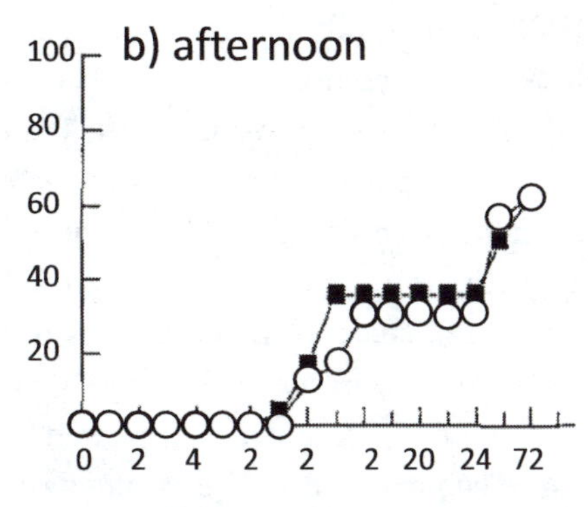

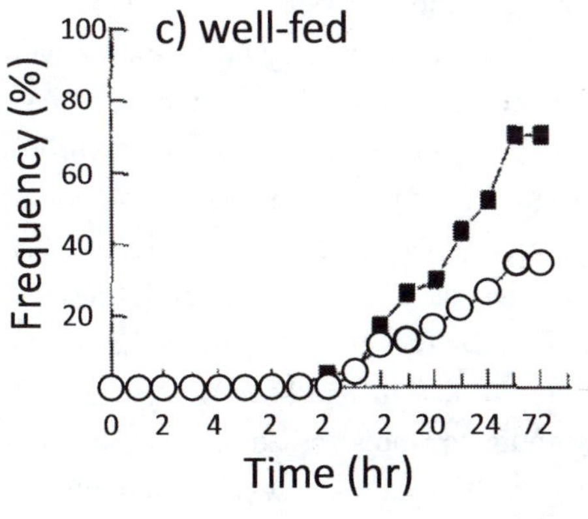

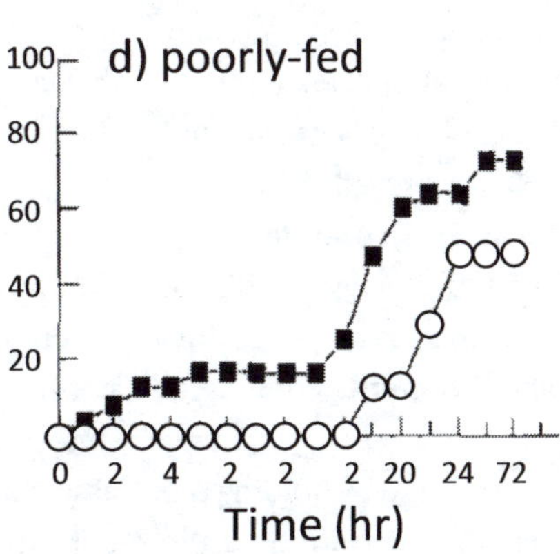

Fig. 8.6. The presence of prey (four calliphorid flies) in the same container induced orb construction by *Zygiella x-notata* (AR). Acclimated spiders were introduced into large wooden frames in the morning (*a*) or near the beginning of the usual web-building period (18:00) (*b*), and then checked every half hour thereafter for 24 hours, then once at 48 hrs and once more at 72 hrs. With flies present, spiders were more likely to build during daylight hours, and more of them had built orbs at the end of 72 hrs (*a*). These effects were lost if the spiders were placed in the containers in the afternoon (at 18:00) (*b*) rather than the morning. When spiders had been especially well-fed (*c*), the daytime building effect vanished but the overall accumulation effect persisted; if they were poorly fed (*d*), both daytime building and total accumulation of webs persisted (after Pasquet et al. 1994).

Tetragnatha elongata (Gillespie and Caraco 1987). The webs of *Z. x-notata* built in the presence of prey had reduced diameters and smaller spaces between sticky spiral lines (Pasquet et al. 1994; Thévenard et al. 2004), but in *A. keyserlingi* there were no design differences (Herberstein et al. 2000c). The cues used by these spiders to sense the prey were not determined, and the degree of ecological realism in the fly experiments is uncertain (*Calliphora vomitoria* flies were enclosed in a limited, web-sized space with the *Z. x-notata* spiders). Both Pasquet et al. (1994) and Herberstein et al. (2000c) speculated that spiders sensed airborne vibrations from the prey, but chemical cues or vibrations of draglines that spiders had produced while walking in the cage prior to orb construction were not ruled out. Chemical cues from the remains of prey captured and consumed previously by adult females were probably responsible for inducing one-month-old juveniles (about 3 mm long) of *Cyrtophora citricola* (AR) to stay when they were introduced into the webs of these females than when juveniles were introduced into webs from which all prey remains had been removed (Mestre and Lubin 2011).

Direct detection of living prey is probable in a few species that have reduced webs and prey on ants, including the theridiids *Dipoena* spp. (Umeda et al. 1996; Gastreich 1999), *Steatoda fulva* (Hölldobler 1979), and the deinopids *Deinopis* spp. (Robinson and Robinson 1971). The presence of ants along trails and near nests is highly predictable, and a spider may thus be able to evaluate website quality before she builds her web. The prey of *Dipoena banksii* is the ant *Pheidole bicornis*, which nests in the plant species *Piper obliquum*; this spider has been collected only from this plant (Gastreich 1999), so it may use cues from the plant. The prey of *Dipoena punctisparsa* is also very restricted (all of 23 prey collected belonged to two species in the ant genus *Lasius*), so it may use cues associated with certain ant species (Umeda et al. 1996). On the other hand, *Yaginumena castrata* and (especially) *Phycosoma mustelinum* preyed on a wide variety of ant species (Umeda et al. 1996). Another theridiid, *Chrosiothes tonala*, hunted termites by building a few non-sticky horizontal lines 10–100 cm above the forest floor and dropping from above onto the ephemeral foraging columns of the termites in the leaf litter (Eberhard 1991). Dropping attacks were elicited by holding the tips of a forceps coated with crushed termite near the spider, so this species apparently used chemical cues from her prey to trigger the attacks. In one case at least 13 spiders were found feeding in one m^2 site that had two termite trails nearby (WE), suggesting that sometimes spiders may be able to locate the nest entrances of the termites. The closely related theridiid *Janula* nr. *erythrophthalma* arrived in 1–5 min after aerial nests of the termite *Nasutitermes ephratae* were broken open, and captured soldiers there (Marshall et al. 2015), suggesting that the spiders lingered near nests and were attracted by termite alarm pheromones.

8.4.3.2.7 PREEXISTING WEBS

The presence of a conspecific's web at a site is probably a good indicator that the site is habitable, and it stimulates building behavior in some species. Individuals of the araneid *Cyrtophora citricola* remained longer and built sooner in a semi-captive situation when placed experimentally in small trees that contained conspecific webs; experimental removal of web owners had no effect, so the preference was due to the webs not to the spiders that built them (Rao and Lubin 2010). In contrast, field and lab data from *Zygiella x-notata* (AR), which often occurs in aggregations on human constructions such as fences and window frames on buildings (Ramousse and Le Guelte 1984), gave mixed results in similar tests.

Unrestrained adult females of *Z. x-notata* were aggregated rather than being distributed randomly in each of two years in a population living on the outside frames of windows (Leborgne and Pasquet 1987). The possibility that the clumping was due to the presence of conspecifics per se, rather than to additional variables (e.g., sites to which to attach lines, presence of prey), was supported by the fact that the windows were uniform in size and design, and that different portions of the same windows were occupied in the two years.

On the other hand, in laboratory experiments with the same species, in which individually-marked spiders were placed in large wooden frames with and without a conspecific, there was no effect of the second spider (Thévenard et al. 2004). Orb size and design were also unaltered by the presence of another spider in these experiments, although the presence of conspecifics in the unrestrained populations on windows was sometimes associated with reduced web sizes. Removal of the spider with the larger web in pairs whose orbs shared frame lines led to an increase in the web size of the remaining spider, so conspecific aggression may induce reductions in web size (Leborgne and Pasquet 1987).

The tendency to build in groups sometimes varies. In *Cyclosa mulmeinensis* (AR), windy conditions increased the tendency to build together (Blamires et al. 2010). Aggregations of *Leucauge mariana* (TET) were almost exclusively of adults; juveniles occasionally occurred in groups of adults, but they never formed their own aggregations even though they lived in the same subhabitats (Valerio and Herrero 1976). The proportion of colonial rather than solitary spiders increased during the windy dry season in the Valle Central of Costa Rica, but there was no tendency for colonies to occur at windier sites, and some colony websites were completely sheltered from the wind (Valerio and Herrero 1976).

Philoponella semiplumosa (UL) showed still different patterns (Lahmann and Eberhard 1979). Colonies included individuals of all sizes, and all sizes of spider were equally likely to be solitary or colonial. Colonial individuals had smaller orbs, and added stabilimenta to their webs less often (perhaps an indication of poorer feeding, or reduced danger from predators—see section 3.3.4.2). Spiders in an open, weedy habitat were much more likely to be in colonies (90% of 692 individuals) than those in a low forest of *Acacia costaricensis* trees (36% of 89 individuals). Solitary and colonial individuals were equally likely to be feeding in a sample of 442 individuals in the open habitat. In contrast, food may play a role in the tendency to share web lines in *Nephila clavipes* (Rypstra 1985). Four females in a group of six abandoned the group over a period of 8 days in which most prey were experimentally removed from the web once/day (at 14:00), and another group of four individuals all disbanded over a period of 10 days in which previous supplementation of prey (10–15 *Drosophila*/day/female) was suspended.

There are several additional reasons why building a new web adjacent to a previous web can be advantageous. Owners of aggregated orbs in the araneids *Gasteracantha minax* and *Metepeira incrassata* suffered lower predation rates (Lloyd and Elgar 1997; Uetz 1986), and were also able to use the strong support lines of neighbors to support their own webs. Some species reinforced their neighbors' lines, giving the entire group access to spaces that would not be feasible for a single individual, as in *Metabus ocellatus* (TET) (Buskirk 1975), *Cyrtophora citricola* (AR) (Blanke 1972), and *Philoponella semiplumosa* (UL) (Lahmann and Eberhard 1979). Using lines from other spiders' webs may also reduce the need for otherwise costly exploration behavior (Hesselberg 2015). Still another possible advantage of sharing lines is that prey that have just encountered and bounced off of or escaped from a nearby web may be easier to stop and retain (Uetz 1989); Wu et al. (2013) speculated that this "ricochet" effect might be more important at windy sites. Ricochets might occur more often if prey have greater difficulty controlling their flight paths in stronger wind, but this is speculation at the moment.

Having a conspecific's web nearby can also be disadvantageous. First and foremost is the presence of the occupant herself (at least if she is larger); her web and her defense of it can exclude the spider from otherwise useful space (Leborgne and Pasquet 1987), and she may also invade and steal struggling prey (Buskirk 1975; Lahmann and Eberhard 1979). The defensive behavior of neighbors also delayed web construction of some individuals in colonial *Metabus ocellatus* (TET) (Buskirk 1975) and *Leucauge mariana* (TET) (WE). In *Nephila edulis* (NE) sites with a conspecific were likely to have greater numbers of kleptoparasites (Elgar 1989). The tendency of *N. edulis* to nevertheless build adjoining webs suggests an overall benefit from attaching to lines of conspecifics.

Web construction is also probably sometimes induced by the webs of other species. Large webs with strong lines, such as the webs of large-bodied individuals of species such as *Nephila* spp. (NE) and *Cyrtophora* spp. (AR) or colonies of *Metepeira incrassata* (AR), are used as attachment sites by various orb weavers, including *Leucauge venusta* (TET) (Hénaut and Machkour-M'Rabet 2010), *L. mariana* (WE), *Micrathena gracilis* (AR) and *Gasteracantha cancriformis* (AR) (Hénaut and Machkour-M'Rabet 2010), and *Philoponella* spp. (UL) (Fincke 1981; Lubin 1986; WE). When marked individuals of the commensal *Philoponella vicina* (UL) were placed on tree buttresses where the webs of their commensal *Tengella radiata* were nearby, their webs were more likely to be attached to the zoropsid lines than to the nearby vegetation when they subsequently built webs in the vicinity (Fincke 1981).

The cues that spiders use to choose sites with preexisting lines are not known, but may be mechanical rather than chemical. Some species build in the webs of several different host species; for example, *P. vicina* (UL) built in the webs of both the zoropsid *Tengella radiata* and the theridiid *Tidarren* sp. (Fincke 1981; WE), and *L. mariana* (TET) built in webs of *N. clavipes*, *L. mariana*, and the recently arrived exotic *Cyrtophora citricola* (AR) (WE).

In sum, the decisions in facultatively gregarious species to aggregate at a given website were complex. Although a few species are induced to build by the presence of lines from other spiders, there was no obvious, simple overall pattern.

8.4.3.2.8 ISOLATION

The distance between a given habitat patch and other inhabited patches probably influences the likelihood that the patch will receive new colonists. Data on the distributions of a pair of kleptoparasitic argyrodine theridiids that inhabit the webs of *Nephila clavipes* showed an effect of a physical connection between host webs, but did not show any effect of the distance between isolated host webs: adding isolation to a multiple linear regression of agyrodid abundance on host web size did not have any effect (Agnarsson 2010). Many spiders can disperse long distances relative to their body sizes, and possibly the distances between host webs were too small (most were <5 m apart) compared with the dispersal capabilities of the theridiids for an effect to be evident.

8.4.3.2.9 THE PRESENCE OF PREDATORS

Natural selection would favor an ability to avoid websites where the danger from predators is high (e.g., Blamires et al. 2007b). Selection on website choice exercised by predation is difficult to study, however. A large variety of animals have been observed attacking web-building spiders, including asilid flies, damselflies, dragonflies, birds, lizards, vespid, sphecid, ichneumonid, and pompilid wasps, emesine bugs, and preying mantids (summaries in Eberhard 1973a, 1990a; Herberstein and Tso 2011; Blackledge et al. 2011; also Soley and Taylor 2013). At least some (perhaps all) of these animals learn from their own experience, and may adjust where they search, what they search for, and how to overcome the defenses of their prey (e.g., Coville and Coville 1980 and Araújo and Gonzaga 2007 on spider-hunting sphecid wasps, Soley and Taylor 2013 on an emesine bug). Even different species in a given taxonomic group can utilize different attack tactics (Rayor 1996; Blackledge and Pickett 2000). Thus, characterizing the danger at a given site to a given spider from a given predator species can be very difficult. Web-building spiders may also be relatively insensitive to one important type of potential cue—airborne odors from predators (Jackson et al. 1993), though emigration by the theridiid *Phylloneta impressa* was higher from sites bearing chemical cues from *Lasius* and *Formica* ants (Mestre et al. 2014). In addition, it is technically difficult to make direct measurements of predation in the field: the disappearance of a spider from a site cannot be assumed to be due to predation (e.g., Craig et al. 2001) without ruling out the possibility that the spider escaped a predator's attack by abandoning her web (Blackledge and Pickett 2000) or simply emigrated.

Some attempts to assess the general predation risks at different sites in the field, for instance on the basis of how "closed" the habitat is (Blamires et al. 2007b), are thus open to considerable doubt. Perhaps the most confident distinctions are special cases comparing otherwise similar islands that do or do not have particular predators present (Lubin 1974; Kerr 1993; Toft and Schoener 1983).

In sum, it seems intuitively likely that habitat selection (and also the timing of web construction—section 8.10) involves a trade-off between predation risk vs. the chances of prey capture in some web-building species; but empirical proofs are generally lacking.

8.4.3.2.10 HUMIDITY

Some web spiders probably prefer to build in more humid conditions. Many are found only near open water; a small sample includes the tetragnathids *Tetragnatha* spp. (Bradley 2013; Gillespie and Caraco 1987) and *Metabus ocellatus* (Buskirk 1975), the anapid *Conculus lyugadinus* (Shinkai and Shinkai 1988), and the theridiosomatids *Epilineutes* spp. and *Wendilgarda* spp. (Coddington and Valerio 1980; Coddington 1986a). Other species, such as for example the theridiids *Helvibis longicauda* and *Chrysso intervales*, and the hahniid *Neoantistea riparia* occur only at very humid sites (Gonzaga et al. 2006; Barrantes 2007; Eberhard 2018b). The araneids *Poecilopachys australasia* and *Pasilobus* sp. are thought to build their highly derived moth orbs only on nights with high humidity (Stowe 1986). The selective advantages of these associations probably vary. The webs of *Wendilgarda* and *Conoculus* were attached directly to water surfaces. In the theridiids *H. longicauda* and *C. intervales*, and *Theridion evexum* the adhesiveness of the long sticky lines may last longer in a more humid atmosphere (Gonzaga et al. 2006; Garcia and Japyassú 2005). The glue on the sticky spiral lines of orbs is often stickier at higher humidities (e.g., Amarpuri et al. 2015). Droplets of water condensed on the sheet web of the hahniid *N. riparia* may help restrain prey (Eberhard 2018b). In *T. elongata* the spider itself is thought to need high humidity to avoid dessication (Gillespie and Caraco 1987). These advantages are not mutually exclusive, and could be combined with greater prey availability at more humid sites. The specific cues used by spiders to select more humid websites are unknown.

8.4.3.2.11 PLANT SPECIES

Some spiders may use the identities of plants when choosing websites (Masumoto and Okuma 1995). Bristowe (1939) noted some preferences that may represent general habitat choice, such as *Mangora acalypha* (AR) for heather and *Phylloneta sisyphia* (THD) for holly and gorse, and *Anelosimus decaryi* for pitcher plants. One strict association was that of the Brazilian *Alpaida quadrilorata* (AR), which built its orbs in the relatively small central area of the rosette-type array of leaves of *Paepalanthus bromelioides* (Romero and Vasconcellos-Neto 2007). When disturbed, the spider fell from the hub and submerged herself in the pool of water that was typically present at the bases of the leaves. This spider may also derive benefit from the rigid supports provided by the leaves, the defense provided by their sharp tips, the micro-climate, and the insects that arrive at or leave this water (Romero and Vasconcellos-Neto 2007).

Another spider-plant association, between the sister nocturnal araneids *Eustala illicita* and *E. oblonga* and *Acacia* trees in central Panama, is probably due to the defense provided by the *Pseudomyrmex* ants that inhabit these trees (Hesselberg and Triana 2010; Garcia and Styrsky 2013; Styrsky 2014). Experimental removal of ants from trees showed that they defended against several potential predators, such as birds, crab spiders, jumping spiders, predacious bugs and beetles, and other ants; spider populations gradually declined in these trees (Styrsky 2014) (it was not clear whether reductions resulted from increased mortality or immigration). Individuals of *E. oblonga* were more abundant on *Acacia melanocerus* trees than other nearby plants of similar size and structure (Styrsky 2014); they "hid in plain sight" during the day, crouching immobile on compound leaves and small branches at sites that were frequently patrolled by the ants. Their immobility served as a defense against attacks by the ants (Garcia and Styrsky 2013).

Other, more tentative associations occurred between the domed-sheet weaving linyphiid *Laetesia raveni* and a pair of thorny plants (in different families) (Hormiga and Scharff (2014), and between *Eustala sagana* and *E. taquara* and largely dead portions of several plants (da Silva-Souza et al. 2015). The choice of dead vegetation may afford these species camouflage against visually hunting sphecid wasps, as both species show strikingly wide intra-specific variation in color patterns (da Silva-Souza et al. 2015).

A more general association is that between species in the pholcid genus *Modisimus* and a variety of plants that have medium to large, horizontal leaves in humid tropical forests (fig. 8.7; Eberhard and Briceño 1985;

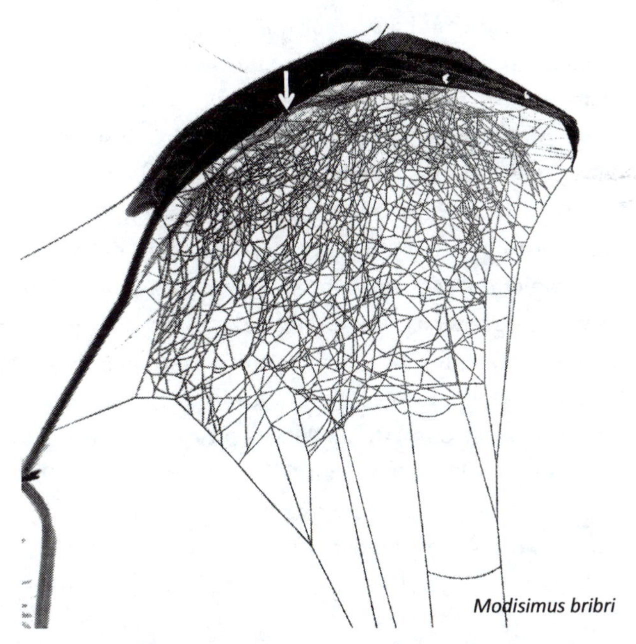

Fig. 8.7. This lateral view of the web of a mature male and female of the pholcid *Modisimus bribri* (the male is indicated by the arrow) shows a dome-shaped sheet and a small tangle above that are just beneath a horizontal leaf. It illustrates the loose, non-specific association that this species has with plants; nearly all webs were under leaves (in diverse taxa) that were more or less horizontal and at least 7 cm long.

Deeleman-Reinhold 1986; WE). Their webs consist of a domed sheet with a small tangle above, and are consistently located with the top of the dome near the underside of a leaf. Here the spider's green color and small body size, and the usual backlighting when viewed from below, combine to make her very difficult to see (at least for humans). The leaf also protects the spider from predation from above. The dictynids *Dictyna foliacea* (Comstock 1967) and *Mallos hesperius* (WE) build their webs just off the surfaces of large, slightly curled, rigid leaves.

8.4.3.2.12 WIND AND OTHER FACTORS THAT MAY BIAS INSECT MOVEMENTS

Although wind damage to webs is probably of little importance at some sites (e.g., Moore 1977 on *Nephila clavipes* [NE]), winds that are persistent and relatively constant in direction could affect website choice in several ways. Wind has several likely negative effects: reduced retention times (Turner et al. 2011); reduced attack speed (Turner et al. 2011); increased damage from movements of surrounding vegetation; reduced insect activity; increased damage from detritus moving more rapidly when it strikes the web; and increased stress from drag. It could also limit the directions in which a spider can float lines and find attachment sites to initiate construction. On the other hand, wind could also favor prey capture: insects flying into the wind would have lower ground speeds and would thus be easier to stop; and turbulence in air flow would make webs more difficult to avoid as they move erratically in the wind, and could also reduce an insect's ability to execute evasive maneuvers (section 3.2.11). Design traits that affect an orb's drag and movements in the wind include the lengths and diameters of lines, the number and spacing of sticky spiral loops, the angle with prevalent winds, the angle of the web's plane with respect to gravity, and the tensions on web lines. The relative weights of these different factors are generally undocumented, and probably vary for different species, different ontogenetic stages, and different sites.

The importance of these considerations will vary with details of the website as well as the general windiness of the habitat; for instance, sites lower in the vegetation near the ground tend to have lower wind velocities (Geiger 1965). In addition, brief changes of intensity or direction, which cause the web to "snap" in the wind, generate high momentary stresses (Langer 1969), making webs under especially low tensions more likely to be damaged. The patterns of eddies and swirls around large objects such as tree trunks can be complex, and will vary with changes in wind direction and velocity. Stresses on webs from the wind differ from those from prey impacts in that they are usually not applied at particular locations in the web, but over large areas (Langer 1969; section 3.2.11). They may also differ in producing additive damage; when the wind ripped small holes in the webs of *Uloborus diversus* (UL) that were not repaired by the spider, they soon grew into larger holes (Eberhard 1971a).

Avoidance of damage could favor orbs whose planes were parallel to the prevailing wind. Several araneids with approximately vertical orbs, *Metepeira daytona*, *Araneus gemmoides*, and *Argiope trifasciata*, tended to bias web orientation so that the orb's plane was approximately parallel to the prevailing winds at relatively windy sites (Schoener and Toft 1983; Hieber 1984; Ramírez et al. 2003 respectively). *Araneus diadematus* (AR) also reduced the area of its orb in response to wind (Hieber 1984). Smaller individuals of *M. daytona* failed to orient their orbs with respect to the wind, perhaps because they built lower in the vegetation, where wind speeds were probably lower (Schoener and Toft 1983). The webs of *M. daytona* tended to occur on the lee rather than on the windward sides of small islands (Schoener and Toft 1983). These trends suggest that avoiding web damage may sometimes be more important in website choice than differences in prey abundance. Choice of densely vegetated habitats in immature *Argiope aurantia* (AR), and increased website desertion when surrounding vegetation was removed (without damaging the web), might be explained by preference for low-wind websites (Enders 1976b).

Differences in web designs and website choice also suggest selection to avoid wind damage. The draglines of a species from a windy seashore habitat, *Cyclosa mulmeinensis* (AR), were thicker, had higher ultimate tension and breaking energies, and were more extensible than those of the related forest species, *C. ginnaga* (Liao et al. 2009) (average wind velocities were 9.0 m/s at the seashore site and 1.1 m/s at the forest site). Individuals of *C. mulmeinensis* made smaller orbs with fewer and more widely spaced sticky spiral loops when subjected experimentally to wind, and direct measurements showed that these webs had less drag in the wind (Liao et al. 2009). At

Fig. 8.8. Despite its dense mesh and lax lines, this horizontal orb of *Maymena* sp. (MYS) (#2168) did not suffer wind damage in the nearly windless sub-habitat where it was built—in the leaf litter on the floor of a tropical forest. The dense array of sticky lines above the orb sagged deeply under the coat of powder in these lateral (*a*) and dorsal views (*b–d*). The sticky lines above the orb sagged even before they were coated. The lateral view (*a*) shows that only a few (about 12) among the large number of radii supported the orb; the sticky spiral lines were apparently not attached to many of the other radii (*c*, *d*). The spider rested at the hub (*b*), where she had broken and then reattached the radii.

the other end of the wind spectrum, the long, lax sticky lines of tiny symphytognathoid spiders (figs. 7.9, 8.8) remain distinct and are often not entangled in the leaf litter of dense tropical forests, where there is very little air movement (Geiger 1965).

The nocturnal araneid *Larinioides sclopetarius* showed a different response to prevailing winds (Herberstein and Heiling 2001). Spiders in captivity subjected to experimental changes in wind direction consistently oriented themselves so their dorsal sides were directed toward the wind (when the web was inclined, the spider tended to rest on the lower side, irrespective of wind direction). The side of the webs on which prey impact occurred did not affect predatory success, and Herberstein and Heiling (2001) argued instead that the bias in web orientation protected spiders from being dislodged by strong gusty winds. This hypothesis needs testing; I have never seen an orb weaver blown off of an orb.

In the Negev desert in Israel, very young juveniles (about 1 mm long) of *Cyrtophora citricola* (AR), which built horizontal sheets in the midst of tangles, were more likely to leave more exposed trees; colonies of this facultatively colonial species tended to be on the lee sides of trees, rarely extending to the windward side (Johannesen et al. 2012). Another desert species, *Uloborus diversus* (UL), also showed several patterns that may be related to wind damage. Of 296 webs built during a period of consistent, light winds from the S and SW, the webs tended to be more common on the downwind sides of the approximately 1 m high nests of sticks and detritus of the packrat *Neotoma lepida* (Eberhard 1971a). Several additional trends suggested that this bias was due to wind damage. The most fragile webs (of the smallest immatures) were closer to the ground (one-third of them were <15 cm above the ground) where wind speeds were likely lower (Geiger 1965), and they were also more nearly horizontal (offering less resistance to horizontal airflow; wind directions were not tested, however). The webs of intermediate-sized spiders on the windward sides of the nests also tended to be more nearly horizontal than those on the leeward sides. Caged spiders subjected to approximately 11 kmph winds during the previous day and night were less likely to build webs than when they did not experience wind during the previous day and night. Finally, the biased distribution of spiders around the packrat nests changed when wind conditions changed. After several weeks of moderate S and SW winds during the day and little wind at night, there were three days with violent evening thunderstorms with estimated winds of 45–80 kmph. On the three days immediately following the storms (two nearly windless, the third with the usual steady SSW wind), the bias had disappeared, and the distribution of webs around the nests was random (Eberhard 1971a).

In sum, wind could theoretically affect website choice and web design in several ways. Data from several species suggest that avoidance of wind damage explains some variations in both websites and web designs.

8.4.3.2.13 HEIGHT ABOVE THE GROUND

The heights of orb webs above the ground vary in different species (table O8.1), and the arrays of both potential prey and potential predators surely vary in complex ways at different distances above the surface of the ground. Some extremes are easily distinguishable. For instance, the araneids *Pozonia* spp. (Levi 1993; Moya et al. 2010) and *Araneus cingulatus* and *A. niveus* (Coddington 1987) seem to be restricted to the canopy of forest trees. The horizontal webs of symphytognathoids, on the other hand, are almost always close to the ground, usually in the leaf litter (an exception occurs in cloud forests high in the Andes, where the dense accumulations of moss and other small plants on tree trunks far above the ground harbor species typical of leaf litter communities, such as *Mysmena* sp. (MYS) and *Ogulnius* sp. (TSM) (Valderrama 2000). The orbs of some araneids, including *Pronous* spp., *Gea heptagon*, and *Alpaida leucogramma*, are consistently built within 15 cm of the ground. These websites may help defend these species, as the spiders all appear to be especially prone to drop from their hubs when disturbed to the ground below, where the spider hides effectively under leaf litter, etc. (Sabath 1969; WE); *G. heptagon* also immediately changed her abdomen color to dark brown, increasing the effectiveness of this defense (Sabath 1969). Nielsen (1932) observed a similar association between websites close to the ground and defensive dropping in the linyphiid *Microlinyphia* (= *Linyphia*) *pusilla*. The short distances that spiders fell made their arrival on the ground almost instantaneous, and it was difficult to follow the spiders visually.

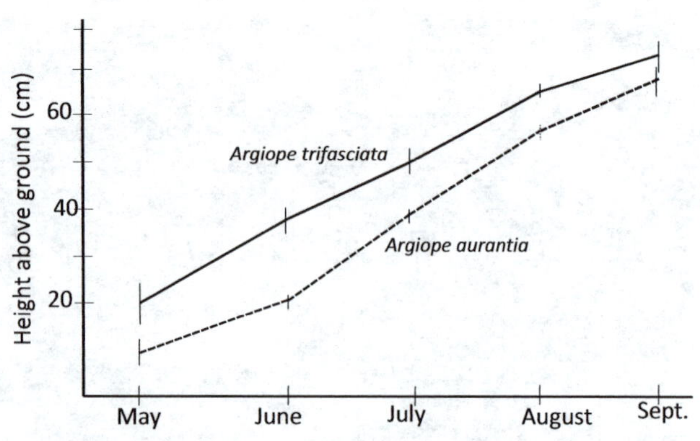

Fig. 8.9. Competitive exclusion, due to competition either for prey or for websites, was suggested by the vertical stratification of the webs of *Argiope aurantia* and *A. trifasciata* throughout the summer in a relatively simple and uniform stand of plants (mostly the exotic species *Lespedeza cuneata*) (Enders 1974) (vertical lines are standard errors). But this impression was misleading. In the first place, the density of the spiders was not high (there was only about one mature female of each species/5 m²). Because the webs of mature females of these species are generally <50 cm in diameter (Witt et al. 1968), there were almost certainly many physically inhabitable but unoccupied websites in this uniform habitat. More importantly, experimental manipulations of the densities of these two species in more natural conditions (Horton and Wise 1983) showed that they did not compete for prey, even at greater densities (on the order of one spider/1–3 m²).

In some species, such as the araneids *Argiope trifasciata* and *A. aurantia* (Enders 1974) (fig. 8.9), *Uloborus diversus* (UL) (Eberhard 1971a), and *Leucauge venusta* (TET) (Hénaut et al. 2006), the height of the web increased as the spiders grew. Samples of potential prey from artificial sticky traps at different heights above the ground (for problems with this technique, see section 3.2.5.4 and appendix O3.2.3) suggested that larger individuals of *L. venusta* built orbs higher above the ground because larger prey were available there, even though the overall abundance of prey was lower (Hénaut et al. 2006). On the other hand, the lower heights of the webs of smaller individuals of *U. diversus* (especially of the very smallest individuals, which were only 0–15 cm above the ground) were taken to represent defense against damage from wind (Eberhard 1971a) (no attempt was made to measure prey abundance, however). As illustrated in table O8.1, the heights of the webs of the adults of some species such as *Cyclosa caroli* (AR), *Micrathena pilaton* (AR), and *Naatlo sutila* (TSM) were relatively constant in different habitats, while those of others such as *Azilia* sp. (TET) varied with the habitat.

Although it is more difficult to measure precisely, the top of the herb layer is probably often a more biologically important reference point for the height of a web than its height with respect to the ground. This is because the typical velocities of prey flying within the herb layer are probably lower than those of prey flying above it, and because webs within the herb layer will also be less easily seen by prey due to the complex visual background against which they are viewed (section 3.3.3.2). Prey that are flying through dense tangles of vegetation probably also tend to be smaller, though I know of no data on this point. In sum, orbs within the herb layer are probably less often called upon to stop high-energy prey. It is possible that these effects are diluted to some extent by other high-energy prey such as grasshoppers and some homopterans that hop rather than fly; further empirical data are needed (table O10.1).

The adults of some orb weavers, such as the ladder web spider *Tylorida* sp. (TET) (Robinson and Robinson 1972), and *Nephila pilipes* (NE) (Robinson and Robinson 1973a) tended to build their webs above the herb layer. In others, such as the araneids *Mangora* spp., *Araneus expletus*, and *Uloborus trilineatus* (UL), webs were often within a dense herb layer (WE). Adult female *Leucauge argyra* (TET) often built at or just above the upper edge of the herb layer in oil palm plantations, while juveniles of this species built within the well-defined herb layer (WE) (quantitative measurements are lacking, however).

The cues that spiders used to select websites at certain heights above the ground remain unknown. Direct height measurements, by descending to make contact with the ground, seem possible in some species, especially those with webs built close to the ground or over relatively bare ground (apparent searches by making descents to the ground were made by *Micrathena duodecimspinosa* in the process of hiding her egg sac package in leaf litter—Eberhard 2015). It seems likely, however, that it would be difficult for a spider to reliably reach the ground at sites with complex, dense vegetation; spiders high above the herb layer also seem unlikely to check the ground directly.

A field experiment in which three species of linyphiids were released on the lowest branches of fir trees 0.1–0.2 m above the ground and then checked in their webs the next day (Herberstein 1997a) showed two trends: *Neriene radiata* built webs closer to the ground (mean = 0.79 ± 0.22 m) than did *Frontinella frutetorum* or *Linyphia triangularis* (respective means of 1.32 ± 0.3 m and 1.52 ± 0.40 m). And removal of the understory vegetation within 1 m of the edge of the tree resulted in all three species building webs somewhat lower (a mean of 17.3%). Possible causes of these responses mentioned by Herberstein (1997a) include microclimatic cues and the availability of open spaces with sufficient anchor sites.

8.4.3.2.14 FOOD QUALITY AND SATIATION

Foraging theory predicts that investments in foraging will be balanced against those in other activities such as growth, reproduction, and defense against predation; the relative advantage of foraging decreases when the animal is better nourished (fig. 8.2). Thus web-building spiders are predicted to reduce their investments in webs when they have previously fed well. This prediction has been confirmed in several groups, including both non-orb weavers such as the eresids *Stegodyphus lineatus* (Pasquet et al. 1999) and *Seothyra henscheli* (Lubin and Henschel 1996), and the orb weavers *Larinioides cornutus* (AR) (Sherman 1994), *Argiope keyserlingi* (AR) (Herberstein et al. 2000a), and *Nephila clavipes* (NE) (Higgins 1990). Food-supplemented *S. lineatus* behaved especially conservatively: they reduced the sizes of their webs, and many stopped building altogether (Pasquet et al. 1999). One implication for website selection is that the size of a spider's preferred website will increase when the spider has fed more poorly. I know of no tests of this hypothesis.

Prey species and thus food quality as well as quantity may also vary (Toft 2013). Along the increasingly urbanized length of a 16 km river in Puerto Rico, increasing pollution reduced the populations of some potential prey species populations (such as stoneflies) and increased those of others (such as chironomid midges). Adult females of the pan-tropical *Tetragnatha bogotensis* (TET) showed dramatic web differences along this river; the orbs at the more urbanized end had on average only about 64% as many sticky spiral loops and radii that were 52% as long (Sanchez-Ruiz et al. 2017) (surprisingly, radius numbers were not changed—contrast with fig. 3.29). Spider body condition (mass/cephalothorax width^3) also decreased, as did several additional factors, so attributing cause-effect relations can only be tentative; the large differences merit further study.

8.4.3.2.15 SEASON OF THE YEAR

Groups of *Leucauge mariana* (TET), in which orb frame lines and nearby sparse tangles were shared,

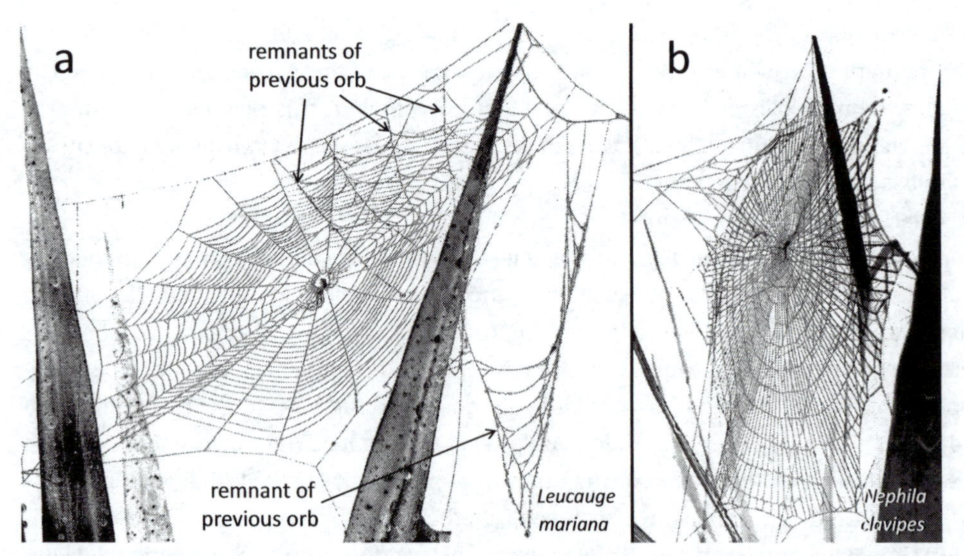

Fig. 8.10. Two solutions to problems resulting from fog are illustrated by these webs, which are coated with tiny water droplets. The penultimate instar spiderling of *Leucauge mariana* (TET) (*a*) that rested at the hub of a newly built orb on a foggy afternoon had just lost a "bet" on the weather. Radii and other remnants of its previous orb, loaded with larger water droplets that had accumulated during previous hours of fog, hung at the edges of the new orb. The fog did not lift, however, and the new web, which was only about 30 min old in this photo, had already begun to accumulate water droplets. These water droplets probably had multiple disadvantages: they loaded the web and probably reduced its ability to absorb the impact of high-energy prey; they made it more visible and easier for prey to avoid; they presumably diluted the sticky material on the spiral lines and made it less strongly adhesive; and they probably made it difficult (or impossible) for the spider recycle her web by reingesting it (see sections 2.5.2, 7.9). Two other spiderlings of *L. mariana* on this same plant were resting on old, tattered webs; they had made more conservative, winning "bets" by not building a new web (but of course they also had less chance of capturing prey). Another conservative strategy was employed by a spiderling (estimated fourth instar) of *Nephila clavipes* (NE) (*b*), which built all of the non-sticky lines in its orb and then waited at the hub until the fog lifted before adding the sticky spiral. It is not known whether orb weavers can estimate likely weather conditions when they decide whether or not to build (to different scales).

tended to occur more often in the early dry season than in the wet season in Costa Rica (Valerio and Herrero 1976).

8.4.3.2.16 POSSIBILITY OF DAMAGE (?)

It may sometimes be possible for spiders to predict the likelihood of some types of damage and choose websites accordingly. One type of damage that varies between even nearby sites and that might be predictable is that from light rainfall; for instance, incidental observations showed that webs of *Leucauge mariana* exposed to the open sky usually suffered more damage from brief, light rains than did those sheltered by trees (WE). The behavior of spiders themselves gave clear indications that rain is damaging. Mature female *Leucauge mariana* (TET) and *Allocyclosa bifurca* (AR) sometimes responded to the approach of a rain shower (thunder, sudden lowering of the air temperature, brief gusts of breeze) by removing and ingesting the radii and sticky spiral of their orbs before a single drop fell (WE). *Nephila pilipes* (Robinson and Robinson 1973a) and *N. clavipes* (NE) (WE) as well as *Micrathena duodecimspinosa* (AR) (WE) responded to early morning showers by postponing sticky spiral construction until the rain had stopped, while *Neoscona nautica* and *Tetragnatha* sp. refrained from building on evenings when storms threatened (Hingston 1920). The most important damage from rain is probably to the sticky spiral, causing many lines to adhere to their neighbors and at least partially washing off the water-soluble glue; these effects have never, however, been documented carefully.

Another type of damage that it might be possible to predict in the short term is from fog. Fog causes small droplets of water to form on web lines, and can impose a heavy mechanical load on a web: the droplets must make it less able to absorb the impact of high-energy prey, more visible to prey (figs. 8.10, 8.11), redistribute or remove the water-soluble aggregate gland glue on sticky spiral lines, and probably make it more difficult for spiders to reingest their webs (extra-oral silk-digesting enzymes would be strongly diluted by the water drops). Fog and light rain are common in some tropical cloud forest habitats, and often come and go repeatedly during the day. Web spiders are nevertheless common at such sites (e.g., Valderrama 2000).

Intermittent fog and light rains present an orb weaver with a conundrum. If she builds a new web during an interlude, but the fog or rain soon returns, she will have lost her investment. If, on the other hand, she refrains from renewing her web and the interlude is long, she will have lost a chance to capture prey. As noted in fig. 8.10, orb weavers such as *Leucauge mariana* (TET) may be under selection to predict whether the interludes will be long enough to make it profitable to build a new orb. Some *L. mariana* renewed their webs while water drops from a previous light rain were still falling from the nearby leaves (WE). Others refrained from building, and instead removed the previous orb and waited on a resting web of radii and frame lines, ready to build quickly when conditions improved (WE). Another option is to remove water droplets: following a light rain, a *Cyclosa monteverde* (AR) spiderling on a small resting web repeatedly gave the web small, sharp jerks which caused some of the droplets to fall, and walked around and flicked others off of the lines with her anterior legs (WE).

8.4.3.2.17 ANT NESTS

Three British species of linyphiids (in the genera *Acartauchenius*, *Evansia*, and *Thyreosthenius*) and an agelenid (*Tetrilus*) are known only from inside ant nests, and *Brontosauriella melloleitaoni* (THD) is known only from within termite nests (Bristowe 1939).

8.5 A GENERAL CORRELATION BETWEEN WEBSITE SELECTIVITY AND WEB DESIGN FLEXIBILITY?

Spiders use many different criteria in selecting sites for their webs. The size of the range of conditions under which a spider will build her orb can be termed her degree of selectivity. In some cases reduced selectivity may be correlated with a greater ability to adjust web designs to different conditions (the spider's adaptive flexibility), though such a correlation is not logically necessary. A non-selective species might build orbs of the same shape at the different sites, regardless of differences in their shapes.

Selectivity and flexibility were correlated in a comparison of two similarly sized araneids, *Eustala illicita* and *Cyclosa caroli*. The orbs of *E. illicita* were tightly associated with ant acacia trees (*Acacia collinsii*), where the spiders probably obtained protection from their predators provided by patrolling commensal *Pseudomyrmex* ants (Hesselberg and Triana 2010; Garcia and Styrsky 2013). The orbs of *E. illicita* were built among the tree's leaves and many small branches, had relatively short anchor lines (Hesselberg and Triana 2010; Hesselberg 2013), and occupied spaces with a variety of sizes and shapes (Hesselberg 2010). Perhaps not surprisingly, this species proved to be relatively non-selective when caged in containers of different sizes and shapes; spiders built with similar frequencies in all of them (Hesselberg 2013). More significantly, the shapes and sizes of *E. illicita* webs built in these containers varied substantially. The variation was adaptive, in that the shapes of the webs tended to conform closely to the shapes of their cages and to fill them.

In contrast, the orbs of *C. caroli* in nature were anchored to forest understory vegetation of various types. Their webs often spanned relatively large spaces, and their supports, especially for the upper frame line, were often relatively far apart (Craig 1989; Hesselberg 2013; WE). Using the proportion between the total of the length of all anchor lines and the mean diameter of the capture zone as a measure of the relative degree of confinement (Hesselberg 2013), the mean for 17 *C. caroli* webs was 4.69 ± 0.94, while that for 45 *E. illicita* webs was 1.83 ± 0.50. When *C. caroli* were offered the same containers that were offered to *E. illicita*, they were less likely to build in small and odd-shaped cages, but equally likely to build in the large, "control" cages. In addition, overall estimates showed that the shapes of the few webs that *C. caroli* were built in the confined cages appeared to be less closely adjusted to the shapes of the cages and did not fill them as completely (frame lines tended to be some distance from the wall of the container) (Hesselberg 2013; T. Hesselberg pers. comm.).

This experiment raises an interesting general question. Is the correlation seen in these two species between greater physical separation of the web from the substrate and both a higher degree of selectivity and a lower degree of flexibility a general phenomenon in web-building spiders?

8.5.1 POST-BUILDING SELECTIVITY AND CUES

Like other foraging animals, a web-building spider is repeatedly faced with the decision whether to remain at the current site or to move away in search of a better site. Some trends in the sites at which spiders and their webs are found in nature are undoubtedly due to stimuli that induce the spider to abandon a site. Janetos (1986) argued that many cases of apparent selectivity in website choice are the result of relatively non-selective choices of websites, followed by highly selective abandonment of sites.

Website abandonment rates in nature are sometimes surprisingly high. On any given day, about one in five individuals of *Cyclosa argenteoalba* (AR) abandoned her website (18–21% /web/day) (Nakata and Ushimaru 1999); *Tetragnatha elongata* (TET) moved, on average, once every 1.3 days in a lakeside habitat (Gillespie and Caraco 1987) (see table 8.1). Several studies indicate that abandonment of websites is selective rather than random, and that various cues were probably used to assess the "quality" of a website.

8.5.1.1 Prey capture success

Previous prey capture success is an obvious cue that spiders could use to decide whether to abandon a site. Turnbull's classic study (1964) showed that when previous capture success was better, the chance that the house spider *Parasteatoda* (= *Achaearanea*) *tepidariorum* (THD) would stay at a given site increased. When spiders were released in a simple habitat (a room with a very uneven distribution of escaped *Drosophila* flies), those spiders that built their webs in parts of the room that had lower concentrations of prey were more likely to relocate. Over a period of days the spiders gradually accumulated near a window where the flies were more abundant (Turnbull 1964). Similar associations between a greater tendency to leave a website following poorer feeding success have been demonstrated in several other species (Turnbull 1966; Dąbrowska-Prot et al. 1973; Olive 1982; Rypstra 1983, 1985; Gillespie 1987a,b; McNett and Rypstra 1997; Nakata and Ushimaru 1999; Chmiel et al. 2000; Mestre and Lubin 2011). There are possible exceptions, however (Enders 1976b; Spiller 1992).

The "leave if the hunting is poor" criterion seems simple and obvious. But if there is a large temporal variance in short-term feeding rates at a given site, and if the spatial variance in prey availability is also large between sites within the subhabitat, it may be difficult for a spider to distinguish a good quality site from a poor one (Gillespie and Caraco 1987; Nakata and Ushimaru 1999; Edwards et al. 2009). This difficulty in ranking sites may explain the unexpected pattern in website tenacity shown by *T. elongata* in two contrasting habitats. Spiders near a creek, with relatively few prey on average, changed websites much less often (once every 17.9 days on average) than did those near a lake (once every 1.3 days on average), where prey were more abundant (Gillespie and Caraco 1987). Cross-transfers of individuals between habitats indicated that phenotypic plasticity rather than genetic differences between the populations at the two sites were responsible for this difference. A model associating a fixed-site strategy with riskproneness, and a mobile strategy with risk-aversion, showed that the spiders could best avoid reproductive failure in generally poor sites by using the fixed-site tactic, and by using the mobile tactic at generally good sites (Gillespie and Caraco 1987). Incidentally, this species made another post-building decision that is unusual. When fed to satiation in captivity, spiders consistently removed their webs entirely (Gillespie and Caraco 1987). Presumably removal enabled the spider to conserve silk that would be lost if the web was destroyed.

A subsequent study of *T. elongata* at the same site concluded that the low website tenacity in the rich habitat was due to higher spider density and aggression, not to risk-aversive foraging (Smallwood 1993). This study is difficult to evaluate because it employed a flawed technique (artificial traps) to estimate prey available (section 3.2.5.4, appendix O3.2.3), and prey richness was quantified in terms of prey numbers rather than mass. Similar uncertainties due to artificial traps make it difficult to evaluate another conclusion, that in *Gasteracantha fornicata* (AR) the variance in prey captures was so high that the spider did not have enough information to evaluate any given website with respect to others in her habitat (Edwards et al. 2009). A further technical complication is that the technique of prey supplementation cannot be

Table 8.1. Mean durations of site residencies (days or nights) in the field for species with different types of webs; probably none of the distributions of residence times had normal distributions, and median values were probably smaller than those reported here (adapted from Smith 2009).

	Mean length of residency (days or nights)	*Reference*	*Notes*
Orbs			
Araneidae			
Allocyclosa bifurca	>57[1,2]	WE	On the side of a house
Argiope aurantia	2.6–4.5[2,3]	Enders 1975, 1976b	June dates, 5th instar
Argiope aurantia	3.4–7.7[3,4]	Enders 1976b	Controls in field experiments
Argiope aurantia	5.3–14.3[3]	Enders 1975, 1976b	August, prob. subadult females
Argiope trifasciata	8.7[2,5]	McNett and Rypstra 1997	Control replicates
Cyclosa argenteoalba	4.3–5.6[2]	Nakata and Ushimaru 1999, 2004	Controls from separate field expts.
Cyclosa octotuberculata	26.3[2,6]	Nakata and Ushimaru 2004	
Cyclosa octotuberculata	16.9[6]	Nakata and Ushimaru 2004	Juveniles, perhaps penultimate instars
Micrathena gracilis	6.7–8[2,5]	Hodge 1987a,b	Controls; informal estimate of longest residency was "weeks"
Poltys noblei	24.9–40.8	Smith 2009	
Tetragnathidae			
Tetragnatha elongata	3.8[2,5]	Smallwood 1993	Low density of spiders in prey-poor habitat
Tetragnatha elongata	17.9[2,6]	Gillespie and Caraco 1987	Low density of spiders in prey-poor habitat
Nephilidae			
Nephila clavipes	16[5]	Vollrath 1985	Spiderlings in prey-poor enclosure
Nephila clavipes	58.8[7]	Vollrath 1985	Spiderlings in prey-rich enclosure; longest recorded residence >41 days
Uloboridae			
Uloborus diversus	about 5.6	Eberhard 1971a	Estimated from mean desertion rate
Tangle with gumfoot lines			
Theridiidae			
Latrodectus revivensis	17.9[4]	Lubin et al. 1993[8]	
Latrodectus revivensis	44.1[2,5]	Lubin et al. 1993[8]	
Funnel webs			
Agelenidae			
Agelena limbata	143-permanent[2,5]	Tanaka 1989	No adults changed sites during study

[1] 26 of 29 spiders marked individually on 4 Nov. were still at the same sites 57 days later.
[2] Primarily or exclusively mature females.
[3] Original figures were for website tenacity (probability of spiders remaining per day); mean residence was calculated as = 1/probability of turnover.
[4] Includes both juveniles and adult females.
[5] Original figures were for mean days per residence.
[6] Original figures were for turnover (probability of spider leaving per day); mean residence was calculated as = 1/probability of turnover.
[7] No calculation was made in original paper; extrapolated from figures in text and based on a few actual spider movements.
[8] Possibly affected by marks applied to animals.

expected to produce changes in behavior in a habitat in which prey are already abundant (Spiller 1992). The basic point, that it is sometimes difficult for spiders to evaluate prey richness when there is high temporal variation in prey captures (Blackledge et al. 2011), seems certain.

The tendency to respond to poor prey capture success by leaving probably varies between species. For example, the likelihood of leaving a site in captivity in which previous feeding success had been good but where prey were no longer present was greater in *Metazygia rogenhoferi* (AR) than in the uloborid *Zosis geniculata* (Kawamoto and Japyassú 2008). The authors linked this difference to the relatively lower costs of web and building behavior in *M. rogenhoferi*, and it is also possible that *Z. geniculata*, which builds in highly sheltered sites (e.g., inside buildings), is more limited with respect to appropriate subhabitats in nature, and may thus have a higher threshold for moving (see also Blackledge and Wenzel 2001b on the dictynid *Dictyna volucripes*).

Foraging theory predicts that reduced feeding success will also be associated with a greater tendency to renew the web. This prediction was confirmed in *Argiope trifasciata* (AR): spiders in the field that were fed more abundantly (three 10–15 mm grasshoppers—probably about one-half to one-third the spider's body length) built webs less frequently (Tso 1999). In *A. aurantia*, however, experimental manipulations of feeding had no effect on website tenacity in either the field or captivity (though here the sample was very small) (Enders 1976b). Another option in some species is to remove and replace only part of an orb. Although prey capture success was not checked in connection with web renewal, larger individuals of *Nephila clavipes* (NE) were more likely to renew their webs (typically about half of the orb) than were smaller ones (Higgins 2006).

8.5.1.2 Learning *how* to adjust

Some spiders may learn from practice to adjust the size and shape of their webs to unusual websites. When da Cunha-Nogueira and Ades (2012) confined *Argiope argentata* (AR) in vertical frames whose sides were only 5 cm wide until they had built two webs, and then laid the frames horizontally and checked the subsequent webs that the spiders built, they found that the spiders made a diversity of horizontal webs ranging from tangles with no perceptible pattern of lines, to typical orbs (fig. 7.48). When the frames were returned to a vertical position for two more webs, and then once again laid horizontally, the spiders built recognizable orbs much more quickly than they had the first time, and that the designs of these orbs were more similar to those of normal orbs. In contrast, *Eustala illicita* showed no improvement (or even a trend in this direction) when adjusting to a different type of change in frame shapes (Hesselberg 2013).

I know of no further work on the interesting implication that a spider's adjustment to a new site is sometimes influenced by a combination of relatively innate and learned responses. Orb-weaving spiders are well known to be able to learn in other contexts, such as attacks on prey (Bays 1962; Rodríguez and Gamboa 2000; Nakata 2013; Hénaut et al. 2014), so the apparent learning by *A. argentata* could be of general significance.

8.5.1.3 Web damage

If web damage is common, non-randomly distributed among sites, and difficult to predict beforehand (e.g., movements of branches or leaves in the wind later in the day, lack of shelter from direct rainfall), an abandonment response could be advantageous. Spiders with webs that have relatively low costs and can be built quickly, such as many orbs and small pholcid sheets (figs. 5.2, 5.21, Eberhard 1992b; box O5.1), might be expected to be especially likely to abandon websites following damage. This expectation may be fulfilled in at least some groups that move very frequently (section 8.6), but the role of web destruction in these species has yet to be documented.

In several species physical damage to the web increases the likelihood that the spider will move away from a site (Enders 1976b; Riechert and Łuczak 1982; Hodge 1987a; Leclerc 1991; Chmiel et al. 2000). Web destruction induced *Micrathena gracilis* (AR) to abandon websites the next day, and destruction of the bridge thread had a stronger effect than removing the sticky spiral and leaving most radii intact (a type of damage that seldom if ever occurs in nature) (Biere and Uetz 1981). Total web removal in *A. aurantia* increased the frequency with which spiders abandoned websites, from 9% in controls to 67% (Enders 1976b). Major (but not minor) damage to the orbs of *Argiope keyserlingi* (AR) resulted in spiders building smaller orbs with larger stabilimenta (Walter and Elgar 2011). In the desert uloborid *Uloborus diversus* a series of three storms on four evenings with heavy wind and rain that probably destroyed most if not all orbs was followed by website changes by a substantial portion the population (Eberhard 1973a). There are probably taxonomic differences on spiders' responsiveness to web destruction; it would be expected to be lower when the costs of relocating are higher. The desert eresid *Stegodyphus lineatus*, which is thought to suffer especially large costs in searching for a new website and in building a new web (Henschel and Lubin 1992), was remarkably impervious to web damage; spiders showed no tendency to abandon websites when their webs were destroyed in the morning on each of four days during an eight-day experiment (Pasquet et al. 1999).

The tendency to abandon sites and move to long distances appeared to be greater in very young and adults of *Araneus diadematus* (AR) (Colebourn 1974). Of adult females placed on 50 intact patches of heath (*Calluna vulgaris*) or on 50 similar patches in which a circular space 50 cm in diameter and 10 cm deep had been cut, a total of only 18% remained in the patch the next day. When small, 3rd instar juveniles were placed at similar sites (N = 50 in both), only 10% were on orbs in the area the next day. In contrast, similar numbers of intermediate juveniles (instar 5) were more likely to stay, with a total of 53% remaining on webs the following day.

"Moving away" in other studies sometimes involved only small distances. Small scale movements of *Argiope keyserlingi* (AR) occurred in a 1 m wide cage when one side of an orb was damaged; the spider tended to shift the hub of the next orb away from the side that had been damaged (Chmiel et al. 2000). The cumulative distance moved back and forth in these cages over seven days when webs were damaged daily was larger when the spiders were not fed. It is difficult to translate behavior under these conditions, in which longer distance abandonment was not possible, to behavior in nature.

The frequencies with which webs are damaged and the extent of the damage in nature are poorly documented. Craig (1987a) thought that smaller and more fragile webs, like those of *Epilineutes globosus* (THM) and *Leucauge globosa* (TET), were more frequently damaged in an all-or-none manner than were larger webs like those of the araneids *Micrathena schreibersi*, *Mangora pia*, or *Cyclosa caroli*. There were differences, however, in their websites and web designs, so further comparisons are needed.

Some webs are quite durable. While most orbs are only up for a day or less, because the spider herself removes them (Carico 1986), some non-orb webs last much longer. In sheltered sites, such as cellars or caves, webs of filistatids, zoropsids, agelenids, loxoscelids, and pholcids are only slowly degraded, even after the spider leaves (WE). The strong, tightly meshed sheet-plus-tangle webs of *Cyrtophora moluccensis* (AR) may be the longest-lasting aerial webs known from exposed natural sites: an adult female typically builds only one web during her 3.5 month lifetime, and repairs amounted to replacing only about 10% of the web every 18 days (Lubin 1973). One abandoned web of this species in New Guinea showed no appreciable damage from the heavy rains it experienced over the space of three weeks (Lubin 1973). Webs of the theridiids *Nihonhimea tesselata* and *Anelosimus* nr. *studiosus* also remain more or less intact even after strong rains (WE), and these species also seem to stay at the same sites for many days.

8.5.1.4 Material fatigue in silk lines?

A further, unexplored possible influence on web replacement is the possibility of material fatigue in silk lines. Fatigue is the progressive and localized structural damage that occurs when a material is subjected to cyclic loading. In general, the greater the range of applied stresses, the shorter the probable life of the material before it fails. Fatigue damage is cumulative and materials do not recover when rested. Some materials do not suffer fatigue if the amplitude of cyclic stresses is below a critical value (the "fatigue limit"). The fact that fatigue originates on and is promoted by irregularities in the material brings into focus the possible significance of the shiny uniformity of the surfaces of silk lines, even in the SEM (fig. 2.4) (Lehmensick and Kullmann 1957; Kullmann 1975).

Spider webs that billow in the wind or are attached to objects that are flexible and move could be particularly susceptible to fatigue. Fatigue could select in favor of periodic abandonment or rebuilding of webs, especially of longer-lived aerial non-orbs, such as those of many theridiids and linyphiids, or the araneid *Cyrtophora* and relatives (Lubin 1973). I know of no studies that test for possible fatigue under natural conditions; the relatively frequent renewal of frame lines by *Uloborus diversus* (UL) in moderate winds in the field (where frame lines were only seldom visibly damaged) was linked (speculatively) to possible material fatigue (Eberhard 1971a) (see also Breed et al. 1964 on *Araneus diadematus*).

8.5.1.5 Kleptoparasites

Web-building spiders are plagued by various kleptoparasitic spiders and insects (e.g., Wiehle 1928; Robinson and Robinson 1973a; Vollrath 1984; Eberhard et al. 1993; Agnarsson 2010), and host species have a wide variety of web designs that range from funnel-web diplurids and zoropsids to orb weavers and aerial sheet weavers. Although some kleptoparasitic species parasitize multiple types of host webs (Larcher and Wise 1985; Whitehouse 1986), they generally specialize on certain hosts; it is not clear why they are often associated with some host species but not others with more or less similar webs in the same habitat.

Some kleptoparasites can cause substantial reductions in host feeding rates; for example, the nephilid *Nephila plumipes* experienced a 55% reduction in weight gain when accompanied by spider parasites (Grostal and Walter 1997; see also Rypstra 1981 on *N. clavipes*). Some kleptoparasites also feed on the web itself, or on the host spider (Whitehouse 1986, 1987; Wise 1982; summary in Larcher and Wise 1985). It is thus no surprise that the emigration rates of some host spiders increased when there were kleptoparasites in their webs. For instance, the theridiid *Argyrodes trigonum* induced more frequent movements to new sites in both *Metepeira labyrinthea* (AR) and *Neriene radiata* (LIN) (Larcher and Wise 1985).

The cues that spiders use to trigger emigration in response to kleptoparasites are not well known. Experiments with the nephilid *Nephila plumipes*, which relocated to new websites 4.5 times more often in a greenhouse when their webs contained the kleptoparasite *Argyrodes antipodianus* (THD), showed that neither food theft nor web damage alone explained the increased relocation rates (Grostal and Walter 1997). The opposite experimental strategy, of adding extra prey to the hosts *M. labyrinthea* and *N. radiata*, supported this same conclusion: extra feeding failed to reduce the emigration induced by *A. trigonum* (Larcher and Wise 1985). Perhaps some host spiders can sense the kleptoparasites themselves, or their silk. The degree of damage caused by *A. trigonum* to its host was greater when its size with respect to the host was greater; perceptible effects on emigration occurred only when the parasite's weight was greater than 10% that of the host (Larcher and Wise 1985).

8.5.1.6 "Pilot" webs—a risk-minimizing tactic

Previous experience can provide information on the quality of a website, but it comes at the cost of building a web there. One risk-minimizing strategy is to first build a cheap, pilot web to test the quality of a new site, and then make more substantial investments in subsequent webs if the site passes this first test (Riechert and Gillespie 1986; Nakata and Ushimaru 1999; Zschokke and Vollrath 2000). The pilot web hypothesis predicts that the first orb that a spider builds at a site will tend to be cheaper (e.g., smaller, with fewer sticky spiral loops, possibly with non-optimal planarity or inclination) (Zschokke and Vollrath 2000). This prediction was fulfilled when *Araneus diadematus* (AR) were introduced into rectangular cages. Their first webs had fewer loops of sticky spiral, and were less planar than subsequent webs at the same site (Zschokke and Vollrath 2000). Observations of first and subsequent webs in the field showed that they were also less planar, at least by some criteria. Similar trends for first webs to be smaller than second webs when spiders arrived at a new website (e.g., were brought into captivity) occurred in several other araneids, including *Argiope argentata* (da Cunha-Nogueira and Ades 2012), *Zygiella x-notata* (Leborgne and Pasquet 1987), *Cyclosa argenteoalba* (Nakata and Ushimaru 1999), and *Larinioides cornutus* (field vs. lab comparisons in Sensenig et al. 2010b). There are exceptions to this pilot web trend, however (Nakata and Ushimaru 2004 on *C. octotuberculata*, Hesselberg 2014 on *Eustala illicita*). When a spider makes smaller pilot webs, her evaluation of the new website should ideally take into account the expected lower interception success due to her smaller web (Nakata and Ushimaru 2004), but current understanding of payoffs is not refined enough to test this idea.

A second prediction of the pilot web hypothesis is that the criteria for abandoning a site where the spider has recently arrived may differ from those that are used to decide whether to abandon long-inhabited sites. The heroically labor-intensive experiment of Nakata and Ushimaru (1999) tested this idea with free-ranging marked *C. argenteoalba* (AR) in a botanical garden. When they removed all the prey from the spiders soon after the prey were captured (within 3 min), those spiders that had just shifted websites the night before were more likely to leave the next night than were spiders that had been at their websites for several previous nights. The rates of website abandonment for both groups were greater than those of control spiders that did not have prey removed from their webs. The implication is that the spider remembered her recent past. The behavior of *Nephila clavipes* (NE), in contrast, suggested that they had no memories of this sort (Vollrath and Houston 1986), so this tactic may vary.

Zschokke and Vollrath (2000) argued that spiders using the pilot web technique might be expected to have lower frequencies of website abandonment, because abandonment would be more costly due to the construction of a less than optimal first web at the next website. I think that this argument can be inverted: the process of searching for a good new website will be cheaper if a species makes pilot webs, because the losses at inappropriate new sites will be reduced. If more is gained (in terms of reduced costs) from a pilot web's cheaper design than is lost due to the pilot web's reduced effectiveness in prey capture (in terms of the lost chances for prey capture), then the pilot web strategy will be favored. The pilot web tactic may be more effective for evaluating factors that are relatively constant at a site (physical stability of the supports, damage from detritus such as falling flowers or seeds, temperature and wind regimes) than likely more variable factors such as prey arrival in the web.

8.6 WEBSITE TENACITY, WEB DURABILITY, AND RECYCLING

The variables just discussed (and probably others) can induce a spider to abandon a website and search for another. How frequently do spiders in nature abandon websites? The answers vary widely for different species, and even for members of the same species with different ages and under different ecological conditions (table 8.1; see also Smith 2009). The extremes range about four orders of magnitude: about 50% of the websites were abandoned each day by several species of pholcids with domed sheet webs, including *Physocyclus globosus* (Eberhard 1992a), *Holocnemus pluchei* (Jakob et al. 2000), and *Modisimus guatuso* (WE); in contrast, only about 0.01% were abandoned each day by the sand dune spider *Seothyra henscheli* (Lubin and Henschel 1990), and there were no observed abandonments by adult females of the funnel-web agelenid *Agelena limbata* (Tanaka 1989). In perhaps the most extensive study ever, involving 3237 spider days of the orb-weaver *Nephila pilipes* (NE), the durations of website occupation averaged more than 12 days, and ranged from <1 to 7–8 weeks; patterns included an increased tendency to leave after having laid eggs, and lack of increases in leaving associated with poor prey-capture success or with the numbers of associated kleptoparasitic spiders (Robinson and Robinson 1976b).

An undisturbed, free-ranging population of the pholcid *Physocyclus globosus* in a cave-like environment (smooth-walled empty tunnels) where environmental damage to webs was minimal nevertheless showed very low website tenacity. The webs themselves were left intact when spiders moved and, in contrast with most other studies, it was possible to determine the initiation as well as the termination of the occupancy of each website by marking the walls and thus obtain more precise data on durations. The websites of mature females had a mean total occupancy time of only 2.6 ± 2.7 days (not counting females guarding eggs or spiderlings, which moved less); nearly half of the spiders stayed at a new website for only a single day. Penultimate females had similarly short residence times (2.5 ± 1.6 days) (Eberhard 1992a). Other, distantly related species in which webs were left intact when spiders moved away also showed surprisingly low website tenacity. In captivity, the oecobiid *Oecobius annulipes* routinely abandoned intact, functional webs only 2–4 days after building them (Glatz 1967). Many of the sicariids *Loxosceles reclusa* and *L. laeta* in captivity shifted refugia frequently. Of 20 *L. laeta* kept in containers for 30 days, 25% stayed at the same site, 40% shifted 1–8 times, and 35% shifted >9 times (*L. reclusa* showed similar frequencies) (Vetter and Rust 2008) (see also Levi and Spielman 1964).

Linked to low site tenacity was evidence of unexpectedly high rates of website colonization. Student field projects that I have led over more than 30 years have consistently shown that when most or all of the individuals of two small pholcid species (probably in the genus *Modisimus*) and a tiny mysmenid (*Mysmena* sp.) that live near the leaf litter in lowland tropical forests in Costa Rica were removed from 10–15 1 m^2 plots (after locating all webs by dusting the litter lightly with corn starch), the next day there were often similar and sometimes even greater numbers of spiders of the same species in the same plots (see Eberhard et al. in prep.). Population densities were on the order of only five webs or less/m², and nearly all webs on the second day were at different websites within the plots, so direct competitive exclusion (Smallwood 1993) seems unlikely.

Rapid recolonization also occurred in several other species. Artificial websites in a sugar beet field in England that were created by making 85 cm² diameter cylindrical pits in the soil with an auger were rapidly colonized: 40% of the pits had a new web of either *Lepthyphantes tenuis* or *Bathyphantes gracilis* (both Linyphiidae) after only a single day (Thornhill 1983). High rates of movements between websites were also suggested in several cases in which researchers have had problems with frequent natural recolonizations in experiments in which they attempted to keep plots free of the web spiders of a given species; examples include *Linyphia triangularis* (LIN) (especially young juveniles) (Toft 1986), *Nephila clavipes* (NE) (Vollrath 1985), *Parasteatoda* (=*Achaearanea*) *tepidariorum* (THD) (Riechert and Cady 1983), the agelenid *Coelotes montanus* (Riechert and Cady 1983), and *Argiope aurantia* and *A. trifasciata* (AR) (Horton and Wise 1983). Even in the extremely specialized single-line theridiid *Dipoena banksii*, whose prey is the ant *Pheidole bicornis*, some sites were colonized within only a day after a previous spider had been removed (Gastreich 1999).

These high rates of movement between websites are surprising, because movements to new sites seem likely to be risky (e.g., Vollrath 1985, Lubin et al. 1993). In addition, the expected difficulty for the spider to determine the quality of a website without long-term experience there (Blackledge et al. 2011) reduces the chances that the spider will be able to choose a new site that will be an improvement. Local depletion of prey due to previous predation seems unlikely (I know of no tests of this idea, however), and thus unlikely to be the cause of most movements. In some species, such as the desert theridiid *Latrodectus revivensis* (Lubin et al. 1991) and the desert eresids *Seothyra henscheli* (Henschel and Lubin 1992, 1997) and *Stegodyphus lineatus* (Ward and Lubin 1993), as well as in a tropical wet forest orb weaver *Nephila clavipes* (Vollrath 1985), the disadvantages of movements are thought to be large enough to explain why some spiders remain at websites even when they are unsuitable.

Several variables correlate with reductions in website tenacity: poor feeding success in *Nephila clavipes* (NE) (Vollrath 1985); difficulty in finding a new site with appropriate camouflage in *Poltys noblei* (AR) (Smith 2009); higher frequency of web damage from falling leaves in *Lepthyphantes flavipes* (LIN) (Leclerc 1991); violent thunderstorms in *Uloborus diversus* (UL) (Eberhard 1971a); greater age of the spider in several species (table 8.1); and lower mortality while moving between sites in *Latrodectus revivensis* (THD) (Lubin et al. 1993). Henschel and Lubin (1992, 1997) found especially high mortality associated with moves in the eresid *Seothyra henscheli*, which presented extremely high website tenacity (table 8.1). This species had several traits likely to promote tenacity: high cost of building a new web; high danger from heat, dessication, and predation during moves between sites; and highly unpredictable resources at new sites. Individual *Tengella radiata* remained at the same websites with protected retreats for up to 16 weeks, despite the disturbance associated with being removed weekly from their webs to be weighed and then returned to their webs (Santana et al. 1990). Incidental observations of some non-orb spiders such as filistatids and diplurids that inhabit small cavities or retreats in the substrate suggest that these species also move infrequently (WE). A relative lack of mobility in spiders with retreats in the substrate could be related to the difficulty of finding another appropriate cavity. Species that abandon rather than recycle their webs might be expected to move less readily, but I know of no comparative data.

8.7 WEB DURABILITY

Spider webs show a wide range of durabilities in the field. At one extreme, some non-orb webs like those of filistatids, pholcids, dictynids, and loxoscelids built in very sheltered sites such as cellars, undoubtedly last intact for months, even when unoccupied (Opell 1999a). The majority of occupied webs of the large tropical funnel web mygalomorph *Linothele* sp. in wet lowland forest in Colombia lasted at least 4–6 months (Paz 1988). At the opposite extreme are webs like those of the gradungulid *Progradungula carraiensis* (Gray 1983), the theridiids *Dipoena banksii* (Gastreich 1999) and *Phoroncidia* spp. (Eberhard 1981b), the uloborids in the genera *Hyptiotes* and *Miagrammopes* (Wiehle 1927, 1929;

Marples and Marples 1937; Lubin et al. 1978), and the deinopids such as *Deinopis* (Coddington and Sobrevila 1987; Getty and Coyle 1996) that are often largely or completely destroyed during the capture of a single prey. The gumfoot lines (but apparently not the tangle portion of the web) of the theridiid *Achaearanea globispira* usually disappeared during the day (perhaps due to wind), and were rebuilt each night (Henschel and Jocqué 1994). The sheets of a *Linothele macrothelifera* were largely destroyed daily by only moderate rains (Eberhard and Hazzi 2013). The webs of the theridiid *Theridion evexum* in the field had only 5–25 long sticky lines in the field, but they grew to include up to 63 sticky lines when the spiders were sheltered in the lab (Barrantes and Weng 2007b), suggesting that damage to sticky lines is common in the field.

Orb webs undoubtedly tend toward the ephemeral end of this spectrum (Opell 1999a). I know of few quantitative measures of the rates at which orbs accumulate damage in nature. Incidental observations of web damage while searching for minimally damaged orbs to photograph suggest that some orb webs, including those of *Hypognatha* sp., *Verrucosa* spp., and *Bertrana striolata*, accumulate substantial damage within only an hour or two and are often removed within a few hours by the spider (WE). A more typical rhythm of replacement is one orb/day (e.g., Carico 1986; see table 8.2). Some orb weavers routinely replace their webs more frequently. For instance, *Leucauge mariana* and *L. argyra* (TET) typically build a new orb at 4–5 AM, tear it down and replace it around noon, and then often build a third orb early in the evening; they also replace webs damaged by rain or fog (fig. 8.10) (WE). During intermittent rain showers *Acacesia hamata* (AR) built up to three orbs/night; the spider removed each previous web prior to the next rain shower (Carico 1986).

In contrast, several very large orb weavers such as *Nephila pilipes* (Robinson and Robinson 1973a), *N. clavipes* (Peters 1955) (NE), and *Caerostris darwini* (AR) (which builds at extremely exposed sites suspended over rivers) (Agnarsson et al. 2010; Gregorič et al. 2011), have especially strong lines and probably experience low rates of general web destruction. Their rates of repair and renewal vary. In a sample of 3237 web days, 75.1% of the webs of adult female *N. pilipes* were renewed or repaired after only a single day; comparable numbers were 18.2% after two days, 4.4% after three, 1.0% after four, and 1.3% after 5–9 days (Robinson and Robinson 1973a). Orb designs without sticky spirals, as in *Cyrtophora citricola* (AR), make webs virtually impervious to some sources of damage such as fog (fig. 8.11). Building orbs at very protected sites may have a similar effect of prolonging their useful life, and some smaller species such as *Cryptaranea* (= *Araneus*) *atrihastula* (AR) (Forster and Forster 1985) and *Zosis geniculata* (UL) (WE) often leave their orbs up for several days at such sites. In the desert uloborid *Uloborus diversus*, which built in the early morning (Eberhard 1972a), 29% of 191 orbs at more or less exposed sites on pack rat nests had 10% or more of their area missing by late afternoon on the day when they were built; webs of this species built at more sheltered sites deep in dense vegetation and inside buildings in an urban setting seemed to accumulate damage less rapidly. Marking lines with small accumulations of powder showed that some sectors of an orb of an unrestrained mature female in the lab were more more than 1 month old. At especially sheltered sites, uloborids like *U. diversus* and *Z. geniculata* sometimes built especially large orbs, which they needed two nights to fill with sticky spiral, or which they extended on subsequent nights with additional, "repair" sectors (e.g., fig. 6.19, section 7.3.3.2); *U. diversus* webs with multiple repairs were rare at exposed sites in the field (WE). The sheet webs of the giant linyphiid *Orsonwelles* spp. were also durable, and often had substantial repairs (Hormiga 2002).

The cost of a web in both material and construction behavior (section 3.3.3.3.2) may covary with web durability. Tanaka (1989) considered these costs to be the major cost of web relocation for the funnel-web agelenid *Agelena limbata*, which moves only seldom (table 8.1). A second complicating variable may be a species' overall life history strategy. In subtropical Okinawa *Nephila pilipes* (NE) grew to a larger adult size than *N. clavata* during a shorter developmental period, and thus required greater foraging success (Miyashita 2005). Associated with these differences, *N. pilipes* residence times were shorter, and supplementary feeding of *N. pilipes* resulted in longer residence times; *N. pilipes* appeared to be more prone to take the risk of moving in search of especially high-quality sites (Miyashita 2005; see also Robinson and Robinson 1976b).

Table 8.2. A simplified sampler of times (undoubtedly overly typological) at which orb webs are built, operated, and removed by species in seven families, illustrating the wide variety of tactics. Most data are from my own incidental, unquantified observations at sites where I spent substantial amounts of time both day and night searching for webs, and for which I thus feel I knew (at least approximately) both construction times (from direct observations of buiding behavior or consistent times at which I observed fresh, relatively undamaged webs) and destruction times. Other data are from species that I have kept in substantial numbers in captivity, or from published accounts in which it was clear that the authors checked webs both day and night for both construction and destruction. I characterized building times in the field on the basis of direct observations of spiders building, or of finding fresh webs only at certain times of the day. Behavior on days when it rained and caused spiders to postpone construction or to destroy their orbs is not included, nor are cases in which spiders built a web soon after a previous web had been damaged. The time periods (approximate) are the following: "evening" = dusk and following hour or two; "late at night" = hours between evening and early morning; "early morning" = within about 1–2 hrs preceding first perceptible light; "dawn" = up to first hour following first perceptible light; "mid-morning" = hours following "dawn"; "day" = daylight hours. During the time between construction and removal, spiders operated the orb (attacking prey that it trapped) from the hub or from a retreat. The coverage of published studies is incomplete, and was biased to increase the taxonomic range of the table. Although this table is thus idiosyncratic in its coverage (as are, of course, previous studies themselves, which are probably biased against short-lived webs, especially those built later at night), general patterns are evident (see text).

Time of day web built	*Time of day web removed*	*Species and references*
Webs operated mainly at night		
1. Mainly in the evening	Dawn	AR: *Acacesia hemata* (Carico 1986), *Scoloderus tuberculifer* (Eberhard 1975), *Poltys* spp. (Smith 2006), *Wixia* sp. (#573) (WE), *Eustala anastera* (Carico 1986), *E. fuscovittata* (WE), *Hypognatha* sp. (#1714)[1] (WE), *Eriophora transmarina* (Herberstein and Elgar 1994)
2. Mainly in evening	Next morning or day (sometimes from a retreat)	AR: *Eriophora edax* (WE)[2], *Verrucosa arenata* (Carico 1986)
3. Late at night, then operated it both day and night	Not removed (web durable)	AR: *Cyrtophora moluccensis* (Lubin 1973, 1980, Lubin and Dorugl 1982), *C. citricola* (WE)
4. Late at night	Around dawn or late at night (?)	AR: *Scoloderus* spp. (Stowe 1978, 1986; WE), *Alpaida* sp. (WE), *Tylorida* sp. (Robinson and Robinson 1972)
5. Evening	Abandoned web intact (though often in tatters) at dawn, maintained no contact with any line running directly to orb during the day	AR: *Metazygia gregalis* (WE)[3,4], *Larinioides* (= *Nuctenea*) *cornutus*[1] (WE) TET: *Tetragnatha* spp. (WE)
Webs operated mainly during the day		
6. Early morning, and also built and operated others later in the day and/or in the evening	Mid-day, afternoon, night	TET: *Leucauge mariana*, *L. argyra*, *L. venusta* (WE)
7. Early morning	Abandoned web intact in the evening (maintained no contact with lines that ran directly to orb during the night)	UL: *Philoponella republicana*[5] (Lubin 1980)
8. Early or mid-morning	Removed most or the entire web in the evening	AR: *Araneus legonensis*[6,7] (Grasshoff and Edmunds 1979), *Micrathena duodecimspinosa*[8], *M. horrida*, *Gasteracantha cancriformis*[9] (WE), *G.* sp. (#2036, 2038-9)
9. Early morning (or late at night) and operated web both the following day and night (unless it was destroyed by rain)	Late night or early morning; sometimes (especially if not destroyed by rain, etc.) web was not removed, and was operated for multiple days	AR: *Azilia* sp. 591 (WE), *Phonognatha* sp. (Hormiga et al. 1995) NE: *Nephila pilipes* (=*maculata*) (Lubin and Dorugl 1982; Herberstein and Elgar 1994), *N. clavipes*[10] (Higgins 1987), *Herennia ornatisima* (WE) TET: *Chrysometa saladito*, *Dolichognatha pentagona* (WE)

Table 8.2. Continued

Time of day web built	*Time of day web removed*	*Species and references*
		UL: *Octonoba sybotides* (Watanabe 1999b, 2001), *Uloborus diversus, U. glomosus, Zosis geniculata*[11] (WE) AN: *Anapisona simoni* (WE) MYS: *Mysmena* sp. (WE)
10. Early morning	Removed hub and at least some portions of the capture area in the evening, leaving some radii, frames, and anchors intact during the night (skeleton)	AR: *Cyclosa turbinata* (Carico 1986), *C. argenteoalba*[12] (Nakata et al. 2003), *Allocyclosa bifurca* (WE), *Arachnura higginsi* (McKeown 1952), *Arachnura melanura* (Robinson and Lubin 1979a)
11. Early morning (operated web during the day via a signal line while resting in a retreat)	*A. expletus* often removed web incompletely early in the evening or later at night (timing of destruction, if it occurs, was not clear in *Z. x-notata*)	AR: *Zygiella x-notata* (Ramousse and Le Guelte 1984; Venner et al. 2000; Thévenard et al. 2004), *Araneus expletus* (WE)
12. All times of the day, but especially often late in the afternoon (about 4 PM)	Sometimes webs abandoned in tatters (at different times of the day) after only a few hours of use	AR: *Caerostris darwini*[13] (Gregorič et al. 2011) AN: *Conoculus lyugadinus* (Shinkai and Shinkai 1988) TET: *Metabus ocellatus*[14] (Buskirk 1975) THDS: *Wendilgarda galapagensis* (Eberhard 1989a)
13. Early morning, then early evening	Very fresh webs seen at 5–7 AM and 7–8 PM; no webs present during the rest of the day	AR: *Wagneriana tauricornis*[15] (WE)

[1] Webs were generally already very tattered early in the evening.

[2] Spider hid in a retreat but attacked prey during the day.

[3] Spiders often built two or three orbs during the course of a night, then abandoned the last orb and hid in a small cavity or a silken retreat; usually there was no line running directly to the orb, so the orb was not operated during the day.

[4] First web very early in the evening.

[5] At night, spiders of this colonial species abandoned their orbs in the periphery of the colony and clustered at a central, presumably more protected position.

[6] Built between 04:00 and 07:00.

[7] Spiders removed orbs when they were partially damaged during the day; on days when no web was built (21% of 795 observations), the web from the previous day was not left up (Grasshoff and Edmunds 1979).

[8] Some spiders removed the first web and built a second web during the morning if the first was damaged by rain. Of eight spiders checked on a foggy night (when even very thin lines were easily visible, two were on a single horizontal line long enough to be an upper frame, and six were on a set of three lines in the form of a "Y" that might have been the initial stage of orb construction. Other, very fine, lax lines sagged from these lines (produced during attempts to float lines on the breeze? produced by other spiders?).

[9] Muma (1971) saw one unusual female build two webs in one day.

[10] The complete range of building times was from 22:00 to 11:00, with a major peak from 03:00 to 07:30 in the unusually complete study of Higgins (1987).

[11] Building time varied substantially indoors.

[12] "At sunset it ceases foraging activity and start(s) consuming web silk" (p. 373 of Nakata et al. 2003). This gives the impression that the removal behavior is gradual, probably like that of *C. turbinata*, leaving the frame lines intact (Carico 1986).

[13] 74% of building took place between 16:00 and 18:00; other building times were not specified.

[14] Spiders evidently rested in a retreat at the edge of the stream at night, as the author describes how they emerged from the retreat around daybreak.

[15] The majority of early morning orbs were removed completely before noon, leaving only one or two non-sticky lines; in some cases the spider first rested for a period during the morning on the substrate at the edge of the orb rather than at the hub, then removed the orb.

Fig. 8.11. The fine droplets of water from fog loaded the webs in this colony of *Cyrtophora citricola* (AR), probably making them more easily visible to prey and perhaps also less able to resist high-energy impacts. All the lines were non-adhesive silk, however, and the webs were apparently undamaged when the fog dissipated and the water dried off (for more damaging effects of fog on orb webs see fig. 8.10, section 2.5.2). Fog commonly comes and goes repeatedly in tropical cloud forests. Orb weavers are nevertheless common there (e.g., Valderrama 2000).

8.8 LIMITED BY WEBSITES? POSSIBLE COMPETITION FOR PREY AND WEBSITES

Evaluating the possible importance of inter-specific competition for websites and for food is potentially crucial for understanding the evolution of differences and similarities in choices of sites and web designs, as it could promote divergence in website preferences. I will argue in this section (based largely on the work of David Wise summarized in his classic 1995 book) that inter-specific competition for food is generally weak or absent among web-building spiders, at least regarding the limited number of cases that have been studied. Intra-specific competition for websites may, in contrast, be common in some species that have highly specific website requirements, but is probably generally not of major importance, because websites are usually not limiting. The common behavioral observation that web spiders threaten and drive other spiders from their webs probably represents defense of the web itself (and avoidance of the robbery of any prey that it may intercept), not of the website per se.

8.8.1 INTER-SPECIFIC COMPETITION

The web designs, websites, phenology, and body sizes of different web-building species that coexist in the same habitat often differ. The growth and fecundity of web spiders is often limited by prey, so it is tempting to conclude that these differences represent niche-partitioning that has resulted from selection against niche overlap (summary Wise 1995; also Herberstein 1997a,b). For instance, inter-specific differences between web designs and websites result in differences in the prey that coexisting species capture. It does not follow automatically, however, that these differences are the result of selection to avoid inter-specific competition for prey. Nor does the coexistence of different species that consume similar prey at a given site (Foelix 2011) imply that they are competing for prey there. For example, syntopic species may be limited not by food, but by predators (e.g., by lizards—see Spiller and Schoener 2008), or by egg parasites, or by abiotic factors such as floods or wind damage to their webs. Their population levels may be low enough that the prey captured by one species has no effect on prey available to the other.

Wise (1995) reviewed and carefully evaluated evidence regarding possible inter-specific competition for prey among species of web spiders, and found support for the following general conclusions. Their population densities were generally so low that their impacts on the populations of their prey tended to be so low that the abundance of prey for a given spider was generally not affected by the density of other web-building species (fig. 8.12). These low population densities were, at least in some species, the result of predation by other animals such as lizards and birds (Schoener and Toft 1983). Wise's balanced and critical discussions are particularly convincing (and admirable) in view of the fact that they led him to reject central conclusions from his own PhD thesis.

Experimental manipulations of the population densities of the araneids *Argiope aurantia* and *A. trifasciata* in early second growth (Horton and Wise 1983), and of *Metepeira labyrinthea* and *Mecynogea lemniscata* in forest habitats (Wise 1981, 1983; Wise and Barata 1983), provided the strongest evidence regarding lack of competition for prey. Removal of *A. aurantia* from experimental plots where the two species coexisted did not increase spider size, rate of prey capture, or the rate of growth

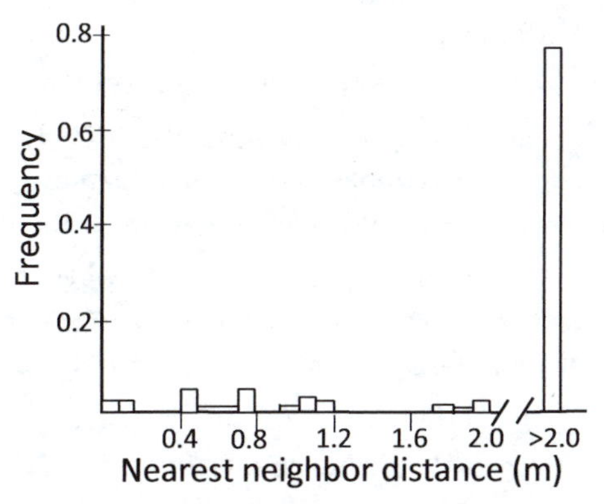

Fig. 8.12. The large nearest neighbor distances between adult females of *Metepeira labyrinthea* (AR) in natural vegetation argue against the importance of competitive exclusion via either reduction in prey captures due to the presence of conspecifics, or to aggressive exclusion by conspecifics (web diameters were generally <0.5 m) (section 8.8) (from Wise 1983).

of *A. trifasciata*; nor did reciprocal experiments, in which *A. trifasciata* were removed, have positive effects on *A. aurantia*. Nor did a further pair of experiments in which the density of one species was increased have negative effects on the other (Horton and Wise 1983).

Generalizing from these results to other spiders is not simple, because some other experimental manipulations of spider densities in the araneids *Metepeira grinnelli* and *Cyclosa turbinata* suggested inter-specific competition for websites (Spiller 1984a,b). But these other experiments were performed, however, in an ecologically simple habitat, the open (unforested) edge of a man-made levee bordering a salt marsh, where predators such as pompilid and sphecid wasps were scarce, perhaps because the ground was too hard for nesting (Spiller 1984a,b). After critically analyzing these and other data suggesting inter-specific competition, Wise concluded that field experiments "have shown that food shortages act as a density-independent factor for many web-building spiders," and that "spiders do not often compete with other spiders for prey . . . because spider densities are below competitive levels" (Wise 1995, pp. 286–287).

8.8.2 INTRA-SPECIFIC COMPETITION

The evidence regarding competition with conspecifics for food and websites is mixed. A reanalysis of the finding of negative density dependence in feeding rates in the sheet-weaving linyphiid *Neriene radiata* in juniper bushes (Wise 1975) found no indication of competition for food (Wise 1995). Intra-specific competition for food and for websites occurred only in the spring but not the summer in *Metepeira grinnelli* (AR), but in both spring and summer in *Cyclosa turbinata* (AR) (Spiller 1984a,b). The effects were weaker when, as is biologically appropriate for spiders (Venner and Casas 2005; Blackledge 2011), data on prey biomass rather than prey numbers were analyzed.

Competition for websites (rather than food) does seem to occur in several species. I will describe two illustrations. The funnel webs of the agelenid *Agelenopsis aperta* had sheltered tunnel retreats where the spider lurked during the day. In the harsh desert grassland environment in an old lava bed in central New Mexico, sheltered refuges that were approximately the diameter of the spider and were cool enough to allow the spider a long enough period of daylight hours to be active on her web to capture prey, and that were also near vegetation where insects were thought to be more abundant (at least according to the possibly flawed technique of sticky trap catches–see section 8.9.2), were in short supply (Riechert 1982, 1985, 1988; Riechert and Gillespie 1986). Spiders defended not only their own webs, but also the areas around them against conspecific wanderers. Experiments with sticky traps suggested that by excluding neighbors, a spider increases her expected prey capture. In a more prey-rich and less harsh riparian environment, the defended territories were smaller and the battles over territories were less intense, in accord with this hypothesis.

Riechert argued (1978a,b, 1986) that territoriality of this sort is widespread in spiders. But the evidence she cited regarding regular dispersion patterns and the direct observations of web usurpations do not constitute strong support. Regular distribution patterns have alternative explanations (Wise 1995). In addition, data on regular dispersion patterns in the orb weaver *Araneus marmoreus* (AR) (Pasquet 1984) involved population densities that were so low that one would have to make the unlikely supposition that the spiders were defending areas substantially larger than their own orbs even though no lines extended this far. Such territory defense also seems unlikely because an invading spider could easily break connecting lines and become "invisible" to a territory holder. The very low rates of displacements of marked spiders by other individuals in the desert theridiid *Latrodectus revivensis* (1 of 88 web relocation events in Lubin et al. 1993), and the tropical araneid *Allocyclosa bifurca* (never observed in >100 daily visits to webs of marked mature females—WE) also argue against this sort of territoriality. Six species of araneids and tetragnathids showed little inter-specific segregation in spatial distribution or timing of orb construction in the same hedges (Ward and Lubin 1992).

Web usurpation occurs in several species; *M. grinnelli* commonly invaded *C. turbinata* webs (AR) (Spiller 1984a,b). It was rare in other species, such as *Argiope* spp. (AR) (Enders 1974; Horton and Wise 1983; Wise 1995), and by conspecifics in *Uloborus diversus* (UL) (Eberhard 1972a). But neither web usurpation nor rapid colonization of artificial websites is a reliable indicator of food limitation. Usurpation could be due, for instance, to robbery of preexisting webs for prey capture (as in mature males—(Eberhard et al. 1978; Vollrath 1987b), or to using preexisting webs as indicators of good quality websites (Vollrath 1987c), rather than to competition for prey.

One further, previously unappreciated type of evidence indicates that intra-specific web usurpation is uncommon. When mature males usurped webs from younger conspecifics, the spiderling was often found at the edge of the same web in *Metazygia gregalis* (AR) (webs stolen by mature males of *M. gregalis*, *Larinia directa*, *Eustala fuscovittata*, and *Tetragnatha* sp.), *Leucauge argyra* and *L. mariana* (web stolen by conspecific males) (Eberhard et al. 1978; WE). If spiderlings and mature females were displacing each other, similar observations of displaced spiders lurking at the edge of the orb would presumably occur. Although I have never made careful counts, I am sure from extensive experience in the field with thousands of webs of *L. mariana*, *L. argyra* (TET), *M. gregalis* (AR), *Allocyslosa bifurca* (AR), *Micrathena duodecimspinosa* (AR), and *Uloborus diversus* (UL) that such cases, with two individuals on a single orb, are extremely rare. There is general accord that in web-building spiders in general there is only one spider per web (e.g., Bristowe 1954; Witt et al. 1968; Foelix 2011) (except of course, certain colonial and social taxa—Avilés 1997); this implies that web usurpation is uncommon.

Testing for website limitation and competition by removing spiders and then checking their websites for recolonization has the weakness that if no spiders arrive to colonize new websites, it may be due to a lack of migrating individuals in the population rather than a lack of website limitation. An additional simple, but apparently seldom used technique to address questions regarding website limitations that avoids some of the drawbacks of manipulations is to check the same habitat year after year to see whether spiders occupy the same or different websites there. Long-term surveys of this sort seem to be rare. I have never made careful counts, but am quite certain that some species were present only sporadically at particular discrete and easily surveyed sites around my residence near San Antonio de Escazu in Costa Rica and on the campus of the Universidad de Costa Rica during >40 years. In particular *Allocyclosa bifurca* was present nearly continuously, but only sporadically occupied particular plants and particular windows of the house. Although it could be argued that the surroundings of any website are never strictly constant (the branches of nearby trees or bushes will have grown, or died back, nearby weeds may be larger or smaller, etc.), the implication is that supports per se were not in limited supply.

On the other hand, incidental observations of the year-to-year consistency with which some particular shelters near the same house were occupied by the tunnels of the funnel-web zoropsid *Tengella radiata*, was suggestive of website limitations, perhaps of appropriate retreat cavities in the substrate (see fig. 10.7). The evolution of aerial webs from substrate-bound webs may have liberated spiders from some kinds of constraints (Blackledge et al. 2009c). Better-structured, more extensive multi-year surveys might give interesting results.

In sum, it is likely that for most species intra-specific competition for websites is weak, because websites are not limiting. There are exceptions to this trend, however. The common behavioral observation that web spiders drive others (both conspecifics and others) from their webs may be due to defense of the web itself (and the prey that it may capture), rather than of the website per se.

8.9 PROBLEMS IN ATTEMPTS TO STUDY CUES THAT GUIDE WEBSITE CHOICES

The associations between particular species and habitats undoubtedly often arise due to dispersing spiders using particular cues to make decisions about whether or not to build at a given site (section 8.4.3.2). Nevertheless, the specific traits of sites that spiders use to make these decisions are poorly known. Experimental studies to determine the cues that are used, especially in the early stages of the process, are very uncommon. Several reasons for this ignorance and difficulties in making biologically appropriate measurements are discussed in this section.

In general it is not clear how to deduce from the general habitat preferences of a spider exactly which cues are being used. For example, the conclusion that the araneid *Argiope keyserlingi* chooses "closed habitats" (Blamires et al. 2007b) leaves open which cues are used by spiders to guide website selection (possible candidates include illumination at night, chemical cues from different types of objects, and typical velocities of air movement). I believe our current ignorance of these topics is profound, and suspect that one of the major problems may be that the cues that are used by spiders are not necessarily the same as the variables that are commonly measured by researchers. Perhaps more emphasis on direct observations of spiders while they are in the process of making choices, and less on simply examining the final results, is needed.

8.9.1 EXPERIMENTAL TESTS NEED CONTROLS: HOW TO COUNT UNOCCUPIED SITES?

One of the largest hurdles to overcome in devising experimental tests of website preferences is the difficulty of defining an uninhabited "website." This is especially true for aerial webs. Recognizing uninhabited sites is a necessary step that is missing in many field studies that attempt to discover preferences for one subhabitat (say shaded sites) rather than another (say sunny sites). Establishing a "preference" for a given subhabitat requires comparison of the fraction of available sites in that subhabit that are inhabited with the fraction of inhabited websites that are available in other subhabitats; counting uninhabited websites is crucial. Thus finding more spiders/m^2 in shaded sites is not sufficient to establish that the spiders prefer shade, because their greater numbers there could be due to greater numbers of possible websites in the shade than in the sun. But counting unoccupied websites is difficult.

Take, for instance, a tangle of weedy growth and bushes at the edge of the grassy yard near my house. I know that *Leucauge mariana* (TET) often builds orbs in such sites, and have seen them at the edge of the yard in the past. I also know that the plants to which their webs are attached varied from year to year, and also during a given season, and that some plants in a given square meter that were not occupied in one year nevertheless had webs in other years. Although I know that this species generally builds more or less horizontal orbs in spaces between the distal rather than the basal portions of plants, and that they attach their orbs to objects with a certain degree of rigidity but seem to avoid extremely flexible objects such as the tips of the long flowering heads of some grass species, I have no idea of how to determine the precise number of websites present.

One technique for counting unoccupied websites is to quantify geographically defined areas, such as square meter plots, and then repeatedly count the numbers of spiders present over long periods of time; the maximum number of spiders observed gives the minimum number of websites, some of which will have been unoccupied during some counts; but the technique is blind to unoccupied websites. Another is to stretch horizontal lines at different heights above the ground and count the number of plants in which spiders might build touched by each line (da Silva-Souza et al. 2015). But of course such units do not necessarily define a single website; a single plant might contain several, or none of the appropriate size. Another is to characterize several aspects of a typical website, and then count the number of sites with such a combination of variables. But, as discussed by Schoener and Toft (1983), this too has serious limitations such as inadvertent omission of crucial variables. Still another possible solution is to experimentally force choices on spiders. One can provide spiders with standardized websites (e.g., a standard cage) and then vary some particular stimulus (e.g., wind) and check whether web-building frequency is affected. This technique can involve situations of only limited biological reality, however. To take an extreme experiment as an example, *L. argyra* (TET) will build orbs when confined in tiny, 5.6 cm diameter containers (fig. 7.42, Barrantes and Eberhard 2012), but this probably has no relevance to website choice in nature. Perhaps a more effective solution would be to build artificial websites that vary in specific ways (e.g., Robinson 1981), and then leave them in the field in replicated pairs, to see if one type is colonized more than the other.

8.9.2 MEASURING HABITAT RICHNESS: STICKY TRAPS DO NOT MIMIC SPIDER WEBS

The distribution patterns of web spiders are surely affected by the richness of prey available in different habitats. There is a long tradition in web spider ecology (and ecology in general) of attempting to measure the numbers of prey in order to determine the richness of a habitat. Unfortunately, things are often not as simple as they might seem (Eberhard 1990a).

Most importantly, the prey that exist at a given site are not all equally available to a spider; web spiders capture only a biased sample of all the prey present. For instance, prey that fly directly from leaf to leaf are available to species building webs near leaves, but less to those that build in large spaces between leaves. Prey that have effective defenses against adhering to webs such as oily surfaces, detachable scales, long legs held ahead of the body, or a slow, tentative slow flight pattern (e.g., Craig 1986, 2003; Blackledge and Zevenbergen 2006), that fly at times of the day when the spider does not have a web up, or that have effective chemical defenses, may be unavailable to spiders even though they are abundant in exactly the same subhabitat. From the point of view of a spider, any potential prey that she cannot capture and eat are not available to her, and the presence or absence of such prey is irrelevant with respect to the habitat quality. Such prey are "available" only in the evolutionary sense that future changes in the spider or her web might enable the species to feed on it.

Unfortunately, all human techniques for collecting insects have intrinsic biases (Southwood 1978), and none of these biases match precisely the biases of the spiders (section 3.2.5.4). Web spiders would seem to offer a substantial advantage over other predators such as lizards or birds in this respect, because their webs determine (or at least affect) their biases, so a trapping technique that mimics a web should overcome these problems. Instead of having to resort to such poorly correlated biases as that between the insects captured by an insectivorous bird and a sweep net, web spider ecologists can use sticky sheets or other web mimics to obtain estimates of the available prey.

I was once an enthusiastic proponent of such techniques, and was proud of my own artificial web orb design (Eberhard 1977c). This optimism evaporated, however, when José Castillo and I performed a calibration study that I still believe constitutes the most thorough job of mimicking orb webs with traps ever performed. We mimicked many aspects of the orbs of mature females of the nocturnal araneid *Metazygia gregalis* with unusual precision, including their visibility, height above the ground, weather conditions, time of day, microhabitat, and orientation with respect to the wind (Castillo and Eberhard 1983). Incredibly, even with all these factors pretty much under control and with large sample sizes (238 prey captured by spiders, 654 by the traps), the correlation between the identities and numbers of prey was only mediocre: only 23% of the mean squared variation in spider captures was explained by the trap captures. The single most common prey in the traps (a seemingly palatable green leaf hopper) was not fed on by a single spider! How much trust can one place in other artificial trap studies in which most or all of these factors were not controlled? As is clear, however, from the continued use of sticky traps and other similar collecting techniques in recent studies to estimate prey availability (Harwood et al. 2001, 2003; Guevara and Avilés 2009; Guevara et al. 2011; Rao and Lubin 2010), other workers have not found these reservations convincing.

There is one labor-intensive (and less often employed) way to taken into account the biases of webs and spider attack behavior: to inspect webs in the field at frequent intervals, and count how often spiders are found feeding (e.g., Buskirk 1975; Howell and Ellender 1984). If the rates at which food is ingested from different prey sizes and during the course of a given feeding bout are uniform (I suspect they are not), this could give better estimates of prey available in different habitats. Another tactic would be to estimate the masses of prey captured by the spiders from their lengths (e.g., Venner and Casas 2005); this technique also has serious limitations (Eberhard 2013a), but nevertheless seems likely to be better than artificial traps. Even these modifications cannot correct, however, for the substantial differences in nutritional value of different kinds of prey (Toft 2013).

In sum, practical problems such as the difficulty of counting the numbers of unoccupied websites, and of measuring the relative quantities and qualities of prey available to web spiders at different sites, have impeded progress in understanding the cues used by spiders to select their websites.

8.10 TIME OF DAY: DAY WEBS VS. NIGHT WEBS

It appears that most non-orb webs are built during the night. I speculate that ancestral web spiders attacked prey both day and night, and that one of the original advantages of prey capture webs was to enable the spider to remain protected in her retreat during the day while she could nevertheless sense and attack prey. Spiders in many modern non-orb weaving groups that rest hidden in retreats during the day are often responsive to prey on their webs during the day. For instance, the pisaurid *Architis tenuis* built its modified funnel web (fig. 1.3) at night or in the early morning, but also attacked prey both day and night (Nentwig 1985c). Daytime attempts to feed such "nocturnal" spiders were generally successful in the filistatid *Filistata insidiatrix* (except at very high temperatures) (Nørgaard 1951), the zoropsid *Tengella radiata*, the agelenid *Melpomene* sp., the theridiids *Nihonhimea tesselata*, *Latrodectus* spp., and *Steatoda* spp., the therophosids *Psalmopoeus reduncus* and *Aphonopelma* sp. (Barrantes and Eberhard 2007), the lycosid *Aglaoctenus castaneus* (Eberhard and Hazzi 2017), and the diguetid *Diguetia canities* (WE).

The evolution of relatively delicate but rapidly constructed orb webs brought with it both the need to rebuild these webs frequently, and also the opportunity for the spider to coordinate the timing of web construction with peaks in the numbers of available prey, so that her orb would be fresh and in optimal condition when more prey were active. Table 8.2 demonstrates that there is great variety in temporal patterns of orb construction. There is a spider that builds at any time of the day or night, though the early afternoon seems to be the least popular. The data are somewhat biased because arachnologists (with some heroic exceptions, such as Sherman 1994 and Thévenard et al. 2004) have shown only a limited disposition to alter their own diurnal cycles of activity and make extensive observations at night.

The extremes for high specificity in timing seem to be *Cyrtarachne inaequalis* (AR), a nocturnal moth specialist in Japan (Baba et al. 2014), and *Araneus diadematus* (AR) in captivity when an increase in temperature and a simultaneous switch from dark to light induced initiation of web construction by 100 spiders in the same 2 min interval (!) (Rawlings and Witt 1973). At the other extreme, web construction by *Zosis geniculata* (UL) has been observed in captivity from 9 PM to 10 AM (WE). Overall, spiders show two peaks of web building: around dawn (I will call these "diurnal" webs), and in the early evening (I will call these and those built later at night "nocturnal" webs) (e.g., fig. 8.13).

The huge majority of species build under low light conditions, even when the web is operated during the day. Presumably darkness provides protection from their many visual predators. Building around dawn, when the humidity is especially high, could also be beneficial for araneoids, because it would increase the ability of the glue on the sticky lines to take up moisture from the air and thus its stickiness (Edmonds and Vollrath 1992; Amarpuri et al. 2015); ecologically realistic differences in humidity resulted in substantial differences in retention times of a prey in *Araneus marmoreus* (Opell et al. in prep.).

The temporal distribution of prey activity is complex. Several insect trapping techniques have demonstrated one peak of flying insects around dawn, and a second around dusk (Southwood 1978). Presumably these peaks

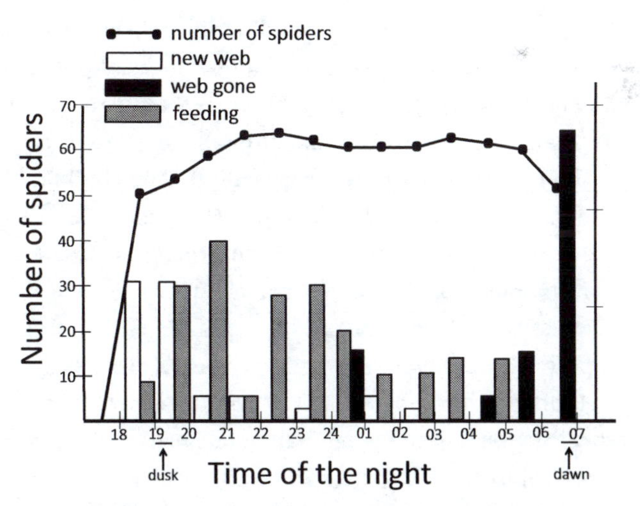

Fig. 8.13. In this unusually complete study of the nightly cycles of orb construction, web removal, and prey consumption during repeated surveys on three nights in the nocturnal tropical araneid *Eriophora edax* (AR), most webs were built early in the evening. Spiders fed throughout the night, but most of the 55 prey were captured before midnight; 67.7% were Lepidoptera. The reasons why some orbs were taken down long before dawn, and why some were built long after dusk are not clear (after Ceballos et al. 2005).

of insect activity have resulted from selection on the insects to avoid predation by many of the same visually orienting predators of orb weavers. The tetragnathid *Metabus ocellatus* constitutes a particularly dramatic exception to the dawn or dusk pattern in orb weavers, and supports the probable importance of prey abundance. The peak of web construction occurred in the late afternoon, at the same time that there was a peak of insect abundance just above the surface of the tropical streams where this species built its webs (Buskirk 1975). Construction of orbs also peaked in the late afternoon in other, distantly related species that build orbs just above streams, including *Conculus lyugadinus* (AN) in Japan (Shinkai and Shinkai 1988), *Caerostris darwini* (AR) in Madagascar (Gregorič et al. 2011), and *Wendilgarda galapagensis* (TSO) on Cocos Island (Eberhard 1989a). Other local patterns of prey availability may also have important effects on diurnal building patterns. In hedges near ponds in Israel there was a large early evening peak of midges, and the younger spiderlings of four species of araneids and tetragnathines showed a preference for building early in the evening as opposed to the rest of the night (Ward and Lubin 1992).

An indication of the importance of avoiding predators is that some spiders fail to respond during the day to prey that becomes entangled in the remnants of their nocturnal webs (e.g., Harmer and Herberstein 2010). Indeed, there are scattered observations of visual predators attacking spiders that had broken their usual patterns and were on their webs during the day: a salticid spider captured an *Araneus legonensis* that was building late in the morning following a rain (Edmunds 1986); a *Momotus momota* bird captured an *Eriophora* sp. (AR) that emerged from her retreat to attack a prey on her web (Carlos Valerio pers. comm.); and an unidentified frog captured a *Tetragnatha* sp. building at early dusk (WE).

Probably other factors also act on the timing of orb construction. Edmunds and Edmunds (1986) speculated that temperate zone species are less likely to have strictly nocturnal webs because nights are shorter during the portion of the year in which spiders are active. The adhesive properties of some araneoid webs are greater when the web is built and operated under more humid conditions (Edmonds and Vollrath 1992), and the reduced orb species *Cyrtarachne bufo*, *C. akirai*, and *C. nagasakiensis* (AR), which depend on the unusual stickiness of their lines (Cartan and Miyashita 2000) to capture their moth prey, preferred to build when humidity was higher (Baba et al. 2014) (the related *Pasilobus* sp. appeared to have a similar preference [Robinson and Robinson 1975]).

There appear to be taxonomic patterns in the timing of orb construction (table 8.3). For instance, all of the *Argiope*, *Micrathena*, *Gasteracantha*, *Cyclosa*, and *Uloborus* species with which I am familiar build early morning orbs; in contrast, all the araneids observed in the genera *Cyrtarachne*, *Metazygia*, and *Poltys* build at night (table 8.2). As usual, there are also exceptional groups with substantial variation, such as *Alpaida* (AR), *Araneus* (AR), *Caerostris* (AR), *Neoscona* (AR), *Miagrammopes* (UL), and some tetragnathids (table 8.3).

A survey of building behavior (Eberhard 1982) incidentally revealed two somewhat poorly defined groups that build in the early evening. The araneids *Metazygia gregalis* and *Hypognatha* sp. both tended to begin orb construction while there was still enough light to see the spider at work. At least *M. gregalis* began construction with little or no exploration and, perhaps as a result of incomplete reconnaissance, some orbs were built in constrained or irregularly shaped spaces. In contrast, several species of *Eustala* as well as *Wagneriana tauricornis* at the same sites did not begin to build until darkness had fallen completely; the spiders' behavior was never visible without additional illumination. Presumably the payoff for dusk webs like those of *M. gregalis* was access to a larger portion of the dusk peak of insect flight; *Hypognatha* sp. orbs were often in tatters by the time darkness fell.

An additional distinction among night webs is that some nocturnal species tended to build late at night, rather than near dawn or dusk. As shown in table 8.2, some of these are species are thought to specialize on moths (on the basis of their orb designs). This timing of web construction fits the general activity pattern of moths, which tend to fly throughout the night, rather than being concentrated at dawn and dusk.

Many of the characterizations in tables 8.2 and 8.3 are probably overly simple, especially for species whose diurnal cycles have not been studied in a structured manner. Intra-specific variation seems to be the rule among the better-studied species. The tetragnathid *Metabus ocellatus* built at least some webs at every hour from 6 AM to 6 PM (Buskirk 1975). The adults of *Larinioides cornutus*

Table 8.3. A sampler of genera for which there are data for at least three species, contrasting in which the timing of orb construction was either relatively constant (A), or in which it was variable (B). Many of the limitations mentioned in table 8.2 also apply here.

A. Genera in which different species appear to build at approximately the same time

Araneidae

Argiope (early morning): *argentata* (WE), *flavipalpis* (Ewer 1972), *trifasciata* (Edmunds and Edmunds 1986), *anasuja* (WE), *Chrysometa* (early morning): sp. (#'s O-3, 249, 2810) (WE)
Cyclosa (early morning): *argenteoalba* (Nakata and Ushimaru 1999), sp. nr. *atrata*[1], *caroli, insulana, jose, monteverde, tapetifaciens, triquetra*, spp. (#1867, 1892, 1961) (WE), *turbinata* (Carico 1986; Moore et al. 2016; WE)
Gasteracantha (early morning): *cancriformis*, sp. (#2036), *curvispina* (Edmunds and Edmunds 1986), *fornicata* (Hauber 2002), sp. (#2036, 2038-9) (WE)
Mangora (early morning): *melanocephala, pia, vito* (WE)
Metazygia (very early evening): *gregalis*[2], *lopezi, wittfeldae, yobena* (WE)
Metepeira (early morning): *incrassata* (Jakob et al. 1998), sp. (WE), *gressa* (WE)
Micrathena (early to mid-morning): *gracilis* (B. Opell pers. comm.), *acuta, horrida, molesta, plana, schreibersi, sexspinosa*, sp. nr. *lucasi, undecimspinosa* (WE)

Tetragnathidae

Leucauge (early morning): at least 9 species[3] (WE)
Tetragnatha (dusk or early evening): *kaestneri* (Crome 1954), *elongata* (Ceballos et al. 2005), spp. (#2190, 1837) (WE)

Uloboridae

Uloborus (early morning): *diversus, glomosus, trilineatus* (WE)

B. Genera in which different species build at different times

Araneidae

Araneus:
Evening: *niveus, cingulatus* (Coddington 1987), *cereolus* (Edmunds and Edmunds (1986), *sexpunctatus* (Savory 1952)
Early morning: *diadematus* (Witt et al. 1968), *cucurbitanus* (Savory 1952), *legonensis* (Grasshoff and Edmunds 1979), *expletus* (WE)
Both late in evening and around dawn: *marmoreus* (B. Opell pers. comm.)
Eustala:
Evening: *fuscovittata, guttata, scutigera*, spp. (#'s 508, 784, 806, 2345) (WE)
Early morning: *conformans, perfida* (Messas et al. 2014)
Eriophora:
Evening: *edax, ravilla* (Bradley 2013), *transmarina* (Herberstein and Elgar 1994)
Early morning: *nephiloides* (WE)
Alpaida:
Evening: sp. (Parrita, Costa Rica), *veniliae* (WE)
Early morning: *leucogramma, rhodomela, truncata, tuonabo* (WE)
Neoscona
Early evening: *nautica* (Hingston 1920), *hentzi*[4] (Carico 1986), *crucifera* (B. Opell pers. comm.)
Early morning: sp. (#1947) (WE)

[1] This species, from Tamilnadu, India, was identified on the basis of its unsual color and the form of its abdomen using Shinkai and Takano 1984.
[2] Some webs built later at night.
[3] Some (perhaps all) also build other webs later in the day.
[4] Webs generally left intact during the day while the spider was in a retreat; they were generally removed at dusk and a new web was built immediately.

initiated webs from 9:30 PM to 4:30 AM, with a peak around midnight (Sherman 1994). Different individuals of the araneid *Eriophora edax* at the same site built and tore down orbs at different times during the same nights (fig. 8.13) (Ceballos et al. 2005). *Metepeira incrassata* (AR) often paused during construction when clouds appeared (Jakob et al. 1998). *Gasteracantha cancriformis* (AR) often built at or soon after dawn; but a few individuals built as late as 11 AM, even on mornings when it did not rain (WE).

Possible reasons for such individual variation have hardly been explored. During the rainy season (with most of the rain falling in the afternoon or evening) younger individuals of the tropical colonial tetragnathid *Metabus ocellatus* tended to build later in the morning than adult females, perhaps to avoid having their webs damaged or being chased away from websites by larger individuals, which tended to build their webs earlier in the morning (Buskirk 1975). A similar pattern of larger spiders building first occurred in the colonial araneid *M. incrassata*, but experimental manipulations of colony composition revealed that, at least in terms of proximate causes, the differences in timing were not due to interactions among different-sized spiders (Jakob et al. 1998). Solitary *Zygiella x-notata* (AR) were more likely to build earlier in the evening when kept in the presence of live *Calliphora vomitoria* flies in captivity (Pasquet et al. 1994; Thévenard et al. 2004). Some variations in table 8.3 may be due to some individual spiders building more than a single orb each 24-hour cycle.

Two probable mechanisms that spiders could use to build specific times are endogenous circadian rhythms, and cues from the environment. Endogenous rhythms are known to occur in orb weavers. Locomotor activity of captive *Cyclosa turbinata* (AR) showed two peaks, one within an hour after lights-off, and a broader peak soon before lights-on (when orbs were normally built). The rhythm was endogenous and exceptionally short (about 19 hours); it persisted under conditions of constant dark and temperature (Moore et al. 2016). Web construction, in contrast, was sporadic in constant darkness, so construction behavior may also be driven by exogenous factors (Moore et al. 2016).

The most obvious external cue is the diurnal light-dark cycle. For instance, displacing the light-dark cycle, so that dawn occurred at about noon, altered web construction times in the uloborids *Uloborus diversus* and *Philoponella vicina*, *Leucauge mariana* (TET) and *Nephila clavipes* (NE) (Eberhard 1972a, 1990b). Another possible cue, at some sites in some seasons, is temperature. Typically the lowest temperatures occur in the early morning before dawn (Geiger 1965). Mature females of the typically dusk- and dawn-building *Araneus diadematus* kept in constant darkness but with a diurnal cycle of temperature changes tended to build while the temperature was decreasing; when the temperature was held constant, building times became irregular (Spronk 1935). When light but not temperature or humidity cycled, spiders showed typical dusk and dawn peaks; when light was also held constant, web destruction (which had little effect when light levels changed on a 24-hr cycle) elicited construction (Le Guelte and Ramousse 1979).

There can be limits in the use of these cues. In southern Louisiana in the US, where the transition between winter and spring is often erratic and warm spells are interspersed with cold periods, I have seen the usually quickly-moving species *Leucauge venusta* (TET) slowly plodding through the process of building an orb in plain daylight at 9 to 10 AM on cold mornings when the temperature was only 10–12°C. Later, in the spring, this species built earlier, before dawn. Low temperatures may be a problem at some other sites. At a moderately warm site in Mexico at 1000 m during the summer, spiders of all sizes in colonies of *M. incrassata* (AR) apparently built later in the morning when the day was colder (Jakob et al. 1998). Building times also varied with the season of the year in *Araneus diadematus* (Spronk 1935). Similarly, *L. mariana* (TET) often postponed orb construction until later in the morning in tropical Costa Rica on mornings when it rained or when there was heavy fog (WE) (fig. 8.10). I do not know whether the peak of insect flight activity is also displaced until later on cold, rainy, or foggy mornings.

8.10.1 MULTIPLE ORBS IN A SINGLE DAY

Part of the fascination of orb webs is their exquisite economy in materials and construction behavior. In half an hour or so, a spider can set up a fresh new trap. Some orb weavers exploit this economy and routinely build two or three orbs every day in the field even when their webs are not damaged by catastrophes such as rain, falling detritus, or large prey (Carico 1986; Craig 1989). For instance, individuals of the tetragnathids *Leucauge mariana* and *L. argyra* probably usually build three orbs each day—one in the early morning before dawn, a second around noon, and still another early in the evening (WE). As in many other species, all but the frame and anchor lines of the previous web are ingested (Carico 1986). The timing of web replacement varies in different groups. *Neoscona nautica* and *Singa lucina* (AR) sometimes built one orb before dawn and another at dusk (Edmunds and Edmunds 1986; Ward and Lubin 1992); *Metazygia gregalis* (AR) consistently built two or three orbs during a single night, the first at nightfall, and the others late at night (W. Eberhard, M. Barreto, and W. Pfizenmaier unpub.). Young individuals of *Poltys* often renewed their webs several times in a single night, but renewal was apparently less common in adults (Smith 2006). *Micrathena duodecimspinosa* normally built a single web in the morning, after dawn, but some individuals built a second orb around noon when there was a brief rain later in the morning (WE).

Drawing a line between the spiders that "spontaneously" tear down and replace their orbs, and those which respond to relatively minor web damage by replacing the entire web is not easy. The timing of the "spontaneous" noon replacement webs in *L. mariana* was moved forward by removing the radii and sticky lines of the spider's first web at 8 AM; she generally built a new web between 9:30 and 11 AM (Eberhard 1988a). Some species show intra-specific variations. For instance, most *Araneus expletus* (AR) built only one orb/day (usually in the hours just before dawn); but in 4 of 60 cases in which the spider was checked in the early evening, she had built a second orb (WE).

Craig (1989) linked more frequent web replacement with smaller spider size in five Neotropical species; web replacement rates by adult females (means of three, and up to five webs/day) were higher in the two smallest species, *Epilineutes globosus* (TSM) (0.8 mg) and *Leucauge globosa* (TET) (2.7 mg) than in the two larger species *Micrathena schreibersi* (146 mg) and *Mangora pia* (AR) (21 mg). Estimated web "breakdown" rates prior to replacement were also higher in the smaller species. Craig speculated that small orb weavers may be able to tolerate high rates of web "loss" because of their high rates of prey intake, but that larger spiders may not be able to tolerate high rates of web loss. This correlation was based however on only adult females and only five species, and did not take into account the possible effects of differences other than spider size such as orb design, websites, mechanical propertiers of silk, or phylogeny. In addition, the trend did not occur in other species. For instance, adult females of two other species of *Leucauge* (*mariana* and *argyra*) and the araneid *Metepeira gregalis* all weigh on the order of 40–60 mg and are thus "large," but behave as did the small species of Craig's study, building about three webs/day. As Craig noted, a more powerful test would be to check the web renewal rates of spiderlings of different sizes of the same species.

8.11 SUMMARY

This chapter links web-building behavior with the ecological interactions between spiders and their environments via studies of website selection. Not surprisingly, the topic is complex. Included among the many different variables that influence website selection are the sizes and shapes of available spaces in which to build, the presence of smooth rather than rough surfaces, the rigidity of available supports, dead (as opposed to living) supports, temperature, humidity, fog, rain, wind, lighting conditions, the presence of prey, previous history of prey captures, the presence of predators, previous web destruction, the degree of isolation, the height above the ground, the presence of kleptoparasites, the presence of conspecifics, the distance from the natal web, and the presence of egg sacs and of abandoned lines from previous webs.

All of these variables can function as potential indicators of the single, crucial, but intrinsically unpredictable aspect of a potential website—the biomass of the prey that a web there would be able to capture. The accuracy with which spiders adjust their behavior to the balances between different variables undoubtedly differs greatly, and the precision of their estimates must always be limited by day-to-day variation in prey at any given site. The magnitude of this variation undoubtedly varies for different prey in different habitats. For example, the number of prey that will encounter a web of the theridiid *Dipoena banksii* that is built next to the entrance to a *Pheidole bicornis* ant nest (Gastreich 1999), or of a *Deinopis longipes* that is built over an ant trail (Robinson and Robinson 1971), is likely to vary less than the numbers that will encounter the webs of many typical orb weavers from one day to the next.

Limitations imposed by spiders' senses probably have important consequences for how they choose and evaluate potential websites. Although there is little field data on searching behavior, spiders physically explore sites prior to construction. Exploration, which is to some extent subject to vagaries such as wind directions, is used to verify, through still poorly understood mechanisms, that there are attachment sites that are appropriately distant from each other and have appropriately sized open spaces between them. Spiders are probably unable to compare different websites before building and apparently do not retrace their steps during searches for websites (I know of no field data, however); rather they probably employ "satisficing" techniques that utilize minimum and maximum thresholds to decide whether or not to build at a particular site. One apparently risk-aversive tactic in some orb weavers is to make only a reduced web at a new site (a "pilot web"), and then make larger webs subsequently if they decide to stay. This tactic seems more likely to be useful in evaluating relatively invariable website factors such as physical stability of supports, temperature, humidity, and the degree of wind stress, than more intrinsically variable factors such as prey availability or damage from falling detritus.

After choosing a given website, a spider's likely degree of commitment to that site (her website tenacity) also varies widely, and is probably strongly influenced by web designs (though comparative data are fragmentary). The level of tenacity varies from a long-term commitment to brief, one-night stands in species whose webs are built quickly. Tenacity may be greater in species in which web construction is more costly. The derivation of low commitment from higher commitment webs probably enabled spiders to explore more extensively and make more precise discriminations of website quality. Some species at the long-term commitment end of the spectrum, such as the sand dune eresid *Seothyra henscheli*, prefer to remain at a site that has proven suboptimal rather than run the risk and make the investment that would be necessary to find and occupy a new site. This conservative eresid is probably an extreme case, but probably many other species that invest heavily in the web before it is fully functional, that run large risks when migrating to new websites, and that do not reingest their webs when abandoning them are also likely to often occupy suboptimal websites.

Problems determining website quality plague not only spiders, but also arachnologists attempting to resolve the apparently simple but ultimately difficult question of website quality. Nearly all techniques of quantifying insect abundance have intrinsic biases, and in no case are these biases the same as the biases of spiders and their webs. Many published estimates of website quality must therefore be treated with caution. Unfortunately, the most convincing techniques (such as making and analyzing video recordings of thousands of hours of activity at webs, or making frequent surveys of whether spiders are feeding on prey over substantial periods of time) are also the most time- and labor-intensive.

Another question is whether the numbers of appropriate websites limit spider populations or predatory opportunities in nature. Again, the answers seem to vary in different species. Website limitation is suggested in some species in which some sites are occupied from one year to the next by different spiders (though data are as yet very preliminary). Another type of evidence is that gregarious species like the colonial orb weavers *Metabus ocellatus* (TET), *Leucauge mariana* (TET), *Philoponella semiplumosa* (UL), and *Parawixia bistriata* (AR) frequently fight over websites. A similar conclusion is suggested by the rapid colonization by some other orb weavers when new supports for webs are placed experimentally in the field. On the other hand, there are several indications of a lack of limitation in aerial web spiders. Counts of possible websites in a structurally simple but densely populated habitat suggested that only about 10% of the possible sites were occupied. Less structured observations of sites occupied from one year to the next give a similar impression that only a small fraction of potential sites are occupied at any given moment. In addition, when some pholcids, a mysmenid, and linyphiids that are abundant near the floor of tropical forests were removed from experimental plots, other individuals moved in the next day but mostly occupied websites different from those occupied the day before.

Another aspect of website selection involves the time of day when webs are built. Many spiders build during the night; presumably this timing is ancestral, and functions as a defense against diurnal predators. The timing of construction probably also functions in some species, especially those with more ephemeral webs such as orbs, to have the web fresh and undamaged at the time of day when it is most likely to intercept prey. In several groups of orb weavers, the timing of orb construction corresponds with the dusk and dawn peaks of insect flight activity. In contrast, others whose webs are probably specialized to capture moths, build their webs later at night when some moth species tend to fly. Still another set of species build their webs just above streams at about 4:00 PM in the afternoon, when there is a peak of insects flying just above the stream.

One unsolved mystery concerning the timing of different activities involves the question of why some orb weavers remove their webs many hours before they build a new web. I speculate that this behavior is related to the neglected question of whether spiders sleep. By removing her previous web (rather than leaving its more or less tattered remains intact), the spider may reap the benefits (whatever they may be) of uninterrupted sleep.

It is not always easy to distinguish cause and effect in relations between the designs of the webs that a species is capable of building and its biases in website choice.

Many species adjust at least some aspects of their web designs to the sites that are available. There is preliminary evidence that spiders may learn how to make such adjustments, and that species that build in large open spaces in nature (and thus normally have to make fewer adjustments of this sort) may be less capable of adjusting. Further data are needed on both points. Have evolutionary changes in website selection caused evolutionary changes in web design during the course of evolution, or vice versa? There are not enough data yet to identify general trends, and it seems likely that both types of cause and effect relations occur. One especially dramatic case is the Australian araneid *Telaprocera maudae*, in which experiments showed that website choice criteria are the cause rather than the result of the striking evolutionary vertical elongation of its orb.

9

EVOLUTIONARY PATTERNS

AN ANCIENT SUCCESS THAT HAS PRODUCED HIGH DIVERSITY AND RAMPANT CONVERGENCE

9.1 INTRODUCTION

This chapter on evolutionary patterns in the designs of the webs of different species reveals a very different pattern compared with the uniformity in behavioral acts and guiding stimuli seen in chapters 6 and 7: there is instead an extreme diversity of web designs both in orbs and in non-orbs. I will attempt to convey a sense of this diversity by using several techniques: many photos; Whitman-like lists of variants in the text; long tables; and mini-essays (boxes 9.1–9.4). My objective is less to give a coherent series of logical arguments than to to pile up examples and eventually overwhelm the reader with elaborate, exotic, and beautiful variations. I will also sketch a second major general pattern—of multiple convergences on a number of different designs, many of which can be explained by their probable functions. I will argue that these two large-scale patterns, along with a third, an abundance of intermediate forms, can be explained by recognizing that web construction is both an ancient and a very successful tactic, and that spiders have made an extensive evolutionary exploration over a long history of the possible ways in which webs can be exploited. The combination of the typical "theme and variations" pattern seen in evolutionary comparisons and the unexpected inventiveness of many of the variations is a source of aesthetic pleasure in spider webs.

Before beginning, it is important to reemphasize the problems of overly typological statements. Descriptions of "the" web of different species and groups in this chapter suffer from important limitations. Orb designs vary substantially even from one web of a given individual to the next (e.g., figs. 7.38, 7.41, 10.15; summaries by Heiling and Herberstein 1999; Herberstein and Tso 2011; Harmer and Herberstein 2009; Barrantes and Eberhard 2012; Quesada 2014). Intra-specific variation can also be substantial in non-orb webs (fig. 10.7; see for example Eberhard and Hazzi 2013 and Jörger and Eberhard 2006 on a diplurid and a theridiid). The true range of web designs may not be known for any species.

An additional problem is that discussions of convergence and divergence depend, in the end, on words. The words available in English do not begin to portray the subtle differences that occur in web designs. Using photos rather than words to illustrate given webs can compensate, but only to a limited extent.

9.2 PATTERNS IN THE DIVERSITY OF WEBS

9.2.1 HIGH DIVERSITY

Spider webs show a seemingly endless variety of forms, as illustrated in the following run-on lists. Table 9.1 summarizes the large number of web types on which multiple taxa have converged, table 9.2 documents a similar pattern in the single family Theridiidae, and table 9.3 classifies 54 different web designs built by Japanese spiders (Shinkai 1989). Spiders have explored many of the different dimensions in which webs can vary. Consider the following examples. Some spiders are extremely economical of silk (*Miagrammopes* spp., *Phoroncidia* spp., *Chrosiothes tonala, Episinus* spp.) (e.g., Shinkai 1989), while others are profligate, such as the mygalomorph *Diplura* sp. (Viera et al. 2007), to the point of having webs so dense that ants sometimes nest in the web and drive the spiders away from some prey (Fowler and Venticinque 1996). Some webs are designed to be built in highly specific habitats, including *Wendilgarda* spp. (TSM) (fig. 10.29) and *Metabus ocellatus* (TET) (figs. 2.19) just over the surfaces of tropical streams, and *Telaprocera maudae* (AR) against the trunks of large trees (Harmer 2009). Many other webs in a variety of families (e.g., *Kukulcania, Segestria, Ariadna, Liphistius*) are built around retreats or burrows that have very specific sizes, and sometimes also occur in specific substrates. In contrast, many aerial webs can seemingly be hung almost anywhere.

A variety of forms function only as extensions of the spider's tactile sense, and do not to retain prey—for example, *Liphistius, Atypus* (Comstock 1967; Coyle 1986), *Ariadna* (Griswold et al. 2005), *Hersilia* sp. (Williams 1928), *Uroctea durandi* (Kullmann and Stern 1981), *Ocrepeira ectypa* (AR) (Stowe 1978), and *Modisimus culicinus* (figs. 9.1, 9.2, 10.5). Others that serve only as resting sites for mature males also have a variety of forms (fig. 9.4, appendix O9.2). Those of the salticid *Portia fimbriata* function as lookouts from which spiders sight potential prey visually (Jackson 1986), while the vertical rectangular or trapezoidal "sail" webs of another salticid, *Pellenes arciger* (fig. 9.3), have no known function (Lopez 1986). Some diversity currently has no functional explanation (fig. 9.5). Some sticky lines are highly adhesive, and are clearly traps to retain prey, while the sticky lines of other species, including several linyphiids (Benjamin et al. 2002; Benjamin and Zschokke 2004), *Anelosimus pacificus* (THD), *Theridion hispidum* (THD), the pholcid *Modisimus* (Briceño 1985), and the synotaxids *Synotaxus* spp. (Eberhard et al. 2008b; Barrantes and Triana 2009), have such small, sparse glue droplets (fig. 2.8) that are so weakly adhesive that their function is still uncertain.

Most webs are anchored to the substrate, but some, such as *Mastophora, Dichrostichus, Cladomelea* (Yeargan 1994), can be swung by the spider, and others are brought down onto the prey or thrust into its path (*Deinopis, Menneus*—Akerman 1926; Clyne 1967; Robinson and Robinson 1971; Coddington and Sobrevila 1987); one species of *Wendilgarda* sometimes drags its reduced web (a single line with a sticky tip) back and forth across the surface of a lake to capture water striders (Eberhard 2001a). Some trapping webs consist of sticky lines on the substrate itself (*Oecobius concinnus*—Solano-Brenes et al. 2018) (fig. 9.6), and phyxelidids *Vidole* and *Xevioso* build sticky substrate webs surrounding their retreats (Griswold et al. 2005). Some others, such as the filistatids *Kukulcania hibernalis* (Comstock 1967), and *Filistata insidiatrix* (Griswold et al. 2005), the amaurobiid *Callobius* sp. (Griswold et al. 2005), and the sicariid *Loxosceles* (Vetter 2015), have sticky lines on or just above the substrate, or non-sticky lines that are largely attached directly to the substrate and serve as extensions of the spider's tactile sense, as in the liphistiids of the genus *Liphistius* (Bristowe 1932, 1976), the diplurid *Linothele macrothelifera* (Eberhard and Hazzi 2013), the linyphiid *Drapetisca* (Kullmann 1964), and the oecobiid *Uroctea durandi* (Kullmann 1975; Kullmann and Stern 1981). These "substrate" or "appressed" webs (Griswold et al. 2005) are thought to function in capturing walking prey, though this may be an oversimplification, since the substrate webs of the filistatid *Pritha* (= *Filistata*) *nana* often (31%) captured nematocerous flies (mostly chironomids and limoniids) that usually fly rather than walk (Nentwig 1982b). Other webs (with a great variety of forms) are far from the substrate and facilitate capture of flying prey ("aerial" webs), but some aerial webs such as gumfoot webs have secondarily become adapted to capture walking prey; here too there is substantial variation, even among "star" webs (fig. 9.7). Some aerial trapping webs are flimsy and retain prey only just momentarily, as in the tetragnathid *Leucauge mariana* (Zschokke et al. 2006; Briceño and Eberhard 2011), while others (both substrate

Table 9.1. Web designs and the ways spiders operate them have frequently converged; separate entries in the "taxa" column are, on the basis of their taxonomic status, likely to be independent. There are several limitations, however, that prevent definitive statements regarding precise numbers of independent evolutionary derivations: some web categories grade gradually into each other; phylogenies in many groups are currently uncertain; some homologies are uncertain (see Blackledge et al. 2009c; Eberhard and Hazzi 2017); and the behavior used to build different types of webs is frequently unknown. Even so, convergences are clearly rampant (there are 45 non-orb and 50 orb entries and even larger numbers of probable convergences in the table). There several reasons to believe that the numbers of convergences are probably underestimated: the webs of most species are not yet known; some apparently similar web designs are built using different behavior patterns (see, for example, Eberhard and Hazzi 2017 on "brushed" sheet webs); and additional convergences in the family Theridiidae (table 9.2, Eberhard et al. 2008b) were not included. "Dense" sheets were distinguished from "sparse" sheets in that individual lines in dense sheets could not be easily distinguished with the naked eye (these include the "brushed" sheets of Blackledge et al. 2009c; see section 10.5.2).

Web trait	*Taxa*	*References*
A. Non-orbs		
1. Non-sticky signal lines radiate from edge of tunnel mouth or the spider's resting place on or very near the substrate	*Liphistius* (Liphistiidae)	Bristowe 1930, Murphy and Murphy 2000 (fig. 10.5)
	Hexathele (Hexathelidae), *Atrax* (Atracidae)	Coyle 1986
	Cteniza, others (Ctenizidae)	Coyle 1986
	Missulena (Actinopodidae)	Coyle 1986
	Ariadna spp. (Dysderidae)	Comstock 1967, Griswold et al. 2005
	Hersilia (Hersiliidae)	Williams 1928 (fig. 9.1)
	Modisimus culicinus (Pholcidae)	F. Cargnelutti pers. comm. (fig. 9.1)
	Uroctea durandi (Oecobiidae)	Kullmann and Stern 1981 (fig. 10.5)
2. Rigid objects (leaves, twigs) are attached to and radiate from the edge of tunnel mouth, and serve as signal lines	*Atypoides*, *Aliatypus* (Antrodiaetidae)	Coyle 1986
	Psalistops (Barychelidae)	Coyle 1986
	Many Ctenizidae	Coyle 1986
3. Silken tubes or chambers above ground are used to extend sense of touch	*Atypus*, *Sphodros* (Atypidae)	Comstock 1967, WE (fig. 10.28)
	Dyarcyops (Ctenizidae)	Coyle 1986
	Microhexura[1] (Dipluridae)	Coyle 1981a, 1981b
	Eresus niger[2] (Eresidae)	Nielsen 1932
	Paraplectanoides crassipes (Araneidae)	Hickman 1975 (fig. 10.28)
4. Round, cup-like silk structure (hub) attached to substrate, from which spider monitors web (orbs and non-orbs)	*Thaumasia* sp. (Pisauridae)	Eberhard 2007a
	Herennia, *Clitaetra* NE	Kuntner et al. 2008a,b
5. Few, long, non-sticky lines	*Ariamnes attenuatus*, *A. cylindrogaster*, *A.* sp. THD[3]	Eberhard 1979a, Shinkai and Shinkai 1981, Bradoo 1971
	Thwaitesia sp. THD	Agnarsson 2004
6. Web reduced to few strands; rests on lines and senses chemical cues to find social insects (ants or temites) or their trails	*Steatoda fulva* THD	Hölldobler 1970, 1979
	Dipoena banksii[4] THD	Gastreich 1999
	Chrosiothes tonala[3] THD	Eberhard 1991
	Euryopsis funebris THD	Carico 1978
7. Single sticky horizontal line	*Phoroncidia* spp. THD	Marples 1955, Eberhard 1981b, Shinkai and Takano 1984, Kariko 2014
	Miagrammopes spp.[5] UL	Akerman 1932, Lubin et al. 1978, Lubin 1986 (fig. 3.1)

Table 9.1. Continued

Web trait	*Taxa*	*References*
8. Spider reels in line and holds it tense with legs IV, then releases it when prey strikes this line, causing it to sag	*Miagrammopes* spp. UL	Akerman 1932, Lubin et al. 1978
	Phoroncidia spp. THD	Marples 1955, Eberhard 1981b
9. Distal ends of gumfoot lines split	*Neottiura* sp. THD	Eberhard et al. 2008b
	Nesticus sp., *Eidmanella pallida* (Nesticidae)	Coddington 1986b, Griswold et al. 1998
10. Star gumfoot webs	*Chrysso spiniventris* THD	Benjamin and Zschokke 2003
	Achaearanea spp. THD, *Theridion* sp. THD	Eberhard et al. 2008b, Agnarsson 2004 (fig. 9.7)
11. Tangle with gumfoot lines[6]	Many genera THD, Nesticidae	summary Eberhard et al. 2008b
	Physocyclus globosus (Pholcidae)	Japyassú and Macagnan 2004, Escalante and Masís-Calvo 2014
12. Asterisk web (short, non-sticky lines radiate directly to substrate from central, aerial point)	*Ocrepeira ectypa* AR[7]	Stowe 1978 (fig. 9.2)
	Comaroma simonii AN[6]	Kropf 1990
	Unidentified mysmenid	M. Ramirez and P. Michalik pers. comm. (fig. 9.2)
13. Sticky silk in aerial tangle	*Anelosimus pacificus* THD	Eberhard et al. 2008b
	Nihonhimea tesselata THD	Barrantes and Weng 2006b
	Tidarren sisyphoides THD	Madrigal-Brenes and Barrantes 2009
	Linyphia triangularis, *Microlinyphia pusilla* LIN	Benjamin et al. 2002
	Diplothyron simplicatus LIN	WE
14. Sheet with cribellate silk applied nearly directly to substrate (supporting non-sticky lines only barely above the substrate)	*Oecobius* spp.[8] (Oecobiidae)	Solano-Brenes et al. 2018,
	gen. nov. (Filistatidae)	G. Barrantes pers. comm. (fig. 9.6)
	Mallos hesperius, *Emblyna* sp. (Dictynidae)	WE
15. Horizontal "unit" dense sheet near substrate with vestigial tangle or none at all	*Labulla thoracica* LIN	Nielsen 1932, Hormiga and Eberhard in prep. (fig. 1.2)
	Neoantistea agilis, *N. riparia* (Hahniidae)[9]	Comstock 1967, Opell and Beatty 1976, Eberhard 2018b
16. Horizontal dense sheet that lies on or barely above the substrate (some split into many sub-sheets), no tangle above	*Eperigone tridentata*, *Sphecozone* sp.[10] LIN	Hormiga and Eberhard in prep.
	Tapinopa longidens[11] LIN	Nielsen 1932, Shinkai 1979, Hormiga and Eberhard in prep.
	Linothele macrothelifera (Dipluridae)[12]	Eberhard and Hazzi 2013 (fig. 10.6)
	Ischnothele spp.[12] (Dipluridae)	Coyle 1986, 1995
17. Aerial dense horiz. sheet near or above substrate[13], with tubular retreat (funnel)	*Linothele*[12] spp., *Diplura* spp. (Dipluridae)	Paz 1988, Viera et al. 2007, Nielsen 1932, Coyle 1986
	Agelenopsis, *Agelena naevia*, *Melpomene* sp., *Tegenaria*[12] (Agelenidae)	Emerton 1902, Riechert and Gillespie 1986, Rojas 2011, McCook 1889, Job 1974
	Aulonia albimanus, *Sosippus* spp., and *Aglaoctenus* spp. (Lycosidae)[12,13]	Brady 1962, Santos and Brescovit 2001, Viera et al. 2007, Eberhard and Hazzi 2017 WE (fig. 5.5)
	Tengella radiata (Zoropsidae)[12]	Eberhard et al. 1993 (fig. 10.7)

Table 9.1. Continued

Web trait	*Taxa*	*References*
	Architis spp. [14] (Pisauridae)	Nentwig 1985c, Davies 1982, Viera et al. 2007 (fig. 1.3)
	Dendrolycosa[15] (Pisauridae)	Doran et al. 2001
	Hickmania troglodytes (Austrochilidae)	M. Ramirez and P. Michalik pers. comm., fig. 10.8
18. Aerial dense, horizontal, flat sheet near or just above substrate, no tubular retreat	*Labulla contortipes*, *Floronia bucculenta*, several genera LIN	Shinkai and Takano 1984 Nielsen 1932, Comstock 1967, Hormiga and Eberhard in prep.,
	Several genera (Pimoidae)[10]	Griswold et al. 1998, Hormiga and Eberhard in prep.
	Calymmaria cavicola[12,16] (Agelenidae)	Shear 1986, pers. comm.
	Scytodes intricata, *S.* sp. (Scytodidae)[17]	Eberhard 1986c, Li et al. 1999
	Tapinillus longipes (Oxyopidae)	Griswold 1983
	Hygropoda (?) sp. (Pisauridae)	WE (fig. 5.14)
19. Domed horizontal sheet with tangle above	*Modisimus*, *Mesabolivar*, *Bolivar*, *Spermaphora* (Pholcidae)	Eberhard 1992b, Eberhard and Briceño 1983, Huber and Kwapong 2013, WE (figs. 1.3, 5.2, 5.21, 8.7)
	Pahoroides whangarei (Synotaxidae)	Griswold et al. 1998
	Neriene (= *Prolinyphia*) *longipedella*, *yunohamensis*, many other genera LIN	Shinkai 1979, Shinkai and Takano 1984, Nielsen 1932, Emerton 1902, Kaston 1948, Homiga and Eberhard in prep.,
	Ochryocera cachote, *O.* sp. (Ochyroceratidae)	Hormiga et al. 2007, G. Barrantes and W. Eberhard unpub. (fig. 2.13)
	several genera THD	Eberhard et al. 2008b
	Tekella unisetosa (Cyatholipidae)	Griswold et al. 1998
20. Tangle with two horizontal sheets (spider under denser upper sheet)	*Linyphia communis* LIN	Comstock 1967, Nielsen 1932
	Tidarren sisyphoides THD	Madrigal-Brenes and Barrantes 2009 (fig. 9.15)
	Steatoda sp. THD	Eberhard et al. 2008b (fig. 9.14)
21. Aerial, sparse to somewhat dense sheet far above the substrate, sheet with downward "dimples" and upward-slanting edges, tangle above and below	*Nihonhimea tesselata*, *Cryptachaea* sp. nr. *porteri* THD[10]	Eberhard 1972b, Jörger and Eberhard 2006 (fig. 9.16)
	Theridion japonicum[18] THD	Shinkai and Takano 1984
	Chrosiothes sp.[10] THD	Eberhard et al. 2008b (fig. 9.16)
	Steatoda spp.[10] THD	Viera et al. 2007, Eberhard et al. 2008b
	Diguetia spp.[10,19] (Diguetidae)	Cazier and Mortenson 1962, Eberhard 1969, Bentzien 1973
	Cyrtophora. *Manogea*, *Mecynogea* AR[10,20]	Wiehle 1927, Kullmann 1958, Lubin 1973, Shinkai 1979, WE (fig. 1.7)
22. Aerial "sparse" horizontal sheet with no dimples	*Psechrus* spp.[10] (Psechridae)	Dippenaar-Schoeman and Jocqué 1997, Eberhard 1987a, Griswold et al. 2005 (fig. 5.20)
	Tapinillus longipes (Oxyopidae)	Griswold 1983
	Diaea sp. (Thomisidae) (verbal descr., no photo)	Jackson et al. 1995

Table 9.1. Continued

Web trait	*Taxa*	*References*
23. More or less flat, usually only moderately dense, horizontal sheet with sticky lines in the sheet	*Thaida peculiaris* (Austrochilidae)	Lopardo et al. 2004
	Psechrus spp. (Psechridae)	Dippenaar-Schoeman and Jocqué 1997, Eberhard 1987a, Griswold et al. 2005 (fig. 5.20)
	Chrysso sp. nr. *nigriceps*, sp. nr. *cambridgei* THD	Eberhard et al. 2008b
	Tengella radiata (Zoropsidae)	Eberhard and Hazzi 2017 (fig. 10.7)
	Dubiaranea lubugris LIN	Hormiga and Eberhard in prep.
	Metaltella simoni (Amphinectidae)	I. Escalante pers. comm.
	Hickmania trodglodytes (Austrochilidae)	Doran et al. 2001, M. Ramirez and P. Michalik pers. comm.
24. Flimsy sheet with sparse sticky lines bearing small, widely spaced droplets of glue	*Synotaxus* spp. (Synotaxidae)	Eberhard 1977a, 1995, Eberhard et al. 2008b (figs. 5.3, 9.19)
	Belisana (Pholcidae)	Deeleman-Reinhold 1986, Huber 2005 (fig. 9.19)
	Theridion hispidum, *T.* sp. nr. *melanosticum* THD	Eberhard et al. 2008b
	Achaearanea sp. THD	G. Hormiga unpub. (fig. 9.12)
25. Cupped sheet at bottom of tangle	*Anelosimus* spp. THD	Eberhard et al. 2008b
	Tidarren sisyphus THD	Eberhard et al. 2008b
	Achaearanea wau THD	Lubin 1982
26. Planar, open-meshed sheet to which vertical lines with adhesive are attached	*Latrodectus* spp. THD	Eberhard et al. 2008b
	Achaearanea nr. *tepidariorum* THD	Eberhard et al. 2008b
27. Sheet web in which reduced reliance on sticky lines for prey capture is apparently derived	*Tengella radiata* (juveniles) (Zoropsidae)	Barrantes and Madrigal-Brenes 2008
	Neolana pallida (Neolanidae)	Opell 1999a, 2013
28. Web in an indentation or cavity in substrate and sheltered from outside by a strong silk wall	*Achaearanea rostrata* THD	Eberhard et al. 2008b
	Theridion melanurum THD	Nielsen 1932
29. Non-adhesive "space web" tangle built in sheltered site (e.g., under rocks)	*Plectreurys* sp. (Plectreuridae)	WE
	Maniho sp. (Amphinectidae)	Griswold et al. 2005
30. Radially organized non-sticky lines that converge on retreat and bear zig-zag sticky lines	*Badumna longiqua* (Desidae)	Griswold et al. 2005
	Dictyna sublata (Dictynidae)	Comstock 1967
31. Lines bear tiny sticky droplets (difficult and sometimes impossible to see with the naked eye)[21]	*Modisimus guatuso* (Pholcidae)	Briceño 1985
	Synotaxus spp (Synotaxidae)	Eberhard et al. 2008b (figs. 2.8, 5.3, 9.12)
	Linyphia triangularis, *Microlinyphia pusilla*, *Diplothyron* sp., *Neriene coosa* LIN	Benjamin et al. 2002, WE
	Nihonhimea tesselata THD	Barrantes and Weng 2006b
	Anelosimus sp THD	Eberhard et al. 2008b
	Theridion spp. THD	Eberhard et al. 2008b

Table 9.1. Continued

Web trait	*Taxa*	*References*
32. Sticky lines in aerial tangle	*Argyrodes antipodiana* THD	Whitehouse 1986
	Theridion spp. THD	Eberhard et al. 2008b
	Theridula sp. THD	Eberhard et al. 2008b
	Achaearanea spp. THD	Eberhard et al. 2008b
	Chrysso spp. THD	Eberhard et al. 2008b
	Philoponella oweni UL	WE (fig. 1.6)[22]
33. Sticky lines in wide-meshed, more or less planar array	*Synotaxus* spp. (Synotaxidae)	Eberhard 1977a, 1995
	Chrysso spp. THD	Eberhard et al. 2008b
	Latrodectus spp. THD	Eberhard et al. 2008b
	Theridion spp. THD	Eberhard et al. 2008b
	Helvibis longicauda THD	Viera et al. 2007
34. Long vertical lines sticky along nearly entire length below small non-sticky tangle	*Chrysso ecuadorensis* THD	Eberhard et al. 2008b
	Gaucelmus calidus (Nesticidae)	Eberhard et al. 2008b
35. Sticky lines across depression in a large leaf	*Chrysso vallensis*[23] THD	Eberhard et al. 2008b
	Neottiura sp.[23] THD	Eberhard et al. 2008b
36. Tangle of sticky lines with no obvious pattern on and close to the substrate surrounding retreat	*Loxosceles* spp. (Sicariidae)	Viera et al. 2007
	Vidole capensis (Phyxelididae)	Griswold et al. 2005
	Callobius sp., *Amaurobius* (Amaurobiidae)	Griswold et al. 2005
	Neoramia sana (Agelenidae)	Griswold et al. 2005
37. Abandon webs, prey on other spiders (see also #38 in the second section of this table)	*Enoplognatha ovata* THD	Bristowe 1958
	Theridion tinctum, *T. sterninotatum*, *T. adamsoni* THD	Jones 1983, Shinkai 1988d
	Rhomphaea spp. THD, *Faiditus* spp. THD	Vollrath 1979, Whitehouse 1987, WE
	Thwaitesia spp. THD	Agnarsson 2004
	Menosira ornata TET[22]	Shinkai 1998b
38. Abandon webs, steal silk from other spiders	*Argyrodes cylindratus* THD	Shinkai 1988c
	Leucauge mariana (mature males) TET	WE
39. Abandon webs, often to steal prey from other spiders (see also #38 in second section of this table)	*Enoplognatha ovata*, *Argyrodes* spp., *Neospintharus*, *Faiditus* spp. THD	Bristowe 1958, Kullmann 1959, Vollrath 1979, Agnarsson 2004, Eberhard et al. 2008b
	Parasteatoda tepidariorum THD	Kullmann 1959
	Neoscona neotheis AR (facultative)	Armas and Alayón 1987
	Archaeodictyna ulova (Dictynidae)	Griswold and Meikle-Griswold 1987
	Oedothorax spp. LIN	Alderweireldt 1994
40. Approximately radial lines converge at an aerial retreat	*Nihonhimea tesselata* THD	Jörger and Eberhard 2006
	Achaearanea apex THD	Eberhard et al. 2008b
	Latrodectus pallidus THD	Szlep 1965
	Neottiura sp. THD	Eberhard et al. 2008b
41. Build mats of sand grains (to which to attach lines) by stabbing with long, extensible AL spinnerets that are crowned by long, divergent piriform spigots	*Seothyra henscheli* (Eresidae)	Peters 1992 (figs. 9.30, 9.31)
	Leucorchestris arenicola (Heteropodidae)	Peters 1992

Table 9.1. Continued

Web trait	*Taxa*	*References*
42. Stiff silk (?) "poles" to elevate lines above the substrate	*Ariadna bicolor*[24] (Dysderidae); *Uroctea durandi*[24] (Oecobiidae)	Comstock 1967; Kullmann and Stern 1981 (fig. 10.5)
43. Small web that contracts after being applied to the prey	*Deinopis* sp. (Deinopidae)	Coddington and Sobrevila 1987 (fig. 3.1)
	Scytodes thoracica (Scytodidae)	Stratton and Suter 2009
44. "Lampshade" web built against underside of planar surface	*Crossopriza cylindrogaster* (Pholcidae)	Huber 2009
	Hypochilus gertschi (Hypochilidae)	Shear 1969
45. Domed sheet immediately below horizontal planar surface (leaf)	*Modisimus bribri* (Pholcidae)	WE (fig. 8.7)
	Spermaphora akwamu (Pholcidae)	Huber and Kwapong 2013
B. Orbs		
1. Tangle of non-sticky lines built alongside a roughly vertical orb	*Nephila* spp., *Nephilingis* spp. NE	Wiehle 1931, Peters 1954, Robinson and Robinson 1973*a*, 1973*b*, Shinkai 1985, Kuntner et al. 2008a,
Spider rests at hub	*Allocyclosa bifurca* AR	Eberhard 2003, WE
	Argiope spp. (but not *A. flavipalpis*) AR	Wiehle 1928, Kaston 1948, Peters 1955, Edmunds and Edmunds 1986
Spider rests in retreat (in tangle, curled leaf, or off the web)	*Arachnura* spp. AR	McKeown 1952, Robinson and Lubin 1979a, WE
	Isoxya penizoides AR	Edmunds and Edmunds 1986
	Metepeira spp. AR	Comstock 1967, Peters 1955, Levi 1977, Viera et al. 2007, Eberhard 2014, WE (fig. 3.7)
	Araneus omnicolor, *A. expletus*, *Araneus* (= *Epeira*) *pegnius* AR	Gonzaga and Sobczak 2007, Kaston 1948, WE
	Phonognatha spp. AR	McKeown 1952, Hormiga et al. 1995, Fahey and Elgar 1997, Kuntner et al. 2008a
2. Tangle of non-sticky lines below horizontal orb	*Leucauge* spp. TET	Comstock 1967, Maroto 1981, Triana-Cambronero et al. 2011, WE (fig. 10.3)
	Philoponella spp. UL	WE
	Edricus sp. AR	WE
	Uloborus barbipes, *U.* spp UL[25]	Lubin 1986, Lubin et al. 1982
	Tangaroa sp. UL	Davies unpub. cited in Lubin et al. 1982
3. Tangle with some radially arranged lines above horizontal orb, pulls the orb upward at the hub to form a cone	*Spilasma* spp.[26] AR	Coddington 1986*b*, Eberhard 1986a, WE
	Paraneus cyrtoscapus, *P. spectator* AR	Edmunds 1978
	Leucauge argyra (imm.) TET	Triana-Cambronero et al. 2011
	Several genera AN, SYM, MYS	Lopardo et al. 2011 (figs. 1.14, 3.12, 7.9)
	Philoponella oweni, *P. arizonica*, *P. sempiplumosa*, *P. variegatus*[27] UL	Peters 1955, Smith 1997, Lahmann and Eberhard 1979, WE

Table 9.1. Continued

Web trait	*Taxa*	*References*
4. "Orb plus cone"—more or less conical, orb-like tangle of non-sticky lines on one side of or below the orb where spider rests (radial lines converge)	*Nephila clavipes, N. pilipes* (immatures) NE	Kuntner et al. 2008a, Gonzaga et al. 2010, Robinson and Robinson 1973*a*, Murphy and Murphy 2000 (fig. 10.32)
	Conifaber spp. UL	Lubin et al. 1982, Lubin 1986, Grismado 2008
	Uloborus spp. UL	Lubin 1986 (fig. 10.37)
	Allocyclosa bifurca AR	Eberhard 2003, WE
	Argiope argentata, A. trifasciata AR	Peters 1955, WE
5. One sector lacks sticky spiral around a radius that carries egg sacs or detritus	*Arachnura feredayi, A. higginsi* AR	Forster and Forster 1973, McKeown 1952
	Allocyclosa bifurca AR	WE
6. Open sector with signal line to where the spider rests in tangle or retreat off the orb	*Parawixia kochi* AR (= *Turckheimia morabelli*)	Hingston 1932
	"*Meta*" sp. TET	Hingston 1932
	Chrysometa spp. TET	Levi 1986, Eberhard 1986a, WE
	Dolichognatha sp. TET	WE
	Zygiella (s.l.) x-notata, Z. atrica, Z. keyserlingi, montana, thorelli, stroemi, AR	Wiehle 1927, Kaston 1948, Shinkai and Takano 1984, Shinkai 1989, Gregorič et al. 2015
	Araneus legonensis, A. fuscocolorata, A. mitificus, A. pentagrammicus, A. expletus[28] AR	Grasshoff and Edmunds 1979, Shinkai 1979, Shinkai and Takano 1984, H. W. Levi, pers. comm. cited in Grasshoff and Edmunds 1979, Gregorič et al. 2015, WE
	Aculepeira escazu (?) AR	WE (fig. 3.24)
	Guizygiella sp. AR	Gregorič et al. 2015
	Metazygia dubia (= *Epeira morabelli*), *M.* sp. AR	Hingston 1932, WE
	Lubinella morobensis UL	Lubin 1986
	Metepeira spp. AR	Comstock 1967, WE
	Eustala illicita, E. sp. AR	Hesselberg and Triana 2010
	Meta menardi AR (sometimes)	Yoshida and Shinkai 1993
	Philoponella tingena, P. undulata UL	Lubin 1986
7. Signal line out of plane of intact orb (no free sector) to resting site	*Metepeira* sp. AR	WE
	Araneus rufipalpis, A. expletus, A. diadematus (occasional), *A. pegnius* AR	Bristowe 1958, Kaston 1948, Grasshoff and Edmunds 1979, WE
	Enacrosoma anomalum AR	WE
	Dolichognatha spp. TET	WE
8. Multiple radii attached directly to the substrate without a frame line	*Tetragnatha* (= *Eucta*) *kaestneri, T. lauta, T.* sp. TET	Crome 1954, Shinkai 1988b (fig. 9.10)
	Cyrtognatha TET	Eberhard 1986a, WE
	Paraneus cyrtoscapus AR[29]	Edmunds 1978
	Anapisona spp. AN	Platnick and Shadab 1979, Eberhard 2007b (fig. 3.12)

Table 9.1. Continued

Web trait	*Taxa*	*References*
9. Sticky lines attached to water surface	*Wendilgarda* spp. TSM	Coddington and Valerio 1980, Coddington 1986a, Eberhard 1989a, 2000a (fig. 10.29)
	Conoculus lyugadinus AN	Shinkai and Shinkai 1988
10. Short, non-sticky radial lines direct to substrate, no sticky lines ("asterisk" webs)[30]	*Menosira ornata* TET	Shinkai 1998b
	Ocrepeira (= *Wixia*) *ectypa* AR	Stowe 1978 (fig. 9.2)
	unidentified MYS	M. Ramírez and P. Michalik, pers. comm. (fig. 9.2)
11. Orb spans single leaf with edges that curled upward or downward[31]	*Cyclosa jose*[32] AR	WE (fig. 3.34)
	Araneus displicatus, A. niveus[33] AR	Comstock 1967, Coddington 1987
12. Twig runs through hub and spider rests under it	*Tetragnatha* spp. TET	Marples 1955b, Eberhard 1986a, WE (fig. 10.24)
	Uloborus eberhardi UL	WE (fig. 10.25)
	Polenecia producta UL	Peters 1995a (fig. 3.44)
	Poltys noblei AR	Smith 2006
13. Ladder webs (orbs extremely elongate vertically)	*Tylorida* sp. TET	Robinson and Robinson 1972
In open space	*Scoloderus tuberculifer, S. cordatus* AR	Eberhard 1975, Stowe 1978 (fig. 3.4)
	Cryptaranea (= *Araneus*) *atrihastula* AR	Forster and Forster 1985
Against a tree trunk	*Herennia* spp., *Clitaetra* spp. NE	Robinson and Lubin 1979a, Kuntner et al. 2008a,b) (fig. 4.3)
	Telaprocera maudae, T. joanae AR	Harmer and Herberstein 2009, Harmer and Framenau 2008
	Eustala perfida, E. conformans, E. sp. AR	Messas et al. 2014, WE (fig. 4.3)
14. Stabilimentum at or near hub or other resting place	*Eriophora sagana* AR	Nakata 2008
	Araneus expletus[35], *A.* spp. AR	Eberhard 2008, Levi 2001 (fig. 3.49)
Silk (section 3.3.4.2)[34]	*Metepeira* spp. AR	Piel 2001
	Molinaranea AR	Levi 2001
	Allocyclosa bifurca, Cyclosa spp. AR	Hingston 1927, 1932, Marson 1947b, Eberhard 2003
	Polenecia producta UL	Peters 1995a
	8 other independent derivations	Herberstein et al. 2000b
Detritus (section 3.3.4.1)	*Landana* sp., *Dolochignatha*[35] TET	Hingston 1932, WE (fig. 3.43)
	Leucauge mariana (rare) TET[36]	WE
	Allocyclosa bifurca AR	Eberhard 2003
	Arachnura AR	WE
	Uloborus scutifaciens UL	Hingston 1927, 1932
15. Multiple objects (detritus) dangle from radii or frame in horizontal web; spider when disturbed hangs on or amongst them	*Uloborus* spp. UL	Lubin 1986, G. Hormiga pers. comm. (fig. 3.42)
	Enacrosoma anomalum, E. sp. AR[37]	G. Hormiga pers. comm., WE (fig. 3.42)
	Dolichognatha sp. TET	Eberhard 1986a, WE (fig. 4.35)
	Crytophora citricola (occasionally) AR	WE
16. "Disc" stabilimentum at center of hub	*Zosis geniculata* UL	WE (fig. 3.46)

Table 9.1. Continued

Web trait	*Taxa*	*References*
	Argiope spp. AR	Hingston 1927, 1932, Ewer 1972, Robinson and Robinson 1974, Herberstein et al. 2000b, Shinkai and Takano 1984, Edmunds 1986 (figs. 3.38, 3.49)
	Allocyclosa bifurca AR	Eberhard 2003
17. Arms of stabilimentum formed by zig-zag swaths of white (probably aciniform) silk	*Argiope* spp. AR	Hingston 1927, 1932, Marson 1947a, Robinson and Robinson 1974, Herberstein et al. 2000b, Shinkai and Takano 1984, Edmunds 1986 (fig. 3.45, 3.47)
	Allocyclosa bifurca AR	Eberhard 2003
18. Yellow silk in orb	*Nephila* spp.[38] NEP	Craig 2003, Austin and Anderson 1978, Kuntner et al. 2008a
	Araneus expletus[39] AR	WE
	Araneus legonensis AR	Grasshoff and Edmunds 1979
	Araneus mitificus AR	Shinkai and Takano 1984
	Aspidolasius branicki AR	Calixto and Levi 2006
	Cyclosa spp. AR	Levi 1999, WE
	Micrathena schreibersi AR	WE
	Neoscona sp. AR	Kuntner et al. 2008a
	Poltys illepidus AR[40]	Smith 2006
19. Move web (throw, sag, trawl, etc.) when sense prey	*Mastophora* spp. AR (sling)	Eberhard 1980a, Yeargan 1994 (fig. 3.1)
	Deinopis, *Menneus* DEI (spread web, lunge)	Akerman 1926, McKeown 1952, Austin and Blest 1979 (fig. 3.1)
	several genera of TSM (spring orb)	McCook 1889, Coddington 1986a, WE
	Hyptiotes, *Miagrammopes* UL (spring)	Comstock 1967, Marples and Marples 1937, Marples 1955, Akerman 1932, McKeown 1952, Shinkai and Takano 1984 (figs. 3.1, 10.23)
	Phoroncidia spp. THD (sag)	Eberhard 1981b
Spontaneous	*Cladomelea*, *Ordgarius*, *Dichrostichus* AR (twirl)	McKeown 1952
	Wendilgarda sp. TSM (trawl)	Eberhard 2001a
20. Tense hub lines while wait at hub, and release tension when a prey hits the web (immediately)[41]	*Micrathena* spp., *Gasteracantha* sp. AR	WE (fig. 3.37), Peters 1955
	Wagneriana sp. AR	WE
	Pozonia nigroventris AR	G. Barrantes, pers. comm. (fig. 3.35)
21. Tense orb (or orb sector) with a spring line and abruptly release this tension when prey strikes web[42]	*Hyptiotes* spp. UL	McCook 1889, Wiehle 1927, Marples and Marples 1937
	Theridiosoma and related genera TSM	Coddington 1986a (summary)

Table 9.1. Continued

Web trait	*Taxa*	*References*
22. Retain temporary spiral in finished orb	*Nephila, Herrenia,* relatives NE	Kuntner et al. 2008a (summary)
	Phonognatha spp. AR	McKeown 1952, Hormiga et al. 1995, Kuntner et al. 2008a
	Scoloderus tuberculifer, S. nigriceps AR	Eberhard 1975, Stowe 1978
	Tylorida sp. TET	Robinson and Robinson 1972
	Cyrtophora, Mecynogea, Kapogea, Manogea AR	Wiehle 1927, Kullmann 1958, Levi 1980, WE
	Uloborus, Zosis (erratic)[43] UL	Eberhard 1972a, WE
	Polenecia (= *Sybota*) *producta* UL	Wiehle 1931, Peters 1995a
23. Omit temporary spiral	*Tetragnatha* (= *Eucta*) *kaestneri* TET	Crome 1954
	Anapisona simoni AN	Eberhard 1982, 1987e
	Mysmena sp. MYS	Eberhard 1982
24. Zig-zag attachments of the hub spiral or temporary spiral to radii[44]	*Uloborus* spp., *Philoponella* spp.[45,46] UL	Lubin 1986
	Tetragnatha sp.[46] TET	WE
	Nephila spp.[47] NE	Wiehle 1931, Peters 1955, Kuntner et al. 2008a
	Leucauge mariana[46] (occasionally) TET	WE
	Phonognatha sp.[47] (occasionally) AR	Hormiga et al. 1995, Kuntner et al. 2008a
	Cyrtophora, Mecynogea, Manogea[47] AR	Wiehle 1927, Kullmann 1958, 1975, WE
25. Sticky spiral frequently laid for a short distance along each radius, forming a zig-zag pattern	*Uloborus* spp., *Philoponella* spp. UL[48]	Eberhard 1972a, Lubin et al. 1978, Lubin 1986
	Miagrammopes sp.[49] UL	WE, Wiehle 1931, Peters 1995a
	Polenecia (= *Sybota*) *producta* UL	Eberhard 1986a, Shinkai and Takano 1984
	Cyrtarachne spp., *Pasilobus* sp. AR	Robinson and Robinson 1975,
26. Very high number of sticky spiral loops that are very closely spaced, relatively low numbers of radii; nocturnal web (for moths)	*Acacesia hemata* AR (fig. 3.5)	Eberhard 1976a, Carico 1986
	Deliochus sp. AR	Kuntner et al. 2008a (fig. 3.5)
	Poltys spp. AR[50]	Smith 2006, WE
	Pozonia nigroventris AR	G. Barrantes unpub. (fig. 3.35)
27. Spider faces upward rather than downward while sitting at hub	*Anepsia rhomboides* AR	Marples 1955
	Verrucosa spp. AR	Levi 1976, WE
	Cyclosa octotuberculata AR	Nakata and Zschokke 2010, WE
28. Nearly perfectly vertical webs[51] (low variance in slant) (see table O3.1)	*Cyclosa* (some species) AR	Nakata and Zschokke 2010
	Allocyclosa bifurca AR	WE
	Aspidolasius branicki AR	Calixto and Levi 2006
	Wagneriana spp. AR	WE
	Tylorida sp?[52] TET	Robinson and Robinson 1972
	Scoloderus tuberculifer, S. nigriceps AR	Eberhard 1975, Stowe 1978
	Eriophora spp. AR	WE

Table 9.1. Continued

Web trait	*Taxa*	*References*
29. Orb nearly often close to perfectly horizontal (see table O3.1)	*Uloborus* sp. UL	WE
	Dolichognatha sp. TET	WE
	Metabus ocellatus TET	Buskirk 1975, WE
	Gibbaranea (= *Araneus*) *abcissus* AR	Shinkai 1989
	Neoscona nautica AR[53]	Edmunds 1978
	Azilia affinis AR	WE
	Enacrosoma sp. AR	WE
	Cyrtarachne spp. AR	Cartan and Miyashita 2000, Miyashita et al. 2001
	Paraneus cyrtoscapus (imm.; seldom adults) AR	Edmunds 1978
	Several genera AN, SYM, MYS	Lopardo et al. 2011, Ramírez et al. 2004, Shinkai and Shinkai 1988, WE
30. Typical prey capture orbs built by mature males	Several genera AN	Eberhard 2007b, Ramírez et al. 2004, Lopardo et al. 2011 (fig. 9.4f)
	Wendilgarda sp. TSM	Eberhard 2001a
	Cyrtophora, *Mecynogea* AR	Blanke 1972, J. Vasconcellos-Neto, pers. comm., WE
31. Hub-like resting web built by mature male (non-sticky lines only—hub, radii, frames)	*Naatlo splendida* TSM	WE (fig. 9.4d)
	Epeirotypus brevipes TSM	WE (fig. 9.4e)
	Allocyclosa bifurca AR	Eberhard 2003, WE (fig. 9.4a)
	Zosis geniculata UL	WE (fig. 3.48)
	Argiope argentata, *A. trifasciata* AR	Kaston 1948, WE (fig. 9.4b)
	Nephila clavipes NE[54]	WE
32. Bend radii to fill over-sized spaces between radii	*Uloborus diversus*, *U.* sp. UL	Eberhard 1972a, 1986 (fig. 3.31)
	Scoloderus tuberculifer, *S. nigriceps* AR	Eberhard 1975, Stowe 1978 (fig. 3.4)
	Eustala sp. AR	Eberhard 1975
	Metepeira sp. AR	Eberhard 2014
	Nephila spp NE	Eberhard 2014
33. High symmetry of sticky spiral lines (usually no turnbacks)	*Micrathena duodecimspinosa* AR	WE
	Hypognatha sp. AR	WE
34. "Egg sac" orbs[55]	*Miagrammopes* nr. *unipus* UL	Lubin et al. 1978
	Philoponella sp. UL	WE
	Uloborus diversus UL	WE (fig. 10.16)
	Zosis geniculata UL	WE (fig. 10.35)
35. Web reduced to one or a few long sticky lines	*Miagrammopes* spp. UL[56]	Akerman 1932, Lubin et al. 1978, Shinkai and Shinkai 1981
	Wendilgarda galapagensis TSM	Eberhard 2001a
	Phoroncidia spp. THD[56,57]	Marples 1955, Shinkai and Shinkai 1981, Shinkai 1988a, Eberhard 1981b
36. Reduction or loss of orb associated with predation on moths	*Kaira*, *Pycnacantha tribulus*[58] AR	Stowe 1978, Dippenaar-Schoeman and Leroy 1996
	Pasilobus, *Cyrtarachne* AR	Robinson and Robinson 1975, Miyashita et al. 2001
	Poecilopachys australasia	Clyne 1973
37. Complement web with prey attractant[59]	*Kaira alba* AR	Stowe 1985, 1986
	Mastophora and related genera AR	Stowe 1986 (summary)
	Phoroncidia sp.	Eberhard 1981b

Table 9.1. Continued

Web trait	*Taxa*	*References*
38. Complete loss of orb for prey capture	*Chorizopes nipponicus, C.* sp. (araneophages) AR	Eberhard 1983, Shinkai and Takano 1984
	Neoarchemorus, Archemorus, Oarces AR	Stowe 1986 (summary), Dimitrov et al. 2016
	Arkys sp. (Arkyidae)	Dimitrov et al. 2016
	Taczanowskia sp. AR	Eberhard 1981c
	Tetragnatha spp. TET	Gillespie 1991a
	Tetragnatha squamata[60] TET	Shinkai and Takano 1984
	Tetragnatha (=*Eucta*) *kaestneri*[61] TET	Crome 1954
	Doryonychus raptor TET	Gillespie 1991b
	Mysmenopsis spp., *Isela* sp. MYS	Eberhard et al. 1993, Lopardo et al. 2011
	Curimagua bayano SYM	Lopardo et al. 2011
	Sofanapis antillanca AN	Lopardo et al. 2011
	Malkara sp. (Malkaridae)	Dimitrov et al. 2016
	Mimetidae (largely araneophagic)	Dimitrov et al. 2016
	Holarchaeidae (now AN)	Dimitrov et al. 2016
39. Replace hub during construction with multiple direct connections between radii	*Uloborus conus*[62] UL	Lubin et al. 1982
	Mysmena sp. MYS	Eberhard 1987e
	several genera TSM	McCook 1889, Coddington 1986*a*
40. Sticky spiral line laid only along radii and frame lines	*Eustala* sp. (adults) AR	Eberhard 1985
	Polenecia producta UL	Wiehle 1931, Peters 1995a
41. Build orbs (or stay) in tangle webs of other species	*Philoponella vicina, P. tingena, P.* spp. UL	Fincke 1981, Opell 1979, WE
	Uloborus terokus UL	Bradoo 1989
	Leucauge venusta[63] TET	Zschokke et al. 2006
42. Sticky material limited to central portion of each segment of sticky spiral	*Sybota atlantica* UL	Grismado 2001
	Poecilopachys Australasia, Pasilobus sp., *Cyrtarachne bufo, Paraplectana tsushimensis* AR	Clyne 1973, Robinson and Robinson 1975, Stowe 1986, Shinkai and Takano 1984
43. Reacquire ancestral trait of adding to webs over the space of several days	*Trogloneta granulum*[64] AN	Hajer 2000
	Zosis, Uloborus, Octonoba[64] UL	WE
44. Reacquire ancestral two-dimensional orb derived from three-dimensional orb	*Symphytognatha, Patu* SYM	Lopardo et al. 2011
	Several genera AN	Lopardo et al. 2011
45. Tubular dry leaf suspended vertically as a retreat, opens just above the hub	*Acusilas coccineus, A.* sp. AR	Shinkai and Takano 1984, Blackledge et al. 2011
	Phonognatha spp. AR	McKeown 1952, Hormiga et al. 1995
	Hingstepeira folisecens AR	Levi 1995
46. Build webs most frequently in the afternoon[65]	*Metabus ocellatus* TET	Buskirk 1975
	Conculus lyugadinus AN	Shinkai and Shinkai 1988
	Wendilgarda galapagensis TSM	Eberhard 1989a
	Caerostris darwini AR	Gregorič et al. 2011
47. Hole in center of hub is large and more or less circular[66]	several genera TET	Wiehle 1927, Yoshida and Shinkai 1993, Shinkai 1989
	Micrathena spp., *Pronous* spp. AR	Comstock 1967, Emerton 1902, Eberhard 1986a, Levi 1985, 1995, Coddington 1986b

Table 9.1. Continued

Web trait	*Taxa*	*References*
	Araneus niveus AR	Coddington 1987
	Neoscona fuscocolorata AR	Shinkai 1989
	Cyrtophora citricola AR	WE
48. Add fine radial (or non-radial) lines after sticky spiral and other lines are in place	*Patu* sp. MYS	Hiramatsu and Shinkai 1993, Lopardo et al. 2011
	Maymena MYS	Eberhard 1987e, Lopardo et al. 2011
	Uloborus spp.[67] UL	Szlep 1961, Eberhard 1977d, Lubin 1986
	Conifaber parvus UL	Lubin et al. 1982
49. Cues from leg oIV guide sticky spiral construction	*Nephila* spp. (and other genera) NE	Eberhard 1982, Kuntner et al. 2008a
	Micrathena spp. AR	Peters 1954
		Eberhard 1982, 2012
50. Only cues from the distance traveled along the radius guide sticky spiral construction	*Hyptiotes*[68] UL	Peters 1954
	Several genera TSM, AN, SYM, MYS	Eberhard 1982

[1] Prey capture was not observed, but web consists of flattened tubes under mats of moss, with only a shallow funnel-shaped opening at the tube edge.

[2] The tough, elongate horizontal cylinder was an extension of the vertical burrow below; it had an oval opening at one end, and expanded into a small sheet.

[3] The widely different uses of these lines in *Ariamnes* and *Chrosiothes* (landing sites for prey, and paths to the webs of host spiders from which prey are stolen, versus monitoring posts from which to sense the odor of termite prey) intimate possibly independent origins.

[4] The concordance between the geographical range of another species of *Dipoena* and its ant prey species (Davidson 2011 on *D. torva*) suggest that other species may also use cues from ant prey to site their webs.

[5] Some lines horizontal, some vertical, some at other angles. In two species with multiple long sticky lines (up to 5), there were also occasional very thin, lax lines of unknown significance (Lubin et al. 1978).

[6] Lines more or less perpendicular to the substrate with sticky material only near the tip. Those of theridiids and pholcids were apparently also fastened to the substrate with special, weak attachment discs, but those of the anapid may not break away readily (Eberhard et al. 2008b).

[7] Secondarily derived from orb web.

[8] Unidentified spiders that Hingston (1920) called Urocteidae (= Oecobiidae) laid silk directly onto the substrate; he did not note whether or not this silk was cribellum silk.

[9] Genus not specified by Comstock 1967; probably it was *Hahnia* or *Neoantistea*.

[10] Spider moved on undersurface of the sheet.

[11] Differed from all other known spider webs in being covered with glistening "slime."

[12] Spider moved on upper surface of the sheet.

[13] Although some were near the substrate, these sheets were "aerial" in generally being attached to the substrate only at their edges; they graded into substrate-bound sheets (previous category).

[14] Orientation of the sheet with respect to horizontal varied widely.

[15] Sheets were generally not horizontal.

[16] The sheet was near the roof of the cave rather than the floor. The tangle of lines was below, and was funnel-shaped; it narrowed at the top, possibly to channel prey upward toward the sheet.

[17] Sheet was very sparse and sometimes not easily distinguished.

[18] Dimples uncertain.

[19] Sheet of *D. canities* was occasionally close to the ground (Cazier and Mortenson 1962).

[20] Sheet was an orb web in which the temporary spiral was tightly spaced and not removed; sticky spirals were absent.

[21] Incidental observations indicated that these droplets provide only very small adhesive forces.

[22] Facultative; built multiple web designs.

[23] Some webs.

[24] Elevated lines were trip lines in both species.

[25] It is uncertain whether the uloborid derivations were independent, but in both genera the lines beside the orb formed a cone, and were built using behavior typical of orb construction.

[26] Cylindrical retreat hung in small tangle above the orb.

[27] The species name in the publication, *vicina*, was probably incorrect (handwritten note on reprint from H. Peters).

Table 9.1. Continued

[28] Sometimes the signal line was in an open sector, sometimes it was out of the plane of the orb and there was no open sector (WE).
[29] Mostly in webs of juveniles built low in grass.
[30] Apparently functioned to signal the presence of prey walking nearby, which the spider attacked without other help from the web.
[31] See Eberhard et al. 2008b on *Achaearanea apex* and *A.* sp. nr. *tepidariorum* for possible convergences in non-orbs.
[32] Leaf edges bent upward.
[33] Leaf edges bent downward.
[34] As far as is known, the convergence involved both the type of silk used (from the aciniform glands) as well as its placement on the web (few data are available on glandular origins, however; and in *P. producta* cribellum silk was also included in the stabilimentum [Peters 1995a]).
[35] Stabilimentum was away from the hub, generally at the edge of the web or beyond, where the spider rested during the day (Eberhard 2008).
[36] Very occasionally the body of a large prey captured previously was fastened near the hub; this may only represent storing the prey until it could be completely consumed, but illustrates a possible first step in the evolution of detritus stabilimenta.
[37] Lubin (1986) mentions a similar web as being built by the tetragnathid *Dolichognatha*; her identification of the spider as an araneid and her description of the site and design of the web suggest she was referring to *Enacrosoma*.
[38] Only sticky spiral and the top (strong) frame line were yellow in *N. edulis* (Austin and Anderson 1978); only sticky spiral lines were tested for color in *N. clavipes* by Craig et al. (1996).
[39] The sticky spiral and also strong frame lines were yellow; wrapping silk, retreat silk, and stabilimentum silk were white (Eberhard 1986a, 2008).
[40] The bridge line, which remained in place during the day, was yellow; the orb was only up at night.
[41] The spider tightened hub lines by holding her legs flexed in *M. duodecimspinosa*, and her first response in video recordings of the impact of prey was to release this tension.
[42] Used legs IV to reel in line in *Hyptiotes*, but legs I in the theridiosomatids.
[43] Uncertain whether temporary spiral segments were left intact by error or by design.
[44] The effect of these attachments was to tighten the radii (see section 4.4.2).
[45] Some but not all attachments; see also figs. 3.31 and 7.42d,e of the uloborids *Uloborus* sp. and *Zosis geniculata*.
[46] Hub spiral but not temporary spiral.
[47] Temporary spiral.
[48] In both orbs and some egg sac webs.
[49] On radial lines around egg sac.
[50] Orb was far from supports. Smith (2006) mentioned bridge lines up to 4 m long, and that "moths are the most frequent prey" and that other insects were also preyed upon in this genus; but she did not give empirical data or references.
[51] Movement on the web is difficult for the spider because her own weight causes her to sag into other portions of the orb, and also damages the orb; this may be a derived trait.
[52] Description was not completely clear on this point.
[53] On different days webs of the same individual sometimes varied from horizontal to vertical.
[54] Also tangle on both sides of the "orb" (with radii, temporary spiral but no sticky spiral) plus a sheet with detritus on the side with the spider, less dense planar shield on the other side in the midst of the tangle.
[55] Webs built only around egg sacs, but in at least *M.* nr. *unipus* the spider captures prey trapped in this web (which was built only at night). It is not certain whether or not the derivations in these uloborids are independent. In at least some (*Zosis geniculata*) cribellate silk was also laid on the external wall of the sac itself (WE), indicating that some portions of egg sac webs function as protection against enemies of the eggs, rather than to capture prey.
[56] In both genera, some species built a sticky line that was nearly horizontal, while others built nearly vertical sticky lines.
[57] Distant ancestors probably built orbs.
[58] Specialization on moths was documented only on the basis of prey accepted in captivity, rather than on prey captured in the field as in other species.
[59] Evidence for prey attraction was generally indirect, based largely on the highly biased array of prey captured. In some cases (e.g., *Phoroncidia* spp.) the data are only fragmentary.
[60] Occasionally.
[61] Mature females both built orbs and hunted without orbs (Crome 1954).
[62] In the cone below the horizontal orb.
[63] Especially in webs of *Nephila clavipes*.
[64] Websites were deep in stable piles of loose rocks, and inside buildings.
[65] Peak in web construction coincided with the greatest abundance of prey just above the stream surface, at least in *Metabus ocellatus*.
[66] There are several uncertainties in classification, including the relative size of the hole: should a relatively small hole, such as that of *Neoscona fuscocolorata* (Shinkai 1989), be included? When draglines across the hole were built subsequently while the spider moved around on her web, they were not taken to indicate a lack of a hole.
[67] Spiderlings newly emerged from the egg sac, also mature males.
[68] Based on the fact that the spaces between sticky spiral loops were considerably larger than the entire length of the spider.

Table 9.2. This sampler illustrates the many convergent derivations of web traits in different genera of Theridiidae and closely related families, and also intra-generic diversity. Derivations tentatively considered independent are separated by ";", and indicated by separate underlining. Intra-generic diversity is emphasized by "some" (incomplete sampling of species in all genera underestimates this diversity) (after Eberhard et al. 2008b, where references are given).

Derived web trait	*Taxa*
Sticky silk lost from tangle/sheet	some *Steatoda*; some *Latrodectus*; some *Chrosiothes*; *Thwaitsia*; *Selkirkiella*; some *Argyrodes*; *Cephalobares*; *Meotipa*; some *Ameridion*; some *Archaearanea*
Lost sticky lines attached to the substrate	Synotaxidae; some *Theridion*; some *Achaearanea*
Lost gumfoot lines	some *Latrodectus*; *Phoroncidia*; some *Chrysso*; *Rugathodes*; some *Theridion*; some *Achaearanea*
Sticky line to substrate lost	some Nesticidae
Star gumfoot web	*Chrysso spiniventris*; some *Achaearanea*; some *Theridion*
Sticky silk in tangle	*Argyrodes antipodiana*; some *Theridion*; some *Nihonhimea*; some *Chrysso*; *Theridula* sp.
Sticky silk in sheet or other more or less planar array of lines	*Synotaxus* spp.; some *Chrysso*; some *Latrodectus*; some *Theridion*
Small isolated balls of sticky material on lines	*Synotaxus* spp.; *Theridion hispidum* and *T.* nr. *melanosticum*
Tightly meshed horizontal sheet with extensive tangle above and below	*Achaearanea tesselata* and *A.* nr. *porteri*; *Chrosiothes portalensis*
Cupped sheet at bottom of a tangle	*Anelosimus* spp.; *Tidarren* sp.; *Achaearanea wau*
Planar, open-meshed sheet to which vertical adhesive lines are attached below	some *Latrodectus*; *Achaearanea* nr. *tepidariorum*
Few, non-sticky lines	*Ariamnes attenuatus*[1]; *Thwaitesia* sp.; *Chrosiothes tonala*[2]
Retreat at or beyond edge of prey capture web	*Theridion sisyphium*; some *Latrodectus*
Curl living leaf to form retreat	some *Chrysso*; *Enoplognatha*; *Latrodectus bishopi*; some *Theridion*
Spider rests against unmodified leaf	*Synotaxus* spp.; *Chrysso* spp.; some *Theridion*
Strong, dense sheet at outer limit of 3-D web that is probably protective	*Theridion melanurum*; *Achaearanea rostrata*
Distal ends of gumfoot lines split	*Neottiura* sp.; *Nesticus* sp., *Eidmanella pallida*

[1] Landing site for prey.

[2] From which spider hunts termite prey.

Table 9.3. This classification of different web types of Japanese spiders by E. Shinkai (1989) is probably the most complete ever proposed. It illustrates, with a still different set of species, the themes in this chapter of high diversity and convergence. Despite the large number of categories, numerous additional web designs that are mentioned in the text are not included. Shinkai used photographs of the webs of the species marked with "*" to illustrate different categories.

1. Orb Web

- I. Closed hub (center removed and replaced)
  - A. Complete orb
    - (1) Normal orb-*Araneus, Neoscona melloteei**, *Araniella, Larinioides, Atea, Zilla, Yaginumia, Singa, Hyposinga sanguinea**, *Larinia, Poltys, Gasteracantha, Argiope, Mangora herbeoides* (horizontal: *Araniella, Araneus viperifer, A. abcissus**, *Neoscona adiantum*)
    - (2) Orb with retreat line—*Araneus macacus, A ishisawai, A. ventricosus, A. diadematus, A. marmoreus, A. pinguis, A. variegatus**, *Neoscona fuscocolorata, N. subpullata, Larinioides cornutus, Yaginuma sia, Atea* sp.
    - (3) Orb with stabilimentum—*Araneus macacus, Zilla astridae, Z. sachalinensis, Argiope aemula, A. amoena, A. bosenbergi, A. bruennichii**, *A. aetherea, A. minuta**, *A. ohsumiensis*
  - B. Free sector orb
    - (1) *Zygiella* web—*Z. montana**, *Araneus mitificus, A. pentagrammicus, Arachnura logio**
    - (2) Secondary free-sector web—*Araneus variegatus, Neoscona fuscocolorata**, *N. subpullata*
    - (3) Camouflaged free-sector web—*Acusilas coccineus, Arachnura logio**
- II. Hub orb (center left intact)
  - A. Camouflaged orb
    With stabilimentum—*Cyclosa argenteoalba, C. camelodes, C. ginnaga**, *C. insulans, C. japonica, C. laticauda, C. octotuberculata**, *C. sedeculata**
  - B. Horizontal camouflaged free sector orb—*Cyclosa vallata**, *C. mulmeinensis*
- III. Horseshoe orb
  - A. Horseshoe orb—*Nephila clavata**
  - B. Incomplete horseshoe orb—*N. pilipes (= maculata)*
- IV. Mesh web
  - A. Dome mesh web—*Cyrtophora moluccensis**
  - B. Horizontal mesh—*C. exanthematica**, *C. unicolor*
- V. Open hub-orb (center removed)
  - A. Horizontal orb—*Meta, Metleucauge yunohamensis**, *Leucauge magnifica**, *L. subgemmea, Tetragnatha praedonia**, *Tylorida striata, Menosira ornata* (spiderling)
  - B. Square horizontal orb—*Tetragnatha lauta*
- VI. Quasi-concentric circular web
  - A. Quasi-concentric web—*Cyrtarachne bufo, C. inaequalis**, *C. induta, C. nagasakiensis, C. nigra, C. yunoharuensis, Paraplectana sakaguchii, P. tsushimensis*
- VII. No hub orb (entire hub removed)
  - A. Ray-form orb—*Theridiosoma epeiroides**
  - B. Frameless orb (no hub)—*Ogulnius pullus**
  - C. Spherical orb—*Mysmena jobi**
  - D. Irregular orb
    - (1) Floating web—*Conculus lyugadinus**
    - (2) Horizontal orb—*Conculus lyugadinus**
  - E. Disk orb—symphytognathid*
- VIII. Hackled-band orb
  - A. Hackled-band orb—*Uloborus varians**, *U. sybotides, U. sinensis**, *U. yesoensis, U. walckenaerius, Zosis geniculata*
  - B. Filmy orb—*Uloborus varians* (spiderling)

2. Irregular web

- I. Irregular web
  - A. Irregular web
    - (1) Tangle web—*Achaearanea, Theridion, Anelosimus, Enoplognatha margarita, Nesticus, Pholcus, Spermaphora*
    - (2) Basket web—*Achaearanea tepidariorum**, *A. culicivora, A. angulithorax* (spiderling), *Theridion yunohamense* (spiderling)
    - (3) Irregular web with bell—*A. angulithorax, A. ferrumequinum**, *A. culicivora, A. riparia, A. tabulata*
    - (4) Tent-like irregular web—*Theridion chikunii, T. latifolium, Anelosimus, Pholcus, Spermaphora, Physocyclus, Crossopriza*

Table 9.3. Continued

B. Reticulate irregular web—*Theridion pinastri**

II. Sheet and tangle web—*Achaearanea japonica**, *Chrosiothes subabides*

III. Irregular web with retreat-*Steatoda albomaculata*, *S. cavernicola*, *S. grossa*, *S. erigoniformis*, *Enoplognatha dorsinotata*, *E. japonica*, *E. transfersifoveata*

IV. Vertical thread web—*Chrysso venusta**, *C. punctifera*, *C. argyrodiformis*, *Theridion rapulum*, *T. nigrolimbata*, *Theridula gonigaster**, *Coleosoma blandum*, *Comaroma nakahirai*, *C. maculosum*

V. X-shaped web—*Episinus affinis*, *E. nubilus*, *E. kitazawai*, *E. mirabilis*

3. Sheet web

I. Sheet web

A. Sheet web—*Arcuphantes tamaensis**, *Bathyphantes*, *Centromerus*, *Doenitzius*, *Drephanotylus*, *Floronia*, *Labulla*, *Lepthyphantes cericeus**, *Linyphia montana**, *Meioneta*, *Ostearius*, *Porrhomma*, *Taranucnus*, *Aprifrontalia muscula**, *Asperthorax*, *Cresomatoneta*, *Gonatium*, *Nematogmus*, *Walckenaeria*, *Ummeliata*, *Solenysa*, *Tmetius*, *Caviphantes*

B. Dome web—*Prolinyphia limbatinella*, *P. emphana*, *P. longipedella**, *P. radiata*, *P. marginella*, *P. yunohamensis**

C. Hammock web—*Neolinyphia fusca**, *N. japonica**, *N. Nigripectoris*, *N. angulifera*

II. Sack-like web—*Tapinopa longidens**

III. Filmy sheet web—*Leptoneta nippara**, *Telema nipponica*

4. Funnel web

I. Funnel web

A. Platform web—*Agelena limbata**, *A. opulenta*, *A. labyrinthica*, *Coelotes corasides*, *C. antri*, *C. insidiosus*, *Litisedes shrahamensis*, *Macrothele* sp.

B. Funnel web—*Cicurina japonica*, *Coelotes luctuosus**, *Litisedes shirhamensis*, *Tengenaria domestica*, *Dolomedes saganus* (spiderling)

II. Filmy funnel web—*Hahnia corticora*, *Neoantistea quelpartensis*

5. Signal-line web

I. Signal-line web

A. Tunnel and signal-line web—*Cybaeus nipponicus**

B. Flat-retreat and catching-thread web—*Uroctea compactilis**

II. Catching-thread web

A. Tube and catching thread web—*Ariadna lateralis**, *A. insulicola*, *Segestria nipponica*

B. Flat-retreat and catching-thread web—*Oecobius annulipes*

6. Naruko web

I. Naruko web—*Wendilgarda* sp.*

7. Thread web

I. Single-line web (sticky thread)—*Phoroncidia pilula*

II. Non-sticky thread web—*Ariamnes cylindrogaster*

III. Hackled-band thread web—*Miagrammopes orientalis**

8. Triangular web

I. Triangular web—*Hyptiotes affinis**

II. Triangular horizontal web—*Pasilobus bufoninus*

9. Lace web

I. Lace web

A. Lace web—*Filistata marginata*, *F. longiventris*, *Amaurobius flavidorsalis*, *Callobius hokkaido*, *Cybaeopsis typica*, *Ixeuticus robustus*, *Titanoeca albofasciata*, *Lathys humilis**, *L. punctosparsa*, *Dictyna felis*, *D. foliicola*, *D. arundinacea*, *D. uncinata*

B. Tent web—*Dictyna felis**, *D. foliicola*, *D. arundinacea*, *D. uncinata*

II. Crimpy web—*Loxosceles rufescens*

10. Tube web (purse web)

I. Tube web—*Atypus karschi*

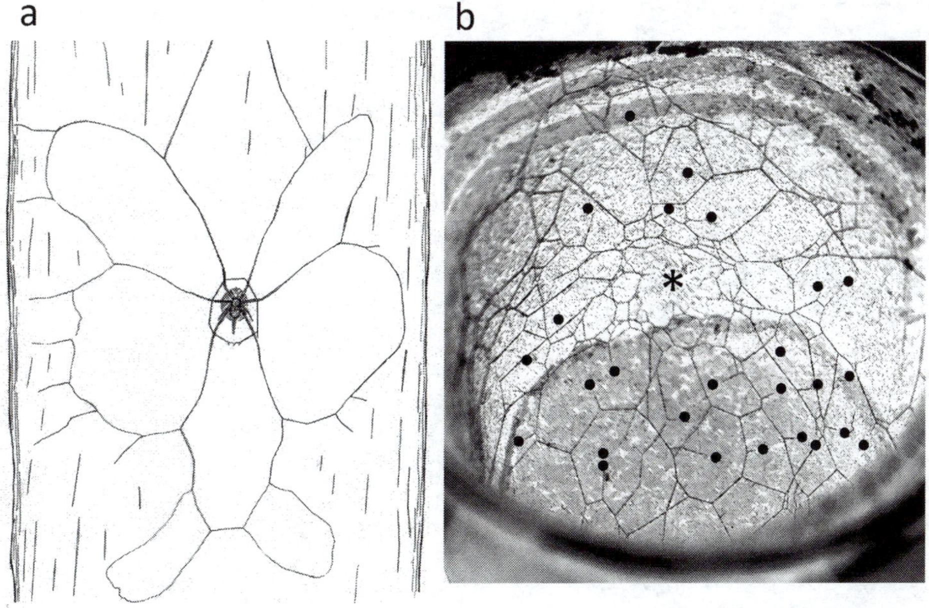

Fig. 9.1. The non-sticky webs of the trunk-dwelling hersiliid *Hersilia* sp. from the Philippines (*a*) and the tiny pholcid *Modisimus culicinus* that was built at the bottom of a cylindrical container (*b*) illustrate the convergences in several distantly related spiders on a radial design that may function to extend the spider's sense of touch while it waits at the center (table 9.1) (* marks the spider's approximate resting site near the floor in *b*). Most of the *M. culicinus* lines were slightly above the substrate; the black dots mark some of the short lines that connected them to the substrate. Both of these species run very rapidly (*a* from Williams 1928; *b* courtesy of Franco Cargnelutti).

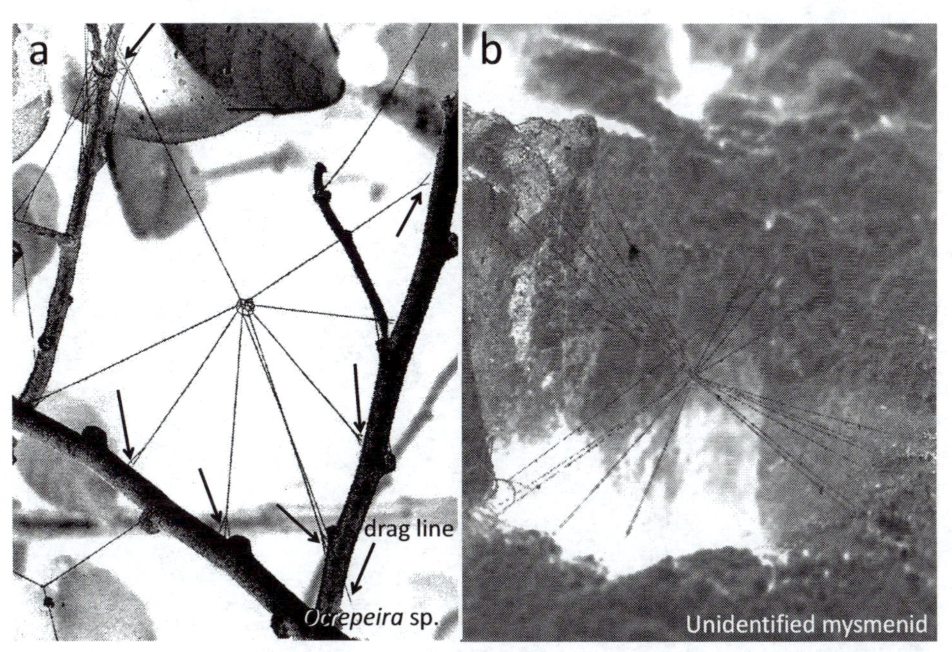

Fig. 9.2. These "asterisk" webs of non-sticky lines, built by (*a*) a tropical immature New World *Ocrepeira* sp. (#2115) (AR) on a tree branch several meters above the ground and by (*b*) a tiny, unidentified mysmenid in a small cavity in the floor of an Australian cave, both evolved from orb-building ancestors. The spiders rested at the hub (note the dragline in left *a* when the spider was removed); the convergent designs probably promote interception and vibration transmission (note the multiple attachments to the substrate indicated by arrows in *a* that may promote interception of walking prey; the mysmenid web apparently lacked split radii). The species *Ocrepeira ectypa* attacked prey that contacted a radius by running rapidly to the branch and wrapping them (Stowe 1978) (mysmenid photo courtesy of Martin Ramírez, Museo Argentino de Ciencias Naturales, and Peter Michalik, University of Greifswald).

Fig. 9.3. The function of this vertical, sail-like web of the tiny jumping spider *Pellenes arciger* is unknown. Webs are almost unknown in this large family of visual hunters. The lower edge of the web was against the ground; "R" marks the spider's retreat. Lopez (1983–1986), who saw hundreds of these webs, speculated that they were not prey capture devices, but rather resting places or observation posts that might transmit prey vibrations to their highly visual owners. Some spiders hid near their webs, while others rested on the sheet (from Lopez 1983–1986).

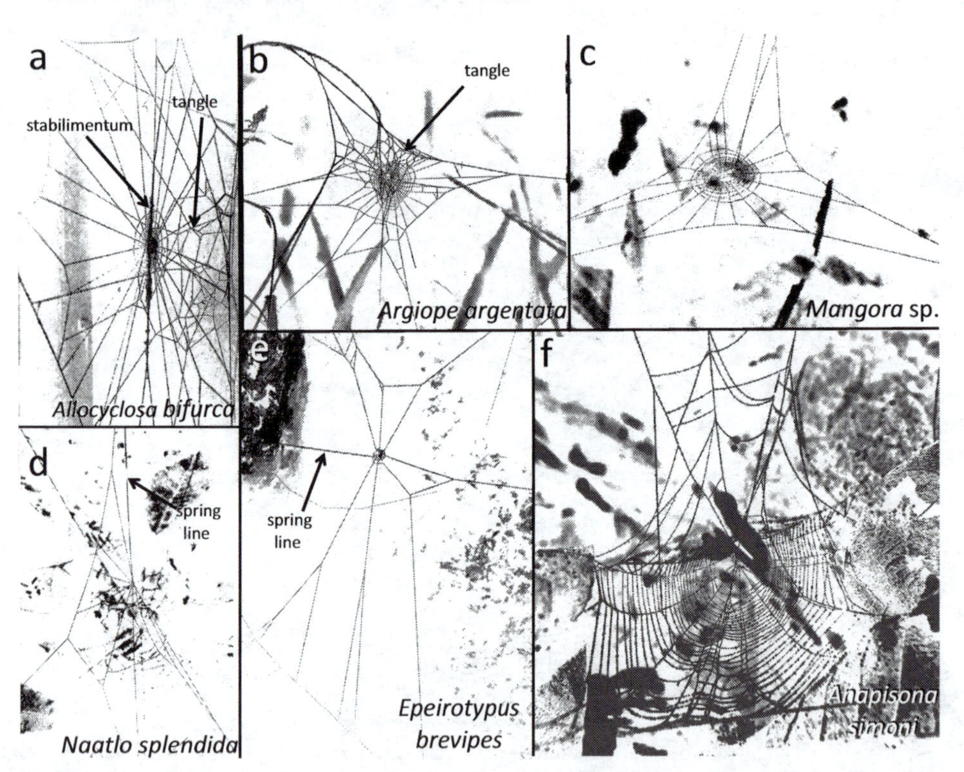

Fig. 9.4. Although the webs of mature male orb weavers are probably mostly utilized only for defensive purposes, such as resting sites protected from substrate-bound predators like ants, their designs include traits similar to those in the prey capture orbs built by conspecific females and spiderlings; they thus reflect probable taxonomic constraints on web designs. Males of some species (*a*, *b*) built a small tangle on the side of the web where the spider rested, probably protecting them against flying or jumping predators (as in molting webs—see figs. 10.32, O9.1). The dense hub of a male *Mangora* sp. (*c*) resembled the hubs of the orbs of mature female *Mangora* spp. (figs. 3.13e, 3.36b, 3.39). In two theridiosomatids (*d*, *e*) the male held a radial "spring" line tight and then released it suddenly when disturbed, causing him to snap back several mm with the web. The rapid movement may protect him from aerial predators, rather than functioning to capture prey as in female and immature conspecifics. The mature male anapid *Anapisona simoni* (*f*) constitutes a striking exception: mature males of this species built prey capture orbs that were indistinguishable from those of conspecific adult females (as in other anapids; see Lopardo and Hormiga 2008 on *Acrobleps hygrophilus*).

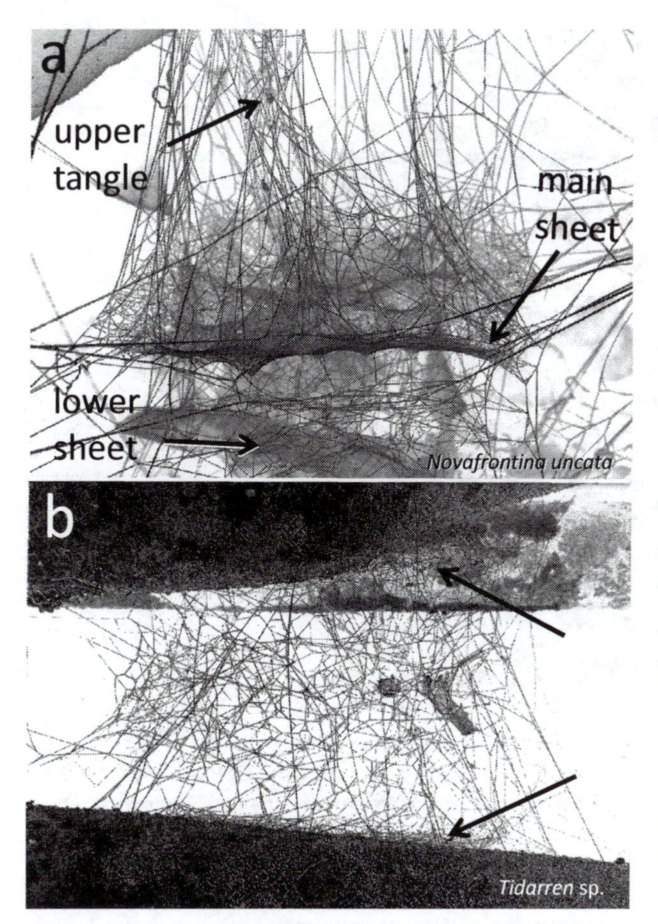

Fig. 9.5. The usual explanation for a double sheet like that of the linyphiid *Novafrontina uncata* (*a*) is that the upper sheet is for prey capture, and the lower, less extensive one is to protect the spider from attacks from predators; this also seems likely for the similar double sheets of *Tidarren sisyphoides* (see fig. 9.15b). This tidy interpretation does not fit easily, however, with the puzzling web of another species of *Tidarren* (*b*) from Fanies Island of South Africa, which apparently included small sheets next to the substrate at the very top and the very bottom (arrows), but no sheet in the middle (photographs by Gustavo Hormiga).

and aerial webs) have many strong and treacherously sticky lines that retain prey securely, as in some dictynids (WE), the amphinectid *Metaltella simoni* (Escalante et al. 2015), the filistatid *Kukulcania hibernalis* (Barrantes and Eberhard 2007), and *Micrathena* spp. (AR) (WE). Some webs are ephemeral, and are destroyed by the capture of a single prey, as in *Deinopis* spp. (Robinson and Robinson 1971; Coddington and Sobrevila 1987) and the progradungulid *Progradungula carraiensis* (Gray 1983), while others (at protected sites) can last up to months, as in the filistatid *K. hibernalis* and the pholcids *Pholcus phalangioides* and *Physocyclus globosus* (WE).

There is also diversity in the properties of homologous silk lines, which differ much more than was previously appreciated (section 3.2.9). The aggregate gland material that coats the sticky spiral lines of the araneoid orb weaver *Cyrtarachne bufo* and *C. nagasakiensis* adheres especially powerfully to moths, whose bodies are covered with deciduous scales (Cartan and Miyashita 2000), and is correlated with their highly reduced orbs that have only a few, widely spaced sticky lines (Stowe 1986; Shinkai 1982, 1989; Cartan and Miyashita 2000), as also occurs in the related genera *Pasilobus* (Robinson and Robinson 1975), *Poecilopachys* (Clyne 1973), and *Paraplectana* (Chigira in Stowe 1986) (fig. 9.23). The non-sticky lines (ampullate silk? aciniform silk?) are stiff rather than flexible in a few groups, and are used to raise the web's lines off the substrate in the oecobiid *Uroctea durandi* (fig. 10.5, Kullmann and Stern 1981) and the filistatid *Kukulcania hibernalis* (WE), and to raise the spider's tent-like silk shelter above the substrate in other oecobiids (Glatz 1967; Kullmann and Stern 1981).

Some of the most spectacular diversity involves the attachment discs made with piriform silk. Subtle modifications have evolved for different functions: attaching lines to water (in some theridiosomatids and anapids—fig. 10.29, section 2.2.9); "break away" attachments designed to fail with only moderate stress (section 2.2.9); some are to the substrate (the gumfoot lines of theridiids) and others to other lines (sticky lines to radii in cyrtarachnine araneids) (fig. 9.23c). Still others are designed to fail only partially when stressed, allowing sticky lines to slide past radii without breaking free (in many araneids) (fig. 3.33, section 3.3.3.1.6).

Still another type of diversity results from active manipulations of tensions by the spider. Several groups probably increase the retention abilities of their webs by reducing the tensions on sticky lines by pulling out more line than is needed to reach from one attachment point to the next during construction; these include many cribellates that build visibly slack sticky lines, as in *Tengella radiata* just above the sheet (Eberhard et al. 1993), *Loxosceles* sp. (Knight and Vollrath 2002), filistatids (figs. 5.10, 10.12), as well as orb weavers in general that pull out sticky spiral lines with the hind legs until they are slack or nearly slack (Eberhard 1982, fig. 7.9, appendices O6.1, O6.2). Some groups also make slack non-sticky lines that probably serve to entangle prey, as in the pholcid "screw

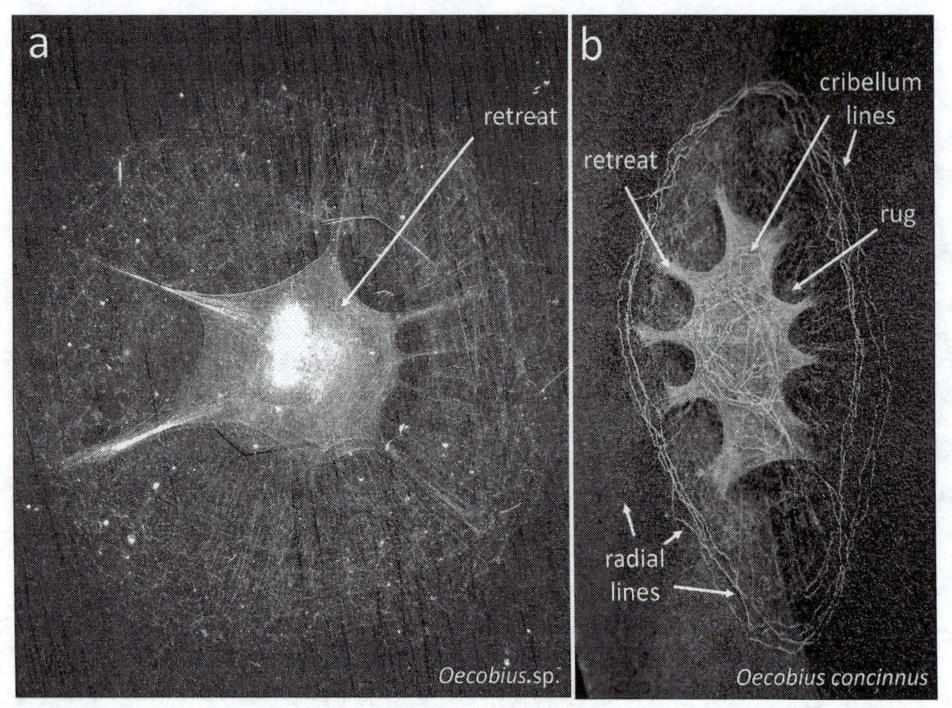

Fig. 9.6. Field and lab webs of *Oecobius* show unexpected similarities with orb webs. The web of *O.* sp. in *a* was built in a small indentation on a vertical black wooden fence post, and had accumulated dust and fine detritus, revealing a dense array of radial lines visible in the circular mat just beyond the margins of the white star-shaped retreat roof (white eggs are visible inside the retreat). The web of *O. concinnus* in *b* was built in the angle where the floor (right) of a small petri dish that had been painted black met the vertical wall (left). It also had many radial lines, as well as cribellum lines (thicker white lines) that formed circular patterns on the radial lines around the retreat and an irregular pattern on the retreat itself; the "rug" to which the radial lines were attached is also visible (the web was coated with white talcum powder, and then the dish was jarred repeatedly until nearly all of the powder had fallen from all non-sticky lines). The sticky lines on the retreat may have a defensive rather than prey capture function. Direct observations of *O. concinnus* building behavior (Solano-Brenes 2018) confirmed that spiders made radial trips while laying non-sticky lines, and circular trips around the retreat while laying cribellum lines (*b* photo by Gilbert Barrantes, from Solano et al. in prep.).

lines" of *Pholcus phalangioides* (Kirchner 1986). *Chrysiothes* sp. (THD) and *Cyrtophora citricola* (AR) built loose, entangling lines just above the dense horizontal sheet (fig. 1.7; WE). Others, such as many theridiids, build tense sticky lines, but attach them only weakly to the substrate, so that a struggling prey will break the attachment and lift the prey away from the substrate (Bristowe 1958; Eberhard et al. 2008b; Sahni et al. 2012). Convergent "break-away" attachments occur in other reduced orbs, as in *Pasilobus* sp. and *Poecilopachys australasia*, where the weak attachments apparently facilitate tethering moths without damaging the rest of the more or less horizontal web (Robinson and Robinson 1975; Clyne 1973).

An additional, recently discovered technique allows spiders like the mygalomorph *Linothele macrothelifera* (Eberhard and Hazzi 2013) and the lycosid *Aglaoctenus castaneus* (Eberhard and Hazzi 2017) to lay slack lines: the spider spread the multiple non-sticky fine lines emerging from each posterior lateral spinneret to form a band or sheet (instead of in a multi-strand cable) by attaching the swaths of lines at only a fraction of the attachment sites (fig. 5.17, section 5.6). This increased the area the lines covered, and presumably increased prey retention times.

Other webs function to capture prey in still other creative ways. The funnel-web agelenid *Coelotes terrestris* regularly suspended many small pebbles or grains of soil on short thin lines, even at sites where there were no such particles nearby; their swinging movements resulting from web vibrations may facilitate prey detection (Tretzel 1961). *Ariamnes* (=*Argyrodes*) *attenuatus* and *A.* sp. (THD) built only a few long lines that constituted perching or resting sites for potential prey such as

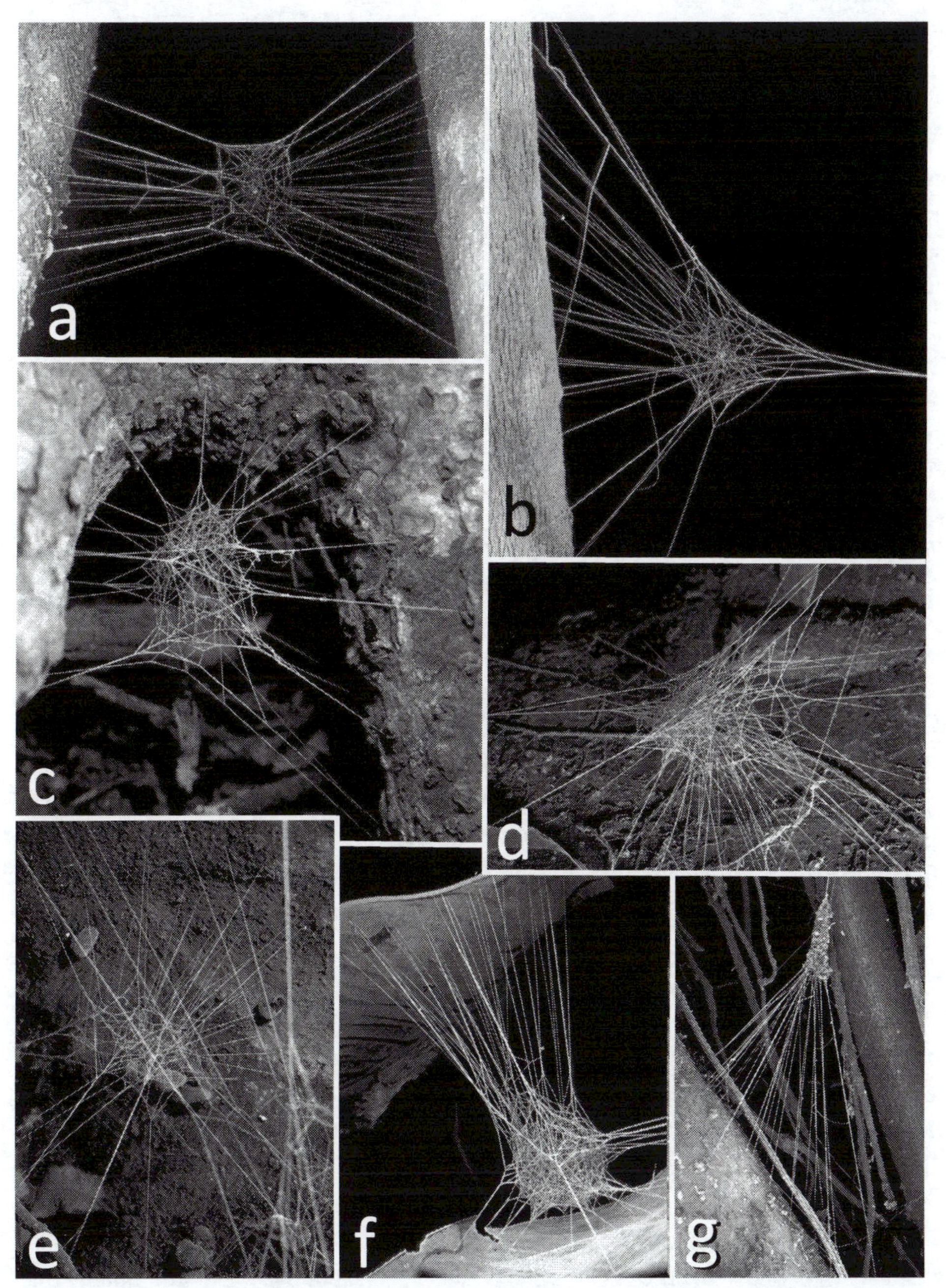

Fig. 9.7. The familiar "theme and variations" pattern in biological diversity is illustrated by the "star webs" in the large theridiid genus *Achaearanea* (see also fig. 10.31). All had a small, more or less spherical aerial tangle where the spider (visible in *a* and *b*) rested, and one or more bundles of long, more or less parallel gumfoot lines that ran directly from the tangle to the substrate. But the details varied. Some groups of gumfoot lines were oriented in only one direction (*b*, *g*), others in two directions (*a*, *f*), and still others radiated widely from the central tangle (*c*, *d*, *e*). The central tangle in some was relatively dense (*f*), in others sparse (*a*, *b*, *e*), and others intermediate (*c*). In one the central tangle was replaced with a retreat built from detritus (*g*). Some tangles were far from the substrate with long and nearly parallel gumfoot lines (*f*), while others were close to the substrate and had relatively short gumfoot lines (*d*). The gumfoot lines in some were sparse (*c*), but abundant in others (*d*, *f*). It is not known how much of this variation is due to intra-specific flexibility, and how much to inter-specific differences. Other species in this large genus have quite different web designs (e.g., figs. 5.16, 9.12); perhaps the *Achaearanea* species with star webs form a distinct taxonomic subgroup, but this only speculation (photographs by Gustavo Hormiga).

insects or spiders; the spider approached perched visitors slowly and stealthily along the line until she was in range to loop sticky wrapping silk onto them (Bradoo 1971; Eberhard 1979a). Other skimpy arrays of non-sticky lines have different functions. A few thin lines of the kleptoparasite *Faiditus* (=*Argyrodes*) spp. (THD) transmit vibrations from the host spider's web, allowing the parasite to steal the host's prey (Vollrath 1979; Whitehouse 1986). The sparse, approximately horizontal lines of the theridiid *Chrosiothes tonala* provide only aerial runways about 30–100 cm above the forest floor, along which the spider moves in order to drop onto columns of termites foraging below when she senses olfactory cues (Eberhard 1991).

A different kind of minimalism occurs in the long, thick-walled camouflaged closed tubes or sacs ("purse webs") of *Atypus* and *Sphodros* that extend along the surface of the ground or a tree trunk. The spider remains protected inside at all times, and depends on prey walking onto the tube's external surface; she uses her long, thin fangs to stab the prey through the wall (Comstock 1967; Coyle 1986; WE). The araneid *Paraplectanoides crassipes* has converged a similar enclosed web (fig. 10.28), and may employ a similar tactic to capture cockroaches (Hickman 1975). Converging on a similar design but with a different function, some female pholcids such as *Holocnemus pluchei* (Sedey and Jakob 1998) and *Modisimus bribri* (WE) altered the normal domed sheet design of their prey capture webs to enclose themselves in approximately spherical webs that probably functioned to protect new spiderlings. A few highly visual salticids built "watch towers" from which spiders spied out potential prey at a distance (*Portia*—Jackson 1986).

There is also striking diversity among variations on a given web design such as "sheet" webs (fig. 9.8). The hahniid *Neoantistea* spp. built small, densely meshed delicate aerial sheets that completely lacked tangles, spanning small depressions in moss or soil (Opell and Beatty 1976; Eberhard 2018b); the pisaurid *Hygropoda* sp. (#3656) made a minimalist small sheet spanning a single leaf with up-curled edges (fig. 5.14), while the small sheet of the dictynid *Mallos hesperius* also spanned a single leaf, but included many sticky lines (WE). The phyxelidid *Phyxelida* built barely discernable sheet-like central horizontal portions that consisted of slightly greater concentrations of lines in the midst of loosely meshed tangles. The sticky sheets of titanoecids *Titanoeca albomaculata*, *T. nipponica*, and *T. nigrella* were slightly raised from the substrate; the spider walked on the upper surface of the web (Szlep 1966a; Shinkai 1979; Griswold et al. 2005). The agelenids *Agelena*, *Agelenopsis*, *Melpomene*, and *Tegenaria* built densely meshed horizontal non-sticky sheets somewhat off the substrate but which were connected with a peripheral, protected tube retreat ("funnel web") and had tangles above to intercept flying prey and knock them onto the sheet where the spider could attack them (Bristowe 1958; Rojas 2011); some lycosids, such as *Sosippus* and *Aglaoctenus*, have converged on the same designs (Brady 1962; Viera et al. 2007; González et al. 2015; Eberhard and Hazzi 2017). The stiphidiid *Stiphidion facetum* and the zoropsid *Tengella radiata* also built densely meshed funnel webs in sheltered sites, but they included sticky silk in the sheet and the tangle just above it (Eberhard et al. 1993; Griswold et al. 2005). The stiphidiid had a central rather than a peripheral retreat, and the sheets in some webs were vertical rather than horizontal (fig. 204E in Griswold et al. 2005); in *T. radiata* the retreat was central in juveniles but peripheral in adults (Barrantes and Madrigal-Brenes 2008).

The family Linyphiidae has had a separate, extensive diversification of sheets, that includes the classic, densely meshed horizontal aerial sheets suspended in extensive tangles that in some cases contain weakly sticky lines (Emerton 1902; Nielsen 1932; Comstock 1967; Benjamin and Zschokke 2004; Benjamin et al. 2002). These sheets vary in form, in number, in having or not having associated tangles above or below, in the sites where they are built, and in the patterns of the lines in the sheets (fig. 9.9, G. Hormiga and W. Eberhard in prep.). In the end, the diversity of "sheets" is such that it is difficult to decide what a "sheet" is (section 9.4). Another family that is small but that nevertheless has divergent web forms that include sheets is Synotaxidae (fig. 9.8) (Agnarsson 2003b).

Then there are the loosely meshed sheets. Some have sticky lines, such as the nicodamid *Megadictyna thilenii* (Griswold et al. 2005), the austrochilids *Thaida peculiaris* (Lopardo et al. 2004) and *Hickmania troglodytes* (Griswold et al. 2005), the psechrids *Psechrus* spp. and *Fecenia* spp. (Robinson and Lubin 1979b; Shear 1986; Zschokke and Vollrath 1995b; Eberhard 1987a; Griswold et al. 2005), the theridiid *Chrysso* nr. *nigriceps* (Eberhard

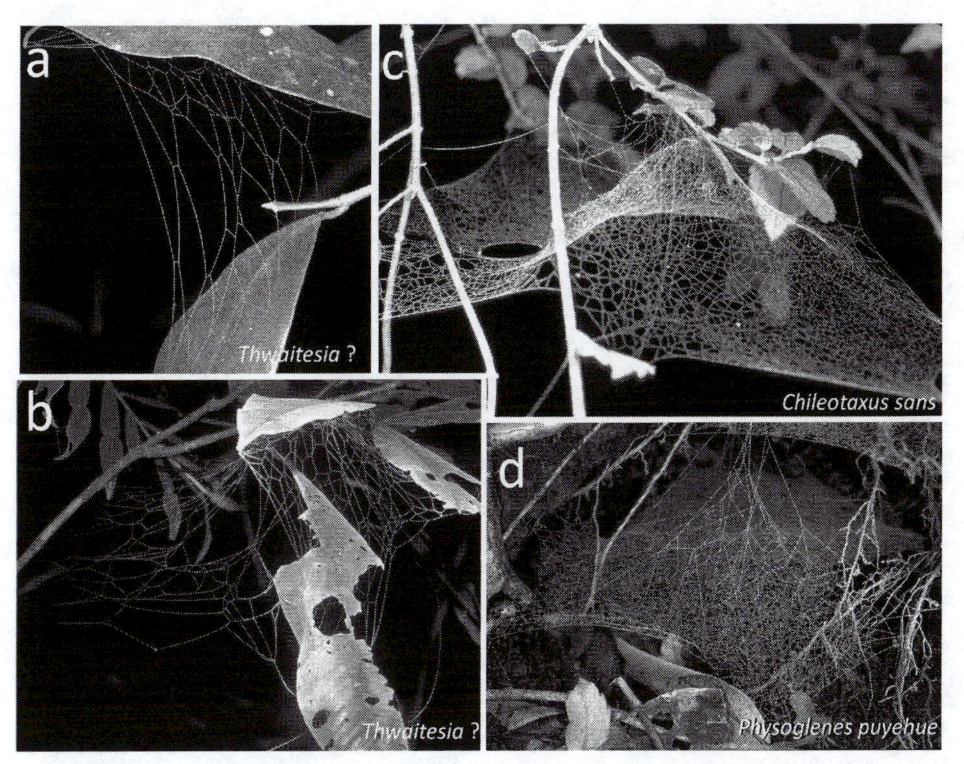

Fig. 9.8. These webs illustrate further convergences and diversity. The sparse planar arrays of an unidentified Australian theridiid (perhaps near *Thwaitesia*) (*a*, *b*) resembled the sparse, planar "rectangular orbs" of several species of the New World genus *Synotaxus* (figs. 5.3, 5.4, 9.12a), but lacked the sets of parallel lines in the rectangular orbs. The other two species belong to Physoglenidae, a group formerly associated with synotaxids, but they had quite different webs. *Chileotaxus sans* (*c*) built a sparsely-meshed horizontal sharply-domed sheet that spread away from a sparse tangle above the central area near the underside of a leaf where the spider rested, resembling the webs of the pholcid *Modisimus bribri* (fig. 8.7); *Physoglenes puyehue* built a nearly flat, more finely-meshed sheet that was not associated with leaves (*d*) (photographs by Gustavo Hormiga).

et al. 2008b), and the amphinectid *Metaltella simoni* (I. Escalante pers. comm.) (these are skeletons of non-sticky lines with cribellum silk added, not "sheets of cribellum silk" as described by Griswold et al. 2005). Others lack sticky lines, such as some *Latrodectus* and *Steatoda* species (Eberhard et al. 2008b). The dictynid *Mallos hesperius* adds both sticky lines and small swaths of thin non-sticky lines to the planar skeleton formed by non-sticky spanning lines (WE). Some other dictynid webs are described as multiple small sheets in different planes (not horizontal) around a central funnel-like retreat (Griswold et al. 2005; see also photo of *Dictyna sublata* in Comstock 1967). The desids *Matachia* sp. and *Badumna longiqua* make almost orb-like aerial, approximately planar, sparse sheets with central or peripheral retreats, approximately radial non-sticky lines, and regular patterns of zig-zag sticky lines that are attached to the radial lines (fig. 1.3) (Opell 1999a; Griswold et al. 2005). Some "sheets" consist of only two non-sticky lines supporting multiple sticky lines strung between them, as in some parts of colonial *Stegodyphus* sp. webs (Eberhard 1987a) and the pholcid *Belisana* (fig. 9.19). Thus, at close range, the patterns of lines in sheets are also diverse (fig. 9.9).

There is also diversity in how spiders use their sheets. In many species, the spider generally stays on the underside of her sheet; but in several, including *Psechrus argentatus* (Robinson and Lubin 1979b), *P.* sp. (Eberhard 1987a), and the diguetid *Diguetia canities* (Eberhard 1967), the spider inserts her legs through the sheet during attacks and presses the prey to her chelicerae. In many others the spider normally runs on the upper surface of the sheet.

Several types of webs appear to function only to intercept prey and transmit vibrations, as in the scytodid *Scytodes longipes* (Nentwig 1985a); the oecobiid *Uroctea durandi* (fig. 10.5; Kullmann and Stern 1981); *Ariadna* spp. (Shinkai 1989; Griswold et al. 2005); *Liphistius* spp. (Bristowe 1976); and *Ocrepeira* (= *Wixia*) *ectypa* (Stowe 1978).

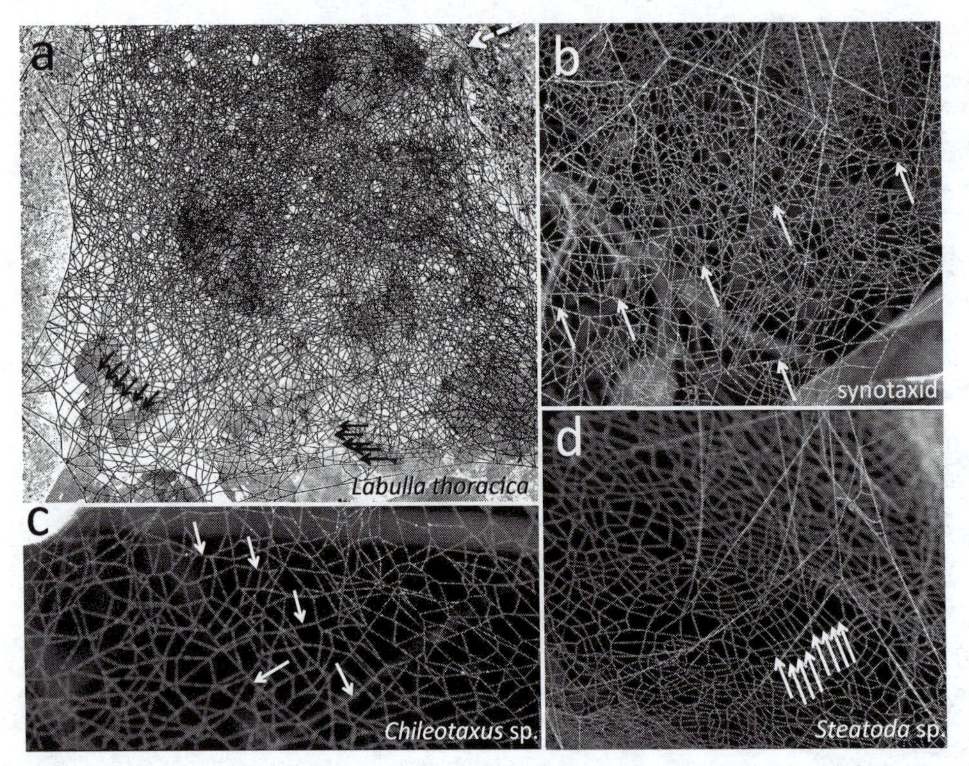

Fig. 9.9. There are as yet no careful comparative studies of the patterns of lines in the sheets of many taxa, but this sampler suggests that there may be interesting differences. The sheet of *Labulla thoracica* (LIN) (*a*) had both converging lines in one corner (dashed white arrow at the upper right) that radiated from the probable retreat of the spider as in *Agyneta* sp. (LIN) (fig. 1.3a), and small groups of nearly parallel lines in other areas (a few lines are marked with smaller black arrows at the bottom left). In two synotaxid sheets (*b*, *c*) there were possible skeleton lines (arrows) but some were not straight, intimating (if they were indeed laid early during web construction) that tensions on other lines pulled them out of line. The sheet of *Steatoda* sp. (THD) (*d*) resembled *a* in lacking signs of skeleton lines; instead there were relatively uniform spaces between lines, and lines in some areas were approximately parallel to each other (arrows) (*a–c* photographs by Gustavo Hormiga).

Other webs function only in retention: *Progradungula carraiensis* (fig. 10.9; Gray 1983); *Mastophora* spp. and other bolas spiders (Yeargan 1994); and *Deinopis* spp. (Robinson and Robinson 1971; Coddington and Sobrevila 1987). Some have costs so small that the spider readily discards the web entirely and leaves when disturbed: the theridiid *Phoroncidia* (= *Ulesanis*) *pukeiwa* simply released its single line web and walked away when disturbed (Marples 1955). Many webs have combinations of functions, but probably gave special importance to one: prey retention was probably the major function of the large numbers of sticky lines in the convergent "moth" webs of the araneids *Poltys*, *Acacesia*, *Deliochus*, and *Pozonia* (figs. 3.5, 3.35), the ladder webs of *Tylorida* (Robinson and Robinson 1972) and *Scoloderus* spp. (fig. 3.4) (Eberhard 1975; Stowe 1978; Traw 1995). The special adhesives of *Cyrtarachne* (Cartan and Miyashita 2000) and perhaps the related genera *Poecilopachys* (Clyne 1973) and *Pasilobus* (Robinson and Robinson 1975) probably also functioned mostly in prey retention.

Some functions are entirely lacking in some general web types; for instance, absorbing the momentum of prey (as opposed to retention) is likely of little or no importance in most substrate webs, and of reduced importance for trunk orbs (section 4.3.3.3). In still others particular functions may be important for some prey but not others; for instance, greater visibility in a substrate web probably does not affect the chances that walking nocturnal prey will contact it, but may reduce the chances of capturing some diurnal prey. Vertically elongate "trunk" orbs of *Herennia multipuncta* (= *ornata*) (Robinson and Lubin 1979a; Kuntner et al. 2008a), *Clitaetra irenae* (Kuntner et al. 2008b), *Eustala perfida* (Messas et al. 2014), and *Telaprocera maudae* (Harmer and Herberstein 2009), are built against tree trunks, and are especially difficult to see.

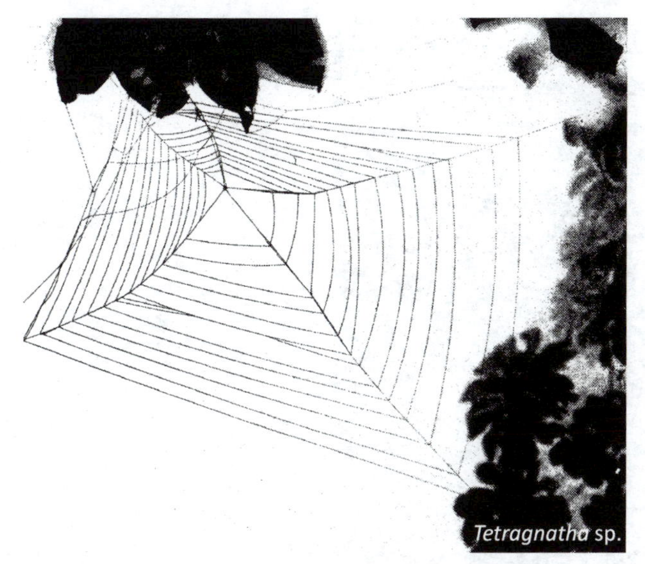

Fig. 9.10. This large (maximum diameter about 47 cm), highly simplified 5-radius orb of *Tetragnatha* sp. (#1277) was about 20 m from the edge of a lowland tropical river in Costa Rica. The web's plane was approximately 30° with horizontal (the photo was taken looking upward into the night sky). The hub was essentially non-existent, and there may have been no frame lines. The multiple lines of the radii at 11:00 (where the spider is visible), 2:00 and 4:00 were presumably laid when the spider walked away from the hub. Shinkai (1988b) observed a similarly reduced orb in the Japanese species *Tetragnatha lauta*.

Finally, at some taxonomic levels, there is diversity in whether or not there is diversity. For instance, hub designs are quite diverse in some orb weaver genera such as *Alpaida* (AR) (fig. 3.15) and *Tetragnatha* (TET) (figs. 4.6, 9.10, 10.24), but appear to be relatively uniform in others such as *Micrathena* (AR) (fig. 3.14). Striking intra-generic diversity in overall web design is rampant in Theridiidae (Eberhard et al. 2008b), while web designs are relatively unform at the genus level in Nephilidae (Kuntner et al. 2010b). Some species, like *Uaitemuri rupicola* (UL), show mixes of divergent and conservative web traits (fig. 9.11, Santos and Gonzaga 2017).

To end this section, it is interesting to note that there may also be limits to web diversity. Substrate-bound webs almost never have uniform spacing between sticky lines (despite repeated convergences on such regularity in aerial webs—next section) (the filistatid *Misionella mendensis* and the dictynid *Emblyna* sp. are exceptions, however—figs. 10.12, 10.14). Perhaps prey approach aerial webs from widely different angles, whereas substrate-bound prey tend to encounter substrate webs at their edges, rendering internal regularity in the arrangement of their sticky lines less relevant in prey capture. It is also puzzling that spiders have never evolved the obviously successful hanging-curtain type of web designs of larvae of the nematocerous fly family Keroplatidae (fig. 9.28, section 9.7) (*Wendilgarda* spp. [TSM] attached to water surfaces are an exception).

In sum, spider webs show extraordinary diversity and evolutionary inventiveness. Further examples concerning tangle webs associated with orbs, orbs built by mature males, and orb slants with respect to gravity are given in appendices O9.1–O9.3.

9.2.2 FREQUENT CONVERGENCE

Widespread convergence (e.g., figs. 9.1, 9.12, 9.13, 9.14, 9.15, 9.16) constitutes a second strong general pattern in both orbs and non-orbs. The long lists in tables 9.1 and 9.2 give semi-quantitative estimates (17 traits in theridiids, 42 other non-orb traits, and 50 orb traits). In some cases up to five or more different groups have converged independently on the same trait. Additional cases undoubtedly occur, though they are not indicated specifically, in table 9.4 of Japanese spiders and their web types. Although these numbers are surely imprecise (see the table headings for some limitations), convergences are clearly rampant.

To illustrate, consider the diguetids *Diguetia* spp. (Cazier and Mortenson 1962; Eberhard 1967; Bentzien 1973), theridiids in the genera *Achaearanea*, *Nihonhimea* (Jörger and Eberhard 2006; Eberhard et al. 2008b), and *Steatoda* (Viera et al. 2007; Eberhard et al. 2008b), *Chrosiothes portalensis* (Eberhard et al. 2008b), and *Theridion melanurum* (Eberhard et al. 2008b) (some of these theridiid groups are only distantly related to each other—Agnarsson 2004; Arnedo et al. 2004), and the araneids in the related genera *Cyrtophora*, *Manogea*, *Kapogea*, and *Mecynogea* (Wiehle 1927; Blanke 1972; Levi 1980; WE). All built densely woven horizontal aerial sheets that slanted upward at the edges that had tangles above and below and a few downward projecting dimples midway between the edge and the center (figs. 1.7 and 9.16). The sheets' horizontal positions in aerial tangles, their high densities of lines, and their upward curls at the edge all presumably function to stop and retain prey, as documented in *C. moluccensis* (Lubin 1973; Blamires et al. 2011). The downward directed, widely spaced dimples

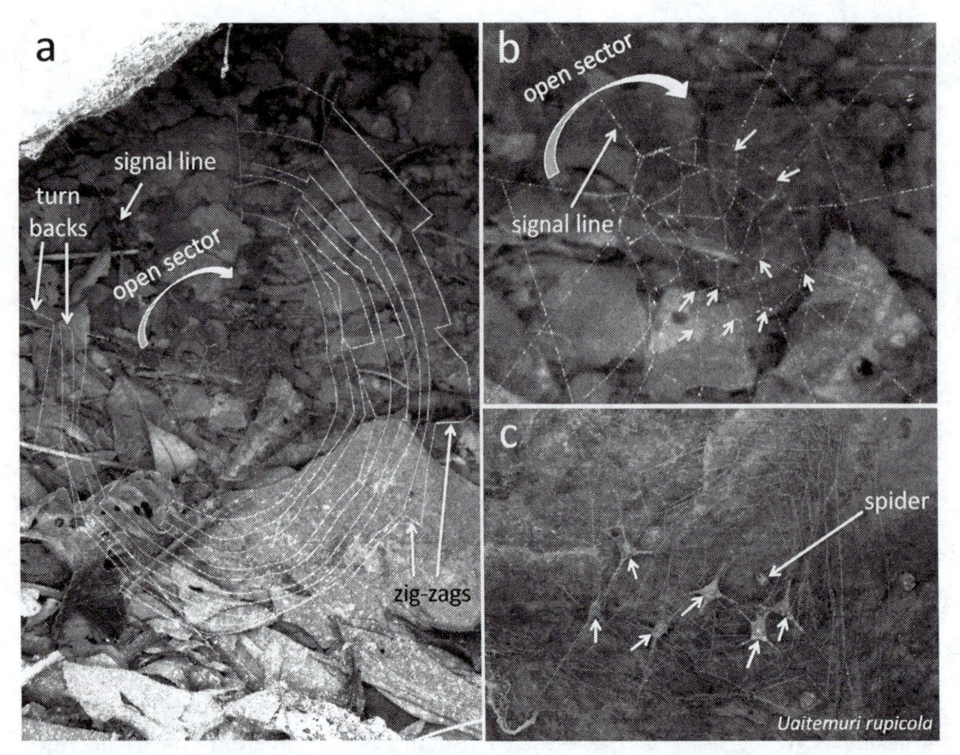

Fig. 9.11. The webs of this species in a newly discovered uloborid genus, *Uaitemuri rupicola*, illustrate two major themes of this chapter: convergence and taxon-specificity. During the day the spider rested near the substrate, at the far end of the signal line that ran through an open sector that lacked sticky spiral lines; the sticky spiral turned back at the edges of the open sector (*a*). These are both unusual traits for a uloborid, and represent convergences with the webs of araneids like *Zygiella x-notata* (table 9.1). The low numbers of radii, hub loops (*b*), and sticky spiral loops, the relatively abrupt transition from hub to temporary spiral, and the addition of apparently protective sticky loops to the lines circling the egg sacs and the resting site (*c*) are also unusual in uloborids. In contrast, the short zig-zag segments where the sticky spiral runs along a radius near the edge of the web (*a*), and the double attachments of the hub spiral to each radius that pull the radius out of line (small arrows in *b*), are typical of other uloborids (e.g., figs. 3.31, 6.11, O9.1). Egg sacs also occurred near the resting site (unlabeled arrows in *c*), and the sticky lines that encircled the resting site may represent transfer of the egg sac web construction behavior module to a new context (see figs. 10.16a,b and 10.35 of *Uloborus* spp. and *Zosis geniculata*) (from Santos and Gonzaga 2017; photos courtesy of Marcelo Gonzaga).

probably serve to tense the sheet while minimizing the number of lines impeding the spider as she runs under the sheet to attack prey (Lubin 1980; Barrantes and Weng 2006a).

A second illustration of convergence concerns steep-sided dome-shaped sheets with extensive tangles above, found in the pholcid *Mesabolivar* sp. (fig. 9.15), *Tidarren sisyphoides* (THD) (fig. 9.15) (Madrigal-Brenes and Barrantes 2009), and several linyphiids (Emerton 1902; Comstock 1967; Shinkai and Takano 1984). An example of convergence in orb webs is the split radii in strongly asymmetric webs (fig. 9.17).

Some groups have converged on non-orb webs that have geometrically regular arrays of sticky lines that are attached to and supported by more or less regular arrays of non-sticky lines. They include independent derivations in two orbicularian families, Synotaxidae and Theridiidae (fig. 9.12), as well as the non-orbicularians Amaurobiidae (fig. 10.14), Pholcidae (fig. 9.19), Psechridae (fig. 10.14, box 9.1), and Desidae (fig. 1.3). The convergences on orb-like web traits extends to the tapping behavior in *Titanoeca albomaculata* that resembles the inner loop localization behavior of orb weavers (Szlep 1966a). Even minor details, such as widely spaced tiny droplets of glue on short, lax lines that are interspersed with non-sticky lines, occur in widely separated groups such as the synotaxids *Synotaxus* spp. (Eberhard et al. 2008b; Barrantes and Triana 2009) and the pholcids *Belisana* spp. (Huber 2005).

The theme of diversity continues even in these convergences, because different species have converged on different aspects of orbs. For instance, the psech-

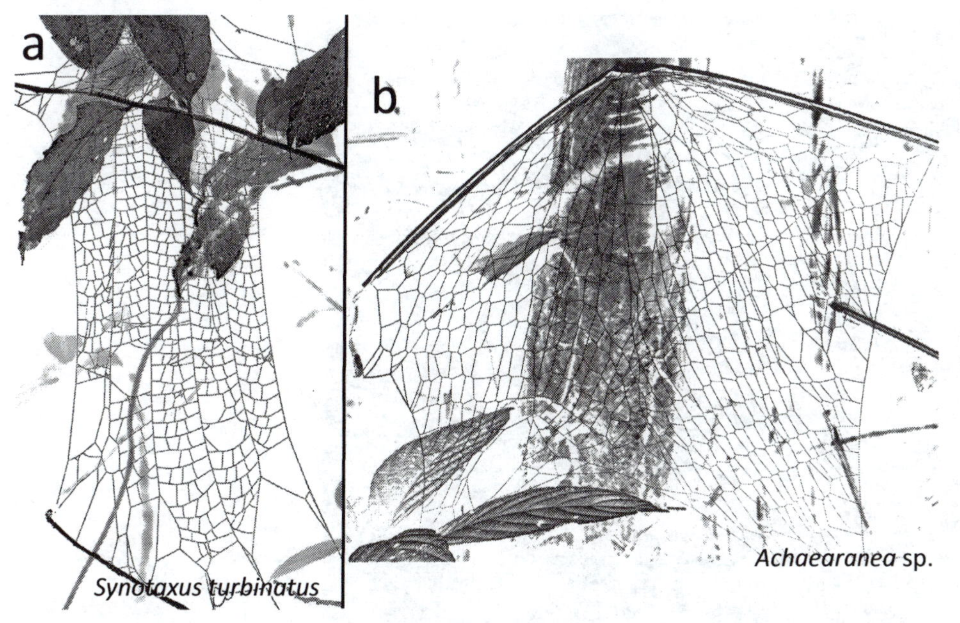

Fig. 9.12. The "rectangular orb" of the neotropical synotaxid *Synotaxus turbinatus* (*a*), and the eerily similar vertical "chicken wire" web of the theridiid *Achaearanea* sp. from Thailand (*b*) illustrate the wide-ranging pattern of convergences in spider web designs. The *Synotaxus* web was organized in vertical rectangles or sections that were delimited by nearly straight vertical lines running from top to bottom that were built early in web construction, and had a substantial tangle above. New sections were added in succession. In contrast, the web of *A.* sp. had no straight vertical lines; instead it had more or less parallel horizontal lines. Construction behavior has never been observed in this species (photographs by Gustavo Hormiga).

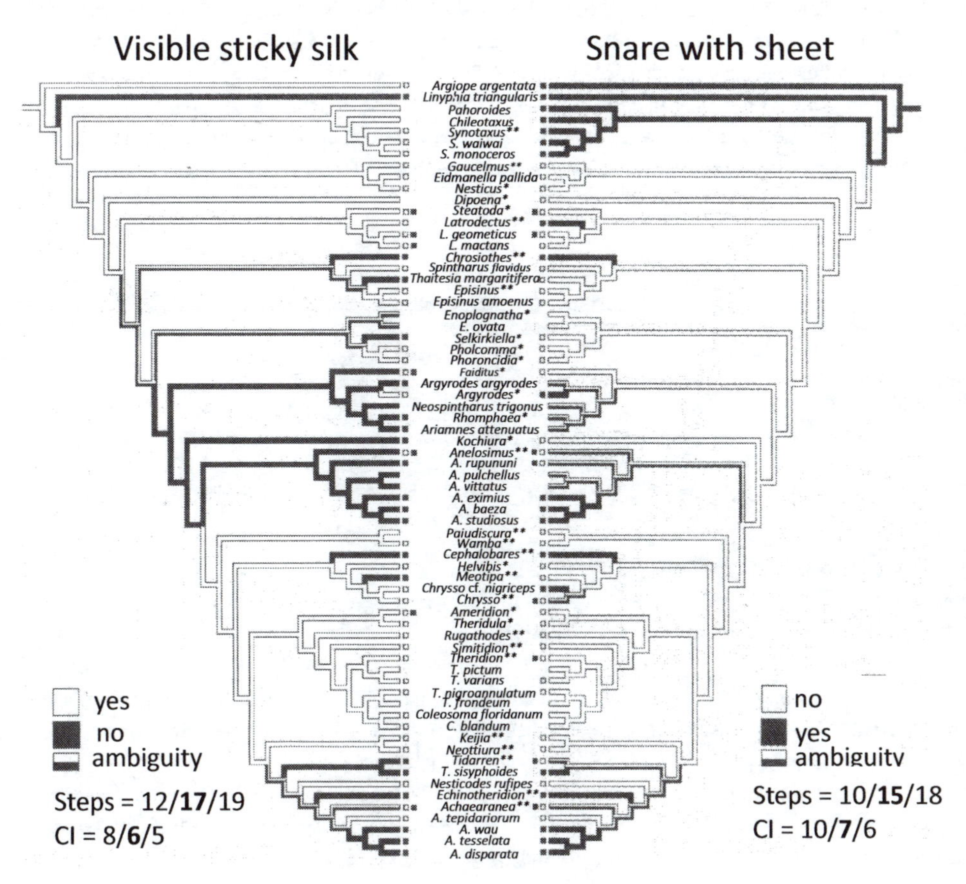

Fig. 9.13. The evolution of two web traits in Theridiidae, inclusion of visibly sticky lines and inclusion of a planar sheet, illustrate the common pattern of multiple convergences in web designs in this family. The traits are plotted on a consensus phylogeny derived from molecular, behavioral, and morphological data. Both sticky lines and sheets evolved and disappeared multiple times (polymorphism is shown by associating more than one box with a taxon, ambiguity by branches that are bi-colored). Three methods were used to count steps: $minimum_1$ (ignoring polymorphism); $minimum_2$ (counting polymorphism but ignoring status in composite taxa); and maximum (counting polymorphism). The consistency index (CI) was calculated according to each step-counting method; the bold numbers were the preferred values (from Eberhard et al. 2008b).

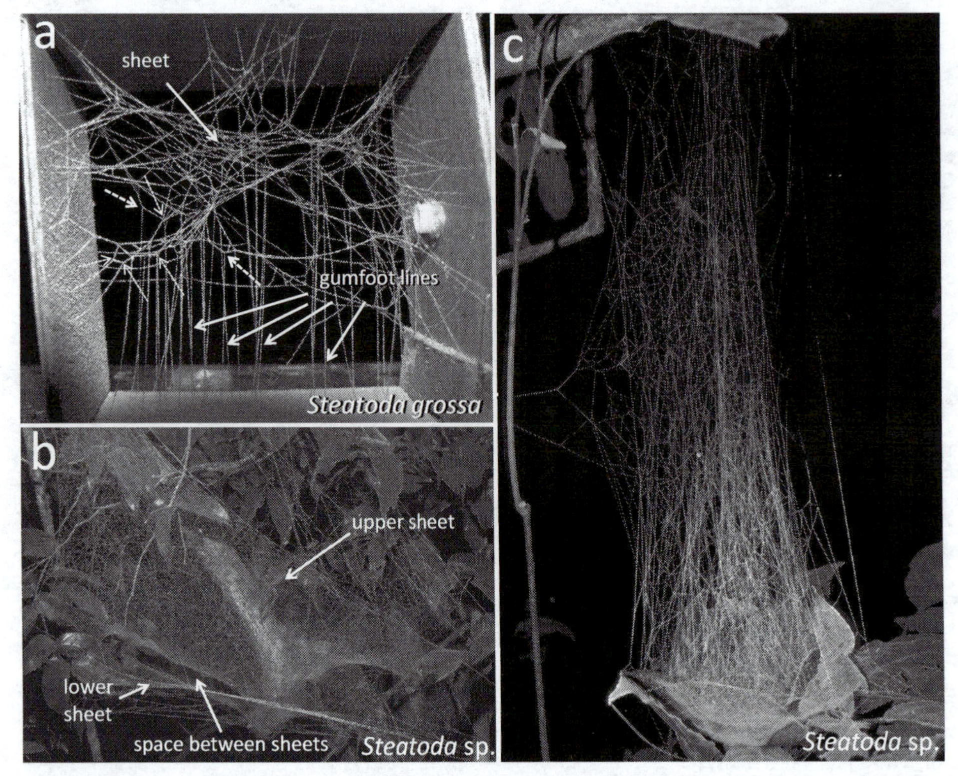

Fig. 9.14. The intra-generic diversity in web designs in *Steatoda* illustrates the exuberant diversity and convergences typical of theridiid genera. Some *Steatoda* species as *S. grossa* built webs in sheltered sites, with a tangle that included a sparse, weakly delimited horizontal platform and numerous gumfoot lines that were strung between the substrate below and approximately horizontal lines above (unlabeled dashed arrows) (*a*). The webs of some other species (*b, c*) were more exposed and lacked gumfoot lines. In *S.* sp. from South Africa (*b*) the sparse platform was replaced with a moderately dense horizontal sheet, with a relatively regular mesh size (see fig. 5.7) that lacked obvious long skeleton lines like those in pholcid sheets (figs. 5.2, 5.21, 9.19); there was a second sparse sheet just below, and a sparse tangle above. In *S.* sp. from Central America (*c*), the sheet was dense, there was apparently also a second sheet below, and the tangle above was strikingly extended (see also fig. 9.9d). The web in *c* represents a convergence with some linyphiids) (*a* from Eberhard et al. 2008b, *b* and *c* photographs by Gustavo Hormiga).

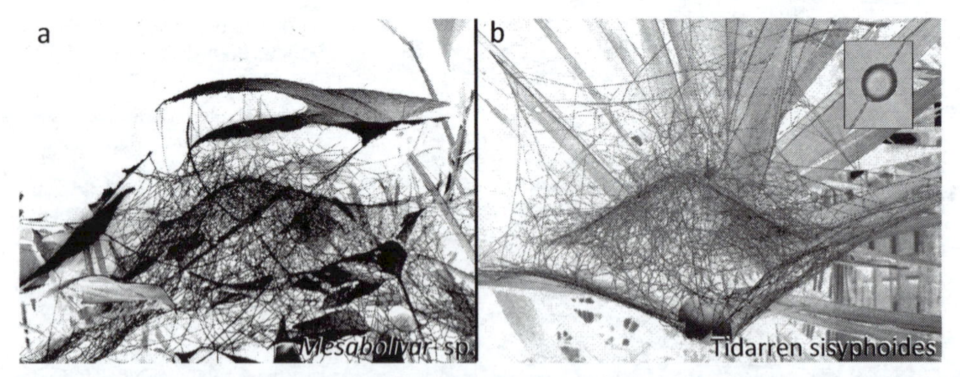

Fig. 9.15. The webs of the distantly related pholcid *Mesabolivar* sp. (#1384) (*a*) and *Tidarren sisyphoides* (THD) (*b*) illustrate two common evolutionary patterns, "theme and variations" and rampant convergence. The *Mesabolivar* web resembled the webs of other pholcids like *Modisimus guatuso* (fig. 5.2) and *M. bribri* (fig. 8.7) in having a domed horizontal sheet; but it differed in having a more extensive tangle above (in which *Philoponella* sp. (UL) often built its orbs), and in seldom being associated with large objects such as tree trunks or large leaves. It also resembled the web of *T. sisyphoides* in having a domed sheet that was moderately sparsely meshed, and an extensive tangle above. The *T. sisyphoides* web (in which *P.* sp. also built orbs) differed in having a cup-like sheet at the bottom, below the dome, and sparse sticky droplets (inset) (*b* from Madrigal-Brenes and Barrantes 2009, courtesy of Gilbert Barrantes).

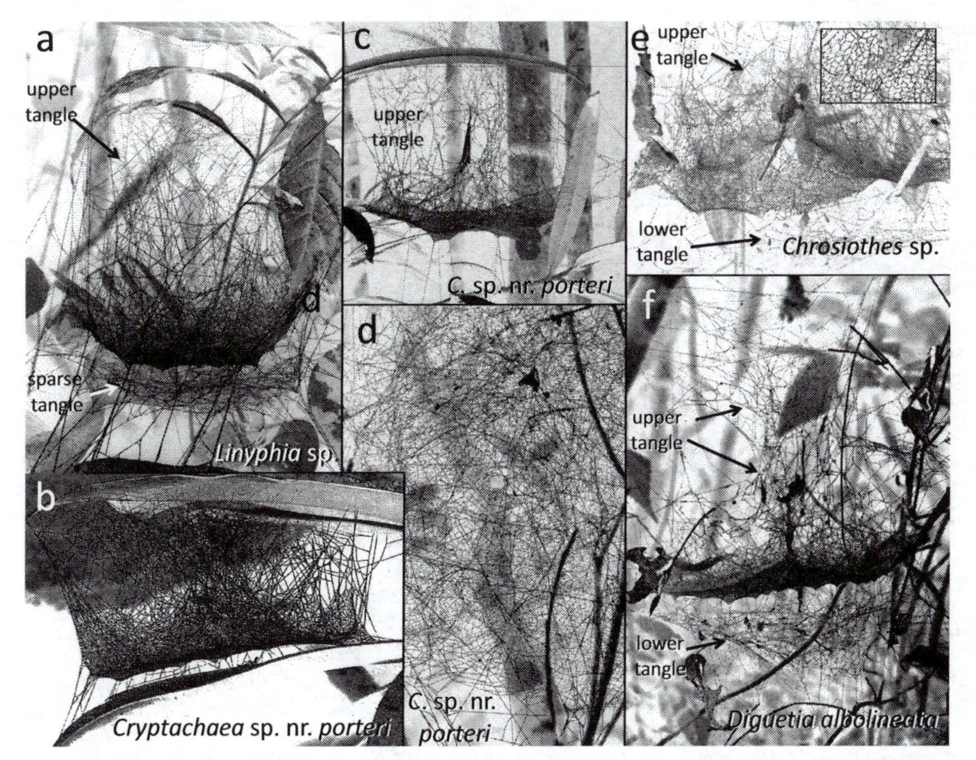

Fig. 9.16. There were variations even in superficially similar aerial sheet webs. The classic web of *Linyphia* sp. (LIN) (#3370) (*a*) had a cup-shaped sheet with downward-projecting dimples, a dense tangle above, an open space just below, and a smaller, more sparse tangle below. The sheet of the australian *Cryptachaea* sp. nr. *porteri* (THD) (#3696) (*b–d*) also had downward directed dimples, an open space just below, and a tangle above; it differed in having a sheet that was more nearly flat rather than curved, having a curled leaf retreat suspended in the upper tangle, and almost lacking a tangle below; some sheets were built just above the upper surface of a leaf (*b*), but not others (*c*). The pattern of lines in the sheets of both these species was irregular, though the sheet of *C.* sp. (*d*) was slightly more sparse near the edge (lower margin of *d*) and had many curved lines that did not form "V's" at the edge (see, in contrast, the sheets of the pholcids *Modisimus* spp. in figs. 5.2d, 8.7). The sheet of *Chrosiothes* sp. (THD) (*e*) (seen from slightly below) had still another combination of traits: a slightly domed shape that curled upward at the edges; sparse downward dimples; an only moderately dense upper tangle with a curled leaf retreat and a sparse lower tangle; a large open space below the sheet; and, most striking, a very regular, fish-net pattern of lines in the sheet (inset) (see also Eberhard et al. 2008b). The sheet of the distantly related diguetid *Diguetia albolineata* (*f*) also had a shallow dome and curled upward at the edges as in the *Chrosiothes* web, a tangle above with suspended retreat (in this case made from small pieces of detritus fastened together rather than a leaf), a large space immediately below the sheet, and abundant downward dimples. The mesh of the sheet was irregular. In sum, the common category of horizontal "aerial sheet web" used in some recent studies of web evolution clearly hides substantial diversity (see also the domed sheets in fig. 9.15) (*b-d* from Eberhard et al. 2008b).

rid *Fecenia* (box 9.1; fig. 9.18; Agnarsson et al. 2013), the desid *Matachia* (Griswold et al. 2005), and the synotaxid *Synotaxus* (Eberhard 1977a; Agnarsson 2003b) had tense, strong frame lines supporting multiple parallel or radial lines that were in some cases only slightly deflected by the lines attached to them. Just as clearly, the titanoecid *Titanoeca albomaculata* completely lacked frame lines. *Synotaxus turbinatus* constructed non-sticky support lines that served as bridges (Eberhard 1977a), like the temporary spiral of orbs weavers, while *T. albomaculata* lacked a temporary spiral (Szlep 1966a). *Titanoeca*, *Matachia*, and *Fecenia* built clear hub-like concentrations of non-sticky lines, while the webs of *Synotaxus* spp., and two independently derived theridiids, *Achaearanea* sp. (fig. 9.12) and *Theridion hispidum* (Eberhard et al. 2008b), lacked any hint of a hub. The "spanning" lines that supported geometrically regular array of zigzag sticky lines in webs of the dictynid *Mallos hesperius* were parallel to each other, running from one edge of the leaf to the other, rather than radial (fig. 10.14).

The likely adaptive explanations for these convergences are diverse. Take, for example, the convergent loss of webs in the tetragnathids the spiny clade of *Tetragnatha* species in Hawaii (Gillespie 1999) and *T. squamata*

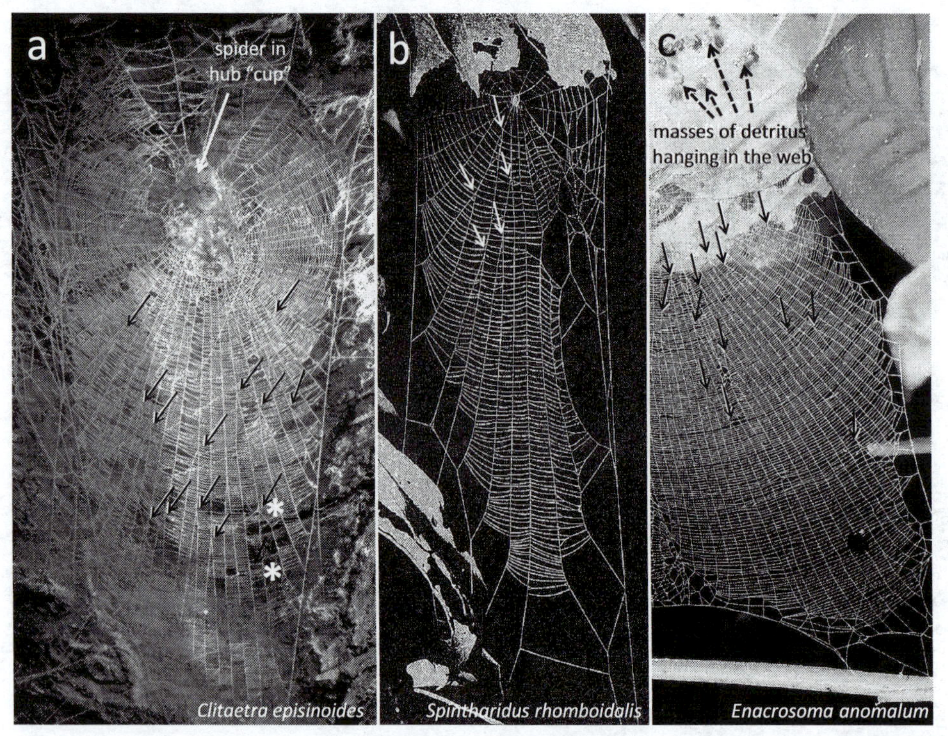

Fig. 9.17. The split, "tertiary" radii (whose origins are marked with arrows) probably evolved independently in these three highly asymmetrical orbs. Most of the tertiary radii were similar in that they diverged gradually rather than suddenly from the radii where they originated (a few exceptions, which may have originated on the temporary spiral, are marked with with "*" in *a*). They differed, however, in some other respects. The tertiary radii originated in the middle portion of the vertical, greatly expanded lower portion of the "intermediate trunk orb" of the nephilid *Clitaetra episinoides* (NE) (*a*) (as also occurred in *Nephila* spp.—see fig. 4.4); but they originated nearer the hub in both the vertical web of *Spintharidus rhomboidalis* (AR) (*b*) and the horizontal web of *Enacrosoma anomalum* (AR) (*c*) (seen from below). The outer portions of both the secondary and tertiary radii in the lower portion of the *S. rhomboidalis* web curved sharply downward (*b*), while the radii were nearly straight in the other two species. The photo of the *E. anomalum* web shows only the newly repaired portion of a larger web; the shadows of the masses of detritus that dangled on short lines below the web are indicated with dashed arrows) (*a* courtesy of Matjaž Kuntner; *b* from Levi 2008, courtesy of Jonathan Coddington).

in Japan (Shinkai and Takano 1984), and in the araneid genera *Archemorus* (Robinson and Robinson 1980), *Neoarchemorus* and *Arkys* (Stowe 1986). All have parlayed quick reactions and long spiny legs into an ability to snatch up passing prey. Older spiderlings and adults of *Pachygnatha* (Bristowe 1929; Martin 1978; Platnick et al. 1991) and the Japanese tetragnathid *Metellina ornata* (also with long spiny legs—Shinkai and Takano 1984) also snatch prey out of the air, even though young spiderlings still build horizontal orbs (Martin 1978; Shinkai 1998b). In contrast, web loss in *Celaenia* (AR) was associated with attracting moth prey using a chemical that mimics their long-distance sexual attractant pheromones (McKeown 1952). Convergence on web loss in the specialized spider predator *Chorizopes* sp. (AR) is associated with a still different set of traits—oversized chelicerae, a fast-acting venom, and the otherwise suicidal tendency to walk onto the webs of larger spiders and threaten them, luring their victims into striking range (Eberhard 1983). The entire family Mimetidae may have discarded orbs (its phylogenetic position is uncertain) to specialize on capturing other spiders in their webs (Bristowe 1954; L. Benavides pers. comm.), though one unidentified spider rested on the occupied sheet web of a pholcid and rapidly attacked prey that fell onto the sheet (while the smaller owner was immobile) (WE). Web loss also apparently evolved repeatedly as part of a kleptoparasitic lifestyle (Lopardo et al. 2011) in the small symphytognathoids *Sofanapis antillanca* (AN) (Griswold et al. 2005), *Curimagua bayano* (SYM) (Vollrath 1978), and *Mysmenopsis* spp. (MYS) (Coyle and Meigs 1989, 1992; Eberhard et al. 1993) and in the theridiid genera *Argyrodes*, *Faiditus*, and

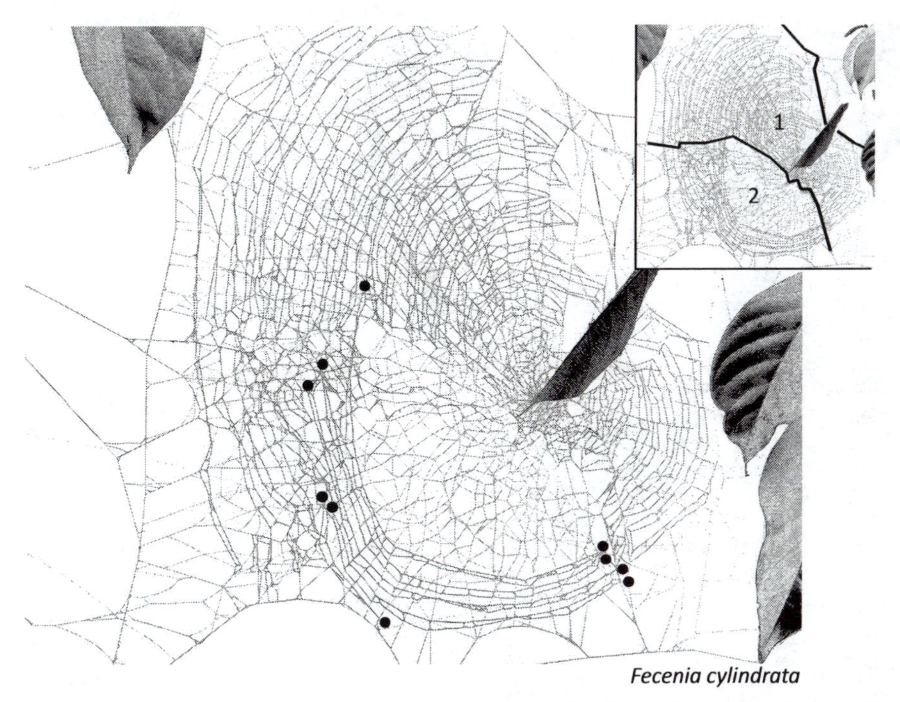

Fig. 9.18. This "pseudo-orb" web of the psechrid *Fecenia cylindrata* from China differs dramatically from the web of *F. ochracea* from New Britain (frontispiece). Some of the differences are probably due to the *F. cylindrata* web having a repaired sector (labeled 2 in the inset) in which sticky spiral production was not finished (probable turnbacks of the sticky lines in section 2, identified by the greater retention of powder on sticky lines, are marked with dots). The non-sticky lines in both the presumed repair sector and other portions of the web give the impression of a disorderly, loosely-meshed sheet rather than a radial organization. Puzzlingly, non-sticky lines seem to have two thicknesses (judging by their coverings with powder). There is no sign of the sheet-like modifications of spiral lines in the inner portion as in *F. ochracea*. This surprising degree of intra-generic divergence, along with the indication that these spiders can execute highly organized web repair behavior (in some ways repair seems more challenging behaviorally than building an web from scratch—see section 6.3.8, appendix O6.3.8.3) suggest that further study of the webs and building behavior of this extraordinary genus would be rewarding (photo from Bayer 2011, courtesy of Jeremy Miller; this photo was published under a Creative Commons Attribution License, which permits unrestricted use, distribution, and reproduction provided the source is cited).

Rhomphaea (Vollrath 1979; Whitehouse 1986; Eberhard et al. 2008b), which steal prey from the webs (or even the jaws!) of other spiders. The orb weaver *Neoscona neoteis* (AR) facultatively robs webs from *Argiope argentata* (AR) and also builds orbs of its own (Armas and Alayón 1987). The scytodid *Scytodes* sp. facultatively hunts both with and without a web (a three-dimensional tangle of non-sticky lines over a sparse sheet close to a curled leaf) (Li et al. 1999).

In other groups webs have convergently been reduced but not lost. Kaston (1964) and Shear (1986) argued that there is a pattern for spiders with reduced webs to tend to manipulate their webs more actively. There are many examples that support this idea, but a few that contradict it. For instance, *Deinopis* and *Menneus* actively move their reduced orbs, pressing them onto their pedestrian prey or expanding them into the paths of prey flying nearby (Akerman 1926; Robinson and Robinson 1971; Coddington and Sobrevila 1987); the gradungulid *Progradungula carraiensis* springs her probably reduced web (it is so small and sophisticated in design [fig. 10.9] that it seems likely to be derived from a larger, less organized ancestral design), using her legs III to catch the prey that she scoops upward with her legs I and II (Gray 1983); *Mastophora, Dichrostichus,* and *Cladomelea* swing a large sticky ball toward nearby prey (Akerman 1923; Eberhard 1980a; Yeargan 1994); *Theridiosoma* spp. spring their entire orbs (possibly reduced with respect to other orbs) toward prey flying nearby (McCook 1889; Wiehle 1929; Shinkai and Shinkai 1985); and *Wendilgarda* sp. sometimes tows her undoubtedly reduced single line web skating erratically below her across a water surface to capture water striders (Eberhard 2001a). In several other groups with reduced webs, such as *Hyptiotes*

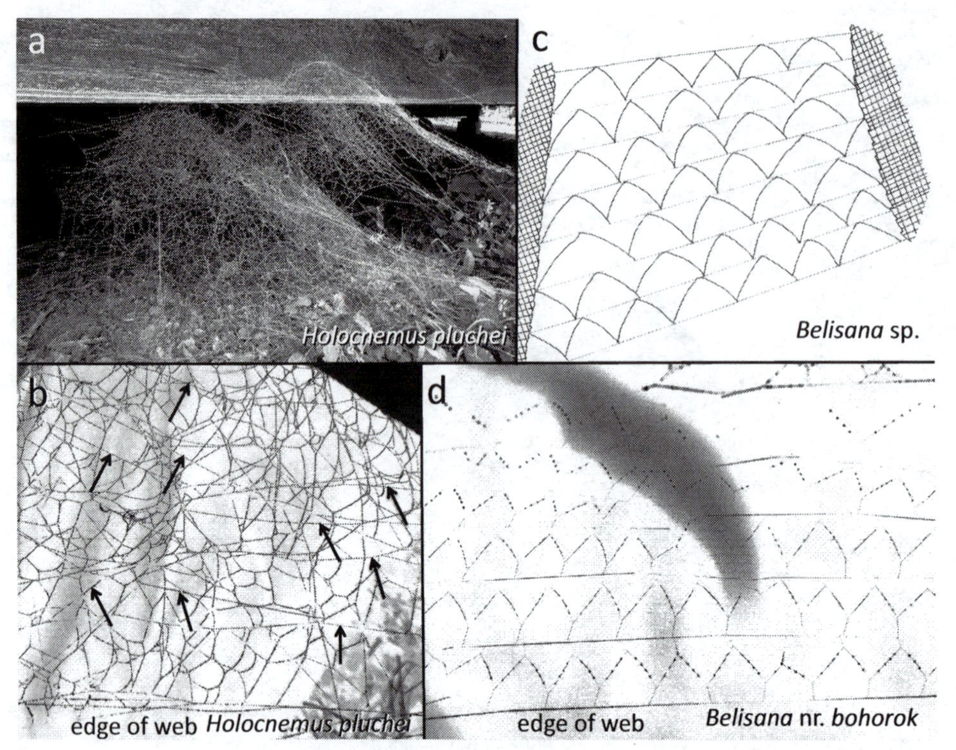

Fig. 9.19. Two apparent trends in pholcid webs are illustrated here. Webs often have asymmetrically domed sheets, as in *Holocnemus pluchei* (*a*). The sheet in a closeup view near the edge (*b*) appears to have a combination of relatively long "skeleton" web lines (arrows; see also figs. 5.2, 9.9b) and shorter, lax lines (*b*). In *Modisimus guatuso* the lax lines in a domed sheet carried sparse sticky droplets, and were laid with wrapping movements of legs IV onto a skeleton of more taut lines (fig. 5.2, Briceño 1985; Eberhard 1992b). The small, very sparse sheets of *Belisana* sp. (*c*) and *B.* nr. *bohorok* (*d*) were approximately planar rather than having a dome, but also had an apparent skeleton of long, nearly parallel and apparently taut lines that supported shorter, apparently lax and in this case geometrically regular lines. At least in *B.* nr. *bohorok* these lines carried small, evenly spaced balls of adhesive along part of their length (*d*) (*c* from Deeleman-Reinhold 1986; *d* courtesy of Bernhard Huber).

(Wiehle 1927; Marples and Marples 1937), *Miagrammopes* (Akerman 1932; Lubin et al. 1978), and *Phoroncidia* (= *Ulesanis*) spp. (Marples 1955; Eberhard 1981b; Shinkai 1988a), the spider sags her web actively when a prey strikes it. In contrast, the theridiid *Argyrodes attenuatus*, whose web of a few non-sticky lines serves only as a landing site or perch for insect and spider prey, stalks its prey very slowly and carefully (for up to five min) before launching a rapid wrapping attack with sticky silk (Eberhard 1979a). Convergent web reduction in the theridiids *Dipoena* spp. (Gastreich 1999; Davidson 2011) and *Steatoda fulva* (Hölldobler 1970) and complete web loss in *Euryopis funebris* (THD) (Carico 1978) are all associated with frequenting sites where their ant prey are concentrated, but not with web manipulation.

The combination of high diversity and repeated convergence in web form also occurs within smaller taxonomic groups. The large family Theridiidae is the queen in this respect (box 9.2). Widespread diversity, convergence, and intermediate forms constitute such an evolutionary tangle in Theridiidae that the title of a recent publication that mapped the web forms of just over 150 species onto a relatively well-resolved phylogeny mentioned extracting "chaos from order" (Eberhard et al. 2008b) (fig. 9.13, box 9.2). Of 22 web traits examined, 14 varied intra-specifically, and 16 evolved convergently in at least two lineages (there were up to nine independent origins for a single trait). And this study undoubtedly gave severe underestimates of diversity and convergences, because web data were available for only 7% of the known species in the family. Of the 22 web traits, five had a "consistency index" or CI (the minimum possible number of steps in the phylogenetic tree/the observed number of steps) of less than 0.15 (for comparison, only

15 of 242 morphological traits in the same spiders had values this low) (Eberhard et al. 2008b). Theridiid web traits had a much lower mean CI (0.30 ± 0.17) than those of orb weavers (0.63 ± 0.26) (Kuntner 2005, 2006) or Nephilidae (0.49) (Kuntner et al. 2008a). This index was also lower than the behavioral and morphological traits in 22 taxa, including insects, arachnids, shrimp, and vertebrates (respectively, 0.84 ± 0.14 for behavior and 0.84 ± 0.12 for morphology) (de Queiroz and Wimberger 1993).

Some of these convergences involved behavior patterns that are not trivially simple, such as reorganizing the web to include or omit sheets, to include or not include sticky lines, or to alter the form of the sheet (Eberhard et al. 2008b). Fragmentary information on the much more poorly known families Synotaxidae and Pholcidae suggests that they resemble theridiids in also having wide intra-familial web diversity (figs. 9.8, 9.9, 9.19; Deeleman-Reinhold 1986; Japyassú and Macagnan 2004; Eberhard et al. 2008b).

In summary, convergences have occurred in a wide variety of web traits. Convergences in web loss show the same pattern of a wide diversity of selective reasons for web loss. Further examples of convergences with respect to tangle webs accompanying orbs, orbs built by mature males, and orb angles with horizontal are presented in appendices O9.1–O9.3.

9.2.3 ABUNDANT INTERMEDIATE FORMS AND A SUMMARY

A third major pattern in web diversity is the widespread existence of multiple intermediate forms. This is perhaps no surprise, given the first two patterns, but it is not a necessary consequence. If a given design is advantageous and multiple lineages converge on this design, selection might have eliminated intermediates. But intermediate web designs are often common. Take, for example, the web design in which regularly spaced sticky lines are laid on a scaffold of strong, non-sticky lines, as occurs in orb webs. At least partial convergences occurred in other, non-orb groups (table 9.1). including *Matachia* (fig. 1.3), *Fecenia* (fig. 9.18), *Titanoeca* (fig. 10.14), *Achaearanea* (fig. 9.12), *Synotaxus* (fig. 9.12), *Mallos*, and *Theridion* (table 9.1). Despite the likely intense and sustained selection that in the orbicularian lineage resulted in the elaborate set of traits used to produce the highly uniform and delicately adjusted spacing between sticky lines in orb webs (chapters 6 and 7), selection on these other lineages has allowed some with only very approximate uniformity of spacing to persist.

Summing up, the spectacular diversity of web designs is combined with a bewildering array of multiple, convergent reinventions and partial reinventions of many different designs (tables 9.1, 9.2). These multiple convergences suggest that there are only a limited number of ways to make effective traps for prey, and that these have been repeatedly discovered in different evolutionary lineages. The combination of high diversity and rampant convergence suggests that modern web spiders are the result of an especially ancient and thorough exploration of the adaptive space associated with webs. Spider webs are ancient, probably dating back 245 million years (Vollrath and Selden 2007); orbs may have arisen about 190 Mya, and the orb-weaving genus *Nephila* (NE) is at least 165 million years old (Selden et al. 2011). The old hope of tracing web evolution by enumerating intermediate designs (e.g., Kaston 1964; Kullmann 1972) has faded and is now gone. The problem for tracing histories is now an overabundance rather than a lack of intermediates. In addition, our improved understanding of phylogenetic relations (section 1.6) has failed to conform to many former simplistic schemes of web evolution.

One probable consequence of high diversity and abundant intermediate forms with varying combinations of traits is that it has been difficult to make valid generalizations regarding web evolution and function (table 9.4). Many apparently logical potential generalizations run up against clear exceptions. This makes it more difficult to understand the causes of web evolution, but helps us steer clear of overly simplistic conclusions.

9.2.4 INTRA-SPECIFIC ALTERNATIVE WEB DESIGNS

Arachnologists have probably generally done a poor job of documenting variation in web construction behavior and designs. I suspect that many other species will eventually resemble the history of the zoropsid *Tengella radiata*, in which an original description of "the" web (Eberhard et al. 1993) was followed by subsequent observations of quite different web designs in juveniles (Barrantes and Madrigal-Brenes 2008) and adults (fig.

Table 9.4. Beautiful theories and cruel data: many apparently convincing theoretical expectations about orb webs are contradicted by "inconvenient" facts. This table gives a sampler of such disappointments (with my own failures perhaps over-emphasized). This "contrariness" of spiders may be the result of their especially thorough exploration over evolutionary time of the possible ways to use silken webs to capture prey (see text).

THEORETICALLY

Larger prey are more important nutritionally, and also tend to have more kinetic energy: therefore the stopping function of an orb is so important that its swamps out the importance of other functions (Sensenig et al. 2010a; Blackledge 2011).

BUT

There are multiple examples of evolution of webs that have evolved to have designs with *reduced* stopping abilities, such as many theridiosomatids, tetragnathids; and patterns in spacing of sticky spiral lines testify to the selective importance of retention.

THEORETICALLY

Orbs built in situations in which prey velocities are known to be relatively low should be relatively radius-poor (WE).

BUT

Trunk webs are not radius-poor.

THEORETICALLY

The highly regular geometry of orb webs demonstrates the strong selective advantage of regular over irregular geometries (Witt 1965; many others).

BUT

Numerous groups show only intermediate levels of geometric regularity, and the irregularity itself shows consistent patterns, and is thus not simply due to construction errors (section 4.3).

THEORETICALLY

Website tenacity will be lower in groups in which spiders re-ingest old webs, because re-ingestion reduces losses when a spider moves (WE).

BUT

The lowest tenacities yet measured occur in species of Pholcidae that abandon their old webs intact.

THEORETICALLY

Radius-rich orbs, a design that improves the stopping function of an orb, should tend to be vertical, because prey velocities will be greater when they are moving horizontally (WE).

BUT

There are many groups with approximately horizontal orbs that build radius-rich orbs, such as *Mangora* (AR) and *Uloborus* (UL).

THEORETICALLY

Asymmetrical elongations of orbs should be correlated with the direction in which the spider faces while resting at the hub and with gravity, because she can attack prey more rapidly in the direction she was facing and when moving downward (Masters and Moffat 1983; Nakata and Zschokke 2010; Zschokke and Nakata 2015).

BUT

The predicted correlation occurs in some groups, such as *Cyclosa* spp. and *Neoscona*; but it does not occur in others, such as *Azilia*, *Uloborus*, and *Verrucosa*.

THEORETICALLY

Sticky spiral–rich orbs that are apparently designed to retain moths ought to be vertical (improves retention of prey) (WE).

BUT

Some are consistently vertical, such as *Scoloderus*, *Deliochus*, and *Pozonia*; but some are not, such as *Poltys*, and *Acacesia* (table O3.1).

Table 9.4. Continued

THEORETICALLY

In those groups in which orbs have become reduced, the spider's active participation in prey capture (via behavior or chemical attraction) is greater (Kaston 1964).

BUT

The correlation occurs in several groups, such as the uloborids *Miagrammopes*, *Hyptiotes*, the araneids *Mastophora* and allies, Cyrtarachneae, *Chorizopes*, and the theridiids *Phoroncidia* and *Chrosiothes tonala*; but it is lacking in others, such as the theridiosomatids *Wendilgarda* spp., *Ogulnius* spp., and the theridiids *Dipoena* spp.

THEORETICALLY

The invisibility of webs built and operated at night should be correlated with design differences such as more dense sticky spiral lines that are disadvantageous during the day because of their greater visibility (Craig and Freeman 1991; Craig 2003).

BUT

No consistent differences in design have been documented between nocturnal and diurnal orbs (except for the lack of stabilimenta in nocturnal orbs).

THEORETICALLY

There are four reasonable theories to explain stabilimentum function (table 3.7).

BUT

None of the theories escapes serious empirical contradictions (section 3.3.4.2, tables O3.3-O3.5).

THEORETICALLY

Orb designs might differ in different species in order to avoid competition for the same prey (section 3.2.6).

BUT

The few sets of convincing experimental data that are available strongly indicate that species do not generally compete for prey.

THEORETICALLY

Orb web geometry represents the peak of web evolution toward increasingly effective and efficient trap designs (Witt 1965; Shear 1986).

BUT

Secondary loss of orb webs has occurred repeatedly, and one loss led to the most successful evolutionary radiation (in terms of numbers of species) in spiders (e.g., Garrison et al. 2016; sections 1.6, 10.10).

THEORETICALLY

Vertical orbs are better able to intercept prey (Chacón and Eberhard 1980), and vertical orbs evolved from horizontal orbs and then radiated widely due to this advantage (Bond and Opell 1998).

BUT

In some contexts spider web mimics that were less nearly vertical intercepted more prey (Bishop and Connolly 1992), and evolutionarily successful species-rich taxa such as *Tetragnatha* and *Leucauge* build approximately horizontal orbs, despite the fact that there is ample variation in these groups on which selection could act to favor more vertical orbs. In addition, horizontal orbs have arisen repeatedly from ancestors with vertical orbs, for instance in the araneids *Mangora*, *Enacrosoma*, some *Eustala*, and most symphytognathoids.

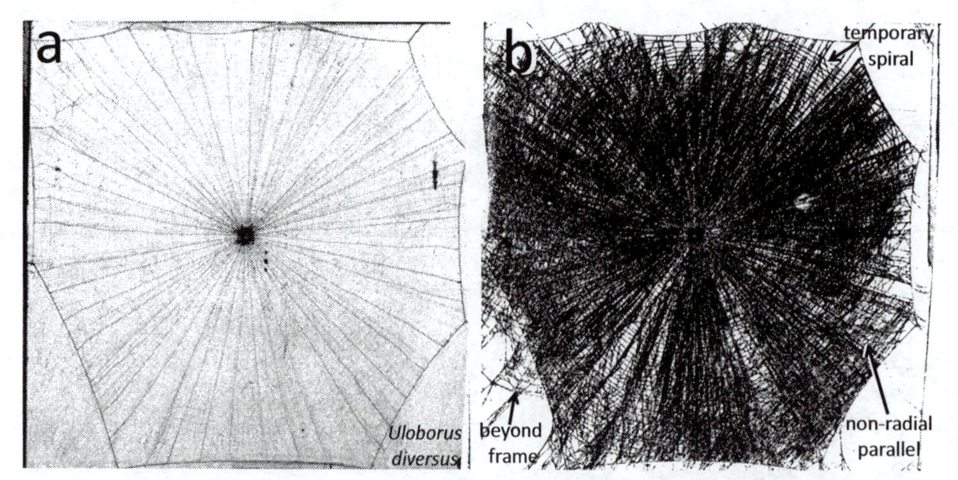

Fig. 9.20. Two views of the same web of a second instar *Uloborus diversus* (UL) spiderling, which is thought to lack a functional cribellum. When illuminated strongly from all sides against a black background, the uncoated web appeared to consist mainly of densely arranged fine radial lines (*a*). But it was revealed to be much more complex when powdered with cornstarch (*b*), and had several unique traits: relatively tightly spaced temporary spirals that were presumably non-sticky (the spiderlings are said to lack a functional cribellum, though this detail was not checked); many supplementary radial lines that were built following completion of the temporary spiral; and many very fine, non-radial supplementary lines that sometimes extended beyond the frame lines. The glandular origins of the fine supplementary lines are unknown. It is a mystery how the many non-radial lines were laid, because during direct observations of construction behavior spiders moved almost exclusively in radial directions, and never moved beyond frame lines (from Eberhard 1977d). The webs of mature males of *Uloborus* spp., *Philoponella* spp. (UL), and *Zosis* sp. n. (UL) had similar designs (fig. 9.21).

10.7). I suspect that the apparent rarity of discrete alternative designs in spider webs is an artifact of lack of attention to variation by observers. I will describe a few extreme and especially interesting cases to illustrate.

Alternative designs appear to be common in the orb-weaving uloborids. When the second instar spiderling emerges from the egg sac, the spiderling lacks a functional cribellum, and produces a "juvenile" web, with a dense mat of very fine lines that are laid onto a skeleton orb that contains anchor, frame, hub, radii, and temporary spiral lines (fig. 9.20); juvenile webs have been seen in *Uloborus* spp. (Wiehle 1927; Szlep 1961; Eberhard 1977d; Opell 1979; Lubin 1986), *Philoponella* spp. (WE), and *Zosis geniculata* (WE). *Conifaber parvus* continues to build juvenile orbs when it is an adult female (Lubin et al. 1982) (but, surprisingly, the congeneric *C. yasi* builds typical sticky spiral loops—Grismado 2008). Mature males of some uloborid species also build juvenile webs (fig. 9.21, Eberhard 1977d). When the spider molts to the third instar, it acquires a cribellum and can make orbs with sticky spiral lines. But the transition to typical orbs is sometimes incomplete, and can vary in different individuals of the same species; later instars of *Uloborus* and *Philoponella*, for instance, sometimes build webs that include more or less extensive mats of very fine lines as well as sticky spiral lines (Lubin 1986; WE), as do "senile" *Uloborus* females (fig. 6.22).

Interestingly, two derived uloborid genera, *Hyptiotes* and *Miagrammopes*, which have independently evolved to build secondarily reduced webs (fig. 1.13), have both discarded juvenile webs (Lubin 1986). The spiderlings emerge from the egg sac in the second instar, but molt again in the near vicinity after several days without building a prey capture web. In the third instar they have functional cribella, and disperse to build their typical, derived webs (Marples and Marples 1937; Lubin et al. 1978; Opell 1982; Lubin 1986; Opell 2001). The reasons for the lack of a cribellum in second instars in uloborids is a mystery. This lack and the accompanying distinctive juvenile web may be derived traits, as I know of no report of juvenile webs that lack cribellum silk in other cribellate families such as Dictynidae, Desidae, or Psechridae that also build aerial webs (though admittedly they are poorly studied). The funnel web zoropsid *Tengella radiata* only began to build cribellum lines several instars after having emerged from the egg sac (Barrantes and Madrigal-Brenes 2008). The extremely fine lines of the uloborid juvenile web mat are easily broken, and retain

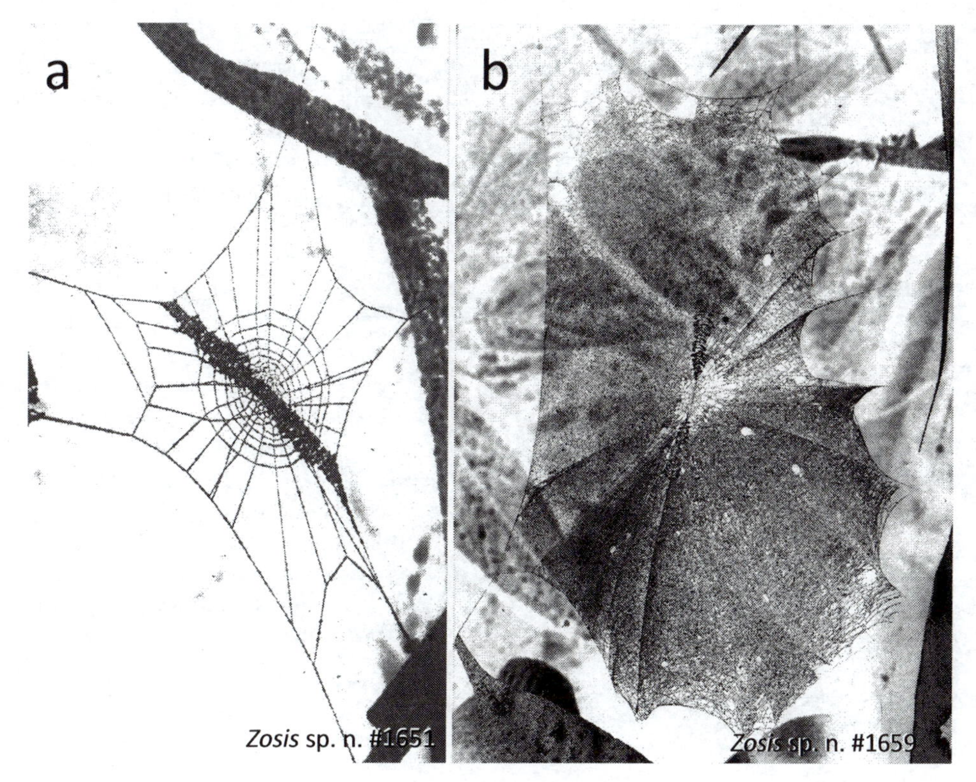

Fig. 9.21. These field webs of mature males in an undescribed species of *Zosis* sp. n. (UL) (#1659) represent a pair of alternative phenotypes. The web in *a* is a resting web, with short radii, no prey capture lines, and a silk stabilimentum (presumably for visual defense). The other form (*b*) is for prey capture, with a dense mat of fine lines just like those in the orbs of the second instar spiderlings (fig. 9.20) and the mature males of other uloborids (fig. 10.36c); only a few radial lines supported the rest of the nearly horizontal mat.

prey poorly (Lubin 1986; WE). The tiny spiderlings resting at the hub are difficult to see from above, and one possible advantage of juvenile webs is to hide the spiderling (Lubin 1986); another possible advantage is to avoid the costly process of combing out cribellum silk.

Tretzel (1961) found that the webs of young *Coelotes terrestris* (Agelenidae) associated with moss consisted of especially narrow, shallow, and highly branched tubes, and lacked the usual prey capture sheet of older individuals; he did not describe the method of prey capture.

There are other intra-specific alternative designs not associated with either ontogeny or with differences in the types of silk. The facultative "stick orbs" of the anapid *Anapisona simoni* (Eberhard 2011) and *Tetragnatha* sp. (fig. 10.26) probably represent adaptations for hiding the spider at the hub. The typical water webs, the low land, and the high land webs of the theridiosomatid *Wendilgarda galapagensis* (Eberhard 1989a) were, in contrast, associated with differences between subhabitats with respect to available water surfaces to which to attach lines. The two webforms of the mature male webs of *Zosis* sp. n. (#1651) may serve as defense and for prey capture (fig. 9.21). The araneid *Cryptaranea atrihastula* facultatively omitted the top half of its ladder web when its retreat was at the top of the cage (Forster and Forster 1985). Webs of the theridiid *Achaearanea hieroglyphica* (THD) varied in several respects (fig. 9.22) (Agnarsson and Coddington 2006).

9.2.5 BEHAVIORAL BRICKS AND BUILDINGS

While diversity and convergence are the rule in web designs, the distribution of the behavior patterns used to build different designs suggests much stronger evolutionary conservatism—for instance, in which legs are used to hold which lines at the moment when one line is attached to another, or which legs explore to find which lines. This conservatism is clearest in the eight orb-weaving families in Orbiculariae (where data are most abundant) (Eberhard 1982) (observations are too scarce in other taxa to draw confident conclusions—table 5.2). Particular variants are conserved over relatively large taxonomic groups, even though the designs of their finished webs vary substantially (e.g., Eberhard and Barrantes 2015). It is as if the behavioral "bricks" that orb weavers use vary much less than the "buildings" (webs) that they construct using these bricks. This pattern is described in more detail in sections 10.8 and 10.9.

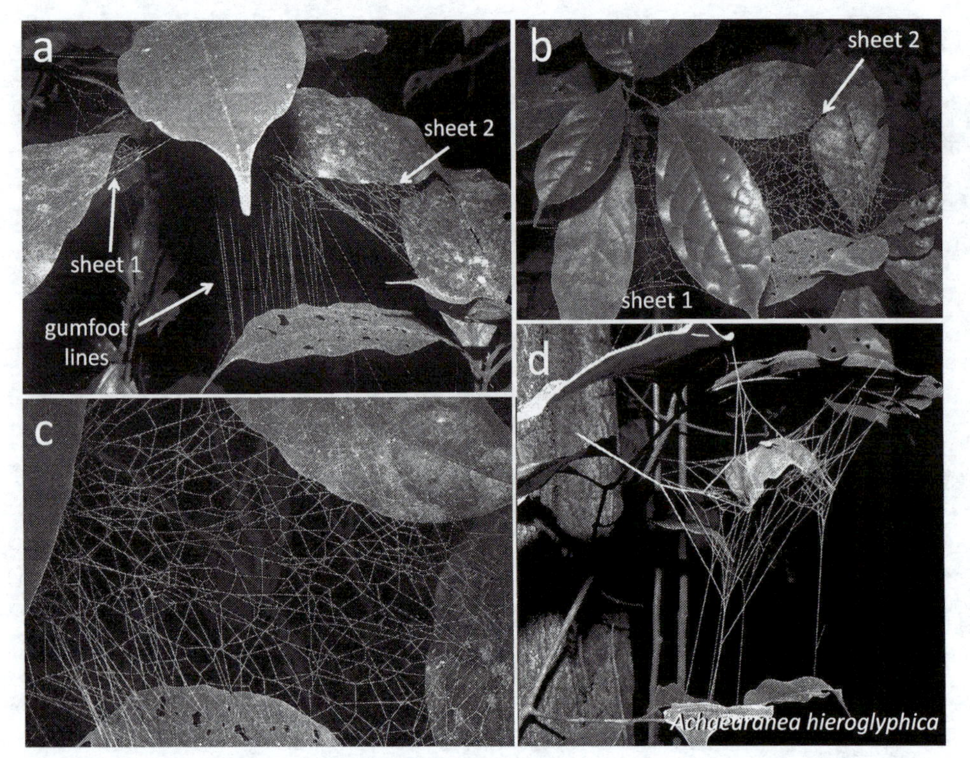

Fig. 9.22. The diversity of web designs extends to within-species differences in *Achaearanea hieroglyphica* (THD). One web (*a–c*) was centered under a broad leaf (in lateral view in *a*; from above in *b* and *c*). Two slanting, wide-meshed planar sheets extended beyond opposite sides of a loose tangle retreat under the leaf; many additional, nearly parallel gumfoot lines were attached to a flat leaf directly below the retreat (*a*). A second web (*d*) differed in not being under a leaf, in lacking a sheet, and in having only a sparse array of gumfoot lines below; instead it had a suspended, curled leaf where the spider rested (from Agnarsson and Coddington 2006, courtesy of Ingi Agnarsson).

9.2.6 ADAPTIVE CHEMICAL DIVERSITY OF SILK

New discoveries in the currently very active field of the biomechanical properties of silk have documented substantial diversity in an additional dimension—differences in the chemical composition and the physical properties of silk lines (section 3.2.9). Previous discussions tended to discuss "the" properties of dragline silk or sticky spiral silk, but recent work suggests that silk properties vary substantially, and are adapted to different needs in different species. I will describe one especially striking and well-studied example, and give a run-on list of other possible cases to give an idea of the potential scope of this evolutionary flexibility.

The mechanical properties of silk in the araneid genus *Cyrtarachne* (Shinkai 1982; Eberhard 1986a; Cartan and Miyashita 2000) and related groups, such as *Pasilobus* sp. (Robinson and Robinson 1975), *Paraplectana tsushimensis* (Chigira 1978 in Stowe 1986), and *Poecilopachys australasia* (Clyne 1973), have evolved in conjunction with unusual web designs. The webs of most species in this relatively basally derived group conserve recognizable orb-like traits (fig. 9.23) but are strongly reduced. Like the related bolas spiders (Scharff and Coddington 1997), most (but not all) of the prey of cyrtarachnines are moths (Suginaga 1963; Robinson 1975 and Shinkai et al. 1985 cited in Cartan and Miyashita 2000; Stowe 1986). Bolas spiders specialize on certain prey (male moths) using chemical mimics of their long-distance attractant pheromones (Stowe 1986; Stowe et al. 1987; Yeargan 1994). But chemical prey attraction seems to be ruled out in *Pasilobus* sp. and *Cyrtarachne* spp., because of the variety of moth species that the spiders of an individual species captured (up to 19 genera in 5 families in *C. inaequalis*), and because they capture both male and female moths in approximately equal numbers (Miyashita et al. 2001). The sticky lines in cyrtarachnine webs are so far apart and the numbers of radii are so low that often only a single sticky line intercepts a moth (Clyne 1973; Robinson and Robinson 1975; Cartan and Miyashita 2000).

After correcting for the effects of spider body size, Cartan and Miyashita (2000) found that the morphological and mechanical properties of sticky lines in *Cyrtarachne* spp. differed from those of other araneids that build typical orb webs in having sharply larger sticky droplets, thicker axial lines, greater strength (ability to absorb kinetic energy), and greater stickiness (fig. 9.24). The stickiness of the droplets also differed in being much more short-lived (as is also true in bolas spiders), disappearing completely only 3–11 hr after construction; in com-

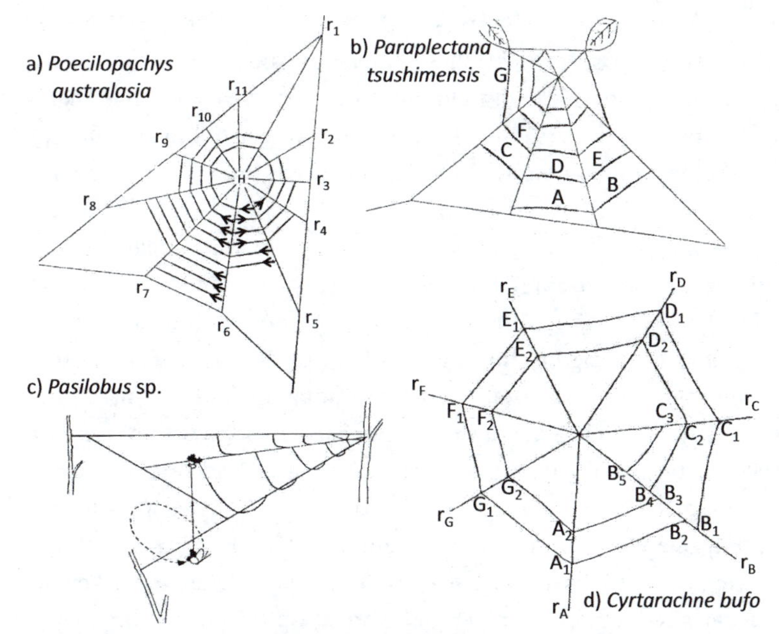

Fig. 9.23. The web construction behavior of these closely related araneid genera reveals both affinities and differences with the behavior of typical orb weavers (after Stowe 1986). *a*) The sticky lines in the web of *Poecilopachys australasia* were built from the outside in, but sticky line construction was repeatedly interrupted and then reinitiated (arrows). The first 3 sticky spiral loops were begun on radius r_6, the next two on r_5, the next three on r_6, and the innermost loop on r_5. None of the sticky lines continued past r_1, and most ended earlier; the innermost four loops were started twice, first going in one direction and then in the other. There was sticky material only in the middle of each segment of the sticky spiral (originally after Clyne 1973). *b*) The lettered segments of sticky spiral in the web of *Paraplectana tsushimensis* were also built from the outside in (in the order A to G), and only had sticky material on the central portion of each segment (originally after Chigira 1978). *c*) The two sector web of *Pasilobus* sp. is seen in oblique view, just after a moth contacted and broke one segment of sticky spiral line (at a specialized, low-shear junction) and the spider traveled down the central line and began to reel in the moth (originally after Robinson and Robinson 1975). *d*) The letters in the central portion of this web of *Cyrtarachne bufo* (frame lines are omitted) indicate how the segments of sticky spiral were built from the outside in, as approximate circles rather than as a spiral. The order in which the segments were built was B1–C1, C1–D1, around to A1–B2; then B3–A2, A2–G2, around to C2–B4; then finally B5–C3 (originally from Suginaga 1963). At least some of these web designs appear to be specialized to capture moths (see Stowe 1986; Cartan and Miyashita 2000).

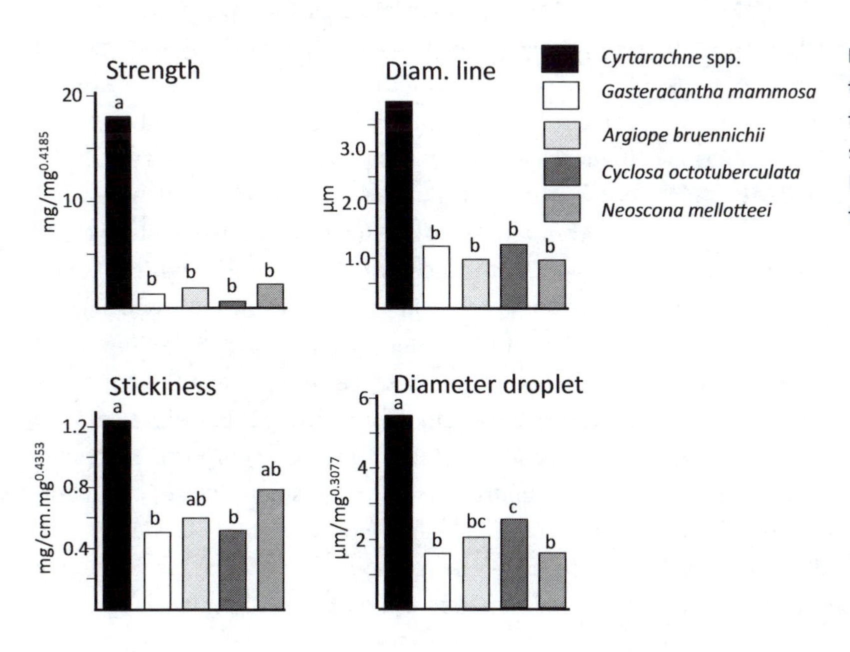

Fig. 9.24. The properties of the sticky lines in the webs of two species of *Cyrtarachne* (AR) that specialize on moths differed sharply in several respects from those of the sticky spiral lines of four other araneid species that build typical orbs (after Cartan and Miyashita 2000).

parison, the stickiness of spiral lines in the typical orb of *Larinioides cornutus* had not even begun to decrease (Cartan and Miyashita 2000). The more liquid nature of *Cyrtarachne* droplets may allow the adhesive to flow past the moth's detachable scales and reach its cuticle below. Differences in adhesive properties were accompanied by morphological differences in both the droplets and the surface of the axial lines when they were dried and viewed in the SEM. The sticky lines were widely spaced and lax, and sagged substantially under their own weight (fig. 9.23). In addition, some sticky spiral attachments to radii in *Cyrtarachne* (Suginaga 1963 in Cartan and Miyashita 2000) and also in its relatives (Clyne 1973; Robinson and Robinson 1975) were quite weak; they broke much more easily than those in typical orbs, leaving only one end of the sticky line attached to a radius (fig. 9.23c).

This combination of web design and unusual silk properties allows a *Cyrtarachne* and relatives to capture a moth with only a single sticky line, which breaks free from the web at one end and tethers the moth in flight until the spider reels it in (fig. 9.23c). The spiders are not large, but they make relatively large webs up to around 1 m in diameter (Stowe 1986), which do not represent disproportionally large investments in sticky silk compared with typical orbs (Cartan and Miyashita 2000). The wide spaces between sticky spiral lines and the horizontal orientation of the orb probably reduce the chances that other sticky lines will be damaged by a moth's struggles. These spiders may represent an intermediate stage between typical araneid orbs and the bolas spiders (Mastophoreae), which are thought to be closely related to them (Scharff and Coddington 1997). The capture webs of bolas spiders are often reduced to a single line with a sticky ball, or lost completely (Robinson and Robinson 1975; Eberhard 1980a; Stowe 1986; Cartan and Miyashita 2000).

A brief list of other correlations between altered properties of silk lines and web designs includes the following. In the uloborid genera *Miagrammopes* and *Hyptiotes*, the web is highly reduced, and the adhesiveness of the cribellum silk (corrected for differences in the spider's weight) was higher than that in five other typical orb-weaving genera in the same family. The webs of *Miagrammopes* (some of which are only a single line) are more highly reduced than those of *Hyptiotes* (which are sectors of an orb), and the stickiness of a *Miagrammopes* sticky line was 1.8 times that of a *Hyptiotes* sticky line (Opell 1999a). These differences are not known to be associated with chemical differences in the lines themselves (though I know of no studies of this possibility); rather the stickier *Miagrammopes* lines are composed of larger numbers of fine cribellum fibers, which in turn are due to larger numbers of spigots on the cribellum (Opell 1996, 2013).

The subtle correlation between silk stiffness and web design in the oecobiid *Uroctea durandi* is described in box 9.4. The strength and toughness of the major ampullate gland silk of both pseudo-orb weavers (*Fecenia*) and true orb weavers (orbicularia) have independently improved compared with that of relatives, presumably due to selection to improve the stopping abilities of their webs (fig. 9.25) (Blackledge et al. 2012a). In *Nephila clavipes*, dragline and sticky spiral silk are bright yellow in well-illuminated webs, a property that is thought to make the lines less visible to prey, and they are less intensely yellow at more shaded sites (Craig and Freeman 1991; Craig 2003). The huge araneid *Caerostris darwini* from Madagascar builds bridge lines of unequalled lengths (up to 25 m!) that sometimes span small rivers (Gregorič et al. 2011) and have unprecedented toughness (high strength combined with high elasticity) (Agnarsson et al. 2010). Their orbs per se are also large (up to 2.7 m^2), but have relatively unremarkable designs, with moderately low numbers of radii and relatively widely spaced sticky spiral loops (Blackledge et al. 2011; Gregorič et al. 2011). When spider size was controlled, spiders of several species that spun "lower quality silk" (judged by several traits with different weights, derived from a principal components analysis) compensated by building more dense orb designs (Sensenig et al. 2010a). The adhesiveness of the glue from aggregate glands varies among araneoids (Blackledge et al. 2011), and generally increased with spider size (Opell and Hendricks 2009). The strength of sticky spiral baselines in these species was elegantly adjusted to the strength of adhesion of the glue on these lines: the strength of adhesion never exceeded the strength of their baselines (fig. 3.32) (Agnarsson and Blackledge 2009). The complex morphology of the glue droplets on sticky spiral lines of some araneoids (see Opell and Hendricks 2010; Blackledge et al. 2011) intimates there may also be substantial taxonomic variation in their morphology.

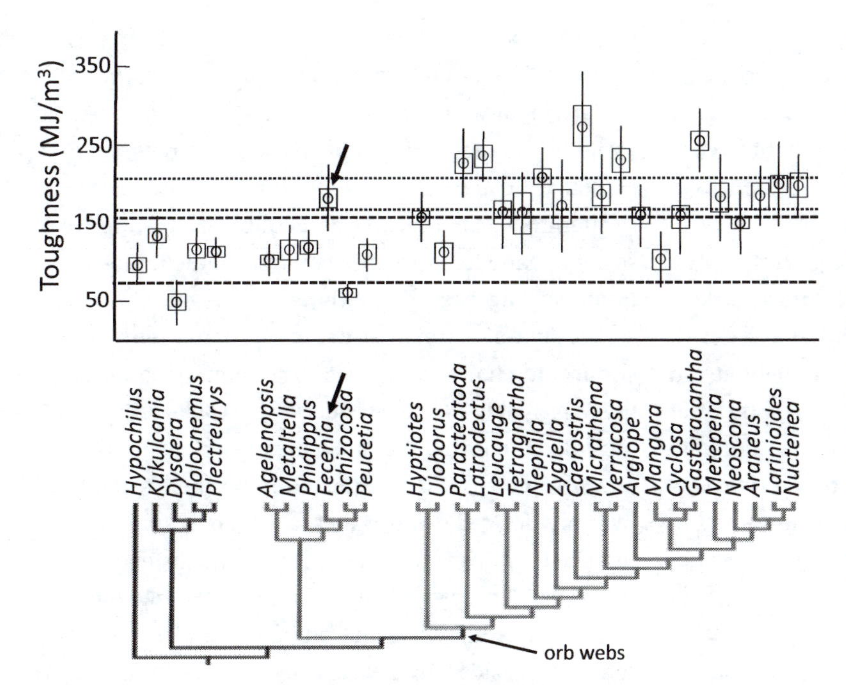

Fig. 9.25. The toughness of the major ampullate gland silk in the "pseudo-orb" genus *Fecenia* (arrow) is greater than that of related taxa and resembles that of orbicularians. The major ampullate silk of orb weavers and *Fecenia* is also stronger (with higher "true stress" values) than their relatives. The independent improvements in these two lineages may have evolved in association with the convergence in their web designs (boxes represent ± SEM, and tails represent ± SD for individual species; gray lines in the phylogeny indicate RTA clade and orb weavers; the horizontal dashed and dotted lines represent, respectively, the 95% confidence limits for mean performance of silk from the RTA clade and from orb weavers) (after Blackledge et al. 2012a, courtesy of Todd Blackledge).

9.2.7 DIFFERENCES BETWEEN CONSPECIFICS: ARE THERE "INDIVIDUAL STYLES" OF WEB DESIGN?

In some animal species, different individuals show consistent, repeated behavioral differences compared with other conspecific individuals of the same size, sex, and ontogenetic stage. An example in spiders is the funnel web spider *Agelenopsis aperta,* in which some individuals were consistently more aggressive to conspecifics than others (Riechert and Hedrick 1993). Some individuals of the black widow spider *Latrodectus hesperus* (THD) consistently tended to build webs with more gumfoot lines and also to respond more to vibration stimuli by attacking even when they were satiated, possibly aspects of an alternative foraging strategy (DiRienzo and Montiglio 2016).

It is not clear whether individual web weavers also present such consistent inter-individual variation in different aspects of the designs of their webs. There are some anecdotal descriptions of possible "individual styles." For instance, Witt et al. (1968) mention that two individuals of *Zygiella x-notata* (AR) differed consistently: one built an anchor line more or less directly below the hub of each of her vertical webs, while another built the two lowest anchor lines at the lateral edges of the web; this difference persisted even when the two spiders were placed in each other's cages. Another individual of this species always left one band of oversized spaces between the loops of the sticky spiral in the middle portion of her orb during about one month of her life in captivity (later the open space gradually disappeared). Different individuals of *Argiope argentata* (AR) showed different patterns of edge-to-hub difference in the diameters of sticky spiral baseline fibers (Blackledge et al. 2005a).

The possibility that genetic differences were responsible for web differences was suggested by finding that the means of several web traits differed in different groups of offspring of different female *Araneus diadematus* (AR) (Rawlings and Witt 1973) (in none of these cases, however, were other possible causes, such as differences in previous feeding, leg injuries, etc., checked systematically).

In sum, evidence suggesting that consistent differences occur in the orb designs among different individuals of a species is only anecdotal at the moment.

9.3 CONSEQUENCES OF THE FAILURE OF THE PREY SPECIALIST HYPOTHESIS FOR UNDERSTANDING DIVERSITY AND CONVERGENCE

One commonly mentioned idea is that differences in designs represent fine tuning of webs to capture different arrays of prey. But (with exceptions) most web spiders are apparently not prey specialists, but are instead prey generalists (section 3.2.5). With few exceptions, the bulk of the evidence for most groups argues against prey specialization (though there are lingering doubts concerning selection from rare, life-altering experiences—section 3.2.5.3). "Specialization" seems more likely to have occurred at more general levels (e.g., walking vs. flying prey) (section 3.2.5.9). Wise and his coworkers demonstrated that even closely related species with very similar web designs in the same subhabitat did not compete for prey (section 8.8, Wise 1995), and that most previous studies that purported to demonstrate inter-specific competition for food among web weavers (including his own) were not convincing. This is not to say that inter-specific competition for prey never occurs (Spiller 1984a,b, 1986). But competition for prey has probably had only a limited role in explaining evolutionary changes in web design; when inter-specific competition does occur, it may play out via physical displacements from websites, rather than in differences in web design (Wise 1983). Thus, even if it were true that different orbs were fine-tuned to capture different prey, one of the expected advantages of evolving divergent designs, avoiding inter-specific competition, has probably not been generally important.

Another puzzle is that the lack of prey specialization does not fit with the pattern of multiple convergences. Convergences suggest strong selection favoring particular designs on which different groups have converged. The lack of prey specialization suggests, in contrast, only very diffuse selection on web design to capture general classes of prey. The most reasonable solution I can see for this dilemma is to suppose that there is weak selection favoring some web forms over others, and that, acting over many repetitions (a new web every day or even several/day) and over long evolutionary time spans, different web forms differ enough in prey capture to favor changes. The many evolutionary convergences may have resulted not from intense selection to specialize tightly on specific prey, but from relatively weak but nevertheless long-sustained selection that favored the ability to capture certain general prey classes.

This hypothesis is challenged by cases in which evolutionary changes in web design seem to have been relatively rapid. In the theridiids, for example, web design varies *within* nearly all genera for which more than two or three species' webs are known (box 9.2, Eberhard et al. 2008b) (see also fig. 3.15 illustrating the diversity of designs in the araneid genus *Alpaida*). But *intra*-specific variation in seemingly quite basic web traits such as presence or absence of sticky lines, is also common in theridiids (Eberhard et al. 2008b), suggesting weak selection, and this variation could facilitate rapid evolutionary changes (section 10.13).

The abundance of intermediate designs seems, on the surface, to challenge a basic idea about evolution, but this is probably a false puzzle. Darwin (1859) argued that intermediate forms that arise during the evolution of a new, adaptive trait are expected to be rapidly replaced by more derived forms, in which the new trait is more perfected. If this is true, then why the persistence of the decidedly "imperfect," intermediate orb-like designs of several groups, like the amaurobiid *Titanoeca* (fig. 10.14), the desids *Matachia* (fig. 1.3), the psechrid *Fecenia* (frontispiece, fig. 9.18), and *Badumna* (one member of which, the invasive Australian *B. longiqua*, is currently sweeping across the New World and Europe—Griswold et al. 2005; Kielhorn and Rödel 2011)? Why haven't orb weavers, with their supposedly optimal sticky line spacing, not replaced these groups? Probably this is a false puzzle resulting from oversimplifications regarding other dimensions of the niches of these spiders.

The modularity of behavior patterns involved in web construction may have also predisposed web form to evolutionary change (section 10.13). Reshuffling and recombining behavior units, common in the evolution of web construction behavior (Eberhard 2018a), have played a large role in evolutionary change in many other organisms (West-Eberhard 2003). The substantial flexibility of orb designs adjusting to different local conditions (Heiling and Herberstein 1999; Harmer and Herberstein 2009; Herberstein and Tso 2011; chapter 7) could also promote evolutionary divergence.

9.4 WHAT IS A SHEET WEB? PROBLEMS INHERITED FROM PREVIOUS IMPRECISION

The term "sheet web" has been applied to various web forms built by many groups (table 9.3). An undoubtedly incomplete list includes the following: diplurids (fig. 10.6; Coyle 1986), austrochilids (fig. 10.8, Lopardo et al. 2004), diguetids (fig. 9.16, Eberhard 1967), pholcids (figs. 5.2, 8.7,

9.15, Eberhard 1992b), ochyroceratids (fig. 2.13, Hormiga et al. 2007), scytodids (Eberhard 1986a,c), agelenids (McCook 1889; Bristowe 1958; Rojas 2011), lycosids (figs. 5.5, 8.5, González et al. 2015; Eberhard and Hazzi 2017), pisaurids (fig. 5.14, Nentwig 1985c), zoropsids (fig. 10.7, Eberhard et al. 1993), psechrids (fig. 5.20, Robinson and Lubin 1979b), dictynids (Griswold et al. 2005), linyphiids (figs. 1.2, 1.3, 9.16) (multiple web types) (Benjamin et al. 2002), pimoids (Eberhard et al. 2008b), oxyopids (Griswold 1983), synotaxids (fig. 5.7, Eberhard et al. 2008b; Hormiga and Griswold 2014), theridiids (figs. 5.7, 9.16) (multiple web types) (Eberhard et al. 2008b), and cyrtophorine araneids (fig. 1.7, Blanke 1972; Shinkai 1989; Murphy and Murphy 2000). No systematic surveys are available, but the photographs just cited illustrate substantial variation in both the overall shapes and orientations of the sheets (e.g., horizontal, vertical, flat, domed, cup-shaped, flat with downward projecting dimples), their placement in the habitat (e.g., against the ground, above the ground, with an extensive tangle above and/or below), and the patterns of the lines in the sheet itself (e.g., long straight lines, tight and geometrically regular meshes, meshes so tight that individual lines cannot be readily distinguished, uniform line diameters, multiple line diameters, lines laid under tension, lines that are slightly lax, lines that are very lax).

One basic problem is that our language does not have a sufficient diversity of commonly used words to describe the diversity of designs in spider webs. Even within a single relatively homogenous category of the webs there are serious problems, as illustrated by the webs of lycosids and their close relatives. Murphy et al. (2006) included as "sheets" such diverse structures as the tents that *Dolomedes* spp. build around their egg sacs to protect their spiderlings (Bristowe 1958; Comstock 1967), and the extensive sheets and silk tubes of *Aulonia albimana* (Job 1968, 1974), *Hippasa olivacea* (Hingston 1920), and sosippine lycosids (González et al. 2015; Eberhard and Hazzi 2017). Rejection of the homologies implied by this use of the word *sheet* could strengthen the case for considering funnel webs as derived rather than ancestral for Lycosidae.

Blackledge et al. (2009c) made a welcome first step in recognizing the "sheet" web problem. They proposed four categories of sheet webs: brushed; irregular ground; irregular aerial; and stereotyped aerial. These categories are unfortunately difficult to distinguish. Criteria were lacking, for instance, for how distant a web must be from the ground to be classified as an "aerial" rather than a "ground" web, or how to distinguish "stereotyped" from "irregular." Other axes not mentioned by Blackledge et al. (2009c) also seem likely to be important: with vs. without a tangle above the sheet; with vs. without a tangle below; with vs. without sticky lines (in the sheet, in the tangle above, in the tangle below); the geometric form of the sheet (flat and horizontal, flat but with edges curling up, domed, cup); the degree of uniformity of thread diameters; the degree of inter-connectedness of lines; the density of lines in sheet; the regularity of arrangements of lines in sheet; the degree of independence from or elevation above the substrate; spider walks on top vs. on the bottom of the sheet (Eberhard and Hazzi 2017) (fig. 9.16 illustrates several of these variables).

Pholcids offer an example of how precision may be improved. Although the webs of many pholcid groups remain to be studied, their sheets appear to have a typical set of traits. Often sheets were domed rather than flat, with the spider resting centrally, under the top of the dome; usually there was only a small tangle above and none below (figs. 5.2, 8.7, 9.19, Huber 2000). The sheets themselves were composed of two types of lines: longer, apparently more tense lines (direct observations in two species of *Modisimus* showed that these were built first, forming a skeleton web); and lax lines that bore sticky droplets, at least in *M. guatuso* and 10–20 species of *Belisana* (Huber 2005) (fig. 9.19). In two species of *Modisimus* the sticky lines were added with wrapping movements of legs IV as the spider walked under the skeleton web(Eberhard 1992b; WE). Other species of *Belisana* had other designs: at least one had a typical pholcid dome; another had a "curtain" with parallel vertical lines carrying droplets (B. Huber pers. comm.).

One preliminary step would be to reserve the term "sheet" for more or less horizontal planar structures. This will exclude groups such as many dictynids (e.g., Nielsen 1932 and Blackledge and Wenzel 2001b on *Dictyna* spp.), eresids (*Stegodyphus* spp.) (Griswold et al. 2005), pisaurid nursery webs (Murphy et al. 2006), and salticid "sail" webs (fig. 9.3, Lopez 1986). As noted by Blackledge et al. (2009c), webs with isolated trip lines radiating from a retreat or a burrow entrance (e.g., *Ariadna* sp., *Uroctea durandi*, *Oecobius annulipes*) are best excluded from

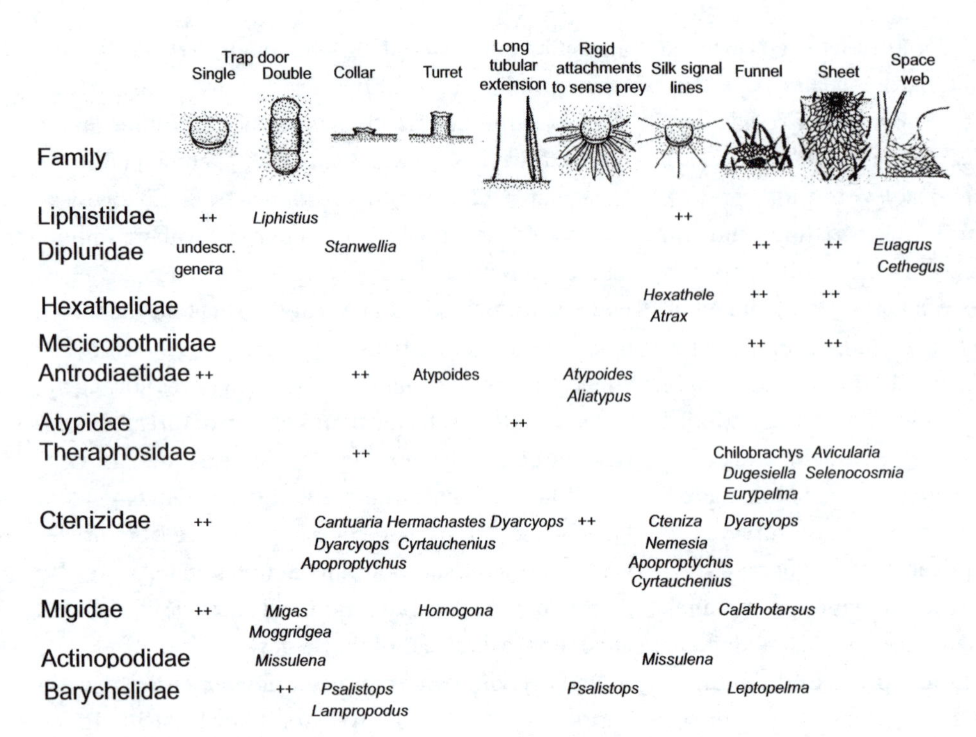

Fig. 9.26. This summary of the webs and other structures of mygalomorph spiders shows that many of their designs have evolved more than once in different lineages; most of them probably function to increase the spider's ability to sense nearby prey ("++" indicates that the structure is common within the family; each genus name indicates that the structure is built by at least one species within that genus) (some genus names have changed since) (after Coyle 1986).

the category "sheet," even when their lines all occur in close to the same horizontal plane. One unfortunate consequence of previous lax use of the word "sheet" by arachnologists is that many published verbal characterizations of the webs of a given taxonomic group as "sheets" convey little information, except in the few cases in which "sheet" was carefully defined, or when good photographs were provided. One basic lesson is the importance of photographs in describing webs; a picture can be worth even more than a thousand words for documenting some traits.

9.5 MYGALOMORPHS: SIMILAR PATTERNS OF DIVERSITY AND RAMPANT CONVERGENCE IN A DIFFERENT WORLD

Most mygalomorph spiders live in burrows or other retreats, and many build structures of silk or silk plus debris that increase their ability to sense prey near the mouth of the retreat ("entrance constructs"). An extensive review of the highly scattered literature (Coyle 1986) showed the same patterns of diversity and convergences (fig. 9.26) in this smaller taxonomic world that are seen in araneomorphs elsewhere in this chapter. Many families build various types of entrance constructs. Intra-generic and even intra-specific variation also occurred, for instance with respect to below-ground vs. above-ground or arboreal retreats, trap doors vs. lack of trap doors, entrance number, sheet-web development, double- vs. single-door retreats, and obligatory vs. facultative twig lining.

In addition, mygalomorphs have apparently repeatedly converged on the same types of entrance constructs, such as radial signal lines (at least 6 times), funnel plus sheet webs (5 times), trap doors, and incorporation of plant material into the collar (fig. 9.26). Some of these traits have also been lost repeatedly.

Persistence of supposed intermediate forms is also common in some mygalomorph families, such as Hexathelidae. The fact that even the shortest extensions of an entrance rim can improve prey capture (Coyle 1986) leaves unexplained why these extensions are so tiny in some groups, as for instance in the antrodiaetid *Aliatypus*.

Although entrance constructs at the burrow mouth per se can serve to hide and physically shield the spider from predators, and turrets may sometimes defend against flooding, it is likely that many (probably most) constructs evolved as a result of selection to increase predatory success, mainly via transmitting prey vibrations to the spider from farther away and thus extending her sense of touch (Coyle 1986). In a few cases such as the hinges on some trap doors that physically impede strikes at prey, selection for predatory and protective functions has probably acted in opposite directions.

9.6 DIVERSITY OF RELATIONS WITH INSECTS

The interactions between web-building spiders and insects are also diverse. Spider webs sometimes provide shelter and even feeding sites for insects. They are used by flies, wasps, and small moths as resting sites that afford protection from substrate-bound predators such as ants and other spiders (Robinson and Robinson 1976c; Eberhard 1980b; Lahmann and Zúñiga 1981); as sources of prey by empidid flies and plokiophyllid bugs (Nentwig and Heimer 1987; Eberhard et al. 1993); and as sources of discarded cadavers by detritovorous moth larvae (Robinson 1977). Some tiny nematocerous flies cluster in large numbers on the webs of the pholcid *Modisimus* sp. (fig. 9.27) (perhaps deriving selfish herd benefits against predators?), while other milichiid flies ride

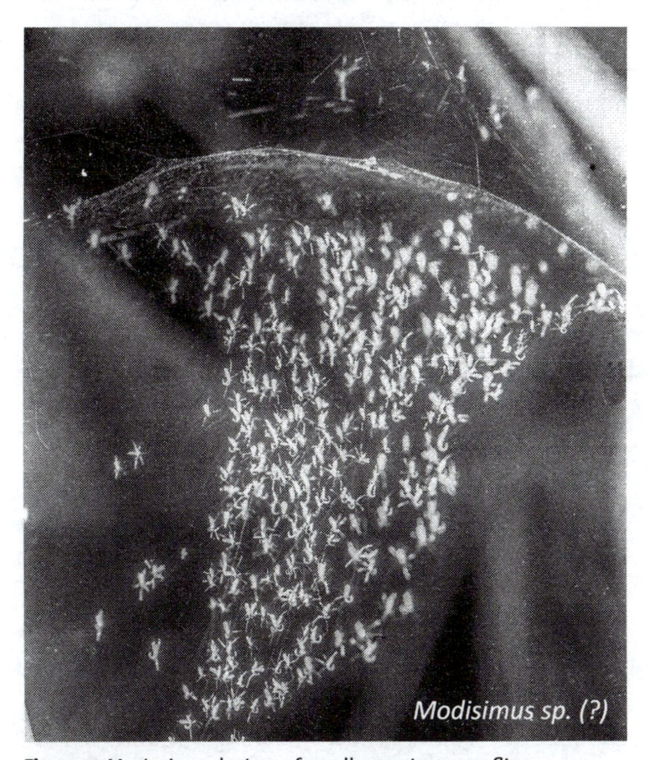

Fig. 9.27. Mysterious clusters of small nematocerous flies sometimes occur in the webs of pholcid spiders like this *Modisimus* sp. in Colombia. When the web was jarred gently, the flies flew up, then settled again on the web. Neither the function of the flies' grouping behavior nor the reason why the spiders failed to respond to them is known.

on the spider herself and steal liquefied prey from her as she feeds (Robinson and Robinson 1977). And then there are assassin bugs (Soley et al. 2011) and spiders in at least four different families (Mimetidae, Theridiidae, Salticidae, and Araneidae) (Bristowe 1958; Vollrath 1979; Eberhard 1983; Jackson and Whitehouse 1986; Jackson 1990c) that are specialized web spider predators that use the spider's web to stalk their prey; other mimetids (Jackson and Whitehouse 1986; WE) and salticids opportunistically steal prey from webs (and sometimes fall prey themselves) (Jackson 1985a,b).

It is not known how it is that some insects are able to land on particular non-sticky lines in host webs with such precision that they do not become entangled; some even land on spider lines in the dark (Lahmann and Zúñiga 1981). Some of these visitors may exercise selection on the web designs of their victims. Webs of the pholcid *Trichocyclus arawari* that were built farther from supporting objects made predation more difficult for the emesine bug *Stenolemus giraffa* (Soley et al. 2011; Soley and Taylor 2013).

9.7 LACK OF MINIATURIZATION EFFECTS

Does the evolution of miniature body size (and thus a small brain) have negative effects on an animal's behavioral capabilities? The symphytognathoid orb weavers, which are derived from larger-bodied ancestors (fig. 1.14), include probably the tiniest adult spiders known (young spiderlings of *Anapisona simoni* [AN] weighed only about 0.005 mg). Small body sizes also occur during development in other groups: even the spiderlings of the large *Nephila clavipes* (NE) (adult female body weight of about 2000 mg) weighed only about 0.6 mg when they built their first orbs (fig. 1.5). Orb-weaving spiders thus span more than five orders of magnitude in weight, offering the chance to compare the performance of very similar behavioral tasks among animals of very different sizes (Eberhard 2007b).

Tiny animals confront special problems in the designs of their nervous systems that may impose behavioral limitations (Eberhard and Wcislo 2011). The general trend for smaller animals to have relatively larger brains ("Haller's Rule"), which holds in spiders (Quesada et al. 2011), combined with the relatively high metabolic cost of building and maintaining nervous tissue, is likely to result in especially intense selection on tiny animals favoring either reduced behavioral capacities (the "size-limited behavior" hypothesis), increased economies in nervous system designs, or increased tolerance of energy limitations (Chittka and Niven 2009; Eberhard and Wcislo 2011). The brains of tiny orbicularian spiders were smaller than the brains of insects of the same body weight (Quesada et al. 2011; Eberhard and Wcislo 2011). The likely intensity of selection to limit their relatively large size (that results from Haller's Rule) is illustrated by the fact that in the tiniest individuals, the lobes of the brain did not fit in the cephalothorax and extended into the coxae (Quesada et al. 2011).

One way to test the size-limited behavior hypothesis is to quantify the behavioral precision of tiny vs. more normal-sized animals (their ability to repeat the same behavior precisely). Misunami et al. (2004) made the logical speculation that behavioral precision is reduced in animals with smaller brains. There are several possible reasons to expect greater imprecision in tiny animals. They may have less complete sensory information, due to lower numbers of sense organs; they may make less thorough or precise analyses of sensory information due to lower numbers of inter-neurons, fewer dendrites, or fewer or more noisy connections between them (Eberhard and Wcislo 2011); they may have less extensive internal feedback mechanisms that correct for the imprecision that results from nervous system noise (Eberhard 1990c, 2000a); or they may have reduced motor precision, due to smaller numbers of motor axons or less feedback information. In an animal like a web-building spider that uses movements and positions of its legs to provide sensory information (chapters 6, 7), reduced precision in leg movements or analyses of these movements could also reduce sensory precision.

Sticky spiral construction is especially well-suited for analyzing these predictions regarding behavioral precision: the spider repeatedly executes the same process of sensing the location of the inner loop of sticky spiral, measuring the distance at which the current loop should be attached, and then attaching it to the radius. Even very small differences in the results of this process are recorded in the sites where attachments are made, and the high degree of repetition (often hundreds of times in a web) permits especially powerful, intra-individual comparisons. Control of the spacing between sticky spiral loops is complex, and each attachment decision may be affected by >10 different variables, but nearly all of these factors can be held very close to constant by making appropriate comparisons.

The prediction of the size-limited behavior hypothesis is that very tiny orb weavers will show greater imprecision (greater differences in the spacing between adjacent loops of sticky spiral in the same part of the orb). The variations between neighboring spaces constitute a realistic measure of imprecision, because the nearly constant spaces between successive loops of sticky spiral in orb webs imply that a particular distance between loops is advantageous, at least in a given area of the orb (section 4.3). When the indices of precision were compared for spiders whose weights varied over about four orders of magnitude, including both second instar spiderlings and adults from three different species, however, the predicted trend was absent (Eberhard 2007b). The spacing in the orbs of the smallest individuals (spiderlings of the smallest species, *Anapisona simoni* [AN], which weighed about 0.005 mg and whose bodies were 0.6 mm long) was no less precise than that in the webs of larger ones. The conclusion was that the relatively large brains of tiny spiders (Haller's Rule) compensated sufficiently for their small size to produce equally precise behavior.

This test could be weak, because determining sticky spiral spacing may be relatively undemanding in terms of behavioral capacities. Subsequent studies, of *Anapisona simoni* (AN) (fig. 7.9, Eberhard 2011) and *Leucauge argyra* (TET) (Quesada 2014) that examined behavior that is probably more challenging confirmed that miniature spiders have undiminished behavioral capabilities. The most striking data are from spiderlings and adults of the moderate-sized *L. argyra* (adult females weigh 60–100 mg). Adults built about 60 cm diameter orbs in the field. But when confined in small spaces (with diameters <1/10 the normal web span), they built highly modified, tiny orbs with many design features adjusted to the limited space (fig. 7.42, section 7.8.2). Many changes were apparently independent of each other (Barrantes and Eberhard 2012). Early instar *L. argyra* spiderlings are tiny, weighing less than 1/100 of the adult weight, but they showed the same extraordinary flexibility: they built miniature orbs with similarly adaptively altered designs when enclosed in proportionally small spaces. The precisions of their adjustments (residuals of regressions on total orb area) did not differ from those of adults. Preliminary estimates intimated that the smaller neurons and the relatively over-sized CNS of the spiderlings resulted in similar numbers of neurons in spiderlings and adults. This is the most challenging comparison of the behavioral capabilities of tiny vs. larger spiders yet performed (indeed, it may be the most detailed behavioral comparison ever for any animal this small).

In sum, several types of data from orb weavers lead to the surprising conclusion that tiny individuals are not behaviorally limited in web construction compared with much larger conspecifics or relatives. The evolution of tiny body sizes in orb weavers thus did not entail reductions in their web-spinning capabilities. Not enough is known about the functional anatomy of orb weaver brains to ask whether those portions of the brain that are involved in web construction are disproportionately well developed in tiny individuals. The best data come from localized lesions of the supra-esophageal ganglion of *Araneus diadematus* (AR), which produced severe and persistent alterations in several aspects of orb construction that varied widely among individuals (Witt 1969; Le Guelte and Witt 1971). But variations in the sites in the brain where damage was produced, uncertainties regarding the extent of damage around each site where the laser beam that produced the damage was focused (Witt 1969; Le Guelte and Witt 1971), and the inherent difficulties of interpreting ablation experiments precluded detailed resolution of the functions of different parts of the brain.

9.8 PATHS NOT FOLLOWED: ALTERNATIVE WEB FORMS IN OTHER ANIMALS

I will close this discussion of the diversity of spider web designs by mentioning web designs and ways of organizing building behavior that spiders could have adopted but did not (fig. 5.13). The prey capture webs of the larvae in the nematocerous fly family Keroplatidae are an example. Their webs consist of one or at most a few rows of vertical sticky lines, each of which hangs free below a more or less horizontal non-sticky line (fig. 9.28b; Richards 1960; Eberhard 1980b). Building behavior in one species involved independent construction of each of the different vertical lines in the web, and quick replacement of lines that were damaged during prey capture (WE). Immediate replacement of lines damaged during prey capture is very rare in spiders (I have seen it only in the nesticid *Gaucelmus* sp. and the theridiid *Phoroncidia* sp.). Free-hanging sticky lines are also very rare in spiders, occurring only in occasional individuals of *Mastophora* spp. (AR) (Eberhard 1980a; Yeargan 1994; R. Bradley pers. comm.) and *Ordgarius hobsoni* (AR) (Shinkai and Takano 1984; Murphy and Murphy 2000) bolas spiders that hang one or more short bolas lines with a sticky ball at the tip from a horizontal non-sticky line. Probably a major limitation of the beaded curtain design is that it is only effective at sites such as caves or close to the ground deep in dense forests where there is little or no wind, but keroplatid webs are common in some tropical forest habitats.

Another function for which spider webs have never been shown to be designed is to sieve small particles such as pollen grains, fungal spores, or other detritus from the air or water. Aerial webs of this type of design are exemplified by the dense silk sheet webs of the larvae of mycetophylid fly *Leptomorphus* spp. that filter fungal spores (fig. 9.28a) (Eberhard 1970b), and those of the larvae of caddis flies in the family Hydropsychidae that function as filters in aquatic environments (fig. 9.28c), (Grimaldi and Engel 2005). Specialization for sieving is not known to occur in any spider, though the dense sheet webs of the linyphiids *Frontinella communis* and *Tennesseellum formica* do trap pollen in corn fields for short periods during certain times of the year, and the spiders ingested corn pollen soon after it was experimentally dusted onto their webs (Peterson et al. 2010). In contrast, *Pinus* pollen was apparently not utilized as food by the linyphiid sheet weaver *Frontinella pyramitela* (Carrel et al. 2000), and *Leucauge venusta* (TET) gathered up several balls of silk lines that were coated with tree pollen (probably *Quercus* sp., judging by the trees in the vicinity) while renewing their webs, but then flicked them away without feeding on them (WE). Blackledge et al. (2011) speculated that the dense orbs of tiny symphytognathid spiders function to capture pollen or spores, but gave no supporting evidence; in my experience, these webs show no sign of being preferentially built under fungal fruiting bodies as occurs with those of the fungus gnats. Perhaps spiders have not exploited this feeding strategy for lack of the necessary enzymes to digest spores without grinding them (the likely importance of grinding was suggested

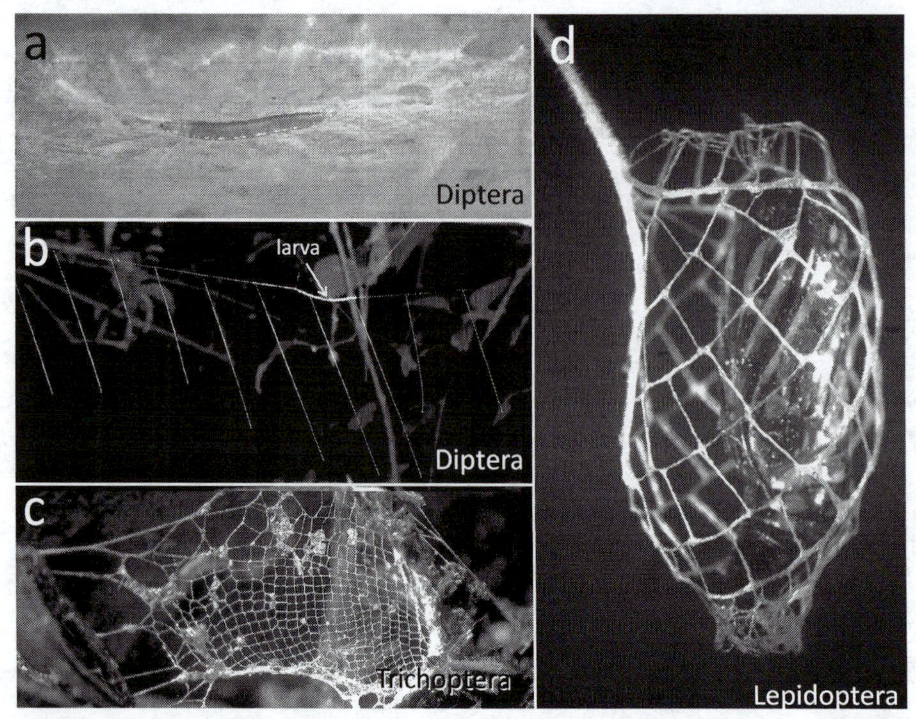

Fig. 9.28. Three functional designs seen in the silk webs of insects that spiders have apparently never discovered are illustrated here (they are labeled with the order of insect). *a*) Dense, sheet-like filters to capture fungal spores were built by larvae of fungus gnats in the genus *Leptomorphus* on the undersurfaces of fungal fruiting bodies. The larva, which stayed on the undersurface of the sheet, periodically cut a hole in the sheet and ingested it and the spores that it had collected, then repaired the sheet. Spiders may occasionally ingest spores or pollen grains captured incidentally (section 9.8) but show no signs in their web-site selection of specializing on such food. *b*) Larvae of an unidentified keroplatid fly (possibly *Neoditomyia* sp.) built traps for flying prey that consisted of a horizontal, non-sticky line along which the larva moved that supported a sparse curtain of free-hanging sticky lines, each weighted at its tip with a large droplet of adhesive. These webs had the disadvantage that even slight breezes caused the free-hanging lines to swing, and they often eventually adhered to each other. Nevertheless, they obviously captured enough prey to enable these larvae to survive and grow. A third design, based on a rectangular mesh, occurred in both the sieve web used by the larva of the caddisfly *Leptonema* sp. to strain detritus from a stream (*c*), and the pupal cocoon of an undetermined moth (probably an yponomeutid) (*d*) (the more open mesh of thinner lines at the top allowed the adult moth to escape). Rectangular designs may have been more difficult for spiders to evolve than for holometabolous insect larvae because of their different body designs: spiders do not have their spinnerets at the end of a long flexible, cylindrical body; in contrast, the insect larvae can lift one end of the body and swing the head with its silk spigots back and forth over relatively long distances and in different directions (*c* courtesy of Danny Vasquez).

by mouthpart morphology in some but not other spore-feeding staphylinoid beetles—Betz et al. 2003).

Still further designs that capture tiny aquatic food items exist in the ocean. Polychaete worms in the genus *Chaetopterus* make sieves out of mucous (Ruppert and Fox 1988), while the sea butterfly snails *Limacina* use sticky mucous sheets up to 5 cm wide. Like the *Leptomorphus* larvae, these snails periodically ingest the sheet with its tiny trapped prey, and produce another. In these groups the web functions as a sticky sieve that accumulates relatively large numbers of tiny food items. Perhaps the density of small nutritious items is generally greater in the water than in the air, where it is usually not sufficient (except near fungal fruiting structures) to provide sustained sustenance.

The rectangular meshes of the sieve webs of some hydropsychid caddisflies (fig. 9.28c), a design that also occurs in the cocoons of some yponomeutid moths (fig. 9.28d; Grimaldi and Engel 2005), is surprisingly rare in spider webs. With the exception of *Cyrtophora* sheets and the non-sticky lines in *Synotaxus* spp. webs (figs. 1.7, 5.3, 9.8) I know of no certain rectangular meshes in spider webs (and the sticky lines are not parallel in *Synotaxus*) (a photo of the linyphiid *Arcuphantes tamaensis* on its slanting sheet [Shinkai 1979] suggests such a design, but could be an artifact of lighting). Geometrical regularity in many spider webs involves radial rather than rectangular patterns. The long, flexible bodies of many insect larvae, which they can lift and extend and also swing widely from side to side, may predispose them to build more rectangular meshes.

In sum, spider web designs have diversified extensively over a long span of evolutionary time; but they have also missed a few opportunities that have been utilized by other animals.

9.9 SUMMARY AND A NEW SYNTHESIS

There are three major patterns in this chapter: the spectacular diversity of web designs; the rampant evolutionary convergences on similar web designs in different groups; and the existence of many intermediate forms connecting extreme modifications with more typical forms. I propose that newly recognized consequences of several insights mentioned in previous chapters explain these three patterns, and form part of a single overall vision. In the first place, building silken webs to capture prey is both an extremely ancient (at least 245 million years old) and a very successful and widespread trait in spiders. Orb webs had evolved before many major groups such as birds and flowering plants even first appeared in the fossil record (section 1.6). For a very long time, there have been large numbers of web-building spider species evolving web-based predatory strategies. Secondly, it appears that inter-specific competition between web spiders for prey may be uncommon (sections 3.2.5, 8.8). Orb designs probably seldom represent adaptations to capture specific prey (section 3.2.4), and inter-specific competition for prey may have rarely selected for divergent web designs; divergence may tend to result instead from selection to gain access to different general classes of prey. If the lack of inter-specific competition for prey proves to be a general phenomenon, it could explain the evolution, persistence, and co-occurrence of independently convergent web types, the continued coexistence of intermediate as well as derived web forms, and some of the exceptions to general rules (table 9.4). The long periods of time and the large numbers of species increase the probability of all three patterns—diversity, convergence, and intermediates; modern web spiders are the product of an especially ancient and thorough exploration of the adaptive space associated with webs.

Whatever the reasons may have been for the evolution of the high diversity of webs, the radiation of web designs has been accompanied by a bewildering array of multiple reinventions and partial reinventions of many different designs (tables 9.1, 9.2). One implication of these multiple convergences is that there are apparently only a limited number of ways to make effective traps for prey, and that these designs have been repeatedly discovered in different evolutionary lineages. This interpretation fits with the phylogenetic studies that place the origin of orbs (and of course, of the webs preceding orbs) early, at least 165 million years ago (fig. 1.8, section 1.6). It also fits with the trend for intermediate web forms to be common, and with the fact that even some groups that branched off very early, like progradungulids, build highly derived web designs.

The different combinations of mechanical traits in different web designs suggest that selection has favored quite different mixes of relative abilities to perform the basic trapping functions of intercepting, stopping, and retaining prey. This diversity of functions is in accord with the argument (e.g., section 4.3) that no single function has consistently been more important than the others in the evolution of spider webs.

The same three trends of diversity, convergence, and intermediate forms are repeated in the webs of the family Theridiidae (box 9.2), the tangle webs built alongside orbs (appendix O9.1), and the webs built by mature males (appendix O9.2), and even in the family Linyphiidae (which was previously said to lack diversity) (Hormiga and Eberhard in prep.). Perhaps the most spectacular convergence is the "pseudo-orbs" in the psechrid *Fecenia* (box 9.1). The webs of the eresid *Seothyra henscheli* (box 9.3) represent an extreme example of a divergent design, morphology, and behavior adapted to an unusual environment.

Our present incomplete knowledge, especially of non-orbs, in combination with a typological bias to describe "the" web design of a species and to fail to recognize intra-specific variation, implies that all three trends are probably underestimated. The additional chemical and mechanical properties of homologous silk lines in different webs suggest that there is substantial additional diversity in these traits, that has hardly begun to be explored.

Box 9.1 The most spectacular convergence of all: *Fecenia*

Members of the psechrid "pseudo-orb" weaver genus *Fecenia* form a very modest-sized group, with only five known species ranging from southern India to the Solomon Islands (Bayer 2011); the only other genus in this family is *Psechrus*, with 46 species (Platnick 2014). There is general agreement that psechrids are deeply embedded in the RTA clade (figs. 1.8–1.11), and are only distantly related to true orb weavers (Blackledge et al. 2012a; Agnarsson et al. 2012). Patching together sketchy information from direct behavioral observations of *F.* sp. (from New Guinea) (Robinson and Lubin 1979b) and *F. protensa* (Zschokke and Vollrath 1995b) (this tentative identification is based on the geographic ranges of different species in Bayer 2011; the specimens were evidently collected in Singapore—S. Zschokke pers. comm.), and from web photos of *F.* sp. (from Malasia) (Murphy and Murphy 2000), *F.* sp. or spp. (Blackledge et al. 2012a; Agnarsson et al. 2012), *F. cylindrata* (fig. 9.18) (Bayer 2011), and *F. ochracea* (frontispiece) (Agnarsson et al. 2012), the pseudo-orbs of *Fecenia* spp. resembled orbicularian orb webs ("orbs" in what follows) in multiple ways. They all had more or less radially arranged non-sticky lines that converged on the lower end of a curled leaf suspended in the web; the non-sticky lines in any single portion of the web were built before the sticky lines; at least some of the radial lines ended at the outer edge of the web on long, non-sticky frame lines rather than being attached directly to the substrate; some frame lines were attached to other frame lines, as in the secondary and tertiary frames of orbicularian orbs; the area where the radial lines converged had many, tightly meshed non-sticky lines (they were too dense to determine whether they had any hint of the circular patterns of the hubs of orbs); all of the lines were in approximately the same plane; the more or less circular, evenly-spaced sticky lines were attached to the radial lines; and the spider started building the sticky lines at the edge of the web and worked her way inward (cribellum silk production was tentatively distinguished from that of non-sticky silk in *F. protensa* on the basis of movement patterns and speeds, which were about 0.66 mm/sec for what was probably sticky silk and around 5 mm/s for what was probably non-sticky silk) (S. Zschokke pers. comm.). The convergences with true orbs extended to some properties of *Fecenia* major ampullate gland dragline silk. It resembled dragline silk in orbicularians in being both tougher and stronger than the homologous silk of other, related RTA-clade groups (fig. 9.25). But it lacked a major ampullate spidroin protein of orbicularians, MaSp2, and was stiffer and less extensible than the major ampullate silk of orb weavers (Blackledge et al. 2012a).

Fecenia pseudo-orbs also appeared to differ from orbs in several basic ways. There was no temporary spiral in the webs of *F. protensa* (Zschokke and Vollrath 1995b) or *F.* sp. (Robinson and Lubin 1979b) (it was also lacking in *F. cylindrata* if my interpretation of the web in fig. 9.18 is correct). Some split radii diverged abruptly from each other, however, just beyond the split, and this pattern in *N. clavipes* and *N. edulis* radii is produced by temporary spiral lines (see figs. 4.4, 7.31, O6.8) (Hesselberg and Vollrath 2012; Eberhard 2014), so non-radial lines of some sort may connect at least some *Fecenia* radii. The temporal pattern of construction differed: both the radii and the sticky spiral lines were built in two bursts rather than one in one web of *F. protensa*: the radii and then the sticky spiral were built first on one side of the retreat, and then on the other (fig. 9.29) (Zschokke and Vollrath 1995b); W. Piel observed similar sector-by-sector construction (W. Piel pers. comm.). The patterns in the *F.* sp. webs in figs. 2a and 2d of Agnarsson et al. 2012, however, suggest single bouts of sticky spiral construction. Radii also appeared to fuse (e.g., arrow at 08:30 in fig. 9.18), a pattern that is very rare in orbs. The sticky spiral lines of *F. ochracea* from New Britain and *F. protensa* (but none of the other *Fecenia*) (nor in any other spider) spread out to form a continuous sheet in the inner portion of the web (frontispiece, fig. 9.18). Even in the most orb-like *Fecenia* web (of *F. ochracea*), the "radial" lines seemed to be more highly interconnected in some sectors (e.g., the 4:00—6:00 sector) than others, and did not simply tend to split at greater distances from the hub as in the webs of araneoids like *Cyrtophora* and *Nephila* (fig. 1.5, 1.7, 4.4). Agnarsson et al. (2012) noted two additional, less basic differences with many but not all orb weavers: webs are more long-lived; and the sticky spiral doubled back repeatedly in a zig-zag pattern rather than circling the entire web. The ontogenetic change from a cone to a planar pseudo-orb observed in *F.* sp. from New Guinea (Robinson and Lubin 1979b) has not been noted in any other species.

The similarities between the pseudo-orbs of *Fecenia* and the "true" orbs of orbicularians likely evolved convergently, as other psechrids built open-meshed horizontal sheets (fig. 5.20). The mechanical properties of the dragline silk in both

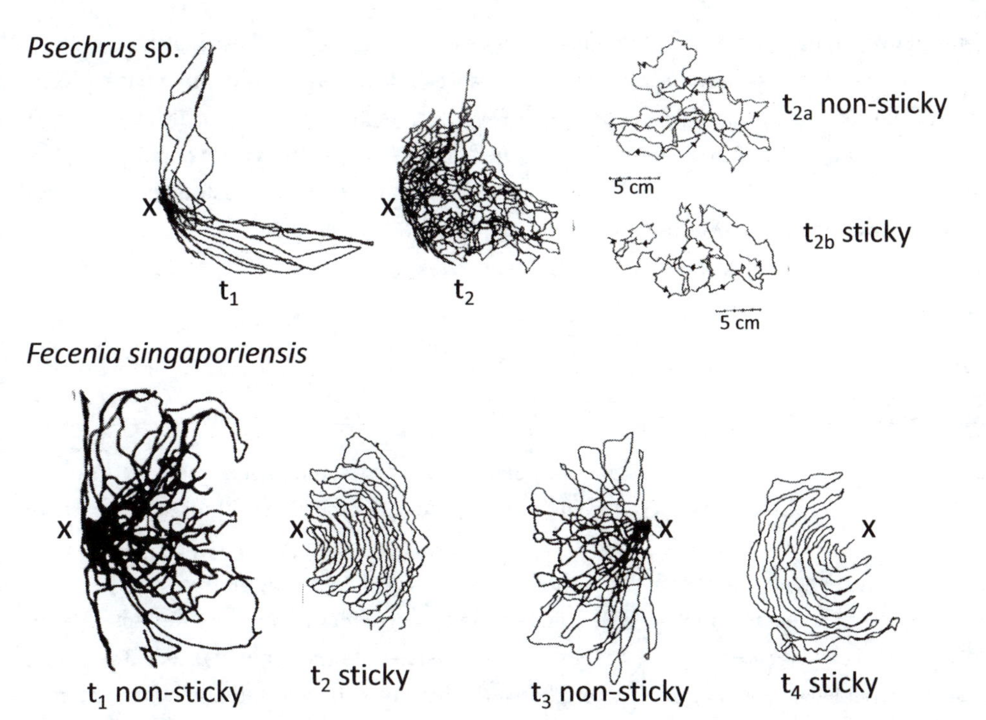

Fig. 9.29. The paths taken by two species of non-orb weaving psechrid spiders during web construction were traced from video recordings. *Psechrus* sp. began by building long, more or less straight lines radiating from her retreat ("X") (t_1). Some paths were very similar to secondary radius construction by orb weavers (fig. 6.5i) (Zschokke and Vollrath 1995b). Subsequently the spider filled the area covered by the previous lines with additional, less consistently oriented lines (t_2). Some of these were sticky while others were not, and she alternated between periods of more rapid movements (mean = 21.8 cm/min; partial record in t_{2a}), and slower ones (mean = 4.3 cm/min) (partial record in t_{2b}). The two speeds probably corresponded, respectively, to laying non-sticky dragline silk and sticky cribellum silk, but this was not verified directly. The pseudo-orb weaver *Psechrus singaporensis* also built her web around a central retreat ("X") but showed more discrete stages of building non-sticky and sticky lines. First (t_1) she made approximately radial non-sticky lines on the right side of the retreat, followed by curving sticky lines attached to these radii (t_2). Then she made a second set of non-sticky radial lines to the left (t_3), and added curving sticky lines to these (t_4) (after Zschokke and Vollrath 1995b).

groups improved over those of their relatives, presumably due to selection on orb properties. But the inferior mechanical properties of *Fecenia* draglines compared with those of orbicularians were probably associated with a lack of MaSp2, the product of a gene which apparently arose in the orbicularian lineage after it split from the RTA-clade lineage an estimated 225 Mya (Blackledge et al. 2012a).

Fragmentary observations of other psechrids in the genus *Psechrus* suggest possible similarities with pseudo-orb construction. These spiders built approximately horizontal, open-meshed sheets, either with a dome where the spider rested during the day, as in *P. protensa* (Murphy and Murphy 2000), or with a tubular retreat at one edge, as in *P. argentatus* (Robinson and Lubin 1979b), *P.* sp. from India (Eberhard 1987a), and *F. protensa* (Zschokke and Vollrath 1995b). There are fragmentary observations of the construction behavior of the latter two species (Eberhard 1987a; Zschokke and Vollrath 1995b). Construction began with the spider building a skeleton of long non-sticky lines, some of which were at least vaguely radially oriented. Next, the spider "filled in" with additional, shorter non-sticky lines, perhaps especially in the central area. Finally she added sticky lines in approximately circular patterns. In the species from India, the spider initiated the sticky line in four of five webs near the outer margin of the sheet, moving in a path more or less parallel to its edge (fig. 5.20), and the sticky lines traced a zig-zag path produced when the spider swung her abdomen slowly from side to side as she walked. No zig-zag patterns were discernable, however, in the sticky line of the *F. protensa* (Zschokke and Vollrath 1995b) (t_{2b} in fig. 9.29). Similarities between *Psechrus* and *Fecenia* included the loosely meshed, approximately planar skeleton of non-sticky lines, the vaguely radial arrangement of these lines, and sticky line construction that began near the edge and ran nearly parallel to it.

These descriptions are only preliminary. The information on *Fecenia* and *Psechrus* webs and web-building behavior is sparse (some of the statements above are based on only a single observation). Interpretation of the photos of pseudo-orbs taken in the field is also difficult, because some lines may have been added after the web was finished, rather than during the original construction.

In sum, *Fecenia* pseudo-orbs have converged with orbicularian orbs in several traits; a few of these may have evolved as prior, more ancestral convergences in more basal psechrids. The discovery of pseudo-orbs has conclusively resolved the old orb web monophyly debate in favor of polyphyly; but this second origin occurred in an previously unexpected portion of the phylogenetic tree of spiders.

Box 9.2 The most spectacular divergence of all: Theridiidae

With >2300 described species, the family Theridiidae is one of the largest families of spiders. Phylogenetic analyses of molecules (Arnedo et al. 2004) and morphology and (to a lesser extent) behavior (Agnarsson 2004) have yielded largely congruent phylogenies for theridiids. When the resulting consensus tree was used to trace the evolution of web designs over 166 species in 37 genera (nearly half of the genera in the family), a chaotic evolutionary story of rampant divergence and convergence emerged. There was substantial within-genus and even within-species diversity in web design; of the 22 web traits discussed, 14 varied intra-specifically in 53 different species. In addition, there were many between-genera convergences (table 9.2; figs. 9.12, 9.13, 9.16).

Take, for example, the spiders in the black widow genus *Latrodectus*. Previous authors had taken this genus to have a single, gumfoot design, termed "the *Latrodectus*-type web" (Benjamin and Zschokke 2003). But web designs in *Latrodectus* were quite variable, differing with respect to the presence of sticky lines, the sites where sticky lines occurred, and the presence and forms of sheet-like structures (Eberhard et al. 2008b).

One might wonder whether perhaps the lack of congruence between web types and phylogeny is the result of flawed taxonomy. Indeed, some of the within-genus diversity occurred in acknowledged "dumping ground" genera, such as *Achaearanea*, *Theridion*, and *Chrysso*. But the species in these genera are probably related, because multiple species in these genera clustered on the same branches. The strong accord between the molecular and morphological phylogenies also argues that the phylogenies are reasonable. The phylogenetic "chaos" in theridiid webs contrasted sharply with evolutionary patterns in orb weavers, where only 2 of 21 orb web traits showed intra-specific variation, and involved only 3 of 32 species analyzed (Kuntner 2005, 2006).

Why are the webs of theridiids so evolutionarily labile compared with those of orb weavers? One possibility is related to how variant behaviors originate. There is probably a relaxation of selection for behavioral consistency in theridiid webs vs. orbs, and "random" intra-specific and intra-individual variance increases due to less intense repression of intrinsic imprecision in the nervous system, thus facilitating the origin of substantially new web designs (Eberhard 1990a). A second possible factor is that, as descendants of orb weavers, theridiids have inherited the capability of making fine manipulations of lines such as agile cut-and-reel behavior, facilitating the evolution of new web forms (fig. 6.3, Eberhard et al. 2008a,b).

Box 9.3 Sand castles: extreme modifications of *Seothyra henscheli* webs to shifting sand

Nearly all silk lines are flexible rather than rigid (for exceptions see section 10.5.6, box 9.4). So perhaps the very last place one would expect to find web-building spiders is the scorching dunes of shifting, fine-grained sand of the Namibian desert, where there are no rigid supports of any kind (and surface temperatures range up to 65–75°C). Nevertheless, the eresid spider *Seothyra henscheli* builds webs here (fig. 9.30), and in fact is limited to this habitat (Henschel and Lubin 1992). Trace fossils of their burrows indicate that *Seothyra* have lived in this habitat for more than 14 million years (Pickford 2000). Their adjustments to this forbidding environment involve extraordinary morphology and behavior.

A spider begins web construction by building a support to which she can anchor her lines, attaching sand grains together to form a mat (a behavior also seen early in tube construction by the funnel-web agelenid *Coelotes terrestris* in Europe, perhaps to facilitate walking across loose substrates—Tretzel 1961). She then uses this mat as a roof, under which she first builds a central chamber and then attaches sticky silk to its edges. She digs a more or less vertical 10–15 cm tunnel under the chamber, into the cooler sand below (<35°C), and runs a signal line from the bottom of the tunnel up to the roof (Henschel and Lubin 1992; Peters 1992).

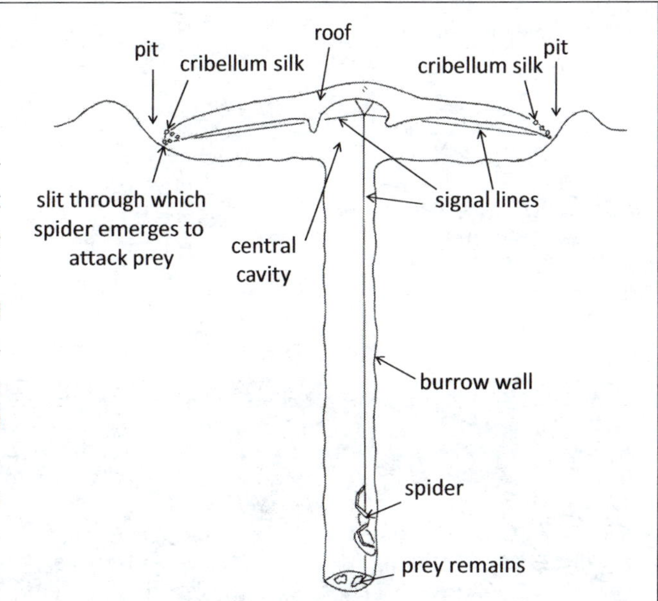

Fig. 9.30. Webs of the eresid *Seothyra henscheli* were built in thermally and mechanically forbidding dunes of loose sand in Namibia. Spiders compensated for the lack of rigid supports by first constructing a mat or "roof" of sand grains bound together with silk (see also fig. 9.31), then used this mat plus a tube lined with silk to form a cavity under the roof; fluffy masses of sticky cribellum silk were placed in pits at the edges of the roof, while signal lines extended to deep in the burrow where the spider rested. Spiders captured insects (mostly ants) walking on the surface, but avoided the scorching (up to 65–75° C) surface temperatures (after drawings and descriptions in Henschel and Lubin 1992 and Peters 1992).

The morphology and behavior used to build the web were unusual. The anterior lateral spinnerets of *S. henscheli* are strongly modified for attaching sand grains together. They are elongate and also extensible, and bear on their tips unusual, mace-like arrays of elongate piriform spigots, each of which has a long, thin, flexible distal prolongation (fig. 9.31d). At the beginning of mat construction, the anterior lateral spinnerets were extended, and swept back and forth across the sand to produce a loose sheet of lines that were attached to sand grains and to each other, apparently with a secretion from the piriform glands. Then the spider literally dove into the sand, turning upside down and grasping the sheet with her legs as she slid under the mat (fig. 9.31b)! Moving beneath the mat, she created a chamber, and strengthened and extended the preliminary mat by adding further grains that she collected with her spinnerets (some from the floor of the cavity) and attaching them to the underside of the roof formed by the mat. She built the vertical tunnel below the chamber over several nights. She reinforced the tunnel walls by repeatedly thrusting her anterior lateral spinnerets into the wall in quick succession, producing very orderly rows of small pits directed away from the tunnel cavity and in which the sand grains were bound to each other by silk lines (fig. 9.31c). The distantly related dune-dwelling sparassid, *Leucorchestris arenicola*, showed a striking convergence, reinforcing its tunnel walls by stabbing them repeatedly with similar long, eversible, anterior lateral spinnerets (Peters 1992).

The mat was extended by *S. henscheli* into lobes in several directions, and sand dug from the chamber and tunnel was pushed into piles located just beyond the tips of the lobes, leaving small pits between the edge of the lobe and the pile (fig. 9.30). Later she attached small mounds of cribellum silk to the upper surfaces of the lobes; this sticky silk retained prey (mostly ants) that fell into the pit. Vibrations from their struggles were transmitted down the signal line to the spider lurking at the bottom of the tunnel. The edge of the lobe sagged downward into the pit, forming a thin slit through which the spider emerged to attack the prey.

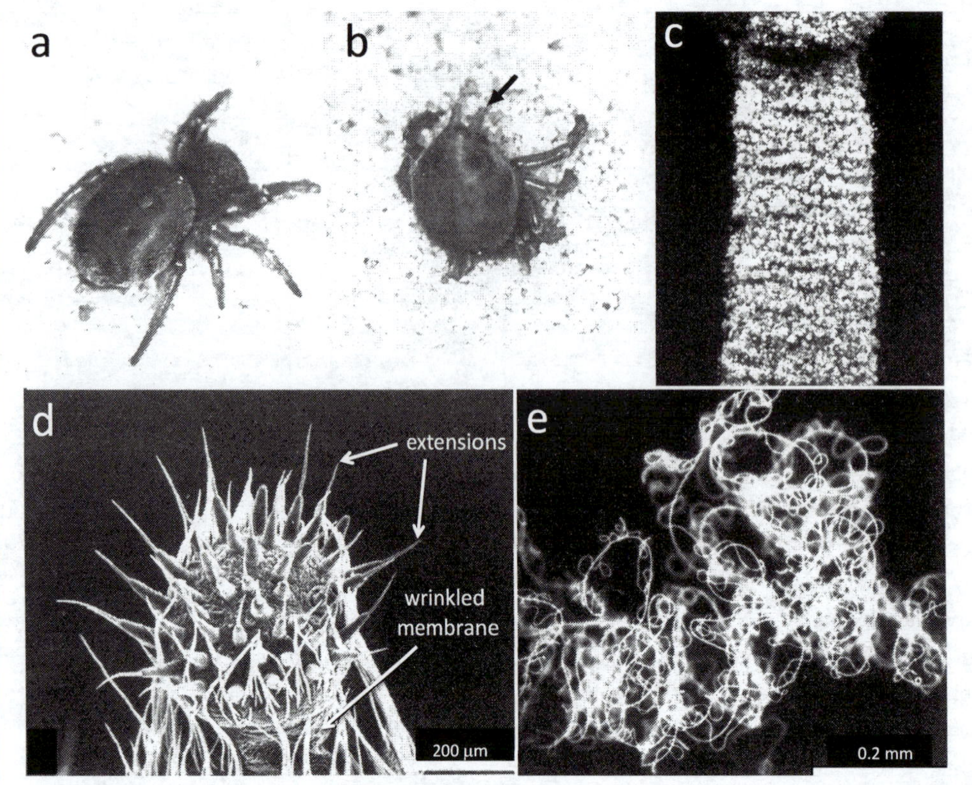

Fig. 9.31. The sand dune eresid *Seothyra henscheli* used specialized behavioral and morphological traits to build webs in a forbidding habitat of loose, shifting sand dunes (see also fig. 9.30). The spider began by using her legs to grasp a loose mat of sand grains bound together with silk lines (*a*) and "dove" (!) under its edge (*b*) to begin building the roof and the central chamber (the arrow marks an extended spinneret adding lines to the mat). The tunnel she built below the mat was lined with silk that bound sand grains together; the extremely regular ridges in this lining (*c*) were composed of large numbers of highly aligned tiny projections, each of which was probably produced by a single thrust of an anterior lateral spinneret. The tip of each anterior lateral spinneret (*d*) was extensible (note the wrinkled membrane at the base of the distal segment), and was equipped with an unusual arrangement of apparent piriform spigots. Each of these bore a long thin extension at its tip, and they projected in many different directions. These spigots differed from those of many other spiders (see figs. 2.11, 2.12) including those of other eresids such as *Stegodyphus* (Griswold et al. 2005). By examining spiders killed while they were building the roof, Peters (1992) showed that these spigots produced the lines laid at this stage. The trapping portion of the web consisted of loose piles of cribellum silk (*e*) in small tangles at the edges of the roof. The spider spent up to 40 s combing out each pile of sticky silk (from Peters 1992).

The derivation of these webs is unclear, as very little is known about web construction in other Eresidae, and nothing about that of *Eresus*, whose web design seems similar to that of *S. henscheli*. Nielsen found that the webs of young *Eresus niger*, which occurred in heather-clad hills in Jutland, had a vertical burrow in the ground that was covered by a strong, broad sheet, and that the sheet could be lifted like a flap, revealing the burrow entrance near the sheet's attachment to the ground (Nielsen 1932). He was uncertain regarding the location of cribellum lines. While the designs of *S. henscheli* and *E. niger* webs show a general resemblance, the contortions that the dune spiders go through in order to build them in loose sand represent truly spectacular innovations.

Box 9.4 Relation between web design and silk properties: stiff silk in *Uroctea durandi*

The web of the small oecobiid *Uroctea durandi*, described in beautiful detail by Kullmann and Stern (1981), gives a taste of coevolution between the mechanical properties of silk and web design. The silk of *U. durandi* is unusual in its stiffness, and its webs exploit this stiffness in two ways. The spider built a central, star-shaped, symmetrical sheet or tent under which she sheltered (fig. 10.5). The stiff tent was dome-shaped in cross-section, even when it was built on a flat surface. There were no supports above or below to hold up the dome, and there were six rigid, arched "portals" at its edges through which the spider ran to attack prey.

Connected directly to the silk floor of the tent were 12 long signal lines (up to 20 cm each); these lines functioned as "trip" lines for prey. They were raised above the substrate by tiny stiff silken posts, thus increasing the chances that walking prey would stumble against them and produce vibrations signaling to the waiting spider that she should attack. Neither the trip lines nor the silken posts were sticky, and the web did not stop or retain prey; these functions were accomplished by the spider's ability to run rapidly and to wrap prey quickly with a thick swath of lines from her long posterior lateral spinnerets (Crome 1957), whose surfaces are packed with up to 1500 aciniform spigots (there are also chemosensory hairs nearby, so the spider may also sense chemical signals from the prey with her rear end!).

The silken posts have still another elegant design trait. The trip lines are long, and thus need to transmit vibrations with a minimum of damping. There were up to 20 or more posts on a single long trip line, and each rigid supporting post could impede this transmission. The damping effect on transmission was reduced by the thin, delicate junction between the trip line and each post (inset in fig. 10.5). This suite of adaptations enabled these spiders to build effective webs with elevated trip lines even at websites on concave and flat rocks that lacked protruding supports.

The recently discovered diversity in the mechanical properties of the silk in different species (section 3.2.9) leads one to suspect that there are probably many other yet to be undiscovered variations in web designs that take advantage of special mechanical properties of silk lines.

10

ONTOGENY, MODULARITY, AND THE EVOLUTION OF WEB BUILDING

The developmental uncoupling between [web] algorithms opens, at the evolutionary level, the possibility of structural novelties in web building. —Japyassú and Ades 1998, p. 932

10.1 INTRODUCTION

Knowledge of the web forms of both orb weavers and non-orb weavers has grown explosively over the last 40 years, thanks in large part to the efforts of taxonomists using the "low tech" technique of photographing webs dusted with corn starch or talcum powder in the field. For instance, Griswold et al. (2005) presented original photos of non-orb web designs for 22 species in 18 families, many of whose web designs were previously completely unknown. Happily, the time when nearly the only non-orbs known were those of a few North Temperate Zone species is now only a distant memory.

The blizzard of new information has, however, complicated evolutionary analyses of webs. Just over 50 years ago Kaston (1964) introduced his review of web evolution by noting that the accumulation of new data made his task immeasurably easier than it had been for Comstock about 50 years earlier (Comstock 1912). I would say that today, 50 years later, the additional knowledge has made the task much harder. Many new and possibly intermediate forms have been discovered, and the huge advances in understanding the phylogeny of spiders discourage making previous simple assumptions regarding evolutionary order, such as presuming that more complex webs are more derived, or attributing intermediate evolutionary status to the webs of extant species (see figures in Kaston 1964; Kullmann 1972; Vollrath 1988b). Orb webs are ancient (at least 165 million years old), and previous portraits of web evolution as a progressive series of steps leading gradually to increasingly complex web forms and culminating in orbs (e.g., Savory 1952; Kaston 1965; Kullmann 1972; Shear 1986) must be discarded. Current uncertainty regarding some phylogenetic relations (section 1.6) adds further difficulties.

This chapter discusses various topics, including possible relations between ontogeny and web evolution (section 10.2), a new hypothesis regarding the origins of webs (section 10.4), several aspects of non-orb evolution (section 10.5), the probable monophyly of orb webs (section 10.8), changes in orbs (section 10.9), post-orb changes that occurred after orbs first originated (section 10.10), and a brief summary concerning the details of how particular behavior patterns may have changed over evolutionary time (sections 10.11, 10.13, 10.14).

10.2 WEB ONTOGENY AND EVOLUTION

The so-called biogenetic law, or the tendency for the ontogenetic changes in a species to occur in the same order as the evolutionary changes in its ancestral lineage, has long been a topic of controversy (Nelson 1985). Most studies have concentrated on morphology rather than behavior. Wenzel (1993) found several apparent cases (and one possible contradiction) of the biogenetic trend in the nest construction behavior of paper wasps; he used, however, a somewhat unusual definition of ontogenetic change that concentrated on the sequence of behavior patterns that are used to build a given structure by different individuals, rather than on changes in the behavior of the individuals themselves.

Web-building spiders are an exception to this neglect. Early workers noted a biogenetic pattern in changes in orb web designs (e.g., Petrusewiczowa 1938; Bristowe 1941); Bristowe (1958) used this pattern to deduce evolutionary history, noting that in some species the orbs of juveniles "are more symmetrical and less specialized, thereby indicating the inheritance of their design from some distant ancestor." Table 10.1 tests this idea with the most extensive (but nevertheless incomplete) summary ever compiled for spiders. It shows, in the first place, that ontogenetic changes in web designs are common. Secondly, it demonstrates a general biogenetic trend for designs of adults to be more derived, but that exceptions are not uncommon. In the sample of 47 species or groups of congeneric species, 71 traits in which ancestral and derived states could be deduced with some certainty differed between juveniles and adults. The juvenile trait was ancestral in 12 (75%) of 16 non-orbs, and 39 (71%) of 55 orbs. Both trends are statistically significant, but their inconsistency makes attempts to use ontogeny to deduce phylogenetic changes risky. Some spiders, incidentally, build highly derived web forms for their entire lives, including *Hyptiotes cavatus* (UL) (Opell 1982), *Miagrammopes* sp. ca. *unipus* (UL) (Lubin et al. 1978; WE), *Comaroma simoni* (AN) (C. Kropf pers. comm.), and *Wendilgarda galapagensis* (TSM) (Eberhard 1989a) and *W.* sp. (P. Palavicini unpub.).

Ontogenetic patterns can nevertheless illuminate possible evolutionary pathways. An especially thorough study of the ontogenetic changes in the webs of the nephilid *Nephilengis* (= *Nephilengys*) *cruentata* revealed the evolutionary independence of different web traits (Japyassú and Ades 1998). The trajectories of the ontogenetic increases in the relative sizes of the upper and lower portions of the orb differed: the lower portion showed a gradual, unidirectional increase, while the upper portions first increased, and later decreased (fig. 10.1). In addition, ontogenetic changes in other web traits (changes in slopes when plotted against body size) occurred at different stages of development (fig. 10.1). In the first transitional period, the numbers of radii, the presence of sticky and temporary spiral loops, the percentage of orbs with an empty sector, and web angles with horizontal all changed; later the horizontal diameter, the radius length, and the percentage of tent-like retreats changed. As noted by Japyassú and Ades (1998), the modularity revealed by this developmental uncoupling could facilitate evolutionary diversification of web designs (section 10.13, Eberhard 2018a). The ontogeny of webs in *Zygiella x-notata* (AR) (fig. 10.2; Le Guelte 1966) and two *Leucauge* species (TET) (Barrantes et al. 2017) showed similarly independent changes in different aspects of web design.

Explanations for trends in behavioral versus morphological ontogeny are likely to differ, because intermediate forms in morphology constitute physical links or bridges between stages. The newly formed zygote in the egg is morphologically very simple and relatively uniform, and this imposes certain types of patterns. In contrast, behavioral traits do not need to be linked physically. In fact, some web traits can change abruptly from one day to the next when the environment changes (fig. 7.42 in response to limited space), or when an individual adopts alternative web designs (figs. 10.25, 10.26, section 9.2.4).

10.2.1 LIMITS OF INTERPRETATIONS

The outgroup technique used in table 10.1 to distinguish ancestral from derived traits has a serious problem regarding which outgroup should be used (Gregorič et al. 2013), because using more or less distantly related outgroup sometimes alters the apparent polarity of a trait. The complete phylogeny is not known for any of the species in table 10.1, so there are uncertainties of this sort throughout the table. On the other other hand (bar-

Table 10.1. A sampler of ontogenetic changes in web design emphasizing species in which it was possible to deduce (at least tentatively) whether adult or juvenile traits are more like to be derived (less certain cases are indicated with "?"). There was a trend for juvenile rather than adult web traits to be plesiomorphic (i.e., ontogenetic changes tended to repeat phylogenetic changes), but there were many exceptions. The polarities of theridiid traits are from Eberhard et al. 2008b; for other groups they were estimated using comparisons with outgroups (usually closely related genera)[1] or the publications cited as references (see text for further limitations on determining polarities). Changes in webs that were probably direct consequences of the spider's growth in size, such as web diameter and spaces between sticky spiral loops, were not included (see also section 7.19 for the possible role of learning).

Family and species	*Juvenile web trait*	*Adult web trait*	*Which stage plesio-morphic?*	*References*
Pholcidae				
Physocyclus globosus	Lack gumfoot lines	Gumfoot lines present	Juv.	Escalante and Masís-Calvo 2014
Filistatidae				
Kukulcania hibernalis	Cribellate silk attached along a non-sticky line	Cribellate silk in piles on non-sticky line	Juv.	Lopardo and Ramírez 2007, G. Barrantes in prep.
Zoropsidae				
Tengella radiata	Retreat first lacking, then in center of sheet	Retreat at edge	?	Barrantes and Madrigal-Brenes 2008
	Lack cribellum lines	Cribellum lines present	Adult	Barrantes and Madrigal-Brenes 2008
	Short tunnel	Long tunnel	?	Barrantes and Madrigal-Brenes 2008
Psechridae				
Fecenia sp.	Web 3-D cone	Web vertical sheet, with retreat in upper portion	?	Robinson and Lubin 1979b
Oxyopidae				
Peucetia viridans	Build web	No web	Juv.?	Kaston 1972, Griswold 1983
Pisauridae				
Architis nitidopilosa	2-D sheet on surface of leaf	3-D sheets, spanning multiple objects	Juv?[2]	Nentwig 1985c
Linyphiidae				
Oedothorax sp.	Small web	No web	Juv.	Alderweireldt 1994
Frontinella pyramitela	Bowl and doily sheet with tangle above and below	Tangle only (mature male)	?	Suter et al. 1987

Table 10.1. Continued

Family and species	*Juvenile web trait*	*Adult web trait*	*Which stage plesio-morphic?*	*References*
Theridiidae				
Steatoda grossa	No sticky lines in tangle	Sticky lines in tangle	Juv.	Eberhard et al. 2008b, Barrantes and Eberhard 2010
Latrodectus, *Steatoda*	Retreat in midst of tangle	Retreat at edge of web	Juv.	Szlep 1965, Barrantes and Eberhard 2010, Eberhard et al. 2008b
Latrodectus spp.	No sheet	Sheet in tangle	Juv.	Eberhard et al. 2008b, Barrantes and Eberhard 2010
Latrodectus sp.	Retreat only silk	Retreat with detritus	Juv.	Szlep 1965, Eberhard et al. 2008b
Latrodectus spp.	Central disc with radial lines	No central disc	Adult?[3]	Barrantes and Eberhard 2010
Latrodectus spp.	Built entire web in a single night	Built web over several nights	Adult[4]	Szlep 1965
Theridion melanurum	No cylindrical sheet	Cylindrical sheet enclosed gumfoot web	Juv.	Nielsen 1932, Eberhard et al. 2008b
Achaearanea (= Theridium) lunata	Gumfoot plus tangle	Tangle only	Juv.	Nielsen 1932, Eberhard et al. 2008b
Enoplognatha ovata	No retreat with leaves fastened together	Retreat with leaves fastened together	Juv.	Nielsen 1932, Eberhard et al. 2008b
Theridion saxatile	No gumfoot lines	Gumfoot lines	Adult	Freisling 1961
Araneidae				
Allocyclosa bifurca	Disc stabilimentum common	Disc stabilimentum uncommon	?[5]	Eberhard 2003
Argiope argentata	Few with tangle	More tangles beside orb	Juv.?	Lubin 1974
Argiope aurantia, *bruennichii*, *trifasciata*	Tangle beside orb more common or more dense	Fewer tangles, less dense[6]	Adult?	Tolbert 1975, Baba and Miyashita 2006, Edmunds and Edmunds 1986
Argiope spp.	Disc stabilimentum common	Stabilimentum usually with arms, not disc	?[7]	Comstock 1967, Peters 1955, Wiehle 1927, 1928, Edmunds 1986, Robinson and Robinson 1970a
Cyclosa insulana	All with stabilimentum	Stabilimentum infrequent	Adult	Marson 1947b
Cyclosa morretes	Stabilimentum cluster of detritus	Stabilimentum line of detritus	?	Gonzaga and Vasconcellos-Neto 2005

Table 10.1. Continued

Family and species	*Juvenile web trait*	*Adult web trait*	*Which stage plesio-morphic?*	*References*
Cryptaranea atrihastula	Less elongate orb	Elongate above and below hub	Juv.[8,9]	Forster and Forster 1985
Cyrtophora citricola	Tangle with only rudimentary orb sheet	Tangle with large orb sheet	Adult	Kullmann 1964
Eriophora fuliginea	Not perfectly planar	Orb planar	Adult[10]	Graf and Nentwig 2001
Eriophora fuliginea	Not close to vertical	Close to vertical	Juv.[10]	Graf and Nentwig 2001
Eriophora fuliginea	Diurnal	Nocturnal	Adult[10]	Graf and Nentwig 2001
Eriophora fuliginea	Horizontal orb on upper side of leaf	Horizontal between plants	Adult[10]	Graf and Nentwig 2001
Eriophora fuliginea	Spider at hub	Spider in curled leaf	Juv.[10]	Graf and Nentwig 2001
Eriophora sagana	Orb larger above than below the hub	Orb more nearly vertically symmetrical	Adult[11]	Nakata 2010*b*
Eustala sp. #2345	Typical orb	Highly asymmetrical, with sticky lines along the frames and radii	Juv.	Eberhard 1985
Eustala illicita	Orb more symmetrical	Orb larger below hub	Juv.[8,12]	Hesselberg 2010
Eustala illicita	Free sector less common	Free sector more common	Juv.	Hesselberg 2010
Eustala perfida	Orb more symmetrical vertically	Orb very asymmetrical vertically (trunk web)	Juv.[8]	Messas et al. 2014
Gasteracantha curvispina	Barrier webs more common	Barrier webs rare	Adult	Edmunds and Edmunds 1986
Larinioides scopetarius	Orb more symmetrical vertically	Orb less symmetrical vertically	Juv.[8]	Heiling and Herberstein 1998
Mastophora dizzydeani	No web[13]	Bolas	Adult	Eberhard 1980*a*
Metepeira labyrinthea, M. gressa, M. sp. (Costa Rica)	Lack retreat, spider at hub	Spider in retreat in tangle near orb	Juv.	Peters 1955, Viera 2008, WE
M. gressa, M. sp. (Costa Rica)	No tangle near orb	Tangle beside orb	Juv.	Viera 2008, WE
M. gressa, M. sp. (Costa Rica)	Approximately symmetrical	Hub near upper edge of orb	Juv.?[8]	Viera 2008, WE
Pararaneus cyrtoscapus	Horizontal	Vertical	Adult	Edmunds 1978
Pararaneus cyrtoscapus	Orb drawn into shallow cone by 2–4 lines	Planar	Adult	Edmunds 1978
Phonognatha graeffei	No curled leaf	Curled leaf suspended in orb and used as retreat	Juv.	McKeown 1952

Table 10.1. Continued

Family and species	*Juvenile web trait*	*Adult web trait*	*Which stage plesio-morphic?*	*References*
Poecilopachys australasia	Interruptions in covering of glue on sticky spiral	No interruptions	Adult	Clyne 1973
Poltys spp.	Hub center left intact	Hub center removed	Adult	Smith 2009
Poltys spp.	Free zone present between hub and sticky spiral	Free zone reduced or absent	Juv.	Smith 2009
Poltys spp.	Circular outline orb	Taller than wide[14]	Juv.[8]	Smith 2009
Poltys spp.	Begin construction before dusk, destroy after dawn	Build and destroy at night	Adult?	Smith 2009
Scoloderus tuberculifer	Approximately symmetrical vertically[15]	Orb greatly elongate vertically	Juv.	Eberhard 1975
Zygiella x-notata	No open sector with signal line	Open sector with signal line	Juv.	Le Guelte 1966
Zygiella x-notata	Retreat at edge rarer[16]	Retreat at edge	Juv.	Le Guelte 1966
Zygiella x-notata	Approximately symmetrical vertically[16]	Less symmetrical vertically	Juv.[8]	Le Guelte 1966
Zygiella x-notata	Hub not displaced toward retreat	Hub displaced toward retreat	Juv.	Le Guelte 1966
Zygiella x-notata	More radii[16]	Fewer radii	?	Le Guelte 1966
	Fewer radii[17]	More radii	?	Le Guelte 1966
Nephilidae				
Clitaetra perroti, irenae	Vertical elongation weak	More vertical elongation	Juv.[8]	Kuntner et al. 2008b, Kuntner and Agnarsson 2009
Herennia multipuncta (= *ornatisima*) and *H. papuana*	Orb more nearly planar	Orb curved laterally to fit the rounded curve of the trunk with "pseudoradii"	Juv.	Robinson and Lubin 1979a, Kuntner 2005
Herennia multipuncta (= *ornatisima*) and *H. papuana*	Orb more symmetrical vertically	Orb very asymmetrical vertically	Juv.[8]	Robinson and Lubin 1979a, Kuntner 2005
Nephila pilipes	Attach radius to frame once	Double attachment of radius to frame	Juv.	Kuntner et al. 2008a

Table 10.1. Continued

Family and species	*Juvenile web trait*	*Adult web trait*	*Which stage plesio-morphic?*	*References*
Nephila clavipes	Barrier web dense[18]	Barrier web less dense or absent	Adult[18]	Peters 1953, Higgins 1992b, Kuntner et al. 2008a
Nephila clavipes	Orb more symmetrical vertically	Orb very asymmetrical vertically[19]	Juv.[8]	Comstock 1967, Brown and Christenson 1983 (but see Hesselberg 2010)
Nephila spp.	Usually replace entire orb at once	Usually replace only about half of orb at once	Juv.	Higgins 1992b, Kuntner et al. 2008a
Nephilingis cruentata	Orb more symmetrical vertically	Orb very asymmetrical vertically	?[8,20]	Japyassú and Ades 1998
Nephilingis cruentata	Variable aerial sites	Near sheltered retreat	Juv.[20]	Japyassú and Ades 1998
Nephilingis cruentata	No retreat	Retreat in silk cylinder[21]	Juv.[20]	Edmunds 1978, 1993, Japyassú and Ades 1998
Nephilingis cruentata	Frame, radial and sticky lines above hub	No frame, radial or sticky lines above hub	Juv.[20]	Japyassú and Ades 1998
Nephilingis cruentata	More nearly horizontal	More nearly vertical	Adult[20]	Japyassú and Ades 1998
Nephilingis cruentata	Hub aerial	Hub adjacent to substrate	Juv.[20]	Japyassú and Ades 1998
Nephilingis cruentata	Build complete orb in one night	Different sections of orb built on different nights	Juv.	Edmunds 1993
Tetragnathidae				
Leucauge argyra	Tangle above orb	Tangle sparse in large juveniles, absent in adults	Adult	Triana-Cambronero et al. 2011
Leucauge mariana	No tangle below orb	Tangle below common	Juv.	I. Moroto unpub., WE
Metellina ornata	Typical horizontal orb	Only non-sticky vertical radial lines	Juv.[22]	Shinkai 1998b
Pachygnatha spp.	Orb web[23]	No web[24]	Juv.	Balogh 1934, Martin 1978 in Levi 1980
Tetragnatha (= *Eucta*) *kaestneri*	Orb web	Sometimes no web	Juv.	Crome 1954
Theridiosomatidae				
Epeirotypus sp.	Hub not close to substrate when tensed	Hub close to substrate when tensed	Juv.	Eberhard 1986b, Coddington 1986a
Epilineutes	Spring line more often present	Spring line less often present	Juv.	Coddington 1986a, Lopardo et al. 2008

Table 10.1. Continued

Family and species	*Juvenile web trait*	*Adult web trait*	*Which stage plesio-morphic?*	*References*
Uloboridae				
Polenecia producta	Lack hub or temporary spiral	With hub, often with temporary spiral	Adult	Peters 1995a
Uloborus diversus	More nearly horizontal	Less horizontal	Juv.?	Eberhard 1973a
Uloborus diversus	More frequent silk stabilimenta	Less frequent silk stabilimenta	?	Eberhard 1973a
Uloborus conus	Rim of orb with few loops of sticky spiral	Rim with more loops of sticky spiral	?	Lubin et al. 1982
Several genera (incl. *Zosis, Uloborus, Philoponella)*	Orb lacks sticky spiral, but has temporary spiral and dense mat of very fine lines	Orb with sticky spiral, no mat	Adult[25]	Szlep 1961, Eberhard 1972a, 1977d, WE
Anapidae				
Conculus lyugadinus	Aerial orb (?)[26]	Sticky spiral attached to water surface	Juv.	Shinkai and Shinkai 1988

[1] Choosing outgroups, and thus the relatedness of the ancestor, is crucial. For instance, the lack of a stabilimentum in adult *Cyclosa insulana* is classified as primitive on the basis of comparisons with other araneid genera. But this trait may be derived in this species if there was a subsequent *Cyclosa* ancestor of *C. insulana* that had a stabilimentum (and most modern *Cyclosa* species do indeed have stabilimenta).

[2] Based on the relative simplicity of the two designs.

[3] Central discs are also present in the webs of adult *L. pallidus*, but are not known in webs of other *Latrodectus* species adults or juveniles or of the adults of the closely related *Steatoda*, so polarization is uncertain (Eberhard et al. 2008b). On the assumption that theridiids are derived from an orb-weaving ancestor, hubs are plesiomorphic at a deeper level.

[4] Pleisomorphic status deduced from table 5.1.

[5] This change has a possible functional explanation: adult females mimic the gray-green egg sacs of similar size and shape that are present above and below the hub; so additional white silk at the hub might make the spider more visible. Juveniles mimic (in color) the detritus itself, which has no fixed size or shape.

[6] Different individuals ceased making barrier webs at different stages; there was also some intra-individual variation (Edmunds and Edmunds 1986).

[7] Disc stabilimenta are generally thought to hide the spider; it is possible that they are abandoned when the spider's body becomes too large to hide this way.

[8] Many authors have supposed that greater symmetry is plesiomorphic, on the basis of closely related genera. For instance, greater symmetry is taken as pleisiomorphic in Nephilidae (Kuntner et al. 2008a). At a deeper level, however, asymmetry may have been plesiomorphic in the original derivation of orb webs (see frontispiece, fig. 9.18).

[9] Many other *Araneus* species, including the closely related New Zealand *A. subcomptus*, make only moderately asymmetrical orbs (Forster and Forster 1985).

[10] Planar, diurnal, asymmetrical, and nearly vertical orbs occur both in other species of *Eriophora* and in *Verrucosa* (WE), its sister genus (Scharff and Coddington 1997).

[11] The reduced upper portion of the orb in mature spiders is associated with a reduced relative speed of attack above the hub, and thus may be due to natural selection rather than the biogenic law (attack speed was not measured, however) (Nakata 2010b).

[12] A similar, but statistically non-significant trend occurred in *Clitaetra espisinoides* (NE).

[13] Ecological factors may be responsible for this difference: evaporation of water from the bolas would render a smaller bolas non-functional relatively quickly, and thus explain the lack of a bolas in smaller spiders.

[14] The upper portion was lengthened by widening sticky spiral spacing, or, in *P. frenchi*, by adding zig-zags of sticky spiral (Smith 2009).

Table 10.1. Continued

[15] The webs of young *S. cordata* are elongate ladder webs (Stowe 1978), so this species apparently does not show ontogenetic changes.

[16] This change occurs in early vs. later webs during the second instar (the first instar spent outside the egg sac).

[17] This change occurs from late second instar to adult.

[18] Barrier webs function to defend against predators, and smaller spiders may be under greater predation pressure; barrier webs decreased at a later age in areas thought to have higher predator pressure (Higgins 1992b).

[19] Hesselberg 2010 found no change in the up-down symmetry of *N. clavipes* webs that were built in captivity, but in the field the orbs of adult females were more asymmetrical than those of second and third instar spiderlings (WE).

[20] Based on the phylogeny of Kuntner et al. 2008a.

[21] Retreats of intermediate instars had intermediate forms (tent, open tube).

[22] Both the horizontal orientation and the presence of sticky spiral lines in the juvenile webs are ancestral traits.

[23] The existence of orbs build by young juveniles was at first controversial (see summary by Levi 1980), but the descriptions by Martin (1978) (in Levi 1980) of horizontal, open-meshed orbs close to the ground resolved the uncertainty.

[24] Adult females of *P. autumnalis* lose their flagelliform and aggregate gland spigots, confirming their inability to produce sticky spiral lines (Platnick et al. 1991).

[25] The derived nature of juvenile webs was deduced from the fact that in other cribellate families the webs contain sticky cribellum silk, including those of juveniles. Given, however, the relative neglect of juvenile webs (section 3.2.14), this must be considered tentative. Dense mats of fine lines like juvenile uloborid webs are, to my knowledge, unknown in any other cribellate family.

[26] Both types of web are built by spiders of all sizes, but it is implied that aerial orbs may be more common in smaller individuals. Floating webs were only built when the web was over water, when the velocity of the current was relatively slow, and when the plane of the web was less than 10 cm above the water surface (Shinkai and Shinkai 1988).

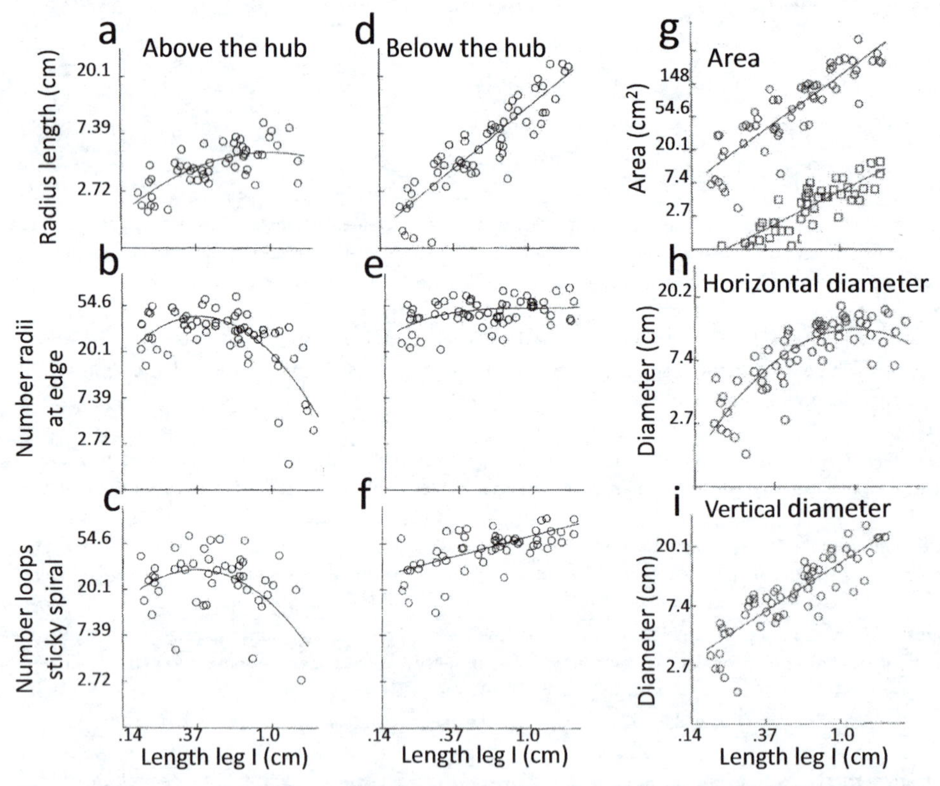

Fig. 10.1. The ontogenetic trajectories of these nine web traits of *Nephilingis cruentata* (NE) differed, suggesting that they are under at least partially independent control (see also fig. 10.2). The numbers of radii and sticky spiral loops above the hub increased somewhat early in life, but later fell off sharply; in contrast, these same numbers were more nearly constant below the hub. Changes in the area of the web (circles in *g*) and the hub (squares in *g*) were similar to those of the vertical diameter (*i*), except that the hub did not increase sharply early in life. The vertical diameter of the orb below the hub increased monotonically, while the horizontal diameter (*h*) increased during early stages and leveled off later. Differences in trajectories suggest that these traits are controlled independently, and can thus presumably evolve independently (after Japyassú and Ades 1998).

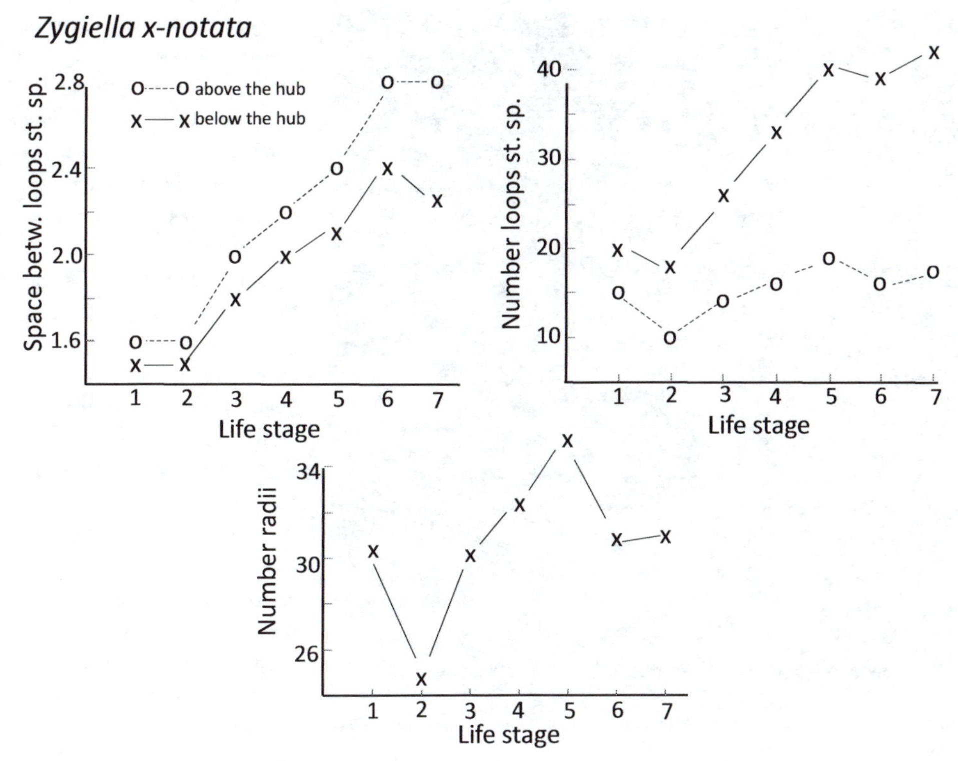

Fig. 10.2. The ontogenetic changes in several features documented extensively in the orbs of *Zygiella x-notata* AR in captivity suggest a modular organization of web construction behavior (see also fig. 10.1). The "life stages" and the numbers of webs measured are the following: 1 = early second instar spiderling (first instar outside egg sac) without a retreat (55 webs); 2 = late second instar spiderling with a retreat beyond the web's edge (78 webs); 3 = third instar spiderling (96 webs); 4 = fourth instar spiderling (89 webs); 5 = fifth instar spiderling (87 webs); 6 = adult female before oviposit (105 webs); 7 = adult female after oviposit (45 webs). The differences between traits indicate a degree of independence between behavioral traits. Especially diamatic changes occurred in the number of radii during the second instar, when the spider changed from resting at the hub (stage 1) to resting in a retreat (stage 2) (after Le Guelte 1966).

ring a possible unconscious bias in choosing outgroups), I do not see any reason to suppose that this type of error would produce the bias in the table toward juvenile traits being ancestral.

Two examples illustrate this problem. The young of *Nephila pilipes* (Kuntner et al. 2010b) and some *N. clavipes* (NE) (WE) built orbs that were more vertically symmetrical than those of the adults (the hub was more nearly midway between the upper and lower edges of the web). Deep in the phylogeny of the ancestors of all nephilids, the strong vertical asymmetry that typifies most nephilid webs today was presumably derived from more typical, moderate vertical asymmetries found in most other orb weavers (Blackledge et al. 2009c). But the one favored phylogeny of Nephilidae suggests that the web of the last common ancestor of this family was a small ladder web, which evolved into a highly specialized arboricolous ladder ("trunk web"), then into a large semi-ladder, and finally, in *Nephila*, into a secondarily derived aerial orb (Kuntner et al. 2008a, 2010b). The ontogeny of *N. pilipes* webs clearly does not follow this trajectory (Kuntner et al. 2010b).

Which ancestor should be used to check for the biogenetic pattern in *Nephila*? I do not have an easy answer. An added complication is that phylogenetic hypotheses are sometimes unstable. I have no reason to doubt the particular sequence just described for nephilid web evolution, but confidence is greater regarding some transitions than others, and changes in branching patterns can change conclusions regarding character polarities. In addition, the ontogenies of the webs of putative ancestors are unknown.

In a second well-documented ontogeny, young *Leucauge argyra* (TET) had more elaborate tangles below their horizontal orbs than did adults, and they also built tangles above their orbs (Triana-Cambronero et al. 2011),

Fig. 10.3. This web of a *Leucauge argyra* (TET) spiderling apparently "broke" the ontogenetic law. The tangle above the conical orb was reduced in the webs of older instars, and absent from those of adult females. Comparisons with other *Leucauge* species and other tetragnthid genera suggest that building a tangle above the orb is probably a derived trait; thus the ontogenetic change in *L. argyra* appears to be the reverse of the usual trend. Presumably one or more of the postulated functions of the tangle (stabilize the web, defend against enemies, knock prey down onto the orb) is more important for juveniles than adults in this species (from Triana-Cambronero et al. 2011; courtesy of Gilbert Barrantes).

a site where conspecific adults and the adults and juveniles of other *Leucauge* species almost never built tangles (fig. 10.3, Barrantes and Eberhard in prep.; WE). In *L. mariana*, however, these trends were inverted. Spiders in early instars of *L. mariana* never had tangles above or below their orbs, while adult females often had a sparse tangle below (Maroto 1981; WE). And in *L. venusta* both juveniles and adults consistently had tangles below but not above the orb (WE). At first glance at least some of these data might appear to be incompatible with the biogenetic law, but interpretation is not simple. Outgroup comparison with other tetragnathid genera, such as *Tetragnatha*, *Metabus*, and *Dolichognatha*, suggests that the common ancestor of all modern *Leucauge* species lacked a tangle: on the long term, the lack of a tangle is likely ancestral. But on the shorter term, the immediate ancestors of *L. argyra* and of *L. mariana* may have had different orb designs. For instance, the *L. mariana* lineage may have acquired a tangle from ancestral *Leucauge* that lacked tangles; the *L. argyra* lineage may have had an ancestor with a well-developed tangle on both sides of the orb. Until the web designs are described for more species in this large genus of 100+ species and an intra-generic phylogenetic hypothesis become available, it will not be possible to test these speculations.

Another possible complication is that no provision was made for possible phylogenetic biases in totaling up the data in table 10.1. For instance, juvenile webs were more vertically symmetrical than adult webs in all four nephilid species, *Herennia multipuncta*, *Nephila pilipes*, *N. clavipes*, and *Nephilengys cruentata*. Some would argue that these four observations should be collapsed into one because of their close phylogenetic relations (e.g., Harvey et al. 1991). I believe that the existence of abundant intra-specific variation in the degree of vertical web symmetry in adults and juveniles of these species argues, however, against lumping them. The presumed phylogenetic constraint seems to be very weak, as there is an ample supply of variants in the degree of vertical symmetry on which natural selection could act and produce changes if they were advantageous.

In sum, ontogenetic changes in webs tend to reflect the likely phylogenetic changes, but with many exceptions. The frequent exceptions, as well as possible imprecisions in the analyses, argue against attempting to deduce phylogenetic changes from ontogeny. Contrary to my previous opinions (Eberhard 1985; Eberhard et al. 2008a), I feel that the biogenetic trend in spider webs can only, at best, be used cautiously to suggest probable phylogenetic patterns.

10.2.2 A NEW HYPOTHESIS FOR ONTOGENIC CHANGES: CONSISTENT SELECTION ASSOCIATED WITH SMALLER SIZE

It is possible that the ontogenetic changes in web design are selectively advantageous, rather than the result of phylogenetic constraints as implied by the biogenetic law. This "adaptive changes" hypothesis would propose that the tendency for juvenile web designs to be more ancestral is due not to phylogenetic constraints on the behavior itself, but to a pattern in natural selection: selection on early stages may differ in consistent ways from that on later stages. In other words, a different and more uniform set of selective pressures may act on younger spiders.

What would be the most likely characteristics of this more uniform adaptive space for juveniles? Major trends in table 10.1 offer some suggestions. The most well-represented and consistent trait was the degree of vertical asymmetry in approximately vertical orbs: in 14 of 15 species adult orbs were less symmetrical, and their greater asymmetry resulted from elongation below rather than above the hub (with the exception of the aberrant ladder webs of *Scoloderus* spp. [AR], which are probably specialized to capture moths) (Eberhard 1975; Stowe 1978, 1986). This ontogenetic change could result from heavier spiders being relatively slower in climbing upward to attack prey. The attack behavior in the extreme case of the giant *Nephila clavipes* (NE) supported this hypothesis. Adult females (which weigh on the order of 1,000–2,000 mg) attacked some prey below the hub nearly instantaneously (the spider took only tenths of a second to descend 15–20 cm to the prey); all horizontal or upward movements in their webs were, in contrast, very slow and painstaking (WE). Similarly, the lighter weights of younger *N. pilipes* may also make their attacks on prey above the hub relatively quicker than those by larger individuals (Kuntner et al. 2010a). The lack of change in the vertical asymmetry in *Epeirotypus* sp. (TSM) (Eberhard 1986b) also fits this idea, because these short-legged spiders attacked all prey slowly (above and below the hub) (WE), relying instead on springing their strongly adhesive webs. Further support comes from the additional nephilid pattern of increasing the orb's curvature and the verticality of the lateral frame lines in later ontogenetic stages in *Herennia multipuncta*, which allows the web to fit against the surfaces of large tree trunks. At the scale of the web of a younger spider, the trunk surface is more nearly flat and less limiting laterally (Kuntner et al. 2010a,b).

A second major trend in table 10.1 was that juveniles tend to have more direct (and presumably more rapid) access to prey by waiting in less protected sites. This also seems to fit the adaptive changes hypothesis. The allometric scaling of brain size on body size means that juveniles (especially very young ones) have relatively oversized, energy-demanding central nervous systems (Quesada et al. 2011). This implies an increased importance of prey capture relative to other traits, such as defense against predators. This could explain the following several ontogenetic changes related to retreats: retreats in the midst of the tangle rather than at the web's edge in *Latrodectus* and *Steatoda* (THD); the lack of a protective sheet at the most exposed edge of the web (which probably excludes some possible prey) in *Theridion melanurum* (THD); the spider waiting at a central location in the web (usually the hub) in *Enoplognatha* (THD), *Metepeira* (AR), *Eriophora* (AR), *Zygiella* (AR), and *Nephilengys* (NE); and failing to locate the web especially near sheltered sites in *Zygiella* (AR), *Nephilengys* (NE), and *Epeirotypus* (TSM).

There are several other possible cases of this sort. Surface-volume relationships may result in the liquid in the sticky balls of small spiderlings of the bolas spider *Mastophora dizzydeani* evaporating too quickly to function for trapping prey (even the balls of the much larger adults became unusable after only about 20 min) (Eberhard 1980a), and thus explain the more derived behavior of the juveniles of hunting without any web at all (table 10.1). The thinner lines in the orbs of juvenile *L. argyra* are probably more susceptible to damage from whole web movements in the wind than are those of adults (fig. 10.4), and could thus benefit more from increased support from a tangle above the orb (fig. 10.3, Triana-Cambronero et al. 2011). The higher frequency of the defensive tangles at the sides of the orbs of smaller individuals of *Nephila clavipes* (NE) is thought to have been favored by the greater predation pressure on smaller spiders of this species, whose giant adults are undoubtedly out of the size range for many generalist invertebrate predators (Higgins 1992b).

There are additional differences in the ways in which smaller spiders interact with their environments whose

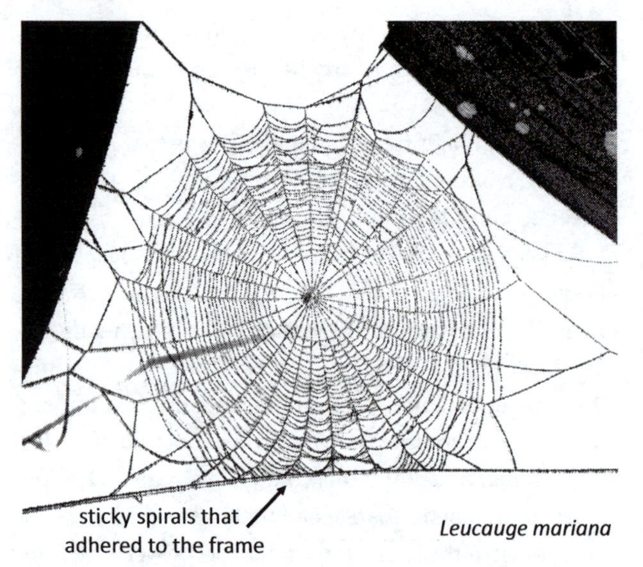

Fig. 10.4. Wind damage was a new problem when aerial webs with sticky lines arose. This nearly horizontal orb of a second or third instar *Leucauge mariana* (TET) spiderling, built on a windy day between a rigid pair of supports (the bases of leaflets at the tip of a large pinnate palm leaf), illustrates how billowing in the wind can result in damage to an orb; 8–10 sticky spiral lines touched and adhered to a frame line, opening holes in the orb's array of sticky lines. This distinctive type of damage may be especially important in smaller spiders because of their relatively weak webs, but data to test this speculation are lacking.

possible consequences for selection on different orb web designs are yet to be explored. For instance, the thinner diameters and the larger surface/volume ratios of their silk lines probably make them more susceptible to breaking under the friction produced by given wind velocities (ontogenetic variation in silk composition, if it exists, could introduce further complications). And the greater relative flexibility of some objects (like leaves) to which webs are attached would be greater under the stresses placed on them by the webs of larger spiders, thus increasing the risks of some types of web damage (fig. 10.4).

The hypothesis fails to explain convincingly, however, several cases in which younger individuals retain a more ancestral web form. For instance, if open sectors for signal lines are advantageous for adults of *Z. x-notata* in facilitating moving quickly from the retreat to the hub, why would they not also be advantageous for younger individuals that have retreats? In fact, young spiderlings of another araneid, *Araneus dalmaticus*, do build open sectors with signal lines just like those of conspecific adults (Kaston 1964). The advantage of ladder webs would only be absent in young *Scoloderus tuberculifer* (AR) if there are fewer small moth species in the same habitat (I know of no data) (see also Eberhard 1985).

Another web trait that has repeatedly been found to undergo ontogenetic change is the stabilimentum. But changes in stabilimentum design are not easy to interpret, because it is difficult to determine directions of evolutionary change, and there is uncertainty regarding possible functions (section 3.3.4.2); the significance of the common ontogenetic changes in stabilimenta is thus uncertain.

10.3 EARLY WEB EVOLUTION

Silk production dates to at least the origin of spiders about 410 Mya, and the morphology of the earliest spinneret known (380–385 Mya) has all the traits "required to produce silk at a level of sophistication paralleling that of some modern spiders" (Shear et al. 1989). There is indirect evidence of trip line webs with the origin of Mesothelae from about 310 Mya (figs. 9.26, 10.5a), and of lines organized into sheets about 245 Mya (Vollrath and Selden 2007). I will focus mostly on only a few major transitions, and will not give detailed historical accounts; current lack of data and phylogenetic uncertainties (section 1.6) make such speculation unprofitable. The reader should keep in mind the paraphyletic nature of the group "non-orbicularians" (figs. 1.9, 1.10).

10.3.1 BURROW ENTRANCES VS. EGG SACS

All spiders have silk glands, spinnerets, and spigots through which silk lines are produced. The proteinaceous nature of silk suggests that it arose from nitrogenous waste products in predatory ancestors (e.g., Craig 2003; Foelix 2011). Spiders in several evolutionary lineages that branched off very early do not use their silk to build prey capture webs. Thus silk probably originally evolved to function as it does in modern representatives of these ancient groups: to cover their eggs (originally proposed by Pocock 1895) (and as also occurs in other arachnids such as pseudoscorpions and whip scorpions); to line the tunnel or retreat where early spiders probably lived (originally proposed by McCook 1889); to facilitate transfer of sperm from the male's gonopore to his palps (below); or to cover the opening of the burrow to protect the spider (Kaston 1964, Shear 1986, and Blackledge et al. 2009c give historical summaries of these hypotheses). Only later, at least 300 million years ago (Vollrath and Selden 2007), did spiders begin to use silk for prey capture. Spiders and silk evolved long before insects evolved the ability to fly, so early webs were for pedestrian rather than aerial prey (Bristowe 1930; Coyle 1986; Grimaldi and Engel 2005; Vollrath and Selden 2007).

The origin of prey capture webs has generally been thought to be from silk that lined burrow openings (Shear 1986; Craig 2003) or from egg sacs (Bristowe 1930, 1941). Of these two possibilities, silk around the burrow opening seems more likely, because of its location. Before webs evolved, ancient spiders likely hunted by waiting at the openings of their silk-lined burrows or retreats for passing prey, as do present-day *Liphistius* spp. (Bristowe 1976; Blackledge et al. 2009c) and most modern mygalomorphs (Coyle 1986). Even a very small extension of the lining at the mouth of a retreat could increase the range of the spider's sense of touch and thus her predatory success (Coyle 1986). In addition, the web might increase the spider's ability to distinguish large, potentially dangerous prey from others and attack them more cautiously (Riechert and Łuczak 1982); and some potentially dangerous prey that were entangled in the spider's silk would also be attempting to free themselves when the spider approached, and thus be less alert to defend themselves. In contrast, eggs are generally laid in highly protected sites (e.g., deep in the burrow), where prey seldom occur.

The sequence proposed by Shear (1986; p. 374) seems likely: "Web evolution has proceeded from the primitive lined retreat, through expanded collars, to sheets along the substrate guyed in vegetation, to aerial sheets." A second possibility is extension of the collar with trip lines rather than a sheet (fig. 9.1, Kaston 1964; Coyle 1986). The course of web evolution has surely included multiple reversals (table 9.1, section 10.6), however, contrary to Shear's supposition (1986) that web evolution tended to go from geometrically irregular to more regular designs.

Because early web weavers produced numerous thin lines from multiple, fine spigots rather than a strong dragline from ampullate glands (Shear et al. 1989; Vollrath and Selden 2007), and because tunnel lining or closing behavior performed outside the tunnel would be sufficient to build a primitive sheet (see section 10.4), the sheet hypothesis is attractive. Nevertheless, many modern mygalomorphs build trip-line structures of silk (probably cables of multiple fine lines) (Coyle 1986). The lack of clear distinctions between these hypotheses is illustrated by segestriid genus *Ariadna*. Several species of this genus build trip lines that radiate from a small tangle at the mouth of the retreat (Comstock 1967; Shinkai and Takano 1984; Griswold et al. 2005), while *A. lateralis* in Japan had a clear sheet around the mouth of the retreat, with radial lines projecting from the edge of the sheet (Shinkai 1989). Sheet-like designs might have originated with structures to cover the mouth of the burrow to protect the spider, when the spider simply filled in the spaces between signal lines, or when the lining of the retreat was extended beyond the mouth of the burrow (Kaston 1964; Coyle 1986; Shear 1986).

Many spiders leave silk lines behind as they walk, and webs could have originated simply as accumulations of such lines made when the spider left the tunnel mouth briefly to attack passing prey or remove detritus (Bristowe 1930, 1976; Shear 1986). I would speculate, on the basis of the behavior of modern sheet-weaving diplurids (Eberhard and Hazzi 2013) and the behavior of filastatids (WE), that lines were deposited using movements of the abdomen and the spinnerets, and that more extensive use of the legs to find attachment sites only appeared later (see section 10.4).

Extensions of the tunnel lining at the mouth of the burrow that function to transmit vibrations, such as lines, tabs, or attachments to rigid objects like twigs and leaves near the mouth, have arisen many times (table 9.1). In mygalomorphs alone, Coyle (1986) mentions a minimum of eight derivations (not counting more elaborate funnels and sheets). The sister group to all other spiders, *Liphistius*, builds classic trip lines radiating from its burrow mouth (fig. 10.5a) (Bristowe 1976; Murphy and Platnick 1981), which are presumably cables of many much finer lines (possibly aciniform lines, though gland homologies are not yet clear). This web design was reinvented, for example, in araneomorphs such as the dysderid *Ariadna bicolor* (Comstock 1967) and the oecobiid *Uroctea durandi* (fig. 10.5, Kullmann and Stern 1981).

Probably one early change in spider webs with trip lines was to raise the lines off of the substrate, thus decreasing the likelihood that prey would simply step over them rather than bumping against them. Elevation can

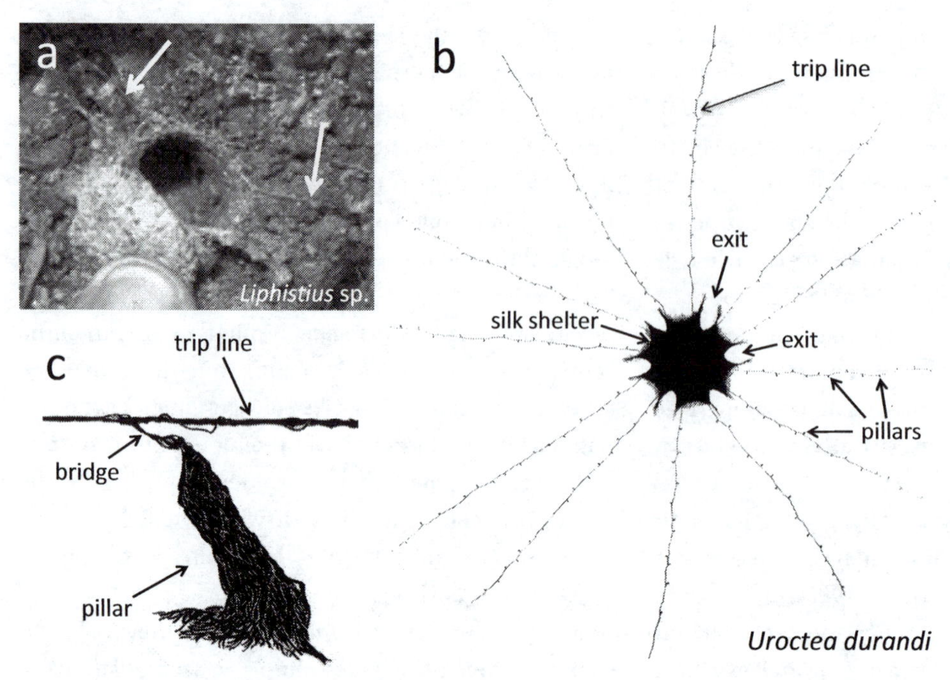

Fig. 10.5. An ancient (*a*) and a newer, "improved" version (*b*, *c*) of the signal line web design that functions to extend the spider's sense of touch. Spiders in the ancient *Liphistius*, sister group to all other spiders, build signal lines (arrows) radiating from the opening of the tunnel in which the spider lurks (*a*). The "tent plus trip lines" web built by *Uroctea durandi* (Oecobiidae) (*b*, *c*) includes a highly symmetrical, stiff, domed sheet that forms a central shelter within which the spider hides, and 12 long trip lines that radiate from their connections with the floor of the shelter (*b*). The trip lines were held above the substrate by large numbers of stiff silken pillars, each about 0.1 mm tall (*c*). Each pillar was joined delicately to the trip line (" bridge" in *c*), thus allowing the trip line to move relatively freely with respect to the pillar, and thus to transmit vibrations to the spider in her retreat with a minimum of damping. Six arched openings at the edges of the shelter (*b*) served as exits through which the spider dashed to attack prey that contacted a trip line (*a* courtesy of John Murphy; *b* and *c* after Kullmann and Stern 1981; *b* is based on a photograph of only a part of one web and verbal descriptions; *c* was traced from a photo).

be accomplished by preferentially attaching trip lines to protruding objects near the retreat. Elevation of this sort occurs in the filistatids *Filistata insidiatrix* (Nørgaard 1951) and *Kukulcania hibernalis* (WE) (fig. 7.52), raising both non-sticky and sticky lines from the substrate. *F. insidiatrix* mostly captured pedestrian rather than flying insects (45 [83%] of 54 prey) (Nørgaard 1951), so raising lines off the substrate probably increased their ability to contact and retain walking prey. A second more derived method is to construct tiny silk "poles" and attach the trip lines to their tips; poles have evolved convergently in the dysderid *Ariadna bicolor* (Comstock 1967; Griswold et al. 2005) and the oecobiid *Uroctea durandi* (fig. 10.5, Kullmann 1972; Kullmann and Stern 1981) (table 9.1) (the silk nature of the poles has not been confirmed; *U. durandi* poles contain multiple cylinders). Elevating lines away from the walls of the retreats of *F. insidiatrix* was also advantageous at especially warm sites on a sunny rock cliff, because the spider could stay out of contact with the hot walls (Nørgaard 1951).

Still another method of laying lines over irregularities in the substrate is to raise the spinnerets high in the air while walking, as in the diplurid *Linothele macrothelifera* (Eberhard and Hazzi 2013). Similar spinneret raising behavior occurred during prey wrapping in a surprising variety of groups (Barrantes and Eberhard 2007), where it probably also serves to spread apart the multiple fine lines; it is probably an ancient aspect of prey wrapping (Barrantes and Eberhard 2007).

I know of no detailed studies of mygalomorph egg sac construction, so cannot judge the alternative possibility that the spinning abilities that were later used to construct prey capture webs originally evolved as part of egg sac construction (Bristowe 1930); this behavior is complex in some araneomorphs (e.g., Gheysens et al. 2005; Moya et al. 2010).

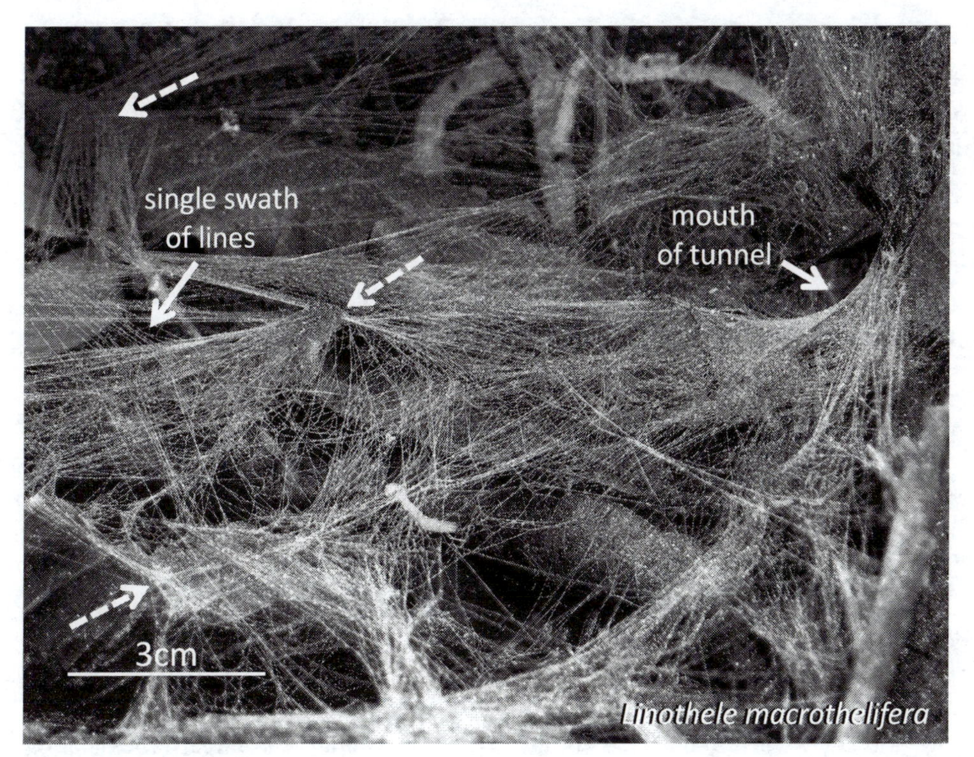

Fig. 10.6. This closeup of a partially built sheet web of the diplurid *Linothele macrothelifera* illustrates a swath of about 15–20 parallel thin lines that was probably produced by a single sweep of one posterior lateral spinneret. The swath's lines were attached to protruding objects in the litter (dotted arrows), especially near the edges of the sheet (from Eberhard and Hazzi 2013).

10.3.2 INTERCEPTION FUNCTION FOR EARLIEST WEBS

Early web weavers probably made very rapid attacks on prey that encountered their webs (Bristowe 1976; Nentwig 1985b), and the earliest prey capture webs probably functioned not as traps, but rather to increase the area in which spiders could sense prey by transmitting vibrations to the waiting spider from ambulatory prey (Bristowe 1976). The advantage of this design is illustrated by subsequent reinventions, such as those in the jumping spiders *Euryattus* sp. and *Simaetha* spp., that sometimes alerted the spider to attack prey that brushed against the small tangles of lines around their silken retreats (Jackson 1985a,b). The behavioral modifications needed to produce an web of this sort would seem to be minimal (even the behavioral change that converts a collar to a trapdoor in mygalomorphs is remarkably small—Coyle and Icenogle 1994).

10.3.3 RETENTION FUNCTION IN EARLY WEBS

A second early transition was the production of loose, fine lines that entangled and detained prey. Retention devices probably also occurred very early. Mats of fine lines that may have this effect occur in some mygalomorph webs (e.g., fig. 10.6; Coyle 1986). A later development that improved retention was production of adhesive lines containing multiple fibrils of cribellum silk (section 2.2.8.1). At first these fibrils formed part of compound cables; later, with the invention of the calamistrum, they were combed into loose piles that were more powerfully adhesive (Michalik et al. 2019). The cribellum was present in the ancestor of the largest spider suborder Labidognatha just under 200 Mya (figs. 1.9, 1.10).

Presumably specialized sticky lines were first added to the substrate, or to lines just off the substrate around the mouth of the spider's retreat, perhaps something like the present day webs of *Kukulcania* (fig. 5.10) and the amaurobiid *Callobius* sp. (fig. 206D of Griswold et al. 2005), or to a sheet-like continuation of the tube lining at the mouth, as in the udubid *Uduba* sp. (fig. 207E of Griswold et al. 2005). The subsequent acquisition of a calamistrum (section 2.2.8.1) allowed spiders to increase the retention abilities of these sticky lines by piling them into masses of loose fibrils. The mysterious ability of cribellate spiders to avoid adhering to their own cribellum silk (section 2.6) likely coevolved with the silk itself, as apparently leg IV never adheres to the lines it pulls from the cribellum. The pattern with which orb weavers add sticky lines, starting at the edge of the web and working inward, may also be ancient, as it occurs in numerous cribellate families, including filistatids, psechrids, dictynids, austrochilids, and eresids (table 5.1, Eberhard 1987a).

A primitive "wrapping" technique may have played a heretofore unappreciated role in the evolution of webs. Some web builders in groups that branched early from the labidognath stem, such as for instance the filistatid *Kukulcania hibernalis* (Barrantes and Eberhard 2007) and the diguetid *Diguetia canities* (WE), do not overpower their prey, but simply seize and hold a projecting extremity such as a leg or an antenna with their chelicerae, and then move backward, dragging the prey through the web and further entangling it. The spider then waits for the poison to take effect. This sort of "indirect wrapping," if it occurred early in the evolution of prey capture webs, could have favored the presence of "snagging" lines in the vicinity of the central retreat, and perhaps also less powerful legs and chelicerae. Entangling prey in silk this way may have allowed spiders to begin to abandon their ancient arachnid predatory tactic of overcoming prey by a combination of speed, strength, and venom.

10.3.4 WEBS WITHOUT RETREATS IN THE SUBSTRATE

Another important transition, which surely occurred multiple times during evolution, was to abandon refuges in cracks or cavities in the substrate such as the retreats of gradungulids (Michalik et al. 2013) and filistatids (Nørgaard 1951; figs. 5.10, 7.49) and the funnel webs of many modern-day groups (fig. 10.7). This change would have had potentially large ecological consequences. Only retreats within a limited range of diameters and depths are acceptable to some spiders. To cite two ancient lineages, the burrow sites of *Liphistius sumatranus* and *L. desultor* were limited to "bare vertical banks which are neither parched nor too damp, and must be in the shade of trees or have a north-west aspect" (p. 5, Bristowe 1976). The retreats of *Progradungula otwayensis* are apparently restricted to the hollows of a single species of tree (Michalik et al. 2013). Probably the retreats used by many species must have particular configurations (e.g., Vetter and Rust 2008 on *Loxosceles* spp.), and in a substrate that has a minimum hardness (though I know of no experimental studies on the latter). Such specificity strongly implies that the availability of appropriate retreat sites in nature (and the costs of finding them) have played important limiting roles in the ecology of some web spiders.

Limitations in the availability of appropriate retreats may also sometimes induce spiders to occupy suboptimal websites or to build suboptimal web designs, just because the sites have good retreat sites (fig. 10.7). In sum, web spiders that became liberated from needing substrate retreats would have been able to occupy new, previously unavailable habitats, and also to increase the consistency with which they built more optimal web designs.

An intermediate degree of liberation may have evolved earlier. For instance, some webs have a retreat, but one that is largely built by the spider and thus depends less on the substrate. A modern example of this sort would be the funnel webs of the Costa Rican lycosid *Sosippus* sp. (presumably *S. agalenoides*, judging by its range—see Brady 1962), which built silk retreat tubes among the branches and leaves of small plants high above the ground (WE). Close relatives such as *S. californicus*, *S. texanus* (Brady 1962), and *Aglaoctenus castaneus* (Santos and Brescovit 2001; Eberhard and Hazzi 2017) utilize only preformed substrate shelters, which were usually only close to or in the ground. The adults and late instar spiderlings of the pisaurid *Architis tenuis* built in a wider variety of sites than earlier instars, because the adults did not have the younger spiders' obligatory association with a hole in a leaf for the retreat (fig. 1.3) (Nentwig 1985c). The widow spiders *Latrodectus geometricus* and *L. revivensis* built tough silk tubes for retreats, or bent tall grass to form a shelter, rather than using substrate cavities as do *L. mactans* (Lamoral 1968) and species in the closely related genus *Steatoda* (Eberhard et al. 2008b; Barrantes and Eberhard 2010). Abandoning protected, preformed retreats may also result in greater exposure to predators, and the balance between the costs and benefits of substrate retreats probably varies.

10.3.4.1 Independence from the substrate is not a qualitative trait

The degree of physical independence from the substrate was used to classify webs in several recent discussions of web evolution. For instance, Blackledge et al. (2009c) contrasted "substrate defined webs" as opposed to more

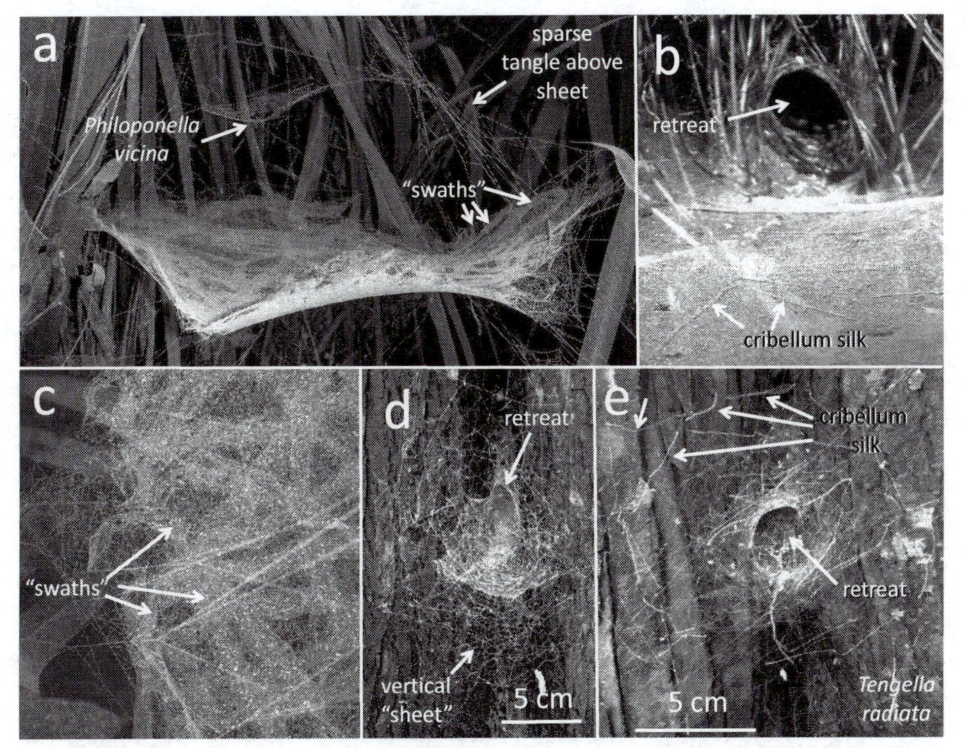

Fig. 10.7. The "typical" sheet web of *Tengella radiata* (Zoropsidae) is depicted in *a–c*; but this typological characterization is ruined by the exceptions in *d* and *e*, which offer important lessons. This species is limited to Costa Rica and western Panama but, despite its puzzlingly small geographic range, occurs in a wide variety of habitats from 5–1500 m in elevation. Their webs have been described as dense horizontal sheets with a sparse tangle above (*a*, *c*), a tubular retreat at one sheltered edge (*b*), and a few scattered, long, loosely suspended cribellum lines laid just above the sheet or in the tangle above (*b*) (Fincke 1981; Eberhard et al. 1993; Griswold et al. 2005; Eberhard and Hazzi 2017). Swaths of very fine lines that are laid on a skeleton of apparently thicker lines could sometimes be distinguished near the edges of newly built sheets (*c*); on subsequent nights the sheet becomes completely covered with fine lines (*b*). This description of "the" web was spoiled, however, by the discovery of occasional "rudimentary" webs (*d*, *e*) that were associated with especially impregnable retreats such as cracks and holes in large tree trunks, but where there were no potential supports for a sheet within 5–20 m (WE). Of 27 rudimentary webs in cracks in tree trunks of this sort, 20 lacked even a single attachment to an object other than the trunk. The only trait that all of these rudimentary webs shared with the typical webs was the tunnel retreat. Some rudimentary webs lacked sheets altogether (*e*), while others had only tiny, deformed, and sometimes vertical sheets with no tangle (*d*). Cribellum lines were sometimes incorporated in a small tangle around the mouth of the tunnel, and sometimes formed sparse, vaguely circular arrays (*e*). The highly protected retreats were apparently valuable enough that they induced the spiders to build these vestigial webs, despite the lack of attachment sites nearby that would have been needed to build a typical sheet web. This spectrum of web designs illustrates the dangers of typology, and documents a striking and previously unappreciated ability to make radical adaptations in web design. One important implication for web evolution is that when web forms arose that did not require appropriately sized cavities in the substrate for retreats, they "liberated" spiders in two ways: previously uninhabitable sites (e.g., those lacking retreat sites) could be occupied (see Blackledge et al. 2009c); and spiders could build more consistent web designs.

aerial webs ("orb webs," "cobwebs," "stereotyped aerial sheets"), and emphasized the advantages that resulted when spiders "shifted ecologies away from constraints of substrate-bound sheet webs" by adopting more aerial forms or by discarding webs altogether. Emphasis on the relation to the substrate is useful in thinking about web evolution, because it is clear that the earliest webs were on or only slightly above the substrate, around retreats or tunnels in the substrate itself. Nevertheless the degree of association is not qualitative (as sometimes implied—Bond et al. 2014; Fernández et al. 2014). Instead it is quantitative, is often difficult to apply in practice, and has clear exceptions. The following is an attempt to refine and extend these ideas.

It is difficult to classify some web designs. How far from the substrate must a web be to be free from its constraints? All webs are attached to the substrate, and thus "constrained" to some degree. And even lines laid directly onto the substrate are often attached to protruding points and span spaces between these points and are thus "aerial." The degree of a web's physical association with the substrate is clearly quantitative, not qualitative, but no quantitative criteria have been proposed. Perhaps extreme cases could be defined in terms of the spider's body. If there is not enough room under a sheet web for the spider to move about and she instead moves on its upper surface, the web is reasonably considered tightly associated with the substrate. Alternatively, if the spider sometimes places her tarsi on the substrate rather than the web's lines when she walks on her web, as in the diplurid *Linothele macrothelifera* (Eberhard and Hazzi 2013) and the filistatid *Kukulcania hibernalis* (WE), then it also seems reasonable to classify the web as substrate-bound. Another possible criterion might be regularity of design, as Blackledge et al. (2009c) argued that substrate-bound webs tend to be relatively irregular in design (e.g., fig. 5.10). But there are clear exceptions to this "constraint," such as the geometrically regular webs of the oecobiids *Uroctea durandi* (fig. 10.5) and *Oecobius concinnus* (fig. 9.6), the filistatid *Misionella mendensis* (fig. 10.12), and the dictynid *Emblyna* sp. (fig. 10.14).

Sometimes substrate constraints are imposed only indirectly. Among species that build funnel webs, *Tengella radiata* represents an extreme substrate-imposed constraint by placing such importance on retreats that it sometimes builds webs at sites with strong retreats that entirely lack appropriate attachment sites for either the usual sheet or the tangle above (fig. 10.7). The lycosid funnel weaver *Sosippus agelenoides* is near the opposite end of this gradient; these spiders built their tubular retreats among the branches and leaves of plants with a variety of forms, often many cm above the ground (WE). The agelenid *Coelotes terrestris* was also somewhat less dependent on preformed retreats, because it often dug tunnels in the substrate (Tretzel 1961). Another lycosid, *Aglaoctenus castaneus*, was moderately substrate-limited with respect to tunnels, but not limited with respect to supports for tangles (Eberhard and Hazzi 2017); it omitted tangles completely if there were no attachment sites available above the sheet. This case argues against the implication from the Blackledge et al. classification scheme that the designs of aerial webs are not constrained by substrate characteristics.

In sum, the topic of substrate-imposed constraints on web design is complicated, and has not been treated adequately in recent studies of web evolution. Substrate effects were likely important in spider web evolution.

10.3.5 SHEETS WITH STICKY LINES AND TANGLES

Judging by their presence in groups that branched very early from all other labidognaths, including the austrochilids *Thaida peculiaris* (Lopardo et al. 2004) and *Hickmania troglodytes* (fig. 10.8), webs that are organized in more or less horizontal sheets of sparsely spaced non-sticky lines with a tubular retreat at one sheltered edge and with similarly sparse and geometrically unorganized sticky lines attached to them, may have been a very early capture web design. The trends to diversity and convergence documented earlier (table 9.1) make this reasoning shaky, but it does help understand the origin of the lampshade type web (Shear 1970; Fergusson 1972) in another ancient group, Hypochilidae, which amounts to rolling a sheet of this sort into a cylinder.

Another change in web design would be to build silk lines above a more or less horizontal substrate to capture previously inaccessible flying prey. Perhaps the earliest aerial design knocked prey down onto a sheet-like web below, as in modern families such as Dipluridae (Paz 1988), Agelenidae (Comstock 1967), Lycosidae (Santos and Brescovit 2001; Viera et al. 2007), Zoropsidae (Eberhard et al. 1993), and Psechridae (Griswold et al. 2005). Another modern design with aerial lines that might have occurred is that of the udubid *Uduba* sp. (Griswold et al. 2005), which has tunnel walls extending above a burrow in the ground, with radiating lines or a sheet around the opening at the upper end. It is not clear whether the apparent lack of modern species with aerial knockdown lines above more or less formless webs close to or on the substrate, such as those of *Loxosceles* spp. (Vetter 2015) and the amaurobiid *Callobius* sp. (Griswold et al. 2005), indicates that such designs did not occur early in evolution. Such a design would be ineffective when the substrate is vertical substrates (e.g., tree trunks).

Construction of aerial lines (whether in the early stages of sheet construction or for tangles above) would

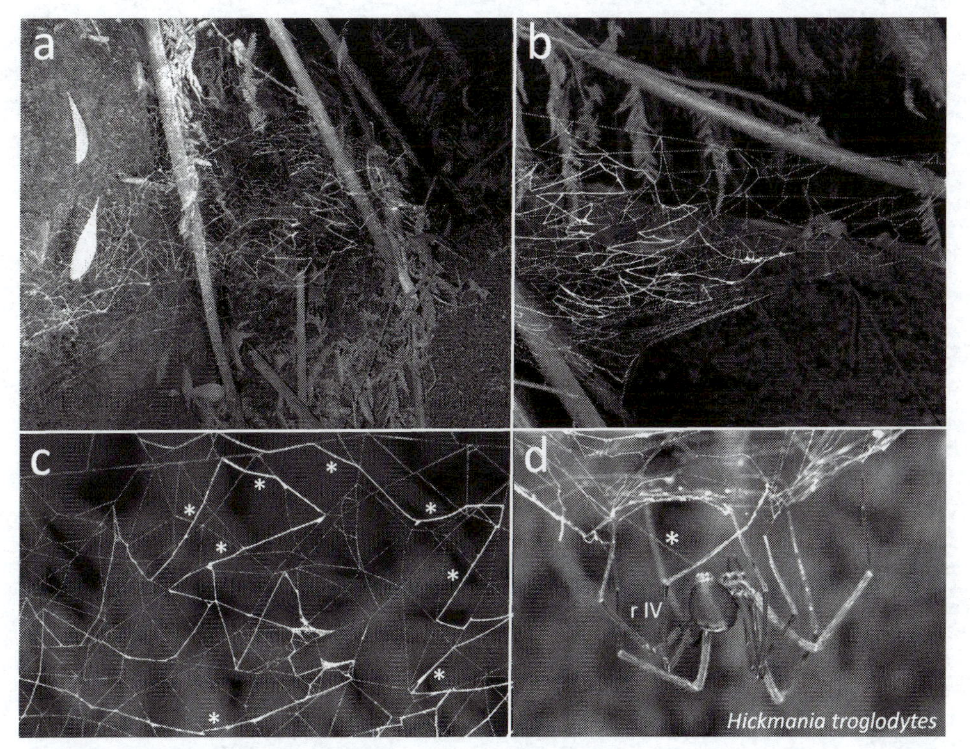

Fig. 10.8. These photos of recently discovered webs of the early-derived Australian austrochilid *Hickmania troglodytes* offer possible insights into early web evolution. Webs (in dorsal view in *a*, lateral-dorsal view in *b*, and closeup in *c*) consisted of a more or less horizontal sheet formed by a sparse, irregular array of sticky lines (the thicker lines) that were supported by wide-meshed, irregularly oriented non-sticky lines (the thinner lines). Direct observations of behavior (*d*) confirmed that spiders added sticky lines (marked with * in *c* and *d*) after non-sticky lines had been built (leg RIVwas supported by LIV on *d* while it combed cribellum fibrils from the calamistrum). The similarity of this basic web design with the webs of another austrochilid, the New World *Thaida peculiaris* (Lopardo et al. 2004), and with *Hypochilus* spp. in the similarly ancient family Hypochilidae, suggests that this sheet design is ancient (see text) (photos courtesy of Martin Ramírez, Museo Argentino de Ciencias Naturales, and Peter Michalik, University of Greifswald).

have obliged the spider to walk along these lines rather than on the substrate. Evolution of such an ability could have far-reaching consequences, including gaining access to previously inaccessible flying and jumping prey, increased ability to overcome relatively larger prey (Nentwig and Wissel 1986), unprecedented accentuations of sexual dimorphism (Moya-Laraño et al. 2002; Brandt and Andrade 2007; see Prenter et al. 2010), and new attack behavior such as shaking the web to cause prey to fall onto the sheet below.

The mygalomorph family Dipluridae illustrates some possible intermediate stages. Diplurid webs vary considerably with respect to their independence from the substrate. The "appressed" simple sheets of *Linothele macrothelifera* were associated with retreats in preexisting cavities, and sheets were built on the litter surface (Eberhard and Hazzi 2013). Disorderly "piles" of such sheets and tunnels that form dense, complex honeycombed arrays on the substrate are also associated with preexisting retreats in several diplurids such as *Euagrus* spp. (Coyle 1986, 1988). A further diplurid design is a single well-defined aerial sheet, as in *Diplura* sp. and *Linothele* sp. (Paz 1988). These sheets sometimes extended up to a meter from a tubular retreat in the substrate, and were up to a meter or more above the ground. Some also had an elaborate tangle of strong lines extending a meter or more above the sheet.

10.3.6 EARLY-BRANCHING ARANEOMORPH LINEAGES WITH DERIVED WEBS

The webs of modern representatives of the earliest branches of labidognaths would seem to be a likely source of information for understanding early evolutionary transitions in this group. Unfortunately, these early

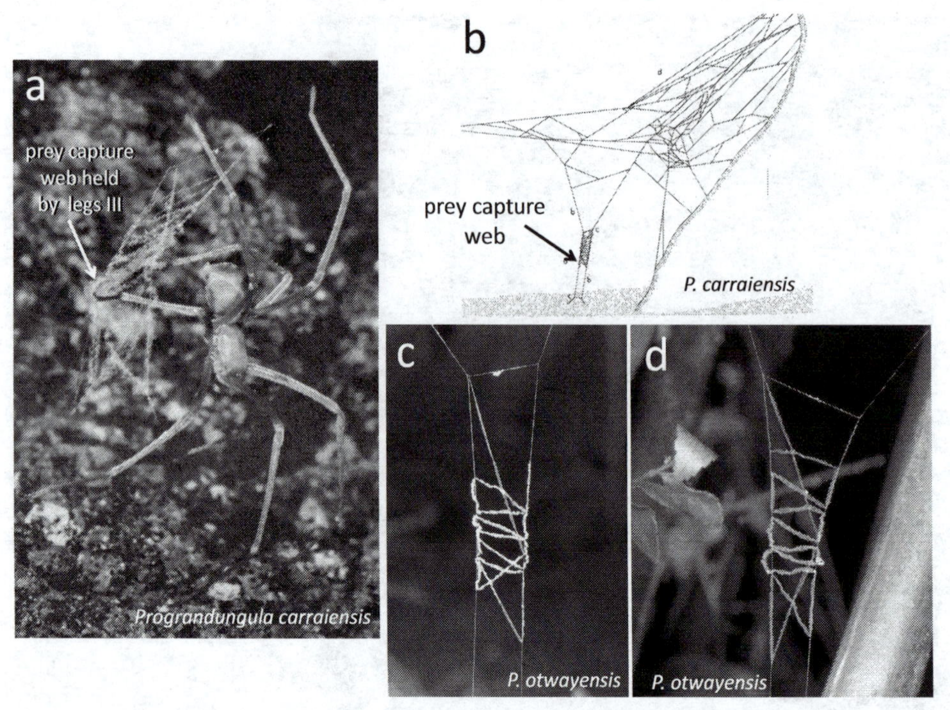

Fig. 10.9. *Progradungula carraiensis* (*a*, *b*) and *P. otwayensis* (*c*, *d*) belong to the small, ancient lineage of Gradungulidae, but their webs are unique and highly derived; they illustrate the frequent finding that web evolution cannot be traced by simply compiling the web designs of ancient lineages. In both species the spider waited on a "catching ladder," a small set of sticky cribellate lines fastened in zig-zag fashion to approximately vertical non-sticky lines near the substrate, with legs III holding the sticky lines away from her body (*a*, *b*). The cave species *P. carraiensis* captured prey by scooping them from the substrate upward and ventrally with her legs I and II (the claws on these legs are modified for this function); at the same time she brought the ladder toward her body to receive the prey. The catching ladder of *P. otwayensis*, which lives in deep forest, was often distant (up to 3 m) from the spider's resting site. The extraordinary similarity of the capture webs in *c* and *d* suggests highly stereotyped construction behavior (see Michalik et al. 2013) (*a* and *b* from Gray 1983; *c* and *d* courtesy of Martin Ramírez, Museo Argentino de Ciencias Naturales, and Peter Michalik, University of Greifswald).

lineages have been little help in this respect, because many build highly derived webs. The most spectacular case of this sort occurs in one of the earliest labidognath branches, the family Gradungulidae. The webs of *Progradungula carraiensis* and *P. otwayensis* are unique among all spiders. They have only a small patch of zig-zag sticky cribellate lines that spans the small space between two more or less vertical lines that are attached above to a tangle of non-sticky lines, and below to the substrate (fig. 10.9, Gray 1983; Michalik et al. 2013). The spider holds the vertical lines near the cribellum lines with her legs III, and scoops walking prey up onto the sticky lines with a quick movement of legs I and II; at the same time she springs the adhesive lines toward the prey with her legs III (Gray 1983). Legs I and II have modified, raptorial tarsal proclaws (Griswold et al. 2005). The derived traits in this early evolutionary branch include a very short-lived capture web, frame lines above and below the patch that contains cribellate silk, active participation of the spider herself in snaring prey, and tensing of the web and springing it to capture prey.

The webs of other early branches are of little more help. That of the hypochilid *Hypochilus gertschi* consisted of a unique, circular, flaring sheet ("lampshade") containing cribellate lines that was attached under a flat surface. The sheet surrounded the spider's resting place on the surface, and was held in place by long radiating guy lines (Shear 1970) (fig. 195B of Griswold et al. 2005). The spider moved on the underside of the sheet (Shear 1970), probably a derived trait. The webs of the austrochilids *Thaida peculiaris* and *Hickmania troglodytes* may be less derived, consisting of extensive but relatively sparse horizontal sheets that contained cribellate lines, with a tubular retreat at one edge (Lopardo et al. 2004; Doran et al. 2001; Griswold et al. 2005). These spiders also moved on the underside of the sheet (Griswold et al. 2005).

10.3.7 SPIDER WEBS AND INSECT FLIGHT

Spider web designs may have coevolved with traits of their insect prey (Vollrath and Selden 2007). For instance, spider webs may have been responsible for major evolutionary changes in insects such as changes in locomotion from walking to jumping and flying, their coverings of detachable scales, and the slow tentative flight patterns of some groups. These insect traits could have in turn selected for modifications of spider webs. Despite the fact that spiders are a major, abundant predatory group that must have strong selective effects on their potential prey, detailed coevolution has yet to be demonstrated. For instance, Vollrath and Selden (2007) concluded that there is no convincing fossil evidence that spiders and their webs drove insects to acquire flight. The extremely fragmentary nature of the fossil record and the difficulty of distinguishing cause and effect from simple temporal correlations will make it difficult to obtain confident conclusions. With the gradual spread of webs into a wider and wider variety of habitats, and with the emergence of short-lived webs that facilitated improvements in choosing superior websites (section 8.5.1.6), the overall ecological and evolutionary impacts of spider webs on prey populations probably increased. In sum, coevolution may have occurred, but details have yet to be demonstrated convincingly.

10.3.8 SUMMARY

As suggested by the large numbers of convergences in web designs (section 9.2.2), some of the transitions described here almost surely occurred more than once. The evolution of webs was not a neat, ascending ladder of progress, but rather a tangled bush of multiple origins and multiple losses, even of the same trait in the same lineage (e.g. the ancestors of modern lycosids like *Sisippus* and *Aglaoctenus* that build webs are descended from ancestors that acquired, then lost, then re-acquired prey capture webs). It has even been speculated that some early stages occurred while spiders still lived under water (Decae 1984), though there is little direct evidence supporting this hypothesis. The earliest webs were likely composed only of non-sticky lines built close to the substrate. Only later did aerial lines and then sticky lines evolve. These developments gave spiders access to flying and jumping prey in the air, but they also brought new problems to be solved (fig. 10.4).

10.4 THE BEHAVIOR PATTERNS USED TO BUILD EARLY WEBS

10.4.1 MALE SPERM WEBS, BURROW CLOSURES, AND THE ORIGIN OF PREY CAPTURE WEBS

Several behavior patterns are employed to manipulate silk during web construction, including precise movements of spinnerets and legs, and coordination of these movements with those of the rest of the body. One hypothesis, not previously mentioned, is that the evolutionary origins of such basic abilities to manipulate lines occurred very early, prior to web construction, and involved mature males. The mature males of most web-building species cease building prey capture webs to search for females. Nevertheless, mature male spiders do make one type of web—the sperm web. Making a sperm web is apparently an ancestral trait in spiders, associated with one of the defining traits of spiders—the use of the male pedipalps to transfer sperm (Eberhard and Huber 2010). In order to transfer sperm from his genital opening to his palps, the male first builds a small silk sheet and deposits a drop of sperm on it; then he changes position to take up the sperm in his pedipalps. Precision in constructing the sperm web is obviously crucial to both successful deposition and successful uptake of sperm, and thus to the male's fitness. Sperm webs must have arisen before about 400 million years ago, long before females and immatures began building prey capture webs. Selection against imprecision in sperm web construction is almost certainly more intense than against similar levels of imprecision in building prey capture webs.

The elaborate nature of sperm web construction, even in spiders that live in tunnels and make no capture webs, is not generally appreciated, so I will describe in detail that of the mygalomorph *Grammostola monticola* and several other Uruguayan theraphosids (Costa and Pérez-Miles 2002; F. Costa pers. comm.). The male spends several hours building a substantial, 10–15 cm diameter sheet in his tunnel or in some other sheltered site such as under a rock. Males evidently judge the sizes and shapes of possible sites in which to build this web, as the male consistently selected a site with a small cavity under the sheet into which he later crawled upside down to deposit his sperm. If no appropriate cavity was available, the male dug one himself (digging behavior was used by Costa and Pérez-Miles as a signal that a male in captivity was about to charge his palps). Using small movements of his abdomen and closely spaced tapping movements with his long spinnerets as he swung them back and forth (but with no direct participation of his legs), the male first lined the cavity with silk, and then built a dense sheet across it. Next he modified one edge of this sheet, apparently pulling and molding lines with his palps and chelicerae to form a thick, distinct edge with a notch in it. Then, crawling under the sheet on his back, he positioned his body so that his sexual opening was near this modified edge (possibly using setae on the ventral surfaces of his coxae to sense the border?), added an additional small oval patch of silk from his epiandrous spigots (just anterior to his genital opening) to the sheet, and then moved slightly forward to deposit the drop of semen. Then he carefully backed out from under the sheet, avoiding contact with the precious sperm droplet, climbed onto the upper surface of the sheet, and dipped his palps into the droplet to slowly absorb it.

Perhaps prey capture webs originally evolved using behavior patterns used by males in building sperm webs. Behavior modules that could have been important include the following: construction of the large dense sheet; the ability to make strong attachments of lines from the sheet to the surrounding supports (the sheet must not sag while the male is crawling out from under it after depositing the sperm droplet); and the ability to evaluate the size and shape of potential building sites.

I know of no similarly careful accounts of sperm web construction in liphistiids or other mygalomorphs, so the possibility that the behavioral complexity just described is derived rather than ancestral cannot be evaluated. The need for fine precision in the movements involved in sperm deposition and uptake seems undeniable, however, and is in accord with the otherwise mysterious persistence of epiandrous glands in spiders (section 2.2.10). Sperm web construction in the relatively basal araneomorph *Kukulcania hibernalis* (Filistatidae) also involved elaborate, fine motor control (fig. 10.10) (Barrantes and Ramírez 2013). The male did not make the "Y" shaped dragline support web typical of most spiders (Foelix 2011), but instead built two or more approximately parallel lines (each of which was probably a cable of finer lines, many from the aciniform spigots—see fig. 10.11). Then, presumably using aciniform silk, he built a small sheet between two lines, using a series of precise triangular movements of the tip of his abdomen while moving his body slowly anteriorly and then posteriorly. After several cycles of forward and rearward movements, he positioned his genital opening against this small sheet and deposited a droplet of sperm.

A second, previously unappreciated context in which dextrous construction behavior occurs is construction of a sheet to cover a burrow entrance, seen in a theraphosid (possibly *Pterinochilus lugardi*) (https://www.youtube.com/watch?v=y5NVfr8kcUU). The entire process occurred with the spider inside or only partially outside the mouth of her burrow. She began construction by facing downward in the tunnel and spinning a sheet across the entrance. Movements included rapid dorsoventral taps with her abdomen and her posterior lateral (PL) spinnerets on the entrance margins, lateral swings of her abdomen to draw swaths of silk lines from one side of the entrance to another and then to fasten them there with ventral movements and taps of the spinnerets, and rotation of her body on its longitudinal axis to reorient her spinnerets to tap on different sides of the burrow entrance. After laying about 5–10 swaths to complete the sheet, she moved deeper into the tunnel, then returned to lay a few more swaths apparently slightly deeper in the tunnel.

Some theraphosids close the burrow with a simple sheet but this spider then re-emerged by breaking through the sheet near one edge, thus forming an asymmetrical collar of silk, and subsequently incorporated lumps of earth from the edge of the burrow entrance that she dug with her chelicerae, pulled onto the collar, and then turned again to use her PL spinnerets to attach silk to the lower surfaces of the lumps and to partially close the sheet again. She repeated this general process over and over, bringing lumps of earth from inside the tunnel or digging them at the edge of the entrance, and thrusting them through the sheet or at its edge, often using her anterior legs, then turning to spin more silk and to attach the lumps to the sheet. Spinning included various PL spinneret movements, sometimes in unison and sometimes not: the directions they moved included distally, dorsally, laterally, and ventrally; late in the process they made some small lateral or approximately cir-

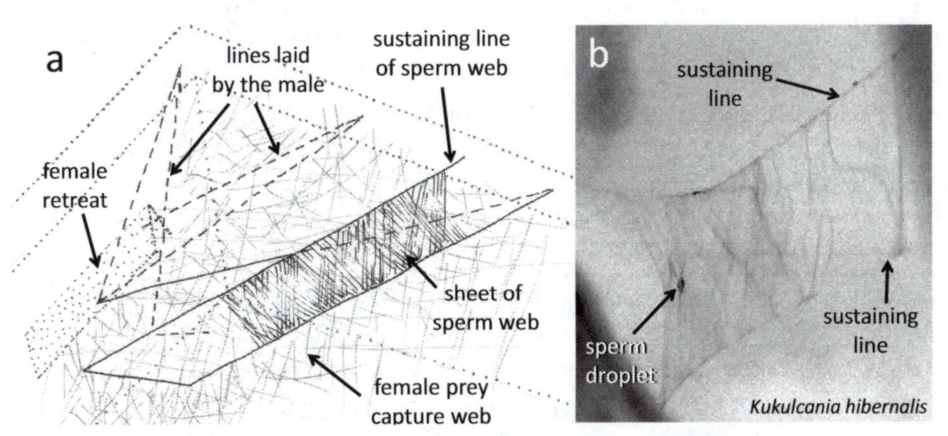

Fig. 10.10. Sperm web construction by a basal araneomorph, the filistatid *Kukulcania hibernalis* (*a*), illustrates the complex and precise male spinning behavior that formed part of the crucial process in which the male charged his palps with sperm; such abilities may have played important roles in the evolution of prey capture webs. The construction process was as follows. The male interrupted courtship behavior to walk across the female's web and its supports, building several scaffolding lines (long solid and dotted lines in *a*); he walked on the substrate, and (less often) along these lines as he moved. Each time he attached his dragline to another line, he grasped the other line with one leg III and touched his spinnerets to it 3–10 mm posterior to this grasp. Two of the lines were long and nearly parallel ("sustaining lines" in *a* and *b*), and here he built a small sheet to receive his sperm. Positioning himself above the sustaining lines (and not holding them with his legs), he rocked his body and abdomen forward and backward to describe precise, slow triangular movements with his abdomen, laying a swath of fine (presumably of aciniform) lines that spanned the space between the sustaining lines. He moved forward and then rearward in a triangular pattern, producing an orderly, finely meshed sheet. Each attachment of the swath was made by grasping a sustaining line with his leading leg III, and touching his spinnerets to the line just posterior to this point. The sheet was under tension, and pulled the sustaining lines toward each other (*b*). When the sheet was complete, the male grasped one sustaining line with his left legs III and IV, and other with his right legs III and IV; legs III were positioned at the anterior edge of the sheet, and the male deposited a droplet of sperm near this edge (*b*). Then he climbed under the sheet, positioning himself under the sustaining lines and facing toward the edge with the sperm droplet, and reached around the edge of the sheet with his palps to take up the sperm. Some of these male abilities to manage silk lines precisely (construction of a pair of nearly parallel lines, holding the lines with leg III to make attachments to them, anterior-posterior and lateral rocking movements of his body and abdomen to lay a swath of fine lines, precise alignment of legs III and IV on these lines, and the ability to orient his body precisely under the small sheet) may have been employed in the origin and evolution of prey capture webs (after Barrantes and Ramírez 2013, and G. Barrantes pers. comm.).

cular "scrubbing" movements on the lip of the tunnel. On occasion the spider curled them over the upper surface of lumps at the entrance margin to fasten a swath of lines to a collar, and then pulled the lumps and the collar so they turned upside down and fastened them to the sheet. Much of the silk was thus on their lower, hidden side. In no case did the spider raise one PL spinneret and lower the other while making an attachment, as was the rule while attaching swaths of silk while wrapping prey or building sheets in other mygalomorphs and araneomorphs (Barrantes and Eberhard 2007; Eberhard and Hazzi 2013, in press). Ancestral spiders may have employed some of these burrow closing operations when they evolved the ability to construct prey capture webs. The variety of designs of mygalomorph tunnels and their internal linings and closures (Main 1976; Coyle 1986) whose building behavior has not been observed, also suggests that early web builders may have had abilities to manage silk that originated in the context of burrow construction.

10.4.2 MOVING UPSIDE DOWN BELOW SILK LINES

Lines in early webs were almost certainly on and very near the substrate, with the spider moving on the web's upper surface rather than hanging below it (footnote O of table 9.1; Bristowe 1930). Few new abilities (if any) would be needed to walk on a silk sheet, except perhaps improved coordination of lifting the tarsal claws when initiating each step to avoid snagging them on the lines in the sheet (section 2.4.1); probably this ability was already used to walk on the substrate or silk burrow linings

(Labarque et al. 2017). The diplurid *Linothele macrotheli-fera* often walked with no apparent problems on both silk sheets and on twigs and leaves in the litter during sheet construction (Eberhard and Hazzi 2013).

Spiders later evolved the ability to grasp single lines with the middle tarsal claw (fig. 2.16, section 2.4). This allowed them to grasp lines more firmly, and probably to move more easily as they hung under single lines. Presumably the ability to walk under sheets was derived from an ancestor that walked on the top of sheets (Bristowe 1930), as, for instance, in many araneomorphs such as amphinectids (Escalante et al. 2015), zoropsids (Eberhard et al. 1993), agelenids and lycosids (Bristowe 1930). Aerial sheet-weaving pholcids and the diguetid *Diguetia canities*, which normally walked under their sheets, sometimes also walked on their upper surfaces during construction (WE); the amphinectid *Metaltella simoni* hung under the roof of her retreat, but moved on the upper surface of her sheet when attacking prey (I. Escalante pers. comm.).

The transition from walking on top of sheets of silk lines to walking under single lines may have required new abilities to find, grasp, and release single lines (or cables of lines). Walking under single lines occurs early in sheet construction, even in the lycosid *Aglaoctenus castaneus*, which otherwise walks on the upper surface of its somewhat elevated finished sheet (Eberhard and Hazzi 2017). Perhaps the abilities that many spiders use to climb single draglines, including seizing a single line with their anterior legs and reeling it up and packing together the mass of loose line that accumulates (Eberhard 1986c), were used in this transition. Even in a species such as the filistatid *Kukulcania hibernalis*, in which individuals probably never hang under lines while they build their webs (WE), spiderlings can climb up their draglines. An immature zoropsid, *Tengella radiata*, a species that nearly always moved on the upper surface of their dense sheets (Eberhard et al. 1993), showed a surprising ability to move underneath lines when attacking a relatively large prey; it cut a hole in the sheet, and moved to the underside to bite the prey from below (presumably to gain protection from the prey's struggles) (Barrantes and Madrigal-Brenes 2008). Young spiderlings of the mygalomorphs *Ummidia* sp. (Ctenizidae) (Coyle 1985; Eberhard 2006), and the atypids *Sphodros rufipes* (WE) and *Atypus affinis* (J. Cox pers. comm.), walked relatively easily under elevated silk lines when dispersing, even though these species probably never make aerial lines the rest of their lives and have only a single pair of claws.

10.4.3 USING LEGS TO MANIPULATE LINES

One pattern documented in tables 5.2 and 10.2 (though their limited taxonomic coverage and frequently incomplete observations leave room for doubt) suggests a derived status for two different ways of using the legs during web construction: grasping the lines to which the spider is about to attach the new line she was producing (hold "old" line; section 5.3.4); and holding the new line that she is attaching at the moment she attaches it (hold "new" lines; section 5.3.3). The diplurid *Linothele megatheloides* lacked both types of behavior (box 5.2), and only manipulated lines using dabbing movements of the spinnerets and the abdomen. Similar dabbing behavior occurred during retreat construction in other non-web groups, such as the anyphaenid *Aysha* sp. and salticids (Eberhard 2007a; WE), and during prey wrapping in many non-orb non-orbicularian species (Barrantes and Eberhard 2007). On the basis of simplicity as well as taxonomic distribution, both "hold old" and "hold new" lines with the legs are probably derived.

In contrast with mygalomorphs, web-building araneomorphs, which generally have relatively short spinnerets, used leg and abdomen movements to manipulate lines during web construction (chapters 5 and 6). Basal araneomorphs such as the austrochilid *Thaida peculiaris* (Lopardo et al. 2004) and several filistatids (Lopardo and Ramírez 2007; WE) held old lines with one leg III, lifting the line into contact with their spinnerets when making an attachment (section 5.3.4); they did not, however, hold new lines. Both the orb and non-orb members of Orbiculariae held both old lines (with ipsilateral legs III and IV), and new lines (with the other leg IV) when attaching (e.g., Eberhard 1982; Jörger and Eberhard 2006; Eberhard et al. 2008a).

Another possible source for line manipulation behavior during web construction is prey wrapping. Pushing silk lines from the spinnerets onto prey with legs IV is probably very ancient. It was absent in mygalomorphs (Barrantes and Eberhard 2007), but has been observed in several basal araneomorph groups, including gradungulids (Gray 1983), hypochilids (Shear 1970), and austrochilids (Lopardo et al. 2004). Filistatids illustrate a possible transition: *Kukulcania hibernalis* used "rearward casting" movements of legs IV in which the leg grasped lines near the spinnerets to wrap prey (Barrantes and Eberhard 2007), and also to extended cribellum lines near the retreat with legs IV during web construction (WE). *Misionella mendensis* made similar casting movements with legs IV to extend cribellum lines during web construction (fig. 10.12; Lopardo and Ramírez 2007).

Holding old lines while making an attachment has been secondarily lost in several species with dense sheets. The pholcids *Modisimus guatuso* (Eberhard 1992b) and *M. bribri* (WE), the agelenid *Melpomene* sp. (Rojas 2011), the lycosid *Aglaoctenus castaneus* (Eberhard and Hazzi 2017), several linyphiids (Benjamin and Zschokke 2004), and *Nihonhimea tesselata* (THD) (Jörger and Eberhard 2006) did not grasp the old lines to which they attached the dragline during the later stages of filling in their densely meshed sheets. Orbicularians, in contrast, were uniform in using the same legs (ipsilateral III and IV) to grasp the old line just anterior to and posterior to the spinnerets when they attached a new line while building other types of web (Eberhard 1982).

10.4.4 MANAGING SWATHS OF FINE LINES

There is a suite of traits, including elongate posterior lateral spinnerets that have rows of spigots along their ventral surfaces, and movements of these spinnerets (section 5.3.8) that result in spreading the swath of lines apart. Two of these same traits are employed to wrap prey, probably with the same function of widening the swath (Barrantes and Eberhard 2007). Because these movements are employed in prey wrapping by some mygalomorph species in the family Theraphosidae, which is thought not to have had any ancestor that built sheet webs (fig. 1.9), these behavioral traits were presumably originally part of prey wrapping behavior, and later evolved to be included in sheet web construction (Eberhard and Hazzi 2017).

10.4.5 DIPLURID BEHAVIOR: A POSSIBLE GUIDE TO ANCESTRAL TRAITS

Although very little is known about the details of mygalomorph web construction behavior, recent preliminary observations of the diplurid *Linothele macrothelifera* (fig. 10.6; Eberhard and Hazzi 2013) (described in box 5.2) give some suggestions that are summarized in table 10.2 regarding the ancestral status of several behavioral details. Tube construction behavior by another mygalomorph, the atypid *Sphodros rufipes*, resembled *L. macrothelifera* behavior in several details: the spider also rocked rearward to attach lines, waved her posterior lateral spinnerets actively while building, and extended her anterior lateral spinnerets ventrally (WE).

The webs of other diplurids provide insight into the probable prey of ancestral web weavers, and thus into some of the selection pressures they experienced. Even in ischnothelines, which make complex 3D arrays of multiple tunnels connecting small sheets, nearly all of the prey in a large sample were walking invertebrates, especially ants (78% of 1324 prey at several sites for five species of *Ischnothele*—Coyle and Kentner 1990). Strong selection to facilitate capture of ambulatory rather than flying prey probably continued long after prey capture webs originated.

10.5 EVOLUTION OF LATER NON-ORB WEBS

10.5.1 CONSEQUENCES OF CRIBELLUM SILK LOSS IN LABIDOGNATHS

The common ancestor of the Araneomorphae, by far the most evolutionarily successful and ecologically important group of spiders that accounts for the large majority of the world's 41,000+ described species (figs. 1.9, 1.10), likely produced sticky cribellum silk. This type of silk was lost in two major lineages of araneomorphs, the orbicularians (>11,000 modern day species, most of which spin webs with an alternative type of sticky silk), and the RTA clade (>21,000 species, most of which do not make webs) (Coddington and Levi 1991; Blackledge et al. 2009c; Fernández et al. 2014; Bond et al. 2014). Blackledge et al. (2009c) argued that discarding cribellum silk constituted a "key innovation" that allowed species in both of these lineages to escape from the multiple constraints imposed by spinning costly cribellum silk, and that this escape resulted in greater fecundity (Blackledge et al. 2009a,c). These ideas merit careful consideration because of the prominence of their authors, their wide-ranging summaries of modern discoveries, the generality of their conclusions, and their general use in subsequent studies of phylogeny (Fernández et al. 2014;

Table 10.2. Tentative polarizations of different types of web construction movements associated with building structures with silk lines, based mostly on tables 5.1 and 5.2 and observations of web construction by the diplurid *Linothele macrothelifera* (Eberhard and Hazzi 2013).

Presumed ancestral state	*Presumed derived state*
1. Return to the retreat frequently during web construction	1. Return to the retreat less often or never
2. Build web gradually over the span of several nights	2. Build entire web in one burst of activity
3. Not use legs to find attachment sites (use spinnerets instead)	3. Explore with legs to locate sites to which to attach
4. Use spinnerets rather than legs to position lines being produced	4. Manipulate lines as they emerge from the spinnerets with tarsus IV
5. Not grasp lines to which attachments are being made	5. Grasp lines to which attachments are being made and bring them to touch the appropriate spinnerets precisely
6. Use movement of the entire abdomen to produce attachments by bringing the spinnerets into contact with the substrate	6. Use movements of long spinnerets to contact the substrate to make attachments[1]
7. Attach by using the spinnerets symmetrically and simultaneously[2]	7. Use asymmetrical and asynchronous spinneret movements[2]
8. Produce many similar lines simultaneously	8. Produce only pairs of lines (e.g., major ampullate draglines)

[1] This is possibly an ancestral character in mygalomorphs, if original spinnerets were long. In labidognaths, ancestral spinnerets were probably relatively short (and became secondarily elongate in groups like Hersiliidae and Hahniidae; in hersiliids movements of the spinnerets as well as the abdomen and the legs result in contact with the substrate).

[2] The posterior lateral spinnerets of *L. macrothelifera* generally moved symmetrically and simultanously during tunnel construction, but often moved asymmetrically during sheet construction. Tunnel construction is presumably more ancient in spiders. This polarization is less certain in labidognaths. The impression that spinnerets in higher labidognaths are moved symmetrically and simultaneously may be an artifact of the difficulty in observing them; in the lycosid *Aglaoctenus castaneus* the posterior lateral spinnerets often moved asymmetrically (Eberhard and Hazzi 2017), and the anterior lateral spinnerets of *Nephila clavipes* (NE) moved in alternation when attaching the sticky spiral to a radius (Eberhard 2010a).

Bond et al. 2014; Garrison et al. 2016). The two major constraints were attributed to cribellum silk: the time and the energy needed to comb the sticky lines from the cribellum with the calamistrum; and the greater freedom from having their webs tightly linked to the relatively variable shapes of the websites where these "substrate defined" webs are built. Evaluating these constraint hypotheses leads to mixed conclusions.

On the physiological side, the time constraint imposed by the slower production of cribellate compared with ecribellate sticky silk is large (up to a factor of 10 or more) and well documented (fig. 6.14, section 10.8.5). The metabolic cost of combing the cribellum silk has never to my knowledge been measured directly, but it is likely to be substantial; the cribellate orb weaver *Zosis geniculata* made on the order of 69,000 combing movements with her legs IV in building a single orb (section 3.3.3.3.2.1). There is, however, another, potentially counter-balancing factor. The physical stickiness of a given amount of cribellum silk is greater than the stickiness of an equal quantity of ecribellate sticky silk; cribellum silk is substantially more efficient (Opell and Schwend 2009). The balance between the metabolic costs of cribellum silk and the gains from its more efficient stickiness is undetermined. Strictly speaking, the cost aspect of the constraint hypothesis is uncertain, though the greater time cost is certain and likely higher metabolic cost of combing silk from the cribellum intimates that cribellum silk is more costly.

The "substrate defined" portion of the constraint argument is less convincing. In the first place, neither the supposed greater design constraint imposed by the substrate nor the effects of such constraints on a spider's survival and reproduction have been documented empirically for any species, with or without a cribellum (see section 10.3). The specific contrast was between a web that is "centered around a pre-existing retreat (crevice), attaches directly to the substrate, and therefore has a shape largely determined by the web's location" (a "substrate defined" web), versus a "web suspended in air away from the substrate that possesses a stereotypical shape independent of the substrate to which it is attached" (a "stereotyped" web) (Blackledge et al. 2009c). But these are quantitative rather than qualitative traits, and no quantitative criteria for thresholds or limits were given (both "substrate defined" and "stereotyped" were treated, however, as presence/absence traits).

The proposed association between geometry and nearness to the substrate is, in fact, inconsistent, and there are cases that do not fit it (at least on the basis of overall subjective impressions). For instance, the cribellate "substrate" webs of *Kukulcania hibernalis* were usually more or less radially symmetrical around the spider's retreat (e.g., fig. 5.10). The webs of the cribellate oecobiids *Oecobius annulipes*, *O. concinnus*, and *Uroctea durandi*, which are also very close to the substrate, were nevertheless also often relatively symmetrical (e.g., figs. 9.6, 10.5, Glatz 1967; Kullmann and Stern 1981; Solano-Brenes et al. 2018). On the other hand, the aerial, presumably "stereotyped" sheet webs of linyphiids, sheet-building theridiids like *Nihonhimea tesselata* (Jörger and Eberhard 2006), the linyphiid sheet-weaver *Novafrontina uncata* (fig. 7.51), and several orb weavers (e.g., fig. 1.6d, Sensenig et al. 2010b) all showed substantial variations in designs that conformed to different shapes imposed by the substrate.

In sum, there are problems with the hypothesis that abandoning cribellum silk was a key innovation that resulted in liberation of web spiders from energetic and web design constraints (Blackledge et al. 2009c), due to uncertainties regarding the proposed constraints. There are also problems with the logic of key innovation hypotheses in general (section 10.5.3) In sum, the cause(s) of the undeniably successful radiations of orbicularian and the RTA clades are not certain.

10.5.2 PROBLEMS CATEGORIZING WEB TYPES IN EVOLUTION

Labels that employ everyday terms to distinguish biological categories sometimes lead to confusion; the limited vocabulary of terms that have been used to describe the geometric forms of spider webs has sometimes led to problems in understanding the evolution. While lumping non-homologous structures is useful in discovering convergence on functionally advantageous designs (e.g., table 9.1), excessive lumping can confuse discussions of evolution. Recent phylogeny studies employed categories like "brushed sheet," "terminal line," "irregular ground," "irregular aerial sheet," "orb," "cob," "stereotyped aerial sheet," and "bolas" webs (Blackledge et al. 2009c; Bond et al. 2014; Fernández et al. 2014; Garrison et al. 2016; Dimitrov et al. 2016). Only two of these categories (orb and bolas) are relatively easy to distinguish, however. The others were not defined precisely (below), and no quantitative criteria were proposed. Criteria were lacking, for instance, for how distant a web must be from the ground to be classified as an "aerial" rather than a "ground" web, or how to distinguish "stereotyped" from "irregular." The useful suggestions of Coddington (1986c) about how to go about categorizing webs were not followed. Discussing web evolution without giving operational definitions of either the categories being considered or the means by which additional webs can be classified runs the risk of reducing the usefulness of hard-earned phylogenies. As pointed out by Viera et al. (2007) and Blackledge et al. (2009c), one especially thorny problem is the term "sheet web," which has been used with a wide variety of different meanings (see section 9.4). To pick just one example, Murphy et al. (2006) included as "sheets" such diverse structures as the tents that *Dolomedes* spp. build around their egg sacs to protect their spiderlings (Comstock 1967; Bristowe 1958), the dense silk retreat of *Pirata* embedded in sphagnum moss with an open doorway from which the spider attacks passing prey (Bristowe 1958), the sparse planar arrays of non-sticky and sticky silk in dictynids (e.g., fig. 200B of Griswold et al. 2005), and the extensive sheets and silk tubes of presumably non-sticky silk in the lycosids *Aulonia albimana* (Job 1968, 1974) and Sosippinae. Indiscriminate use of the single term "sheet" has also obscured substantial structural diversity in the webs of Linyphiidae (G. Hormiga and W. Eberhard in prep.). This problem is especially acute because nearly all classifications to date have been made on the basis of only the superficial appearance of the web, with no data on construction behavior or microscopic study of threads and their attachments in finished webs (see also section 9.4).

Recent authors have begun to work on this "lumping" problem. Blackledge et al. (2009c) deduced from phylogeny, for example, that the "araneoid sheet webs" of Griswold et al. (2005) are not all homologous, and that they probably evolved independently in linyphiids and theridioids. The classification of Blackledge et al. (2009c) (later adopted in other phylogenetic studies) distinguished four categories of sheets: "brushed" ("brushed silk lines are not specifically and repetitively attached to structural silk threads, but rather tend to lie upon them"; "irregular ground" ("relatively complex three-dimensional webs that consist of multiple sheets intersecting at various angles and whose overall form tends to follow closely the contours of the substrate to which the webs are attached"); "irregular aerial" ("webs that are suspended or free standing . . . relatively amorphous and fill available space in the microhabitat location"; and "stereotyped aerial" ("architecturally stereotyped and usually taxonomically distinctive regardless of variation in microhabitat location") (these definitions were given in the supplementary material; they did not exactly parallel the categories recognized in the figure in the text that described web evolution). "Junction" webs constitute the contrast term for brushed webs, giving a name to the synapomorphy of piriform mediated attachments (J. Coddington pers. comm.). Unfortunately, these categorizations included almost no data on construction behavior (documented only in one short abstract concerning a single species) (Coddington 2001), lack quantitative guides for judging traits such as "relatively complex," "tend to follow closely the contours," and "relatively amorphous," and lacked any microscopic observations documenting connections and lack of connections between lines in finished webs.

These distinctions are sometimes very difficult to apply in practice. For instance, it is not clear how to characterize a web as "irregular" rather than "regular" without specifying which aspects of the webs are to be evaluated. In what respect are the domed sheet webs of the pholcids *Modisimus* spp. (figs. 5.2, 8.7) or the horizontal sheets of the diguetid *Diguetia canities* (fig. 9.16f) more "irregular" than the sheets of linyphiids, as suggested by Bond et al. (2014)? How can one decide whether a web does or does not "fill available space"? (See section 8.9 on the difficulty of defining "available" websites.) Many of the distinctions are quantitative, but offer no precise values to guide categorizations.

The distinction between "brushed" and "junction" webs is useful in specifying the derived use of piriform silk to attach lines to each other. But the two categories do not yield clear characterizations of webs. There are "junction" webs, such as that of the pholcid *Modisimus guatuso*, that include lines that fit the "brushed" definition (they lie on others and are not attached specifically to them) (Eberhard 1992b). Presumably these pholcids have reinvented unattached fine lines after having lost them previously; their distinctive method of applying them to the web (using wrapping movements of legs IV) favors this hypothesis. In addition, the "brushed" webs of the lycosid *Aglaoctenus castaneus*, the zoropsid *Tengella radiata*, and the agelenid *Melpomene* sp. include piriform attachments that involve numerous fine lines in the sheet; in the case of *A. castaneus*, direct observations confirmed that spiders dab their spinnerets against the sheet to attach swaths of fine lines during sheet construction (Eberhard and Hazzi 2017). There was also substantial variation even in one of the traits (lack of attachments of fine lines) that defines the category.

Closeup photos show that the mesh details in sheet webs are also diverse, including the following: many long, straight lines crossing over each other without apparent attachments in the small sheets of the pisaurid *Hygropoda* sp. (fig. 5.14) (WE) that span a single curled leaf; long lines converging on a single corner in the linyphiid *Agyneta* sp. (fig. 1.3); short swaths of strictly parallel lines in *Ochyrocera* sp. (fig. 2.13, Hormiga et al. 2007); widely spaced lines that in some places seem to run parallel to neighboring lines in the theridiid *Achaearanea* sp. nr. *porteri* (fig. 9.16); large numbers of fine lines with tiny droplets of glue that are scraped into larger droplets when the lines move against each other in the linyphiids *Diplothyron simplicatus* and *Neriene coosa* (WE); and lines so dense that no pattern can be discerned in the agelenid *M.* sp., the zoropsid *T. radiata*, and the lycosid *A. castaneus*.

Other axes not mentioned by Blackledge et al. (2009c) also seem likely to be useful: with vs. without a tangle above the sheet (see, for example, Nielsen 1932 on abundant lines just above the sheet of *Tegenaria domestica* to hinder prey escape); with vs. without a tangle below; with vs. without sticky lines (in the sheet, in the tangle above, in the tangle below); the geometric form of the sheet (flat and horizontal, flat but with edges curling up, domed, cup-shaped); the degree of uniformity of thread diameters; the degree of inter-connectedness of lines; the density of lines in the sheet; the degree of regularity of arrangements of lines in the sheet; the degree of independence from or elevation above the substrate; whether the spider walks on top vs. on the bottom of the sheet (Eberhard and Hazzi 2017) (fig. 9.16 illustrates several of these variables).

In sum, overly simple classifications of web types hamper understanding of web evolution. It seems clear, for instance, that different types of "sheet" webs need to be distinguished, but we do not yet have satisfactory categories. Other than the suggestion that the criterion of whether the web is dense enough that the spider normally walks on top of it rather than under it may be useful, I do not have easy answers on how to best make the divisions. Until additional studies remedy the current nearly complete absence of relevant behavioral data, useful criteria may be difficult to find. In the end, the validity of phylogenetic analyses of the evolution of spider webs will depend on the rigor in the distinctions that are made between homologies and convergences in web-building behavior; these in turn will require careful observations of the behavior patterns that are used to build different webs, and of the webs themselves. Some recent powerful and sophisticated phylogenetic analyses of spider web evolution were weakened by a lack of rigor in distinguishing different types of web design.

10.5.3. PROBLEMS WITH KEY INNOVATION ARGUMENTS IN GENERAL

"Key innovations" arguments are based on associations between the acquisition of a particular trait (or group of traits) in a phylogenetic tree and the subsequent proliferation of descendant groups. Key innovations arguments run the risk of confusing correlations with cause-and-effect relations. I suspect, for instance, that few would be convinced by the hypothesis that the correlation between the evolutionary loss of external insemination openings in the female genitalia of the mega-genus *Tetragnatha* and its close relatives constituted the "key innovation" that explains its undoubted evolutionary success. The risk of error is especially great when the sample is tiny (in analyses of web evolution, the contrasts are usually between only two lineages). The argument seems even weaker when, as in the hypothesis that cribellum loss explains subsequent radiation, the key innovation represents a loss rather than a gain of a trait and is thus likely to be relatively easy to evolve. Several other key innovations have been proposed in different groups of web spiders. One example is the hypothesis that acquisition of silk proteins with reduced UV reflectance explains the evolutionary success of araneoids in general (Craig et al. 1994). Another is that the vertical orientation of orbs with viscid glue on their sticky spirals explains the success of orb-weaving araneoids compared with uloborids (Bond and Opell 1998) (sections 10.8 and 10.9.1).

There is an important contrast between key innovation arguments of this sort and other evolutionary arguments made elsewhere in this book (e.g., sections 3.3, 4.3), which are based on the mechanical advantages that would result from the evolution of certain traits. For instance, I argued that the evolution of webs built on skeletons of strong non-sticky silk lines would be unlikely to have occurred until spiders had developed the ability to use piriform silk to fasten individual lines firmly to each other and to the substrate. This kind of argument, in the end, concerns the selective advantages of mechanical and physiological designs, and thus the likely adaptive links between different traits. Key innovations hypotheses, in contrast, purport to explain evolutionary success in terms of the numbers of descendant species, with success being measured in terms of the relative numbers of surviving descendant taxa.

10.5.4 SILK GLANDS AND OTHER MORPHOLOGICAL TRAITS

A spider's ability to build webs surely depends on the types of silk she is able to produce, so web evolution was probably related to silk gland evolution. The diversity in known mygalomorphs of silk glands and spigot types, the relatively sparse data, and the uncertainties about gland homologies (Glatz 1972) result in uncertainty regarding the types of mygalomorph silk and their properties during and prior to the evolutionary origin of araneomorphs. Two early acquisitions in araneomorph spiders were probably important for subsequent web evolution: the strong, relatively thick draglines produced by the ampullate glands; and the attachment discs of multiple fine lines and glue produced by the piriform glands (Lopardo et al. 2004, sections 2.2.3, 2.2.9). The apparent presence of both ampullate and aciniform gland silk in the radial and support lines of the webs of *Kukulcania hibernalis* (fig. 10.11), a member of the basal araneomorph family Filistatidae (fig. 1.9), suggests a pos-

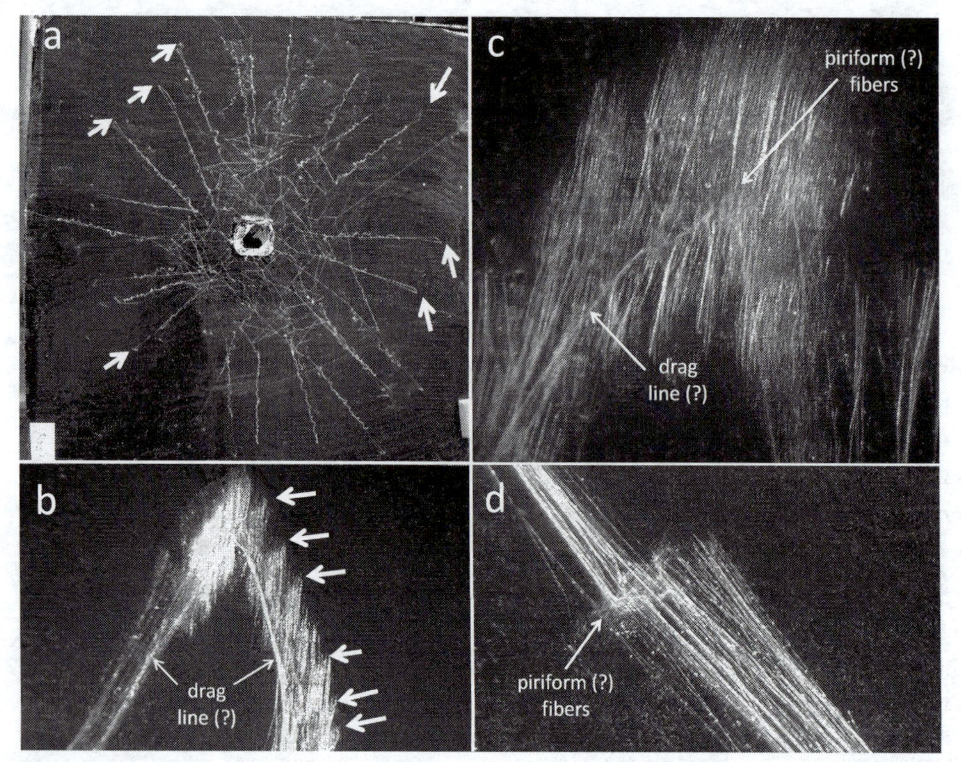

Fig. 10.11. The attachments of radial lines in the web (*a–c*) and of draglines (*d*) to the substrate by the filistatid *Kukulcania hibernalis* may give an insight into intermediate stages in the evolution of attachment discs, between the mygalomorph extreme of applying swaths of presumed aciniform silk directly to the substrate without any piriform silk and araneomorph attachment discs that consist of only piriform silk (fig. 2.10). In the "U"-shaped attachments of the long, non-sticky radial cables in the webs (arrows in *a*) there were many fine, presumably aciniform fibers; these fanned out where they were attached to the substrate (*b*), revealing what appeared to be a central, thicker dragline (*b*, *c*) (from the ampullate glands?). The array of fine lines often had a serrate outline (thick arrows in *b*); each prominence was probably produced by one of the rapid, lateral open and closing movement of the spinnerets that occurred in rapid series when each attachment was made. In some attachments, especially those made while the spider walked across an open surface (*d*), additional, small, more dense masses of what may have been piriform silk were visible near the dragline. The "draglines" of this species differed from the more familiar draglines of orbicularians in containing large numbers of fine fibers (instead of only two major ampullate lines), and in being relatively rigid (WE).

sible intermediate stage; the thickening of the ampullate lines may have been gradual.

One key to building lines increasingly distant from the substrate was undoubtedly the ability to fasten lines firmly to the substrate and to each other (Blackledge et al. 2009c). Earlier spiders probably made relatively weak attachments such as those of modern-day mygalomorphs, which probably depend on the slightly wet surface of aciniform, minor ampullate, and cylindrical silk lines as they emerge from the spigots (Peters and Kovoor 1991; Barrantes and Eberhard 2007; Eberhard 2010a) (section 2.2.8, fig. 10.11). Stronger attachments arose with the acquisition of piriform glands and spigots in the common ancestor of the labidognaths (Blackledge et al. 2009c) (Glatz 1972 mentions "piriform glands" in the mygalomorph *Hexathele hochstetteri*, but their homologies are uncertain).

Again the filistatid *K. hibernalis* offers a possible intermediate condition, combining the ancient technique of many aciniform lines brushed broadly onto the substrate by the posterior lateral spinnerets (as occurs, for example, when a theraphosid wraps a prey—Barrantes and Eberhard 2007) and the use of newer piriform material. The attachments of one non-sticky line to another, of non-sticky lines to the substrate, and of cribellate silk to these lines, all had small masses of material (presumably from the piriform glands, though this has not been verified) (fig. 10.11). But the attachments of the strong radial web lines to the substrate also included an additional, extensive array of many thin lines (presumably from the aciniform glands). The radial lines splayed out into a multitude of fine lines, and the two swaths zigzagged on the substrate (up to 20 times in a single attachment) (fig. 10.11), presumably serving to increase the strength of the attachment by increasing the surface area of the silk in contact with the substrate. The substrate attachments of *K. hibernalis* were relatively weak, and could be ripped away from painted cardboard by pulling on the radial line, though the attachments seldom broke during prey capture; the less extensive attachments of the dragline made as the spider walked across the surface could also be pulled away without breaking the line (WE). The evolution of secure piriform attachments to silk lines was probably associated with a highly precise attachment behavior, in which the spider grasped the previous line, and then positioned it precisely with respect to her spinnerets (sections 2.3, 2.4).

The suite of traits that characterize the common ancestor of labidognaths set the stage for many new design possibilities for webs, and in turn for modifying the simple prey capture formula of previous webs such as those of diplurids (high densities of fine silk lines, rapid responses, and powerful legs and chelicerae) (Coyle 1986; Paz 1988; Eberhard and Hazzi 2013). They facilitated possible further independence from the substrate, particularly with respect to the need for hiding places for the spider, and for the capture of flying as opposed to walking prey (Blackledge et al. 2009c). Perhaps early forms would have been tangles or "space webs" in sheltered sites such as under rocks or fallen logs, as in the plectreurid *Plectreurys tristis* (WE), the desid *Maniho* sp. (Griswold et al. 2005), and, in somewhat less sheltered sites, the hypochilid *Hypochilus gertschi* (Shear 1970, 1986), but this is only speculation.

The evolutionary weakening of the dependence on and the physical connections with retreats that occurred as webs became more aerial was accompanied by a new problem associated with molting. Burrow-dwelling mygalomorphs support themselves while they molt by simply lying on their backs at the bottom of the burrow (Baerg 1958). In contrast, many araneomorph spiders (many of which do not inhabit sheltered sites) dangle at the ends of silk draglines when they molt (e.g., Foelix 2011). This requires that the spiders continue to make silk lines in the moments just preceding and during the process of molting (section 2.7), with lines passing through the new cuticle and emerging from the old spigots (Yu and Coddington 1990). Some of the small, nipple-shaped structures on spinnerets called tartipores (Kovoor 1986; Townley and Tillinghast 2009) may be remnants of these sites in the new cuticle. The function of tartipores has been controversial (Yu and Coddington 1990); as expected under the molting hypothesis, tartipores were absent on ancient Devonian spinnerets (Shear et al. 1989).

The pattern of the evolution of the cribellum can be used as a proxy for the presence of strongly adhesive lines in the webs of some groups (other than in Orbicularia, where loss of the cribellum was associated with the gain of an alternative type of sticky silk). The general pattern is of an ancient origin of cribellum, and repeated, independent losses, some of which were subsequently followed by major radiations (fig. 1.9, 1.8; Spagna and Gillespie 2008; Garrison et al. 2016). This trend seems surprising, given the advantage of retaining prey. Perhaps the high energetic cost of combing lines from the cribellum favored these transitions (see section 10.5.1).

10.5.5 VISIBILITY OF SILK TO PREY

It has been proposed that reducing the UV reflectance of silk was a major transition that allowed a major habitat expansion by orbicularians, from dark forested or otherwise protected sites to brightly lit environments during the day, by making the silk less visible to flying prey (Craig et al. 1994; Bond and Opell 1998; Craig 2003; Miyashita and Shimazaki 2006). The reflectance spectra of silks from one group within Orbicularia, Araneoidea, showed proportionally less UV than those of other non-orbiculians or of Deinopoidea (Craig et al. 1994). Many insects can see UV, so araneoid silk may be less visible to them. This relative invisibility has been cited as a key innovation explaining the evolutionary success of araneoids (Craig et al. 1999).

I find this UV hypothesis difficult to evaluate. In the first place, there are doubts regarding the foraging behavior of some of the groups in which silk reflectance was studied. Members of three of six of the "ancestral" non-orbicularian araneomorphs in the analysis of Craig et al. (1994) do not forage only in "dark" environments as stated. These include *Diguetia canities* (Diguetidae) (which inhabits open scrub on the Sonoran desert) (WE), *Sosippus* sp. (Lycosidae) (also found in open habits)

(Brady 1962), and *Kukulcania hibernalis* (Filistatidae); all attacked prey readily during the day (WE). I have repeatedly observed unstaged prey captures by *K. hibernalis* when an insect flew into a web during the day. It is also true, however, that in all four species the spiders were better positioned to respond to prey (e.g., not hidden in a retreat) at night than during the day. Does this make them them "dark" foragers as classified by Craig et al. (1994)?

Similarly, characterizations of some other, orbicularian groups as foraging in dim light (e.g., Uloboridae, *Leucauge, Mangora*) are overly simplistic. Some species in these groups, such as *Uloborus glomosus, L. mariana, L. argyra,* and *M. melanocephala,* inhabit the upper strata of weeds in very early succession sites (WE) (see also Kato et al. 2008 on *L. blanda*); the two *Leucauge* species are so exposed to the sun that they frequently adopt special positions at the hub to reduce insolation on warm days in the field (WE).

These problems weaken the correlations that were cited in support of the key innovation UV hypothesis (Craig et al. 1994; Bond and Opell 1998). An additional problem is the weakness of confusing correlations with cause-and-effect relations in key innovation hypotheses (section 10.5.3), especially when they involve small numbers of evolutionary origins (the number of derivations here is one). This does not give grounds for rejecting such a hypothesis definitively, but rather for doubting its strength. In sum, the most reasonable conclusion is that the UV hypothesis can be neither confidently accepted nor confidently rejected.

10.5.6 WEB EVOLUTION IN TWO SMALL GROUPS

10.5.6.1 Filistatid webs

One small corner of non-orb web evolution has recently come into better focus in the small family Filistatidae. Filistatids are of particular interest because they apparently branched off from the rest of the araneomorphs relatively early (fig. 1.9, 1.10), and can thus serve as an outgroup for phylogenetic analyses of many other labidognaths. All spiders in this family, as far as is known, build webs with sticky and non-sticky lines around a central retreat that is usually in a sheltered cavity of some sort (fig. 5.10, Comstock 1967; Griswold et al. 2005; Lopardo and Ramírez 2007). The sticky silk of the filistatines *Kukulcania hibernalis* and *Filistata insidiatrix* is laid in two forms, one of which is apparently unique to filistatids. Near the distal ends of the long non-sticky lines that radiate from the mouth of the retreat (figs. 5.10, 10.11a), the spider coils long snaking bands of loose cribellum silk, and attaches them at widely separated points to non-sticky lines (Griswold et al. 2005; Lopardo and Ramírez 2007; fig. 5.10). In *K. hibernalis* webs there were also uncoiled bands of cribellate silk; the spider used one tarsus IV to seize the sticky line band, leaned rearward, and extended this leg posteriorly and laterally, and then ventrally to snag the line on the substrate (or other lines) (arrows in fig. 5.10). These bands formed long, lax lines that usually ran in more nearly circular than radial directions, and that crossed many non-sticky lines without being attached to them.

A similar though more elaborated behavioral patern of attaching cribellum lines to the substrate that also employed rearward extensions of legs IV occurred in the prithine filistatids *Misionella mendensis* (fig. 10.12) and *Pikelinia tambilloi* (and also, judging by their webs, in *Pritha nana*) (Lopardo and Ramírez 2007). After combing out a loose pile of cribellum silk, the spider used each of the two tarsi IV to seize one of the two cribellum swaths that emerged from the split cribellum, and carried it rearward and laterally to press it against the substrate; this spread the two lines laterally, and formed regular triangular patterns of sticky lines (fig. 10.12). The probable derived nature of this technique fits well with the phylogeny for filistatids suggested by morphological characters (Ramírez and Grismado 1997) (fig. 10.12).

The evolutionarily derived nature of the coils of cribellum silk with respect to other araneomorphs is suggested by observations of ontogenetic changes in *K. hibernalis* (G. Barrantes pers. comm.). When second instar spiders were removed from their mother's retreat (where they normally remain and feed on prey captured by the female), they gradually built webs that included more or less radial non-sticky lines around a retreat. These webs resembled those of adult females, except that the bands of cribellum silk on these radial lines were not highly coiled and loose, but were instead more or less straight, apparently attached to the radial, non-sticky lines. This ontogenetic change suggests that both the loose, radially arranged coils, and the uncoiled, more nearly circular lines nearer the retreats of adult webs evolved from a predecessor that built only straight sticky lines attached

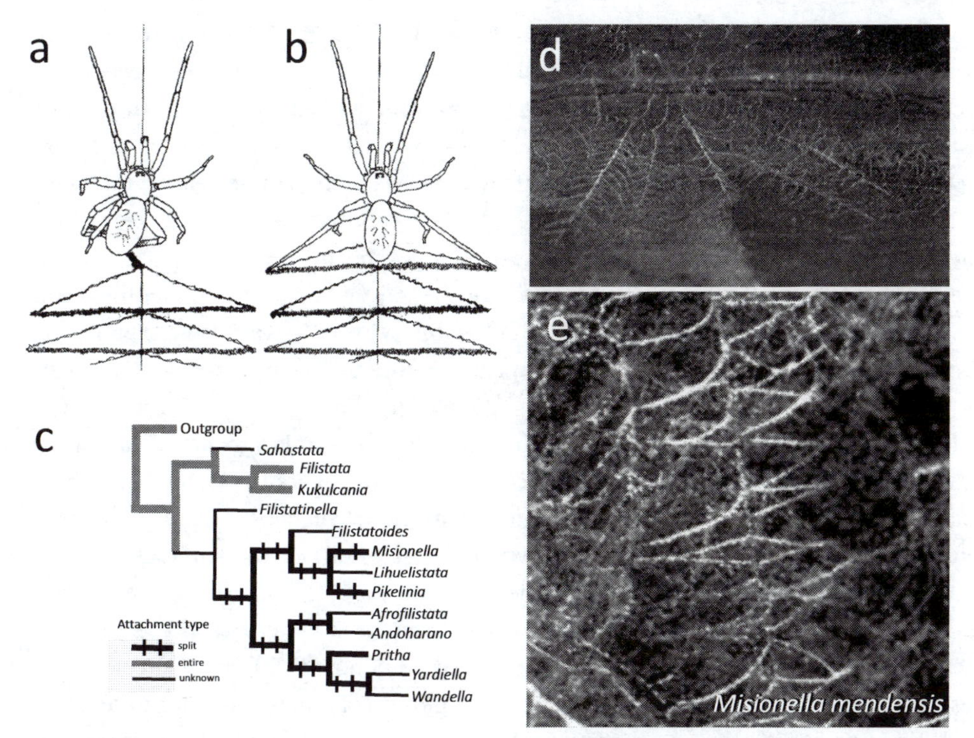

Fig. 10.12. The filistatid *Misionella mendensis* used her legs IV in a unique way (*a*, *b*) to place her sticky cribellum silk in geometrically regular, zig-zag patterns (*d*, *e*), demonstrating that substrate-bound webs are not necessarily geometrically irregular, as has sometimes been thought. Moving slowing inward from the outer edge of her web along a previously laid radial line of non-sticky silk (*a*), the spider combed two swaths of cribellum silk from the paired cribellum plates on her abdomen with the calamistrum of one leg IV. Periodically she paused, grasped each swath near her abdomen with her two tarsi IV, and then spread her legs IV wide, splitting the two swaths apart, and touched the tarsi IV to the substrate where the cribellum silk adhered (*b*); then she attached the cribellum silk to the radial line and continued onward. The phylogeny of filistatids (*c*) suggests that split attachments using this leg IV behavior evolved from an ancestor that laid loose mats of cribellum silk on non-sticky radial lines (see figs. 5.10, 7.52) (from Lopardo and Ramírez 2007, courtesy of Lara Lopardo).

to radial non-sticky lines. It is not known whether these straight cribellum lines were produced without combing behavior, as occurred in the even more basal group *Progradungula* (Michalik et al. 2019).

Because most of the prey captured by a web built close to the substrate like that of *K. hibernalis* probably arrive by walking rather than flying, the lines in the web's periphery are more likely to be contacted by prey than those in its central portion. Thus the addition of sticky lines to the central area after the web has already been largely constructed (e.g., fig. 5.10) might seem paradoxical. The spider's attack behavior on large prey offers a possible explanation. Spiders seized a prey appendage (usually a leg) in their chelicerae, and then further entangled it by dragging it through the web, moving backward toward the retreat. In effect they "wrapped" the prey in web lines closer to the retreat. Such indirect wrapping could make addition of sticky lines in the central portion of the web advantageous; their approximately circular orientation would increase the probability that prey encountered them while being dragged.

10.5.6.2 Interception vs. retention in oecobiid webs

A classic contrast occurs in the web designs of the diminutive spiders (adults generally ≤3 mm) in the small cribellate family Oecobiidae. The webs of *Oecobius annulipes* and *O. concinnus* consist of a more or less circular mat of sticky cribellum silk 2–3 cm in diameter laid directly on the substrate with a central, star-shaped rigid "tent" in the central area where the spider rests (fig. 9.6; Glatz 1967; Shear 1970; Solano-Brenes et al. 2018). The spider can run out from under the tent in any direction through the openings at its edges to attack pedestrian prey (mainly ants) that adhere to the cribellum lines. She attacks prey by rapidly running over and around the prey while she enswathes it with a thick band of fine wrapping lines; her specialized anal setae aid by holding the wrapping lines apart (Glatz 1967).

In *Uroctea durandi*, in contrast, the cribellum has been lost. Associated with this loss in retention function, the prey capture webs of *U. durandi* have increased interception and signaling capabilities, with long, rigid, non-sticky "trip lines" radiating several cm from a similar tent-like structure where the spider waits (fig. 10.5; Kullmann and Stern 1981; box 9.4) (a web photo suggests a similar design in *U. compactilis*) (Shinkai and Takano 1984). The trip lines connect with the floor of the tent, and the spider runs out rapidly to attack when a prey jars a trip line (Crome 1957). Because the cribellum has likely been lost secondarily in *Uroctea*, the improved interception and signaling design is presumably derived from the greater retention design of *Oecobius*-type webs.

Intra-specific variation in *O. annulipes* webs suggests additional changes. When the tent was built near a wall, the side nearest the wall was closed, and the number of entrances fell from up to 7 down to as few as 3; in extreme cases, the tent became a tube with only two entrances (Glatz 1967). These adjustments of web design to the substrate could have served as the original variants that gave rise to the evolution of the tube-shaped mating web built by the male (or vice versa) (Glatz 1967).

10.6 INCONSISTENT EVOLUTIONARY TRENDS IN NON-ORB WEBS

Several authors have attempted to find general trends in web evolution, but with only limited success. Examples and some problems they suffer include the following. Witt (1965) argued that at a large scale, there is a trend for web-building spiders to reduce the spider's direct participation in prey capture and to increase that of the web. Kaston (1964) argued, however, that at the level of orb weavers, there was a trend in the opposite direction, to reduce the web and to increase the direct participation of the spider in prey capture. Nentwig and Heimer (1983) claimed that there was a first evolutionary phase in which the web was gradually enlarged, eventually reaching the third dimension. This was followed by a second phase, leading from three-dimensional to two-dimensional structures (they also made the unlikely claim that "larger webs, such as orb webs or even space webs, are a waste of silk") (p. 35). The repeated derivation of 3-D webs from orbs (fig. 1.14, section 10.10) does not fit this model. Shear (1986) argued that the basic evolutionary sequence was from geometrically irregular to more regular designs, and that the amount of silk invested in a web gradually increased and then decreased, but again there are multiple exceptions, especially in view of the ancient origin of orb webs (fig. 1.8).

Riechert and Łuczak (1982) proposed that there was a general evolutionary trend toward specialization on particular prey. It is true that many of the specialist web designs mentioned in table 9.1 to capture ants, moths, water surface insects, and termites seem not have given rise to secondarily more generalist descendants (phylogenies are still incomplete in many cases). On the other hand, however, the ancestral web form for theridiids and nesticids was probably a gumfoot web that specialized for capturing walking insects (Agnarsson 2004; Arnedo et al. 2004; Eberhard et al. 2008b), but it subsequently gave rise to an explosion of different web forms, many of which probably capture a wide array of flying prey. The derivation in *Wendilgarda galapagensis* (TSM) of webs built far from streams that probably capture a wide array of flying insects rather than only insects associated with the water surface (Eberhard 1989a) is a another, smaller-scale exception.

There are other possible trends, but they also prove to have exceptions. There seems to have been a large-scale shift from "brute force" solutions to problems in trapping prey, such as adding more and more undifferentiated lines (e.g., diplurid sheet webs) to using much fewer but stronger, more firmly attached lines that are more precisely positioned (see Shear 1986). The more derived silk glands (especially the piriform glands, which allow strong, precisely positioned attachments to other lines and to the substrate) were surely important in this set of transitions. There are also reversions, however, such as the derivation of dense sheets in linyphiids and a few theridiids and synotaxids from orb webs, and repeated losses of webs that entailed more forceful struggles with prey (table 9.1).

Another major change, associated with the transition to orb webs, was construction of the entire web in one session, rather than over several nights as is typical in non-orb builders (table 5.1). This "all-at-once" tactic was subsequently lost independently, however, in several groups of orb weavers and their descendants. For instance, gradual additions to their webs are common in theridiids, some linyphiids, and *Cyrtophora citricola* (AR) (WE, fig. 1.7). All of the orb webs with multiple construction episodes are unusually durable, either because the website is especially protected from rain, wind, and detritus, as in *Azilia affinis* (AR) and *Zosis geniculata* (UL) (Eberhard and Barrantes 2015) and the cave-dwelling mysmenid *Trogloneta granulum* (Hajer 2000; Hajer and Řeháková 2003), or because the spider is very large and the web is strong as in *Nephila* spp. (NE) (Wiehle 1931; Robinson and Robinson 1973a; Kuntner et al. 2008a).

Secondary loss of sticky lines has occurred repeatedly in the araneomorphs and has often (but not always) been accompanied by a complete loss of webs. Web loss is implied by comparisons with related spiders (Blackledge et al. 2009c), and by ontogenetic changes within some species in which immatures build webs but mature individuals do not, as illustrated in the pisaurids *Pisaura mirabilis* (Lenler-Eriksen 1969) and *Pisaurina* sp. (Carico 1972, 1985), and the oxyopid *Peucetia viridans* (Kaston 1972). There are also ontogenetic changes in the opposite direction, however, as in the zoropsids *Tengella* spp. that acquire cribellum silk part way through their development (Barrantes and Madrigal-Brenes 2008; Mallis and Miller 2017). Blackledge et al. (2009c) argued that loss of cribellate silk and webs played a crucial role in facilitating spider diversification. In the major RTA clade of araneomorphs, which contains about half of all spider species, more than 90% of "species richness" is associated with losses of cribellate silk and prey capture webs. Some RTA spiders have subsequently re-acquired webs, however, and cause-and-effect deductions from this correlation are risky (sections 10.5.1, 10.5.3).

Evolutionary losses of webs and sticky lines have not been unidirectional. For instance, three or perhaps four reinventions of sticky lines occurred in araneomorphs after the cribellum was lost. One of these lineages, however, the small subgroup *Loxosceles* of the small family Sicariidae (Lehmensick and Kullmann 1957), has been a relative failure. A second independent derivation occurred in pholcids, which convergently evolved both gumfoot sticky lines (Japyassú and Macagnan 2004; Escalante and Masís-Calvo 2014) and the sparsely beaded lines in the aerial webs of *Modisimus guatuso* (Briceño 1985; Eberhard 1992b) and *Belisana* spp. (Deeleman-Reinhold1980; Huber 2005) (fig. 9.19). Sticky droplets also occur on lines in sheet webs in other groups such as the small family Ochryoceratidae (Hormiga et al. 2007). One other reinvention of sticky silk was a spectacular evolutionary success, giving rise to the large majority of the species in the largest group of families, the Orbiculariae. Some orbicularians, in turn, subsequently discarded sticky silk (section 10.10).

In sum, there have been numerous partial trends, but there are clear exceptions to each; none of the trends is truly consistent. The complex evolutionary tangles can be illustrated with an extreme example in the evolutionary sequences leading (independently) to *Chrosiothes tonala* (THD) and to *Argyrodes* spp. (THD): their ancestors acquired webs, then acquired sticky silk, then discarded one type of sticky silk for another, transferred this second type of sticky silk onto new types of lines (gumfoot lines), then lost these sticky lines from their webs (but maintained sticky silk to wrap prey) (fig. 10.13, Bradoo 1971; Eberhard 1979a), and then finally discarded prey trapping webs altogether! The most general conclusion

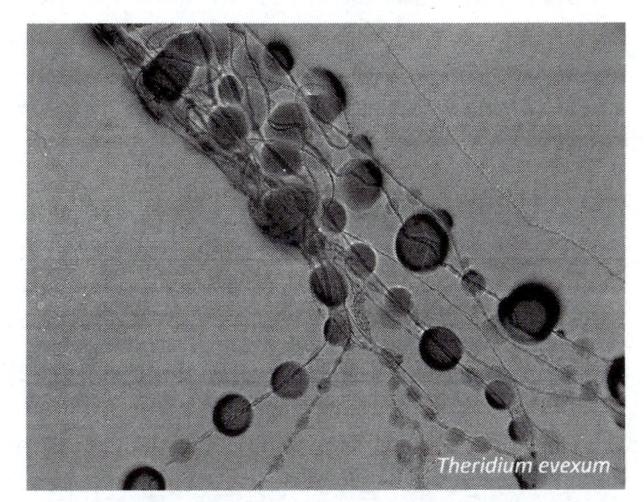

Fig. 10.13. The sticky droplets and multiple silk lines in the wrapping silk of *Theridion evexum* illustrate how functions have sometimes changed. The sticky droplets from the aggregate glands, which probably arose in basal araneoids to produce the sticky spiral in orb webs, persisted in two contexts in Theridiidae after orbs were lost: on gumfoot lines that have sticky droplets placed on relatively non-extensible dry lines (from the major ampullate glands?); and in wrapping silk (from the aciniform glands?). Despite the high diversity of web designs in theridiids, neither the highly extensible, rubbery properties of the flagellum silk of the sticky spiral nor the orb design itself appear to have ever been recovered (information on the properties of non-orb weaver silk is still extremely fragmentary, however—see Blackledge et al. 2011) (from Barrantes and Weng 2007b, courtesy of Gilbert Barrantes).

is that there has not been any clear, consistent trend in the evolution of spider webs (Blackledge et al. 2011; Herberstein and Tso 2011).

10.7 DIVERSITY IN NON-ORBS THAT RESULTS FROM BEHAVIORAL STABILITY

Non-orb webs are strikingly diverse (section 9.2.1, box 9.2). Just as with orb webs, this diversity probably evolved in combination with conservative, unchanged behavior patterns and divergent behaviors (table 5.2). It is too early to give behavioral details for many groups, but a comparison of the behavior of the theridiid *Tidarren sisyphoides* with that of other theridiids that produce other web designs can serve to illustrate the kind of story that is likely to be discovered in many groups. The design of the web of this species is unusual among theridiids, and differs in at least one respect with that of the congeneric *T. haemorrhoidale*. Nevertheless Madrigal-Brenes and Barrantes (2008) showed that nine different, apparently conservative behavioral traits of *T. sisyphoides* are shared with other theridiids; several of these also occurred in other araneoid families. Peters and Kovoor (1991) noted the same "diversity from stability" pattern in the conservative morphological "base" of glands, spinnerets, and spigots that are used to produce a diversity of web forms.

10.8 THE (PROBABLY) MONOPHYLETIC ORIGIN OF ORB WEBS

10.8.1 EVOLUTIONARY ORIGINS WHEN BEHAVIOR IS MODULAR

It is now clear that orb webs, far from being a recent pinnacle of web evolution, are an ancient web form that probably arose at least 165 Mya (Vollrath and Selden 2007), and that they have been repeatedly modified and discarded (e.g., figs. 1.9, 1.10, 3.1). Perhaps the grandfather of controversies in spider web evolution is whether orbs evolved more than once. The intensity of this discussion has been lower than that of the stabilimentum wars described in section 3.3.4.2, but the tides of opinion have ebbed and flowed through the years, and the major players include such icons as McCook (1889), Thorell (1886), Wiehle (1931), Kaston (1964), and Kullmann (1972) (for detailed early histories, see Coddington 1986b and Shear 1986). The most commonly asked question has been whether or not the cribellate orb weavers (Deinopoidea in fig. 1.9) and ecribellate orb weavers (Araneoidea in fig. 1.9) are the descendants of a single orb-weaving ancestor (the monophyly hypothesis), or whether orb webs evolved independently in the two groups.

Before turning to the data, it is helpful to clear up a few more theoretical points. Previous chapters have indicated that orb web construction behavior (and indeed web construction by spiders in general) is organized hierarchically, and that there are often semi-independent modules or subroutines in a given level (sections 10.13, 10.14). The evidence for modularity is varied, and includes the discreteness of subroutines and their repetitive expression in time and space, the evolutionary patterns in which modules have been gained, lost, and shuffled in different species, and the ability shown by parasitoid wasps to selectively elicit or repress particular modules chemically.

These indications of modularity complicate the question of the evolutionary origin of orb webs. Orb web construction can validly be considered to be a single trait, in the sense that it results in the production of a single, recognizable structure: a planar array of non-sticky radial lines that are attached distally to lines that are attached to the substrate, and are connected centrally in a hub, and an evenly spaced, more or less circular array of sticky lines supported by the radial lines. The focus of previous discussions of orb web monophyly has been almost exclusively at this relatively high level. On the other hand, the subroutines used to build an orb can evolve independently, and most likely they did not all originate simultaneously; in this sense, orb construction is not a single trait, but the combination of many.

The "pseudo-orb" web of *Fecenia cylindrata* (fig. 9.18) offers an example of the importance of this possible modular independence for understanding the evolution of orb webs. The sticky lines in this web are orb-like in being evenly spaced and approximately circular in outline; in contrast, many of the non-sticky lines on which they are laid are not radial, but rather have many different orientations and interconnections. An orb-like sticky spiral may thus have evolved before a radial organization of the non-sticky lines in these spiders. The webs of an unidentified "amaurobiid" and (to a lesser extent)

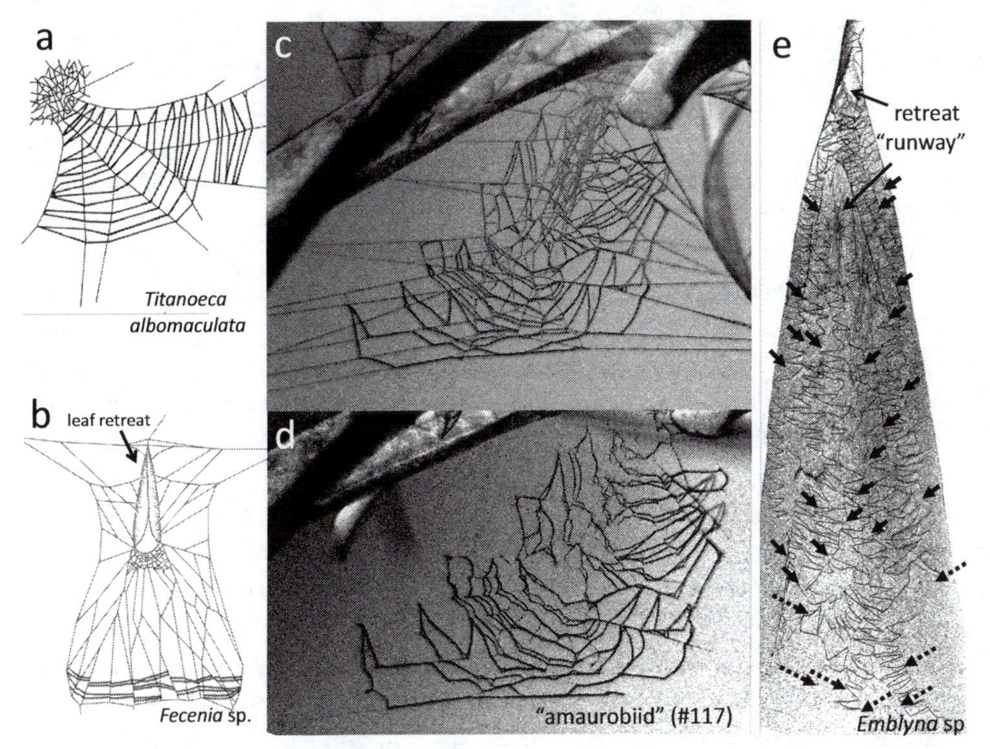

Fig. 10.14. Three independent inventions of "proto" orbs, with arcs of sticky silk, occur in three cribellate groups: two amaurobioids (*a*, *c*, *d*), Psechridae (*b*), and Dictynidae (*e*). In the drawing of a *Titanoeca albomaculata* web (*a*) the sticky lines are drawn thicker; in the unidentified "amaurobiid species" (#117) web all the lines are coated with powder in *c*, while only the sticky cribellum lines are coated in *d*); the web of *Fecenia* sp. (*b*) was incomplete, with only the first 2–3 loops of sticky spiral in place and lacked those that covered the rest of the "pseudo orb" (see fig. 9.18, frontispiece). The web of the dictynid *Emblyna* (*e*) was powdered and then carefully jarred to knock much but not all of the powder off the non-sticky lines, leaving the sticky lines more heavily coated. It was built on the long thin succulent leaf of a *Sanseviera* sp. plant that was slightly twisted was also slightly bent longitudinally. The web had many long, non-sticky spanning lines running across the bend; many of these lines ran downward along the leaf (solid arrows), except those at the bottom of the web where they were more radially oriented (dotted arrows). Many spanning lines were approximately parallel to each other, and most were in approximately the same plane; sticky lines zig-zagged between many pairs of spanning lines. There was a retreat at the top, and a "runway" of non-sticky lines that was a downward extension of the retreat floor (see *c* for a superficially similar structure in the amaurobiid web). Frame lines were absent in the amaurobiid and dictynid webs, but present in *Fecenia* sp. webs. The *E.* sp. web was just off the surface of a leaf, the *T. albomaculata* web was built at a sheltered site, but the web in *c* and *d* was in an aerial bromeliad plant; it was approximately horizontal, while that of *F.* sp. was aerial, close to vertical, and had a curled leaf retreat hanging in its midst (*a* from Szlep 1966a; *b* from Robinson and Lubin 1979b).

the titanoecid *Titanoeca albomaculata* (Szlep 1966a) (fig. 10.14a,c,d) present similar, independent cases of regularly spaced sticky lines on non-sticky lines that are more nearly transverse than radial. The long non-sticky lines of the dictynid *Emblyna* sp. (arrows in fig. 10.14e) also supported sticky lines, but they were mostly parallel, spanning a longitudinal bend in the leaf; the sticky lines were zig-zag rather than circular (the spanning lines of the related *Mallos hesperius* were also often co-planar and supported zig-zag sticky lines, but were usually more nearly transverse, spanning the shallow longitudinal curl of a leaf) (WE). Those of the desid *Matachia livor* (fig. 1.3), the eresid *Stegodyphus* sp. (Eberhard 1987a), and the filistatid *Misionella mendensis* (fig. 10.12) (Lopardo and Ramírez 2007) represent further independent origins of evenly spaced sticky lines, and also at least somewhat radial non-sticky lines. Radially oriented non-sticky lines that are connected to each other where they converge have evolved independently in several other groups (e.g., figs. 9.1, 9.26).

The current lack of direct observations of construction behavior in most of these species precludes using lower levels of behavior to examine questions concerning evolutionary derivations, but they will likely be informative. Observations of *M. mendensis* showed that unique leg IV movements were used to lay sticky lines with uniform spacing, implying an independent origin (fig. 10.12, Lopardo and Ramírez 2007).

Due to the lack of data, the discussion below will be focused mainly at the usual, higher-level question of monophyly. A full understanding of the origin of orb webs will require analyses at multiple levels of behavior. Orbs surely did not originate fully formed in a single step, but rather through a gradual process of evolution of different modules. This recognition does not change the question of whether this series of steps occurred in a single lineage (monophyly) or in different lineages (polyphyly). But it does change the criterion for recognizing an "orb," and suggests that intermediate categories need to be recognized ("partial orbs," "semi-orbs," etc.). The way in which webs are categorized may lead to biases in phylogeny analyses. For instance, coding all alternative web types in unique categories would bias the results toward orb monophyly, because of the way in which current algorithms for ancestral character reconstruction work (I. Agnarsson pers. comm.).

10.8.2 MORPHOLOGY, MOLECULES, AND BEHAVIOR

It has long been clear that cribellate and ecribellate orb weavers employ similar sequences of similar behavior patterns (chapter 6). This evidence favoring monophyly was summarized effectively more than a century ago by Emerton (1883): "It is highly improbable that the making of such complicated webs of the same kind should be acquired separately by *Uloborus* and by the Epeiridae" (p. 205). Most of the early controversy revolved around apparently contradictory morphological evidence, especially from the cribellum and the associated calamistrum. Realization that the cribellum is an ancestral trait in araneomorphs (Lehtinen 1967; Platnick and Gertsch 1976), and that there may be construction constraints (Coddington 1986b) that impose certain details (for instance, spiral lines cannot be built without first building radial lines to support them), helped make monophyly seem more probable. More recent analyses have largely involved molecular data. Until the last few years, a general consensus seemed to be emerging that favored the monophyly hypothesis (Griswold et al. 1998; Garb et al. 2006; Blackledge et al. 2009c; Dimitrov et al. 2011). Based on additional morphological and molecular data and rigorous cladistic analyses, orb weavers were included in a single taxon, the Orbiculariae, in which orb construction was a synapomorphy (Coddington 2005).

Recently, however, molecular analyses that attempted to correct for several potential problems, including artificial inflation of support due to missing data, unequal rates of evolution in different lineages, compositional heterogeneity, and heterotachy (Bond et al. 2014; Fernández et al. 2014; Garrison et al. 2016), found support for linking the deinopoids more closely with a large RTA-clade of about 40 non-orb weaving (and largely webless) families, rather than with araneoids. If this grouping holds up under future tests (see section 1.6), it invalidates the group "Orbicularia," and implies either separate derivations of orb webs in cribellates and ecribellates, or a single, even more ancient derivation of orbs and their subsequent loss in the ancestor of the large RTA clade (the preferred hypothesis of both Bond et al. 2014 and Garrison et al. 2016).

Recent discovery of the strong resemblance between the cues that are used by cribellate and ecribellate spiders to guide their orb construction behavior (Eberhard and Barrantes 2015; fig. 10.15, tables 7.2, 10.3) also favors the ancient monophyly hypothesis. If orbs evolved only once, then modern species would be expected to conserve both the cues that were used by the common ancestor of all orb weavers, and the behavioral responses that these cues trigger. This prediction is not trivial, because the geometric regularity of orb webs means that it would be possible to use many different aspects of orbs to guide construction because of their strong correlations with each other. For a concrete illustration of this problem, recall the successful simulation of temporary spiral construction that failed to utilize the cue that is used by the spiders (section 7.15).

Although the behavioral data in table 10.3 are not complete, they were clearly in accord with the prediction of similar cues and responses (Eberhard and Barrantes 2015). For instance, even distantly related descendants of orb weavers in the families Anapidae and Mysmenidae, which build tiny orbs and non-orbs in small spaces, showed the same trend to reduce or eliminate frame lines (Kropf 1990; Hajer 2000; Hajer and Řeháková 2003; Eberhard 2007b) seen in experiments with tetragnathids and uloborids in small spaces (Eberhard and Barrantes

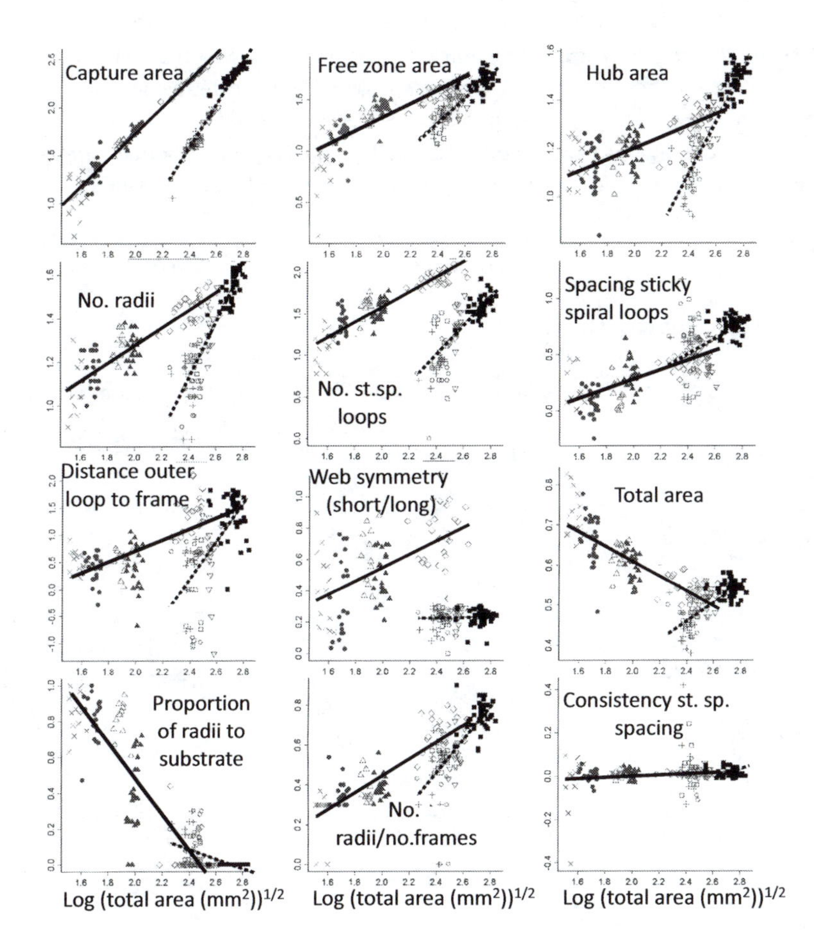

Fig. 10.15. The multiple similarities in the relationships between different aspects of orb design and web area in orbs that were built by mature females of the distantly related species *L. argyra* (TET) (dotted lines) and *Zosis geniculata* (UL) (solid lines) that were confined in small containers (different symbols represent webs built in different-sized containers) constitute evidence favoring the "orb monophyly" hypothesis—that all modern orb weavers are descended from a single, orb-weaving ancestor. Many modifications of the orbs built in especially small spaces appear to be extensions of trends in normal orbs in the field. For instance, the positive relationships of the number of radii and of the spacing of sticky spiral loops to web area in the *L. argyra* webs in small cylinders in captivity were similar to those in normal orbs in the field (see also fig. 7.42, tables 7.2and 7.3) (from Eberhard and Barrantes 2015).

2015). Overall, there are 11 cues for which there are comparable data from both uloborids and araneoids; in ten of these both groups seem to respond to the same cue and their responses are similar (table 10.3).

One might argue that, as with the order of behavioral stages in orb construction (section 6.3.1), some of these similarities result from design constraints imposed by the geometry of orb designs. For instance, it would seem obvious that to achieve regular spacing between loops of sticky spiral the spider would need to locate the previous loop laid (e.g., by touching it—the first trait in table 10.3) and use it as a point of reference. But the use of an alternative cue that has evolved in several araneoids, which include the araneid *Poecilopachys* (and probably its relatives such as *Cyrtarachne* and *Pasilobus*), as well as theridiosomatids and anapids in the symphytognathoid families, shows that they are not "constrained": these spiders do not touch the inner loop, and probably rely instead on the TS-IL distance to guide sticky spiral placement. In fact, the use of two different cues (IL site, TS-IL distance) to guide the same sticky spiral attachment decision in both uloborids and araneoids is a striking similarity.

Two other important similarities are the lack of response to radius tensions, and the apparent ability to sense the amount of sticky silk available in their silk glands and to adjust the sticky spiral spacing accordingly (smaller spaces when more silk is available) (Eberhard 1988a; Eberhard and Barrantes 2015). The homology of the sticky silk cue is complex: the araneoids presumably use information from their aggregate or flagelliform glands; uloborids lack these glands, and presumably use information from their cribellum or pseudoflagelliform glands instead.

In sum, there is an extraordinary uniformity in the cues that cribellate and ecribellate spiders use to guide orb web construction. This uniformity is not reasonably attributed to constructional constraints. Combined with evidence from morphology and molecules, this behavioral evidence constitutes strong evidence in favor of orb web monophyly.

Table 10.3. The stimuli and the responses to these stimuli that guide orb construction behavior in uloborids compared with araneids. "—" indicates there is no evidence for or against; "A" indicates evidence from araneoids, "U" indicates evidence from uloborids; "stsp" = sticky spiral (after Eberhard and Barrantes 2015). The original references are the following: 1 = Hingston 1920; 2 = Peters 1954; 3 = Eberhard 1982; 4 = Eberhard 1972a; 5 = Eberhard and Barrantes 2015; 6 = Eberhard and Hesselberg 2013; 7 = Eberhard 1988c; 8 = Barrantes and Eberhard 2012; 9 = Hesselberg 2013; 10 = Peters 1939; 11 = Le Guelte 1966; 12 = Eberhard 2012; 13 = Eberhard 2014.

				Evidence			*References*	
Cues used by araneoids	*Response to cue by araneoids*	*Same cue used by uloborids?*	*Response same in uloborids?*	*Details of behavior*	*Finished webs*	*Expts.*	*Araneoids*	*Uloborids*
Site where prev. inner loop of stsp crossed radius	Used as point of ref. for stsp atts.	Probably yes	Yes	A, U	U	A, U[1]	1–3	3–5
TS-IL dist. during stsp const.	An increase induces larger stsp spacing	Yes?[2]	Probably yes[2]	—	A, U	—	6	5
Amt. sticky silk available in glands	A decrease induces larger stsp spacing	Possibly yes[3]	Yes	—	U	A	7	5
Low tension on radii	NO response	Yes (NOT used)	Yes (lack response)	—	—	A, U	6	5
Smaller space in which to build[4]	Reduce stsp spacing	Yes?[4]	Yes	—	A, U	A, U	8, 9	5
Smaller space in which to build[4]	Reduce num. radii	Yes?[4]	Yes	—	A, U	A, U	8, 9	5
Smaller space in which to build[4]	Reduce num. frames, num. r/frame, increase num. radii without a frame	Yes?[4]	Yes (all)	—	A, U	A, U	8	5
Smaller space in which to build[4]	Rel. larger hub	Yes?[4]	Yes[5]	—	A, U[5]	A, U[5]	8	5
Smaller space in which to build[4]	Outer loop stsp nearer frame (or beyond)	Yes?[4]	Yes	—	—	A, U	8	5
Smaller space in which to build[4]	Reduce rel. size free zone	Yes?[4]	Yes	—	—	A, U	8	5
Smaller space in which to build[4]	Reduce symmetry of orb	No?[4]	No change in symmetry	—	—	A, U	8	5
Edge vs. central area (rel. distance from hub)	stsp spacing varies predictably	?[6]	Yes	—	A, U	—	10–13	4, 5, 13

Table 10.3. Continued

[1] Conclusion is tentative due to small sample size.

[2] The rarity of complete responses in spontaneous "Hingston experiments" and the "compensatory" changes in sticky spiral spacing in *Z. geniculata* are both compatible with the spider using the TSP-IL distance as a cue to guide sticky spiral spacing, as has been shown experimentally to occur in the araneoids *Micrathena duodecimspinosa* and *Leucauge mariana* (Eberhard and Hesselberg 2013).

[3] Both the increased spacing of sticky spiral loops just prior to spontaneous suspension of sticky spiral construction, and the sharply reduced spacing when construction was resumed the following night in *Z. geniculata*, are compatible with the hypothesis that sticky spiral spacing is influenced by the amount of silk available in the silk glands.

[4] The specific cue (or cues) used are not known in either araneoids or uloborids. All data are from finished orbs built under experimental conditions. For reasons to consider the different responses to smaller areas as independent, see the text, table 7.1, and Eberhard and Barrantes 2015.

[5] Data are available for ulobords on relative hub size but not the number of hub loops. In any case, all hub loops in uloborids were built during radius construction (in contrast with many araneids—section 6.3.3.4, appendices O6.3.2.3, O6.3.4.3), so the number of loops may not be biologically independent of radius number in uloborids.

[6] The cue (or cues) are not known for either araneoids or uloborids.

10.8.3 FOSSILS AND POSSIBLE PRECURSOR WEBS

The fossil record is, as far as it goes, in accord with an ancient derivation of orb webs (fig. 1.8; Vollrath and Selden 2007). The record is famously only fragmentary, and in addition one must assume that modern associations between web forms and morphological traits occurred in the past in order to infer the probable web designs of fossil species. Nevertheless different strands of recent evidence give strong evidence that orb webs are ancient, at least 165 million years old (section 1.6).

The most basal orb-weaving cribellates, Deinopidae and Uloboridae, were formerly linked, but have recently been separated in some phylogenies (fig. 1.10; Garrison et al. 2016). A recent study (Selden et al. 2015), based in part on abundant, extremely well-preserved, specimens from about 165 Mya, described a Mesozoic stem assemblage of diverse taxa that appears to be more related to these two groups than to any other extant superfamily. They concluded that the diagnostic deinopoid morphological traits apparently accumulated through a series of stem taxa that are all now extinct, and that the families Uloboridae and Deinopidae likely represent remnants of a former greater diversity in the Mesozoic. The molecular data from orb weavers are in approximate accord with the fossils; they place the date of origin in the lower Jurassic, 187–201 million years ago (Bond et al. 2014; Garrison et al. 2016).

What were the precursors of orbs? There is no obvious, closely related sister group that builds "proto-orbs," so there is no direct answer from modern webs. Instead, there are various more or less orb-like web forms, with more or less radially arranged non-sticky lines ("radii") and more or less perpendicular sticky lines attached to them ("sticky spirals"), scattered among several apparently distantly related groups, even including egg sac rather than prey capture webs (fig. 10.16). Webs with more or less radial lines that converge on the spider's resting site and at least somewhat circular arcs of sticky lines occur in the psechrid *Fecenia* spp. (box 9.1) and the titanoecid *Titanoeca* spp. (fig. 10.14, Szlep 1966a; Shinkai 1979; Griswold et al. 2005); the sticky lines are somewhat less arc-like in the desids *Badumna* and *Matachia livor* (fig. 1.3, Griswold et al. 2005). Arc-like sticky lines also occur on a sheet in the zoropsid *Tengella perfuga* (Mallis and Miller 2017). Other webs that have radial arrangements of non-sticky lines to which sticky lines are attached include even ancient branches, such as the hypochilid *Hypochilus gertschi* (Shear 1970) and the filistatids *Kukulcania* (fig. 5.10, Comstock 1967), *Filistata* (Griswold et al. 2005), and *Misionella* (fig. 10.12) (Lopardo and Ramírez 2007), as well as the phyxelidid *Xevioso orthomeles* (Griswold et al. 2005) and (in the lower portions of the web) the dictynid *Emblyna* sp. (fig. 10.14e). In sum, there is a plethora rather than a scarcity of possible "missing links."

The ancestor from which orb weavers were derived undoubtedly built webs with sticky cribellum silk. Some have speculated that its webs may have been built close to the substrate, capturing mainly walking prey (Blackledge et al. 2009c; Harmer et al. 2011), but I see no reason to assume a strong association with the substrate. Proto-orbs could have been relatively aerial, as for instance those of the desids *Badumna* and *Matachia* (fig. 1.3).

The origin of orbs may have been associated with an increase in the number of spigots on the cribellum. Comparisons of the numbers of cribellum spigots in the

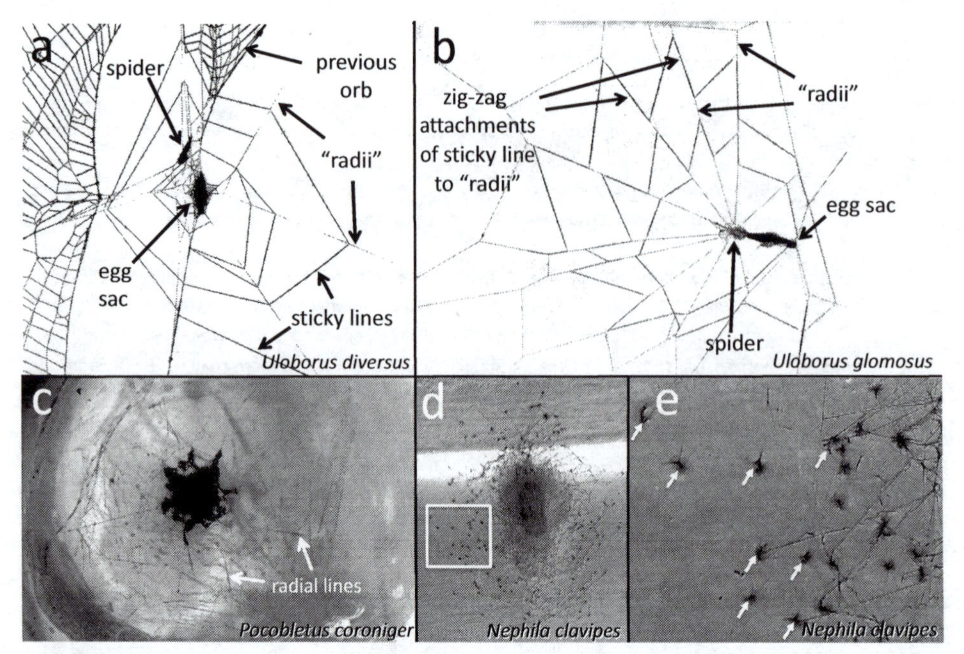

Fig. 10.16. The apparent "radii" in the egg sac webs of the uloborids *Uloborus diversus* (UL) (*a*) and *U. glomosus* (*b*), the linyphiid *Pocobletus coroniger* (*c*), and the egg sac of *Nephila clavipes* (NE) (*d*, and *e*, which is a closeup of the area in the white rectangle in *d*) illustrate how observations of finished products are not always sufficient to evaluate behavioral homologies. Radius-like lines converged on the sac of *U. diversus* (*a*) but direct observations of construction behavior showed that the resemblance to an orb was misleading. The sac was not built at the center of an "orb"; rather the radial lines were built after the sac was finished, during the removal of other lines from the previous large orb on which the sac was built. Radial lines also suspended the sac of *P. coroniger* (*c*), which was adorned with fluffy white silk outside a transparent, silk-walled polyhedron in a curled leaf; but the prey capture web of this spider was not an orb, but a sheet with small tangles above and below. The fluffy sac of *N. clavipes* was held against the side of a house (*d*) with an outer cover, and direct observations of the behavior of a *N. clavipes* female near the end of egg sac construction showed that she repeatedly moved her abdomen in radial directions; but close inspection of the attachments of the sac's outer cover (inset *e*) showed that these lines formed "V" shaped configurations, rather than being straight radial lines, and likely also represent an independent convergence on radial designs. In contrast, the less obviously functional detail of the zig-zag attachments of the sticky line in an egg sac web of *U. glomosus* (*b*) seem more likely to be an evolutionary transfer from orb web construction: similar zig-zags occur in the early loops of sticky spiral of the prey capture orbs of this and other uloborids (figs. 7.24, 9.11, 10.17).

cribella of species of orb-weaving and non-orb weaving cribellate species showed that in general the orb weavers had at least twice as many or more cribellum spigots/mg (the means for five orb weaver and eight non-orb weaver species were 445 and 102 respectively) (Opell 1999a). The stickiness of cribellate silk correlates positively with the number of spigots, and the stickiness of the cribellate silk of the orb weavers, corrected for the weight of the spider, was more than eight times that of the non-orb weaving species (1.83 vs. 0.22 μN/mm/mg). Thus orb web construction is associated with greater stickiness of cribellum silk (Opell 1999a). In contrast to other cribellates, the number of cribellum spigots of orb-weaving species of uloborids scaled with spider mass.

Whether greater stickiness came before orbs and facilitated their evolution, or came afterward as a consequence of selection on sticky lines in orbs, is not clear. The two non-orb species in Opell's sample with the most orb-like arrangement of sticky lines (more widely and regularly spaced), the dictynid *Mexitlia trivittata* and the desid *Matachia livor* (fig. 1.3), had especially sticky cribellum silk (Opell 1999a), suggesting (though weakly) that increased stickiness may have preceded orb webs. Cribellate orb weavers apparently invested more per mm of sticky line. Orb vs. non-orb comparisons are complicated by the fact that sticky lines are piled or looped on each other in the webs of at least some non-orb weavers (see fig. 5.10 of the filistatid *Kukulcania hibernalis*, fig. 1.3 of *M. livor*), presumably increasing adhesion to prey.

Orb webs appear abruptly in "finished" form in

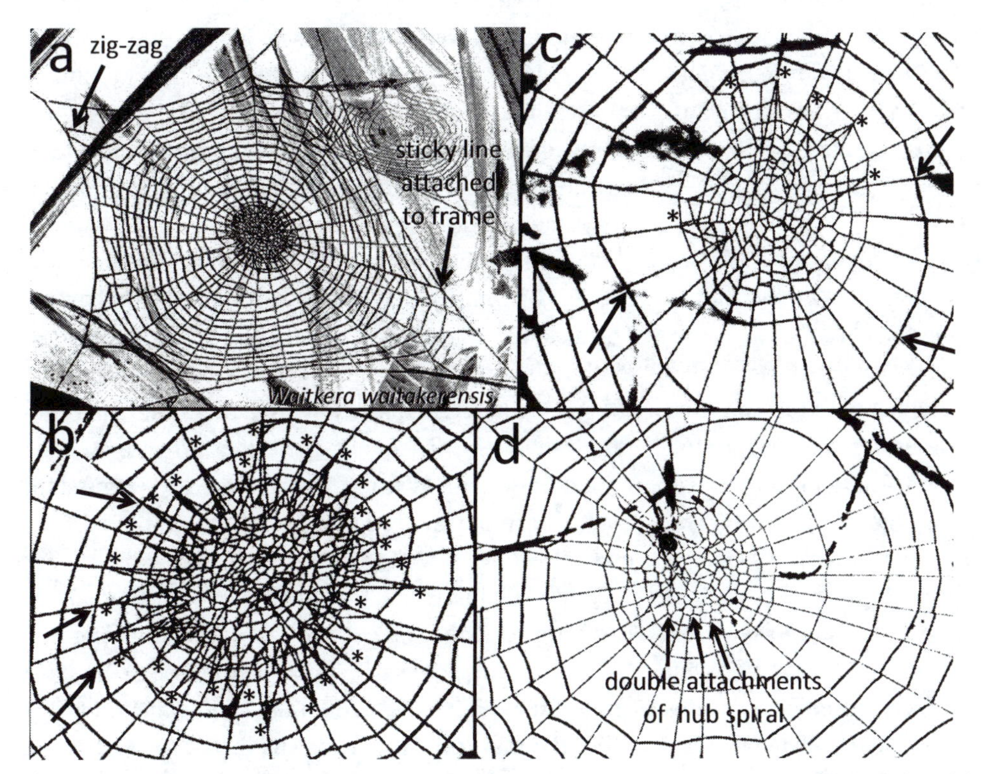

Fig. 10.17. Because the monotypic New Zealand genus *Waitkera* (UL) is apparently the sister of all other uloborids, these orbs of *W. waitakerensis* suggest that four otherwise unique traits that they share with other uloborids probably occurred in the webs of the last common ancestor of the entire family. The outermost loop of sticky spiral was attached in places to frame lines (*a*), and was also sometimes attached twice in succession to the same radius, giving a jagged zig-zag outline to the sticky spiral (*a*). The hub spiral was attached twice to each radius that it crossed, pulling the radii into zig-zag forms (arrows in *d*). And finally, the sticky spiral was not attached to all radial lines it crossed, especially in the inner part of the orb (arrows in *b* and *c*). The *W. waitakerensis* webs had another trait (marked with "*" in *b* and *c*), that is unique to this genus: "V" lines that ran from the outer portion of the hub to radii at or just beyond the outer edge of the hub. The frequency of "V" lines varied from nearly all radii (*b*) to intermediate numbers (*c*) to completely absent (*d*). The behavior used to produce the "V" patterns is unknown (photos courtesy of Brent Opell).

both uloborids and the araneoids. Several uloborid genera build webs that are less elaborate (*Polenecia*, *Miagrammopes* and *Hyptiotes*), but these genera are all thought, on morphological grounds, to be derived rather than ancestral within this family (fig. 1.13) (Opell 1979; Coddington 1990); some less elaborate webs in araneoids are also thought to be derived rather than basal (fig. 1.12, section 10.9).

Comparing the orbs of the most basal uloborid genus, *Waitkera* (fig. 10.17), with those of other uloborid genera suggests that the orbs of the last common ancestor of uloborids had three traits that are absent from (or very rare in) those of araneoid orb-weavers: outer loop or loops of sticky spiral that are attached to frame lines; outer loop or loops of sticky spiral that are attached twice in succession to the same radii, giving a jagged outline to the sticky spiral; and (probably) sticky spiral lines that frequently skip over some radii (especially near the hub) without being attached (there is only one published photo of a *W. waitakerensis* orb, but Brent Opell kindly provided several others for me to check). These may be synapomorphies of uloborids, or traits subsequently lost in araneoids.

One unexplored possibility is that the egg sac webs preceded orbs. In some uloborid species such as *Zosis geniculata* (fig. 10.35), *Uaitemuri rupicola* (fig. 9.11), and *Uloborus* spp. (fig. 10.16), egg sac webs have radial lines in a single plane that converge on the sac and that have more or less spiral sticky lines attached to them. This is only speculation, however, and not supported by the few available observations of building behavior (fig. 10.16). Photos of two *W. waitakerensis* egg sacs show only three converging lines and no certain cribellum lines (B. Opell pers. comm.).

In sum, some orb-like aspects of web designs have repeatedly evolved convergently. Current knowledge does not point to an obvious taxonomic group, nor an obvious web design from which orb webs are especially likely to have been derived.

10.8.4 SPECULATIONS ON THE ORIGINS AND CONSEQUENCES OF CUT AND REEL BEHAVIOR (AND THE POSSIBLE ROLE OF MALES)

Small behavioral units, such as particular leg movements and positions, can resemble bricks that are used to build diverse types of web designs (section 9.2.5), and some web designs may not be feasible (and thus will not evolve) unless the spiders possess certain behavioral modules. Understanding the lower level behavioral mechanisms that are used to produce particular web construction operations may thus improve understanding of evolutionary transitions. Several behavior patterns mentioned in chapter 2 represented increases in the abilities of early spiders to manipulate silk lines and were probably necessary to build more effective webs: grasping individual lines (or cables of fine lines) firmly (section 2.4.1); following behavior of the legs (section 2.4.2.1); and grasping a line and bringing it to the spinnerets so that the new line being produced can be attached to it (section 2.4.3). Only the grasping ability has been observed in mygalomorphs; perhaps the other two abilities did not arise until araneomorphs appeared (very few mygalomorphs have been observed, however).

Two low level behavioral modules that may have been prerequisites in the transition from non-orbs to orbs are "cut and reel" and "reconnecting" behavior; both involve highly coordinated, precise leg movements (fig. 6.3). Orbs are relatively fragile and need to be rebuilt frequently, and an orb-weaver's ability to rapidly utilize superior websites depends on the ability to explore new sites and to build new webs quickly from scratch. I speculate that the ability to cut lines and eliminate and/or reposition them early during construction was a prerequisite for the evolution of orbs. As a spider moves about during the process of exploring a site for an aerial web, she inevitably produces new draglines. Many of the paths that she follows during exploration, and thus the lines that she produces, are highly likely *not* to be in the positions and orientations that are appropriate for the orb that she eventually produces. If all the exploration lines were still in place when definitive construction began, they would "clutter" the area, forming a sparse, three-dimensional tangle. Even in the relatively simple case of *Micrathena duodecimspinosa*, lines were repeatedly cut and removed or repositioned during the early stages of orb construction (section 6.3.2).

Fine precision and coordination of leg movements are used in cut and reel and reconnecting behavior. For instance, the effective packing movements by legs III during cut and reel eliminate loose lines floating in the space where the spider will build that could interfere physically with subsequent operations. Precise coordination between legs is also crucial (fig. 6.3); a misstep could cause the spider to lose her hold on the broken end of the line and fail to reconnect it, or to fall.

This is not to say that building a web with a radial organization depends on cut and reel or reconnecting behavior. Several groups with radial webs close to the substrate, including the filistatid *Kukulcania hibernalis* (fig. 5.10), the pisaurid *Thaumasia* sp., the hersiliid *Hersilia* sp. (fig. 9.1), the oecobiid *Oecobius concinnus* (fig. 9.6) (Solano-Brenes et al. 2018), and the segestriid *Ariadna* sp. (Griswold et al. 2005), probably achieve such organization without either ability. They all probably walk on the substrate, where the spider's path need not be affected by the silk lines already present, even if they are not radially oriented. But the spider is more likely affected by previous lines in building more aerial webs.

A related question concerns the origin of cut and reel and reconnecting behavior, and its distribution among non-orb weavers. The pholcid *Modisimus guatuso* did not cut and reel (Eberhard 1992b; WE), but nevertheless rapidly built organized webs from scratch. Do the extremely long legs of pholcids represent alternative exploration mechanisms to the fishing lines of stubby-legged araneids like *M. duodecimspinosa*? Some theridiids only seldom employed cut and reel behavior in constructing gumfoot and other three-dimensional tangle webs (Eberhard et al. 2008b). Perhaps the more disorganized lines laid during exploration can be incorporated with less problem into finished webs of this type.

The origins of cut and reel behavior seem to never have been studied. One possible ancestral behavior is that employed in floating bridge lines on the wind. Bridge lines are produced during long-distance dispersal, including the movements of mature males searching for females. How widespread are such long-distance movements? Surely the heavy-bodied mature males of mygalomorphs are unlikely to move on single, long aerial lines. Mygalomorph spiderlings (*Ummidia* sp., *Sphodros rufipes*, and *Atypus affinis*) sometimes ballooned, however (section 10.4.2); but none showed signs of break and reel behavior. No behavioral observations are available, however, for many other basal groups with aerial webs, including hypochilids, plectreurids, diguetids, and pholcids, or for the males of these and many other possibly interesting non-orb building entelegynes like eresids, dictynids, oecobiids, oxyopids, and psechrids.

10.8.5 DERIVATION OF ECRIBELLATE STICKY LINES FROM CRIBELLATE STICKY LINES

Perhaps the most functionally important transition that occurred after orb webs had arisen was the acquisition of sticky lines composed of flagelliform and aggregate gland silk (sections 2.2.6, 2.2.8.2) to substitute for cribellum silk (section 2.2.8.1). This transition had occurred by the Jurassic or possibly earlier (Vollrath and Selden 2007), on the order of at least 165 Mya (fig. 1.8). Ecribellate sticky spiral lines are produced by a diagnostic, derived triad of spigots where wet gluey material from the aggregate glands is placed on the axial line from the flagelliform glands (section 2.2.8.2). Orbs with the derived, ecribellate type of sticky silk are now clearly dominant; about 95% of extant orb weaver species use this type of silk (Coddington and Levi 1991; Opell 1998). Bond and Opell (1998) argued, on the basis of phylogenetic correlations, that the transition to wet stickiness, combined with a change in the orientation of the plane of the orb from horizontal to vertical, constituted key adaptations that were responsible for the evolutionary success of araneoids. Key adaptation arguments of this sort, which use correlations to infer particular cause-and-effect relations between selective forces and evolutionary radiations, are risky (section 10.5.3). Nevertheless, there are reasons to suppose that ecribellate sticky silk is superior in some ways, and that it could have facilitated the ecribellate radiation (I will postpone discussion of the second factor, web orientation, until section 10.9.1).

Ecribellate sticky spirals may be advantageous both because they have superior mechanical properties, and because they have reduced costs of production. Two potentially important mechanical properties are extensibility and elasticity. Both cribellate and ecribellate sticky spiral lines are extremely extensible: cribellate lines (including both the pseudoflagelliform axial fibers and the cribellum fibers) extended up to 500% of their original length in *Deinopis spinosa* (and >200% in *Hyptiotes* spp. and *Uloborus diversus*); typical values for ecribellate sticky spiral lines are 300–400% (fig. 10.18) (Blackledge and Hayashi 2006a). The presence of other, relatively thick coiled fibers in *D. spinosa* lines (but not in the others), in addition to the axial and cribellum fibers, may help account for the extreme extensibility of *D. spinosa* lines (coiled fibers also occur in other cribellate families—Eberhard and Pereira 1993; Griswold et al. 2005).

The sticky spiral lines of ecribellates are much more

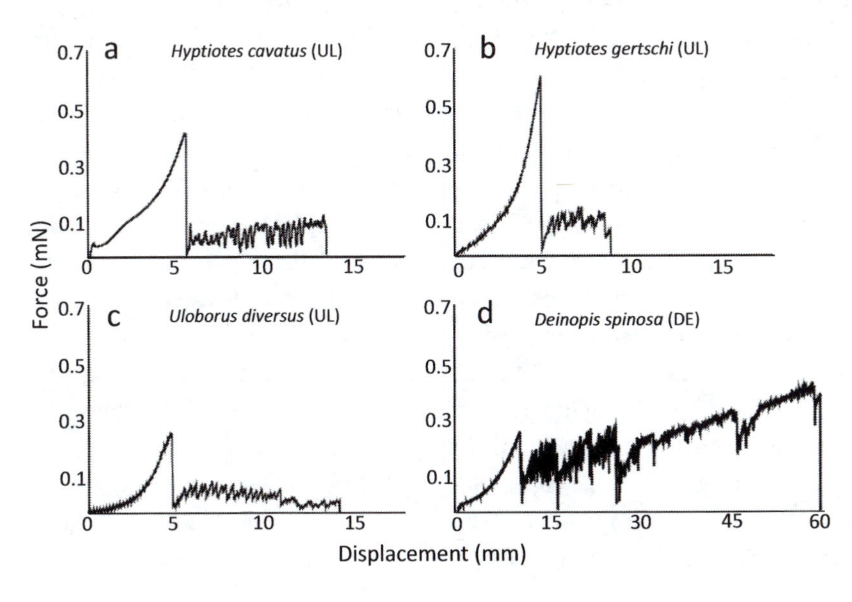

Fig. 10.18. These force-displacement curves for cribellate capture lines (all were initially 10 mm long) were qualitatively similar in that the axial fibers extended and broke (larger peak at left), followed by the gradual breaking of cribellar fibrils. *Deinopis* cribellum fibers, however, extended more than four times as far as those of the uloborids (from Blackledge and Hayashi 2006).

elastic, however, than those of cribellates. Most of the extension of cribellate sticky spiral lines is due to irreversible unfolding of the highly tangled cribellum lines, rather than to extension of their axial pseudoflagellum fibers (which break after extending only 50–100%) (Blackledge and Hayashi 2006b). The cribellum silk lines probably break in groups as the tangle is extended, producing numerous small, rapid increases in the length of the line as the sticky spiral as a whole is extended (fig. 10.18) (Blackledge and Hayashi 2006b); they thus lack elastic recovery. In summary, ecribellate sticky spiral lines are similar in extensibility to those of some cribellate spiders (*Deinopis*) or superior to others (*Uloborus* and *Hyptiotes*), and they are superior in elasticity compared with all known cribellate sticky lines.

The relative costs of cribellate and ecribellate sticky silk include both material and behavioral factors. On the material side, ecribellate sticky silk was thought to be more efficient, because it was thought to produce equivalent stickiness with smaller amounts of material (Bond and Opell 1998; Opell 1998). Nevertheless, a recent cost-effectiveness analysis of the material costs to produce a given degree of stickiness in sticky lines, which corrected a previous systematic computational error in determining cribellar fibril volumes (by a factor of 103), cast doubt on the material cost argument. In terms of the adhesive forces that can be generated to capture prey, the gross material economy was greater in the cribellum threads of several uloborid species and of the related cribellate neolandid *Neolana pallida* than in the viscous sticky lines of several araneids (Opell and Schwend 2009).

Other possible mechanical shortcomings of cribellar threads include the following. Increases in adhesion are probably achieved with less efficiency. That is, the only way to increase stickiness is to make more cribellar fibrils but only the fibers on the surface of the mat of fibrils that is exposed to the prey can contribute to stickiness. Fibrils inside the mat and on the other sides tend to be "wasted" (though they probably make contributions to the unreeling behavior described above) (see fig. 5 of Opell and Schwend 2009). In contrast the viscoelastic glycoprotein core of an ecribellate glue droplet flattens when it contacts the prey, increasing the area of contact so that less material is "wasted." Secondly, the stiffness of the cribellar thread's axial lines means that when a segment of sticky line lies in contact with a surface and then pulls away, most adhesion is generated towards the ends of the contact zone; thus the stickiness of cribellar thread does not increase with the length of thread contact as rapidly as it does with ecribellate sticky lines (fig. 8 of Opell and Hendricks 2007; Opell and Schwend 2009).

The differences in behavioral costs are especially clear: cribellum silk must be combed out with repeated movements of the hind legs, which are probably metabolically expensive, while viscous sticky lines are largely just pulled from the spider's body as she moves. The metabolic costs of the combing movements have never been measured, but the extremely high number of repetitions that a spider needs to build a single orb suggests that they may be substantial: the uloborid *Zosis geniculata* made about 69,000 combing movements/orb, and *Uloborus conus* an estimated 54,000/orb (see section 3.3.3.3.2.1.2).

In sum, behavioral costs, perhaps in combination with material costs, could have favored the transition from cribellate to viscous sticky spiral lines in the evolution of orb webs (Lubin 1986). The energetic costs of the synthesis of the silk in the two types of sticky lines are also unknown.

Another, previously unemphasized behavioral advantage that makes ecribellate sticky spiral lines likely to be cheaper is that the time needed to build an equivalent orb is substantially shorter. The rate at which the spider pulls out sticky lines and attaches them to radii is generally substantially more rapid in ecribellates. A mature female of the araneid *Micrathena duodecimspinosa* takes on average about 1.6 s/attachment near the edge of her web, and <1 s/attachment nearer the hub (see fig. 6.13); in contrast, the uloborid *Zosis geniculata* at the edge of her web takes more than 10 times longer between attachments (mean 31.6 s/attachment) (WE). Some orb designs with especially dense sticky spirals, such as those of the moth specialist araneids *Scoloderus* spp. (fig. 3.4), *Acacesia hamata*, *Deliochus* sp. (fig. 3.5), *Poltys* spp., and *Pozonia nigroventris* (fig. 3.35), may not be feasible in cribellates. The spiders made several thousand sticky line attachments instead of the 500 or so in a *Z. geniculata* orb (WE), and direct observations showed that *Scoloderus tuberculifer* and *Acacesia hamata* worked even more rapidly than does *M. duodecimspinosa* (WE). Thus an uloborid might take nearly 50–100 times longer to build the same dense web (she might not be able to finish building before it was time to take it down again!). In addition, incidental observations of species in all four major orb weaving families suggest that attack times to reach prey that strike the web are much slower while the spider is laying the sticky spiral than when the orb is finished and she is resting at the hub. So even the portion of an orb that has already been built is partially out of action during sticky spiral construction, suggesting an additional advantage for the shorter duration of ecribellate sticky spiral construction.

There could also be other, more indirect consequences of the possible metabolic savings associated with ecribellate sticky silk. Kawamoto and Japyassú (2008) found that in a comparison between the cribellate uloborid *Zosis geniculata* and the ecribellate araneid *Metazygia rogenhoferi*, the cribellate was less likely to adjust to food stress by moving to other sites. They argued that "a less costly and more quickly built web, along with the lower [website] tenacity" (p. 419) would give ecribellate orb weavers an advantage. I am not sure that reduced website tenacity is a logical consequence of building cheaper orbs, unless appropriate new sites are easier to find for cheaper webs or a cheaper web is more easily recovered when a spider abandons a site (and of course a sample of only two species does not necessarily represent the overall differences between cribellate and ecribellate orbicularians). But this example illustrates the possibility that there are indirect payoffs from metabolic savings that deserve further testing.

Opell and Schwend (2009) listed several challenges that need to be resolved in order to understand how a transition from cribellate to ecribellate sticky lines could have occurred in orb weavers. The basic problem is that stickiness needs to have been maintained throughout the process (an orb without sticky lines would be nonfunctional). There are no modern groups known that produce both cribellate and ecribellate sticky lines, so discussions must be speculative (this lack of intermediates stands in contrast to the abundance of possible intermediates in many other possible evolutionary transitions) (see above, chapter 9). In one recently proposed hypothetical transitional form (Opell et al. 2011c), wet sticky material was added to cribellate silk. An experimental combination of the viscous glue from the sticky spiral of the araneids *Argiope aurantia* and *Araneus marmoreus* and the cribellum silk from *Hyptiotes cavatus* (UL) was more adhesive than the sticky lines of either type alone. A possible limitation is that highly derived viscous silk was used, whose properties probably differed from those of transitional forms.

Blackledge et al. (2011) and Sahni et al. (2014) proposed the alternative hypothesis that an early intermediate cribellate added hygrophyllic salts (such as those present in modern aggregate gland glue) to cribellate silk, and that this facilitated adhesion because the nodules on the individual fibrils allowed for additional hygroscopic adhesion. Even slight increases in adhesion associated with these salts could "set up a tipping point that favored the origin of the large viscoelastic glue droplets and glycoproteins" of araneoid sticky silk (Sahni et al. 2014). This step could have resembled the transition in cribellates to nodules on the cribellum fibers, which may have increased the spiders' reliance on capillary forces for adhesion (Hawthorn and Opell 2002, 2003).

A tantalizing observation (Blackledge et al. 2009c) is that genes for the derived, rubbery flagelliform silk in the axial lines of ecribellate sticky lines are also expressed in cribellate spiders (Garb et al. 2006); their role in cribellates is unknown (see Blackledge and Hayashi 2006b).

10.8.6 SUMMARY REGARDING ORB MONOPHYLY

Many uncertainties remain regarding the evolutionary origins of orb webs, but the weight of the evidence favors a molophyletic origin. The earlier claim of "robust" support for monophyly from molecular data (Blackledge et al. 2009c) was criticized by Dimitrov et al. (2011); but when these authors then added more molecular data of their own, they found increased support for monophyly. In turn, this support was judged to be "weak" by Bond et al. (2014). Recent analyses have included data from transcriptomes as well as from the sequences of traditionally analyzed genes (Garrison et al. 2016), and further anchored enrichment analyses are planned. Some of the molecular techniques have serious weaknesses (discussed in Garrison et al. 2016). Evidence on the cues used during construction and the spiders' responses to them also favors monophyly.

There is also a logical problem for the monophyly hypothesis regarding the possible lack of adaptive intermediate stages; sticky cribellum silk must not have been lost until after ecribellate sticky silk had evolved. There are possible solutions: perhaps an intermediate ancestor combined both types of sticky silk in the same lines (section 10.8.5); or perhaps an ancestor placed different types of sticky lines in different portions of webs, like those of New Zealand endemic *Neolana pallida*, which builds a sticky sheet and also a separate planar platform where the spider rests (B. Opell pers. comm.). But these are only speculations. Even if there was a single origin of orb webs, different behavioral traits that go into making an orb may have evolved independently in the cribellate and ecribellate lineages (Eberhard 1990a).

Ironically, just as the orb monophyly controversy in orbicularians seems to be resolving itself in favor of the single origin hypothesis, incontrovertible evidence of a second origin of orb-like webs has recently emerged, in the very distantly related "pseudo-orb" psechrid *Fecenia* (box 9.1). Many questions about the construction behavior of these poorly known spiders remain unanswered.

10.9 EVOLUTIONARY CHANGES IN ORB DESIGNS

A recent overview of araneoid phylogeny and web evolution based on molecular evidence (Dimitrov et al. 2016) found that uncertainties regarding the branching patterns at the interfamilial level precluded confident resolution of web architecture evolution within Araneoidea (additional difficulties stemmed from the incomplete fossil record, and uncertainties in the dating and the systematic circumscription of some of the oldest known orb-weaver fossils). Several general trends emerged, however. The ancient nature of orbs was confirmed. The general web architectures similar to those that characterize the extant species of non-orb groups of araneoids were already diverse at the time of the spectacular Cretaceous diversification of holometabolous insects (primarily Hymenoptera, Diptera, and Lepidoptera) that coincided with the angiosperm radiation.

10.9.1. HORIZONTAL VS. VERTICAL AND NEARLY VERTICAL ORBS

The two major lineages of orb weavers form a contrast in evolutionary success: the modest cribellate families Deinopidae and Uloboridae have only about 300 described species together; the highly successful ecribellate Araneoidea has >12,000 described species in seven families (Coddington and Levi 1991). Can this difference be explained by these spiders' webs? Opell and Bond (Opell 1998; Bond and Opell 1998; Opell and Bond 2001) argued that two key innovations were responsible for the araneoid success: replacement of cribellate sticky silk with ecribellate sticky silk; and replacement of horizontally slanted orbs with vertical orbs. I remain unconvinced. As just discussed, the relative advantages of ecribellate over cribellate silk are less certain than previously thought. More importantly, the horizontal vs. vertical contrast is inconclusive.

Opell and Bond noted that araneoid sticky lines have lower tensile strengths and also a lower Young's modulus than cribellum sticky spiral lines, and are thus better equipped to dissipate the force of prey impact by stretching (Opell and Bond 2001). They argued that this ability could be more advantageous in vertical orbs, which are thought to intercept faster moving prey (Opell and Bond 2001). This seems intuitively appealing, though I know of no data on the relative velocities of prey moving in horizontal and vertical directions (the kinetic energy of a flying insect is a function of its velocity squared, so imprecisions in determining velocities could have large consequences). It appears, however, that sticky spiral lines in araneid orbs play a relatively minor role in absorbing prey momentum (Sensenig et al. 2012; see section 3.3.2.1). Thus their special properties probably did not evolve to resist high-energy prey impacts.

A second problem is that although it is true that web slants in the two groups differ on average, there is substantial overlap between the two groups (table O3.1). In addition, and probably more importantly, there is large intra-specific variation in slants in various species of both uloboids and araneoids (table O3.1). Among uloborids, in two species of *Philoponella* with samples of >10 webs in the field, *P. tingena* had mean slopes of 65 ± 14°, and *P. vittata* 47 ± 11°. The maximum slant in this sample of only 30 webs was nearly perfectly vertical (85°). The slopes of the webs of *Uloborus trilineatus* (in a sample only 4 individuals) also ranged up to 60° (WE), and Hingston (1932) mentions an orb of *U. filidentatus* as being "vertical." On the araneoid side, some species in the araneid genera that are classified as having "vertical" orbs have mean slants that hardly differ from those of the uloborids. These include 69° for *Alpaida leucogramma*,

64° for *Argiope argentata*, 60° for *Alpaida veniliae*, 48° for *Metazygia lopez*, and 58° for *Cyclosa jose* (table O3.1). Some species of *Cyclosa* tended to build nearly perfectly vertical orbs (table O3.1), but *C. serena* occurred on horizontal orbs (Levi 1999). The slants of *Araniella cucurbitina* (Zschokke and Vollrath 1995b) and *Micrathena gracilis* (Opell et al. 2006) varied from vertical to nearly horizontal. Other examples of araneid species with occasional individuals building webs with slants of <45° include *Micrathena sexspinosa*, *Gasteracantha cancriformis*, *Metazygia gregalis*, *Neoscona moreli*, *Hypognatha mozamba*, and *Witica crassicauda* (WE). Intra-specific variation also ranged from vertical to nearly horizontal in the tetragnathids *Meta ovalis* and *Tetragnatha laboriosa* (Bradley 2013).

Nearly horizontal orbs are typical of some araneid species, including *Araniella cucurbitina* (Zschokke and Vollrath 1995b), *Araneus niveus* (Coddington 1987), *Azilia* sp. (Hingston 1932; WE), *Mangora melanocephala*, *Enacrosoma anomalum* and *E. quizarra* (WE); and tetragnathids as a family tend to have more nearly horizontal orbs. Still another probable derivation of horizontal orbs in araneoids occurs in the symphytognathoid families (Lopardo et al. 2011). These families may in fact build orbs that are more consistently close to horizontal than those of the uloborids.

In sum, this particular pair of key innovations seems to me a poor candidate to explain the undoubtedly major difference in the evolutionary success of cribellate and ecribellate orb weavers. The two traits are not convincingly associated functionally, and web slants are not consistently different. Slants also show ample intraspecific variation (and thus material on which natural selection could work), implying that phylogenetic constraint is particularly weak, and araneoids have reverted to the supposedly selectively inferior state several times. Finally, there is the major problem of the intrinsic weakness of "key innovation" arguments (section 10.5.3). In sum, I believe that the causes of the relative success of araneoids compared with deinopoids remain uncertain.

10.9.2 SMALL DERIVED LINEAGES: LADDER AND TRUNK WEBS

"Ladder" and "trunk" orbs, which are extremely elongate above or below the hub (or both), have evolved convergently in three or four genera of araneids that may not be closely related (Blackledge et al. 2011), and also in three other families (table 9.1). Some ladder webs probably function to trap moths, while others allow the spider to build orbs against tree trunks, and are better termed "trunk webs" (fig. 4.3; Kuntner et al. 2008a). One phylogeny of Nephilidae indicates that both increases (in *Herennia*) and decreases (in *Nephila*) have occurred in top-bottom asymmetry (fig. 10.19). The striking increase in symmetry in orbs built by the trunk web species *Telaprocera maudae* (AR) when spiders were confined on a symmetrical support (Harmer and Herberstein 2009) suggests that the evolutionary transition to a trunk web in this group resulted mostly from changes in website selection, rather than in orb construction per se. Other trunk webs (in Nephilinae) had additional modifications that represent adjustments to building on a trunk, such as silk retreat cups built on the trunk and "pseudo-radii" that help hold the orb away from the trunk in *Herennia* spp. (Robinson and Lubin 1979a; Kuntner 2005). Using trunks as websites does not always result in design changes: *Argiope ocyaloides* (AR) preferred tree trunks but its orbs resembled those of other congeneric species that live in other habitats (Robinson and Lubin 1979a) (some may be elongate—A. Harmer pers. comm.); *Allocyclosa bifurca* (AR) occasionally built typical orbs on large vertical tree trunks (WE).

10.9.3 DERIVATION OF DEINOPID WEBS

Deinopidae is a small family of only two genera and 57 named species. Their unique webs appear to be uniform throughout the family in being bilaterally symmetrical sectors of orbs with three radii on each side (Coddington et al. 2012). The webs are very small in relation to the spider (only about the size of her own body), and are actively manipulated by the spider to capture prey with behaviors not known in other spiders. The spider's long legs I and II hold the corners of the dense array of sticky silk (fig. 3.1), and the spider spreads these legs apart and expands the web or thrusts it downward to envelop prey that are flying or walking nearby. Operation of webs is facilitated by the unusually extensible dragline and the cribellum silk in these webs (sections 2.2.3, 2.2.8.1). Perhaps the extensible draglines are produced by the unique supernumerary major ampullate gland spigots.

Deinopids clearly share a number of derived behav-

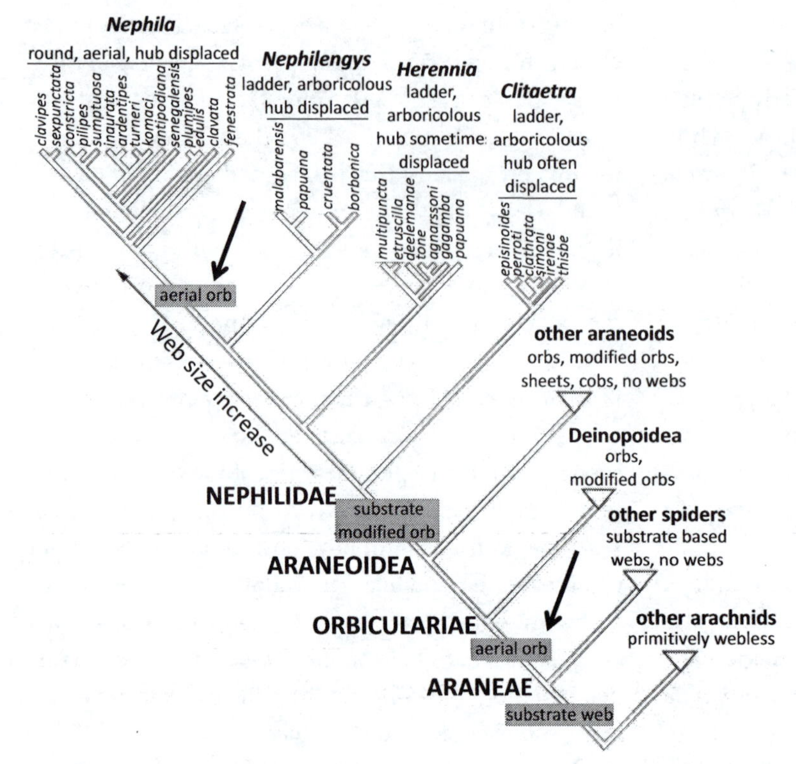

Fig. 10.19. The acquisition, loss, and then re-acquisition of an aerial orb web (arrows) in this phylogeny of nephilid spiders is a possible example of the common lack of consistent, unidirectional trends in web evolution (from Kuntner et al. 2010b; more recent information suggests that ladder webs may have evolved more than once—M. Kuntner pers. comm.).

ioral traits with typical orb weavers (fig. 10.20, Coddington 1986c, summary in Coddington et al. 2012). Some ancestral traits have become reduced, several have been transferred in the construction sequence so that they are expressed in new contexts, and others have gone from being only occasionally expressed in the ancestor to being consistent parts of web construction (fig. 10.20). One such "fixation" of an ancestral trait probably results from the spider's great size compared with the size of her web: she was reliably able to establish new frame attachments by walking along the substrate, because the sites of the two lower anchors to the substrate were only on the order of two body lengths from the site of the hub (Coddington 1986c). Despite the web's highly modified final form, deinopid construction behavior involved relatively few new behavioral traits (Coddington 1986c).

10.9.4 THERIDIOSOMATIDS AND THEIR ALLIES

Orb designs are strikingly diverse in the approximately 450 described species of the small-bodied orb weavers of the five families that include Theridiosomatidae and its allies (the "symphytognathoids") (fig. 1.14) (table 10.4). These families are probably derived with respect to other orb-weaving araneoids (figs. 1.9, 1.10). The changes in their orbs vary widely, and often include lines out of the plane of the orb. Some anapids, mysmenids, and symphytognathids add many additional "supplemental" radii to the orb after the sticky spiral is finished (Eberhard 1987e; Hiramatsu and Shinkai 1993; Lopardo et al. 2011); this trait sometimes varies within a genus (Ramírez et al. 2004 on *Crassanapis*) and even intra-specifically (Ramírez et al. 2004 on *Sheranapis villarrica*). Some theridiosomatids (e.g., *Theridiosoma*) and mysmenids have converged independently on removing the hub entirely (fig. O6.9, table 10.4); other theridiosomatids (e.g., *Epeirotypus*), anapids, symphytognathids, and mysmenids break all of the radii near the hub after completing the sticky spiral and then build a new hub, where they reattach all or some of these radii.

Some other changes involve non-sticky lines that are added out of the plane of the orb and are probably homologous to radii. These include the following: non-sticky spring lines with which the spider pulls the web into a cone and then snaps it back to snare prey flying

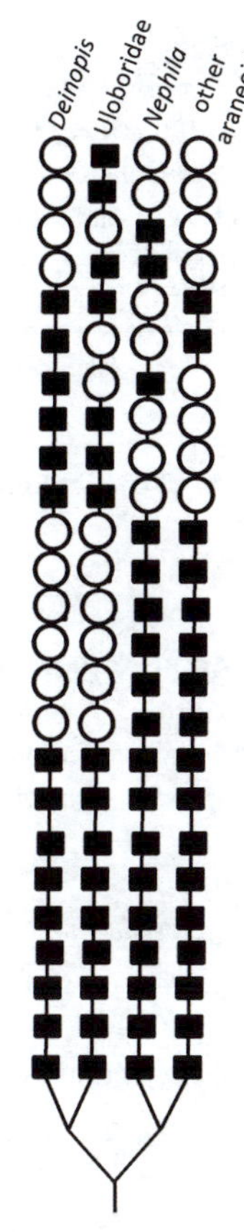

Fig. 10.20. Different details of web design ("*") and construction behavior ("x") of *Deinopis* were shared with orb weavers. At least one (#7) was probably related to the spider's great size compared with the mesh of its web, rather than phylogeny, but others may be taxonomically useful. Those (#17–24) that occurred in all orb weavers argue for an orb-weaving ancestor for deinopids (after Coddington 1986c).

nearby in several theridiosomatid genera (summary Coddington 1986a); non-sticky lines radiating from the hub above a more or less horizontal orb that often pull the hub upward, producing a conical rather than a planar orb in anapids and mysmenids (summary in Lopardo et al. 2011); and sticky lines out of the orb plane that are probably homologous to sticky spiral lines (Platnick and Shadab 1978, 1979; Eberhard 1987e, 2011; Lopardo and Hormiga 2008; Lopardo et al. 2011). Sticky lines running to a water surface directly below the web that are built just above the surfaces of streams evolved twice independently, in the horizontal orb of *Conculus lyugadinus* (AN) (Shinkai and Shinkai 1988) and in the highly reduced orb of *Wendilgarda* (THM) (Coddington and Valerio 1980; Coddington 1986a; Eberhard 1989a; Shinkai and Shinkai 1997).

One highly modified design is the "spherical orb" web of the *Mysmena* spp. and three other genera of Mysmeninae, which includes a dense "cloud" or "rind" of sticky lines that surround the central area where the three-dimensional array of non-sticky radii converge and the spider waits; the hub itself is removed (fig. 10.21) (Coddington 1986a; Eberhard 1987e; Lopardo et al. 2011). The webs in the theridiosomatid genus *Ogulnius* are also three-dimensional, but are very sparse rather than dense; they consist of a few, widely spaced vaguely spiral sticky lines that are attached to a few (5–10) non-sticky lines that do not converge on a single hub (fig. 10.22).

Some symphytognathoid webs are so modified that it is difficult to determine homologies. The few radii of the micropholcommatine anapid *Comaroma simoni* converge on an irregular "hub" that lacks radial or circular lines; the distal tips of the radii have short, sticky branches that are attached to the substrate (Kropf 1990, 2004; pers. comm.). The "sheet" web of the mysmenid *Trogloneta granulum* appears to have both radial and transverse sticky lines, and is so modified and variable that it has proven difficult to describe and categorize (Hajer 2000; Lopardo et al. 2011). The synaphrid *Synaphris lehtineni* builds a small thin sheet under stones (Marusik et al. 2005 in Lopardo et al. 2011). Most spectacular are the recently discovered "quilt" orbs of the anapid *Tasmanapis strahan*, in which several different tightly woven co-planar orbs are joined along their edges to form a single horizontal sheet (fig. 10.27, Lopardo et al. 2011).

Still another unusual aspect of symphytognathoid webs is that mature males construct "normal" prey capture orbs indistinguishable from those of conspecific females (Table 10.5) (the mysmenid *Trogloneta granulum* is a possible exception—Lopardo et al. 2011); this contrasts with the nearly complete lack of prey capture orbs in the mature males of other araneoids (see appendix O9.2). Symphytognathoid males generally retain the triad of aggregate and flagelliform spigots for making sticky lines, a trait unknown in the mature males of all other orbicularians except a single species (the linyphiid *Orsonwelles malus*) (Ramírez et al. 2004).

Another symphytognathoid tendency is kleptoparasitism in the webs of other spiders, which has evolved at least three times independently (one species each in Anapidae and Symphytognathidae, three genera in Mys-

Table 10.4. Multiple independent transitions from more or less typical orb webs in the five symphytognathoid families. Despite the fact that this is not a particularly speciose group (at total of about 450 described species), and that there is relatively little information on the web forms of these small to very small, mostly tropical and under-studied spiders, it is clear that they have been unusually flexible evolutionarily compared with large araneoid groups such as Tetragnathidae and Araneidae. For possible phylogenetic relations among symphytognathoids, see fig. 1.10.

Taxa	*Basal traits*	*Derived traits*	*References*
Theridiosomatidae			
Epeirotypus, Naatlo	Orb, spring line from hub out of web plane	Hub removed, radii attached directly ("anastomize")	Coddington 1986a
Theridiosoma, Epilineutes[1]			Many authors
Wendilgarda		"Curtain" of sticky lines attached to water surface (homologies uncertain)	Fig. 10.29
Ogulnius[2]		Very low number "radial" lines	Coddington 1986a
Baalzebub		Complete hub, no spring line	Coddington 1986a
Plato[2]		Radii anastomize, lack spring line	Coddington 1986a
Synaphridae			
Synaphris lehtineni		Small sparse sheet web	Marusik et al. 2005
Anapidae			
Acrobleps, Anapis, Anapisona, Crassanapis, Chasmocephalon, Sheranapis	Low number of radii and sticky spiral lines above a more or less horizontal orb	"Cobweb" like theridiid	Lopardo et al. 2011
Comaroma simonii			Kropf 1990
Conculus lyugadinus		"Aquatic orb" attached in places to stream surface	Shinkai and Shinkai 1988
Sofanapis antillanca		Kleptoparasite, with irreg. tangle superimposed on host's sheet	Ramírez et al. 2004
Elanapis		Horizontal orb lacking radii above plane	Ramírez et al. 2004
Tasmanapis strahan		Several contiguous orbs combined, lack lines above	Lopardo et al. 2008
Micropholcommatines		Irregular "sheets"	Hickman 1944, 1945, Forster 1959
Undescribed genus from Madgascar		Horizontal sheet	G. Hormiga unpub. in Lopardo et al. 2008
Symphytognathidae			
Patu spp., *Anapistula* spp.	Horizontal orb lacking lines above	Accessory radii that are later broken near the hub	Coddington 1986b, Lopardo et al. 2008
Symphytognatha (= *Patu*) sp.			Griswold and Yan 2003
Anapistula sp.		Sheet web	Cardoso and Scharff 2009
Symphytognatha globosa		Few irregular lines in more or less horizontal plane	Hickman 1931 in Forster and Platnick 1977

Table 10.4. Continued

Taxa	*Basal traits*	*Derived traits*	*References*
Curimagua bayano		Kleptoparasite lacking any web	Vollrath 1978, Lopardo et al. 2008
Mysmenidae			
Maymena	Horizontal orb with radii above that pull hub upward (similar to anapids)	Kleptoparasites lacking any web	Lopardo et al. 2008
Isela spp., *Mysmenopsis*			Lopardo et al. 2008
Trogloneta granulum		Tiny 3-D irregular web, variable shapes with sticky radial and transverse lines	Hajer 2000, Schütt 2003
Mysmena, Microdipoena, Simaoa, Gaolingonga		3-D "spherical orb"	Coddington 1986b, Eberhard 1987e, Lopardo et al. 2008

[1] Spring lines only present in some juvenile webs.

[2] May sometimes tense and spring a radius (WE, Coddington 1986a).

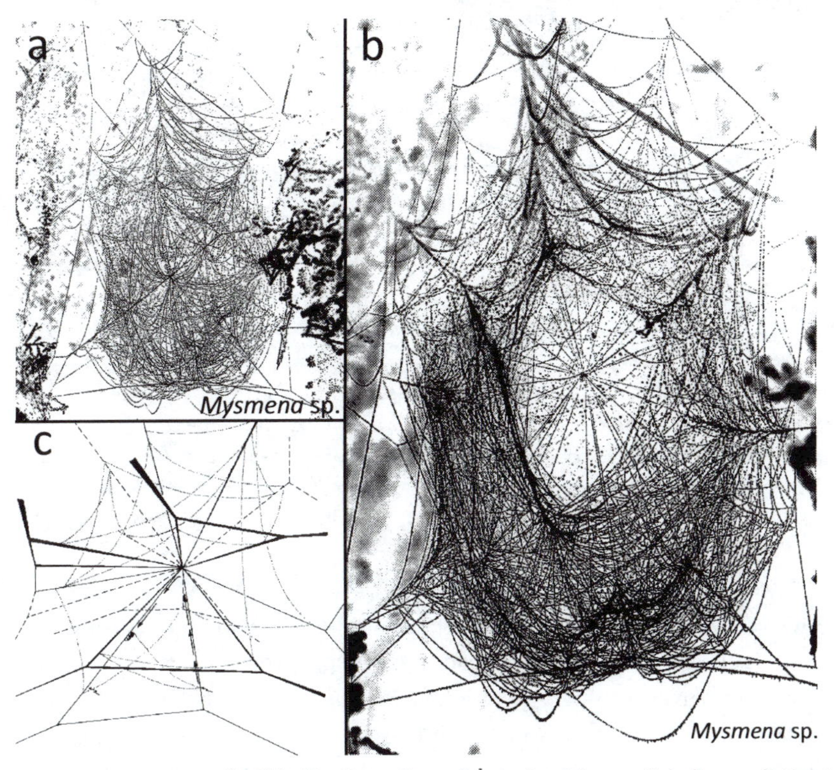

Fig. 10.21. *Mysmena* sp. (MYS) built a three-dimensional orb, with non-sticky lines radiating in all directions from a central hub area that was vaguely distinguishable in the intact web (*a*) and more obvious when a portion was removed (*b*). During sticky spiral construction (*c*) the spider walked inward to the hub and then back out another radius (arrows along the radii) (the dotted lines are sticky; the thicker lines are closer to the viewer, and the dashed lines farther away). The sticky lines hung loose, in catenaries. Consecutive attachments to radii were not on opposite sides of the web (this would have produced sticky lines running near the hub), but no other pattern was discerned. The spaces between the sticky lines attached to a given radius were often larger than the length of the spider's body, so attachment points were presumably determined on the basis of the distances the spider walked inward to the hub and back outward, as in theridiosomatids and some other symphytognathoids (fig. 7.7, appendix O6.3.6.5), and not by contact with the inner loop of sticky spiral as in other araneoids and uloborids (*b* courtesy of J. Coddington, *c* from Eberhard 1987e).

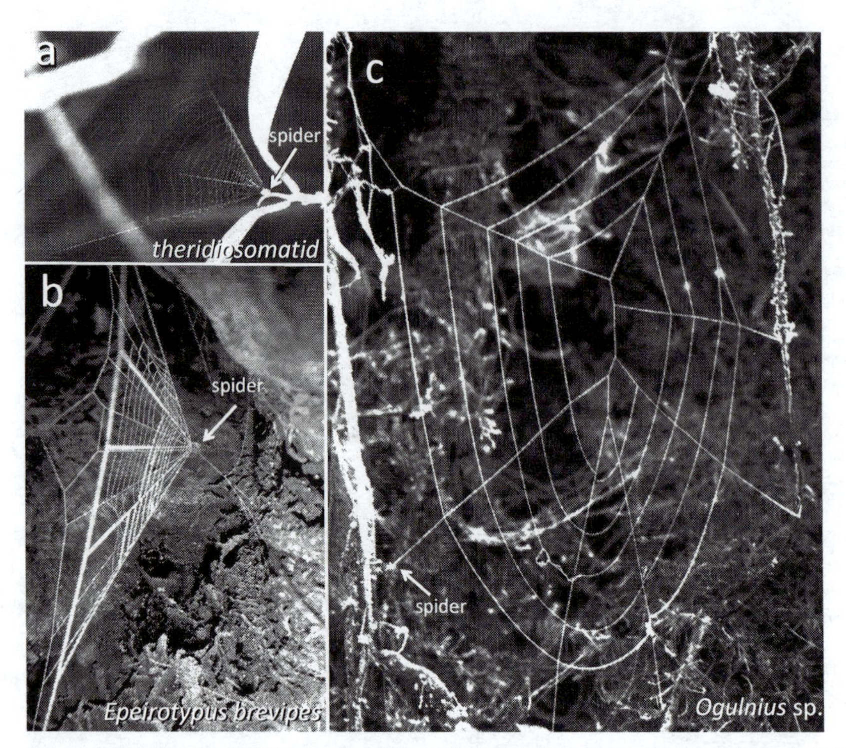

Fig. 10.22. These webs depict several puzzling variations in three typical traits of the webs in the family Theridiosomatidae: a "spring line" (*a*, *b*) extending out of the orb's plane that the spider reels in to hold the web tense, and then releases suddenly to make web spring back into a plane and snare prey when she senses a humming noise nearby (e.g., when a vibrating tuning fork was brought near); low numbers of radii and sticky spiral loops; and lax, sagging sticky lines. In both an unidentified species (#967) (*a*) (unpowdered) and in *Epeirotypus brevipes* (*b*) the spider rested at the hub and pulled the orb into a cone by reeling up a spring line. The movement in the web in *a* was substantial, probably more than 10 times the length of the spider's body: the frame and anchor lines were long and thus especially extensible; the spring line was perpendicular to the orb's plane; and the spider pulled it tight, forming a sharp angle. The webs of *E. brevipes*, in contrast, were less well designed to spring such large distances: the frames and anchors were relatively short, the spring line was often far from perpendicular to the orb plane, and the spider only pulled the orb into a more obtuse angle. The web of *Ogulnius* sp. (#802) (*c*), accentuated these puzzlingly ineffective traits. The spider produced almost no movement of the web as a whole (which lacked frame lines entirely) when she tensed a radius slightly and then released the tension abruptly in response to humming sounds (WE). The low numbers of radii and sticky spiral loops and the low tensions on sticky lines were also more extreme in the *O.* sp. web. The evolution of such sparse webs from more typical orbs runs counter to predictions of the "rare large prey hypothesis" (section 4.2.2), which is based on the assumption that the advantage of stopping prey that have greater momentum will consistently override other functions. The low tensions on sticky lines may be related to a tendency (which has never, however, been specifically tested) for theridiosomatids to build in sheltered sites.

menidae) (table 10.4). These have lost their own webs entirely or (in the case of *Sofanapis antillanca*) make an irregular tangle over the host's sheet web (Lopardo et al. 2011). The hosts were all relatively large spiders with sheet webs (fig. 10.7); perhaps the small size of the symphytognathids predisposed them to go unperceived by their hosts.

Why did this spectacular burst of divergent webs occur in these small spiders? Craig (1987a, 2003) argued that an absence of selective advantages for different architectures in the low-energy orbs of such spiders allowed this diversity to evolve. Selective neutrality of web design seems highly unlikely, however, taking into account the likely effects of designs on prey capture (section 3.3), the spider's dependence on her web for prey capture, and the substantial differences in designs. Craig's argument was derived from an analysis of only five species (Craig 1987a), in which the effects of different web designs may have been confounded with the effects of different body sizes (section 3.2.5.1). The nearly complete ignorance of the prey captured by these tiny spiders (are their impacts truly "low energy"?), and the addition of sometimes very abundant secondary radial lines in some orbs whose likely effect is to improve energy-absorbing properties

Table 10.5. A sample of species in which mature males build webs to capture prey. The list is undoubtedly incomplete, but serves to make the point that construction of prey capture webs by mature males has evolved repeatedly in distantly related taxa; this behavior appears to be especially common in symphytognathoids.

Taxonomic group	*Type of web*	*Reference*
Dipluridae		
Linothele sp.	Funnel web[1]	Paz 1988
Ischnothele sp.	Sheet on substrate[2]	WE
Pholcidae		
Modisimus spp.	Domed aerial sheet in tangle	Eberhard and Briceño 1985
Physocyclus globosus	Domed aerial sheet in tangle	Eberhard 1992a, I. Escalante in prep.
Holocnemus pluchei	Domed aerial sheet in tangle	Jakob 1991, 1994, WE
Pholcus phalangioides	Tangle with sheet	Kirchner 1986
Ochyroceratidae		
Ochyrocera cachote[3]	Sheet	Hormiga et al. 2007
Lycosidae		
Aglaoctenus castaneus	Funnel web	Eberhard and Hazzi 2017
Pisauridae		
Architis nitidopilosa	Double funnel web	Nentwig 1985c
Linyphiidae		
Frontinella pyramitela	Sheet with tangle	Suter 1984, Hirscheimer and Suter 1985
Uloboridae[4]		
Uloborus spp.	Orb with sheet of thin lines instead of sticky spiral	Eberhard 1977d, Lubin 1986
Philoponella spp.	(Same)	Lubin 1986, WE
Araneidae		
Cyrtophora citricola	Non-sticky "orb-sheet"[1]	Blanke 1972, WE
Mecynogea sp.	(Same)[1]	J. Vasconcellos-Neto pers. comm., WE
Manogea porracea	(Same)	Sobczak et al. 2009, J. Vasconcellos-Neto pers. comm.
Metepeira gressa	Orb[5]	Viera 1988
Theridiosomatidae		
Wendilgarda sp.	Sticky lines attached to water	Eberhard 2001a
Symphytognathoidea		
Acrobleps hygrophilus	Orb[1]	Lopardo and Hormiga 2008
Anapisona simoni	Orb[1]	Eberhard 2007b
A. kethleyi	Orb[1]	Lopardo et al. 2011
Crassanapis chilensis	Orb	Ramírez et al. 2004
Elanapis aisen	Orb[1]	Ramírez et al. 2004
Gertschanapis schantzi	Orb[6]	Ramírez et al. 2004
Sheranapis villarrica	Orb[6]	Ramírez et al. 2004
Tasmanapis strahan	Quilt orb[1]	Lopardo et al. 2011

Table 10.5. Continued

Taxonomic group	*Type of web*	*Reference*
Mysmenidae		
Mysmena spp.	3-D orb[1]	Lopardo et al. 2011, WE
Maymena spp.	Orb with tent above[1]	Lopardo et al. 2011

[1] Webs of mature males smaller than those of mature females, but not distinguishable in design.
[2] After 3–4 days in captivity in a container with moist soil on the bottom, a mature male had built a small, irregular sheet about 7–8 cm in diameter on the surface of the soil around a central, sheltered resting place. There were few or no lines at the bases of the walls of the container, suggesting that this silk was not laid incidentally while the spider attempted to escape.
[3] Web production deduced from presence of specialized spinneret structures.
[4] Mature males of *Zosis geniculata* that were kept for several months in captivity made only resting webs (Fig. 3.48), but a male of another species of *Zosis* made a prey capture web in the field (fig. 9.21).
[5] Strongly altered web with poorly defined hub, long loose loops of sticky line, and barely recognizable radii.
[6] Some mature male orbs were like those of females, but others (contiguous with a female orb) were much more sparse.

(e.g., Hiramatsu and Shinkai 1993 on *Patu*), represent additional weaknesses in the neutral argument.

10.9.5 THE REDUCED WEBS OF *HYPTIOTES* AND *MIAGRAMMOPES*

Hyptiotes and *Migrammopes* are relatively derived sister genera within Uloboridae (fig. 1.13) that independently evolved reduced webs: one or a few long sticky lines in *Miagrammopes* (Akerman 1932; Lubin et al. 1978; Lubin 1986); and a triangular sector of an orb with four radii in *Hyptiotes* (fig. 10.23) (McCook 1889; Emerton 1902; Wiehle 1927; Marples and Marples 1937; Opell 1982). Associated with these web reductions are decreases in the total length of sticky lines in their webs, and compensatory increases in the stickiness/mm of sticky line, probably due to their greater numbers of spigots on the cribellum (Opell 1996, 2013). The cribella of *Miagrammopes* species have about twice as many spigots as the cribella of similar sized orb-weaving uloborids (Opell 1999a). Consequently, the total stickiness in each web is, relative to spider weight, very similar. Spiders in both genera tense the web by reeling in a line, and release this reeled in silk abruptly when a prey strikes the web, causing the web to go slack and probably improving both stopping and retention (Wiehle 1927; Marples and Marples 1937; Lubin et al. 1978). Spiders often subsequently lengthened (and thus loosened) this line further as they approached to attack. In *Hyptiotes* this probably further improved retention, as it collapsed the web further around the prey (Wiehle 1927; Marples and Marples 1937).

These tension manipulations are convergent with tensing and loosening behavior in some orb-weaving araneids and theridiosomatids (table 9.1). Tension manipulations in uloborids apparently arose independently, as none of the other, orb-weaving uloborids are known to manipulate web tensions. Inspired by *Hyptiotes* and *Miagrammopes* (as well as the bolas spiders), Kaston (1972) argued that orb web reduction and modification is consistently associated with more active attack behavior by the spider herself. Many exceptions are now known, however, so this idea is no longer defensible (section 10.10).

10.9.6 "TWIG ORBS": AN OBJECT PROJECTS THROUGH THE HUB

A spider waiting at the hub of her orb is exposed to many types of visually orienting predators, including birds, lizards, frogs, damselflies, sphecid and ichneumonid wasps, asilid flies, and salticid spiders. One partial remedy that has been discovered independently in several lineages is to include objects (detritus and egg sac stabilimenta) near the hub, with which the spider can camouflage herself (section 3.3.4.1). A second solution, adopted by some theridiids (Eberhard et al. 2008b) and several sheet weavers (fig. 9.16) as well as some pseudo-orb weavers (fig. 9.18), has been to place a curled leaf or conical retreat near the hub, within which the spider is hidden while she waits for prey. Still another probable defense against visual predators that has evolved independently several times is to build the orb with an object (often a stick) at the site of the hub ("twig" webs). Species with twig orb webs include *Anapisona simoni* (AN) (fig.

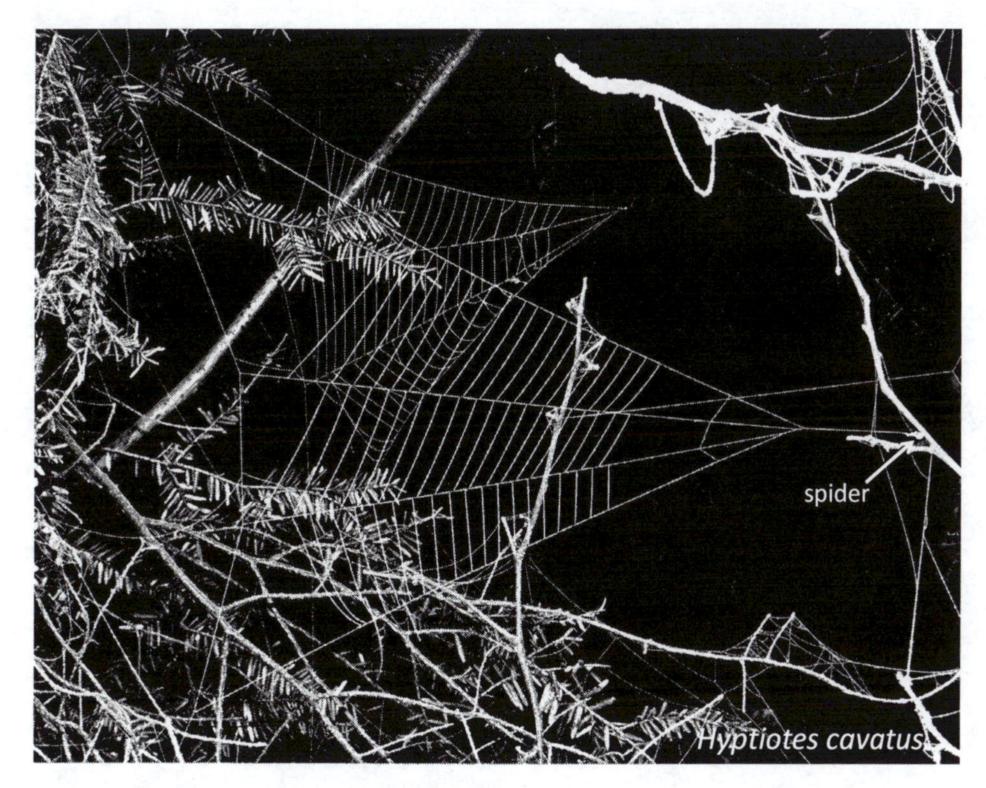

Fig. 10.23. The reduced, "triangle" webs of *Hyptiotes cavatus* (UL) (a mature female and her web are in the foreground, the web of an immature male in the background) illustrate the apparent (though never carefully documented) preference of this species for rigid supports such as the dead conifer twigs in this photo (from Opell 1982, courtesy of Brent Opell).

10.25b, Eberhard 2011), *Uloborus eberhardi* and *U.* sp. #1324 (UL), perhaps six species of *Tetragnatha* (TET) from Western Samoa (Marples 1955), southern India (#2043–2053), northeastern Australia (#3674), western Mexico (#3514), mountains near Escazu, Costa Rica, and the western United States (likely *T. elongata*) (figs. 1.6d, 10.24), and one with a modified orb, *Polenecia producta* (UL) (fig. 3.1) (Wiehle 1931; Peters 1995a). Presumably twig webs evolved from ancestors in which the hub was very near one edge of the web (e.g., *Tetragnatha* sp. #1474 in fig. 10.26) (similar orbs, with the hub very near a linear support, also occurred occasionally in *Metazygia gregalis* [AR] [#1369]).

Twig webs are interesting in several respects. In two species, *Tetragnatha* sp. (fig. 10.26) and *A. simoni* (fig. 10.25), twig webs and complete orbs were alternative designs produced by individuals of the same species (single individuals of *A. simoni* built both designs) (Eberhard 2011). The twig must sharply alter resonant frequencies and the patterns of vibrations in an orb, so these spiders' ability to build both web types are in accord with other evidence in arguing that these possible cues are not used to guide construction behavior (e.g., sections 7.3.2.6, 7.4.4). A second pattern seen in several twig orb species, including *Tetragnatha* sp. #2043–2053 from India (fig. 10.24), *T.* sp. (#3514) from western Mexico, and *P. producta* (Peters 1995a), was that there was substantial intra-specific variation with respect to whether the spiders built complete orbs or only sectors of orbs (fig. 10.24). As far as I know, no free hub species in *Tetragnatha* builds only a sector of an orb (though *T.* sp. #1474 built a highly asymmetric, seemingly intermediate web [fig. 10.24]). Peters (1995a) argued that the variation in *P. producta* resulted from limitations in appropriate websites, and thus the need to adjust to a wide variety of supports. Perhaps related to this, the twig orbs of *Tetragnatha* spp. from Colorado and Mexico (#3514) were often not in a single plane, but sloped downward away from the twig.

Finally, twig webs raise the question of the speed of evolutionary loss of behavior. In the species of *Tetragnatha* in India, twig webs seem to have become fixed. All of the 20+ orbs seen in the field were twig webs. In a few webs (fig. 10.24) the hub of the orb was built beside but not precisely under the twig; the spider did not remove the center of the hub, despite the opportunity to do so (see also fig. 1.6d). This contrasts with the widespread open hole in the center of the hub in other *Tetragnatha*

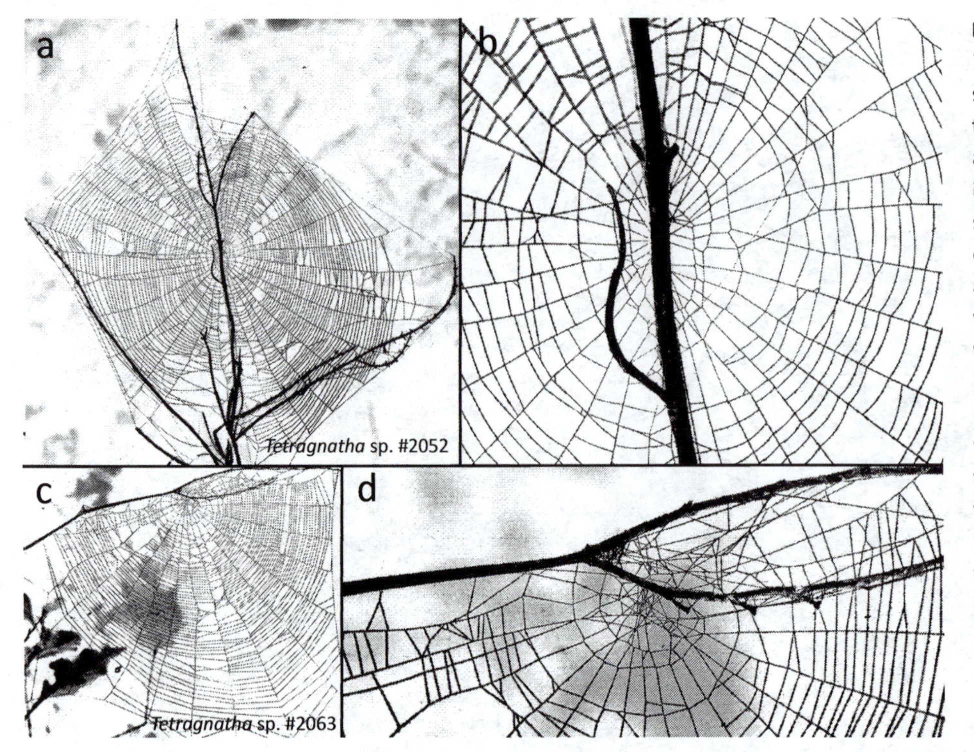

Fig. 10.24. These webs of a *Tetragnatha* (TET) species from southern India in which only twig orbs were found, illustrate a common pattern also seen in other twig web species: sometimes spiders built entire orbs (*a*, *b*), and other times only a section of an orb (*c*, *d*) (similar variation occurred in the twig webs of *Polenecia producta* [UL]—Peters 1995a). Usually the radii in the center of the hub were attached directly to the twig (*b*) (see also fig. 10.25c,d of twig webs of *Uloborus eberhardi* [UL]), but in a few most or all of the hub was built slightly to the side of the twig (*d*). The lines in the center of the hub were left intact, rather than being removed as is typical in most *Tetragnatha* species (figs. 3.18, 10.26).

webs (e.g., figs. 4.6, 10.26, Kaston 1948; Comstock 1967; Blackledge and Gillespie 2004), and raises the question of whether this twig orb species has "forgotten" how to remove the center of the hub over evolutionary time.

10.10 "POST-ORB" WEB EVOLUTION IN ORBICULARIAE

The likely resolution of the old monophyly controversy (section 10.8) creates new problems of interpretation. No longer does the orb web, with its geometric regularity and supposedly optimal spacing of sticky lines, represent the pinnacle of web evolution. Instead it is a very old, highly successful intermediate form, which over a long evolutionary timespan has repeatedly given way to a variety of other, less geometrically "perfect" alternative designs (Vollrath and Selden 2007; Blackledge et al. 2009c, 2011; Herberstein and Tso 2011). Some "post-orb" families of Orbiculariae, in particular Theridiidae (next section) and Linyphiidae, are very large and abundant and make a wide variety of web forms. The transitions to these webs from orbs have been little studied.

As early as about 150 Mya (using molecular dating), some orbs had already been transformed into significantly different architectures such as the sheet webs of linyphioids and the tangle webs of theridiids. The ancient diversification of araneoid families occurred in a relatively short period of time (Dimitrov et al. 2016). The orb-weaving families themselves show several independent transitions from orbs to other types of webs (see table 9.1, figs. 1.13, 1.14). As usual, there is no shortage of convergent and intermediate forms. To take a single example of convergence, the sticky lines in the webs of both the uloborid *Polenecia producta* (Wiehle 1931; Peters 1995a) and the later ontogenetic stages of the araneid *Eustala* sp. (Eberhard 1985) have changed from a spiral to a radial pattern. A similar design of approximately radial sticky lines also evolved convergently in some theridiids, such as some *Chrysso* and *Theridion evexum* (Barrantes 2007; Barrantes and Weng 2007b; Eberhard et al. 2008b). Sheet webs, 3-D tangle webs, and kleptoparasitism with concomitant web loss have also evolved several times, along with extravagant new web forms, such as the "spherical orbs" of at least four related genera of Mysmenidae, and the "quilt" orbs of *Tasmanapis strahan* (AN) (fig. 10.27). Perhaps the strangest are the completely closed ovoid sheets of the Tasmanian araneid *Paraplectanoides crassipes* into which detritus is incorporated; the spider rested inside, on an internal, central "hub" with a few short radii, and the web entirely lacked sticky lines (fig. 10.28,

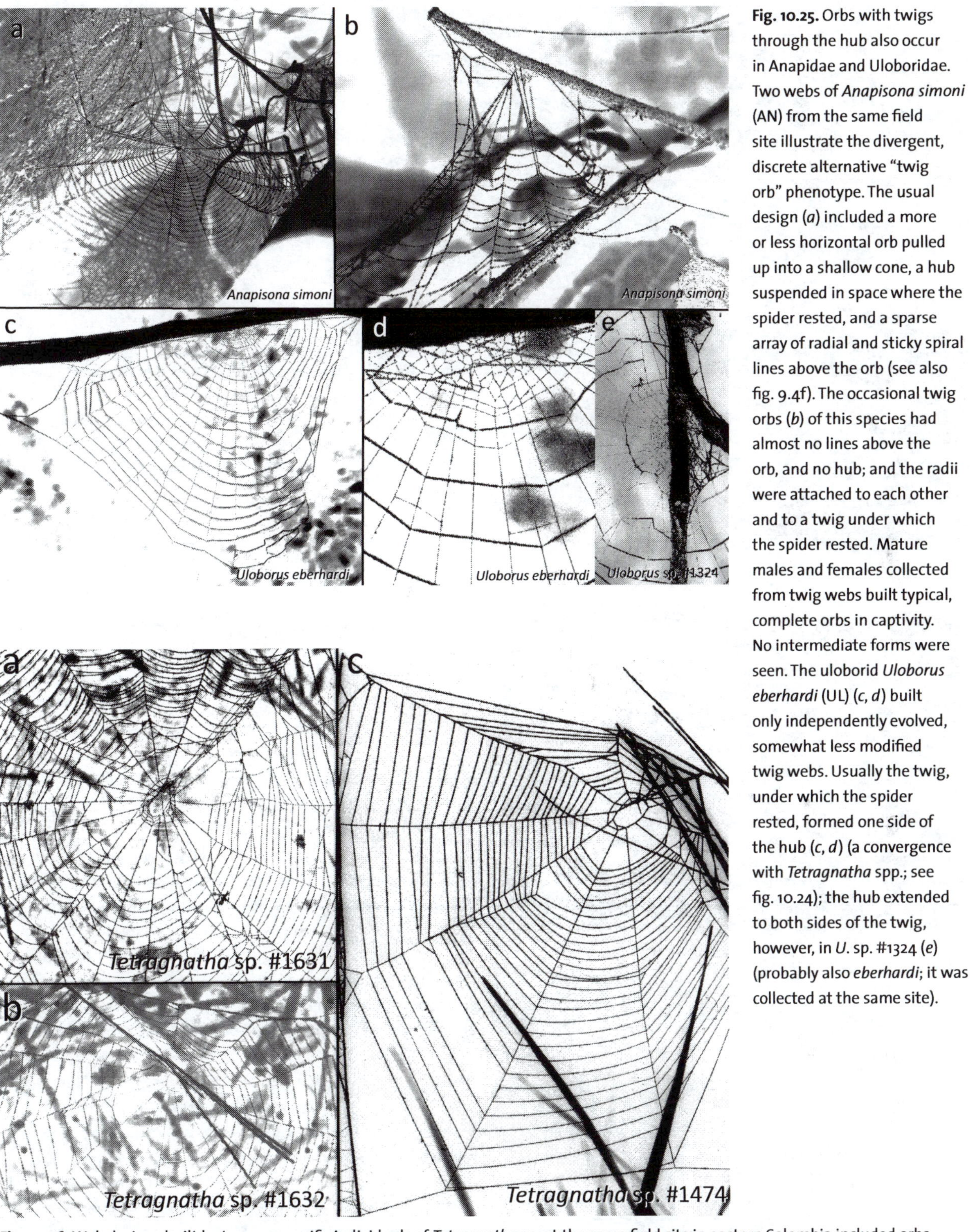

Fig. 10.25. Orbs with twigs through the hub also occur in Anapidae and Uloboridae. Two webs of *Anapisona simoni* (AN) from the same field site illustrate the divergent, discrete alternative "twig orb" phenotype. The usual design (*a*) included a more or less horizontal orb pulled up into a shallow cone, a hub suspended in space where the spider rested, and a sparse array of radial and sticky spiral lines above the orb (see also fig. 9.4f). The occasional twig orbs (*b*) of this species had almost no lines above the orb, and no hub; and the radii were attached to each other and to a twig under which the spider rested. Mature males and females collected from twig webs built typical, complete orbs in captivity. No intermediate forms were seen. The uloborid *Uloborus eberhardi* (UL) (*c*, *d*) built only independently evolved, somewhat less modified twig webs. Usually the twig, under which the spider rested, formed one side of the hub (*c*, *d*) (a convergence with *Tetragnatha* spp.; see fig. 10.24); the hub extended to both sides of the twig, however, in *U.* sp. #1324 (*e*) (probably also *eberhardi*; it was collected at the same site).

Fig. 10.26. Web designs built by two conspecific individuals of *Tetragnatha* sp. at the same field site in eastern Colombia included orbs with "free hubs" which had no direct attachments to any objects (as is typical in this genus) (*a*), and "twig orbs" (*b*) in which a grass stem crossed the web in the plane of the orb, the hub had fewer loops, and the spider hid from sight under the stem. These designs represented extremes in a continuum, and other orbs had intermediate, highly asymmetrical designs (*c*) in which a grass stem ran along the orb's edge, and the hub was complete but very close to the stem.

Fig. 10.27. This unique "quilt" web of *Tasmanapis strahan* epitomizes the modularity of orb construction behavior, at the level of the orb itself. The spider was at the hub of the central orb, around which three additional orbs were built. The planes of the small orbs at the left and right were apparently somewhat different from that of the central orb, and they partially overlapped it. The middle peripheral orb shared at least some frame lines with the other orbs. Presumably each orb was built separately, but construction behavior of these spiders has never been observed (photo from Lopardo et al. 2011, courtesy of Lara Lopardo).

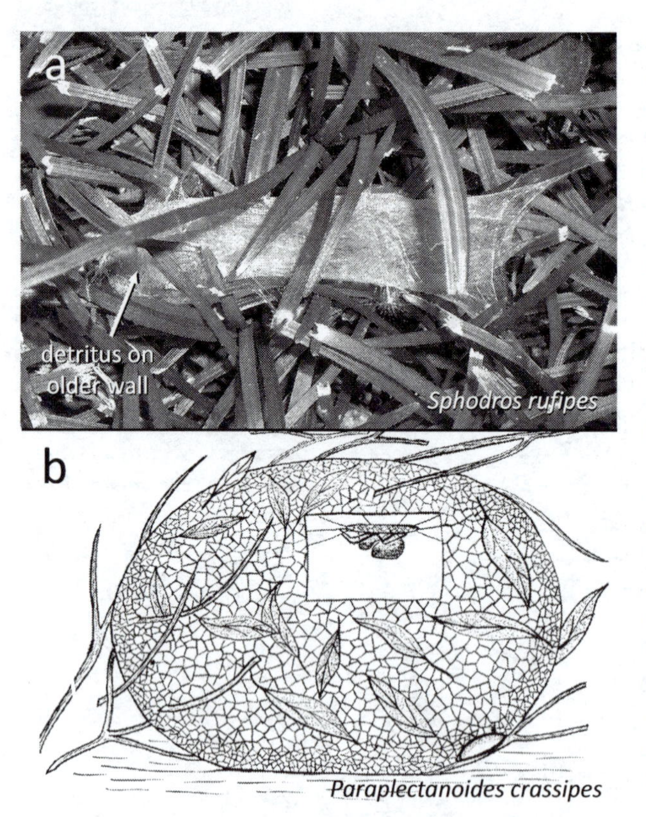

Fig. 10.28. These two species, at opposite extremes in the phylogenetic tree of spiders, have apparently converged on the unlikely web design of enclosing the spider in a sealed chamber to capture prey. As in other atypids, the web of the mygalomorph purse web spider *Sphodros rufipes* (Atypidae) (seen from above in *a*) was a closed cylinder; the lower end was in an underground tunnel at the left, while the upper end was a more or less horizontal tube built among the bases of grass plants in an open yard (other atypids make vertical tubes attached to tree trunks). Finished *S. rufipes* webs were covered with detritus, but the web in the photo had recently been mostly rebuilt after it was damaged when the lawn was mowed. The spider remained inside the tube, and stabbed prey that walked on its outer surface with her long cheliceral fangs. The web of the Tasmanian araneoid *Paraplectanoides crassipes* (*b*), which a recent molecular analysis placed as a sister group for classical Nephilidae (or Nephilinae) (N. Scharff pers. comm.), was also closed. It formed an ovoid whose walls were covered with detritus that was thick enough to seriously hinder observation of the spider inside, which rested suspended under a small horizontal hub. There were no sticky lines. The webs of mature females in the field were littered with the remains of cockroaches, and observations of spiders in captivity showed that they remained inside the ovoids and attacked prey that walked on outer surface of their walls (Hickman 1975). This species showed signs of low rates of prey capture, taking three years to mature and living perhaps up to nine years or more (*b* from Hickman 1975).

Hickman 1975). The sheet may function like the purse webs of atypid mygalomorphs (Coyle 1986), as a surface on which prey (cockroaches) walk and are attacked (Hickman 1975). In most of these species the intermediate steps in the transformations are unknown. I discuss below two cases in which at least reasonable hypotheses can be proposed.

Current lack of divergence dates generally precludes assessing rates of evolution of construction behavior. Two special cases in which maximum ages could be established because the species are endemic to oceanic islands of known ages suggest relatively rapid changes. *Wendilgarda galapagensis* (TSM) is endemic to Isla del Coco (Eberhard 1989a), which has existed for about 2 million years; and *Tetragnatha* spp. (TET) are endemic on several Hawaiian islands that emerged between 3.7 and 0.7 million years ago (Blackledge and Gillespie 2004). The most striking changes occurred in *W. galapagensis*, which profoundly reorganized sticky line production. Classically, *Wendilgarda* sticky lines are laid in strict order from outward to inward on each radial line (fig. 8.1e). During construction of one web type, *W. galapagoensis* visited the central area after laying each sticky line, and she often changed radii (fig. 8.1f); in building a second type of web some individuals paused in the center and initiated new radial lines, broke and shifted the attachments of previous sticky lines to radial non-sticky lines, descended up to 20–30 cm and climbed back up, or broke and shifted attachments of non-sticky lines to the substrate before laying the next sticky line (Eberhard 1989a). Some Hawaiian *Tetragnatha* species have discarded webs altogether; the changes to the orb designs of others were less profound, involving overall web sizes and relative densities of radii and sticky spiral loops (Blackledge and Gillespie 2004).

10.10.1 POSSIBLE DERIVATION OF OTHER WEB TYPES FROM ORBS

10.10.1.1 Gumfoot webs

Orb webs trap prey in the air, but they cannot capture the many kinds of substrate-bound prey. Several orbicularian groups have independently replaced orbs with webs designed to capture substrate-bound prey. The largest of these groups is the non-orb building family Theridiidae, in which gumfoot webs (figs. 5.9, 5.19, 9.7, 9.22, Wiehle 1927) are apparently ancestral (Eberhard et al. 2008b). The ancestors of Theridiidae and Nesticidae (which also builds gumfoot webs) probably evolved from an ancestor that built orbs (fig. 1.10). It is not clear whether gumfoot webs evolved convergently in these two families. Theridiids are the sister group for all other araneoids and are widely separated from Nesticidae in some recent molecular analyses (e.g., Garrison et al. 2016), but they are, however, linked morphologically (Coddington 1986b, 1989, 1990). Given the ubiquity of ants, the effectiveness of gumfoot webs in capturing ants, and the many theridiids that specialize on ants (Bristowe 1939; Cushing 2012), it is tempting to speculate that gumfoot webs evolved in order to gain access to ants. A further factor is the effectiveness of theridiid wrapping attacks that utilize sticky silk in subduing even powerful, dangerous prey such as ants.

The same sequence from orbs to webs with sticky material at the ends of lines to the substrate also occurred independently in two genera of orb-weaving families, *Wendilgarda* (TSM) (Coddington and Valerio 1980; Coddington 1986a; Eberhard 1989a), and *Comaroma simoni* (AN) (Kropf 2004) (fig. 10.29j). An additional convergence is *Ocrepeira ectypa* (AR), which also captures walking prey, using branched, non-sticky lines to reach the substrate and to trigger rapid attacks (Stowe 1978). Two other possible but less studied convergences on radially-arranged sticky lines that were derived from an orb occur in the mysmenid *Trogloneta granulum* (Hajer 2000; Hajer and Řeháková 2003) and an undetermined cave mysmenid (fig. 9.2b). The homologies of the sticky lines in *Wendilgarda* and *C. simonii* webs with the lines in orbs seem relatively clear, thus offering possible insights into the process by which gumfoot webs evolved. The construction behavior of these two groups, when combined with recent discoveries of ontogenetic changes in the web design of basal theridiids in the genus *Latrodectus* (black widows), suggest possible details of the sequence of events that occurred as theridiid gumfoot webs evolved from orbs.

The early juvenile webs of *Latrodectus* (THD) differed from those of adults in having several features that resembled those of orbs (fig. 10.30), and there was a hub-like platform in the star web of a penultimate *Parasteatoda* or *Achaearanea* (THD) (fig. 10.31). If these traits are ancestral (section 10.2.2), they suggest several possible homologies with orb webs (table 10.6) (for a detailed dis-

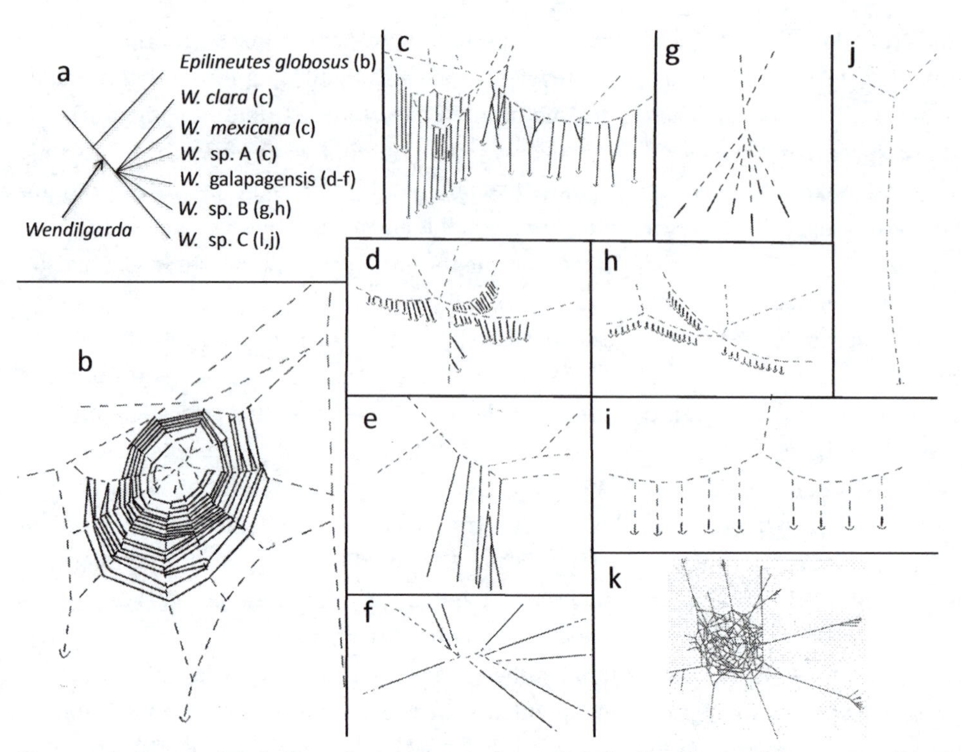

Fig. 10.29. Independent derivations of substrate webs from orbs in the theridiosomatid genus *Wendilgarda* (*a–i*) and the anapid *Comaroma simoni* (*k*) have converged on designs similar to those of the better known gumfoot webs of theridiids. The close relative of *Wendilgarda*, *Epilineutes globosus* (*b*) built orbs similar to those of other theridiosomatids but that were sometimes attached to the surface of water (attachments to water are indicated with small half-circles; thicker lines represent sticky lines; dashed lines represent non-sticky lines). The typical, derived design seen in several species of *Wendilgarda* is to have multiple vertical lines that are sticky at their lower ends attached to the surface of water below one or a few horizontal non-sticky lines (*c*, *d*, *h*); in one species the number of vertical lines was sometimes reduced to only one (*j*). Two species facultatively built, in addition to typical water webs, webs attached to solid objects, and these had additional forms (*e*, *f*, *g*) that in some cases resembled gumfoot webs. The independently derived gumfoot web of *C. simoni* differed in having multiple small branched lines at the tips of radial lines that were sticky at their tips (*k*) (C. Kropf 2004, and pers. comm.) (theridiosomatid webs after Shinkai and Shinkai 1997, and Eberhard 2000a; anapid web after Kropf 1990; from Eberhard 2000a).

cussion of the homology claims, and possible details of the transition, see Eberhard et al. 2008a). The supposed homologies with orb lines on which table 10.6 is based are relatively certain for *Wendilgarda* spp. and *C. simonii*, but less so for the theridiids.

There are extensive similarities in derived traits of juvenile *Latrodectus* webs and those of *Wendilgarda* webs and also (to a lesser extent, perhaps because of less complete data) those of *C. simonii* webs. The multiple similarities in the changes in *W.* spp. and *C. simonii* as they evolved from orb weaving ancestors to capture substrate-bound prey favor the hypotheses that the orb weaving ancestor of theridiids converted its sticky spiral lines into gumfoot lines, its radii into the lines at the edge of the tangle that sustain the gumfoot lines, and its hub into the central disc seen in the webs of *Latrodectus* spiderlings. Tellingly, 7 of the 12 traits that are shared only between young *L. geometricus* and *Wendilgarda* but not with the orb weavers ("*" and in bold in table 10.6) could be explained by probable selection resulting from a single convergence—entry into the new adaptive zone of capturing substrate-bound prey. The importance of observing the webs of juvenile in addition to those of adult *L. geometricus* is emphasized by the fact that many of the 11 points of similarity with orb weavers (especially those involving radial lines) disappear in adult webs (table 10.6).

Additional changes occurred in the sticky silk itself. Comparing the aggregate gland products of the gumfoot lines of *Latrodectus hesperus* (THD) and *Larinioides cornutus* (AR) (Sahni et al. 2011), the theridiid glue behaved like a viscoelastic liquid rather than like a visco-

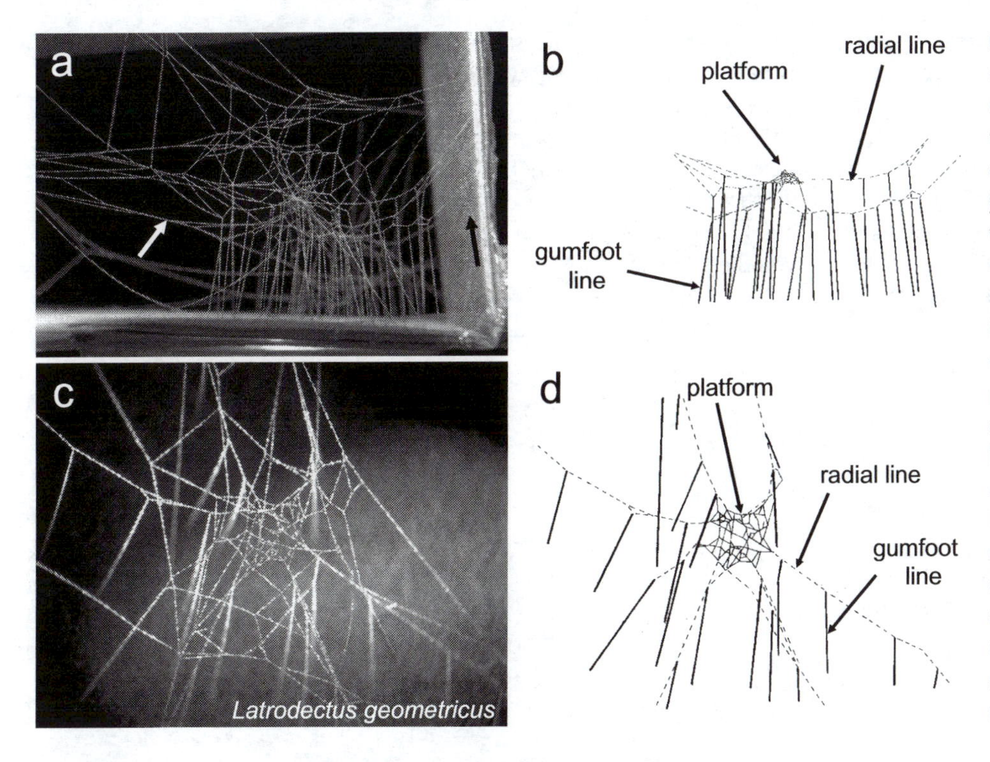

Fig. 10.30. These photographs (*a*, *c*) and interpretive drawings (*b*, *d*) (in which many lines in the tangle were omitted) of a web of a second instar *Latrodectus geometricus* (the first instar outside the egg sac) illustrate a possible (though uncertain) homology with the hubs of orb webs. Seen in a lateral view of the entire web (*a*, *b*), the spider rested at a central platform or "hub," where several approximately horizontal radial lines (dotted lines in the drawings *c* and *d*) converged. The radial lines had adhesive gumfoot lines (thicker solid lines in the drawings) attached to them. The webs of adult female *L. geometricus* had a more extensive, loosely-meshed sheet rather than a central platform (after Eberhard et al. 2008a).

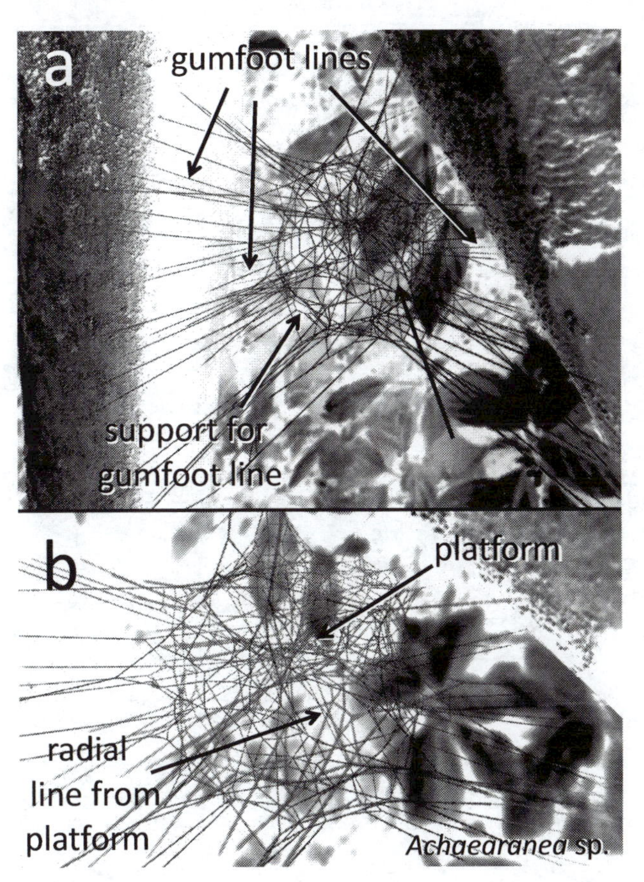

Fig. 10.31. The unusually sparse nature of this "star" web built by an immature *Parasteatoda* or *Achaearanea* (THD) (the genus name is controversial) revealed traits that hark back to the presumed orb design of the distant ancestors of theridiids. As is common in star webs (see fig. 9.7), numerous gumfoot lines radiated from a more or less spherical central area (*a*). At greater magnification (*b*), a small, round, domed platform under which the spider rested was visible in the central portion of the sphere. Numerous more or less radially oriented lines (one is indicated in *b*) connected the edges of this platform with the sparse wall of the sphere. Some lines in the spherical portion supported multiple gumfoot lines (*a*). The hub-like form of the central platform and the radial lines attached to it might be homologous with hub and radial lines in orbs, or (more likely?) may be independently derived (spider identification courtesy of Ingi Agnarsson).

Table 10.6. Similarities and differences between typical araneoid orb weavers, young *Latrodectus geometricus* THD, *Wendilgarda* spp. TSM, and *Comaroma simonii* AN, with respect to their sticky and radial lines and their attachments. The suppositions on which this table is based are that the gumfoot lines are homologous with the sticky spiral of an orb; that the more or less horizontal and radial lines to which the gumfoot lines are attached are homologous with the radii of an orb; and that the central disc where *Latrodectus* spiderlings rest is homologous with the hub of an orb. The resemblances between young *L. geometricus* and *Wendilgarda* or *C. simonii* that are not shared with orb weavers are probably convergences. Traits are classified as "line/behavior" according to whether they involve properties of the lines themselves (L) or of the spider's behavior (B). Resemblances that are not shared with the orb weavers and that may be convergences are marked in bold. Data on *L. geometricus* are from Eberhard et al. (2008a,b); those on *W.* spp. are from Coddington and Valerio (1980) and Eberhard (1989a, 2000a, and 2001a); those on *C. simoni* are from Kropf (1990, 2004, and pers. comm.).

Trait	*Typical orbs*	*Latrodectus geometricus*	*Wendilgarda sp.*	*Comaroma simoni*	*Line or behavior*
Radial lines					
1. Are only lines supporting sticky lines	Yes	Yes	Yes	Yes	L
2. Converge on single point or area	Yes	Yes	Approximate	Yes	L
3. There are "hub" lines at the intersection	Yes	Yes	No	Yes	L
4. Interior of hub with radial organization	Yes	Yes/**No**	—	**No**	L
5. End on frame lines	Yes	**No** (yes in *L. pallidus*)	**No**	**No**	L
6. Strict order (other lines first, then radii)	Yes	**No**	**No**	?	B
7. Number of radii	High	**Low**	**Low**	**Low**	L
8. Cut and reel before sticky line(s)	Yes/No	No?	Yes	?	B
9. Construction of radii and hub/disc lines alternates	Yes	Yes (?)	—	?	B
Sticky lines					
10. Orientation of sticky lines	Spiral	**Vertical**[1]	**Vertical**[1]	Radial[2]	L
11. Begin sticky lines at edge and work inward	Yes	Yes	Yes	?	B
12. Uniform spaces between sticky lines	Yes	More or less	Yes	Yes (approximately)	L
13. Cut and reel during sticky line construction	No	**Yes**	**Yes**	?	B
14. Push sticky line with leg IV as attach it	Yes	Yes	**No**	?	B
15. Reel out sticky line with IV before attach it	Yes (some)	**No**	**No**	?	B
16. Site of glue on sticky line	Entire line	**Tip**	**Tip**	Entire line[2]	L
17. Sticky line from aggregate gland	Yes	Yes (?)[3]	Yes (?)[3]	Yes (?)[3]	L
18. Baseline of sticky line is highly elastic	Yes	No	No[4]	?	L
19. Form of attachment of sticky line	(Other)	**Star**	**Star**	?	L
20. Strict order (non-sticky lines, then sticky lines)	Yes	**No**	**No**	?	B
21. Sticky lines built in bursts	Yes	**No**	**No**	?	B
22. Always turn toward hub after attach	Yes	Yes	Yes	?	B
23. Construct hub/disc alternates with sticky line	No	**Yes**	—	?	B
24. Weak attachment of sticky line to substrate	—	**Yes**	**Yes**	No[5]	L

Table 10.6. Continued

[1] Prior to and during addition of glue to line.
[2] Multiple short lines near the tip of the signal line are sticky. The web of *C. simoni* is highly variable and the orientation of the signal lines ending in sticky capture threads depends probably only on the available substrate. There were webs with the "hub" on a vertical structure, in which case the signal lines and capture threads are oriented obliquely downwards (i.e., no lines leading upwards, see fig. 5 in Kropf 2004); other webs had almost horizontal and more or less radial signal lines and a "hub" on a horizontal surface, while still others were intermediate (see fig. 1 in Kropf 1990) (C. Kropf pers. comm.).
[3] Documentation is incomplete (as is typical for most glandular attributions—see section 1.7).
[4] Report of 300% extension by Coddington and Valerio (1980) is probably wrong. The apparent contraction of the sticky line of a different species of *Wendilgarda* was due largely to the thread being taken up windlass-fashion in a large ball of glue (as also occurs in *Araneus*—Edmunds and Vollrath 1992) (G. Barrantes unpub.). The apparent contraction of very long sticky lines of still another species (Eberhard 2000a, 2001a) when they broke free from the water surface was much smaller (WE), as expected if the line "contracted" due to being taken up into the ball of glue rather than to the line being highly elastic.
[5] C. Kropf pers. comm.

elastic solid. Moreover, the gumfoot glue was much more humidity-resistant; its elasticity and adhesion were constant at different humidities, and showed only weak volume-dependence, in contrast to strong humidity effects in the araneid glue, which expanded an order of magnitude and showed a monotonous reduction in elasticity under increased humidity, and optimum adhesion at intermediate humidities.

This evolutionary account is only speculative. Crucial questions involve the validity of the proposed homologies. The very behavioral modularity that may help explain rapid divergence in the evolution of theridiid webs (Eberhard et al. 2008b) makes it difficult to judge possible homologies. How is one to decide, to take a hypothesis that seems to be particularly weak, whether the radial organization of lines in the central disc of young *L. geometricus* is homologous with the radii of orbs? Radial organization occurs in many other contexts, such as the occasionally radial organization of lines in the retreats that these same spiders build at the edges of their webs, the radially organized protective tangles of some nephilids (fig. 10.32), the egg sac and protective tangles of some uloborids such as *Zosis geniculata* (fig. 10.35) and *Uaitemuri rupicola* (fig. 9.11) (Santos and Gonzaga 2017), the radial patterns of lines in the retreats of *Araneus expletus* (AR) (Eberhard 2008), and webs of distantly related filistatid (fig. 5.10), segestriid, hersiliid, and pisaurid spiders (fig. 5.11); they are almost certainly not all homologous. In difficult cases like the theridiids or webs associated with egg sacs (fig. 10.33), searching for homologies in behavior feels uncomfortably similar to searching for hidden meanings in dreams.

Gumfoot lines connected to an aerial tangle have also evolved convergently in the distantly related family Pholcidae (Japyassú and Macagnan 2004; Escalante and Masís-Calvo 2014), and were surely not derived from an orb-weaving ancestor.

10.10.1.2 Other web types

A recent study of symphytognathoids (Lopardo et al. 2011) showed that these tiny spiders, comprising only a few hundred species, most in four families (Platnick 2014), have been particularly inventive evolutionarily. The common ancestor of this group is thought to have built three-dimensional orbs (with at least one radial line out of the orb plane [e.g., fig. 10.22a,b]) (many species build sticky lines and multiple radial lines out of the orb plane—fig. 7.9), and they reverted to strictly planar orbs without additional lines twice (in Symphytognathidae and some anapids). Sheet webs evolved from 3-D orbs twice (in Synaphridae and *Trogloneta* [MYS]), tangle webs evolved from 2-D orbs once (in *Comaroma simoni* [AN]), and a unique "spherical orb" (fig. 10.21) evolved from 3-D orbs in Mysmeninae. The reason for this unusual evolutionary flexibility is uncertain (section 10.9.4).

10.10.2 WEBS COMBINED WITH PREY ATTRACTANTS

Spiders in at least three web weaving lineages are known to have apparently acquired the ability to attract prey chemically (table 9.1). In all three cases the web became reduced or eliminated: in the araneid lineages of *Kaira* spp. (Stowe 1986); the Cyrtarachneae tribe of 10 genera, which includes the bolas spiders (Akerman 1923; McKeown 1952; Stowe et al. 1987; Stowe 1988; Eberhard 1980a, 1990a; Yeargan 1994); and the theridiid genus

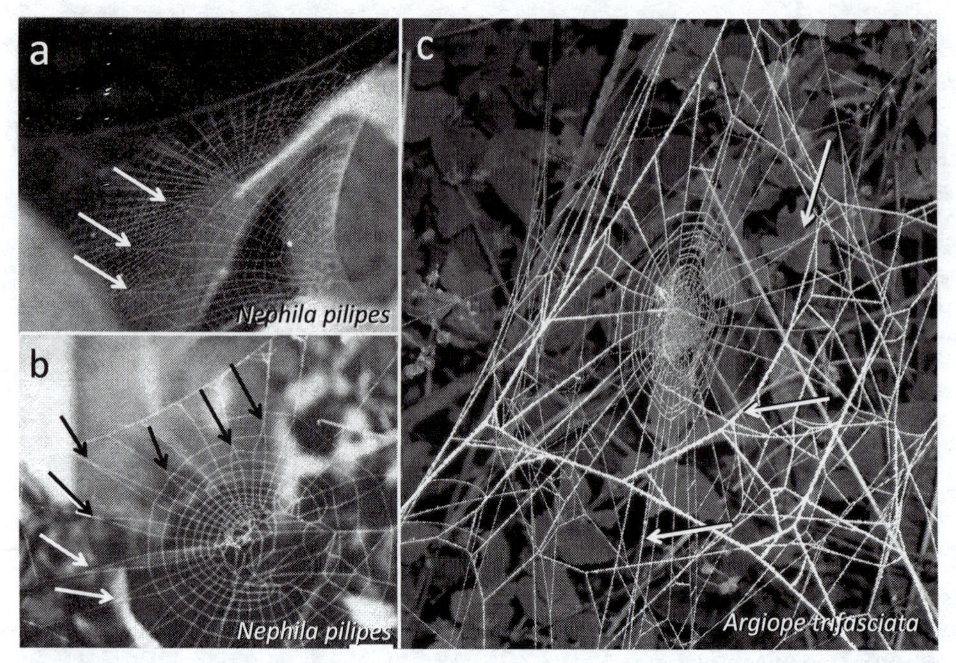

Fig. 10.32. The barrier web alongside the orb of *Nephila pilipes* (NE) (*a*), the resting webs of this species (*b*), and the resting web of *Argiope trifasciata* (AR) (*c*) all included modules of orb web construction and modifications of these modules. All had the general form of a circular hub with radial lines that resembled the hubs of prey capture webs of these species (similar radial organization was generally absent in the barrier webs of *N. clavipes*, however—WE). The radii in the *N. pilipes* barrier web differed from those in orbs in being attached to frame lines at only a single point but then splitting apart (arrows in *b*), rather than the usual double attachments to the frame in *Nephila* spp. orbs (Eberhard 1982). The "hub and radii" pattern in *c* also resembled the prey capture webs of *A. trifasciata* in having an adjacent tangle, but differed in that the radii were doubled rather than single (three are marked with arrows); the spider evidently failed to break and reel on the return trip to the hub when building these radii (*a* and *b* from Robinson and Robinson 1973).

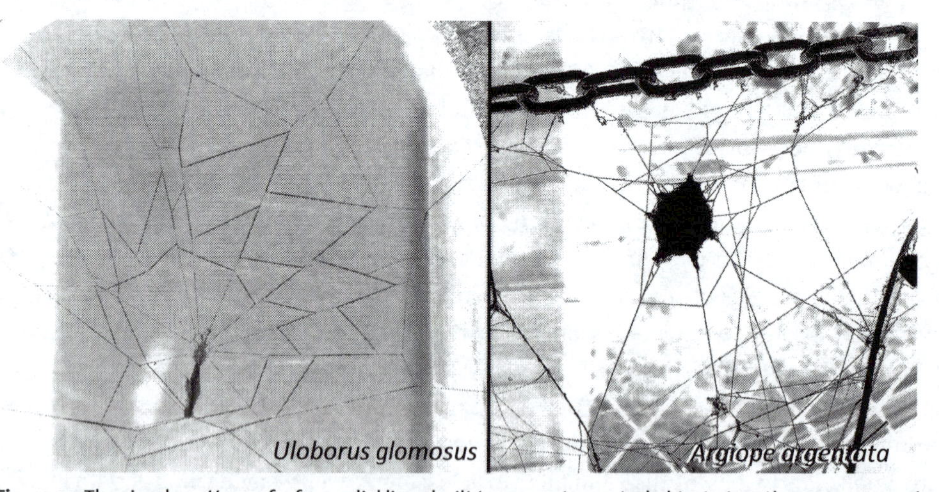

Fig. 10.33. The simple pattern of a few radial lines built to support a central object gives these egg sac webs of *Uloborus glomosus* (UL) and *Argiope argentata* (AR) the appearance of possible orb-web precursors; the similarity in *U. glomosus* was increased by the approximately circular loops of sticky lines (thicker lines; presumably they defend the eggs against enemies such as parasitic wasps). Derivation of orb-like patterns in prey capture webs from egg sac webs is logically possible, but discussions of orb origins have centered instead on derivations from other prey capture web designs. In the end, there is a plethora rather than a scarcity of possible "missing links" to orbs.

Phoroncidia (Eberhard 1981a; Shinkai 1988a). In two of the three groups the reduced web may disperse attractant, but in Cyrtarachneae the attractant substance apparently comes from the spider's body (Yeargan 1994). Intra-specific variation in the webs of bolas spiders in the genus *Mastophora* links them to species that build more elaborate but nevertheless highly modified orbs like *Pasilobus*, and also to species with further web reductions like *Celaenia* and *Taczanowskia* (Eberhard 1980a, 1981c; Yeargan 1994).

In both araneid lineages the spider played an active role in prey capture, twirling or swinging a sticky ball (bolas spiders) or snatching the prey from the air with her legs. In *M. cornigera* different individuals produced different blends of attractant substances to draw males of different moth species (Stowe et al. 1987). But *Phoroncidia* attacked only slowly, and relied on her sticky line to intercept and retain the relatively delicate nematocerous flies that were her prey (Eberhard 1981a; Shinkai 1988a). Both of the araneid groups seem to be relatively scarce wherever they occur, and one of them, the bolas spiders, also has the classic "tiny male" adaptation that characterizes groups with widely dispersed females (Ghiselin 1974). *Phoroncidia*, however, seemed not to be rare at the two sites where I have observed it in Colombia and Brazil.

The transition to using an attractant and a reduced web can be more complex than it might originally seem, because the prey species attracted by adults are likely to be much too large for young immatures to deal with (a *Mastophora dizzydeani* spiderling recently emerged from the egg sac weighed about 1 mg, while the moths attracted by adult females weighed about 60 mg) (Eberhard 1980a). The smallest prey (in a sample of 460) of *M. hutchinsoni* were about 10 mg (Yeargan 1994), again far too large for an early spiderling to subdue. It seems likely that the transition to reduced webs must have involved intermediate stages, in which web design varied through ontogeny, and reduction depended on the acquisition of different chemical lures for prey of appropriate sizes to support different life stages of the spiders. Perhaps there is some chemical relationship between the long-distance pheromones of female moths (mostly noctuids and pyralids) on which adult prey and those of psychodid and ceratopogonid flies trapped by small spiderlings (Yeargan 1994) that facilitated the evolutionary transition from attracting one type of prey to attracting the others. There is evidence suggesting that adult spiders use multiple attractants, compatible with the hypothesis of ontogenetic changes among multiple attractants.

The association between prey attractants and web reduction remains tentative, because if there are species that emit attractants but do not have obviously reduced webs, their use of an attractant would be less likely to be noticed. An attractant with sufficiently general effects might also go unperceived. For instance, 14 of the 29 prey of the single line uloborid *Miagrammopes* nr. *unipus* were winged ants, and one of four of the identifiable prey of *M. simus* were also winged ants. Could it be that the highly adhesive webs of this genus (Opell 1994, 2013), whose reduction to only one or a few sticky lines is otherwise mysterious, are associated with production of an attractant that also incidentally captures an additional smattering of other prey, misleading researchers into deciding it does not produce an attractant (Lubin et al. 1978)?

10.11 COEVOLUTION BETWEEN ATTACK BEHAVIOR AND WEB DESIGN (AND ITS LACK)

As emphasized by numerous authors (e.g., Burgess and Witt 1976; Olive 1980; Miyashita 1997; Hénaut et al. 2001; Zschokke et al. 2006), a spider and her web are complementary parts of a single overall tactic for capturing prey, and the functionality of a spider's web is influenced by her own morphology and attack behavior. Attack behavior, including its flexibility in adjusting to different prey, has clearly evolved in concert with web-building behavior (Japyassú and Caires 2008), and certain web designs make particular details of attack behavior feasible or necessary. Examples include the following. Aerial (as opposed to substrate) webs make it advantageous for the spider to wrap prey in at least a few lines and suspend it in the web, to facilitate attacks on additional prey (Eberhard 1967; Rovner and Knost 1974). Placing sticky lines near the retreat in filistatids is associated with wrapping prey "indirectly" by dragging them through the web toward the retreat (Barrantes and Eberhard 2007). Shaking the web to cause prey to fall is likely an adaptation associated with webs with a tangle above a horizontal sheet that has evolved repeatedly; it occurred in the diguetids *Diguetia* spp. (Eberhard 1967; Bentzien 1973), the araneids *Cyrtophora* spp. (Wiehle 1927; Lubin 1973), the theridiid *Nihonhimea tesselata* (Eberhard 1972b; Barrantes and Weng 2006a), and several linyphiids (Nielsen 1932; Bristowe 1958; Eberhard 1969). Thrusting legs I through a horizontal sheet to press the prey against the spider's chelicerae has also probably arisen convergently in several sheet weavers that attack from underneath their sheets (Eberhard 1967; Bentzien 1973; Blanke 1972).

There are other very general confirmations of web and attack coevolution. Judging by mygalomorph attack behavior (e.g., Coyle 1986), the starting point for spider web builders was a suite of prey capture structures and tactics that emphasized speed and power—quick, short lunges or strikes using thick, powerful legs and chelicerae. When prey capture webs became more elaborate, spiders had the opportunity to modify these tactics. Because the web extended the spider's tactile sense beyond her own body, it must have favored the ability to run short distances to the prey more rapidly. The ability of webs to retain prey could also have favored attack behavior preceded by preliminary sensory examination of the prey (section 3.2.12). Araneomorph web builders in general fit the expectations of these ideas, as they have legs that are relatively longer and less powerful (thinner), and chelicerae that are less robust than those of mygalomorphs.

Hypotheses linking web design to attack behavior receive only a confusing mix of apparent confirmations (table 3.4) and contradictions from data on finer levels of comparisons between species with different types of webs (sections 3.2.10, 4.8). For instance, comparison of four co-occurring species of web builders in Mexico showed some correlations between the speed of attack and the length of time that the web was able to retain prey (Zschokke et al. 2006), but similar comparisons of three species of *Cyclosa* in Japan showed the opposite trend, with the species having the weakest web being the slowest to attack (Miyashita 1997). Another possible example of coevolution was the possible reduction in prey capture repertoire size comparing orb weavers with theridiids (Japyassú and Jotta 2005); in this case, however, the webs of theridiids vary so much (Eberhard et al. 2008b) that one could argue that the general lack of changes in attack behavior during the substantial divergence of web forms in theridiids represents a *lack* of coevolution). I see no reason to expect a priori that all species should make some standard total investment (web traits + attack behavior) in the ability to capture prey; there are many other dimensions in which trade-offs could occur.

It is worth examining, however, one easily quantified set of variables that could be used to estimate running speed—the relative lengths and diameters of the spider's legs. While some long-legged animals move slowly (e.g., *Hypochilus gertschi*—Shear 1969), a stubby-legged animal cannot move rapidly; thus stubby legs are probably reliable indicators of relatively slow attack behavior. It is striking that none of the groups with especially densely spaced sticky lines, such as *Acacesia* (fig. 3.5), *Deliochus* (fig. 3.5), and *Pozonia* (fig. 3.35), are especially stubby-legged (though *Scoloderus tuberculifer* tends in this direction) (fig. 3.4). At the opposite extreme, especially radical reductions in sticky line density were built by stubby-legged spiders (*Phoroncidia, Hyptiotes, Cyrtarachne* and related genera, *Ogulnius*). Some of these hold the web tensed while waiting for prey (*Phoroncidia, Hyptiotes*, theridiosomatids), and their short thick legs may have evolved to provide strength. These are only qualitative impressions, however; more rigorous analyses that take phylogenetic relations into account are needed.

10.12 WHAT *DIDN'T* HAPPEN, POSSIBLE SYNAPOMORPHIES FOR ORB WEAVERS, AND FURTHER PUZZLES

Several possible synapomorphies are brought to light by considering possible evolutionary transitions that failed to occur. One such "path not taken" can be illustrated using the mistakes in some popular accounts of frame construction in orbs. Frame lines built without utilizing a previously built hub, such as in the sequence *a* in fig. 10.34, have never been reported, even though it is mechanically feasible to produce frame lines prior to producing a hub. This suggests a possible synapomorphy of all orb weavers: frame construction that includes use of a previously constructed hub.

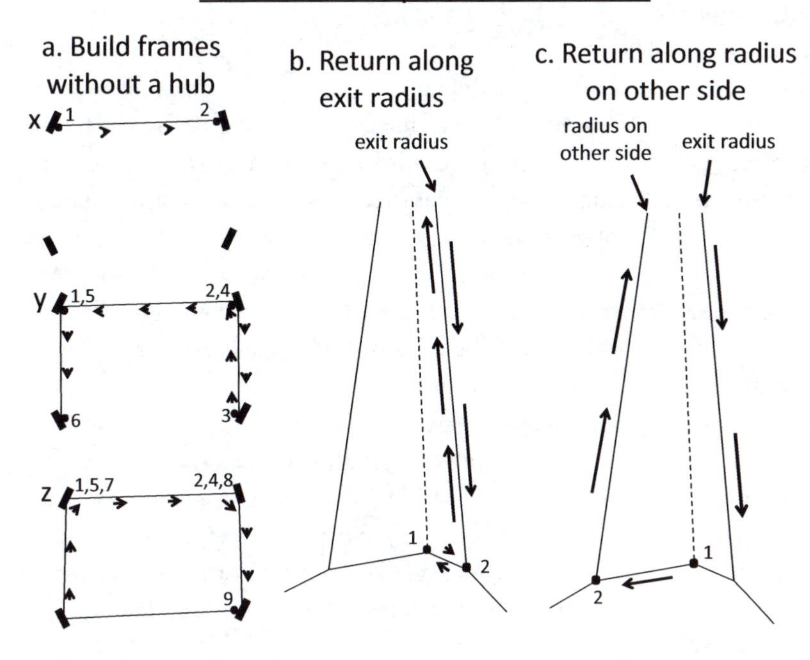

Fig. 10.34. The existence of patterns of radius and frame construction behavior that are feasible but are never utilized by any known orb-weaving spider supports the hypothesis of a monophyletic origin for orbs. For instance, there are mistaken descriptions of frame construction without radii (*a*) in the popular literature (e.g., Dugdale 1969) but this behavior sequence is unknown in spiders. It would also be possible during construction of a new radius (dotted lines in *b* and *c*) for the spider to return to the hub by walking along the original exit radius (*b*) or along the radius on the other side of the empty sector (*c*) (dots represent new attachments following previous drawing). Instead, this never occurs: all of the 100–200 species that have been observed return along the new radius that was just laid. Returning to the hub along the new radius, and laying radii as part of frame construction are thus possible synapomorphies that link all known orb weavers.

Another feasible but unknown behavior is to return to the hub while laying a new radius by walking back along the original exit radius, rather than the newly laid radial line (*b* in fig. 10.34). Nevertheless, not a single species among >100 that have been observed (Eberhard 1982) utilized this alternative. Still another possible alternative that has never been observed would be to return to the hub along the radius that is on the other side of the sector in which the new radius is being placed (*c* in fig. 10.34). Thus returning to the hub along the newly laid radius may be another synapomorphy for orbicularians. Similar behavior, returning to the central area along newly-laid lines attached to the substrate, also occurred, but only inconsistently, in non-orb spiders such as the theridiid *Latrodectus geometricus* (Eberhard et al. 2008a) and the pholcid *Modisimus guatuso* (Eberhard 1992b).

This "return-along-the-latest-line-attached-at-the-edge" trait was surely not present early in web evolution when spiders built only lines close to the substrate, and thus walked on the substrate rather than on their own lines. It was also absent in some modern non-orb species, such as the pisaurid *Thaumasia* sp. (Eberhard 2007a) (figs. 5.11), but present in others such as the oecobiid *Oecobius concinnus* (Solano-Brenes et al. 2018). I speculate that the return-along-the-latest-line behavior arose at some point in the evolution of aerial webs (or perhaps several different times independently), and that it was already present or evolved at the same time when orbs evolved. More detailed studies of non-orb construction are needed to refine and test these hypotheses.

Still another feasible but never utilized pattern in radius construction is illustrated by the "repeated bisection of angles" pattern of addition of radial lines in the web of the filistatid *Kukulcania hibernalis* (fig. 5.10) (see also fig. 5.11 of the pisaurid *Thaumasia* sp.). No orb weaver is known to use this tactic. Instead, orb radii are laid at "final angles" (section 6.3.3) that (with a few exceptions such as *Nephila* spp.—see appendix O6.3.2) are nearly never subdivided by subsequent radii. This "final angle" pattern would seem to be more demanding mentally than the alternative of simply subdividing angles over and over until they are small enough. The difficulty is that a given spider builds orbs with widely differing numbers of radii in different-sized spaces (see section 7.7), so she must adjust radial angles beginning very early in orb construction, possibly using information regarding the projected final size of the orb as a guide. This striking "new-radii-at-final-angles" trait may also be a synapomorphy of orb weaving orbicularians.

Another missing transition concerns the highly derived rubbery property of the flagelliform gland axial lines of aranoid sticky spirals. This property is apparently highly adaptive (Opell and Bond 2001); nevertheless, although the glands themselves are still present in non-orb orbicularians. Unquantified observations suggest that the high elasticity does not occur in the sticky lines of theridiids, linyphiids, or the synotaxid *Synotaxus* (fig. 5.3) (WE; the data are, however, scarce and only preliminary).

Still another puzzle concerns convergence in reduced orbs on interruption of sticky line production. In some species, such as in the uloborid *Hyptiotes* (Marples and Marples 1937; Eberhard 1982), the theridiosomatids *Wendilgarda* spp. (Coddington and Valerio 1980; Shinkai and Shinkai 1997; and Eberhard 1989a), the araneids *Poecilopachys australasia* (Clyne 1973), *Pasilobus* sp. (Robinson and Robinson 1975), and *Eustala* sp. (Eberhard 1985), interruption is highly stereotyped. These interruptions contrast sharply with the strong trend in orb weavers to continue sticky spiral construction once it has begun until it is finished, as illustrated by the failure of *Neoscona nautica* to interrupt sticky spiral construction even in the face of experimental changes that made it urgent for the spider to repair non-sticky lines to avoid producing a highly irregular array of sticky lines (Hingston 1920). Why have most orb weavers been so evolutionarily conservative in not altering the order of production of sticky and non-sticky lines; and why are the symphytognathoids exceptions, having evolved such alterations multiple times?

Another puzzling lack of a transition is that none of the thousands of orbicularian species that abandoned orbs returned to making substrate webs built around a retreat and including ecribellate sticky silk placed on or very near the substrate (e.g., figs. 9.6, 10.7). Once ecribellates with viscous glue abandoned orbs, they never re-evolved substrate webs, even in the hyper-diverse family Theridiidae. The reason is not because these spiders do not utilize retreats associated with the substrate. Numerous substrate retreats exist in theridiids (Eberhard et al. 2008b), and they also occur (though less frequently) in some araneids, tetragnathids, nephilids, and linyphiids (table 9.1). Nor is it because substrate webs are not adaptive—witness the many cribellates with substrate webs.

10.13 MODULARITY AND ADAPTIVE FLEXIBILITY

Much of this chapter discussed the different aspects of webs that have changed, and the order in which the changes occurred. Lurking behind these questions are other, more mechanistic questions. The internal organization of building behavior and its flexibility, in terms of the behavioral units that can and cannot be triggered by particular "decisions," could bias the frequency with which different variants will appear and be available for natural selection to act upon (West-Eberhard 2003). Both adaptive flexibility (the ability of an organism to manifest adaptive alternative behavioral phenotypes), and the degree of independence of the controls (both genetic and neuronal) of expression of different behavior patterns, are likely to bias which variants will appear, and the evolutionary results of selection on them.

10.13.1 MODULARITY IS A CENTRAL PATTERN IN WEB CONSTRUCTION

A high degree of modularity is one of the basic patterns in observations of web construction behavior (chapters 5 and 6). I will concentrate here mostly on data from finished webs that further document modularity, and the role of modules in the evolution of web designs and of the behavior used to build them.

Before beginning, I need to clarify my use of the term "modularity." The more independent a trait is with respect to its control and the more consistent it is in its expression in different contexts, the greater its degree of modularity; modularity is thus a quantitative rather than a qualitative trait, and the phrase "degree of modularity" is more appropriate than the black-and-white classification implied by the single word modular. The degree of modularity of a trait can change over evolutionary time or even over the lifetime of an individual, and the components of a given module can be made up of sub-modules that have different degrees of modularity (West-Eberhard 2003). I will largely ignore these subtleties in the discussion that follows.

10.13.1.1 Direct observations of behavior

Modularity was a central theme in the stages of construction orb web behavior such as radius construction, tem-

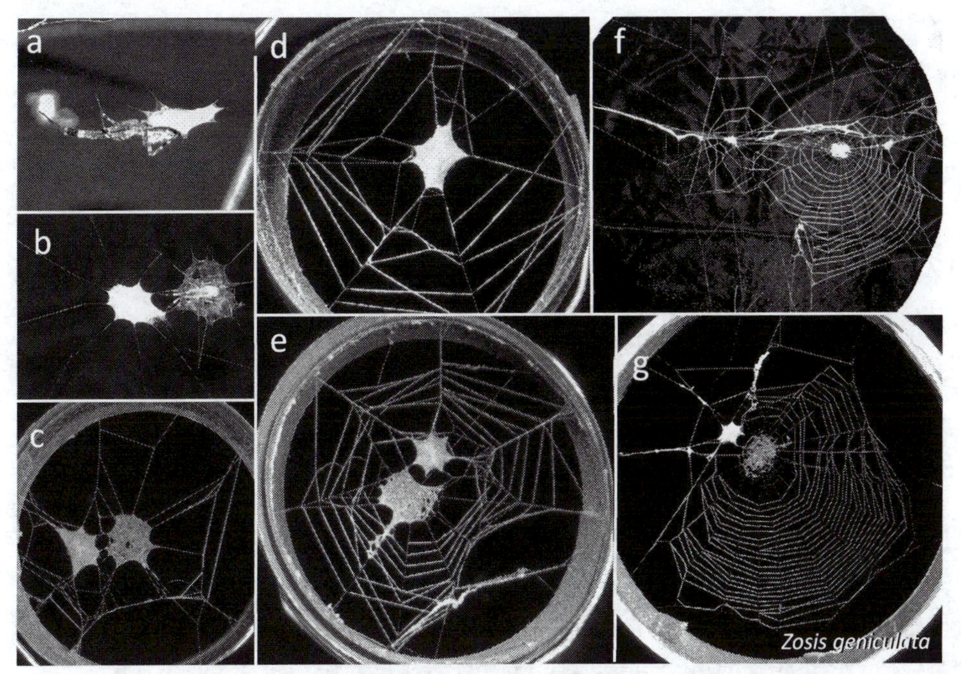

Fig. 10.35. These combinations of web designs, all made in captivity by mature female *Zosis geniculata* (UL) in association with the production of a new egg sac, represent different combinations of similar web construction modules: a sac with no additional web, with the spider resting under it (*a*); a sac with a small hub nearby covered with a disc stabilimentum where the spider rested (spider is present in *b*, absent in *c*); a sac with a few loops of sticky lines attached to the radial lines supporting the sac (*d*); a sac with nearby hub and stabilimentum, with loops of sticky lines attached both to the radii of the hub and also to the support lines of the sac (*e*); a sac with a few loops of sticky lines, as well as an adjacent complete orb (*f*); and a sac with no associated loops of sticky lines but with an adjacent complete orb (*g*). Some individuals made different web designs with different egg sacs. The sizes of the containers (note the sizes of egg sacs in the different photos) may have influenced the web designs accompanying them.

porary spiral construction, sticky spiral construction, etc.: during each stage the spider repeated particular behavior patterns over and over (sections 6.3.2–6.3.5). There are numerous examples of evolutionary transfers from one context to another of both the behavior patterns and the web structures (e.g., figs. 10.35–10.37, table 10.7).

Modular units also occurred at lower levels of analysis, such as the details of leg movements (section 6.5, appendix O6.2). Here, however, the pattern of distribution of modules during the different stages of construction was sometimes quite different. For example, one modular unit, the movements that the spider uses to attach the line that she is currently producing to another line by grasping the previous line with ipsilateral legs III and IV and holding it against her spinnerets (Eberhard 1982), was used at several stages of construction by all orb weavers. These movements were used by orb weavers in exploration, radius construction, temporary spiral construction, sticky spiral construction, and hub construction. Similarly, *Argiope aurantia* (AR) made alternate wrapping movements with legs IV while applying a disc stabilimentum to the hub (WE) that closely resembled (and were presumably derived from) the prey wrapping movements of these same legs (Robinson and Olazarri 1971). Other modules at low levels involved following behavior (section 2.4), and retrolateral vs. prolateral tapping movements (section 6.8.1). Some of these may constitute "atoms" of building behavior; a preliminary analysis of a few aspects of a small, taxonomically scattered set of species suggested very widespread uniformity (table 2.3).

Modularity probably also occurs in non-orb webs. For instance, Mayo (1988) found that the behavior used by the mygalomorph spiders *Antrodiaetus*, *Aliatypus* (both Antrodiaetidae), and *Hebestatis* (Helonoproctidae) to build structures at the mouths of their burrows (collars, turrets, doors) consisted of only five behavior patterns.

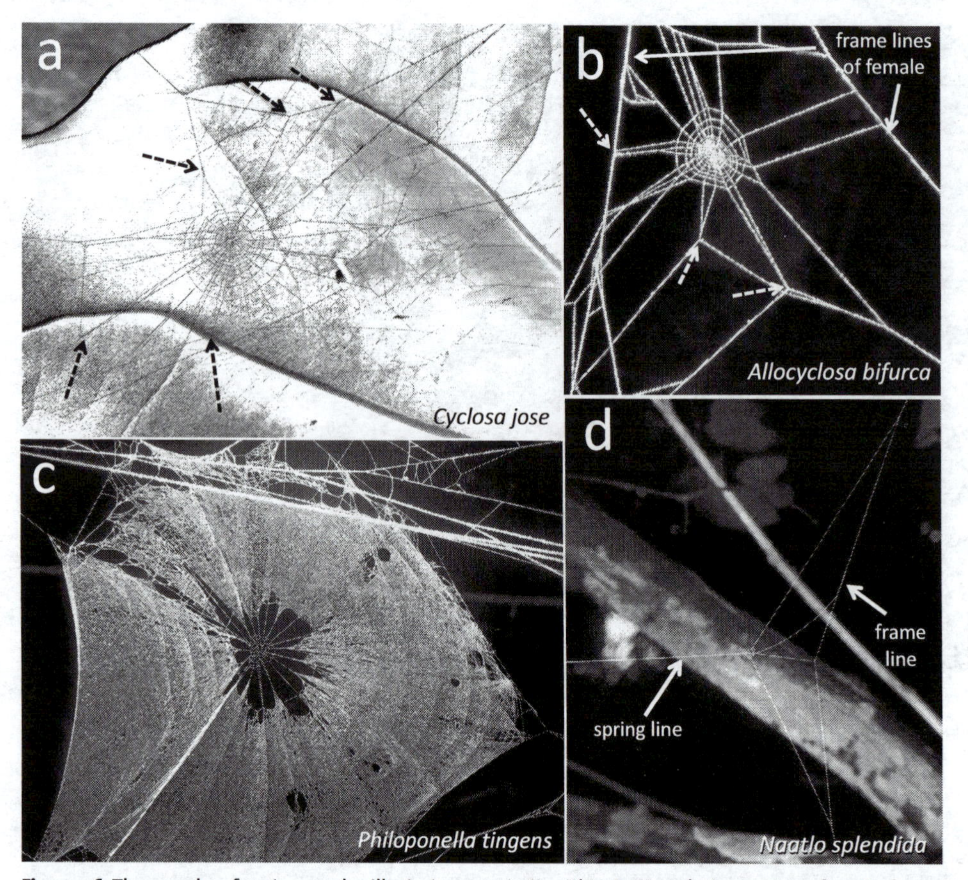

Fig. 10.36. These webs of mature males illustrate some trait-within-trait evolutionary transfers. Each species transferred the ancestral prey-capture orb web pattern—non-sticky radial lines attached peripherally to frame lines and converging on a hub with tightly spaced circular loops—to the new context. Ancestral components were altered in various ways. In the small resting webs of *Cyclosa jose* (AR) (*a*) and *Allocyclosa bifurca* (AR) (*b*) (which was attached to the frame lines of an adult female's orb), some radii were strengthened by being doubled during radius construction (dashed arrows in *a* and *b*), omitting the break-and-reel behavior of normal radius construction (sections 6.3.3, O6.2.2). The *A. bifurca* male also omitted some frame lines. The prey-capture web of a male *Philoponella tingena* (UL) (*c*) combined hub and temporary spiral lines similar to those adult females with a mat composed of huge numbers of very fine lines in various orientations that resembled the prey capture web of a newly emerged spiderling (fig. 9.20) except that it lacked reduced spaces between loops of temporary spiral. Judging by the rounded holes at the inner edge of this mat, the spider may have built the hub at the end of construction, after the fine lines had been built, thus adding a new hub construction stage. The pattern in which the mat of fine lines sagged indicated that it was supported by a temporary spiral (the round holes were probably sites where prey or other objects struck the web). In the *Naatlo splendida* (TSM) web (*d*), only the radii of a single sector were present (in reduced numbers), and they were attached to a long frame line that made the web more extensible. The male tensed this web by reeling in the spring line as he rested at the hub (fig. 10.22); he released the spring line abruptly when disturbed, causing him to spring rearward a cm or more (to the right in the figure).

Very simple reorganizations of these behavior patterns were associated with differences in the final structures. For instance, when the behavior patterns that were used by one species to build a turret were re-oriented in another species so that they were concentrated on only one side of the burrow mouth instead of being distributed equally on all sides, a door was produced.

It might be thought that the divisions of web construction into modules are arbitrary human constructs. Experimental tests showed that this doubt can be discarded in some cases, suggesting that it also does not apply in others where it has not been tested. Hingston (1920) eliminated the entire temporary spiral of *Neoscona nautica* (AR) after she had finished the temporary spiral construction stage, while she was building the sticky spiral, and found that the spiders were not able to return to temporary spiral construction. As a result of the modularity of these two stages, the spiders built highly deviant, irregular arrays of sticky spiral lines because they lacked the bridges needed to walk from one

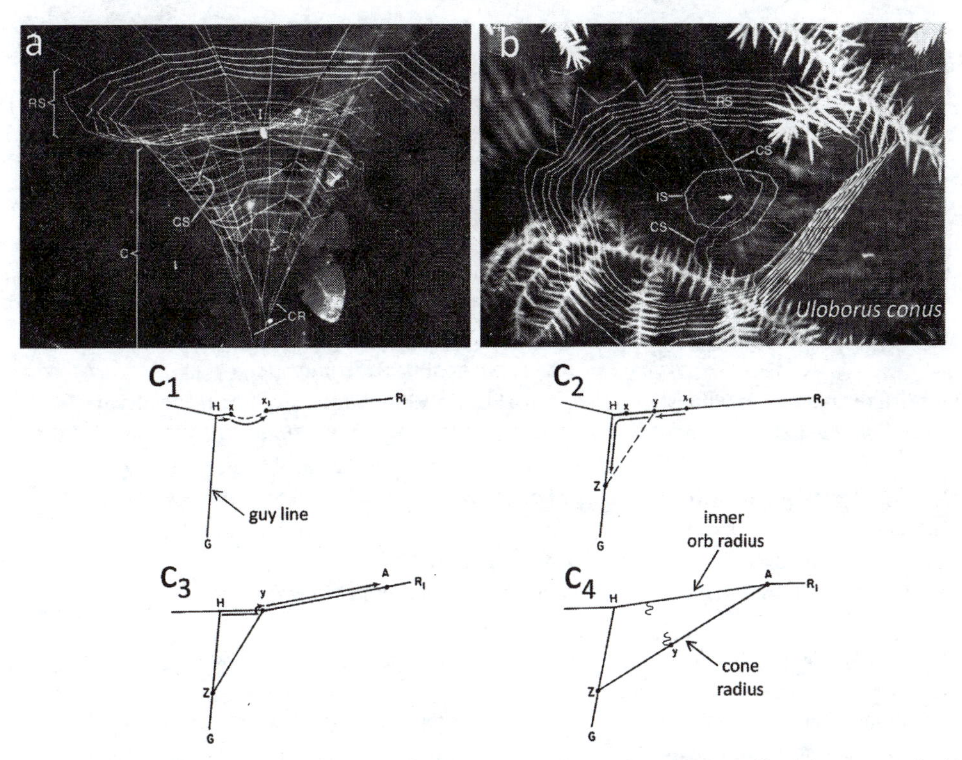

Fig. 10.37. To produce her orb plus cone web (seen from the side in *a*, from above in *b*), *Uloborus conus* (UL) built, in essence, two orbs by employing the same basic orb construction behavior used by other uloborids for both. First, to make the cone, she built a planar orb with a hub, radii, and temporary spiral; she added several sticky spiral loops in its outer portion and then a rapidly spiraling sticky line in its inner portion (RS and CS in *a* and *b*). She then added a guy line below the hub, and with a complex series of movements (c_1–c_4) loosened each radius and pulled its inner portion (CR in *a*) into a cone (C in *a*). After forming the cone, she cut most of the other temporary inner orb radii, collapsing the hub into a small bit of silk (at H in c_4) to which a few temporary radii were attached, and then replaced these temporary inner radii using behavior very similar to normal orb initiation (appendix O6.3.2.3). She attached each new radius at or just short of the innermost sticky loop in the rim (forming the inner radius H–A in c_4) and returned to the center. She built the hub spiral during the process of replacing other temporary inner radii and adding further inner radii, as in radius construction in a typical orb (see appendices O6.3.4.3, O6.3.5.3). Each new inner radius was attached near the inner edge of the rim spiral by walking across the temporary spiral of the cone. When the spider had finished these new inner radii, she added a temporary spiral (IS) (but no sticky lines) to this "second" orb-like portion of the web (after Lubin et al. 1982, courtesy of Y. D. Lubin).

radius to the next. At a finer level, parasitoid wasps chemically manipulate their host spiders, causing them to perform particular subroutines of web construction behavior and selectively repress many other aspects of orb construction, isolating even behavioral traits that normally occurred together very consistently (Eberhard 2000b, 2010b, 2013b; Sobczak et al. 2009; Korenko et al. 2014). Manipulations involved even very fine details, such as the omission of cut and reel behavior in otherwise normal radius and frame line construction.

10.13.1.2 Finished structures

The designs of finished webs also point to evolutionary shuffling of relatively intact units at several different levels (table 10.7). The clearest cases are in orb weavers, where the units are easier to perceive and better studied. A simple but graphic illustration is the newly discovered anapid *Tasmanapis strahan* (Lopardo et al. 2011), whose "quilt" web is composed of several orbs (the ancestral module) that are attached at or near their edges, forming a large, composite sheet (fig. 10.27). The egg sac webs of *Zosis geniculata* (UL) (fig. 10.35) showed different combinations of several modules, including hub, stabilimentum, radial lines, and sticky spiral lines that were centered on the sac, on the hub, or both. The mysmenid *Trogloneta granulum* also appeared to make (at least in captivity) several unit webs that gradually accumulated and were tied together by a gradually increasingly dense

Table 10.7. An undoubtedly incomplete list of "transfers" (mostly cases in which a modified version of an ancestral module is expressed in a new context) in which a higher-level behavior program evolved in one context, and later evolved to be expressed relatively unchanged in a new context. These examples argue for a relatively high degree of modularity in web construction. Further examples involving relatively fine details of behavior are given in table O6.2.

Original structure or behavior	*New structure or behavior*	*Taxa and references*
Orb web	Protective tangle with radial and spiral organization next to orb[1]	*Nephila* spp. (NE) Robinson and Robinson 1973a,b, Kuntner et al. 2008a; *Conifaber parvus* (UL) Lubin et al. 1982; *Allocylosa bifurca* (AR) WE; *Uloborus* spp. (UL) Lubin et al. 1982, Lubin 1986
Complete orb with sticky spiral on supporting non-sticky lines	Orb-like resting web lacking sticky lines built by mature male	*Zosis* (UL) Lubin 1986; *Naatlo splendida*, *Epeirotypus brevipes* (TSM) WE; *Argiope trifasciata*, *A. aurantia* (AR) Kaston 1948, Eberhard 2013b
Orb with sticky spiral	Molting or resting web with small, orb-like array of non-sticky lines nearby[1]	*Uloborus trilineatus* (UL) Lubin et al. 1982; *Nephila clavipes* (NE), *Leucauge mariana* (TET), *Argiope trifasciata* (AR) WE
"Sheet orb" webs of second instar spiderlings[2]	"Sheet orb" of mature male[2]	*Uloborus* spp. (UL) Lubin 1986
"Sheet orb" webs of second instar spiderlings	"Sheet orb" of adult female	*Conifaber parvus* (UL) Lubin et al. 1982
"Sheet orb" webs of second instar spiderlings	Peripheral "sheet orb" with typical orb central portion pulled into a cone underneath	Unidentified uloborid Australia Agnarsson et al. 2012
Orb web	Multiple orbs joined in "quilt"	*Tasmanapis strahan* (Lopardo et al. 2011)
Remove proto-hub	Destroy and replace entire hub after sticky spiral finished	Anapids, mysmenids, symphytognathids Eberhard 1987e, Lopardo et al. 2011
Reduce radius tension by break and reel during radius construction	Break and lengthen radii after sticky spiral is finished	Anapids, mysmenids, symphytognathids Eberhard 1987e, Lopardo et al. 2011
Tangle below approximately horizontal orb to protect spider (section 3.3.3.3.4)	Tangle above orb[3]	*Leucauge argyra* (TET) Triana-Cambronero et al. 2011
Construction of early radii in orb	Construction radial lines to which egg sac is added[4]	*Zosis geniculata* (UL) fig. 10.35

[1] Most if not all of the genera listed represent independent, convergent derivations.

[2] Direction of derivation of this web form is not certain (webs of mature males could have preceded those of second instar spiderlings). Both second instar spiderlings and mature males lack a functional cribellum and thus cannot build the orb webs with sticky spirals that are typical of all other developmental stages.

[3] There is a tangle of lines both above and below the small orb-like portion of some resting webs of *L. mariana*.

[4] The homology is less certain, because of possible constructional constraints: a radially organized set of lines converging on the object to be supported is mechanically effective and efficient way to support the object (the egg sac); the sticky line may protect the egg sac from predators or parasitoids, and have no other site available for support other than the radial lines.

tangle of sticky lines (Hajer 2000). Intra-specific alternative web forms also occur in non-orb weavers, such as *Bathyphantes simillius* (LIN) (Rybak 2007).

Still other examples come from the resting webs of mature males (fig. 10.36), orb plus cone web of *U. conus* (fig. 10.37), and the molting webs (fig. 10.38) of several groups. In each case the webs resembled small versions of the normal prey capture orbs of conspecific females in some aspects (e.g., hub-like structures), but also omitted others (e.g., sticky spiral). The male webs all lacked sticky lines, but had relatively normal though small frames, radii, and anchor lines. Hub designs differed among

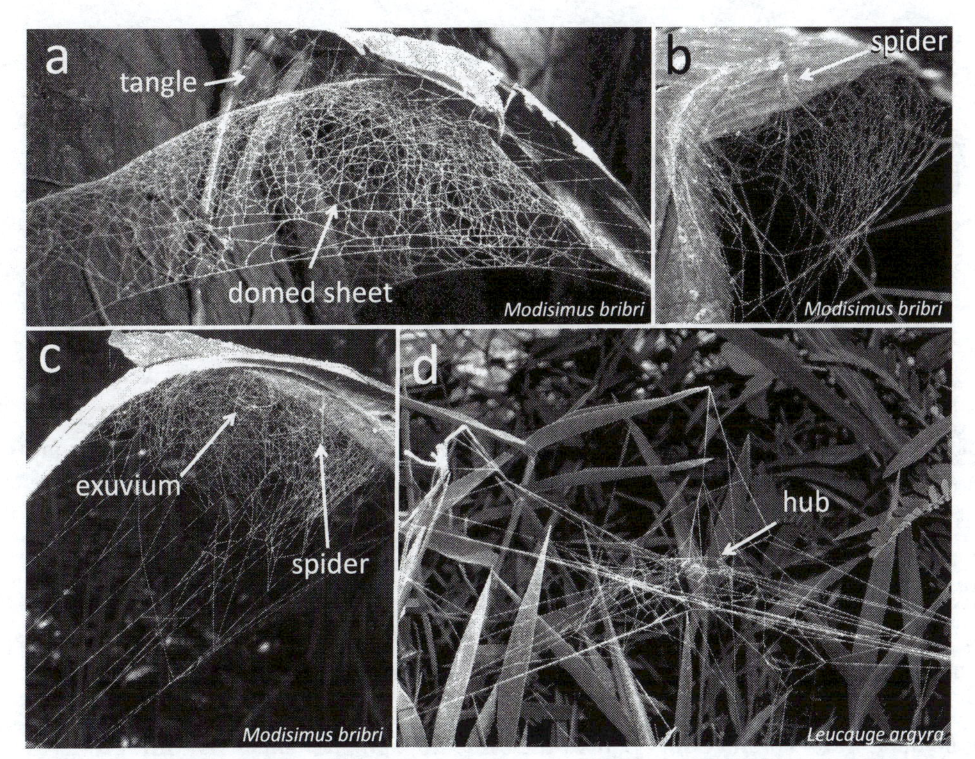

Fig. 10.38. The molting webs of two distantly related species both included aspects of their species' prey capture webs. In the pholcid *Modisimus bribri*, the normal dome-shaped sheet of a prey capture web built against the underside of a leaf (*a*) was reduced in size and the sides were more nearly vertical so that they nearly closed below, thus protecting the spider when it molted under the leaf (*b*, *c*). A molting web (*d*) of the orb weaver *Leucauge argyra* (TET) included several traits of conspecific prey capture webs, including a hub with a few hub loops and an open center, more or less horizontal radii, and a tangle below (which was often more extensive than those seen with prey capture webs, e.g., figs. 8.10, 10.3); both the temporary and the sticky spirals of prey capture webs were lacking (*a*, *b* courtesy of Kevin Sagastume).

species in ways that reflected the differences among the hubs of prey capture webs. In addition, the radii of some were modified in ways likely to increase the strength of the web (e.g., *A. bifurca* and *Z. geniculata* in fig. 10.36). Still other modular transfers occurred in the orb-like barrier webs built beside some orbs to protect the spider at the hub (figs. 10.32, 10.37, table 9.1).

Another probable but more subtle case involved shuffling high-level units in the "rectangular orbs" of the synotaxids *Synotaxus* spp. The webs of *S. turbinatus* contain multiple units that are combined in somewhat variable ways into a single vertical sheet (fig. 9.12, Eberhard 1977a; Agnarsson 2003b; Eberhard et al. 2008b). The strong similarity between the web of *S. ecuadorensis* and a single unit in the web of *S. turbinatus* (fig. 5.3) suggests that *S. turbinatus* behavior is organized in modules (Eberhard 1995).

Modular losses also occurred in prey capture webs. For instance, the temporary spiral was lost independently in two theridiosomatid genera (*Ogulnius* and *Wendilgarda*), and a third time in anapids (table 10.6; Eberhard 1982; Coddington 1986a). Frame construction was also lost in these theridiosomatid genera, and it was reduced in *Anapisona simoni* (AN) (Coddington 1986a; Eberhard 2007b). Frame loss was induced in *Leucauge argyra* (TET) and *Zosis geniculata* (UL) by severely reducing the space available to the spider in which to build her orb (fig. 7.42, Barrantes and Eberhard 2012; Eberhard and Barrantes 2015), so frame loss may have originated in anapids in association with the small spaces in the leaf litter where these spiders build their orbs. A possible modular *acquisition* occurred in the theridiosomatid *Baalzebub*: its intact hub was probably secondarily recovered following a previous loss of hubs in an ancestor (Coddington 1986a).

Modular elements themselves sometimes also changed. For example, the removal of the central portion of the hub after the sticky spiral is complete, typical of araneids and tetragnathids, may have been transformed in Theridiosomatidae and their symphytognathoid allies to include removal of the entire hub. Radius construction was altered in resting webs of *Cyclosa jose*, *Allocyclosa bifurca*, and *Argiope trifasciata* (AR) to provide increased strength (figs. 10.32, 10.36), and the radius attachments to the frame in the barrier and resting webs of *Nephila pilipes* (NE) reverted to single rather than double attachments (fig. 10.32).

10.13.1.3 Ontogenetic and experimentally induced changes

Ontogenetic changes in web design also testify to the modularity of web design. In *Nephilengys cruentata* (NE), for example, the patterns of change in radius length, the number of radii at the edge of the web, and the number of sticky spirals differ over the spider's lifetime and in different parts of the orb (fig. 10.1) (Japyassú and Ades 1998). Differences of the same sort occurred during the ontogeny in *Zygiella x-notata* (AR) (Le Guelte 1966) (fig. 10.2).

Experiments in which spiders were confined in different-sized containers produced several relatively independent changes in different aspects of orb design, including numbers of sticky spiral loops, capture area, numbers of hub loops, and hub area (Barrantes and Eberhard 2012; Hesselberg 2013; Eberhard and Barrantes 2015). The different combinations of effects induced by psychoactive drugs (fig. 7.46, Witt and Reed 1965, section 7.13), and by behavior-modifying parasitic wasp larvae (section 10.14), also suggest independent controls of different aspects of orb construction.

10.13.1.4 Summary

A high degree of modularity is a central aspect of orb web construction behavior, and may well also be important in constructing non-orbs. While the inter-connectedness of different aspects of phenotypes often imposes important limits on the usefulness of the concept of a module (West-Eberhard 2003), the evidence for the relative independence of many different aspects of web construction behavior is strong, at least at some levels of analysis. The high degree of modularity in the organization of behavior is evident in spider web evolution. The fact that orb web construction behavior is organized in modules is of course no surprise. Behavioral modularity has long been known from observations of inter-specific hybrids in vertebrates and insects (summary in Franck 1974).

10.14 MODULES AND EVOLUTIONARY TRANSITIONS IN WEB-BUILDING BEHAVIOR

One common problem in understanding the evolution of behavior is that it is usually difficult to know details of how ancestors behaved. This problem is largely solved for some spiders whose web designs were derived from orb webs, because 1) the details of orb construction behavior and the cues that guide it are relatively well known, 2) they are evolutionarily very conservative (sections 6.5, 7.21, 9.2.5), and 3) the construction behavior of some derived webs has been described in detail. In addition, various taxa build web forms similar enough that the difficult task of determining homologies is simplified.

Making decisions regarding homologies requires careful specification of the level of analysis, because homologies may differ for the same behavior depending on the level at which it is analyzed. A behavioral pattern can at the same time be both ancestral with respect to the movements that are employed in its execution, and derived with respect to the context in which it occurs. In the words of Coddington (1986c; p. 66) in a discussion of homologies in *Deinopis*, "The question is whether the presence of a probably plesiomorphic action pattern (cutting and reeling) in an apomorphic sequence (frame and radius construction) should be considered apomorphic or plesiomorphic." The answer will vary, depending on the level of analysis; this is especially important when using details of web construction behavior as characters in taxonomic analyses (next section).

A summary of 51 behavioral transitions in 11 species (Eberhard 2018) revealed that just over two-thirds involved transfers of modules or the cues directing them. Transfers included the following types (in descending frequency): transfer an ancestral behavior more or less intact to a new context (23.4%); lose an ancestral trait (10.9%); change the relative frequency of an ancestral trait (including "fixation" of a previously optional ancestral trait) (9.1%); increase or reduce the repetition of an ancestral trait (9.1%); lose one ancestral cue but conserve another with the same function (5.5%); accentuate an ancestral trait (3.6%); use a new stimulus or point of reference to guide an ancestral response (3.6%); and use an ancestral stimulus to trigger a new behavior (1.8%) (the remaining 32.7% of the transitions involved acquisition of a new trait that had no clear homology with ancestral traits; many of the behavior patterns that were classified as "new" would probably, when examined at lower levels of analysis, be seen to be lower-level modules that were derived from ancestral behavior) (Eberhard 2018).

The most general conclusion is that many different kinds of modular changes have occurred during the evolution of web-building behavior. Evolutionary changes have not been limited to one or just a few types of transition. A second general conclusion is that many transitions involved some sort of modification of an ancestral behavior. This fits with the argument that orb construction is largely composed of semi-independent modules. Of course, evolutionary change classically involves the transformation of ancestral traits into new traits, so the frequent traces of ancestral behavior in derived building behavior is no surprise. Nevertheless, the frequently modular nature of these traces is striking.

It must be noted that it is possible that categorizing the behavior patterns in the discussions above may overestimated their modularity. A concrete example illustrates both problems and possible solutions. Observations of behavior and completed structures can give an impression of apparent homologies, such as in the orb-like egg sac web of *Uloborus diversus* (UL) in fig. 10.16a. The star-shaped sac itself was built not at the center of an orb-like set of radial lines, but rather in the midst of a large, intact sector of a previous orb. When it was finished, the spider repeatedly attached a line to the sac, moved away from the sac breaking and collapsing a portion of the old web as she went, attached her dragline to the web, and returned directly to the sac. These radial trips differed from typical radius construction behavior, in lacking lateral movements at the end of a radial excursion, lacking any sign of hub construction between radial excursions, shifting from one exit radius to another during the same movement away from the center, in some cases breaking and reeling up the exit line as she moved away from the sac, and in sometimes climbing onto the top of the sac and briefly onto the tops of the radial lines already in place.

Part of the problem of homologies revolves around deciding whether or not particular variant behaviors constitute independent traits. One assay for independence is to use the chemical manipulations (probably involving ecdysone—see Kloss et al. 2017; Eberhard and Gonzaga 2019) that have been evolved by parasitoid ichneumonid wasps to alter the behavior of their host spiders (Gonzaga et al. 2017). In at least some cases, a wasp larva elicits very specific fragments of normal orb construction behavior from its spider host. For instance, the larva of *Hymenoepimecis argyraphaga* induced *Leucauge argyra* (TET) to perform one subroutine of one type of frame construction (Eberhard 1990b) behavior over and over. This particular subroutine is thus a distinct module of behavior in the spider's repertoire, which can be elicited independently of other behavior, and could potentially evolve independently. The wasp larvae justified the decision in the original published description of frame construction behavior to resist the temptation to lump different subroutines of frame construction into one behavioral category (as was urged by one referee). Of course, to give the referee his due, no two behavioral acts are identical; some degree of lumping is always necessary; there is no "magic" cut-off criterion (Eberhard 2018).

10.14.1 USE OF WEB CONSTRUCTION BEHAVIOR IN TAXONOMY

10.14.1.1 Historical successes and failures

Arachnologists have long used behavioral traits to determine the relationships among web-building spiders (e.g., Thorell 1886 in Coddington 1986b; Emerton 1902; Wiehle 1931; Fukumoto 1981; Eberhard 1982, 1988b; Shinkai 1982), and modern studies (Kuntner et al. 2008a; Kuntner and Agnarsson 2009; Blackledge et al. 2009c) continue to use characters at both higher and lower levels of analysis. Recent quantitative analyses show a mixture of both failures (in theridiids at the level of genera and groups of genera—Eberhard et al. 2008b) and successes (in araneoids at the level of families and groups of families—Kuntner et al. 2008a; Lopardo et al. 2011; in cribellates at the level of families—Griswold et al. 2005). Although behavioral characters comprised only 15% of the 231 characters in an analysis of Nephilidae, they contributed 47% of the 15 unambiguous synapomorphies (Kuntner et al. 2008a).

Useful traits in orb weavers include the identities of the legs that are used to locate the inner loop during sticky spiral construction, the legs that hold different lines when attachments are made, whether the spider maintains contact with the temporary spiral lines early in sticky spiral construction, the order of thread placement during radius construction, whether the hub is removed, whether the transition between hub and temporary spiral construction is abrupt, whether the temporary spiral is continuous or broken into circles or omitted entirely, and which cues are used to guide construction behavior and how the spiders respond to them (Eberhard and Barrantes 2015).

Perhaps the greater consistency of behavioral traits in orb weavers was due (at least in part) to the lower level of behavioral details analyzed. The theridiid observations did not include lower-level behavioral details, but rather general aspects of web design such as the sites of sticky lines. And among orb weavers, some web design characters (such as stabilimenta, open hubs, greater or lesser vertical symmetry, relative densities of radii or sticky spiral lines) showed substantial variation and frequent homoplasy, even among congeneric species (or sometimes members of the same species) (figs. 1.6, 3.15, 10.24–10.26; section 3.3.4.2.2.1; Kuntner et al. 2008a). Another possible explanation is that orb web construction is more highly conserved because of its demands on the spider's abilities to orient precisely and consistently without errors.

In sum, low- and medium-level behavioral characters sometimes but not always provide useful phylogenetic signals for grouping orb weavers.

10.14.1.2 Implications for orb monophyly

The orb monophyly hypothesis would predict that the operations in building orbs and the order in which they are performed should be similar in deinopoid and araneoid spiders. As has been known for a long time (McCook 1889; Emerton 1902; Comstock 1912; Wiehle 1931), this prediction is clearly confirmed. Although subsequent studies have shown that the early stages of uloborid orb construction associated with proto-hub construction and destruction differ from those of araneoids (appendix O6.3.2.3), the overall pattern of operations in orb construction is quite uniform. The differences within araneoids (appendix O6.3.2.1) seem greater than those between araneoids and uloborids. The tiny theridiosomatoid araneoids differ especially clearly, with respect to the timing of radius construction (both before and after sticky spiral construction), the presence and form of the temporary spiral (missing in some, circular rather than spiral in others), and the destruction of the entire hub and relaxing radial lines just after sticky spiral construction is finished and in some cases rebuilding it. One could argue that constructional constraints may account for some general similarities (e.g., the spider must build at least some non-sticky radial lines before she can build any sticky spiral lines attached to them—Coddington 1986b). But the existence of these different theridiosomatoid behaviors shows that many other divergent processes are feasible, and thus makes the lack of such uloborid-araneoid differences striking.

10.15 SUMMARY

This chapter focused on the evolutionary history of web-building behavior in spiders. It continued the theme of chapter 9 that spider webs are ancient, and that their long evolutionary history has resulted in an extensive evolutionary exploration of feasible functional designs. It began by discussing the hypothesis that a common pattern of ontogenetic change can aid in tracing evolutionary trajectories (the "biogenic law"): the pattern is for the order in which web designs change as a spider matures to reflect the order in which they evolved. Web construction behavior thus often shows the same "ontogeny repeats phylogeny" pattern that is common (though by no means universal) in morphology. There are numerous exceptions in web construction behavior, however, and ontogeny is only an inconsistent predictor of evolutionary history. I speculate that the biogenic trend may derive not from developmental constraints, but from uniformity in the differences between selection on larger and smaller individuals in different species.

The questions discussed subsequently related to the order of events during the evolution of web building. Topics varied widely, and included the following. The behavioral traits used to build male sperm webs may have been involved in the origins of prey capture webs. The earliest prey capture webs probably served only to increase interception (by extending the spider's sensory range); stopping and retaining functions probably evolved later. Several general trends can be seen in web evolution (e.g., coevolution between attack behavior and web design), but all have multiple exceptions, and no truly consistent direction of evolutionary change has been demonstrated. Evolutionary acquisitions of new types of silk (especially piriform sillk) probably played important roles in web evolution. Similarly, one set of behavioral capabilities involved in "cut and reel" and "reattachment" behavior may have been important in setting the stage for evolution of orb webs. Sticky cribellum silk is ancient, and has been lost repeatedly, but arguments that its loss was a "key innovation" that permitted subsequent radiations are not conclusive. Recent categorizations of how webs evolved in phylogenetic studies are oversimplified; more behavioral data are needed. Questions about traits that did *not* evolve (and speculations as to why) helped focus attention on previously unrecognized probable synapomorphies of orb weavers.

Data from morphology and behavior indicate that orb webs probably evolved only once; currently available molecular data are also compatible with monophyly. Key innovation arguments regarding the origin of ecribellate orbs from cribellate orbs were discussed critically. Orbs have been repeatedly modified and lost. A new hypothesis for the derivation of some gumfoot webs (surely polyphyletic) from orbs was proposed. The reduction and loss of webs is apparently associated with the use of prey attractant substances (also polyphyletic, and relatively unsuccessful evolutionarily).

There are several patterns in the evolution of the behavioral acts that are associated with web construction. Web-building spiders tend to manipulate lines more dextrously as they build, and to walk under rather than above lines. Ancestral use of movements of the abdomen and the often elongate spinnerets to lay and attach lines were replaced, to a greater or lesser extent in different groups, with use of the legs to manipulate lines, combined with small but very precise movements of the shorter spinnerets. A newly recognized set of behavioral acts functions to manage swaths of lines rather than single draglines. There is abundant evidence that the building behavior of orb weavers is organized in semi-independent modules, that during building the spider makes repeated decisions concerning the use of alternative modules, and that this modular organization has played an important role in the evolution of web construction behavior. These historical accounts highlighted the possibility of using web designs and building behavior as taxonomic characters in phylogenetic studies. Minor details of orb construction behavior have generally proven to be more conservative, and thus better indicators of higher-level relationships than have web designs.

Two threads running through this chapter are the complex concept of homologies, and the existence of semi-independent behavioral modules that can evolve and be the objects of natural selection. Just as in comparative analyses of morphological traits, there are limits and complexities in the interpretations of homologies in analyses of web construction behavior. Nevertheless, the multiple criteria available to evaluate homologies and the often abundant data that are relevant to these criteria, mean that many homology hypotheses can be evaluated in multiple, independent tests, providing relative confidence in many cases. There is a high degree of modularity in web construction, and web evolution has probably been strongly influenced by this modularity (Eberhard 2018).

REFERENCES

Adams, M. R. 2000. Choosing hunting sites: web site preferences of the orb weaver spider, *Neoscona crucifera*, relative to light cues. Journal of Insect Behavior 13:299–305.

Ades, C. 1986. A construção da teia geométrica como programa comportamental. Ciência e Cultura 38:760–775.

———. 1991. Memória e instinto no comportamento de predação da aranha Argiope argentata. Instituto de Psicologia, Universidade de São Paulo, São Paulo.

———. 1995. A construcão da teia geométrica enquanto instinto: primeira parte de um argumento. Psicologia Universidade de Sao Paulo 6:145–172.

Ades, C., and E. N. Ramires. 2002. Asymmetry of leg use during prey handling in the spider *Scytodes globula* (Scytodidae). Journal of Insect Behavior 15:563–570.

Ades, C., S. Cunha, and K. Tiedemann. 1993. Experience-induced changes in orb-web building. 23 International Ethological Conference. Torremolinos, España.

Agnarsson, I. 2003a. Spider webs as habitat patches: the distribution of kleptoparasites (*Argyrodes*, Theridiidae) among host webs (*Nephila*, Tetragnathidae). Journal of Arachnology 31:344–349.

———. 2003b. The phylogenetic placement and circumscription of the genus *Synotaxus* (Araneae: Synotaxidae), a new species from Guyana, and notes on theridioid phylogeny. Invertebrate Systematics 17:719–734.

———. 2004. Morphological phylogeny of cobweb spiders and their relatives (Araneae, Araneoidea, Theridiidae). Zoological Journal of the Linnean Society 141:447–626.

———. 2010. Habitat patch size and isolation as predictors of occupancy and number of argyrodine spider kleptoparasites in *Nephila* webs. Naturwissenschaften 98:163–167.

Agnarsson, I., and T. A. Blackledge. 2009. Can a spider web be too sticky? Tensile mechanics constrains the evolution of capture spiral stickiness in orb-weaving spiders. Journal of Zoology 278:134–140.

Agnarsson, I., and J. A. Coddington. 2006. Notes on web and web plasticity and description of the male of *Achaearanea hieroglyphica* (Mello-Leitão) (Araneae, Theridiidae). Journal of Arachnology 34:638–641.

Agnarsson, I., L. Avilés, J. A. Coddington, and W. P. Maddison. 2006a. Sociality in theridiid spiders: repeated origins of an evolutionary dead end. Evolution 60:2342–2351.

Agnarsson, I., G. Barrantes, and L. J. May-Collado. 2006b. Notes on the biology of *Anelosimus pacificus* Levi, 1963 (Theridiidae, Araneae)—evidence for an evolutionary reversal to a less social state. Journal of Natural History 40:2681–2687.

Agnarsson, I., C. Boutry, and T. A. Blackledge. 2008. Spider silk aging: initial improvement in a high performance material followed by slow degradation. Journal of Experimental Zoology Part A: Ecological Genetics and Physiology 309A:494–504.

Agnarsson, I., A. Dhinojwala, V. Sahni, and T. A. Blackledge. 2009a. Spider silk as a novel high performance biomimetic muscle driven by humidity. Journal of Experimental Biology 212:1990–1994.

Agnarsson, I., C. Boutry, S. C. Wong, A. Baji, A. Dhinojwala, A. T. Sensenig, and T. A. Blackledge. 2009b. Supercontraction forces in spider dragline silk depend on hydration rate. Zoology 112:325–331.

Agnarsson, I., M. Kuntner, and T. A. Blackledge. 2010. Bioprospecting finds the toughest biological material: extraordinary silk from a giant riverine orb spider. PLOS ONE 5:e11234.

Agnarsson, I., M. Gregorič, T. A. Blackledge, and M. Kuntner. 2012. The phylogenetic placement of Psechridae within Entelegynae and the convergent origin of orb-like spider webs. Journal of Zoological Systematic Evolution Research 51:100–106.

Agnarsson, I., J. A. Coddington, and M. Kuntner. 2013. Systematics: progress in the study of spider diversity and evolution. Pages 58–111 *in* D. Penney and N. I. Platnick, eds. Spider Research in the 21st Century: Trends and Perspectives. Siri Scientific Press, Manchester.

Aisenberg, A., and G. Barrantes. 2011. Sexual behavior, cannibalism, and mating plugs as sticky traps in the orb weaver spider *Leucauge argyra* (Tetragnathidae). Naturwissenschaften 98:605–613.

Akerman, C. 1923. A comparison of the habits of a South African spider, *Cladomelea*, with those of an Australian *Dicrostichus*. Annals of the Natal Museum 5:83–88.

———. 1926. On the spider, *Menneus camelus* Pocock, which constructs a moth catching expanding snare. Annals of the Natal Museum 5:411–422.

———. 1932. On the spider *Miagrammopes* sp. which constructs a single-line snare. Annals of the Natal Museum 7:137–143.

Alam, M. S., and C. H. Jenkins. 2005. Damage tolerance in naturally compliant structures. International Journal of Damage Mechanics 14:365–384.

Alcock, J. 2013. Animal Behavior: An Evolutionary Approach. 10th ed. Sinauer Associates, Sunderland, MA.

Alderweireldt, M. 1994. Prey selection and prey capture strategies of linyphiid spiders in high-input agricultural fields. Bulletin of the British Arachnological Society 9:300–308.

Allmeling, C., C. Radtke, and P. M. Vogt. 2013. Technical and biomedical uses of nature's strongest fiber: spider silk. Pages 475–490 *in* W. Nentwig, ed. Spider Ecophysiology. Springer-Verlag, Berlin, Heidelberg.

Álvarez-Padilla, F., and G. Hormiga. 2011. Morphological and phylogenetic atlas of the orb-weaving spider family Tetragnathidae (Araneae: Araneoidea). Zoological Journal of the Linnean Society 162:713–879.

Amarpuri, G., C. Zhang, C. Diaz, B. D. Opell, T. A. Blackledge, and A. Dhinojwala. 2015. Spiders tune glue viscosity to maximize adhesion. ACS Nano 9:11472–11478.

Anderson, C. M., and E. K. Tillinghast. 1980. GABA and taurine derivatives on the adhesive spiral of the orb web of *Argiope* spiders, and their possible behavioural significance. Physiological Entomology 5:101–106.

Anderson, J. F. 1970. Metabolic rates of spiders. Comparative Biochemistry and Physiology 33:51–72.

———. 1974. Responses to starvation in the spiders *Lycosa lenta* Hentz and *Filistata hibernalis* (Hentz). Ecology 55:576–585.

Anotaux, M., J. Marchal, N. Châline, L. Desquilbet, R. Leborgne, C. Gilbert, and A. Pasquet. 2012. Ageing alters spider orb-web construction. Animal Behaviour 84:1113–1121.

Aoyanagi, Y., and K. Okumura. 2010. Simple model for the mechanics of spider webs. Physical Review Letters 104: 038102.

ap Rhisiart, A., and F. Vollrath. 1994. Design features of the orb web of the spider, *Araneus diadematus*. Behavioral Ecology 5:280–287.

Araújo, M. S., and M. O. Gonzaga. 2007. Individual specialization in the hunting wasp *Trypoxylon* (Trypargilum) *Albonigrum* (Hymenoptera, Crabronidae). Behavioral Ecology and Sociobiology 61:1855–1863.

Argintean, S., J. Chen, M. Kim, and A. M. F. Moore. 2006. Resilient silk captures prey in black widow cobwebs. Applied Physics A: Materials Science & Processing 82:235–241.

Armas, L. F., and G. Alayón. 1987. Observaciones sobre la ecología trofica de una población de *Argiope trifasciata* (Araneae: Araneidae) en el sur de La Habana. Poeyana 344:1–18.

Arnedo, M. A., J. A. Coddington, I. Agnarsson, and R. G. Gillespie. 2004. From a comb to a tree: phylogenetic relationships of the comb-footed spiders (Araneae, Theridiidae) inferred from nuclear and mitochondrial genes. Molecular Phylogenetics and Evolution 31:225–245.

Austin, A. D., and D. T. Anderson. 1978. Reproduction and development of the spider *Nephila edulis* (Koch) (Araneidae: Araneae). Australian Journal of Zoology 26:501–518.

Austin, A. D., and A. D. Blest. 1979. The biology of two Australian species of dinopid spider. Journal of Zoology 189:145–156.

Avilés, L. 1986. Sex-ratio bias and possible group selection in the social spider *Anelosimus eximius*. American Naturalist 128:1–12.

———. 1997. Causes and consequences of cooperation and permanent-sociality in spiders. Pages 476–498 *in* Jae C. Choe and Bernard J. Crespi, ed. The Evolution of Social Behavior in Insects and Arachnids. Cambridge University Press, Cambridge.

Ayoub, N. A., J. E. Garb, A. Kuelbs, and C. Y. Hayashi. 2013. Ancient properties of spider silks revealed by the complete gene sequence of the prey-wrapping silk protein (AcSp1). Molecular Biology and Evolution 30:589–601.

Baba, Y. G., and T. Miyashita. 2006. Does individual internal state affect the presence of a barrier web in *Argiope bruennichii* (Araneae: Araneidae)? Journal of Ethology 24:75–78.

Baba, Y. G., M. Kusahara, Y. Maezono, and T. Miyashita. 2014. Adjustment of web-building initiation to high humidity: a constraint by humidity-dependent thread stickiness in the spider *Cyrtarachne*. Naturwissenschaften 101:587–593.

Babb, P. L., N. F. Lahens, S. M. Correa-Garhwal, D. N. Nicholson, E. J. Kim, J. B. Hogenesch, M. Kuntner, et al. 2017. The *Nephila clavipes* genome highlights the diversity of spider silk genes and their complex expression. Nature Genetics advance online publication.

Babu, K. S. 1973. Histology of the neurosecretory system and neurohaemal organs of the spider, *Argiope aurantia* (Lucas). Journal of Morphology 141:77–97.

Baerg, W. J. 1958. The Tarantula. University of Kansas Press, Lawrence, KS.

Balogh, J. I. 1934. Vorlaufige Mitteilung iiber radnetzbauende Pachygnathen. Folia Zool. Hydrobiol. Riga 6:94–6.

Baltzer, F. 1930. Ueber die Orientierung der Trichterspinne *Agelena labyrinthica* (Cl) nach der Spannung des Netzes. Revue Suisse de Zoologie 37:363–369.

Barghusen, L. E., D. L. Claussen, M. S. Anderson, and A. J. Bailer. 1997. The effects of temperature on the web-building behaviour of the common house spider, *Achaearanea tepidariorum*. Functional Ecology 11:4–10.

Barrantes, G. 2007. Detritus camouflage in webs of *Helvibis longicauda* (Araneae: Theridiidae). Arachnology 14:59–60.

Barrantes, G., and W. G. Eberhard. 2007. The evolution of prey-wrapping behaviour in spiders. Journal of Natural History 41:1631–1658.

———. 2010. Ontogeny repeats phylogeny in *Steatoda* and *Latrodectus* spiders. Journal of Arachnology 38:485–494.

———. 2012. Extreme behavioral adjustments by an orb-web spider to restricted spaces. Ethology 118:438–449.

Barrantes, G., and R. Madrigal-Brenes. 2008. Ontogenetic changes in web architecture and growth rate of *Tengella radiata* (Araneae, Tengellidae). Journal of Arachnology 36:545–551.

Barrantes, G., and M. J. Ramírez. 2013. Courtship, egg sac construction, and maternal care in *Kukulcania hibernalis*, with information on the courtship of *Misionella mendensis* (Araneae, Filistatidae). Arachnology 16:72–80.

Barrantes, G., and E. Triana. 2009. Characteristics of the capture threads of *Synotaxus* sp. (probably *S. turbinatus*) webs (Araneae, Synotaxidae). Arachnology 14:349–352.

Barrantes, G., and J. L. Weng. 2006a. The prey attack behavior of *Achaearanea tesselata* (Araneae, Theridiidae). Journal of Arachnology 34:456–466.

———. 2006b. Viscid globules in webs of the spider *Achaearanea tesselata* (Araneae: Theridiidae). Journal of Arachnology 34:480–482.

———. 2007a. Carrion feeding by spiderlings of the cob-web spider *Theridion evexum* (Araneae, Theridiidae). Journal of Arachnology 35:557–560.

———. 2007b. Natural history, courtship, feeding behaviour

and parasites of *Theridion evexum* (Araneae: Theridiidae). Bulletin of the British Arachnological Society 14:61–65.

Barrantes, G., Triana, E. and Sanchez-Quiroz, C. 2017. Functional changes in web design along the ontogeny of two orb-weavers. Journal of Arachnology 45:152–159.

Barrows, W. M. 1915. The reactions of an orb-weaving spider, *Epeira sclopetaria* Clerck, to rhythmic vibrations of its web. Biological Bulletin 29:316.

Barth, F. G. 1985. Insects and Flowers. The Biology of a Partnership. Princeton University Press, Princeton, NJ.

———. 1993. Sensory guidance in spider pre-copulatory behaviour. Comparative Biochemistry and Physiology Part A: Physiology 104:717–733.

———. 2002. A Spider's World: Senses and Behavior. Springer Verlag, Berlin.

Bayer, S. 2011. Revision of the pseudo-orbweavers of the genus *Fecenia* Simon, 1887 (Araneae, Psechridae), with emphasis on their pre-epigyne. ZooKeys 153:1.

Bays, S. 1962. A study of the training possibilities of *Araneus diadematus* Cl. Cellular and Molecular Life Sciences 18:423–424.

Becker, N., E. Oroudjev, S. Mutz, J. P. Cleveland, P. K. Hansma, C. Y. Hayashi, D. E. Makarov, and H. G. Hansma. 2003. Molecular nanosprings in spider capture-silk threads. Nature Materials 2:278–283.

Bednarski, J., H. Ginsberg, and E. M. Jakob. 2009. Competitive interactions between a native spider (*Frontinella communis*, Araneae: Linyphiidae) and an invasive spider (*Linyphia triangularis*, Araneae: Linyphiidae). Biological Invasions 12:905–912.

Bel-Venner, M. C., and S. Venner. 2006. Mate-guarding strategies and male competitive ability in an orb-weaving spider: results from a field study. Animal Behaviour 71:1315–1322.

Benjamin, S. P., and S. Zschokke. 2002. Untangling the tangle-web: web construction behavior of the comb-footed spider *Steatoda triangulosa* and comments on phylogenetic implications (Araneae: Theridiidae). Journal of Insect Behavior 15:791–809.

———. 2003. Webs of theridiid spiders: construction, structure and evolution. Biological Journal of the Linnean Society 78:293–305.

———. 2004. Homology, behaviour and spider webs: web construction behaviour of *Linyphia hortensis* and *L. triangularis* (Araneae: Linyphiidae) and its evolutionary significance. Journal of Evolutionary Biology 17:120–130.

Benjamin, S. P., M. Düggelin, and S. Zschokke. 2002. Fine structure of sheet-webs of *Linyphia triangularis* (Clerck) and *Microlinyphia pusilla* (Sundevall), with remarks on the presence of viscid silk. Acta Zoologica (Stockholm) 83:49–59.

Bentzien, M. 1973. Biology of the spider *Diguetia imperiosa* (Araneida: Diguetidae). Pan-Pacific Entomologist 49:110–123.

Berry, J. W. 1987. Notes on the life history and behavior of the communal spider *Cyrtophora moluccensis* (Doleschall) (Araneae, Araneidae) in Yap, Caroline Islands. Journal of Arachnology 15:309–319.

Betancourt, J. L., T. R. Van Devender, and P. S. Martin. 1990. Packrat Middens: The Last 40,000 Years of Biotic Change. University of Arizona Press, Tucson, AZ.

Betz, O., M. K. Thayer, and A. F. Newton. 2003. Comparative morphology and evolutionary pathways of the mouthparts in spore-feeding Staphylinoidea (Coleoptera). Acta Zoologica 84:179–238.

Bianchi, F. A. 1945. Notes on the abundance of the spiders *Latrodectus mactans*, *L. geometricus* and *Argiope avara*, and of their parasites on the Island of Hawaii. Proceedings of the Hawaii Entomological Society 12:245–247.

Biere, J. M., and G. W. Uetz. 1981. Web orientation in the spider *Micrathena gracilis* (Araneae: Araneidae). Ecology 62:336–344.

Bilde, T., and Y. Lubin. 2011. Group living in spiders: cooperative breeding and coloniality. Pages 275–306 *in* M. E. Herberstein, ed. Spider Behaviour: Flexibility and Versatility. Cambridge University Press, Cambridge.

Bishop, L., and S. R. Connolly. 1992. Web orientation, thermoregulation, and prey capture efficiency in a tropical forest spider. Journal of Arachnology 20:173–178.

Bittencourt, D., K. Dittmar, R. V. Lewis, and E. L. Rech. 2010. A MaSp2-like gene found in the Amazon mygalomorph spider *Avicularia juruensis*. Comparative Biochemistry and Physiology Part B: Biochemistry and Molecular Biology 155: 419–426.

Bjorkman-Chiswell, B. T., M. M. Kulinski, R. L. Muscat, K. A. Nguyen, B. A. Norton, M. R. Symonds, G. E. Westhorpe, et al. 2004. Web-building spiders attract prey by storing decaying matter. Naturwissenschaften 91:245–248.

Blackledge, T. A. 1998a. Stabilimentum variation and foraging success in *Argiope aurantia* and *Argiope trifasciata* (Araneae: Araneidae). Journal of Zoology 246:21–27.

———. 1998b. Signal conflict in spider webs driven by predators and prey. Proceedings of the Royal Society of London B: Biological Sciences 265:1991–1996.

———. 2011. Prey capture in orb weaving spiders: are we using the best metric? Journal of Arachnology 39:205–210.

———. 2012. Spider silk: a brief review and prospectus on research linking biomechanics and ecology in draglines and orb webs. Journal of Arachnology 40:1–12.

———. 2013. Spider silk: molecular structure and function in webs. Pages 267–281 *in* W. Nentwig, ed. Spider Ecophysiology. Springer-Verlag, Berlin, Heidelberg.

Blackledge, T. A., and C. M. Eliason. 2007. Functionally independent components of prey capture are architecturally constrained in spider orb webs. Biology Letters 3:456–458.

Blackledge, T. A., and R. G. Gillespie. 2004. Convergent evolution of behavior in an adaptive radiation of Hawaiian web-building spiders. Proceedings of the National Academy of Sciences of the United States of America 101:16228–16233.

Blackledge, T. A., and C. Y. Hayashi. 2006a. Silken toolkits: biomechanics of silk fibers spun by the orb web spider *Argiope argentata* (Fabricius 1775). Journal of Experimental Biology 209:2452–2461.

———. 2006b. Unraveling the mechanical properties of composite silk threads spun by cribellate orb-weaving spiders. Journal of Experimental Biology 209:3131–3140.

Blackledge, T. A., and K. M. Pickett. 2000. Predatory interactions between mud-dauber wasps (Hymenoptera, Sphecidae) and *Argiope* (Araneae, Araneidae) in captivity. Journal of Arachnology 28:211–216.

Blackledge, T. A., and J. W. Wenzel. 1999. Do stabilimenta in orb webs attract prey or defend spiders? Behavioral Ecology 10:372–376.

———. 2000. The evolution of cryptic spider silk: a behavioral test. Behavioral Ecology 11:142–145.

———. 2001a. Silk mediated defense by an orb web spider against predatory mud-dauber wasps. Behaviour 138:155–171.

———. 2001b. State-determinate foraging decisions and web architecture in the spider *Dictyna volucripes* (Araneae Dictynidae). Ethology Ecology & Evolution 13:105–113.

Blackledge, T. A., and J. M. Zevenbergen. 2006. Mesh width influences prey retention in spider orb webs. Ethology 112: 1194–1201.

———. 2007. Condition-dependent spider web architecture in the western black widow, *Latrodectus hesperus*. Animal Behaviour 73:855–864.

Blackledge, T. A., J. A, Coddington and R. G. Gillespie. 2003a. Are three-dimensional spider webs defensive adaptations? *Ecology Letters* 6:13–18.

Blackledge, T. A., G. J. Binford, and R. G. Gillespie. 2003b. Resource use within a community of Hawaiian spiders (Araneae: Tetragnathidae). Annales Zoologici Fennici 40:293–303.

Blackledge, T. A., R. A. Cardullo, and C. Y. Hayashi. 2005a. Polarized light microscopy, variability in spider silk diameters, and the mechanical characterization of spider silk. Invertebrate Biology 124:165–173.

Blackledge, T. A., A. P. Summers, and C. Y. Hayashi. 2005b. Gumfooted lines in black widow cobwebs and the mechanical properties of spider capture silk. Zoology 108: 41–46.

Blackledge, T. A., J. E. Swindeman, and C. Y. Hayashi. 2005c. Quasistatic and continuous dynamic characterization of the mechanical properties of silk from the cobweb of the black widow spider *Latrodectus hesperus*. Journal of Experimental Biology 208:1937–1949.

Blackledge, T. A., J. A. Coddington, and I. Agnarsson. 2009a. Fecundity increase supports adaptive radiation hypothesis in spider web evolution. Communicative & Integrative Biology 2:459–463.

Blackledge, T. A., C. Boutry, S. C. Wong, A. Baji, A. Dhinojwala, V. Sahni, and I. Agnarsson. 2009b. How super is supercontraction? Persistent versus cyclic responses to humidity in spider dragline silk. Journal of Experimental Biology 212:1981–1989.

Blackledge, T. A., N. Scharff, J. A. Coddington, T. Szüts, J. W. Wenzel, C. Y. Hayashi, and I. Agnarsson. 2009c. Reconstructing web evolution and spider diversification in the molecular era. Proceedings of the National Academy of Sciences 106:5229–5234.

Blackledge, T. A., M. Kuntner, and I. Agnarsson. 2011. The form and function of spider orb webs: evolution from silk to ecosystems. Advances in Insect Physiology 40:175–262.

Blackledge, T. A., M. Kuntner, M. Marhabaie, T. C. Leeper, and I. Agnarsson. 2012a. Biomaterial evolution parallels behavioral innovation in the origin of orb-like spider webs. Scientific Reports 2: 833. DOI:10.1038/srep00833.

Blackledge, T. A., J. Pérez-Rigueiro, G. R. Plaza, B. Perea, A. Navarro, G. V. Guinea, and M. Elices. 2012b. Sequential origin in the high performance properties of orb spider dragline silk. Scientific Reports 2: 782. DOI:10.1038/srep00782

Blamires, S. J. 2010. Plasticity in extended phenotypes: orb web architectural responses to variations in prey parameters. Journal of Experimental Biology 213:3207–3212.

Blamires, S. J., D. F. Hochuli, and M. B. Thompson. 2007a. Does decoration building influence antipredator responses in an orb-web spider (*Argiope keyserlingi*) in its natural habitat? Australian Journal of Zoology 55:1–7.

Blamires, S. J., M. B. Thompson, and D. F. Hochuli. 2007b. Habitat selection and web plasticity by the orb spider *Argiope keyserlingi* (Argiopidae): do they compromise foraging success for predator avoidance? Austral Ecology 32:551–563.

Blamires, S. J., D. F. Hochuli, and M. B. Thompson. 2008. Why cross the web: decoration spectral properties and prey capture in an orb spider (*Argiope keyserlingi*) web. Biological Journal of the Linnean Society 94:221–229.

Blamires, S. J., Y.-H. Lee, C.-M. Chang, I.-T. Lin, J.-A. Chen, T.-Y. Lin, and I.-M. Tso. 2010a. Multiple structures interactively influence prey capture efficiency in spider orb webs. Animal Behaviour 80:947–953.

Blamires, S. L., I.-C. Chao, and I.-M. Tso. 2010b. Prey type, vibrations and handling interactively influence spider silk expression. Journal of Experimental Biology 213:3906–3910. DOI:10.1242/jeb.046730.

Blamires, S. J., Y. C. Chao, C. P. Liao, and I. M. Tso. 2011. Multiple prey cues induce foraging flexibility in a trap-building predator. Animal Behaviour 81:955–961.

Blamires, S. J., C. Hou, L. F. Chen, C. P. Liao, and I. M. Tso. 2013. Three-dimensional barricading of a predatory trap reduces predation and enhances prey capture. Behavioral Ecology and Sociobiology 67:709–714.

Blanke, R. 1972. Untersuchungen zur Okophysiologie und Okethologie von *Cyrtophora citricola* Forskal (Araneae, Araneidae) in Andalusien. Forma et functio 5:125–206.

———. 1975. Das Sexualverhalten der Gattung *Cyrtphora* als Hilfsmittel fur phylogenetisch e Aussagen. Proceedings of the 6th International Arachnological Congress 6:116–119.

Bond, J. E., and B. D. Opell. 1998. Testing adaptive radiation and key innovation hypotheses in spiders. Evolution 52:403–414.

Bond, J. E., N. L. Garrison, C. A. Hamilton, R. L. Godwin, M. Hedin, and I. Agnarsson. 2014. Phylogenomics resolves a spider backbone phylogeny and rejects a prevailing paradigm for orb web evolution. Current Biology 24:1765–1771.

Bonny, A. P. 1980. Seasonal and annual variation over 5 years in contemporary airborne pollen trapped at a Cumbrian lake. Journal of Ecology 68: 421–441.

Bott, R. A., W. Baumgartner, P. Bräunig, F. Menzel, and A.-C. Joel. 2017. Adhesion enhancement of cribellate capture threads by epicuticular waxes of the insect prey sheds new light on spider web evolution. Proceedings of the Royal Society of London B: Biological Sciences 284:20170363.

Boutry, C., and T. A. Blackledge. 2008. The common house spider alters the material and mechanical properties of cobweb silk in response to different prey. Journal of Experimental Zoology Part A: Ecological Genetics and Physiology 309A:542–552.

———. 2009. Biomechanical variation of silk links spinning plasticity to spider web function. Zoology 112:451–460.

———. 2010. Evolution of supercontraction in spider silk: structure–function relationship from tarantulas to orb-weavers. Journal of Experimental Biology 213:3505.

———. 2013. Wet webs work better: humidity, supercontraction and the performance of spider orb webs. Journal of Experimental Biology 216:3606–3610.

Boutry, C., M. Rezak, and Blackledge, T. A. 2011. Plasticity in major ampullate silk production in relation to spider phylogeny and ecology. PLOS ONE 1–8.

Bradley, R. A. 2013. Common Spiders of North America. University of California Press, Berkeley.

Bradoo, B. L. 1971. Some observations on the habits of *Ariamnes* sp. (Araneae: Theridiidae) from Kerala (India). Entomologist's Monthly Magazine.

———. 1972. Some observations on the ecology of social spider *Stegodyphus sarasinorum* Karsch (Araneae: Eresidae) from India. Oriental Insects 6.

———. 1989. Advantages of commensalism in *Uloborus ferokus* Bradoo (Araneae: Uloboridae). Journal of the Bombay Natural History Society 86:323–328.

Brady, A. R. 1962. The spider genus *Sosippus* in North America, Mexico, and Central America (Araneae, Lycosidae). Psyche: A Journal of Entomology 69:129–164.

Brandt, Y., and M. Andrade. 2007. What is the matter with the gravity hypothesis? Functional Ecology 21:1182–1183.

Brandwood, A. 1985. Mechanical properties and factors of safety of spider drag-lines. Journal of Experimental Biology 116:141.

Breed, A. L., V. D. Levine, D. B. Peakall, and P. N. Witt. 1964. The fate of the intact orb web of the spider *Araneus diadematus* Cl. Behaviour 23:43–60.

Briceño, R. D. 1985. Sticky balls in webs of the spider *Modisimus* sp. (Araneae, Pholcidae). Journal of Arachnology 13:7.

Briceño, R. D., and W. G. Eberhard. 2011. The hub as a launching platform: rapid movements of the spider *Leucauge mariana* (Araneae: Tetragnathidae) as it turns to attack prey. Journal of Arachnology 39:102–112.

———. 2012. Spiders avoid sticking to their webs: clever leg movements, branched drip-tip setae, and anti-adhesive surfaces. Naturwissenschaften 99:337–341.

Bristowe, W. S. 1929. The mating habits of spiders, with special reference to the problems surrounding sex dimorphism. Proceedings of the Zoological Society of London 99:309–358.

———. 1930. Notes on the biology of spiders. I. The evolution of spiders' snares. Annals and Magazine of Natural History 6:334–342.

———. 1932. The Liphistiid Spiders. Proceedings of the Zoological Society of London 102:1015–1057.

———. 1939. The Comity of Spiders Volume I. The Ray Society, London.

———. 1941. The Comity of Spiders Volume II. The Ray Society, London.

———. 1954. The chelicerae of spiders. Endeavour 13:42–49.

———. 1958. The World of Spiders. Collins Press, London.

———. 1976. A contribution to the knowledge of liphistiid spiders. Journal of Zoology 178:1–6.

Brown, S. G., and T. E. Christenson. 1983. The relationships between web parameters and spiderling predatory behavior in the orb-weaver, *Nephila clavipes*. Zeitschrift für Tierpsychologie 63:241–250.

Bruce, M. J., M. E. Herberstein, and M. A. Elgar. 2001. Signalling conflict between prey and predator attraction. Journal of Evolutionary Biology 14:786–794.

Bruce, M. J., A. M. Heiling, and M. E. Herberstein. 2004. Web decorations and foraging success in *'Araneus' eburnus* (Araneae: Araneidae). Annales Zoologici Fennici 41:563–575.

Buchli, H. H. R. 1969. Hunting behavior in the Ctenizidae. American Zoologist 9:175–193.

Burgess, J. W., and P. N. Witt. 1976. Spider webs: design and engineering. Interdisciplinary Science Reviews 1:322–335.

Bush, A. A., D. W. Yu, and M. E. Herberstein. 2008. Function of bright coloration in the wasp spider *Argiope bruennichi* (Araneae: Araneidae). Proceedings of the Royal Society of London B: Biological Sciences 275:1337–1342.

Buskirk, R. E. 1975. Coloniality, activity patterns and feeding in a tropical orb-weaving spider. Ecology 56:1314–1328.

———. 1986. Orb-weaving spiders in aggregations modify individual web structure. Journal of Arachnology 14:259–265.

Caine, L. A., and C. S. Hieber. 1987. Web orientation in the spider *Mangora gibberosa* (Hentz) (Araneae, Araneidae). Journal of Arachnology 15:263–265.

Calixto, A., and H. Levi. 2006. Notes on the natural history of *Aspidolasius branicki* (Araneae: Araneidae) at Tinigua National Park, Colombia, with a revision of the genus. Bulletin of the British Arachnological Society 13:314.

Caraco, T., and R. G. Gillespie. 1986. Risk-sensitivity: foraging mode in an ambush predator. Ecology 67:1180–1185.

Carico, J. E. 1972. The nearctic spider genus *Pisaurinae* (Pisauridae). Psyche: A Journal of Entomology 79:295–310.

———. 1978. Predatory behavior in *Euryopis funebris* (Hentz) (Araneae: Theridiidae) and the evolutionary significance of web reduction. Symposia of the Zoological Society of London 42:51–58.

———. 1985. Description and significance of the juvenile web of *Pisaurina mira* (Walck.) (Araneae: Pisauridae). Bulletin of the British Arachnological Society 6:295–296.

———. 1986. Web removal patterns in orb-weaving spiders. Pages 306–318 *in* W. A. Shear, ed. Spiders: Webs, Behavior, and Evolution. Stanford University Press, Stanford, CA.

Carrel, J. E. 1978. Behavioral thermoregulation during winter in an orb-weaving spider. Symposia of the Zoological Society of London 42:41–50.

Carrel, J. E., H. K. Burgess, and D. M. Shoemaker. 2000. A test of pollen feeding by a linyphiid spider. Journal of Arachnology 28:243–244.

Cartan, C. K., and T. Miyashita. 2000. Extraordinary web and silk properties of *Cyrtarachne* (Araneae, Araneidae): a possible link between orb-webs and bolas. Biological Journal of the Linnean Society 71:219–235.

Castillo, L., and W. G. Eberhard. 1983. Use of artificial webs to determine prey available to orb weaving spiders. Ecology 64:1655–1658.

Castro, T. J. 1995. Estudio comparativo del comportamiento reproductor, en las arañas del género *Leucauge* [Araneae, Tetragnathidae], del Soconusco Chiapas. (Licenciatura). Instituto de la Ciencias y Artes de Chiapas, México.

Cazier, M., and M. Mortenson. 1962. Analysis of the habitat, web design, cocoon and egg sacs of the tube weaving spider *Diguetia canities* (McCook) (Aranea, Diguetidae). Bulletin of the Southern California Academy of Sciences 61:65–88.

Ceballos, L., Y. Hénaut, and L. Legal. 2005. Foraging strategies of *Eriophora edax* (Araneae, Araneidae): a nocturnal orb-weaving spider. Journal of Arachnology 33:509–515.

Cerveira, A., and R. R. Jackson. 2002. Prey, predatory behaviour, and anti-predator defences of *Hygropoda dolomedes* and *Dendrolycosa* sp. (Araneae: Pisauridae), web-building pisaurid spiders from Australia and Sri Lanka. New Zealand Journal of Zoology 29:119–133.

Chacón, P., and W. G. Eberhard. 1980. Factors affecting numbers and kinds of prey caught in artificial spider webs, with considerations of how orb webs trap prey. Bulletin of the British Arachnological Society 5:29–38.

Champion de Crespigny, F. E., M. E. Herberstein, and M. A. Elgar. 2001. The effect of predator-prey distance and prey profitability on the attack behaviour of the orb-web spider *Argiope keyserlingi* (Araneidae). Australian Journal of Zoology 49:213–221.

Chapin, K. J., and E. A. Hebets. 2016. The behavioral ecology of amblypygids. Journal of Arachnology 44:1–14.

Chaw, R. C., S. M. Correa-Garhwal, T. H. Clarke, N. A. Ayoub, and C. Y. Hayashi. 2015. Proteomic evidence for components of spider silk synthesis from black widow silk glands and fibers. Journal of Proteome Research 14:4223–4231.

Chaw, R. C., Y. Zhao, J. Wei, N. A. Ayoub, R. Allen, K. Atrushi, and C. Y. Hayashi. 2014. Intragenic homogenization and multiple copies of prey-wrapping silk genes in *Argiope* garden spiders. BMC Evolutionary Biology 14:31.

Cheng, R. C., and I. Tso. 2007. Signaling by decorating webs: luring prey or deterring predators? Behavioral Ecology 18:1085.

Cheng, R. C., E. C. Yang, C. P. Lin, M. E. Herberstein, and I. M. Tso. 2010. Insect form vision as one potential shaping force of spider web decoration design. Journal of Experimental Biology 213:759.

Chittka, L. and J. Niven 2009. Are bigger brains better? Current Biology 19:995–1008.

Chigira, Y. 1978. A note of the web of *Paraplectana tsushimensis*. Atypus 72:19–24.

Chmiel, K., M. E. Herberstein, and M. A. Elgar. 2000. Web damage and feeding experience influence web site tenacity in the orb-web spider *Argiope keyserlingi* Karsch. Animal Behaviour 60:821–826.

Chou, I.-C., P.-H. Wang, P.-S. Shen, and I.-M. Tso. 2005. A test of prey-attracting and predator defence functions of prey carcass decorations built by *Cyclosa* spiders. Animal Behaviour 69:1055–1061.

Christiansen, A., R. Baum, and P. N. Witt. 1962. Changes in spider webs brought about by mescaline, psilocybin and an increase in body weight. Journal of Pharmacology and Experimental Therapeutics 136:31–37.

Clements, R., and D. Li. 2005. Regulation and non-toxicity of the spit from the pale spitting spider *Scytodes pallida* (Araneae: Scytodidae). Ethology 111:311–321.

Clyne, D. 1967. Notes on the construction of the net and sperm-web of a cribellate spider *Dinopus subrufus* (Koch) (Araneidae: Dinopidae). Australian Zoology 14:189–197.

———. 1973. Notes on the web of *Poecilopachys australasia* (Griffith & Pidgeon, 1833) (Araneidae, Argiopidae). Australian Entomological Magazine 1:23–29.

———. 1979. The Garden Jungle. Collins Press, London.

Coddington, J. A. 1986a. The genera of the spider family Theridiosomatidae. Smithsonian Contributions to Zoology 422:1–96.

———. 1986b. The monophyletic origin of the orb web. Pages 319–363 *in* W. A. Shear, ed. Spiders: Webs, Behavior and Evolution. Stanford University Press, Stanford, CA.

———. 1986c. Orb webs in "non-orb weaving" ogre-faced spiders (Araneae: Dinopidae): a question of genealogy. Cladistics 2:53–67.

———. 1987. Notes on spider natural history: the webs and habits of *Araneus niveus* and *A. cingulatus* (Araneae, Araneidae). Journal of Arachnology 15:268–270.

———. 1989. Spinneret silk spigot morphology: evidence for the monophyly of orbweaving spiders, Cyrtophorinae (Araneidae), and the group Theridiidae plus Nesticidae. Journal of Arachnology 17:71–95.

———. 1990. Ontogeny and homology in the male palpus of orb-weaving spiders and their relatives, with comments on phylogeny (Araneoclada: Araneoidea, Deinopoidea). Smithsonian Contributions to Zoology 496:1–52.

———. 2001. Whence sheet webs? Web construction in *Thaida peculiaris* Karsch (Araneae: Austrochilidae). Abstracts, International Congress of Arachnology. South Africa.

———. 2005. Phylogeny and classification of spiders. Pages 18–24 *in* D. Ubick, P. Paquin, P. E. Cushing, and V. D. Roth, eds. Spiders of North America: An Identification Manual. American Arachnological Society.

Coddington, J. A., and H. W. Levi. 1991. Systematics and evolution of spiders (Araneae). Annual Review of Ecology and Systematics 22:565–592.

Coddington, J. A., and C. Sobrevila. 1987. Web manipulation and two stereotyped attack behaviors in the ogre-faced spider *Deinopis spinosus* Marx (Araneae, Deinopidae). Journal of Arachnology 15:213–225.

Coddington, J. A., and C. G. Valerio. 1980. Observations on the web and behavior of *Wendilgarda* spiders (Araneae: Theridiosomatidae). Psyche: A Journal of Entomology 87: 93–106.

Coddington, J. A., H. D. Chanzy, C. L. Jackson, G. Raty, and K. C. Gardner. 2002. The unique ribbon morphology of the major ampullate silk of spiders from the genus *Loxosceles* (Recluse spiders). Biomacromolecules 3:5–8.

Coddington, J. A., M. Kuntner, and B. D. Opell. 2012. Systematics of the spider family Deinopidae, with a revision of the genus *Menneus*. Smithsonian Contributions to Zoology 636:1–61.

Colebourn, P. 1974. The influence of habitat structure on the distribution of *Araneus diadematus* Clerck. Journal of Animal Ecology 43:401–409.

Colin, M. E., D. Richard, and S. Chauzy. 1991. Measurement of electric charges carried by bees: evidence of biological variations. Journal of Bioelectricity 10:17–32.

Colinvaux, P. A., P. E. De Oliveira, and M. B. Bush. 2000. Amazonian and Neotropical plant communities on glacial time-scales: the failure of the aridity and refuge hypotheses. Quaternary Science Reviews 19:141–169.

Collatz, K.-G. 1987. Structure and function of the digestive tract. Pages 229–238 *in* W. Nentwig, ed. Ecophysiology of Spiders. Springer-Verlag, Berlin, Heidelberg.

Comstock, J. H. 1967. The Spider Book. Revised and edited by W. J. Gertsch. Cornell University Press, Ithaca.

———. 1912. The evolution of the webs of spiders. Annals of the Entomological Society of America 5:1–10.

Coslovsky, M., and S. Zschokke. 2009. Asymmetry in orb-webs: an adaptation to web building costs? Journal of Insect Behavior 22:29–38.

Costa, F. G., and F. Pérez-Miles. 2002. Reproductive biology of Uruguayan theraphosids (Araneae, Mygalomorphae). Journal of Arachnology 30:571–587.

Coville, R. E., and P. L. Coville. 1980. Nesting biology and male behavior of *Trypoxylon* (Trypargilum) *enoctitlan* in Costa Rica (Hymenoptera: Sphecidae). Annals of the Entomological Society of America 73:110–119.

Cox, J. 2015. *Atypus affinis* spiderling dispersal on the Great Orme, North Wales. Newsletter British Arachnological Society 134:6–7.

Coyle, F. A. 1971. Systematics and natural history of the mygalomorph spider genus *Antrodiaetus* and related genera (Araneae: Antrodiaetidae). Bulletin of the Museum of Comparative Zoology 141:269–402.

———. 1981a. Notes on the behaviour of *Ummidia* trapdoor spiders (Araneae; Ctenizidae): burrow construction, prey capture, and the functional morphology of the peculiar third tibia. Bulletin of the British Arachnological Society 5:159–165.

———. 1981b. The mygalomorph spider genus *Microhexura* (Araneae, Dipluridae). Bulletin of the American Museum of Natural History 170:64–75.

———. 1985. Ballooning behavior of *Ummidia* spiderlings (Araneae, Ctenizidae). Journal of Arachnology 13:137–138.

———. 1986. The role of silk in prey capture by nonaraneomorph spiders. Pages 269–305 *in* W. A. Shear, ed. Spiders: Webs, Behavior and Evolution. Stanford University Press, Stanford, CA.

———. 1988. A revision of the American funnel-web mygalomorph spider genus *Euagrus* (Araneae, Dipluridae). Bulletin of the American Museum of Natural History 187: 203–292.

———. 1995. A revision of the funnelweb mygalomorph spider subfamily Ischnothelinae (Araneae, Dipluridae). Bulletin of the American Museum of Natural History 226:1–333.

Coyle, F. A., and W. R. Icenogle. 1994. Natural history of the Californian trapdoor spider genus *Aliatypus* (Araneae, Antrodiaetidae). Journal of Arachnology 22:225–255.

Coyle, F. A., and N. D. Kentner. 1990. Observations on the prey and prey capture behaviour of the funnelweb mygalomorph spider genus *Ischnothele* (Araneae, Dipluridae). Bulletin of the British Arachnological Society 8:97–104.

Coyle, F. A., and T. E. Meigs. 1989. Two new species of kleptoparasitic *Mysmenopsis* (Araneae, Mysmenidae) from Jamaica. Journal of Arachnology 17:59–70.

———. 1992. Web co-habitants of the African funnelweb spider, *Thelechoris karschi* (Araneae, Dipluridae). Journal of African Zoology 106:289–295.

Coyle, F. A., R. Dellinger, and R. Bennet. 1992. Retreat architecture and construction behaviour of an East African idiopine trapdoor spider (Araneae, Idiopidae). Bulletin of the British Arachnological Society 9:99–104.

Craig, C. L. 1986. Orb-web visibility: the influence of insect flight behaviour and visual physiology on the evolution of web designs within the Araneoidea. Animal Behaviour 34:54–68.

———. 1987a. The ecological and evolutionary interdependence between web architecture and web silk spun by orb web weaving spiders. Biological Journal of the Linnean Society 30:135–162.

———. 1987b. The significance of spider size to the diversification of spider-web architectures and spider reproductive modes. American Naturalist 129:47–68.

———. 1988. Insect perception of spider orb webs in three light habitats. Functional Ecology 2:277–282.

———. 1989. Alternative foraging modes of orb web weaving spiders. Biotropica 21:257–264.

———. 1990. Effects of background pattern on insect perception of webs spun by orb-weaving spiders. Animal Behaviour 39:135–144.

———. 1991. Physical constraints on group foraging and social evolution: observations on web-spinning spiders. Functional Ecology 5:649–654.

———. 1992. Aerial web-weaving spiders: linking molecular and organismal processes in evolution. Trends in Ecology & Evolution 7:270–273.

———. 1994a. Limits to learning: effects of predator pattern and colour on perception and avoidance-learning by prey. Animal Behaviour 47:1087–1099.

———. 1994b. Predator foraging behavior in response to perception and learning by its prey: interactions between orb-spinning spiders and stingless bees. Behavioral Ecology and Sociobiology 35:45–52.

———. 1997. Evolution of arthropod silks. Annual Review of Entomology 42:231–267.

———. 2003. Spiderwebs and Silk: Tracing Evolution from Molecules to Genes to Phenotypes. Oxford University Press, New York.

Craig, C. L., and G. D. Bernard. 1990. Insect attraction to ultraviolet-reflecting spider webs and web decorations. Ecology 71:616–623.

Craig, C. L., and K. Ebert. 1994. Colour and pattern in predator–prey interactions: the bright body colours and patterns of a tropical orb-spinning spider attract flower-seeking prey. Functional Ecology 8:616–620.

Craig, C. L., and C. R. Freeman. 1991. Effects of predator visibility on prey encounter: a case study on aerial web weaving spiders. Behavioral Ecology and Sociobiology 29:249–254.

Craig, C. L., and M. Lehrer. 2003a. Insect color vision is a potential selective factor on the evolution of silk chromatic properties and web design. Pages 108–122 *in* C. L. Craig, ed. Spider Webs and Silk. Oxford University Press, New York.

———. 2003b. Insect spatial vision is a potential selective factor on the evolution of silk achromatic properties and web architecture. Pages 84–107 *in* C. L. Craig, ed. Spider Webs and Silk. Oxford University Press, New York.

Craig, C. L., A. Okubo, and V. Andreasen. 1985. Effect of spider orb-web and insect oscillations on prey interception. Journal of Theoretical Biology 115:201–211.

Craig, C. L., G. D. Bernard, and J. A. Coddington. 1994. Evolutionary shifts in the spectral properties of spider silks. Evolution 48:287–296.

Craig, C. L., R. S. Weber, and G. D. Bernard. 1996. Evolution of predator-prey systems: spider foraging plasticity in response to the visual ecology of prey. American Naturalist 147:205–229.

Craig, C. L., M. Hsu, D. Kaplan, and N. E. Pierce. 1999. A comparison of the composition of silk proteins produced by spiders and insects. International Journal of Biological Macromolecules 24:109–118.

Craig, C. L., S. G. Wolf, J. L. D. Davis, M. E. Hauber, and J. L. Maas. 2001. Signal polymorphism in the web-decorating spider *Argiope argentata* is correlated with reduced survivorship and the presence of stingless bees, its primary prey. Evolution 55:986–993.

Cranford, S. W., A. Tarakanova, N. M. Pugno, and M. J. Buehler. 2012. Nonlinear material behaviour of spider silk yields robust webs. Nature 482:72–76.

Crews, S. C., and B. D. Opell. 2006. The features of capture threads and orb-webs produced by unfed *Cyclosa turbinata* (Araneae: Araneidae). Journal of Arachnology 34:427–434.

Crome, V. W. 1954. Beschreibung, Morphologie und Lebensweise der *Eucta kaestneri* sp. n. (Araneae, Tetragnathidae). Zoologische Jahrbucher 82:425–452.

———. 1956. Bemerkungen zur Biologie der Kreuzspinne *Araneus sclopetarius* Clerck. Deutschen Entomologischen Gesellschaft 15:45–46.

———. 1957. Bau und Funktion des Spinnapparates und Analhügels, Ernährungsbiologie und allgemeine Bemerkungen zur Lebensweise von *Uroctea durandi* (Latreille) (Araneae, Urocteidae). Zoologische Jahrbücher, Abteilung für Systematik 85:501–672.

Cushing, P. E. 1989. Possible eggsac defense behaviors in the spider *Uloborus glomosus* (Araneae: Uloboridae). Psyche: A Journal of Entomology 96:269–277.

———. 2012. Spider-ant associations: an updated review of myrmecomorphy, myrmecophily, and myrmecophagy in spiders. Psyche: A Journal of Entomology. DOI:10.1155/2012/151989

Cushing, P. E., and B. D. Opell. 1990a. Disturbance behaviors in the spider *Uloborus glomosus* (Araneae, Uloboridae): possible predator avoidance strategies. Canadian Journal of Zoology 68:1090–1097.

———. 1990b. The effect of time and temperature on disturbance behaviors shown by the orb-weaving spider *Uloborus glomosus* (Uloboridae). Journal of Arachnology 18:87–93.

da Cunha-Nogueira, S. S., and C. Ades. 2012. Evidence of learning in the web construction of the spider *Argiope argentata* (Araneae: Araneidae). Revista de Etología 11:23–36.

da Silva-Souza, H., Y. F. Messas, M. de Oliveira Gonzaga, and J. Vasconcellos-Neto. 2015. Substrate selection and spatial segregation by two congeneric species of *Eustala* (Araneae: Araneidae) in southeastern Brazil. Journal of Arachnology 43:59–66.

Dąbrowska-Prot, E., and J. Łuczak. 1968. Spiders and mosquitoes of the ecotone of alder forest (Carici elongate-Alnetum) and oakpine forest (Pino-Quercetum). Ekologia Polska Ser. A 16:461–483.

Dąbrowska-Prot, E., J. Łuczak, and Z. Wójcik. 1973. Ecological analysis of two invertebrate groups in the wet alder wood and meadow ecotone. Ekologia Polska Ser. A 21:753–812.

Dalton, S. K. 1975. Borne on the Wind: The Extraordinary World of Insects in Flight. Reader's Digest Press, New York.

Darchen, R. 1965. Ethologie d'une araignee sociale, *Agelena consociata* Denis. Biologia Gabonica 1:117–146.

Darwin, C. 1859. On the Origin of Species by Means of Natural Selection. J. Murray, London.

Davidson, M. 2011. Some observations on the wood ant spider (*Dipoena torva*). www.woodants.org.uk, retrieved on 5 February 2014.

Davies, V. T. 1982. *Inola* nov. gen., a web-building pisaurid (Araneae: Pisauridae) from northern Australia with descriptions of three species. Memoirs of the Queensland Museum 20:479–487.

de Queiroz, A., and P. H. Wimberger. 1993. The usefulness of behavior for phylogeny estimation: levels of homoplasy in behavioral and morphological characters. Evolution 47:46–60.

Decae, A. E. 1984. A theory on the origin of spiders and the primitive function of spider silk. Journal of Arachnology 12:21–28.

Deeleman-Reinhold, C. L. 1986. Leaf-dwelling Pholcidae in Indo-Australian rain forests. Pages 45–48 *in* W. G. Eberhard, Y. D. Lubin, and R. Robinson, eds. Proceedings of the Ninth International Congress of Arachnology, Panama 1983. Smithsonian Institution Press, Washington, DC.

Denny, M. 1976. The physical properties of spider's silk and their role in the design of orb-webs. Journal of Experimental Biology 65:483–506.

Dimitrov, D., and G. Hormiga. 2010. Mr. Darwin's mysterious spider: on the type species of the genus *Leucauge* White, 1841 (Tetragnathidae, Araneae). Zootaxa 2396:19–36.

Dimitrov, D., L. Lopardo, G. Giribet, M. A. Arnedo, F. Álvarez-Padilla, and G. Hormiga. 2011. Tangled in a sparse spider web: single origin of orb weavers and their spinning work unravelled by denser taxonomic sampling. Proceedings of the Royal Society of London B: Biological Sciences. DOI:10.1098/rspb20112011.

Dimitrov, D., L. R. Benavides, M. A. Arnedo, G. Giribet, C. E. Griswold, N. Scharff, and G. Hormiga. 2016. Rounding up the usual suspects: a standard target-gene approach for resolving the interfamilial phylogenetic relationships of ecribellate orb-weaving spiders with a new family-rank classification (Araneae, Araneoidea). Cladistics 1–30. DOI:10.1111/cla.12165.

Diniz, S., J. Vasconcellos-Neto, and V. Stefani. 2017. Orb inclination in Uloboridae spiders: the roles of microhabitat structure and prey capture. Ethology Ecology & Evolution 29:474–489.

Dippenaar-Schoeman, A. S., and R. Jocqué. 1997. African Spiders: An Identification Manual. Plant Protection Research Institute Handbook. ARC Plant Protection Research Institute, Pretoria.

Dippenaar-Schoeman, A. S., and A. Leroy. 1996. Notes on the biology of *Pycnacantha tribulus*, another araneid without an orbweb (Araneae: Araneidae). Revue Suisse de Zoologie Hors Série 165–171.

DiRienzo, N., and P.-P. Montiglio. 2016. Linking consistent individual differences in web structure and behavior in black widow spiders. Behavioral Ecology 1–8. DOI:10.1093/beheco/arw048.

Doran, N. E., A. M. M. Richardson, and R. Swain. 2001. The reproductive behaviour of the Tasmanian cave spider *Hickmania troglodytes* (Araneae: Austrochilidae). Journal of Zoology 253:405–418.

Dugdale, B. 1969. The weaving of an engineering masterpiece, a spider's orb, done at Fryson Lakes, NJ, August 8, 1942, as observed and reported by BE Dugdale, structural engineer. Natural History 78:36–41.

Dukas, R. 2004. Causes and consequences of limited attention. Brain, Behavior and Evolution 63:197–210.

Duncan, W. 1949. Webs in the Wind. Ronald Press, New York.

Eberhard, W. G. 1967. Attack behavior of diguetid spiders and the origin of prey wrapping in spiders. Psyche: A Journal of Entomology 74:173–181.

———. 1969. Computer simulation of orb-web construction. American Zoologist 9:229.

———. 1970a. The predatory behavior of two wasps, *Agenoideus humilis* (Pompilidae) and *Sceliphron caementarium* (Sphecidae), on the orb weaving spider, *Araneus cornutus* (Araneidae). Psyche: A Journal of Entomology 77:243–251.

———. 1970b. The natural history of the fungus gnats *Leptomorphus bifasciatus* (Say) and *L. subcaeruleus* (Coquillett) (Diptera: Mycetophilidae). Psyche: A Journal of Entomology 77:361–384.

———. 1971a. The ecology of the web of *Uloborus diversus* (Araneae: Uloboridae). Oecologia 6:328–342.

———. 1971b. Senile web patterns in *Uloborus diversus* (Araneae: Uloboridae). Developmental Psychobiology 4:249–254.

———. 1972a. The web of *Uloborus diversus* (Araneae: Uloboridae). Journal of Zoology 166:417–465.

———. 1972b. Observations on the biology of *Achaearanea tesselata* (Araneae: Theridiidae). Psyche: A Journal of Entomology 79:209–212.

———. 1973a. Stabilimenta on the webs of *Uloborus diversus* (Araneae: Uloboridae) and other spiders. Journal of Zoology 171:367–384.

———. 1973b. The ability of a spitting spider, *Scytodes hebraica*, to capture flying prey. *Entomological News* 84:156.

———. 1975. The "inverted ladder" orb web of *Scoloderus* sp. and the intermediate orb of *Eustala* (?) sp. Araneae: Araneidae. Journal of Natural History 9:93–106.

———. 1976a. Photography of orb webs in the field. Bulletin of the British Arachnological Society 3:200–204.

———. 1976b. Physical properties of sticky spirals and their connections: sliding connections in orb webs. Journal of Natural History 10:481–488.

———. 1977a. "Rectangular orb" webs of *Synotaxus* (Araneae: Theridiidae). Journal of Natural History 11:501–507.

———. 1977b. Aggressive chemical mimicry by a bolas spider. Science 198:1173–1175.

———. 1977c. Artificial spider webs. Bulletin of the British Arachnological Society 4:126–128.

———. 1977d. The webs of newly emerged *Uloborus diversus* and of a male *Uloborus* sp. (Araneae: Uloboridae). Journal of Arachnology 4:201–206.

———. 1979a. *Argyrodes attenuatus* (Theridiidae): a web that is not a snare. Psyche: A Journal of Entomology 86:407–414.
———. 1979b. Rates of egg production by tropical spiders in the field. Biotropica 11:292–300.
———. 1980a. The natural history and behavior of the bolas spider, *Mastophora dizzydeani* sp. n. (Araneae). Psyche: A Journal of Entomology 87:143–170.
———. 1980b. Spider and fly play cat and mouse. Natural History 56–61.
———. 1981a. Construction behaviour and the distribution of tensions in orb webs. Bulletin of the British Arachnological Society 5:189–204.
———. 1981b. The single line web of *Phoroncidia studo* Levi (Araneae: Theridiidae): a prey attractant? Journal of Arachnology 9:229–232.
———. 1981c. Notes on the natural history of *Taczanowskia* sp. (Araneae: Araneidae). Bulletin of the British Arachnological Society 5:175–176.
———. 1982. Behavioral characters for the higher classification of orb-weaving spiders. Evolution 36:1067–1095.
———. 1983. Predatory behaviour of an assassin spider, *Chorizopes* sp. (Araneidae), and the defensive behavior of its prey. Journal of the Bombay Natural History Society 79:522–524.
———. 1985. The "sawtoothed" orb web of *Eustala* sp. (Araneae, Araneidae), with a discussion of ontogenetic changes in spiders' web-building behavior. Psyche: A Journal of Entomology 92:105–118.
———. 1986a. Effects of orb-web geometry on prey interception and retention. Pages 70–100 *in* W. Shear, ed. Spiders: Webs, Behaviour and Evolution. Stanford University Press, Palo Alto.
———. 1986b. Ontogenetic changes in the web of *Epeirotypus* sp. (Araneae, Theridiosomatidae). Journal of Arachnology 14:125–128.
———. 1986c. Trail line manipulation as a character for higher level spider taxonomy. Pages 49–51 *in* W. G. Eberhard, D. Lubin, and B. Robinson, eds. Proceedings of the Ninth International Congress of Arachnology, Panama 1983. Smithsonian Institution Press, Washington, DC.
———. 1987a. Construction behaviour of non-orb weaving cribellate spiders and the evolutionary origin of orb webs. Bulletin of the British Arachnological Society 7:175–178.
———. 1987b. Effects of gravity on temporary spiral construction by *Leucauge mariana* (Araneae: Araneidae). Journal of Ethology 5:29–36.
———. 1987c. How spiders initiate airborne lines. Journal of Arachnology 15:1–9.
———. 1987d. Hub construction by *Leucauge mariana* (Araneae, Araneidae). Bulletin of the British Arachnological Society 7:128–132.
———. 1987e. Web-building behavior of anapid, symphytognathid and mysmenid spiders (Araneae). Journal of Arachnology 14:339–356.
———. 1988a. Behavioral flexibility in orb web construction: effects of supplies in different silk glands and spider size and weight. Journal of Arachnology 295–302.
———. 1988b. Combing and sticky silk attachment behaviour by cribellate spiders and its taxonomic implications. Bulletin of the British Arachnological Society 7:247–251.
———. 1988c. Memory of distances and directions moved as cues during temporary spiral construction in the spider *Leucauge mariana* (Araneae: Araneidae). Journal of Insect Behavior 1:51–66.
———. 1989a. Niche expansion in the spider *Wendilgarda galapagensis* (Araneae, Theridiosomatidae) on Cocos Island. Revista de Biología Tropical 37:163–168.
———. 1989b. Effects of orb web orientation and spider size on prey retention. Bulletin of the British Arachnological Society 8:45–48.
———. 1990a. Function and phylogeny of spider webs. Annual Review of Ecology and Systematics 21:341–372.
———. 1990b. Early stages of orb construction by *Philoponella vicina*, *Leucauge mariana*, and *Nephila clavipes* (Araneae, Uloboridae and Tetragnathidae), and their phylogenetic implications. Journal of Arachnology 18:205–234.
———. 1990c. Imprecision in the behavior of *Leptomorphus* sp. (Diptera, Mycetophilidae) and the evolutionary origin of new behavior patterns. Journal of Insect Behavior 3: 327–357.
———. 1991. *Chrosiothes tonala* (Araneae, Theridiidae): a web-building spider specializing on termites. Psyche: A Journal of Entomology 98:7–20.
———. 1992a. Notes on the ecology and behavior of *Physocyclus globosus* (Araneae, Pholcidae). Bulletin of the British Arachnological Society 9:38–42.
———. 1992b. Web construction by *Modisimus* sp. (Araneae, Pholcidae). Journal of Arachnology 20:25–34.
———. 1995. The web and building behavior of *Synotaxus ecuadorensis* (Araneae, Synotaxidae). Journal of Arachnology 23:25–30.
———. 2000a. Breaking the mold: behavioral variation and evolutionary innovation in *Wendilgarda* spiders (Araneae Theridiosomatidae). Ethology Ecology & Evolution 12:223–235.
———. 2000b. The natural history and behavior of *Hymenoepimecis argyraphaga* (Hymenoptera: Ichneumonidae) a parasitoid of *Plesiometa argyra* (Araneae: Tetragnathidae). Journal of Hymenoptera Research 9:220–240.
———. 2001a. Trolling for water striders: active searching for prey and the evolution of reduced webs in the spider *Wendilgarda* sp. (Araneae, Theridiosomatidae). Journal of Natural History 35:229–251.
———. 2001b. Under the influence: webs and building behavior of *Plesiometa argyra* (Araneae, Tetragnathidae) when parasitized by *Hymenoepimecis argyraphaga* (Hymenoptera, Ichneumonidae). Journal of Arachnology 29:354–366.

———. 2003. Substitution of silk stabilimenta for egg sacs by *Allocyclosa bifurca* (Araneae: Araneidae) suggests that silk stabilimenta function as camouflage devices. Behaviour 140:847–868.

———. 2006. Dispersal by *Ummidia* spiderlings (Araneae, Ctenizidae): ancient roots of aerial webs and orientation? Journal of Arachnology 34:254–257.

———. 2007a. Radial organisation in the web of a pisaurid, *Thaumasia* sp. (Araneae: Pisauridae). Bulletin of the British Arachnological Society 14:49–53.

———. 2007b. Miniaturized orb-weaving spiders: behavioural precision is not limited by small size. Proceedings of the Royal Society of London B: Biological Sciences 274:2203–2209.

———. 2008. *Araneus expletus* (Araneae, Araneidae): another stabilimentum that does not function to attract prey. Journal of Arachnology 36:191–194.

———. 2010a. Possible functional significance of spigot placement on the spinnerets of spiders. Journal of Arachnology 38:407–414.

———. 2010b. Recovery of spiders from the effects of parasitic wasps: implications for fine-tuned mechanisms of manipulation. Animal Behaviour 79:375–383.

———. 2011. Are smaller animals behaviourally limited? Lack of clear constraints in miniature spiders. Animal Behaviour 81:813–823.

———. 2012. Cues guiding placement of the first loop of the sticky spiral in orbs of *Micrathena duodecimspinosa* (Araneidae) and *Leucauge mariana* (Tetragnathidae). Arachnology 15:224–227.

———. 2013a. The rare large prey hypothesis for orb web evolution: a critique. Journal of Arachnology 41:76–80.

———. 2013b. The polysphinctine wasps *Acrotaphus tibialis*, *Eruga* ca. *gutfreundi*, and *Hymenoepimecis tedfordi* (Hymenoptera, Ichneumonidae, Pimplinae) induce their host spiders to build modified webs. Annals of the Entomological Society of America 106:652–660.

———. 2014. A new view of orb webs: multiple trap designs in a single structure. Biological Journal of the Linnean Society 111:437–449.

———. 2015. How *Micrathena duodecimspinosa* (Araneae: Araneidae) uses the elasticity of her dragline to hide her egg sac. Journal of Arachnology 43:417–418.

———. 2017. How orb-weavers find and grasp silk lines. Journal of Arachnology 45:145–151.

———. 2018a. Modular patterns in behavioural evolution: webs derived from orbs. Behaviour DOI:10.1163/1568539X-00003502.

———. 2018b. The webs of *Neoantistea riparia* (Araneae: Hahniidae): are dew drops useful in prey capture? Arachnology 17:491–496.

———. 2018c. Hunting behavior of the wasp *Polysphincta gutfreundi* and related polysphinctine wasps (Hymenoptera, Ichneumonidae). Journal of the Kansas Entomological Society 91:177–191.

———. 2019a. Radius construction by *Micrathena duodecimspinosa* (Araneae: Araneidae): a puzzle within a puzzle. Journal of Arachnology 47:57–62.

———. 2019b. Adaptive flexibility in cues guiding spider web construction and its possible implications for spider cognition. Behaviour 156:331–362.

Eberhard, W. G., and G. Barrantes. 2015. Cues guiding uloborid construction behavior support orb web monophyly. Journal of Arachnology 43:371–387.

Eberhard, W. G., and R. D. Briceño. 1983. Chivalry in pholcid spiders. Behavioral Ecology and Sociobiology 13:189–195.

———. 1985. Behavior and ecology of four species of *Modisimus* and *Blechroscelis* (Araneae, Pholcidae). Revue Arachnologique 6:29–36.

Eberhard, W. G., and M. O. Gonzaga. 2019. Evidence that *Polysphincta*-group wasps (Hymenoptera: Ichneumonidae) use ecdysteroids to manipulate the web-construction behaviour of their spider hosts. Biological Journal of the Linnean Society 20:1–43,

Eberhard, W. G., and N. A. Hazzi. 2013. Web construction of *Linothele macrothelifera* (Araneae: Dipluridae). Journal of Arachnology 41:70–75.

———. 2017. Webs and building behavior of *Aglaoctenus castaneus* (Araneae: Lycosidae: Sosippinae). Journal of Arachnology 45:177–197.

Eberhard, W. G., and T. Hesselberg. 2012. Cues that spiders (Araneae: Araneidae, Tetragnathidae) use to build orbs: lapses in attention to one set of cues because of dissonance with others? Ethology 118:610–620.

Eberhard, W. G., and B. A. Huber. 2010. Spider genitalia. Pages 249–284 *in* J. Leonard and A. Córdoba-Aguilar, eds. The Evolution of Primary Sexual Characters in Animals. Oxford University Press, Oxford, New York.

Eberhard, W. G., and F. Pereira. 1993. Ultrastructure of cribellate silk of nine species in eight families and possible taxonomic implications (Araneae: Amaurobiidae, Deinopidae, Desidae, Dictynidae, Filistatidae, Hypochilidae, Stiphidiidae, Tengellidae). Journal of Arachnology 21:161–174.

Eberhard, W. G., and W. T. Wcislo. 2011. Grade changes in brain-body allometry: morphological and behavioural correlates of brain size in miniature spiders, insects, and other invertebrates. Advances in Insect Physiology 60:155–214.

Eberhard, W. G., M. Barreto, and W. Pfizenmaier. 1978. Web robbery by mature male orb-weaving spiders. Bulletin of the British Arachnological Society 4:228–230.

Eberhard, W. G., N. I. Platnick, and R. T. Schuh. 1993. Natural history and systematics of arthropod symbionts (Araneae; Hemiptera; Diptera) inhabiting webs of the spider *Tengella radiata* (Araneae, Tengellidae). American Museum Novitates 3065:1–17.

Eberhard, W. G., G. Barrantes, and R. Madrigal-Brenes. 2008a. Vestiges of an orb-weaving ancestor? The "biogenetic law" and ontogenetic changes in the webs and building behavior of the black widow spider *Latrodectus geometricus* (Araneae Theridiidae). Ethology Ecology & Evolution 20:211–244.

Eberhard, W. G., I. Agnarsson, and H. W. Levi. 2008b. Web forms and the phylogeny of theridiid spiders (Araneae: Theridiidae): chaos from order. Systematics and Biodiversity 6:415.

Edmonds, D. T., and F. Vollrath. 1992. The contribution of atmospheric water vapour to the formation and efficiency of a spider's capture web. Proceedings of the Royal Society of London B: Biological Sciences 248:145–148.

Edmunds, J. 1978. The web of *Paraneus cyrtoscapus* (Pocock, 1989) (Araneae: Araneidae) in Ghana. Bulletin of the British Arachnological Society 4:191–96.

———. 1986. The stabilimenta of *Argiope flavipalpis* and *Argiope trifasciata* in West Africa, with a discussion of the function of stabilimenta. Pages 61–72 *in* W. G. Eberhard, Y. D. Lubin, and B. Robinson, eds. Proceedings of the Ninth International Congress of Arachnology, Panama 1983. Smithsonian Institution Press, Washington, DC.

———. 1993. The development of the asymmetrical web of *Nephilengys cruentata* (Fabricius). Memoirs of the Queensland Museum 33:503–506.

Edmunds, J., and M. Edmunds. 1986. The defensive mechanisms of orb weavers (Araneae: Araneidae) in Ghana, West Africa. Pages 73–90 *in* W. G. Eberhard, Y. D. Lubin, and B. Robinson, eds. Proceedings of the Ninth International Congress of Arachnology, Panama 1983. Smithsonian Institution Press, Washington, DC.

Edwards, W., P. A. Whytlaw, B. C. Congdon, and C. Gaskett. 2009. Is optimal foraging a realistic expectation in orb-web spiders? Ecological Entomology 34:527–534.

Eggs, B., and D. Sanders. 2013. Herbivory in spiders: the importance of pollen for orb-weavers. PLOS ONE 8(11):e82637. DOI:10.1371/journal.pone.0082637.

Ehlers, M. 1939. Untersuchungen über Formen aktiver Lokomotion bei Spinnen. Zoologische Jahrbücher, Abteilung für Systematik 72:373–499.

Eisner, T., and S. Nowicki. 1983. Spider web protection through visual advertisement: role of the stabilimentum. Science 219:185–187.

Eisner, T., R. Alsop, and G. Ettershank. 1964. Adhesiveness of spider silk. Science, New Series 146:1058–1061.

Elettro, H., S. Neukirch, F. Vollrath, and A. Antkowiak. 2016. In-drop capillary spooling of spider capture thread inspires hybrid fibers with mixed solid–liquid mechanical properties. Proceedings of the National Academy of Sciences 113:6143–6147.

Elgar, M. A. 1989. Kleptoparasitism: a cost of aggregating for an orb-weaving spider. Animal Behaviour 37:1052–1055.

Elices, M., G. R. Plaza, M. A. Arnedo, J. Pérez-Rigueiro, F. G. Torres, and G. V. Guinea. 2009. Mechanical behavior of silk during the evolution of orb-web spinning spiders. Biomacromolecules 10:1904–1910.

Emerit, M. 1968. Contribution à l'étude de la biologie et du développement de l'araignée tropicale *Gasteracantha versicolor* (Walck.) (Argiopidae): Note préliminaire. . . . Bulletin de la Societe Zoologique de France 93:49–68.

Emerton, J. H. 1883. The cobwebs of *Uloborus*. American Journal of Science 3:203–205.

———. 1902. The Common Spiders of the United States. Athanaeum Press, Boston.

Enders, F. 1974. Vertical stratification in orb-web spiders (Araneidae, Araneae) and a consideration of other methods of coexistence. Ecology 55:317–328.

———. 1975. Change of web site in *Argiope* spiders (Araneidae). American Midland Naturalist 94:484–490.

———. 1976a. Clutch size related to hunting manner of spider species. Annals of the Entomological Society of America 69:991–998.

———. 1976b. Effects of prey capture, web destruction and habitat physiognomy on web-site tenacity of *Argiope* spiders (Araneidae). Journal of Arachnology 3:75–82.

Endo, T. 1988. Patterns of prey utilization in a web of orb-weaving spider *Araneus pinguis* (Karsch). Researches on Population Ecology 30:107–121.

Escalante, I., and M. Masís-Calvo. 2014. The absence of gumfoot threads in webs of early juveniles and adult males of *Physocyclus globosus* (Pholcidae) is not associated with spigot morphology. Arachnology 16:214–218.

Escalante, I., A. Aisenberg, and F. G. Costa. 2015. Risky behaviors by the host could favor araneophagy of the spitting spider *Scytodes globula* on the hacklemesh weaver *Metaltella simoni*. Journal of Ethology 33:125–136.

Esquivel, C. 2006. Dragonflies and damselflies of Middle America and the Caribbean. Editorial INBio, San José, Costa Rica.

Evans, H. E., and M. J. West-Eberhard. 1971. The Wasps. University of Michigan Press, Ann Arbor.

Evans, S. C. 2013. Stochastic Modeling of Orb-Web Capture Mechanics Supports the Importance of Rare Large Prey for Spider Foraging Success and Suggests How Webs Sample Available Biomass. Master's thesis, University of Akron, Ohio.

Ewer, R. F. 1972. The devices in the web of the West African spider *Argiope flavipalpis*. Journal of Natural History 6:159–167.

Fabre, J. H. 1912. Social Life in the Insect World. Fisher Unwin, London.

Fahey, B. F., and M. A. Elgar. 1997. Sexual cohabitation as mate-guarding in the leaf-curling spider *Phonognatha graeffei* Keyserling (Araneoidea, Araneae). Behavioral Ecology and Sociobiology 40:127–133.

Fan, C.-M., E. C. Yang, and I.-M. Tso. 2009. Hunting efficiency and predation risk shapes the color-associated foraging traits of a predator. Behavioral Ecology 20:808–816.

Fergusson, I. C. 1972. Natural history of the spider, *Hypochilus thorelli* Marx (Hypochilidae). Psyche: A Journal of Entomology 79:179–199.

Fernández-Campón, F. 2007. Group foraging in the colonial spider *Parawixia bistriata* (Araneidae): effect of resource levels and prey size. Animal Behaviour 74:1551–1562.

Fernández, R., G. Hormiga, and G. Giribet. 2014. Phylogenomic analysis of spiders reveals nonmonophyly of orb weavers. Current Biology 24:1772–1777.

Finck, A. 1972. Vibration sensitivity in an orb-weaver. American Zoologist 12:539–543.

Finck, A., G. M. Stewart, and C. F. Reed. 1975. The orb-web as an acoustic detector. Journal of the Acoustical Society of America 57:753–754.

Fincke, O. M. 1981. An association between two Neotropical spiders (Araneae: Uloboridae and Tengellidae). Biotropica 13:301–307.

———. 1992. Behavioural ecology of the giant damselflies of Barro Colorado Island, Panama (Odonata: Zygoptera: Pseudostigmatidae). Pages 103–113 *in* D. Quintero and A. Aiello, eds. Insects of Panama and Mesoamerica: Selected Studies. Oxford University Press, Oxford.

Fischer, F. G., and J. Brander. 1960. Eine Analyse der Gespinste der Kreuzspinne. Hoppe-Seyler´s Zeitschrift für physiologische Chemie 320:92–102.

Foelix, R. F. 1970. Structure and function of tarsal sensilla in the spider *Araneus diadematus*. Journal of Experimental Zoology 175:99–123.

———. 2011. Biology of Spiders. Oxford University Press, New York.

Foelix, R. F., B. Erb, and B. Rast. 2013a. Alleged silk spigots on tarantula feet: electron microscopy reveals sensory innervation, no silk. Arthropod Structure & Development 42:209–217.

———. 2013b. Feinstruktur der Spinnspulen von Vogelspinnen. Arachne 3:4–10.

Ford, M. J. 1977. Energy costs of the predation strategy of the web-spinning spider *Lepthyphantes zimmermanni* bertkau (Linyphiidae). Oecologia 28:341–349.

Forster, L. M., and R. R. Forster. 1985. A derivative of the orb web and its evolutionary significance. New Zealand Journal of Zoology 12:455–465.

Forster, R. R., and L. M. Forster. 1973. New Zealand Spiders, an Introduction. Collins Press, Auckland.

Fowler, H. G., and J. Diehl. 1978. Biology of a Paraguayan colonial orb-weaver, *Eriophora bistriata* (Rengger) (Araneae, Araneidae). Bulletin of the British Arachnological Society 4:241–250.

Fowler, H. G., and E. M. Venticinque. 1996. Interference competition and scavenging by *Crematogaster* ants (Hymenoptera: Formicidae) associated with the webs of the social spider *Anelosimus eximius* (Araneae: Theridiidae) in the Central Amazon. Journal of the Kansas Entomological Society 69:267–269.

Franck, D. 1974. The genetic basis of evolutionary changes in behaviour patterns. Pages 119–140 *in* J. H. F. van Abeelen, ed. The Genetics of Behaviour. American Elsevier, New York.

Freisling, J. 1961. Netz und Netzbauinstinkte bei *Theridium saxatile* Koch. Z. wiss. Zool 165:396–421.

Frings, H., and M. Frings. 1966. Reactions of orb-weaving spiders (Argiopidae) to airborne sounds. Ecology 47:578–588.

Frohlich, C., and R. E. Buskirk. 1982. Transmission and attenuation of vibration in orb spider webs. Journal of Theoretical Biology 95:13–36.

Fukumoto, N. 1981. Notes on the web-weaving activity. Atypus 78:17–20.

Gallistel, C. R. 1980. From muscles to motivation: three distinct kinds of elementary units of behavior combine to form complex units whose coordination by selective potentiation and depotentiation yields motivated behavior. American Scientist 68:398–409.

Gálvez, D. 2011. Web decoration of *Micrathena sexpinosa* (Araneae: Araneidae): a frame-web-choice experiment with stingless bees. Journal of Arachnology 39:128–132.

Garb, J. E., T. DiMauro, V. Vo, and C. Y. Hayashi. 2006. Silk genes support the single origin of orb webs. Science 312:1762.

Garb, J. E., T. DiMauro, R. V. Lewis, and C. Y. Hayashi. 2007. Expansion and intragenic homogenization of spider silk genes since the Triassic: evidence from Mygalomorphae (Tarantulas and their kin) spidroins. Molecular Biology and Evolution 24:2454–2464.

Garb, J. E., N. A. Ayoub, and C. Y. Hayashi. 2010. Untangling spider silk evolution with spidroin terminal domains. BMC Evolutionary Biology 10:243.

Garcia, L. C., and H. F. Japyassú. 2005. Estereotipia e plasticidade na seqüência predatória de *Theridion evexum* keyserling 1884 (Araneae: Theridiidae). Biota Neotropica 5:27–43.

Garcia, L. C., and J. D. Styrsky. 2013. An orb-weaver spider eludes plant-defending acacia ants by hiding in plain sight. Ecological Entomology 38:230–237.

Garrido, M. A., M. Elices, C. Viney, and J. Pérez-Rigueiro. 2002. Active control of spider silk strength: comparison of drag line spun on vertical and horizontal surfaces. Polymer 43:1537–1540.

Garrison, N. L., J. Rodriguez, I. Agnarsson, J. A. Coddington, C. E. Griswold, C. A. Hamilton, M. Hedin, et al. 2016. Spider phylogenomics: untangling the Spider Tree of Life. PeerJ 4:e1719.

Gaskett, A. C. 2007. Spider sex pheromones: emission, reception, structures, and functions. Biological Reviews 82:27–48.

Gastreich, K. R. 1999. Trait-mediated indirect effects of a theridiid spider on an ant-plant mutualism. Ecology 80:1066–1070.

Gatesy, J., C. Hayashi, D. Motriuk, J. Woods, and R. Lewis. 2001. Extreme diversity, conservation, and convergence of spider silk fibroin sequences. Science 291:2603.

Geiger, R. 1965. The Climate near the Ground. Harvard University Press, Cambridge, MA.

Gennard, D. 2012. Forensic Entomology: An Introduction. 2nd ed. John Wiley and Sons, Chichester, UK.

Getty, R. M., and F. A. Coyle. 1996. Observations on prey capture and anti-predator behaviors of ogre-faced spiders (*Deinopis*) in southern Costa Rica (Araneae, Deinopidae). Journal of Arachnology 24:93–100.

Gheysens, T., L. Beladjal, K. Gellynck, E. Van Nimmen, L. Van

Langenhove, and J. Mertens. 2005. Egg sac structure of *Zygiella x-notata* (Arachnida, Araneidae). Journal of Arachnology 33:549–557.

Ghiselin, M. T. 1974. The Economy of Nature and the Evolution of Sex. University of California Press, Berkeley, CA.

Gilbert, C., and L. S. Rayor. 1985. Predatory behavior of spitting spiders (Araneae: Scytodidae) and the evolution of prey wrapping. Journal of Arachnology 13:231–241.

Gillespie, R. G. 1987a. The mechanism of habitat selection in the long-jawed orb-weaving spider *Tetragnatha elongata* (Araneae, Tetragnathidae). Journal of Arachnology 15:81–90.

———. 1987b. The role of prey availability in aggregative behaviour of the orb weaving spider *Tetragnatha elongata*. Animal Behaviour 35:675–681.

———. 1991a. Hawaiian spiders of the genus *Tetragnatha*: I. Spiny leg clade. Journal of Arachnology 19:174–209.

———. 1991b. Predation through impalement of prey: the foraging behavior of *Doryonychus raptor* (Araneae: Tetragnathidae). Psyche: A Journal of Entomology 98:337–350.

———. 1999. Comparison of rates of speciation in web-building and non-web-building groups within a Hawaiian spider radiation. Journal of Arachnology 27:79–85.

Gillespie, R. G., and T. Caraco. 1987. Risk-sensitive foraging strategies of two spider populations. Ecology 68:887–899.

Gillespie, R. G., H. B. Croom, and S. R. Palumbi. 1994. Multiple origins of a spider radiation in Hawaii. Proceedings of the National Academy of Sciences 91:2290–2294.

Glatz, L. 1967. Zur Biologie und Morphologie von *Oecobius annulipes* Lucas (Araneae, Oecobiidae). Zeitschrift für Morphologie der Tiere 61:185–214.

———. 1972. Der Spinnapparat haplogyner Spinnen (Arachnida, Araneae). Zeitschrift für Morphologie der Tiere 72:1–25.

Gold, A. 1973. Energy expenditure in animal locomotion. Science 181:275–276.

Gómez, J. E., J. Lohmiller, and A. Joern. 2016. Importance of vegetation structure to the assembly of an aerial web-building spider community in North American open grassland. Journal of Arachnology 44:28–35.

Gonzaga, M. O., and J. F. Sobczak. 2007. Parasitoid-induced mortality of *Araneus omnicolor* (Araneae, Araneidae) by *Hymenoepimecis* sp. (Hymenoptera, Ichneumonidae) in southeastern Brazil. Naturwissenschaften 94:223–227.

Gonzaga, M. O., and J. Vasconcellos-Neto. 2005. Testing the functions of detritus stabilimenta in webs of *Cyclosa fililineata* and *Cyclosa morretes* (Araneae: Araneidae): do they attract prey or reduce the risk of predation? Ethology 111:479–491.

Gonzaga, M. O., N. Leiner, and A. Santos. 2006. On the sticky cobwebs of two theridiid spiders (Araneae: Theridiidae). Journal of Natural History 40:293–306.

Gonzaga, M. O., J. F. Sobczak, A. M. Penteado-Dias, and W. G. Eberhard. 2010. Modification of *Nephila clavipes* (Araneae Nephilidae) webs induced by the parasitoids *Hymenoepimecis bicolor* and *H. robertsae* (Hymenoptera Ichneumonidae). Ethology Ecology & Evolution 22:151–165.

Gonzaga, M. O., T. G. Kloss, and J. F. Sobczak. 2017. Host behavioral manipulation of spiders by ichneumonid wasps. Pages 417–437 *in* C. Viera and M. O. Gonzaga, eds. Behaviour and Ecology of Spiders: Contributions from the Neotropical Region. Springer, New York.

González, M., F. G. Costa, and A. V. Peretti. 2015. Funnel-web construction and estimated immune costs in *Aglaoctenus lagotis* (Araneae: Lycosidae). Journal of Arachnology 43:158–167.

Gorb, S. N., and F. G. Barth. 1996. A new mechanosensory organ on the anterior spinnerets of the spider *Cupiennius salei* (Araneae, Ctenidae). Zoomorphology 116:7–14.

Gorb, S. N., S. Niederegger, C. Y. Hayashi, A. P. Summers, W. Vötsch, and P. Walther. 2001. Silk-like secretion from tarantula feet. Science 291:2603–2605.

Görner, P. 1988. Homing behavior of funnel web spiders (Agelenidae) by means of web-related cues. Naturwissenschaften 75:209–211.

Görner, P., and C. Zeppenfeld. 1980. The runs of *Pardosa amentata* (Araneae, Lycosidae) after removing its cocoon. Pages 243–248 *in* Proceedings of the 8th International Congress of Arachnology. Vienna.

Gosline, J. M., P. A. Guerette, C. S. Ortlepp, and K. N. Savage. 1999. The mechanical design of spider silks: from fibroin sequence to mechanical function. Journal of Experimental Biology 202:3295–3303.

Gotts, N. M., and F. Vollrath. 1991. Artificial intelligence modelling of web-building in the garden cross spider. Journal of Theoretical Biology 152:485–511.

———. 1992. Physical and theoretical features in the simulation of animal behavior: the spider's web. Cybernetics and System 23:41–65.

Gould, S. J., and R. C. Lewontin. 1979. The spandrels of San Marco and the Panglossian paradigm: a critique of the adaptationist programme. Proceedings of the Royal Society of London B: Biological Sciences 205:581–598.

Graf, B., and W. Nentwig. 2001. Ontogenetic change in coloration and web-building behavior in the tropical spider *Eriophora fuliginea* (Araneae, Araneidae). Journal of Arachnology 29:104–110.

Grasshoff, M., and J. Edmunds. 1979. *Araneus legonensis* n. sp. (Araneidae: Araneae) from Ghana, West Africa, and its free sector web. Bulletin of the British Arachnological Society 4:303–309.

Grawe, I., J. O. Wolff, and S. N. Gorb. 2014. Composition and substrate-dependent strength of the silken attachment discs in spiders. Journal of the Royal Society Interface 11:20140477.

Gray, M. R. 1983. The male of *Progradungula carraiensis* Forster and Gray (Araneae, Gradungulidae) with observations on the web and prey capture. Proceedings of the Linnean Society of New South Wales 107:51–58.

Green, P. T., K. E. Harms, and J. H. Connell. 2014. Nonrandom, diversifying processes are disproportionately strong in the smallest size classes of a tropical forest. Proceedings of the National Academy of Sciences 111:18649–18654.

Gregorič, M., R. Kostanjšek, and M. Kuntner. 2010. Orb web features as taxonomic characters in *Zygiella* s.l. (Araneae: Araneidae). Journal of Arachnology 38:319–327.

Gregorič, M., I. Agnarsson, T. A. Blackledge, and M. Kuntner. 2011. Darwin's bark spider: giant prey in giant orb webs (*Caerostris darwini*, Araneae: Araneidae)? Journal of Arachnology 39:287–295.

Gregorič, M., H. C. Kiesbüy, S. G. Q. Lebrón, A. Rozman, I. Agnarsson, and M. Kuntner. 2013. Optimal foraging, not biogenetic law, predicts spider orb web allometry. Naturwissenschaften 100:263–268.

———. 2015. Phylogenetic position and composition of Zygiellinae and *Caerostris*, with new insight into orb-web evolution and gigantism. Zoological Journal of the Linnean Society 175:225–243.

Griffiths, B. V., G. I. Holwell, M. E. Herberstein, and M. A. Elgar. 2003. Frequency, composition and variation in external food stores constructed by orb-web spiders: *Nephila edulis* and *Nephila plumipes* (Araneae: Araneoidea). Australian Journal of Zoology 51:119–128.

Grimaldi, D., and M. S. Engel. 2005. Evolution of the Insects. Cambridge University Press, New York.

Grismado, C. J. 2001. Notes on the genus *Sybota* with a description of a new species from Argentina (Araneae, Uloboridae). Journal of Arachnology 29:11–15.

———. 2008. Uloboridae. Pages 97–103 *in* L. E. Claps, G. Debandi, and J. Roig-Juñent, eds. Biodiversidad de Artrópodos Argentinos. Editorial Sociedad Entomológica Argentina, Mendoza.

Griswold, C. E. 1983. *Tapinillus longipes* (Taczanowski), a web-building lynx spider from the American tropics (Araneae: Oxyopidae). Journal of Natural History 17:979–985.

Griswold, C. E., and T. Meikle-Griswold. 1987. *Archaeodictyna ulova*, new species (Araneae: Dictynidae), a remarkable kleptoparasite of group-living eresid spiders (*Stegodyphus* spp., Araneae: Eresidae). American Museum Novitates 2897:1–12.

Griswold, C. E., and H. M. Yan. 2003. On the egg-guarding behavior of a Chinese symphytognathid spider of the genus *Patu* Marples, 1951 (Araneae, Araneoidea, Symphytognathidae). Proceedings of the California Academy of Sciences 54:356–360.

Griswold, C. E., J. A. Coddington, G. Hormiga, and N. Scharff. 1998. Phylogeny of the orb-web building spiders (Araneae, Orbiculariae: Deinopoidea, Araneoidea). Zoological Journal of the Linnean Society 123:1–99.

Griswold, C. E., M. J. Ramírez, and J. A. Coddington. 2005. Atlas of phylogenetic data for entelegyne spiders (Araneae: Araneomorphae: Entelegynae) with comments on their phylogeny. Proceedings of the California Academy of Sciences 56:1–324.

Grostal, P., and D. E. Walter. 1997. Kleptoparasites or commensals? Effects of *Argyrodes antipodianus* (Araneae: Theridiidae) on *Nephila plumipes* (Araneae: Tetragnathidae). Oecologia 111:570–574.

Guevara, J., and L. Avilés. 2007. Multiple techniques confirm elevational differences in insect size that may influence spider sociality. Ecology 88:2015–2023.

———. 2009. Elevational changes in the composition of insects and other terrestrial arthropods at tropical latitudes: a comparison of multiple sampling methods and social spider diets. Insect Conservation and Diversity 2:142–152.

Guevara, J., M. O. Gonzaga, J. Vasconcellos-Neto, and L. Avilés. 2011. Sociality and resource use: insights from a community of social spiders in Brazil. Behavioral Ecology 22:630.

Guinea, G. V., M. Elices, J. Pérez-Rigueiro, and G. R. Plaza. 2005. Stretching of supercontracted fibers: a link between spinning and the variability of spider silk. Journal of Experimental Biology 208:25.

Gutiérrez-Fonseca, P. E., and L. Ortiz-Rivas. 2014. Substrate preferences for attaching gumfoot lines in *Latrodectus geometricus* (Araneae: Theridiidae). Entomological News 123:371–379.

Hajer, J. 2000. The web of *Trogloneta granulum* Simon (Araneae, Mysmenidae). Bulletin of the British Arachnological Society 11:334–338.

Hajer, J., and D. Řeháková. 2003. Spinning activity of the spider *Trogloneta granulum* (Araneae, Mysmenidae): web, cocoon, cocoon handling behaviour, draglines and attachment discs. Zoology 106:223–231.

Hallas, S. E. A., and R. R. Jackson. 1986. Prey-holding abilities of the nests and webs of jumping spiders (Araneae, Salticidae). Journal of Natural History 20:881–894.

Harmer, A. M. T. 2009. Elongated orb-webs of Australian ladder-web spiders (Araneidae: Telaprocera) and the significance of orb-web elongation. Journal of Ethology 27:453–460.

Harmer, A. M. T., and V. W. Framenau. 2008. *Telaprocera* (Araneae: Araneidae), a new genus of Australian orb-web spiders with highly elongated webs. Zootaxa 1956:59–80.

Harmer, A. M. T., and M. E. Herberstein. 2009. Taking it to extremes: what drives extreme web elongation in Australian ladder web spiders (Araneidae: *Telaprocera maudae*)? Animal Behaviour 78:499–504.

———. 2010. Functional diversity of ladder-webs: moth specialization or optimal area use? Journal of Arachnology 38:119–122.

Harmer, A. M. T., T. A. Blackledge, J. S. Madin, and M. E. Herberstein. 2011. High-performance spider webs: integrating biomechanics, ecology and behaviour. Journal of the Royal Society Interface 8:457–471.

Harmer, A. M. T., H. Kokko, M. E. Herberstein, and J. S. Madin. 2012. Optimal web investment in sub-optimal foraging conditions. Naturwissenschaften 99:65–70.

Harvey, P. H., M. D. Pagel, R. M. May, and P. H. Harvey. 1991. The Comparative Method in Evolutionary Biology. Oxford Studies in Ecology and Evolution. Oxford University Press, Oxford, UK.

Harwood, J. D., K. D. Sunderland, and W. O. C. Symondson.

2001. Living where the food is: web location by linyphiid spiders in relation to prey availability in winter wheat. Journal of Applied Ecology 38:88–99.
———. 2003. Web-location by linyphiid spiders: prey-specific aggregation and foraging strategies. Journal of Animal Ecology 72:745–756.
Hauber, M. E. 2002. Conspicuous colouration attracts prey to a stationary predator. Ecological Entomology 27:686–691.
Hawthorn, A. C., and B. D. Opell. 2002. Evolution of adhesive mechanisms in cribellar spider prey capture thread: evidence for van der Waals and hygroscopic forces. Biological Journal of the Linnean Society 77:1–8.
———. 2003. van der Waals and hygroscopic forces of adhesion generated by spider capture threads. Journal of Experimental Biology 206:3905–3911.
Hayashi, C. Y., and R. V. Lewis. 1998. Evidence from flagelliform silk cDNA for the structural basis of elasticity and modular nature of spider silks. Journal of Molecular Biology 275: 773–784.
———. 2000. Molecular architecture and evolution of a modular spider silk protein gene. Science 287:1477.
———. 2001. Spider flagelliform silk: lessons in protein design, gene structure, and molecular evolution. BioEssays 23:750–756.
Hayashi, C. Y., T. A. Blackledge, and R. V. Lewis. 2004. Molecular and mechanical characterization of aciniform silk: uniformity of iterated sequence modules in a novel member of the spider silk fibroin gene family. Molecular Biology and Evolution 21:1950.
Hazzi, N. A. 2014. Natural history of *Phoneutria boliviensis* (Araneae: Ctenidae): habitats, reproductive behavior, postembryonic development and prey-wrapping. Journal of Arachnology 42:303–310.
Heiling, A. M. 1999. Why do nocturnal orb-web spiders (Araneidae) search for light? Behavioral Ecology and Sociobiology 46:43–49.
———. 2004. Effect of spider position on prey capture success and orb-web design. Acta Zoologica Sinica 50:559–565.
Heiling, A. M., and M. E. Herberstein. 1998. The web of *Nuctenea sclopetaria* (Araneae, Araneidae): relationship between body size and web design. Journal of Arachnology 26:91–96.
———. 1999. The importance of being larger: intraspecific competition for prime web sites in orb-web spiders (Araneae, Araneidae). Behaviour 136:669–677.
———. 2004. Predator-prey coevolution: Australian native bees avoid their spider predators. Proceedings of the Royal Society of London B: Biological Sciences 271:S196–S198.
Hénaut, Y., and S. Machkour-M'Rabet. 2010. Interspecific aggregation around the web of the orb spider *Nephila clavipes*: consequences for the web architecture of *Leucauge venusta*. Ethology Ecology & Evolution 22:203–209.
Hénaut, Y., J. Pablo, G. Ibarra-Nuñez, and T. Williams. 2001. Retention, capture and consumption of experimental prey by orb-web weaving spiders in coffee plantations of Southern Mexico. Entomologia Experimentalis et Applicata 98:1–8.
Hénaut, Y., J. A. García-Ballinas, and C. Alauzet. 2006. Variations in web construction in *Leucauge venusta* (Araneae, Tetragnathidae). Journal of Arachnology 34:234–240.
Hénaut, Y., S. Machkour-M'Rabet, P. Winterton, and S. Calmé. 2010. Insect attraction by webs of *Nephila clavipes* (Araneae: Nephilidae). Journal of Arachnology 38:135–138.
Hénaut, Y., S. Machkour-M'Rabet, and J.-P. Lachaud. 2014. The role of learning in risk-avoidance strategies during spider-ant interactions. Animal Cognition 17:185–195.
Henschel, J. R. 1998. Predation on social and solitary individuals of the spider *Stegodyphus dumicola* (Araneae, Eresidae). Journal of Arachnology 26:61–69.
———. 2002. Long-distance wandering and mating by the dancing white lady spider (*Leucorchestris arenicola*) (Araneae, Sparassidae) across Namib dunes. Journal of Arachnology 30:321–330.
Henschel, J. R., and R. Jocqué. 1994. Bauble spiders: a new species of *Achaearanea* (Araneae: Theridiidae) with ingenious spiral retreats. Journal of Natural History 28:1287–1295.
Henschel, J. R., and Y. D. Lubin. 1992. Environmental factors affecting the web and activity of a psammophilous spider in the Namib Desert. Journal of Arid Environments 22:173–189.
———. 1997. A test of habitat selection at two spatial scales in a sit-and-wait predator: a web spider in the namib desert dunes. Journal of Animal Ecology 66:401–413.
Herberstein, M. E. 1997a. Niche partitioning in three sympatric web-building spiders (Araneae: Linyphiidae). Bulletin of the British Arachnological Society 10:233–238.
———. 1997b. The effect of habitat structure on web height preference in three sympatric web-building spiders (Araneae, Linyphiidae). Journal of Arachnology 25:93–96.
———. 2000. Foraging behaviour in orb-web spiders (Araneidae): do web decorations increase prey capture success in *Argiope keyserlingi* Karsch, 1878? Australian Journal of Zoology 48:217–223.
Herberstein, M. E., and M. A. Elgar. 1994. Foraging strategies of *Eriophora transmarina* and *Nephila plumipes* (Araneae: Araneoidea): nocturnal and diurnal orb-weaving spiders. Australian Journal of Ecology 19:451–457.
Herberstein, M. E., and A. M. Heiling. 1999. Asymmetry in spider orb webs: a result of physical constraints? Animal Behaviour 58:1241–1246.
———. 2001. Positioning at the hub: does it matter on which side of the web orb-web spiders sit? Journal of Zoology 255:157–163.
Herberstein, M. E., and I.-M. Tso. 2011. Spider webs: evolution, diversity and plasticity. Pages 57–98 *in* M. E. Herberstein, ed. Spider Behaviour: Flexibility and Versatility. Cambridge University Press, Cambridge.
Herberstein, M. E., K. E. Abernethy, K. Backhouse, H. Bradford, F. E. Crespigny, P. R. Luckock, and M. A. Elgar. 1998. The effect of feeding history on prey capture behaviour in the orbweb spider *Argiope keyserlingi* Karsch (Araneae: Araneidae). Ethology 104:565–571.

Herberstein, M. E., C. L. Craig, and M. A. Elgar. 2000a. Foraging strategies and feeding regimes: web and decoration investment in *Argiope keyserlingi* Karsch (Araneae: Araneidae). Evolutionary Ecology Research 2:69–80.

Herberstein, M. E., C. L. Craig, J. A. Coddington, and M. A. Elgar. 2000b. The functional significance of silk decorations of orb-web spiders: a critical review of the empirical evidence. Biological Reviews 75:649–669.

Herberstein, M. E., A. C. Gaskett, D. Glencross, S. Hart, S. Jaensch, and M. A. Elgar. 2000c. Does the presence of potential prey affect deb design in *Argiope keyserlingi* (Araneae, Araneidae)? Journal of Arachnology 28:346–350.

Hergenröder, R., and F. G. Barth. 1983. Vibratory signals and spider behavior: how do the sensory inputs from the eight legs interact in orientation? Journal of Comparative Physiology 152:361–371.

Hesselberg, T. 2010. Ontogenetic changes in web design in two orb-web spiders. Ethology 116:535–545.

———. 2013. Web-building flexibility differs in two spatially constrained orb spiders. Journal of Insect Behavior 26:283–303.

———. 2014. The mechanism behind plasticity of web-building behavior in an orb spider facing spatial constraints. Journal of Arachnology 42:311–314.

———. 2015. Exploration behaviour and behavioural flexibility in orb-web spiders: a review. Current Zoology 61:313–327.

Hesselberg, T., and E. Triana. 2010. The web of the acacia orb-spider *Eustala illicita* (Araneae: Araneidae) with notes on its natural history. Journal of Arachnology 38:21–26.

Hesselberg, T., and F. Vollrath. 2004. The effects of neurotoxins on web-geometry and web-building behaviour in *Araneus diadematus* Cl. Physiology & Behavior 82:519–529.

———. 2012. The mechanical properties of the non-sticky spiral in *Nephila* orb webs (Araneae, Nephilidae). Journal of Experimental Biology 215:3362–3369.

Hickman, V. V. 1975. On *Paraplectanoides crassipes* Keyserling (Araneae: Araneidae). Bulletin of the British Arachnological Society 3:166–174.

Hieber, C. 1984. Orb-web orientation and modification by the spiders *Araneus diadematus* and *Araneus gemmoides* (Araneae: Araneidae) in response to wind and light. Zeitschrift für Tierpsychologie 65:250–260.

Higgins, L. E. 1987. Time budget and prey of *Nephila clavipes* (Linnaeus) (Araneae, Araneidae) in southern Texas. Journal of Arachnology 15:401–417.

———. 1990. Variation in foraging investment during the intermolt interval and before egg-laying in the spider *Nephila clavipes* (Araneae: Araneidae). Journal of Insect Behavior 3:773–783.

———. 1992a. Developmental plasticity and fecundity in the orb-weaving spider *Nephila clavipes*. Journal of Arachnology 20:94–106.

———. 1992b. Developmental changes in barrier web structure under different levels of predation risk in *Nephila clavipes* (Araneae: Tetragnathidae). Journal of Insect Behavior 5:635–655.

———. 1993. Constraints and plasticity in the development of juvenile *Nephila clavipes* in Mexico. Journal of Arachnology 21:107–119.

———. 1995. Direct evidence for trade-offs between foraging and growth in a juvenile spider. Journal of Arachnology 23:37–43.

———. 2004. Silk spinners. Journal of Experimental Biology 207:4339–4340.

———. 2006. Quantitative shifts in orb-web investment during development in *Nephila clavipes* (Araneae, Nephilidae). Journal of Arachnology 34:374–386.

Higgins, L. E., and R. E. Buskirk. 1992. A trap-building predator exhibits different tactics for different aspects of foraging behaviour. Animal Behaviour 44:485–499.

Higgins, L. E., and K. McGuinness. 1991. Web orientation by *Nephila clavipes* in Southeastern Texas. American Midland Naturalist 125:286–293.

Higgins, L. E., and M. A. Rankin. 1999. Nutritional requirements for web synthesis in the tetragnathid spider *Nephila clavipes*. Physiological Entomology 24:263–270.

———. 2001. Mortality risk of rapid growth in the spider *Nephila clavipes*. Functional Ecology 15:24–28.

Higgins, L. E., M. A. Townley, E. K. Tillinghast, and M. A. Rankin. 2001. Variation in the chemical composition of orb webs built by the spider *Nephila clavipes* (Araneae, Tetragnathidae). Journal of Arachnology 29:82–94.

Hill, D. E. 2006. Jumping spider feet (Araneae, Salticidae), version 3. Peckhamia Epublications. Online resource (http://peckhamia.com/epublications.html), accessed 16 April 2012.

Hingston, R. W. 1920. A Naturalist in Himalaya. H. F. & G. Witherby, London.

———. 1922a. The snare of the giant wood spider (*Nephila maculata*). Part I. Journal of the Bombay Natural History Society 28:642–649.

———. 1922b. The snare of the giant wood spider (*Nephila maculata*). Part II. Journal of the Bombay Natural History Society 28:911–917.

———. 1922c. The snare of the giant wood spider (*Nephila maculata*). Part III. Further lessons of the *Nephila*. Journal of the Bombay Natural History Society 28:917–923.

———. 1925. Nature at the desert's edge. H. F. & G. Witherby, London.

———. 1927. Protective devices in spiders' snares, with a description of seven new species of orb-weaving spiders. Proceedings of the Zoological Society of London 97:259–293.

———. 1929. Instinct and intelligence. Macmillan Company, New York.

———. 1932. A naturalist in the Guiana forest. Annals of the Entomological Society of America 25:654.

Hiramatsu, T., and A. Shinkai. 1993. Web structure and web-building behaviour of *Patu* sp. (Araneae: Symphytognatidae). Acta Arachnologica 42:181–185.

Hirscheimer, A. J., and R. B. Suter. 1985. Functional webs built by adult male bowl and doily spiders. Journal of Arachnology 13:396–398.

Hodge, M. A. 1987a. Factors influencing web site residence time of the orb weaving spider, *Micrathena gracilis*. Psyche: A Journal of Entomology 94:363–371.

———. 1987b. Macrohabitat selection by the orb weaving spider, *Micrathena gracilis*. Psyche: A Journal of Entomology 94:347–361.

Hogan, N., and D. Sternad. 2013. Dynamic primitives in the control of locomotion. Frontiers in Computational Neuroscience 7.

Hogue, C. L., and S. E. Miller. 1981. Entomofauna of Cocos Island, Costa Rica. Atoll Research Bulletin 250:1–29.

Holl, A., and M. Henze. 1988. Pigmentary constituents of yellow threads of *Nephila* webs. Page 350 *in* XI. Europa isches Arachnologisches Colloquium. Technische Universitat Berlin Dokumentation Kongresse und Tagungen, Berlin.

Hölldobler, B. 1970. *Steatoda fulva* (Theridiidae), a spider that feeds on harvester ants. Psyche: A Journal of Entomology 77:202–208.

———. 1979. Territoriality in ants. Proceedings of the American Philosophical Society 123:211–218.

Holzapfel, M. 1934. Die nicht-optische Orientierung der Trichter-spinne *Agelena labyrinthica* (Cl.). Zeitschrift für Vergleichende Physiologie 20:55–116.

Hormiga, G. 2002. *Orsonwelles*, a new genus of giant linyphiid spiders (Araneae) from the Hawaiian Islands. Invertebrate Systematics 16:369–448.

Hormiga, G., and C. E. Griswold. 2014. Systematics, phylogeny, and evolution of orb-weaving spiders. Annual Review of Entomology 59:487–512.

Hormiga, G., and N. Scharff. 2014. The strange case of *Laetesia raveni* n. sp., a green linyphiid spider from Eastern Australia with a preference for thorny plants (Araneae, Linyphiidae). Zootaxa 3811:83.

Hormiga, G., W. G. Eberhard, and J. A. Coddington. 1995. Web-construction behavior in Australian *Phonognatha* and the phylogeny of Nephiline and tetragnathid spiders (Araneae, Tetragnathidae). Australian Journal of Zoology 43:313–364.

Hormiga, G., F. Alvarez-Padilla, and S. P. Benjamin. 2007. First records of extant Hispaniolan spiders of the families Mysmenidae, Symphytognathidae, and Ochyroceratidae (Araneae), including a new species of Ochyrocera. American Museum Novitates 3577:1–21.

Horton, C. C. 1979. Apparent attraction of moths by the webs of araneid spiders. Journal of Arachnology 7:88.

———. 1980. A defensive function for the stabilimenta of two orb weaving spiders (Araneae, Araneidae). Psyche: A Journal of Entomology 87:13–20.

Horton, C. C., and D. H. Wise. 1983. The experimental analysis of competition between two syntopic species of orb-web spiders (Araneae: Araneidae). Ecology 64:929–944.

Howell, F. G., and R. D. Ellender. 1984. Observations on growth and diet of *Argiope aurantia* Lucas (Araneidae) in a successional habitat. Journal of Arachnology 12:29–36.

Hu, X., J. Yuan, X. Wang, K. Vasanthavada, A. M. Falick, P. R. Jones, C. La Mattina, et al. 2007. Analysis of aqueous glue coating proteins on the silk fibers of the cob weaver, *Latrodectus hesperus*. Biochemistry 46:3294–3303.

Huber, B. A. 1996. On the distinction between *Modisimus* and *Hedypsilus* (Araneae, Pholcidae), with notes on behavior and natural history. Zoologica Scripta 25:233–240.

———. 2000. New World pholcid spiders (Aranea: Pholcidae): a revision at generic level. Bulletin of the American Museum of Natural History 254:1–348.

———. 2005. High species diversity, male-female coevolution, and metaphyly in Southeast Asian pholcid spiders: the case of *Belisana* Thorell 1898 (Araneae, Pholcidae). Zoologica 155:1–126.

———. 2009. Life on leaves: leaf-dwelling pholcids of Guinea, with emphasis on *Crossopriza cylindrogaster* Simon, a spider with inverted resting position, pseudo-eyes, lampshade web, and tetrahedral egg-sac (Araneae: Pholcidae). Journal of Natural History 43:2491–2523.

Huber, B. A., and P. Kwapong. 2013. West African pholcid spiders: an overview, with descriptions of five new species (Araneae, Pholcidae). European Journal of Taxonomy 59:1–44.

Huber, B. A., B. Petcharad, C. Leh Moi Ung, J. K. Koh, and A. R. Ghazali. 2016. The Southeast Asian *Pholcus halabala* species group (Araneae, Pholcidae): new data from field observations and ultrastructure. European Journal of Taxonomy 190:1–55.

Humpreys, W. F. 1992. Stabilimenta as parasols: shade construction by *Neogea* sp. (Arneae, Argiopinae) and its thermal behavior. Bulletin of the British Arachnological Society 9:47–52.

Hutchinson, C. E. 1903. A bolas throwing spider. Scientific American 89:172.

Jackson, R. R. 1973. Nomenclature for orb web thread connections. Bulletin of the British Arachnological Society 2:125–126.

———. 1974. Effects of d-Amphetamine sulfate and diazepam on thread connection fine structure in a spider's web. Journal of Arachnology 2:37–41.

———. 1985a. The biology of *Euryattus* sp. indet., a web-building jumping spider (Araneae, Salticidae) from Queensland: utilization of silk, predatory behaviour and intraspecific interactions. Journal of Zoology 1:145–173.

———. 1985b. The biology of *Simaetha paetula* and *S. thoracica*, web-building jumping spiders (Araneae, Salticidae) from Queensland: co-habitation with social spiders, utilization of silk, predatory behaviour and intraspecific interactions. Journal of Zoology (B) 1:175–210.

———. 1986. Web building, predatory versatility, and the evolution of the Saltiddae. Pages 232–268 *in* W. A. Shear, ed. Spiders: Webs, Behavior, and Evolution. Stanford University Press, Stanford, CA.

———. 1990a. Ambush predatory behaviour of *Phaeacius malayensis* and *Phaeacius* sp. indet., spartaeine jumping

spiders (Araneae: Salticidae) from tropical Asia. New Zealand Journal of Zoology 17:491–498.
———. 1990b. Comparative study of lyssomanine jumping spiders (Araneae: Salticidae): silk use and predatory behaviour of Asemonea, Goleba, Lyssomanes, and Onomastus. New Zealand Journal of Zoology 17:1–6.
———. 1990c. Predator-prey interactions between jumping spiders (Araneae, Salticidae) and *Phokus phalangioides* (Araneae, Pholcidae). Journal of Zoology 220:553–559.
Jackson, R. R., and R. J. Brassington. 1987. The biology of *Pholcus phalangioides* (Araneae, Pholcidae): predatory versatility, araneophagy and aggressive mimicry. Journal of Zoology 211:227–238.
Jackson, R. R., and A. M. Macnab. 1989a. Display, mating, and predatory behaviour of the jumping spider *Plexippus paykulli* (Araneae: Salticidae). New Zealand Journal of Zoology 16:151–168.
———. 1989b. Display behaviour of *Corythalia canosa*, an ant-eating jumping spider (Araneae: Salticidae) from Florida. New Zealand Journal of Zoology 16:169–183.
Jackson, R. R., and S. D. Pollard. 1990. Web-building and predatory behaviour of *Spartaeus spinimanus* and *Spartaeus thailandicus*, primitive jumping spiders (Araneae, Salticidae) from South-east Asia. Journal of Zoology 220:561–567.
Jackson, R. R., and M. E. A. Whitehouse. 1986. The biology of New Zealand and Queensland pirate spiders (Araneae, Mimetidae): aggressive mimicry, araneophagy and prey specialization. Journal of Zoology 210:279–303.
Jackson, R. R., E. M. Jakob, M. B. Willey, and G. E. Campbell. 1993. Anti-predator defences of a web-building spider, *Holocnemus pluchei* (Araneae, Pholcidae). Journal of the Zoological Society of London 229:347–352.
Jackson, R. R., P. W. Taylor, A. S. McGill, and S. D. Pollard. 1995. The web and prey-capture behaviour of *Diaea* sp., a crab spider (Thomisidae) from New Zealand. Records of the Western Australian Museum 52:33–37.
Jacobi-Kleemann, M. 1953. Über die Lokomotion der Kreuzspinne Aranea diadema beim Netzbau (nach Filmanalysen). Zeitschrift für Vergleichende Physiologie 34:606–654.
Jaffé, R., W. G. Eberhard, C. De Angelo, D. Eusse, A. Gutierrez, S. Quijas, A. Rodríguez, et al. 2006. Caution, webs in the way! Possible functions of silk stabilimenta in *Gasteracantha cancriformis* (Araneae, Araneidae). Journal of Arachnology 34:448–455.
Jäger, P. 2012. Observations on web-invasion by the jumping spider *Thyene imperialis* in Israel (Araneae: Salticidae). Arachnologische Mitteilungen 43:63–65.
Jakob, E. M. 1991. Costs and benefits of group living for pholcid spiderlings: losing food, saving silk. Animal Behaviour 41:711–722.
———. 2004. Individual decisions and group dynamics: why pholcid spiders join and leave groups. Animal Behaviour 68:9–20.
Jakob, E. M., G. W. Uetz, and A. H. Porter. 1998. The effect of conspecifics on the timing of orb construction in a colonial spider. Journal of Arachnology 26:335–341.
Jakob, E. M., J. A. Blanchong, M. A. Popson, K. A. Sedey, and M. S. Summerfield. 2000. Ontogenetic shifts in the costs of living in groups: focal observations of a pholcid spider (*Holocnemus pluchei*). American Midland Naturalist 143: 405–413.
Janetos, A. C. 1982. Foraging tactics of two guilds of web-spinning spiders. Behavioral Ecology and Sociobiology 10:19–27.
———. 1986. Web-site selection: are we asking the right questions. Pages 9–22 *in* W. A. Shear, ed. Spiders: Webs, Behavior, and Evolution. Stanford University Press, Stanford, CA.
Japyassú, H. F., and C. Ades. 1998. From complete orb to semi-orb webs: developmental transitions in the web of *Nephilengys cruentata* (Araneae: Tetragnathidae). Behaviour 135:931–956.
Japyassú, H. F., and R. A. Caires. 2008. Hunting tactics in a cobweb spider (Araneae-Theridiidae) and the evolution of behavioral plasticity. Journal of Insect Behavior 21:258–284.
Japyassú, H. F., and E. G. Jotta. 2005. Forrageamento em *Achaearanea cinnabarina* Levi 1963 (Araneae, Theridiidae) e evoluçiao da caça em aranhas de teia irregular. Biota Neotropica 5:53–67.
Japyassú, H. F., and C. R. Macagnan. 2004. Fishing for prey: the evolution of a new predatory tactic among spiders (Araneae, Pholcidae). Revista de Etologia 6:79–94.
Job, W. 1968. Das Röhrengewebe von *Aulonia albimana* (Walckenaer) (Araneida: Lycosidae) und seine systematische Bedeutung. Zoologischer Anzeiger 180:403–409.
———. 1974. Beitrage zur Biologie der fangnetzbauenden Wolfsspinne *Aulonia albimana* (Walckenaer 1805) (Arachnida, Araneae, Lycosidae, Hippasinae). Zoologische Jahrbücher, Abteilung für Systematik, Okologie und Geographie der Tiere.
Joel, A.-C., P. Kappel, H. Adamova, W. Baumgartner, and I. Scholz. 2015. Cribellate thread production in spiders: complex processing of nano-fibres into a functional capture thread. Arthropod Structure & Development 44:568e573.
Johannesen, J., J. T. Wennmann, and Y. Lubin. 2012. Dispersal behaviour and colony structure in a colonial spider. Behavioral Ecology and Sociobiology 66:1387–1398.
Jones, D. 1983. The Country Life Guide to Spiders of Britain and Northern Europe. Country Life Books, Feltham, UK.
Jörger, K. M., and W. G. Eberhard. 2006. Web construction and modification by *Achaearanea tesselata* (Araneae, Theridiidae). Journal of Arachnology 34:511–523.
Kaestner, A. 1968. Invertebrate Zoology: Arthropod Relatives, Chelicerata, Myriapoda. Interscience Publishers, New York.
Kariko, S. J. 2014. The glitterati: four new species of *Phoroncidia* (Araneae: Theridiidae) from Madagascar, with the first description of the male of *P. aurata* O. Pickard—Cambridge, 1877. Arachnology 16:195–213.
Kaston, B. J. 1948. Spiders of Connecticut. Rev ed. State

Geological and Natural History Survey of Connecticut, Bulletin No. 70.

———. 1964. The evolution of spider webs. American Zoologist 4:191–207.

———. 1965. Some little known aspects of spider behavior. American Midland Naturalist 73:336–356.

———. 1970. Comparative biology of American black widow spiders. Transactions of the San Diego Society of Natural History 16:33–82.

———. 1972. Web making by young *Peucetia viridans* (Hentz) (Araneae: Oxyopidae). Notes of the Arachnologists of the Southwest 3:6–7.

Kato, N., M. Takasago, K. Omasa, and T. Miyashita. 2008. Coadaptive changes in physiological and biophysical traits related to thermal stress in web spiders. Naturwissenschaften 95:1149–1153.

Katz, M. J., R. J. Lasek, and J. Silver. 1983. Ontophyletics of the nervous system: development of the corpus callosum and evolution of axon tracts. Proceedings of the National Academy of Sciences 80:5936–5940.

Kavanagh, E. J., and E. K. Tillinghast. 1979. Fibrous and adhesive components of the orb webs of *Araneus trifolium* and *Argiope trifasciata*. Journal of Morphology 160:17–31.

Kawamoto, T. H., and H. F. Japyassú. 2008. Tenacity and silk investment of two orb weavers: considerations about diversification of the Araneoidea. Journal of Arachnology 36:418–424.

Kawamoto, T. H., F. de A. Machado, G. E. Kaneto, and H. F. Japyassú. 2011. Resting metabolic rates of two orbweb spiders: a first approach to evolutionary success of ´ecribellate spiders. Journal of Insect Physiology 57:427–432.

Keiser, C. N., A. E. DeMarco, T. A. Shearer, J. A. Robertson, and J. N. Pruitt. 2015. Putative microbial defenses in a social spider: immune variation and antibacterial properties of colony silk. Journal of Arachnology 43:394–399.

Kerr, A. M. 1993. Low frequency of stabilimenta in orb webs of *Argiope appensa* (Araneae: Araneidae) from Guam: an indirect effect of an introduced avian predator? Pacific Science 47:328–337.

Kielhorn, K.-H., and I. Rödel. 2011. Badumna longinqua nach Europa eingeschleppt (Araneae: Desidae). Arachnologische Mitteilungen 42:1–4.

Kikuchi, D. W., and D. W. Pfennig. 2013. Imperfect mimicry and the limits of natural selection. Quarterly Review of Biology 88:297–315.

Kirchner, W. 1986. Das Netz der Zitterspinne (*Pholcus phalangioides* Fuesslin) (Araneae: Pholcidae). Zoologischer Anzeiger 216:151–169.

Klärner, D., and F. G. Barth. 1982. Vibratory signals and prey capture in orb-weaving spiders (*Zygiella x-notata*, *Nephila clavipes*; Araneidae). Journal of Comparative Physiology A: Neuroethology, Sensory, Neural, and Behavioral Physiology 148:445–455.

Kloss, T. G., M. O. Gonzaga, L. L. de Oliveira, and C. F. Sperber. 2017. Proximate mechanism of behavioral manipulation of an orb-weaver spider host by a parasitoid wasp. PLOS ONE 12:e0171336.

Knight, D. P., and F. Vollrath. 2002. Spinning an elastic ribbon of spider silk. Philosophical Transactions of the Royal Society of London B: Biological Sciences 357:219–227.

Ko, F. K., and J. Jovicic. 2004. Modeling of mechanical properties and structural design of spider web. Biomacromolecules 5:780–785.

König, M. 1951. Beiträge zur Kenntnis des Netzbaus orbiteler Spinnen. Zeitschrift für Tierpsychologie 8:462–492.

Korenko, S., M. Isaia, J. Satrapová, and S. Pekár. 2014. Parasitoid genus-specific manipulation of orb-web host spiders (Araneae, Araneidae). Ecological Entomology 39:30–38.

Kovoor, J. 1977a. Donnees histochimiques sur les glandes sericigenes de la veuve noire *Latrodectus mactans* Fabr. (Araneae, Theridiidae). Annales des Sciences Naturelles Zoologie.

———. 1977b. La soie et les glandes séricigènes des arachnides. Année Biologique 16:97–171.

———. 1986. Affinités de quelques Pholcidae décelables à partir des caractères de l'appareil sericigene. Mém. Soc. T. Belge Ent 33:111–118.

———. 1987. Comparative structure and histochemistry of silk-producing organs in arachnids. Pages 160–186 *in* W. Nentwig, ed. Ecophysiology of Spiders. Springer-Verlag, Berlin, Heidelberg.

Kovoor, J., and A. Lopez. 1983. Composition et histologie de l'appareil sèricigène des *Argyrodes* relations avec le comportement de ces araignèes (Theridiidea). Revue Arachnologique 5:29–43.

Kovoor, J., and L. Zylberberg. 1972. Histologie et infrastructure de la glande chélicérienne de *Scytodes delicatula* Sim. (Araneidae, Scytodidae). Annales des Sciences Naturelles Zoologie 14: 333–388.

———. 1980. Fine structural aspects of silk secretion in a spider (*Araneus diadematus*). I. Elaboration in the pyriform glands. Tissue and Cell 12:547–556.

Krakauer, T. 1972. Thermal responses of the orb-weaving spider, *Nephila clavipes* (Araneae: Argiopidae). American Midland Naturalist 88:245–250.

Krink, T., and F. Vollrath. 1999. A virtual robot to model the use of regenerated legs in a web-building spider. Animal Behaviour 57:223–232.

———. 2000. Optimal area use in orb webs of the spider *Araneus diadematus*. Naturwissenschaften 87:90–93.

Kropf, C. 1990. Web construction and prey capture of *Comaroma simoni* Bertkau (Araneae). Acta Zoologica Fennica 190: 229–233.

———. 2004. Eine interessante Kleinspinne: *Comaroma simonii* Bertkau 1889 (Arachnida, Araneae, Anapidae). Denisia 12, zugleich Kataloge der OÖ. Landesmuseen 14:257–270.

Kropf, C., D. Bauer, T. Schläppi, and A. Jacob. 2012. An organic coating keeps orb-weaving spiders (Araneae, Araneoidea, Araneidae) from sticking to their own capture threads.

Journal of Zoological Systematics and Evolutionary Research 50:14–18.

Kullmann, E. 1958. Beobachtung des Netzbaues und Beiträge zur Biologie von *Cyrtophora citricola* Forskål (Araneae: Araneidae). Zoologische Jahrbuecher, Abteilung fuer Anatomie und Ontogenie der Tiere 86:181–216.

———. 1959. Beobachtungen und Betrachtungen zum Verhalten der Theridiide *Conopistha argyrodes* Walckenaer (Araneae). Mitteilungen aus dem Museum für Naturkunde in Berlin. Zoologisches Museum und Institut für Spezielle Zoologie (Berlin) 35:275–292.

———. 1964. Neue Ergebnisse über den Netzbau und das Sexualverhalten einiger Spinnenarten (*Cresmatoneta mutinensis, Drapetisca socialis, Lithyphantes paykullianus, Cyrtophora citricola*) als Beitrag zur Frage der Bedeutung besonderer Verhaltensmerkmale für die Systematik1. Journal of Zoological Systematics and Evolutionary Research 2:41–122.

———. 1971/1972. Bemerkenswerte Konvergenzen im Verhalten cribellater und ecribellater Spinnen. Freunde Kolner Zoo 13:123–150.

———. 1972. Evolution of social behavior in spiders (Araneae; Eresidae and Theridiidae). American Zoologist 12:419.

———. 1975. The production and function of spider threads and spider webs. Pages 318–378 *in* K. Bach, B. Burkhardt, R. Graefe, and R. Raccanello, eds. Nets in Nature and Technics. Druckerei Heinrich Fink KG., Stuttgart.

Kullmann, E., and H. Stern. 1981. Leben am seidenen Faden. Kindler Verlag, Munich.

Kullmann, E., S. Nawabi, and W. Zimmermann. 1971. Neue Ergebnisse zur Brutbiologie cribellater Spinnen aus Afghanistan und der Serengeti (Araneae, Eresidae). Zeitschrift des Kölner Zoos 14:87–108.

Kuntner, M. 2005. A revision of *Herennia* (Araneae: Nephilidae: Nephilinae), the Australasian "coin spiders." Invertebrate Systematics 19:391–436.

———. 2006. Phylogenetic systematics of the Gondwanan nephilid spider lineage Clitaetrinae (Araneae, Nephilidae). Zoologica Scripta 35:19–62.

———. 2007. A monograph of *Nephilengys*, the pantropical "hermit spiders" (Araneae, Nephilidae, Nephilinae). Systematic Entomology 32:95–135.

Kuntner, M., and I. Agnarsson. 2009. Phylogeny accurately predicts behaviour in Indian Ocean *Clitaetra* spiders (Araneae: Nephilidae). Invertebrate Systematics 23:193–204.

Kuntner, M., J. A. Coddington, and G. Hormiga. 2008a. Phylogeny of extant nephilid orb-weaving spiders (Araneae, Nephilidae): testing morphological and ethological homologies. Cladistics 24:147–217.

Kuntner, M., C. R. Haddad, G. Aljančič, and A. Blejec. 2008b. Ecology and web allometry of *Clitaetra irenae*, an arboricolous African orb-weaving spider (Araneae, Araneoidea, Nephilidae). Journal of Arachnology 36:583–594.

Kuntner, M., M. Gregorič, and D. Li. 2010a. Mass predicts web asymmetry in *Nephila* spiders. Naturwissenschaften 97:1097–1105.

Kuntner, M., S. Kralj-Fišer, and M. Gregorič. 2010b. Ladder webs in orb-web spiders: ontogenetic and evolutionary patterns in Nephilidae. Biological Journal of the Linnean Society 99:849–866.

Labarque, F. M., J. O. Wolff, P. Michalik, C. E. Griswold and M. J. Ramírez. 2017. The evolution and function of spider feet (Araneae: Arachnida): multiple acquisitions of distal articulations. Zoological Journal of the Linnean Society 20:1–34.

Lahmann, E. J., and W. G. Eberhard. 1979. Factores selectivos que afectan la tendencia a agruparse en la araña colonial *Philoponella semiplumosa* (Aranae; Uloboridae). Revista de Biología Tropical 27:231–240.

Lahmann, E. J., and C. M. Zúñiga. 1981. Use of spider threads as resting places by tropical insects. Journal of Arachnology 9:339–341.

Lai, C.-W., S. Zhang, D. Piorkowski, C.-P. Liao, and I.-M. Tso. 2017. A trap and a lure: dual function of a nocturnal animal construction. Animal Behaviour 130:159–164.

Lamoral, B. H. 1968. On the nest and web structure of *Latrodectus* in South Africa, and some observations on body colouration of *L. geometricus* (Araneae: Theridiidae). Annals of the Natal Museum 20:1–14.

Landolfa, M. A., and F. G. Barth. 1996. Vibrations in the orb web of the spider *Nephila clavipes*: cues for discrimination and orientation. Journal of Comparative Physiology A: Neuroethology, Sensory, Neural, and Behavioral Physiology 179:493–508.

Langer, R. M. 1969. Elementary physics and spider webs. American Zoologist 9:81–89.

Langer, R. M., and W. G. Eberhard. 1969. Laboratory photography of spider silk. American Zoologist 9:97.

Larcher, S. F., and D. H. Wise. 1985. Experimental studies of the interactions between a web-invading spider and two host species. Journal of Arachnology 13:43–59.

Lassen, H., and E. Toltzin. 1940. Tierpsychologische Studien an Radnetzspinnen. Zeitschrift für Vergleichende Physiologie 27:615–630.

Lawrence, B. A., C. A. Vierra, and A. M. Moore. 2004. Molecular and mechanical properties of major ampullate silk of the black widow spider, *Latrodectus hesperus*. Biomacromolecules 5:689–695.

Le Guelte, L. 1966. Structure de la toile de *Zygiella x-notata* Cl. (Araignées, Argiopidae) et facteurs qui régissent le comportement de l'Araignée pendant la construction de la toile. Université de Nancy, Faculté des Sciences, France.

———. 1967. La structure de la toile et les facteurs externes modifiant le comportement de *Zygiella-x-notata* Cl. (Araignées, Argiopidae). Revue du Comportement Animal 1:23–70.

———. 1968. Sur le comportement des Araignées libérées du cocon avant la date normale. Comptes Rendus de l'Académie des Sciences 266:382–383.

Le Guelte, L. and R. Ramousse. 1979. Effets de facteurs environnementaux sur le rythme du comportement constructeur chez l'araignée *Araneus diadematus* Cl. Biology of Behaviour 4:289–302.

Le Guelte, L. and P. N. Witt. 1971. Consequences histologiques et comportementales de lesions (par laser) du ganglion sus-oesophagien de l'araignee (*Araneus diadematus* Cl). Revue Comportement Animal 5:19–26.

Leborgne, R., and A. Pasquet. 1987. Influences of aggregative behaviour on space occupation in the spider *Zygiella x-notata* (Clerck). Behavioral Ecology and Sociobiology 20:203–208.

Leborgne, R., Y. Lubin, and A. Pasquet. 2011. Kleptoparasites influence foraging behaviour of the spider *Stegodyphus lineatus* (Araneae, Eresidae). Insectes Sociaux 58:255–261.

Leclerc, J. 1991. Optimal foraging strategy of the sheet-web spider *Lepthyphantes flavipes* under perturbation. Ecology 72:1267–1272.

Lehmensick, R., and E. Kullmann. 1956. Uber den Feinbau der Spinnenfaden. Pages 307–309 *in* Proceedings of the Stockholm Conference on Electron Microscopy.

———. 1957. Über den Feinbau der Fäden einiger Spinnen. Zoologischer Anzeiger, Supplementband 20:123–129.

Lehtinen, P. T. 1967. Classification of the cribellate spiders and some allied families, with notes on the evolution of the suborder Araneomorpha. Annales Zoologici Fennici 4:199–467.

Lenler-Eriksen, P. 1969. The hunting-web of the young spider *Pisaura mirabilis*. Journal of Zoology 157:391–398.

Levi, H. W. 1976. The orb-weaver genera *Verrucosa*, *Acanthepeira*, *Wagneriana*, *Acacesia*, *Wixia*, *Scoloderus* and *Alpaida* north of Mexico (Araneae: Araneidae). Bulletin of the Museum of Comparative Zoology 147:351–391.

———. 1977. The orb-weaver genera *Metepeira*, *Kaira* and *Aculepeira* in America north of Mexico (Araneae: Araneidae) [new taxa]. Bulletin of the Museum of Comparative Zoology 148:185–238.

———. 1978. Orb-webs and phylogeny of orb-weavers. Symposia of the Zoological Society of London 42: 1–15.

———. 1980. The orb-weaver genus *Mecynogea*, the subfamily Metinae and the genera *Pachygnatha*, *Glenognatha* and *Azilia* of the subfamily Tetragnathinae north of Mexico (Araneae: Araneidae). Bulletin of the Museum of Comparative Zoology 149:1–74.

———. 1981. The American orb-weaving genera *Dolichognatha* and *Tetragnatha* north of Mexico (Araneae: Araneidae, Tetragnathinae). Bulletin of the Museum of Comparative Zoology 149:271–318.

———. 1985. The spiny orb-weaver genera *Micrathena* and *Chaetacis* (Araneae: Araneidae). Bulletin of the Museum of Comparative Zoology 150:429–618.

———. 1986. The Neotropical orb-weaver genera *Chrysometa* and *Homalometa* (Araneae: Tetragnathidae). Bulletin of the Museum of Comparative Zoology 151:91–215.

———. 1993. The Neotropical orb-weaving spiders of the genera *Wixia*, *Pozonia*, and *Ocrepeira* (Araneae: Araneidae). Bulletin of the Museum of Comparative Zoology 153:47–141.

———. 1995. Orb-weaving spiders *Actinosoma*, *Spilasma*, *Micrepeira*, *Pronous*, and four new genera (Araneae: Araneidae). Bulletin of the Museum of Comparative Zoology 154:153–213.

———. 1999. The Neotropical and Mexican orb weavers of the genera *Cyclosa* and *Allocyclosa* (Araneae: Araneidae). Bulletin of the Museum of Comparative Zoology 155:299–379.

———. 2001. The orbweavers of the genera *Molinaranea* and *Nicolepeira*, a new species of *Parawixia*, and comments on orb weavers of temperate South America (Araneae: Araneidae). Bulletin of the Museum of Comparative Zoology 155:445–475.

———. 2008. The South American genus *Spintharidius* (Araneae, Araneidae). Journal of Arachnology 36:216–217.

Levi, H. W., and L. R. Levi. 1968. Spiders and Their Kin. St. Martin's Press, New York.

Levi, H. W., and A. Spielman. 1964. The biology and control of the South American brown spider, *Loxosceles laeta* (Nicolet), in a North American focus. American Journal of Tropical Medicine and Hygiene 13:132–136.

Li, D. 2005. Spiders that decorate their webs at higher frequency intercept more prey and grow faster. Proceedings of the Royal Society of London B: Biological Sciences 272:1753.

Li, D., and W. S. Lee. 2004. Predator-induced plasticity in web-building behaviour. Animal Behaviour 67:309–318.

Li, D., and M. L. Lim. 2005. Ultraviolet cues affect the foraging behaviour of jumping spiders. Animal Behaviour 70:771–776.

Li, D., R. R. Jackson, and A. T. Barrion. 1997. Prey preferences of *Portia africana*, *P. labiata* and *P. schultzi*, araneophagic, web-building jumping spiders (Araneae: Salticidae) from Kenya, Sri Lanka and Uganda. New Zealand Journal of Zoology 24:333–349.

———. 1999. Parental and predatory behaviour of *Scytodes* sp., an araneophagic spitting spider (Araneae: Scytodidae) from the Philippines. Journal of Zoology 247:293–310.

Li, D., S. H. Yik, and W. K. Seah. 2002. Rivet-like nest-building and agonistic behaviour of *Thiania bhamoensis*, an iridescent jumping spider (Araneae: Salticidae) from Singapore. Raffles Bulletin of Zoology 50:143–152.

Li, D., L. M. Kok, W. K. Seah, and M. L. M. Lim. 2003. Age-dependent stabilimentum-associated predator avoidance behaviours in orb-weaving spiders. Behaviour 140:1135–1152.

Li, G., and A. Shojaei. 2012. A viscoplastic theory of shape memory polymer fibres with application to self-healing materials. Proceedings of the Royal Society of London A 468:2319–2346.

Liao, C.-P., K.-J. Chi, and I.-M. Tso. 2009. The effects of wind on trap structural and material properties of a sit-and-wait predator. Behavioral Ecology 20:1194–1203.

Lien, O. J., and G. J. Fitzgerald. 1973. Several factors influencing web-spinning activity in the common house spider

Achaearanea tepidariorum Koch. Animal Learning and Behavior 1:290–292.

Liesenfeld, F. J. 1956. Untersuchungen am Netz und über den Erschütterungssinn von *Zygiella x-notata* (Cl.) (Araneidae). Zeitschrift für Vergleichende Physiologie 38:563–592.

Lin, L. H., D. T. Edmonds, and F. Vollrath. 1995. Structural engineering of an orb-spider's web. Nature 373:146–148.

Lissner, J., R. Bosmans, and J. Hernández-Corral. 2016. Description of a new ground spider from Majorca, Spain, with the establishment of a new genus *Chatzakia* n. gen. (Araneae: Gnaphosidae). Arachnology 17:142–146.

Liu, Y., Z. Shao, and F. Vollrath. 2008. Elasticity of spider silks. Biomacromolecules 9:1782–1786.

Lloyd, N. J., and M. A. Elgar. 1997. Costs and benefits of facultative aggregating behaviour in the orb-spinning spider *Gasteracantha minax* Thorell (Araneae: Araneidae). Australian Journal of Ecology 22:256–261.

Lopardo, L., and G. Hormiga. 2008. Phylogenetic placement of the Tasmanian spider *Acrobleps hygrophilus* (Araneae, Anapidae) with comments on the evolution of the capture web in Araneoidea. Cladistics 24:1–33.

Lopardo, L., and M. Ramírez. 2007. The combing of cribellar silk by the prithine *Misionella mendensis*, with notes on other filistatid spiders (Araneae: Filistatidae). American Museum Novitates 3563:1–14.

Lopardo, L., M. J. Ramírez, C. Grismado, and L. A. Compagnucci. 2004. Web building behavior and the phylogeny of austrochiline spiders. Journal of Arachnology 32:42–54.

Lopardo, L., G. Giribet, and G. Hormiga. 2011. Morphology to the rescue: molecular data and the signal of morphological characters in combined phylogenetic analyses—a case study from mysmenid spiders (Araneae, Mysmenidae), with comments on the evolution of web architecture. Cladistics 27:278–330.

Lopez, A. 1986. Construction de toiles en "voile de bateau" par une araignee salticid languedocine. Bull. Soc. Arch. Beziers (ser. 6) 2:65–68.

———. 1987. Glandular aspects of sexual biology. Pages 121–132 *in* W. Nentwig, ed. Ecophysiology of Spiders. Springer-Verlag, Berlin, Heidelberg.

Lubin, Y. D. 1973. Web structure and function: the non-adhesive orb-web of *Cyrtophora moluccensis* (Doleschall) (Araneae: Araneidae). Forma et functio 6:337–358.

———. 1974. Stabilimenta and barrier webs in the orb webs of *Argiope argentata* (Araneae, Araneidae) on Daphne and Santa Cruz Islands, Galapagos. Journal of Arachnology 2:119–126.

———. 1978. Seasonal abundance and diversity of web-building spiders in relation to habitat structure on Barro Colorado Island, Panama. Journal of Arachnology 31–51.

———. 1980. The predatory behavior of *Cyrtophora* (Araneae: Araneidae). Journal of Arachnology 8:159–185.

———. 1982. Does the social spider, *Achaearanea wau* (Theridiidae), feed its young? Zeitschrift für Tierpsychologie 60:127–134.

———. 1986. Web building and prey capture in the Uloboridae. Pages 132–171 *in* W. A. Shear, ed. Spiders: Webs, Behavior, and Evolution. Stanford University Press, Stanford, CA.

Lubin, Y. D., and S. Dorugl. 1982. Effectiveness of single-thread webs as insect traps: sticky trap models. Bulletin of the British Arachnological Society 5:399–407.

Lubin, Y. D., and J. R. Henschel. 1990. Foraging at the thermal limit: burrowing spiders (Seothyra, Eresidae) in the Namib desert dunes. Oecologia 84:461–467.

———. 1996. The influence of food supply on foraging behaviour in a desert spider. Oecologia 105:64–73.

Lubin, Y. D., W. G. Eberhard, and G. G. Montgomery. 1978. Webs of *Miagrammopes* (Araneae: Uloboridae) in the Neotropics. Psyche: A Journal of Entomology 85:1–23.

Lubin, Y. D., B. D. Opell, W. G. Eberhard, and H. W. Levi. 1982. Orb plus cone-webs in Uloboridae (Araneae), with a description of a new genus and four new species. Psyche: A Journal of Entomology 89:29–64.

Lubin, Y., M. Kotzman, and S. Ellner. 1991. Ontogenetic and seasonal changes in webs and websites of a desert widow spider. Journal of Arachnology 19:40–48.

Lubin, Y. D., S. Ellner, and M. Kotzman. 1993. Web relocation and habitat selection in desert widow spider. Ecology 74:1916–1928.

Łuczak, J. 1970. Behaviour of spider populations in the presence of mosquitoes. Ekologia Polska 18:625–634.

Ludy, C., and A. Lang. 2006. Bt maize pollen exposure and impact on the garden spider, *Araneus diadematus*. Entomologia Experimentalis et Applicata 118:145–156.

Madrigal-Brenes, R. 2012. Substrate selection for web-building in *Cyrtophora citricola* (Araneae: Araneidae). Journal of Arachnology 40:249–251.

Madrigal-Brenes, R., and G. Barrantes. 2009. Construction and function of the web of *Tidarren sisyphoides* (Araneae: Theridiidae). Journal of Arachnology 37:306–311.

Madsen, B., and F. Vollrath. 2000. Mechanics and morphology of silk drawn from anesthetized spiders. Naturwissenschaften 87:148–153.

Madsen, B., Z. Z. Shao, and F. Vollrath. 1999. Variability in the mechanical properties of spider silks on three levels: interspecific, intraspecific and intraindividual. International Journal of Biological Macromolecules 24:301–306.

Main, B. Y. 1976. Spiders. Collins Press, London.

Mallis, R. E., and K. B. Miller. 2017. Natural history and courtship behavior in *Tengella perfuga* Dahl, 1901 (Araneae: Zoropsidae). Journal of Arachnology 45:166–176.

Maroto, I. 1981. Cambios ontogenéticos en la tela de la araña *Leucauge* sp. (cf. venusta) (Araneae, Araneidae) en el Valle Central, Costa Rica. (Licenciatura en Biología). Universidad de Costa Rica, San José, Costa Rica.

Marples, B. J. 1955. Spiders from Western Samoa. Journal of the Linnean Society of London, Zoology 42:453–504.

———. 1962. Notes on spiders of the family Uloboridae. Ann. Zool. Agra 4:1–11.

———. 1967. The spinnerets and epiandrous glands of spiders. Journal of the Linnean Society of London, Zoology 46:209–222.

———. 1969. Observations on decorated webs. Bulletin of the British Arachnological Society 1:13–18.

Marples, M. J. 1935. Notes on *Argiope bruennichi* and other Pyrenean spiders. Journal of the Linnean Society of London, Zoology 39:195–202.

Marples, M. J., and B. J. Marples. 1937. 18. Notes on the spiders *Hyptiotes paradoxus* and *Cyclosa conicn*. Proceedings of the Zoological Society of London A107:213–221.

Marshall, S. A., A. Borkent, I. Agnarsson, G. W. Otis, L. Fraser, and D. d'Entremont. 2015. New observations on a Neotropical termite-hunting theridiid spider: opportunistic nest raiding, prey storage, and ceratopogonid kleptoparasites. Journal of Arachnology 43:419–421.

Marson, J. E. 1947a. Some observations on the ecological variation and development of the cruciate zigzag camouflage device of *Argiope pulchella* (Thor.). Proceedings of the Zoological Society of London 117: 219–227.

———. 1947b. Some observations on the variations in the camouflage devices used by *Cyclosa insulana* (Costa). Proceedings of the Zoological Society of London 117:598–605.

Martin, D. 1978. Zum Radnetzbau der Gattung *Pachygnatha* Sund. (Araneae: Tetragnathidae). Mitteilungen aus dem Museum für Naturkunde in Berlin. Zoologisches Museum und Institut für Spezielle Zoologie (Berlin) 54:83–95.

Marusik, Y. M., V. A. Gnelitsa, and M. M. Kovblyuk. 2005. A new species of *Synaphris* (Araneae, Synaphridae) from Ukraine. Bulletin of the British Arachnological Society 13:125–130.

Masters, W. M., and H. Markl. 1981. Vibration signal transmission in spider orb webs. Science 213:363–365.

Masters, W. M., and A. J. Moffat. 1983. A functional explanation of top-bottom asymmetry in vertical orbwebs. Animal Behaviour 31:1043–1046.

Masumoto, T., and C. Okuma. 1995. Specific web building on eucalyptus trees in *Herennia ornatissima* (Araneae: Tetragnathidae). Acta Arachnologica 44:171–172.

Maughan, O. E. 1978. The ecology and behavior of *Pholcus muralicola*. American Midland Naturalist 100:483–487.

Mayer, G. 1952. Untersuchungen über Herstellung und Struktur des Radnetzes von *Aranea diadema* und *Zilla x-notata* mit besonderer Berücksichtigung des Unterschiedes von Jugend- und Altersnetzen. Zeitschrift für Tierpsychologie 9:337–362.

Mayntz, D., S. Toft, and F. Vollrath. 2009. Nutrient balance affects foraging behaviour of a trap-building predator. Biology Letters 5:735.

Mayo, A. B. 1988. Door Construction Behavior of the Mygalomorph Spider Family Antrodiaetidae and One Member of the Family Ctenizidae (Araneae, Mygalomorphae). Master's thesis, Western Carolina University, Cullowhee, NC.

Mayr, E. 1982. The Growth of Biological Thought: Diversity, Evolution, and Inheritance. Harvard University Press, Cambridge, MA.

McCook, H. C. 1889. American Spiders and Their Spinning Work. I. Webs and Nests. Coachwhip Publications, Landisville, Philadelphia.

McKeown, K. C. 1952. Australian Spiders: Their Lives and Habits. 2nd ed. Sirius Books, Sydney.

McNett, B. J., and A. L. Rypstra. 1997. Effects of prey supplementation on survival and web site tenacity of *Argiope trifasciata* (Araneae, Araneidae): a field experiment. Journal of Arachnology 25:352–360.

———. 2000. Habitat selection in a large orb-weaving spider: vegetational complexity determines site selection and distribution. Ecological Entomology 25:423–432.

Messas, Y. F. 2014. História natural e ecologia populacional de *Eustala perfida* Mello-Leitão, 1947 (Araneae, Araneidae) na Serra do Japi, Jundiaí, São Paulo—Brasil. Universidade Estadual de Campinas, Campinas, Brazil.

Messas, Y. F., H. S. Souza, M. O. Gonzaga, and J. Vasconcellos-Neto. 2014. Spatial distribution and substrate selection by the orb-weaver spider *Eustala perfida* Mello-Leitão, 1947 (Araneae: Araneidae). Journal of Natural History 48:2645–2660.

Mestre, L., and Y. Lubin. 2011. Settling where the food is: prey abundance promotes colony formation and increases group size in a web-building spider. Animal Behaviour 81:741–748.

Mestre, L., R. Bucher, and M. H. Entling. 2014. Trait-mediated effects between predators: ant chemical cues induce spider dispersal. Journal of Zoology 293:119–125.

Michalik, P., L. Piacentini, E. Lipke, and M. Ramírez. 2013. The enigmatic Otway odd-clawed spider (*Progradungula otwayensis* Milledge, 1997, Gradungulidae, Araneae): natural history, first description of the female and micro-computed tomography of the male palpal organ. ZooKeys 335:101–112.

Michalik, P., D. Piorkowski, T. A. Blackledge, and M. J. Ramírez. 2019. Functional trade-offs in cribellate silk mediated by spinning behavior. Scientific Reports 9:9092 https://doi.org/10.1038/s41598-019-45552-x.

Miller, J. A. 2007. Review of *Erigonine* spider genera in the Neotropics (Araneae: Linyphiidae, Erigoninae). Zoological Journal of the Linnean Society 149:1–263.

Misunami, M., F. Yokohari, and M. Takahata. 2004. Further exploration into the adaptive design of the arthropod "microbrain": I. Sensory and memory-processing systems. Zoological Science 21:1141–1151.

Mittelstaedt, H. 1985. Analytical cybernetics of spider navigation. Pages 298–316 *in* P. D. F. G. Barth, ed. Neurobiology of Arachnids. Springer-Verlag, Berlin, Heidelberg.

Miyashita, T. 1992. Food limitation of population density in the orb-web spider *Nephila clavata*. Researches on Population Ecology 34:143–153.

———. 1997. Factors affecting the difference in foraging success in three co-existing *Cyclosa* spiders. Journal of Zoology 242: 137–149.

———. 2005. Contrasting patch residence strategy in two species of sit-and-wait foragers under the same environment: a constraint by life history? Ethology 111: 159–167.

Miyashita, T., and A. Shimazaki. 2006. Insects from the grazing food web favoured the evolutionary habitat shift to bright environments in araneoid spiders. Biology Letters 2:565–568.

Miyashita, T., and A. Shinkai. 1995. Design and prey capture ability of webs of the spiders *Nephila clavata* and *Argiope bruennichii*. Acta Arachnologica 44:3–10.

Miyashita, T., Y. Sakamaki, and A. Shinkai. 2001. Evidence against moth attraction by *Cyrtarachne*, a genus related to bolas spiders. Acta Arachnologica 50:1–4.

Monaghan, P., A. Charmantier, D. H. Nussey, and R. E. Ricklefs. 2008. The evolutionary ecology of senescence. Functional Ecology 22:371–378.

Moon, M. J., and E. K. Tillinghast. 2004. Silk production after mechanical pulling stimulation in the ampullate silk glands of the barn spider, *Araneus cavaticus*. Entomological Research 34:123–130.

Moore, C. W. 1977. The life cycle, habitat and variation in selected web parameters in the spider, *Nephila clavipes* Koch (Araneidae). American Midland Naturalist 98:95–108.

Moore, D., J. C. Watts, A. Herrig, and T. C. Jones. 2016. Exceptionally short-period circadian clock in *Cyclosa turbinata*: regulation of locomotor and web-building behavior in an orb-weaving spider. Journal of Arachnology 44:388–396.

Morse, D. H. 2007. Predator upon a Flower: Life History and Fitness in a Crab Spider. Harvard University Press, Cambridge, MA.

Mortimer, B., and F. Vollrath. 2015. Diversity and properties of key spider silks and webs. Research and Knowledge 1:32–42.

Mortimer, B., S. D. Gordon, C. Holland, C. R. Siviour, F. Vollrath, and J. F. C. Windmill. 2014. The speed of sound in silk: linking material performance to biological function. Advanced Materials 26:5179–5183.

Moya, J., R. Quesada, G. Barrantes, W. Eberhard, I. Escalante, C. Esquivel, A. Rojas, et al. 2010. Egg sac construction by folding dead leaves in *Pozonia nigroventris* and *Micrathena* sp. (Araneae: Araneidae). Journal of Arachnology 38:371–373.

Moya-Laraño, J., J. Halaj, and D. H. Wise. 2002. Climbing to reach females: Romeo should be small. Evolution 56:420–425.

Moya-Laraño, J., D. Vinković, C. Allard, and M. Foellmer. 2007. Gravity still matters. Functional Ecology 21:1178–1181.

Moya-Laraño, J., D. Vinković, E. De Mas, G. Corcobado, and E. Moreno. 2008. Morphological evolution of spiders predicted by pendulum mechanics. PLOS ONE 3:e1841.

Moya-Laraño, J., D. Vinković, C. M. Allard, and M. W. Foellmer. 2009. Optimal climbing speed explains the evolution of extreme sexual size dimorphism in spiders. Journal of Evolutionary Biology 22:954–963.

Müller, G. B., and G. P. Wagner. 1991. Novelty in evolution: restructuring the concept. Annual Review of Ecology and Systematics 22:229–256.

Muma, M. H. 1971. Biological and behavioral notes on *Gasteracantha cancriformis* (Arachnida: Araneidae). The Florida Entomologist 54:345–351.

Murphy, F., and J. Murphy. 2000. An Introduction to the Spiders of South East Asia. Malaysian Nature Society, Kuala Lumpur, Malaysia.

Murphy, J. A., and N. J. Platnick. 1981. On *Liphistius desultor* (Araneae, Liphistiidae). Bulletin of the American Museum of Natural History 170:46–56.

Murphy, N. P., V. W. Framenau, S. C. Donnellan, M. S. Harvey, Y. C. Park, and A. D. Austin. 2006. Phylogenetic reconstruction of the wolf spiders (Araneae: Lycosidae) using sequences from the 12S rRNA, 28S rRNA, and NADH1 genes: implications for classification, biogeography, and the evolution of web building behavior. Molecular Phylogenetics and Evolution 38:583–602.

Nakata, K. 2007. Prey detection without successful capture affects spider's orb-web building behaviour. Naturwissenschaften 94:853–857.

———. 2008. Spiders use airborne cues to respond to flying insect predators by building orb-web with fewer silk thread and larger silk decorations. Ethology 114:686–692.

———. 2009. To be or not to be conspicuous: the effects of prey availability and predator risk on spider's web decoration building. Animal Behaviour 78:1255–1260.

———. 2010a. Attention focusing in a sit-and-wait forager: a spider controls its prey-detection ability in different web sectors by adjusting thread tension. Proceedings of the Royal Society of London B: Biological Sciences 277:29–33.

———. 2010b. Does ontogenetic change in orb web asymmetry reflect biogenetic law? Naturwissenschaften 1–4.

———. 2012. Plasticity in an extended phenotype and reversed up-down asymmetry of spider orb webs. Animal Behaviour 83:821–826.

———. 2013. Spatial learning affects thread tension control in orb-web spiders. Biology Letters 9:20130052.

Nakata, K. and Y. Mori. 2016. Cost of complex behavior and its implications in antipredator defence in orb-web spiders. Animal Behaviour 120:115–121.

Nakata, K., and A. Ushimaru. 1999. Feeding experience affects web relocation and investment in web threads in an orb-web spider, *Cyclosa argenteoalba*. Animal Behaviour 57:1251–1255.

———. 2004. Difference in web construction behavior at newly occupied web sites between two *Cyclosa* species. Ethology 110:397–411.

Nakata, K., and S. Zschokke. 2010. Upside-down spiders build upside-down orb webs: web asymmetry, spider orientation and running speed in *Cyclosa*. Proceedings of the Royal Society of London B: Biological Sciences 277:3019–3025.

Nakata, K., A. Ushimaru, and T. Watanabe. 2003. Using past experience in web relocation decisions enhances the foraging efficiency of the spider *Cyclosa argenteoalba*. Journal of Insect Behavior 16:371–380.

Neet, C. 1990. Function and structural variability of the stabilimenta of *Cyclosa insulana* (Costa) (Araneae, Araneidae). Bulletin of the British Arachnological Society 8:161–164.

Nelson, G. 1985. Outgroups and ontogeny. Cladistics 1:29–45.

Nelson, X. J., and R. R. Jackson. 2011. Flexibility in the foraging strategies of spiders. Pages 31–56 *in* M. E. Herberstein, ed. Spider Behaviour: Flexibility and Versatility. Cambridge University Press, Cambridge.

Nentwig, W. 1982a. Why do only certain insects escape from a spider's web? Oecologia 53:412–417.

———. 1982b. Beutetieranalysen an cribellaten Spinnen (Araneae: Filistatidae, Dictynidae, Eresidae). Entomologische Mitteilungen aus dem Zoologischen Museum Hamburg 7:233–244.

———. 1983a. The non-filter function of orb webs in spiders. Oecologia 58:418–420.

———. 1983b. The prey of web-building spiders compared with feeding experiments (Araneae: Araneidae, Linyphiidae, Pholcidae, Agelenidae). Oecologia 56:132–139.

———. 1983c. An association of earwigs (Dermaptera) and bugs (Heteroptera) in a spider's (Araneae) web? Journal of Arachnology 11:450.

———. 1985a. Feeding ecology of the tropical spitting spider *Scytodes longipes*. Oecologia 65:284–288.

———. 1985b. Prey analysis of four species of tropical orb-weaving spiders (Araneae: Araneidae) and a comparison with araneids of the temperate zone. Oecologia 66:580–594.

———. 1985c. *Architis nitidopilosa*, a Neotropical pisaurid with a permanent catching web (Araneae, Pisauridae). Bulletin of the British Arachnological Society 6:297–303.

Nentwig, W., and T. E. Christenson. 1986. Natural history of the non-solitary sheetweaving spider *Anelosimus jocundus* (Araneae: Theridiidae). Zoological Journal of the Linnean Society 87:27–35.

Nentwig, W., and S. Heimer. 1983. Orb webs and single-line webs: an economic consequence of space web reduction in spiders. Journal of Zoological Systematics and Evolutionary Research 21:26–37.

———. 1987. Ecological aspects of spider webs. Pages 211–225 *in* W. Nentwig, ed. Ecophysiology of Spiders. Springer-Verlag, Berlin, Heidelberg.

Nentwig, W., and H. Rogg. 1988. The cross stabilimentum of *Argiope argentata* (Araneae: Araneidae)—nonfunctional or a nonspecific stress reaction? Zoologischer Anzeiger 221:248–266.

Nentwig, W., and H. Spiegel. 1986. The partial web renewal behaviour of *Nephila clavipes* (Araneae: Araneidae). Zoologischer Anzeiger 216:351–356.

Nentwig, W., and C. Wissel. 1986. A comparison of prey lengths among spiders. Oecologia 68:595–600.

Nielsen, E. 1932. The Biology of Spiders: With Especial Reference to the Danish Fauna. 2 vols. Levin and Munksgaard, Copenhagen.

Nijhout, H. F. 1991. The Development and Evolution of Butterfly Wing Patterns. Smithsonian Institution Press, Washington, DC.

Nørgaard, E. 1951. Notes on the biology of *Filistata insidiatrix* (Forsk.). Entomologiske Meddelelser 26:170–184.

Nyffler, M., and K. Birkhofer. 2017. An estimated 400–800 million tons of prey are annually killed by the global spider community. Science of Nature 104:30–42.

Olive, C. W. 1980. Foraging specialization in orb-weaving spiders. Ecology 61:1133–1144.

———. 1982. Behavioral response of a sit-and-wait predator to spatial variation in foraging gain. Ecology 63:912–920.

Opell, B. D. 1979. Revision of the genera and tropical American species of the spider family Uloboridae. Bulletin of the Museum of Comparative Zoology 148:443–549.

———. 1982. Post-hatching development and web production of *Hyptiotes cavatus* (Hentz) (Araneae, Uloboridae). Journal of Arachnology 10:185–191.

———. 1987a. Changes in web-monitoring forces associated with web reduction in the spider family Uloboridae. Canadian Journal of Zoology 65:1028–1034.

———. 1987b. The influence of web monitoring tactics on the tracheal systems of spiders in the family Uloboridae (Arachnida, Araneida). Zoomorphology 107:255–259.

———. 1987c. The new species *Philoponella herediae* and its modified orb-web (Araneae, Uloboridae). Journal of Arachnology 15:59–63.

———. 1992. Web-monitoring force exerted by the spider *Waitkera waitakerensis* (Uloboridae). Journal of Arachnology 20:146–147.

———. 1994. Increased stickness of prey capture threads accompanying web reduction in the spider family Uloboridae. Functional Ecology 8:85–90.

———. 1996. Functional similarities of spider webs with diverse architectures. American Naturalist 148:630–648.

———. 1997. The material cost and stickiness of capture threads and the evolution of orb-weaving spiders. Biological Journal of the Linnean Society 62:443–458.

———. 1998. Economics of spider orb-webs: the benefits of producing adhesive capture thread and of recycling silk. Functional Ecology 12:613–624.

———. 1999a. Changes in spinning anatomy and thread stickiness associated with the origin of orb-weaving spiders. Biological Journal of the Linnean Society 68:593–612.

———. 1999b. Redesigning spider webs: stickiness, capture area and the evolution of modern orb-webs. Evolutionary Ecology Research 1:503–516.

———. 2001. Cribellum and calamistrum ontogeny in the spider family Uloboridae: linking functionally related but separate silk spinning features. Journal of Arachnology 29:220–226.

———. 2002. How spider anatomy and thread configuration shape the stickiness of cribellar prey capture threads. Journal of Arachnology 30:10–19.

———. 2013. Cribellar thread. Pages 303–318 *in* W. Nentwig, ed. Spider Ecophysiology. Springer-Verlag, Berlin, Heidelberg.

Opell, B. D., and J. A. Beatty. 1976. The Nearctic Hahniidae (Arachnida: Araneae). Bulletin of the Museum of Comparative Zoology 147:393–433.
Opell, B. D., and J. E. Bond. 2000. Capture thread extensibility of orb-weaving spiders: testing punctuated and associative explanations of character evolution. Biological Journal of the Linnean Society 70:107–120.
———. 2001. Changes in the mechanical properties of capture threads and the evolution of modern orb-weaving spiders. Evolutionary Ecology Research 3:567–581.
Opell, B. D., and W. G. Eberhard. 1984. Resting postures of orb-weaving uloborid spiders (Araneae, Uloboridae). Journal of Arachnology 11:369–376.
Opell, B. D., and M. L. Hendricks. 2007. Adhesive recruitment by the viscous capture threads of araneoid orb-weaving spiders. Journal of Experimental Biology 210:553.
———. 2009. The adhesive delivery system of viscous capture threads spun by orb-weaving spiders. Journal of Experimental Biology 212:3026.
———. 2010. The role of granules within viscous capture threads of orb-weaving spiders. Journal of Experimental Biology 213: 339.
Opell, B. D., and D. C. Konur. 1992. Influence of web-monitoring tactics on the density of mitochondria in leg muscles of the spider family Uloboridae. Journal of Morphology 213:341–347.
Opell, B. D., and H. S. Schwend. 2007. The effect of insect surface features on the adhesion of viscous capture threads spun by orb-weaving spiders. Journal of Experimental Biology 210: 2352–2360.
———. 2009. Adhesive efficiency of spider prey capture threads. Zoology 112:16–26.
Opell, B. D., G. Roth, and P. E. Cushing. 1990. The effect of *Hyptiotes cavatus* (Uloboridae) web-manipulation on the dimensions and stickiness of cribellar silk puffs. Journal of Arachnology 18:238–240.
Opell, B. D., J. E. Bond, and D. A. Warner. 2006. The effects of capture spiral composition and orb-web orientation on prey interception. Zoology 109:339–345.
Opell, B. D., G. K. Lipkey, M. L. Hendricks, and S. T. Vito. 2009. Daily and seasonal changes in the stickiness of viscous capture threads in *Argiope aurantia* and *Argiope trifasciata* orb-webs. Journal of Experimental Zoology Part A: Ecological Genetics and Physiology 311:217–225.
Opell, B. D., S. E. Karinshak, and M. A. Sigler. 2011a. Humidity affects the extensibility of an orb-weaving spider's viscous thread droplets. Journal of Experimental Biology 214:2988–2993.
Opell, B. D., H. S. Schwend, and S. T. Vito. 2011b. Constraints on the adhesion of viscous threads spun by orb-weaving spiders: the tensile strength of glycoprotein glue exceeds its adhesion. Journal of Experimental Biology 214:2237.
Opell, B. D., A. M. Tran, and S. E. Karinshak. 2011c. Adhesive compatibility of cribellar and viscous prey capture threads and its implication for the evolution of orb-weaving spiders. Journal of Experimental Zoology Part A: Ecological Genetics and Physiology 315A:376–384.
Opell, B. D., S. E. Karinshak, and M. A. Sigler. 2013. Environmental response and adaptation of glycoprotein glue within the droplets of viscous prey capture threads from araneoid spider orb-webs. Journal of Experimental Biology 216:3023–3034.
Opell, B. D., S. F. Andrews, S. E. Karinshak, and M. A. Sigler. 2015. The stability of hygroscopic compounds in orb-web spider viscous thread. Journal of Arachnology 43:152–157.
Opell, B. D., K. E. Buccella, M. K. Godwin, M. X. Rivas, and M. L. Hendricks. 2017. Humidity-mediated changes in an orb spider's glycoprotein adhesive impact prey retention time. Journal of Experimental Biology 220:1313–1321.
Ortega-Jiménez, V. M., and R. Dudley. 2013. Spiderweb deformation induced by electrostatically charged insects. Scientific Reports 3:2108.
Ortlepp, C. S., and J. M. Gosline. 2004. Consequences of forced silking. Biomacromolecules 5:727–731.
———. 2008. The scaling of safety factor in spider draglines. Journal of Experimental Biology 211:2832.
Osaki, S. 1989. Seasonal change in color of spiders' silk. Acta Arachnologica 38:21–28.
Osaki, S., and M. Osaki. 2011. Evolution of spiders from nocturnal to diurnal gave spider silks mechanical resistance against UV irradiation. Polymer Journal 43:200–204.
Otto, F., R. Raccanello, and F. Gröbner. 1975. Fundamentals and classification—general survey and classification of net structures. Pages 10–45 *in* K. Bach, B. Burkhardt, R. Graefe, and R. Raccanello, eds. Nets in Nature and Technics. Druckerei Heinrich Fink KG., Stuttgart.
Pan, Z. J., C. P. Li, and Q. Xu. 2004. Active control on molecular conformations and tensile properties of spider silk. Journal of Applied Polymer Science 92:901–905.
Pasquet, A. 1984. Predatory-site selection and adaptation of the trap in four species of orb-weaving spiders. Biology of Behaviour 9:3–19.
Pasquet, A., A. Ridwan, and R. Leborgne. 1994. Presence of potential prey affects web-building in an orb-weaving spider *Zygiella x-notata*. Animal Behaviour 47:477–480.
Pasquet, A., R. Leborgne, and Y. Lubin. 1999. Previous foraging success influences web building in the spider *Stegodyphus lineatus* (Eresidae). Behavioral Ecology 10:115.
Pasquet, A., J. Cardot, and R. Leborgne. 2007. Wasp attacks and spider defence in the orb weaving species *Zygiella x-notata*. Journal of Insect Behavior 20:553–564.
Paz, N. 1988. Ecologia y aspectos del comportamiento en *Linothele* sp. (Araneae, Dipluridae). Journal of Arachnology 16:5–22.
Peakall, D. B. 1964. Effects of cholinergic and anticholinergic drugs on the synthesis of silk fibroins of spiders. Comparative Biochemistry and Physiology 12:465–470.
———. 1965. Regulation of the synthesis of silk fibroins of spiders at the glandular level. Comparative Biochemistry and Physiology 15:509–515.

———. 1969. Synthesis of silk, mechanism and location. Integrative and Comparative Biology 9:71–79.

———. 1971. Conservation of web proteins in the spider, *Araneus diadematus*. Journal of Experimental Zoology 176:257–264.

Peakall, D. B., and P. N. Witt. 1976. The energy budget of an orb web-building spider. Comparative Biochemistry and Physiology Part A: Physiology 54:187–190.

Peretti, A., W. G. Eberhard, and R. D. Briceño. 2006. Copulatory dialogue: female spiders sing during copulation to influence male genitalic movements. Animal Behaviour 72:413–421.

Pérez-Miles, F., and D. Ortíz-Villatoro. 2012. Tarantulas do not shoot silk from their legs: experimental evidence in four species of New World tarantulas. Journal of Experimental Biology 215:1749–1752.

Pérez-Miles, F., A. Panzera, D. Ortiz-Villatoro, and C. Perdomo. 2009. Silk production from tarantula feet questioned. Nature 461:E9.

Pérez-Rigueiro, J., M. Elices, G. Plaza, J. I. Real, and G. V. Guinea. 2005. The effect of spinning forces on spider silk properties. Journal of Experimental Biology 208:2633–2639.

Peters, H. M. 1931. Die Fanghandlung der Kreuzspinne (*Epeira diademata* L.). Zeitschrift für Vergleichende Physiologie 15:693–748.

———. 1933a. Weitere Untersuchungen über die Fanghandlung der Kreuzspinne (*Epeira diademata* Cl.). Zeitschrift fur vergleichende Physiologie 19:47–67.

———. 1933b. Kleine Beitrage zur Biologie der Kreuzspinne *Epeira diademata* Cl. Zeitschrift fur Morphologie und Okologie der Tiere 26:447–468.

———. 1936. Studien am Netz der Kreuzspinne (*Aranea diadema*.) I. Die Grundstruktur des Netzes und beziehungen zum Bauplan des Spinnenkörpers. Zeitschrift für Morphologie und Ökologie der Tiere 32:613–649.

———. 1937. Studien am Netz der Kreuzspinne (*Aranea diadema* L.). II. Uber die Herstellung des Rahmens, der Radialfäden und der Hilfsspirale. Zeitschrift für Morphologie und Ökologie der Tiere 33:128–150.

———. 1939. Probleme des Kreuzspinnen-Netzes. Pages 179–266 *in* Probleme des Kreuzspinnennetzes. Springer-Verlag, Berlin, Heidelberg.

———. 1953. Beiträge zur vergleichenden Ethologie und ökologie tropischer Webespinnen. Zeitschrift für Morphologie und Ökologie der Tiere 42:278–306.

———. 1954. Estudios adicionales sobre la estructura de la red concéntrica de las arañas. Comunicaciones 3:1–18.

———. 1955. Contribuciones sobre la etología y ecología comparada de las arañas tejedoras tropicales. Comunicaciones 4:37–46.

———. 1969. Maturing and coordination of web-building activity. Integrative and Comparative Biology 9:223–227.

———. 1982. Wie Spinnen der Familie Uloboridae ihre Beute einspinnen und verzehren. Verhandlungen des Naturwissenschaftlichen Vereins in Hamburg 25:147–167.

———. 1984. The spinning apparatus of Uloboridae in relation to the structure and construction of capture threads (Arachnida, Araneida). Zoomorphology 104:96–104.

———. 1990. On the structure and glandular origin of bridging lines used by spiders for moving to distant places. Acta Zoologica Fennica 190:309–314.

———. 1992. On the burrowing behaviour and the production and use of silk in *Seothyra*, a sand-inhabiting spider from the Namib Desert (Arachnida, Araneae, Eresidae). Verhandlungen naturwissenschaften vereins Hamburg (NF) 33:191–211.

———. 1993a. Functional organization of the spinning apparatus of *Cyrtophora citricola*. Zoomorphology 113:153–163.

———. 1993b. Über das Problem der Stabilimente in Spinnennetzen. Zoologische Jahrbücher, Abteilung für allgemeine Zoologie und Physiologie der Tiere 97:245–264.

———. 1995a. *Polenecia producta* and its web: structure and evolution (Araneae, Uloboridae). Zoomorphology 115:1–9.

———. 1995b. Ultrastructure of orb spiders' gluey capture threads. Naturwissenschaften 82:380–382.

Peters, H. M., and J. Kovoor. 1991. The silk-producing system of *Linyphia triangularis*. Zoomorphology 111:1–17.

Peters, H. M., Witt, P. N., and Wolff, D. 1950. Die Beeinflussung des Netzbaues der Spinnen durch neurotrope Substanzen. Zeitschrift fur vergleichende Physiologie 32:20–44.

Peters, P. J. 1970. Orb web construction: interaction of spider (*Araneus diadematus* Cl.) and thread configuration. Animal Behaviour 18:478–484.

Peters, R. 1967. Vergleichende Untersuchungen über Bau und Funktion der Spinnwarzen und Spinnwarzenmuskulatur einiger Araneen. Zoologische Beitrage 13:29–119.

Peterson, J. A., S. A. Romero, and J. D. Harwood. 2010. Pollen interception by linyphiid spiders in a corn agroecosystem: implications for dietary diversification and risk-assessment. Arthropod-Plant Interactions 1–11.

Petrusewiczowa, E. 1938. Beobachtungen ber den Bau des Netzes der Kreuzspinne. Travaux de l'Institut de Biologie Gemirale de l'Universite de Vilno 9:1–25.

Pickford, M. 2000. Fossil spider's webs from the Namib Desert and the antiquity of *Seothyra* (Araneae, Eresidae). Annales de Paléontologie 86:147–155.

Piel, W. H. 2001. The systematics of Neotropical orb-weaving spiders in the genus *Metepeira* (Araneae: Araneidae). Bulletin of the Museum of Comparative Zoology 157:1–92.

Platnick, N. I. 2014. The World Spider Catalog, version 15. American Museum of Natural History.

Platnick, N. I., and W. J. Gertsch. 1976. The suborders of spiders: A cladistic analysis (Arachnida, Araneae). American Museum Novitates 2607:1–15.

Platnick, N. I., and M. U. Shadab. 1978. A review of the spider genus *Anapis* (Araneae: Anapidae) with a dual cladistic analysis. American Museum Novitates 2663:1–23.

———. 1979. A review of the spider genera *Anapisona* and *Pseudanapis* (Araneae, Anapidae). American Museum Novitates 2672:1–20.

Platnick, N. I., J. A. Coddington, R. R. Forster, and C. E. Griswold. 1991. Spinneret evidence and the higher classification of the haplogyne spiders (Araneae, Araneomorphae). American Museum Novitates 3016:1–73.

Pocock, R. I. 1895. On a new and natural grouping of some of the Oriental genera of Mygalomorphæ, with descriptions of new genera and species. Annals and Magazine of Natural History 15:165–184.

Polis, G. A. 1990. The Biology of Scorpions. Stanford University Press, Standford, CA.

Popper, K. R. 1970. Normal science and its dangers. Pages 51–59 *in* I. Lakatos and A. Musgrave, eds. Criticism and the Growth of Knowledge. Cambridge University Press, Cambridge.

Prenter, J., D. Pérez-Staples, and P. W. Taylor. 2010. The effects of morphology and substrate diameter on climbing and locomotor performance in male spiders. Functional Ecology 24:400–408.

Prestwich, K. N. 1977. The energetics of web-building in spiders. Comparative Biochemistry and Physiology Part A: Physiology 57:321–326.

Prokop, P. 2006. Prey type does not determine web design in two orb-weaving spiders. Zoological Studies 45:124–131.

Qin, Z., B. G. Compton, J. A. Lewis, and M. J. Buehler. 2015. Structural optimization of 3D-printed synthetic spider webs for high strength. Nature Communications 6:7038.

Quesada, R. 2014. Una prueba de plasticidad en arañas diminutas: construcción de telas orbiculares por ninfas de *Leucauge argyra* (Tetragnathidae) en espacios reducidos. Master's thesis, Universidad de Costa Rica.

Quesada, R., E. Triana, G. Vargas, J. K. Douglass, M. A. Seid, J. E. Niven, W. G. Eberhard, W. T. Wcislo. 2011. The allometry of CNS size and consequences of miniaturization in orb-weaving and cleptoparasitic spiders. Arthropod Structure & Development 40:521–529.

Raff, R. A. 1996. The Shape of Life: Genes, Development, and the Evolution of Animal Form. University of Chicago Press, Chicago.

Ramírez, M. G., E. A. Wall, and M. Medina. 2003. Web orientation of the banded garden spider *Argiope trifasciata* (Araneae, Araneidae) in a California coastal population. Journal of Arachnology 31:405–411.

Ramírez, M. J. 2014. The morphology and phylogeny of dionychan spiders (Araneae: Araneomorphae). Bulletin of the American Museum of Natural History 390:1–374.

Ramírez, M. J., and C. Grismado. 1997. A review of the spider family Filistatidae in Argentina (Arachnida, Araneae), with a cladistic reanalysis of filistatid genera. Insect Systematics & Evolution 28:319–349.

Ramírez, M. J., Lopardo, L., and N. I. Platnick. 2004. Notes on Chilean anapids and their webs. American Museum Novitates 3428:13pp.

Ramousse, R., and L. Le Guelte. 1984. Statégies de construction de la toile chez deux espèces d'araignées. Revue Arachnologique 5:255–265.

Rao, D. 2010. Stingless bee interception is not affected by variations in spider silk decoration. Journal of Arachnology 38:157–161.

Rao, D., and Y. Lubin. 2010. Conditions favoring group living in web-building spiders in an extreme desert environment. Israel Journal of Ecology and Evolution 56:21–33.

Rao, D., and G. Poyyamoli. 2001. Role of structural requirements in web-site selection in *Cyrtophora cicatrosa* Stoliczka (Araneae: Araneidae). Current Science 81:678–679.

Rao, D., K. Cheng, and M. E. Herberstein. 2007. A natural history of web decorations in the St Andrew's Cross spider (*Argiope keyserlingi*). Australian Journal of Zoology 55:9–14.

———. 2008. Stingless bee response to spider webs is dependent on the context of encounter. Behavioral Ecology and Sociobiology 63:209–216.

Rao, D., O. Ceballos Fernandez, E. Castañeda-Barbosa, and F. Diaz-Fleischer. 2011. Reverse positional orientation in a neotropical orb-web spider, *Verrucosa arenata*. Naturwissenschaften DOI:10.1007/s00114-011-0811-2.

Rawlings, J. O., and P. N. Witt. 1973. Appendix: preliminary data on a possible genetic component in web building. Behavioral Genetics Simple Systems, Symposium Held at the University of Colorado. J. R. Wilson, ed.

Rayor, L. S. 1996. Attack strategies of predatory wasps (Hymenoptera: Pompilidae; Sphecidae) on colonial orb web-building spiders (Araneidae: Metepeira incrassata). Journal of the Kansas Entomological Society 69:67–75.

Reed, C. F. 1969. Cues in the web-building process. American Zoologist 9:211–221.

Reed, C. F., P. N. Witt, and R. L. Jones. 1965. The measuring function of the first legs of *Araneus diadematus* CL. Behaviour 25:98–118.

Reed, C. F., P. N. Witt, and M. B. Scarboro. 1969. The orb web during the life of *Argiope aurantia* (Lucas). Developmental Psychobiology 2:120–129.

Reed, C. F., P. N. Witt, M. B. Scarboro, and D. B. Peakall. 1970. Experience and the orb web. Developmental Psychobiology 3:251–265.

Řezáč, M., T. Krejči, T. S. Goodacre, C. R. Haddad, and V. Řezáčová. 2017. Morphological and functional diversity of minor ampullate glands in spiders from the superfamily Amaurobioidea (Entelegynae: RTA clade). Journal of Arachnology 45:198–208.

Richards, A. M. 1960. Observations on the New Zealand glow-worm *Arachnocampa luminosa* (Skuse) 1890. Transactions of the Royal Society of New Zealand 88: 559–574.

Riechert, S. E. 1982. Spider interaction strategies: communication vs. coercion. Pages 281–315 *in* P. N. Witt and J. S. Rovner, eds. Spider Communication: Mechanisms and Ecological Significance. Princeton University Press, Princeton, NJ.

———. 1985. Decisions in multiple goal contexts: habitat selection of the spider, *Agelenopsis aperta* (Gertsch). Zeitschrift für Tierpsychologie 70:53–69.

Riechert, S. E., and A. B. Cady. 1983. Patterns of resource use and tests for competitive release in a spider community. Ecology 64:899–913.

Riechert, S. E., and R. G. Gillespie. 1986. Habitat choice and utilization in web-building spiders. Pages 23–48 *in* W. A. Shear, ed. Spiders: Webs, Behavior, and Evolution. Stanford University Press, Stanford, CA.

Riechert, S. E., and A. V. Hedrick. 1993. A test for correlations among fitness-linked behavioural traits in the spider *Agelenopsis aperta* (Araneae, Agelenidae). Animal Behaviour 46:669–675.

Riechert, S. E., and J. Łuczak. 1982. Spider foraging: behavioral responses to prey. Pages 353–385 *in* P. N. Witt and J. S. Rovner, eds. Spider Communication: Mechanisms and Ecological Significance. Princeton University Press, Princeton, NJ.

Riechert, S. E., and C. R. Tracy. 1975. Thermal balance and prey availability: bases for a model relating web-site characteristics to spider reproductive success. Ecology 56:265–284.

Riechert, S., R. Roeloffs, and A. C. Echternacht. 1986. The ecology of the cooperative spider *Agelena consociata* in equatorial Africa (Araneae, Agelenidae). Journal of Arachnology 14:175–191.

Risch, P. 1977. Quantitative analysis of orb web patterns in four species of spiders. Behavior Genetics 7:199–238.

Rittschof, C. C. 2012. The effects of temperature on egg development and web site selection in *Nephila clavipes*. Journal of Arachnology 40:141–145.

Rittschof, C. C., and K. V. Ruggles. 2010. The complexity of site quality: multiple factors affect web tenure in an orb-web spider. Animal Behaviour 79:1147–1155.

Robinson, B., and M. H. Robinson. 1974. The biology of some *Argiope* species from New Guinea: predatory behaviour and stabilimentum construction (Araneae: Araneidae). Zoological Journal of the Linnean Society 54:145–159.

———. 1978. Developmental studies of *Argiope argentata* (Fabricius) and *Argiope aemula* (Walckenaer). Symposia of the Zoological Society of London 42:31–40.

Robinson, J. V. 1981. The effect of architectural variation in habitat on a spider community: an experimental field study. Ecology 62:73–80.

Robinson, M. H. 1969a. Defenses against visually hunting predators. Evolutionary Biology 3:5–59.

———. 1969b. Predatory behavior of *Argiope argentata* (Fabricius). American Zoologist 9:161–173.

———. 1975. The evolution of predatory behaviour in araneid spiders. Pages 292–312 *in* C. Baerends, C. Beer, and A. Manning, eds. Function and Evolution in Behaviour. Clarendon Press, Oxford.

———. 1977. Symbioses between insects and spiders: an association between lepidopteran larvae and the social spider *Anelosimus eximius* (Araneae: Theridiidae). Psyche: A Journal of Entomology 84:225–232.

Robinson, M. H., and Y. D. Lubin. 1979a. Specialists and generalists: the ecology and behavior of some web-building spiders from Papua New Guinea. I. *Herennia ornatissima*, *Argiope ocyaloides* and *Arachnura melanura* (Araneae: Araneidae). Pacific Insects 21:97–132.

———. 1979b. Specialists and generalists: the ecology and behavior of some web-building spiders from Papua New Guinea. II. *Psechrus argentatus* and *Fecenia* sp. Pac. Insects 21:133–164.

Robinson, M. H., and H. Mirick. 1971. The predatory behavior of the golden-web spider *Nephila clavipes* (Araneae: Araneidae). Psyche: A Journal of Entomology 78:123–139.

Robinson, M. H., and J. Olazarri. 1971. Units of behavior and complex sequences in the predatory behavior of *Argiope argentata* (Fabricius):(Araneae: Araneidae). Smithsonian Contributions to Zoology 65:1–36.

Robinson, M. H., and B. Robinson. 1970a. The stabilimentum of the orb web spider, *Argiope argentata*: an improbable defense against predators. Canadian Entomologist 102: 641–655.

———. 1970b. Prey caught by a sample population of the spider *Argiope argentata* (Araneae: Araneidae) in Panama: a year's census data. Zoological Journal of the Linnean Society 49:345–358.

———. 1971. The predatory behavior of the ogre-faced spider *Dinopis longipes* F. Cambridge (Araneae: Dinopidae). American Midland Naturalist 85:85–96.

———. 1972. The structure, possible function and origin of the remarkable ladder-web built by a New Guinea orb-web spider (Araneae: Araneidae). Journal of Natural History 6:687–694.

———. 1973a. Ecology and behavior of the giant wood spider *Nephila maculata* (Fabricius) in New Guinea. Smithsonian Contributions to Zoology 149:1–76.

———. 1973b. The stabilimenta of *Nephila clavipes* and the origins of stabilimentum-building in araneids. Psyche: A Journal of Entomology 80:277–288.

———. 1975. Evolution beyond the orb web: the web of the araneid spider *Pasilobus* sp., its structure, operation and construction. Zoological Journal of the Linnean Society 56:301–313.

———. 1976a. Discrimination between prey types: an innate component of the predatory behaviour of araneid spiders. Zeitschrift für Tierpsychologie 41:266–276.

———. 1976b. The ecology and behavior of *Nephila maculata*: a supplement. Smithsonian Contributions to Zoology 218: 1–22.

———. 1976c. A tipulid associated with spider webs in Papua New Guinea. Entomologist's Monthly Magazine 112:1–3.

———. 1977. Associations between flies and spiders: bibiocommensalism and dipsoparasitism? Psyche: A Journal of Entomology 84:150–157.

———. 1978. Thermoregulation in orb-web spiders: new descriptions of thermoregulatory postures and experiments on the effects of posture and coloration. Zoological Journal of the Linnean Society 64:87–102.

———. 1980. Comparative studies of the courtship and

mating behavior of tropical araneid spiders. Pacific Insects Monographs 36:1–218.

Robinson, M. H., H. Mirick, and O. Turner. 1969. The predatory behavior of some araneid spiders and the origin of immobilization wrapping. Psyche: A Journal of Entomology 76:487–501.

Robinson, M. H., B. Robinson, and W. Graney. 1971. The predatory behavior of the nocturnal orb web spider *Eriophora fuliginea* (CL Koch) (Araneae: Araneidae). Revista Peruana de Entomología 14:304–315.

Rodríguez, R. L., and S. E. Gamboa. 2000. Memory of captured prey in three web spiders (Araneae: Araneidae, Linyphiidae, Tetragnathidae). Animal Cognition 3:91–97.

Rodríguez, R. L., R. D. Briceño, E. Briceño-Aguilar, and G. Höbel. 2015. *Nephila clavipes* spiders (Araneae: Nephilidae) keep track of captured prey counts: testing for a sense of numerosity in an orb-weaver. Animal Cognition 18:307–314.

Rojas, A. 2011. Sheet-web construction by *Melpomene* sp. (Araneae: Agelenidae). Journal of Arachnology 39:189–193.

Romero, G. Q., and J. Vasconcellos-Neto. 2007. Aranhas sobre plantas: dos comportamentos de forrageamento às associações específicas. Pages 67–87 *in* M. O. Gonzaga, A. J. Santos, and H. F. Japyassú, eds. Ecologia e Comportamento De Aranhas. Interciência, Rio de Janeiro, Brazil.

Rovner, J. S. 1976. Detritus stabilamenta on the webs of *Cyclosa turbinata* (Araneae, Araneidae). Journal of Arachnology 4:215–216.

———. 1978. Adhesive hairs in spiders: behavioral functions and hydraulically mediated movement. Symposia of the Zoological Society of London 42:99–108.

Rovner, J. S., and S. J. Knost. 1974. Post-immobilization wrapping of prey by lycosid spiders of the herbaceous stratum. Psyche: A Journal of Entomology 81:398–415.

Ruppert, E. E., and R. S. Fox. 1988. Seashore Animals of the Southeast: A Guide to Common Shallow-Water Invertebrates of the Southeastern Atlantic Coast. University of South Carolina Press, Columbia.

Rybak, J. 2007. Structure and function of the web of *Bathyphantes simillimus* (Araneae: Linyphiidae) in an isolated population in the Stołowe Mountains, SW Poland. Bulletin of the British Arachnological Society 14:33–38.

Rypstra, A. L. 1981. The effect of kleptoparasitism on prey consumption and web relocation in a Peruvian population of the spider *Nephila clavipes*. Oikos 37:179–182.

———. 1982. Building a better insect trap: an experimental investigation of prey capture in a variety of spider webs. Oecologia 52:31–36.

———. 1983. The importance of food and space in limiting web-spider densities: a test using field enclosures. Oecologia 59:312–316.

———. 1985. Aggregations of *Nephila clavipes* (L.) (Araneae, Araneidae) in relation to prey availability. Journal of Arachnology 13:71–78.

Rypstra, A. L., and G. J. Binford. 1995. *Philoponella republicana* (Araneae, Uloboridae) as a commensal in the webs of other spiders. Journal of Arachnology 23:1–8.

Sabath, L. E. 1969. Color change and life history observations of the spider *Gea heptagon* (Araneae: Araneidae). Psyche: A Journal of Entomology 76:367–374.

Sabath, M. D., L. E. Sabath, and A. M. Moore. 1974. Web, reproduction and commensals of the semisocial spider *Cyrtophora moluccensis* (Araneae: Araneidae) on Guam, Mariana Islands. Micronesica 10:51–55.

Sahni, V., T. A. Blackledge, and A. Dhinojwala. 2011. Changes in the adhesive properties of spider aggregate glue during the evolution of cobwebs. Scientific Reports 1:41. DOI:10.1038/srep00041.

Sahni, V., J. Harris, T. A. Blackledge, and A. Dhinojwala. 2012. Cobweb-weaving spiders produce different attachment discs for locomotion and prey capture. Nature Communications 3:1106.

Sahni, V., T. Miyoshi, K. Chen, D. Jain, S. J. Blamires, T. A. Blackledge, and A. Dhinojwala. 2014. Direct solvation of glycoproteins by salts in spider silk glues enhances adhesion and helps to explain the evolution of modern spider orb webs. Biomacromolecules 15:1225–1232.

Salomon, M. 2007. Western black widow spiders express state-dependent web-building strategies tailored to the presence of neighbours. Animal Behaviour 73:865–875.

Sanchez-Ruiz, J. A., A. Ramírez, and S. P. Kelly. 2017. Decreases in the size of riparian orb webs along an urbanization gradient. Journal of Arachnology 45:248–252.

Sandoval, C. P. 1994. Plasticity in web design in the spider *Parawixia bistriata*: a response to variable prey type. Functional Ecology 8:701–707.

Santana, M., W. G. Eberhard, G. Bassey, K. N. Prestwich, and R. D. Briceño. 1990. Low predation rates in the field by the tropical spider *Tengella radiata* (Araneae: Tengellidae). Biotropica 22:305–309.

Santos, A. J., and A. D. Brescovit. 2001. A revision of the South American spider genus *Aglaoctenus* Tullgren, 1905 (Araneae, Lycosidae, Sosippinae). Andrias 15:75–90.

Santos, A. J., and M. O. Gonzaga. 2017. Systematics and natural history of *Uaitemuri*, a new genus of the orb-weaving spider family Uloboridae (Araneae: Deinopoidea) from south-eastern Brazil. Zoological Journal of the Linnean Society 180:155–174.

Savory, T. H. 1952. The Spider's Web. Frederick Warne, London.

Schaeffer, M. 1978. Some experiments on regulation of population density in the spider *Floronia bucculenta* (Araneidae: Linyphiidae). Symposia of the Zoological Society of London 42:203–210.

Scharf, I., Y. Lubin, and O. Ovadia. 2011. Foraging decisions and behavioural flexibility in trap-building predators: a review. Biological Reviews 86:626–639.

Scharff, N., and J. A. Coddington. 1997. A phylogenetic analysis of the orb-weaving spider family Araneidae (Arachnida, Araneae). Zoological Journal of the Linnean Society 120: 355–434.

Schildknecht, H., P. Kunzelmann, D. Krauß, and C. Kuhn. 1972. Über die Chemie der Spinnwebe, I. Naturwissenschaften 59:98–99.

Schneider, J. M. 1995. Survival and growth in groups of a subsocial spider (*Stegodyphus lineatus*). Insectes Sociaux 42:237–248.

Schneider, J. M., and F. Vollrath. 1998. The effect of prey type on the geometry of the capture web of *Araneus diadematus*. Naturwissenschaften 85:391–394.

Schoener, T. W. 1980. Length-weight regressions in tropical and temperate forest-understory insects. Annals of the Entomological Society of America 73:106–109.

Schoener, T. W., and D. A. Spiller. 1992. Stabilimenta characteristics of the spider *Argiope argentata* on small islands: support of the predator-defense hypothesis. Behavioral Ecology and Sociobiology 31:309–318.

Schoener, T. W., and C. A. Toft. 1983. Spider populations: extraordinarily high densities on islands without top predators. Science 219:1353–1355.

Schütt, K. 1996. Wie Spinnen ihre Netze befestigen. Mikrokosmos 85:274–278.

Seah, W. K., and D. Li. 2002. Stabilimentum variations in *Argiope versicolor* (Araneae: Araneidae) from Singapore. Journal of Zoology 258:531–540.

Sedey, K. A., and E. M. Jakob. 1998. A description of an unusual dome web occupied by egg-carrying *Holocnemus pluchei* (Araneae, Pholcidae). Journal of Arachnology 26:385–388.

Selden, P. A., W. A. Shear, and P. M. Bonamo. 1991. A spider and other arachnids from the Devonian of New York, and reinterpretations of Devonian Araneae. Palaeontology 241–281.

Selden, P. A., C. Shih, and D. Ren. 2011. A golden orb-weaver spider (Araneae: Nephilidae: Nephila) from the Middle Jurassic of China. Biology Letters 7:775–778.

———. 2013. A giant spider from the Jurassic of China reveals greater diversity of the orbicularian stem group. Naturwissenschaften 100:1171–1181.

Selden, P. A., D. Ren, and C. Shih. 2015. Mesozoic cribellate spiders (Araneae: Deinopoidea) from China. Journal of Systematic Palaeontology 14:49–74.

Sensenig, A. T., I. Agnarsson, and T. Blackledge. 2010a. Behavioural and biomaterial coevolution in spider orb webs. Journal of Evolutionary Biology 23:1839–1856.

Sensenig, A. T., I. Agnarsson, T. M. Gondek, and T. A. Blackledge. 2010b. Webs in vitro and in vivo: spiders alter their orb-web spinning behavior in the laboratory. Journal of Arachnology 38:183–191.

Sensenig, A. T., I. Agnarsson, and T. Blackledge. 2011. Adult spiders use tougher silk: ontogenetic changes in web architecture and silk biomechanics in the orb-weaver spider. Journal of Zoology 285:28–38.

Sensenig, A. T., K. A. Lorentz, S. P. Kelly, and T. A. Blackledge. 2012. Spider orb webs rely on radial threads to absorb prey kinetic energy. Journal of the Royal Society Interface 9:1880–1891.

Sewlal, J. A. N. 2016. Possible functions of the detritus stabilimentum and hanging detritus in webs of *Azilia vachoni* (Araneae: Tetragnathidae). Arachnology 17:1–6.

Seyfarth, E. A., and F. G. Barth. 1972. Compound slit sense organs on the spider leg: mechanoreceptors involved in kinesthetic orientation. Journal of Comparative Physiology 78:176–191.

Seyfarth, E. A., R. Hergenröder, H. Ebbes, and F. G. Barth. 1982. Idiothetic orientation of a wandering spider: compensation of detours and estimates of goal distance. Behavioral Ecology and Sociobiology 11:139–148.

Shear, W. A. 1969. Observations on the predatory behavior of the spider *Hypochilus gertschi* Hoffman (Hypochilidae). Psyche: A Journal of Entomology 76:407–417.

———. 1970. The evolution of social phenomena in spiders. Bulletin of the British Arachnological Society 1:65–76.

———. 1986. The evolution of web-building behavior in spiders: a third generation of hypotheses. Pages 364–400 *in* W. A. Shear, ed. Spiders: Webs, Behavior, and Evolution. Stanford University Press, Stanford, CA.

Shear, W. A., J. M. Palmer, J. A. Coddington, and P. M. Bonamo. 1989. A Devonian spinneret: early evidence of spiders and silk use. Science 246:479–481.

Shelly, T. E. 1984. Prey selection by the Neotropical spider *Micrathena schreibersi* with notes on web-site tenacity. Proceedings of the Entomological Society of Washington 86:493–502.

Sherman, P. M. 1994. The orb-web: an energetic and behavioural estimator of a spider's dynamic foraging and reproductive strategies. Animal Behaviour 48:19–34.

Shettleworth, S. J. 2010. Cognition, Evolution, and Behavior. Oxford University Press, New York.

Shinkai, A. 1982. Web structure and construction behaviour of *Cyrtarachne yunoharuensis* Strand. Is *Cytarachne* web an ordinary web? Atypus 100:4–12.

———. 1985. Comparison in the web structure between *Nephila clavata* L. Koch and *Nephila maculata* (Fabricius) (Araneae: Araneidae), and the origin of genus *Nephila*. Acta Arachnologica 34:11–22.

———. 1988a. Single line web of *Phoroncidia pilula* (Karsch), and its prey insects. Atypus 92:37–39.

———. 1988b. Web structure of *Tetragnatha lauta* Yaginuma. Kishidaia 56:15–18.

———. 1988c. A note on the web silk theft by *Argyrodes cylindratus* (Thorell) (Araneae: Theridiidae). Acta Arachnologica 36:115–119.

———. 1988d. Predatory behavior of "spider eat spider": *Theridion sterninotatum*. Kiyosumi 12:1–5.

———. 1998a. Orb web repairing of *Cyclosa confusa*. Kishidaia 74:7–10.

———. 1998b. The web structure and the predatory behavior of *Menosira ornata* Chikuni (Araneae: Tetragnathidae). Acta Arachnologica 47:53–58.

Shinkai, A., and E. Shinkai. 1985. The web-building behavior and predatory behavior of *Theridiosoma epeiroides* Bösenberg et

Strand (Araneae: Theridiosomatidae) and the origin of the ray-formed web. Acta Arachnologica 33:9–17.
———. 1988. Web structure of *Conoculus lyugadinus* Komatsu (Araneae: Anapidae). Acta Arachnologica 37:1–12.
———. 1997. The web structure and predatory behavior of *Wendilgarda* sp. (Araneae: Theridiosomatidae) [in Japanese]. Acta Arachnologica 46:53–60.
Shinkai, A., M. Sadamoto, K. Suzuki, and E. Sinkai. 1985. Notes on the web and the prey of *Cyrtarachne*. Atypus 86:9–15.
Shinkai, E. 1979. Spiders of Hachioji City, Tokyo. 2. Classification of life-forms and webs. Memoirs Educational Institute of Private Schools of Japan 64:1–14.
———. 1989. Classification of web types in weaving spiders of Japan. Arachnological Papers Presented to Takeo Yaginuma on the Occasion of His Retirement, Osaka Arachnologists' Group, Osaka 153–179.
Shinkai, E., and A. Shinkai. 1981. The single line web of three Japanese spiders. Anima Magazine of Natural History 102:50–56.
Shinkai, E., and S. Takano. 1984. A Field Guide to the Spider of Japan. Tokai University Press, Kanagawa, Japan.
Simon, E. 1892. Histoire Naturelle des Araignées. Roret, Paris.
Smallwood, P. D. 1993. Web-site tenure in the long-jawed spider: is it risk-sensitive foraging, or conspecific interactions? Ecology 74:1826–1835.
Smith, D. R. 1997. Notes on the reproductive biology and social behavior of two sympatric species of *Philoponella* (Araneae, Uloboridae). Journal of Arachnology 25:11–19.
Smith, H. M. 2006. The systematics and biology of the genus *Poltys* (Araneae: Araneidae) in Australasia (unpublished). University of Sydney, Australia.
———. 2009. The costs of moving for a diurnally cryptic araneid spider. Journal of Arachnology 37:84–91.
Smith, J. N., D. J. Emlen, and D. E. Pearson. 2016. Linking native and invader traits explains native spider population responses to plant invasion. PLOS ONE 11:e0153661.
Smith, R. B., and T. P. Mommsen. 1984. Pollen feeding in an orb-weaving spider. Science 226:1330–1332.
Sobczak, J. F., A. P. S. Loffredo, A. M. Penteado-Dias, and M. O. Gonzaga. 2009. Two new species of *Hymenoepimecis* (Hymenoptera: Ichneumonidae: Pimplinae) with notes on their spider hosts and behaviour manipulation. Journal of Natural History 43:2691–2699.
Solano-Brenes, D. 2018. Web building behavior in a wall spider (Araneae: Oecobiidae) supports a closer relationship with orb web spiders. Thesis Lic., Universidad de Costa Rica.
Solano-Brenes, D., X. Miranda, and G. Barrantes. 2018. Making the invisible visible: methods to enhance features of tiny spider webs. Journal of Arachnology 46:538–540.
Soler, A., and R. Zaera. 2016. The secondary frame in spider orb webs: the detail that makes the difference. Scientific Reports 6.
Soley, F. G., and P. W. Taylor. 2013. Ploys and counterploys of assassin bugs and their dangerous spider prey. Behaviour 150:397–425.
Soley, F. G., R. R. Jackson, and P. W. Taylor. 2011. Biology of *Stenolemus giraffa* (Hemiptera: Reduviidae), a web invading, araneophagic assassin bug from Australia. New Zealand Journal of Zoology 38:297–316.
Southwood, T. R. 1978. Ecological Methods: With Particular Reference to Study of Insect Populations. 2nd ed., rev. Chapman and Hall, New York.
Spagna, J. C., and R. G. Gillespie. 2008. More data, fewer shifts: molecular insights into the evolution of the spinning apparatus in non-orb-weaving spiders. Molecular Phylogenetics and Evolution 46:347–368.
Spiller, D. A. 1984a. Seasonal reversal of competitive advantage between two spider species. Oecologia 64:322–331.
———. 1984b. Competition between two spider species: experimental field study. Ecology 65:909–919.
———. 1986. Interspecific competition between spiders and its relevance to biological control by general predators. Environmental Entomology 15:177–181.
———. 1992. Relationship between prey consumption and colony size in an orb spider. Oecologia 90:457–466.
Spiller, D. A., and T. W. Schoener. 1994. Effects of top and intermediate predators in a terrestrial food web. Ecology 75:182–196.
———. 2008. Climatic control of trophic interaction strength: the effect of lizards on spiders. Oecologia 154:763–771.
Sponner, A., B. Schlott, F. Vollrath, E. Unger, F. Grosse, and K. Weisshart. 2005. Characterization of the protein components of *Nephila clavipes* dragline silk. Biochemistry 44:4727–4736.
Spronk, F. 1935. Die Abhängigkeit der Nestbauzeiten der Radnetzspinnen *Epeira diademata* und *Zilla-X-notata* von verschiedenen Aussenbedingungen. Zeitschrift für Vergleichende Physiologie 22:604–613.
Starks, P. T. 2002. The adaptive significance of stabilimenta in orb-webs: a hierarchical approach. Annales Zoologici Fennici 39:307–315.
Stellwagen, S. D., B. D. Opell, and K. G. Short. 2014. Temperature mediates the effect of humidity on the viscoelasticity of glycoprotein glue within the droplets of an orb-weaving spider's prey capture threads. Journal of Experimental Biology 217:1563–1569.
Stellwagen, S. D., B. D. Opell, and M. E. Clouse. 2015. The impact of UVB radiation on the glycoprotein orb-weaving spider capture thread. Journal of Experimental Biology 218:2675–2684.
Stowe, M. K. 1978. Observations of two nocturnal orbweavers that build specialized webs: *Scoloderus cordatus* and *Wixia ectypa* (Araneae: Araneidae). Journal of Arachnology 6:141–146.
———. 1985. Capture methods of spiders of the family Araneidae, genera *Kaira* and *Mastophora*. Research Reports 463.
———. 1986. Prey specialization in the Araneidae. Pages 101–131 *in* W. A. Shear, ed. Spiders: Webs, Behavior, and Evolution. Stanford University Press, Stanford, CA.

Stowe, M. K., J. H. Tumlinson, and R. R. Heath. 1987. Chemical mimicry: bolas spiders emit components of moth prey species sex pheromones. Science 236:964.

Stratton, G. E., and R. B. Suter. 2009. Water repellent properties of spiders: topographical variations and functional correlates. Functional Surfaces in Biology 1:77–95.

Stropa, A. A. 2010. Effect of architectural angularity on refugia selection by the brown spider, *Loxosceles gaucho*. Medical and Veterinary Entomology 24:273–277.

Styrsky, J. D. 2014. An orb-weaver spider exploits an ant-acacia mutualism for enemy-free space. Ecology and Evolution 4:276–283.

Suginaga, A. 1963. Web-building and predation behavior of *Cyrtarachne bufo* and *C. inaequalis*. Atypus 31:13–15.

Suter, R. B. 1978. *Cyclosa turbinata* (Araneae, Araneidae): prey discrimination via web-borne vibrations. Behavioral Ecology and Sociobiology 3:283–296.

———. 1981. Behavioral thermoregulation: solar orientation in *Frontinella communis* (Linyphiidae), a 6-mg spider. Behavioral Ecology and Sociobiology 8:77–81.

———. 1984. Web tension and gravity as cues in spider orientation. Behavioral Ecology and Sociobiology 16:31–36.

Suter, R. B., and G. E. Stratton. 2009. Spitting performance parameters and their biomechanical implications in the spitting spider, *Scytodes thoracica*. Journal of Insect Science 9:1–15.

Suter, R. B., A. J. Hirscheimer, and C. Shane. 1987. Senescence of web construction behavior in male *Frontinella pyramitela* (Araneae, Linyphiidae). Journal of Arachnology 15:177–183.

Sutherland, T. D., J. H. Young, S. Weisman, C. Y. Hayashi, and D. J. Merritt. 2010. Insect silk: one name, many materials. Annual Review of Entomology 55:171–188.

Swanson, B. O., T. A. Blackledge, A. P. Summers, and C. Y. Hayashi. 2006. Spider dragline silk: correlated and mosaic evolution in high-performance biological materials. Evolution 60:2539–2551.

Swanson, B. O., S. P. Anderson, C. DiGiovine, R. N. Ross, and J. P. Dorsey. 2009. The evolution of complex biomaterial performance: the case of spider silk. Integrative and Comparative Biology 49:21–31.

Szlep, R. 1958. Influence of external factors on some structural properties of the garden spider (*Aranea diademata*) web. Folia Biologica 6:287–299.

———. 1961. Developmental changes in the web spinning instinct of Uloboridae: construction of the primary-type web. Behaviour 17:60–70.

———. 1965. The web-spinning process and web-structure of *Latrodectus tredecimguttatus*, *L. pallidus* and *L. revivensis*. Proceedings of the Zoological Society of London 145:75–89.

———. 1966a. Evolution of the web spinning activities: the web spinning in *Titanoeca albomaculata* Luc. (Araneae, Amaurobidae). Israel Journal of Zoology 15:83–88.

———. 1966b. The web structure of *Latrodectus variolus* Walckener and *L. bishop* Kaston. Israel Journal of Zoology 15:89–94.

Takada, M., T. Kobayashi, A. Yoshioka, S. Takagi, and I. Washitani. 2013. Facilitation of ground-dwelling wolf spider predation on mirid bugs by horizontal webs built by *Tetragnatha* spiders in organic paddy fields. Journal of Arachnology 41:31–35.

Tan, E. J., and D. Li. 2009. Detritus decorations of an orb-weaving spider, *Cyclosa mulmeinensis* (Thorell): for food or camouflage? Journal of Experimental Biology 212:1832–1839.

Tan, E. J., S. W. Seah, L. M. Y. Yap, P. M. Goh, W. Gan, F. Liu, and D. Li. 2010. Why do orb-weaving spiders (*Cyclosa ginnaga*) decorate their webs with silk spirals and plant detritus? Animal Behaviour 79:179–186.

Tanaka, K. 1989. Energetic cost of web construction and its effect on web relocation in the web-building spider *Agelena limbata*. Oecologia 81:459–464.

Tew, E. R., and T. Hesselberg. 2017. The effect of wind exposure on the web characteristics of a tetragnathid orb spider. Journal of Insect Behavior 30:273–286.

Tew, E. R., A. Adamson, and T. Hesselberg. 2015. The web repair behaviour of an orb spider. Animal Behaviour 103:137–146.

Théry, M., and J. Casas. 2009. The multiple disguises of spiders: web colour and decorations, body colour and movement. Philosophical Transactions of the Royal Society B: Biological Sciences 364:471–480.

Thévenard, L., R. Leborgne, and A. Pasquet. 2004. Web-building management in an orb-weaving spider, *Zygiella x-notata*: influence of prey and conspecifics. Comptes Rendus Biologies 327:84–92.

Thorell, P. T. 1886. XXVIII.—On Dr. Bertkau's classification of the order araneœ, or spiders. Annals and Magazine of Natural History 17:301–326.

Thornhill, W. A. 1983. The distribution and probable importance of linyphiid spiders living on the soil surface of sugar-beet fields. Bulletin of the British Arachnological Society 6:127–136.

Tietjen, W. J. 1986. Social spider webs, with special reference to the web of *Mallos gregalis*. Pages 172–206 *in* W. A. Shear, ed. Spiders: Webs, Behavior, and Evolution. Stanford University Press, Stanford, CA.

Tillinghast, E. K., and T. Christenson. 1984. Observations on the chemical composition of the web of *Nephila clavipes* (Araneae, Araneidae). Journal of Arachnology 12:69–74.

Tillinghast, E. K., and E. J. Kavanagh. 1977. The alkaline proteases of *Argiope* and their possible role in web digestion. Journal of Experimental Zoology 202:213–222.

Tillinghast, E. K., and M. A. Townley. 1987. Chemistry, physical properties, and synthesis of araneidae orb webs. Pages 203–210 *in* W. Nentwig, ed. Ecophysiology of Spiders. Springer-Verlag, Berlin, Heidelberg.

Tilquin, A. 1942. La Toile Géométrique Des Araignées. Presses University de France, Paris.

Toft, C. A., and T. W. Schoener. 1983. Abundance and diversity of orb spiders on 106 Bahamian islands: biogeography at an intermediate trophic level. Oikos 41:411–426.

Toft, S. 1986. Field experiments on competition among web spiders. Pages 328 *in* W. G. Eberhard, Y. D. Lubin, and B. C. Robinson, eds. Proceedings of the 9th International Congress of Arachnology, Smithsonian Institution Press, Washington, DC.

———. 1987. Microhabitat identity of two species of sheet-web spiders: field experimental demonstration. Oecologia 72:216–220.

———. 1989. Interactions among two coexisting *Linyphia* spiders. Acta Zoologica Fennica 190:367–372.

———. 2013. Nutritional aspects of spider feeding. Pages 373–384 *in* W. Nentwig, ed. Spider Ecophysiology. Springer-Verlag, Berlin, Heidelberg.

Toft, S., and D. H. Wise. 1999a. Growth, development, and survival of a generalist predator fed single- and mixed-species diets of different quality. Oecologia 119:191–197.

———. 1999b. Behavioral and ecophysiological responses of a generalist predator to single- and mixed-species diets of different quality. Oecologia 119:198–207.

Tolbert, W. W. 1975. Predator avoidance behavior and web defensive structures in the orb weavers, *Argiope aurantia* and *Argiope trifasciata* (Araneae, Araneidae). Psyche: A Journal of Entomology 82:29–52.

———. 1979. Thermal stress of the orb-weaving spider *Argiope trifasciata* (Araneae). Oikos 32:386–392.

Toscani, C., R. Leborgne, and A. Pasquet. 2012. Behavioural analysis of web building anomalies in the orb-weaving spider *Zygiella x-notata* (Araneae, Araneidae). Arachnologische Mitteilungen 43:79–83.

Townley, M. A., and E. K. Tillinghast. 1988. Orb web recycling in *Araneus cavaticus* (Araneae, Araneidae) with an emphasis on the adhesive spiral component, GABamide. Journal of Arachnology 16:303–319.

———. 2009. Developmental changes in spider spinning fields: a comparison between *Mimetus* and *Araneus* (Araneae: Mimetidae, Araneidae). Biological Journal of the Linnean Society 98:343–383.

———. 2013. Aggregate silk gland secretions of araneoid spiders. Pages 283–302 *in* W. Nentwig, ed. Spider Ecophysiology. Springer-Verlag, Berlin, Heidelberg.

Townley, M. A., D. T. Bernstein, K. S. Gallagher, and E. K. Tillinghast. 1991. Comparative study of orb web hygroscopicity and adhesive spiral composition in three araneid spiders. Journal of Experimental Zoology 259:154–165.

Townley, M. A., E. K. Tillinghast, and N. A. Cherim. 1993. Moult-related changes in ampullate silk gland morphology and usage in the araneid spider *Araneus cavaticus*. Philosophical Transactions of the Royal Society of London B: Biological Sciences 340:25.

Townley, M. A., E. K. Tillinghast, and C. D. Neefus. 2006. Changes in composition of spider orb web sticky droplets with starvation and web removal, and synthesis of sticky droplet compounds. Journal of Experimental Biology 209: 1463–1486.

Trabalon, M. 2013. Chemical communication and contact cuticular compounds in spiders. Pages 125–140 *in* W. Nentwig, ed. Spider Ecophysiology. Springer: New York.

Traw, M. B. 1995. A revision of the Neotropical orb-weaving spider genus *Scoloderus* (Araneae: Araneidae). Psyche: A Journal of Entomology 102:49–72.

Tretzel, E. 1961. Biologie, Ökologie und Brutpflege von *Coelotes terrestris* (Wider) (Araneae, Agelendidae) Teil I: Biologie und Ökologie. Zeitschrift fur Morfologie Ökologie de Tiere 49:658–745.

Triana-Cambronero, E., G. Barrantes, E. Cuyckens, and A. Camacho. 2011. Function of the upper tangle in webs of young *Leucauge argyra* (Araneae: Tetragnathidae). Journal of Arachnology 39:519–522.

Triana, E., G. Barrantes, and P. Hanson. 2012. Incidence of parasitoids and predators on eggs of seven Therididae species (Araneae). Bulletin of the British Arachnological Society 15: 293–298.

Tseng, H. J., R.-C. Cheng, S.-H. Wu, S. J. Blamires, and I.-M. Tso. 2011. Trap barricading and decorating by a well-armored sit-and-wait predator: extra protection or prey attraction? Behavioral Ecology and Sociobiology 65:2351–2359.

Tseng, L., and I. M. Tso. 2009. A risky defence by a spider using conspicuous decoys resembling itself in appearance. Animal Behaviour 78:425–431.

Tso, I. M. 1996. Stabilimentum of the garden spider *Argiope trifasciata*: a possible prey attractant. Animal Behaviour 52:183–191.

———. 1998a. Isolated spider web stabilimentum attracts insects. Behaviour 135:311–319.

———. 1998b. Stabilimentum-decorated webs spun by *Cyclosa conica* (Araneae, Araneidae) trapped more insects than undecorated webs. Journal of Arachnology 26:101–105.

———. 1999. Behavioral response of *Argiope trifasciata* to recent foraging gain: a manipulative study. American Midland Naturalist 141:238–246.

———. 2004. The effect of food and silk reserve manipulation on decoration-building of *Argiope aetheroides*. Behaviour 141:603–616.

———. 2013. Insect view of orb spider body colorations. Pages 319–332 *in* W. Nentwig, ed. Spider Ecophysiology. Springer-Verlag, Berlin, Heidelberg.

Tso, I. M., and L. Severinghaus. 1998. Silk stealing by *Argyrodes lanyuensis* (Araneae: Theridiidae): a unique form of kleptoparasitism. Animal Behaviour 56:219–225.

Tso, I. M., P.-L. Tai, T.-H. Ku, C.-H. Kuo, and E.-C. Yang. 2002. Colour-associated foraging success and population genetic structure in a sit-and-wait predator *Nephila maculata* (Araneae: Tetragnathidae). Animal Behaviour 63:175–182.

Tso, I. M., H.-C. Wu, and I.-R. Hwang. 2005. Giant wood spider *Nephila pilipes* alters silk protein in response to prey variation. Journal of Experimental Biology 208:1053–1061.

Tso, I. M., S.-Y. Chiang, and T. A. Blackledge. 2007. Does the giant wood spider *Nephila pilipes* respond to prey variation by altering web or silk properties? Ethology 113:324–333.

Turillazzi, S. 2013. The Biology of Hover Wasps. Springer-Verlag, Berlin, Heidelberg.

Turnbull, A. L. 1960. The prey of the spider *Linyphia triangularis* (Clerck) (Araneae, Linyphiidae). Canadian Journal of Zoology 38:859–873.

———. 1964. The search for prey by a web-building spider *Achaearanea tepidariorum* (C. L. Koch) (Araneae, Theridiidae). Canadian Entomologist 96:568–579.

———. 1966. A population of spiders and their potential prey in an overgrazed pasture in Eastern Ontario. Canadian Journal of Zoology 44:557–583.

Turner, J. S., F. Vollrath, and T. Hesselberg. 2011. Wind speed affects prey-catching behaviour in an orb web spider. Naturwissenschaften 98:1063–1067.

Uetz, G. W. 1986. Web building and prey capture in communal orb weavers. Pages 207–231 *in* W. A. Shear, ed. Spiders: Webs, Behavior, and Evolution. Stanford University Press, Stanford, CA.

———. 1989. The "ricochet effect" and prey capture in colonial spiders. Oecologia 81:154–159.

Uetz, G. W., and S. P. Hartsock. 1987. Prey selection in an orb-weaving spider: *Micranthend gracilis* (Araneae: Araneidae). Psyche: A Journal of Entomology 94:103–116.

Uetz, G. W., A. D. Johnson, and D. W. Schemske. 1978. Web placement, web structure, and prey capture in orb-weaving spiders. Bulletin of the British Arachnological Society 4:141–148.

Uetz, G. W., C. S. Heiber, E. M. Jakob, R. S. Wilcox, D. Kroeger, A. McCrate, and A. M Mostrom. 1994. Behavior of colonial orb-weaving spiders during a solar eclipse. Ethology 96:24–32.

Uhl, G. 2008. Size dependent occurrence of different types of web decorations and a barrier web in the tropical spider *Argiope argentata* (Fabricius 1775) (Araneae Araneidae). Tropical Zoology 21:97–108.

Umeda, Y., A. Shinkai, and T. Miyashita. 1996. Prey composition of three *Dipoena* species (Araneae: Theridiidae) specializing on ants. Acta Arachnologica 45:95–99.

Vaknin, Y., S. Gan-Mor, A. Bechar, B. Ronen, and D. Eisikowitch. 2000. The role of electrostatic forces in pollination. Plant Systematics and Evolution 222:133–142.

Valderrama, A. C. 2000. Vertical distribution of orb-weaving spiders in a Colombian cloud forest. Pages 140–142 *in* N. M. Nadkami and N. T. Wheelwright, eds. Ecology and Conservation of a Tropical Cloud Forest. Oxford University Press, New York.

Valerio, C. E., and M. V. Herrero. 1976. Tendencia social en adultos de la araña *Leucauge* sp. (Araneae: Araneidae) en Costa Rica. Brenesia 10:69–76.

Venner, S., and J. Casas. 2005. Spider webs designed for rare but life-saving catches. Proceedings of the Royal Society of London B: Biological Sciences 272:1587–1592.

Venner, S., A. Pasquet, and R. Leborgne. 2000. Web-building behaviour in the orb-weaving spider *Zygiella x-notata*: influence of experience. Animal Behaviour 59:603–611.

Venner, S., L. Thevenard, A. Pasquet, and R. Leborgne. 2001. Estimation of the web's capture thread length in orb-weaving spiders: determining the most efficient formula. Annals of the Entomological Society of America 94:490–496.

Venner, S., M. C. Bel-Venner, A. Pasquet, and R. Leborgne. 2003. Body-mass-dependent cost of web-building behavior in an orb weaving spider, *Zygiella x-notata*. Naturwissenschaften 90:269–272.

Venner, S., I. Chadès, M.-C. Bel-Venner, A. Pasquet, F. Charpillet, and R. Leborgne. 2006. Dynamic optimization over infinite-time horizon: web-building strategy in an orb-weaving spider as a case study. Journal of Theoretical Biology 241: 725–733.

Vetter, R. S. 2015. The Brown Recluse Spider. Cornell University Press, New York.

Vetter, R. S., and M. K. Rust. 2008. Refugia preferences by the spiders *Loxosceles reclusa* and *Loxosceles laeta* (Araneae: Sicariidae). Journal of Medical Entomology 45:36–41.

———. 2010. Influence of spider silk on refugia preferences of the recluse spiders *Loxosceles reclusa* and *Loxosceles laeta* (Araneae: Sicariidae). Journal of Economic Entomology 103:808–815.

Vetter, R. S., L. M. Penas, and M. S. Hoddle. 2016. Laboratory refugia preferences of the brown widow spider, *Latrodectus geometricus* (Araneae: Theridiidae). Journal of Arachnology 44:52–57.

Viera, C. 1986. Comportamiento de captura de *Metepeira* sp. (Araneae: Araneidae) sobre *Acromyrmex* sp. (Hymenoptera, Formicidae) en condiciones experimentales. A (Araneae, Araneidae) sobre Acromyrmex sp.(Hymenoptera, Formicidae) en condiciones experimentales. Aracnología 6:1–8.

———. 1988. An irregular orb-like web built by an adult male of *Metepeira* sp. A (Aranea, Araneidae). Journal of Arachnology 16:387–388.

———. 2008. Spatial and temporal variability in webs of *Metepeira gressa* (Keyserling, 1892) (Araneae, Araneidae): a year field study. Anales de Biología 25:13–20.

Viera, C., H. F. Japyassú, A. J. Santos, and M. O. Gonzaga. 2007. Teias e forrageamiento. Pages 45–65 *in* M. O. Gonzaga, A. J. Santos, and H. F. Japyassú, eds. Ecologia e comportamiento de aranhas. Editora Interciencia, Rio de Janeiro.

Vollrath, F. 1978. A close relationship between two spiders (Arachnida, Araneidae): *Curimagua bayano* Synecious on a *Diplura* species. Psyche: A Journal of Entomology 85: 347–354.

———. 1979. Behaviour of the kleptoparasitic spider *Argyrodes elevatus* (Araneae, Theridiidae). Animal Behaviour 27:515–521.

———. 1984. Kleptobiotic interactions in invertebrates. Pages 61–94 *in* C. J. Barnard, ed. Producers and Scroungers: Strategies of Exploitation and Parasitism. Coom Helm, London.

———. 1985. Web spider's dilemma: a risky move or site dependent growth. Oecologia 68:69–72.

———. 1986. Gravity as an orientation guide during web-construction in the orb spider *Araneus diadematus* (Araneae, Araneidae). Journal of Comparative Physiology A: Neuroethology, Sensory, Neural, and Behavioral Physiology 159:275–280.

———. 1987a. Altered geometry of webs in spiders with regenerated legs. Nature 328:247–248.

———. 1987b. Kleptobiosis in spiders. Pages 274–286 *in* W. Nentwig, ed. Ecophysiology of Spiders. Springer-Verlag, Berlin, Heidelberg.

———. 1987c. Growth, foraging and reproductive success. Pages 357–370 *in* W. Nentwig, ed. Ecophysiology of Spiders. Springer Verlag, Berlin, Heidelberg.

———. 1988a. Spiral orientation of *Araneus diadematus* orb webs built during vertical rotation. Journal of Comparative Physiology A: Neuroethology, Sensory, Neural, and Behavioral Physiology 162:413–419.

———. 1988b. Untangling the spider's web. Trends in Ecology & Evolution 3:331–335.

———. 1992. Analysis and interpretation of orb spider exploration and web-building behavior. Advances in the Study of Behavior 21:147–199.

———. 2003. Web masters: spiders' silky skills hold the key to their evolutionary success. Nature 426:121–122.

———. 2010. Silk evolution untangled. Nature 466:319.

Vollrath, F., and D. Edmonds. 1989. Modulation of normal spider silk by coating with water. Nature 340:305–307.

Vollrath, F., and A. Houston. 1986. Previous experience and site tenacity in the orb spider *Nephila* (Araneae, Araneidae). Oecologia 70:305–308.

Vollrath, F., and D. P. Knight. 2001. Liquid crystalline spinning of spider silk. Nature 410:541–548.

Vollrath, F., and T. Köhler. 1996. Mechanics of silk produced by loaded spiders. Proceedings of the Royal Society of London B: Biological Sciences 263:387–391.

Vollrath, F., and W. Mohren. 1985. Spiral geometry in the garden spider's orb web. Naturwissenschaften 72:666–667.

Vollrath, F., and F. Samu. 1997. The effect of starvation on web geometry in an orb-weaving spider. Bulletin of the British Arachnological Society 10:295–298.

Vollrath, F., and P. Selden. 2007. The role of behavior in the evolution of spiders, silks, and webs. Annual Review of Ecology, Evolution, and Systematics 38:819–846.

Vollrath, F., and E. K. Tillinghast. 1991. Glycoprotein glue beneath a spider web's aqueous coat. Naturwissenschaften 78:557–559.

Vollrath, F., M. Downes, and S. Krackow. 1997. Design variability in web geometry of an orb-weaving spider. Physiology & Behavior 62:735–743.

Vollrath, F., B. Madsen, and Z. Shao. 2001. The effect of spinning conditions on the mechanics of a spider's dragline silk. Proceedings of the Royal Society of London B: Biological Sciences 268:2339–2346.

Walter, A., and M. A. Elgar. 2011. Signals for damage control: web decorations in *Argiope keyserlingi* (Araneae: Araneidae). Behavioral Ecology and Sociobiology 65:1909–1915.

———. 2012. The evolution of novel animal signals: silk decorations as a model system. Biological Reviews 87:686–700.

Walter, A., M. A. Elgar, P. Bliss, and R. F. Moritz. 2008. Wrap attack activates web-decorating behavior in *Argiope* spiders. Behavioral Ecology 19:799.

Walter, A., P. Bliss, M. A. Elgar, and R. F. Moritz. 2009. *Argiope bruennichi* shows a drinking-like behaviour in web hub decorations (Araneae, Araneidae). Journal of Ethology 27:25–29.

Ward, D., and Y. Lubin. 1992. Temporal and spatial segregation of web-building in a community of orb-weaving spiders. Journal of Arachnology 20:73–87.

———. 1993. Habitat selection and the life history of a desert spider, *Stegodyphus lineatus* (Eresidae). Journal of Animal Ecology 62:353–363.

Watanabe, T. 1999a. The influence of energetic state on the form of stabilimentum built by *Octonoba sybotides* (Araneae: Uloboridae). Ethology 105:719–725.

———. 1999b. Prey attraction as a possible function of the silk decoration of the uloborid spider *Octonoba sybotides*. Behavioral Ecology 10:607–611.

———. 2000a. Life history and seasonal change in the frequency of dimorphic stabilimenta of the orb-web spider, *Octonoba sybotides* (Uloboridae). Acta Arachnologica 49:1–12.

———. 2000b. Web tuning of an orb-web spider, *Octonoba sybotides*, regulates prey-catching behaviour. Proceedings of the Royal Society of London B: Biological Sciences 267: 565–569.

———. 2001. Effects of web design on the prey capture efficiency of the uloborid spider *Octonoba sybotides* under abundant and limited prey conditions. Zoological Science 18:585–590.

Wehner, R. 1981. Spatial vision in arthropods. Pages 287–616 *in* H. Autrum, ed. Comparative Physiology and the Evolution of Vision in Invertebrates. C: Invertebrate Visual Centers and Behavior II. Springer, New York.

Weissmann, M., and F. Vollrath. 1998. The effect of leg loss on orb-spider growth. Bulletin of the British Arachnological Society 11:92–94.

Weng, J. L., G. Barrantes, and W. G. Eberhard. 2006. Feeding by *Philoponella vicina* (Araneae, Uloboridae) and how uloborid spiders lost their venom glands. Canadian Journal of Zoology 84:1752–1762.

Wenzel, J. W. 1993. Application of the biogenetic law to behavioral ontogeny: a test using nest architecture in paper wasps. Journal of Evolutionary Biology 6:229–247.

West-Eberhard, M. J. 2003. Developmental Plasticity and Evolution. Oxford University Press, New York.

Whitehouse, M. E. A. 1986. The foraging behaviours of *Argyrodes antipodiana* (Theridiidae), a kleptoparasitic spider from New Zealand. New Zealand Journal of Zoology 13:151–168.

———. 1987. "Spider Eat Spider": the predatory behavior of *Rhomphaea* sp. from New Zealand. Journal of Arachnology 15:355–362.

Wiederman, S. D., and D. C. O'Carroll. 2013. Selective attention in an insect visual neuron. Current Biology 23:156–161.

Wiehle, H. 1927. Beiträge zur Kenntnis des Radnetzbaues der Epeiriden, Tetragnathiden und Uloboriden. Zoomorphology 8:468–537.

———. 1928. Beiträge zur Biologie der Araneen, insbesondere zur Kenntnis des Radnetzbaues. Zeitschrift für Morphologie und Ökologie der Tiere 11:115–151.

———. 1929. Weitere Beiträge zur Biologie der Araneen, Insbesondere zur Kenntnis des Radnetzbaues. Zeitschrift für Morphologie und Ökologie der Tiere 15:262–308.

———. 1931. Neue Beiträge zur Kenntnis des Fanggewebes der Spinnen aus den Familien Argiopidae, Uloboridae und Theridiidae. Zoomorphology 22:349–400.

Williams, F. X. 1928. The natural history of a Philippine nipa house with descriptions of new wasps. Philippine Journal of Science 35:53–118.

Williams, G. C. 1966. Adaptation and Natural Selection: A Critique of Some Current Evolutionary Thought. Princeton University Press, Princeton.

———. 1998. The Pony Fish's Glow; and Other Clues to Plan and Purpose in Nature. Basic Books, New York.

Wilson, R. S. 1962. The structure of the dragline control valves in the garden spider. Journal of Cell Science s3-103:549–555.

Wirth, E., and F. Barth. 1992. Forces in the spider orb web. Journal of Comparative Physiology A: Neuroethology, Sensory, Neural, and Behavioral Physiology 171:359–371.

Wise, D. H. 1975. Food limitation of the spider *Linyphia marginata*: experimental field studies. Ecology 56:637–646.

———. 1981. Inter- and intraspecific effects of density manipulations upon females of two orb-weaving spiders (Araneae: Araneidae). Oecologia 48:252–256.

———. 1982. Predation by a commensal spider, *Argyrodes trigonum*, upon its host: an experimental study. Journal of Arachnology 10:111–116.

———. 1983. Competitive mechanisms in a food-limited species: relative importance of interference and exploitative interactions among labyrinth spiders (Araneae: Araneidae). Oecologia 58:1–9.

———. 1995. Spiders in Ecological Webs. Cambridge University Press, Cambridge.

Wise, D. H., and J. L. Barata. 1983. Prey of two syntopic spiders with different web structures. Journal of Arachnology 11:271–281.

Witt, P. N. 1962. Effects of atropine on spider's web building behaviour and thread production. Federation Proceedings 21:180.

———. 1963. Environment in relation to behavior of spiders. Archives of Environmental Health: An International Journal 7:4–12.

———. 1965. Do we live in the best of all worlds? Spider webs suggest an answer. Perspectives in Biology and Medicine 8:475–487.

———. 1969. Behavioral consequences of laser lesions in the central nervous system of *Araneus diadematus* Cl. American Zoologist 9:121–132.

———. 1971. Drugs alter web-building of spiders: a review and evaluation. Behavioral Science 16:98–113.

Witt, P. N., and R. Baum. 1960. Changes in orb webs of spiders during growth (*Araneus diadematus* Clerck and *Neoscona vertebrata* McCook). Behaviour 16:309–318.

Witt, P. N., and C. F. Reed. 1965. Spider-web building. Science 149:1190–1197.

Witt, P. N., C. F. Reed, and D. B. Peakall. 1968. A Spider's Web: Problems in Regulatory Biology. Springer-Verlag, New York.

Witt, P. N., J. O. Rawlings, and C. F. Reed. 1972. Ontogeny of web-building behavior in two orb-weaving spiders. American Zoologist 12:445.

Witt, P. N., M. B. Scarboro, R. Daniels, D. B. Peakall, and R. L. Gause. 1977. Spider web-building in outer space: evaluation of records from the skylab spider experiment. Journal of Arachnology 4:115–124.

Wolff, J. O., and M. E. Herberstein. 2017. Three-dimensional printing spiders: back-and-forth glue application yields silk anchorages with high pull-off resistance under varying loading situations. Journal of the Royal Society Interface 14:20160783.

Wolff, J. O., J. M. Schneider, and S. N. Gorb. 2014. How to pass the gap—functional morphology and biomechanics of spider bridging threads. Pages 165–177 *in* T. Asakura and T. Miller, eds. Biotechnology of Silk, Biologically-Inspired Systems. Springer Netherlands.

Wolff, J. O., I. Grawe, M. Wirth, A. Karstedt, and S. N. Gorb. 2015. Spider's super-glue: thread anchors are composite adhesives with synergistic hierarchical organization. Soft Matter 11:2394–2403.

Wolff, J. O., A. van der Meijden, and M. E. Herberstein. 2017a. Distinct spinning patterns gain differentiated loading tolerance of silk thread anchorages in spiders with different ecology. Proceedings of the Royal Society of London B: Biological Sciences 284:20171124.

Wolff, J. O, M. Řezáč, T. Krejčí, and S. N. Gorb. 2017b. Hunting with sticky tape: functional shift in silk glands of araneophagous ground spiders (Gnaphosidae). Journal of Experimental Biology 220: 2250–2259. DOI:10.1242/jeb.154682.

Work, R. W. 1977. Dimensions, birefringences, and force-elongation behavior of major and minor ampullate silk fibers from orb-web-spinning spiders—the effects of wetting on these properties. Textile Research Journal 47:650–662.

———. 1978. Mechanism for the deceleration and support of spiders on draglines. Transactions of the American Microscopical Society 97:180–191.

———. 1981. A comparative study of the supercontraction of major ampullate silk fibers of orb-web-building spiders (Araneae). Journal of Arachnology 9:299–308.

Work, R. W., and N. Morosoff. 1982. A physico-chemical study of the supercontraction of spider major ampullate silk fibers. Textile Research Journal 52:349–356.

Wright, C. M., C. N. Keiser, and J. N. Pruitt. 2016. Colony personality composition alters colony-level plasticity and magnitude of defensive behaviour in a social spider. Animal Behaviour 115:175–183.

Wright, S., and S. L. Goodacre. 2012. Evidence for antimicrobial activity associated with common house spider silk. BMC Research Notes 5:326.

Wu, C.-C., S. J. Blamires, C.-L. Wu, and I.-M. Tso. 2013. Wind induces variations in spider web geometry and sticky spiral droplet volume. Journal of Experimental Biology 216:3342–3349.

Wunderlich, J. 2004. Fossil spiders in amber and copal: conclusions, revisions, new taxa and family diagnoses of fossil and extant taxa. Beitraege zur Araneologie 3A-B:1–1908.

Xavier, G. M., R. R. Moura, and M. O. Gonzaga. 2017. Orb web architecture of *Wixia abdominalis* O. Pickard-Cambridge, 1882 (Araneae: Araneidae): intra-orb variation of web components. Journal of Arachnology 45:160–165.

Yamanoi, T., and T. Miyashita. 2005. Foraging strategy of nocturnal orb-web spiders (Araneidae: Neoscona) with special reference to the possibility of beetle specialization by *N. punctigera*. Acta Arachnologica 54:13–19.

Yanoviak, S. P., R. Dudley, and M. Kaspari. 2005. Directed aerial descent in canopy ants. Nature 433:624–626.

Yeargan, K. V. 1994. Biology of bolas spiders. Annual Review of Entomology 39:81–99.

Yoshida, M. 1987. Predatory behavior of *Tetragnatha praedonia* (Araneae: Tetragnathidae). Acta Arachnologica 35:57–75.

———. 1989. Predatory behavior of three Japanese species of *Metleucauge* (Araneae, Tetragnathidae). Journal of Arachnology 17:15–25.

Yoshida, M., and A. Shinkai. 1993. Predatory behavior and web structure of *Meta menardi* (Araneae: Tetragnathidae). Acta Arachnologica 42:21–25.

Young, A. M. 1980. Feeding and oviposition in the giant tropical damselfly *Megaloprepus coerulatus* (Drury) in Costa Rica. Biotropica 12:237–239.

Yu, L., and J. A. Coddington. 1990. Ontogenetic changes in the spinning fields of *Nuctenea cornuta* and *Neoscona theisi* (Araneae, Araneidae). Journal of Arachnology 18:331–345.

Zaera, R., A. Soler, and J. Teus. 2014. Uncovering changes in spider orb-web topology owing to aerodynamic effects. Journal of the Royal Society Interface 11:20140484.

Zevenbergen, J. M., N. K. Schneider, and T. A. Blackledge. 2008. Fine dining or fortress? Functional shifts in spider web architecture by the western black widow *Latrodectus hesperus*. Animal Behaviour 76:823–829.

Zhang, S., T. H. Koh, W. K. Seah, Y. H. Lai, M. A. Elgar, and D. Li. 2011. A novel property of spider silk: chemical defence against ants. Proceedings of the Royal Society of London B: Biological Sciences 279:1824–1830.

Zimmer, C., and D. J. Emlen. 2013. Evolution: Making Sense of Life. Roberts, New York.

Zschokke, S. 1993. The influence of the auxiliary spiral on the capture spiral in *Araneus diadematus* Clerck (Araneidae). Bulletin of the British Arachnological Society 9:169–173.

———. 1996. Early stages of orb web construction in *Araneus diadematus* Clerck. Revue Suisse de Zoologie 2:709–720.

———. 1997. Factors influencing the size of the orb web in *Araneus diadematus*. Proceedings of the 16th European Colloquium of Arachnology 329:329–334.

———. 1999. Nomenclature of the orb-web. Journal of Arachnology 27:542–546.

———. 2000. Radius construction and structure in the orb-web of *Zilla diodia* (Araneidae). Journal of Comparative Physiology A: Neuroethology, Sensory, Neural, and Behavioral Physiology 186:999–1005.

———. 2002. Form and function of the orb-web. Pages 99–106 *in* S. Toft and N. Scharff, eds. European Arachnology 2000. Aarhus University Press, Aarhus.

———. 2011. Spiral and web asymmetry in the orb webs of *Araneus diadematus* (Araneae: Araneidae). Journal of Arachnology 39:358–362.

Zschokke, S., and A. Bolzern. 2007. Erste Nachweise sowie Kenntnisse zur Biologie von *Cyclosa oculata* (Araneae: Araneidae) in der Schweiz. Arachnologische Mitteilungen 33.

Zschokke, S., and K. Nakata. 2010. Spider orientation and hub position in orb webs. Naturwissenschaften 97:43–52.

———. 2015. Vertical asymmetries in orb webs. Biological Journal of the Linnean Society 114:659–672.

Zschokke, S., and F. Vollrath. 1995a. Unfreezing the behaviour of two orb spiders. Physiology & Behavior 58:1167–1173.

———. 1995b. Web construction patterns in a range of orb-weaving spiders (Araneae). European Journal of Entomology 92:523–541.

———. 2000. Planarity and size of orb-webs built by *Araneus diadematus* (Araneae: Araneidae) under natural and experimental conditions. Ekológia (Bratislava) 19:307–318.

Zschokke, S., Y. Hénaut, S. P. Benjamin, and J. A. García-Ballinas. 2006. Prey-capture strategies in sympatric web-building spiders. Canadian Journal of Zoology 84:964–973.

INDEX

The letter *t* following a page number denotes a table, the letter *f* denotes a figure, the letter *b* denotes a box, and the letter O denotes online-only material.